Lecture Notes in Mechanical Engineering

Lecture Notes in Mechanical Engineering (LNME) publishes the latest developments in Mechanical Engineering - quickly, informally and with high quality. Original research reported in proceedings and post-proceedings represents the core of LNME. Volumes published in LNME embrace all aspects, subfields and new challenges of mechanical engineering. Topics in the series include:

- Engineering Design
- Machinery and Machine Elements
- Mechanical Structures and Stress Analysis
- Automotive Engineering
- Engine Technology
- Aerospace Technology and Astronautics
- Nanotechnology and Microengineering
- Control, Robotics, Mechatronics
- MEMS
- Theoretical and Applied Mechanics
- Dynamical Systems, Control
- Fluid Mechanics
- Engineering Thermodynamics, Heat and Mass Transfer
- Manufacturing
- Precision Engineering, Instrumentation, Measurement
- Materials Engineering
- Tribology and Surface Technology

To submit a proposal or request further information, please contact the Springer Editor in your country:

China: Li Shen at li.shen@springer.com
India: Dr. Akash Chakraborty at akash.chakraborty@springernature.com
Rest of Asia, Australia, New Zealand: Swati Meherishi at swati.meherishi@springer.com
All other countries: Dr. Leontina Di Cecco at Leontina.dicecco@springer.com

To submit a proposal for a monograph, please check our Springer Tracts in Mechanical Engineering at http://www.springer.com/series/11693 or contact Leontina.dicecco@springer.com

Indexed by SCOPUS. The books of the series are submitted for indexing to Web of Science.

More information about this series at http://www.springer.com/series/11236

Nizar Aifaoui · Zouhaier Affi ·
Mohamed Slim Abbes ·
Lassad Walha · Mohamed Haddar ·
Lotfi Romdhane · Abdelmajid Benamara ·
Mnaouar Chouchane · Fakher Chaari
Editors

Design and Modeling of Mechanical Systems - IV

Proceedings of the 8th Conference on Design and Modeling of Mechanical Systems, CMSM'2019, March 18–20, Hammamet, Tunisia

Springer

Editors
Nizar Aifaoui
Preparatory Institute of Engineering Studies of Monastir (IPEIM)
Monastir, Tunisia

Zouhaier Affi
National Engineering School of Monastir (ENIM)
Monastir, Tunisia

Mohamed Slim Abbes
National School of Engineers of Sfax (ENIS)
Sfax, Tunisia

Lassad Walha
National School of Engineers of Sfax (ENIS)
Sfax, Tunisia

Mohamed Haddar
National School of Engineers of Sfax (ENIS)
Sfax, Tunisia

Lotfi Romdhane
National Engineering School of Sousse (ENISO)
Sousse, Tunisia

Abdelmajid Benamara
National Engineering School of Monastir (ENIM)
Monastir, Tunisia

Mnaouar Chouchane
National Engineering School of Monastir (ENIM)
Monastir, Tunisia

Fakher Chaari
Department of Mechanical Engineering
National School of Engineers of Sfax (ENIS)
Sfax, Tunisia

ISSN 2195-4356 ISSN 2195-4364 (electronic)
Lecture Notes in Mechanical Engineering
ISBN 978-3-030-27145-9 ISBN 978-3-030-27146-6 (eBook)
https://doi.org/10.1007/978-3-030-27146-6

This Springer imprint is published by the registered company Springer Nature Switzerland AG
The registered company address is: Gewerbestrasse 11, 6330 Cham, Switzerland

Preface

The eighth edition of the "International Congress Design and Modelling of Mechanical Systems" CMSM'2019 was held in Hammamet, Tunisia, from March 18 to 20, 2019. This congress was organized jointly between two Tunisian research laboratories: the Laboratory of Mechanical Engineering of the National School of Engineers of Monastir and the Laboratory of Mechanics, Modelling and Production of the National school of Engineers of Sfax.

This book is the fourth volume of the LNME book series "Design and Modelling of Mechanical Systems". It contains 101 selected from papers presented during the CMSM'2019 congress. More than 350 participants discussed during the 3 days of the congress, latest advances in the field of design and modeling of mechanical systems. Similar to the past editions, the topics presented and discussed during the congress were very broad covering of research topics of mechanical modeling and design.

The organizers of the conference were honored by the presence of six keynote speakers, who are experts in the field of modeling of mechanical systems, namely:

- Prof. Maurice Pillet, SYMME Laboratory, University of Savoie Mont Blanc, France
- Prof. Pierre-Olivier Mattei, CNRS Deputy Director of LMA, University of Marseille, France
- Prof. Lucas F. M. da Silva, Faculty of Engineering of the University of Porto, Portugal
- Prof. David Daney, Inria Laboratory, Sofia Antipolis, Bordeaux, France
- Prof. Jean Yves Choley, Laboratory QUARTZ, SUPMECA, Paris, France
- Prof. Faida Mhenni, Laboratory QUARTZ, SUPMECA, Paris, France

Each chapter included in this book was rigorously reviewed by three referees. Our gratitude goes to all members of the scientific committee for their valuable efforts to reach a high quality of contributions. We would like also to thank all authors, presenters, and participants of eighth edition of CMSM'2019. Finally, special thanks go to Springer for their continuous support of this conference.

Hammamet, Tunisia
March 2019

Nizar Aifaoui
Zouhaier Affi
Mohamed Slim Abbes
Lassad Walha
Mohamed Haddar
Lotfi Romdhane
Abdelmajid Benamara
Mnaouar Chouchane
Fakher Chaari

Contents

From Assembly Planning to Secondary Assembly's Lines Identification

Imen Belhadj(✉), Moez Trigui, Nizar Aifaoui, and Abdelmajid Benamara

Mechanical Engineering Laboratory, National Engineering School of Monastir (LGM_ENIM), University of Monastir, 5 Av. Ibn Eljazzar, 5019 Monastir, Tunisia
Imenne.belhadj@gmail.com

Abstract. With the new technologies of the product design, the assembly automation lines has expatriated to a new assembly strategy characterized by an assembly modularization. This strategy consists, in a first time, in preparing subassemblies on secondary lines and then assembled with each other on the main line in a second time. The choice has been made to automate certain sections of the main line and to leave the subassembly lines in a manual way. The aim of this work is to identify the subassemblies of a mechanical product from its CAD model in order to predict secondary's assembly lines. For better discussing and explaining all the steps of the proposed approach a CAD assembly of an industrial complex product is presented in all sections of this paper.

Keywords: Assembly automation lines · Modular product · Sub-assembly · CAD model · Base part · Contact matrices · Interference matrices

1 Introduction

The organization of an assembly line is an important issue of the industrialization of complex products. As example, for the case of aircraft structure, up to 80% of the final cost is determined during the design phase, while up to 30% is due to assembly operations [1]. The relevance of both: the product design phases and the assembly line design are pointed out by two relevant cost-related facts. Nowadays, the assembly automation lines has transformed on a new assembly strategy characterized by an assembly modularization. This strategy consists, in a first time, in preparing sub-assemblies on secondary assembly lines and then assembled with each other on the main assembly line in a second time. The choice has been made to automate certain sections of the main assembly line and to leave the subassembly lines in an automatic or manual way.

Regarding the literature, both concepts: Assembly Line (AL) and Assembly/Disassembly Sequence Planning (ASP/DSP) are treated independently. AL

N. Aifaoui et al. (Eds.): CMSM 2019, LNME, pp. 1–11, 2020.
https://doi.org/10.1007/978-3-030-27146-6_1

is a flow production system and is usually composed of a set of sequentially linked workstations in which a set of tasks is performed by operators [2]. While the early AL was utilized by Henry Ford and his colleagues in 1913, the design and industrialization of AL have been important problems for practitioners and academics [3]. Several authors have approved the need to develop specific methods, guidelines, and software applications to support AL designers using the product database [4–6]. The assembly line balancing problem is basically the difficulty, not only to allocate assembly tasks in the ASP of a product but also to arrange the workstations with respect to the precedence relations and optimized performance.

In the other hand, Ullah et al. [7] proposed an ASP generation approach. The developed tool named "Computer Aided Assembly Sequence Plan" is composed of two mains steps. The first step is the extraction of the CAD assembly data. The second one deals with the identification of all possible free directions during the part displacement in order to generate feasible ASP. Trigui et al. [8] presented an approach based on a mobility matrix defined for every part as mobility constraints. An original model of disassembly directions based on geometric and assembly CAD data has been proposed and implemented under Open Cascade© platform. In order to avoid the ASP/DSP combinatory problem of complex mechanism having important number of parts, Issaoui et al. [9] proposed a DSP approaches based on genetic and ant colony algorithms. The developed methods allow the generation of an optimal and feasible DSP of a complex product from its CAD data. In these approaches, several criteria have been considered such as disassembly directions, part volume, tool change, and the maintainability of usury part. Kheder et al. [10] proposed a new approach to identify subassemblies in order to transform a product into a modularization representation. Based on the new representation of the product, an optimal and feasible DSP can be generated easily. The proposed tool is implemented using an automatic coupling between Solidworks© and Matlab© softwares.

Kheder et al. [11] presented an integrated framework for Assembly-Oriented Product Design and Optimization (AOPDO), which integrates ASP process and production simulation, supporting each other in a single framework. They used a technique of producing "function models", which is a structured representation of the functions, activities, or processes within the modeled system or subject area. The user imports a CAD model to the AOPDO tool and the system evaluates the model and provides redesign suggestions, and also generates optimal ASP. Belhadj et al. [12] developed a highly interactive computer-aided decision support system for ASP in an assembly-oriented environment in order to facilitate the design of the associated assembly lines. This approach is attractive, however, each aspect is represented by a separately framework. Su and Smith [13] described a design system framework for an automotive assembly process. Their system is a computer-aided intelligent system, which is able to automatically generate the optional ASP for a best dimensional quality.

As could be seen, it is clear that none of cited approaches have considered simultaneously the AL and the ASP/DSP problems. In few references, the two aspects are considered but using separately frameworks [14, 15]. The aim of this paper is to develop a coupled approach between the product design and the assembly line design (main and secondary lines). It aims to identify the subassemblies of a complex industrial product from its CAD model in order to predict its secondary's assembly lines. For better discussing and explaining all the steps of the proposed approach a CAD assembly of a complex industrial product is presented in all sections of this paper.

2 Proposed Approach

Figure 1 shows the strategy of the proposed approach. The CAD assembly attributes using the developed CADLab© plug-in are firstly extracted. These data should be treated subsequently by Matlab© to generate the set of subassemblies (Subi). Based on the obtained (Subi) set, the secondary's assembly lines and the main assembly line are identified in order to arrange the workstations with respect to the anteriority relations between components.

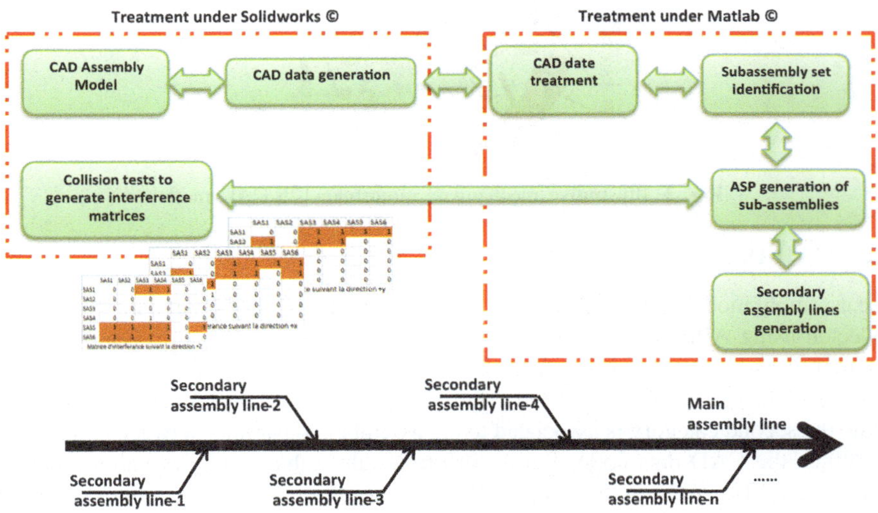

Fig. 1 Background of the proposed approach

2.1 Illustrative Mechanism

In order to detail the proposed approach, an industrial illustrative example is chosen. It is an internationally recognized motorbike-125. It consists of 152 parts joined by 74 connector parts (screws, washers, nuts, rings, etc.) (Fig. 2).

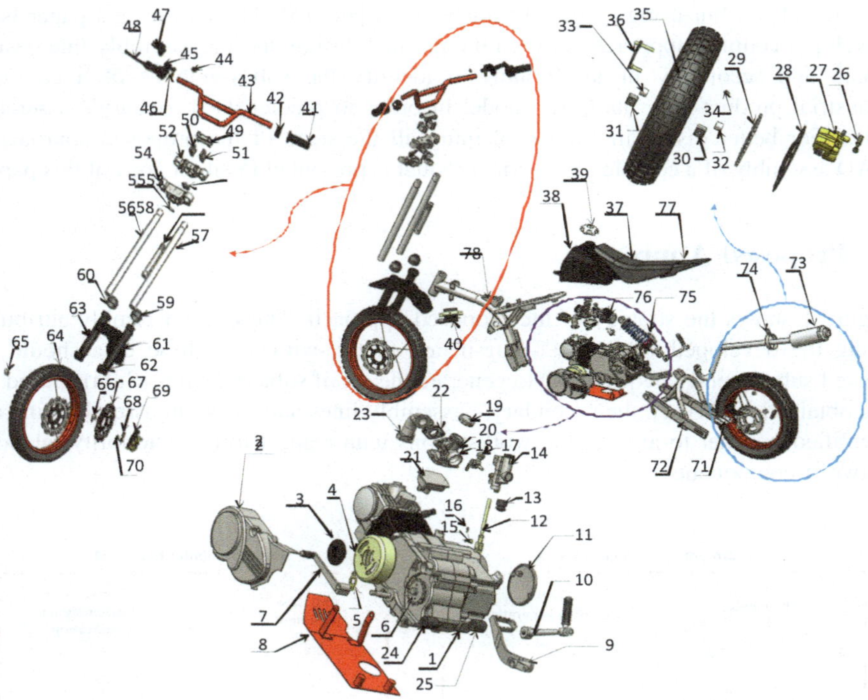

Fig. 2 Illustrative mechanism: motorbike-125 CAD model

2.2 CAD Data

The assembly model produced by CAD software contains a lot of smart data which can be collected using API (Application Programming Interfaces) and used to search the sub-assemblies and the optimal and feasible ASP. Two main categories of data should be identified. The first category is related to the part (topological and geometrical information) while the second category is associated to the assembly constraints between components.

When the CAD data collection is performed, the subassembly research procedure can be starts. This procedure is formed by three stages: creation of the contact matrices; reduction of the contact matrices size by removing of all connector parts from the contact matrices; the base parts identification and the research of subassemblies.

a *Creation of the contact matrices*

The Contact matrix [CM] in the direction of the (k) axe is a square and symmetric matrix, its size is equal to n where n represents the number of parts. The CM (i, j) element, representing a probable contact between the two parts i and j and can have three possible values as follows:

- CM (i, j) = 1 if there is a contact between i and j;
- CM (i, j) = 0 if there is no contact between i and j;
- CM (i, j) = 0 if i = j.

b *Reduction of the contact matrices size*

The size of [CM] can by strongly reduced by suppressing all connector parts. These connector parts are identified directly from the feature manager given by the CAD software. In the case of the illustrative example, all parts from item 79 to item 152 are suppressed from the contact matrices along (x, y, z) directions and the Reduced Contact Matrix [RCM].

c *Base parts research*

The Base Parts (BP) research can be started when the [RCM] matrices are identified. A BP can be defined as a vital part on which some parts will be assembled. In order to identify the BP list, an evaluation function is proposed and associated to each part of the mechanism. For a part i, the score of the evaluation function is given by Eq. (1).

$$EF_i = u.Tc + v.\frac{Vo_i}{Vo} + z.\frac{Su_i}{Su} \quad (1)$$

where

- u, v and z: represent the weight parameters,
- Tc: represents the total contact between part i and other parts in the assembly;
- Vo_i: represents the volume of part i;
- Vo: represents the total volume of parts;
- Su_i: represents the border surface of part i;
- Su: represents the total surface of parts;

Considering the motorbike mechanism, Table 1 shows the obtained score of the evaluation function of each part. Based on these results, the obtained BP list is: {(**1**), (**71**), and (**78**))}.

Table 1 Score of each part of the motorbike mechanism

Item	Score	Item	Score	Item	Score
1	5.62	28	3.76	54	1.82
2	1.63	29	1.62	55	1.21
3	1.00	30	0.60	56	1.23
4	2.01	31	0.60	57	1.23
5	2.00	32	1.20	58	1.01
6	1.80	33	0.86	59	1.20
7	180	34	1.20	60	0.80
8	1.61	35	1.66	61	0.63
9	1.20	36	0.60	62	1.22

(*continued*)

Table 1 (*continued*)

Item	Score	Item	Score	Item	Score
10	2.00	37	0.02	63	1.42
11	1.80	38	1.87	64	1.15
12	1.20	39	0.60	65	1.40
13	0.60	40	0.60	67	1.63
14	1.80	41	0.61	68	0.60
15	1.80	42	0.63	69	1.21
16	0.60	43	1.82	70	1.20
17	0.60	44	1.27	71	5.03
18	0.60	45	0.62	72	0.60
19	0.60	46	1.4	73	1.22
21	0.60	47	2.31	74	0.40
22	0.60	48	0.60	75	1.01
23	1.20	49	1.20	76	1.21
24	0.60	50	1.00	77	0.21
25	1.40	51	1.20	78	5.87
26	0.60	52	2.81		
27	1.02	53	1.00		

d Subassemblies research

The subassemblies (subi) procedure begins by reading all connections between all BP and removing them. Then, it browses all connections with other parts to identify (subi). The output of the subassembly research algorithm (Fig. 3) is a list of subassemblies. For the treated example, the list of the identified subassemblies is represented by Fig. 4.

2.3 Assembly Plan Generation

In this work, the ASP is considered as the inverse of the DSP. For that, DSP is firstly generated.

When the Subi are recognized, the DSP generation process of Subi can be started. The DSP algorithm begins by generating the interference matrices of Subi $[IM_k]$ in the direction of (k) axis, which are automatically extracted from CAD software. The size of $[IM_k]$ is equal to $(m \times m)$ where m represents the numbers of Subi. The $IM_k(i, j)$ element related to the k-axis is as follows:

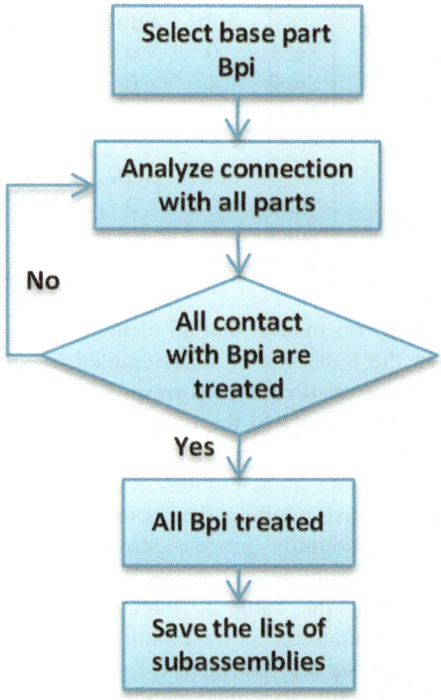

Fig. 3 Subassembly research algorithm

- If the Subi does not interfere with another Subj in the direction of the k-axis, the Subi can be disassembled freely from the Subj in the direction of the k-axis; in this case the value of $IM_k(i, j) = 0$.
- If the Subi that moves in the direction of the k-axis has interference with another Subj, then the Subj that moves in the direction of the k-axis will, in turn, has interference with the Subi; in this case the value of $IM_k(i, j) = 1$.

The $[IM_k]$ in the direction of the (x, y, z) axis of the motorbike mechanism is given as follow.

$$[IMx] = \begin{bmatrix} & \text{Sub1} & \text{Sub2} & \text{Sub3} \\ \text{Sub1} & 0 & 1 & 1 \\ \text{Sub2} & 1 & 0 & 0 \\ \text{Sub3} & 1 & 0 & 0 \end{bmatrix}$$

$$[IMy] = \begin{bmatrix} & Sub1 & Sub2 & Sub3 \\ Sub1 & 0 & 0 & 0 \\ Sub2 & 1 & 0 & 0 \\ Sub3 & 1 & 1 & 0 \end{bmatrix}$$

$$[IMz] = \begin{bmatrix} & Sub1 & Sub2 & Sub3 \\ Sub1 & 0 & 0 & 0 \\ Sub2 & 1 & 0 & 0 \\ Sub3 & 1 & 1 & 0 \end{bmatrix}$$

It is noticed, from the [IMy] matrix, that all elements of the line corresponding to Sub1 are equal to 0 then the Sub1 can be disassembled freely in the direction of the (+y) axis. Similarly, all elements of column corresponding to Sub3 are equal to 0 then the Sub3 can be disassembled freely in the direction of the (−y) axis. The decision, disassembling Sub1 or Sub3, is carried out by comparing the scores of the evaluation function corresponding to the both subassemblies: Sub1 and Sub3. Table 2 presents the score of each subassembly corresponding the motorbike mechanism. In fact, in the industrial practices, the disassembly process starts by dismantling parts or sub assemblies having the smaller volume or surfaces i.e. having the smaller value of the evaluation function. Regarding the Table 2, Sub3 is disassembled firstly because it have the smaller EF. After that, the corresponding line and column to Sub3 are removed from $[IM_k]$. A new square matrix $[IM_k]$ (size $(m - 1)$) is, then, generated. This procedure is repeated until all subassemblies are treated. In conclusion the identified disassembly sequence for the motorbike is

$$Sub3 \rightarrow (-y) \rightarrow Sub2(-y) \rightarrow Sub1(\pm y)$$

Table 2 Score of each subassembly of the motorbike mechanism

Subassembly item	Score
Sub1	10.76
Sub2	6.82
Sub3	3.41

Based on the DSP given in Fig. 5, the ASP is obtained by the inversion of the previous DSP. In the case of the treated example, the ASP is: Sub1 → (±y) → Sub2 (−y) → Sub1 (−y).

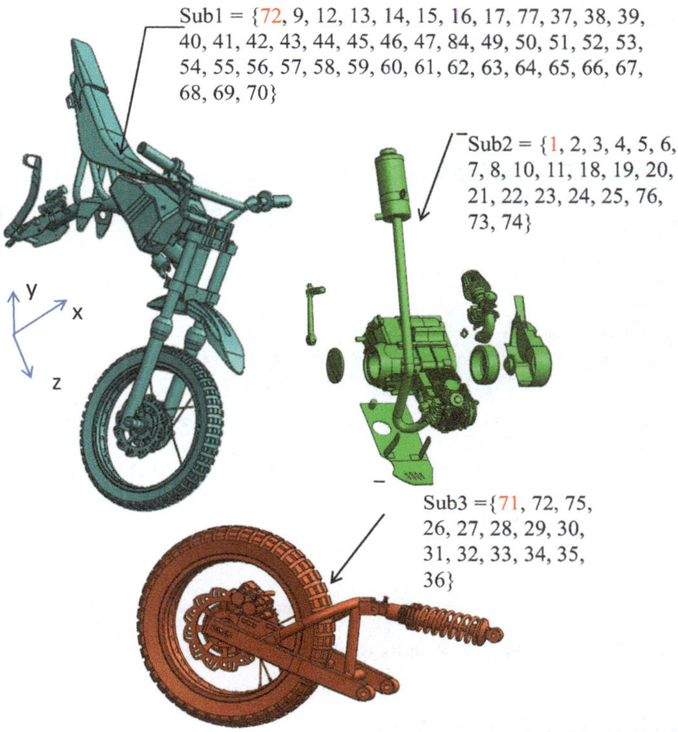

Fig. 4 List of subassemblies of the motorbike mechanism

3 Data Implementation

The data-processing implementation of the developed approach was carried out, using the SolidWorks © CAD system with its API and Matlab © software.

The tool interface is shown in Fig. 5. The results of the running process are presented in four aspects: a list of base parts, a list of subassemblies, a DSP report containing the disassembly plan of subassemblies and the associated senses and directions and the associated ASP report containing the assembly plan of subassemblies and the associated senses and directions. In the other hand a proposed secondary's assembly lines is proposed on order to perform the proposed ASP of subassemblies.

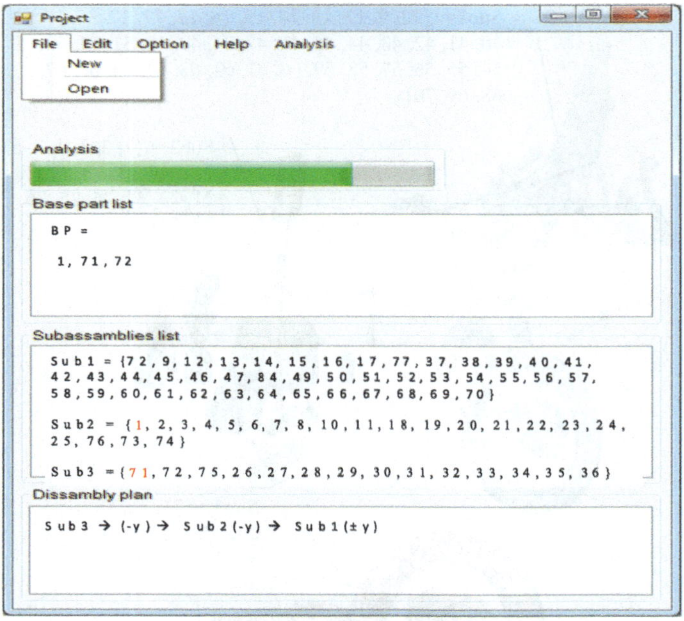

Fig. 5 Results of the running process

4 Conclusion and Future Work

In this paper, an original approach to generate secondary's assembly lines of complex products is proposed. It is based on the identification of subassemblies from a CAD model. The main objective is to elaborate an optimal and feasible ASP based not only on a geometric reasoning but also on the modality to assemble the industrial product i.e. secondary and main assembly lines.

The developed approach starts by generating all smart CAD data useful to execute the proposed approach. Then, the identification of all subassemblies with respect to the assembly' constraints is given based on the list of the BP. Using an interference matrices, along the three directions (x, y, z), of the identified subassemblies, a simplification method was performed in order to generate the DSP of subassemblies then the associated ASP. Consequently, a proposed assembly line is performed in order to optimally execute the proposed ASP. As a result of the implemented approach, several points can be conducted:

- The developed approach is in a permanent connection with CAD system. In a real time, it sends and receives data related to the treated CAD model of a mechanism (contact matrices, interference matrices, part attributes, subassembly groups, etc.).
- The ASP research is greatly facilitated by using the subassembly concept.
- All identified subassemblies and the sequences found in ASP are feasible in the context of industrial practices.
- The secondary's assembly lines are optimized due to the identified subassemblies.

– The generation of secondary's assembly lines is automated based on the identified subassemblies.

In the future work, workstations will be defined and organized based on the identified subassemblies in order to optimally execute the assembly plan.

References

1. Raju JA (2003) Conceptual design and cost optimisation methodology. In: 44th AIAA/ASME/AHS structures, structural dynamics and materials conference, Norfolk, Virginia; April 2003
2. Kucukkoc I, Zhang DZ (2015) Balancing of parallel U-shaped assembly lines. Comput Oper Res. https://doi.org/10.1016/j.cor.2015.05.014
3. Ozdemir RG, Ayag Z (2000) An integrated approach to evaluating assembly-line design alternatives with equipment selection. Prod Plan Control 22(2):194–206
4. Wirtorsson M, Andersson T, Broman M (2000) A note on the specification of assembly systems. Int J Prod Res 38:3999–4004
5. Khan AD (2002) A knowledge based design methodology for manufacturing assembly lines. Comput Ind Eng 41:441–467
6. Barnes CJ (2000) A methodology for the concurrent design of products and their assembly sequences. PhD Thesis. Cranfield University, UK
7. Ullah S, Zailin G, Li Z, Baoxi W, Jahanzaib M (2017) Multi-objective artificial bee colony algorithm for order oriented simultaneous sequencing and balancing of multi-mixed model assembly line. J Intell Manuf. https://doi.org/10.1007/s10845-017-1316-4
8. Trigui M, BenHadj R, Aifaoui N (2015) An interoperability CAD assembly sequence plan approach. Int J Adv Manuf Technol 79:1465–1476
9. Issaoui L, Aifaoui N, Benamara A (2015) Solution space reduction of disassembly sequences generated automatically via computer aids. Proc Inst Mech Eng Part C J Mech Eng Sci 22–16:2977–2986
10. Kheder M, Trigui M, Aifaoui N (2017) Optimization of disassembly sequence planning for preventive maintenance. Int J Adv Manuf Technol. https://doi.org/10.1007/s00170-016-9434-2
11. Kheder M, Trigui M, Aifaoui N (2016) Disassembly sequence planning based on genetic algorithm. Mech Eng Sci 229(12):2281–2290
12. Belhadj I, Trigui M, Benamara A (2016) Subassembly generation algorithm from a CAD model. Int J Adv Manuf Technol 87:2829–2840
13. Su Q, Smith SF (2003) An integrated framework for assembly oriented product design and optimization. J Ind Technol 19(2)
14. Barnes CJ, Jared GEM, Swift KG (2004) Decision support for sequence generation in an assembly oriented design environment. Robot Comput-Integr Manuf 20:289–300
15. Chen G, Zhou J, Cai WC, Lai X, Lin Z, Menassa R (2006) A framework for an automotive body assembly process design system. Comput Aided Des 38:531–539

CAD Tolerancing Integration: A Tool for Optimal Tolerance Allocation

Maroua Ghali(✉), Mehdi Tlija, and Nizar Aifaoui

Laboratory of Mechanical Engineering, National School of Engineers of Monastir (LGM_ENIM), University of Monastir, 5 Av. Ibn Eljazzar, 5019 Monastir, Tunisia
maroua.ghali3@gmail.com, tlija.mehdi@gmail.com, nizar.aifaoui@gmail.com

Abstract. Tolerance allocation approaches serve as effective tools for design engineers to reduce the total manufacturing cost of mechanisms as well as to improve the product quality. In every mechanical design, the major task of design engineer is to allocate tolerances and clearances to the studied dimensions and joints, respectively, in a mechanism assembly. This paper presents an optimum tolerance allocation tool, based on manufacturing difficulty quantification using tools for the study and analysis of reliability of the design or the process, as the Failure Mode, Effects and Criticality Analysis (FMECA) and Ishikawa diagram. The proposed method is performed to produce, economically and accurately, allocated tolerances according to difficulty requirements. For this, an integrated CAD model is developed using Graphical User Interface (GUI) in MATLAB to expose diverse tolerance allocation approaches that respect the functional and manufacturing requirement. Many examples can be executed using the established GUI in order to highlight the advantages of the proposed approach.

Keywords: Tolerance allocation · Manufacturing difficulty · CAD tool · Quality technique

1 Introduction

Tolerancing have become the focus of improved activity as well as manufacturing industries strive to raise productivity and increase the quality of their products. As a matter of fact, Fig. 1 demonstrates that the tolerancing effects are far-reaching. Consequently, the tolerances affect not only the capability to assemble the final product, but also the production cost, process selection, tooling, setup cost, operator skills, inspection and gaging, and scrap and rework. Besides, tolerances directly touch engineering performance and design robustness. Products that are characterized by poor performance, i.e. lesser quality and excess cost, will eventually lose out in the marketplace [1]. In addition, Fig. 1 proves that tolerancing brings the functional requirements and production ability together in a well-understood CAD tolerance model.

N. Aifaoui et al. (Eds.): CMSM 2019, LNME, pp. 12–22, 2020.
https://doi.org/10.1007/978-3-030-27146-6_2

1.1 Literature Review

Optimal tolerance allocation approach is a trade-off between functional and quality requirements and manufacturing cost. In fact, various manufacturing cost-tolerance models have been proposed as in Dong et al. [2].

- **Overview on optimal tolerance allocation approach**

Various optimization methods are established to optimal tolerance allocation as developed in [3, 5]. The genetic algorithm, colony algorithm, teaching-learning-based optimization algorithm, particle swarm optimization and bat algorithm are used in many works [6–13] in order to optimize tolerance allocation. In this context and based on the investigation of fuzzy factors, various methods have been exposed in many literatures [14–16]. Liu et al. [17] presented a method of tolerance grading allocation based on the uncertainty analysis of the remanufacturing assembly.

- **Overview on dimension transfer approach**

Regarding dimension transfer, two methods are generally used to determine Manufacturing Dimensions (MD): Wade [18] and Bourdet [19–21] methods. Comparison studies of these methods are established in [22–24]. More details are given in Ghali et al. [25].

1.2 Synthesis and Research Objectives

The major disadvantages of Traditional dimension Transfer (TT) and the tolerance allocation approaches are the following: The classical dimension transfer methods are performed by double transfer involving tolerance reduction and without software assistance in industrial practice. The implementation of the failure mode and effects analysis concept is neglected in order to quantify the difficulty of manufacturing operation and to optimize tolerance allocation. The traditional approaches of dimension transfer and tolerance allocation do not improve the concurrent engineering environment. Based on the advantages and inconveniences of the above methods, this paper proposes a new approach that led to direct transfer of Functional Requirement (FR) to MDs without using Part Dimensions (PD)s. Besides, the tolerance allocation is achieved considering difficulty coefficient β The Difficulty Coefficient Computation (DCC) is established based on FMECA tool and Ishikawa diagram. Therefore, the originality of the proposed approach is the coupling between the Unique Transfer (UT) and DCC in the tolerance allocation to enhance the concurrent engineering environment (Fig. 1).

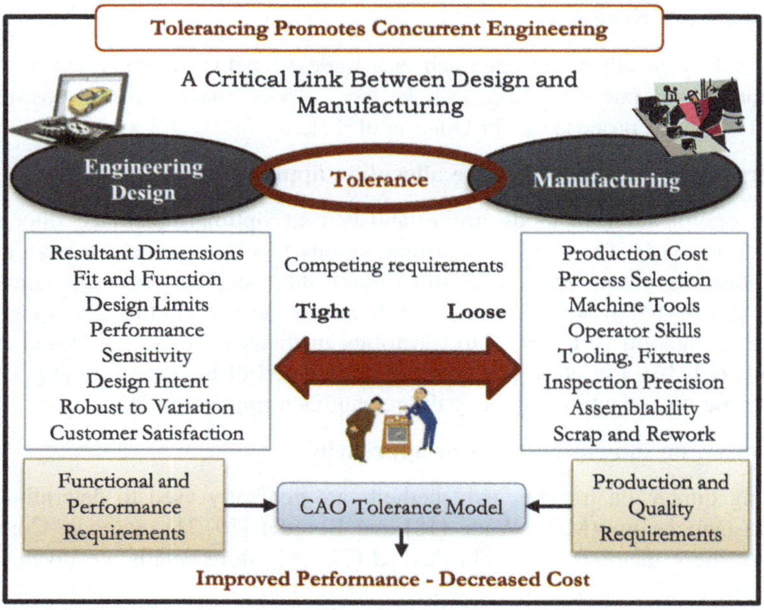

Fig. 1 Tolerancing impact towards concurrent engineering

2 Proposed Approach

This work proposes a coupling between unique transfer methodology avoiding the determination of PDs [25] and DCC technique [26]. Therefore, the proposed approach allows directly the transfer of FR into MD and integrates subsequently the β coefficients in tolerance allocation. The Fig. 2 elucidates the flowchart of the whole proposed approach. The main steps of DCC are the flowing: Determine the difficult Manufacturing Operation (MO); Establish the Ishikawa diagram, Calculate the Risk Priority Number (RPN), Compute the difficulty coefficient β. Fill the proposed FMECA Worksheet as established in Ghali et al. [26]. The dimension tolerances t_i are computed as shown in Table 1 according to Worst Case (WC) and Root Sum Square (RSS) methods, where, α is the influence coefficient and β is the difficulty coefficient.

Table 1 Tolerance formulas

Approaches	Formulas
WC	$t_i = \beta_i \times \frac{t_{FR}}{\sum_i \lvert\alpha_i\rvert \times \beta_i}$
RSS	$t_i = \beta_i \times \frac{t_{FR}}{\sqrt{\sum_i \alpha_i^2 \times \beta_i^2}}$

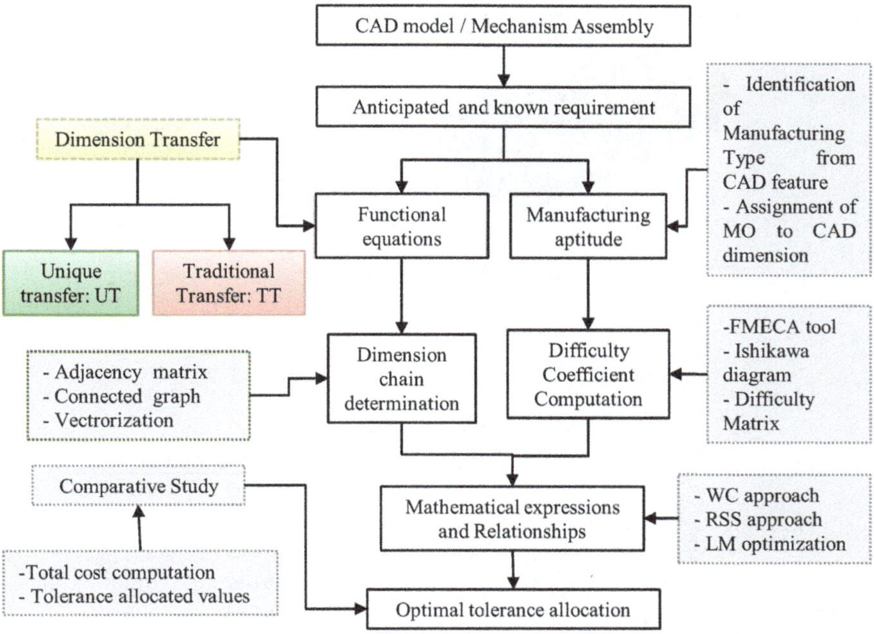

Fig. 2 Flowchart of the whole proposed approach

3 Case Study

3.1 Studied Example

In this paper, the rotor key base assembly is the applied case study. The mechanism is chosen to use the tolerance cost model proposed by Sampath et al. [28, 29]. The rotor key base is composed by two parts a and b as shown in Fig. 3. The FR is between the faces a3 and b5: FR = a3b5. A tolerance of 1.016 mm is required: $t_{a3b5} = 1.016$ mm. The tolerance cost model is given in Table 2. The Total Manufacturing cost (C_{Tm}) is given in Eq. 1. Thus, the Quality Loss (QL) [27] is given as Eq. 2. The summation of C_{Tm} and QL give the Total Cost C_T.

Table 2 Cost model parameters

Cost model	t_{a13}	t_{a15}	t_{a25}	t_{b12}	t_{b25}
C_0	27.84	431.5	431.5	27.84	66.43
C_1	3.661	17.64	17.64	3.661	2.738

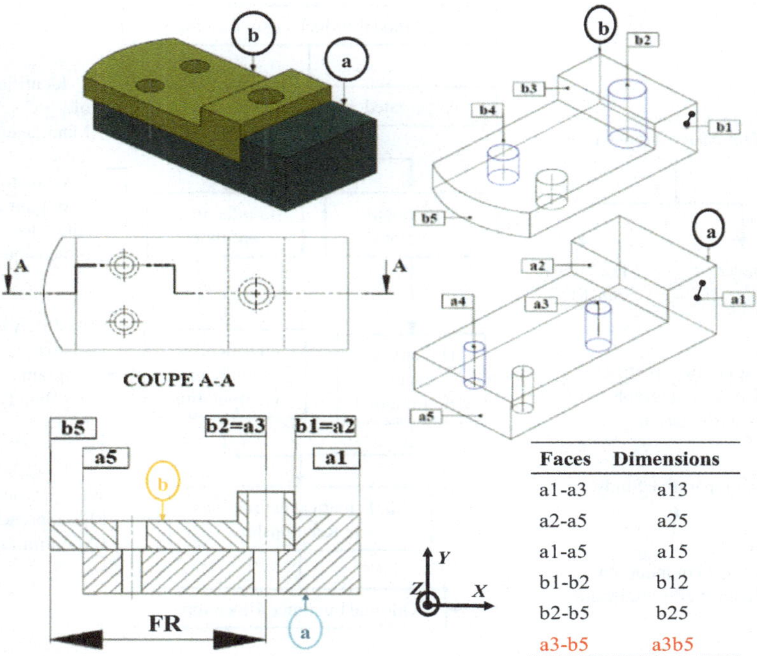

Faces	Dimensions
a1-a3	a13
a2-a5	a25
a1-a5	a15
b1-b2	b12
b2-b5	b25
a3-b5	a3b5

Fig. 3 Rotor Key base assembly

$$C_{Tm} = \beta_{a13}C_{13}(t_{a13}) + \beta_{a15}C_{15}(t_{a15}) + \beta_{a25}C_{25}(t_{a25}) + \beta_{b12}C_{12}(t_{b12}) + \beta_{a25}C_{25}(t_{b25}) \quad (1)$$

$$QL = \frac{A}{36t_{a3b5}^2}\left(t_{a13}^2 + t_{a15}^2 + t_{a25}^2 + t_{b12}^2 + t_{b25}^2\right) \quad (2)$$

3.2 Implementation of Proposed Model and Compared Approaches

The Fig. 4 presents the welcome user interface and the steps of the proposed tool. The choice of example is performed using the popup menu indicated in Fig. 4.

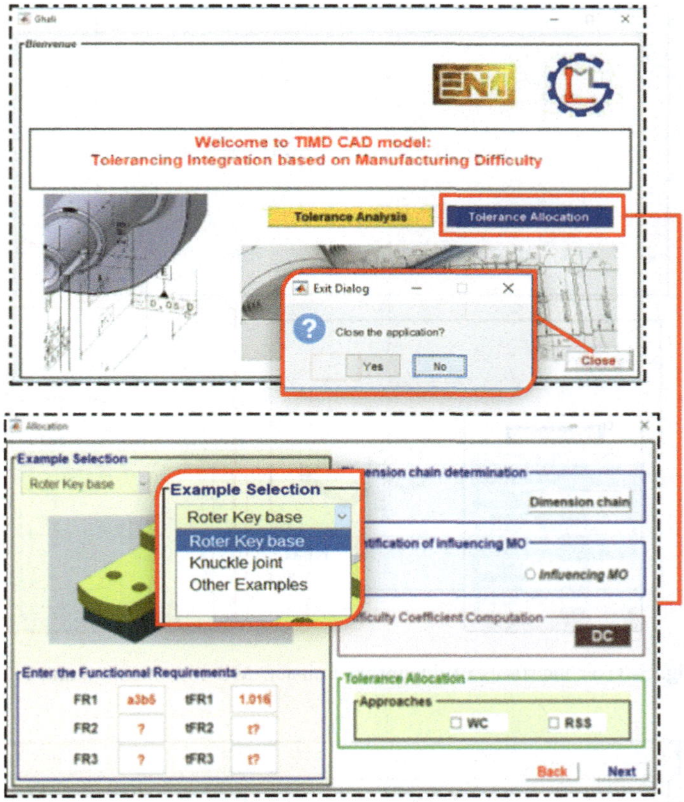

Fig. 4 Welcome and steps user interfaces of the proposed tool

Moreover, a click on the button 'Next' leads to open the user interface of next steps as dimension transfer, MO identification and DCC (Fig. 5).

The user interface of Fig. 6 leads to achieve tolerance allocation according to four tolerance allocation alternatives, which are the compared approaches (1-TT using Uniform Allocation (TTUA), 2-UT using UA (UTUA), 3-TT using DC (TTDC) and 4-UT using DCC (UTDC)). In addition, a click on the radio-button 'Total cost' induces C_T of each alternative. Warning box are done to select the choice of tolerance allocation approaches and to enter the cost model (Fig. 6).

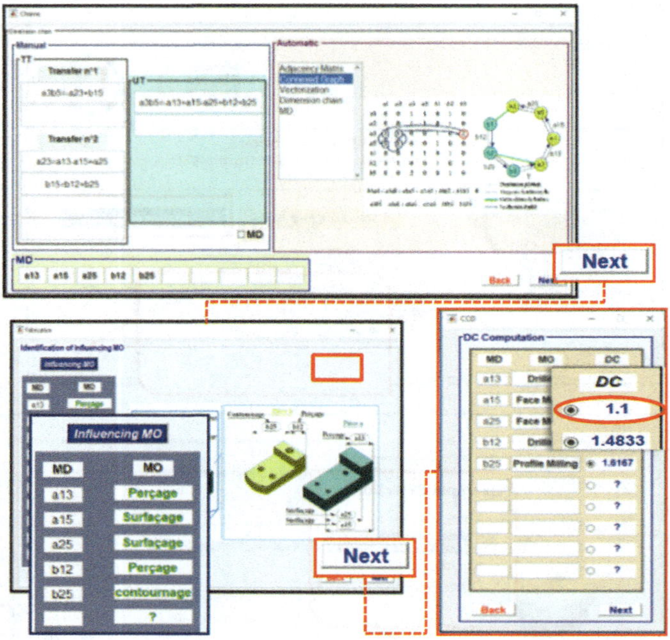

Fig. 5 User interfaces of dimension transfer, MO identification and DCC

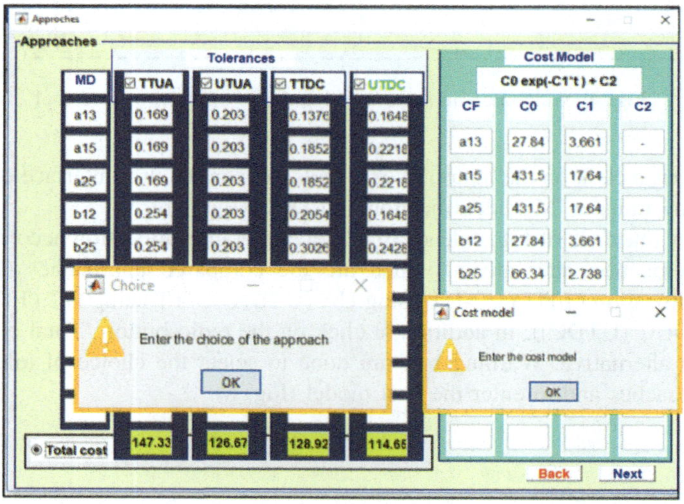

Fig. 6 User interface of tolerance allocation approaches

4 Results and Discussion

4.1 Transfer and Tolerance Results

The Transfer diagrams of double transfer (TT) and UT as well as, tolerance relationships are developed in Ghali et al. [25]. The allocated tolerance according to TT and UT approaches using UA are resumed in Table 4. In addition, the MOs affecting MDs are presented in Table 3.

Table 3 MDs tolerances of TT and UT and associated influencing MO

$t_{\cdot MD}$	TT	UT	MO
t_{a13}	0.169	0.203	Drilling
t_{a15}	0.169	0.203	Face milling
t_{a25}	0.169	0.203	Face milling
t_{b12}	0.254	0.203	Drilling
t_{b25}	0.254	0.203	Turning

4.2 DCC and Tolerance Results

The β values are computed after achieving DCC procedure steps and completing FMECA worksheet of MO affecting MD [26]. The β values are obtained as elucidated in Table 4. Whence, the tolerance results according to compared approaches are resumed in Table 4. Based on allocated tolerance analysis, the proposed UTDC approach induces the most suitable tolerances by broadening tolerance of difficult MDs. For example a15, which has $\beta_{a15} = 1.48$, is more difficult than a13 which has $\beta_{a13} = 1.10$. Thus, the new obtained t_{a15} is upper than t_{a13} ($t_{a15} = 0.222$ mm > $t_{a13} = 0.165$ mm) as demonstrated in the Table 4. This fact guarantees consequently optimal quality and cost. In this respect, the proposed UTDC gives tolerance fluctuation perfectly proportional to β variation.

Table 4 β values and allocated tolerances according to compared approaches

MD	β notation	β values	Tolerances (mm)			
			TTAU	UTAU	TTDC	UTDC
a13	β_{a13}	1.10	0.169	0.203	0.138	0.165
a15	β_{a15}	1.48	0.169	0.203	0.185	0.222
a25	β_{a25}	1.48	0.169	0.203	0.185	0.222
b12	β_{b12}	1.10	0.254	0.203	0.205	0.165
b25	β_{b25}	1.62	0.254	0.203	0.303	0.243

4.3 Cost Results

- **Cost comparison of compared approach**

The Table 5 summarizes the tolerance total cost according to TTUA, UTUA, TTDC and the proposed UTDC approaches.

Table 5 Comparison of tolerance total cost

Approach	TTUA	UTUA	TTDC	UTDC
C_T (€)	147.410	126.763	129.004	114.741

Based on the analysis of Table 5, the UTDC is the most economical and efficient. Indeed, the proposed UTDC promotes a Monetary Gain (MG) per assembly:

- MG = (147.410 – 114.741) × 100/147.410 = 22.162% compared to TTUA,
- MG = (126.763 – 114.741) × 100/126.763 = 9.484% compared to UTUA,
- MG = (129.004 – 114.741) × 100/129.004 = 11.056% compared to TTDC.

- **Comparison with the author** [28]

Figure 7 presents comparison of allocated tolerance and total cost with Sampath et al. [28]. The total cost computation proves that the proposed UTDC approach is more economical than the author's tolerance allocation approach. In fact, a gain of 43, 46% is provided by UTDC per assembly due to combine UT and DCC simultaneous.

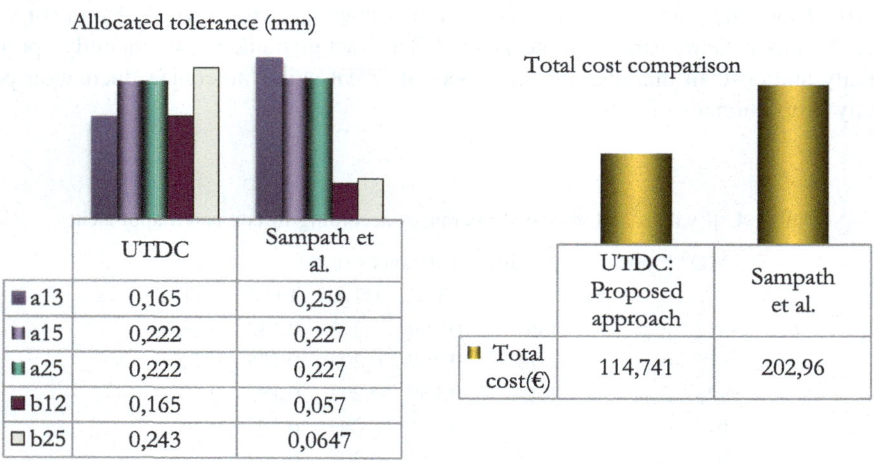

Fig. 7 Comparison of allocated tolerance and total cost with Sampath et al. [28]

Therefore, from the obtained results analysis and based on the established comparisons, the proposed approach creates an economical cost achievement and guarantees privileges to concurrent engineering environment by coupling of UT and DCC approaches. Whence, a common meeting ground is established where design and manufacturing can interact and appraise the effects of their requirements. Thus, such an efficient tolerancing approach promotes concurrent engineering and offers a tool for improving performance and decreasing cost in an early stage of development product cycle.

5 Conclusion

This paper proposes a new tool for tolerance allocation and dimension transfer based on DCC and UT approaches simultaneous. The proposed UTDC lead to quantify manufacturing dimension difficulty and minimize the tolerance cost. This fact promotes accordingly the concurrent engineering environment. Thus, the proposed tolerance allocation tool is both economical and successful. Future works will focus on optimization algorithms and geometrical tolerances. In addition, the implementation of the proposed approach in manufactories is also among the desired outlooks.

References

1. Chase KW, Magleby SP, Glancy CG (1998) A comprehensive system for computer-aided tolerance analysis of 2-D and 3-D mechanical assemblies. In: ElMaraghy HA (ed) Geometric design tolerancing: theories, standards and applications
2. Dong Z, Hu W, Xue D (1994) New production cost tolerance models for tolerance synthesis. J Eng Ind Trans ASME 116(2):199–206
3. Zong Y, Mao J (2015) Tolerance optimization design based on the manufacturing costs of assembly quality. Proc CIRP 27:324–329
4. Ramesh Kumar L, Padmanaban KP, Balamurugan C (2016) Least cost tolerance allocation based on Lagrange multiplier. Concur Eng Res Appl 1–14
5. Rao RV, More KC (2014) Advanced optimal tolerance design of machine elements using teaching-learning-based optimization algorithm. Prod Manuf Res 2(1):71–94
6. Muthu P, Dhanalakshmi V, Sankaranarayanasamy K (2009) Optimal tolerance design of assembly for minimum quality loss and manufacturing cost using metaheuristic algorithms. Int J Adv Manuf Technol 44(11):1154–1164
7. Huang MF, Zhong YR, Xu ZG (2005) Concurrent process tolerance design based on minimum product manufacturing cost and quality loss. Int J Adv Manuf Technol 25(7–8):714–722
8. Lu C, Zhao WH, Yu SJ (2012) Concurrent tolerance design for manufacture and assembly with a game theoretic approach. Int J Adv Manuf Technol 62(1):303–316
9. Cheng KM, Tsai JC (2011) Optimal statistical tolerance allocation of assemblies for minimum manufacturing cost. Appl Mech Mater 52–54:1818–1823
10. Cheng KM, Tsai JC (2013) Optimal statistical tolerance allocation for reciprocal exponential cost-tolerance function. Proc Inst Mech Eng Part B J Eng Manuf 227(5):650–656

11. Prabhaharan G, Asokan P, Rajendran S (2005) Sensitivity-based conceptual design and tolerance allocation using the continuous ants colony algorithm (CACO). Int J Adv Manuf Technol 25(5):516–526
12. Sivakumar K, Balamurugan C, Ramabalan S (2010) Concurrent multi-objective tolerance allocation of mechanical assemblies considering alternative manufacturing process selection. Int J Adv Manuf Technol 53(5–8):711–732
13. Ramesh L, Padmanaban KP, Ganesh Kumar S, Balamurugan C (2016) Design and optimization of concurrent tolerance in mechanical assemblies using bat algorithm. J Mech Sci Technol 30(6):2601–2614
14. Ji S, Li X, Ma Y, Cai H (2000) Optimal tolerance allocation based on fuzzy comprehensive evaluation and genetic algorithm. Int J Adv Manuf Technol 16:461–468
15. Kumar A, Choi SK, Goksel L (2011) Tolerance allocation of assemblies using fuzzy comprehensive evaluation and decision support process. Int J Adv Manuf Technol 55:379–391
16. Wang Y, Zhai WJ, Yang LP, Wu WG, Ji SP, Ma YL (2007) Study on the tolerance allocation optimization by fuzzy-set weight-center evaluation method. Int J Adv Manuf Technol 33:317–322
17. Mingzhou Liu, Liu Conghu, Xing Lingling, Mei Fadong, Zhang Xi (2016) Study on a tolerance grading allocation method under uncertainty and quality oriented for remanufactured parts. Int J Adv Manuf Technol 87:1265–1272
18. Wade OR (1967) Tolerance control in design and manufacturing. Industrial Press, New York
19. Bourdet P (1975) Chaînes de cotes de fabrication (méthode des delta L): deuxième partie, le mode opératoire. L'Ingénieur et Technicien de l'Enseignement Technique 191:15–23
20. Bourdet P, Schneider F (2007) Détermination des chaînes de cotes unidirectionnelles et quatification des tolérances. Spécification géométrique des produits, cotation et tolérancement ISO. Dunod, Paris, pp 269–290
21. Fainguelernt D, Weil R, Bourdet P (1986) Computer aided tolerancing and dimensioning in process planning. Ann CIRP Manuf Technol 35(1):381–386
22. Lehtihet EA, Ranade S, Dewan P (2000) Comparative evaluation of tolerance control chart models. Int J Prod Res 38(7):1539–1556
23. Pairel E, Goldschmidt E, Adragna PA et al (2011) The pilot dimensions method: reconciling steering and conformity in workshops. Int J Prod Res 49(19):5943–5956
24. Goldschmidt E (2009) Gammes et cotation pour le réglage des machines-outils de décolletage. Thesis (PhD). University of Savoy, France
25. Ghali M, Tlija M, Pairel E, Aifaoui N (2018) Unique transfer of functional requirements into manufacturing dimensions in an interactive design context. Int J Interact Des Manuf (IJIDeM). https://doi.org/10.1007/s12008-018-0472-x
26. Maroua Ghali, Mehdi Tlija, Nizar Aifaoui, Eric Pairel (2017) A CAD method for tolerance allocation considering manufacturing difficulty based on FMECA tool. Int J Adv Manuf Technol 91:2435–2446
27. Taguchi G et al (2005) Taguchi's quality engineering handbook quality loss function. Wiley, Part III, pp 171–198
28. Sampath KR, Alagumurth N, Ramesh R (2009) Calculation of total cost, tolerance based on Taguchi's, Asymmetric quality loss function approach. Am J Eng Appl Sci 2(4): 628–634
29. Sampath KR, Alagumurth N, Ramesh R (2010) Integrated optimization of machining tolerance and Asymmetric quality loss cost for Rotor key base assembly. In: International conference on recent advances in Mechanical Engineering (ICRAME2010), at Noorulislam University, Kanyakumari-629180, India

A Computer Aided Tolerancing (CAT) Tool of Non-rigid Cylindrical Parts Assemblies

Anis Korbi[1(✉)], Mehdi Tlija[1], Borhen Louhichi[2], and Abdelmajid BenAmara[1]

[1] Mechanical Engineering Laboratory (LGM), National Engineering School of Monastir (ENIM), University of Monastir, Monastir, Tunisia
anis.Korbii@gmail.com, tlija.mehdi@gmail.com, benamara.enim@gmail.com

[2] Mechanical Engineering Laboratory (LMS), National Engineering School of Sousse (ENISo), University of Sousse, Sousse, Tunisia
louhichi.ca@gmail.com

Abstract. The consideration of manufacturing defects in CAD tools is an important goal of industrials. Several studies have been proposed to take into account the geometrical and dimensional tolerances on the CAD modeler. However, the assumption of the rigid body is used and the deformations of non-rigid parts are neglected. In this context, the aim of this paper is to present a new Computer Aided Tolerancing (CAT) tool for cylindrical parts assemblies by taking into account both of geometrical and dimensional defects as well as deformations of components already generated during the system operation. The worst case tolerancing and the Small Displacements Torsor (SDT) are applied to model parts with dimensional and geometrical tolerances. The deformations of non-rigid cylindrical components are determined basing on the Finite Elements (FE) simulation. The realistic assembly is obtained by the update of mating constraints between couples of Rigid (R) and Non-Rigid (NR) realistic parts.

Keywords: CAD · Geometrical and dimensional tolerances · Cylindrical parts · Deformations · FE · Mating constraints

1 Introduction

The realistic modeling of mechanical components with manufacturing defects, in the Digital Mock-up (DMU), is one of the major interests of manufacturers in mechanics. These defects are of two categories: dimensional and geometrical defects already resulted from the machining stage as well as the deformations of non-rigid parts generated during the product operation. In this regard, this paper shows a new CAT tool for the tolerance analysis of non-rigid cylindrical parts assemblies by taking into account both of the above defects. This paper is organized as follows. Section 2 presents a literature review of the tolerance analysis methods approaches of rigid and non-rigid parts assemblies. The main steps of the novel tolerancing CAT tool are presented in Sect. 3. Sub-algorithms to update mating constraints between couples of Rigid/Rigid (R/R), Rigid/Non-Rigid (R/NR) cylindrical parts are illustrated.

N. Aifaoui et al. (Eds.): CMSM 2019, LNME, pp. 23–32, 2020.
https://doi.org/10.1007/978-3-030-27146-6_3

The Functional Condition (FR) of the realistic CAD assembly is controlled in the end. The conclusion and perspectives of this work are highlighted in Sect. 4.

2 State of the Art

2.1 Tolerance Analysis Approaches of Rigid and Non-rigid Parts Assemblies

Several methods are developed for the modeling of the toleranced geometry [2, 13] and are assisted others researches to the establishment of tolerancing approaches to consider the functional aspect of the product and the tolerances stuck-up [11]. Despite their advantages, the contact evolution between the assembly components and the assembly sequence (AS) are not taken into account and during the product operation. In this regard, some researchers such as Louhichi et al. [8] and Ansemetti et al. [1] contributed to the development of tolerance analysis methods considering the above aspects. In the above tolerancing approaches, the hypothesis of the rigid part is just considered and the deformations of components subjected to external and internal loads are neglected.

To overcome the limitations of the above tools, other research works are proposed in the literature and allow the consideration especially of the deformations of non-rigid parts in the tolerance analysis and synthesis process. Camelio et al. established the Method of Influence Coefficients (MIC) for multi-station compliant sheet metal assembly lines while taking into account geometrical variations into components, tools and fixture [3]. Mazur et al. proposed a CAD tool called Process Integration and Design Optimization platform (PIDO) for the tolerance analysis and synthesis of assemblies under external and internal loads [9]. Hermansson et al. analyzed the geometrical deviations of thin cables and hoses, by the computation of the optimal tolerance envelopes for each part [4]. Söderberg et al. developed a method to study the impact of the spot welding position variation on the final quality of sheet metal assemblies [12]. Pierre et al. considered thermos-mechanical strains in the strategy of tolerance analysis using the FE computation [10]. Korbi et al. established a CAD tool for the tolerance analysis of mechanical assembly considering both of orientation and positional defects as well as deformations of non-rigid planar parts [7].

3 The Main Steps of the Developed Tolerance Analysis Model

The developed method uses several engineering tools such as CAD, the tolerancing and the FE simulation. Thus, this model is a new CAT tool allowing the tolerance analysis of rigid and non-rigid parts assembly in the DMU. The flow chart in (Fig. 1) shows the main steps of the developed CAT model, as following:

- Extraction of CAD data (Inputs of the model).
- Modeling of rigid and non-rigid parts with defects.
- Determination of the realistic CAD assembly configurations.
- Control the Functional condition (FR) of the realistic assembly.

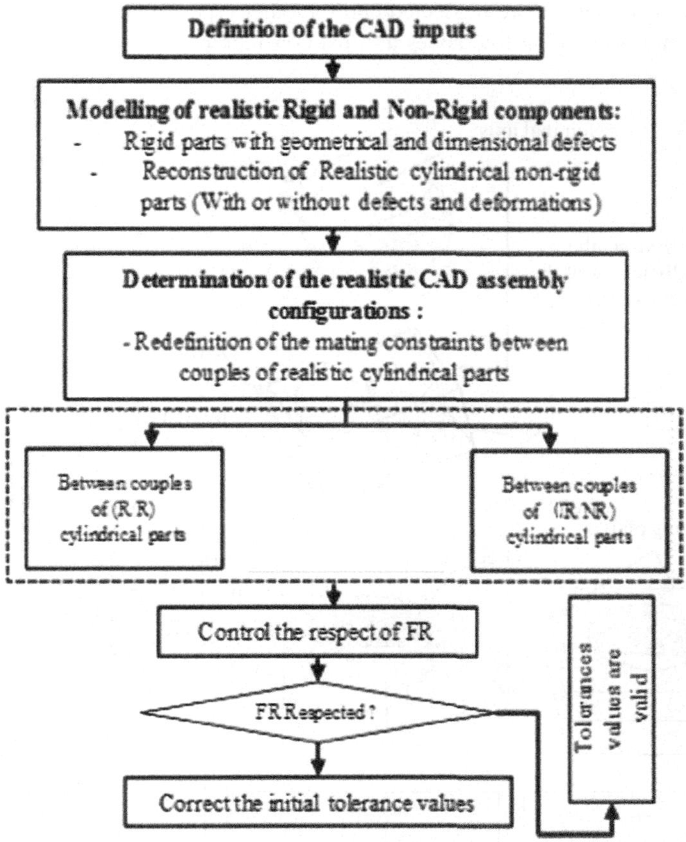

Fig. 1 Flow chart of the developed model

3.1 Extraction of CAD Data

The inputs data of the developed CAT tool are determined directly from the 3D nominal assembly: The Assembly Sequence (AS) is identified from mates already specified in the tree feature manager of the CAD software. The tolerances type, datum, value and toleranced features as well as FR are deduced from the assembly and components models.

3.2 Modeling of Realistic Cylindrical Components with Defects

In this sub-section, the methods developed in [8] are exploited to model the rigid components with dimensional and geometrical tolerances. The dimensional deviations of parts are taken into account and two configurations are considered: the maximum and the minimum material limits. To model the assembly components with geometrical defects, the worst case tolerancing concept and SDT parameters are used. As assumption, form defects are neglected and the other types of geometrical defects such as position and orientation tolerances are just considered.

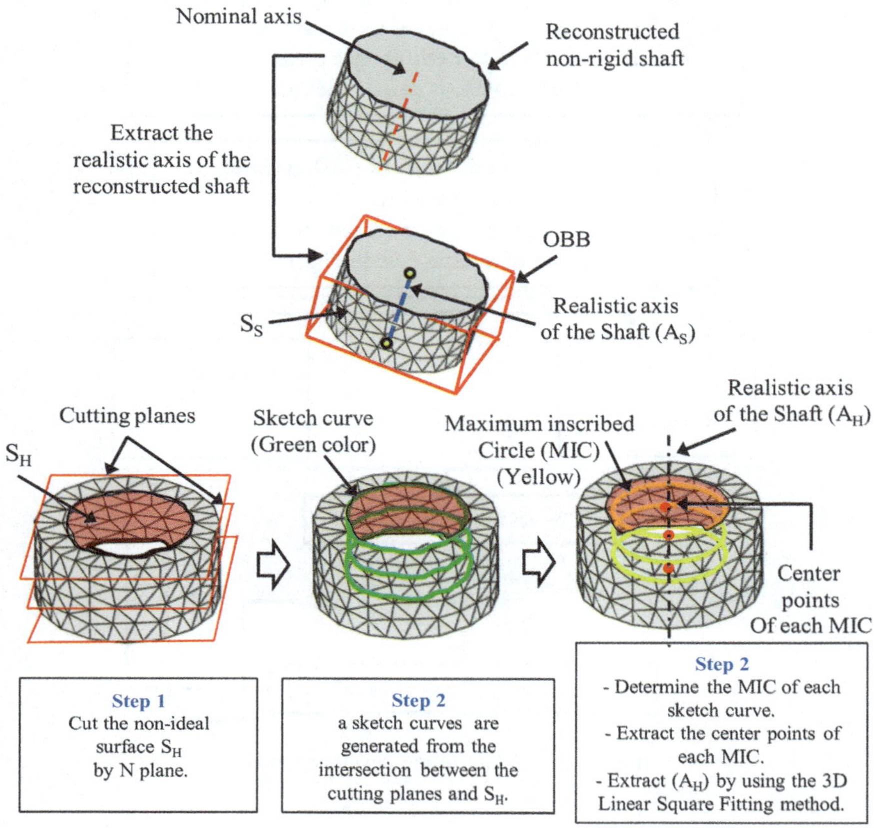

Fig. 2 The determination of the realistic axes of non-rigid cylindrical surfaces S_S and S_H of a Shaft and a Hole

Each non-rigid component with geometrical and dimensional defects is isolated from the 3D assembly into process for the FE calculation. As result, the mesh nodes and the nodal displacements of each face of non-rigid part already subjected to loads are extracted. After that, the points cloud of the above deformed faces of non-rigid components are triangulated using the Digitized Shape Editor (DSE) of CATIA V5 and reference surfaces are created to obtain the final configuration of the deformed non-rigid components with defects. In order to construct the realistic axis (A_S) of a realistic cylindrical surface (S_S) of the shaft (S) (Fig. 2), the Oriented Bounding Box (OBB) is used [6]. Another method based on the feature operation principle of ISO–GPS and the 3D regression line algorithm is used to obtain the realistic axis (A_H) of a realistic non-rigid cylindrical surface (S_H) of a hole (H). More precisely, a Matlab Code is applied to determine the MIC of each sketch curve.

After that, the 3D square fitting algorithm already developed by Jacquelin [5] is used to obtain A_H from the center points of each MIC.

3.3 Modeling of the CAD Assembly with Realistic Parts

To obtain all the realistic configurations of the CAD assembly, the mating constraints between couples of Rigid/Rigid (R/R) and Rigid/Non-Rigid (R/NR) parts should be updated based on an Objective Function of the Assembly (OFA). The OFA is deduced from the initial nominal assembly and contains the following conditions:

- The joints types between the assembly components are defined for each assembly sequence.
- The assembly order is identified from the initial mating constraints between parts.
- The interferences between parts should not be detected.
- After updating the mating constraints, the initial contact between couples of rigid and non-rigid parts should be conserved.
- For each assembly sequence, the first part according to the AS order, is considered as fixed component and the second is movable.

Sub-algorithms based on the API of Solidworks are used to update the assembly mating constraints. In the case of couples of rigid planar parts, the algorithms developed by Louhichi et al. [8] are applied.

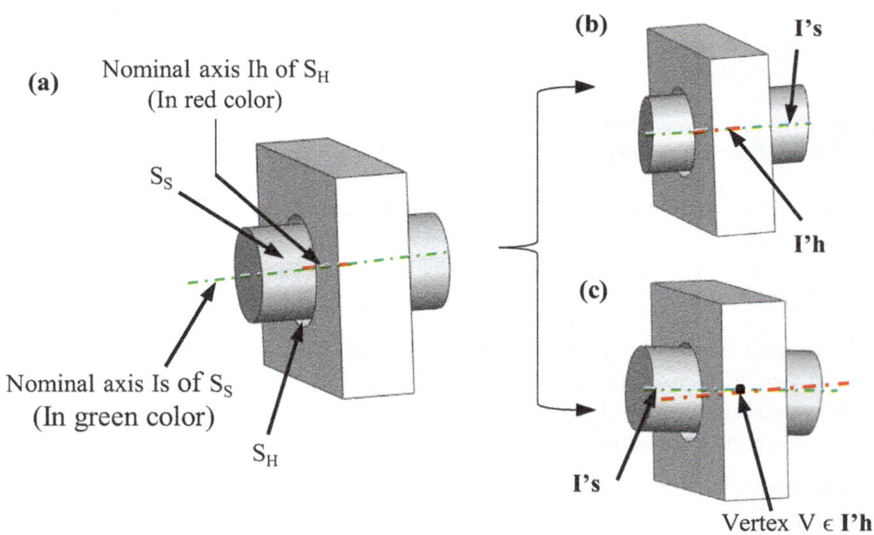

Fig. 3 **a** The coincident constraint between the nominal axes Ih and Is; **b** The coincident constraint between two realistic axes I's and I'h of the Shaft and the Hole surfaces respectively; **c** The coincident constraint between the realistic axis I's and a vertex of I'h

In the case of (R/R) cylindrical joint (Fig. 3a), the initial coincident constraint (Co: Is & Ih) between two axes Is and Ih of a Shaft surface (S_S) and a Hole surface (S_H) respectively, is replaced in the realistic configurations of parts by one of the following constraints according to the OFA:

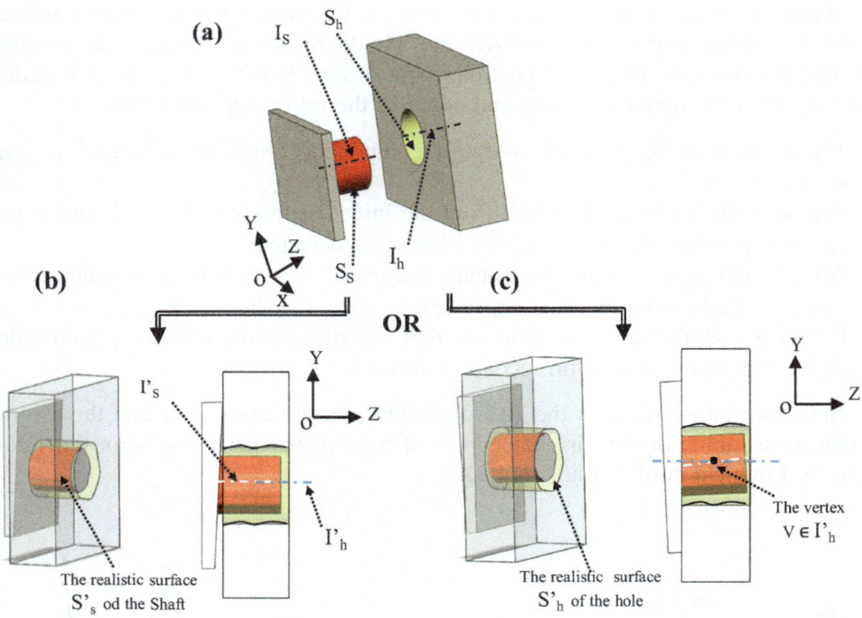

Fig. 4 **a** The initial coincident constraint (Co: Is & Ih); **b** The realistic coincident constraint (Co: I's & I'h); **c** The realistic coincident constraint (Co: I's & Vertex V ϵ I'h)

- Coincident constraint (Co: I's & I'h) between the realistic axes I's and I'h of the cylindrical shaft and hole surfaces respectively (Fig. 3b)
- Coincident constraint (Co: I's & Vertex of I'h) between I's and a vertex of I'h (Fig. 3c).

In the case of (R/NR) cylindrical joint, the initial coincident constraint (Co: Is & Ih) between two ideal axes Is and Ih of cylindrical shaft and hole surfaces S_S and S_H respectively (Fig. 4a), is substituted in the realistic configuration of parts by one of the following mating constraints according to the OFA:

- Coincident constraint (Co: I's & I'h) between the realistic axes I's and I'h of the cylindrical shaft and hole surfaces S'_S and S'_H respectively (Fig. 4b). The realistic axis I's is determined based on the OBB tool (as described in the Sect. 3.2) if S'_S is non-rigid. I's is the realistic axis of S'_S just with geometrical defect, in the rigid configuration. Moreover, I'h is extracted using the 3D least square fitting method if S'_H is non-rigid. In the rigid configuration, I'h is the realistic axis of S'_H with geometrical defect.

- Coincident constraint (Co: I's & Vertex of I'h) between I's and a vertex of I'h (Fig. 4c). In this case, I's or I'h can be rigid or non-rigid.

3.4 Control of the FR: Tolerances Analysis

In order to validate the initial tolerances values already attributed to each assembly parts in the nominal configuration, the FR of the realistic assembly is controlled and checked. If FR is respected, then the initial tolerances values are correct. However, if FR is not satisfied, an iterative optimization algorithm is used to correct the tolerances specifications. This task is based on the sensitivities analysis method (identify and classify tolerances according to their impact on the FR) and the tolerances cost. Generally, the less expensive tolerances are selected first to be optimized.

4 Case Study

To validate the approach already developed, a case study is illustrated in this section.

A rod and crankshaft mechanical system is used to validate the different steps of the CAT model already developed (Fig. 5), and composed of the following components with geometrical defects:

- The crank part (3), considered as non–rigid with positional tolerance TP1 = 0.3 mm.
- The connecting rod (4), considered as rigid with positional tolerance TP2 = 0.1 mm.
- The support crank (1), considered as rigid with positional tolerance TP3 = 0.2 mm.

The other components (0), (2), (5), (6) of the assembly are rigid without defects. The FR of the assembly consists of keeping the clearance G1, G2, G3 and G4 between the piston (6) and the cylinder (2) (as presented in Fig. 6).The mating constraints and the AS between the assembly parts in the nominal and the 6th realistic configurations are shown in Table 1.

Table 1 AS and constraints

(Nominal configuration) AS N°. Constraint	(6th realistic configuration) AS N°. Constraint
1. Co: A3,1 & A1,1	1. Co: A3,1 & A'1,1
2. Co: F1,3 & F3,3	2. Co: Edge (E) of F1,3 & F3,3
3. Co: A4,1 & A3,2	3. Co: A4,1 & $A3,2_{OBB}$
4. Co: A4,2 & A6,1	4. Co: A'4,2 & A6,1
5. Co: A5,1 & A6,1	5. Co: A5,1 & A6,1
6. Co: F5,3 & F6,3	6. Co: F5,3 & F6,3
7. Co: A6,2 & A2,1	7. Co: 2 edges of F6,2 & F2,3

A'i,i are the realistic axes of Ai,i; $A3,2_{OBB}$ is the realistic axis of A3,2 created using the OBB tool

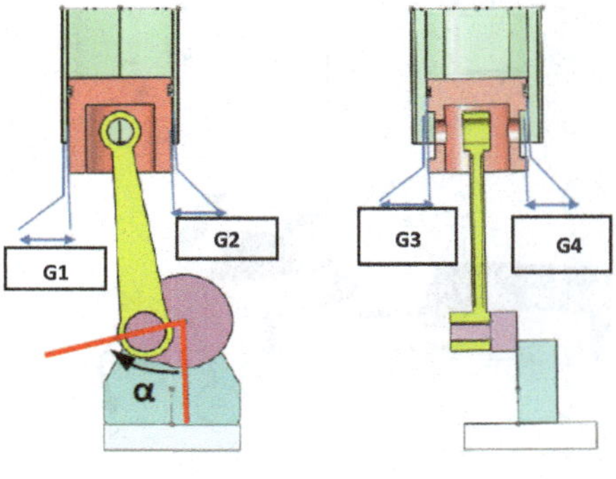

Fig. 6 The clearances G1, G2, G3 and G4 between the piston and the cylinder

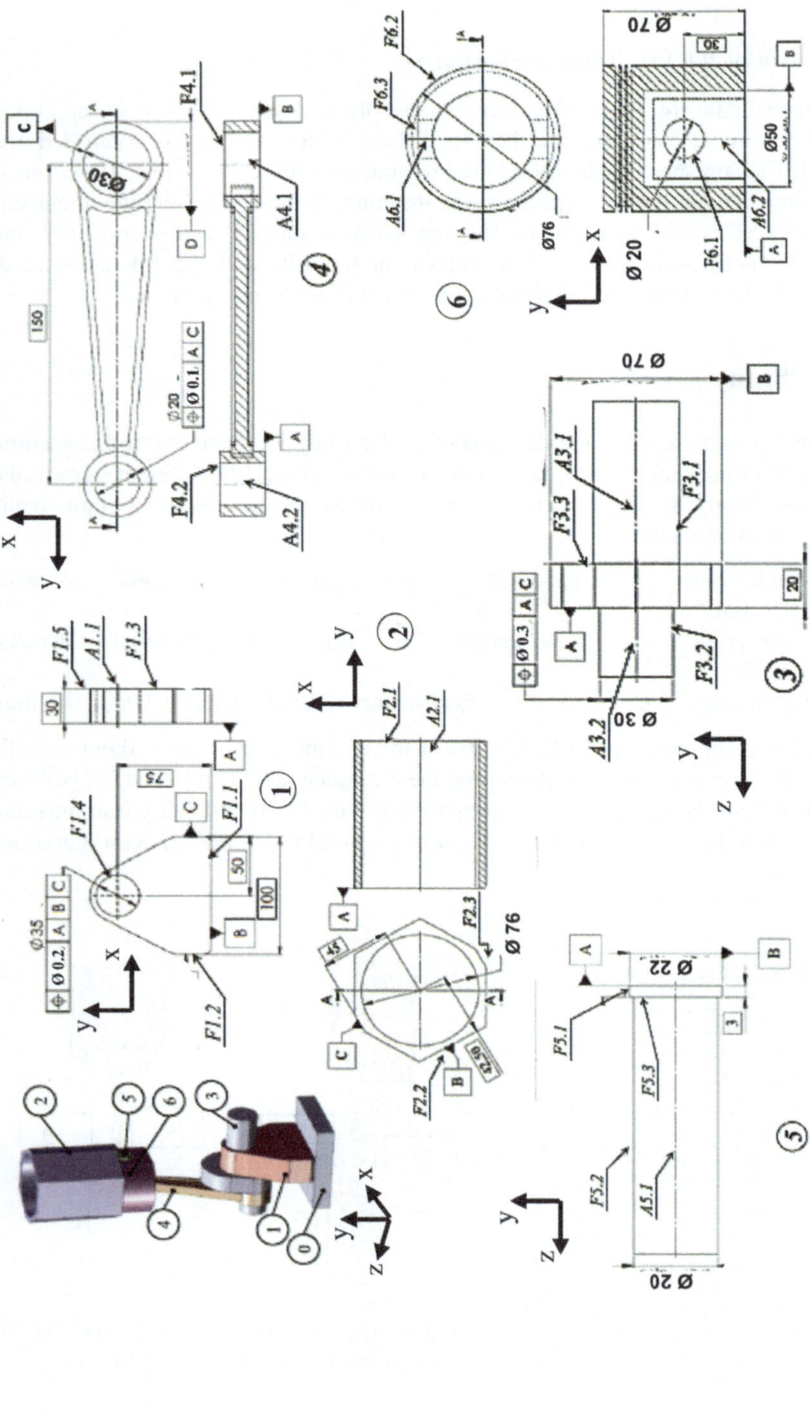

Fig. 5 The drawing of the rod and crankshaft mechanical system

The non-rigid crank part (3) is isolated from the assembly and subjected to the FE simulation in order to reconstruct the surface F3,2 of axis A3,2. A sub-algorithm based on the API of Solidworks® is used to run the FE computation while considering the material of (3) as (AISI 1020), the faces F3,1 and F3,3 as fixed features and a Force (**F**) = P × S = 0.5 × 4537 = 1361 N according to ($-\overrightarrow{y}$), where P: on the piston, S: is the section of the piston.

Table 2 Control of FR with initial tolerances values TP1 = 0.3 mm, TP2 = 0.1 mm, TP3 = 0.2 mm

Realistic Configuration	Value of clearances for a rotation angle of the crank α = 0°				Value of clearances for a rotation angle of the crank α = 60°				Value of clearances for a rotation angle of the crank α = 230°				FR
	G1	G2	G3	G4	G1	G2	G3	G4	G1	G2	G3	G4	
1	0	0.758	0.479	0.336	0.345	0.465	−0.001	−0.002	0.002	0.523	0.552	0.004	No
2	0.482	0.413	0	0.661	0.023	0.540	0.348	0.552	0.482	0.481	0	0	No
3	0.486	0.502	−0.001	−0.002	0.489	0.497	−0.012	−0.003	0.481	0.518	0.003	0.013	No
4	−0.001	−0.001	0.002	0.023	0.491	0.508	0.071	0.001	0.003	0.603	0.581	0.268	No
5	−0.002	−0.001	0	0	0.241	0.502	0.119	0.108	0.671	0.152	0.055	0.931	No
6	0.466	0.526	0	0.001	0.254	0.115	0	0.558	0.302	0.430	0.012	0	Ok

Table 3 Control of FR with optimal tolerances values $TP1_{OPT}$ = 0.23 mm, TP2 = 0.1 mm, TP3 = 0.2 mm

Realistic configuration	Value of clearances for a rotation angle of the crank α = 0°				Value of clearances for a rotation angle of the crank α = 60°				Value of clearances for a rotation angle of the crank α = 230°				FR
	G1	G2	G3	G4	G1	G2	G3	G4	G1	G2	G3	G4	
1	0.018	0.755	0.471	0.332	0.341	0.460	0.031	0.027	0.029	0.551	0.573	0.035	Ok
2	0.511	0.441	0.021	0.710	0.051	0.610	0.396	0.586	0.531	0.502	0.021	0.031	Ok
3	0.527	0.535	0.035	0.028	0.521	0.549	0.028	0.037	0.516	0.559	0.045	0.052	Ok
4	0.037	0.042	0.038	0.064	0.528	0.547	0.112	0.039	0.046	0.642	0.625	0.319	Ok
5	0.051	0.047	0.034	0.033	0.278	0.544	0.162	0.147	0.713	0.188	0.086	0.964	Ok
6	0.512	0.571	0.027	0.038	0.305	0.162	0.036	0.583	0.344	0.455	0.047	0.029	Ok

- **Results analysis**

The tolerance analysis using the initial tolerance values (TP1 = 0.3 mm, TP2 = 0.1 mm, TP3 = 0.2 mm), shows that FR is not respected for all the realistic configurations of the assembly (Only 6 realistic configurations are presented to facilitate the illustration) (Table 2). This result justify the initial assumption of our work about the deformations impact on the choice of tolerances. By using the iterative sub-algorithm of tolerances optimization, the tolerance TP1 is chosen to be corrected by an increment $\varphi = 0.01$ mm. The new tolerance value $TP1_{OPT}$ = 0.23 mm allows to respect FR with consideration of the deformation factor and the positional defects (Table 3). In this case, the correction is equal to 0.07 mm for the tolerance TP1. The other tolerances values TP2 and TP3 are conserved.

5 Conclusion

A novel CAD tool for the tolerance analysis of non-rigid cylindrical parts assemblies is presented in this paper. The parts with defects are modeled in the DMU. The non-rigid cylindrical parts are reconstructed using the FE computation. The final realistic assembly is obtained after the updating of the mating constraints between (R/R) and (R/NR) realistic components. The FR is controlled at the end to validate the accuracy of the initial tolerance values. Our future works will focus on the consideration of form defects as well as (NR/NR) cylindrical joints in the tolerancing step.

References

1. Anselmetti B (2006) Generation of functional tolerancing based on positioning features. Comput-Aided Des 38:902–919
2. Benichou S (2012) Integration of the effects of thermal expansion in tolerancing. Dissertation. The Normal Superior School of Cachan, France
3. Camelio J, Hu SJ, Ceglarek D (2003) Modeling variation propagation of multi-station assembly systems with compliant parts. J Mech Des 1257:59–790
4. Hermansson T, Carlson J, Björkenstam S, Söderberg R (2012) Geometric variation simulation and robust design for flexible cables and hoses. In: Proceedings of 12th CIRP conference on computer aided tolerancing, Huddersfield, UK
5. Jacquelin J (2011) 3-D linear regression. https://fr.scribd.com/doc/31477970/Regressionsettrajectoires-3D
6. Jbira I, Tlija M, Louhichi B, Tahan A (2017) CAD/tolerancing integration: mechanical assembly with form defects. Adv Eng Softw 114:312–324
7. Korbi A, Tlija M, Louhichi B, Benamara A (2018) A CAD Model for Tolerance Analysis of non-rigid Planar parts assemblies. The 28th CIRP design conference. Nantes, France, pp 126–131
8. Louhichi B, Tlija M, BenAmara A, Tahan A (2015) An algorithm for CAD tolerancing integration: generation of assembly configurations according to dimensional and geometrical tolerances. Comput-Aided Des 62:259–274
9. Mazur M, Leary M, Subic A (2011) Computer aided tolerancing (CAT) platform for the design of assemblies under external and internal forces. Comput-Aided Des 43:707–719
10. Pierre L, Teissandier D, Nadeau JP (2009) Integration of thermo-mechanical strains into tolerancing analysis. Int J Interact Des Manuf (IJIDeM) 3:247–263
11. Sellakh R, Sellem E, Riviere A (2000) specification and simulation of geometric Imperfections. In: CAD/CAM. mechanics & industry, pp 365–372
12. Söderberg R, Wärmefjord K, Lindkvist L, Berlin R (2012) The influence of spot weld position variation on geometrical quality. CIRP Ann Manuf Technol 6:13–16
13. Tsai JC (2007) Stiffness variation of compliant devices due to geometric tolerancing. In: The 10th international conference on computer aided tolerancing (CIRP), Erlangen, Germany

Why and How to Move from SPC (Statistical Process Control) to APC (Automated Process Control)

Thomas Muller[1], Maurice Pillet[2(✉)], Jean Luc Maire[2], and Davy Pillet[1]

[1] Ellistat, Annecy, France
{Thomas.muller,davy.pillet}@ellistat.com
[2] Laboratoire SYMME, Université Savoie Mont-Blanc, Chambéry, France
{maurice.pillet,jean-luc.maire}@univ-smb.fr

Abstract. SPC (Statistical Process Control) is widely used to monitor processes. However, recent developments in Industry 4.0 and improvements of machine tools question the relevance of SPC. Using SPC in conjunction with EPC (Engineering Process Control) can partly improve the performance of processes. In this article, we define a new meaning of APC (Automated Process Control) which is based on three main innovations: the use of machine learning to control processes more accurately, the simultaneous consideration of all available characteristics in multidimensional processes and finally the dissociation between conformity and control for these characteristics.

Keywords: Process control · SPC · EPC · APC · Machine learning · Industry 4.0

1 Introduction

Reducing variability is one of the main objective of product quality control. Since Shewhart's proposals [18], significant improvements have been obtained, especially in the part industries, with the using of SPC (Statistical Process Control). However, the technological transformations related to the deployment of Industry 4.0 question the relevance of this approach. Indeed, several evolutions are challenging the traditional SPC, notably in mechanical companies using Computer Numerical Control (CNC) machines. For example:

- The increasing complexity of these machines now makes it possible to produce a whole part with a single operation.
- The complexity of the functional shapes to be realized, which are no longer the combination of elementary shapes (plans/cylinders…).
- The increase of cycle times resulting from these complexifications makes it difficult to control the machines from a single sample.

N. Aifaoui et al. (Eds.): CMSM 2019, LNME, pp. 33–40, 2020.
https://doi.org/10.1007/978-3-030-27146-6_4

- The existence of dependencies between several characteristics makes difficult the calculation of appropriate adjustments.
- The reduction in batch sizes requires the first part to be compliant.
- The reduction of product life cycle time implies frequent changes of processes

All these changes challenge traditional SPC charts. Manufacturers must find new approaches in order to control their processes by avoiding the use of sampling. Ideally, the control should be done directly from the measurement of a single part.

EPC (Engineering Process Control) has often been combined with SPC to improve its control performance. Jiang and Farr [10] provides a review of various control techniques using this method. For example, [2] compares SPC and EPC regarding when and where they could be useful for monitoring the process. Janakiram and Keats [9] point out several limits of SPC/EPC regarding sampling rate, fault detection, real-time compensation and process control. Akram et al. [1] detailed different strategies for combining the two approaches. Hu and Cheng [8] proposes an integrated control policy based on SPC and EPC to minimize the process fluctuation, stabilize the output quality, and finally improve the process reliability.

In the case of CNC machines with relatively long cycle times, these approaches may not be appropriate. In this paper, we propose to show that an approach using machine learning algorithms combined with multi-criteria control is perfectly adapted to the control of CNC machines, even with several hundred characteristics dependent on each other.

2 The Challenges of Process Control in Mechanical Production

The evolution of CNC machines combined with the increasing complexity of mechanical parts often leads to an increase of cycle times. Now it is often possible to carry out all machining operations with a single step, which significantly improves quality by eliminating repositioning. However, the process control must then take into account several constraints such as:

1. Dependency between characteristics: since the part is realized in a single step, adjusting one dimension may cause another to be maladjusted. Figure 1 gives a schematic example of a part. An action on the corrector T3a can correct the dimension c and disturb the dimension e. This is the first challenge.
2. Because of the reduction of batch sizes, the process must be adjusted as accurately as possible from the first part. The second challenge is therefore to find more relevant approaches than SPC and EPC in order to control the process with the data obtained from a single part. This eliminates some EPC approaches such as those using PID (Proportional Integral Derivative) methods.
3. The 3D scanning technology gives a knowledge about the geometry skin of the parts. The last challenge is to use this entire data rather than calculating dimensions from it, which can led to a loss of information.

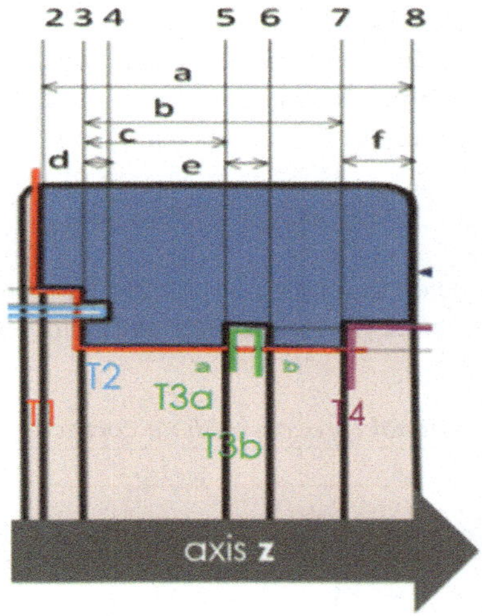

Fig. 1 Example of mechanical part operations

The three challenges, which are adapted to CNC machines, are the foundations for APC. We define it as *"a method of process control that adjusts automatically the dimensions of the part on the target from a limited number of data (one part) using all available information. APC's performance is the result of the joint use of mathematical, statistical and machine learning techniques."*

3 Optimum Control from a Single Part

The first challenge is the ability to control a process with information from the last measured part. Process adjustment is critical for short production runs. Therefore some researchers have aimed at adapting SPC control cards for small batches such as Lill [13] Pillet [14]. Quesenberry [16] and Castillo et al. [3] proposed specific charts notably in the case of start-up processes and short runs. Other researchers proposed solutions about the adjustment of the process just after detecting a significant deviation. Trietsch [19], Del Castillo [4] and Pan [17] provided an extension of the Grubbs [7] rules, who was the first to introduce a statistical approach for the adjustment of machines. Kibe et al. [11] focus on adjustment of the tools using the in situ measurement.

Difficulties in finding the best adjustment when the value of the deviation from a given target is estimated often lead to a spread of the process. These difficulties arise more frequently with the use of autonomous machines using automatic monitoring. Duret and Pillet [5] proposes a Bayesian approach to estimate the best correction. Liao et al. [12] have focused on the benefit of using neural networks to learn how to grasp the behavior of the process and stabilize it.

Machine learning approaches undeniably have remarkable properties to focus a process on its target. We present here a comparative study between a Shewhart control chart approach to individual values and an approach based on a machine learning. For this study, we use a Monte Carlo method with a simulation of a production subject to drifts and maladjustments. The comparison of the two approaches is based on the ability to obtain in the long term a standard deviation as close as possible to the short-term standard deviation. Therefore, the stability indicator of the process is:

$$SR = \sigma_{short\_term} / \sigma_{long\_term} \tag{1}$$

The simulation was done on a sample of 1,000,000. Figure 2 gives the three signals of the simulation on 1,000 parts (for readability and clarity).

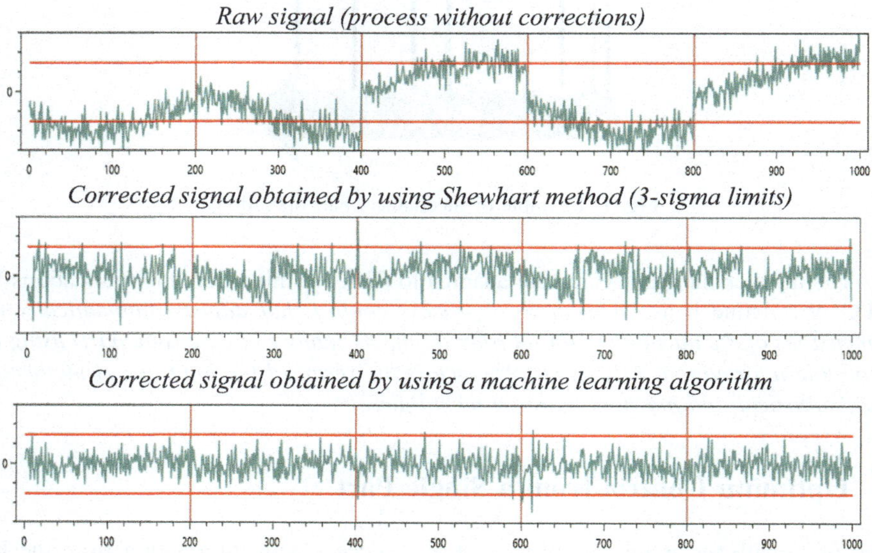

Fig. 2 The three signals of the simulation (1,000 parts)

Machine learning provides a significant gain in quality with a long-term standard deviation very close to the short-term standard deviation (Table 1).

Table 1 Result of the simulation (1,000,000 parts)

	Raw signal	Shewhart method	Machine learning
σ_{long_term}	2.58	1.50	1.09
σ_{short_term}	1	1	1
SR	**0.3876**	**0.6667**	**0.9174**

4 Multi-criteria Control

Figure 1 shows a process with dependencies between characteristics. In the case of CNC machines, the dependencies are analytical. We can find the matrix [15] representing the relationships between the correctors and the dimensions.

$$\begin{bmatrix} \Delta a \\ \Delta b \\ \Delta c \\ \Delta d \\ \Delta e \\ \Delta f \end{bmatrix} = \begin{bmatrix} -1 & 0 & 0 & 0 \\ -1 & 0 & 0 & 1 \\ -1 & 0 & 1 & 0 \\ -1 & 1 & 0 & 0 \\ 0 & 0 & -1 & 0 \\ 0 & 0 & 0 & -1 \end{bmatrix} \begin{bmatrix} T1 \\ T2 \\ T3a \\ T3b \\ T4 \end{bmatrix} \tag{2}$$

$$[\Delta C] = [A][T] \tag{3}$$

The calculation of the corrections is directly carried out from the previous relationship by using the pseudoinverse matrix of the model (inertial steering method).

$$[T] = \left[(A^t A)^{-1} A^t\right][\Delta C] \tag{4}$$

A generalization of the machine learning approach to each dimension ΔC gives the optimal setting for the whole part.

To show the results obtained, we also carried out a Monte Carlo simulation to compare the multi-criteria control associated with machine learning against a multi-criteria approach without machine learning:

- The first simulation is done with the Shewhart rules: deviations ΔC are only considered if the characteristic is out of control
- The second simulation is done by using a machine learning algorithm to identify the most likely maladjustment to be corrected

In both cases, the corrections are calculated with the formula (4).

Figure 3 gives the three signals of the multi-criteria simulation (the raw signal, the signal obtained with standard SPC rules and the signal obtained with Machine Learning rules).

The use of machine learning rules provides significant overall improvement of the process.

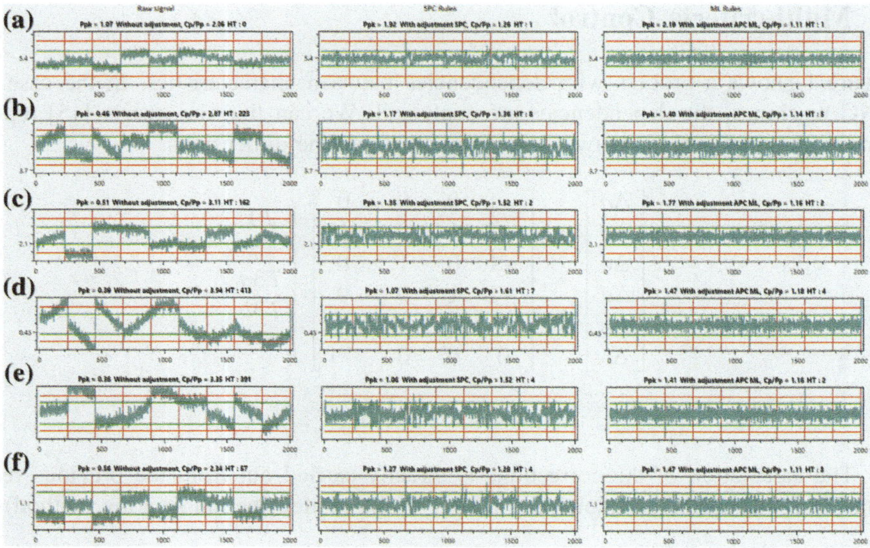

Fig. 3 The three signals of the multi-criteria simulation (6 characteristics)

5 APC: A Dissociation Between Conformity and Control

The third challenge concerns the control of complex parts, better defined by point cloud (surfaces) than by dimensional characteristics. Figure 4 shows a part defined from only three specifications in common zone.

Each surface is totally specified by a shape tolerance defined, which results of the calculation of the inertia I on all the scanned points (5).

$$I = \sqrt{\sum deviation^2/n} \tag{5}$$

The conformity of the part is reached if the values of its characteristics (three characteristics for the example) belong to the control limits. The adjustment should not be calculated based on the conformity characteristics. Indeed, there is a loss of information when we move from information of the point cloud to the reduced information on the three characteristics. It is therefore necessary to use all the measured points to adjust the process. Conformity and control should be dissociated.

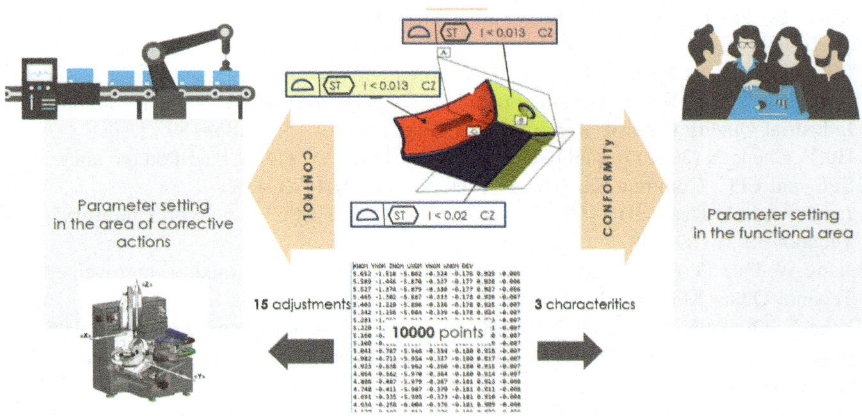

Fig. 4 An example of a part defined by only three specifications

6 Conclusion

The increasing complexity of machine tools combined with the development of the digital chain leads to change practices in process control. SPC, combined or not with EPC, is no longer sufficient to achieve stability of a process. In this paper, we have discussed the benefit of APC, which leads to three main improvements:

- An optimum control from the information of a reduced number of parts produced, ideally from a single part. We discussed the benefit of using machine-learning algorithms to find the right adjustments obtained from this reduced information.
- A simultaneous consideration of all characteristics to calculate the best fit. Indeed, if there are dependencies between the characteristics, only a global approach works.
- The dissociation between conformity and control in order to avoid a loss of information. To ensure the best possible fit, the control of the process should not be based on the parameterized conformity information but on the raw information.

These three improvements are already implemented in a software called Ellisetting [6]. This software is not an evolution of SPC but an innovative approach for process control.

References

1. Akram MA, Saif AWA, Rahim MA (2012) Quality monitoring and process adjustment by integrating SPC and APC: a review. Int J Ind Syst Eng 11(4):375–405
2. Box GE, Coleman DE, Baxley RV Jr (1997) A comparison of statistical process control and engineering process control. J Qual Technol 29(2):128–130
3. Castillo D, Enrique M, Douglas C (1994) Short-run statistical process control: Q-chart enhancements and alternative methods. Qual Reliab Eng Int 10(2):87–97
4. Del Castillo E (1998) A note on two process adjustment models. Qual Reliab Eng Int 14 (1):23–28

5. Duret D, Pillet M (2015) Process adjustment by a Bayesian approach. Cogent Eng 2 (1):1096999
6. Ellisetting (2019). https://ellistat.com/ellisetting/
7. Grubbs FE (1983) An optimum procedure for setting machines or adjusting processes. Industrial Quality Control, Reprinted in J Qual Technol 15(4):186–189
8. Hu Y, Cheng X (2012) A quality control approach of PVC gloves based on the integration of SPC and EPC. Commun Stat-Theory Methods 41(24):4470–4483
9. Janakiram M, Keats JB (1998) Combining SPC and EPC in a hybrid industry. J Qual Technol 30(3):189–200
10. Jiang W, Farr JV (2007) Integrating SPC and EPC methods for quality improvement. Qual Technol Quant Manag 4(3):345–363
11. Kibe Y, Okada Y, Mitsui K (2007) Machining accuracy for shearing process of thin-sheet metals—Development of initial tool position adjustment system. Int J Mach Tools Manuf 47 (11):1728–1737
12. Liao XP, Xie HM, Zhou YJ, Xia W (2007) Adaptive adjustment of plastic injection processes based on neural network. J Mater Process Technol 187(188):676–679
13. Lill H, Chu Y, Chung K (1991) Statistical set-up adjustement for low volume manufacturing. In: Statistical process control in manufacturing - Dekker, pp 23–38
14. Pillet M (1996) A specific SPC chart for small-batch control. Qual Eng 8(4):581–586
15. Pillet M, Pairel E (2011) Determination of an adjusting rule in the case of multi-criteria inertial piloting. Int J Metrol Qual Eng 2(1):51–59
16. Quesenberry CP (1991) SPC Q charts for start-up processes and short or long runs. J Qual Technol 23(3):213–224
17. Pan R (2002) Statistical process adjustement methods for quality control in short-run manufacturing, Thesis in Industrial Engineering, Pennsylvania State University, August 2002
18. Shewhart WA (1931) Economic control of quality of manufactured products. MacMillan, New York: Van Nostrand, London, 501 p. (1980 by American Society for Quality)
19. Trietsch D (1998) The harmonic rule for process setup adjustment with quadratic loss. J Qual Technol 30(1):75–84

Proposal of a New Based Scenarios Eco-Manufacturing Methodology on CAD Phase

Hadhami Ben Slama[1], Raoudha Gaha[2](✉),
and Abdelmajid Benamara[1]

[1] Laboratoire de Génie Mécanique de Monastir, Ecole Nationale D'ingénieurs de Monastir, Université de Monastir, Monastir, Tunisie
abdel.benAmara@enim.rnu.tn

[2] Laboratoire de Mécanique de Sousse, Ecole Nationale D'Ingénieurs de Sousse, Université de Sousse, Sousse, Tunisie

Abstract. The integration of environmental aspects in the design process is getting more and more covered by the scientific research. Owing to the issued regulations, industries tend to reduce their environmental impacts (EIs) while keeping a minimal manufacturing cost. The process plan operations order has an influence on the EIs and cost of a product. In this paper, we propose a new based scenarios methodology by using feature technology (FT) which allows the non-environmental expert designer to choose the most optimal manufacturing scenario with the best compromise EIs/Cost. First, we present an overview of works using the FT in CAD phase to reduce EIs and cost of a manufacturing process. Secondly, we propose a new methodology based on both FT and scenarisation which helps inexperienced eco-designers. Finally, a case study is considered to validate the proposed approach.

Keywords: CAD · Eco-design · Eco-manufacturing · Environmental impact · Cost/scenarios

1 Introduction

Face to the growth of environmental awareness, industrials aim to improve the quality of their products by respecting all regulations and environmental legislation. Therefore, eco-friendly manufacturing is widely studied. In this paper, we start with a state of art showing the necessity of scenarisation in designer's decision support of manufacturing process of his product. Secondly, we present our new methodology based on FT and integrated systems which gives the optimal compromise EIs/Cost. Finally, we apply our methodology into a case study for validation.

N. Aifaoui et al. (Eds.): CMSM 2019, LNME, pp. 41–46, 2020.
https://doi.org/10.1007/978-3-030-27146-6_5

2 State of Art

The generation of various manufacturing scenarios is used in former works based on FT. Sheng in [1] verified that the order of machining operations has an influence on the EIs as well as cost [1] (Fig. 1). This part is consisting of three types of features: a pocket, a hole and a finished planar face. He assumed that this part can be realized in three different sequences:

- Sequence 1: End Mill → Drill → Face Mill
- Sequence 2: Drill → End Mill → Face Mill
- Sequence 3: Face Mill → Drill → End Mill

Sheng studied the process outputs for each sequence. He finds out that the first sequence provides the optimal compromise between energy, process time and weighted mass. Therefore, features are useful for selecting the most ecological manufacturing sequence from a set of possible scenarios. Indeed, among the works found in the literature is that of Cao [2] where they propose a multi-objective decision making optimization model in order to develop a software for generating possible machining sequences and choose the most ecological one. In 2006, Tan introduces a new sustainable production procedure based on expert systems, case-based reasoning and feature technology to design possible component manufacturing process scenarios [3]. Zhao proposed a method for environmentally conscious process planning [4]. This method starts with an existing process plan, determines the impactful steps and associated design features, in terms of EIs and manufacturing cost. These features can achieve alternative processes which can be used in order to generate alternative process plans. In 2015, Gaha et al. proposed a new methodology based on FT to generate scenarios in computer-aided process plan (CAPP). This methodology allows us to evaluate the environmental impact of a product manufacturing process from geometric modelling phase by a life cycle assessment (LCA) tool and to choose the greenest

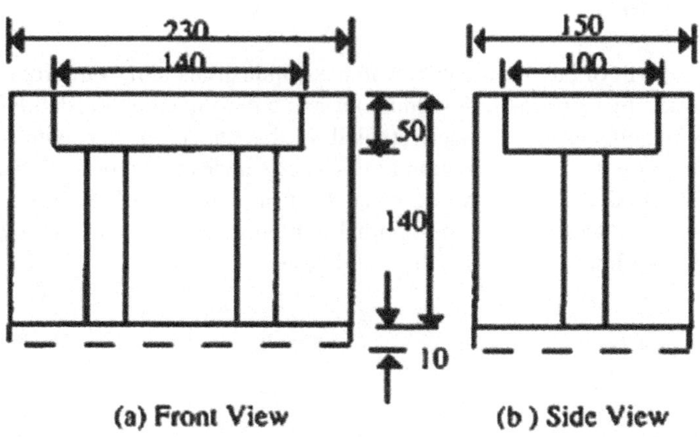

Fig. 1 A sample part used by Sheng to prove the influence of machining sequences order on the EIs [1]

manufacturing scenario [5]. A case study also is developed showing a comparison of environmental impacts of possible manufacturing scenarios of a sample part, as shown in Fig. 2. Thus, scenarisation presents a necessity to reach a green manufacturing based in geometric modelling [6, 7].

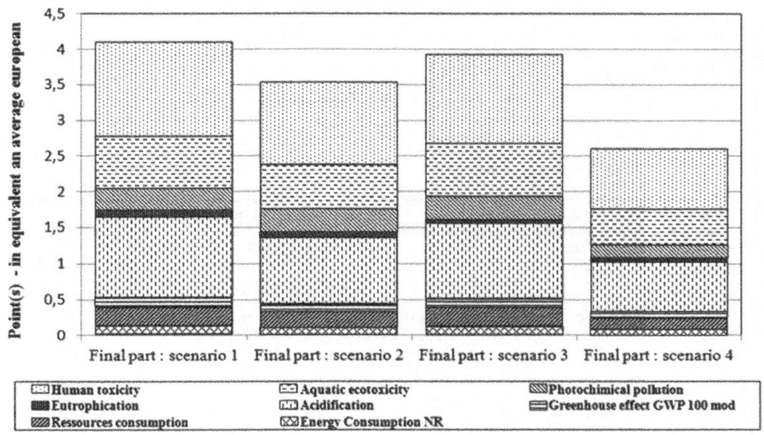

Fig. 2 Comparison of environmental impacts of possible manufacturing scenarios for the sample part [5]

3 Proposed Approach

In our methodology, we explore the feature technology to evaluate EIs and manufacturing cost of a product from its geometric modelling phase. The methodology presented is an algorithm that allows the designer; non-expert in terms of environment; to see the difference of machining environmental impact and manufacturing cost for each possible scenario and allows him to choose the optimal compromise EIs/Cost. Our approach is composed of several steps as shown in Fig. 3.

In order to estimate EIs, our approach allows us to process in two different ways: The first method consists in creating a Database (DB). This DB relates each feature or set of features to their EIs. Our developed program extracts features consisting the final part, then, generates the different possible alternative machining scenarios and calculates the EIs for each one using the DB created in advance. The second method we can use is based on CAD/LCA systems integration. After the generation of different possible manufacturing scenarios, we extract the necessary data for realizing LCA by means of CAD/LCA systems integration. In order to estimate manufacturing cost, our approach needs the creation of a DB which relates every feature or a set of features to

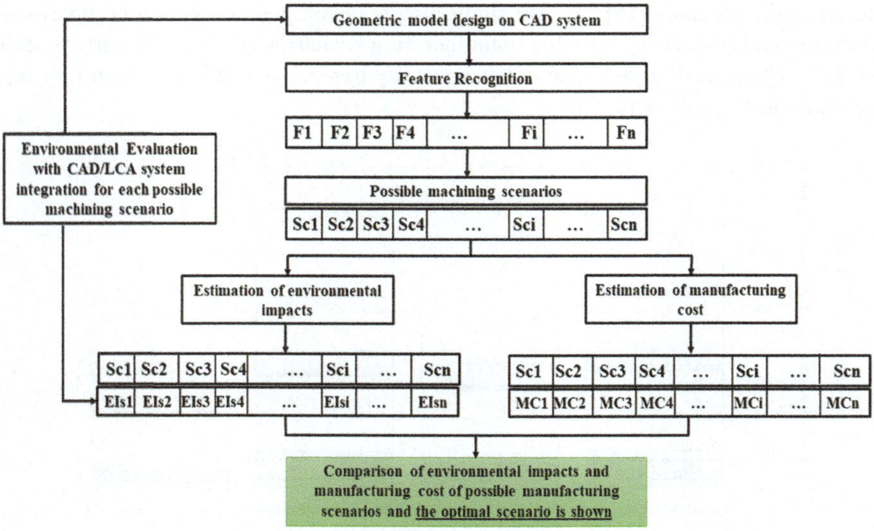

Fig. 3 Proposed methodology for selecting the optimal compromise EI/Cost based on scenarisation

their cost. Like for the estimation of EIs, our system extracts the features that constitute our part, generates possible machining scenarios and calculates the total manufacturing cost of each one.

Finally, our system shows to the non-environmental or economy expert designer comparison of EIs and cost of each scenario and the optimal one.

4 Case Study

We further illustrate the proposed methodology with a case study. We consider the part shown in Fig. 4 which is a part made of the steel S235JR. This part represents the mantle of a metal cabinet that serves to protect the electrical installations in different climatic conditions.

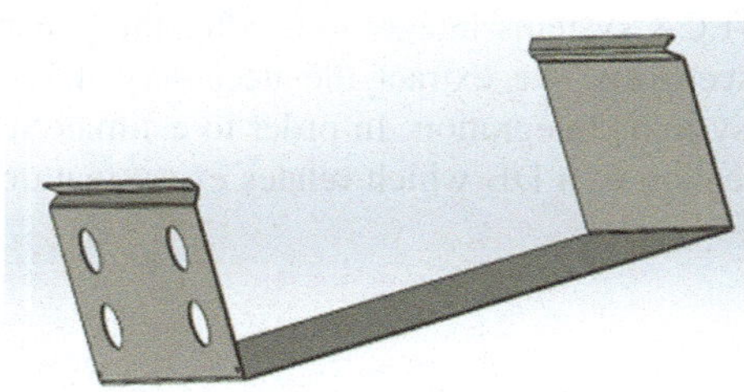

Fig. 4 The sample part

At this stage, based on features, the identification of geometric forms of the considered part allows to generate possible scenarios to realize it. For our case, two machining scenarios are proposed:

- Scenario 1: '*Cutting*', '*Notching*', '*Punching*', '*Embossing*' and '*Folding*'.
- Scenario 2: '*Cutting*', '*Notching*', '*Punching*' for four times, 'Embossing' and '*Folding*'.

Table 1 shows a comparison between carbon emission (kg eq. CO_2) provided by both manufacturing scenarios. Table 2 shows a comparison between manufacturing cost for both manufacturing scenarios. Our system gives also much details about the different costs of manufacturing operations. The comparison between EIs and manufacturing costs which highlights that *scenario 1* is the most ecological and economic one.

Table 1 Carbon emission for both manufacturing scenarios

	Scenario 1	Scenario 2
CO_2 emission (kg eq. CO_2)	20.95	21.85

Table 2 The different manufacturing costs of both scenarios displayed by "ECO fabCAD"

	Cost (Dinars)	
	Scenario 1	Scenario 1
Folding cost	1.200	1.200
Embossing cost	0	0
Notching cost	0.360	0.360
Punching cost	0.525	1.200
Cutting cost	0.450	0.450
Manufacturing cost	2.535	3.210
Material cost	10.505	10.505
Total cost	**13.040**	**13.715**

5 Conclusion

This paper presents the necessity of scenarisation to choose the greenest and the most economical manufacturing scenario which helps the designer in his decision making from CAD phase. Our proposed methodology explores both FT and integrated systems (CAD/CAM/CAPP/LCA) to evaluate environmentally possible manufacturing processes of a product. It also allows the estimation of manufacturing costs using features consisting the product. A case study is also developed to show the necessity of scenarisation in the determination of the optimal compromise EIs/Cost.

References

1. Sheng P, Srinivasan M, Kobayashi S (1995) Multi-objective process planning in environmentally conscious. Department of Mechanical Engineering, University of California, Berkeley, CA, USA
2. Cao HJ, Liu F, He Y, Zhang H (2002) Study on model set based process planning strategy for green manufacturing. Jisuanji Jicheng Zhizao Xitong/Comput Integr Manuf Syst (China) 8 (12):978–982
3. Tan X, Liu F, Dacheng L, Li Z, Wang H, Zhang Y (2006) Improved methods for process routing in enterprise production processes in terms of sustainable development II. Tsinghua Sci Technol 11(6):693–700
4. Zhao F, Murray VR, Ramani K, Sutherland JW (2012) Toward the development of process plans with reduced environmental impacts. Front Mech Eng 1–16
5. Gaha R, Yannou B, Benamara A (2015) Selection of green manufacturing process based on CAD-features. Int J Adv Manuf Technol. https://doi.org/10.1007/s00170-015-7499-y
6. Gaha R, Benamara A, Yannou B (2014) Eco-design of a basin mixer in geometric modeling phase. In: Key engineering materials, vol 572. Trans Tech Publications, pp 7–11
7. Gaha R, Benamara A, Yannou B (2013) A feature-based methodology for eco-designing parts on detail phase. In: Design and modeling of mechanical systems. Springer, Berlin, pp 645–654

Experimental Study of Vehicle Noise and Traffic Pollution

Sana Abbes[1], Hichem Hassine[2](✉), Maher Barkallah[2], Jamel Louati[2], and Mohamed Haddar[2]

[1] National School of Engineers of Sousse, Sousse, Tunisia
abbessana06@gmail.com
[2] Laboratory of Mechanics, Modeling and Manufacturing (LA2MP), Sfax, Tunisia
hassinehichem@yahoo.fr

Abstract. Transport and road traffic in particular are highly emitting noise pollution. These nuisances are considered one of the first threats to the quality of life and health. Indeed, noise pollution will continue to increase in magnitude and severity as a result of population growth, urbanization and growth associated with automobile use. In this paper, we propose an experimental study of vehicle noise based on measurement noise in three different sites. The main goal of this study is to determine the influence of different parameters, related to traffic and vehicle characteristics, in noise pollution emitted by vehicles. The experimental procedure was applied in three different sites used in order to integrate the influence of the infrastructure in noise comportment of vehicles. Results demonstrate that vehicle characteristics, traffic flow and infrastructure have an important influence in noise level of road traffic.

Keywords: Sustainable development · Noise · Traffic parameters · Vehicle characteristics

1 Introduction

Sustainable development is defined as "a development that meets the needs of the present without compromising the ability of future generations to meet their own needs." Brundtland Commission [2]. This development is based on three pillars: social, ecologic and economic pillar. Transport is one of the most contributor on environmental degradation. The traffic noise is the major environmental problem that causes annoyance and disturbance World Health Organization [16, 17]. It has adverse effects on the quality of life and the health of individuals. In fact, prolonged and excessive exposure to noise not only harms the hearing instrument but also the whole organism.

Generally, vehicle noise is mainly due to the four major types of sources: mechanical noise or powertrain noise (intake, exhaust and engine), tire/road contact noise. Sources related to aerodynamic noise or the interaction between the vehicle surface and the flow of air and other sources related to brakes, horns, etc [10].

The noise pollution of road traffic depends mainly on certain factors that can be grouped into two families: road traffic and road infrastructure. For the first family, these

N. Aifaoui et al. (Eds.): CMSM 2019, LNME, pp. 47–54, 2020.
https://doi.org/10.1007/978-3-030-27146-6_6

factors are known as explanatory variables for the acoustic models for assessing the impact of noise on the environment. The principle factors are: Vehicle types, vehicle speed, the acceleration, vehicle flow. Then, for the second family, we have the following factors: The pavement, Slope of the road, The presence of buildings.

Recent work in this area of research has been devoted to the theoretical and empirical modeling of vehicle noise [5, 12, 11]. Acoustic models generally are experimental models, which from the measurements try to calculate the respective influence of the various explanatory parameters selected. The emission law is constructed by measurements made for vehicles taken microscopically (or individually). These measurements are taken either in situ or on a test track using a sound level meter. Generally, measurement and modeling are two complementary methods of evaluating environmental noise. The main acoustic empirical models of sound level calculation at different points of the environment are: The German model: RLS 90, Model C.R.T.N (Calculation of Road Traffic Noise) (Delany et al. 1988), Rawat model [13], Subramani model [14], Griffith and Langdon model [8], Cvetkovic model [4], Sukeerth model [15], Burgess model [3], FHWA traffic noise level [7, 9].

For the theoretical ones, [6] as regards the traffic noise forecast model suggests using the French model named "new noise forecasting method" or simply NMPB-1996. This method makes it possible to calculate the acoustic level and the various terms at the origin of the attenuation of the sound wave (geometrical divergence, atmospheric absorption, diffraction, effect of the ground) under the conditions favorable to its propagation or homogeneous conditions [1].

The developed studies presented some methodologies and approach to quantify noise pollution. But we can conclude that each one is based on its own law of emission and uses other factors. This paper deals with the vehicle noise measurements in order to prepare a data base allowing the study of the impact of different parameters on vehicle noise and minimize this environmental problem.

The remaining content of this paper is organized as follows. In Sect. 2, methodology and the experimental procedure are described. The measurements results and discussion are detailed in Sect. 3. Finally, main conclusions are summarized in Sect. 4.

2 Methodology and Experimental Procedure

Assessing the vehicle noise is important for improving the quality of life, therefore, noise models are widely used for monitoring and assessing the impact of noise on the environment. These models were needed and built to predict the noise level based on field measurement of noise level and traffic parameters. In the same context, this section aims to quantitatively assess the impact of road traffic related parameters on noise emissions. The methodology of measurement noise can be described using the following Fig. 1.

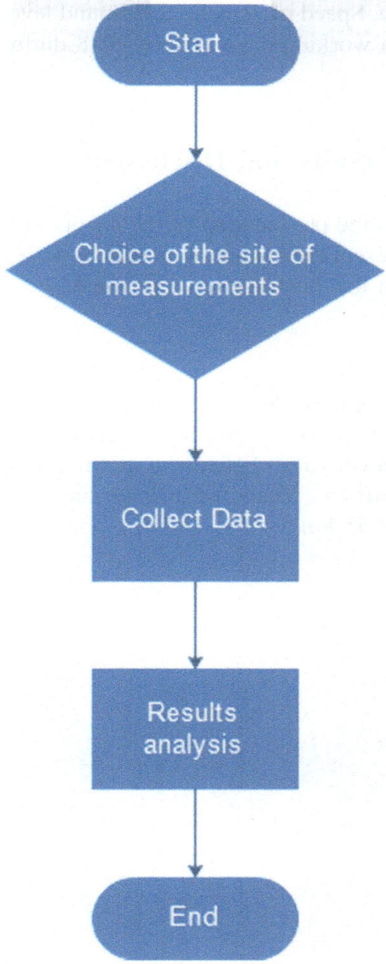

Fig. 1 Flowchart of the experimental procedure

To reach our goal, we started with the choice of sites, the identification of instruments, measurement procedures, parameters to be measured and then, the analysis of noise variation over time and various traffic parameters. In our case study, we choose 3 different sites: the first represent a rural segment, the second is an urban segment and the third site is the highway. These sites were selected on the basis of a set of factors influence the traffic noise such as the presence of large reflective surface (buildings, parked vehicles, traffic signs), the type of vehicle, nature of flow (continuous or not), the type of road, the condition of the roadway.

The sound level is measured using a portable digital sound level meter TES-1352A. The noise measurements were made simultaneously with the recording of road traffic flow with the help of a camera. The data was recorded on each work site: Number of vehicles for each category, Speed of vehicles and Sound level. The measurements were carried out during certain working days in June 2018 during a peak hour and an off-peak hour of road traffic.

3 Measurements Results and Discussion

This section is devoted to the presentation of the results of noise measurements in the three studied sites. We present the evolution of vehicle noise and the influence of traffic and vehicle parameters in this environmental problem.

3.1 Site 1

The following figure shows noise level measurements over time during a peak hour.

By exploiting Fig. 2, two noise peaks are observed that exceed 90 dB (A). However, between the noise level varies between a minimum value of 53.5 dB (A) and a maximum value of 87.3 dB (A). Then, the average speed of vehicles traveling on this site varies between 8 and 55 km/h.

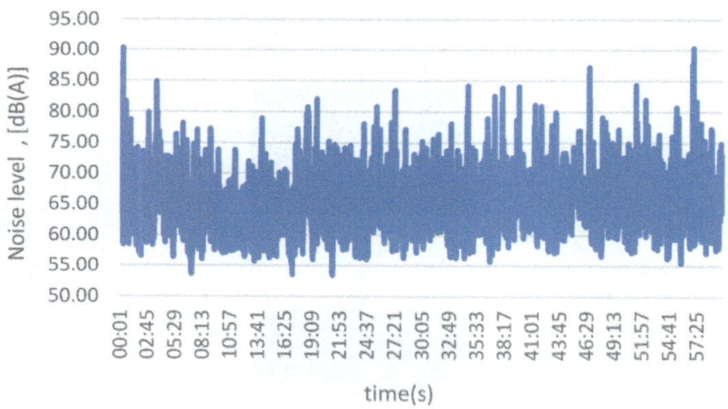

Fig. 2 Evolution of noise level in site 1 during a peak hour

In the other hand, during an off-peak hour, results show the presence of several variant noise peaks between a minimum value of 52.2 dB (A) and a maximum value of 89.5 dB (A) (Fig. 3).

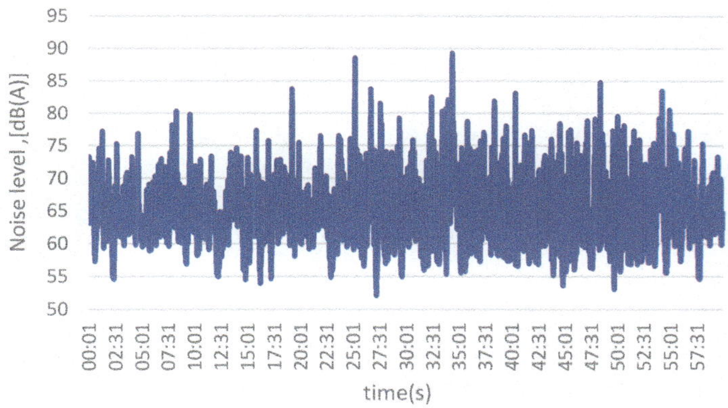

Fig. 3 Evolution of the noise level in site 1 during an off-peak hour

3.2 Site 2

The total number of vehicles crossing this zone is between 1600 and 1800 vehicles/hour. During a peak hour, there are about 1796 vehicles/hour, while about 1668 vehicles/hour are recorded during an off-peak road traffic.

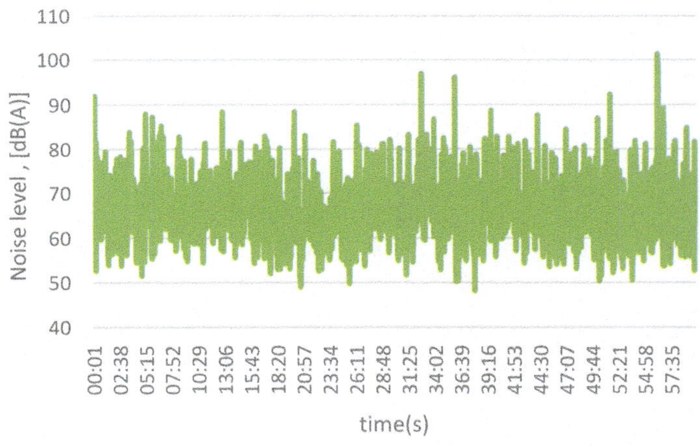

Fig. 4 Evolution of the noise level in site 2 during a peak hour

Results shows the appearance of a loud noise peak of 92.2 dB (A) in the first few minutes. Then, over time, several peaks ranging from 48.4 dB (A) to 83.6 dB (A) and in the last minutes, the presence of about two peaks of noise exceed 97 dB (A) and a peak of 101.5 dB (A) (Fig. 4).

However, it is observed that most noise peaks vary between a minimum value of 52.7 dB (A) and a maximum value of 85.9 dB (A) (Fig. 5).

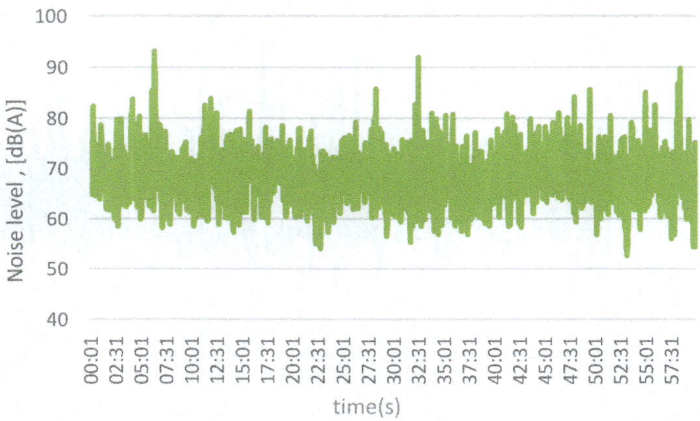

Fig. 5 Evolution of the noise level in site 2 during an off-peak hour

3.3 Site 3

For this site, traffic on the motorway is characterized by a traffic flow between 900 and 1200 vehicles/hour. However, it is observed that during a peak hour about 1133 vehicles and about 913 during an off-peak hour of hollow. The following figure shows the profile of the sound level over time and depending on the noise influence factors in this site. Several noise peaks exceeding 80 dB (A) are observed. However, the noise level in the motorway during the peak period is between a minimum value of 50.9 dB (A) and a maximum value of 87.8 dB (A) (Fig. 6).

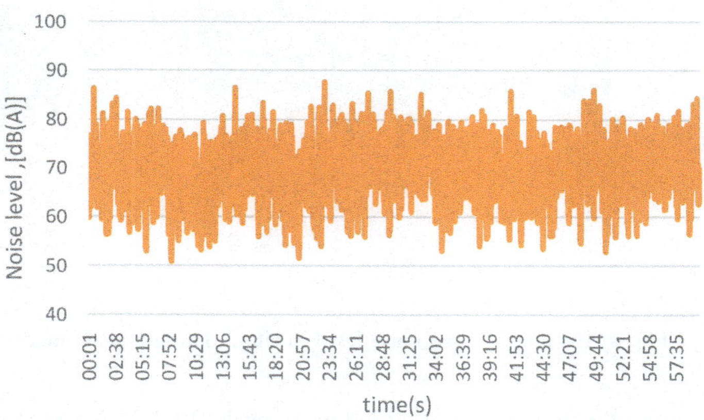

Fig. 6 Evolution of the noise level in site 3 during a peak hour

The noise level during an off-peak hour varies between a minimum value of 50.6 dB (A) and a maximum value of 97.5 dB (A).

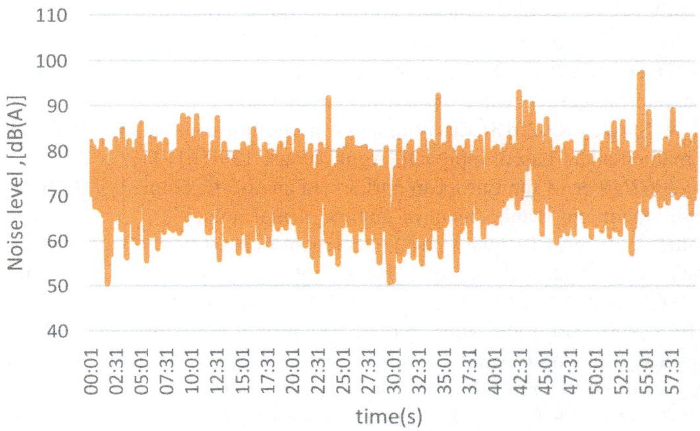

Fig. 7 Evolution of the noise level in site 3 during an off-peak hour

Measured data demonstrate an increase in noise level when the speed is greater than 40 km/ h. however, it is observed that from 25 to 40 km/ h, the noise level between 55 and 80 dB (A), gold, from 40 to 110 km/ h, the noise level between 60 and 90 dB (A) (Fig. 7).

Based on these results, it is important to indicate that the sound level profile varies from site to site. The change in noise level generally due to temporal changes in road traffic, the average speed of vehicles and the total number of vehicles as well as the infrastructure of the study site.

4 Conclusion

In this paper an experimental procedure to evaluate measurement noise is presented. We proposed the evolution of vehicle noise in three sites selected in Sousse-Tunisia. We started by collecting data related to traffic flow and vehicle characteristic simultaneously with the recording of noise level in each site. Then, we treated data to develop the evolution of noise level according to the selected parameters. Results demonstrate the influence of the vehicle speed, the infrastructure and the total number of vehicles in the noise level. In future work, this data base constructed will serve to develop a predictive model to estimate vehicle noise.

References

Journal article

1. Aballéa F-E, René P-J, Cosandey L (2010) Road map: model comparison emissions. In: 10ème Congrès Français d'Acoustique Lyon, 12–16 Avril 2010
2. Brundtland Commission (1987) Report of the world commission on environment and development, United Nations

3. Burgess MA (1977) Noise prediction for urban traffic conditions-related to measurements, Sydney metropolitan area, Applied acoustics, England, 10, 7p
4. Cvetokiv D et al (1997) NAISS-model for traffic noise prediction, Working and living environmental protection, vol 1, no 02, pp 73–81
5. Debnath A, Singh PK (2018) Environmental traffic noise modelling of Dhanbad township area—a mathematical based approach. Appl Acoust 129:161–172
6. Directive 2002/49/ec of the European parliament and of the council of 25 June 2002 relating to the assessment and management of environmental noise
7. Fhwa traffic noise model (fhwa tnm®)—technical manual. U.S. Department of Transportation Research and Special Programs Administration (1998)
8. Griffiths ID, Langdon FJ (1968) Subjective response to road traffic noise. J Sound Vib 8:16
9. Hamet JF (1996) Bruit de trafic routier-Abaques d'emission du modèle de prévision américain: FHWA Traffic Noise model, INRETS, Rapport MMA no 9603, 33 p
10. Hugrel C, Joumard R, Transport routier-parc, usage et émissions des véhicules en France de 1970 à 2025. INRETS Bron cedex, France. Rapport LTE no 0420, Septembre 2004
11. Morel J (2012) Caractérisation physique et perceptive du bruit routier urbain pour la définition d'indicateurs de gêne sonore en situation de mono-exposition et de multi-exposition en présence de bruit industriel, 311 p
12. Quartieri J, Mastorakis NE, Iannone G, Guarnaccia C, D'Ambrosio S, Troisi A, Lenza TLL (2009) A Review of traffic noise predictive models, Recent advances in applied and theoretical mechanics, Puerto De La Cruz, Tenerife, Canary Islands, Spain, 14–16 December 2009
13. Rawat K, Katiyar K, Pratibha V (2009) Mathematical modeling of environmental noise impact. Indian J Biomech
14. Subramani T, Kavitha M, Sivaraj KP (2012) Modelling of traffic noise pollution. IJERA 2:3175–3182
15. Sukeerth G, Munilakshmi N, Anilkumarreddy C (2017) Mathematical modeling for the prediction of road traffic Noise levels in Tirupati town. Int J Eng Dev Res 5(2), 8
16. World Health Organization (2009) Night noise guidelines for Europe, WHO Regional Office for Europe
17. World Health Organization (2011) Burden of disease from environmental noise: quantification of healthy life years lost in Europe. WHO Regional Office for Europe. http://www.euro.who.int

Design of an Electronic Throttle Body Based on a New Knowledge Sharing Engineering Methodology

Mouna Fradi[1,2(✉)], Raoudha Gaha[1(✉)], Abdelfattah Mlika[1], Faïda Mhenni[2], and Jean Yves Choley[2]

[1] Laboratoire de Mécanique de Sousse, Ecole Nationale d'Ingénieurs de Sousse, Université de Sousse, BP 264 Sousse Erriadh 4023, Sousse, Tunisie
mounafradi@yahoo.com, raoudha.gaha@gmail.com, abdelfattah.mlika@gmail.com

[2] QUARTZ EA7393, SUPMECA-Paris, 3 Rue Fernand Hainaut, Saint-Ouen 93407, France
{Faida.mhenni,jean-yves.choley}@supmeca.fr

Abstract. To minimize the number of iterations and correction returns while designing a system, sharing crucial parameters and data between different actors is needed. Since mechatronic systems are considered complex because of their multi-disciplines, their design requires collaborative work to ensure the sharing of parameters between contributors from different domains. This paper proposes a new methodology based on the collaborative design to choose the architecture of a mechatronic system. This methodology is structured around three main phases: the pre-collaboration phase, the collaboration phase and the post-collaboration phase. The proposed methodology has been validated by applying it on a mechatronic system called Electronic Throttle Body (ETB). In order to share the different established activities and capitalize knowledge, KARREN (Knowledge Acquisition and reuse for Robust Engineering) a collaborative tool from DPS (Digital Product Simulation) a French company, is used in this project to help us to choose the appropriate architecture of the Electronic Throttle Body.

Keywords: Collaborative design · Mechatronic system · Methodology · ETB · KARREN

1 Introduction

Mechatronics is defined as a synergic integration of mechanical, electronics, information systems and computer systems in the design of a product in order to increase and optimize the functionality. This alliance requires a simple and fast data exchange between different disciplines to obtain a product of high quality, low price and a short time. For this reason, the concept of collaborative engineering has replaced sequential engineering to improve communication between project actors and to ensure robust and efficient sharing of knowledge. In this context, knowledge Based-Engineering (KBE) is a technology able to capture and reuse knowledge to reduce costs and time of development. This technology has large application such as knowledge reuse and

N. Aifaoui et al. (Eds.): CMSM 2019, LNME, pp. 55–63, 2020.
https://doi.org/10.1007/978-3-030-27146-6_7

collaboration support [12]. In this direction, this paper proposes a new methodology based on a collaborative design to choose the most appropriate architecture of a mechatronic system using KBE application to support collaboration between actors.

The paper is organized in several sections. Section 2 presents a review of related works. In Sect. 3, our methodology is presented in details. In Sect. 4, the proposed methodology is illustrated on a case study. Section 5 gives some results and Sect. 6 summarizes and concludes the paper.

2 State of the Art

2.1 Mechatronic Systems Development Methodologies

As previously mentioned, mechatronic systems are known with their multidisciplinary nature, which increases the complexity of the design. In order to decrease this complexity, various groups of researchers to develop mechatronic systems with high quality and low cost have done significant works. Four types of development methodologies are discussed in this section.

First, the methodologies based on the optimization with integration of control were presented as methodologies to develop mechatronic systems. In this context, Zhang et al. [21] proposed an integrated approach for mechatronic design to facilitate the control system. To improve this work [10] developed a design framework called Control For Design (DFC). This approach emphasizes the importance of designing the control parameters simultaneously with the structural parameters even though it is possible to change the controller parameters after the system is built. Another methodology has been proposed by Villarreal-Cervantes et al. [20] to formulate the mechatronic design problem of the 5R 2Dof parallel robot and its Proportional-Integral-Derivative (PID) controller. They integrated a nonlinear dynamic optimization problem. This work concerns parallel planar robots. Nevertheless, some applications require high performance positioning systems with a spatial parallel mechanism. In this framework, the contribution of Lara-Molina et al. [9] takes into account the structure and control design variables simultaneously. They proposed an optimal design methodology for Stewart-Gough robot. This robot is a three-dimensional mechanism where its movable platform is connected to the fixed base by six legs. This methodology aims at maximizing the accuracy of positioning to optimize the overall performance of the robot. In these previous works, effective approaches have been introduced. However, the improvement of the system performance based on the possibility of modifying controller parameters was overlooked, which represents an important criteria in design evaluation.

Methodologies based on the design evaluation are another type, which aims at developing mechatronic systems. Moulianitis et al. [16] proposed an evaluation index that includes the complexity, flexibility and intelligence of mechatronic systems. In addition, Hammadi et al. [8] proposed a multi-criteria performance indicator called Mechatronic Design Indicator (MDI) as a neuronal network of radial basis function. This indicator provides accurate information for decision making while decreasing design time. However, this methodology overlooks other mechatronic design requirements such as risk assessment, cost, etc. As an improvement to this work, Mohebbi et al. [15] developed a new multi-criteria profile. This profile includes five criteria, which are cost, flexibility, complexity, intelligence and reliability of mechatronic systems. These criteria can be determined using a specific mathematical function. This methodology is applied to the conceptual design of a 6-Dof robotic manipulator. Following this work, Moulianitis et al. [17] proposed a new index includes seven criteria, which are configurability, adaptability, dependability, decisional autonomy, motion ability, perception ability and interaction ability. In formulating this index, the cost and the complexity are taken into account. The proposed index is applied to the design of firefighting robot in order to evaluate its design solutions. However, all these approaches are not extended for use in the detailed design phase.

Due to their flexibility in design representation, bond graphs and evolutionary algorithms are considered as an efficient method for system representation in detailed phases as well as preliminary phases. Seo et al. [19] developed a unified and automated methodology for the design of mechatronic systems. The proposed approach consists in combining Bond-Graphs (BG) and Genetic programming (GP). However, this work is oriented to the structural part in mechatronic systems and lacks the controller part that is always present in these systems. In order to achieve the best topology and parameters of a mechatronic system, Behbahani and De Silvia [5] proposed a new evolutionary algorithm. This algorithm makes it possible to explore efficiently the design space. In this work, the control part has been taken into account. Another methodology combining (BG) and (GP) is presented by Samarakoon et al. [18]. This approach was implemented in an industrial fish cutting machine. Nevertheless, the nonlinearity of the machine has been overlooked in this work.

Finally, although the structure of mechatronic systems has been treated during these previous works, the control part has been overlooked in some of them. This absence can influence the reality of these systems since the control is an essential part in mechatronic products. In this context, several works to model mechatronic systems have been done using the object-oriented modeling language Modelica and Dymola software. This tool makes it possible to control mechatronic systems while modeling their structures in the same environment. Modelica has been used by Ferretti et al. [7] to model and simulate the behavior of a mechatronic machining center. Another work was done by Mcharek et al. [13]. In this work, a mechatronic system called Electronic Throttle Body (ETB) was modelled with Modelica language. The contribution of this work is represented by taking into account the integration of the fail-safe technique in the ETB model.

2.2 Methodologies Based on Collaborative Design and Knowledge Sharing

Over many decades, the complexity of products is in increase. This increase implies also an increase in the number of people needed to develop these systems. Actors need to interact and exchange crucial data to obtain an ideal product. This can be achieved by following collaborative design. Such a strategy enable companies to reduce costs, improve quality and reduce time to market while innovating their products. In this context, researchers developed several methodologies in order to exchange and share knowledge. First, methodologies based on Product Lifecycle Management (PLM) have been used to exchange knowledge between actors. Chen et al. [6] developed an approach of dividing the mechatronic system into components and classifying them into two parts an electrical and a mechanical one. After this classification, the constraints between components are defined and a table summarizing the components and the connection between them is established. This methodology reduces the complexity of mechatronic systems but the classification of components can cause problems while developing these systems. Alexopoulos et al. [2] proposed a methodology for integrating data mechatronic system into the PLM system. In this approach, Automation Markup Language (AutomationML) has been used as an intermediate format to exchange data between different tasks. This language facilitates the integration and exchange of data. However, there are not many translation tools available because AutomationML is a new format. Another interesting study is proposed by Abid et al. [1] based on the use of behavioral and structural diagrams of system Modeling Language (SysML) to model structure of a complex product in a PLM system. The PLM in this case makes it possible to coordinate the exchange between designers. This methodology is validated with a 3D robot. All these methodologies have shown their efficiency in the exchange of data between actors. However, they do not allow an easy access to data embedded in different models. Second, in the field of simulation and modeling, the goal is to provide an integrated environment for engineers in order to achieve optimal multi-domain design. This can be satisfied using technology called Process Integration and Design Optimization (PIDO). There are several tools to apply this technology such as Model Center [14] and Isight. In this context, Azzouzi et al. [3] proposed a methodology aims at improving system modeling by applying PIDO technology. This approach has been validated by applying it to a hybrid vehicle using an optimization tool (Model Center). This tool is used to identify unknown parameters of a system. An improvement of this methodology can be made by minimizing the error between model results and experimental data. However, PIDO technology is used in the optimization of a single architecture not to share knowledges in order to choose the most appropriate architecture among several ones of a system. Finally, Knowledge Based-Engineering (KBE) tools are used in collaborative design since they allow knowledge capitalization and reuse. Several methodologies were done using KBE tools. Badin [4], during his PhD research work, developed a new methodology based on configuration management of crucial knowledge for product design and numerical simulation. This approach allows engineers to capitalize, reuse and maintain coherently knowledges between different domains. As an improvement of this methodology, decision support approaches to guide designers can be integrated. Mcharek et al. [12], also proposed a new methodology based on KBE concept. This approach is characterized by its ability to support designers during design phases as well as its ability to reuse design results. The contribution of Mcharek et al. [12] is illustrated by its application to the design of a mechatronic system called Electronic Throttle Body (ETB). We will consider the same system during the application of the methodology. Therefore, our contribution in this paper is illustrated by the development of a new methodology, which will help us not only to collaborate between different actors of one architecture of a mechatronic system but also to compare between several architectures and to choose the most appropriate one.

3 Proposed Methodology

The interaction between several companies working at the same time for the same project has become a standard for manufacturers and more practically in the aeronautical and automotive fields. These fields are known by their complexity because of the integration of different disciplines. Therefore, each actor must have access to the right information, at the right time and with the right information format. Mechatronic systems are one of complex systems, which are characterized by the combination of mechanical, computer systems and electronics aided by control strategies. These systems can be represented by various architectures or solutions, which make their design very tedious. The design process is composed of seven steps as shown in Fig. 1. In this context, our contribution aims at developing a new methodology that helps engineers to choose the best architecture of mechatronic system among several ones in the conceptual phase. This methodology is divided into three main phases. The first phase of our methodology starts with an idea of designing a system in order to satisfy one or more specific needs. Once the function of the system is defined, a search of different design solutions must be established. If the different solutions have been selected, we can create the models for each solution. These models have different forms such as 3D, fluid, procurement and dynamic models. During the second phase, all project actors will collaborate with each other. This phase represents the core of our methodology since at this stage the actors can share their work to detect the inconsistencies between them and reduce correction returns. This phase starts with the extraction of crucial data from different established models and from the requirements imposed on the system. Once important data are extracted, they can be shared in the collaboration support. Indeed, only the data characterizing the interdependence between disciplines need to be shared. The last step of the second phase consists in visualizing and managing conflicts either in one model or between several models. In the third phase, a comparison between the different solutions or architectures of the system must be done and the best solution can be chosen based on this comparison. Figure 2 illustrates the global process of our methodology.

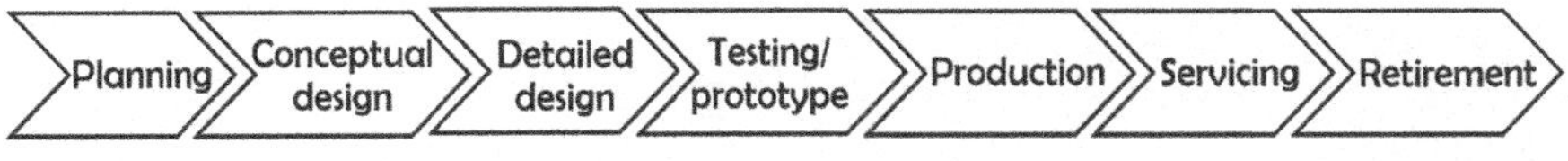

Fig. 1 Product development process [11]

4 Case Study

The proposed approach is used for the choice of the best architecture among three of an electronic Throttle Body (ETB). The ETB adjusts the airflow entering to the combustion engine. Following the methodology steps, we apply it to the ETB. This application starts with finding the different solutions.

For the ETB we choose three different architectures one with a DC motor, one with a brushless motor and another one with a stepper motor. After choosing these

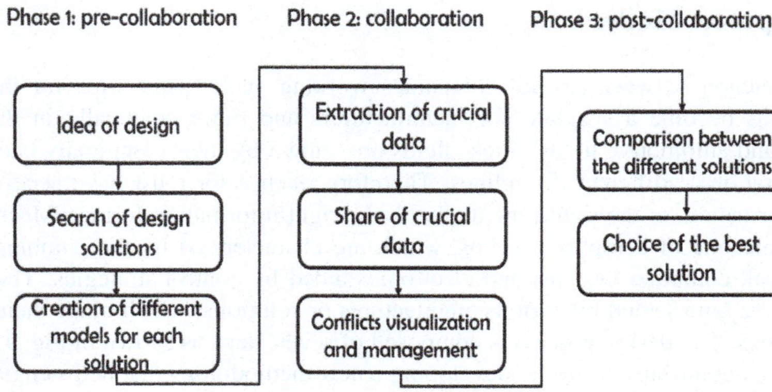

Fig. 2 The global process of the proposed methodology

architectures, the different models for each architecture must be established. For this application, three models will collaborate with each other: a dynamic model using the environment Dymola and the object-oriented language Modelica, a 3D model using CATIA and procurement model, which consists in finding the most appropriate motors conforming within simulation results. The ETB model with a DC motor in Dymola environment is illustrated in Fig. 3. The DC motor is replaced by the brushless motor in the second alternative and by the stepper motor in the third one. After creating these dynamic models with Dymola software, the performances of each motor such as torque, power, voltage, etc. must be visualized. In the procurement models, three motors have be chosen conforming within visualized performances. Indeed, a single 3D model must be created since the difference between the three alternatives concerns motors type not the geometry.

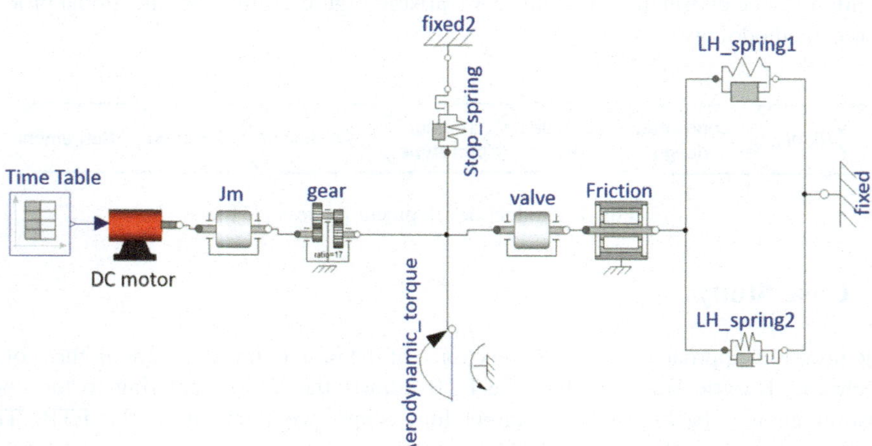

Fig. 3 ETB model with a DC motor in Dymola environment

Once all models have been created, crucial data can be extracted and shared between actors. Sharing data consists in organizing it in Information Core Entities (ICEs), which are generic entities aims to organize and capitalize crucial data, as presented in Table 1. In the end of this application, the existing conflicts are visualized and managed and the best architecture is chosen.

Table 1 Organization of crucial data

Parameters	Unit	Description	Constraints
ICE_performance motor			
P	W	Power of motor	–
C	N.m	Torque of motor	<1.5 N.m
ICE_dimension motor			
L	mm	Motor length	–
D	mm	Motor diameter	–

5 Results and Discussion

Our methodology was tested in KARREN platform. Figure 4 summarizes the obtained results. Only the chosen DC motor in procurement model can be implemented in the 3D model. The other chosen motors have incompatible dimensions with the 3D model. Therefore, we choose the first alternative of the ETB with a DC motor.

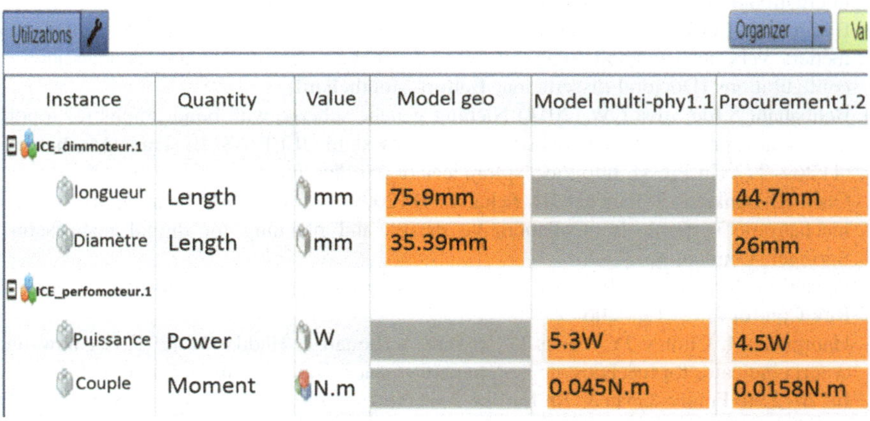

Fig. 4 Obtained results in KARREN platform

6 Conclusions

This paper has presented a new methodology based on collaborative design and knowledge sharing to choose the best architecture among several ones of a mechatronic system. This methodology was applied to an Electronic Throttle Body. In this application, we considered three alternatives each one with a different type of motors (with a DC motor, brushless motor and stepper motor) and we created three models for each alternative which are 3D model, procurement model and dynamic model.

In our future work, we will add another domain such as safety system to improve our methodology.

Acknowledgements. The authors would like to express their acknowledgments to the mechanical laboratory of Sousse (Tunisia) and QUARTZ laboratory in SUPMECA-Paris.

References

1. Abid H, Pernelle P, Noterman D, Campagne JP, Ben Amar C (2015) SysML approach for the integration of mechatronics system within PLM systems. Int J Comput Integr Manuf 28 (9):972–987
2. Alexopoulos K, Makris S, Xanthakis V, Chryssolouris G (2011) A web-services oriented workflow management system for integrated digital production engineering. CIRP J Manuf Sci Technol 4(3):290–295
3. Azzouzi EM, Iuliano M, Hammadi M, Plateaux R, Marchand F (2016) Process integration and design optimization technologies for modelling improvement. In: 2016 IEEE international symposium on systems engineering (ISSE). IEEE, pp 1–5
4. Badin J (2011) Ingénierie hautement productive et collaborative à base de connaissances métier: vers une méthodologie et un méta-modèle de gestion des connaissances en configurations (Doctoral dissertation, Belfort-Montbéliard)
5. Behbahani S, de Silva CW (2014) Niching genetic scheme with bond graphs for topology and parameter optimization of a mechatronic system. IEEE/ASME Trans Mechatron 19 (1):269–277 (In Press). http://ieeexplore.ieee.org/xpl/login.jsp
6. Chen K, Bankston J, Panchal JH, Schaefer D (2009) A framework for integrated design of mechatronic systems. In: Collaborative design and planning for digital manufacturing. Springer, London, pp 37–70
7. Ferretti G, Magnani G, Rocco P (2004) Virtual prototyping of mechatronic systems. Ann Rev Control 28(2):193–206
8. Hammadi M, Choley JY, Penas O, Riviere A, Louati J, Haddar M (2012) A new multi-criteria indicator for mechatronic system performance evaluation in preliminary design level. In: 2012 9th France-Japan & 7th Europe-Asia congress on mechatronics (MECATRONICS), and 2012 13th international workshop on research and education in mechatronics (REM). IEEE, pp 409–416
9. Lara-Molina FA, Dumur D, Koroishi EH (2016) Structure-control optimal design of 6-DOF fully parallel robot. In: Robotics. Springer, Cham, pp 247–266
10. Li Q, Zhang WJ, Chen L (2001) Design for control-a concurrent engineering approach for mechatronic systems design. IEEE/ASME Trans Mechatron 6(2):161–169

11. Masoudi A, You H, Suh SH (2012) A guidance system for selecting an appropriate eco-design checklist in the early stages of product development. In: 3rd International conference on green and sustainable innovation, May 2012
12. Mcharek M, Azib T, Hammadi M, Choley JY, Larouci C (2018) Knowledge sharing for mechatronic systems design and optimization. IFAC-PapersOnLine 51(11):1365–1370
13. Mcharek M, Hammadi M, Choley JY, Azib T, Larouci C (2016) Modeling and multi-objective optimization of an Electronic Throttle in open-loop. In: 17th international conference on research and education in mechatronics (REM) 2016 11th France-Japan & 9th Europe-Asia congress on mechatronics (MECATRONICS). IEEE, pp 331–335
14. Model Center (2012) Phoenix Integration. www.phoenix-int.com
15. Mohebbi A, Achiche S, Baron L (2014) Mechatronic multicriteria profile (MMP) for conceptual design of a robotic visual servoing system. In: 12th Biennial conference on engineering systems design and analysis, ASME 2014. American Society of Mechanical Engineers, p V003T15A015
16. Moulianitis VC, Aspragathos NA, Dentsoras AJ (2004) A model for concept evaluation in design—an application to mechatronics design of robot grippers. Mechatronics 14(6): 599–622
17. Moulianitis VC, Zachiotis GA, Aspragathos NA (2018) A new index based on mechatronics abilities for the conceptual design evaluation. Mechatronics 49:67–76
18. Samarakoon BL, Gamage LB, de Silva CW (2016) Design evolution of engineering systems using bond graphs and genetic programming. Mechatronics 33:71–83
19. Seo K, Fan Z, Hu J, Goodman ED, Rosenberg RC (2003) Toward a unified and automated design methodology for multi-domain dynamic systems using bond graphs and genetic programming. Mechatronics 13(8–9):851–885
20. Villarreal-Cervantes MG, Cruz-Villar CA, Alvarez-Gallegos J (2009) Structure-control mechatronic design of the planar 5r 2dof parallel robot. In: 2009 IEEE international conference on mechatronics, ICM 2009. IEEE, pp 1–6
21. Zhang WJ, Li Q, Guo LS (1999) Integrated design of mechanical structure and control algorithm for a programmable four-bar linkage. IEEE/ASME Trans Mechatron 4(4): 354–362

Optimization Design of the Sewing Mechanism Using Multi-criteria Colonial Competitive Method

Najlawi Bilel[1(✉)], Nejlaoui Mohamed[1], Affi Zouhaier[1], and Romdhane Lotfi[2]

[1] LGM, ENIM, University of Monastir, Monastir, Tunisia
najlawibilelali@gmail.com, nejlaouimohamed@gmail.com, zouhaier.affi@enim.rnu.tn
[2] Department of Mechanical Engineering, American University of Sharjah, Sharjah, United Arab Emirates
lotfi.romdhane@gmail.com

Abstract. In the textile manufacturing, development of sewing mechanisms with optimal performances is very significant. Obviously, the quality of sewing, decrease the sewing machine vibration and the optimal mechanical advantage are greatly dependent on the design criteria of the sewing mechanism. Therefore, this paper presents a multi-objective design optimization of a sewing machines. The needle jerk, the coupler tracking error and the transmission angle index are minimized simultaneously. The aim is to reduce the sewing machine vibration, guarantee a proper stitch formation and ensure the most effective transmission of motion, respectively. The multi-objective colonial competitive algorithm (MOCCA) is used to perform the design optimization. The obtained results confirm improvement of the required design criteria of the sewing mechanism in this study. It is also concluded that the optimized mechanism has 35% less needle Jerk and 20% reduction in term of the transmission angle index (TA) than the one for the Juki machine.

Keywords: Optimization · Colonial competitive method · Vibration

1 Introduction

Nowadays, the application of several engineering fields in textile industry has shown considerable influence. In this context, development of industrial sewing machines with optimal performance is of great importance. In many manufacturing processes, sewing is commonly used for joining different parts. Consequently, small improvements in the design and development of sewing machines would result in significant economic gain [10]. Nevertheless, there is still a clear gap in synthesizing industrial sewing

N. Aifaoui et al. (Eds.): CMSM 2019, LNME, pp. 64–73, 2020.
https://doi.org/10.1007/978-3-030-27146-6_8

mechanisms. For example in this context, vibrations reduction of sewing machines by eccentric balancing of crank mechanism is developed in Sidlof and Votrubec [11]. El Gholmy and El Hawary [2] analyzed the sewing needle vibrations by the Gorman's Eigen values technique. Payvandy and Ebrahimi [10] used the colonial competitive algorithm (CCA) to optimize the needle jerk of the needle bar and thread take up lever (NBTTL) mechanism. Sidlof and Votrubec [11] reduced the sewing machine vibration through an optimization approach based on MathModelica software. Najlawi et al. [6] develop an optimization approach, based on the CCA algorithm, in order to minimize the needle jerk of the NBTTL mechanism.

With a few exception, these optimization approaches are mono-objective ones, which tries to minimize the sewing needle vibrations. However, most of the optimized sewing machine, obtained through mono-objective optimization, has poor performances with respect to other characteristics. The transmission angle index (TA) is one of these characteristics that has to be considered to obtain a practical sewing machine. In fact, the optimization of TA improves the Kinetic performance and practical usability of the machine [4]. Najlawi et al. [5] have presented an analytical expressions for the TA of four-bar mechanism by the application of vector loop equations. Kapse and Handa [3] proposes an approach to identify the effect of change in link length on the TA performance of planar mechanism, using relative velocity method. Khorshidi et al. [4] used an optimization approach to optimize the four-bar linkages by considering the optimization of the transmission angle index.

To the best knowledge of the authors, no specific study has been carried out on this subject. Therefore, in this paper, a multi-objective optimal design of the NBTTL mechanism is developed. The aim is to reduce the sewing machine vibration, guarantee a proper stitch formation and ensure the most effective transmission of motion, respectively. For this purpose, the Multi-objective Colonial Competitive Algorithm (MOCCA) [9] is used as the optimization strategy to find the optimal link lengths of the NBTTL mechanism.

2 The NBTTL System

The NBTTL mechanism, used in sewing machines, consists of a slider crank mechanism and a four-bar mechanism driven by the same crank (Fig. 1).The thread take-up lever mechanism is the four-bar linkage OABC. During the formation of a loop, the take-up lever eye D pulls the upper thread vertically [8]. OEF represents the slider-crank mechanism in which point F denotes the needle. The function of the needle, which is fixed to the needle bar, is to penetrate the fabric. The displacement of the needle bar is represented by the distance ‘S’.

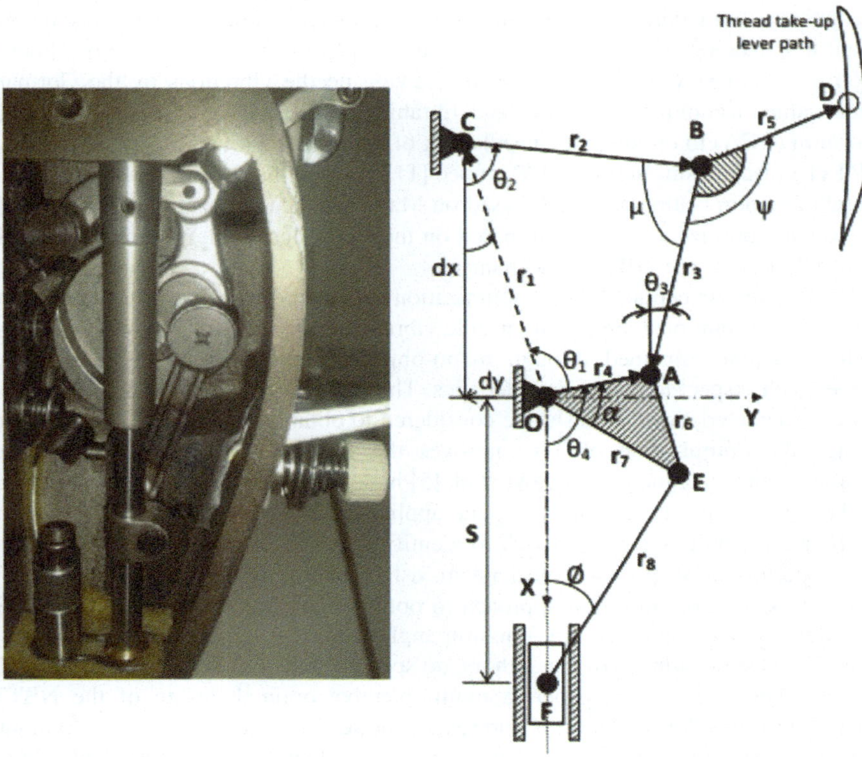

Fig. 1 The NBTTL of a sewing machine

3 The NBTTL Mechanical Performances

In this work, three criteria are considered simultaneously, i.e., the vibration, the input-output torque and the quality of the stitch formation. These criteria are described analytically by the needle Jerk index, transmission angle index and the coupler tracking error, respectively.

3.1 The Needle Jerk (NJ)

The function of the needle (Fig. 1) is to penetrate the fabric and carry the thread under the stitch plate [7]. The needle displacement is given by:

$$s = r_7 \cos(\theta_4 - \alpha) + r_8 \cos\left(\sin^{-1}\left(\frac{r_7}{r_8}\sin(\theta_4 - \alpha)\right)\right) \tag{1}$$

Thus, to decrease the vibration of the sewing machine, we reduce the needle jerk index given as follows:

$$NJ = \int_0^{2\pi} |\dddot{s}| d\theta_4 \tag{2}$$

NJ defines the first objective function to be minimized.

3.2 The Transmission Angle (TA)

The transmission angle index TA denotes the quality of motion transmission in the NBTTL mechanism. TA is given by Najlawi et al. [5]:

$$TA = \left[\left(\mu_{max} - 90^{\circ} \right)^2 + \left(\mu_{min} - 90^{\circ} \right)^2 \right] \tag{3}$$

$$\mu = \cos^{-1}\left(\frac{r_2^2 + r_3^2 - r_4^2 - r_1^2 + 2r_1 r_4 \cos\theta_1}{2r_2 r_3} \right) \tag{4}$$

A mechanism designed with an ideal transmission angle index ($\mu = 90^{\circ}$) leads to a minimum forces acting along the coupler and on the bearings [4, 5].

TA defines the second objective function to be minimized.

3.3 The Coupler Traking Error (TE)

The role of the coupler point D is to ensure appropriate thread feeding and to guide the thread motion. As shown in Fig. 1. The position of the coupler point is given as follows:

$$\begin{cases} r_{Dx} = r_4 \cos\theta_4 - r_3 \cos\theta_3 + r_5 \cos(\psi - \theta_3) \\ r_{Dy} = r_4 \sin\theta_4 + r_3 \sin\theta_3 + r_5 \sin(\psi - \theta_3) \end{cases} \tag{5}$$

To ensure a proper stitch forming, the vertical position of the coupler point D has to follow the prescribed path. Moreover, the horizontal position of the coupler point D has not shown any influence on the stitch quality [5]. Therefore, the tracking error of the coupler point (TE) represents the third objective function, given by:

$$TE = \int_0^{2\pi} |r_{Dd}(\theta_4) - r_{Dx}(\theta_4)| d\theta_4 \tag{6}$$

where $r_{Dd}(\theta_4)$ denotes the desired position of point D. This desired position is given by a sewing machine Juki DDL 8700 taken as a reference. The Juki design parameters are shown in Najlaoui et al. [7].

4 Multi-objective Optimization Design of the NBTTL Mechanism

The optimization problem can be defined as minimizing simultaneously TA, NJ and TE. Therefore, the multi-objective optimization problem can be formulated as:

$$\text{Minimize}\begin{cases} TA = \left[(\mu_{max} - 90^\circ)^2 + (\mu_{min} - 90^\circ)^2\right] \\ NJ = \int_0^{2\pi} |\dddot{s}| d\theta_4 \\ TE = \int_0^{2\pi} |r_{Dd}(\theta_4) - r_{Dx}(\theta_4)| d\theta_4 \end{cases} \tag{7}$$

$$\text{Subject to}: \begin{cases} 2[\max(r_1, r_2, r_3, r_4) + \min(r_1, r_2, r_3, r_4)] \\ < r_1 + r_2 + r_3 + r_4 \\ \theta_4^i - \theta_4^{i+1} < 0 \\ 30\,mm \le S \le 62\,mm \\ x_i \in [x_{i\,min}, x_{i\,max}], \;\; x_i \in X \end{cases} \tag{8}$$

The search domains of the design variables are shown in Table 1.

Table 1 Range of the design variables

X_i	dx(mm)	dy(mm)	r_i(mm)	$\psi(^\circ)$	$\alpha(^\circ)$
$[x_{i\,min}, x_{i\,max}]$	[0 60]	[0 60]	[0 60]	[0 360]	[0 360]

To solve the optimization problem (Eqs. 7 and 8), the Multi-objective colonial competitive algorithm (MOCCA) will be used.

The MOCCA algorithm is an evolutionary optimization method developed recently by Najlawi et al. [9] and is based on the CCA method [1]. Figure 2 shows the flowchart of the MOCCA algorithm. The MOCCA starts with a random initial population where each individual of the population represents a country. The cost of a country is determined by evaluating the objective functions. The most powerful country is called "imperialist" and the remaining countries are considered as "colonies". Each initial empire is composed of one imperialist and several colonies. After forming initial empires, colonies start moving towards the relevant imperialist. In each empire, the crossover and the random mutation operators enhance the colonies with a new one that can have more power and eventually constitute imperialists. In this case, the colony and the imperialist permute positions. Then, the non dominated imperialists are kept in an archive based on the fast non dominated sorting approach (Najlawi et al. [9]. This archive represents the Pareto front. Based on their power, any empire that does not improve in imperialist competition will be diminished. As a result, the imperialistic competition will increase the power of great empires and weaken the frail ones. Thus, weak empires will collapse. Finally, The MOCCA algorithm stops when only one empire remains. This condition can be satisfied when all the empires collapse, except the most powerful one.

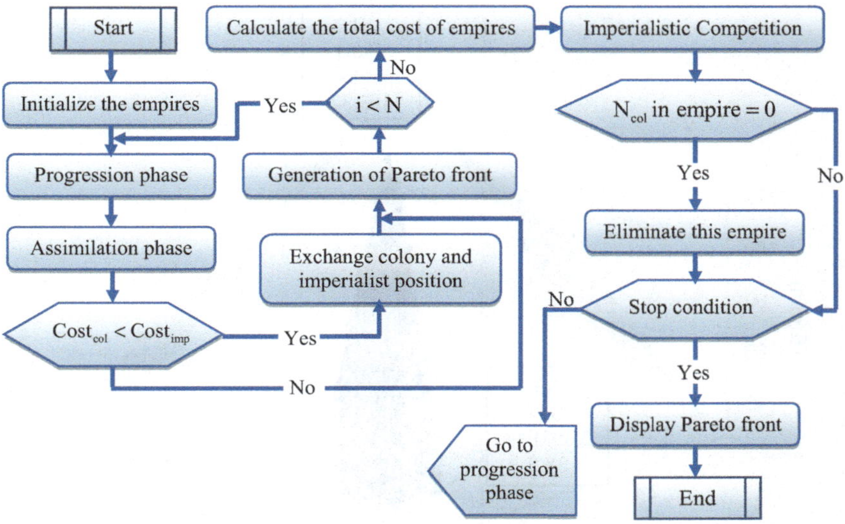

Fig. 2 The MOCCA flowchart

5 Results and Discussion

In what follows, we will present the results of the multi-objective optimization of the problem defined by Eq. (7) under the constraint presented by Eq. (8) with the search domain given by Table 1. Using the MOCCA algorithm, the Pareto front can be determined to find the optimal non dominated solutions (Fig. 3).

In Fig. 3, each non-dominated solution, given by the minimization of the three objective functions, represents an optimal design vector of the NBTTL mechanism. Three optimal solutions are selected from the Pareto front (Fig. 3) and their corresponding design vectors are shown in Table 2. For comparison reason, we give in the same table the design vector of the known industrial sewing machine Juki 8700 [7].

Table 2 Design parameters of the selected optimal solutions

Design parameters	S1	S2	S3	Juki 8700
dx(mm)	36.35	32.14	43.25	38.64
dy(mm)	29.28	**34.78**	**32.98**	33.22
r_2(mm)	28.30	41.64	45.36	44.80
r_3(mm)	38.73	28.17	32.97	30.31
r_4(mm)	6.74	7.09	8.24	7.68
r_6(mm)	28.66	27.45	31.03	40.26
ψ(°)	114.96	129.74	130.65	122
α(deg)	21.27	15.8	19.18	18
r_7(mm)	12.39	13.31	12.87	11.57
r_8(mm)	44.89	45.67	43.09	49.82

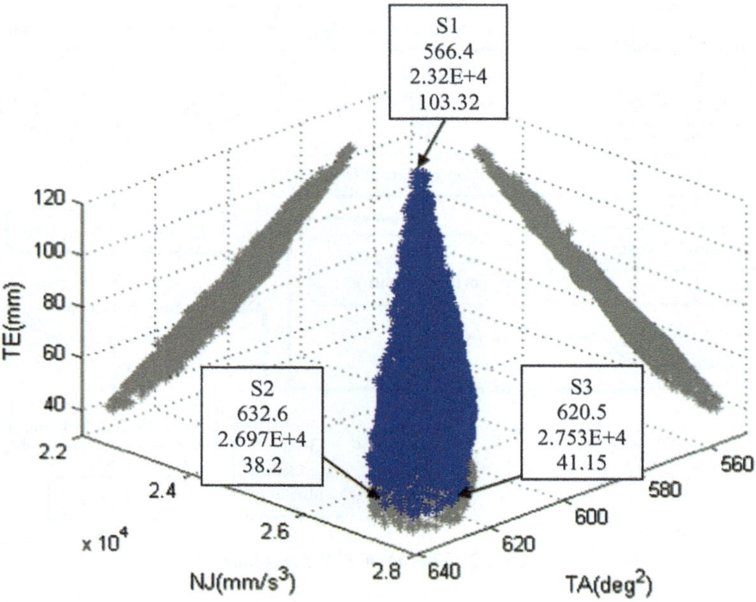

Fig. 3 Pareto front of optimal solutions

Figure 4 shows the evolution of the needle Jerk, for the extreme optimal solutions S1, S2 and S3 as a function of the input link angle. We present also, in the same figure, the evolution of the needle jerk, obtained by the sewing machine Juki 8700. One can note that by using the S1 design vector, we decrease the needle jerk down by 35% compared to the one for the Juki 8700.

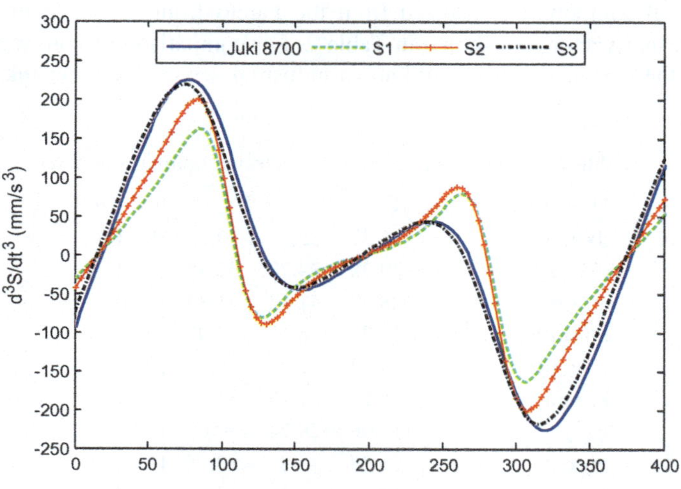

Fig. 4 Evolution of the needle jerk

In Fig. 5, we show the optimal coupler displacement, obtained for S1, S2 and S3 solutions, as a function of the input link angle θ_4. For the comparison raison, we present in the same figure the desired coupler displacement. One can note that the percentage error between all the obtained optimal solutions compared to the Juki 8700 sewing machine (taken as a reference) does not exceed 3%.

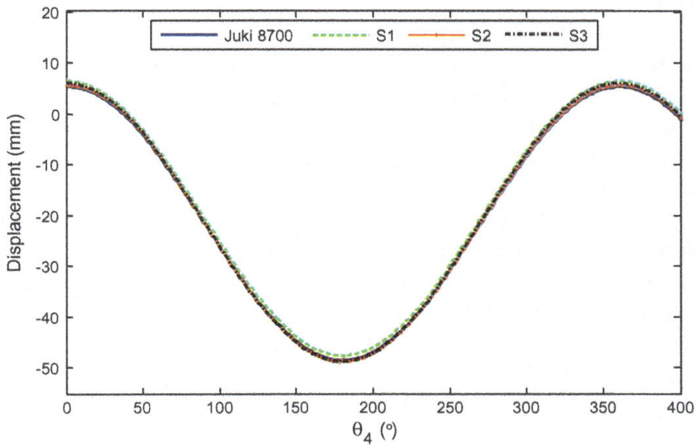

Fig. 5 Variation of the coupler point position

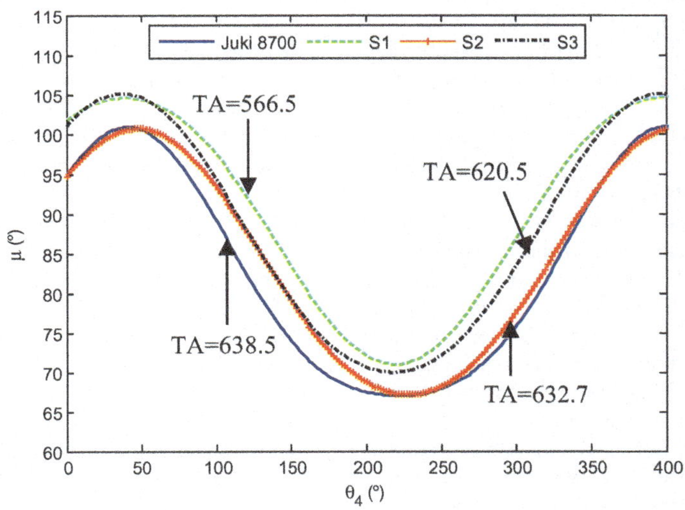

Fig. 6 Variation of the transmission angle

From Fig. 6 one can note that, compared to Juki 8700, we can reduce the TA by about 20% by using the optimal S1 solution.

Based on the previous results, we can conclude that the optimized mechanism has a clear superiority over the reference sewing machine Juki 8700. In fact, all the obtained optimal designs reduce the needle Jerk (NJ) and the transmission angle index (TA) while keeping the same coupler tracking error (TE) (not exceed 3%).

6 Conclusion

The study presented here was dealing with a multi-objective design optimization of the needle bar and thread take-up lever (NBTTL) mechanism used in a typical lockstitch sewing machine. Therefore, the optimization problem was constructed to reduce the sewing machine vibration, minimize the input to output torques ratio and ensure a proper stitch forming by minimizing, simultaneously, the needle jerk (NJ), the transmission angle index (TA) and the coupler tracking error (TE). The multi-objective colonial competitive algorithm (MOCCA) was utilized for this purpose. According to the obtained results, the optimized mechanism had clear superiority in comparison to the industrial machine Juki 8700, in terms, of the needle Jerk (NJ) and the transmission angle index (TA) while keeping the same coupler tracking error (TE).

References

1. Atashpaz-Gargari E, Lucas C (2007) Imperialist competitive algorithm: an algorithm for optimization inspired by imperialistic competition. IEEE Cong Evol Comp 4661–4667
2. El Gholmy SH, El Hawary IA (2016) The application of Gorman's Eigen values to the industrial sewing machine's needle vibration. Alex Eng J 55:983–989
3. Kapse AK, Handa CC (2012) A generalized approach for measurement of performance of planar mechanism using relative velocity method. Int J Eng Res Appl (IJERA), 2:1871–1573
4. Khorshidi M, Soheilypour M, Peyro M, Atai A, Shariat Panahi M (2011) Optimal design of four-bar mechanisms using a hybrid multi-objective GA with adaptive local search. Mech Mach Theory 46:1453–1465
5. Najlawi B, Nejlaoui M, Affi Z, Romdhane L (2015) Optimal design of the needle bar and thread take up lever mechanism using a multi-objective imperialist competitive algorithm. In: 2015 10th International symposium on mechatronics and its applications (ISMA), Sharjah. https://doi.org/10.1109/ISMA.2015.7373463
6. Najlawi B, Nejlaoui M, Affi Z, Romdhane L (2017) Multi-objective robust design optimization of a mechatronic system with uncertain parameters, using a polynomial chaos expansion method. Proc IMechE Part I J Syst Control Eng 1–11. https://doi.org/10.1177/0959651817722333
7. Najlaoui B, Nejlaoui M, Affi Z, Romdhane L (2017) Mechatronic design optimization of the mechanism in a sewing machine. Proc IMechE Part C J Mech Eng Sci. https://doi.org/10.1177/0954406216687786
8. Najlawi B, Nejlaoui M (2018) Analytical modeling of needle temperature in an industrial sewing machine. Heat Transf Res 1–14. https://doi.org/10.1615/HeatTransRes.2018018051
9. Najlawi B, Nejlaoui M, Affi Z, Romdhane L (2018) An efficient evolutionary algorithm for engineering design problems. Soft Comput 1–17. https://doi.org/10.1007/s00500-018-3273-z

10. Payvandy P, Ebrahimi S (2015) Optimization of thread take-up lever mechanism in lockstitch sewing machine using the imperialistic competitive algorithm. J Text Polym 3:12–19
11. Sidlof P, Votrubec V (2011) Vibrations reduction of industrial sewing machines. In: Vibration problems, ICOVP 2011

Investigating the Inline Design Measure in Existing Pressurized Steel Piping Systems

Mohamed Fersi[1(✉)] and Ali Triki[2]

[1] Department of Mechanics, National Engineering School of Sfax, University of SFAX, B.P. 1173, 3038 Sfax, Tunisia
mohamedfersi@yahoo.com
[2] Research Unit: Mechanics, Modelling Energy and Materials M2EM, National Engineering School of Sfax, University of SFAX, Sfax, Tunisia
ali.triki@enis.rnu.tn

Abstract. This paper examined the effectiveness of the inline re-design strategy used to mitigate the cavitating flow into an existing steel piping system. This strategy is based on substituting a short-section of the transient sensitive region of the existing main pipe by another one made of (**HDPE**) or (**LDPE**) plastic material. The (**1−D**) pressurized pipe flow model based on the Ramos formulation was used to describe the flow behavior, along with the fixed grid Method of Characteristics being used for numerical computations. From the case studied, it was shown that such a technique could mitigate the undesirable cavitating flow onset. Besides, this strategy allowed positive-surge magnitude attenuation. It was also found that pressure rise or drop attenuation was slightly more important for the case using an (**LDPE**) inline plastic short-section than that using an (**HDPE**) one. Furthermore, results evidenced that other factors influencing the surge attenuation rate were related to the short-section dimensions.

Keywords: Design · **HDPE** · **LDPE** · Method of characteristics · Plastic material · Ramos formulation · Viscoelasticity · Water-Hammer

1 Introduction

Water-hammer is a common phenomenon that hydraulic designers and engineers have to face in pressurized piping systems. This phenomenon generates pressure–rise and-drop and even sub-atmospheric pressure, which can produce the collapse of the system depending on the conditions of the installation.

Among the various available classical design tools taken to control water-hammer surges, we distinguish: (i) the "passive measures", which are based on the selection of pipe-wall material, pressure classes and thicknesses, according to the ultimate allowable transient pressure. Albeit, this alternative allows good hydraulic performances, it may significantly increase the cost of the piping systems, if used separately; (ii) the "active measures" which influences the operational procedures of the system by equipping the hydraulic systems with protective devices to absorb excessive pressure rise or drop [3, 12, 14, 21, 24]. Aside from the analysis of the used technologies in the

N. Aifaoui et al. (Eds.): CMSM 2019, LNME, pp. 74–82, 2020.
https://doi.org/10.1007/978-3-030-27146-6_9

available protective devices, it must be delineated that none of these technologies is a panacea that can be used inherently for all piping system; the adequate protective strategy is specific for each system case and depends upon the initiating transient event type. Oftentimes, a combination of multiple devices may prove to be the most desirable [22]. Nonetheless, this method exhibits some drawbacks arising from technical implementation constraints and the requirement of system shutdowns or partial load operation of hydraulic machineries ([16, 20]). In addition, due to the complex nature of the transient behavior, a device intended to attenuate a transient condition could even result in the worsening of another condition if the device is not adequately selected and/or located in the system [15].

In this regards, several researches (e.g.: Ghilardi et al. [7–9, 15–20]) addressed the inline-based design strategy, in order to address the forgoing drawbacks. Namely, the authors examined the efficiency of adding a plastic short-section in-line to the sensitive region of the original piping system (or substituting a short-section of the sensitive region of existing steel-piping system by another one made of plastic material) to attenuate both positive and negative hydraulic-head surge. The authors observed that the employed plastic short-section reduced the first pressure-surge peak and crest. Nonetheless, the authors noticed increasingly effect of pressure-wave oscillations period into the protected piping system; which may have adverse effect on the operational procedure of the hydraulic system. Physically, this result is attributed to the viscoelastic behavior of plastic materials which have a retarded deformation component in addition to the immediate one, observed in the case of elastic materials [1, 4, 6, 11, 25].

Considering the aforementioned discussion, the main intention of this paper is to explore the efficiency of the inline strategy to mitigate a cavitating flow onset induced into an existing steel piping system. Special focus is given for the hydraulic-head attenuation rate and the wave oscillation period spreading effects.

The next section outlines the numerical procedure used for solving the transient flow problem.

2 Materials and Methods

According to Ramos et al. [13], the one dimensional (**1−D**) transient flow equations accounting for pipe-wall viscoelasticity and unsteady friction effects, may be expressed as follows:

$$\frac{\partial H}{\partial t} + \frac{a_0^2}{gA}\frac{\partial Q}{\partial x} = 0 \tag{1}$$

$$\frac{1}{A}\frac{\partial Q}{\partial t} + g\frac{\partial H}{\partial x} + g\left(h_{f_s} + \frac{1}{gA}\left(k_{r1}\frac{\partial Q}{\partial t} + k_{r2}a_0\,Sgn(Q)\left|\frac{\partial Q}{\partial x}\right|\right)\right) = 0 \tag{2}$$

where, H is the hydraulic-head; Q is the flow discharge; A is the cross sectional area of the pipe; g is the gravity acceleration; a_0 is the wave speed; x and t are the longitudinal coordinate along the pipeline axis and the time, respectively; the quasi-steady head loss component per unit length h_{f_s} is computed for turbulent and laminar flow, respectively, as follows: $h_{f_s} = RQ|Q|$ and $h_{f_s} = 32\nu Q/(gD^2A)$ where, $R = f/(2DA)$ is the pipe resistance; ν is the Poisson ratio; $k_{r1} = 0.003$ and $k_{r2} = 0.04$ are two decay coefficients [13].

The numerical solution of momentum and continuity Eqs. (1) and (2), within a multi-pipe system framework, is typically performed using the Fixed-Grid Method of Characteristics (**FG-MOC**) [5, 8, 10, 13, 15, 19].

Briefly, the compatibility equations, corresponding to the finite difference discretization of Eqs. (1) and (2) along the set of characteristic lines of the computational grid, may be expressed as follows:

$$\mathbf{C}^j_{\pm}: \frac{dQ^j}{dt} + \frac{gA^j}{\alpha^{\pm} \times (1 + k_{r1})} \frac{dH^j}{dt} + \frac{f^j}{2d^j(1 + k_{r1})} Q^j |Q^j| = 0 \text{ along } \frac{dx^j}{dt} = \alpha^j_{\pm} \tag{3}$$

where,

$$\alpha^j_{\pm} = 2a^j_0(1 + k_{r1}) \Big/ \left(-\mathbf{Sgn}\left(Q\frac{\partial Q}{\partial x} \right) k_{r2} \pm (2 + k_{r1}) \right).$$

The relationships between the hydraulic-head and discharge parameters along the characteristics lines may be deduced directly from Eq. (3) as follows:

$$\begin{cases} \mathbf{C}^j_+ : Q^j_{i,t} = c'^j_p - c''^j_p H^j_{i,t} \\ \mathbf{C}^j_- : Q^j_{i,t} = C'^j_n + c''^j_n H^j_{i,t} \end{cases} \tag{4}$$

in which: $c'^j_p = Q^j_{i-1,t-1} + c''^j_p H^j_{i-1,t-\Delta t} - c'''^j_p \Big/ 1 + k_{r1}$; $c''^j_p = gA^j / B^j \alpha^j_+$; $c'''^j_p = \{(f^j \Delta t)/(2D^j)\} Q^j_{i-1,t-1} \left| Q^j_{i-1,t-1} \right|$; $c'^j_n = Q^j_{i+1,t-1} + c''^j_n H^j_{i+1,t-\Delta t} + c'''^j_p \Big/ 1 + k_{r1}$; $c''^j_n = gA^j / B^j \alpha^j_-$; $c'''^j_n = \{(f^j \Delta t)/(2D^j)\} Q^j_{i+1,t-1} \left| Q^j_{i+1,t-1} \right|$; $B^j = gA^j / a^j_0$; the indices i denotes the section index of the jth pipe $\left(1 \le i \le n^j_s\right)$; n^j_s designates the number of sections of the jth pipe, respectively; the indices $i \pm 1$ refer to the characteristics nodes, in the characteristics grid, at the left and right sides of node i; Δt and Δx correspond to the time and space-step increments, respectively.

The numerical procedure, outlined above, allows hydraulic parameters computation, for a single-phase flow. For the cavitating flow onset, the discrete gas cavity model (**DGCM**) may be included in the (**MOC**) procedure assuming that cavities are lumped at the computing sections [10, 26, 27].

Using the perfect gas law, the isothermic evolution of each isolated gas cavity can be written as:

$$\forall^t_{gi}\left(H^t_i - z_i - H_v\right) = (H_0 - z_i - H_v)\alpha_0 A \Delta t \tag{5}$$

where H_0 is the reference piezometric-head; α_0 the void fraction at H_0; z_i the pipe elevation; and H_v the gauge vapor hydraulic-head of the liquid.

The equation calculating cavity volume $\forall_g$, at a given cross-section, is derived from the discretization of local continuity equation using the **FG-MOC**:

$$\forall_{gi}^{t} = \forall_{gi}^{t-2\Delta t} + \left[\psi\left(Q_i^t - Q_{ui}^t\right) - (1-\psi)\left(Q_i^{t-2\Delta t} - Q_{ui}^{t-2\Delta t}\right)\right]2\Delta t \quad (6)$$

where, $\forall_{gi}$ and $\forall_{gi}^{t-2\Delta t}$ correspond to the cavity volumes at the current time step and at $2\Delta t$ time steps earlier, respectively, and ψ is a weighting factor, chosen in the range: $0.5 \leq \psi \leq 1$ [2].

It is worth noting that the cavity collapses inasmuch as $\forall_g \prec 0$. In this case, and hence, the liquid phase is re-established and the one-phase water-hammer solver is valid again.

Series Junction

A common hydraulic grade-line elevation and no flow storage assumptions are made for calculating the flow parameters at the series section [15, 17, 23].

$$Q_{ns^j,t}^{j-1} = Q_{1,t}^{j} \text{ and } Q_{ns^j,t}^{j-1} = Q_{1,t}^{j} \quad (7)$$

where, the right and left hands of Eq. (7) correspond to the hydraulic parameters estimated at the up- and down-stream sides of the junction.

3 Application, Results and Discussion

The original hydraulic system layout, considered in this study, consists of a sloping pipe system connecting two pressurized tanks and equipped with a ball valve at its inlet. The main steel pipeline characteristics are: $E_0^{\text{steel}} = 210$ GPa; $D = 44.1$ mm; $L = 100$ m; and $a_0 = 1302.5$ m/s. The downstream pipe axis is taken as the horizontal datum level ($z_d = 0$ m) and the upstream reservoir level is $z_u = 2.03$ m. The gauge saturated hydraulic-head of the liquid is equal to: $H_g = -10.2$ m. The initial steady-state regime was established for a constant flow velocity a static hydraulic-head values set equal to $V_0 = 1.04$ m/s and $H_0^{T_2} = 21.4$ m, respectively; prior to a transient event corresponding to the fast and full closure of the upstream valve. The boundary conditions associated with such an event may be expressed as follows.

$$Q_{|x=0} = 0 \text{ and } H_{|x=L} = H_0^{T_2} \ (t \succ 0) \quad (8)$$

In such a situation, the inline technique consists in substituting an upstream short-section of the main steel pipe by another one made of plastic pipe-wall material (Fig. 1).

As a starting step, the inline plastic short-section length and diameter are selected equal to: $l_{\text{short-section}}^{\text{plastic}} = 5\,\text{m}$ and $d_{\text{short-section}}^{\text{plastic}} = 50.6\,\text{mm}$, respectively.

The input parameters of the **FG-MOC** embedding the **DGCM** procedure are: time step $\Delta t = 0.034\,\text{s}$; Courant numbers $c_r^{\text{steel-pipe}} = 0.9841$ and $c_r^{\text{plastic short-section}} = 1$, corresponding to the main steel-pipe and the plastic short-section; and $\psi = 0.5$.

Figure 2 compares the upstream hydraulic-head signals, versus time, involved by the hydraulic systems with and without applying the inline technique. Jointly, the main features of the wave curves, plotted in Fig. 2, are enumerated in Table 1.

As can be seen from Fig. 2, such a transient event leads to the occurrence of the unfavorable cavitation phenomenon, into the original system case. However, if instead a plastic short-section inline technique is implemented, the cavitation phenomenon may be palliated.

A detailed analysis of Fig. 2 and Table 1 reveals that, for the original system case, the change in the upstream boundary condition triggered a series of positive and negative surge waves. In addition, the hydraulic-head profile, corresponding to this case, illustrates a short-duration pulses resulting from the superposition of surge wave involved by the valve-closure and the wave generated by the collapse of the vapor cavity. Besides, this hydraulic-head pulses exhibit a downward gradual attenuated trend due to friction losses.

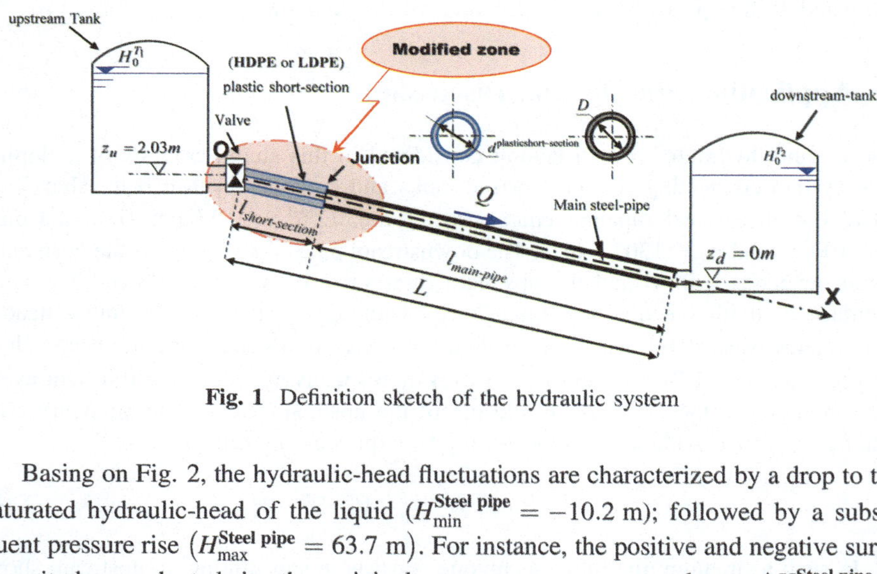

Fig. 1 Definition sketch of the hydraulic system

Basing on Fig. 2, the hydraulic-head fluctuations are characterized by a drop to the saturated hydraulic-head of the liquid ($H_{\min}^{\textbf{Steel pipe}} = -10.2$ m); followed by a subsequent pressure rise $\left(H_{\max}^{\textbf{Steel pipe}} = 63.7 \text{ m}\right)$. For instance, the positive and negative surge magnitudes evaluated in the original system case are equal to: $\Delta H_{\text{up-surge}}^{\textbf{Steel pipe}} = H_{\max}^{\textbf{Steel pipe}} - H_0 = 47.1$ m and $\Delta H_{\text{down-surge}}^{\textbf{Steel pipe}} = H_0 - H_{\min}^{\textbf{Steel pipe}} = 32.2$ m, respectively, above the initial steady-state value.

Nonetheless, analysis of the hydraulic-head signals depicted into the protected hydraulic system, shows a hydraulic-head rise or drop magnitude equal to: $\Delta H_{\text{up-surge}}^{\textbf{HDPE}} = 11.1$ m or $\Delta H_{\text{downsurge}}^{\textbf{HDPE}} = 20.9$ m for the case involving an (**HDPE**) plastic short-section (Table 1). Moreover, a lower hydraulic-head rise or drop is observed for the case involving a (**LDPE**) plastic short-section ($\Delta H_{\text{up-surge}}^{\textbf{LDPE}} = 5.4$ m or

$\Delta H^{\mathbf{HDPE}}_{\mathrm{downsurge}} = 14.7$ m). This in return implies that the positive or negative pressure attenuation ratio obtained using a short section made of (**HDPE**) are equal to $\eta H^{\mathbf{HDPE}}_{\mathrm{up-surge}} = \Delta H^{\mathbf{HDPE}}_{\mathrm{up-surge}} \big/ \Delta H^{\mathbf{steel}}_{\mathrm{up-surge}} = 9.27\%$ or $\eta H^{\mathbf{HDPE}}_{\mathrm{downsurge}} = \Delta H^{\mathbf{HDPE}}_{\mathrm{downsurge}} \big/ \Delta H^{\mathbf{steel}}_{\mathrm{downsurge}} = 26.15\%$, respectively; and more important ratios are involved by the (**LDPE**) short-section material case (i.e.: $\eta H^{\mathbf{LDPE}}_{\mathrm{up-surge}} = \Delta H^{\mathbf{LDPE}}_{\mathrm{up-surge}} \big/ \Delta H^{\mathbf{steel}}_{\mathrm{up-surge}} = 64.91\%$ or $\eta H^{\mathbf{LDPE}}_{\mathrm{downsurge}} = \Delta H^{\mathbf{LDPE}}_{\mathrm{downsurge}} \big/ \Delta H^{\mathbf{steel}}_{\mathrm{downsurge}} = 45.65\%$, respectively).

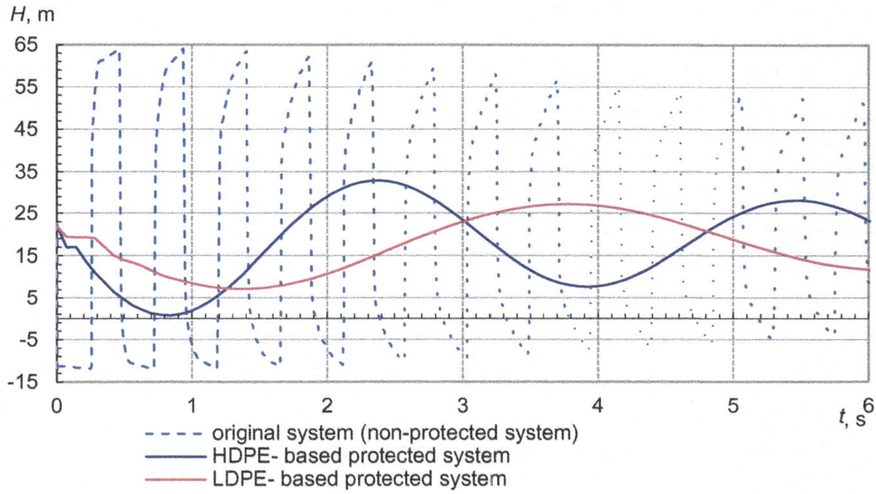

Fig. 2 Comparison of hydraulic-heads at the upstream valve section versus time for the hydraulic system with and without implementation of the protection procedure

Table 1 Characteristics of water-hammer waves in Fig. 2

Parameters		Steel main-pipe **HDPE**	Plastic short-section (**LDPE**)	
$H_{\max}$: 1st hydraulic-head peak	(m)	63.7	33.1	27.4
$H_{\min}$: 1st hydraulic-head crest	(m)	−10.2	1.1	7.3
T_1: period of the 1st cycle of wave oscillation	(s)	0.41	1.37	3.73

On the other side, based on Fig. 2 and Table 1, it is remarkable to point out that the periods of the first cycle of hydraulic-head oscillations, predicted into a (**HDPE**) or (**LDPE**) plastic short-section-based protected system case are: $T_1^{\mathbf{HDPE}-\mathrm{penstock}} = 1.37$ s or $T_1^{\mathbf{LDPE}-\mathrm{penstock}} = 3.37$ s; while the corresponding period associated with the original system case is equal to $T_1^{\mathbf{steel-pipe}} = 0.41$ s. Thereupon, the inline technique based on a (**HDPE**) or (**LDPE**) short-section induces a spreading of the wave oscillations period equal to: $\delta T^1_{\mathbf{HDPE}} = \left|T^1_{\mathbf{steel}} - T^1_{\mathbf{HDPE}}\right| = |0.41 - 1.37| = 0.96$ s or $\delta T^1_{\mathbf{LDPE}} = \left|T^1_{\mathbf{steel}} - T^1_{\mathbf{LDPE}}\right| = |0.41 - 3.73| = 3.32$ s as compared with that involved by the original system case. Furthermore, the period spreading induced by the (**LDPE**) short-section–based inline technique relatively to the **HDPE** short-section–based inline technique is equal to: $\delta' T^1_{\mathbf{LDPE}} = \left|T^1_{\mathbf{HDPE}} - T^1_{\mathbf{LDPE}}\right| = 3.73 - 1.37 = 2.36$ s).

The preceding discussions argue that the (**HDPE**)-based inline technique allows better trade-off, between the attenuation of hydraulic-head peak (and crest) and the limitation of spreading of hydraulic-head oscillation period, as compared with the (**LDPE**)-based one.

All the results presented thus far were obtained for a specific short-section size (i.e.: $d^{\text{plastic}}_{\text{short-section}} = 50.6$ mm and $l^{\text{plastic}}_{\text{short-section}} = 5$ m). Additional results with respect to the magnitude sensitivity of the first hydraulic-head crest to the size of the replaced plastic short-section are reported in Fig. 3a and b, for several diameter and length values of the employed short-section: $d^{\text{plastic}}_{\text{short-section}} = \{45; 50.6; 75 \text{ and } 100 \text{ mm}\}$ and $l^{\text{plastic}}_{\text{short-section}} = \{5;\ 7.5;\ \ 10;\ \ 12.5\text{and } 15 \text{ m}\}$, respectively.

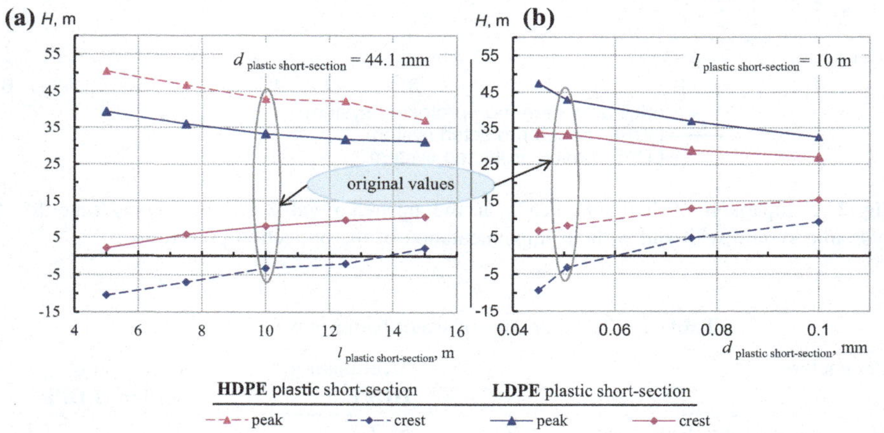

Fig. 3 Variation of hydraulic-head peaks and crests at the downstream valve section depending on the plastic short-section: **a**—length, **b**—diameter

Figure 3a and b suggest that for length and diameter values beyond $l^{\text{plastic}}_{\text{short-section}} \geq 10$ m and $d^{\text{plastic}}_{\text{short-section}} \geq 44.1$ mm, respectively, the variation of the first transient pressure crests is slightly affected. Consequently, the diameter value $d^{\text{plastic}}_{\text{short-section}} = 44.1$ mm and the length value $l_{\text{short-section}} = 10$ m may be considered as the optimal values of the plastic short-section size for the case studied herein.

4 Conclusion

In summary, the present study highlighted that the employed technique provides a large damping of the first pressure peak and crest associated to a transient initiating event. In addition, this pressure damping is observed to be more pronounced when using a (**LDPE**) plastic material for the added short-section than an (**HDPE**) material. However the former technique induces more important wave period spreading as compared with the latter one. Furthermore, it is also shown that other factors contributing to the hydraulic-head attenuation rate depend upon the short-section size (i.e. length and diameter). Specifically, examination of the sensitivity of the pressure peak or crest magnitude, with the short-section length and diameter being the controlling variables, verifies that significant volumes of the short-section provide important hydraulic-head attenuation. However, this correlation is not significant beyond a near-optimum diameter and length values.

One intends that such a technique may greatly enhance the reliability and improve the cost-effectiveness of industrial hydraulic utilities, while safeguarding operators. Future test configurations including pipe networks may represent an extension to this study.

References

1. Aklonis JJ, MacKnight WJ, Shen M (1972) Introduction to polymer viscoelasticity. Wiley-Interscience, Wiley
2. Bergant A, Simpson AR, Tijsseling A (2006) Waterhammer with column separation: a historical review. J Fluids Struct 22(2):135–171. https://doi.org/10.1016/j.jfluidstructs.2005.08.008
3. Besharat M, Tarinejad R, Ramos H (2015) The effect of water hammer on a confined air pocket towards flow energy storage system. J Water Supply Res Technol Aqua 65(2):116–126. https://doi.org/10.2166/aqua.2015.081
4. Brinson HF, Brinson LC (2008) Polymer engineering science and viscoelasticity: an introduction. Springer
5. Chaudhry MH (2014) Applied hydraulic transient. Van Nostrand Reinhold Company
6. Ferry JD (1970) Viscoelastic properties of polymers, 2nd edn. Wiley, New York
7. Fersi M, Triki A (2018) Investigation on redesigning strategies for water-hammer control in pressurized-piping systems. J Press Vessel Technol Trans ASME. https://doi.org/10.1115/1.4040136
8. Fersi M, Triki A (2019) Alternative design strategy for water-hammer control in pressurized-pipe flow. In: Fakhfakh T, Karra C, Bouaziz S, Chaari F, Haddar M (eds) Advances in acoustics and vibration II, ICAV 2018. Applied Condition Monitoring, vol 13, 135–144, Springer, pp 157–165. https://doi.org/10.1007/978-3-319-94616-0_16
9. Ghilardi P, Paoletti A (1986) Additional viscoelastic pipes as pressure surge suppressors. In: Proceedings of 5th international conference on pressure surges, Cranfield (UK), pp 113–121
10. Ghidaoui MS, Zhao M, Duncan AM, David HA (2005) A review of water-hammer theory and practice. Appl Mech Rev 58:49–76. https://doi.org/10.1115/1.1828050
11. Güney MS (1983) Water-hammer in viscoelastic pipes where cross-section parameters are time dependent. In: Proceedings of 4th international conference on pressure surges, BHRA, Bath, U.K, pp 189–209

12. Moussou P, Gibert RJ, Brasseur G, Teygeman C, Ferrari J, Rit JF (2010) Relief instability of pressure valves in water pipes. Press Vessel Technol 132(4):041308. https://doi.org/10.1115/1.4002164
13. Ramos H, Covas D, Borga A, Loureiro D (2004) Surge damping analysis in pipe systems: modelling and experiments? J Hydraul Res 42(4):413–425. https://doi.org/10.1080/00221686.2004.9641209
14. Rosselló JM, Urteaga R, Bonetto FJ (2018) A novel water hammer device designed to produce controlled bubble collapses. Exp Therm Fluid Sci 92:46–55. https://doi.org/10.1016/j.expthermflusci.2017.11.016
15. Triki A (2016) Water-hammer control in pressurized-pipe flow using an in-line polymeric short-section. Acta Mech 227(3):777–793. https://doi.org/10.1007/s00707-015-1493-13
16. Triki A (2017) Water-Hammer control in pressurized-pipe flow using a branched polymeric penstock. J Pip Syst Eng Pract ASCE 8(4):04017024. https://doi.org/10.1061/(ASCE)PS.1949-1204.0000277
17. Triki A (2018) Further investigation on water-hammer control inline strategy in water-supply systems. J Water Suppl Res Technol AQUA 67(1): 30–43. https://doi.org/10.2166/aqua.2017.073
18. Triki A (2018) Dual-technique based inline design strategy for Water-Hammer control in pressurized-pipe flow. Acta Mech 229(5):2019–2039. https://doi.org/10.1007/s00707-017-2085-z
19. Triki A, Fersi M (2018) Further investigation on the Water-Hammer control branching strategy in pressurized steel-piping systems. Int J Press Vessels Pip 165(C):135–144. https://doi.org/10.1016/j.ijpvp.2018.06.002
20. Triki A, Chaker MA (2019) Compound technique -based inline design strategy for water-hammer control in steel pressurized-piping systems. Int J Press Vessel Pip 169C:188–203. https://doi.org/10.1016/j.ijpvp.2018.12.001
21. Wan W, Huang W (2011) Investigation on complete characteristics and hydraulic transient of centrifugal pump. J Mech Sci Technol 25:2583. https://doi.org/10.1007/s12206-011-0729-9
22. Wan W, Li F (2016) Sensitivity analysis of operational time differences for a pump-valve system on a water hammer response. J Press Vessel Technol Trans ASME 138(1):011303. https://doi.org/10.1115/1.4031202
23. Wan W, Huang W (2018) Water hammer simulation of a series pipe system using the MacCormack time marching scheme. Acta Mech 229(7):3143–3160 https://doi.org/10.1007/s00707-018-2179-2
24. Wan W, Zhang B (2018) Investigation of water hammer protection in water supply pipeline systems using an intelligent self-controlled surge tank. Energies 11(6):1450. https://doi.org/10.3390/en11061450
25. Weinerowska-Bords K (2006) Viscoelastic model of waterhammer in single pipeline—problems and questions. Arch Hydro-Eng Environ Mech 53(4):331–351. ISSN 1231–3726
26. Wylie EB, Streeter VL (1993) Fluid transients in systems. Prentice Hall, Englewood Cliffs, NJ
27. Zang B, Wan W, Shi M (2018) Experimental and numerical simulation of water hammer in gravitational pipe flow with continuous air entrainment. Water 10(7):928. https://doi.org/10.3390/w10070928

Exploring the Performance of the Inline Technique-Based Water-Hammer Design Strategy in Pressurized Steel Pipe Flows

Mohamed Amir Chaker[1(✉)] and Ali Triki[2]

[1] Department of Mechanics, National Engineering School of Sfax, University of SFAX, B.P. 1173, 3038 Sfax, Tunisia
medamir.chaker@gmail.com
[2] Research Unit: Mechanics, Modelling Energy and Materials M2EM, National Engineering School of Sfax, University of SFAX, Sfax, Tunisia
ali.triki@enis.rnu.tn

Abstract. This paper explored the effectiveness of the inline technique-based redesign strategy in terms of pressure rise and drop attenuation and wave oscillation period spreading. Basically, this technique consists in replacing an inline short-section of the sensitive zone of the existing steel piping system by another one made of plastic pipe-wall material. Firstly, the **1-D** unconventional water-hammer model combined with the Kelvin-Voigt and the Vitkovsky et al. formulations was solved by the Method of Characteristics. Secondly, the inline technique was implemented in a reservoir pipe valve hypothetical system. The plastic materials mentioned in this paper included high- and low-density polyethylene (**HDPE**) and (**LDPE**). Results illustrated the reliability of the proposed technique in attenuating excessive high- and low-pressure surges. However, they evidenced that this technique induced excessive period spreading, thus affecting negatively the operational procedures of the hydraulic system. Lastly, this study provided an estimate of the near-optimal values of the short-section diameter and length.

Keywords: Control · Design · HDPE · Kelvin-Voigt · LDPE · Plastic material · Pressurized-pipe · Steel · Viscoelasticity · Vitkovsky · Water-Hammer

1 Introduction

Pressurized piping systems are unavoidably subjected to water-hammer phenomenon; which is triggered due to either normal or accidental processes (e.g.: improper setting of valves, hydraulic parts or machinery breakdowns). This phenomenon displays a series of positive- and negative-pressure waves (i.e., unsteady pressure fluctuations), which may be of magnitude large enough to induce severe conditions such as excessive noise, fatigue, stretch or rupture of the pipe-wall and even cause major problems potential risky for operators or users [8]. Consequently, the control of such a transient in pressurized water supply systems constitutes a major concern for design engineers and pipe system managers to ensure safe and efficient operation while providing the adequate service level.

N. Aifaoui et al. (Eds.): CMSM 2019, LNME, pp. 83–91, 2020.
https://doi.org/10.1007/978-3-030-27146-6_10

In this regard, water-hammer control strategies typically include (i) changes within the distribution system (e.g.: pipe diameter or profile, pipe-wall thickness, alignment and other hydraulic components), (ii) wave speed reduction, (iii) optimal operational procedures and (iv) dedicated device installation (e.g.: automatic control valves, surge tanks, and air chambers). Commonly, a combination of numerous strategies are employed by design to soften water-hammer surges at sensitive locations of the hydraulic system; however, the cumulative effect of several types of devices may reversely affects water hammer courses, due to substantial inconsistency between the embedded multiple devices [8].

Alternatively to classical design measures and basing on the ability of plastic materials to extenuate high- and low-transient pressure, recent researches have entailed the use of plastic short-section in order to upgrade existing steel-piping system [4–7, 11–16]. In particular, there is a recent renewed interest on the inline design strategy concept; which is based on the replacement of a short-section of the sensitive zone of the existing steel-piping system by another made of plastic material [10], Gally et al. 1979, [4, 11–13]. In summary, previous investigations on the inline design strategy recognized that this strategy could be an effective tool for attenuation of excessive hydraulic-head–rise and drop in pressurized hydraulic systems; however, this strategy induced large spreading of wave oscillation period during transient flow course. Physically, the viscoelastic behavior attenuates the pressure fluctuations and delays in time the travelling of pressure wave [3].

Classically, safety criteria include a minimum pipe size, extreme pressure-head values, and maximum allowable flowrate. However, other fundamental design parameter such as and the duration of water-hammer event, should also be considered. Specifically, the latter parameter is mainly embedded in the operational procedures of the hydraulic system; such as the admissible critical time for valve closure. Yet, it would be a challenging issue to probe into the conflict between the hydraulic-head attenuation and wave oscillation spreading factors. Accordingly, we planned in this paper to comprehensively assess the effectiveness of the inline strategy concept with regard to the forgoing two factors.

In the following, the one-dimensional (**1-D**) pressurized-pipe flow model is briefly described.

2 Materials and Methods

The one-dimensional (**1-D**) water-hammer model combined with the Kelvin-Voigt and Vitkovsky et al. formulations, may be expressed following form [1, 2, 14, 16–18]:

$$\frac{\partial h}{\partial t} + \frac{a_0^2}{gA}\frac{\partial q}{\partial x} + 2\frac{a_0^2}{g}\frac{d\varepsilon}{dt} = 0 \tag{1}$$

$$\frac{1}{A}\frac{\partial q}{\partial t} + g\frac{\partial h}{\partial x} + g\left(h_{f_s} + h_{f_u}\right) = 0 \tag{2}$$

where, h and q are the instantaneous hydraulic-head and flow-rate; A is the cross sectional area of the pipe; g is the gravity acceleration; a_0 is the wave speed; ε is the radial strain; h_{f_s} is the quasi-steady head-loss component per unit length; h_{f_u} is the unsteady friction loss modelled by the Vitkovsky et al. [17] formula: $h_{f_u} = (k_v/gA)\{(\partial q/\partial t) + a_0\,\mathbf{Sgn}(q)|\partial q/\partial x|\}$, in which, $k_v = 0.03$ is the Vitkovsky decay coefficient, $\mathbf{Sgn}(q) = +1$ or -1 for $q \geq 0$ or $q \prec 0$, respectively; x and t are the coordinates along the pipe axis and time, respectively.

Briefly, the numerical treatments steps of Eqs. (1) and (2) using the Method of Characteristics (**MOC**), built upon a fixed time step mesh, may be outlined as follows (further detailed analysis can be found in e.g. [18]):

The compatibility equations associated with the flow model, written in a finite-difference form, are:

$$\mathbf{C}^{j\pm}: \frac{dh}{dt} \pm \frac{a_0^j}{gA^j}\frac{dq}{dt} + \frac{2a_0^2}{g}\left(\frac{\partial\varepsilon}{dt}\right) \pm a_0^j\left(h_{f_s}^j + h_{f_u}^j\right) = 0 \text{ along } \frac{\Delta x^j}{\Delta t} = \pm\frac{a_0^j}{c_r^j} \tag{3}$$

where the superscript j refers to the pipe number $(1 \leq j \leq np)$ and the lower subscript i refers to the section index of the jth pipe $\left(1 \leq i \leq n_s^j\right)$ n_s^j is the number of the jth pipe sections and np is the number of pipes and Δt is the time- step increment.

The radial strain ε may be calculated using the linear viscoelastic Kelvin-Voigt model, as follows [1]:

$$\varepsilon(x,t) = \sum_{k=1}^{N_{kv}} \varepsilon_k = \sum_{k=1}^{N_{kv}} \rho g \frac{\alpha D}{2e} \int_0^s [h(x,t) - h_0(x)] \frac{J_k}{\tau_k} e^{-\frac{s}{\tau_k}} ds \tag{4}$$

where, J_0 is the elastic creep compliance; J_k and τ_k $(k = 1 \cdots n_{kv})$ denote the creep-compliance and the retardation-time coefficients associated with kth Kelvin-Voigt element, respectively; n_{kv} is the number of Kelvin-Voigt elements; and the subscript "0" stands for the initial steady value.

Given Eqs. (3) and (4), the hydraulic-head and discharge parameters can be expressed, in time domain, as follows:

$$\mathbf{C}^{\pm}: \begin{cases} q_{i,t}^j = c_p^j - c_{a-}^j h_{i,t}^j \\ q_{i,t}^j = c_n^j + c_{a+}^j h_{i,t}^j \end{cases} \text{along } \frac{\Delta x^j}{\Delta t} = \pm\frac{a_0^j}{c_r^j} \tag{5}$$

where, $c_p^j = \left(q_{i-1,t-1}^j + (1/B^j)\,h_{i-1,t-\Delta t}^j + c_{p1}^{\prime\prime j} + c_{p1}^{\prime\prime\prime j}\right) \Big/ \left(1 + c_p^{\prime j} + c_{p2}^{\prime\prime j} + c_{p2}^{\prime\prime\prime j}\right)$; $B = a_0/(gA)$; $c_n^j = \left(q_{i+1,t-1}^j + (1/B^j)h_{i+1,t-\Delta t}^j + c_{n1}^{\prime\prime} + c_{n1}^{\prime\prime\prime}\right) \Big/ (1 + c_n^{\prime j} + c_{n2}^{\prime\prime j})$; $c_{a+}^j = 1 + c_{p2}^{\prime\prime\prime j} \Big/ \left(B^j\left(1 + c_{p2}^{\prime j} + c_{p2}^{\prime\prime j}\right)\right)$; $c_p^{\prime j} = R^j \Delta t \left|q_{i-1,t-1}^j\right|$; $c_n^{\prime j} = R^j \Delta t \left|q_{i+1,t-1}^j\right|$; $R^j = f^j/2D^jA^j$; $c_{p_1}^{\prime\prime j} = k_v \theta q_{i,t-1}^j - k_v(1-\theta)\left(q_{i-1,t-1}^j - q_{i-1,t-2}^j\right) - k_v\mathbf{sgn}\left(q_{i-1,t-1}^j\right)\left(q_{i,t-1}^j - q_{i-1,t-1}^j\right)$; $c_{n_1}^{\prime\prime j} = k_v \theta q_{i,t-1}^j - k_v(1-\theta)\left(q_{i+1,t-1}^j - q_{i+1,t-2}^j\right) - k_v\mathbf{sgn}\left(q_{i+1,t-1}^j\right)\left(q_{i,t-1}^j - q_{i+1,}\right.$

$t-1^j)$; $c_{p_1}^{IIIj}=-c_{n_1}^{IIIj}=-2a_0^jA^j\Delta t\sum_{k=1}^{n_{kv}}\left[\varepsilon_k^j(x,t)/\partial t\right]$; $c_{p_2}^{IIIj}=c_{n_2}^{IIIj}=2a_0^jA^jc_0\gamma\sum_{k=1}^{n_{kv}}J_k^j$ $\left(1-e^{-(\Delta t/\tau_k)}\right)$; $\varepsilon_{k,i,t-\Delta t}^j=J_k^jc_0\left\{\left[h_{i,t-\Delta t}^j-h_{i,0}^j\right]-e^{-(\Delta t/\tau_k)}\left[h_{i,t-2\Delta t}^j-h_{i,0}^j\right]-\tau_k\left(1-e^{-(\Delta t/\tau_k)}\right)\right.$ $\left.\left[h_{i,t-\Delta t}^j-h_{i,t-2\Delta t}^j\right]/\Delta t\right\}+e^{-(\Delta t/\tau_k)}\varepsilon_{k,i,t-2\Delta t}$ $c_{p_2}^{IIj}=c_{n_2}^{IIj}=k_v\theta$; ($\theta=1$ is a relaxation coefficient); and $c_0=\alpha\gamma D^j/2e^j$.

– *Series Connection of Multi Pipes*:

The discharge and the pressure-head may be computed at the series connection, assuming no flow storage and a common hydraulic grade-line elevation at this section [18], yields:

$$q_{ns^j,t}^{j-1}=q_{1,t}^j \text{ and } h_{ns^j,t}^{j-1}=h_{1,t}^j \tag{6}$$

in which, the right and left hands of Eq. (6) correspond to the hydraulic parameters evaluated at the up- and down-stream sides of the connection.

Next section is devoted to explore the effectiveness of the inline strategy with respect to pressure-head attenuation and limitation of wave oscillation period spreading.

3 Application, Results and Discussion

The case studied in this paper relates to a steel-piping system supplied by an upstream reservoir and equipped with a valve at its downstream extremity. The initial steady-state flow regime is characterized by a discharge: $q_0=1.01\,l/s$ and a constant pressure-head level maintained in the upstream reservoir ($h_0^R=80\,\mathrm{m}$). The transient relates to a sudden and full valve closure of the downstream. Such a condition may be expressed as follows:

$$q_{|x=L}=0 \text{ and } h_{|x=0}=h_0^R (\text{for } t\succ 0) \tag{7}$$

For this particular test case, the inline technique consists in substituting a downstream short-section of the main steel piping system by another one made of **(HDPE)** or **(LDPE)** plastic material (Fig. 1). The creep compliance coefficients of the generalized Kelvin-Voigt linear viscoelastic mechanical behaviour associated to **(HDPE)** or **(LDPE)** material are: $J_0^{\mathbf{HDPE}}=1.057\ \mathrm{GPa}^{-1}$, $\{J/\tau\}_{1\cdots5}^{\mathbf{HDPE}}(\mathrm{GPa}^{-1}/\mathrm{s})=\{1.057/0.05;\ 1.054/0.5;\ 0.905/1.5;\ 0.262/5.0;\ 0.746/10.0\}$[9]; and $J_0^{\mathbf{LDPE}}=1.54\ \mathrm{GPa}^{-1}$; $\{J/\tau\}_{1\cdots5}^{\mathbf{LDPE}}(\mathrm{GPa}^{-1}/\mathrm{s})=\{7.54/(89\times10^{-6});\ 10.46/0.022;\ 0.262/1.864\}$ [6], respectively.

Preliminary investigations relate to a short-section length and diameter values equal to: $l_{\text{short-section}}^{\mathbf{plastic}}=5$ m and $d_{\text{short-section}}^{\mathbf{plastic}}=50.6$ mm, respectively.

Figure 2 compares, respectively, the evolution of downstream hydraulic-heads, versus time, predicted into the non-protected hydraulic system case, alongside the corresponding profiles predicted into the protected system cases using an **(HDPE)** or **(LDPE)** configuration- based inline technique. The main wave features displayed in Fig. 2 are listed in Table 1.

Figure 2 clearly confirms the reliability of the inline technique in attenuating excessive hydraulic-heads rise and drop. However, as expected, this technique induces a spreading of hydraulic-heads oscillation period.

In order to assort these two last parameters, the pressure wave attenuation versus the period traces are illustrated in Fig. 3 for each protected system case.

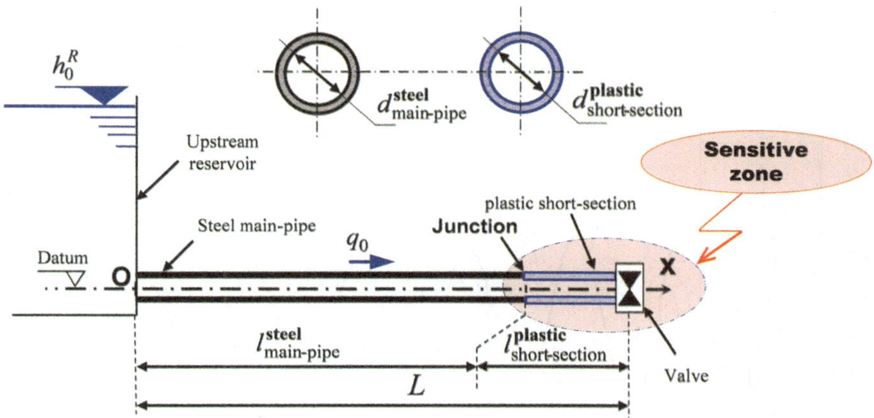

Fig. 1 Definition sketch of implementation of the inline technique

Table 1 Characteristics of applied pipelines

Parameters	Steel main-pipe	Plastic short-penstock	
		(HDPE)	**(LDPE)**
Length L (m)	100.0	5.0	
Inside diameter D (mm)	50.6	50.6	
Pipe-wall thickness e (mm)	3.35	2.41	3.40
Young modulus E_0 (GPa)	1369.7	404.9	263.9

Particularly, for the original system case, the change in the downstream boundary condition triggered both expansion and compression pressure waves. Based on Fig. 2, the hydraulic-head firstly rises to $h_{max}^{\text{Steel pipe}} = 85.6$ m, and subsequently drops to $\left(h_{min}^{\text{Steel pipe}} = 5.4 \text{ m}\right)$; which corresponds to an positive- and negative- surge magnitudes equal to: $\Delta h_{positive-surge}^{\text{Steel pipe}} = h_{max}^{\text{Steel pipe}} - h_0 = 40.6$ m and $\Delta h_{negative-surge}^{\text{Steel pipe}} = h_0 - h_{min}^{\text{Steel pipe}} = 39.6$ m, respectively, above the initial steady state value; nonetheless, this phenomenon is palliated in the protected systems based on the (**HDPE**) or (**LDPE**) configuration of the inline technique.

For example, Fig. 2 and Table 1 show that the positive- and negative- surge magnitudes involved by a (**HDPE**) short-section based protected system are equal to: $\Delta h^{\mathbf{HDPE}\ \text{short-section}}_{\text{positive-surge}} = 40.6\ \text{m}$ and $\Delta h^{\mathbf{HDPE}\ \text{short-section}}_{\text{negative-surge}} = 39.6\ \text{m}$, respectively. This indicates that the positive- and negative-surge attenuation obtained by the **HDPE** inline short-section protected system case as compared with the original system case are equal to: $\delta h^{\mathbf{HDPE}\ \text{short-section}}_{\text{positive-surge}} = h^{\mathbf{steel-pipe}}_{\text{max.}} - h^{\mathbf{HDPE}\ \text{short-section}}_{\text{max.}} = 85.6 - 77.9 = 32.9\ \text{m}$ and $\delta h^{\mathbf{HDPE}\ \text{short-section}}_{\text{negative-surge}} = h^{\mathbf{steel-pipe}}_{\text{min}} - h^{\mathbf{HDPE}\ \text{short-section}}_{\text{min}} = 5.4 - 17.7 = -27.3\ \text{m}$, respectively (Table 2).

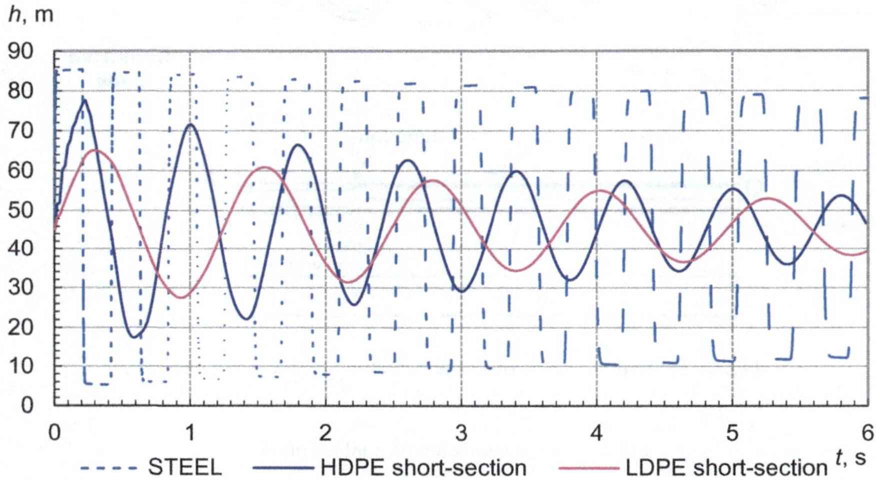

Fig. 2 Comparison of the downstream hydraulic-head signals predicted into the original hydraulic system and the inline strategy-based protected systems

Table 2 Characteristics of water-hammer waves in Fig. 2

Parameters		Steel main-pipe	Plastic short-section	
			HDPE	**LDPE**
h_{max}: hydraulic-head peak	(m)	85.6	77.9	65.3
h_{min}: hydraulic-head crest	(m)	5.4	17.7	27.5
T: period of the first cycle of wave oscillation	(s)	0.40	0.80	1.24

On the other hand, it may be also observed from Fig. 2 and Table 1 that the period of the first cycle of wave oscillation involved by the (**HDPE**) short-section based protected system case is more important than that and performed by the non-protected system case. Precisely, the phase-shift observed between the wave curves involved by the two forgoing cases is: $\Delta T_1 = \left| T_1^{\textbf{Steel pipe}} - T_1^{\textbf{HDPE}\ \text{short-section}} \right| = |0.40 - 0.80| = 0.4$ s.

Analogously, Fig. 2 and Table 1 indicate that the positive- and negative- surge magnitudes provided by the protected system case based upon a (**LDPE**) short-section are equal to: $\Delta h_{\text{positive-surge}}^{\textbf{LDPE}\ \text{short-section}} = 20.3$ m and $\Delta h_{\text{negative-surge}}^{\textbf{LDPE}\ \text{short-section}} = 17.5$ m, respectively. In other words, the positive- and negative-surge attenuation obtained by the **HDPE** inline short-section protected system case as compared with the original system case are equal to: $\delta h_{\text{positive-surge}}^{\textbf{LDPE}\ \text{short-section}} = 20.3\text{m}$ and $\delta h_{\text{negative-surge}}^{\textbf{LDPE}\ \text{short-section}} = -22.1$ m, respectively. In addition, referring to Fig. 2 and Table 1, the phase shift observed between the first cycle of wave oscillation involved by the (**HDPE**) short-section based protected system case and its counterpart performed by the non-protected system case is equal to: $\Delta' T_1 = \left| T_1^{\textbf{Steel pipe}} - T_1^{\textbf{LDPE}\ \text{short-section}} \right| = |0.40 - 1.24| = 0.84$ s.

Consequently, the inline technique allows a significant attenuation of the first pressure peak and crest as compared with those involved in the original system. Furthermore, this effect is slightly more important for the case using a **LDPE** short-section than the one based on a **HDPE** short-section. However, the employed technique induces a spreading effect of wave oscillation period as compared with the original system case. This effect is more important for the case based on a **LDPE** short-section than that build upon a **HDPE** short-section.

In order to evaluate properly the two forgoing effects, the attenuation of the first pressure peak and crest versus the wave oscillation period are plotted in Fig. 3, for each system case. This Figure evidences that the (**HDPE**) short-section based protected system allows more important hydraulic-head attenuation and lower spreading of wave oscillation period as compared with the (**LDPE**) short-section based case. Indeed, the examination of different slops of the curves traced in Fig. 3 shows that the most

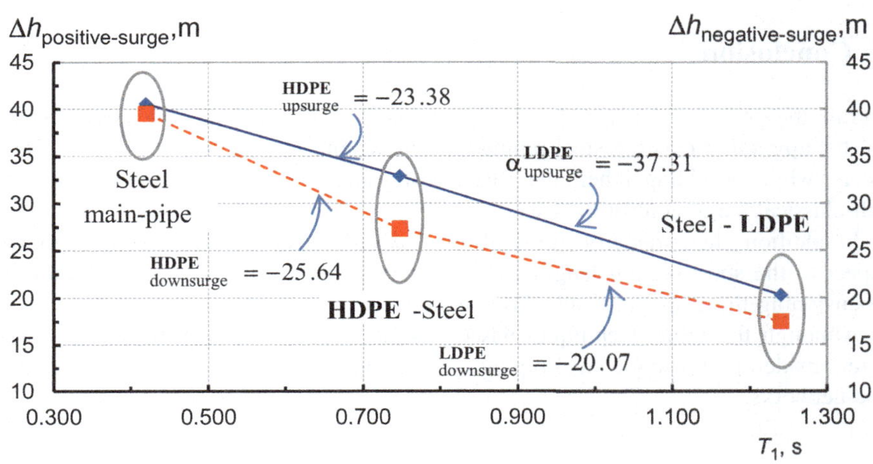

Fig. 3 positive- and negative-hydraulic-head magnitudes versus the period of the 1st cycle of wave oscillations

important slop is located between non protected system case and the (**HDPE**) setup of the protected system.

The second part of this work is dedicated to investigating the relevance of the short-section length and diameter on the first hydraulic-head peak and crest values. The foregoing task is reported in Fig. 4. As per this figure, steepness changes of the different curves can be noticed around the primitive values of the short-section diameter and length ($d^{plastic}_{short-section} = 50.6$ mm and $l^{plastic}_{short-section} = 5$ m). Hence, these values may be considered as the near-optimal diameter and length values of the plastic short-section.

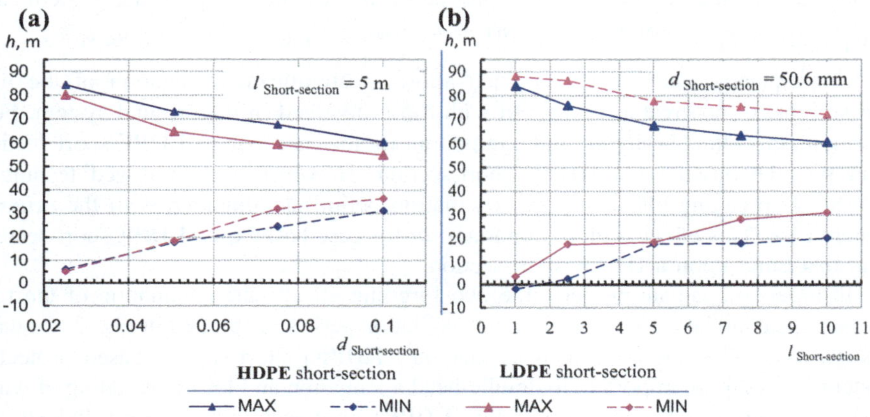

Fig. 4 Variation of downstream hydraulic-head peak and crest depending on the short-section: **a** length, **b** diameter

4 Conclusion

Overall, the findings of this study confirmed that the inline technique is a useful design tool for upgrading existing steel piping systems facing to water-hammer severe surge effects, while providing inherently free maintenance and/or testing merits compared with classical surge control devices.

In addition, the parametric study of the hydraulic-head peak (or crest) values with respect to the short-section length and diameter identified the near-optimal value for dimensioning the compound sub short-section.

Although the inline technique- based design strategy was applied to a hydraulic system made up of a single steel-piping system, this technique may also be applied for pipe networks.

References

1. Aklonis JJ, MacKnight WJ, Shen M (1972) Introduction to polymer viscoelasticity. Wiley-Interscience, Wiley
2. Ferrante M, Capponi C (2017) Viscoelastic models for the simulation of transients in polymeric pipes. J Hydraul Res 55(5):599–612. https://doi.org/10.1080/00221686.2017.1354935
3. Ferry JD (1970) Viscoelastic properties of polymers, Second edn. Wiley, New York
4. Fersi M., Triki A (2018) Investigation on redesigning strategies for water-hammer control in pressurized-piping systems, J. Pressure Vessel Technol. - Transactions of the ASME https://doi.org/10.1115/1.4040136
5. Fersi M, Triki A (2019) Alternative design strategy for water-hammer control in pressurized-pipe flow. In: Fakhfakh T, Karra C, Bouaziz S, Chaari F, Haddar M (eds) Advances in acoustics and vibration II. ICAV 2018. Applied Condition Monitoring, vol 13, 135–144, Springer, pp 157–165. https://doi.org/10.1007/978-3-319-94616-0_16
6. Güney MS (1983) Water-hammer in viscoelastic pipes where cross-section parameters are time dependent. In: Proceedings of 4th international conference on pressure surges, BHRA, Bath, U.K, pp 189–209
7. Ghilardi P, Paoletti A (1986) Additional viscoelastic pipes as pressure surge suppressors. In: Proceedings of 5th international conference on pressure surges, Cranfield (UK), pp 113–121
8. Ghidaoui MS, Zhao M, Duncan AM, David HA (2005) A review of Water-hammer theory and practice. Appl Mech Rev 58:49–76. https://doi.org/10.1115/1.1828050
9. Keramat A, Haghighi A (2014) Straightforward transient-based approach for the creep function determination in viscoelastic pipes. J Hydraul Eng 140(12) https://doi.org/10.1061/(asce)hy.1943-7900.0000929
10. Massouh F, Comolet R (1984) Étude d'un système anti-bélier en ligne-study of a water-hammer protection system in line. La Houille Blanche 5:355–362. https://doi.org/10.1051/lhb/1984023
11. Triki A (2016) Water-hammer control in pressurized-pipe flow using an in-line polymeric short-section. Acta Mech 227(3):777–793. https://doi.org/10.1007/s00707-015-1493-13
12. Triki A (2017) Water-Hammer control in pressurized-pipe flow using a branched polymeric penstock. J Pip Syst Eng Pract ASCE 8(4):04017024. https://doi.org/10.1061/(asce)ps.1949-1204.0000277
13. Triki A (2018) Further investigation on water-hammer control inline strategy in water-supply systems. J Water Suppl Res Technol AQUA 67(1):30–43. https://doi.org/10.2166/aqua.2017.073
14. Triki A (2018) Dual-technique based inline design strategy for Water-Hammer control in pressurized-pipe flow. Acta Mech 229(5):2019–2039. https://doi.org/10.1007/s00707-017-2085-z
15. Triki A, Fersi M (2018) Further investigation on the Water-Hammer control branching strategy in pressurized steel-piping systems. Int J Press Vessels Pip 165(C):135–144. https://doi.org/10.1016/j.ijpvp.2018.06.002
16. Triki A, Chaker MA (2019) Compound technique-based inline design strategy for water-hammer control in steel pressurized-piping systems. Int J Press Ves Pip 169C:188–203. https://doi.org/10.1016/j.ijpvp.2018.12.001
17. Vitkovsky JP, Lambert MF, Simpson AR, Bergant A (2000) Advances in unsteady friction modelling in transient pipe flow. In: The 8th international conference on pressure surges BHR, The Hague, The Netherlands
18. Wylie EB, Streeter VL (1993) Fluid transients in systems. Prentice Hall, Englewood Cliffs, NJ

Investigating the Removal of Hydraulic Cavitation from Pressurized Steel Piping Systems

Mohamed Amir Chaker[1,2(✉)] and Ali Triki[2]

[1] Department of Mechanics, National Engineering School of Sfax, University of Sfax, B.P. 1173, 3038 Sfax, Tunisia
medamir.chaker@gmail.com

[2] Research Unit: Mechanics Modelling Energy and Materials M2EM, Gabes, Tunisia
ali.triki@enis.rnu.tn

Abstract. This paper assessed the branching strategy capacity to mitigate the cavitating flow regime induced into an existing steel piping system. This strategy was based on adding a ramified high or low density polyethylene ((**HDPE**) or (**LDPE**)) short penstock to the transient sensitive regions of the existing piping system. The **1-D** water-hammer model combined with the Vitkovsky et al. and Kelvin-Voigt formulations was used to describe the hydraulic behavior, along with the fixed grid Method of Characteristics, being used for numerical computations. From the case studied, it was found that such a technique could palliate the cavitating flow regime. In addition, this strategy allowed an acceptable first hydraulic-head peak and crest attenuation. Specifically, positive and negative surge magnitude attenuation was slightly more important for the case of a short penstock made of (**LDPE**) material than that using an (**HDPE**) material. Ultimately, it was observed that other factors contributing to the attenuation rate depended upon the short-penstock length and diameter.

Keywords: Cavitation · Design · Kelvin-Voigt · Plastic · **LDPE** · Method of characteristics · **HDPE** · Viscoelasticity · Vitkovsky

1 Introduction

Hydraulic piping systems can be severely affected by the cavitating flow regime, whether generated by normal or accidental maneuvers (e.g. from valve setting; start or stop of pumps; or hydraulic parts breakdowns). Specifically, these maneuvers trigger a series of pressures-rise and-drop (i.e.: positive and negative surges) of sharp magnitudes. In some situations, the pressure drops to its vapor value causing the cavitating flow onset; which leads to negative impacts on hydraulic systems (e.g.: reduction in flow capacity, disruption of the flow, reduced pump and turbine efficiency, effects on pipe materials and pipeline structure) or risks for users A useful review of hydraulic systems damages due to the cavitating flow regime is produced by [2].

N. Aifaoui et al. (Eds.): CMSM 2019, LNME, pp. 92–101, 2020.
https://doi.org/10.1007/978-3-030-27146-6_11

It is quite evident that water-hammer surge cannot be avoided completely. Consequently, certain design measures are commonly taken to mitigate effectively the severe impact of these waves to a desirable extent [17, 15].

Except for employing classical surge control devices, several innovative design measures based on the use of plastic short-section have been enumerated recently in the literature, in order to upgrade existing steel-piping system [4, 5, 7–13]. In particular the authors demonstrated that the branching-strategy, which is based on inserting a ramified plastic short-section at sensitive regions of the original steel-piping system, offers several gains for upgrading existing installations without extensive modifications. Specifically, this strategy allows significant attenuation of first pressure peak and crest values. Nonetheless, the authors observed that this benefit was accompanied with several undesirable effects resulting from the spreading of the wave oscillation period; which may affect negatively the operational procedures of hydraulic utilities; such as the increase of the critical time for valve closure [4, 11, 12]. Physically, the low wave-speed value of viscoelastic material is assumed be the major reason for attenuation of pressure-rise and-drop; however, the spreading of wave propagation period is attributed to the mechanical behaviour of the employed viscoelastic material, which has a retarded response component in coexisting with the immediate one observed in elastic material case [3, 10].

Accordingly, the primary objective of this paper is to explore the reliability of the branching strategy with focuses on the allowed pressure attenuation rate and the induced spreading of the wave oscillation period.

The following section presents the flow model considered in this paper.

2 Materials and Methods

Shortly, the **(1-D)** unconventional water-hammer governing equations embedding the Vitkovsky et al. and Kelvin-Voigt formulations to account for unsteady friction losses and pipe-wall viscoelastic behaviour may be writeen as follows (detailed derivations are reported in [10]:

$$\frac{\partial h}{\partial t} + \frac{a_0^2}{gA}\frac{\partial q}{\partial x} + 2\frac{a_0^2}{g}\frac{d\varepsilon_r}{dt} = 0 \tag{1}$$

$$\frac{1}{A}\frac{\partial q}{\partial t} + g\frac{\partial h}{\partial x} + g\left(h_{f_s} + h_{f_u}\right) = 0 \tag{2}$$

where, h denotes the pressure head, q designates the flow discharge, A is the cross sectional area of the pipe, g is the gravity acceleration, a_0 corresponds to the wave speed; h_{f_s} designates the quasi-steady head-loss component; h_{f_u} corresponds to the unsteady friction losses evaluated using the Vitkovsky et al. [14] formula: $h_{f_u} = (k_v/gA)\{(\partial q/\partial t) + a_0 Sgn(q)|\partial q/\partial x|\}$, in which, $k_v = 0.03$ is a decay coefficient, x and t denote the coordinates along the pipe axis and time, respectively.

The retarded hoop strain ε_r may be evaluated according to the linear-viscoelastic Kelvin-Voight model [1]:

$$\varepsilon_r(x,t) = \sum_{k=1}^{N_{kv}} \varepsilon_{rk} = \sum_{k=1}^{N_{kv}} \frac{\alpha D}{2e} \rho g \int_0^s [h(x,t) - h_0(x)] \frac{J_k}{\tau_k} e^{-\frac{s}{\tau_k}} ds \tag{3}$$

where: J_k and $\tau_k (k = 0 \cdots n_{kv})$ denote the creep-compliance and the retardation-time coefficients associated with the springs and the dashpots of the kth Kelvin-Voigt element, respectively, n_{kv} is the number of Kelvin-Voigt elements; and h_0 designates the initial hydraulic-head value.

In the following, the Method Of Characteristics (**MOC**) procedure selected for transient flow predictions is outlined briefly (a detailed derivation of general algorithm is reported in e.g. Triki [11].

The compatibility equations performed by the (**MOC**) procedure, using a fixed time-step rectangular mesh grid, corresponding to the discretization of Eqs. (1) and (2) along series pipeline parts are given by:

$$\mathbf{C}^{j\pm}: \frac{dh}{dt} \pm \frac{a_0^j}{gs^j} \frac{dq}{dt} + \frac{2a_0^2}{g} \left(\frac{\partial \varepsilon_r}{dt} \right) \pm a_0^j h_f^j = 0 \text{ along } \frac{\Delta x^j}{\Delta t} = \pm \frac{a_0^j}{c_r^j} \tag{4}$$

in which, the superscript j refers to the pipe number ($1 \leq j \leq np$; np is the number of pipes), Δt denotes the time-step increment and c_r designates the Courant number associated with the spatial-discretization of the jth pipe.

Thereby, the unknown flow parameters can be directely integrated from Eq. (5) for each section i of the jth pipe ($1 \leq i \leq n_s^j$; n_s^j is the number of sections of the jth pipe), as follows:

$$\mathbf{C}^{\pm}: \begin{cases} q_{i,t}^j = c_p^j - c_{a-}^j h_{i,t}^j \\ q_{i,t}^j = c_n^j + c_{a+}^j h_{i,t}^j \end{cases} \text{along } \frac{\Delta x^j}{\Delta t} = \pm \frac{a_0^j}{c_r^j} \tag{5}$$

where, $c_p^j = \left(q_{i-1,t-1}^j + (1/B^j) h_{i-1,t-\Delta t}^j + c_{p1}^{\prime\prime j} + c_{p1}^{\prime\prime\prime j} \right) \Big/ \left(1 + c_p^{\prime j} + c_{p2}^{\prime\prime j} + c_{p2}^{\prime\prime\prime j} \right)$; $B = a_0/ .(gA)$; $c_n^j = \left(q_{i+1,t-1}^j + (1/B^j) h_{i+1,t-\Delta t}^j + c_{n1}^{\prime\prime} + c_{n1}^{\prime\prime\prime} \right) \Big/ (1 + c_n^{\prime j} + c_{n2}^{\prime\prime j})$; $c_{a+}^j = 1 + c_{p2}^{\prime\prime\prime j} \Big/ (B^j \left(1 + c_{p2}^{\prime j} + c_{p2}^{\prime\prime j} \right))$; $c_p^{\prime j} = R^j \Delta t \left| q_{i-1,t-1}^j \right|$; $c_n^{\prime j} = R^j \Delta t \left| q_{i+1,t-1}^j \right|$; $R^j = f^j / 2D^j A^j$; $c_{p_1}^{\prime\prime j} = k_v \theta \, q_{i,t-1}^j - k_v (1 - \theta) \left(q_{i-1,t-1}^j - q_{i-1,t-2}^j \right) - k_v \mathbf{sgn} \left(q_{i-1,t-1}^j \right) \left(q_{i,t-1}^j - q_{i-1,t-1}^j \right)$; $c_{n_1}^{\prime\prime j} = k_v \theta \, q_{i,t-1}^j - k_v (1 - \theta) \left(q_{i+1,t-1}^j - q_{i+1,t-2}^j \right) - k_v \mathbf{sgn} \left(q_{i+1,t-1}^j \right) \left(q_{i,t-1}^j - q_{i+} 1, t - 1^j \right)$; $c_{p_1}^{\prime\prime\prime j} = -c_{n_1}^{\prime\prime\prime j} = -2a_0^j A^j \Delta t \sum_{k=1}^{n_{kv}} \left[\varepsilon_{rk}^j (x,t) / \partial t \right]$; $c_{p_2}^{\prime\prime\prime j} = c_{n_2}^{\prime\prime\prime j} = 2a_0^j A^j c_0 \gamma \sum_{k=1}^{n_{kv}} J_k^j \left(1 - e^{-(\Delta t/\tau_k)} \right)$; $\varepsilon_{rk,i,t-\Delta t}^j = J_k^j c_0 \Big\{ \left[h_{i,t-\Delta t}^j - h_{i,0}^j \right] - e^{-(\Delta t/\tau_k)} \left[h_{i,t-2\Delta t}^j - h_{i,0}^j \right] -$

$\tau_k\left(1 - e^{-(\Delta t/\tau_k)}\right)\left[h^j_{i,t-\Delta t} - h^j_{i,t-2\Delta t}\right]/\Delta t\} + e^{-(\Delta t/\tau_k)}\varepsilon_{rk,i,t-2\Delta t}$; $c''^j_{p_2} = c''^j_{n_2} = k_v\theta$ ($\theta = 1$ is a relaxation coefficient); and $c_0 = \alpha\gamma D^j/2e^j$.

It is worth noting that the (**MOC**) procedure, outlined above, relates to a one-phase flow regime. For a cavitating flow occurrence, the approximate flow solution may be achieved by including the discrete gas cavity model (**DGCM**) into the conventional (**MOC**) solution.

Basically, the (**DGCM**) assumes that void cavities are lumped at the computing sections and a homogeneous distribution free gas mixture throughout the liquid. In addition, upon cavity formation at a computational section, a constant absolute hydraulic-head value, equal to the gauge pressure of the liquid (i.e. $h^* = h^*_g$), is imposed at this section [16].

The discretization of the perfect gas law for an isothermic evolution of each isolated gas cavity yields:

$$\forall^j_{gi,t}\left(h^j_{i,t} - z^j_i - h_v\right) = \left(h_0 - z^j_i - h_g\right)\alpha_0 A\Delta t \tag{6}$$

in which, h_0 is the hydraulic-head reference, α_0 the void fraction at h_0, z^j_i the pipe axis elevation and h_g the gauge pressure head of the liquid.

The cavity volume calculation is based on the discretization of the continuity equation applied for the cavity control volume:

$$\forall^j_{gi,t} = \forall^j_{gi,t-2\Delta t} + \left[\psi\left(q^j_{di,t} - q^j_{ui,t}\right) - (1-\psi)\left(q^j_{di,t-2\Delta t} - q^j_{ui,t-2\Delta t}\right)\right] \tag{7}$$

where, q_u and q_d are the flowrates, computed at the upstream and downstream sides of the cavity interface, using $\mathbf{C}^+$ and $\mathbf{C}^-$ characteristic Eq. (5), respectively, and ψ is a weighting factor, chosen in the range: $0.5 \le \psi \le 1$ [16].

It is interesting to delineate here that the flow regime is regarded as cavitating type for: $\forall^j_{gi,t} \prec 0$. Otherwise, it is considered as a one-phase type.

Branch connection of multi pipes:

The flow rate and the hydraulic head at the branch connection section are expressed assuming no flow storage and a common hydraulic grade-line elevation at this section [9, 16]:

$$q^{j-1}_{x=L} = q^j_{x=0} + q^d \text{ and } h^{j-1}_{x=L} = h^j_{x=0} = h^d \tag{8}$$

in which, the superscript d designates the device parameters computed using the lumped capacitance model.

Next section is devoted to assess the reliability of the branching technique to control water-hammer waves accompanied with cavitating flow.

3 Application, Results and Discussion

The hydraulic system, considered herein initially consists of a sloping steel pipe connecting two pressurized-tanks. The up-and down-stream axis elevations of the steel-piping system are: $z_d = 0\,\text{m}$ and $z_u = 2.03\,\text{m}$, respectively. The gauge saturated hydraulic-head of the liquid is equal to: $h_g = -10.29\,\text{m}$. The initial steady-state flow regime corresponds to a constant flow velocity: $V_0 = 0.3\,\text{m/s}$ and a constant hydraulic-head maintained at the downstream pressurized-tank: $h_0^{T_2} = 21.4\,\text{m}$. The transient regime is provoked by the sudden and full closure of the upstream valve. Thereupon, the boundary conditions, associated to such a scenario, may be expressed as follows:

$$q_{|x=0} = 0 \text{ and } h_{|x=L} = h_0^{T_2}(t \succ 0) \tag{9}$$

In such a scenario, the branching technique consists in inserting a ramified plastic short-penstock at the downstream extremity of the steel main piping system (Fig. 1). The piping system and the plastic short-penstock specifications are listed in Table 1. The creep compliance coefficients of the generalized Kelvin-Voigt linear viscoelastic mechanical behaviour associated to **(HDPE)** or **(LDPE)** material type are: $J_0^{\mathbf{HDPE}} = 1.057\,\text{GPa}^{-1}$, $\{J/\tau\}_{1\cdots5}^{\mathbf{HDPE}}(\text{GPa}^{-1}/\text{s}) = \{1.057/0.05; 1.054/0.5; 0.905/1.5; 0.262/5.0; 0.746/10.0\}$; and $J_0^{\mathbf{LDPE}} = 1.54\,\text{GPa}^{-1}$; $\{J/\tau\}_{1\cdots5}^{\mathbf{LDPE}}(\text{GPa}^{-1}/\text{s}) = \{7.54/(89 \times 10^{-6}); 10.46/0.022; 0.262/1.864\}$ [6], respectively.

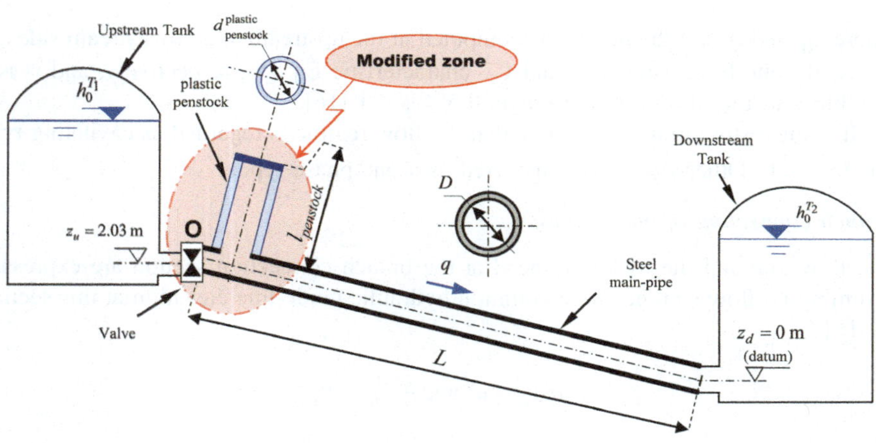

Fig. 1 Definition sketch of the hydraulic system

Figure 2 compares the upstream hydraulic-head, versus time, predicted into the original piping system case along with their counterparts involved by the protected system using a **(HDPE)** or **(LDPE)** ramified short-penstock. Jointly, Table 2 summarizes the main characteristics of wave oscillation curves shown in Fig. 2.

Table 1 Characteristics of applied pipelines

Parameters	Steel main-pipe	Plastic short-penstock	
		(HDPE)	(LDPE)
Length L [m]	143.7	10.0	
Inside diameter D [mm]	50.6	50.6	
Pipe-wall thickness e [mm]	3.35	2.41	3.40
Young modulus E_0 [GPa]	1142.5	404.9	263.9

Figure 2 illustrates a hydraulic-head attenuation effect of the first crest accompanied with the spreading of the hydraulic-head oscillations period observed in the protected system cases. Results evidence, for the first cycle of hydraulic-head oscillations, the cavitating flow onset into the original system case ($h^{\min}_{\mathbf{steel-pipe}} = -10.2\text{m}$); while this phenomenon is palliated when implementing the branching technique using **(HDPE)** or **(LDPE)** plastic short-penstock ($h^{\min}_{\mathbf{HDPE}-\text{penstock}} = -5.6\,\text{m}$ or $h^{\min}_{\mathbf{LDPE}-\text{penstock}} = 7.3\,\text{m}$). In other words, the positive-surge attenuation obtained using **(HDPE)** or **(LDPE)** short-penstock relatively to the original system case is: $\delta\, h^{-}_{\mathbf{HDPE}-\text{penstock}} = h^{\min}_{\mathbf{HDPE}-\text{penstock}} - h^{\min}_{\mathbf{steel-pipe}} = 5.6 + 10.2 = 15.8\,\text{m}$ or $\delta\, h^{-}_{\mathbf{LDPE}-\text{penstock}} = h^{\min}_{\mathbf{LDPE}-\text{penstock}} - h^{\min}_{\mathbf{steel-pipe}} = 7.3 + 10.2 = 17.5\,\text{m}$, respectively. Furthermore, Fig. 2 illustrates that the branching technique allows also the attenuation of positive-surge (i.e.: $\delta\, h^{+}_{\mathbf{HDPE}-\text{penstock}} = 19.7\,\text{m}$ $\delta\, h^{+}_{\mathbf{LDPE}-\text{penstock}} = 31.4\,\text{m}$). Consequently, it may be concluded that the branching technique allows a significant attenuation of the first pressure peak and crest as compared with the original system case. In addition, this attenuation is slightly more important for the case using a **(LDPE)** short-penstock than the one obtained using **(HDPE)** short-penstock.

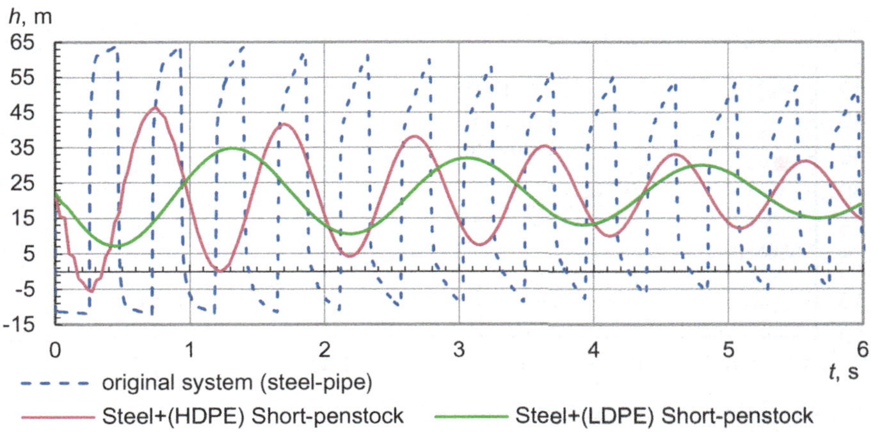

Fig. 2 Comparison of downstream hydraulic-heads versus time for the hydraulic system with and without implementation of the branching procedure

On the other side, based on Fig. 2 together with Table 1, it is remarkable to point out that the periods of the first cycle of hydraulic-head oscillations, predicted into a **(HDPE)** or **(LDPE)** plastic short-penstock-based protected system case are: $T_1^{\mathbf{HDPE-}\text{penstock}} = 0.986\,\text{s}$ or $T_1^{\mathbf{LDPE-}\text{penstock}} = 1.799\,\text{s}$; while the corresponding period, for the piping system without protection (i.e. steel main pipeline), is equal to $T_1^{\mathbf{steel-pipe}} = 0.4\,\text{s}$.

Table 2 Characteristics of water-hammer waves in Fig. 2

Parameters			Steel main-pipe	Plastic short-penstock
			(HDPE)	**(LDPE)**
h_{max}: 1st hydraulic-head peak	[m]	66.03	46.65	34.93
h_{min}: 1st hydraulic-head crest	[m]	−10.2	−5.56	7.29
T: period of the 1st cycle of wave oscillation	[s]	0.828	0.986	1.799

In order to provide a direct comparison between the two employed plastic material types the wave dampening versus the period traces are illustrated in Fig. 3 for protected system cases using a **(HDPE)** or **(LDPE)** short-penstock. It may be conclude from this figure that the protected system case based on a **(HDPE)** short-penstock provides better trade-off between hydraulic-head attenuation and wave oscillation period spreading, as compared with the case based on **(LDPE)** short-penstock.

A further factor that may affect the attenuation of first hydraulic-head peak and crest relates to the ramified plastic short-penstock size. For completeness, this attenuation is reported in Fig. 4a and b depending on the short-penstock diameter and length, respectively.

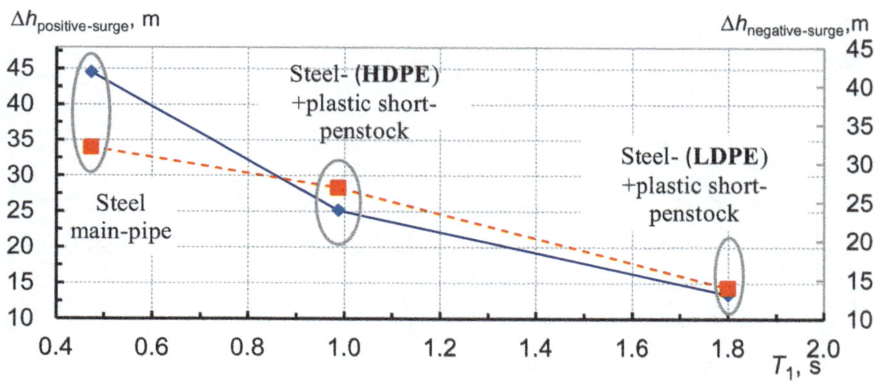

Fig. 3 Comparison of attenuation of hydraulic-head peak and crest depicted into the original system case and the protected system cases using a (HDPE) or (LDPE) ramified plastic short-penstock

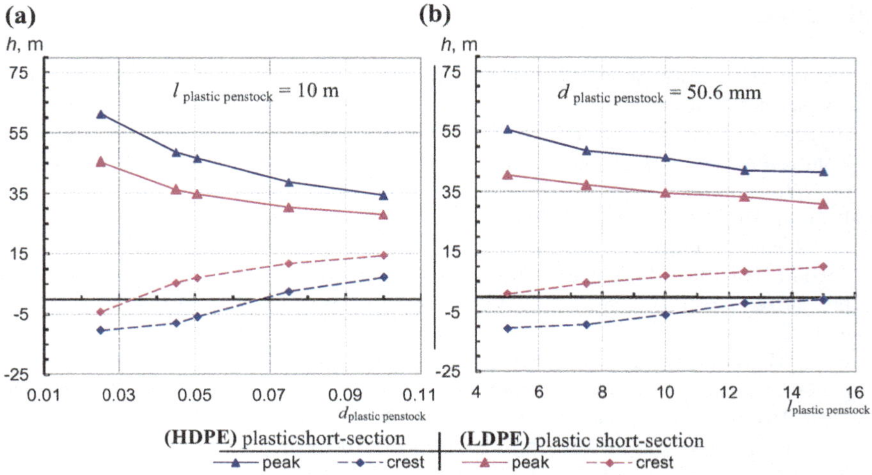

Fig. 4 Variation of downstream hydraulic-head peak and crest, depending on the plastic short-penstock: **a** diameter, and **b** length

As expected, these graphs argue that as the ramified short-penstock volume increases, the associated attenuation effect of the first hydraulic-head peak and crest increase. Furthermore, Fig. 4a and b clearly illustrate that the variation of the first transient pressure peak and crest are slightly affected for the length and diameter values beyond $l_{\text{penstock}}^{\text{plastic}} \geq 7$ m and $d_{\text{penstock}}^{\text{plastic}} \geq 75$ mm, respectively. Thereby, $l_{\text{penstock}}^{\text{plastic}} = 7$ m and $d_{\text{penstock}}^{\text{plastic}} = 75$ mm may be considered as the optimal values of the plastic short section diameter and length.

4 Conclusion

In summary, the present study proved that the branching technique could remove cavitation from transient flow. Furthermore, this technique provides large attenuation of first hydraulic-head peak and crest associated to a transient initiating event. In particular, the protected system case based on a (**HDPE**) ramified plastic short-penstock provides better trade-off between hydraulic-head attenuation and wave oscillation period spreading as compared with the case based on (**LDPE**) short-penstock. Furthermore, examination of the sensitivity of the pressure peak and crest magnitudes, with respect to the plastic short-penstock length and diameter evidenced that significant short-penstock volume allows important hydraulic-head peak and crest attenuation. Nevertheless, this correlation is not significant beyond a near-optimal diameter and length values.

References

1. Aklonis JJ, MacKnight WJ, Shen M (1972) Introduction to polymer viscoelasticity. Wiley-Interscience-Wiley
2. Bergant A, Simpson AR, Tijsseling A (2006) Waterhammer with column separation: a historical review. J Fluids Struct 22(2):135–171. https://doi.org/10.1016/j.jfluidstructs.2005.08.008
3. Brinson HF, Brinson LC (2008) Polymer engineering science and viscoelasticity: an introduction. Springer
4. Fersi M, Triki A (2019a) Investigation on redesigning strategies for water-hammer control in pressurized-piping systems. J Press Vessel Technol Trans ASME. https://doi.org/10.1115/1.4040136
5. Fersi M, Triki A (2019b) Alternative design strategy for water-hammer control in pressurized-pipe flow. In: Fakhfakh T, Karra C, Bouaziz S, Chaari F, Haddar M (eds) Advances in acoustics and vibration II. ICAV 2018. Applied condition monitoring, 13 135–144, Springer, pp 157–165 https://doi.org/10.1007/978-3-319-94616-0_16
6. Keramat A, Haghighi A (2014) Straightforward transient-based approach for the creep function determination in viscoelastic pipes. J Hydraul Eng 140(12). https://doi.org/10.1061/(asce)hy.1943-7900.0000929
7. Pezzinga G, Scandura P (1995) Unsteady flow in installations with polymeric additional pipe. J Hydraul Eng **121**(11):802–811. https://doi.org/10.1061/(asce)0733-9429(1995)121:11(802)
8. Triki A (2017) Water-hammer control in pressurized-pipe flow using a branched polymeric penstock. J Pipeline Syst Eng Pract ASCE 8(4):04017024. https://doi.org/10.1061/(asce)ps.1949-1204.0000277
9. Triki A (2016) Water-hammer control in pressurized-pipe flow using an in-line polymeric short-section. Acta Mech 227(3):777–793. https://doi.org/10.1007/s00707-015-1493-13
10. Triki A (2018a) Further investigation on water-hammer control inline strategy in water-supply systems, J Water Suppl Res Technol AQUA 67(1):30–43. https://doi.org/10.2166/aqua.2017.073
11. Triki A (2018b) Dual-technique based inline design strategy for water-hammer control in pressurized-pipe flow. Acta Mech 229(5):2019–2039. https://doi.org/10.1007/s00707-017-2085-z
12. Triki A, Chaker MA (2019) Compound technique-based inline design strategy for water-hammer control in steel pressurized-piping systems. Int J Pres Ves Pip 169(C):188–203. https://doi.org/10.1016/j.ijpvp.2018.12.001
13. Triki A, Fersi M (2018) Further investigation on the water-hammer control branching strategy in pressurized steel-piping systems. Int J Press Vessel Pip 165(C):135–144. https://doi.org/10.1016/j.ijpvp.2018.06.002
14. Vitkovsky JP, Lambert MF, Simpson AR, Bergant A (2000) Advances in unsteady friction modelling in transient pipe flow. In: The 8th international conference on pressure surges bhr the hague the Netherlands
15. Wan W, Huang W, Li C (2014) Sensitivity analysis for the resistance on the performance of a pressure vessel for water hammer protection. J Press Vessel Technol Trans ASME 136(1):011303. https://doi.org/10.1115/1.4025829

16. Wylie EB, Streeter VL (1993) Fluid transients in systems. Prentice Hall, Englewood Cliffs NJ
17. Yao E, Kember G, Hansen D (2015) Analysis of water hammer attenuation in applications with varying valve closure times. J Eng Mech 141(1):04014107. https://doi.org/10.1061/(asce)em.1943-7889.0000825

Flow Velocity Effect on the Hygrothermal Behavior of the Polyester/Glass Fiber Composite

Mohamed Ounaies[1(✉)], Manel Harchay[2], Fakhreddine Dammak[3], and Hachmi Ben Daly[1]

[1] Laboratory of Mechanics of Sousse, National Engineering School of Sousse, University of Sousse, Sousse, Tunisia
ounaies_m@yahoo.fr
[2] Higher Institute of Applied Sciences and Technology of Kairouan, University of Kairouan, Kairouan, Tunisia
[3] Engineering Production Mechanics and Materials Unit (UGPM2), National Engineering School of Sfax, University of Sfax, B.P W3038 Sfax, Tunisia

Abstract. This is a study of the hygrothermal behavior of the polyester/glass fiber composite material. The composite material is exposed on a lateral face to different humidity velocity under different temperatures. The kinetics of water absorption follows the Fick's second law. The water concentration is higher on the surface of the material and it decreases continuously toward the core of the material. The increase of the humidity velocity induced a rise of the diffusion coefficient (D) and the amount of absorbed humidity at the saturation (M_m). Also, the augmentation of the temperature facilities the diffusion process. The fall of the absorption curve is due to the propagation of cracks in the material and the detachment of certain molecular chains. The finite-element software "Abaqus" is used to simulate the humidity diffusion through the composite material. The comparison between the experimental and numerical results shows that the model can predict the hygrothermal behavior of the polyester/glass fiber composite.

Keywords: Velocity · Diffusion · Simulation · Hygrothermal · Composite

1 Introduction

The use of composite material in the industrial fields is growing up over the years. It is due to the advantages that a composite material offers. Although during its utilization, composite material is exposed to extreme environmental condition like temperature, pressure and humidity. The humidity absorption causes a degradation of the material proprieties and a decrease in its life duration, which can be fatal in the safety side.

The humidity diffusion through composite material is characterized by the free volume theory: it is linked to the number and the size of the free volume existing in the material, which determines the amount of absorbed humidity [3, 8]. Also, it can be described by the molecular approach, which is attached to the hydrophilic nature of the material [3]. The humidity absorption causes a decrease of the transition temperature

N. Aifaoui et al. (Eds.): CMSM 2019, LNME, pp. 102–109, 2020.
https://doi.org/10.1007/978-3-030-27146-6_12

(Tg) [4, 15], a plasticization and swelling of the matrix [8] and a decrease of the fibers resistance. Also, we note a debonding in the fiber/matrix interface due to the increase of the osmotic pressure [2, 14]. The humidity absorption is characterized by the one dimensional Fick's second law:

$$\frac{\partial C}{\partial t} = \frac{\partial^2 (D * C)}{\partial x^2} \tag{1}$$

where C is the concentration of the diffusing, t is the time, x is the distance through the sample thickness, and D is the diffusion coefficient. The diffusion process must respect the following boundary conditions:

$$\begin{aligned} C &= C_i; 0 < x < h; t \leq 0 \\ C &= C_m; x = 0, x = h, t > 0 \end{aligned} \tag{2}$$

where C_i: the initial concentration, C_m: the concentration at the saturation and h: the sample thickness. The solution of Eq. (1) is presented in Eq. (3) [10]:

$$\frac{C - C_i}{C_m - C_i} = 1 - \left(\frac{4}{\pi} * \sum_{n=0}^{\infty} \frac{1}{2n+1} * \sin\left(\frac{(2n+1)\pi x}{h} \right) * e^{\left(-\left(\frac{(2n+1)^2 \pi D_x t}{h^2} \right) \right)} \right) \tag{3}$$

The weight gain (G) is obtained by integrating equation (3) through the thickness:

$$G = \frac{m - m_i}{m_m - m_i} = 1 - \frac{8}{\pi} * \sum_{n=1}^{\infty} \frac{1}{(2n+1)^2} * e^{\left(-(2n+1)^2 * \pi^2 * \left(\frac{(D_x * t)}{h^2} \right) \right)} \tag{4}$$

where m_i: the initial material weight, m_m: the material weight at the saturation and m: the material weight at time t.

Many authors have been interested in the humidity diffusion process: [11, 5] proved that, for small sample, the humidity absorption follows the Fick's law. Likewise, [14] confirmed that, at low temperature, the Fickian model represents well the humidity diffusion through glass fiber-reinforced plastic composite. Also, the impact of the temperature on the humidity diffusion was a subject of interest for several researchers: [6, 12] validated that the temperature augmentation offered more energy for the diffusing molecules, which elevated the diffusion coefficient (D). Moreover, researchers studied the effect of the humidity velocity on the humidity absorption: [13] confirmed that the amount of absorbed humidity increases with the augmentation of the humidity velocity. Also, [9] proved that the humidity rate increases with the rise of the flow velocity for the glass fiber/epoxy composite.

2 Materials and Experimental Methods

The experimental results are extracted from [7]. A polyester resin reinforced with unidirectional E-glass fibers was used. It contains fillers additives Aluminum Silicate Pigments (ASP 400) which are a 4.8 μm size clay powder. The composite material is made by the pultrusion technique. The dimensions of the sample are 20 * 8 * 6.3 mm. The composition of the polyester/glass fiber composite is presented in Table 1.

Table 1 Composition of the polyester/glass fiber material [7]

Material	% ASP 400	% fiber
Polyester/glass fiber composite	20%	51.5

Figure 1 presents the experimental device: the container is filled with water, heated to the desired temperature by means of a heating resistance. The flow velocity is fixed by the handle and displayed on the manometer. Periodically, the diffusion process is stopped and the sample is taken from the experimental device. Then, the sample is dried with an absorbed paper and weighed with a Gibertini E-42-B electronic scale with an accuracy of 0.1 mg. Then, the sample is put back in the assembly. The humidity content is calculated by the following formula:

$$M(\%) = \left(\frac{M_t - M_0}{M_0}\right) * 100 \tag{5}$$

where M_t is the material weight at instant t and M_0 is the initial material weight.

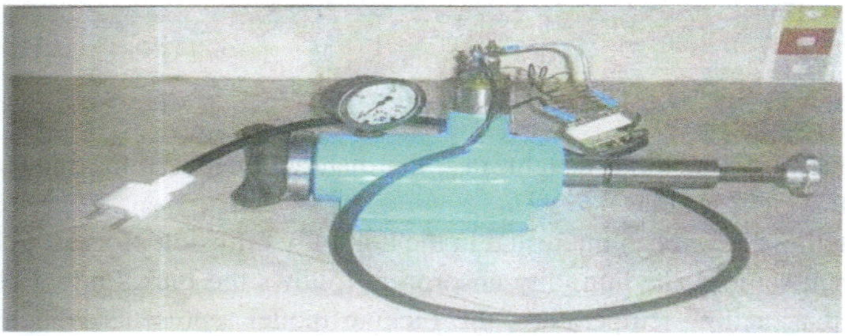

Fig. 1 Designed device for absorption process

3 Modeling and Validation

The humidity diffusion has been modeled thanks to the finite-element software 'Abaqus', through the mass diffusion analysis. The humidity diffusion is assured by the Fickian law, which is based on the hypothesis that the diffusion is governed by a concentration gradient. The flux of the diffusing phase is written in Eq. (6):

$$J = -s * D * \left(\nabla \left(\frac{C}{s} \right) + K_s * \nabla(\ln(\theta - \theta^z) + K_p \nabla(P) \right) \quad (6)$$

where: s: the solubility, D: the diffusion coefficient, g the concentration gradient, Θ: the temperature, P: the pressure, K_s: the sorter effect and K_p: pressure stress factor. The permeability law allows us to modify Eq. (6):

$$J = -s * D * \left(\nabla \left(\frac{C}{s} \right) + \frac{K_s}{\theta - \theta^z} * \nabla(\theta) + K_p * \left(\frac{\eta}{k} * V \right) \right) \quad (7)$$

where V: the flow velocity, k: the water permeability and η: the water viscosity. The diffusion coefficient (D), defined using the Arrhenius law, is presented in Eq. (8):

$$D = D_0 * e^{-\frac{E}{R*T}} \quad (8)$$

where E: the activation energy, D_0: the Arrhenius constant, R: the perfect gas constant and T: the absolute temperature.

The sample is exposed to the humidity flow on a lateral side. The absorption through others sides is assumed to be equal to zero and the initial humidity concentration in the specimen is presumed to be zero. In the analysis, a three dimensional DC3D8 element is utilized, using the heat transfer/mass diffusion element library. The results are normalized with regard to thickness. The thermo-diffusion coupling is done through a UTEMP subroutine. Figures 2 and 3 show a comparison of the experimental and numerical results of humidity absorption through the polyester/glass fiber composite.

For a temperature equal to 45 °C, the numerical model models the Fickian aspect of the absorption curve: a first linear part, while an equilibrium of the concentration gradient happens on the sample surface and a second constant part, during which there is a physical and chemical equilibrium between the diffusing wet molecules and the molecular chains of the matrix. Similarly for a temperature equal to 65 °C, the numerical model reproduces the fall of the absorption curve. Indeed, the continuous humidity absorption increases the size of the free volumes in the sample. When the osmotic pressure in the free volumes exceeds a critical value, the stress concentration at a point induces the propagation of cracks in the material. After a long duration, the cracks coalesce, causing the detachment of certain molecular chains. Consequently, the mass sample falls [1].

The numerical result considers the effect of increasing the flow velocity which gives more speed to the wet molecules. As a result, the time required to reach saturation decreases and the diffusion coefficient (D) increases. Also, the numerical model takes into account the effect of the temperature whose rise offers more mobility to the diffusing molecules and consequently the amount of absorbed humidity rises.

- 45°C:

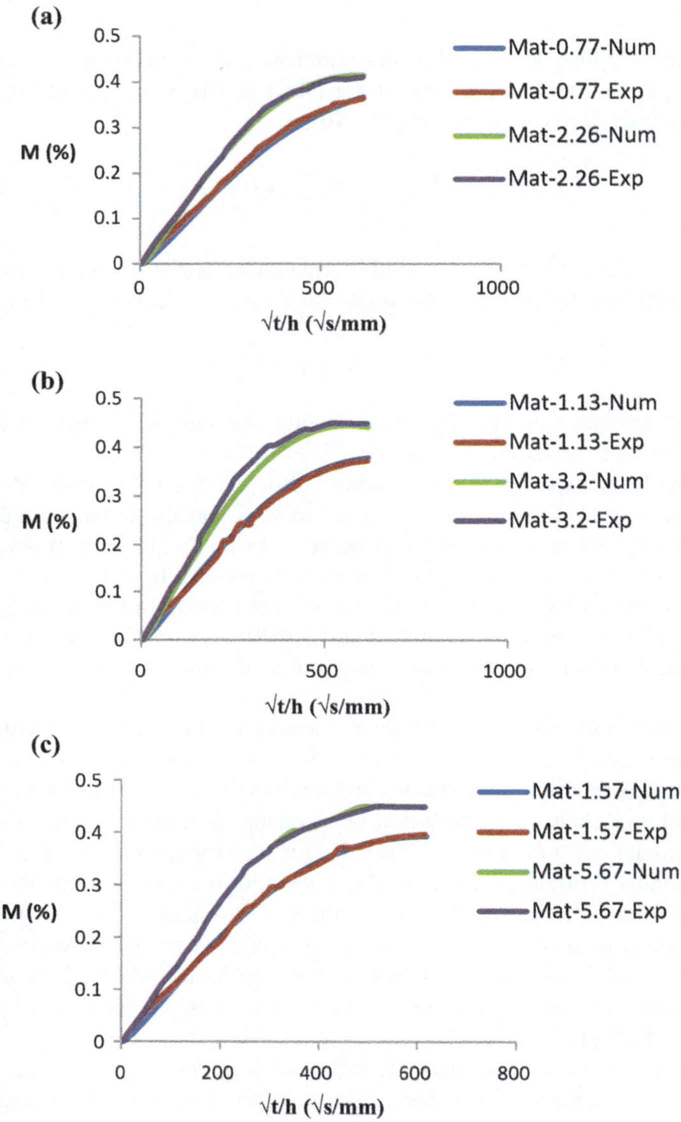

Fig. 2 Humidity absorption as a function of time, normalized to sample's thickness for: **a** 0.77/2.26 m/s, **b** 1.13/3.2 m/s, **c** 1.57/5.67 m/s

- 65°C :

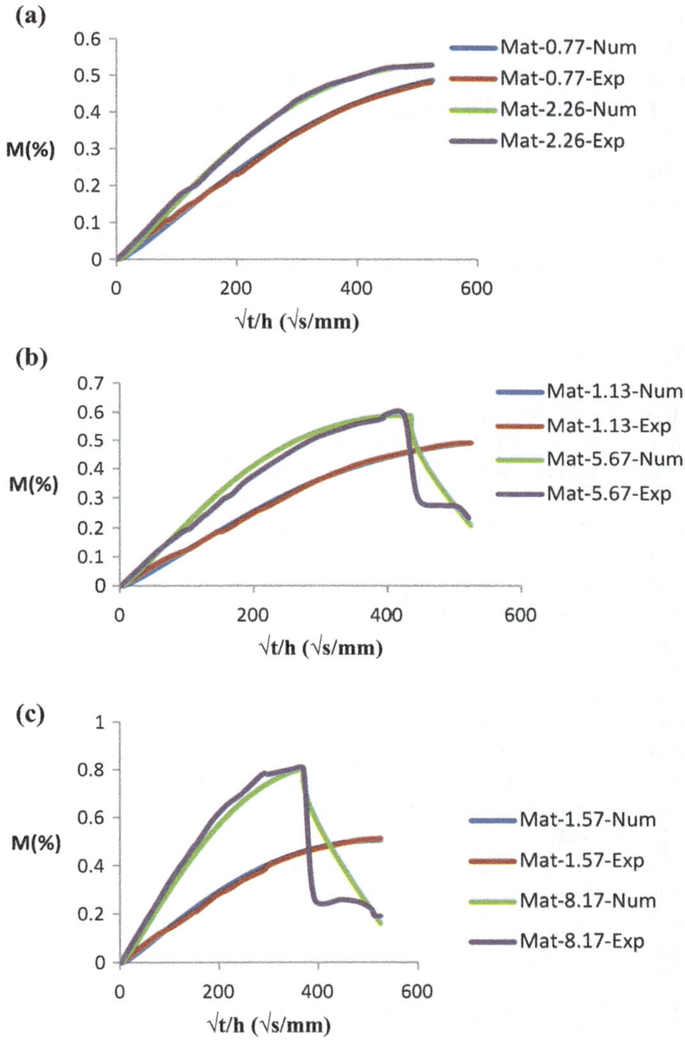

Fig. 3 Humidity absorption as a function of time, normalized to sample's thickness for: **a** 0.77/2.26 m/s, **b** 1.13/5.67 m/s, **c** 1.57/8.17 m/s

4 Conclusion

The objective is to present a numerical model of the hygrothermal behavior of the polyester/glass fiber composite under different flow velocity. The humidity diffusion has been tested at two different temperature 45 and 65 °C.

The introduction of the diffusion coefficient (D) at each temperature is made using the Arrhenius law and the initial concentration of humidity in the sample is assumed to be zero. After the introduction of the material parameters, a flow velocity is exposed on a lateral side, where the absorption is only limited. A three dimensional element CD3D8 was used. The coupling of the thermo-diffusion phenomenon has been ensured by a UTEMP subroutine. The numerical model reproduced the behavior of the experimental result. Also, it considers the effect of the temperature and the flow velocity on the humidity absorption: the increase of the two parameters engenders a rise of the diffusion coefficient (D) and the amount of absorbed humidity. The numerical model makes it possible to predict the hygrothermal behavior of the polyester/glass fiber composite, taking into account the flow velocity.

References

1. Ashbee KHG, Frank FC, Waytt RC (1967) 300 Water damage in polyester resins
2. Barjasteh E, Nutt SR (2012) Moisture absorption of unidirectional hybrid composites. Compos Part A Appl Sci Manuf 43(1):158–164. https://doi.org/10.1016/j.compositesa.2011.10.003
3. Bonniau P, Bunsell AR (1981) A comparative study of water absorption theories applied to glass epoxy composites. J Compos Mater 15(3):272–293
4. Broudin M et al (2015) Water diffusivity in pa66: experimental characterization and modeling based on free volume theory. Eur Polym J 67:326–334. https://doi.org/10.1016/j.eurpolymj.2015.04.015
5. Eslami S, Honarbakhsh-Raouf A, Eslami S (2015) Effects of moisture absorption on degradation of E-glass fiber reinforced vinyl ester composite pipes and modelling of transient moisture diffusion using finite element analysis. Corros Sci 90:168–175. https://doi.org/10.1016/j.corsci.2014.10.009
6. Ben Daly H, Ben Brahim H, Hfaied N, Harchay M, Boukhili R (2007) Investigation of water absorption in pultruded composites containing fillers and low profile additives. Polym Compos 28:355–364
7. Harchay M (2009) Contribution À La Caracterisation Du Comportement Hygrothermique Non Equilibré Dans Les Systemes Polyester/fibre de Verre
8. Huang G, Sun H (2007) Effect of water absorption on the mechanical properties of glass/polyester composites. Mater Des 28(5):1647–1650
9. Humeau C, Davies P, Jacquemin F (2016) Moisture diffusion under hydrostatic pressure in composites. Mater Des 96:90–98. https://doi.org/10.1016/j.matdes.2016.02.012
10. Jost W (1960) Diffusion in solids, liquids, gases. Academic Press, New York
11. Kumosa L, Benedikt B, Armentrout D, Kumosa M (2004) Moisture Absorption Properties of Unidirectional Glass/polymer Composites Used in Composite (Non-Ceramic) Insulators. Compos A Appl Sci Manuf 35(9):1049–1063
12. Ounaies, Mohamed, Manel Harchay, Fackhreddine Dammak, and Hachmi Ben Daly. 2018. "Prediction of Hygrothermal Behavior of Polyester/glass Fiber Composite in Dissymmetric Absorption." *Journal of Composite Materials*
13. Pilli SP, Smith LV (2012) The Effect of Pressure on Moisture Diffusion in Polymer Matrix Composites. Int J Thermophys 33(8–9):1715–1725

14. Shiva Eslami, Fathollah Taheri-Behrooz, and Farid Taheri. 2012. "Effects of Aging Temperature on Moisture Absorption of Perforated GFRP." *Advances in Materials Science and Engineering*
15. Wu L, Murphy K, Karbhari VM, Zhang JS (2002) Short-Term Effects of Sea Water on E-Glass/vinylester Composites. J Appl Polym Sci 84(14):2760–2767

Dynamics of the Flow Field Induced by Multiple Elevated Jets in Crossflow

Amina Radhouane[1](✉), Nejla Mahjoub Said[2], Hatem Mhiri[1], and Philippe Bournot[3]

[1] National Engineering School of Monastir, Route of Ouardanine, 5000 Monastir, Tunisia
radhouane_amina@yahoo.fr

[2] LGM, ENIM, Preparatory Institute for Engineering Studies, University of Monastir, Monastir, Tunisia
mahjoub_nejla@yahoo.f

[3] Aix Marseille University, CNRS, IUSTI, Marseille, France

Abstract. The present study considers multiple tandem jets in cross flow under an injection ratio less than 1. The jets are emitted through 60°-inclined, 8 mm-diameter cylindrical nozzles that are razed at different levels from the ground of the working wind tunnel. The understanding of this configuration is likely to provide a good support for the comprehension of more complicated and then real situations. The main objective of this paper consists of the exploration of the different flow structures induced by the emitted jets with the oncoming mainstream in one hand, and with each other and the different domain boundaries on the other hand. A particular attention is dedicated to the established flow field and the induced vortical structures. It is mainly observed that an injection rate inferior to 1 promotes the jets' flattening and even more the rear jet. A higher injection height, on the other hand, operates differently by providing the jets with a further impulse to cross deeper and higher the mainstream and stay away from the ground attachment effect.

Keywords: Multiple jets · Cross-flow · Injection height · Flow dynamics · Attachment effect

1 Introduction

Multiple jets in cross-flow are commonly found in several applications and domains. An efficient control of the different phenomena they induce is consequently highly recommended. In fact, a hazardous behavior is any application brings for sure a high risk over the application security and efficiency.

Multiple jets in cross-flow are involved in the industry where the control of jets' mixing is highly recommended to obtain rapid and stable thermal or non isothermal jet mixing. Such a configuration is also observed in environmental applications where the control of wastewater jets' discharge is extremely recommended. The latter are generally turbulent buoyant jets that undergo some chemical treatment before being diffused into coastal water through multiport diffusers. Multiple jets in cross-flow are of

N. Aifaoui et al. (Eds.): CMSM 2019, LNME, pp. 110–118, 2020.
https://doi.org/10.1007/978-3-030-27146-6_13

particular interest in the medical field as well in applications like blood injection during hemodialysis through one or more holes at the tip of a catheter, typically positioned at the superior vena cava [1].

Chronologically, the consideration of single jets in cross-flow was prior to that of multiple jets due to the complexity of the problem. In fact, the number of the emitted jets in addition to their characterizing parameters (inclination, arrangement disposition in rows, injection, temperature and buoyancy ratios, etc.) contributes to the dependence of the resulting flow-field to several factors. Exploring all of them directly at once may be complicated and inefficient; while introducing them progressively could be of a great deal in enlightening the impact of each of them.

The earliest single jets in cross flow-studies available in the literature were conducted by Jordinson et al. in [2] and Gordier in [3]; while the earliest multiple jets in cross-flow studies date back to 1971 with Ziegler and Wooler [4, 5] as pioneers in the field.

Since that time, several papers explored both configurations; and some of them conducted comparison between both models in order to deduce the multiple jets' model behavior from the single jet one; majorly the induced vortical structures; or to test the validity of some single jet model correlations in predicting some multiple jets' models phenomena, like mixing and penetration processes in Walker and Kors's paper [6], for example, a row of high circular jets in cross-flow were considered under variable orifice sizes, spacing and temperature ratios. Optimal mixing was obtained for both cold and hot jets at a spacing between the orifices equal to one half of the duct height. In that case, the orifice size changed only the mean exit temperature level but did not significantly alter the shape of the distributions. As to the single jet in crossflow correlations, they were proved to not adequately describe the multiple jets configurations.

Since the correlations are not necessarily and automatically adapted from single to multiple jet configurations, the latter were more extensively considered experimentally. In this context we can mention the papers of Sterland and Hollingsworth in [7], Isaac and Jakubowski in [8] and Ligrani and Lee in 1994 [9] that were based upon hot wire anemometry measurements. Yu et al. used [10] PIV and LIF experiments (Laser Induced Fluorescence techniques, Chen et al. [11] depicted schlieren and CCD images, and Radhouane et al. [12] CCD images and PIV measurements.

Studies on multiple jets in crossflow are also organized according to their orientation towards the oncoming crossflow and the supporting domain; which is a further highly interesting parameter in terms of impact on the resulting flow field dynamics, structures and heat and mass transfers. Placing the jets in one or more rows, tandem, adjacent, staggered or opposed, on a flat plat or radial, in free field or impinging on a wall, normal or inclined, etc. these are all parameters that can help control the characteristics of the resulted flow, but only when extensively explored in order to sort out their proper and combined contributions.

With reference to the abovementioned literature, it is obvious to see that the single jet in cross-flow model was extensively studied, while the prediction of multiple jets in cross-flow is a still not fully resolved problem due to the complexity of its proper progression and mixing processes as well as its interaction with the surrounding oncoming flow. It is even much less documented when it comes to consider a single arrangement; namely the tandem/aligned arrangement. The latter was first examined by

Briggs [13, 14] in a try to compute the resulting plume rise of jets sent from two tandem stacks with the same height and emission conditions. This model was then generalized by Anfossi et al. [15] to include twin stacks with different heights. Radhouane et al. [16] handled later similar twin aligned jets in order to evaluate the dynamic and thermal behavior of the resulting flow field under variable parameters; namely the injection inclination [17], jets and crossflow temperature difference [18, 19], injection height etc. Configurations of more than two aligned jets in cross flow remain still not fully resolved; and this is precisely what we intend to contribute to through this work dedicated to the examination of some developed vortical structures under a variable injection height an under injection ratio inferior to 1.

2 Experimental Set-up

The experiments were carried out in a wind tunnel at the university institute of the industrial thermal systems (IUSTI), a joint research unit between the University of Provence (Aix-Marseille) and the Mediterranean University in Marseille, France.

The wind tunnel contains an upstream inlet section allowing the introduction of an oncoming crossflow with a variable velocity. On the bottom plate were placed three tandem variably elevated injection nozzles, 60°-inclined with reference to the ground and 3 great diameters spaced since the injection cross-section is elliptical. The little diameter, d, being equivalent to 8 mm. Herein, it is to note that the rear injection nozzle was placed 20 diameters from the wind tunnel inlet section in order to consider measurements in a well-established surrounding crossflow, far from the boundary layer influence as illustrated in previous papers with a twin jet in cross-flow model [19] (Fig. 1).

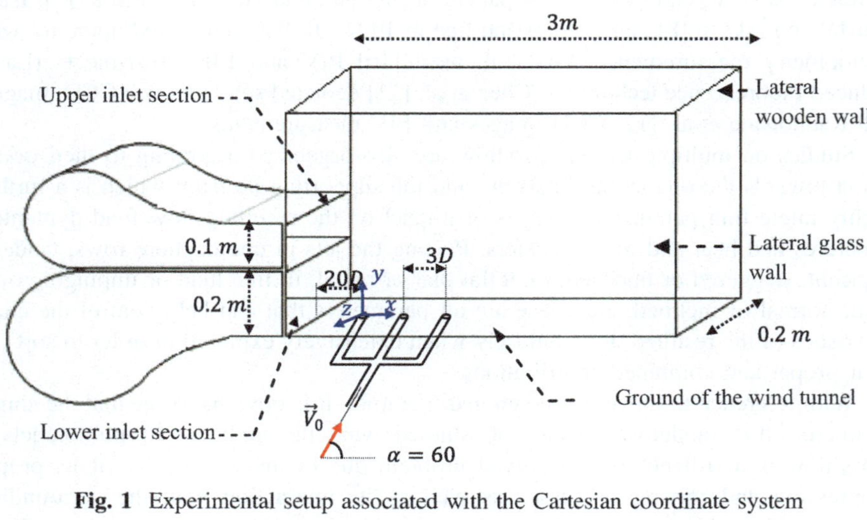

Fig. 1 Experimental setup associated with the Cartesian coordinate system

Measurements were depicted by means of Coupled Charge Device (CCD) images together with a two-component Particle Image Velocimetry (PIV) technique, in the symmetry plan z = 0; which allowed a non intrusive, instantaneous and mean bi-dimensional dynamic diagnosis of the resulting flow field (velocity components and turbulence quantities' development) under an injection ratio under one.

3 Results and Discussion

The dynamics of the resulting flow field are majorly defined by the interaction of the discharged jets with the oncoming crossflow and the surrounding boundaries either of the domain (particularly the ground of the wind tunnel) or the variably elevated emitting nozzles. These interactions take place in the form of vortical structures that develop on the different borders of the jets (shear layer/leading edge vortices on the top

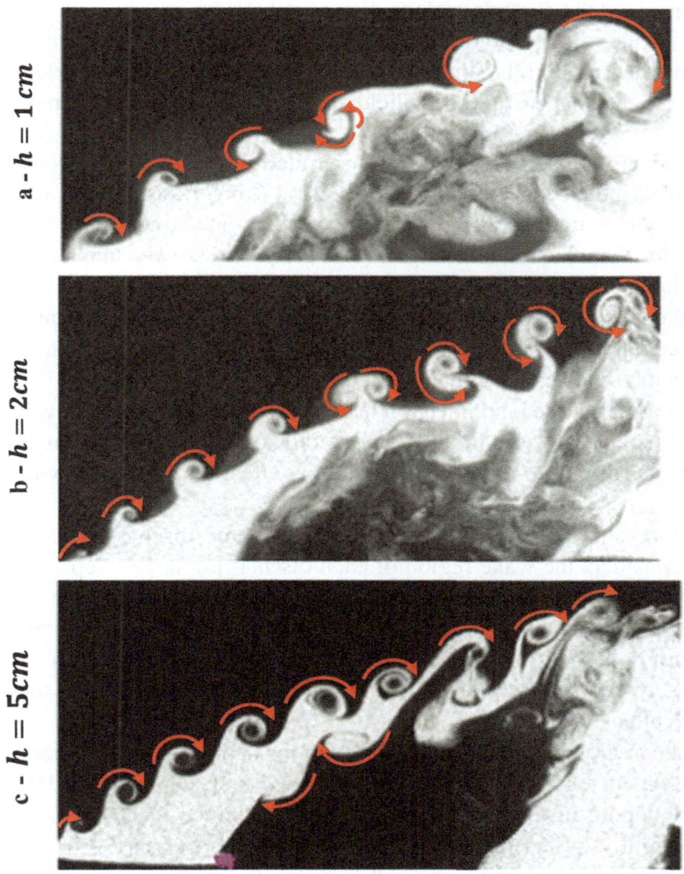

Fig. 2 Visualization of the shear layer vortices on the symmetry plane (z = 0) under R < 1

edge [20]; wake/upright vortices on lower edge of the jets [21] and horseshoe vortices windward of the jets [22] and within them counter-rotating vortices [20]). The order of occurrence of these interactions is closely dependent on the flow regime, dependent on its turn to the jets-to-mainstream velocity ratio, also called injection ratio. In the present paper, concentration will be dedicated to an injection ratio inferior to one (R < 1), which suggests the predominance of the mainstream towards the emitted jets. This reasonable assumption is comforted with the observations depicted from Fig. 2 illustrating the development of the shear layer vortices.

It's obvious to note the dominant clockwise sense of rotation of the shear layer vortices under all injection heights; but is also relevant to note their clearer and more proper development from higher emitting nozzles. In fact, discharging the jets farther from the ground releases them from the attachment effect and provides the jets with a stronger impulse to cross the mainstream. We particularly considered the progression of the rear jet since this is the one to be the most consistently flattened by the mainstream due to its prior position in the mainstream path.

The abovementioned attachment and flattening effects are enlightened in Fig. 3 where we provided the evolution of the discharged jets among the oncoming cross-flow in terms of mean velocity contours and vectors. The capture of the images was particularly tightened around the emitting section of the discharging nozzles; however we can easily note the shape of the flow field streamtraces and the corresponding velocity levels.

As the jets are emitted higher from the ground the rear jet undergoes a decreasing flattening effect indicated by its higher dispersion (longer arrow from a to c in Fig. 3). This observation is tightly related to the decreasing attachment effect downstream of the emitted nozzles, illustrated in Fig. 3 with the progressively smaller encircled zone. While the latter contained stagnant points that attached the surrounding flow and induced wake vortices as illustrated in the case of twin jets in a previous paper [16], it henceforth contains simply a low velocity flow that does not contain any consistent vortical structures; instead it contains regular streamtraces.

Downstream of the second injection nozzle, we observe a progressively slower flow zone as well (smaller vectors in the dotted parallelograms in Fig. 3); as this location is double-protected from the mainstream flattening effect by the two previous discharged jet columns.

Figure 4 reconsiders the CCD images of the three discharged jets in crossflow; however it considers the wake region of each discharging nozzle separately in order to check and validate the abovementioned observations. The wake of the first emitted jet is clearly more consistent when emitted close to the ground as signaled by the encircling rectangles in Fig. 4a; it even ends by fading away when sent at the highest injection case relative to $h = 5cm$. The rear jet's elevation however, seems to be not significantly affected by the jets' elevation.

The wake region of the second jet column is progressively liberated from the ground attachment effect as well. In fact, the two first jets combine later due to less consistent flattening of the first jet, implying a postponed reaching of the second jet; higher and slightly farther downstream of the emitting section level. The wake of the combined jets is consequently reduced and similarly postponed higher and farther downstream (Fig. 3b).

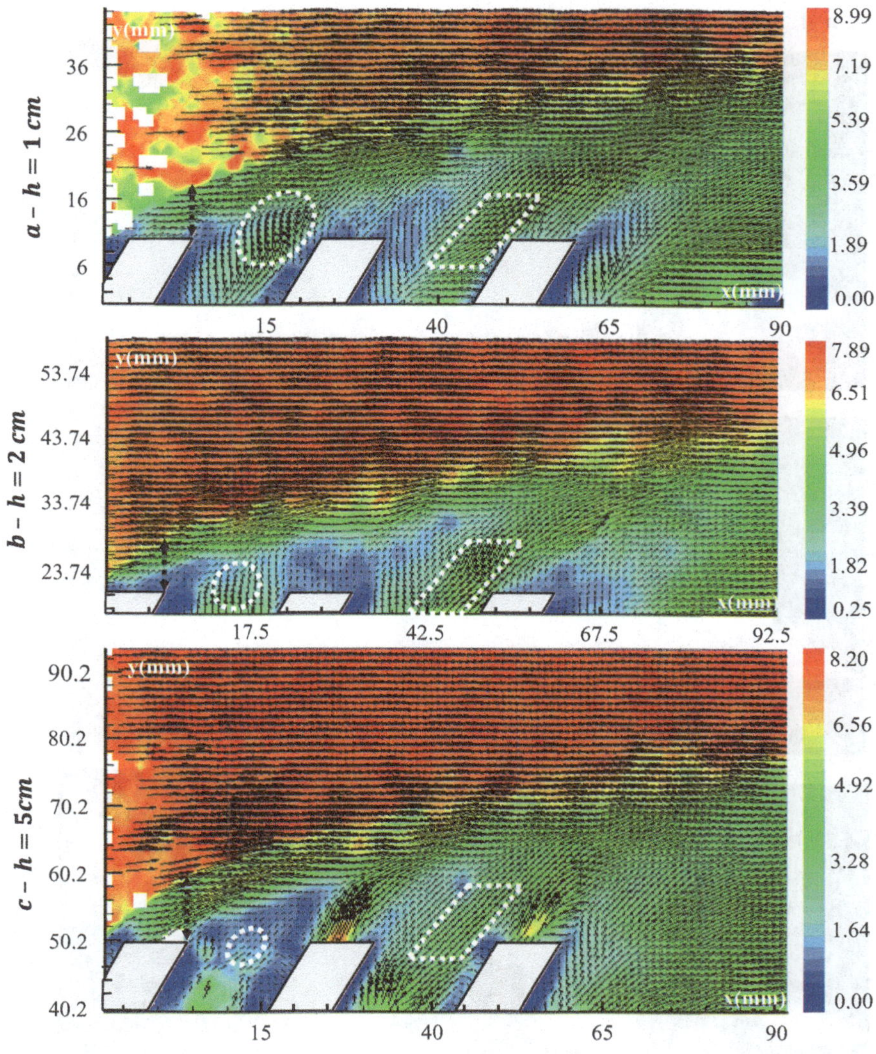

Fig. 3 Establishment of the mean velocity vectors and contours on the symmetry plane (z = 0) under R < 1

The wake region of the third discharging nozzle is consequently and understandably going to be more significant in the case of $h = 1cm$; as it is going to receive jets that were seriously flattened and then as greatly directed downstream. The resulting combined jet plume is consequently well developed, in terms of jet height and curvature, and wake region expansion (Fig. 4c-I). As the jets are sent from a higher

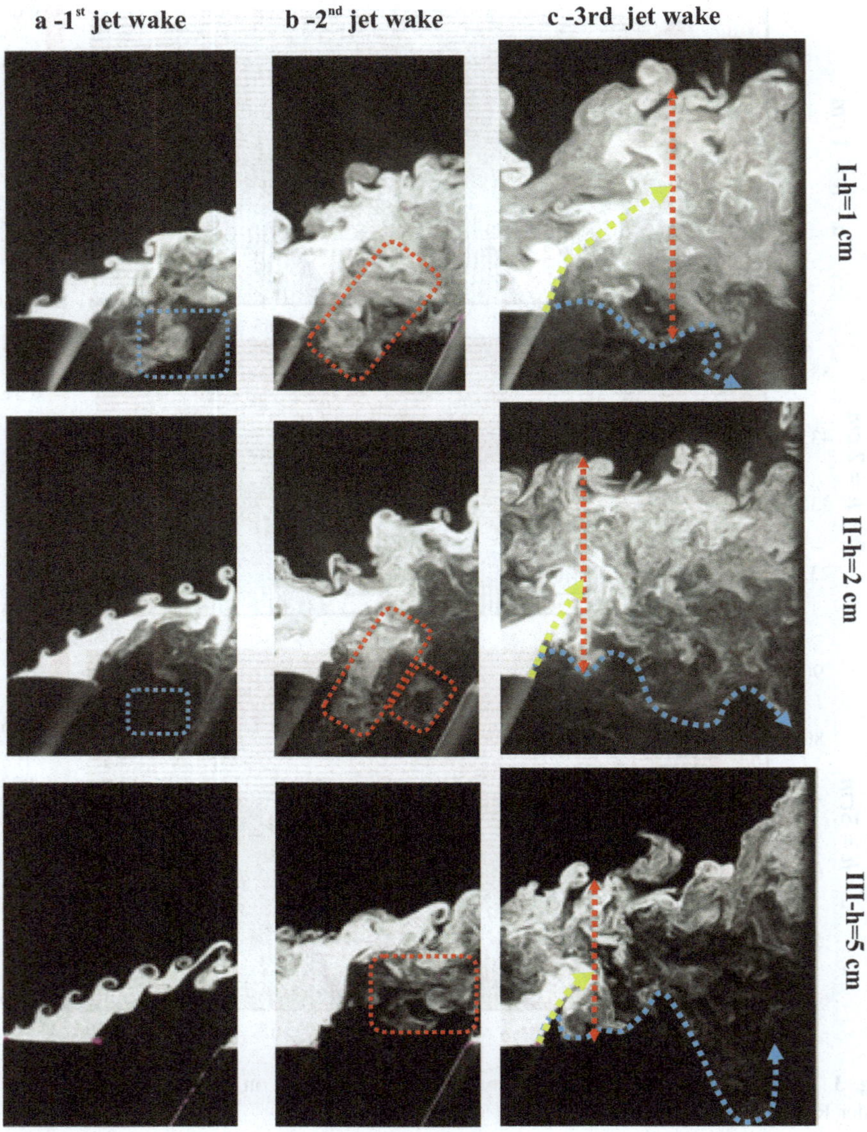

Fig. 4 Effect of the injection height over the different jets' wake regions for R < 1 in the symmetry plane (z = 0)

discharging section, they lose some of their consistency, and when they combine downstream of the third nozzle they result in a weaker combined plume: shorter and more curved; that consequently rapidly fades away (Fig. 4c-III).

4 Conclusion

An experimental study was conducted in order to consider the dynamics of the resulting flow field of multiple tandem jets with an oncoming cross-flow. The CCD images together with PIV measurements provided close images of the vortical structures developed on the symmetry plane (z = 0), consisting mainly of clockwise leading edge vortices (R < 1) that evolve more clearly and regularly as the jets are emitted farther from the ground (increasing h). This is due to a decreasing flattening effect that manifests as well in terms of slower flow zones downstream of the discharging nozzles (wake zones), as progressively deprived of the wake/uprights vortices. The downstream jet columns undergo a decreasing flattening effect due to their shielding by the rear jet. Combined together, such mechanisms are likely to decrease the quality of the mixing near the ground and enhance it rather far among the domain.

References

1. Froust J, Rockwell D (2007) Flow structure associated with the multiple jets from a generic catheter tip. Exp Fluids 42:513–530. https://doi.org/10.1007/s00348-006-0249-z
2. Jordinson R (1958) Flow in a jet directed normal to the wind. R and M No. 3074, Brit. A R C
3. Gordier RL (1959) Studies on fluid jets discharging normally into moving liquid. St Anthony Falls Hyd Lab, Tech Paper, No 28, Series B
4. Ziegler H, Wooler PT (1971) Multiple jets exhausted into a crossflow. J Aircr 8(6):414–420
5. Ziegler H, Wooler PT (1973) Analysis of stratified and closely spaced jets exhausting into a crossflow. National Aeronautics and space Administration, Washington, DC
6. Walker R E, Kors D L, (1973) Multiple Jet Study. Technical Report, NASA-CR-121217; NAS 1.26:121217
7. Sterland PR, Hollingsworth MA (1975) An experimental study of multiple jets directed normally to a cross-flow (for turbojet afterburner flameholder design). J Mech Eng Sci 17:117–124
8. Isaac KM, Jakubowski AK (1985) Experimental study of the interaction of multiple jets with a cross flow. AIAA Journal 23(11):1679–1683. https://doi.org/10.2514/3.9151
9. Ligrani PM, Lee JS (1996) Film Cooling from a Single Row of Compound Angle Holes at High Blowing Ratios. Int J Rotating Mach 2:259–267
10. Yu D, Ali MS, Lee JHW (2006) Multiple tandem jets in cross-flow. J Hyd Eng 132:971–982
11. Chen P, Li S, Luo S, Ni Z (2014) Flow visualization on lateral multiple jet interaction with freestream. J Flow Control Meas Visulization 2:7–11. https://doi.org/10.4236/jfcmv.2014.21002
12. Radhouane A, Mahjoub Saïd N, Mhiri H, Bournot Ph, Le Palec G (2016) Twin inclined jets in crossflow: experimental investigation of different flow regimes and jet elevations. Environ Fluid Mech 16:45–67. https://doi.org/10.1007/s10652-015-9410-7
13. Briggs GA (1974) Plume rise from multiple sources. In: Proceedings of cooling tower environment held at University of Maryland. pp 161–179
14. Briggs GA (1975) Plume rise predictions. Lectures on air pollution and environmental impact analysis, pp 59–I 11
15. Anfossi D, Bonino G, Bossa F, Richiardone R (1978) Plume rise from multiple sources: a new model. Atmos Environ 12:1821–1826

16. Radhouane A, Bournot H, Mahjoub Saïd N, Mhiri H, Le Palec G (2009) Numerical and experimental study of a double jet inclination variation on its dynamic evolution within a crossflow. Heat Mass Transf J 45(12):1597–1616
17. Radhouane A, Mahjoub Saïd N, Mhiri H, Le Palec G, Bournot P (2009) Impact of the initial streamwise inclination of a double jet emitted within a cool crossflow on its temperature field and pollutants dispersion. Heat Mass Transf J 45(6):805–823
18. Radhouane A, Mahjoub Saïd N, Mhiri H, Le Palec G, Bournot P (2010) Impact of the temperature gradient between twin inclined jets and an oncoming crossflow on their resulting heat transfer. J Comput Therm Sci (CTS-1071) 2(6):pp. 487–499
19. Radhouane A, Bhouri Baouab I, Mahjoub Saïd N, Mhiri H, Le Palec G, Bournot P (2013) Temperature impact on the turbulence generated by the interaction of twin inline inclined jets in crossflow. Heat Mass Transf 49:629–656
20. Kelso RM, Lim TT, Perry AE (1996) An experimental study of round jets in cross-flow. J Fluid Mech 306:111–144
21. Fric TF, Roshko A (1994) Vortical structure in the wake of a transverse jet. J Fluid Mech 279:1–47
22. Andreopoulos J, Rodi W (1984) Experimental investigation of jets in a crossflow. J Fluid Mech 138:93–127

Transient Flow Study and Fault Detection in Polymeric Pipelines Inverse-Transient-Based Leak Detection Algorithm

Oussama Choura[1(✉)], Sami Elaoud[1], and Bruno Brunone[2]

[1] Laboratory of Applied Fluid Mechanics, Process Engineering and Environment, National School of Engineering of Sfax, Sfax, Tunisia
Oussema.choura@enis.tn

[2] Water Engineering Laboratory, Department of Civil and Environmental Engineering, University of Perugia, Perugia, Italy

Abstract. This paper presents a technique for detection and location of faults in polymeric pipelines, by means of transient analysis, of water flows. The method uses transient pressure waves initiated by the sudden closure of a downstream shut-off valve. The presence of faults in pipes partially reflects these pressure waves and allows for the location of any type of fault e.g. leaks. Pressure waves are governed by two coupled non-linear, hyperbolic partial differential equations. The generalized Kelvin-Voigt model was adopted to model the viscoelastic behavior of the polymeric pipes. The fluid pressure head and flow rate are considered as two principal dependent variables. To locate the leak, the mathematical formulation has been solved by the method of characteristics (MOC) of specified time intervals along with Nelder-Mead optimization algorithm for the estimation of the flaw parameters (size and location). The numerical obtained results have shown a good agreement with the experimental data for the detection and the location of leaks.

Keywords: Transient flow · Inverse transient analysis · Polymeric pipes · Leak detection · MOC · Nelder-Mead optimization algorithm

1 Introduction

Water covers 70% of Earth. However, only 3% of the Earth's water is freshwater. Nowadays, the lack of fresh water results in 1.1 billion people on Earth lack access to fresh water. As stated the International Water Management Institute (IWMI), 45 countries (33% of the world population) will suffer from water scarcity by 2025 [1]. According to the United Nations (UN) [2], the global population is estimated to reach 9.3 billion in four decades. Along with such growth, an increase in water demands became inevitable. This inescapable rise is due not only to the daily needs of drinking, health, and sanitation but also to the energy production, food and other services and goods that require water to be delivered. The insufficiency of such a need became a major problem. Rationing water use seemed to be the perfect solution for a while but knowing that up to 30% of water is lost in transportation (about 20% in France in 2014, 40% in Canada in 2014) [3], another more sufficient redeem must surface. This loss

N. Aifaoui et al. (Eds.): CMSM 2019, LNME, pp. 119–133, 2020.
https://doi.org/10.1007/978-3-030-27146-6_14

ratio can reach the 50% of the production in some old networks (in Bulgaria, 1996) [4]. In fact, transmission pipes are subject to numerous types of incidences such as traffic, overloads and water hummer that may result in leaks. Most water distribution networks suffer from water leaks which may occur because of corrosion, aging pipelines or excessive pressure resulting from operational errors. In addition, bad workmanship can be a root for pipelines faults. Leaks vary in size and significance. Nevertheless, when combined, they result in major losses. Water supply systems in North America suffer a range of leaks from 0.5 to 1.4 measurable leaks per mile of pipe [5] and up to 3 leaks per mile in some cases. Based on an investigation by the Asia development bank, the water losses in the Asian region ranged from 8% in Singapore to 62% and Bangladesh [6]. In Europe, for the purpose of limiting leaks as much as possible, 60% of drinking water networks were renovated and reconstructed out of High-Density Polyethylene (HDPE) pipes owing to their ability to resist higher pressures and their considerable durability comparing with steel or concrete pipes. Besides their particular mechanical characteristics (visco-elastic behavior), these pipes are also known by for the lower price and the easy manufacturing and manipulation. However, they are, in some cases, susceptible to leaks and even bursts. Based on statistics above, detecting the defective pipe whether polymeric or rigid, providing an accurate and exact position of a leak and quantifying this default plays a vital role in the management and distribution of water as it reduces losses and lessens the negative impact on humans and the surrounding environment.

2 Inverse Analysis Method

2.1 Concept

The Inverse Methods is the resolution of a system in an inverse mode. The objective of such method is not the prediction of the behavior for given set of parameters and boundary conditions but the estimation of these dependent variables based on the physical-data knowledge. In a forward problem, the system response (pressures and flows) is calculated based on the characteristics of the pipe and outlet/inlet conditions. In an inverse problem, estimated parameters are determined based on the pressures and flows measured in the system for specified positions and time-scale.

2.2 Inverse Transient Analysis

Inverse Transient Analysis (ITA) Methods are a super-set of pressure-flow derivation methods. These techniques are widely used in water distribution systems. Pudar and Ligett [7] were the first to present an inverse-analysis for leak detection in a network. Based on the steady-state, they modeled a leak at the node (junction) of the pipes. An objective function which was the sum of the squares of the differences in measurable properties such as pressures and flows is to be minimized. The authors found that the size and location were too sensitive to the measurement accuracy and the parameters of the pipe of interest. For that, it was concluded that this approach can serve only as a supplement to leak survey. Limiting their consideration to a pipeline system transporting incompressible fluids, Mukherjee and Narasimhan [8] followed a similar approach but for an non-node position of the leak. They stated that the method ability to detect leaks is less than 80% for magnitude up to 10% of the flow. Ligett and Chen [9] saw that the technology progress such as the development of the speed of pressure transducers and computers for data logging and solving equations and problems in an interesting time can facilitate the real time monitoring and data acquisition for the inverse-transient method. Nevertheless, they coupled the unsteady-state analysis with the steady-state approach for the calibration of the inverse analysis model. Both Pudar and Ligett [7], and Ligett and Chen [9] defined the same objective function and used a method-of-characteristics-based algorithms to create a calibrated model of the transient.

3 Mathematical Model

3.1 Momentum Equation

The momentum equation is derived from Reynold transport theorem [10]:

$$gA\frac{\partial H}{\partial x} + \frac{dQ}{dt} + h_f = 0 \tag{1}$$

Where g is the gravitational acceleration (m/s^2), A is the cross-sectional area (m^2), H is the pressure head (m), a is the wave velocity (m/s), Q is the flow discharge and h_f is the loss friction term. This friction can be divided into a steady-state and an unsteady-state component. The unsteady friction is neglected since it has no major effect on the pressure wave compared to the steady state friction which can be determined as a function of Darcy-Weisbach friction factor [10].

$$h_f = f\frac{Q|Q|}{8gAD^2} \tag{2}$$

The delimitation of the Darcy-Weisbach friction factor depends on the flow regime. It is calculated in turbulent flow (Re >3000) using the iteratively-solved Colebrook formula [11]:

$$\frac{1}{\sqrt{f}} = -2\log_{10}\left(\frac{k}{3.7D} + \frac{2.51}{\mathrm{Re}\sqrt{f}}\right) \tag{3}$$

Where k is the pipe wall roughness (m), and in laminar flow using Hagen-Poiseuille formula: $f = 64/Re$.

3.2 Continuity Equation

To derive the continuity equation, the law of conservation of mass is applied along with certain assumptions:

(i) The flow is one-dimensional,
(ii) the fluid is slightly compressible,
(iii) the pipe material is homogeneous and isotropic,
(iv) the conduit walls are linearly visco-elastic for small strains,
(v) its mechanical behavior depends on the creep function only (Poisson Ration μ is constant),
(vi) the fluid velocity is neglected compared to the wave velocity, which yields the following expression [12]:

$$gA\frac{\partial H}{\partial t} + a^2\frac{\partial Q}{\partial x} + 2Aa^2\frac{\partial \varepsilon_r}{\partial t} = 0 \tag{4}$$

Where ε_r is the retarded strain (*m/m*). The third term represents the retarded strain derivative while the instantaneous elastic strain was included in the wave velocity expression and the head pressure time derivative. As a matter of fact, polymers have an immediate-elastic response and a retarded-viscous response. Consequently, strain, ε, can be decomposed into an immediate-elastic strain, ε_e, and a retarded-viscous strain, ε_r, expressed as follows [13]:

$$\varepsilon(t) = \int_{-\infty}^{t} \frac{\partial \sigma(t')}{\partial t'} J(t - t')dt' = J_0\sigma(t) + \int_{0}^{t} \sigma(t - t')\frac{\partial J(t')}{\partial t'}dt' \tag{5}$$

Such behavior can be modeled using a Standard Linear Solid model (SLS) where five Kelvin-Voigt elements are considered ($n_{kv} = 5$). In this case, the creep function J is written as follows:

$$J(t) = J_0 + \sum_{k=1}^{n_{kv}} J_k(1 - e^{-t/\tau_k}) \tag{6}$$

Where J_0 is the creep compliance of the first spring the SLS model which is equal to the inverse of its Young Modulus ($J_0 = 1/E_0$) and $\tau_k = \eta_k/E_k$ is the retardation time as the fraction of the viscosity of the dashpot and the elasticity of the spring of k-Kelvin-Voigt element. The hoop stress related to the fluid pressure on the pipe wall is equal to $c_1\rho gD\Delta H/2e$ where c_1 is the pipe-wall constraint coefficient (anchoring coefficient) which yield the following strain expression (Fig. 1):

$$\varepsilon(t) = J_0\frac{c_1 D\rho g}{2e}(H(t) - H_0) + \int_{0}^{t} \frac{c_1 D\rho g}{2e}(H(t - t') - H_0)\frac{\partial J(t')}{\partial t'}dt' \tag{7}$$

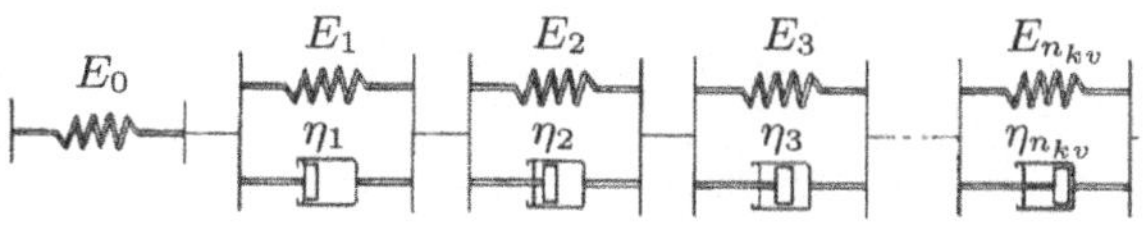

Fig. 1 Standard Linear Solid visco-elastic model [14]

3.3 Leak Modeling

Leak outflow depends mainly on the leak effective area $A_E = C_l A_l$, defined as the product of the discharge coefficient, C_l, and the leak area, A_l, and the pressure head difference ΔH. Assuming that the discharge is into ambient air, a leak outflow is quantified as follows [15]:

$$Q_l = aH^b \tag{8}$$

This form includes the Torricelli formula where $a = C_l A_l \sqrt{2g}$ and $b = 1/2$:

$$Q_l = A_E\sqrt{2gH} \tag{9}$$

The leak is assumed small for a pressure drop generation.

4 Numerical Resolution: Method of Characteristics

The continuity (Eq. 4) and momentum (Eq. 1) equations form a pair of quasilinear, hyperbolic, partial differential equations in terms of two dependent variables, flow rate, Q, and hydraulic head pressure, H, and two independent variables, time, t, and distance along the pipe, x. The method of characteristics is one among numerical analysis schemes. This method has proven its reliability comparing to the other methods because of its unconditional stability, computing-time and complexity.

4.1 Finite-Difference Equations

These governing equations are transformed into two ordinary differential equation by means of the method of characteristics:

$$C^{\pm} : \frac{dH}{dt} \pm \frac{a}{gA}\frac{dQ}{dt} + \frac{2a^2}{g}\frac{\partial \varepsilon_r}{\partial t} \pm h_f = 0 \tag{10}$$

4.2 Retarded Strain

Following the SLS model and according to Boltzmann superposition theorem for small strains, the retarded strain and its derivative can be calculated as the sum of the retarded strains and their derivatives of each Kelvin-Voigt element which means that, in this case:

$$\varepsilon_r(x,t) = \sum\nolimits_1^{n_{kv}} \varepsilon_{k,r} \tag{11a}$$

$$\frac{\partial \varepsilon_r}{\partial t}(x,t) = \sum\nolimits_1^{n_{kv}} \frac{\partial \varepsilon_{r,k}}{\partial t} \tag{11b}$$

Where the time derivative of the retarded strain on a k-Kelvin-Voigt element can be expressed as follows [12]:

$$\frac{\partial \varepsilon_{k,r}}{\partial t}(x,t) = -\frac{\varepsilon_{k,r}(x,t)}{\tau_k} + \frac{J_k}{\tau_k}F(x,t) \tag{12}$$

Covas et al. [12] defined an auxiliary function $F(x,t)$ as:

$$F(x,t) = \frac{c_1 D\rho g}{2e}(H(x,t) - H(x,0)) \tag{13}$$

and proposed the following expression of the elementary retarded strain $\varepsilon_{k,r}(x,t)$:

$$\begin{aligned} \varepsilon_{k,r}(x,t) = {} & J_k F(x,t) + J_k e^{-\frac{\Delta t}{\tau_k}} \mathrm{F}(x,t-\Delta t) \\ & - J_k \tau_k \left(1 - e^{-\frac{\Delta t}{\tau_k}}\right) \frac{F(x,t) - F(x,t-\Delta t)}{\Delta t} + e^{-\frac{\Delta t}{\tau_k}} \varepsilon_{k,r}(x,t-\Delta t) \end{aligned} \tag{14}$$

4.3 Boundary Conditions

Inlet Condition: Upstream Reservoir. Theoretically, the head pressure inside the tank is considered to be constant contrary to the experimental data that showed a variation of this parameter in time while a transient occurs. Thus, the variation of the pressure head at the reservoir was introduced as an inlet condition although it is quite small but affective for an accurate simulation.

Outlet Condition: Downstream Valve. The closure of the valve downstream the Reservoir-Pipe-Valve system is supposed to be instantaneous.

In-between Condition: Leak. In order to simulate such singularity, the pipe was considered as the junction of two pipes where the leak is located (See Fig. 2). No pressure change in the junction was considered. Thus, the condition can be written as follows:

$$H_1(L1,t) = H_2(0,t) \tag{15}$$

Where $L1$ is the location of the leak counting from the reservoir to the valve, the indexes 1 and 2 refers to the conduit number. The law of mass conservation yields the following equation:

$$Q_1(L1,t) = Q_2(0,t) + Q_l \tag{16}$$

The combination of the four equations (Eqs. 10, 15 and 16) gives a second degree polynomial equation where the variable of interest is the square root of the head pressure at the junction.

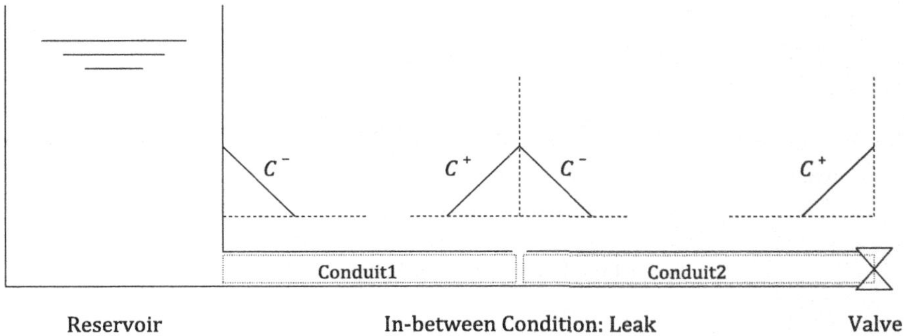

Fig. 2 Boundary conditions explicative chart

5 Experimental Setup

Experiments were carried out in the Water Engineering Laboratory (WEL) at the University of Perugia, Italy on an approximately 168.5 *m*-length HDPE pipe. The setup comprised a single 110 *mm*-nominal diameter pipe connecting an upstream pressurized tank to a pneumatic actuated butterfly valve. The wall thickness of the pipe of interest is 8.1 *mm*. The machined leak is located at 63.2 m from the reservoir with an effective area identified numerically equal to 33.6 m^2. The flow measurement was ensured by the use of an electromagnetic flow meter ML210 - ISOIL Industria mounted at 19.8 *m* (Noted FM in Fig. 3) from the reservoir and an ultrasonic Doppler veloci-meter DOP1000 - Signal Processing mounted few meters before the valve (See Fig. 3). Six transducers were installed for pressure measurement in several measurement sections (see Tab.1).

All measurement instruments whether for pressure or flow were connected to a National Instruments Compact Data Acquisition chassis cDAQ-9188 where 8 Analog Input Modules (NI-9218) were mounted as adapters. The physical data collected was saved into a laptop using a LabVIEW interface. Six test where carried out for 3 different flows (regulated by a gate valve downstream the system) for the two states of the pipe (intact/leaky). Physical data collected from an undamaged pipe test are used for the adjustment of the numerical model (Table 1).

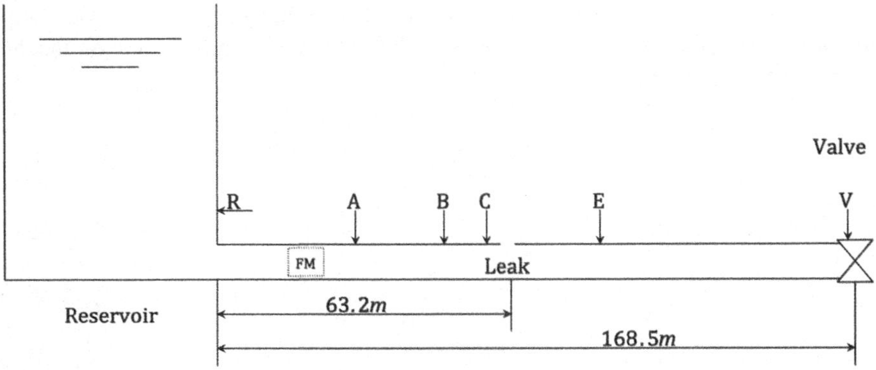

Fig. 3 Experimental Setup sketch

Table 1 Measurement sections position and notation

Notation	R	A	B	C	E	V
Position (*m*)	0	30.1	54.1	62.6	91.8	166.5

6 Model Calibration and Adjustment

Physical data collected from an undamaged pipe test are used for the adjustment of the numerical model.

6.1 Wave Velocity Delimitation

The elastic wave velocity a is a function of the creep compliance of the first spring of the SLS model and therefore its Young Modulus.

$$a = \sqrt{\frac{K}{\rho\left(1 + \frac{K}{E_0}\frac{D}{e}c_1\right)}} \tag{17}$$

Where K is the bulk modulus of the fluid and ρ is its density. D is the internal diameter of the pipe.

This parameter can be determined taking into account the extrema (Maximum/Minimum) of the acquired signal (351 *m/s*), taking into account the first characteristic period (361 *m/s*) and taking into account the first characteristic period of the signal acquired from the transducer next the valve (356 *m/s*). However, these methods are quite inaccurate because of there dependence on the operator. Thus, the wave velocity was determined numerically as 385 *m/s* corresponding to E_0 = 1.7 *GPa*.

6.2 Optimization Problem

Problem-Stating The consistency of the numerical code with the experimental data can be evaluated by the assessment of the accuracy (Relative Error) and the least square error (LSE) between measured and numerically-simulated pressures in different measurement section which can be mathematically presented as follows:

$$E = \sum \left(100 \times \frac{H_x^s - H_x^m}{H_x^m} \right) \tag{18}$$

and,

$$E = \sum \left(H_x^s - H_x^m \right)^2 \tag{19}$$

Where m stands for measured pressure heads, s for simulated and x for the measurement position. Three measurement section were considered; V (166.4 m), E (91.7 m) and B (54.1 m). The decision variables of the optimization problem are the creep function coefficients and the pipe wall roughness k.

$$E = f(J_0, J_1, J_2, J_3, J_4, J_5, k) \tag{20}$$

The retardation times were fixed according to Covas et al. [12] model.

Optimization Algorithm: Nelder-Mead Algorithm The method is used for the minimization of a function of n variables (7 in this case), which depends on the comparison of objective function values at the (n + 1) vertices of a general simplex (polygon). Accordingly, the vertex with the highest value is replaced by another point. The simplex changes its size and shape at every iteration, and contracts when meeting the final minimum. Also called Downhill search method, It is a simple algorithm to search for local minima and applicable for multidimensional optimization applications. Unlike classical gradient methods, this algorithm does not have to calculate derivatives. Instead it creates a geometric simplex and uses its movement to guide its convergence. A simplex is defined as a geometrical figure which is formed by vertices, where is the number of variables of an optimization function, and vertices are points selected to form a simplex. In each iteration, the simplex method will calculate a reflected vertex of the worst vertex through a centroid vertex. According to the function value at this new vertex, the algorithm will do all kinds of operations as reflection or extension, contraction, or shrink to form a new simplex. In other words, the function values at each vertex will be evaluated iteratively, and the worst vertex with the highest function value will be replaced by a new vertex which has just been found. Otherwise, a simplex will be shrunk around the best vertex, and this process will be continued until a desired minimum is met or no further improvement of the objective function can be done [16].

Results Several Attempts of calibration were carried out for the using the mentioned above objective function (Eq. 20) and taking into account the retardation times for different sampling times ranging from 2 to 20 seconds. However, using Nelder-Mead algorithm, the derivative free optimization tool, the best fitted creep function coefficients and pipe wall roughness are determined as in the following table where the overall mean error (for the three considered positions) reached 0.5%. The wave velocity determined after the calibration process, is equal to 374 *m/s* corresponding to an initial Young Modulus equal to E_0 = 1.6 *GPa*. The results showed a mean error equals to 0.5 % for all Sects. (0.065 m^2 At the V-position, 0.049 m^2 at the B-position and 0.039 m^2 for the E-position) with a maximum of 10.5%, 16.8% and 17.7% at respectively the section V, B and E. The determination coefficient R^2 reached 0.990 at the valve, 0.989 at the section E and 0.971 at the section B (see Fig. 4). These errors of consistency between numerical and experimental data is explained by the fact that some phenomenon were not included in the simulation. As shown in the plot below (Fig. 4, zoomed detail), a pressure peaks can be distinguished which are due to the effect of bends on the wave velocity on flow parameters. Besides of the previously mentioned effect, the negligence of the unsteady state friction plays a part in the generation of such error even if its effect is minor compared to the steady-state loss friction and visco-elasticity of the pipe since all of these three factors contribute in the damping and energy loss [12]. (Table 2)

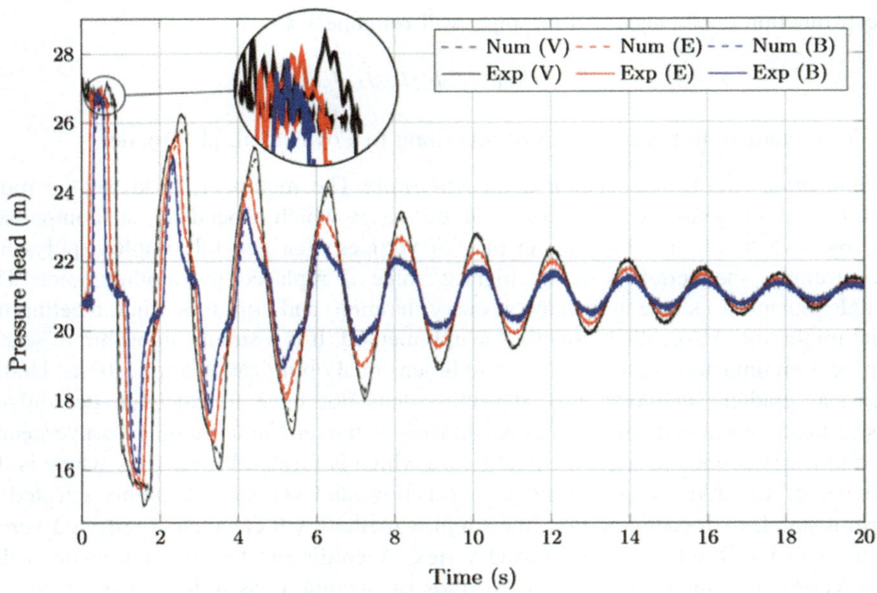

Fig. 4 Calibrated head pressures for the best fitted creep compliance coefficients and pipe wall roughness compared to experimental data of an intact pipe

Table 2 Best fitted creep compliance coefficients and pipe wall roughness for 20 s sampling time

Creep compliance (10 − 10 Pa − 1)		Retardation time (s)	
J_0	6.239		
J_1	0.824	τ_1	0.05
J_2	1.176	τ_2	0.5
J_3	0.087	τ_3	1.5
J_4	0.024	τ_4	5.0
J_5	0.083	τ_5	10.0
Pipe wall roughness (10^{-2} mm)		1.023	

7 Leak Quantification

Following the same previous attempts for creep coefficient calibration, the objective function is the relative error between numerical and experimental data where the decision variables are the leak parameters rather than the creep function which has been already calibrated to an intact pipeline. In this case, the objective function is a function of the effective area of the leak, A_E, and, Pl, its position:

$$\begin{array}{c} E = f(A_E, P_l) \\ 0 < P_l < 1 \end{array} \tag{21}$$

7.1 Numerical to Numerical Calibration

As a first step, a numerically simulated leak was introduced as a reference signal to the solver for the delimitation of its position and size.

The simulated leak is localized at 63 *m* from the reservoir which means 38% of the total length. Its effective area is equal to 3.10^{-5} Square Meters equivalent to 30 Square Millimeters. The initial flow is set to 1.2 *L/s*. Since both leak simulation and the solver are calibrated to the experimental data, the first trial was to run the solver for the sizing and localization of the simulation to be sure the feasibility of the task numerically. The test involved, at first, 20 s as sampling time. In this case, the solver showed an acceptable accuracy when it come to the leak effective area but it was incapable to localize it. The optimization algorithm used is Nelder-Mead Simplex technique. As an initial guess, the input was a localization assumed as the middle of the pipe (50% included to the algorithm as 0.5) and a nil effective area (no leak). The reason behind the incapability of the solver to localize the leak is that the algorithm is able to find only the closest local minima to the initial guess and not the global minima. The accuracy of the leak estimation in this case seemed to be inversely proportional to the sampling time. When considered as 20 s, the solver was incapable of neither locating nor sizing the leak. However, minimizing this parameter enhanced the detectability of the algorithm to find a best fitted result. This divergence of results is due to the presence of local minima near the initial guess which can be reduced by the increase of the sampling time. On these conditions, the solver was able to detect the leak position and size. To investigate this theory, several sampling times were used (same as the calibration of the J-coefficients previously: 20, 10, 5, 2 s). The algorithm showed quite a good accuracy for the determination of the effective area and position for a 10 s sampling time. However, for lower sampling times it was not able to locate the singularity. The optimization for leak parameter estimation for 20 s as a sampling time showed as previously mentioned no accuracy for both the localization and sizing. As output to the algorithm, the total error between numerical results issued from the solver and the numerical simulation reached 0.25% which seemed to be good mathematically speaking but not quite enough since the estimation of the leak position showed an error of 32.9% (50.9% corresponding to 84.7 *m* compared to 38% which corresponds to 63 *m*) and the leak effective area estimation error reached 9.2% (27.2 compared to 30.0 mm^2). For the next step, an optimization process was set to only 10 s. The total error is quantified to be 0.02% with an accuracy of 1.6% for the positioning and 0.4% for the sizing. The algorithm reached the following results: An effective area of 29.8 mm^2 and a position of 38.6% (64.3 *m*) which is considered as the best fit compared to other sampling times. The results are quite accurate since the numerical codes (for leak simulation and detection) are calibrated with the same functions. The third attempt was using a 5 s long sampling time. The algorithm stopped over 0.03% as an overall error, 2.6% and 0.26% as respectively the positioning and sizing error. It estimated the position of the leak as 37% and its effective area as 30.08 mm^2.

The calibration process is afterwards set to 2 s, the algorithm stopped at an overall error of 0.13% with no capability of locating the leak (20% as a positioning error) but able to determine its effective area quite accurately (0.44% as a sizing error).

The previously presented results show that the sampling time choice is quite important as a parameter in the estimation process for leak quantification.

7.2 Numerical-Experimental Calibration

When comparing numerical to experimental data for leak detection, the solver was unable to detect neither the leak nor its position. This is due to the noise that covers the effect of the leak on the pressure wave. The effect of the curves and bends in experimental setup was not included in the solver. Thus, the fluctuations due to this phenomenon were supposed to be an error to be discarded in the analysis process.

Common Monitoring Methods For data where the experimental setup includes bends and unaccounted for phenomenon, the proposed solution is to calibrate a numerical solver to simulate the behavior without taking into consideration these interferences. A solver calculates the relative error in each time-step, determines the maximum of these errors and creates a low threshold as a minimum effective area with regards to the pressure head. Afterwards, the solver locates threshold pass and, therefore, the leak potential position. This case includes signal with an important SNR (Signal-to-Noise Ratio). In other words, only relatively large leaks can be identified. The detectability of such method depends on the development of the pressure transducers and acquisition systems (filters for noise reducing) along with the environmental conditions (noise sources).

Leak Quantification by Estimation The process is to calibrate and regulate the signals of the pressure head at the valve in time (remove time lags: synchronize procedure) and amplitude (effect of the pressure at the reservoir: compensation of the amplitude difference) for an intact and a damaged pipe. The algorithm calculates, then, the difference between these two curves for the isolation of the leak influence on pressure (drop) and add it to the numerical data for an intact system to obtain a to-be-processed pressure head signal which set to be calibrated to for the leak parameter estimation. The solver was able to locate the leak with an accuracy in positioning of 2.2% and in sizing of 23%. The error in sizing is due to the effect of the bend downstream the leak which its amplitude depends on the flow. For that, that effect was not eliminated completely after the differentiation of the intact and damaged pipe pressure head signals. The solver found the following results: a position of 38.76% relative to 64.5 *m* (experimentally 63.2 *m*: 1.3 *m* of error) and an effective area of 25.8 mm^2 (experimentally: 33.6 mm^2). The algorithm was not able to detect the leak for sampling time higher than the duration of the first half period. The initial guess introduced to the solver as in previous attempts is set to be zero for the effective area (no leak), and 50% ($P_l = 0.5$) as the position of the leak.

8 Conclusion

The leak generated sudden pressure drop in the pressure surge and maintains it until the depression pressure wave reaches the valve. The reflection due to the flaw is dissipated until it disappears as a result of the steady-state friction loss and mechanical behavior of pipe wall. Additionally, the leak increases the natural damping of transient pressures same as a relief valve with no dispersion or delay in time of the pressure wave. The reflection of the wave due to the leak and the damping increase with the leak discharge and the distance form the valve. These features are used in the optimization process for the localization and sizing of the flaw. The visco-elastic behavior and the steady state loss friction share the dissipative effect on the pressure wave. However, the visco-elastic effect is higher that the steady-state friction. Thus, numerical calibration is not quite accurate in describing neither of these parameters physically. Numerically, a number of combinations between these two parameters (the creep function describing the visco-elastic behavior and the pipe wall roughness modeling the steady-state loss friction) results in the same shape of the pressure signal. The leak also shares the same effect. The distinction of each parameter outcome in terms of the energy loss in case of the absence a record of the intact pipe signal is a major challenge (Figure 5).

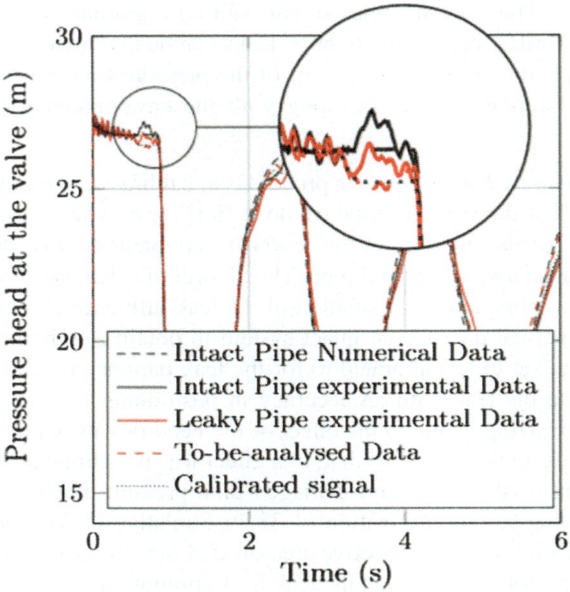

Fig. 5 Regulated (synchronized and regulated for processing) and resulting signals: pressure head at the valve

The delimitation accuracy of a leak size and position depends on the sampling time chosen for the estimation process. It is to note that the task is quite sensitive to such factor. It depends also on the environmental (noise sources) and experimental conditions (bends and curves, temperature, …). Further investigation must take place for a higher accuracy with no relying on the intact system response as a reference for the isolation of the wave and pressure behavior in presence of a leak. Surely, including the bend and curves effect could enhance the algorithm capability for the detection of small leaks. The leak, in this case, is assumed to be rigid and on a specific shape where no crack propagation can occurs and no visco-elastic/dynamic effect is considered. such behavior should be taken into account for the modeling of real leaks in water mains or networks.

References

1. Water Scarcity–World Wild Life. https://thewaterproject.org/water-scarcity/ Accessed 4 May 2018
2. Connor A (2014) Water and energy. United nations educational, scientific and cultural organization, vol 1, Chap 1
3. Le palmars des fuites dans les 101 prfectures de Fance. (French). Danielle Mitterond Fondation - France Liberts. Press release. (2014)

4. Guidara MA (2016) Analyse des conditions de rupture des conduites dadduction deau potable en polythylne, sous leffet dcoulement transitoire, enprsence dun dfaut, (French). PhD thesis. National school of engineering of SFAX
5. Renzetti S, Dupont DP (2013) Buried treasure: the economics of leak. detection and water loss prevention in Ontario
6. Wang XJ, Simpson AR, Lambert MF, Vitkovski JP (2001) Leak detection in pipeline systems using hydraulic methods: a review. Conference on hydraulics in civil engineering, pp 391–400. The Institution of Engineers-Australia
7. Pudar RS, Liggett JA (1992) Leaks in pipe networks. J. Hydraul. Eng. ASCE 118(7): 1031–1046 https://doi.org/10.1061/(asce)07339429(1992)118:7(1031)
8. Mukherjee J, Narasimhan S (1996) Leak detection in networks of pipelines by the generalized likelihood ratio method. In: Indus Eng Chem Res 35(6):18861893 https://doi.org/10.1021/ie950241t
9. Liggett JA, Chen LC (1994) Inverse transient analysis in pipe networks. In: J Hydraul Eng 120(8):934955 https://doi.org/10.1061/(asce)07339429(1994)120:8(934)
10. Chaudry H (2014) Applied hydraulic transients. Springer
11. Colebrook CF (1939) Turbulent flow in pipes, with particular reference to the transition region between the smooth and rough pipe laws. In: J Inst Civil Eng 11(4):133156 https://doi.org/10.1680/ijoti.1939.13150
12. Covas D, Stoianov I, Mano JF, Ramos H, Graham N, Maksimovic C (2005) The dynamic effect of pipe-wall viscoelasticity in hydraulic transients. Part II model development, calibration and verification. In: J Hydraul Res 43(1):5670 https://doi.org/10.1080/00221680509500111
13. Ward W, Sweeney J (2005) Mechanical properties of solid polymers. Wiley, (2004) 14. T. S. Montgomery and W. J. MacKnight. Introduction to Polymer Viscoelasticity. Wiley
14. Meniconi S, Brunone B, Ferrante M, Massari C (2013) Numerical and experimental investigation of leaks in viscoelastic pressurized pipe flow. In: Drink Water Eng Sci 6(1):1116 https://doi.org/10.5194/dwes-6-11-2013
15. Nelder JA, Mead R (1965) A Simplex method for function minimization. Comput. J. 7(4):308313 https://doi.org/10.1093/comjnl/7.4.308

Influence of the Impeller Geometry and the Starting Period on the Hydraulic Performance of a Centrifugal Pump

Fawzi Omri(✉), Sami Elaoud, Noura Bettaib, Issa Chalgoum, and Ezzeddine Hadj Taieb

Laboratory of Applied Fluid Mechanics, Process Engineering and Environment, National School of Engineering of Sfax, 3038 Sfax, Tunisia
{omrifawzi, Elaoudsa, Issa.chalghoum}@yahoo.fr, Noura.bettaieb@enis.tn, Ezed.hadj@enis.rnu.tn

Abstract. In this paper, a numerical study is carried out to investigate transient flow of hydraulic installation during the starting period of a centrifugal pump. The pump impeller angular velocity variation is determined by simulating the dynamic law of the pump electric motor. Transient flow equations are solved by using the method of characteristic of specified time intervals. The computed head and discharge, caused by the starting time of the pump, are presented at some cross sections of the pipe. The results have shown that the evolution of the hydraulic variables is well influenced by impeller geometry and the motor torque starting time.

Keywords: Transient behavior · Centrifugal pump · Startup time · Blade geometry · Motor torque

1 Introduction

Pumps are one of the most complex fluid machine which converts mechanical energy into fluid pressure and kinetic energy. Depending on the specific speed, pumps are divided into three categories: centrifugal, mixed flow and axial flow pumps. In this study, we are interested in investigating the centrifugal pump behavior and its governing parameters. Up to now, many studies have done on the steady state characteristics of turbomachinery. However there are no many advanced studies about pumps under unsteady conditions. The transient performance during starting period of a centrifugal pump has been assumed to be identical with the steady-state performance. However, when the changing rate of the operational point exceeds a certain limit, the pump cannot respond quickly enough to follow along its steady-state characteristics curves, thus resulting in a considerable change in transient behavior [5]. Then, selecting a centrifugal pump to transfer fluid in pipelines is very critical. In fact, when the pump is accelerating very quickly, the pressure and flow rise can be excessively high. This rise can damage the pipelines system. Therefore, we will investigate the transient behavior of a centrifugal pump during its starting period and the parameters that influence on the flow-rate and pressure evolution. It has been demonstrated that there

N. Aifaoui et al. (Eds.): CMSM 2019, LNME, pp. 134–144, 2020.
https://doi.org/10.1007/978-3-030-27146-6_15

are three factors that can influence on the dynamic behavior of a pump during its starting period, which are the mass of water in the pipeline, the valve opening percentage and the starting time [4]. The majority of studies assumed a linear variation of the angular velocity during starting transient. In this work, we will analyze the effect of those parameters based on the actual speed and the motor torque variation considering the parameters of the used hydraulic installation. In addition, we will investigate the effect of the impeller diameter and the blades geometry on the flow-rate and the head evolution.

2 Fundamental Equations

2.1 Fluid Governing Equations

The incompressible flow through the rotating impeller was assumed as a flow through a rotating duct with a diameter equivalent to the mean hydraulic diameter of the impeller passage. The basic equations of continuity and motions, in a reference-rotating frame related to the impeller passage are the following:

Mass conservation equation

$$\frac{\partial \rho}{\partial t} + \rho \frac{\partial w_j}{\partial x_j} = 0 \tag{1}$$

where ρ is the fluid density and $\vec{w}$ the relative velocity. After simplifying the basic continuity equation, we found the following equation [1]:

$$\frac{\partial H}{\partial t} + \frac{C^2}{gA}\frac{\partial Q}{\partial r} = 0 \tag{2}$$

where A is the impeller passage section, C is the pressure wave speed, Q is the fluid discharge and H is the head,

Momentum conservation equation

$$\rho\left(\frac{\partial v_i}{\partial t} + v_j \frac{\partial v_i}{\partial x_j}\right) = -\frac{\partial p}{\partial x_i} + \mu \frac{\partial^2 v_i}{\partial x_j \partial x_j} + \rho fi \tag{3}$$

In this equation $\vec{v}$ represented the absolute velocity of the flow and μ is the dynamic viscosity.

Figure 1 shows the velocity diagram used for determining the absolute velocity at any point of the blade.

$$\vec{v} = \vec{u} + \vec{w} \tag{4}$$

$$u = r.\Omega \tag{5}$$

where Ω is the angular velocity and $\vec{r}$ the position vector from the origin of the rotating frame.

The effect of the Coriolis and centrifugal forces are modeled in this case by:

$$\vec{f} = 2\vec{\Omega} \times \vec{w} + \vec{\Omega} \times (\vec{\Omega} \times \vec{r}) \tag{6}$$

After neglecting unimportant terms and simplifying the equation of motion (3) in a reference frame rotating with the impeller as shown in Fig. 1, we obtain the following equation [1]:

$$\frac{1}{A}\frac{\partial Q}{\partial t} + g\frac{\partial H}{\partial r} + \lambda\frac{Q|Q|}{2D_h A^2} - r\Omega^2 = 0 \tag{7}$$

where λ is the Darcy Weisbach friction coefficient and D_h is the hydraulic diameter of the section A.

It should be noted that the term $r\Omega^2$ of the above equation is an extra term comparing to water hammer equations for conduit flow. This extra term is resulting in the centrifugal force of the rotating impeller. This force will cause the dynamic pressure developed by the pump. In addition, we assumed that the blade angle β of the impeller is equal to 90°.

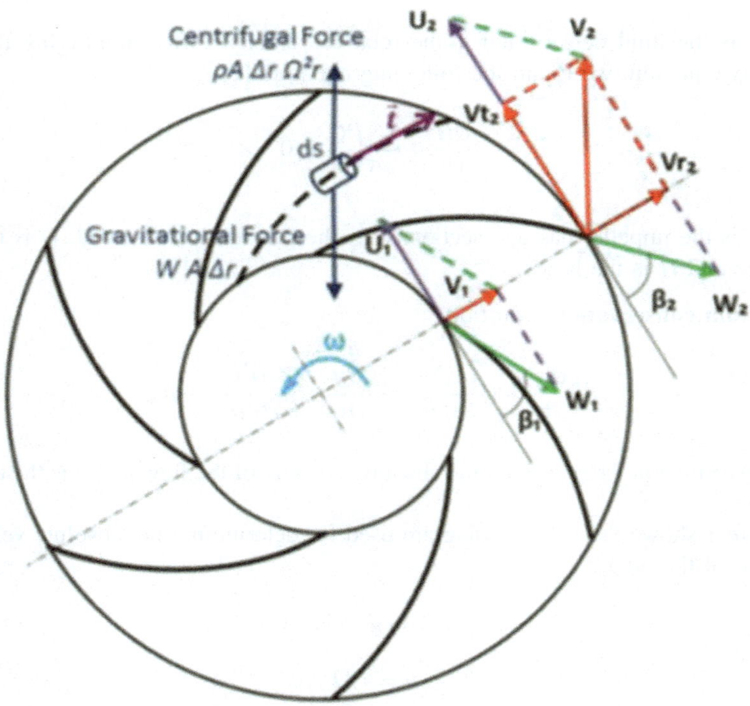

Fig. 1 velocity diagram and applied forces in the rotating impeller

The fluid governing Eqs. (2) and (7) will be used for the numerical computation to study the fluid discharge and head variation during the starting transient in the hydraulic installation described in Fig. 2. For this purpose, we need the impeller angular velocity evolution during the starting period. Therefore, we will use, in the following paragraph, an analytic method to determine the transient flow equation.

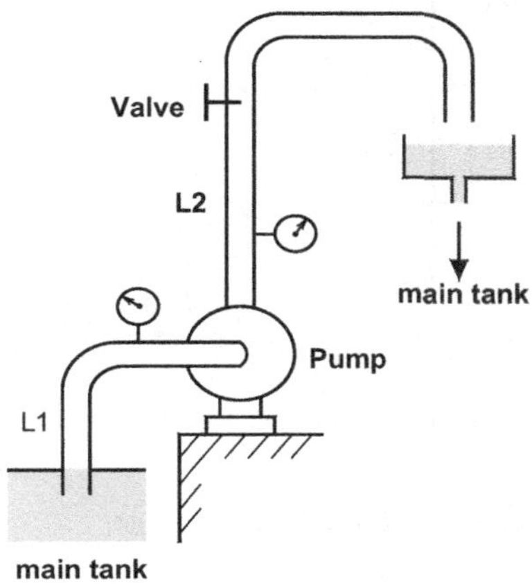

Fig. 2 Hydraulic installation

2.2 Pump Governing Equations

During the starting time, the pump provides besides of the acceleration torque, a torque T_f to overcome the friction of fluid that circulate in the system and the friction of the rotating parts. Reaching the steady state, no torque is needed for acceleration. Torque is needed only to overcome the frictional losses. Therefore, the governing equation for mechanical model can be expressed as following:

$$(I_M + I_p)\frac{d\Omega}{dt} + T_f = T_{em} - T_r \tag{8}$$

where I_M is the inertia of the motor, I_p is the inertia of the propeller including the rotating water inertia, Ω is the propeller angular velocity, *Tem* is the motor torque and *Tr* is the resistant hydrodynamic torque. The resistant hydrodynamic torque depends on the propeller angular speed and the flow rate [4] and given by:

$$T_r = K_1\Omega^2 + K_2\Omega Q, \tag{9}$$

Motor governing equations

The centrifugal pump is connected to a three phase induction motor which converts electrical energy to mechanical energy. The induction motor is modeled using transformation of fixed coordination $_{123}$ to rotating dq coordination [2]. The equivalent circuit diagram is shown in Fig. 3.

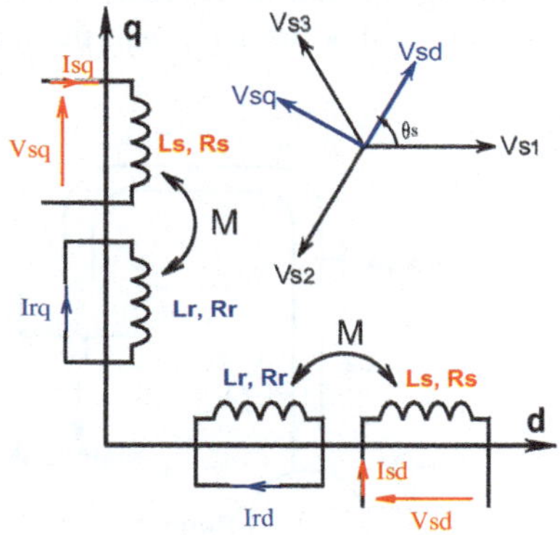

Fig. 3 dq equivalent diagram of induction motor

The induction motor model can be formulated as:

$$\begin{aligned}
V_{sd} &= R_s i_{sd} + L_s \frac{di_{sd}}{dt} + M\frac{di_{rd}}{dt} - M\omega_s(L_s i_{sq} + M i_{rq}) \\
V_{sq} &= R_s i_{sq} + L_s \frac{di_{sq}}{dt} + M\frac{di_{rq}}{dt} + M\omega_s(L_s i_{sd} + M i_{rd}) \\
0 &= R_r i_{rd} + L_r \frac{di_{rd}}{dt} + M\frac{di_{sd}}{dt} - \omega_r(L_r i_{rq} + M i_{sq}) \\
0 &= R_r i_{rq} + L_r \frac{di_{rq}}{dt} + M\frac{di_{sq}}{dt} + \omega_r(L_r i_{rd} + M i_{sd}) \\
p\omega_r &= (1-g)\omega_s
\end{aligned} \tag{10}$$

The motor torque equation is given as:

$$T_{em} = p.M(i_{dr} i_{qs} - i_{qr} i_{ds}) \tag{11}$$

where p is motor pole number, R_s, R_r, L_s, L_r are the resistances and the cyclic inductances of the stator and the rotor, M is the mutual cyclic stator-rotor inductance, g is the sliding coefficient.

To calculate the hydrodynamic torque coefficients, we used the pump power characteristic curve. If the motor torque T_{em} is known as a mathematical function of speed it may be possible to integrate it numerically with the purpose of obtaining the impeller angular velocity variation.

Knowing the motor torque equation and the resistant torque, we can obtain numerically the propeller angular velocity variation. In the following, we will derive ordinary differential equations associated with the flow velocity and the pressure using the characteristic method.

3 Numerical Analysis

In this section, we will derive the characteristic equations in order to reduce our partial differential Eqs. (2) and (7) to a system of ordinary differential equations grouped and identified as C^+ and C^- equations:

$$C^+ \begin{cases} \frac{gA}{C} dH + dQ + JAdt - r\Omega^2 A\, dt = 0 \\ dr = Cdt \end{cases} \tag{12}$$

$$C^- \begin{cases} -\frac{gA}{C} dH + dQ + JAdt - r\Omega^2 Adt = 0 \\ dr = -Cdt \end{cases} \tag{13}$$

where $J = \lambda Q|Q|/2D_h A^2$ is the head loss by impeller length unit.

The ordinary differential Eqs. (12) and (13) are numerically solved in order to determine the flow and the head at any interior point of the pump [1]. In order to get the complete solution, we need to define the boundary conditions in the different junctions. For the pipe lines calculations, we use the same approach with removing the term related to the pump angular velocity.

As has been previously stated, the characteristic equations depend on the pump angular velocity. Therefore, we need the impeller speed variation in order to obtain the head and the discharge variation. A numerical calculation was done using Eqs. (7) and (8) to obtain the motor torque evolution taking into consideration the friction of the system, the total inertia and the hydraulic resistant torque.

4 Results

At the beginning we will consider the example studied by [1] in which they supposed that the speed of the motor driving the centrifugal pump in his starting regime evolves linearly [6] but in the present work we will introduce the real profile of the speed and the engine torque given by the numerical resolution referring to the model of motor and pump.

4.1 Pump Impeller Speed and Motor Torque

The pump impeller speed increases due to the applied motor torque. The rotational speed of the pump encounters some fluctuations at the beginning of the transient response. After the transient time, the angular velocity stabilized at 1460 rpm, Fig. 4.

4.2 Steady State Response Analysis

Considering the rotational speed variation, as shown in Fig. 4 and the motor torque evolution Fig. 5 we will validate the steady state characteristic curve as a first step.

Figures 6 and 7 shows the head and the flow rate variation for different opening positions of the valve.

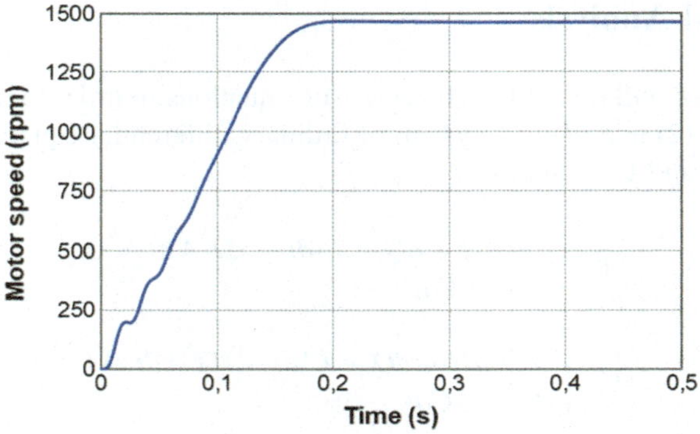

Fig. 4 Pump Impeller speed variation for I = 0.0053 kg.m^2

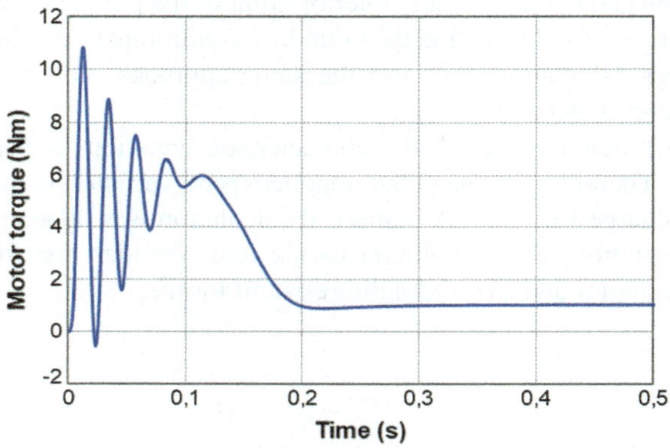

Fig. 5 motor torque variation for I = 0.0053 kg.m^2

Fixing the valve position, the head pressure and the flow rate steady state values correspond to the operating point. By gathering all the operating points we obtain the characteristic curve. Figure 8 shows a comparison between the experimental characteristic curve and numerical modeled curve. The slight difference between the two curves is due to the negligence of the pump volute shape [4].

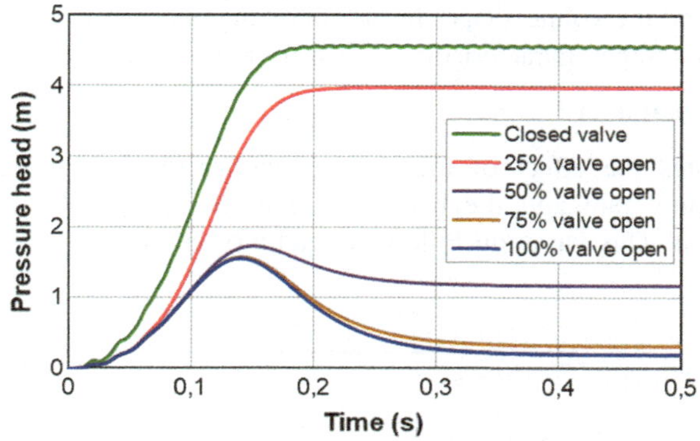

Fig. 6 Pressure head evolution for different valve opening positions

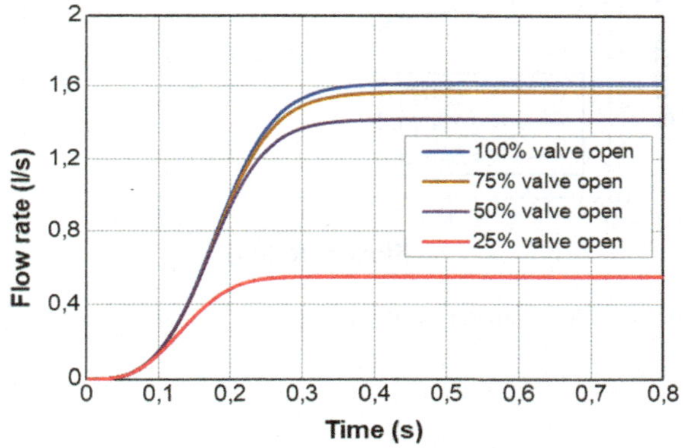

Fig. 7 Discharge evolution for different valve opening positions

As a second step, we will investigate the parameters that affect the pump steady state performance. The steady state curve of the pump depends on the pump angular velocity, the suction and exit diameter, the blade curve, the blade width, the pump

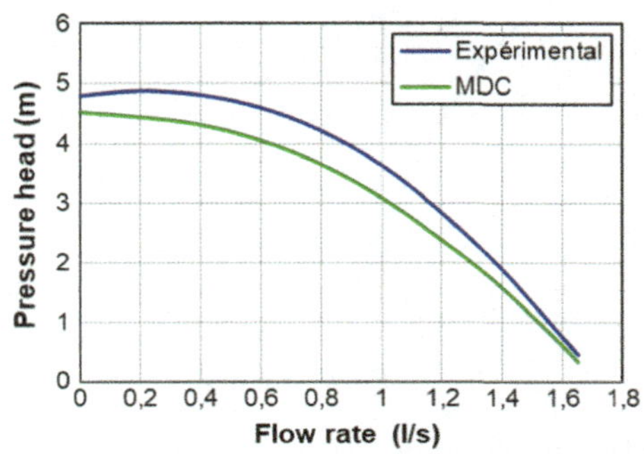

Fig. 8 Comparison between theoretical and experimental characteristic curves for N = 1460 rpm and I = 0.0053 kg.m^2

pressure losses and the volute shape. In the following, we will investigate the effect of some parameters on the characteristic curve from the sited above.

Pump impeller diameter

Figure 9 illustrates the effect of the pump impeller diameter on the steady state behavior. The flow rate and the head variations are proportional to the impeller diameter variation. This variation follows the affinity law [3].

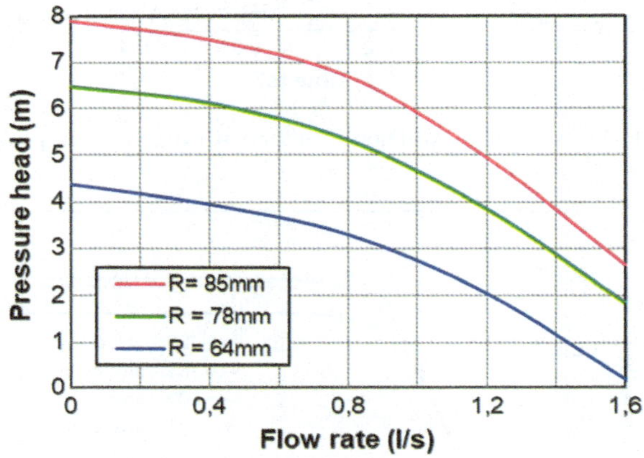

Fig. 9 Effect of the impeller diameter on the pump characteristic curve for N = 1460 rpm and I = 0.0053 kg.m^2

Blade width

Figure 10 shows the effect of the blade width on the pump characteristic curve. If we change the blade width, the flow rate ranges increases while the pressure head remains unchanged.

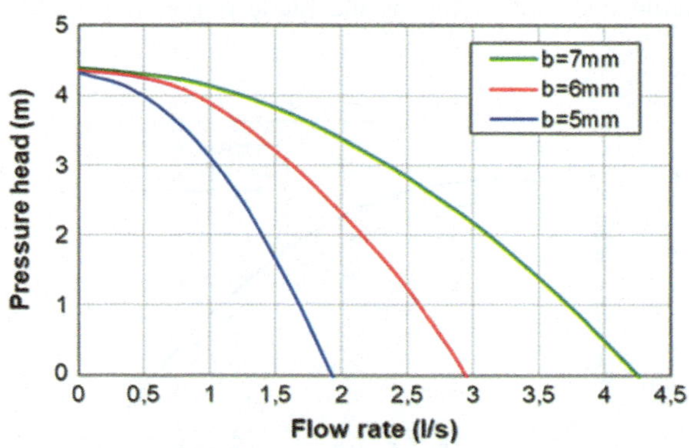

Fig. 10 Effect of the blade width on the pump characteristic curve for N = 1460 rpm and I = 0.0053 kg.m^2

4.3 Transient Response Analysis

During the starting period of the pump, the pressure evolution is affected by the mass of water in the pipe, the opening position of the valve, the impeller speed and the starting time. Using the transient behavior of the angular velocity shown in Fig. 4 and varying the time startup shows that the increase of the startup time causes the decrease of the pressure head during starting transient as demonstrated in Fig. 11.

Finally, we investigate the effect of the width variation on the transient behavior and we conclude that the blade width cause a large increase in the pressure head pick as shown in Fig. 12.

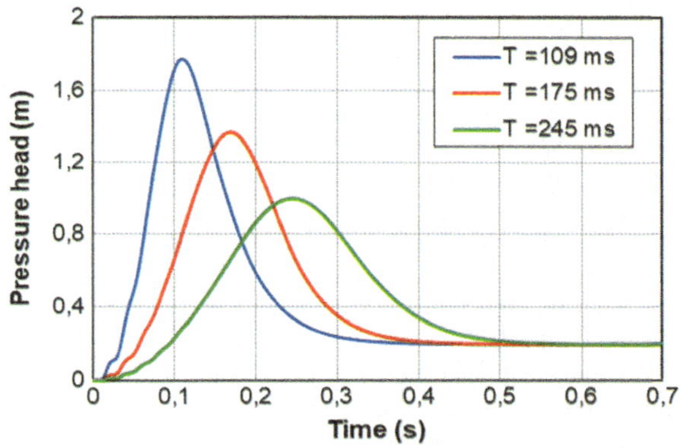

Fig. 11 Effect of the starting time variation on the pressure evolution for N = 1460 rpm

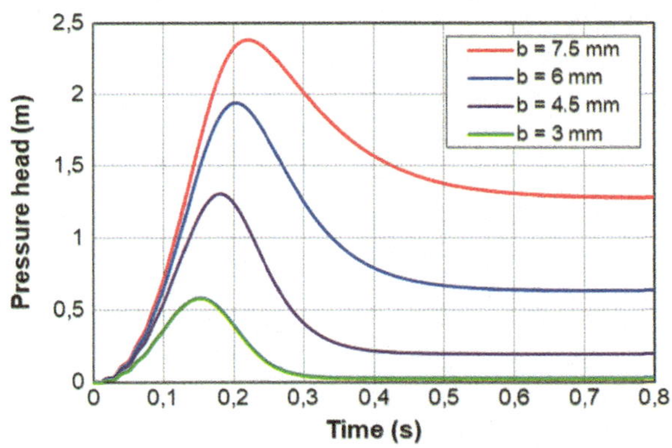

Fig. 12 Effect of the blade width variation on the pressure evolution for N = 1460 rpm

Therefore the optimum configuration is to choose a maximum startup period by fixing the system inertia avoiding the head pick during the transient startup. In fact, to transfer a liquid from a source to a required destination, a fast response is not needed. Also, we need to consider the adequate blade width.

5 Conclusion

In this paper, transient flow through a centrifugal pump has been analyzed. Transient is created due to the variation of the pump motor angular speed from rest to the permanent regime. The motor speed is simulated by introducing the motor induction laws. Transient flow governing equations through the pump are solved numerically by the method of characteristics of specified time intervals. The numerically obtained results have shown that the pressure head evolution and discharge are influenced by the pump starting time and the valve openings. Comparison of the numerical pump characteristic curve and experiment has shown a good agreement. The numerical results have shown too that the pump characteristic curve is well influenced by the impeller geometry.

References

1. Chalghoum I, Elaoud S, Akrout M, Taieb EH (2016) Transient behavior of a centrifugal pump during starting period. Appl Acoust 109(2016):82–89 https://doi.org/10.1016/j.apacoust.2016.02.007
2. Caron JP, et Hautier JP (1995) Modélisation et commande de la machine asynchrone, Edition Technip
3. Menon ES (2014) Transmission pipeline calculations and simulations manual, Prentice 1st edn. Hall, Englewood Cliffs, N.J
4. Saito S (1982) The transient characteristics of a pump during startup, Bull JSME 201 (25):372–379. https://doi.org/10.1299/jsme1958.25.372
5. Tsukamoto H, Ohashi H (1982) Transient characteristics of a centrifugal pump during starting period, ASME J Fluids Eng 104(1). https://doi.org/10.1215/1.3240859
6. Zhifeng Li, Peng Wu, Dazhuan Wu, Leqin Wang (2011) Experimental and numerical study of transient flow in a centrifugal pump during startup. J Mech Sci Technol 25(3):749–757. https://doi.org/10.1007/s12206-011-0107-7

Improvement of Mass Transport at the Surface of an SPR Biosensor Applied in Microfluidics

Yosra Saad[1(✉)], Marwa Selmi[2,3], Mohamed Hichem Gazzah[1], and Hafedh Belmabrouk[3,4]

[1] Quantum and Statistical Physics Laboratory, Faculty of Science of Monastir, University of Monastir, Environment Boulevard, 5019 Monastir, Tunisia
yosra_saad@yahoo.com, hichem.gazzah@gmail.com

[2] Department of Radiological Sciences and Imaging, Faculty of Applied Medical Sciences, Majmaah University, Al Majmaah 11952, Saudi Arabia
marwa_selmi@yahoo.fr

[3] Laboratory of Electronics and Microelectronics, Faculty of Science of Monastir, University of Monastir, Monastir, Tunisia
Ha.Belmabrouk@mu.edu.sa

[4] Department of Physics, College of Science at AlZulfi, Majmaah University, Al Zulfi 11932, Saudi Arabia

Abstract. The main objective of this work is to improve particle transport in a sensitive membrane microchannel in order to reduce the detection time of the biosensor. This requires analyzing the effects to improve the performance of the biosensor. Using numerical simulations, we studied the effect of the application of a magnetic field on the kinetic response of the biosensor. Thus the coupling of diffusion convection phenomenon with the adsorption and desorption reaction of the molecules on the Surface Resonance Biosensor Surface (SPR). These simulations are based on Navier Stokes equations, mass transport equations, chemical kinetics equations, and magnetostatics equations. The results found show an improvement in the fluid velocity and subsequently the diffusion and convection mass transport at the biosensor surface.

Keywords: Mass transport · Biosensor · Microfluidic · Microchannel · Magnetic field

1 Introduction

Since the 1990s, microfluidics has been expanding rapidly. It is a technology at the intersection of various scientific disciplines and that meets the needs in different fields (physics, chemistry, biology) [1–3]. In two words, microfluidics is the science of fluid manipulation at the micrometric scale [4]. Today, it involves tens of thousands of researchers and engineers around the world. This technology is progressively indispensable for the implementation of most bioanalytical protocols [5, 6]. Over the last decade, the improvement of microfabrication techniques has led to the development of integrated fluidic microsystems in which the circulation of the liquid plays a central role [7, 8]. Microsystems are increasingly used in current research and are intended to

N. Aifaoui et al. (Eds.): CMSM 2019, LNME, pp. 145–154, 2020.
https://doi.org/10.1007/978-3-030-27146-6_16

replace many existing systems in our lives. The use of advanced manufacturing techniques has enabled the development and mass production of new miniaturized tools. Particular interest has been given to microsystems used in microfluidics and for applications in the fields of biology or health, which leads to the development of lab-on-chips such as pregnancies, glycemia and DNA tests [7, 9]. Fluidic microsystems are essentially based on the displacement of liquids. This has therefore strongly motivated research to accelerate the study and understanding of fluid dynamics at small size and speed. These advances are supported and complemented by the emergence of biosensors which are the heart of the laboratory on-chip especially biosensors based on surface plasmon resonance [10, 11]. Sikavitsas et al. [12] developed a theoretical model to study the phenomenon of chemical species transport in a microfluidic optical biosensor. This model is used for accurate prediction of kinetic parameters that help to improve the performance of this biosensor. Selmi et al. [13] also studied the electrothermal effect on the biosensor response. This force was generated through the application of an electric field on the microfluidic channel walls to improve the biosensor performance.

The advantages [14] presented by this type of application are numerous. First of all, the reduction of volumes allows to accelerate the chemical reactions and the phenomena of diffusion. This miniaturization approach also leads to a significant reduction in analysis times and the development of highly parallelizable high-speed analysis tools.

At the same time, the use of the magnetic field [15, 16] in biology and medicine is nowadays booming because of its ability to operate on a very small scale. On the other hand, the major advances in the manufacture of magnetic micron sources have enabled the synthesis of magnets of micrometric sizes that can be integrated in microfluidic devices such as labs on a chip, which has been developing rapidly in recent years. It is therefore around this principle of actuation that our work is articulated. It is dedicated to the numerical simulation of the flow in a microchannel of a biosensor. This flow is mixed by a low concentration of the biological analytes which will make a binding reaction with other immobilized particles on one of the walls of this microchannel. We used the Langmuir model to describe the kinetic surface reaction of the biosensor that we coupled with the mass transport mechanisms. A problem that occurs in microfluidic devices is that for high affinity molecules with large association rate constants, the antigen-antibody reaction rate will be limited by mass transport [17]. The mass transport limitation will cause a diffusion boundary layer to develop near the reaction surface. The formation of this layer confines the chemical kinetics and reduces the overall performance of the biosensor. This transport limitation affects the performance of these systems.

In this context, our goal is to improve the particle transport in a sensitive membrane microchannel to reduce the detection time of the biosensor. This requires analyzing the effects to improve the performance of the biosensor. Using numerical simulations, we studied the effect of the application of a magnetic field on the kinetic response of the biosensor. These studies are done in 2D using the finite element method. These different mechanisms and tools for changing the flow are promising for improving the performance of biosensors.

2 Theoretical Consideration

This work therefore consists of studying the microfluidic flow in the SPR biosensor microchannel, following the application of a magnetic field. The Modeling procedure is based on Navier Stokes equations, mass transport equations (convection_ diffusion), chemical kinetics equations and magnetostatic equations.

2.1 The Navier Stokes Equations

This work will be limited to incompressible, homogeneous and Newtonian fluids. The magnetic nanoparticles are supposed to be scattered and move at the velocity of the fluid. This approximation is based on the fact that the time constant for the acceleration phase is too small for the scale of the geometry and the particle size used in the simulation. Thus it can be neglected [15]. The Navier Stokes equations describing the flow in this microchannel are:

$$\nabla . u = 0 \tag{1}$$

$$\rho(u.\nabla)u = -\nabla p + \mu\nabla^2 u + F \tag{2}$$

where ρ is the density of the fluid, u is its velocity, p is the pressure, μ is the dynamic viscosity of the fluid and F is the magnetic force.

2.2 The Binding Reaction

Among the different models of heterogeneous molecular reactions, we chose to rely on the Langmuir model [18]. This model is often used to study the dynamics of molecular reactions in SPR biochips and biosensors. It allows to determine the affinity of a ligand (B) for an analyte (A).

$$[A]_{surf} + [B] \underset{k_{on}}{\overset{k_{off}}{\rightleftharpoons}} [AB] \tag{3}$$

where $[A]_{surf}$ is the analyte concentration at the surface, [B] is the antibody concentration, [AB] is the concentration of the complex. Kon and koff are the association and dissociation rate constants respectively. The B antibodies and the AB complex are supposed to be immobilized on the surface and do not diffuse. This leads us to assume that the binding reaction between the analyte and the immobilized ligand is governed by the first order Langmuir adsorption model. As a result, the reaction rate is described by the following chemical kinetics equation:

$$\frac{\partial [AB]}{\partial t} = k_{on}[A]_{surf}([B_0] - [AB]) - k_{off}[AB] \tag{4}$$

where $[B_0]$ is the concentration of free antibodies.

2.3 Transport Equation

The Navier-Stokes theory is incomplete and requires the consideration of diffusion and convection. For this, it becomes essential to present these transport mechanisms.

$$\frac{\partial[A]}{\partial t} + u.\nabla[A] = D\Delta[A] + G \tag{5}$$

where U is the flow velocity, D is the diffusion coefficient of the analyte and G is the reaction rate. Here, G is zero because there is no reaction in the mass of the fluid.

2.4 Magnetostatic Equations

Because the magnetic part of this problem is static, Maxwell-Ampere's law for the magnetic field H (A/m) and the current density J (A/m2) applies:

$$\nabla \times H = J \tag{6}$$

Moreover, Gauss' law for the magnetic flux density B (Vs/m2) states that:

$$\nabla.B = 0 \tag{7}$$

The following equation describes the relationship between B and H:

$$B = \mu(H + M) \tag{8}$$

Here μ is the magnetic permeability and M is the magnetisation vector. The following equation describes the magnetic vector potential:

$$\nabla \times A = B; \quad \nabla A = 0 \tag{9}$$

Subsequently we obtained the following vector equation after substituting (9) for (6)–(8):

$$\nabla \times \left(\frac{1}{\mu_0\mu_r}\nabla \times A - M\right) = J \tag{10}$$

Simplifying to a 2D problem without perpendicular currents, this equation is reduced to:

$$\nabla \times \left(\frac{1}{\mu_0\mu_r}\nabla \times A - M\right) = 0 \tag{11}$$

As a result, the magnetic force F (Fx, Fy) is written in the following form:

$$F_x = \left(\frac{M_x}{\mu_r}\frac{\partial^2 A_z}{\partial x \partial y}\right) \tag{12}$$

$$F_y = \left(\frac{M_x}{\mu_r}\frac{\partial^2 A_z}{\partial y \partial x}\right) \tag{13}$$

2.5 Boundary Condition

The incoming flow of fluid with magnetic nanoparticles is laminar fully developed, parabolic average velocity at the inlet u = v0 and zero velocity at the walls. The outlet of the microchannel is assumed to be at atmospheric pressure.

The microchannel inlet is set to a constant concentration of analytes [A0]. At the outlet, it is assumed that the analyte leaves the microchannel entirely by convection.

At the reaction surface, the mass flux at the surface is determined by the kinetics of reaction: $-D\left(\frac{\partial[A]}{\partial y}\right)_{surf} = k_a[A]_{surf}\{[B_0 - [AB]]\} - k_d[AB]$

The upper wall rest and the bottom wall are impervious and do not interact with the analytes.

Initially, the concentration of analyte is equal to zero [A] (x, y, 0) = 0, similarly, the initial concentration of the complex is zero: [AB] (x, y, 0) = 0.

A magnetic isolation boundary condition, Az = 0, is applied along a system boundary relatively far from the magnet.

3 Results and Discussion

3.1 Geometric Configuration

The two 2D model geometries are shown in Figs. 1 and 2. The first geometry is composed of a microfluidic microchannel of length L = 80 μm and height H = 10 μm. The microchannel is connected in the center to a circular well of diameter D = 30 μm. On the lower arc of the circular well there is a reaction surface on which the connection reactions take place. The second figure shows the same geometry with the same dimensions but is distinguished by the application of an external permanent rectangular magnet. This one is located on the upper arc of circular well of microfluidic channel. This model is used to study the effect of an external magnetic field on the fluid flow with magnetic nanoparticles.

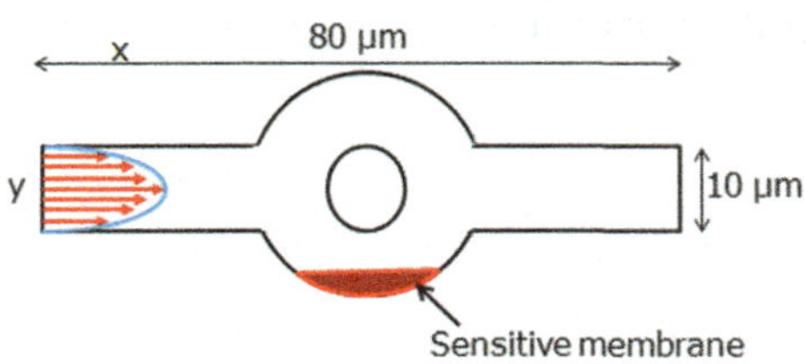

Fig. 1 Geometric configuration of the microchannel of a biosensor studied without the application of a magnetic field

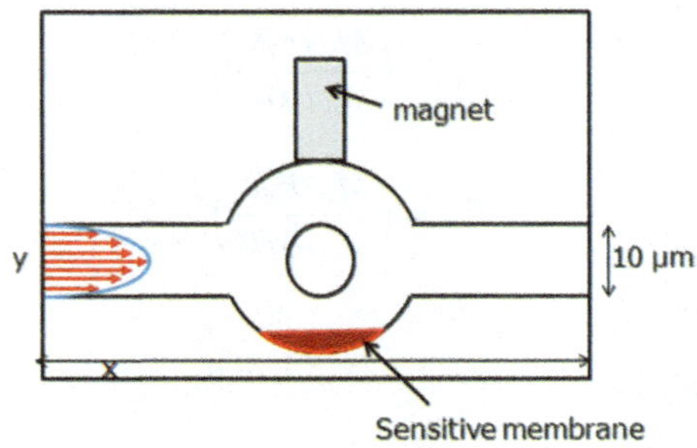

Fig. 2 Geometric configuration of the microchannel of a biosensor studied with the application of a magnetic field

In this work, we adopted the finite element method. This method consists of discretizing the space using simple geometric elements. The finite element method is based on a division of the space according to a mesh.

Both Figs. 3 and 4 describe the mesh used in both cases. We chose a triangular mesh because it is more stable and faster. The small thickness of the concentration limit layer of analytes has strong consequences on numerical computation. It is indeed essential to precisely model the mass transfer in this region. Along the boundary of the reaction surface, the mesh has been refined, in order to accurately model the reaction and the diffusion process that take place there. Indeed, the mesh is tightened near the places of interest for example to the places where the solution will vary a lot as the surface of reaction (Fig. 3) and the zone of the magnet magnet (Fig. 4). The more this mesh is tightened, the more the solution obtained by the finite element method will be precise and close to the "real" solution of the partial differential equation.

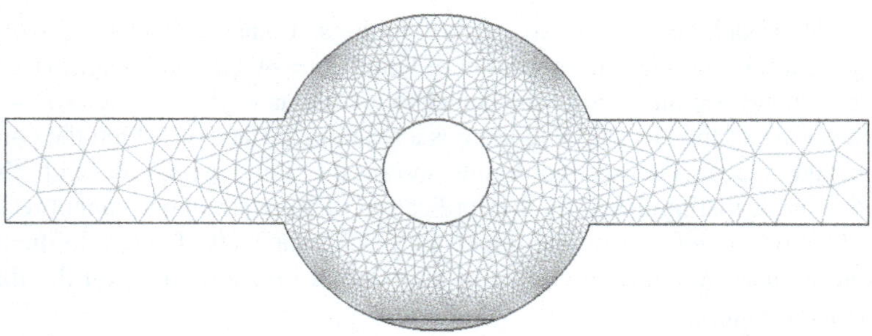

Fig. 3 2D triangular mesh of the first configuration

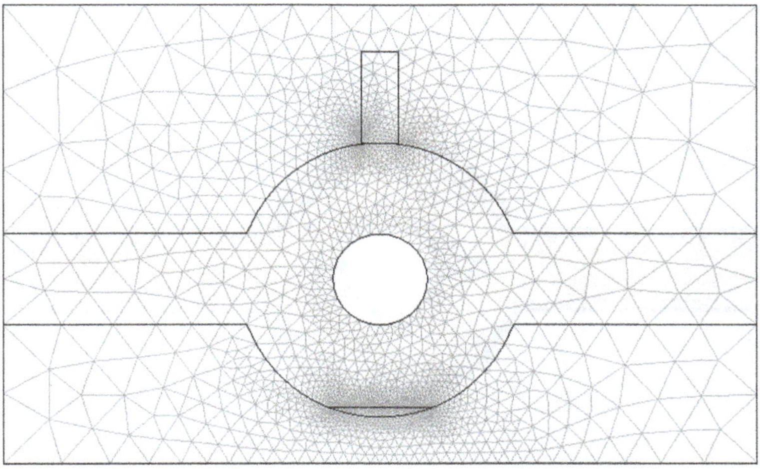

Fig. 4 2D triangular mesh of the second configuration

3.2 Magnetic Field Effect on the Microfluidic Flow Profile

Heterogeneous immunoassays based on the interaction between a target analyte and an immobilized ligand on the reaction surface were well defined and modeled in the literature [19, 20].

This work is devoted to a numerical study of the effect of the magnetic force on the kinetic response of the biosensor. This force is generated following the application of a magnetic field through a permanent rectangular magnet located on one walls of the microchannel. Figure 5 shows the profile and concentration distribution without any applied magnetic field. In this case, the flow is laminar and has a parabolic profile. The effect of the reaction surface on the analyte concentration is visible only in the narrow region near the surface. Compare to it the flow and concentration with an applied magnetic field (Fig. 6). Now the flow profile is far from parabolic: it has two distinct swirls and a narrow region with high flow velocity under the magnet.

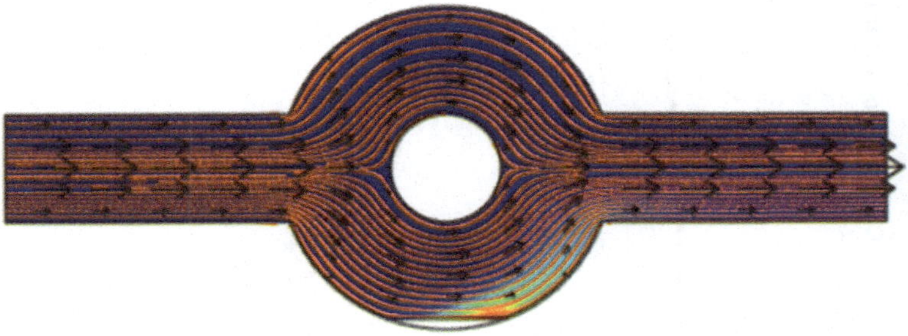

Fig. 5 The flow profile and the concentration distribution without any applied magnetic field

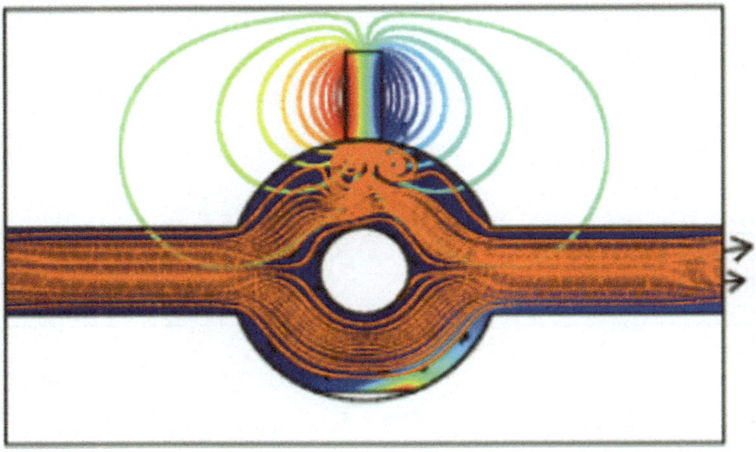

Fig. 6 The flow profile and the concentration distribution with an applied magnetic field

3.3 Magnetic Field Effect on the Kinetic Response

The Fig. 7 shows the time evolution of the AB complex concentration formed at the surface of the reaction without and with application of a magnetic field. The initial concentration of analytes is set at 6.4 µmol/m^3 and the magnetic field value applied is 5 T. It can be observed that the slope of association is more important when we apply a magnetic field that means that the phenomenon of the transport is improved. This improvement is due to the effect of the magnetic force. The role of this force is to agitate the fluid and subsequently reduce the thickness of the diffusion boundary layer.

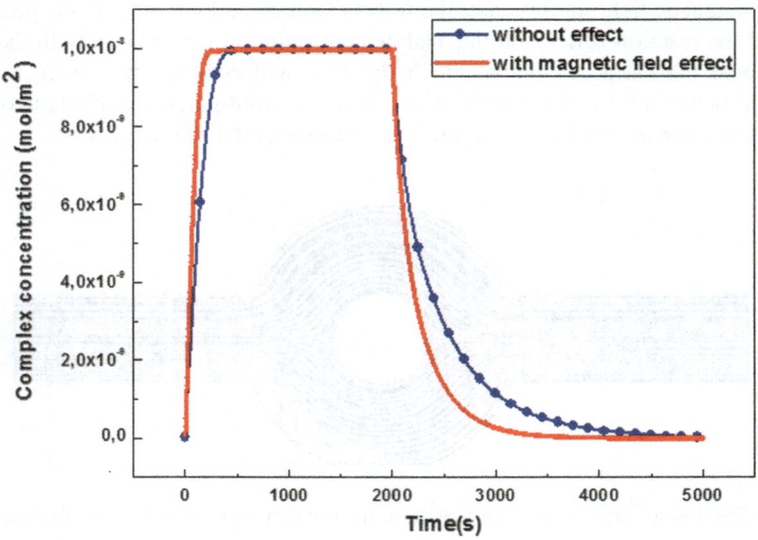

Fig. 7 Temporal evolution of the concentration of the complex without and with the application of the magnetic field

4 Conclusion

In this work, we studied the effect of the application of a magnetic field on the kinetic response of a biosensor. A two-dimensional simulation was performed. We have thus obtained an improvement of the biosensor performance thanks to the magnetic force which improves the transport phenomenon in the microfluidic microchannel.

References

1. Craighead H (2010) Future lab-on-a-chip technologies for interrogating individual molecules. Nanosci Technol Collect Rev Nat Journals 442:330–336
2. Rivet C, Lee H, Hirsch A, Hamilton S, Lu H (2011) Microfluidics for medical diagnostics and biosensors. Chem Eng Sci 66:1490–1507
3. Demello AJ (2006) Control and detection of chemical reactions in microfluidic systems. Nature 442:394–402
4. Ben-Yoav H, Dykstra PH, Bentley WE, Ghodssi R (2012) A microfluidic-based electrochemical biochip for label-free diffusion-restricted DNA hybridization analysis. BiosensS Bioelectronics 38:114–120
5. Haeberle S, Zengerle R (2007) Microfluidic platforms for lab-on-a-chip applications. Lab Chip 7:1094–1110
6. Whitesides GM (2006) The origins and the future of microfluidics. Nature 442:368–373
7. Franke TA, Wixforth A (2008) Microfluidics for miniaturized laboratories on a chip. Chem Phys Chem 9:2140–2156
8. Squires TM, Quake SR (2005) Microfluidics: Fluid physics at the nanoliter scale. Rev Mod Phys 77:977–1026
9. Reyes DR, Iossifidis D, Auroux P-A, Manz A (2002) Micro total analysis systems. 1. introduction, theory, and technology. Anal Chem 74:2623–2636
10. Tsoi PY, Yang M (2002) Kinetic study of various binding modes between human DNA polymerase β and different DNA substrates by surface-plasmon-resonance biosensor. Biochem J 361:317–325
11. Chaitanya RM, Andrea MA (2012) Mass transport effects in suspended waveguide biosensors integrated in microfluidic channels. Sensors 12:14327–14343
12. Sikavitsas V, Nitsche JM, Mountziaris TJ (2002) Transport and kinetic processes underlying biomolecular interactions in the BIACORE optical biosensor. Biotechnol Prog 18:885–897
13. Selmi M, Gazzah MH, Belmabrouk H (2016) Numerical study of the electrothermal effect on the kinetic reaction of immunoassays for a microfluidic biosensor. Langmuir 32:13305–13312
14. Selmi M, Gazzah MH, Belmabrouk H (2017) Optimization of microfluidic biosensor efficiency by means of fluid flow engineering. Sci Rep 7:5721
15. Munir A, Wang J, Zhou HS (2009) Dynamics of capturing process of multiple magnetic nanoparticles in a flow through microfluidic bioseparation system. IET Nano Biotechnol 3:55–64
16. Munir A, Wang J, Li Z, Zhou HS (2010) Numerical analysis of a magnetic nanoparticle-enhanced microfluidic surface-based bioassay. Microfluid Nanofluidics 8:641–652
17. Chaiken I, Rose S, Karlsson R (1992) Analysis of macromolecular interactions using immobilized ligands. Anal Biochem 201:197–210
18. Langmuir I (1918) The adsorption of gases on plane surfaces of Glass, mica and platinum. J Am Chem Soc 40:1361

19. Yang CK, Chang JS, Chao SD, Wu KC (2008) Effects of diffusion boundary layer on reaction kinetics of immunoassay in abiosensor. J Appl Phys 103:084702
20. Hart R, Lec R, Moses Noh H (2010) Enhancement of heterogeneous immunoassays using AC electroosmosis. Sens Actuators B. 147:366–375

Assessing the Inline and Branching Techniques in Mitigating Water-Hammer Surge Waves

Mounir Trabelsi[1,2(✉)] and Ali Triki[2]

[1] Department of Mechanics, National Engineering School of Sfax, University of SFAX, BP 1173, 3038 Sfax, Tunisia
mounirtrabelsi.enis@gmail.com
[2] Research Unit: Mechanics, Modelling Energy and Materials M2EM, Gabes, Tunisia
ali.triki@enis.rnu.tn

Abstract. This paper explored and compared the robustness of the inline and branching re-design techniques used to upgrade existing steel piping systems. These techniques are based on substituting an inline—or adding a ramified plastic short-section at the transient sensitive regions of the steel main pipe. The pressurized pipe flow solver was based on the water hammer model incorporating the Vitkovsky and Kelvin-Voigt formulations; besides, the Method of Characteristics was implemented for numerical computations. The robustness of the proposed protection techniques was tested with regard to a water-hammer event induced into a reservoir pipe valve system. Results demonstrated that both utilized techniques provided a useful tool to mitigate both water-hammer up-and down-surges. Additionally, the attenuation rates of hydraulic-head-rise and-drop were sensitive to the short-section material type and size. Moreover, the branching technique illustrated a marked enhancement compared with the inline one in terms of limitation of wave oscillation period spreading, while providing a surge attenuation rate comparable to that involved by the inline technique. Ultimately, the near-optimal values for the short-section diameter and length were estimated through sensitivity of hydraulic-head peak and crest to the short-section dimension.

Keywords: Branching · Cavitation · Control · **HDPE** · Inline · **LDPE** · Method of characteristics · Plastic material · Water-hammer

1 Introduction

Water-hammer is an undesired hydraulic phenomenon which is typically initiated by changes in flow, either caused by the flow establishment or the flow stoppage event. This phenomenon can potentially have negative impacts on the performance of hydraulic utilities; such as: excessive noise, fatigue and stretch of the pipe wall, and disruption of normal control; and even may causes risks to operators or users [6].

Consequently, to attenuate the severe impact of these surges, several control measures should be considered in the design stage of hydraulic equipment.

In this line, classical design measures may be sorted into three principal categories; including: (i) Alteration of pipeline properties; (ii) Implementation of operational

N. Aifaoui et al. (Eds.): CMSM 2019, LNME, pp. 155–163, 2020.
https://doi.org/10.1007/978-3-030-27146-6_17

control strategies; and (iii) installation of surge control devices at the transient sensitive regions of the hydraulic system [1]. From a design standpoint, a combination of multiple devices is generally employed. Aside from the analysis of the used technologies, it must be delineated that none of them is a panacea that can be used for all system cases; and the adequate technique is specific for each system and operating conditions.

In this context, a surge—control class, based on the use of plastic short-section, has been introduced in the recent literature in order to attenuate excessive pressure–rise or–drop. For example, Massouh and Comolet [8], Triki [10, 12, 13], Triki and Fersi [14], Fersi and Triki [5], and Triki and Chaker [15] introduced the inline technique; which is based on substituting a short-section of the sensitive region of existing steel-piping system by another one made of plastic pipe-wall material. The author evidenced that this technique could significantly reduce both pressure rise and drop magnitudes. However, their studies have shown that such a technique gives rise to a fundamental drawback arising from the amplification of the oscillation period (spreading) of the pressure waves, thus critically increasing the critical time of closing the valve.

Alternatively, recent studies reported by Pezzinga and Scandura [9], Triki [11], Triki and Fersi [14], and Fersi and Triki [4, 5] proposed a second surge-control technique 'known as the branching technique'; based on the implementation of additional ramified plastic short-section at the sensitive region of existing steel-piping system. The authors showed experimentally and numerically that such a technique could significantly reduce unsteady-flow oscillations amplitudes within the up and down-surge initiated water-hammer frameworks. However, the authors emphasized that this technique resulted in spreading of the wave oscillations period. In return, as for the inline technique case, the branching one could delay the subsequent steady regime. Physically, these results are closely related to the viscoelastic mechanical behaviour of plastic materials. Indeed, the reduced modulus of the short-section pipe-wall material and the retarded-strain of the short-section pipe-wall result in a reduced wave speed value (which may attenuate the surge wave amplitude), and a retarded hydraulic transient response (which may expand the wave fluctuation period) [3].

Additionally, the authors concluded that the oscillations attenuation increased with the increase of the ratio between the ramified plastic short-section and the main steel-pipeline volumes.

Overall previous results on plastic short-section-based control techniques highlighted that the use of an inline or ramified plastic short-section are highly promising in terms of limitation of excessive pressure rise or drop; yet, these two technique invoked a spreading of transient wave oscillations period. Consequently, we planned in this paper to explore and compare the reliability of the inline and branching techniques, in terms of attenuation of hydraulic-head rise and drop from one side, and the induced spreading of wave oscillation period from the other side.

Following this introduction, the unconventional water-hammer model along with the Fixed Time-Step Grid—based Method of Characteristics are outlined.

2 Theory and Methods

The (**1-D**) unconventional water hammer model incorporating the Vitkovsky et al. and the kelvin-Voight formulations may be written in the following form [10, 11, 16]; 2018):

$$\frac{\partial H}{\partial t} + \frac{a_0^2}{gA}\frac{\partial Q}{\partial x} + 2\frac{a_0^2}{g}\frac{d\varepsilon}{dt} = 0 \tag{1}$$

$$\frac{1}{A}\frac{\partial Q}{\partial t} + g\frac{\partial H}{\partial x} + g\left(h_{f_s} + h_{f_u}\right) = 0 \tag{2}$$

in which, H and Q designate the hydraulic-head and discharge, respectively; A denotes the pipe cross sectional area; g refers to the gravitational acceleration; a_0 designates the elastic wave speed; x and t denote the spatial coordinates along pipe axis and time, respectively; h_{f_s} describes the quasi-steady-friction loss per unit length; h_{f_u} refers to the unsteady-friction loss approximated according to the Vitkovsky et al. [16] formula: $h_{f_u} = (k_v/gA)\{(\partial Q/\partial t) + a_0\,\mathbf{Sgn}(Q)|\partial Q/\partial x|\}$, in which, $k_v = 0.03$ is the Vitkovsky decay coefficient, $\mathbf{Sgn}(Q) = +1$ or—1 for or $Q \prec 0$, respectively; x and t are the distance along the centerline of the pipe and time, respectively.

The radial-strain ε may be described referring to the linear viscoelastic Kelvin-Voigt model, as follows [2]:

$$\varepsilon(x,t) = \sum_{k=1}^{N_{kv}} \varepsilon_k = \sum_{k=1}^{N_{kv}} \rho g \frac{\alpha D}{2e} \int_0^s [H(x,t) - H_0(x)] \frac{J_k}{\tau_k} e^{-\frac{s}{\tau_k}} ds \tag{3}$$

where, J_0 designates the elastic creep compliance, J_k and τ_k $(k = 0 \cdots n_{kv})$ denote the creep-compliance and the retardation-time coefficients associated with *kth* Kelvin-Voigt element, respectively, n_{kv} refers to the number of Kelvin-Voigt elements.

The numerical solution of Eqs. (1) and (2) may be established using the (**FTSG-MOC**). The numerical procedure outlined below follows from the concept presented in [11, 18].

The discretization of Eqs. (1) and (2) along the characteristic lines leads to:

$$\mathbf{C}^{j\pm}: \frac{dH}{dt} \pm \frac{a_0^j}{gA^j}\frac{dQ}{dt} + \frac{2a_0^2}{g}\left(\frac{\partial \varepsilon}{dt}\right) \pm a_0^j\left(h_{f_s}^j + h_{f_u}^j\right) = 0 \text{ along } \frac{\Delta x^j}{\Delta t} = \pm\frac{a_0^j}{C_r^j} \tag{4}$$

in which, C_r^j refers to the Courant number, being used to allow the grid points to coincide with the intersection of the characteristic curves; the superscript j refers to the pipe number $(1 \le j \le np)$ and the indices i refers to the section number of the *jth* pipe $\left(1 \le i \le n_s^j\right)$; n_s^j designates the number of sections of the *jth* pipe and *np* is the number of pipes; Δt and Δx denote the time-and space-step increments, respectively.

As per Eq. (4), the flow parameters, in the time domain, are linked along the characteristic lines by the following equations set:

$$\mathbf{C}^{\pm}: \begin{cases} Q^j_{i,t} = c^j_p - c^j_{a-} H^j_{i,t} \\ Q^j_{i,t} = c^j_n + c^j_{a+} H^j_{i,t} \end{cases} \text{along } \frac{\Delta x^j}{\Delta t} = \pm \frac{a^j_0}{c^j_r} \tag{5}$$

where, $c^j_p = \left(Q^j_{i-1,t-1} + (1/B^j) H^j_{i-1,t-\Delta t} + c''^j_{p_1} + c'''^j_{p_1}\right) / \left(1 + c'^j_p + c''^j_{p_2} + c'''^j_{p_2}\right)$; $B = a_0/(gA)$; $c^j_n = \left(Q^j_{i+1,t-1} + (1/B^j) H^j_{i+1,t-\Delta t} + c''_{n1} + c'''_{n1}\right) / \left(1 + c'^j_n + c''^j_{n_2}\right)$; $c^j_{a+} = 1 + c'''^j_{p_2} / \left(B^j \left(1 + c'^j_{p_2} + c''^j_{p_2}\right)\right)$; $c'^j_p = R^j \Delta t \left|Q^j_{i-1,t-1}\right|$; $c'^j_n = R^j \Delta t \left|q^j_{i+1,t-1}\right|$; $c''^j_{p_1} = k_v \theta Q^j_{i,t-1} - k_v (1-\theta)\left(Q^j_{i-1,t-1} - Q^j_{i-1,t-2}\right) - k_v \mathbf{sgn}\left(Q^j_{i-1,t-1}\right)\left(Q^j_{i,t-1} - Q^j_{i-1,t-1}\right)$; $c''^j_{n_1} = k_v \theta Q^j_{i,t-1} - k_v (1-\theta)\left(Q^j_{i+1,t-1} - Q^j_{i+1,t-2}\right) - k_v \mathbf{sgn}\left(Q^j_{i+1,t-1}\right)\left(Q^j_{i,t-1} - Q^j_{i+1,t-1}\right)$; $c'''^j_{p_1} = -c'''^j_{n_1} = -2 a^j_0 A^j \Delta t \sum_{k=1}^{n_{kv}} \left[\varepsilon^j_k(x,t)/\partial t\right]$; $c'''^j_{p_2} = c'''^j_{n_2} = 2 a^j_0 A^j c_0 \gamma \sum_{k=1}^{n_{kv}} J^j_k \left(1 - e^{-(\Delta t/\tau_k)}\right)$; ..$c''^j_{p2} = c''^j_{n_2} = k_v \theta$; (is a relaxation coefficient); $R^j = f^j / 2 D^j A^j$; and $c_0 = \alpha \gamma D^j / 2 e^j$.

Inline/Branching Connections:

In the context of series or branching connection of multi-pipes, a common hydraulic grade-line elevation and a no-flow storage assumptions are made at the connection section [17]:

$$H^{j-1}_{ns^j,t} = H^j_{1,t} = H^d_t \text{ and } Q^{j-1}_{ns^j,t} = Q^j_{1,t} + Q_{d,t} \text{ (for inline connection)} \tag{6}$$

$$H^{j-1}_{ns^j,t} = H^j_{1,t} \text{ and } Q^{j-1}_{ns^j,t} = Q^j_{1,t} \text{ (for branching connection)} \tag{7}$$

in the above equations, the indices d refers to the ramified penstock parameters.

3 Application, Results and Discussion

Application relates to the reservoir-pipe-valve system, schematically shown in Fig. 1a. The piping system specifications are: $L = 100\,\text{m}$; $d = 50.6$ mm; $E^{\text{Steel}}_0 = 210\,\text{GPa}$ and $a_0 = 1142.463\,\text{m/s}$. The initial steady-state regime corresponds to constant values of discharge and hydraulic-head at the upstream reservoir equal to: $Q_0 = 0.58$ l/s and $H^R_0 = 45\,\text{m}$, respectively. Transient event is generated by an abrupt and full valve closure; which leads to the following boundary conditions forms:

$$Q_{|x=L} = 0 \text{ and } H_{|x=0} = H^R_0 (t \succ 0) \tag{8}$$

In such a situation, the inline-or the branching-protection techniques consist in handling the downstream extremity of the piping system as sketched in Fig. 1b or c, respectively.

As a starting point of analysis, the inline or ramified plastic short-section length and diameter are selected equal to: $l^{\text{plastic}}_{\text{short-section}} = 5\,\text{m}$ and $d^{\text{plastic}}_{\text{short-section}} = 50.6\,\text{mm}$, respectively. The creep compliance coefficients of the generalized Kelvin-Voigt linear viscoelastic mechanical behavior associated to **(HDPE)** or **(LDPE)** material are: $J_0^{\textbf{HDPE}} = 1.057\ \text{GPa}^{-1}$, $\{J/\tau\}_{1\cdots5}^{\textbf{HDPE}}(\text{GPa}^{-1}/\text{s}) = \{1.057/0.05; 1.054/0.5; 0.905/1.5; 0.262/5.0; 0.746/10.0\}$; or ; $\{J/\tau\}_{1\cdots5}^{\textbf{LDPE}}(\text{GPa}^{-1}/\text{s}) = \{7.54/(89 \times 10^{-6}); 10.46/.0.022; 0.262/1.864\}$, respectively [7].

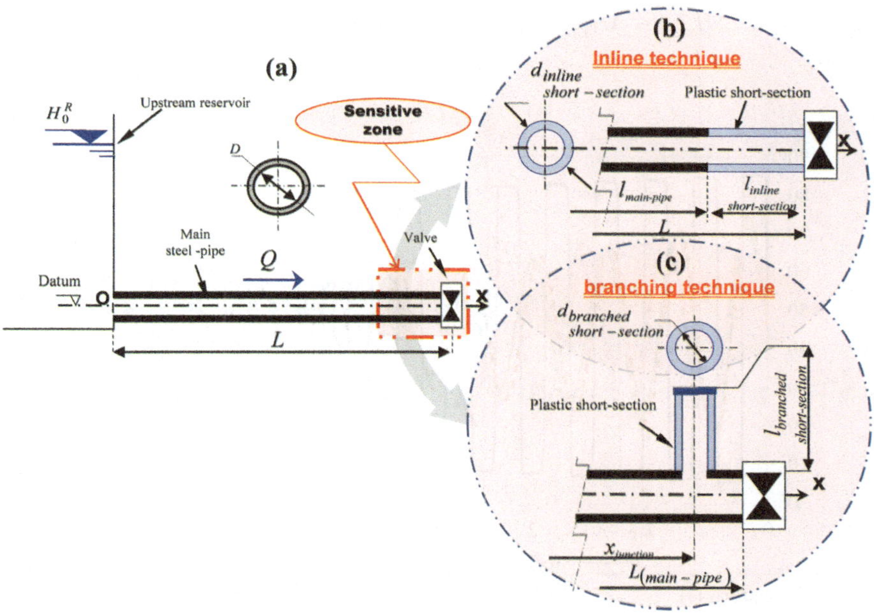

Fig. 1 Definition sketch of the hydraulic system

Figure 2 displays the downstream hydraulic-head signals, as a function of time, predicted into the hydraulic system with and without using the protection techniques. Jointly, Table 1 reports the main characteristics of the wave curve displayed in Fig. 2.

At first sight, Fig. 2 suggests that the wave patterns predicted for the protected systems may be sorted into two wave pattern groups selected depending on the plastic material type of the short-section: (i) **(HDPE)** or (ii) **(LDPE)**.

Moreover, Fig. 2 reveals, for each group of wave patterns, an attenuation effect of the first hydraulic-head peak and crest, along with a spreading effect of the hydraulic-head oscillations period. Particularly, for the cases using **(HDPE)** short-section, the up-surge magnitudes of the first hydraulic-head peaks are: $\Delta H^{+}_{\textbf{HDPE}-\text{branching}} = 30.8\,\text{m}$ or

$\Delta H^{+}_{\mathbf{HDPE}-\text{inline}} = 29.4\,\text{m}$ for the cases applying the branching or the inline techniques, respectively; while the corresponding value is $\Delta H^{+}_{\mathbf{steel}} = 40.6\,\text{m}$ for the original i.e.: non-protected) system case. Consequently, the up-surge hydraulic-head attenuations obtained using the branching- or the inline- protection techniques are, respectively: $\Delta H^{+}_{\mathbf{steel}} - \Delta H^{+}_{\mathbf{HDPE}-\text{branching}} = 9.8\,\text{m}$ or $\Delta H^{+}_{\mathbf{steel}} - \Delta H^{+}_{\mathbf{HDPE}-\text{inline}} = 11.2\,\text{m}$. Similarly, for the cases based on an (**LDPE**) plastic short-section, Fig. 2 illustrates that both techniques also allow a significant attenuation of the first hydraulic-head peak compared with the one predicted into the original system case. More precisely, the up-surge hydraulic-head attenuations obtained using the branching- or the inline- protection techniques are, respectively: $\Delta H^{+}_{\mathbf{steel}} - \Delta H^{+}_{\mathbf{LDPE}-\text{branching}} = 12.3\,\text{m}$ and $\Delta H^{+}_{\mathbf{steel}} - \Delta H^{+}_{\mathbf{LDPE}-\text{inline}} = 14.1\,\text{m}$.

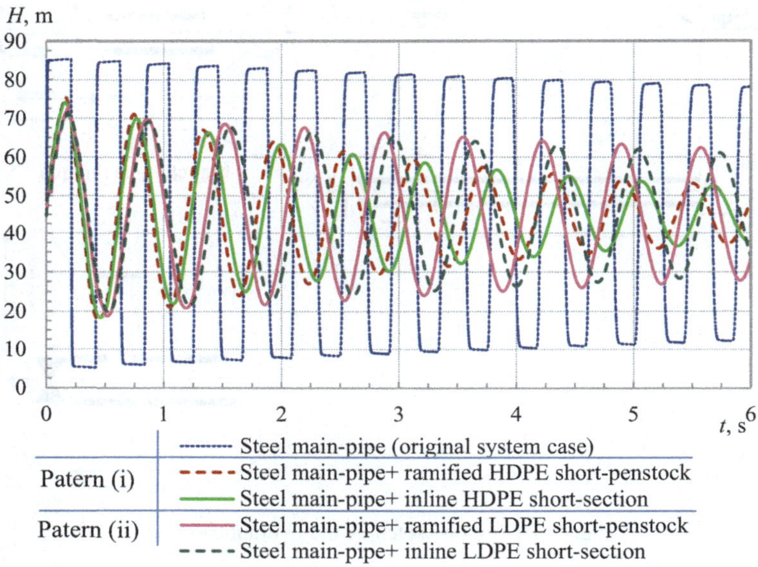

Fig. 2 Comparison of downstream hydraulic-head signals versus time, for the hydraulic system with and without implementation of the protection techniques

As a first result, the employed techniques allow a significant attenuation of the first hydraulic-head peak compared with that predicted for the same transient event initiated into the original system case. Additionally, it important to delineate that, for each control technique, the use of (**LDPE**) short-section provides larger hydraulic-head attenuation than those employing (**HDPE**) material. Similar interpretations may be conducted for the behaviour of the first hydraulic-head crest.

Further analysis of the wave pattern of Fig. 2 shows that the implementation of both control techniques induces a period spreading of transient hydraulic-head oscillations. In particular, this spreading is more important for the cases based on a (**LDPE**) material than those build upon a (**HDPE**) material. For example, the periods of the first cycle of hydraulic-head oscillations, predicted for the branching technique—base protected system cases are: $T^{1}_{\mathbf{HDPE}-\text{branching}} = 0.61\,\text{s}$ or $T^{1}_{\mathbf{LDPE}-\text{branching}} = 0.71\,\text{s}$, for **(LDPE)** or **(HDPE)** plastic ramified short-section. Similarly, the corresponding period values, when employing the inline technique are $T^{1}_{\mathbf{HDPE}-\text{inline}} = 0.62\,\text{s}$ or $T^{1}_{\mathbf{LDPE}-\text{inline}} = 0.7\,\text{s}$, for **(LDPE)** or **(HDPE)** inline short-section; while the period of wave oscillations, predicted in the original piping system, is equal to $T^{1}_{\mathbf{steel}} = 0.42\,\text{s}$.

Table 1 Characteristics of water-hammer waves in Fig. 2

Parameters		Original system (main steel-pipe)	Inline short-section		Branching short-section	
			(HDPE)	**(LDPE)**	**(HDPE)**	**(LDPE)**
h_{max}: 1st hydraulic-head peak	[m]	85.6	74.4	71.5	71.8	73.2
h_{min}: 1st hydraulic-head crest	[m]	5.4	18.5	19.9	18.5	18.8
T^{1} : period of the 1st cycle of wave oscillation	[s]	0.42	0.62	0.70	0.61	0.71

On the side, Fig. 2 illustrates a phase shift between the hydraulic-head traces of each wave pattern groups. In other words, for each short-section material types, a period shift is observed between the inline and the branching techniques. Albeit this period-shift was slightly notable for the first cycle of hydraulic-head oscillations, it becomes more remarkable for the subsequent cycles.

Basing on the above results, it may deduced that the branching technique provides better trade-off between the spreading of wave oscillation period and the attenuation of first hydraulic-head peak (or crest) as compared with the inline technique.

A second task relates to the sensitivity of the first hydraulic-head peak and crest to the short-section size. For completeness, Fig. 3a and b illustrates the forgoing sensitivity analysis within the inline and branching strategies frameworks.

As expected, these curves reveal increasing correlation between the attenuation of fist hydraulic-head peak or crest and the short-section size. More precisely, the attenuation rate of the first hydraulic-head peak or crest is more important for a significant plastic short-section volume. Specifically, Fig. 3a depicts a quite significant decrease effect of the first hydraulic-head peaks or crests for the length values below $l^{\text{plastic}}_{\text{short}-\text{section}} = 5\,\text{m}$. However, this effect is no longer pronounced for the short-section length values beyond $l^{\text{plastic}}_{\text{short}-\text{section}} \geq 5$ m. Similarly, Fig. 3b shows that, as the diameter of the plastic short-section increases from $d^{\text{plastic}}_{\text{short}-\text{section}} = 25$ mm to 50.6 mm, the first hydraulic-head peak or crest is markedly attenuated; however, the last effect is not accentuated for the diameter values beyond $d^{\text{plastic}}_{\text{short}-\text{section}} \geq 75$ mm. Consequently, $l^{\text{plastic}}_{\text{short}-\text{section}} = 5$ m and $d^{\text{plastic}}_{\text{short}-\text{section}} = 75$ mm may be considered as the near-optimal values of the plastic short-section diameter and length.

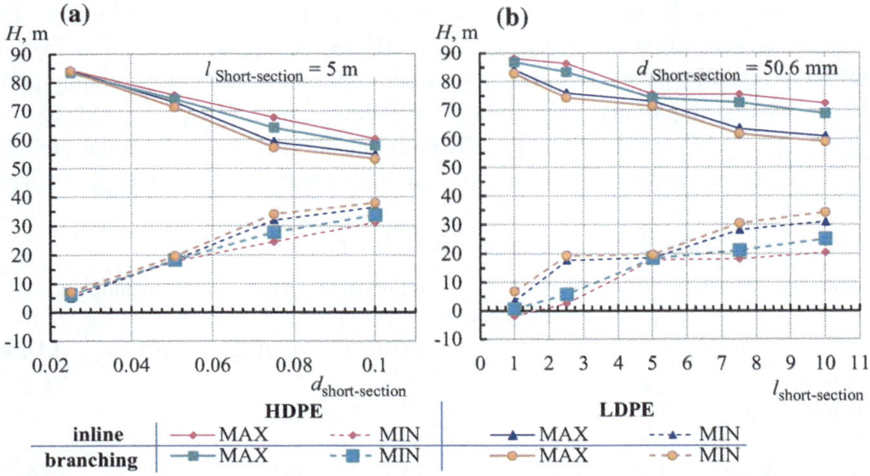

Fig. 3 Variation of first hydraulic-head peaks and crests at the downstream valve section as a function of the short-section **a** diameter and **b** length

4 Conclusion

The present study has demonstrated that the inline and branching water-hammer control techniques are effective to attenuate hydraulic-head peak and crest. In addition, the use of an (**LDPE**) material type for the short-section involves more important attenuation rates compared with the (**HDPE**) one. It is also remarkable to observe that the branching technique illustrates several enhancements to the system reliability as compared with the inline one, in terms of limitation of the spreading of the hydraulic-head oscillation period; while providing comparable hydraulic-head attenuation. Additionally, the sensitivity examination of the first hydraulic-head peaks or crests magnitudes with regard to the short-section length and diameter helped estimate the optimum values of these design parameters.

References

1. Asiaban P, Fathi-Moghaddam M (2018) Flow throttling in surge tanks using porous structures. Int J Pres Ves Pip 168:301–309. https://doi.org/10.1016/j.ijpvp.2018.11.009
2. Aklonis JJ, MacKnight WJ, Shen M (1972) Introduction to polymer viscoelasticity. Wiley-Interscience-Wiley
3. Brinson HF, Brinson LC (2008) Polymer engineering science and viscoelasticity: an introduction. Springer
4. Fersi M, Triki A (2018) Investigation on redesigning strategies for water-hammer control in pressurized-piping systems. J Press Vessel Technol Trans ASME. https://doi.org/10.1115/1.4040136

5. Fersi M, Triki A (2019) Alternative design strategy for water-hammer control in pressurized-pipe flow. In: Fakhfakh T, Karra C, Bouaziz S, Chaari F, Haddar M (eds) Advances in acoustics and vibration II. ICAV 2018. Applied condition monitoring, 13 135–144, Springer, pp 157–165. https://doi.org/10.1007/978-3-319-94616-0_16
6. Ghidaoui MS, Zhao M, Duncan AM, David HA (2005) A Review of Water-hammer Theory and Practice. Appl Mech Rev 58:49–76. https://doi.org/10.1115/1.1828050
7. Keramat A, Haghighi A (2014) Straightforward transient-based approach for the creep function determination in viscoelastic pipes, J Hydraul Eng 140(12). https://doi.org/10.1061/(asce)hy.1943-7900.0000929
8. Massouh F, Comolet R (1984) Étude d'un système anti-bélier en ligne- Study of a water-hammer protection system in line. La Houille Blanche 5:355–362. https://doi.org/10.1051/lhb/1984023
9. Pezzinga G, Scandura P (1995) Unsteady flow in installations with polymeric additional pipe. J Hydraul Eng 121(11):802–811. https://doi.org/10.1061/(asce)0733-9429
10. Triki A (2016) Water-hammer control in pressurized-pipe flow using an in-line polymeric short-section. Acta Mech 227(3):777–793. https://doi.org/10.1007/s00707-015-1493-13
11. Triki A (2017) Water-hammer control in pressurized-pipe flow using a branched polymeric penstock, Journal of Pipeline Systems - Engineering and Practice – ASCE 8(4) 04017024. https://doi.org/10.1061/(asce)ps.1949-1204.0000277
12. Triki A., (2018a) Further investigation on water-hammer control inline strategy in water-supply systems. J Water Suppl Res Technol AQUA 67(1):30–43. https://doi.org/10.2166/aqua.2017.073
13. Triki A (2018b) Dual-technique based inline design strategy for Water-hammer control in pressurized-pipe flow. Acta Mech 229(5):2019–2039. https://doi.org/10.1007/s00707-017-2085-z
14. Triki A., Fersi M. (2018) Further investigation on the Water-hammer control branching strategy in pressurized steel-piping systems. Int. Journal of Pressure Vessels and Piping 165 (C):135–144. https://doi.org/10.1016/j.ijpvp.2018.06.002
15. Triki A, Chaker MA (2019) Compound technique-based inline design strategy for water-hammer control in steel pressurized-piping systems. Int J Pres Ves Pip. 169(C):188–203. https://doi.org/10.1016/j.ijpvp.2018.12.001
16. Vitkovsky JP, Lambert MF, Simpson AR, Bergant A (2000) Advances in unsteady friction modelling in transient pipe flow. In: The 8th international conference on pressure surges BHR, The Hague the Netherlands
17. Wan W, Huang W (2018) Water hammer simulation of a series pipe system using the MacCormack time marching scheme. Acta Mech 229(7):3143–3160. https://doi.org/10.1007/s00707-018-2179-2
18. Wylie EB, Streeter VL (1993) Fluid transients in systems. Prentice Hall, Englewood Cliffs NJ

A Reliability Based Design Method Evaluation for a Coupled Fluid-Structure System

B. Ben Smida[1,2(✉)], E. Mrabet[3], M. Guedri[4], S. Ghanmi[5], and N. Bouhaddi[6]

[1] Laboratoire de Génie Civil, Université de Tunis El Manar, Ecole Nationale d'Ingénieurs de Tunis, BP 37 le Belvédère 1002, Tunis, Tunisia
bbensmida@tvtc.gov.sa

[2] Technical and Vocational Training Corporation, Al Ahsa College of Technology, 7823, 11472 Riyadh, Kingdom of Saudi Arabia

[3] Laboratoire de Mécanique, Ecole Nationale d'Ingénieurs de Sfax, Route Soukra Km 3.5 B.P 1173, 3038 Sfax, Tunisie
mguedri@tvtc.gov.sa

[4] Technical and Vocational Training Corporation, Department of Mechanical Technology, College of Technology at Makkah, Mecca, Kingdom of Saudi Arabia

[5] Technical and Vocational Training Corporation, Tabuk College of Technology, 7823, 11472 Riyadh, Kingdom of Saudi Arabia

[6] Department of Applied Mechanics, University of Bourgogne Franche-Comté, FEMTO-ST Institute, UMR 6174, CNRS/UFC/ENSMM/UTBM, 25000 Besançon, France
noureddine.bouhaddi@femto-st.fr

Abstract. This paper proposes a reliability based designing method study for a coupled fluid-structure system. The main objective is to achieve 90% confidance in our estimate of the failure probability. A First Order Reliability Method (FORM) analysis is performed for a double acoustic cavities numerical example. The evaluation of the (FORM) method is established through a reference Monte Carlo Analysis. The results show that the (FORM) method can approximate the confidence goal with a low computation cost.

Keywords: Reliability · Fluid-structure · FORM · Monte carlo · Model order reduction

1 Introduction

The reliability has shown interest in the design and risk analysis of structures. The concept of Reliability or risk based design is introduced in many codes to predict and optimize the structure behavior [1–3]. The classic determinist design parameters are not realistic. Naturally, there is no perfection in modeling, manufacturing, loads and material specifications. Consequently, all the input parameters are required to be uncertain. The reliability as an output is a result of the uncertain input parameters.

N. Aifaoui et al. (Eds.): CMSM 2019, LNME, pp. 164–172, 2020.
https://doi.org/10.1007/978-3-030-27146-6_18

In the literature we distinguish some popular methods to evaluate reliability [4–6]. The first order reliability methods (FORM) and the second order reliability methods (SORM) are known by their simplicity and accuracy to approximate the reliability with a low computation cost. The Monte Carlo simulations that require a larger number of calculations is used as an accurate reference to evaluate other methods.

This paper evaluates the first order reliability method (FORM) for a coupled vibroacoustic structure. The main objective is to find a low cost and accurate numerical simulation method that estimate the reliability. The quantification of the reliability index β is generally done using iterative [7, 8] and stochastic [9, 10]; Luo et al. [11] methods. To overcome some numerical difficulties, a Model Order Reduction technique with a controlled accuracy proposed by [12] is used.

The evaluation of the presented methodology's performances is illustrated through a numerical double panel example coupled with fluid cavities. This example is commonly used for experimental and numerical fluid structure system behavior prediction [13, 14].

2 Vibroacoustic Damped Model

The fluid structure problem is based on the hypothesis of linear acoustics for a compressible fluid, not viscous, and a small vibration of the fluid flow. For a displacement-pressure formulation (u, p), the damped fluid-structure coupled system with poroelastic material in the rigid walls is based on the following linear equations system:

$$\left(\begin{bmatrix} K_s & -C \\ 0 & K_f \end{bmatrix} - \omega^2 \begin{bmatrix} M_s & 0 \\ \rho_f C^T & M_f \end{bmatrix} + \frac{j\omega}{Z_a(\omega)} \begin{bmatrix} 0 & 0 \\ 0 & A_f \end{bmatrix} \right) \begin{Bmatrix} U \\ P \end{Bmatrix} = \begin{Bmatrix} F_s \\ 0 \end{Bmatrix} \quad (1)$$

where $M_s K_s$ are respectively the mass matrix and the stiffness matrix of the structure, K_s and M_s are respectively the matrices corresponding to the discretization of the kinetic energy and compressibility of the fluid. F_s is the external load applied to the structure and C is the coupling matrix. A_f is the absorbing acoustic matrix, symmetric and depending on the absorbing surface geometry.

In order to damp the fluid cavity resonance, the rigid walls on the source cavity are coated by poroelastic patches characterized by the frequency evolution of the acoustic impedance $Z_a(\omega)$.

3 Resolution Methods

3.1 Direct Method

By using the Eq. (1) and the direct inversion of the dynamic stiffness leads to the computation of the following response:

$$\begin{Bmatrix} U \\ P \end{Bmatrix} = \left(\begin{bmatrix} K_s & -C \\ 0 & K_f \end{bmatrix} - \omega^2 \begin{bmatrix} M_s & 0 \\ \rho_f C^T & M_f \end{bmatrix} + \frac{j\omega}{Z_a(\omega)} \begin{bmatrix} 0 & 0 \\ 0 & A_f \end{bmatrix} \right)^{-1} \begin{Bmatrix} F_s \\ 0 \end{Bmatrix} \tag{2}$$

This method is unachievable for large model size. The inversion calculated at each frequency step is extremely time-consuming and still non possible for various cases. The accuracy is the essential advantage for this method. However, it will be considered as a reference for the comparative study.

3.2 Model Order Reduction Method

The resolution of Eq. (1) is performed using a model order reduction technique presented by [12]. The main objective is to build a reduction modal basis that minimizes the numerical computational cost for predefined prediction accuracy.

The computation of the uncoupled modal basis is built using the following structure and fluid uncoupled eigenvalue problem:

$$\begin{cases} (-\omega^2 M_s + K_s)U = 0 \\ (-\omega^2 M_f + K_f)P = 0 \end{cases} \Rightarrow \quad [T_0] = \begin{bmatrix} T_{s0} & 0 \\ 0 & T_{f0} \end{bmatrix} \tag{3}$$

The structure modal basis is truncated following the frequency criterion $f_s = n \times f_{\max}$, Where $f_{\max}$ is the maximum frequency of interest. Otherwise there is no truncation criterion for the fluid basis. The equation system (2) is projected on the truncated modal basis:

$$\left(-\omega^2 \begin{bmatrix} \tilde{M}_s & 0 \\ \rho_f \tilde{C}^T & \tilde{M}_f \end{bmatrix} + \begin{bmatrix} \tilde{K}_s & -\tilde{C} \\ 0 & \tilde{K}_f \end{bmatrix} \right) \begin{Bmatrix} q_s \\ q_f \end{Bmatrix} = \begin{Bmatrix} \tilde{F}_s \\ 0 \end{Bmatrix} \tag{4}$$

where $\tilde{K}_s = T_{so}^T K_s T_{so}$, $\tilde{M}_s = T_{so}^T M_s T_{so}$, $\tilde{K}_f = T_{fo}^T K_f T_{fo}$ and $\tilde{M}_f = T_{fo}^T M_f T_{fo} \tilde{C}_s = T_{so}^T K_s T_{fo}$, $\tilde{F}_s = T_{so}^T F_s$, $U = T_{so}^T q_s$ and $P = T_{fo}^T q_f$

The computation time is significantly reduced using this superposition method. It is obviously used in structural dynamics. Its main inconvenient for a coupled system, is the non consideration of the coupled modes effect. Consequently, one can obtain a bad convergence of the output result.

The model reduction method used is based on the use of a truncated uncoupled modal basis with a number of chosen n_s^0 structure and n_f^0 fluid modes. The truncation and the kept modes in the basis are selected by an energy criteria presented in the following equations:

$$if \quad E_{s,i} = \frac{T_{so}^T(i) K_s T_{so}(i)}{\sum_{i=1}^{n_s^o} \left(T_{s,sel}^T(i) K_s T_{s,sel}(i) \right)} \geq Tol \quad Then \quad T_{s,i+1}^{sel} = \left[T_{s,i}^{sel} \; T_{so}(i) \right] \tag{5}$$

$$\text{if} \quad E_{f,i} = \frac{T_{fo}^{T}(i) K_f T_{fo}(i)}{\sum_{i=1}^{n_f^o} \left(T_{f,sel}^{T}(i) K_f T_{f,sel}(i) \right)} \geq Tol \quad \text{Then} \quad T_{f,i+1}^{sel} = \left[T_{f,i}^{sel} \; T_{fo}(i) \right] \tag{6}$$

where i represents the current selected mode index.

The basis is accomplished by some vectors to get into account the coupling and the non-considered modes. The vectors $q_{f,i}$ describe the load effects applied by the structure on the fluid. The vectors $q_{s,i}$ describe the load effects applied by the fluid on the structure. They are calculated using the equations:

$$\begin{cases} q_{s,i} = \left(-\omega_c^2 M_s + K_s\right)^{-1} \left(F_s + C T_{f,i}^{sel}\right) \\ q_{f,i} = \left(-\omega_c^2 M_f + K_f\right)^{-1} \omega_c^2 \rho_f C^T T_{s,i}^{sel} \end{cases} \tag{7}$$

The enriched basis T_E is built using the criteria $E_{s,i}$ and $E_{f,i}$ with a Tol tolerance condition:

$$\text{if} \quad \begin{matrix} E_{s,i} \geq Tol \\ E_{f,i} \geq Tol \end{matrix} \quad \text{then} \quad \begin{matrix} \left[T_s^E\right] = \left[T_s^{sel} \; q_{f,1} \cdots q_{f,i} \cdots\right] \\ \left[T_f^E\right] = \left[T_f^{sel} \; q_{s,1} \cdots q_{s,1} \cdots\right] \end{matrix} \tag{8}$$

A modal projection is used to estimate the displacement and pressure responses:

$$\begin{Bmatrix} \hat{U} \\ \hat{P} \end{Bmatrix} = T_E \begin{Bmatrix} q_s \\ q_f \end{Bmatrix} \tag{9}$$

The force residuals, so called error loads, are calculated as follow:

$$\begin{cases} R_{Fs} = \left(K_s - \omega_c^2 M_s\right)^{-1} \hat{U} - C\hat{P} - F_s \\ R_{Ff} = \left(K_f - \omega_c^2 M_f\right)^{-1} \hat{P} - \rho_f \omega_c^2 C^T \hat{U} \end{cases} \tag{10}$$

The displacement residuals are estimated using the force residuals:

$$\begin{cases} R_{Ds} = \left(K_s\right)^{-1} R_{Fs} \\ R_{Df} = \left(K_f\right)^{-1} R_{Ff} \end{cases} \tag{11}$$

The bases T_s^E and T_f^E are fulfilled by the displacement residuals vectors R_{Ds} and R_{Df}. A singular value decomposition (SVD) is therefore necessary.

4 The First Order Reliability Method

The reliability is the probability of a performance function $g(X)$ that is greater than zero. If $X = (x_1, x_2, \ldots, x_n)$ are the random design parameters, they are in the safe region defined by $g(X) > 0$ the probability of failure is evaluated with the integral:

$$p_f = P\{g(X) < 0\} = \int_{g(X)<0} f_x(X)\mathrm{d}x \quad (12)$$

The performance function $g(X)$ is approximated and linearized by the first order Taylor expansion by simplifying the integrand $f_x(X)$. This is the basic of the First Order Reliability Method (FORM).

5 The Reliability Based Design Method Evaluation

The reliability based design method evaluation is presented in four principal steps. The initial one is to select the input design parameters according to their sources and their significance level. Consequently, the simulation time will be significantly reduced. The parameters with weak sensitivity will be considered by their deterministic value.

The second step uses a Monte Carlo (MC) sampling. This technique is the easiest approach for implementation and known by its accuracy. However, the obtained results will be considered as a reference.

The third step uses a first order reliability method (FORM) with the intention to estimate the probability of failure.

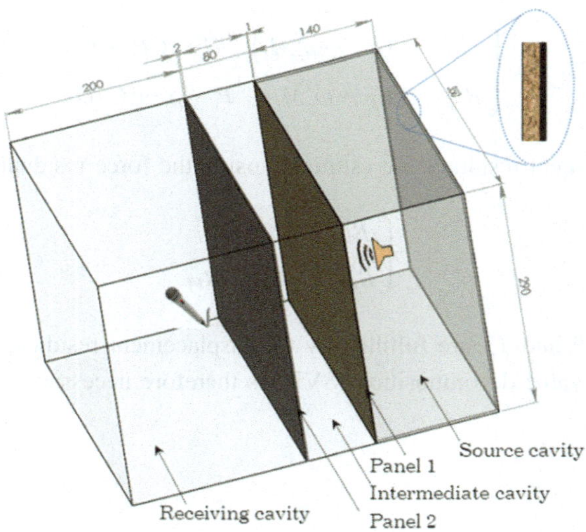

Fig. 1 Geometric model for a three cavities system

The fourth and final step compares the approximated probability of failure results obtained by (FORM) method to the reference.

6 Numerical Simulation

Table 1 Geometric and physical panels properties

Parameters	Young modulus $E(Pa)$	Poisson ratio v	Density $\rho(Kg/m^3)$	Thickness $h(m)$
panel 1	4e9	0,4	1200	2e-3
panel 2	210e9	0,285	7800	1e-3

6.1 Finite Elements Model

The efficiency of the reliability based design method is illustrated through a numerical simulation double panel example (Fig. 1). It is made by three cavities filled with air and coupled through two flexible panels. The system has uniform dimension in the (x, y) plane: 0.35 × 0.29 m. The depth is 0.14 m for the source cavity, 0.08 m for the middle cavity and 0.2 m for the receiving cavity. Poroelastic patches, characterized by acoustic impedance $Z_a(\omega)$, are stacked on the five rigid walls of the source cavity.

The panel 1 is made by epoxy resin and the panel 2 is made by steel. Their physical and geometric properties are presented in the Table 1.

The finite element model consists of 17 × 14 structural quadrilateral elements and 17 × 14 elements acoustic bricks. The finite element model includes about 9600 DDLs degrees of freedom (DOFs) (3108 DOFs of structure, 6480 DOFs of fluid). The mesh size of the finite elements model is based on a criterion of five to six linear elements per mode shape wavelength.

The structural damping factors introduced respectively for the fluid and structure sub-systems are chosen as $\eta_s = 0.02$ and $\eta_s = 0.002$. An external load exciting the system is located in ((x,y,z) = 0.082,0.145,0.12). The vibroacoustic uncertainties analysis is achieved in the frequency range of $[0 - 200Hz]$ containing the first three modes of the coupled system.

The model reduction method is used in the second step; the reduction basis is enriched according to a predefined tolerance ($tol = 10^{-3}$). Singular Value Decomposition (SVD) is useful to guarantee the linear independence of the initial basis modes with the added modes. The model reduction ratio is about 94%.

Table 2 Random design parameters

	Design parameters	Distribution
panel 1	Young modulus $E_1(Pa)$	Deterministic
	Density $\rho_1(Kg/m^3)$	Deterministic
	Thickness $h_1(m)$	Normal Gaussian
Panel 2	Young modulus $E_2(Pa)$	Normal Gaussian
	Density $\rho_2(Kg/m^3)$	Deterministic
	Thickness $h_2(m)$	Normal Gaussian
Intermediate cavity	Width $d(m)$	Deterministic

6.2 The Reliability Based Design Method Evaluation, Results and Discussion

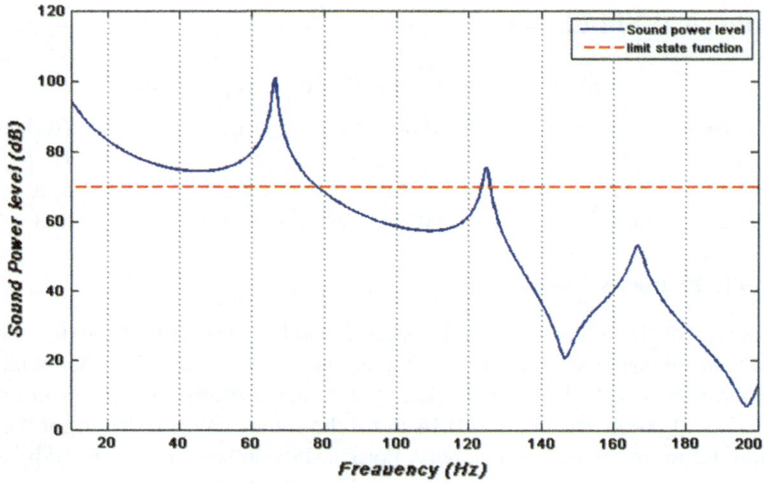

Fig. 2 Acoustic power level and limit state function

Some mechanical properties are considered as random design parameters with normal Gaussian distribution (Table 2). This selection is based on a preliminary design

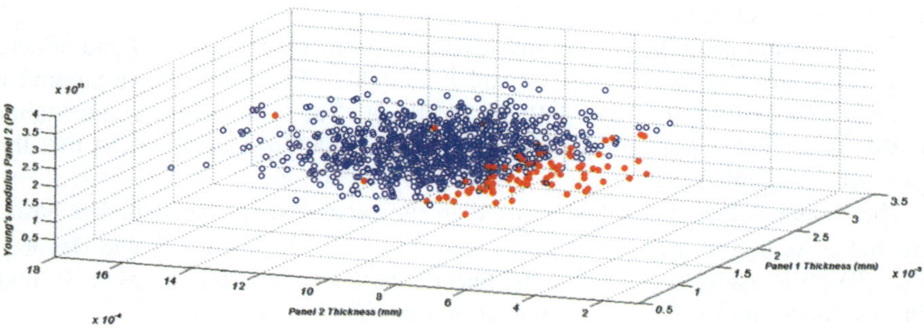

Fig. 3 Failure and Reliability by the design parameters

parameters sensitivities study. The Young modulus E_2 and the thickness h_2 for the panel2 and the thickness h_1 for the panel1 are the more significant design parameters. The deterministic value is used as the nominal value for each variable and the standard deviation is $\sigma = 20\%$.

Table 3 Probability of failure

	MC 1000 runs	FORM 81 runs
p_f	0.0850	0.0851

Figure 2 shows the acoustic power level in the receiving cavity and the limit state function calculated by the deterministic values of the design parameters.

Table 4 CPU Time (seconds)

	MC 1000 runs	FORM 81 runs
CPU (s)	53636.00	29288.04
Ratio (%)	54.6%	

The Fig. 3 presents the fail and the Reliable for 1000 samples of the design parameters. The reliability achieves the initial goal of 90%.

The Table 3 illustrate the probability of failure obtained using the first order reliability method compared to the other one obtained using a reference Monte Carlo simulation. It is clear that the two methods give the same accuracy at the tolerance of 10^{-4}.

The second key performance indicator is the CPU Time that indicate the simulation cost. It is apparent that the MC method is more time consuming than FORM (Table 4).

7 Conclusion and Outlook

In this paper, a reliability based design method (FORM) is compared to the reference Monte Carlo Method. The evaluation is performed throw a vibroacoustic double panel example known by its numerical implementation difficulties related to the model size and the multiphysics problem.

The FORM approximates the probability of failure with accuracy about 10^{-4} and a reduced simulation time more than 50%.

It is well shown that the FORM method can be useful for the coupled fluid structure systems that suffer from excessive computational costs.

As outlook, the second order reliability method (SORM) more accurate than FORM can be used for the same objective.

References

1. Dammak K, El Hami A, Koubaa S, Walha L, Haddar M (2017) Reliability based design optimization of coupled acoustic-structure system using generalized polynomial chaos. Int J Mech Sci 134:75–84. https://doi.org/10.1016/j.ijmecsci.2017.10.003

2. Mrabet E, Guedri M, Ichchou M, Ghanmi S (2014) New approaches in reliability based optimization of tuned mass damper in presence of uncertain bounded parameters. J Sound Vib 355:93–116. http://dx.doi.org/10.1016/j.jsv.2015.06.009
3. Guedri M, Cogan S, Bouhaddi N (2011)Robustness of structural reliability analyses to epistemic uncertainties. Mech Syst Signal Process 28:458–469. https://doi.org/10.1016/j.ymssp.2011.11.024
4. Hasofer AM, Lind NC (1974) An exact and invariant first order reliability format. J Eng Mech Div ASCE 100:111–121
5. Rackwitz RB (1978) Flessler Structural reliability under combined random load sequences. Comput Struct 9(5):489–494
6. Melchers RE (1999) Structural reliability analysis and prediction, 2nd edn, Wiley
7. Der Kiureghian A, Haukaas T, Fujimura K (2006) Structural reliability software at the University of California, Berkeley. StructSaf, 28:44–67. https://doi.org/10.1016/j.strusafe.2005.03.002
8. Periçaro G, Santos S, Matioli A (2015) HLRF–BFGS optimization algorithm for structural reliability Ribeiro. Appl Math Model 39:2025–2035. https://doi.org/10.1016/j.apm.2014.10.024
9. Cheng J, Li QS (2008) Reliability analysis of structures using artificial neural network based genetic algorithms. Comput Methods Appl Mech Eng 197:3742–3750. https://doi.org/10.1016/j.cma.2008.02.026
10. Cheng J (2010) An artificial neural network based genetic algorithm for estimating the reliability of long span suspension bridges. Finite Elem Anal Des 46:658–667. https://doi.org/10.1016/j.finel.2010.03.005
11. Luo X, Li X, Zhou J, Cheng T (2012) A Kriging-based hybrid optimization algorithm for slope reliability analysis. Struct Saf 34:401–406. https://doi.org/10.1016/j.strusafe.2011.09.004
12. Ben Smida B, Majed R, Bouhaddi N, Ouisse M (2012) Investigations for a model reduction technique of fluid-structure coupled systems. Part C J Mech Eng Sci 226:42–54
13. Akrout A, Karra C, Hammami L, Haddar M (2008) Viscothermal fluid effects on vibro-acoustic behavior of double elastic panels. Int J Mech Sci 50:764–773
14. Doutres O, Atalla N (2011) Experimental estimation of the transmission loss contributions of a sound Package placed in a double wall structure. Appl Acoust 72:372–379

Effect of Cylindrical Particle Orientation on the Flow and Temperature Distribution

Hajer Troudi[1(✉)], Moncef Ghiss[1], Mohamed Ellejmi[2], and Zoubeir Tourki[1]

[1] Laboratory of Mechanical of Sousse "LMS", University of Sousse, Bp. 264 Erriadh, 4023 Sousse, Tunisia
hajer.troudi@eniso.rnu.tn, moncef.ghiss@eniso.u-sousse.tn, zoubeir.tourki@mesrs.tn
[2] Alpha Engineering International, AEI, Sahloul III, 4054 Sousse, Tunisia
m.ellejmi@alpha-engineering.com.tn

Abstract. Drag coefficient and average Nusselt number are a critical operating parameters in fluid-particle processes. In this paper, a 3-D computational fluid dynamics (CFD) software is established to investigate the influence of the particle angle orientation on these parameters. A series of particles (spherical and non-spherical) has been developed and corresponding simulations are validated using correlations with reasonable accuracy. The results show that the average Nusselt number increases slowly with the particle angle orientation increasing from 0° to 30°, and rapidly when the angle orientation increases from 45° to 90°. The behavior gives high heat transfer, especially on the upper and front side of the cylinder when the gas velocity is high.

Keywords: Angle orientation · Average Nusselt number · CFD simulation Cylindrical particle · Drag coefficient

1 Introduction

The interaction particle-fluid is of interest in many separation process [15] (e.g., slurry pipeline transportation, fluidized bed systems and multiphase reactors, etc.). Study of different shape, in particular spherical and cylindrical particles in large scale beds remains a challenge, due to lack of accurate measurement correlations. Drag coefficient and average Nusselt number have an important impact on the interaction particle-fluid flows. The drag force is greatly influenced by both the form and the orientation of non-spherical particles. Spherical particles have been well documented in many research projects, e.g., Ellendt et al. [5], Tabata and Itakura [13]. In some cases, there have been studies on the particle drag of ellipsoidal [10], cubic [12], and rectangular prism shapes [1]. However, not much attention has been drawn to the cylindrical particles. Less correlations have been reported to calculate the drag coefficient of cylindrical particles due to their complex form.

Computational fluid dynamics (CFD) has been broadly used as a versatile tool to study the heat transfer and the drag coefficient involving spherical particles and non-spherical particles. Gabitto and Tsouris [7] have successfully developed a new

N. Aifaoui et al. (Eds.): CMSM 2019, LNME, pp. 173–182, 2020.
https://doi.org/10.1007/978-3-030-27146-6_19

correlation to predict drag coefficient of cylindrical particles settling in a liquid medium. They have demonstrated that the error compared with relevant experimental results is very small. Cao and Tafti [2] have studied the drag coefficient in a 3D cylindrical particle of low aspect ratio (0.25) using CFD. In their study, the cylinder was placed on a cubic wall in a flow with *Re* in the range of 10–300. The results show that the angle orientation can make an impressive influence on the leeward side of the cylinder and some different behaviors of the flow development are obtained. However, their results were may not be better enough due to the use of low aspect ratio. Dixon [4] have defined the flow through a hollow cylindrical particle. The authors concluded that the angles 30°–40°, leads to highest thermal performance of the system. Zhang et al. [16] considered particle locations as constant, and analyzed the variation of drag force around the particles. The finite particle method integrated with particle shifting technique (FPM-PST) has been applied to simulate the particles with different temperatures. They concluded that the viscosity of the gas flow could be highly related to the temperature difference, which causes the drag force coefficient to increase for hot particle. The difference becomes more obvious for larger temperature differences and is better for smaller Reynolds number.

To the authors' best knowledge, there are no published studies which systematically vary the angle of orientation of the cylinder for high aspect ratio. Thus, the main aim of this manuscript is to investigate the 3D flow and heat transfer through a cylindrical particle with circular cross section. We have constructed the possible orientation between 0° and 90°, which are good general case studies of cylindrical particles. We have solved the problem using the CFD software ANSYS Fluent 15.

2 Numerical Methodology

In this study, the cylindrical particle is assumed to be stationary and isothermal, and the flow is assumed to be steady and incompressible with heat transfer. With these assumptions, the computational details of the simulation are briefly described below. The governing equations for the gas phase are the locally averaged Navier–Stokes equations. The mass and momentum conservation equations are expressed as:

$$\nabla.\overrightarrow{V} = 0 \tag{1}$$

$$\rho_f \frac{\partial \overrightarrow{V}}{\partial t} + \rho_f \overrightarrow{V}.\nabla \overrightarrow{V} = -\nabla p + \nabla.\tau + \rho_f \overrightarrow{g} \tag{2}$$

$$\rho_f c_p \frac{\partial T}{\partial t} + \rho_f c_p \overrightarrow{V}.\nabla T = k_{eff} \Delta T \tag{3}$$

In the above equations, f refers to the fluid phase. ρ_f, V and p are the density, the velocity vector and the pressure respectively. T is the temperature. c_p and k_{eff} are the specific heat ratio and the thermal conductivity of the fluid f. τ is the stress tensor. Considering that the Reynolds number fluid flow pertaining to this study is in both laminar and turbulent regimes, we used the well-known $k - \varepsilon$ as turbulence model. The drag force belonging to Eqs. (1)–(3), is expressed as follows:

$$F_D = \frac{1}{2}\rho_f A_p C_D \vec{V}\left\|\vec{V}\right\| \quad (4)$$

where p refers to the particle domain. The surface area of the particle A_p is defined as $A_p = (0.5\pi)d_p^2 + \pi l_p d_p$, where (l_p, d_p) are the length and the diameter of the particle p. The heat transfer is calculated based on the difference between the cylinder and inlet fluid temperatures and quantified by local Nusselt number as given in Eq. (5):

$$Nu_{local} = \frac{d_p}{T_{in} - T_p}\left(\frac{\partial T}{\partial n}\right)_s \quad (5)$$

$$Nu_{average} = \frac{\int_S Nu_{local}\, ds}{\int_S ds} \quad (6)$$

where, T_{in} and T_p are the average inlet temperature and the average surface temperature of the particle, respectively. n is the outward normal unit vector on the surface. For the case of cylindrical particle, heat transfer is calculated from the local Nusselt number given on each surface of the cylinder i.e., the curved (c), the front (f) and the rear (r) surface.

2.1 Nusselt Number Correlations

The flow field is characterised by the particle Reynolds number and the Prandtl number, which are commonly defined as $\mathrm{Re}_p = \rho_f V_{in} d_{eq}/\eta_f$ and $\mathrm{Pr}_p = v_f/k_f$. In similar way, the correlations of Nusselt number presented in Table 1 are based on the Re_p and Pr_p. Here V_{in} is the average inlet flow velocity. v_f and k_f are the kinematic viscosity and the thermal diffusivity of the fluid, respectively. d_{eq} is the characteristic size of the particle, e.g. equivalent particle diameter of the cylinder. For assuming "air" as the fluid phase, in this work the Prandtl number, Pr_p, is set to 0.744.

2.2 Drag Coefficient Correlations

As listed in Eq. (4), the drag coefficient C_D is necessary for the calculation of the particle motion. The drag coefficient denotes the hydrodynamic force applied to the particle surface and it is made of particle orientation. Since the arbitrary position or orientation of the particles is the objective of this study, the above mentioned drag coefficients are appropriate to be implemented. For the same reason, these correlations demonstrate high accuracy especially, when the particle orientation was considered.

Table 1 Summary of Nusselt number correlations

	Correlation	Applicability range
Ranz and Marshall [11]	$Nu_p = 2 + 0.4\mathrm{Re}_p^{0.5}\,\mathrm{Pr}_p^{0.33}$	$10 < \mathrm{Re}_p < 10^4$; $\mathrm{Pr}_p > 0.7$
Whitaker [14]	$Nu_p = 2 + \left(0.4\mathrm{Re}_p^{0.5} + 0.06\mathrm{Re}_p^{0.66}\right)\mathrm{Pr}_p^{0.4}\left(\frac{\eta_p}{\eta_s}\right)^{0.25}$	$3.5 < \mathrm{Re}_p < 7.6\,10^4$; $0.7 < \mathrm{Pr}_p < 380$
Feng and Michaelides [6]	$Nu_p = 0.922 + \mathrm{Re}_p^{0.33}\,\mathrm{Pr}_p^{0.33} + 0.1\mathrm{Re}_p^{0.33}\,\mathrm{Pr}_p^{0.33}$	$0 < \mathrm{Re}_p < 2000$; $0 < \mathrm{Pr}_p < 1000$
Richter and Nikrityuk [12]	$Nu_p = 1.76 + 0.55\phi_\perp^{0.075}\sqrt{\mathrm{Re}_p}\,\mathrm{Pr}_p^{0.5} + 0.014\,\mathrm{Pr}_p^{0.33}\,\mathrm{Re}_p^{0.66}\left(\frac{\phi}{\phi_\perp}\right)^{7.2}$	

Table 2 Summary of drag coefficient correlations

	Correlation	Applicability range
Ganser [8]	$C_D = \frac{24}{\mathrm{Re}_p k_1}\left(1 + 0.1118(\mathrm{Re}_p k_1 k_2)^{0.6567}\right) + \frac{0.4305k_2}{1 + [3305/\mathrm{Re}_p k_1 k_2]}$ $k_1 = \left(\frac{1}{3} + \frac{2}{3}\phi^{-0.5}\right)^{-1}$ $k_2 = 10^{1.8148(-\log\phi)^{0.5743}}$	$\mathrm{Re}_p < 3\ 10^5$
Hölzer and Sommerfeld [9]	$C_D = \frac{8}{\mathrm{Re}_p}\frac{1}{\sqrt{\phi_\perp}} + \frac{16}{\mathrm{Re}_p}\frac{1}{\sqrt{\phi}} + \frac{3}{\sqrt{\mathrm{Re}_p}}\frac{1}{\phi^{3/4}} + 0.42\frac{1}{\phi_\perp}10^{0.4(-\log\phi)^{0.2}}$	$\mathrm{Re}_p < 3\ 10^5$
Richter and Nikrityuk [12]	$C_D = 0.21 + \frac{20}{\mathrm{Re}_p}\left(\frac{l_p}{d_p}\right)^{0.58} + \frac{6.9}{\sqrt{\mathrm{Re}_p}}\left(\frac{l_p}{d_p}\right)^{-1.4}$	$10 < \mathrm{Re}_p < 250$

The sphericity ϕ presents the only descriptor of the 3D particle shape and defined as the ratio of the surface area of a sphere having the same volume as the particle to the surface area of the particle. For a spherical particle, the sphericity is $\phi = 1$. For a non-spherical particle, the sphericity is defined by: $\phi = 1.5\left(\pi l_p d_p\right)/A_p$. To characterize the orientation, the so-called crosswise sphericity is utilized: $\phi_\perp = \alpha^{2/3}/\left(\alpha^2 \sin^2\theta_i + \cos^2\theta_i\right)^{1/2}$. Another non-dimensional parameter, the aspect ratio α, is the ratio of the length and the diameter of the particle and defined as $\alpha = l_p/d_p$.

2.3 Solution and Simulation Conditions

The 3D geometries for the problem are shown in Fig. 1. The computational domains are a cylinder tube of uniform aspect ratio 5. The particle is located in the middle and took a uniform aspect ratio 2.1 and different orientation positions. The long axis of the reactor parallels to z-axis which presents the gravity. In this study, the possible contact scenarios between the particle and the gas flow are shown in Fig. 1a. In the first scenario, the particle is orientated in the opposite direction of the streamwise while in the last scenario, the particle is orientated in the streamwise direction. In the remaining scenarios, the orientation of the particle is randomly varied. Figure 1b shows the boundary conditions where a symmetry boundary condition is imposed on the wall. The inflow and outflow boundaries on the z-axis are applied to generate a steady flow in the vertical direction. Around the particle surface, the fluid velocity is damped to zero and an isothermal temperature is attained.

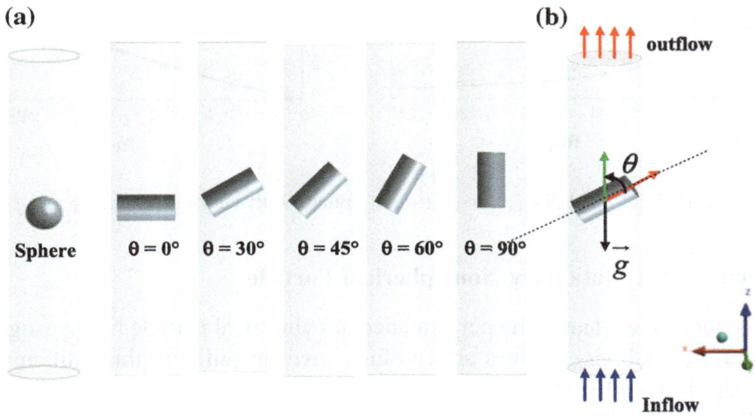

Fig. 1 Spherical and cylindrical particles configurations of aspect ratio 2.1

3 Results and Discussion

First, we carried out the validation on a sphere in Sect. 3.1. Second, the effects of the orientation of cylindrical particles on the drag coefficient and average Nusselt number are investigated in Sect. 3.2. Third, the results of the CFD simulations are presented Sect. 3.3 in the form of temperature contours to illustrate the qualitative features and insight that can be obtained.

3.1 Validation Using Spherical Particle

To validate the numerical results as well as, to test the grid independence, Fig. 2 shows the comparison of the drag coefficient and Nusselt number with correlations illustrated in Tables 1 and 2. Here, we carried out the numerical results of the flow around a sphere at different Reynolds numbers. We operated different grid sizes, e.g., $0.2d_p$, $0.1d_p$, $0.02d_p$ and $0.01d_p$. It is shown that grid 1 might not be accurate where C_D and Nu_{ave} are predicted within a maximum error of 38% and 11%, respectively. Grid 2 presents a slight improvement compared to Grid 1 of 13.4% in C_D and 2.3% in Nu_{ave}. There are no significant difference between the grid 3 and the grid 4 for Nu_{ave} and C_D and the results are consistent with the literature. Thus, only the grid 3 (grid size and total number are =0.75 mm and ≈14 M, respectively) is chosen in the following simulations.

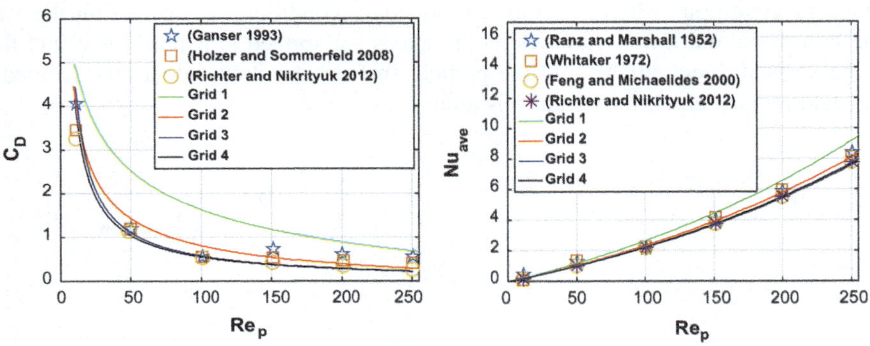

Fig. 2 C_D and Nu_{ave} versus Re_p of sphere with varying particle sizes

3.2 Flow Past a Stationary Non-spherical Particle

In this section, we evaluated the performance of cylindrical particle by investigating the orientation. C_D and Nu_{ave} values are carefully investigated with three different angles $\theta = \{0, 30, 45, 60, 90\}$ and for $\mathrm{Re}_p \in \{5, 15, 20, 100, 200, 250\}$, respectively. We have chosen $\mathrm{Re}_p \leq 250$, because the flow with $\mathrm{Re}_p \geq 250$ do not describe the flow past a 3D cylinder [3]. The flow is both laminar and turbulent. As first insight, the curves in Figs. 3 and 4 exhibit the same trend as the correlations. Figure 3 shows that C_D values decrease as θ increases in all the calculated range. Spheres have the highest C_D. In a similar way to C_D, we evaluated Nu_{ave} as a function of θ and we found that, in contrast to C_D, Fig. 4 demonstrates the increase of Nu_{ave} with an increase of θ. This observation is also analyzed by Ke et al. [10], Richter and Nikrityuk [12] with reference to ellipsoidal particles. But it should be noted that the relationship between C_D decreases more strongly as θ increases than do Nu_{ave}. Since Nu_{ave} depends largely on the temperature gradient, we found that the gradient increases with θ and Re (refer to Figs. 5 and 6).

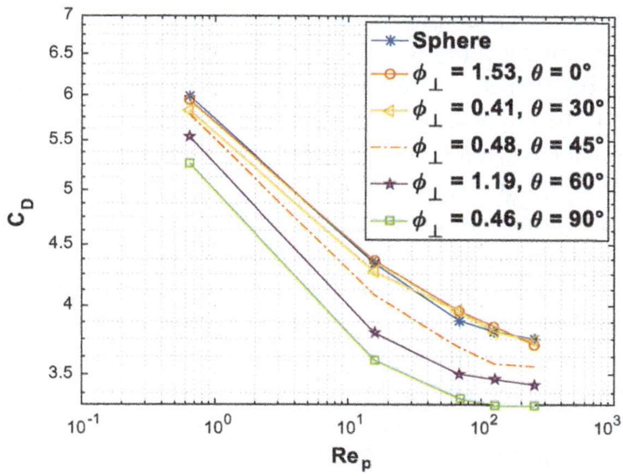

Fig. 3 C_D versus Re_p with varying θ

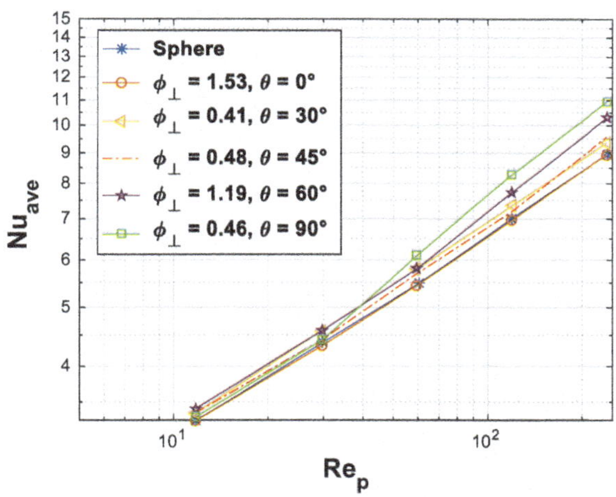

Fig. 4 Nu_{ave} versus Re_p with varying θ

3.3 Hydrodynamic Behavior of Gas and Particles

Figures 5 and 6 depict contour plots of the normalized temperature field at $\mathrm{Re}_p = 6$ and $\mathrm{Re}_p = 90$, respectively along the mid-planes parallel to (yz) and (xz). Clearly, the temperature gradient generates at the bottom of the particle and then are stretched vertically as the gas flowed upwards, until the outlet of the bed. In Fig. 5(a_2) to (g_2), the temperature contour is symmetric in the top surface of the particle along the centerline of the bed. However, for Fig. 5(a_1) to (g_1), the temperature gradient has kept the symmetrical form but the distribution is shown to be non-uniform. The non-uniform temperature distribution suggests that the particle orientation and heat transfer are strongly coupled. Fin layers are presented on the lower region of the curved surface due to the parallel flow of the gas through the particle. When θ increased, the lower circular surface shows a more density of the layers than that in the upper circular surface. For Fig. 6(a_2) and (b_2), the temperature structure of $\theta = 0°$ is similar to that of the sphere. In Fig. 6, with greater Reynolds number (Re = 64), the temperature is becoming no-symmetric, which is a well-known phenomenon. The size of the recirculation wakes becomes bigger above the particle due to the corner effect. In the upper region of the curved surface, the streamlines are very crowded, so that the flow indicates high heat transfer rates. Figure 6(a_2) to (g_2) show that there is a steep temperature gradient with $\theta = 30°$ in the upper curved surface of the particle compared to $\theta = 45°$ and $\theta = 60°$. For $\theta = 90°$, we found that high gradient is observed on the curved and the front surfaces for both $Re_p = 6$ and $Re_p = 90$. Therefore, it can be concluded that effectively high heat transfer and less drag are exerted due to suspension vertical cylinders. As a result, by evaluating the two figures, the heat transfer is enhanced with increasing the fluid velocity and the angle orientation.

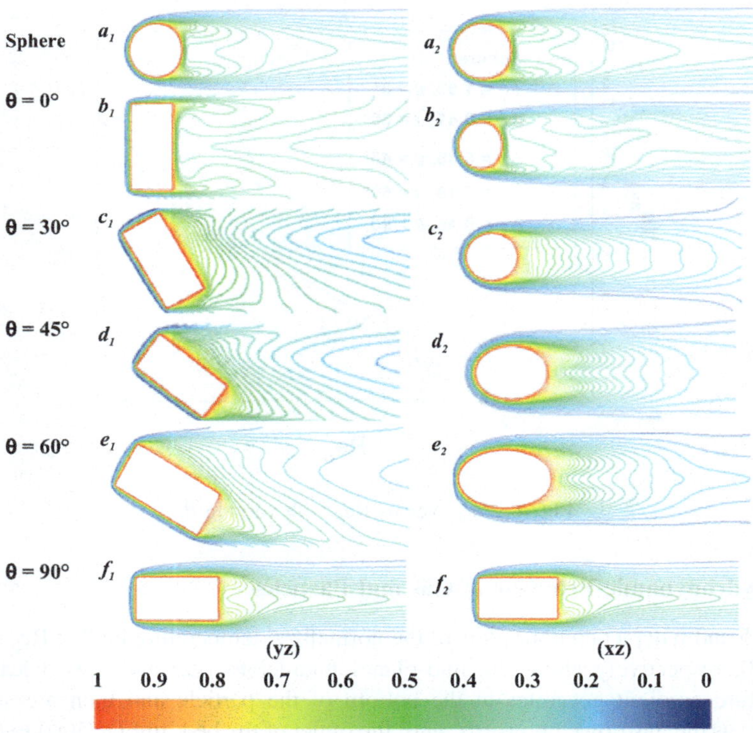

Fig. 5 Contour plots of the isotherms $(T - T_{in})/(T_p - T_{in})$ for $Re_p = 6$

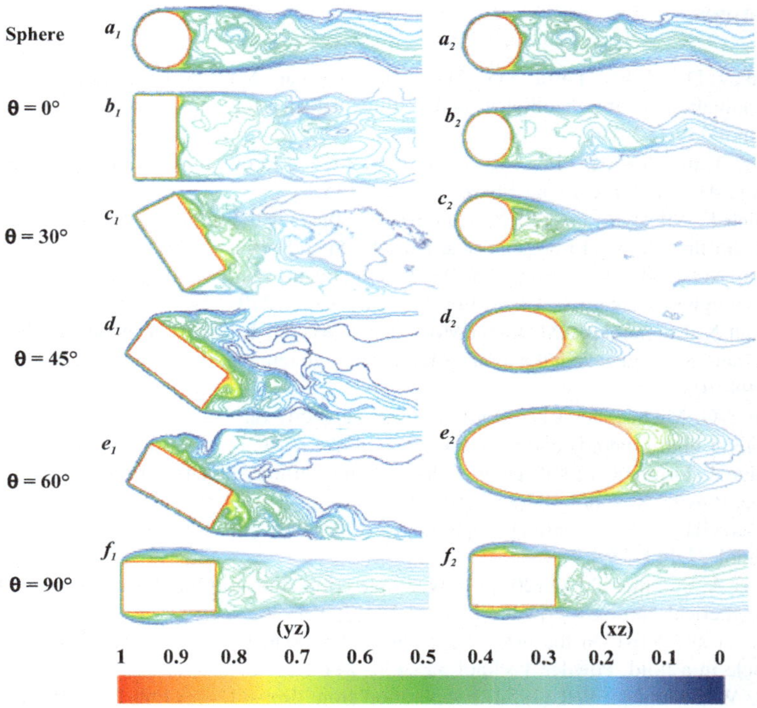

Fig. 6 Contour plots of the isotherms $(T - T_{in})/(T_p - T_{in})$ for $Re_p = 90$

4 Conclusion

Numerical simulation was carried out of steady flow past a cases of cylinders with varying angle orientations. Average Nusselt number and drag coefficient were systematically investigated. By comparing with spherical particles, some conclusions can be summarized as follows: (a) with increasing gas velocity, temperature was altered accordingly and gave rise to different temperature distributions, (b) the temperature is found to be more symmetric for low Reynolds number $Re_p \leq 10$, (c) the increasing of average Nusselt number is found to be high when using the highest angle orientation and (d) the comparison between two different values of Reynolds number shows that the intensity of heat transfer in the upper part of the curved surface of the particle improves as the angle orientation increases.

References

1. Breakey David ES, Farid VG, Masliyah JH, Sean Sanders R (2018) Side-view-only determination of drag coefficient and settling velocity for non-spherical particles. Powder Technol 339:182–191
2. Cao Z, Tafti DK (2018) Investigation of drag, lift and torque for fluid flow past a low aspect ratio (1:4) cylinder. Comput Fluids 177:123–135
3. Dierich F, Nikrityuk PA (2013) A numerical study of the impact of surface roughness on heat and fluid flow past a cylindrical particle. Int J Therm Sci 65:92–103
4. Dixon AG (2014) CFD study of effect of inclination angle on transport and reaction in hollow cylinder catalysts. Chem Eng Res Des 92(7):1279–1295
5. Ellendt N, Lumanglas AM, Imani Moqadam S, Mädler L (2018) A model for the drag and heat transfer of spheres in the laminar regime at high temperature differences. Int J Therm Sci 133:98–105
6. Feng Z-G, Michaelides EE (2000) A numerical study on the transient heat transfer from a sphere at high reynolds and peclet numbers 43:219–29
7. Gabitto J, Tsouris C (2008) Drag coefficient and settling velocity for particles of cylindrical shape. Powder Technol 183(2):314–322
8. Ganser GH (1993) A rational approach to drag prediction nonspherical particles. Powder Technol 77:143–152
9. Hölzer A, Sommerfeld M (2008) New simple correlation formula for the drag coefficient of non-spherical particles. Powder Technol 184(3):361–365
10. Ke C et al (2018) On the drag coefficient and averaged Nusselt number of an ellipsoidal particle in a fluid. Powder Technol 325:134–144
11. Ranz WE, Marshall WR (1952) Evaporation from drops. Chem Eng Prog 48(3):173–180
12. Richter A, Nikrityuk PA (2012) Drag forces and heat transfer coefficients for spherical, cuboidal and ellipsoidal particles in cross flow at sub-critical reynolds numbers. Int J Heat Mass Transf 55(4):1343–1354
13. Tabata M, Itakura K (1998) A precise computation of drag coefficients of a sphere. Int J Comput Fluid Dyn 9(3–4):303–311
14. Whitaker S (1972) Forced convection heat transfer correlations for flow in pipes, past flat plates, single cylinders, single spheres, and for flow in packed beds and tube bundles. AIChE J 18(2):361–371
15. Yang W-C (2003) Handbook of fluidization and fluid-particle systems, New York
16. Zhang ZL et al (2019) A finite particle method with particle shifting technique for modeling particulate flows with thermal convection. Int J Heat Mass Transf 128:1245–1262

Experimental Analysis of Electromyography (EMG) Signal for Evaluation of Isometric Muscle Force

Olfa Jemaa[1(✉)], Sami Bennour[1], David Daney[2], and Lotfi Romdhane[1,3]

[1] Laboratory of Mechanical of Sousse 'LMS', University of Sousse, Bp.264 Erriadh, 4023 Sousse, Tunisia
olfa2jemaa@gmail.com, sami.bennour@gmail.com, lromdhane@aus.edu

[2] National Institute for Research in Computer Science and Automation 'Inria', 33405 Talence, Bordeaux, France
david.daney@inria.fr

[3] College of Engineering, American University of Sharjah, Sharjah, UAE

Abstract. Muscle activation can be quantified by surface Electromyography (EMG) measurements. Indeed, EMG is a technique used to measure the potential charge of muscle fibers during isometric, concentric or eccentric contractions. The aim of this paper is to design an embedded electronic system allowing the detection of the EMG signals. The measured data is then sent through a wireless transmission system to the host computer. Wireless transmission has the advantage of facilitating the capture of the human movement and eliminates electrical wires. The paper is divided into three sections. Firstly, this work will begin with the realization of an embedded system to measure EMG signals. This signal is detected by surface electrodes placed on the muscles. Then, EMG signal analysis will be performed by different processing methods. Finally, these experimental data will be manipulated and used as inputs to musculoskeletal model for estimate isometric muscle forces. As a result, the modelling method presented represents a good way to estimate muscle forces during isometric contraction.

Keywords: Embedded system · Wireless transmission · Electromyography (EMG) · Dynamics contraction · Muscle force

1 Introduction

The human body has more than 650 muscles attached to the skeleton. These muscles constitute about 40% of total body weight [14]. We can distinguish three types of muscle tissues: skeletal muscles, smooth muscles, and cardiac muscles. Skeletal muscles have several functions; it's a key component of the locomotion system [14]. Each skeletal muscle is attached to the bones by two extremity called origin (O) and insertion (I) (See Fig. 1). Furthermore, these muscles are attached to the bones by tendons. Tendons and muscles work jointly to produce the movement of some joint and

N. Aifaoui et al. (Eds.): CMSM 2019, LNME, pp. 183–192, 2020.
https://doi.org/10.1007/978-3-030-27146-6_20

to stabilize another joint. The basic unit of a skeletal muscle is the muscle fibers [15]. These fibers are similar to long cylindrical shapes arranged in parallel. An excitation of the muscle fibers by the nervous system causes the generation of an electric wave, it's action potential (AP) [3]. Following this electrical activation, the muscle generates the mechanical force. This activation can be measured by surface electromyography (EMG). The EMG is an non-invasive technique for assessing the level of muscle activation in human voluntarly contraction [5]. The estimation of the generated force is interesting, not only for biomechanical studies, but also for clinical applications. Indeed, the muscular forces information provides diagnostic assistance and medical treatment. For example, prosthesis can be ordered by analyzing the electrical activities of the agonist and antagonist muscles.

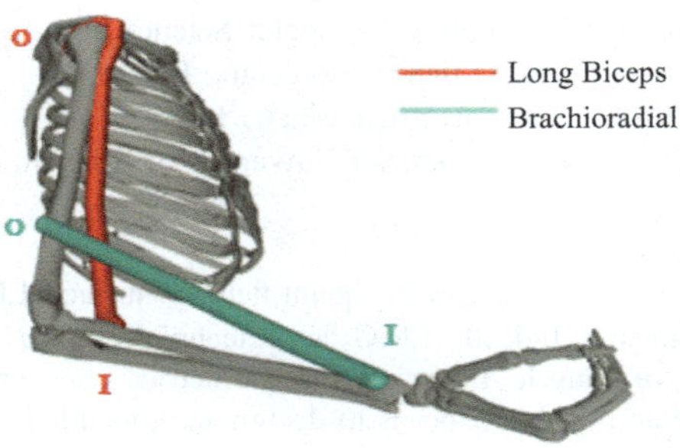

Fig. 1 Origin and insertion of muscle BIClg and BRD modeled by OpenSim software

The main purpose of my paper is to calculate isometric muscle forces using EMG signals. Firstly, we realized an electronic system allowing the acquisition of electromyography signal in real time. Then, the transmission of EMG data is realized by a Bluetooth module to the computer. In the second part, a simplified muscular model will be developed. This Hill muscle model takes into account the muscle geometrical and physiological parameters. Next, an estimation of the isometric effort will be realized by the direct dynamic approach by exploiting the muscle model previously developed. Finally, we will discuss the results obtained by our electronic system, as well as the efforts estimated by the simplified mathematical model of the muscle.

2 Materials and Methods

2.1 Experimental Materials

An embedded system for measuring EMG signals is presented in Fig. 2. This electronic system aims to detect EMG signals using surface electrodes placed directly on the muscles. These electrodes are equipped with electronic sensors that allow acquisition, filtration and rectification of the signal. To facilitate the human movement and eliminate electrical wires, we have developed a new method of data transmission based on wireless transmission (Fig. 2). In general, our system is composed of three main parts which are: (1) acquisition EMG signal, (2) signal processing, (3) display muscle forces using MATLAB GUIs.

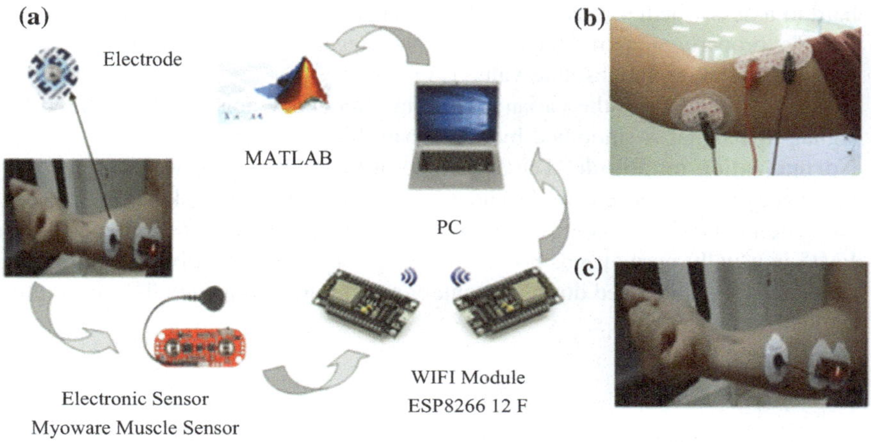

Fig. 2 **a** Experimental procedure to detect EMG, **b** detection with electrical wires, **c** detection without electrical wires

First, the EMG signal is detected by surface electrodes. Indeed, the surface electrodes appear less restrictive and much more adapted to our study. It is important that the electrodes provide good electrode-skin contact, low impedance, low noise and stable behavior over time [8]. The exact placement of the electrodes is parallel to the muscle fibers. Then, the sensor assures the acquisition of the EMG signal. In our project, we chose to work with the sensor "Myoware Muscle Sensor". For the data transmission, we used the wireless transmission technique. This is a very interesting method to avoid wires that could limit the movement of subjects during testing and cause noise. For this, Wifi module ESP8266 12F is used in our project (Fig. 2).

2.2 EMG Acquisition and Processing

EMG aims at detecting signals, which are the electrical manifestations of the excitation process resulting from the propagation of action potentials along the muscle fibers. It is detected by appropriate electrodes (surface or depth sensors). Then, it is amplified, filtered, and finally displayed on a screen [11].

EMG means: Electromyography: Examination, Electromyograph: Device, and Electromyogram: trace obtained by electromyography. The EMG signal has random variability according to several parameters: age, sex, muscle size, etc. Then, EMG signal analysis is generally performed in two domains: time domain and frequency domain. In our work, we are interested in the processing of EMG signals in the time domain. Indeed, time analysis is the most used technique to extract the relevant parameters of muscle activation [11]. This technique involves several treatment steps (Fig. 3), such as:

- **Rectification**: The values of EMG signal are either positive or negative, and it's hard to assess whether the action potential approaches or moves from the electrode [5]. To properly quantify the positive and negative phases, the EMG signal is rectified by taking its absolute value (Fig. 3).
- **Filtering**: To decrease the variance and high frequency noise of the rectified EMG signal, this signal is smoothed by a low pass filter (Fig. 3).
- **Normalization amplitude**: The amplitude of the EMG signal varies depending to the placement of the electrode and the conductivity of the skin. To be able to compare the EMG activation of different subjects in different collection conditions, EMG is usually normalized. For example, the signal can be divided by the maximum amplitude obtained during voluntary maximum contraction. This is amplitude normalization [5].

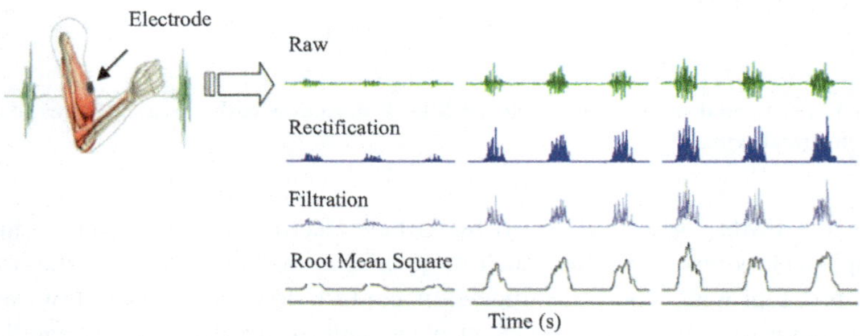

Fig. 3 Different techniques for EMG signal processing

2.3 Development Muscle Model

To measure muscle force from the EMG signal, it is necessary to develop a muscle model which takes into account the properties of skeletal muscles. This mathematical model takes the EMG signal as an input and outputs the isometric muscle force [1, 2]. The calculation procedure is divided into two steps: dynamic activation and dynamic contraction (Fig. 4).

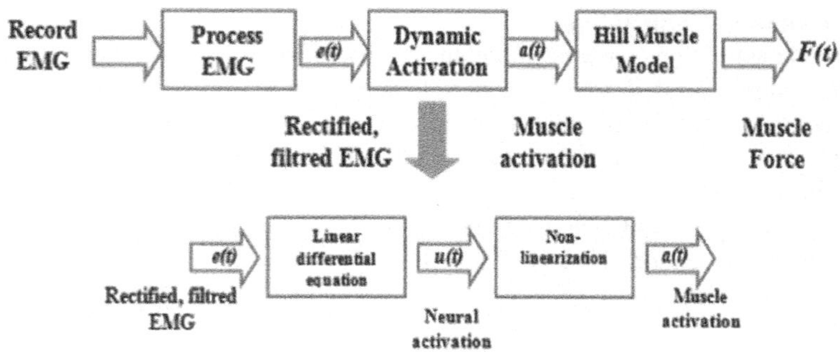

Fig. 4 Forward dynamics approach to estimate muscle force

2.3.1 Dynamic Activation

Dynamic activation is to transform the normalized, rectified and filtered EMG signal $e_i(t)$ of muscle activation $a_i(t)$ [9]. This transformation is realized in two steps [6].

Step 1: First-Order Bilinear Differential Equation

Following electrical excitation, the muscle generates a contraction response. This delayed response can be described by a first order bilinear differential equation [13]:

$$\frac{du}{dt} = (e - u)(K_1 e + K_2) \tag{1}$$

where:

$$K_1 = K_2 + \frac{1}{\tau_{acti}} \tag{2}$$

$$K_2 = \frac{1}{\tau_{desc}} \tag{3}$$

The activation and deactivation time constants: $\tau_{acti} = 0.012\,\text{s}$ and $\tau_{desc} = 0.024\,\text{s}$ [10].

Step 2: Nonlinear Model 'A-Model'

The second step is to transform neural activation $u(t)$ to muscle activation $a(t)$. To obtain a precise value of muscle Force, a non-linear relationship between neural activation and muscle activation should be considered for forces inferior than 40% of maximum voluntary contraction (MVC) and a linear relationship for forces superior than 40% of MVC [10]. This non-linear relationship between neural activation $u(t)$ and muscle activation $a(t)$ is modeled by the following equation:

$$\begin{cases} a = \alpha \ln(\beta u + 1) & 0 \le u \le u_0 \\ a = mu + c & u_0 \le u \le 1 \end{cases} \tag{4}$$

a is the muscle activation, u is the neural activation, α, β, m and c are constants.

2.3.2 Dynamic Contraction

Once muscle activations have been obtained, the next step is to calculate muscle forces. Dynamic contraction is to transform muscle activation in a mechanical force based on Hill model. This model is a muscle simplified mechanical composed of 3 basic elements which are: contractile element (CE), elastic element in series (SEE) and elastic element in parallel (PE). Next, Zajac modified Hill's model by introducing an elastic element to model tendon behavior and an angle φ called pennation angle between muscle fibers and tendon (Fig. 5) (Table 1).

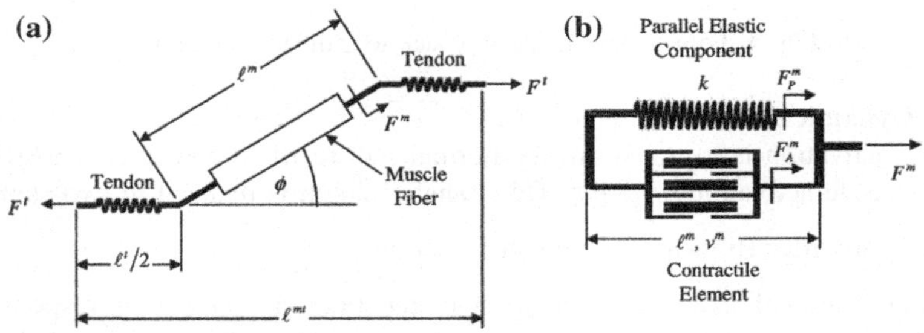

Fig. 5 **a** Schematic of muscle-tendon: muscle fiber and tendon. **b** Schematic of muscle fiber: the contractile element and parallel elastic component [4]

where:

Table 1 Parameters of model musculotendon [7]

Parameter	Description	Parameter	Description
F^T	Tendon force	L^m	Muscle length
F^m	Muscle force	L^{mt}	Muscle-tendon actuator length
ϕ	Pennation angle	L^t	Tendon length
K^T	Tendon stiffness	V^m	Muscle velocity
E^T	Tendon elasticity	V^{MT}	Musclotendon velocity

One significant advantage of Hill-type muscle models is that, in most cases, the dynamics are governed by one differential equation per muscle, making modeling using a system of muscles computationally viable [12]. Therefore, each muscle is modeled as a system of three differential equations (Eq. 5). The first line of Eq. 5 represents the activation dynamics. The second line of Eq. 5 represents the muscle contraction dynamics which is calculated based on muscle activation (Eq. 4). Finally the last line of Eq. 5 represents the muscle length.

$$\begin{bmatrix} \dot{u} \\ \tilde{F}^T(a) \\ \tilde{V}^M(a) \end{bmatrix} = \begin{cases} \frac{du}{dt} = (e-u)(K_1 e + K_2) \\ \frac{d\tilde{F}^T}{dt} = \tilde{K}_T(\tilde{V}^{MT} - \tilde{V}^M \cos(\alpha)) \\ \frac{d\tilde{L}^M}{dt} = \dot{\tilde{L}}^M \end{cases} \tag{5}$$

where u is the neural activation, e the excitation input signal, a the muscle activation, K_1 and K_2 time constants, F^T force tendon, V_{MT} musculotendon velocity, V_M contractile element velocity, α angle pennation and L^M contractile element length.

3 Results and Discussion

After calculating muscle force, we developed a graphical interface using the MATLAB software. This interface is a tool for estimating muscle force of the human upper limb using EMG signals. Our interface is divided in several windows: a main window and secondary windows. The main window contains the coordinates of the person (Name, age, sex, etc.). The secondary windows contain the different muscles of the upper limb, the muscular parameters. In our work, we performed two tests: Maximum Voluntary Contraction (MVC) and Voluntary Contraction (VC).

First test: The MVC test consists to calculate the maximum EMG signal value for each muscle (Biceps long and Brachioradial) to reduce signal noise (Fig. 6).

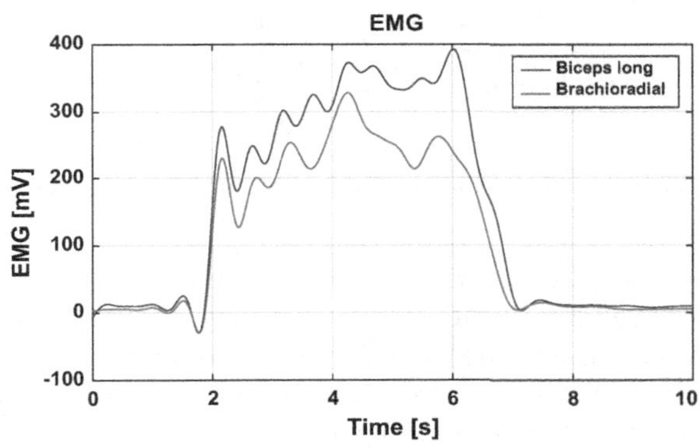

Fig. 6 Determination of the maximum value of EMG signal

Second test: The VC test is used to calculate isometric muscle force. Firstly, all EMG were divided by the peak amplitude EMG for the corresponding muscle. Then, the *e(t)* values for each muscle can then be converted to *u(t)* via Eq. 1, and the *u(t)* can then be converted to *a(t)* using Eq. 4. Finally, the muscle contraction dynamics step requires that we use Eq. 5 to estimate the force in each muscle (Fig. 7).

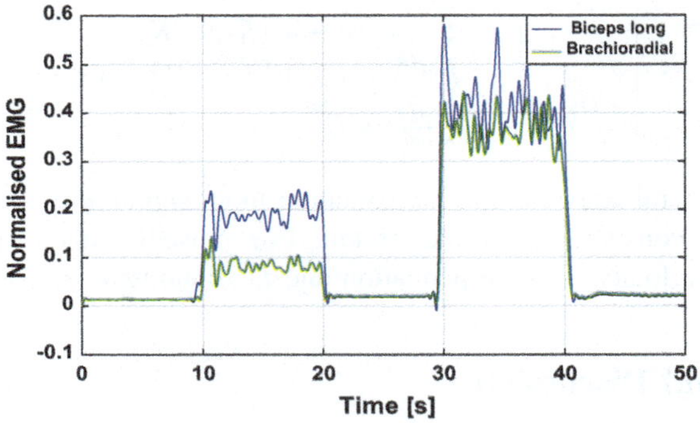

Fig. 7 EMG signals normalized of Voluntary Contraction

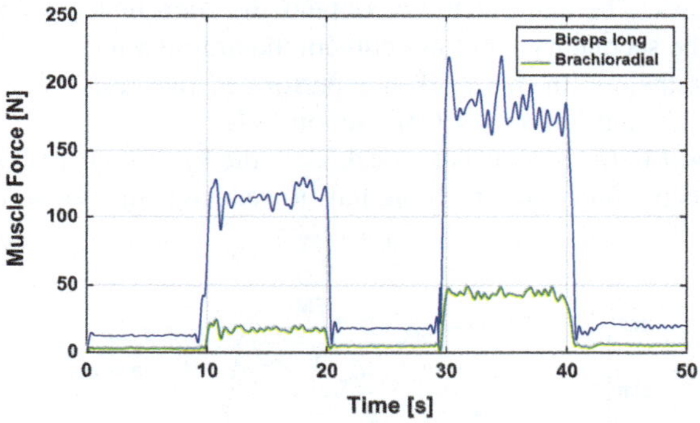

Fig. 8 Isometric muscle forces estimated by forward dynamics approach

In this methodology, we outlined a method to estimate isometric muscle forces of the upper limb obtained from EMG signals and using a forward dynamics approach. This approach uses a Hill-type model that accounts for force-length and force-velocity relationships. In the first step, we used surface electrodes placed directly on the skin. The muscle activation is measured by bipolar electrodes. Figure 2 shows the experimental setup used to obtain the bipolar EMG signals and the isometric force. Data from Biceps long (BIClg) and Brachioradial (BRD) were acquired at a 2 kHz sampling frequency. Then, the model transforms the neural activation, u (t); to muscle activation, a (t). For that, we used a nonlinearity relationship between neural activation and muscle activation. In the final step, forward dynamic approach is used to estimate muscle forces during voluntary contraction (Fig. 8).

4 Conclusion

This paper has addressed a new methodological to estimate isometric muscle forces from EMG signals. Muscle efforts are estimated by a system of differential equations based on a Hill-type muscle model. The procedure contains in two main steps: dynamic activation and dynamic contraction. Firstly, dynamic activation transforms the EMG signals to muscle activation. Then, the dynamic activation was used to estimate isometric muscle forces. To conclude, our obtained results indicate the effectiveness of our developed muscle model. Indeed, the estimated values of the muscle forces are proportional to the measured values of the EMG signal.

References

1. Bennour S et al (2012) Biomechanical model of the ankle to estimate the forces in the tendons an isometric Dorsal/Plantar flexon. J Comput Methods Biomech Biomed Eng 15:167–170
2. Bennour S et al (2013) Validation of Optimisation Model That Estimates the Musculotendinous Forces during an Isometric Extension of Knee. Comput Methods Biomech Biomed Eng 16(supl):167–169
3. Bratty PJA, Peter DL (1978) On modeling the single motor unit action potential. *IEEE Trans Biomed Eng* BME-25(3):236–43
4. Buchanan TS, Lloyd DG, Manal K, Besier TF (2004) Estimation of Muscle Forces and Joint Moments and Movements from Measurements of the Neural Command. J Appl Biomech 20 (4):367–395
5. Cao H, Jean-yves H (2010) Modélisation et Évaluation Expérimentale de La Relation Entre Le Signal EMG de Surface et La Force Musculaire Table Des Matières. *Traitement du Signal*
6. Erdemir A, McLean S, Herzog W, van den Bogert AJ (2007) Model-based estimation of muscle forces exerted during movements. Clin Biomech 22(2):131–154
7. Holzbaur KRS, Murray WM, Delp SL (2005) A model of the upper extremity for simulating musculoskeletal surgery and analyzing neuromuscular control. Ann Biomed Eng 33(6):829–840
8. Kamel M (2014) Paramétrisation Des Potentiels D'action D'une Unité Motrice Détectés Non-Invasivement
9. Lloyd David G, Besier Thor F (2003) An EMG-driven musculoskeletal model to estimate muscle forces and knee joint moments in Vivo. J Biomech 36(6):765–776
10. Manal Kurt, Buchanan Thomas S (2003) A one-parameter neural activation to muscle activation model: estimating isometric joint moments from electromyograms. J Biomech 36 (8):1197–1202
11. Merletti R, Rainoldi A, Farina D (2001) Surface electromyography for noninvasive characterization of exercise and sport sciences reviews. Exerc Sport Sci Rev 29(1):20–25
12. Seth A, Sherman M, Reinbolt JA, Delp SL (2011) OpenSim: a musculoskeletal modeling and simulation framework for in silico investigations and exchange. Procedia IUTAM 2:212–232
13. Piazza Stephen J, Delpt Scott L (1996) The influence of muscles on knee flexion the swing phase of gait. Health Med 6(8):2899–2901

14. Trappe S et al (2003) Single muscle fibre contractile properties in young and old men and women. J Physiol 552(1):47–58
15. Zierath JR, John AH (2004) Skeletal muscle fiber type: influence on contractile and metabolic properties." PLoS Biology 2(10)

Multiscale Approach from Nanoscale to Macroscale to Identify Orthotropic Properties of Trabecular Bone Based on FEM

Houda Khaterchi(✉) and Hédi Belhadjsalah

Laboratoire de Génie Mécanique, Ecole Nationale d'Ingénieurs de Monastir, Université de Monastir, Avenue Ibn Eljazzard, Monastir 5019, Tunisie
houda_enim@yahoo.fr

Abstract. The bone is a hierarchically structured material with mechanical properties depending on its architecture at all scales. It's important to take account the impact of Water which plays a significant role in the bio-mineralization process and serves as a plasticizer, enhancing the toughness of bone. In this study, a trabecular bone multiscale model based on finite element analysis was developed to link scales from nano to macroscale in order to predict the orthotropic properties of bone at different structural level. An inverse identification algorithm is used in order to identify the orthotropic properties. Furthermore, the effect of water is incorporated. Good agreement is found between theoretical and experimental results.

Keywords: Multiscale approach · Finite element analysis · Trabecular bone · Orthotropic material

1 Introduction

Bone tissue is a natural composite material consisting of an organic phase, an inorganic phase and water. The bone is a hierarchically structured material with mechanical properties depending on its architecture at all scales.

In human bodies, the bone consists of three types: Cortical, trabecular and marrow. The mechanical influence of the marrow component is trifling with respect to the bone matrix [1, 3, 4].

The following structural scales can be distinguished in the hierarchical structure of bone: macroscale, mesoscale, microscale, sub-microscale, nanoscale, and sub-nanoscale [6, 7, 11]. It's worth mentioning that the three lower levels of the two types of the bone are the same.

Various analytical and computational models were proposed to predict elastic properties of trabecular bone at these different structural scales. More recent studies incorporated the effect of water [7].

It has been shown that the mechanical properties of the bone vary at different structural scales [13]. Therefore, in order to find the macroscopic properties of the bone, it is important to consider all of these scales [8, 10]. Many researchers have been

N. Aifaoui et al. (Eds.): CMSM 2019, LNME, pp. 193–203, 2020.
https://doi.org/10.1007/978-3-030-27146-6_21

investigated in this field, but the behaviour of the bone materials are modeled at different scales separately.

Structurally, bone is considered like a composite material with a complex structure.

To predict the orthotropic properties at different scales, Hamed et al. [6] used homogenization approach. Elastic properties of trabecular bone are calculated at each structural level, from nanoscale to mesoscale. In the analysis, results from a lower level are used as inputs for a higher level. In the same way, Vaughan et al. [15] proposed a three scale homogenization scheme to estimate the effective properties of trabecular and cortical bone, based on finite element models.

Barkaoui et al. [2] used neural network computation and homogenization method to predict analytically the effective elastic constants of cortical bone by modeling its elastic response at these different scales, ranging from the nanostructural to mesostructural levels.

Khaterchi et al. [11] proposed a multiscale approach for the characterization of mechanical properties of trabecular bone from sub-nanoscale and microscale including the water's effect. To identify the orthotropic properties, an inverse identification algorithm and FE simulation are used.

The objective of this study is to model bone at different scales, identify the orthotropic properties and compare the results obtained from the proposed multiscale approach [11] including the water's effect with experimental and analytical data reported in literature.

The most important originalities of the current study are:

- The implementation of the basic constituents of bone structure (mineral, collagen state, mechanical properties, porosity and water's effect)
- Using of mechanical tests to predict the orthotropic properties of bone at different scales.

The prediction of orthotropic properties of parameterized unit cells using FEM is the advantage of this study and it can be easily incorporated in the finite element code to model the whole bone structure [19].

2 Method Description

In this section, we present the elementary bone composition and the proposed multi-scale approach. At each scales, a description of the proposed EVR model and the method which are used in this approach will be established.

2.1 Bone Composition

We distinguish three elementary bone compositions:

- Collagen: At the lowest hierarchical scale, collagen molecules can be viewed as rods of about 300 nm long and 1.5 nm in diameter.

- Mineral: The mineral phase is composed of impure hydroxyapatite crystals. Those crystals are plate-like shapes. The size of the mineral plates varies according to the bone type.
- And water which represent the third major component in bone.

2.2 Multiscale Model of Trabecular Bone

In human bodies, the bone consists of three types: Cortical, trabecular and marrow. The mechanical influence of the marrow component is trifling with respect to the bone matrix. In this work, we will consider only the trabecular bone.

A multiscale model is used to characterise the orthotropic behaviour of trabecular bone at a number of different length scales [11]. The approach defines a number of representative volume element (RVEs) which describe composition of bone in different scale. Mechanical tests are employed at each scale to link nano and macroscale (Fig. 1).

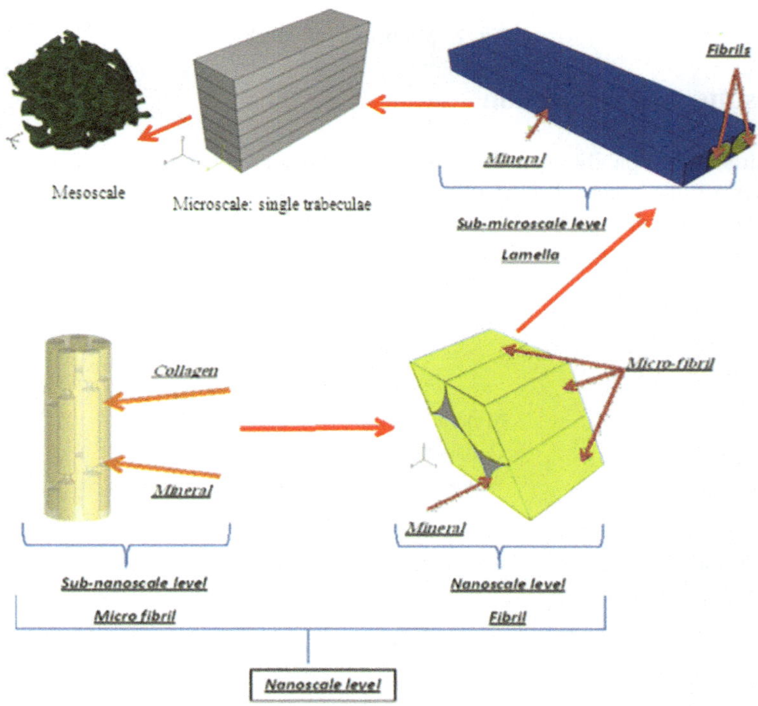

Fig. 1 Hierarchical structure of trabecular bone

2.3 Multiscale Approach

The mutliscale method consists to prepare an algorithm (Fig. 2) composed by different steps which represent the structure of trabecular bone at each scale. The inputs are the mechanical properties of microfibril at sub nanoscale and the outputs are the orthotropic properties of bone at mesoscale.

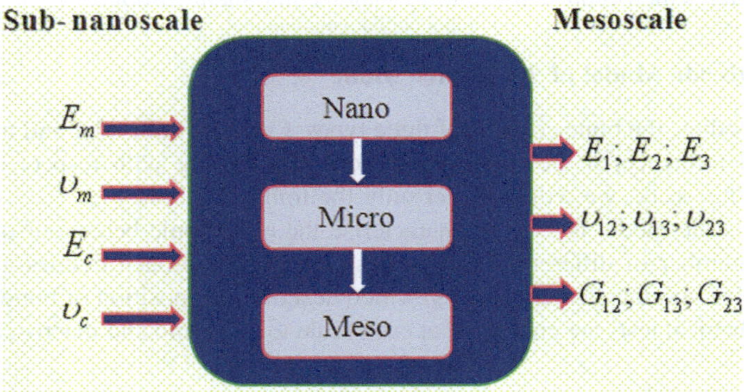

Fig. 2 Multiscale approach

3 Results and Discussion

To illustrate the capabilities of the approach proposed, one combination of mechanical properties of the microfibril (Collagen, mineral and water) is selected which are summarized in Table 1.

Table 1 Selected inputs

Materials	Elastic properties	
	E (GPa)	υ
Mineral	114	0.23
Collagen	1.5	0.28
Water	2.3	0.49

Voigt and Reuss bounds are calculated for collagen and mineral composite using the following relation:

$$E_{CW} = (1 - V_W)E_C + V_W E_W \tag{1}$$

$$\frac{1}{E_{CW}} = \frac{(1 - V_W)}{E_C} + \frac{V_W}{E_W} \tag{2}$$

$$E_{MW} = (1 - V_W)E_M + V_W E_W \tag{3}$$

$$\frac{1}{E_{MW}} = \frac{(1 - V_W)}{E_M} + \frac{V_W}{E_W} \tag{4}$$

With:

E_{CW}, E_{MW} respectively Young Modulus of collagen-water mixture and mineral-water mixture

$V_W = 12\%$, Water volume fraction

E_C, E_M respectively Young Modulus of Collagen and mineral.

The average values of these bounds are selected for the elastic constant of the collagen-water composite and mineral-water composite.

Using the parameters in simulations chosen in Table 1, a FE simulation is generated. The method gives in one step the orthotropic properties of the trabecular bone in different scales (Microfibril, Fibril, lamella, single trabeculae, trabecular network).

3.1 Microfibril

In sub-nanoscale, an ERV is generated [11] and to obtain the orthotropic properties of the microfibrils, the tensile and compression tests are used. Thus, at the sub-nanostructural level, the modulus being determined could be considered acceptable to that determined by Hamed et al. [6] (Fig. 3).

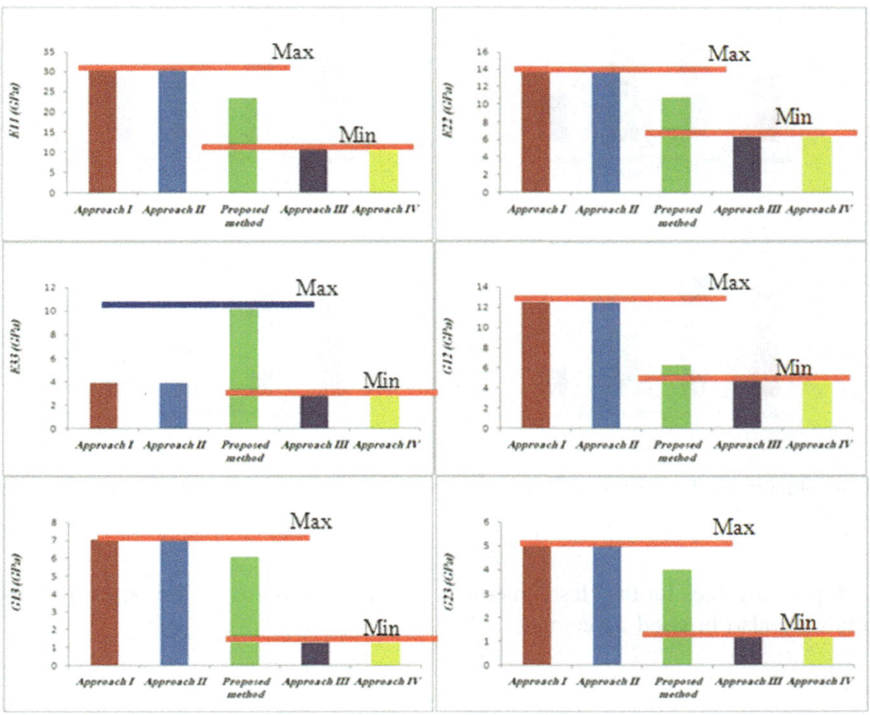

Fig. 3 orthotropic properties at sub nanoscale in comparison with Hamed et al. [6]

3.2 Fibril

The tensile and shear tests are employed to evaluate the orthotropic properties of the fibril [11].

In this scale, the corresponding engineering elastic constants are compared in first hand with data obtained in Hamed et al. [6] (Fig. 4) and in second hand with experimental and analytical data reported in literature (Fig. 5).

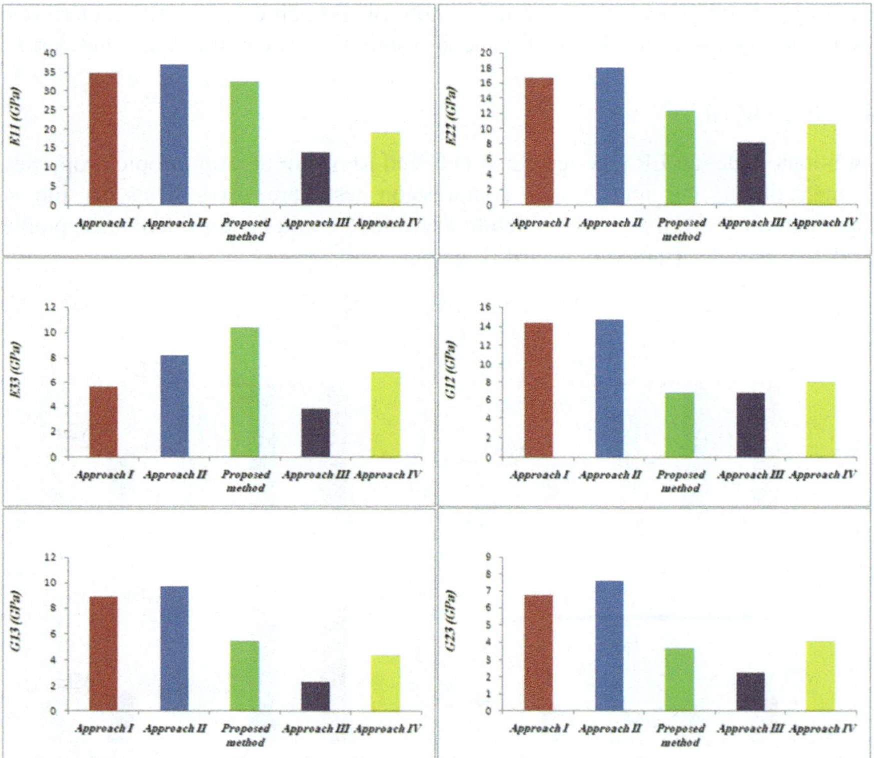

Fig. 4 Orthotropic Properties of fibril in comparison with approach's results listed in Hamed et al. [6]

Reported values for the elastic modulus and shear modulus of fibril level from these studies are also in good agreement.

3.3 Lamella

Mechanical tests are used to evaluate the orthotropic properties of lamella [11]. Table 2 represents the results obtained in this work in comparison with the work of Yoon and Cowin [17, 18]. In Table 3, some experimental results are compared with our work.

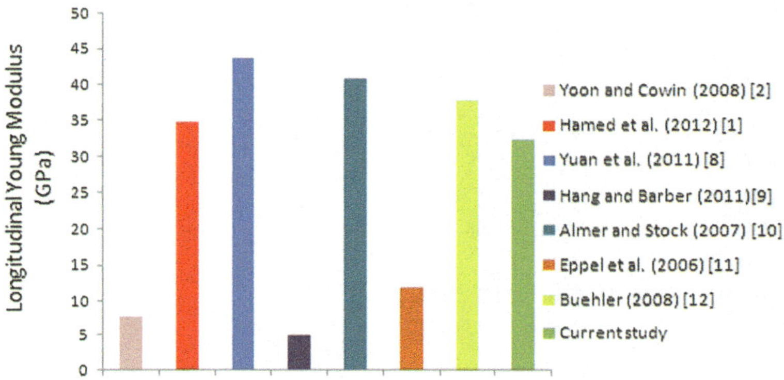

Fig. 5 Orthotropic Properties of fibril in comparison with different study

It can be shown that the value of current study is in good agreement with others results.

Table 2 Orthotropic properties compared with Yoon and Cowin's results [17, 18] (GPa)

	Current study	Yoon and Cowin [17]	Yoon and Cowin [18]
E_1	15,64	19.0	16.9
E_2	15,23	21.3	19.0
E_3	27,7	24 5	22.3
V_{12}	0,39	0.320	0.326
V_{13}	0,12	0.252	0.246
V_{23}	0,19	0.261	0.254
G_{12}	4,8	7.8	7.1
G_{13}	2,87	7.8	7
G_{23}	6,62	9.1	8.3

Table 3 Orthotropic properties compared with results in literature

	Current study	Rho et al. (1997)	Rho et al. [14]	Fan et al. (2002)	Hengsberger et al. (2002)	Hengsberger et al. (2002)
E1	15.64	NR	NR	16.6±1.5	NR	NR
E2	15.23	NR	NR	17.0±2.2	NR	NR
E3	27.7	22.5±1.3	21.8±2.1	25.1±2.1	27.2±3.3	15.8±4.8

	E11 (GPa)	E22 (GPa)	E33 (GPa)
Barkaoui et al. [2]	15,6	16,47	27,6
Martinez et al. [12]	17,2	19,7	22
Yoon and Cowin [17]	16,4	18,7	22,8
Zysset et al. [20]	19,1	–	–
Hamed et al. [7]	17,91	11,88	NA
Hoffler et al. [9]	15,11	–	–
Weiner and Wagner [16]	33,6	24,3	NA
Current study	15,64	15,23	27,7

3.4 Single Trabeculae

In this scale, the Young modulus constants are compared with data obtained in Hamed et al. [6] (Fig. 6). Reported value for the longitudinal elastic modulus of trabeculae in this work is compared with experimental results obtained from literatures and they are also in good agreement (Table 4).

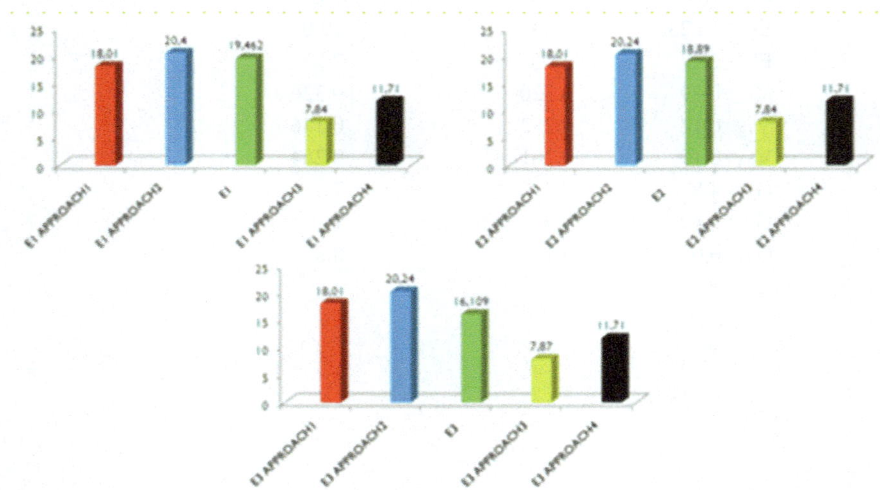

Fig. 6 Young modulus in three directions (1, 2, and 3) compared with data obtained by Hamed et al. [6]

Table 4 Orthotropic properties compared with results from Hamed et al. [6]

	Module d'Young (GPa)				
		Rèsultats expèrimentaux			
Tissu	Prèsent travail	Mesures	Type d'os	tests	Rèfèrence
Trabècule (Travèe)	19.46	11.4 ±5.6	fèmur Humain (wet)	Nanoindentation	Zysset et al. [20]
		12.8 ±1.2	vertèbre d'un Rat (dry)	Nanoindentation	Zysset et al. (1998)
		14.8 ±1.4	tibia Humain (wet)	Ultrason	Rho et al. [13]
		17.5 ±1.12	fèmur Humain	Ultrason	Tumor et al. (1999)
		18.0 ±2.8	fèmur Humain	FEM	Follet et al. (2011)
		18.14 ±1.7	fèmur Humain (dry)	Nanoindentation	Tumor et al. (1999)
		19.9 ±2.5	fèmur Humain (wet)	Ultrason	Jorgensen et al. (2002)
		18.0 ±2.8	Col de fèmur	Test de compression—MEF	Bayraktar et al. (2004)

3.5 Trabecular Network

Same works is done in this scale, the orthotropic properties are obtained using mechanical tests (tensile and shear tests) and they are compared with results obtained from experimental work [5]. Good agreement between the current results for this scale and other studies.

Mechanical properties of trabecualr bone by compression test [5]		
	Module d'Young en compression (MPa)	Contrainte à la rupture (MPa)
Tibia (partie distale)	1.4–500	0.2–45
Fémur (partie distale)	7.6–2942	0.98–22.5
Fémur (partie proximale)	20.68–9800	0.21–16.2
Vertèbre	1.1–428	0.06–15

Orthropic properties of trabecular bone (current study) (GPa)								
E_1	E_2	E_3	G_{12}	G_{13}	G_{23}	υ_{12}	υ_{13}	υ_{23}
6.22	4.855	3.86	3.64	2.818	2.89	0.08	0.03	0.1

4 Conclusion

In this work, a multiscale approach is developed to link between sub-nano, nano, sub-microscale, micro and mesoscale in order to understand the relationship between the mechanical behaviour in different scale including the water's effect. The effective mechanical properties at each scale are compared with experimental and analytical data reported in literature. Good agreement was found between our results and theoretical and experimental results. Finally, some other important variables including geometry and porosity were not included in the present research and will remain interesting future research topics.

References

1. Almer JD, Stock SR (2007) Micromechanical response of mineral and collagen phases in bone. J Struct Biol 157:365–370
2. Barkaoui A, Chamekh A, Merzouki T, Hambli R, Mkaddem A (2013) Multiscale approach including microfibril scale to assess elastic constants of cortical bone based on neural network computation and homogenization method. Int J Numer Meth Biomed Eng. https://doi.org/10.1002/cnm.2604
3. Buehler MJ (2008) Nanomechanics of collagen fibrils under varying cross-link densities: atomistic and continuum studies. J Mech Behav Biomed Mater 1(1):59–67
4. Eppell SJ, Smith BN, Kahn H, Ballarini R (2006) Nano measurements with micro-devices: mechanical properties of hydrated collagen fibrils. J R Soc Interface 3(6):117–121
5. Goldstein SA (1987) The mechanical properties of trabecular bone: dependence on anatomic location and function. J Biomech 20(11–12):1055–1061
6. Hamed E et al (2012) Multiscale modeling of elastic properties of trabecular bone. J R Soc Interface 9(72):1654–1673
7. Hamed E, Lee Y, Jasiuk I (2010) Multiscale modeling of elastic properties of trabecular bone. Acta Mech 213(1–2):131–154
8. Hang F, Barber AH (2011) Nano-mechanical properties of individual mineralized collagen fibrils from bone tissue. J R Soc Interface 8:500–505
9. Hoffler CE, Moore KE, Kozloff K, Zysset PK, Goldstein SA (2000) Age, gender, and bone lamellae elastic moduli. J Orthop Res 18:432–437
10. Khaterchi H, BelHadjSalah H (2015) A comparative study of the mechanical properties of hierarchical trabecular bone with other approaches and existing experimental data. J Biomim Biomater Biomed Eng 23:76–84
11. Khaterchi H, Belhadjsalah H (2013) A three scale identification of orthotropic properties of trabecular bone. In: CMBBE, vol 16, no, si., pp 272–274
12. Martínez-Reina J, Domínguez J, García-Aznar JM (2011) Effect of porosity and mineral content on the elastic constants of cortical bone: a multiscale approach. Biomech Model Mechanobiol 10:309–322
13. Rho JY, Kuhn-Spearing L, Zioupos P (1998) Mechanical properties and the hierarchical structure of bone. Med Eng Phys 20:92–102
14. Rho JY, Zioupos P, Currey JD, Pharr GM (2002) Microstructural elasticity and regional heterogeneity in aging human bone examined by nano-indentation. J Biomech 35:161–165

15. Vaughan T.J, McCarthy C.T, McNamara L.M. 2012. A three scale finite element investigation into the effects of tissue mineralization and lamellar organization in human cortical and trabecular bone. Journal of the mechanical behaviour of biomedical materials, 50–62
16. Weiner S, Wagner HD (1998) The material bone: structure mechanical function relations. Annu Rev Mater Res 28:271–298
17. Yoon YJ, Cowin SC (2008) An estimate of anisotropic poroelastic constants of an osteon. Biomech Model Mechanobiol 7:13–26
18. Yoon YJ, Cowin SC (2008) The estimated elastic constants for a single bone osteonal lamella. Biomech Model Mechanobiol 7:1–11
19. Yuan F, Stock SR, Haeffner DR, Almer JD, Dunand DC, Brinson LC (2011) A new model to simulate the elastic properties of mineralized collagen fibril. Biomech Model Mechanobiol 10:147–160
20. Zysset PK, Guo XE, Hoffler CE, Moore KE, Goldstein SA (1999) Elastic modulus and hardness of cortical and trabecularbone lamellae measured by nanoindentation in the human femur. J Biomech 32:1005–1012

Modeling of a Fatigue Test Performed on a Trans-Tibial Prosthetic Socket Made of Natural Fiber

Mankai Wahbi[1,4](✉), Ben Smida Brahim[1,2], Chafra Moez[3], Ben Cheikh Ridha[4], and Alzoubi Jomah[5]

[1] Al Ahsa College of Technology, Technical and Vocational Training Corporation, Riyadh, Saudi Arabia
mankaiw@act.edu.sa, bbensmida@tvtc.gov.sa

[2] Université de Tunis El Manar, Ecole Nationale d'Ingénieurs de Tunis, Laboratoire de Génie Civil, BP 37 le Belvédère 1002, Tunis, Tunisia

[3] Applied Mechanics and Systems Research Laboratory, Tunisia Polytechnic School, University of Cartage, Tunis, Tunisia
moez.chafra@ept.rnu.tn

[4] Materials, Optimization and Energy for Sustainability, National Engineering School, University of Tunis El Manar, Tunis, Tunisia
ridha.bencheikh@enit.rnu.tn

[5] Department of Mechanical Engineering, College of Engineering, King Faisal University. al-Ahsa, Al Hofuf, Saudi Arabia
Jalzoubi@kfu.edu.sa

Abstract. This current study is considered as a continuation of previous research in which static and cyclic tests were performed in accordance with the ISO 10328 standard on trans-tibial sockets made from natural fiber in order to characterize their mechanical performance. The experimental findings of the fatigue test have revealed a viscoelastic behavior of the composite material of the tested socket. In addition, it has been proposed to model the fatigue test by the Burgers model since it better fits the experimental results presented by fatigue curves. Two main objectives are distinguished. The first one is to identify the parameters of the Burgers model; the second one is to evaluate the natural frequency of the bench structure: Socket, Artificial stump and Test device Assembly used to perform the mechanical tests. The natural frequency, which is deduced from an electro-mechanical analogy method, is then used to optimize the design of the test device.

Keywords: Trans-tibial socket · Fatigue test · Burgers model · Electro-Mechanical analogies · Natural frequency

N. Aifaoui et al. (Eds.): CMSM 2019, LNME, pp. 204–213, 2020.
https://doi.org/10.1007/978-3-030-27146-6_22

1 Introduction

Prostheses allow the amputee to resume his activity and reintegrate social life. In general, trans-tibial prosthesis is composed of three elements: the foot, the pylon and the socket. The socket is the part that connects the prosthesis to the residual limb; it is made of a composite material reinforced with synthetic fibers such as carbon or glass fibers [3]. It is necessary to perform static and cyclic tests on the sockets when new materials are used, such as natural fibers [1] instead of conventional fibers [3]. So far, there is not yet a specific standard for these tests. But the closest and most widely used standard for this purpose is the standard ISO 10328.

In subsequent research work, a cyclic test is performed on an acrylic resin socket reinforced with natural Alfa fibers [2]. The results of this test are represented by curves that follow a deflection according to the number of cycles. The shape of these curves showed a viscoelastic behavior of the socket material tested [7].

In the literature there are several models that describe the viscoelastic behavior of a material such as the Maxwell model or the Voigt-Kelvin model. But the model that best fits the results of the dynamic test is the Burgers model. This model consists of assembling the Maxwell and the Voigt-Kelvin models in series [10].

This study deals with a dynamic test, vibratory phenomena such as resonance could be an interesting issue to be analyzed [8]. Although the standard ISO 10328 did not mention this point in its cyclic test recommendations, it seems important to address this problem. In fact, the vibratory behavior of an insulated socket is totally different from that of a socket in the presence of an external medium such as the test device [6]. A vibration analysis of prosthesis for the through knee amputation has shown that the natural frequency for the first mode of an old prosthesis was 33.2 Hz and for a new designed prosthesis was 46.8 Hz [6].

The main objective of this work is to determine the natural frequency of the test bench used for the cyclic test and to compare it with results of the reference [6] which presents a similarity to our study, in order to propose eventual improvements to the design of the test bench, which could optimize the conditions of the cyclic test on the socket. To achieve this goal, it was necessary to identify the four parameters of the Burgers model and calculate the expression of the mechanical impedance function deduced by an electro-mechanical analogy [4].

2 Cyclic Test of Socket According to ISO 10328

The cyclic test of a socket consists of mounting the latter on a test bench as shown in Fig. 1. The test bench consists of two parts (Fig. 1).

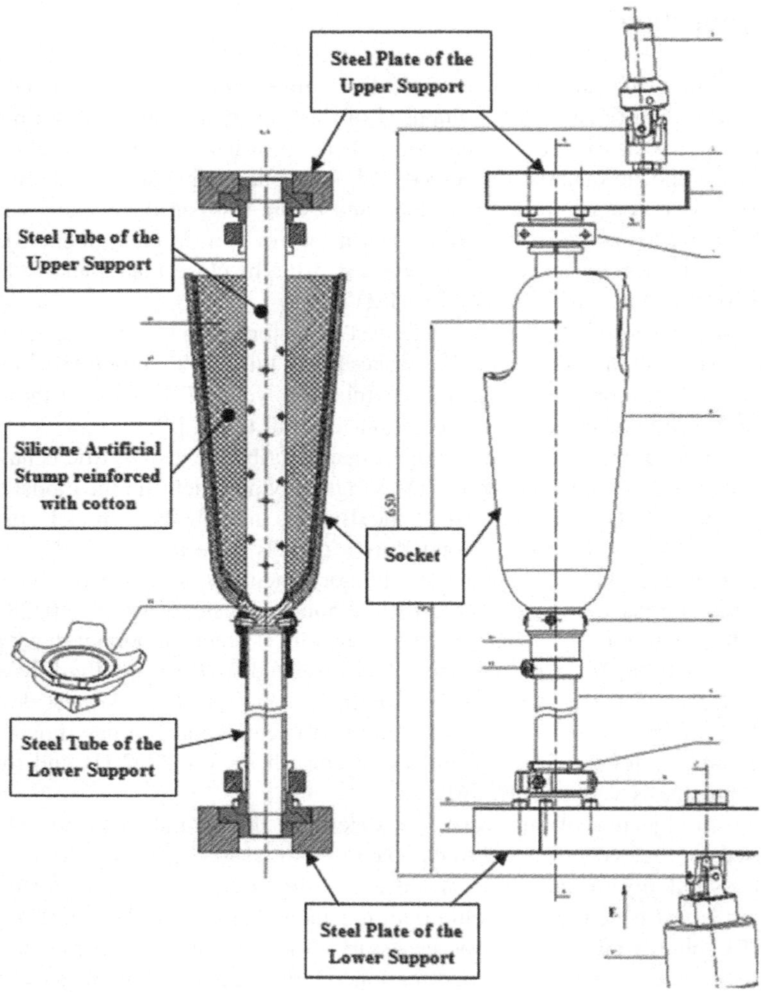

Fig. 1 The test bench

The first one is the upper support formed of a steel plate and a steel tube on which is embedded the artificial stump. The second part is the lower support consisting of a steel plate, a steel tube and the socket to be tested. In all the following, the socket is considered as an element belonging to the test bench. The assembly obtained is, then, fixed by the two jaws of a hydraulic fatigue machine. The upper jaw is equipped with a force sensor and it is unmovable while the lower jaw is connected to a hydraulic jack and is movable. The latter transmit to the test bench the excitation force F (t) whose expression is given by Eq. (1).

$$F(t) = F_m + (F_c/2).sin(\omega.t) \quad (1)$$

where: F_m is the average strength, F_c is the strength variation extent and ω is the pulsation.

This force is transmitted to the socket by means of the artificial stump made of a composite of silicone matrix reinforced with cotton fibers as shown in Fig. 1. Other research used a plaster stump to mount the socket on the bench [9]. It seems to us that the use of an elastic material is better adapted to distribute the load on the whole internal surface of the socket, which makes it possible to better simulate reality. To ensure that this new method does not distort the cyclic test, a theoretical study is conducted to estimate the natural frequency of the test bench where only the artificial stump is considered as an elastic, homogeneous and isotropic material, the other components of test bench being supposed to be rigid bodies, even the socket. This theoretical study is presented in the following section.

The cyclic test is carried out according to the recommendations of the ISO 10328 standard with the loading condition II at the loading level A100 with a frequency equals to 3 Hz (Standard ISO 10328).

3 Theoretical Study

The test bench is schematized by a discrete three-dimensional (3D) modeling where the action of the artificial stump on the socket is replaced by two compression springs, a torsion spring and a ball joint at the bottom of the socket [7]. This 3D model has three degrees of freedom. A preliminary study shows that the loading condition II, used in our case, is favorable to be modeled 1D with an error about 10%. If this error is accepted, the test bench can be modeled with a plane model with a single degree of freedom [7]. Based on this model a dynamic study is conducted which has led to a characteristic equation of motion of a mass-spring system from which the natural frequency $\mathbf{f_n}$, the rigidity **k** and the equivalent mass **M** of the test bench are deduced. All these values are summarized in Table 1.

4 Electro-Mechanical Analogy

Figure 2a presents the results of a cyclic test on an acrylic resin socket reinforced with natural Alfa fibers. The shape of the curves in Fig. 2a shows a visco-elastic behavior of the material constituting the socket. The Burgers model which best describes the obtained experimental data [10] is constituted of four parameters: two spring stiffness values ($\mathbf{k_1}$ and $\mathbf{k_2}$) and two coefficients of viscous damping ($\mathbf{c_1}$ and $\mathbf{c_2}$) [8]. These parameters are insufficient to describe the dynamic behavior of the test bench. It is necessary to add the equivalent mass of the test bench determined by the theoretical study. Thus, the model equivalent to the test bench is shown in Fig. 3.

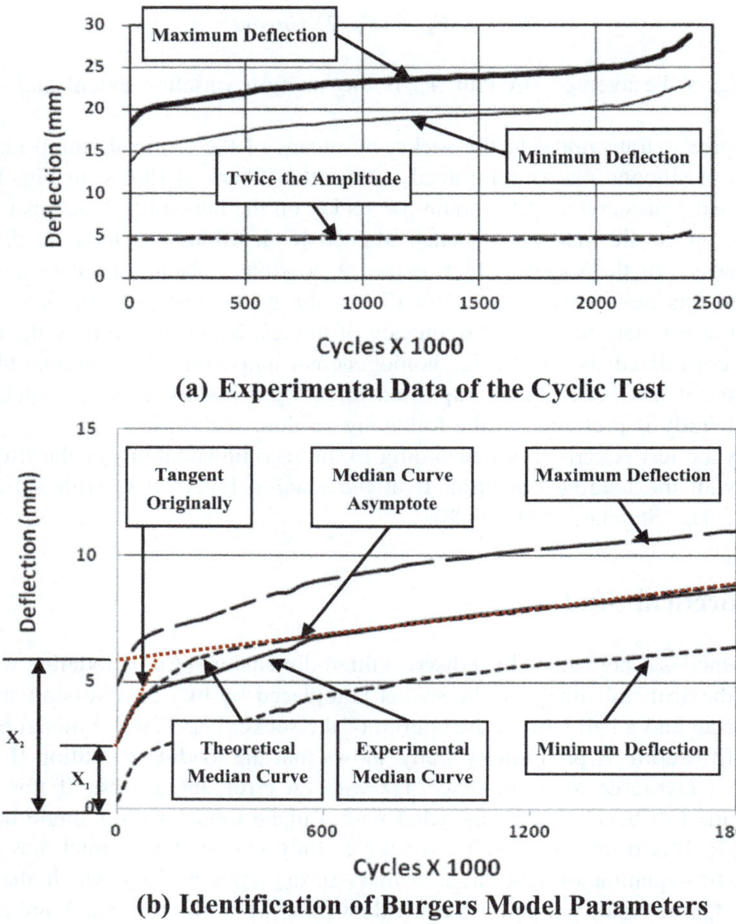

Fig. 2 Cyclic test of a socket made of natural Alfa Fiber

By a direct electro-mechanical analogy the mechanical model of Fig. 3 is replaced by an electric model [4]. The analogy of this study serves not only to determine the natural frequency of the test bench, but also to calculate the expression of the mechanical impedance function **Z** which will be used to calculate the oscillation amplitude of the cyclic test versus the force $\mathbf{F_c}$, the frequency **f**, the equivalent mass **M** and the four parameters of the Burgers model.

5 Identification of Burgers Model Parameters

The identification of the four parameters of the Burgers model is obtained graphically from the median curve of the two curves represented in Fig. 2b. The parameters $\mathbf{k_1}$, $\mathbf{k_2}$, $\mathbf{c_1}$ and $\mathbf{c_2}$ are successively calculated by the respective Eqs. (2), (3), (4) and (5) [10]:

$$k_1 = F_m/x_1 \tag{2}$$

$$k_2 = (k_1.F_m)/(k_1.x_2 - F_m) \tag{3}$$

$$c_1 = F_m/s_1 \tag{4}$$

$$c_2 = (c_1.F_m)/(c_1.s_2 - F_m) \tag{5}$$

where: $\mathbf{x_1}$ is the deflection at t = 0 measured on the median curve, $\mathbf{x_2}$ is the deflection at t = 0 measured on the median curve asymptote, $\mathbf{s_1}$ is the slope of the median curve asymptote and $\mathbf{s_2}$ is the slope of the median curve at t = 0, as presented in Fig. 2b.

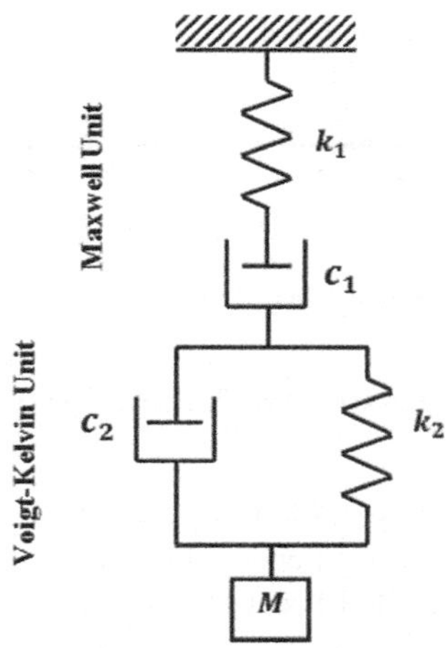

Fig. 3 Burgers Model

Once these four parameters are determined, it is necessary to compare the amplitude of oscillation calculated from the parameters of Burgers and the equivalent mass to that obtained experimentally. When there is a discrepancy, the two experimental curves must be recalibrated and the parameters k1 and k2 taken again to satisfy the equality condition between the calculated and the experimental amplitudes. The retained values of the Burgers model parameters are presented in Table 1.

6 Design Optimization of the Test Bench

6.1 Determination of the Natural Frequency of the Test Bench Based on the Burgers Model

By plotting the inverse of the impedance function, which is known as the Mobility function, versus the frequency, a maximum is reached for a frequency of 23.2 Hz as shown in Fig. 4 by the dashed curve, this frequency corresponds to the natural frequency of the test bench. If we refer to the results of the reference [6] which has a certain similarity with our case study, we note that the natural frequency of the test bench is lower than the two values presented by Mohsen et al. [6]. With the intention of approaching the results of Mohsen et al. [6], the eventual solutions as well as their results to increase the natural frequency are presented in Sect. 6.2, while their consequences on the running of the cyclic test are discussed in Sect. 6.3.

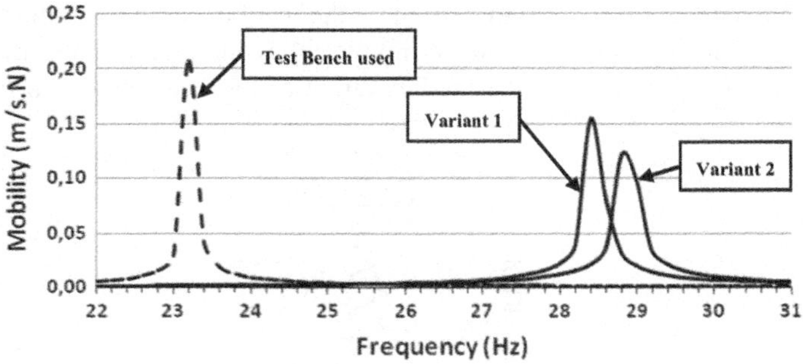

Fig. 4 Determination of the natural frequency of different test benches

6.2 Possible Solutions to Increase the Natural Frequency of the Test Bench

To increase the natural frequency of the test bench there are two alternatives. The first is to increase the stiffness of the test bench while the second is to reduce its equivalent mass. The first proposal is rejected because it consists in stiffening the socket or the artificial stump. As the socket is the tested element, it is unapproachable; while the stiffening of the artificial stump does not bring improvement. Then, only the second alternative, which proposes to reduce the equivalent mass, is a viable alternative.

Table 1 Parameters of theoretical and experimental models

	Theoretical study		Burgers model			
		Test bench used			Variant 1	Variant 2
Parameters	M (kg)	11.41	M (kg)	11.41	7.6	7.38
	k (kN/m)	191000	k_1 (kN/m)	243		
			k_2 (kN/m)	189		
			c_1 (N.s/m)	1.25 E + 11		
			c_2 (N.s/m)	6.58 E + 09		
Natural frequency	f_n (Hz)	652	f_n (Hz)	23.2	28.4	28.8

In fact, apart from the socket and the artificial stump, all the other constituents of the test bench are made of steel. If this material is replaced by another whose density is lower than that of steel, the equivalent mass will be reduced and the natural frequency will increase. We can try a second improvement by reducing the length of the test bench [5].

Referring to Fig. 4, where the mobility is plotted according to the frequency for the three cases, namely the steel test bench, the aluminum test bench (**Variant 1**) and the short aluminum test bench (**Variant 2**), we can see an increase in the natural frequency of the two proposed variants compared to the natural frequency of the test bench used. We also note that Variant 2 did not bring a significant improvement in natural frequency compared to Variant 1. This last observation is quite logical because the equivalent mass of the short test bench has not decreased much compared to the equivalent mass of the long aluminum test bench, as shown in Table 1, where the different natural frequencies are also indicated.

6.3 The Effect of the Reduction of the Equivalent Mass on the Running of the Cyclic Test

In Fig. 5, the amplitude of oscillation is plotted for the low frequencies ranging from 0.2 to 15 Hz for the three cases: the test bench used and the two variants proposed. We note that the two curves of the latter are very close and that they are below the curve of the test bench used. However, they approach the latter and even merge with it, when the frequency tends towards 0 Hz.

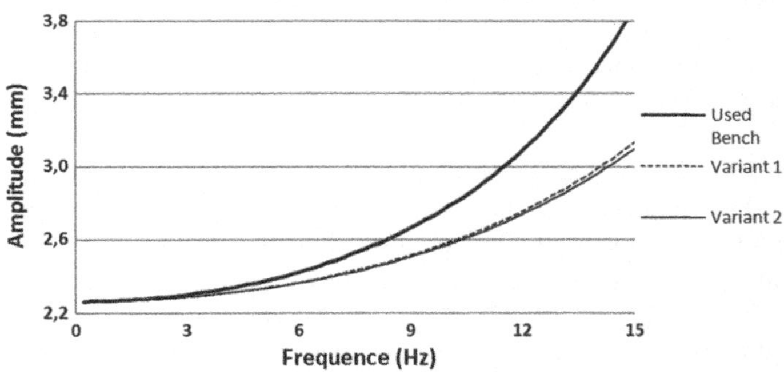

Fig. 5 Amplitude versus low frequencies for the three cases

For different frequencies values, the oscillation amplitudes as well as the relative difference in percentage calculated on the base of the amplitude of the test bench used, are presented in Table 2 for the two variants proposed. By analyzing Table 2, we find that the improvement in the conduct of the cyclic test, based on the amplitude, is not significant at the excitation frequency 3 Hz, but by increasing the frequency it becomes more significant (Table 2).

Table 2 Amplitude improvement

	Frequency (Hz)	3	6	9	12	15
Amplitude (mm)	Used bench	2.3003	2.4236	2.6613	3.0849	3.8786
	Variant 1	2.2873	2.3671	2.5131	2.7507	3.1313
	Variant 2	2.2866	2.3639	2.5051	2.7336	3.0969
Relative difference %	Variant 1	−0.562	−2.33	−5.57	−10.83	−19.27
	Variant 2	−0.595	−2.46	−5.87	−11.39	−20.16

At constant oscillation amplitude, the excitation frequency of the two variants of the test benches is greater than that of the test bench used, as can be verified in Fig. 5, which will result in a decrease in the test duration. This represents a second major improvement achieved by the reduction of the test bench mass.

7 Conclusion

The identification of the parameters of the Burgers model as well as the determination of the equivalent mass of the used test bench made it possible to calculate the natural frequency of the latter evaluated at 23.2 Hz. To increase the natural frequency of the test bench we propose to reduce the equivalent mass of the test bench estimated to 11.41 kg. Two variants are proposed. The equivalent masses and natural frequencies of Variant 1 and Variant 2 are respectively equal to (7.6 kg, 28.4 Hz) and (7.38 kg, 28.8 Hz).

The reduction in mass provided a significant improvement in the running of the cyclic test at low frequencies higher than the frequency of 3 Hz. These improvements have affected two quantities: the reduction of the amplitude of oscillation for a fixed frequency and the increase of the frequency of excitation for a fixed amplitude which will reduce the duration of the test. Although Variant 2 has not added a significant improvement compared to Variant 1, it has the advantage of being less cumbersome.

In this study we presented the results without taking into account the heating of composite materials that may occur when increasing the frequency of excitation. Other tests are needed to further validate the proposed model and compare it to the models obtained from conventional materials used for the manufacture of sockets.

References

1. Begum K, Islam MA (2013) Natural Fiber as a substitute to Synthetic Fiber in Polymer Composites: A Review. Res J Eng Sci 2(3):46–53
2. Ben Brahim S, Ben Cheikh R (2007) Influence of fiber orientation and volume fraction on the tensile properties of unidirectional Alfa-polyester composite. Compos Sci and Technol 67:140–147

3. Campbell AI, Sexton S, Schaschke CJ et al (2012) Prosthetic limb sockets from plant-based composite materials. Prosthet Orthot Int 36(2):181–189
4. Chakroun M, Ghozlen HB (2012) Mechanical impedance of cerebral material. J Modern Phys 3:271–273
5. Gerschutz MJ, Haynes ML, Nixon D et al (2012) Strength evaluation of prosthetic check sockets, copolymer sockets, and definitive laminated sockets. J Rehabil Res Dev 49(3):405–426
6. Jweeg MJ, Jaffar JS (2016) Vibration analysis of prosthesis for the through knee amputation. Al-Nahrain Univ Coll Eng J (NUCEJ) 19(1):46–55
7. Mankai W (2010) Elaboration d'un Protocole d'Essai pour les Emboîtures Tibiales. MRes, Ecole Nationale d'Ingénieurs de Tunis, TU
8. Meirovitch L (2001) Fundamentals of vibrations, 1st edn. Mechanical engineering series. McGraw Hill international Edition
9. Neo LD, Lee PVS, Goh JCH (2000) Principal structural testing of trans-tibial prosthetic assemblies: specimen preparation (Technical note). Prosthet Orthot Int 24:241–245
10. Skrzypek JJ, Ganczarski AW (2015) Constitutive equations for isotropic and anisotropic linear viscoelastic materials. In: Skrzypek J, Ganczarski A (eds) Mechanics of anisotropic materials. Engineering materials. Springer, Cham, pp 57–85

Investigation on the Effect of the Contact-Free Creep Test Loading Conditions on the Human Skin Viscoelastic Parameters

Khouloud Azzez[1,2](✉), Marie-Angèle Abellan[2], Makram Chaabane[1], Jean-Michel Bergheau[2], Hassan Zahouani[2], and Abdelwahab Dogui[1]

[1] Mechanical Engineering Laboratory, University of Monastir, ENIM, LR99ES32, 5000 Monastir, Tunisia
khouloud.azzez@gmail.com

[2] Laboratory of Tribology and Dynamics of Systems, University of Lyon, ENISE, UMR 5513, CNRS, 42023 Saint-Étienne, France

Abstract. The present paper proposes an insight into the objective quantification of internal parameters through experimental tests and numerical simulations of contact-free creep tests. The study investigates effects of applied pressure and dwell-times of contact-free creep tests on the viscoelastic properties of human skin. The skin is seen as a homogeneous, quasi-incompressible, linear, isotropic and viscoelastic material modelled by the Zener's rheological model. A Finite Elements model is developed and an inverse approach is used in order to identify the viscoelastic parameters. Analysis of variance (ANOVA) is carried-out to assess intra- and inter- related effects for two protocols. The analyzed results prove that the imposed pressure exhibits a major effect on the identified viscoelastic properties. However, the dwell time has no remarkable effects. Moreover, the major conclusion establishes that a loading pressure of 1 bar allows to distance objective quantifications from possible protocol influences.

Keywords: Human skin · Viscoelastic parameters · Contact-free creep test · Numerical simulation · ANOVA

1 Introduction

Human skin covers the human body and ensures several functions such as protection and exchange with our environment [4]. Seen as a mechanical material, the skin is a heterogeneous material saturated by a physiological fluid composed of three layers. It has a viscoelastic anisotropic behavior with a natural pre-stress [10, 14].

Various studies proposed an evaluation of mechanical properties of the skin [1, 3, 11] Most of previous works used contact devices which influenced the measurements of the material properties [6]. To overcome this contact drawback, researchers proposed contact-free devices using an air jet indentation system coupled with an OCT system [13]. More recently, [9] developed the Tonoderm® operating with an air jet indentation system. Then, we can cite the WaveSkin© using an air flux to apply pressure on the skin [8].

N. Aifaoui et al. (Eds.): CMSM 2019, LNME, pp. 214–220, 2020.
https://doi.org/10.1007/978-3-030-27146-6_23

In order to identify the viscoelastic mechanical properties, many assumptions were considered. The skin was considered as a linear elastic material [16] or hyperelastic [7]. Extension to viscous phenomenon is proposed in [12]. Furthermore [5] has shown that an inverse procedure associated with an optimization algorithm leads to objective characterization of the viscoelastic parameters of in vivo human skin. Nevertheless, we cannot judge the effect of the experimental parameters such as loading magnitude or velocity on the identified parameters.

To go further, an investigation of the experimental parameters' influence is discussed in the present paper. To this aim, contact-free creep tests were performed on a healthy consenting adult volunteer following two protocols with different creep pressures and different dwell-times. The viscoelastic parameters are identified for each test and an analysis of variance (ANOVA) is conducted in order to discuss the effect of experimental parameters on the identified properties.

This paper presents the method in part 2 i.e. the experimental data, the viscoelastic constitutive law and the numerical model. Part 3 describes and discusses the numerical results and the statistical analysis.

2 Materials and Methods

2.1 Experimental Protocol

The WaveSkin© device is used for in vivo contact-free creep tests [8]. An air flux is imposed onto the external surface of the forearm. The displacement of the surface is recorded using a laser system composed of 800 sensors, with 7 mm of length (Fig. 1a). The WaveSkin© functional diagram is detailed in Fig. 1b.

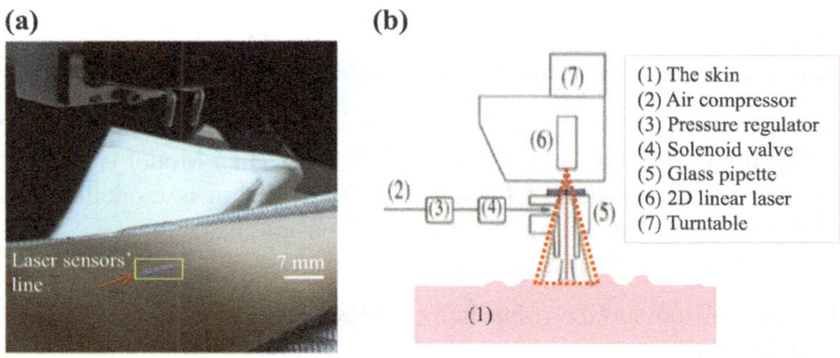

Fig. 1 The WaveSkin© device. With **a** The measurement zone, **b** The functional diagram of the device

Tests were performed in the same zone for a consenting adult following two different protocols.

For Protocol 1, a pressure of 1 bar is applied by the WaveSkin© with: A loading branch from 0 bar to 1 bar and two different dwell times: 200 ms (Test T1) and 400 ms

(Test T2); then the air flux is stopped instantaneously at the end of the dwell time and the undisturbed free-return of the skin is observed during 400 ms.

For Protocol 2, a pressure of 0.5 bar is applied with a loading branch from 0 bar to 0.5 bar with a dwell time of 400 ms (Test T3) then the air flux is stopped instantaneously and the free-return of the skin is observed during 400 ms.

Every test was repeated three times leading to a total of 9 records. The central laser sensor is considered because it corresponds to the maximum of the displacement of the human skin surface.

2.2 Constitutive Law

The mechanical behavior of human skin is defined by Zener's rheological model [5]. The stress-strain law in the Laplace-Carson space is given by (Eq. 1):

$$\boldsymbol{\sigma}^*(p) = 2G^*(p)\boldsymbol{\varepsilon}^{*^{dev}}(p) + K^*(p)Tr(\boldsymbol{\varepsilon}^*(p))\mathbf{I} \tag{1}$$

where: $\boldsymbol{\sigma}$ is the Cauchy stress tensor, $\boldsymbol{\varepsilon}$ is the strain tensor, $\mathbf{I}$ is the identity tensor, * corresponds to the Laplace-Carson transformation, $\boldsymbol{\varepsilon}^{*^{dev}}$ is the deviatoric strain tensor, G and K are respectively the deviatoric and volumetric moduli.

The volumetric modulus is assumed to be elastic. Therefore, we can write: $K^* = K = \frac{E_0}{3(1-2\nu_0)}$ where: E_0 is the instantaneous Young modulus and $\nu_0 = 0.48$ is the Poisson ratio.

The deviatoric modulus is described by: $G^*(p) = G_0(g_\infty + g\frac{p}{p+1/\tau})$,

where: $G_0 = \frac{E_0}{2(1+\nu_0)}$ is the instantaneous shear modulus, g, g_∞ and τ are the Prony series' coefficients; τ is called the characteristic time and E_0 is the instantaneous Young modulus.

2.3 Finite Elements Model and Inverse Problem

The numerical model of the contact-free creep test was developed on ABAQUS© software. The associated axisymmetric Finite Elements (FE) Model is a cylinder with radius 30 mm and a height of 11.11 mm (Fig. 2). The skin is modelled as a homogeneous material with a linear, viscoelastic and isotropic behavior.

The boundary conditions are chosen as follows:

- On the external top surface (pink line); a pressure modelled as a Gaussian distribution is imposed;
- For the bottom surface, no displacement allowed for all the nodes;
- A symmetry along the vertical left boundary is set up;
- The vertical right boundary is free.

A refine mesh (dimensions 4.11 mm × 10 mm) is locally required around the loading zone in order to get a smoother distribution of stress and strain. The mesh is based on bilinear axisymmetric quadratic elements with 4 nodes (CAX4R) with 8485 elements.

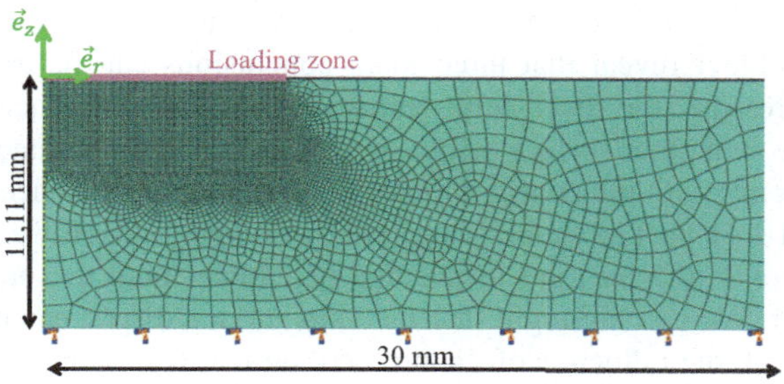

Fig. 2 The FE model

An inverse method is used to identify the three viscoelastic parameters of the skin (E_0, τ and E_∞). The numerical results are optimized numerically referring to the experimental data using the Nelder-Mead Simplex optimization algorithm developed under MATLAB© [5].

2.4 Statistical Analysis

Statistical analyses were carried out using a statistical software program (SPSS for windows). The data sets were subjected to analysis of variance using the general linear model option (Duncan test) in order to investigate differences between samples ($p < 0.05$).

3 Results and Discussion

3.1 Numerical Results

Figure 3 shows the comparison between the experimental displacements and their corresponding optimized numerical displacements for the three different tests.

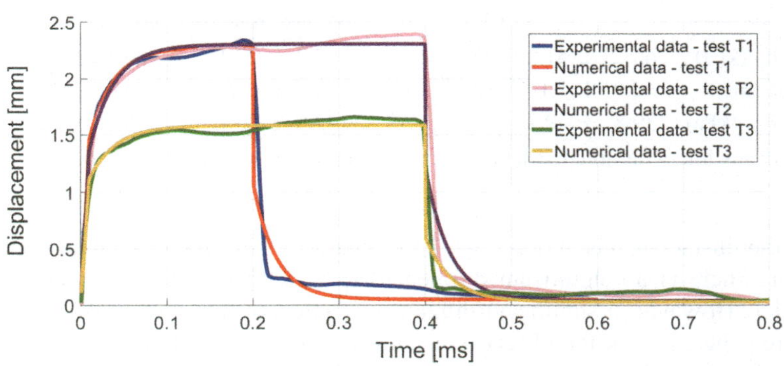

Fig. 3 Comparison between experimental and numerical displacements for the three tests

The WaveSkin© is able to reproduce identically a chosen external load. The results of Fig. 3 and Table 1 reveal that three main conclusions can be drawn. First, for a given loading pressure, the dwell-time seems to have little influence on the order of magnitude of the skin outer surface displacement at the end of the loading branch. This point is in coherence with the contribution of [2]. Second, the displacements of the two tests of Protocol 1 (1 bar) have the same order of magnitude. However, the displacements of Protocol 2 (0.5 bar) are in average half lower than the ones of Protocol 1 (1 bar) (Table 1). We can highlight that the skin surface displacement decreases with the pressure. Third, the influence of the protocol on the displacement fades away after 0.6 s when it becomes equal for all the tests. Hence whatever can be the external load the study retrieves the classical result of [15] that each subject has his own answer for a given solicitation.

Table 1 Experimental and numerical displacements at the end of the dwell-times for all the tests

	Protocol 1			Protocol 2
Displacement (mm)	T1	T2	Means protocol 1	T3
Experimental	2.111 ± 0.057	2.253 ± 0.068	2.182 ± 0.06	1.442 ± 0.049
Numerical	2.172 ± 0.07	2.265 ± 0.027	2.219 ± 0.05	1.571 ± 0.01

A set of the optimized viscoelastic parameters (E_0, τ and E_∞) is identified for each of the 9 experimental data associated to tests (T1, T2 and T3).

3.2 Statistical Analysis Results

Table 2 gives the identified viscoelastic parameters with their associated means and standard deviations for every test.

Table 2 Table of statistical analyses of viscoelastic parameters identified

	Protocol 1			Protocol 2
	T1	T2	Means protocol 1	T3
E_0 (kPa)	$265.2^b \pm 41.2$	$303.1^b \pm 13$	284.2 ± 27.1	$153.5^a \pm 9.3$
τ (ms)	$12.9^a \pm 1.1$	$12.4^a \pm 0.6$	12.7 ± 0.9	$15.4^a \pm 4.2$
E_∞ (kPa)	$139.8^b \pm 40.7$	$145.7^b \pm 12.7$	142.8 ± 26.7	$95.2^a \pm 9.0$

Means line with different letters (a, b) are significantly different ($p < 0.05$)

E_0 the instantaneous Young modulus characterizes the instantaneous material behavior. There is no significant differences ($p > 0.05$) intra-tests and inter-tests for protocol 1. However, a significant difference is relevant inter-protocols. These differences are coherent with the observations already discussed in the previous subsection on the influence of the amplitude of the external load on the answer. Inter-protocols, the means for E_0 are in average lower for Protocol 2 than for Protocol 1. For protocol 1, both tests have the loading branch in common. This explains the results of the ANOVA of Table 2 and corroborates the minimum variations found for E_0 inter-tests. On the contrary, Protocol 2 has a different loading pressure leading to a small change in the slope of the loading branches.

The characteristic time τ a quantification of the time taken by the skin tissues for recovering after an external mechanical solicitation. For this parameter, the ANOVA gives no relevant differences intra-tests and inter-tests for protocol 1 and the amplitude of the external load has no influence on his characteristic time.

When the external load is suppressed instantaneously, the cutaneous tissue is free to eliminate the supplied mechanical energy and to return to its unknown natural state or to search for a new equilibrium. Figure 3 shows that, whatever can be the protocol or the test, after 0.6 s the skin tissues have found a unique final state. At this state of the test, we define the delayed Young modulus E_∞. In Table 2, there is no relevant differences intra-tests and inter-tests for a given protocol. On the other hand, no relevant difference is noticed inter-protocols. Moreover the same tendencies, as the ones observed for the instantaneous Young modulus E_0, are observed here with higher values for Protocol 1 than for the Protocol 2.

4 Conclusion

The present paper has proposed the numerical simulations of contact-free creep tests with characterization of the viscoelastic parameters: E_0 the instantaneous Young modulus, τ the characteristic time and E_∞ the delayed Young modulus. The study is conducted on a healthy adult.

For the experimental records, the measured displacements comply with the observations reported in the literature with respect to their order of magnitudes (between 1.2 and 2.3 mm). For the numerical results and more specifically the estimated mechanical parameters the same conclusions are deduced.

Analyses of variance (ANOVA) were conducted on the numerical results to assess intra- and inter- related effects with a p-value ($p < 0.05$) for statistically significant effects. Based on inter-tests and intra-protocols non relevant differences ($p > 0.05$), the study proposes to conclude that the characterization of the three parameters is not influenced by the dwell-time chosen for the experimental protocol. On the contrary inter-protocols significant differences ($p < 0.05$) lead to the statement that the amplitude of the external load taken in the experimental protocol do have a significant influence.

References

1. Abellan M-A, Feulvarch E, Zahouani H, Bergheau J-M (2013) Numerical simulation of in vivo indentation tests: determination of the mechanical properties of human skin. 21st Congrès Français de Mécanique (CFM'13). Août, Bordeaux, France, pp 26–30
2. Abellan M-A, Ayadh M, Azzez K, Bergheau J-M, Zahouani H (2017) Influence du temps de maintien sur la réponse viscoélastique de la peau humaine jeune in vivo lors d'essais de fluage sans contact. 13e Colloque National en Calculs des Structures (CSMA'17), 15–19 mai, Giens, Var, 2017

3. Agache P, Monneu C, Lévêque J, De Rigal J (1980) Mechanical properties and young's modulus of human skin in vivo. Arch Dermatol Res 269:221–232
4. Agache P (2000) Physiologie de la peau humaine et explorations fonctionnelles cutanées. Lavoisier
5. Azzez K, Chaabane M, Abellan M-A, Bergheau J-M, Dogui A, Zahouani H (2018) Relevance of indentation test to characterize soft biological tissue: application to human skin. Int J Appl Mech 10(7). https://doi.org/10.1142/S1758825118500746
6. Baldwin SA, Larson MJ (2017) An introduction to using Bayesian linear regression with clinical data. Behav Res Ther 98:58–75. https://doi.org/10.1016/j.brat.2016.12.016
7. Bischoff J-E, Arruda E-M, Grosh K (2004) A rheological network model for the continuum anisotropic and viscoelastic behavior of soft tissue. Biomech Model Mechanobiol 3:56–65
8. Bonnet I, Jobeili L, Cadau S, Berthélémy N, Pierrot A, Tesdeschi C, Bardey V, Fargier G, Rival D, Jeanmaire C, Bnnaud-Rosaye C, Chan W, Chavan M, André-Ferei V, Zahouani H, Rousselle P (2017) Collagen XVII: a key interfacial component of the skin architecture. Int J Cosmet Sci 68:35–41
9. Boyer G, Pailler-Mattei C, Molimard J, Pericoia M, Laquieze S, Zahouani H (2012) Non-contact method for in vivo assessment of skin mechanical properties for assessing effect of ageing. Med Eng Phys 34:172–178
10. De Rigal JD, Lévêque J (1985) In vivo measurement of the stratum corneum elasticity. Bioeng Skin 1:13–23
11. Delalleau A, Josse G, Lagarde J-M, Zahouani H, Bergheau J-M (2008) A nonlinear elastic behavior to identify the mechanical parameters of human skin in vivo. Skin Res Technol 14:152–164
12. Flynn C, Taberner A, Nielsen P (2011) Modeling the mechanical response of in vivo human skin under a rich set of deformations. Ann Biomed Eng 39(7):1935–1946
13. Huang Y-P, Zheng Y-P, Wang S-Z, Chen Z-P, Huang Q-H, He Y-H (2009) An optical coherence tomography (OCT)-based air jet indentation system for measuring the mechanical properties of soft tissues. Meas Sci Technol 20(1):1–11. https://doi.org/10.1088/0957-0233/20/1/015805
14. Leveque J-L, De Rigal J, Agache P-G, Monneur C (1980) Influence of ageing on the in vivo extensibility of human skin at a low stress. Arch Dermatol Res 269(2):127–135
15. Wilkes G, Brown I, Wildnauer R (1973) The biomechanical properties of skin. Crit Rev Bioeng 4:453–495
16. Zahouani H, Pailler-Mattei C, Sohm B, Vargiolu R, Cenizo V, Debret R (2009) Characterization of the mechanical properties of a dermal equivalent compared with human skin in vivo by indentation and static friction tests. Skin Res Technol 15:68–76

Effect of Changing Temperature and Wire Cross Section on the Tribological Behavior of the NiTi Alloy

Ines Ben Naceur(✉) and Khaled Elleuch

Department of Materials Engineering and Environment (LGME) ENIS, National Engineering School of Sfax, Sfax, Tunisia
{inesbennaceur, khaled_elleuch}@yahoo.fr

Abstract. Orthodontic therapy generally begins with aligning and leveling phase, wherein a NiTi, superelastic wire is habitually attached, tied and activated to supply the required forces for teeth correction. Deactivation of the archwire following teeth movement will commonly lead to the sliding of wire through the adjoining bracket slots. This study aims to investigate the friction and wear behavior of a deflected NiTi wire sliding against a 316 stainless steel sample with the effect of changing the temperature and the shape of the wire. Superelastic NiTi wires with circular and rectangular cross-section were tested, in Fuzayama artificial saliva and at the temperatures met in the oral cavity, using a modified rotative tribometer. It was found that the tested parameters had a great effect on the tribological behavior of the studied tribo-couple. Indeed, the results showed an increase in the wear rate as the temperature increase because of the rise of the thermal stress within the tested NiTi alloy. It was also found that the wear of the rectangular archwires is more significant than the circular ones, this could be due to the nature of the initial tribo-contact condition.

Keywords: NiTi wires · Friction · Wear · Artificial saliva · Scanning electron microscopy

1 Introduction

NiTi alloy with high spring-back and superelasticity are generally used in the alignment and leveling stages of orthodontic treatment [1]. These properties permit constant forces over large deflections without important permanent deformation. It is believed that the increased friction between the wire and the bracket slot can inhibit sliding of the archwire. Afterwards, teeth correction could be blocked and treatment period might be prolonged. Moreover, this situation is even aggravating because friction will cause the wear of the wire and brackets materials which improves the dissemination of harmful metals such as nickel across the body. Recently, few papers have concentrated on the tribological characterization of this tribo-couple employing a pin on disk sliding tester [2–4]. Nonetheless, these investigations are restricted to the passive disposition where the archwire is entirely parallel to the bracket slot during sliding test. In this search, a rotary tribometer was adapted to examine the tribological characteristics of the NiTi superelastic alloy in artificial saliva medium. This modification afforded the

N. Aifaoui et al. (Eds.): CMSM 2019, LNME, pp. 221–230, 2020.
https://doi.org/10.1007/978-3-030-27146-6_24

feasibility of bending the wire during slidingl tests. The objective was to study the effect of temperature and the shape of the orthodontic wire on the tribological behavior of a deflected NiTi archwire.

2 Materials and Methods

In the present study, we used superelastic NiTi wires with rectangular (0.46 × 0.64 mm^2) and circular (0.46 mm) cross section, which exhibits a fully austenitic structure at room temperature. The friction coefficient produced with each NiTi wire/stainless steel combination were obtained with an adapted rotary sliding friction tester allowing the deflection of the archwire from 0 to 2.5 mm [5] in the presence of the Fusayama artificial saliva (Table 1). The tests was realized with a linear velocity of 68 mm/s (angular velocity of 100 tr/min, which represents the lowest speed of our rotating tribometer) for 12,000 cycles. Divers studies [2, 6] have approximated the normal force enforced by an elastic ligature when constricted the wire into the bracket-slot to 2 N. That's why, the friction tests were conducted against 316 stainless steel flat sample, frequently used as a material for orthodontic brackets, with a fixed applied normal force of 2 N. The chemical composition of this alloy was determined using a spectrometer (Jobin Yvon JY 48) which gives the content of the main elements by weight-percent (Table 2). A heating system adapted to the working environment has been developed in order to be able to perform the friction tests in the temperatures recorded in the oral cavity. This device allows the increase and the maintain of the temperature of the solution at a constant level. The entire system includes a heating element, a thermal probe and a temperature controller.

Table 1 Modified Fusayama artificial saliva used in this Study

Composition	mg/L
NaCl	400
KCl	400
$CaCl_2.2H_2O$	795
$NaH_2PO_4.H_2O$	690
KSCN	300
$Na_2S.9H_2O$	5
Urea	1000

Table 2 Chemical composition of 316 stainless steel

C (%)	Si (%)	Mn (%)	P (%)	S (%)	Cr (%)	Ni (%)	Mo (%)	N (%)
0.084	0.54	1.8	0.018	0.0038	17.28	10.51	2.02	

With a differential scanning calorimeter test (JEOL JSM 5400), the phase transformation temperatures, namely martensite start (M_s) and finish (M_f) temperatures, austenite start (A_s) and finish (A_f) temperatures were determined (Table 3). Quantification of wear was determined geometrically using a scanning electron microscopy, indeed a measurement of the length of the wear track has been carried out. A normalized representation of the wear is then proposed with reference to the test carried out at 25 °C for the non bent wire and with a normal applied force equal to 2 N. At the end of each test, the sample is cleaned with ethanol in an ultrasonic bath to eliminate the non-adherent wear particles. The wear mechanisms will be analyzed based on scanning electron microscope observations of the NiTi alloy.

Table 3 Phase transformation temperatures and room temperature structures

A_s(°C)	A_f(°C)	M_s(°C)	M_f(°C)	Room-temperature structure
11.5	23.5	15	7	Austenite

3 Results and Discussion

3.1 Preliminary Evaluation

- ***Characterization of an unworn NiTi wire***

Micrographic observations made on the surface of an unworn NiTi wire, using the scanning electron microscope, (Fig. 1) shows a non-homogeneous surface with various signs of irregularities marked by the presence of some black precipitates. We also note the presence of pores, scratches and striations in the longitudinal direction.

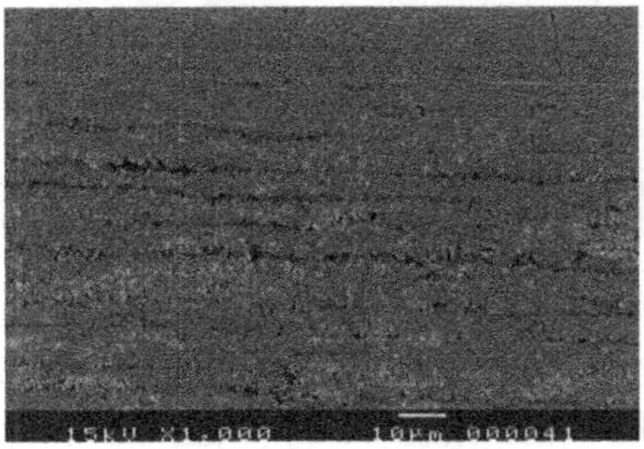

Fig. 1 Micrographics of the surface of an unworn circular NiTi wire

This surface in homogeneity can be attributed to the complex manufacturing process and finishing treatments of this alloy. Surface irregularities of the NiTi alloy wires have also been observed in the work carried out by Krishnan et al. [7] as well as Amini et al. [8]. It should be noted that this diversity of surface condition will obviously influence the friction amount generated by the wires inside the orthodontic brackets.

- ***Characterization of in vivo damaged NiTi wire***

At the alignment and leveling stage of orthodontic treatment, the wires behave elastically over a period of weeks to months. For this reason two superelastic NiTi circular wires used in the mouth for one month (Fig. 2a) and two months (Fig. 2b), given by an orthodontist, were examined using the scanning electron microscope. The aim was to determine the processes and the wear mechanisms involved at the actual clinical conditions.

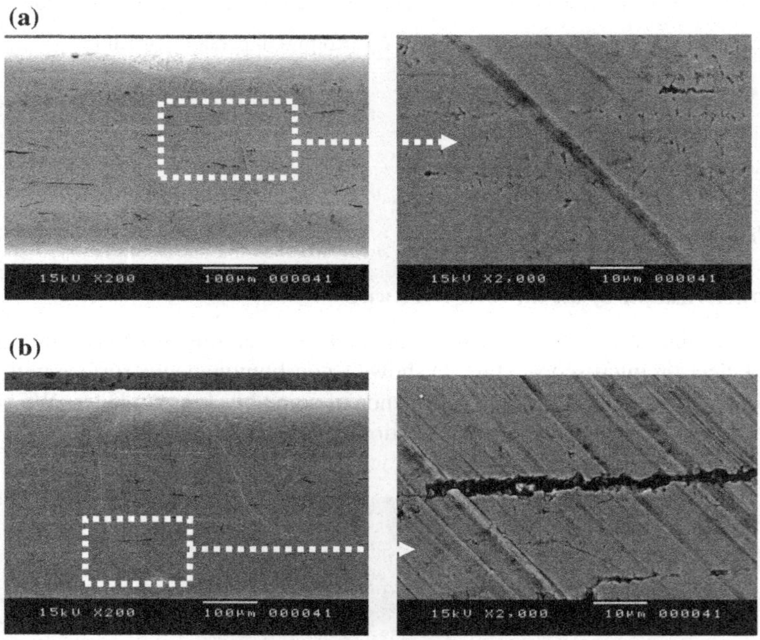

Fig. 2 Wear track of a circular NiTi wire worn in-vivo during **a** 1 month and **b** two months

These observations show the presence of striations and parallel stripes oriented in an inclined direction, which are evidence of abrasive wear. These signs are much more obvious by increasing the oral exposure time. The micrographs obtained from the NiTi wires worn in vivo for two months show the presence of large and deep horizontal cracks as well as extensive crevices (Fig. 2b) which are attributed according to Bourauel et al. [9] to the presence of locally produced tensile and/or compressive forces due to the engagement of the wire through the brackets during the alignment and/or leveling phase of the orthodontic treatment. The observed signs of wear have the typical characteristics of the fatigue wear.

3.2 Tribological Results

Figure 3a and b show typical curves that exhibit, the evolution of the average coefficient of friction with the number of cycles for the circular and rectangular NiTi archwires respectively at 25 and 50 °C. We can notice that the evolution of the friction coefficient over time, for all tested conditions, are similar. Two sliding regimes are observed. The first is characterized by a drop in the coefficient of friction after a highly disturbed transient stage, the second is rather stable and reaches a quasi-stationary value.

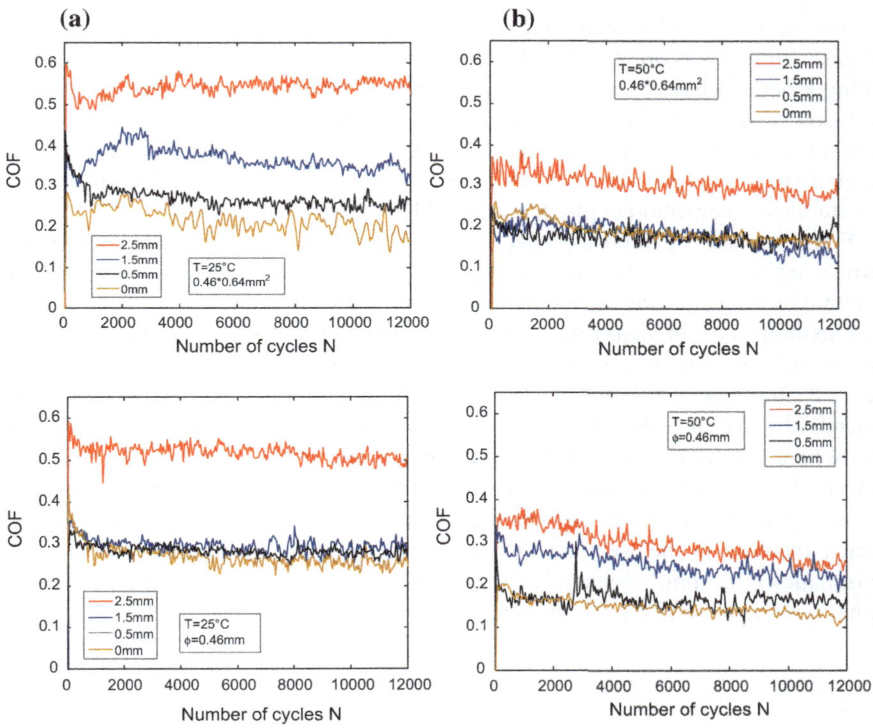

Fig. 3 Coefficient of friction with respect to sliding cycles for different vertical deflections and for rectangular and circular NiTi wires at **a** 25 °C **b** 50 °C

Circular (0.46 mm) and rectangular (0.46 × 0.64 mm^2) superelastic NiTi wires having the most used dimensions in orthodontics were employed in this study. The results shown in Fig. 3 reveal that no noticeable dependency was found between the coefficient of friction and the cross section dimension of the wire. These results are in good agreement with the findings of Drescher et al. [10] in which 0.40 mm NiTi round wire and 0.40 × 0.55 mm^2 rectangular wire were found to practically exhibit the same amount of friction. The authors have explained their findings by the fact that the occluso-gingival dimension (in our configuration the horizontal dimension) of the wire is the most critical dimension affecting friction.

From the DSC measurement (Table 2), it can be noted that all the tested temperatures are greater than the A_f of the tested superelastic NiTi wire. Therefore, the alloy was maintained austenitic at 25 and 50 °C except in the presence of excessive bending deflection. The effect of changing temperature on the friction coefficient at different levels of vertical deflection and for both wire shapes was also illustrated in Fig. 3a and b. A decrease in the coefficient of friction with the increase in the wear testing temperature from 25 to 50 °C can be clearly seen. It is also noteworthy that the average steady state coefficient of friction falls from 0.58 to 0.3 with a deflection of 2.5 mm for the rectangular wire. However, in the case of the non-bent archwire, the coefficient of friction decreases with a much reduced rate from 0.2 to 0.16, when the temperature increases from 25 to 50 °C. Therefore, varying temperature has a strong effect on the sliding kinetic behavior. In fact, with the increase in temperature, the elastic accommodation of the NiTi alloy decreases because of the increase in the austenite elastic modulus and phase transition stress. Moreover, as the temperature increases, the material shear strength decreases, and therefore the coefficient of friction of the NiTi-316 stainless steel tribo-couple decreases [11].

Figures 4 and 5 exhibit the effect of changing wire shape and temperature on the normalized wear rate of the NiTi alloy, as a function of the applied vertical deflection. It should be noted that the rectangular NiTi wire in the austenite state and at low normal force exhibits a better wear resistance than the circular archwire (The initial wear track lengths, at room temperature, for the non deformed rectangular and circular archwires are 1.227 mm and 0.825 mm, respectively). It should be noted that the circular wire generally has a point contact, while rectangular wire has a line contact. Therefore, there would be a greater intensity of the resulting normal force in the case of contact-point and this force could potentially indent the round archwire, which could substantially decrease their wear resistance. On the other hand, the rectangular wire displays a steeper increase in the wear rate when increasing vertical deflection. At the largest deformation (2.5 mm), rectangular and circular wires exhibit about 2 and 1.25 times higher wear rate, respectively, than non-deformed NiTi wires at 25 °C.

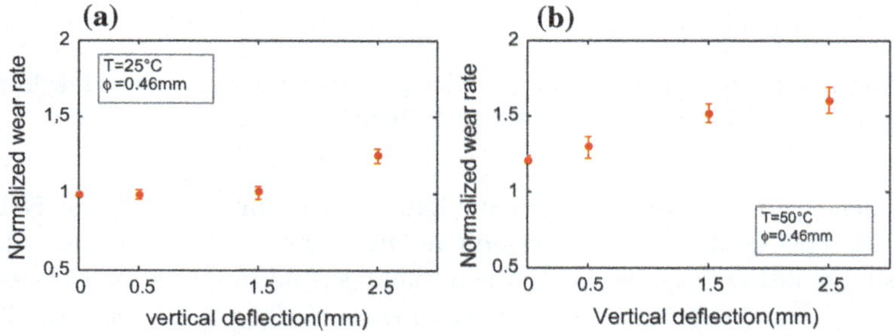

Fig. 4 Normalized wear rate as a function of applied vertical deflections and temperature for the circular NiTi wires **a** 25 °C and **b** 50 °C

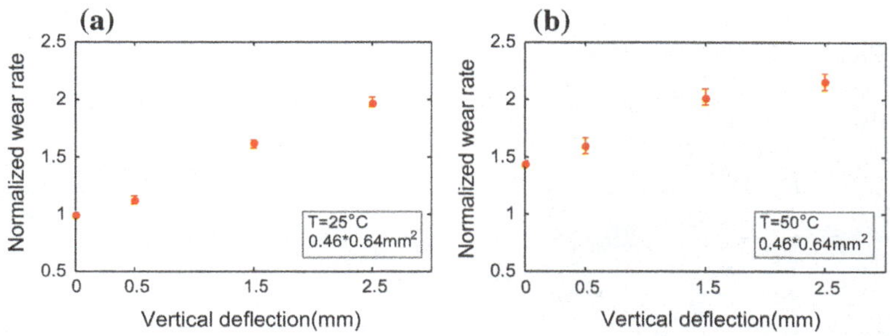

Fig. 5 Normalized wear rate as a function of applied vertical deflections and temperature for the rectangular NiTi wires **a** 25 °C and **b** 50 °C

In addition, the figure indicates a decrease in the wear resistance of the NiTi wire with the increase in the wear testing temperature from 25 to 50 °C. The decrease in the wear resistance at high temperature may be attributed, as already previously mentioned, to the decrease in the elastic accommodation ability of the NiTi and to the amount of the existing phases [11–13] Furthermore, it is well-known that the increase in temperature weakens the materials in contact, leading to an increased thermal stress and, therefore, an increased wear rate [13]. The amount of friction as well as the wear rate of the NiTi alloy are obviously influenced by the changes produced on the rubbing surfaces. Thus, it is appropriate to analysis the wear track to find an explanation for the obtained trends. Therefore, scanning electron microscopy examinations were performed on the wear trace of the tested NiTi samples following 12,000 sliding cycles at 25 and 50°C. The acquired results for different applied vertical deflection are shown respectively in Figs. 6 and 7.

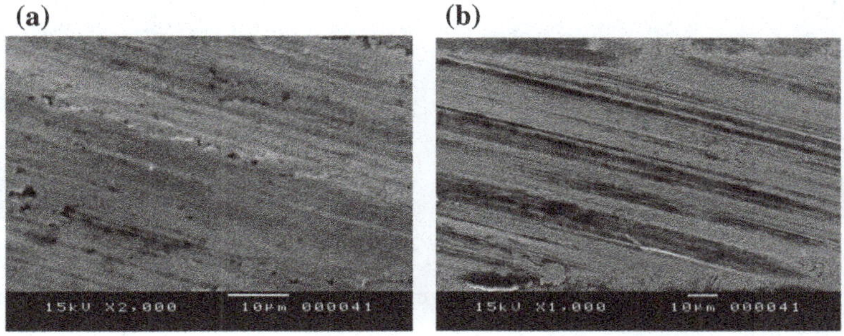

Fig. 6 SEM micrographs of the worn surfaces of NiTi wires tested at 25 °C as a function of vertical deflection after 12,000 sliding cycles: **a** 0.5 mm, **b** 1.5 mm

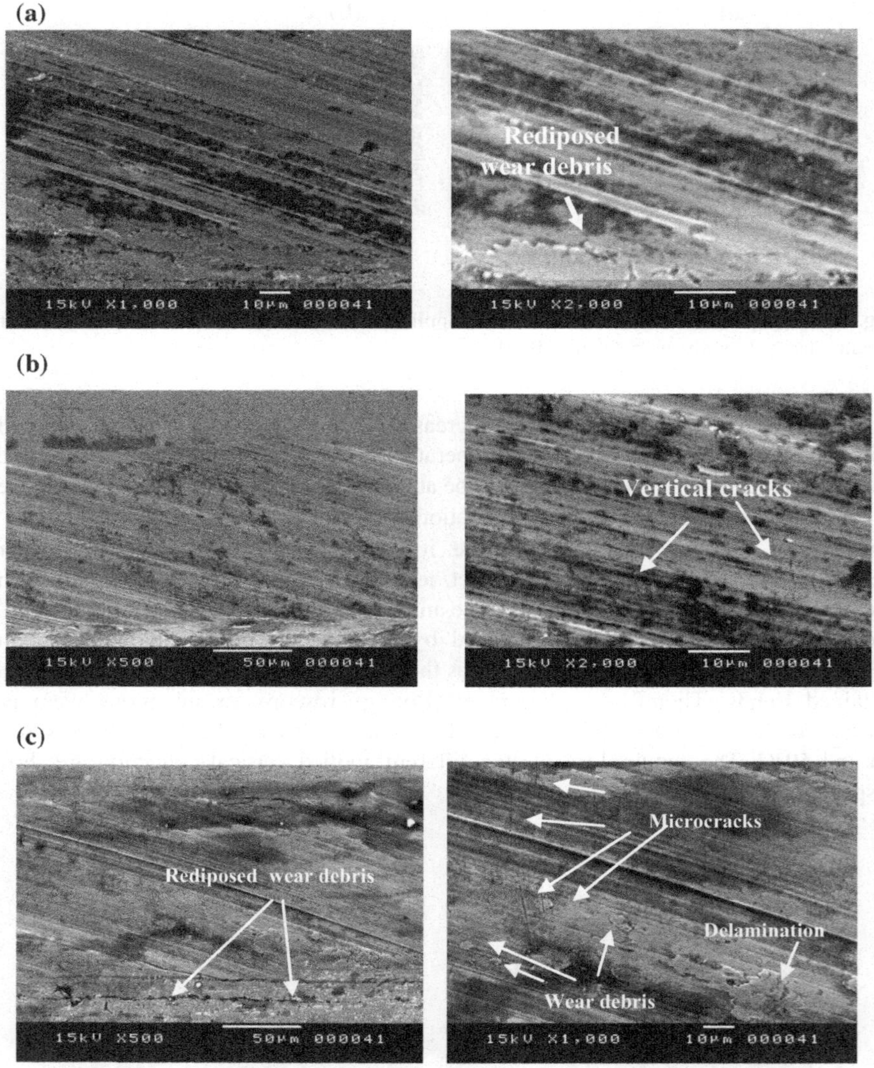

Fig. 7 SEM micrographs of the worn surfaces of NiTi wires tested at 50 °C as a function of applied vertical deflection after 12,000 sliding cycles: **a** 0.5 mm, **b** 1.5 mm and **c** 2.5 mm

We can observe, for the different applied vertical deflection, the presence of accentuated plastic deformations and abrasion striations parallel to the sliding direction. We also note the presence of abundant oxide patches (darker regions) on the friction track reflecting the adhesive wear mechanism. From Fig. 7a and c relating to the wear trace of the circular wire tested at 50 °C, we can clearly discern a detachment and redeposition of wear debris on the friction track following severe plastic deformations. This phenomenon is less observed at room temperature.

The scanning electron microscopy micrographs corresponding to 1.5 mm and 2.5 mm vertical deflections (Fig. 7b and c) reveals, next to the wear mechanisms mentioned previously, the signs of fatigue wear. Indeed, the corresponding friction facies have abundant surface fatigue cracks perpendicular to the sliding direction. It appears that the higher the rate of transformed martensite, the more these cracks are long and abundant. This mode of wear seems to be more accelerated at 50 °C. Likewise, the obtained results show that the increase in temperature seems to favor the abrasive and adhesive components of wear, also the phenomenon of delamination is accentuated, this is well shown in the micrographs of the circular wires represented in Fig. 7.

A compaction phenomenon of detached wear particles following severe plastic deformation on friction trace of the NiTi alloy is very clear in the micrograph of the circular archwires tested at 50 °C. When sliding at higher temperatures, the oxide generation rate of the stainless steel at the interface could increase. The formation of wear debris and the transfer of oxides to the surface of the NiTi alloy and their subsequent compaction can thus form an inter-facial layer formed by a mechanically mixed wear particles dispersed on the surface of the NiTi alloy. The formation of this tribo-layer could reduce the wear damage by acting as a solid lubricant while reducing the metallic contact [13]. However, in the case of the tested NiTi/stainless steel 316 couple this tribological layer, despite having participated in the reduction of the friction coefficient, does not contribute to improving the wear resistance of the NiTi alloy at 50 °C. This leads to consider that this formed layer is not stable and degrades during sliding generating particles of the third body, which in turn feeds the tribological circuit and which is responsible for increasing the wear.

4 Conclusion

In this paper, the friction and the wear behavior of superelastic NiTi orthodontic wires sliding against 316 stainless steel, as well as the morphologies obtained by scanning electron microscopy were analyzed. The aim was to see the effect of the possible changes in clinical conditions as the temperature and the shape of the archwire. The examination of the obtained wear traces validates the results found relating to a significant increase in the wear and friction coefficient by raising the temperature. The damage of the NiTi alloy, for all the combinations tested, is the origin of a complex and interactive mechanisms involving, adhesive, abrasive, delamination and fatigue wear, and it seems that there is an amplification of these phenomena with the increase in temperature leading to an increase in the wear rate of the NiTi alloy.

References

1. Ferčec J, Anžel I, Rudolf R (2014) Stress dependent electrical resistivity of orthodontic wire from the shape memory alloy NiTi. Mater Design 55: 699–706
2. Grosgogeat B, Jablonska E, Vernet JM, Jaffrezic N, Lissac M, Ponsonnet L (2006) Tribological response of sterilized and un-sterilized orthodontic wires. Mat Sci Eng C Mater Biol Appl 26:267–272

3. Alfonso MV, Espinar E. Llamas JM, Rupérez E, Manero JM, Barrera JM, Solano E, Gil FJ (2013) Friction coefficients and wear rates of different orthodontic archwires in artificial saliva. J Mater Sci Mater Med 24:1327–1332
4. Kang T, Huang SY, Huang JJ, Li QH, Diao DF, Duan YZ (2015) The effects of diamond-like carbon films on fretting wear behavior of orthodontic archwire-bracket contacts. J Nanosci Nanotechnol 15:4641–4647
5. Bennaceur, I, Elleuch, K, (2019) Tribological properties of deflected NiTi superelastic archwire using a new experimental set-up: Stress-induced martensitic transformation effect. Tribol. https://doi.org/10.1016/j.triboint.2019.106033
6. Berradja A, Bratu F, Benea L, Willems G, Celis JP (2006) Effect of sliding wear on tribocorrosion behaviour of stainless steels in a Ringer's solution. Wear 261:987–993
7. Krishnan M, Seema S, Tiwari B, Sharma H, Londhe S, Arora V (2015) Surface characterization of nickel titanium orthodontic arch wires. Med J Armed Forces India 71:340–345
8. Amini F, Rakhshan V, Pousti M, Rahimi H, Shariati M, Aghamohamadi B (2012) Variations in surface roughness of seven orthodontic archwires: an SEM-profilometry study. Korean J Orthod 42:129–137
9. Bourauel C, Fries T, Drescher D, Plietsch R (1998) Surface roughness of orthodontic wires via atomic force microscopy, laser specularreflectance, and profilometry. Eur J Orthod 20:79–92
10. Drescher D, Bourauel C, Schumacher HA (1989) Frictional forces between bracket and arch wire. Am J Orthod Dentofacial Orthop 96:397–404
11. Qian LM, Sun QP, Zhou ZR (2005) Fretting wear behavior of superelastic nickel titanium shape memory alloy. Tribology Lett 18:463–475
12. Abedini M, Ghasemi HM, Ahmadabadi MN (2010) Tribological behavior of NiTi alloy against 52100 steel and WC at elevated temperatures. Mater Charact 61:689–695
13. Liu R, Li DY (2000) Experimental studies on tribological properties of pseudoelastic TiNi alloy with comparison to stainless steel 304. Metall Mater Trans 31:2773–2783

The Simulation of Acoustic Cavitation in the Medical Field

Rabeb Bedoui[1(✉)], Aicha Abbassi[1], Hatem Kanfoudi[2], and Ridha Zgolli[2]

[1] Laboratory of Applied Mechanical Research and Engineering (LR-MAI), National School of Engineers of Tunis (ENIT), ELMANAR University, 1002 Tunis, Tunisia
rabeb.bedoui@gmail.com, aicha.abbassi@enit.utm.tn
[2] Laboratory of Hydraulic and Environmental Modeling (LMHE), National School of Engineers of Tunis(ENIT), ELMANAR University, 1002 Tunis, Tunisia
kanfoudih@yahoo.fr, ridha.zgolli@enit.utm.tn

Abstract. In recent years, cavitation is increasingly utilized in a wide range of applications in the biomedical field. Monitoring the temporal evolution of cavitation bubbles is of great significance for efficiency and safety in biomedical applications. Cavitation is characterized by a random phenomenon that causes problems of re-producibility. This could be at the origin of the damage on the vascular walls and the adjacent fabrics of the handled organ. Better control and regulation of the cavitation's activity during the ultrasonic treatment would establish an inescapable way to envisage the development of a therapeutic device. This thesis work aims at developing a model allowing the regulation of the acoustic cavitation. This improvement could have direct applications in the medical field. This paper presents a theoretical study on the modeling of acoustic cavitation in the medical field and a numerical study of the simulation of acoustic cavitation. The study of acoustic cavitation by current CFD numerical simulations is also of great interest. For numerical simulation, we used Open-FOAM which is a toolbox for the dynamics of computer fluids as free software to access codes and algorithms.

Keywords: Cavitation · Acoustic · Medical field · Regulation · OpenFOAM

1 Introduction

This study follows the work on the modeling and simulation of acoustic cavitation bubbles. Cavitation is a complex phenomenon that represents a real challenge for numerical simulations concerning the modeling of the phenomenon itself and the development of powerful numerical methods. Cavitation is a phase change from liquid to vapor by reducing the pressure to a constant temperature. Cavitation describes the birth, the radial oscillation and the implosion of bubbles of gas and vapor in a liquid subjected to a phase of depression. The origin of this depression is related to the variations of a liquid's density subjected to an acoustic wave.

N. Aifaoui et al. (Eds.): CMSM 2019, LNME, pp. 231–239, 2020.
https://doi.org/10.1007/978-3-030-27146-6_25

Pressure waves of high amplitude (ultrasound) propagating in a liquid are likely to create bubbles in the middle, to oscillate and possibly implode: this is the phenomenon of acoustic cavitation. This phenomenon is observed and used in the case of ultrasonic waves for different industrial or medical practices.

In the medical field, acoustic cavitation can be of huge interest because it is the most used in this field like the destruction of kidney stones and so on. The techniques of the use of acoustic cavitation in the medical field are based on the control of the acoustic wave generating this phenomenon, and it is necessary to know the amplitude in direct relation with the pressure of the environment and its frequency. The use of cavitation in therapy can cause damage to adjacent tissues in the treated area, for this, we need to regulate the acoustic cavitation and subsequently model the cavitation.

In addition to the modeling, a simulation of acoustic cavitation is needed.

For this purpose, we used OpenFoam, which is a great tool for numerical simulations in areas such as fluid dynamics, cavitation flows, turbulent flow heat transfer, etc.

2 Mathematical and Numerical Model

2.1 Governing Equations

Cavitation is a multiphase phenomenon. We have several models for the modeling of a multiphase flow [1, 3].

For that we have three main parts: the first part accounts for the presence of the gas-liquid free surface (VOF model), the second part accounts for the presence of the bubbles in the liquid (LPT), whereas the third part accounts for the dynamic of cavitation bubble (the Rayleigh Plesset equation).

(a) **The Volume-Of-Fluid (VOF) method**:

The VOF method was introduced by Hirt and Nichols in [7] which is used for the simulation of flows between several immiscible phases. In this model, the vapor and liquid phases are treated as a single mixture fluid.

The continuity equation for an incompressible flow model is then given by Eq. (1):

$$\frac{\partial u_i}{\partial x_i} = \left(\frac{1}{\rho_l} - \frac{1}{\rho v}\right)\dot{m} \tag{1}$$

where ρ_l is the liquid density, ρ_v is the vapor density and $\dot{m}$ is the inter-phase mass transfer rate due to cavitation.

Furthermore, the Navier–Stokes equations are given by Eq. (2):

$$\frac{\partial(\rho_m u_i)}{\partial t} + \frac{\partial(\rho_m u_i u_j)}{\partial x_j} = \frac{\partial \tau_{ij}}{\partial x_j} + \rho_m u_j \tag{2}$$

where τ is the stress tensor

Here, ρ_m is the mixture density, μ_m is the mixture dynamic viscosity given by (3) and (4):

$$\rho_m = \alpha\rho_l + (1-\alpha)\rho_v \tag{3}$$

$$\mu_m = \alpha\mu_l + (1-\alpha)\mu_v \tag{4}$$

With μ_l is the liquid viscosity, μ_v is the vapor viscosity, and α is the liquid volume fraction that specifies the relative amount of liquid in a control volume, it is defined as follows:

$$\alpha = \frac{Volume\ of\ vapeur}{Total\ volume} \tag{5}$$

(b) **The Lagrangian particle tracking (LPT)**:

In this model, the cavities are treated as discrete Lagrangian bubbles, transported in a continuous ambient Eulerian flow.

A particle p is defined by its position x_p, its diameter d_p, its velocity $u_{p,i}$ and its density ρ_p.

The bubbles are tracked by solving a set of ordinary differential equations along the bubbles trajectories, given by:

$$\begin{cases} \frac{dx_{p,i}}{dt} = u_{p,i} & (6) \\ m_p \frac{du_{p,i}}{dt} = m_p \frac{\left(u_{f,i}-u_{p,i}\right)}{\tau_p} + \left(\rho_p - \rho_f\right) g_i & (7) \\ \tau_p = \frac{3}{4} \frac{\rho_v d_p^2}{\rho_f C_D \left|u_{f,i}-u_{p,i}\right|} & (8) \end{cases}$$

With C_D: particle drag coefficient.

Figure 1 presents the different approaches to treat two-phase flow.

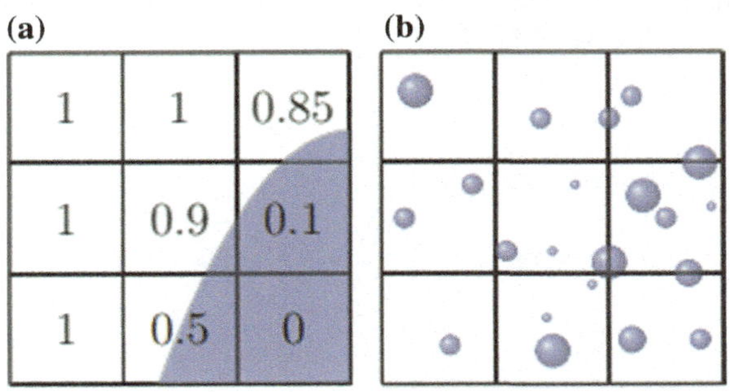

Fig. 1 Different approaches to treat two-phase flow. **a** The Volume of Fluid (VOF) consists of tracking the interface between two fluids. **b** The Lagrangian particle tracking (LPT) methods to track bubbles smaller than the cell size

2.2 Rayleigh-Plesset

The Rayleigh Plesset equation has been proposed by Rayleigh [8] and Plesset [2] for the modeling of bubble dynamics in cavitation phenomena and for multiphase bubble flows.

The generalized Rayleigh-Plesset equation describes the response of the bubble in the following form:

$$R(t)\ddot{R}(t) + \frac{3}{2}\dot{R}^2(t) = \frac{P_g - P_l}{\rho_l} - 4\mu_l \frac{\dot{R}(t)}{R(t)} - \frac{2\sigma_{st}}{\rho_l R(t)} \tag{9}$$

R is the bubble radius, P_l is the fluid pressure, σ_{st} is the surface tension, μ_l is fluid viscosity, and P_g is the gas pressure inside the bubble with,

$$P_g = P_v + P_{g0}\left(\frac{R_0}{R(t)}\right)^{3k} \tag{10}$$

P_{g0} is the equilibrium gas pressure at the equilibrium state with $R = R_0$.

This model assumes a single spherical bubble in an infinite medium, and also the gas content of the bubble is constant.

Eulerian models are limited to the modeling of structures larger than the computational cell and unable to solve cavitation nuclei and bubbles. Also, with Lagrangian models, it is impossible to model non-spherical structures on a large scale, for this, a multi-scale hybrid approach will be used.

2.3 The VOF-LPT Coupling

This new approach consists on coupling between the two VOF and LPT methods. This approach is developed by Vallier [5], in turn inspired by the study of Tomar [4].

The VOF solver is complemented with a LPT solver which accounts for the dispersed phase on a larger grid.

The cavities are classified as Eulerian structures and Lagrangian bubbles. At each time step, the small cavities that are not resolved by the VOF method are transformed into Lagrangian bubbles (see Fig. 2).

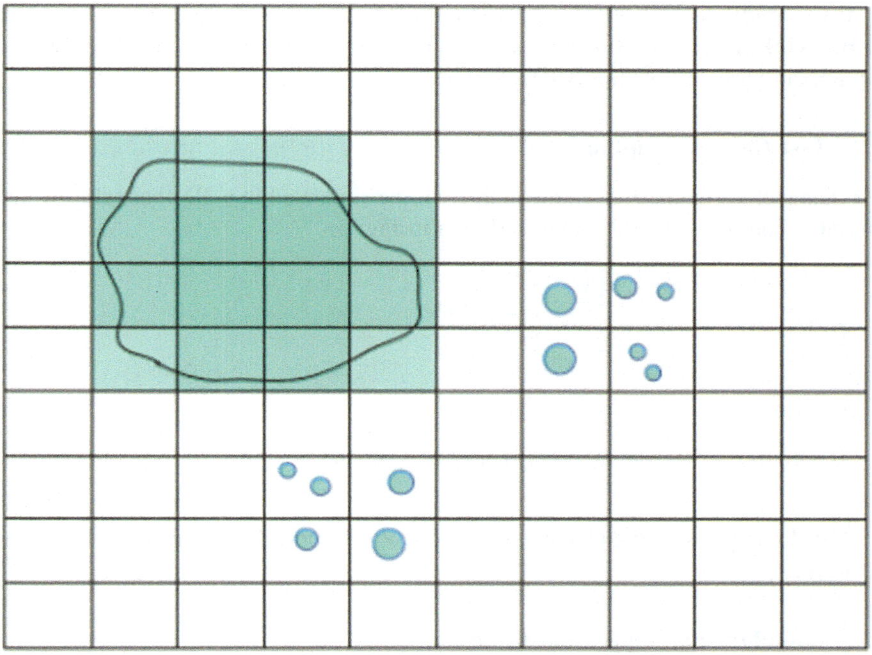

Fig. 2 Figure shows large cavities and small cavities in the fluid

The position, velocity and volume of these bubbles are removed from the Eulerian data and the volume fraction is calculated by the VOF method.

Subsequently the dynamics of small bubbles is solved by a Rayleigh-Plesset equation.

3 Results and Discussions

3.1 The VOF-LPT-RP Solvers

To study the dynamics of a bubble of acoustic cavitation, it is difficult to directly implement the Rayleigh Plesset equation for air bubbles on OpenFoam.

The VOP-LPT solver is then used to treat the air bubbles as solid particles and subsequently implant the RP equation.

The LPT-VOF solver was developed by Vallier [6].

The solver was developed by coupling of interFoam solver with the Lagrangian library.

In this work, a similar solver will be we developed by adding the Lagrangian library to the interPhaseChangeFoam (not the interfoam) solver to make it more suitable for cavitating flows, then we will be implementing the Rayleigh-Plesset equation.

InterphaseChangeFoam is the solver to simulate the cavitation in OpenFOAM.

It is a multiphase solver for 2 incompressible, isothermal immiscible fluids based on the VOF approach. But the Lagrangian library is a compilation of a variety of Lagrangian particle tracking (LPT) libraries.

3.2 Test Case Description

The diameter variation of a single bubble will be simulated in a 2D channel flow under pressure. The initial bubble state is shown in Fig. 3:

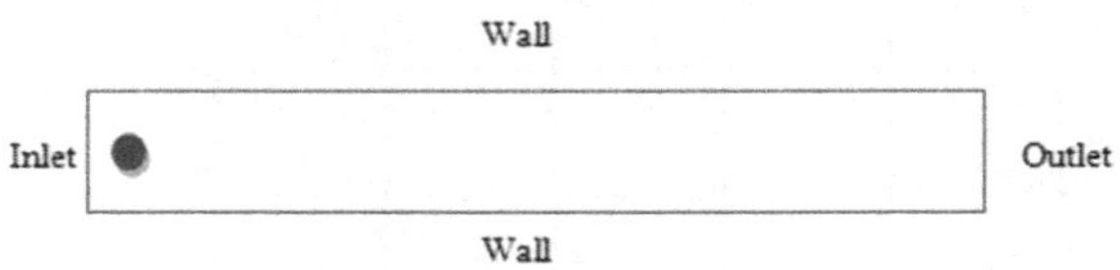

Fig. 3 Bubble in 2D flow

The dimensions of the channel are 10 cm × 0.5 mm (the diameter of the artery is 0.5 mm).

3.3 Boundary and Operation Conditions

We tested our OpenFOAM code with a simple geometry channel, and using water as the fluid.

The inlet pressure is 270 kPa, but the outlet pressure is 70 kPa.

The initial bubble diameter is 4×10^{-4} mm.

The boundary condition for volume fraction has zero-gradient everywhere.

The other boundary conditions are described in Table 1.

Table 1 Boundary conditions

Channel	Boundary
Inlet	Pressure p = 270 kPa,
	Velocity V = zero-gradient
Outlet	Pressure p = 70 kPa
	Velocity V = zero-gradient
Wall	Pressure is zero gradient
	Velocity V = zero
Front and Back	Symmetry planes

The initial volume fraction β is 1(liquid phase). We choose water as the fluid. Its density is 1000 kg/m^3 and its kinematic viscosity is 1E-06. The time step is equal to 0.01 s, and the flow field was solved for 0.65 s.

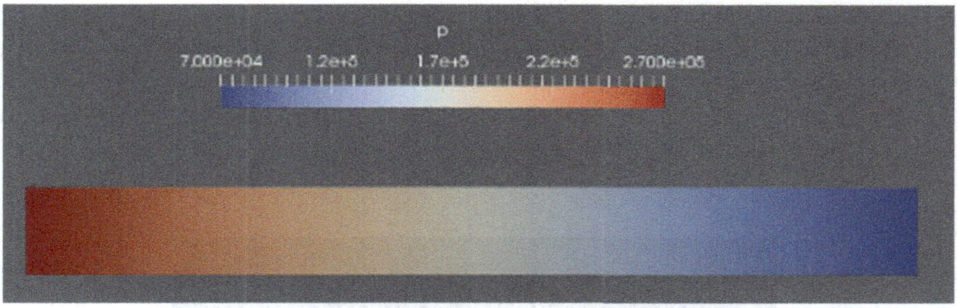

Fig. 4 Numerical simulation: Bubble in channel flow à t = 0 s

3.4 Results

Figure 4 presents the domain and the pressure field before the injection of the bubble.

To visualize the movement of the bubble, we need to open the paraview. The bubble is injected at t = 0.03 s. And because the bubble dimension is so smaller than the dimensions of our channel, we used, in Paraview, 1000 as a scalar factor for making the bubble easier to visualize.

In Figs. 5 and 6, the motion of the bubble inside the channel is depicted for six different time steps. The increase in the size of the bubble during its displacement is due to the decrease in pressure which is visible in Fig. 5. Initially, the diameter of the bubble is 0.4×10^{-4} mm, and its final diameter is about 0.55×10^{-4} mm.

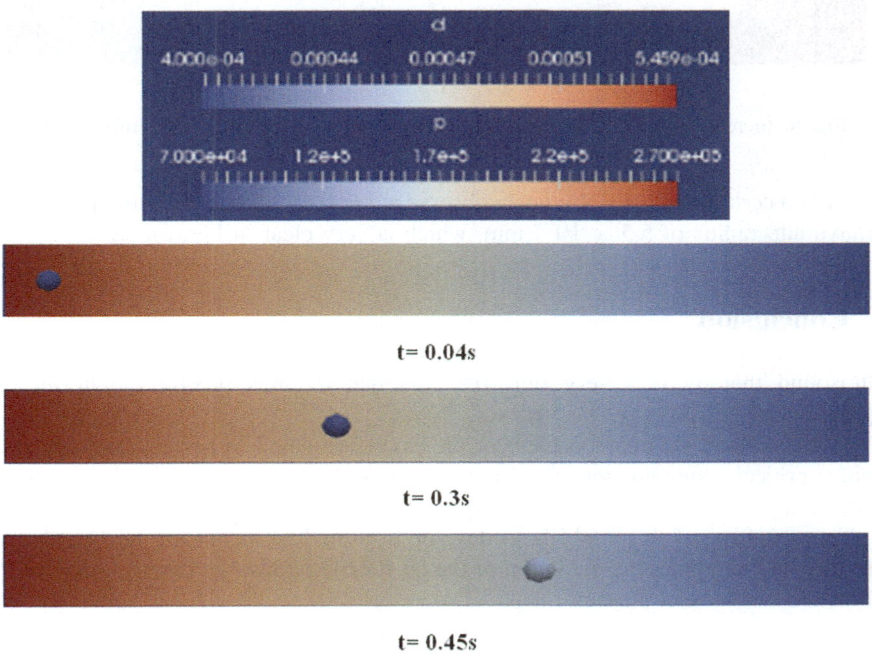

Fig. 5 Bubble size variation with a pressure change in 2D channel flow

The increase in the radius of the bubble was observed as a function of the pressure of the liquid in 3 time steps. There is a minimum radius in the red zone where the pressure is at its peak.

In Fig. 5, the increase in the cavitation bubble during the decrease in pressure over time is clearly observed.

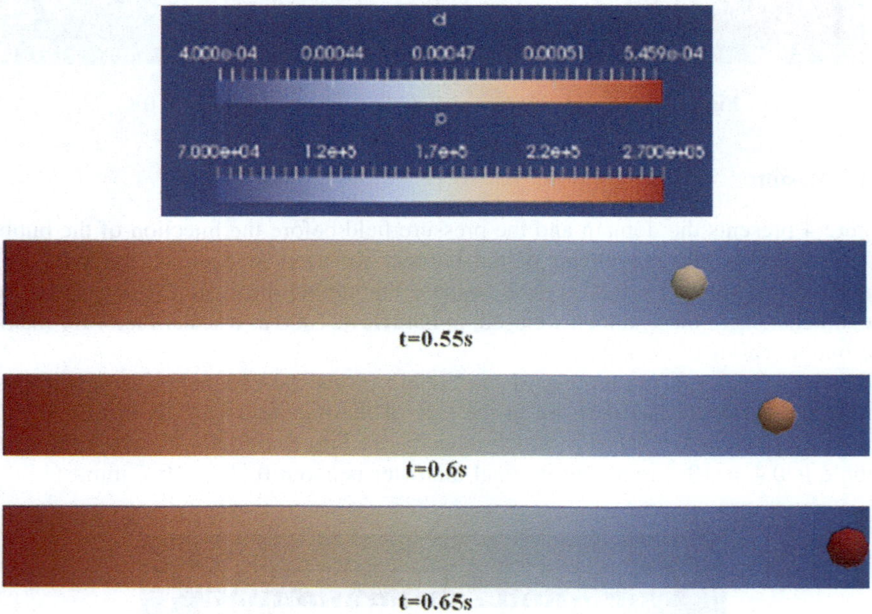

Fig. 6 Increase of the radius of the bubble as a function of the pressure of the liquid

After a certain time, we obviously observe at t = 0,65 s for a minimum pressure for a maximum radius of 5.5×10^{-4} mm, which is very clear in Fig. 6.

4 Conclusion

Ultrasound therapy is a very sensitive area that requires precise monitoring and regulation.

In this context we mention the acoustic cavitation which is a very important and a highly critical phenomenon that requires modeling adequate to the requested application.

The objective of this study is to develop a numerical model on the OpenFOAM interface that simulates the evolution of the air bubble radius (the cavitation bubble) as a function of the inlet pressure.

This server used the Rayleigh Plesset model based on the coupling of the VOF-LPT method which translates the air cavities into Lagrangian particles so that we can solve them with the RP equation.

This study makes it possible to visualize the dynamics of the bubble in a channel of dimension close to that of a blood vessel.

In the next work, we will consider the blood as fluid and we will take into account the pressure of the outer membrane of the blood vessel by studying the parameters that influence cavitation to better propose to another model of this phenomenon.

References

Journal article

1. Ghahramani E (2017) Improvement of the VOF-LPT Solver for Bubbles.
2. Plesset MS (1949) The dynamics of cavitation bubbles. J Appl Mech 16(1949):277–282
3. Singhal AK, Athavale MM, Li H, Jiang Y (2002) Mathematical basis and validation of the full cavitation model. J Fluids Eng 124:617–624. https://doi.org/10.1115/1.1486223
4. Tomar Gaurav, Fuster Daniel, Zaleski Stéphane, Popinet Stéphane (2010) Multiscale simulations of primary atomization. Comput Fluids 39:1864–1874
5. Vallier A (2011) Coupling of VOF with LPT in OpenFOAM
6. Vallier A (2013) The simulations of cavitation from the large vapor structures to the small bubble dynamics, Lund University

Journal article only by DOI

7. Hirt CW, Nicholos BD (1981) Volume of fluid (VOF) method for the dynamics of free boundaries. J Comput Phys 39:201–225. https://doi.org/10.1016/0021-9991(81)90145-5
8. Rayleigh L (1917) Viii. On the pressure developed in a liquid during the collapse of a spherical cavity, The London, Edinburgh, and Dublin Philosophical. Mag J Sci 34 (200):94–98. https://doi.org/10.1080/14786440808635681

Nonlinear Analysis of the Effect of Hydrodynamic Forces on the Stability of an Unbalanced Rigid Rotor

Radhouane Sghir[1,2](✉)

[1] Laboratory of Mechanical Engineering, National Engineering School of Monastir, University of Monastir, Avenue Ibn Eljazzar, 5019 Monastir, Tunisia
sghirradhoineim@yahoo.fr

[2] National Engineering School of Sousse, University of Sousse, Technological Pole of Sousse, Sahloul, 4054 Sousse, Tunisia

Abstract. This paper presents a comparative study between the nonlinear dynamic behavior of an unbalanced rigid rotor in a short hydrodynamic bearing and of an unbalanced rigid rotor in a long hydrodynamic bearing. Two nonlinear mathematical models with two degrees of freedom are used in this investigation to predict the movement of the shaft and its bifurcations. Nonlinearity is introduced into models through analytical expressions of hydrodynamic forces. These analytical expressions are determined by the integration of the oil pressure distribution into the bearing using the short bearing approximation and the long bearing approximation. The numerical integration method is applied to determine the bifurcation diagrams using the rotor speed as a control parameter. In this study, Poincaré sections, frequency spectrum, motion orbits, and bifurcation diagrams are used to characterize the shaft motion. Several nonlinear phenomena such as jumping motion, multi-periodic oscillation, quasi-periodic motion and chaotic motion are predicted. It has been found that the effect of unbalance is very important on the stability threshold speed and on the amplitudes of the oscillations for the case of a short bearing. However, the unbalance effect is negligible in the case of a long bearing.

Keywords: Hydrodynamic forces · Long bearing · Short bearing · Numerical integration · Stability analyses · Nonlinear phenomena

1 Introduction

Hydrodynamic bearings are commonly used in rotating machinery because of their ability to withstand high loads, low friction and high damping compared to ball bearings. However, rotors supported by oil film bearings are susceptible to instability which can lead to wear and premature failure of the bearing.

The movement of a rotor in fluid film bearing is controlled by the hydrodynamic forces. But under proper operating conditions, these efforts can cause instability of the system. This phenomenon is called self-excitation or gyratory movement of oil. It gives rise to large amplitudes of the lateral vibrations. Self-excited vibrations within a hydrodynamic bearing have been described by Newkirk and reported in Refs. [1, 2].

N. Aifaoui et al. (Eds.): CMSM 2019, LNME, pp. 240–248, 2020.
https://doi.org/10.1007/978-3-030-27146-6_26

From the experimental investigations, Newkirk observed that, under certain conditions, the shaft vibrates at a subsynchronous frequency regardless of rotor imbalance. From a study published in Ref. [2], the experimenters noted that the self-excited vibration within a bearing is due to hydrodynamic efforts. In another experimental study [3], two distinct types of vibration called whirl and whip have been reported. Whirl vibrations are characterized by non-violent self-excited oscillations. The frequency of these oscillations is always proportional to the speed of rotation. Whip vibrations are also self-excited vibrations, however, the vibration frequency is constant and independent of the increase in rotor speed. For this type of oscillation, the vibration amplitude is very large and violent, which corresponds to the maximum radial clearance between the shaft and the bearing.

Bearing parameters such as bearing geometry and oil viscosity have a very important effect on the hydrodynamic forces and consequently on the stability threshold of the rotor. Pikus concluded that a high oil viscosity tends to have a stabilizing effect on the rotor bearing systems [4]. Several studies have shown that the variation of the bearing geometry has a very important effect on the stability threshold and the Hopf bifurcation domains (supercritical and subcritical). Myers used the Hopf bifurcation theory to analyze the stability of a rigid rotor symmetrically supported by two infinitely long bearings [5]. He identified the existence of three Hopf bifurcation domains: two subcritical Hopf bifurcation domains, and a supercritical bifurcation domain. A similar analysis, but using the short bearing theory and the numerical continuation method, was published in the Ref. [6]. They indicated the existence of two Hopf bifurcation domains: subcritical and supercritical.

The presence of imbalance defect in addition to the hydrodynamic forces causes a radical change in the rotor movement and in the stability threshold speed. A numerical study reported in the Ref. [7] showed that the rotor imbalance is at the origin of the subharmonic and superharmonic movements. Some articles refer to simple systems such as Jeffcott's rotors [8] to study some examples of non-synchronized periodic movements. This type of movement arises due to unbalance force and non-linear elastic recall forces. Experimental and numerical studies have been carried out in the Refs.[9, 10]. These studies made it possible to identify the paths that lead the movement of a rigid rotor supported by two short hydrodynamic bearings to chaos.

In the present study, we used numerical integration to analyze the effect of hydrodynamique forces on the stability of an unbalanced rigid rotor supported by oil film bearings.

2 Mathematical Modelling

Consider a rigid rotor supported symmetrically by two identical and aligned bearings. A section of a bearing is given in Fig. 1. The position of the journal center O_j relative to the bearing center O_b is defined in polar coordinates by the eccentricity $e = c\varepsilon$ and the attitude angle ϕ, Where ε is the eccentricity ratio and c is the bearing clearance. At a constant rotational speed ω, the forces applied on the journal at each bearing are the hydrodynamics fluid film force components $\overrightarrow{f_\varepsilon}$ and $\overrightarrow{f_\phi}$, the weight of the half of the rotor W and the unbalance force $\overrightarrow{f_b}$.

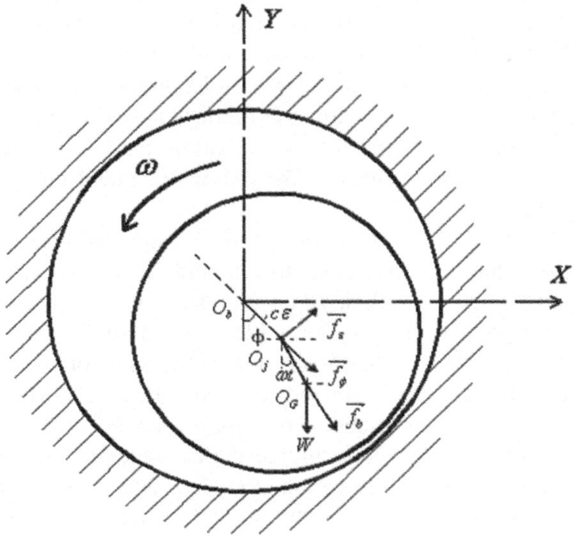

Fig. 1 Bearing section

Applying Newton's second law of the journal center O_j, the following equations are obtained:

$$\begin{cases} \varepsilon'' - \varepsilon\phi'^2 = \frac{\omega^2 a}{c}\cos(\omega t - \phi) + \frac{g}{c}\cos\phi + \frac{f_\varepsilon}{Mc} \\ \varepsilon\phi'' + 2\varepsilon'\phi' = \frac{\omega^2 a}{c}\sin(\omega t - \phi) - \frac{g}{c}\sin\phi + \frac{f_\phi}{Mc} \end{cases} \quad (1)$$

In the above equations, the prime ()' denotes a derivative with respect to time t and M represents the half of the rotor mass.

Hydrodynamic forces are generally determined by the integration of the expression of the pressure field distribution in a bearing. This expression is obtained by solving the Reynolds equation and applying the boundary conditions. The distribution of the pressure field varies according to the bearing geometry. According to the literature, the bearings can be grouped into two families depending on the geometry: the long bearings and the short bearings [11].

When the Half Sommerfeld boundary condition is applied to an infinitely short journal bearing, explicit expressions for the components of the fluid-film force f_ε and f_ϕ in terms of the journal centre position (ε, ϕ) and velocity (ε', ϕ') are obtained:

$$f_\varepsilon = -\frac{\mu RL^3}{2c^2}\left[\pi\varepsilon'\frac{(1+2\varepsilon^2)}{(1-\varepsilon^2)^{5/2}} + 2\varepsilon^2\frac{(\omega-2\phi')}{(1-\varepsilon^2)^2}\right] \tag{2}$$

$$f_\phi = \frac{\mu RL^3}{2c^2}\left[\frac{4\varepsilon'\varepsilon}{(1-\varepsilon^2)^2} + \frac{\pi\varepsilon}{2}\frac{(\omega-2\phi')}{(1-\varepsilon^2)^{3/2}}\right] \tag{3}$$

When the Half Sommerfeld boundary condition is applied to an infinitely long journal bearing, explicit expressions for the components of the fluid-film force f_ε and f_ϕ are obtained:

$$f_\varepsilon = \frac{-12R^3\mu L}{c^2}\left[\frac{\varepsilon'}{(1-\varepsilon^2)^{3/2}}\left(\frac{\pi}{2} - \frac{8}{\pi(2+\varepsilon^2)}\right) + \varepsilon^2\frac{(\omega-2\phi')}{(2+\varepsilon^2)(1-\varepsilon^2)}\right] \tag{4}$$

$$f_\phi = \frac{12R^3\mu L}{c^2}\left[\frac{\pi\varepsilon(\omega-2\phi')}{2(2+\varepsilon^2)(1-\varepsilon^2)^{1/2}} + \varepsilon^2\frac{2\varepsilon\varepsilon'}{(2+\varepsilon^2)(1-\varepsilon^2)}\right] \tag{5}$$

The equation of motion (1) can be written in the following dimensionless form:

$$\begin{cases} \ddot{\varepsilon} = \varepsilon\dot{\phi}^2 + \bar{a}\cos(\tau-\phi) + \frac{1}{\bar{\omega}^2}\cos\phi + \bar{f_\varepsilon} \\ \ddot{\phi} = -\frac{2\dot{\varepsilon}\dot{\phi}}{\varepsilon} + \frac{\bar{a}}{\varepsilon}\sin(\tau-\phi) - \frac{\sin\phi}{\bar{\omega}^2\varepsilon} + \frac{\bar{f_\phi}}{\varepsilon} \end{cases} \tag{6}$$

where, $(\dot{\ }) = d/d\tau = d/\omega dt$ denotes a derivative with respect to nondimensional time τ, and the nondimensional fluid film force components are defined as: $\overline{f_\varepsilon} = f_\varepsilon/Mc\omega^2$, $\overline{f_\phi} = f_\phi/Mc\omega^2$.

When the nondimensional fluid film force components are introduced into the system of Eqs. (6) and using the state variables $x_1 = \varepsilon$, $x_2 = \phi$, $x_3 = \dot{\varepsilon}$, and $x_4 = \dot{\phi}$, the system of Eqs. (6) can be written in the following forms:

$$\dot{x} = f(x, \tau, \bar{\omega}, \Gamma, \bar{a}) \tag{7}$$

$$\dot{x} = f(x, \tau, \bar{\omega}, \lambda, \bar{a}) \tag{8}$$

The above nonlinear systems of equations depends on three nondimensional parameters, the bearing parameter $\Gamma = \mu RL^2/2Mc^{2.5}g^{0.5}$ or $\lambda = \mu R^3L/Mc^{2.5}g^{0.5}$, the dimensionless rotor speed $\bar{\omega} = \omega/\sqrt{g/c}$ and the unbalance parameter $\bar{a} = a/c$, where μ is the lubricant viscosity, R is the journal radius and L is the bearing length.

3 Results

In this study, the numerical integration method is used to determine the rotor center movement, which is defined in polar coordinates ε, ϕ, in the case of a long and a short bearing. To apply numerical integration method, the bearing parameter and the rotor unbalance $\bar{a}$ are selected then, the equations of the nonlinear systems (7) and (8) are integrated by the Runge-Kutta method from a selected initial state. To analyze the effect of unbalance and of hydrodynamic efforts for a low bearing parameter, the nondimensional unbalance value is increased from $\bar{a} = 0.05$ to $\bar{a} = 0.2$ and the variation of the eccentricity ratio is determined at different rotor speeds.

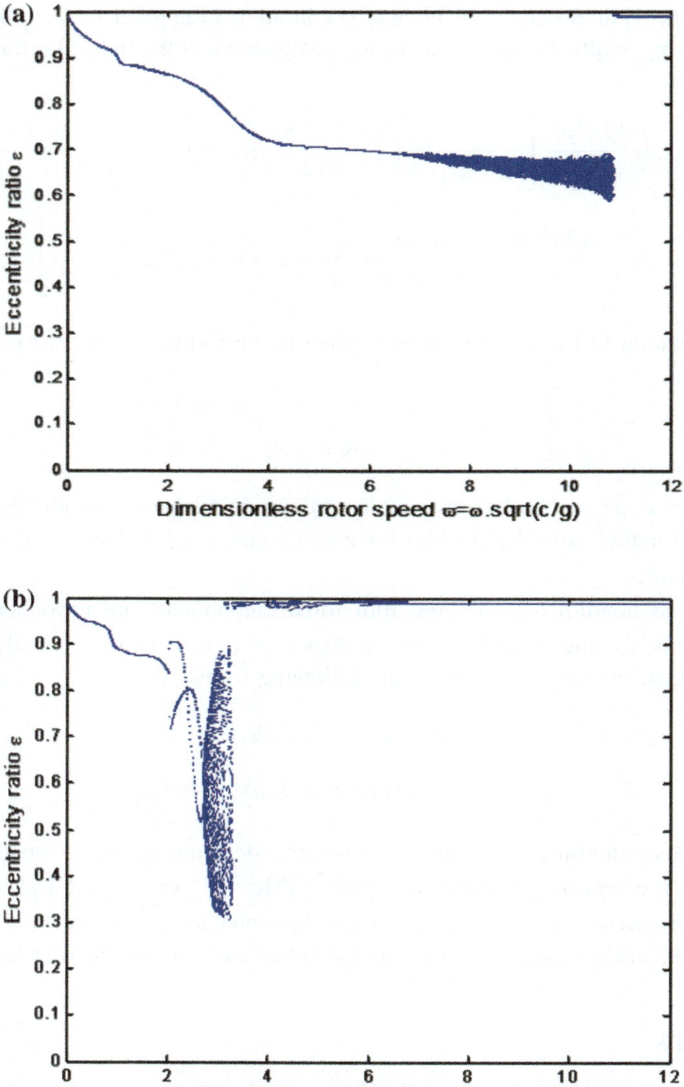

Fig. 2 Effect of increasing unbalance from $\bar{a} = 0.05$ to $\bar{a} = 0.2$ for $\Gamma = 0.02$: **a** Bifurcation diagram for $\bar{a} = 0.05$; **b** Bifurcation diagram for $\bar{a} = 0.2$

Figure 2a shows the bifurcation diagram of a rotor supported by two short bearings for $\Gamma = 0.02$ and $\bar{a} = 0.05$. This diagram indicated that the shaft movement is T-periodic for $\bar{\omega} < 6.74$. For $6.74 < \bar{\omega} < 10.84$ the movement of the rotor center becomes quasi-periodic. The quasi-periodicity of movement is justified by the set of points which forms a closed curve on the Poincaré map and by the unequal spacing between the peaks frequencies in the spectrum, Fig. 3. When the rotor speed $\bar{\omega} > 10.84$, the orbits become 2T-periodic with oscillations of large amplitudes corresponding to $\varepsilon(t) \approx 1$.

For an unbalance value $\bar{a} = 0.2$, the movement of the rotor center transits by varying the speed, from a T-periodic movement to a 2T-periodic motion and then to a quasi-periodic movement before it undergoes a jump towards large oscillations of neighboring amplitude $\varepsilon = 1$ for $\bar{\omega} = 3.24$, Fig. 2b.

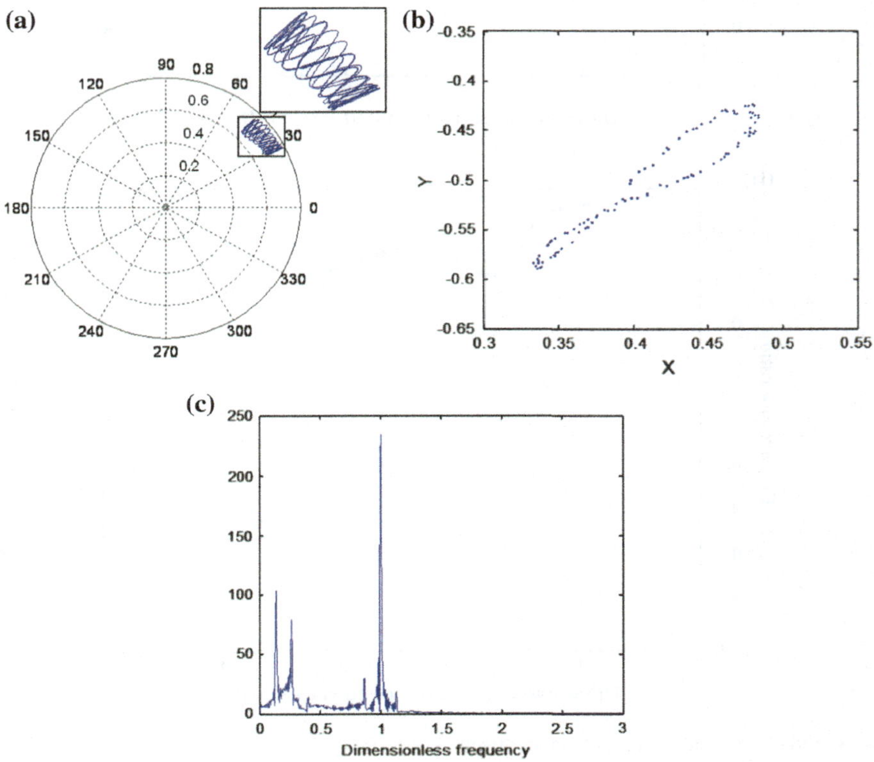

Fig. 3 Quasi-periodic motion for $\Gamma = 0.02$, $\bar{a} = 0.05$ and $\bar{\omega} = 10$: **a** Orbit of rotor center; **b** Poincaré map; **c** Frequency spectrum

In the case of a rotor with short bearings, it can be seen that the increase in the unbalance force has a very important effect on the critical speed of stability and on the movement nature. This effect results in a significant reduction in stability threshold speed. For example, for $\bar{a} = 0.05$, the stability threshold speed is $\bar{\omega} = 10.84$. This

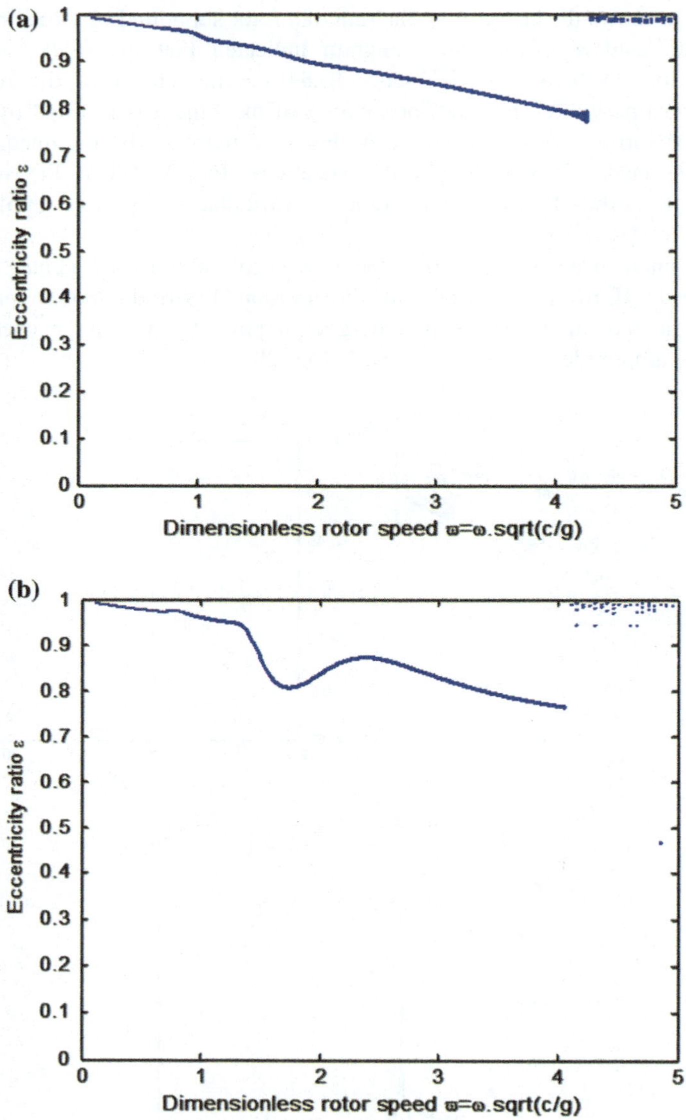

Fig. 4 Effect of increasing unbalance from $\bar{a} = 0.05$ to $\bar{a} = 0.2$ for $\lambda = 0.02$: **a** Bifurtion diagram for $\bar{a} = 0.05$; **b** Bifurcation diagram for $\bar{a} = 0.2$

speed is reduced to $\bar{\omega} = 3.24$, for an unbalance $\bar{a} = 0.2$. This is explained by the fact that for a short bearing the hydrodynamic forces are relatively low as a function of the bearing geometry which is characterized by a negligible length relative to its diameter. These efforts can not control the rotor movement. The latter loses its stability and goes into permanent contact with the bearing under the action of the unbalance force.

Figure 4a show the bifurcation diagram of a rotor with two long bearings for $\lambda = 0.02$ and $\bar{a} = 0.05$. This figure demonstrate that the journal center motion is T-periodic for $\bar{\omega} < 4.24$ and becomes quasi-periodic for $4.24 < \bar{\omega} < 4.33$. The quasi-periodicity of the motion is justified by the closed curve of the Poincaré map in Fig. 5. For $\bar{\omega} > 4.33$, the orbits becomes 2T-periodic with large oscillations corresponding to $\varepsilon(t)$ close to one.

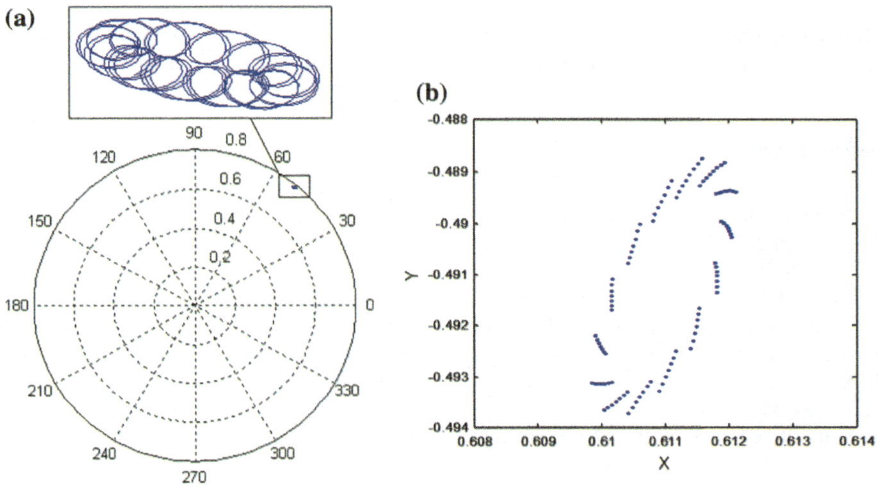

Fig. 5 Quasi-periodic motion for $\lambda = 0.02$, $\bar{a} = 0.05$ and $\bar{\omega} = 4.25$: **a** Orbit of rotor center; **b** Poincaré map

According to Fig. 4a and b, The increase of the unbalance from 0.05 to 0.2 has a weak effect on the amplitude of the motion orbits and on the stability threshold speed. The latter goes from $\bar{\omega} = 4.33$ for $\bar{a} = 0.05$ to $\bar{\omega} = 4.24$ for $\bar{a} = 0.2$. Also, these figures show that the quasi-periodic motion domain narrows progressively with the increase of the unbalance until a total disappearance for $\bar{a} = 0.2$. This result is explained by the fact that for the approximation of a long bearing, the bearing length is quite large relative to its diameter to give a relatively high-pressure field between the shaft and the bearing. Hydrodynamic forces from this pressure field are important enough to control the shaft movement and reduce the effect of the unbalance force.

4 Conclusion

In this paper, two nonlinear mathematical models with two degrees of freedom were used to study the unbalance effect on the shaft movement at different speeds of rotation and for two types of bearing: short bearing and long bearing. The bifurcation diagrams were determined by numerical integration. For a short bearing, the effect of unbalance was found to be very important on the stability threshold speed. On the other hand, for the case of a rotor with long bearings the unbalance effect was found very weak on the critical speed of stability and the amplitudes of movements. This is explained by the dominant role of hydrodynamic forces in a long bearing which can mask the effects of the unbalance defect. This leads us to suggest that the use of long bearings is more adequate than short bearings in the case of a defect. These are very sensitive to unbalance faults, especially for low values of bearing parameters characterized by high rotor mass or low oil viscosity.

Acknowledgements. The authors gratefully acknowledge the helpful comments and suggestions of the reviewers, which have improved the presentation.

References

1. Newkirk BL (1924) Shaft whipping. GE Rev 27:169–178
2. Newkirk BL, Taylor HD (1925) Shaft whipping due to oil action in journal bearings. GE Rev 28:559–568
3. Newkirk BL, Lewis JF (1956) Oil film whirl-an investigation of disturbances due to oil films in journal bearings. Trans ASME 78:21–27
4. Pikus O (1956) Experimental investigation of resonant whip. Trans ASME 78:975–983
5. Myers CJ (1984) Bifurcation theory applied to oil whirl in plain cylindrical journal bearings. J Appl Mech 51:244–250
6. Sghir R, Chouchane M (2015) Prediction of the nonlinear hysteresis loop for fluid-film bearings by numerical continuation. Proc Inst Mech Eng C 229:651–662
7. Ehrich FF (1992) Observations of subcritical superharmonic and chaotic response in rotordynamics. J Vib Acoust 114:93–100
8. Ehrich FF (1991) Some observations of chaotic vibration phenomena in high-speed rotordynamics. J Vib Acoust 113:50–57
9. Adiletta G, Guido AR, Rossi C (1996) Chaotic motions of a rigid rotor in short journal bearings. Nonlinear Dyn 10:251–269
10. Sghir R, Chouchane M (2015) Stability analysis of an unbalanced journal bearing with nonlinear hydrodynamic forces. In: Proceedings of the 9th IFToMM international conference on rotor dynamics. Springer International Publishing, Switzerland, pp 1081–1090
11. Frene J, Nicolas D, Degueurce B, Berthe D, Godet M (1997) Hydrodynamic lubrication bearings and thrust bearings. Elsevier, Amsterdam

Power Losses in a Gearbox Lubricated with Axle Gear Oils

Maroua Hammami[1,2(✉)], Mohamed Slim Abbes[1], Ramiro Martins[3], Jorge H. O. Seabra[2], and Mohamed Haddar[1]

[1] Laboratory of Mechanical, Modelling and Manufacturing, National Engineers School of Sfax (ENIS), BP 1173, 3038 Sfax, Tunisia
hammamimaroi@gmail.com, {slim.abbes,Mohamed.Haddar}@enis.rnu.tn

[2] Faculdade de Engenharia da Universidade do Porto (FEUP), Rua Dr. Roberto Frias s/n, 4200-465 Porto, Portugal
jseabra@fe.up.pt

[3] Institute of Science and Innovation in Mechanical and Industrial Engineering (INEGI), Rua Dr. Roberto Frias 400, 4200-465 Porto, Portugal
rmartins@inegi.up.pt

Abstract. Three axle gear oils (75W90-A, 80W90-A and 75W140-A), available on the market and labeled as "Fuel Efficient", and two candidate products (75W85-B and 75W90-B) were selected and their physical and chemical properties were measured and compared. The friction torque inside rolling bearings lubricated with the five axle gear oils was measured in a dedicated test rig. The measurements and the corresponding numerical simulations indicate that friction torque inside rolling bearings is strongly dependent on the operating conditions and on the axle oil formulations. The model was then applied with success to predict the torque loss in gearbox rolling bearings, in particular, those used in the FZG machine slave and test gearboxes. Gear power loss tests were realized on the FZG gear test machine, using type A10 gears and severe operating conditions. The torque loss measurements and the corresponding numerical simulations clearly pointed out the influence on the base oil type, of the oil viscosity and of the additive package on gear torque loss promoted by different axle oil formulations. An average coefficient of friction between meshing gears was devised from the experimental results. Several aspects regarding the meshing gears power loss were investigated like gear loss factor, the coefficient of friction and the influence of gear oil formulation (axle gear oils). A gearbox total power loss can be predicted by estimating the several power loss sources dissipated in gears, bearings and seals.

Keywords: Axle gear oils · Physical and chemical properties · Friction torque · A10 gears · Power loss

Notation and Units

a	Axis distance [mm]
b	Gear face width [mm]
d_a	Tip diameter [m]

N. Aifaoui et al. (Eds.): CMSM 2019, LNME, pp. 249–260, 2020.
https://doi.org/10.1007/978-3-030-27146-6_27

d_{sh}	Shaft diameter [m]
F_{bt}	Transverse force to tooth flank [N]
H_V	Gear loss factor [–]
H_{VL}	Local gear loss factor using rigid load distribution [–]
K_{FZG}	FZG load stage [–]
m	Gear module [mm]
n	Rotational speed [rpm]
P_{IN}	Input power [W]
P_V	Total power loss [W]
P_V^{exp}	Experimental total power loss [W]
P_{VD}	Seals power loss [W]
P_{VL}	Rolling bearings power loss [W]
ρredC	Equivalent contact radius on the pitch point [mm]
P_{VZ0}	Gears power loss [W]
P_{VZP}	Load dependent gears power loss [W]
R_a	Average surface roughness [m]
T_W	Wheel static torque [Nm]
x	Addendum modification coefficients [–]
$v_{\Sigma c}$	Sum of the rolling velocities on the pitch point [m/s]
X_L	Friction coefficient lubricant parameter [–]
Z	Number of teeth of pinion or gear [–]
α	Pressure angle [°]
β	Gear helix angle [°]
η	Dynamic viscosity [Pas]
ε_γ	Total contact ratio [–]
μ_{bl}	Coefficient of friction in boundary film lubrication [–]
μ_{EHL}	Coefficient of friction in full film lubrication [–]
μ_{mz}	Average friction coefficient along the path of contact [–].

1 Introduction

This work starts with a rolling bearing torque loss model calibrated for RTB (cylindrical roller thrust bearings—81107 TN) and for several axle gear oil formulations and presented in Hammami et al. [1]. The same model was then successfully applied to predict the torque loss in gearbox rolling bearings, in particular those installed in the slave and test gearbox of the FZG machine. This work presents in its second part with the analysis of the power loss in FZG type A spur gears lubricated with the same axle gear oils used in the first part, under oil bath lubrication at 70 °C. With the limitations of the FZG machine in gears geometry, it is not possible to perform tests with hypoid gear set. For that, we selected type A10 spur gears reproducing partially the operating conditions of real hypoid gear such as high temperature, high contact pressure and very high slide-to-roll ratio. Similar tests were found in literature for testing gear oils with high EP [2] and still recommended by ISO Standard—ISO 14635-2 [3].

The power loss tests performed with FZG test machine will be presented and discussed. They allow to validate the lubricant parameter determined for each axle gear oil.

The authors to accurately calculate the meshing gear power loss, they determine an average coefficient of friction along the path of contact. Also, a gear loss factor have a significant influence in the quantification of meshing gears power loss. So, the gear loss factor will be discussed and a method validated for a wide range of gear geometries will be also used.

2 Materials and Methods

The experimental tests were conducted using the FZG gear test machine presented in Fig. 1. In order to provide a fixed torque load to a pair of precision test gears, this machine uses the recirculating power principle. The drive (slave) gearbox was connected to the test wheel and test pinion by two torsional shafts. The shaft connected to test pinion contains a load clutch dividing it in two parts. A locking pin is used to fix one part of the clutch and a load lever with weights are needed to twist the clutch second part.

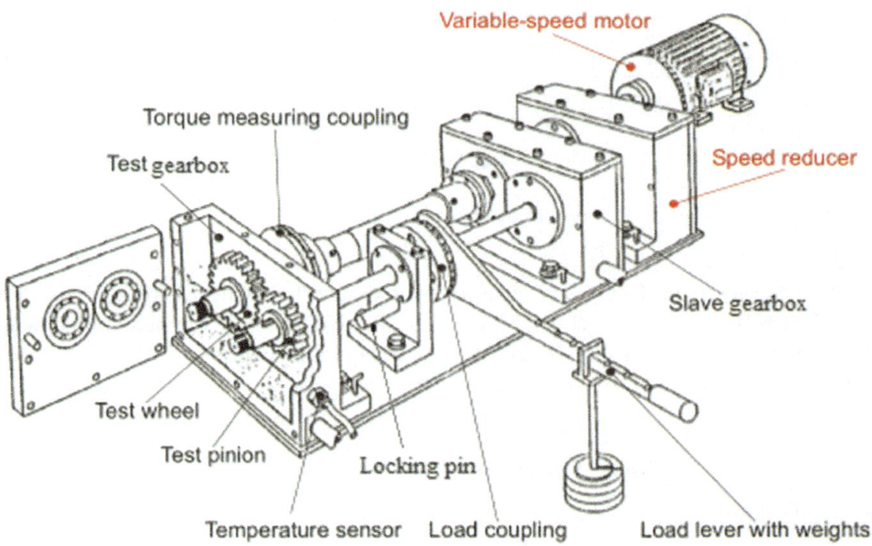

Fig. 1 FZG machine as a main apparatus of the experimental work

For all experimental tests, we are measuring the operating temperatures in several positions of the test rig as presented in Fig. 2 using type K thermocouples.

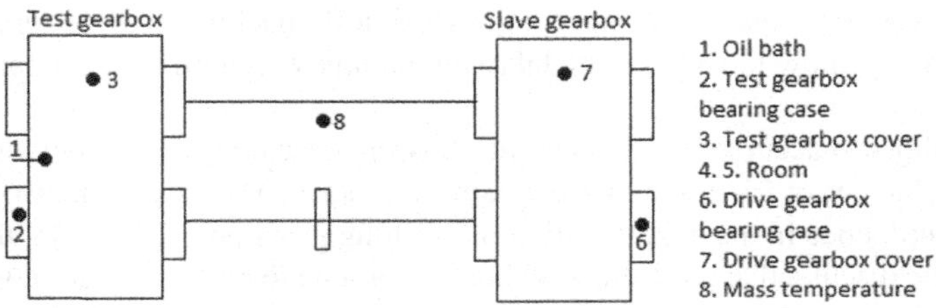

Fig. 2 Thermocouples in specific points of the test rig

An EHT Messtechnik DRDL II torque transducer was mounted in the FZG machine before the slave gearbox as shown in Fig. 3 in order to measure the torque loss (T_L). An input static torque (T_{IN}) was applied on the pinion of test gearbox.

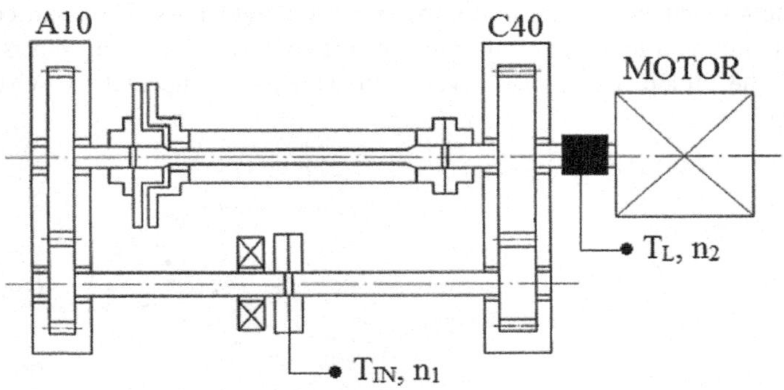

Fig. 3 The system of measuring torque in FZG machine

For these tests the FZG slave gearbox was assembled with a C40 gear set and 4 cylindrical roller bearings (NJ 406MA) while the test gearbox was assembled with a A10 spur gear set (see Table 1) and 4 cylindrical roller bearings (NJ 406 MA) (see Table 2).

Table 1 C40 and A10 FZG gears properties

Gears	Parameters											
		z [/]	m [mm]	a [mm]	α [°]	β [°]	b [mm]	x [/]	da [mm]	ε_Y [/]	H_{VL} [/]	Ra [μm]
FZG C40	Pinion	16	4.5	91.5	20	0	40	0.1817	82.46	1.44	0.1995	0.7
	Wheel	24						0.1715	118.54			
FZG A10	Pinion	16	4.5	91.5	20	0	10	0.8532	73.2	1.33	0.3044	0.3
	Wheel	24					20	−0.5	109.8			

Table 2 Slave and test FZG gearboxes configuration

Gearbox	Slave	Test
Gears	C40	A10
Rolling bearings	4 NJ406 MA	4 NJ406 MA
Seals	2 viton seals (d = 30 mm) + 1 viton seals (d = 26 mm)	2 viton seals (d = 30 mm)

The total power losses (C40 + C40) dissipated in the slave and the test gearboxes were experimentally measured in previous work [4].

The operating conditions used for the torque loss tests were selected based on the conditions of a real axle.

The load stages applied for experimental tests with their corresponding wheel torque are the follows: K1 (T_W = 4.95Nm), K5 (T_W = 104.97 Nm), K8 (T_W = 257.38 Nm) and K11 (T_W = 478.72 Nm).

Table 3 shows the input power for all range speed and load stages.

According to Kolekar study [5] and by taking into account the input motor limitation, the rotational speed range of the input shaft was chosen between 250 rpm and 1750 rpm. Table 3 shows the input power for all range speed and load stages.

A dip lubrication is selected for the test gearbox since it is the distributing way of lubricants in rotating component of axles. In order to decrease the no-load losses, a low lubricant level was introduced in the test gearbox and its volume is fixed to 1.1 l under a temperature of 70 °C (Fig. 3).

A different way of lubrication was used for the slave gearbox which is jet lubrication with an oil flow of 3l/min and under a constant temperature of 80 °C.

Figure 4 presents the experimental tests procedure.

In order to keep the temperature constant during several days, we maintain the oil circulation system working at night and the machine room was equipped with a ventilation system.

The temperature and power loss results shown in this work are the average of the final 30 min of the test since a stabilized environment are achieved.

Table 3 Input power for all operating conditions

Wheel speed [rpm]	Input power [W]			
	K1	K5	K8	K11
250	130	2748	6738	12533
500	259	5496	13476	25066
1000	518	10992	26953	50131
1750	907	19237	47167	87730

The slave gearbox was lubricated with PAO 150 for all torque loss tests while for the test gearbox five axle gear oils were used as follow: 75W90-A, 75W85-B, 80W90-A, 75W90-B and 75W140-A. Those five axle gear oils chemical and physical properties can be found in previous work [6].

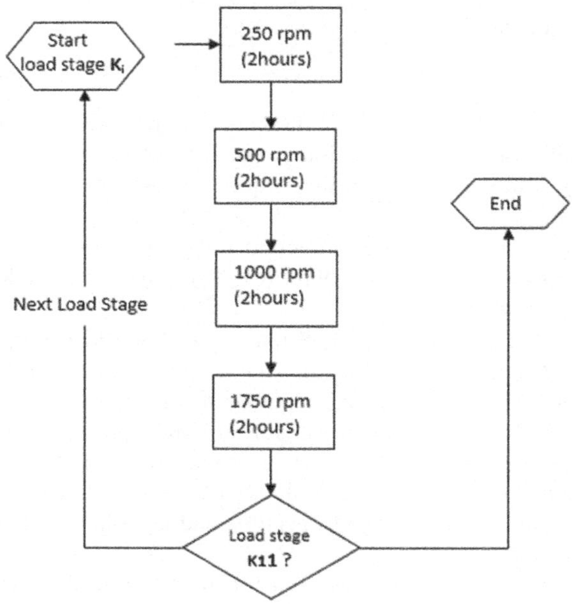

Fig. 4 Experimental tests procedure

3 Power Loss Model with C40/A10 Gears

According to Hohn et al. [7, 8] the gearbox power losses were dissipated mainly in friction loss between the meshing gears, friction loss in the rolling bearings, friction loss in seals and auxiliary losses.

The auxiliary losses are a different dissipative sources which can be generated by oil pumps and they are disregarded in our case. The gearbox power loss takes into account four different power losses as shown in the following Eq. (1).

$$\underbrace{P_V}_{power\,loss} = \underbrace{P_{VZ0}}_{no-load\,gear\,losses} + \underbrace{P_{VZP}}_{load\,dependent\,gear\,loses} + \underbrace{P_{VL}}_{bearings\,losses} + \underbrace{P_{VD}}_{seals\,losses} \quad (1)$$

3.1 Seals Power Loss

The seal manufacturers as e.g. Simrit [9] suggested an unsophisticated equation to predict the seal losses. This simple equation includes only the shaft diameter dsh and the rotational speed n while the oil effect is not considered. The experimental work of Freudenberg summarized in Eq. (2).

$$P_{VD} = 7.69 \times 10^{-6} \cdot d_{sh}^{2} \cdot n \qquad (2)$$

3.2 Rolling Bearing Power Loss

The rolling bearing power loss was calculated using SKF model in previous work [1]. Since the corresponding μ_{bl} and μ_{EHL} coefficients for each axle gear oil lubricating RTB bearings were determined, the power loss of drive gearbox rolling bearings (see Table 2) was calculated as shown in Table 4.

3.3 No-Load Gears Power Loss

The no-load gear losses can be quantified by performing experimental tests under K1 load stage (low input torque, T_W = 4.95 Nm). Based on experimental results and according to Eq. (3), it is possible to estimate this type of power loss.

$$P_{V}^{1\,exp} = P_{VZ0}^{1} + P_{VZP}^{1} + P_{VL}^{1} + P_{VD}^{1} \qquad (3)$$

Table 4 Power loss of drive gearbox rolling bearings P_{VL} [W]

Operating condition	75W85-B	75W90-A	75W90-B	80W90-A	75W140-A
250					
K1	1.83	2.34	2.42	2.29	3.18
K5	5.30	6.65	6.69	6.36	8.73
K8	7.69	9.55	9.48	9.26	12.09
K11	10.41	12.65	12.35	12.09	15.63
500					
K1	5.48	7.17	7.12	6.71	9.83
K5	14.66	19.13	19.02	17.91	26.27
K8	20.53	26.18	25.91	24.61	34.91
K11	26.44	32.92	32.33	29.93	44.02
1000					
K1	15.92	20.76	21.10	19.26	27.95
K5	41.57	54.59	54.96	51.56	73.92
K8	56.11	72.80	72.66	67.33	95.71
K11	66.51	82.15	89.01	67.48	119.39
1750					
K1	37.46	47.62	48.16	45.12	61.81
K5	96.51	123.38	123.62	116.62	162.60
K8	127.42	160.41	162.56	135.14	205.36
K11	327.71	153.65	166.14	139.39	254.83

The no-load gears loss (PVZ0) is determined through Eq. (4), since $P^1_{VZP} = 0$.

$$P_{VZ0} = P_V^{1\,exp} - P^1_{VL} - P_{VD} \tag{4}$$

The corresponding results are displayed in Fig. 5 for the axle gear oils considered.

3.4 Gears Power Loss

The load dependent gear power loss PVZP can be calculated under all load stages using Eq. (5) by subtracting from the total experimental power loss the no-load, the rolling bearings and the seal losses previously calculated.

$$P_{VZP} = P_V^{exp} - (P_{VL} + P_{VD} + P_{VZ0}) \tag{5}$$

Since the meshing gear power loss was calculated, it is possible to determine an experimental average coefficient of friction (μ_{mz}^{exp}) along the path of contact for FZG A10 gears using the Ohlendorf [10] Eq. (6).

$$\mu_{mz}^{exp} = \frac{P_{VZP}}{P_{IN} \cdot H_V} \tag{6}$$

where H_V is the gear loss factor determined in previous work [11] and shown in Table 1.

Table 6 presented the experimental average coefficient of friction (μ_{mz}^{exp}) under all operating conditions and for all axle gear oils.

An important lubricant parameter X_L must be adjusted based on the experimental coefficient of friction between gear teeth in order to take into account the influence of lubricants and their additives. According to Schlenk [12] Eq. (7) and when minimizing the difference between experimental and model values, a lubricant factor X_L can be determined.

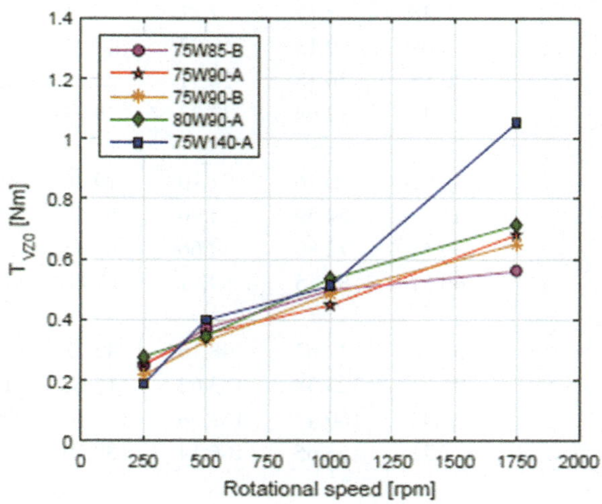

Fig. 5 Load independent torque loss at K1 (T_{VZ0}) for A10 gears

$$\mu_{mZ} = 0.048 \cdot \left(\frac{F_{br}/b}{v_{\Sigma C} \cdot \rho_{redC}}\right)^{0.2} \cdot \eta^{-0.05} \cdot R_a^{0.25} \cdot X_L \tag{7}$$

Table 5 displays the adjusted lubricant parameter for each tested axle gear oil.

Table 5 Lubricant factor X_L for each axle lubricant

Oil	75W85-B	75W90-A	75W90-B	80W90-A	75W140-A
X_L	0.71	0.66	0.71	0.80	0.65

Table 6 Coefficient of friction based on experimental results for A10 gears

(μ_{mz}^{exp})	75W85-B	75W90-A	75W90-B	80W90-A	75W140-A
250					
K5	0.059	0.049	0.068	0.057	0.050
K8	0.057	0.052	0.060	0.060	0.053
K11	0.058	0.061	0.056	0.045	0.049
500					
K5	0.050	0.038	0.054	0.050	0.039
K8	0.049	0.042	0.050	0.056	0.043
K11	0.049	0.051	0.047	0.039	0.041
1000					
K5	0.045	0.029	0.044	0.043	0.037
K8	0.042	0.035	0.042	0.049	0.038
K11	0.039	0.043	0.038	0.037	0.035
1750					
K5	0.044	0.024	0.041	0.037	0.012
K8	0.036	0.030	0.036	0.042	0.022
K11		0.033	0.031	0.028	0.023

From here, the meshing gears power loss P_{VZP} can be calculated using Eq. (8), where the friction coefficient μ_{mz} is only dependent on the test conditions and lubricant parameter (X_L) which is calculated using Schlenk formula Eq. (7).

$$P_{VZP} = P_{IN} \cdot H_V \cdot \mu_{mz} \tag{8}$$

4 Results and Discussion

The model was summarized in Eq. (9) which can provide each power loss source separately in the test gearbox. The several power loss contributions were determined and the friction coefficient was adjusted for each axle lubricant.

$$P_V^i = \underbrace{P_{VZ0}}_{P_V^{1exp} - P_{VL}^1 + P_{VD}} + \underbrace{P_{VZP}^i}_{P_{IN} \cdot H_V \cdot \mu_{mz}(X_L)} + \underbrace{P_{VL}^i}_{NewSK\ FModel} + \underbrace{P_{VD}}_{Simrit\ Equation} \tag{9}$$

Figure 6 presents with lines the experimental torque loss results for the slave gearbox were the type A10 gears were assembled, for K1, K5, K8 and K11 load stages and for different axle gear oil formulations. The different sources of power loss were shown also in Fig. 6 with bars and the sum represent the total power loss of the slave gearbox based on model predictions.

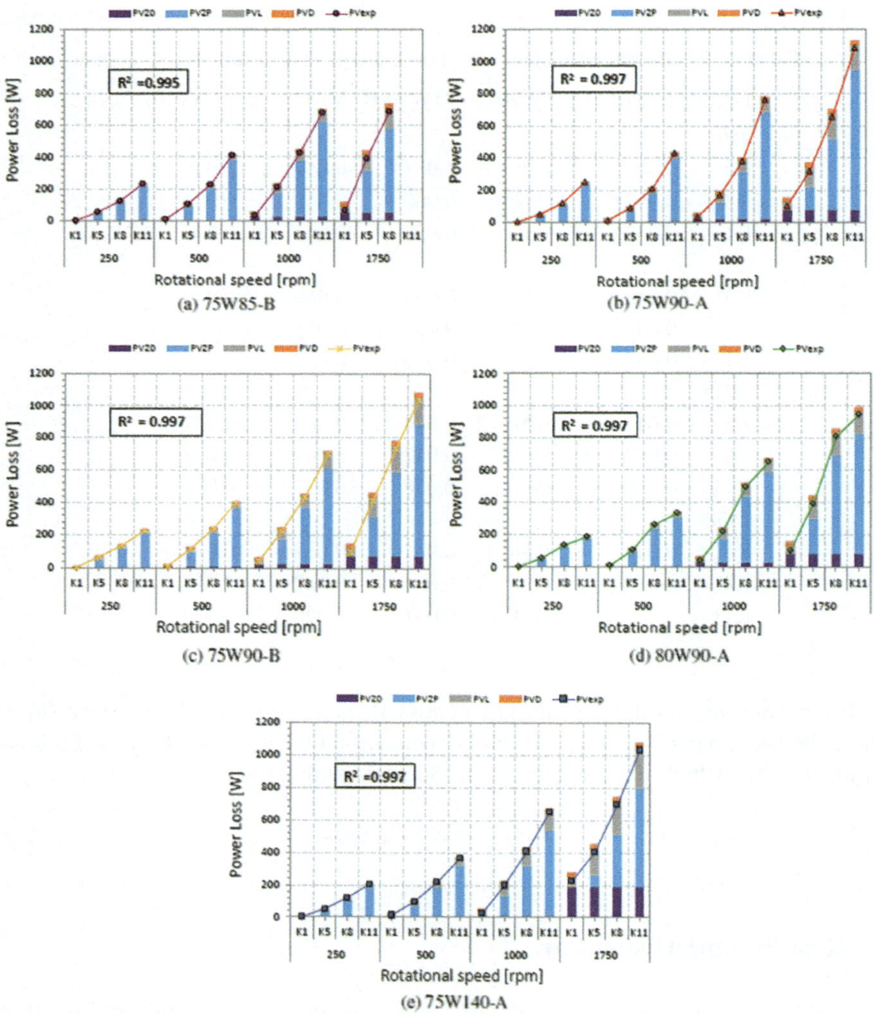

Fig. 6 Model prediction of each source of power loss against experimental test gearbox power loss results (P_{VA10}^{exp}) for all oil formulation

It is observed that whatever the oil selected, the total power loss increases with increasing load and rotational speed.

For all the tests performed, the gears losses (P_{VZP}) are important in the total power loss and after that the bearings losses (P_{VL}) are less important.

Under K1 load stage, the lubricants show a similar behavior since these tests were performed to quantify the no-load losses. For the tests under low rotational speed of 250 and 500 rpm, it seems that the lubricants behavior is affected mainly by the oil coefficient of friction as presented in Table 6. For high rotational speed from 1000 to 1750 rpm, the influence of the physical and chemical properties of the oil appears. These properties can be found in previous work [6].

The two lubricants 75W85-B (Fig. 6a) and 75W90-B (Fig. 6c) with the same PAO base oil, Viscosity

Index and similar additives (B) does not present the same power loss since the 75W90-B oil has the highest viscosity at the operating temperature.

It is noticed that the lubricants 75W90-A (Fig. 6b) and 75W90-B (Fig. 6c) having the same base oil and a close dynamic viscosity but with different additive package (A (Magnesium (Mg), Phosphorus (P) and Sulphur (S)) vs B (Calcium (Ca) and Zinc (Zn)). The 75W90-B generated higher power loss than the 75W90-A oil.

Oils 80W90-A and 75W140-A with the same additive package (A) but with different base oil, mineral base oil for 80W90-A (Fig. 6d) and PAO base oil for 75W140-A (Fig. 6e). It is clear that lubricant with mineral base presented a higher power loss than PAO base oil especially under high load. A very good correlation ($R^2 = 0.995$ to 0.997) is displayed in Fig. 6 between experimental and predicted power loss results which means that we can validated the lubricant parameter, the average coefficient of friction between meshing gears and the gear loss factor used in this work for all lubricants and for all operating conditions except for 1750 rpm because it was not possible to keep the temperature constant (70 °C).

5 Conclusion

The torque loss tests lead as to extract those conclusions concerning axle gear oil formulations: under high load, the 80W90-A oil generated the lowest power loss no matter the rotation speed selected while the 75W90-A oil presented the lowest power loss when lubricating the A10 gear geometry assembled in the test gearbox. The lubricant with mineral base oil and lowest Viscosity Index can reduce at least about 12% when it is compared with the 75W90-A oil.

Those results can be modified under low and medium load stages where the opposite behavior is observed. The 75W90-A lubricant generated the lowest power loss values due to its low viscosity and their additive package (Phosphorus and Sulphur). This oil is selected for its power loss reduction over all other oils which can reach 22%.

The influence of oil dynamic viscosity, Viscosity Index, base oil and additive package was verified through the power loss trend. Replacing lubricant formulation can provide a gearbox efficiency increasing of 0.5%.

The power loss model drawn with its details can predict accurately the power loss sources of each component inside the gearbox since a good agreement between experimental results and model values is observed. For that the lubricant parameter and the friction coefficient of each axle gear oil can be a reference values for future study.

Acknowledgements. The authors gratefully acknowledge the funding supported by National Funds through projects NORTE-01-0145- FEDER-000022—SciTech—Science and Technology for Competitive and Sustainable Industries, co-financed by Programa Operacional Regional do Norte (NORTE2020), and Fundo Europeu de Desenvolvimento Regional (FEDER) and LAETA under the project UID/EMS/50022/2013.

References

1. Hammami M, Martins R, Fernandes C, Seabra J, Abbes MS, Haddar M (2018) Friction torque in rolling bearings lubricated with axle gear oils. Tribol Int 119:419–435. https://doi.org/10.1016/j.triboint.2017.11.018
2. Winter H, Michaelis K (1986) Scoring tests of aircraft transmission lubricants at high speeds and high temperatures. J Synth Lubr 3(2):121–135. https://doi.org/10.1002/jsl.3000030205
3. 14635-2 (2004) I. FZG test procedures–Part 2: FZG step load test A10/16, 6R/120 for relative scuffing load-carrying capacity of high EP oils
4. Fernandes CM, Martins RC, Seabra JH (2014) Torque loss of type C40 FZG gears lubricated with wind turbine gear oils. Tribol Int 70(0):83–93. https://doi.org/10.1016/j.triboint.2013.10.003. ISSN 0301-679X
5. Kolekar AS (2013) Lubrication & efficiency of rear wheel drive axles in road vehicles. PhD. thesis
6. Hammami M, Martins R, Abbes MS, Haddar M, Seabra J (2017) Axle gear oils: Tribological characterization under full film lubrication. Tribol Int 106:109–122
7. Hohn B-R, Michaelis K, Vollmer T (1996) Thermal rating of gear drives: balance between power loss and heat dissipation. AGMA Technical Paper
8. Hohn, B-R, Michaelis K, Hinterstoißer M (2009) Optimization of gearbox efficiency. Goriva imaziva, 48(4):462
9. Simrit (1976) Radialwellendichtringe, Kataog Nr. 100
10. Ohlendorf H (1958) Verlustleistung und Erwärmung von Stirnrädern. PhD. thesis, Dissertation TU München
11. Hammami M (2017) Efficiency and wear in automotive gear transmissions. PhD. thesis, National School of Engineers of Sfax
12. Schlenk L (1994) Unterscuchungen zur Fresstragfähigkeit von Grozahnrädern. PhD. thesis, Dissertation TU Munchen

A Low Cost Uncertainties Propagation Study for a Coupled Fluid Structure System

B. Ben Smida[1,2], M. Guedri[3,4(✉)], and N. Bouhaddi[5]

[1] Université de Tunis El Manar, Ecole Nationale d'Ingénieurs de Tunis, Laboratoire de Génie Civil, BP 37 le Belvédère 1002, Tunis, Tunisia
bbensmida@tvtc.gov.sa

[2] Al Ahsa College of Technology, Technical and Vocational Training Corporation, 7823, Riyadh 11472, Kingdom of Saudi Arabia

[3] Department of Mechanical Technology, College of Technology at Makkah, Technical and Vocational Training Corporation, Makkah, Kingdom of Saudi Arabia
mguedri@tvtc.gov.sa

[4] Preparatory Engineering Institute of Nabeul, 8000 M'Rezgua, Nabeul, Tunisia

[5] Department of Applied Mechanics, University Bourgogne Franche-Comté, FEMTO-ST Institute, UMR 6174, CNRS/UFC/ENSMM/UTBM, 25000 Besançon, France
noureddine.bouhaddi@femto-st.fr

Abstract. In this work, we propose uncertainties propagation study for a coupled fluid structure system. It focuses on the designer needs to predict the vibro-acoustic behavior of systems with uncertainties properties and improve the product quality by mastership the variation sources. The proposed strategy is made by four main steps. The first one is to select the random design parameters according to their sources and their significance level. The second step uses a hypercube latin sampling technique as a reference in the deviation range of Six-Sigma. The third step uses a model order reduction technique. The fourth and last step compares the approximated stochastic responses by a polynomial chaos expansion (PC) to the reference. The computational cost (CPU time) and the accuracies of the proposed strategy are compared and discussed for the extreme statistics and the mean behaviors output results. The performances of the suggested method are established through a double walls numerical example using a stochastic finite element method (SFEM).

Keywords: Uncertainties · Vibro-acoustic · Polynomial chaos · Hypercube latin

1 Introduction

The stochastic behavior study for coupled fluid structure systems faces several implementation problems. The latter relate to the finite element model size and the uncertainty sources' diversity which are time consuming.

Matthies et al. [11] and Van den Nieuwenhof [18] differentiate between three categories of uncertainty sources. The first one emanates the gap between the mathematical model and real phenomena. The second one results from the error between the

N. Aifaoui et al. (Eds.): CMSM 2019, LNME, pp. 261–270, 2020.
https://doi.org/10.1007/978-3-030-27146-6_28

mathematical and numerical model. The third kind of uncertainty sources relates to unknown and not totally identified design parameters in the mathematical model.

In order to deal with uncertainties, we distinguish two approaches which can be used in the context of finite element method (FEM): the parametric and the non-Parametric approach [17]. Mingjie and Qibai [12]. The non-parametric one consists of representing the whole of uncertainties in a single dispersion parameter or a set of parameters [16]. This approach is well adapted to estimate the response variability and uncertainties propagation. However, it has some disadvantages manifested mainly in its inability to estimate the impact of a given design parameter on the global dispersion. The parametric approach is based on the parametric description of random design parameters [8]. The random variables are built by considering the dispersion around their nominal values.

Many recent uncertainties and global sensitivity analysis for vibroacoustic problems have been presented in the literature with various techniques. Monte Carlo (MC) simulation [3], spectral stochastic finite elements method (SFEM) [5] perturbation methods or Neumann-series expansion methods [5, 9]. The MC simulation is the simplest approach known by its accuracy. Nevertheless, it suffers from prohibitive computational costs. Other approaches are based on the global expansion which are more popular. The Karhunen-Loève expansions [10] and the generalized Polynomial Chaos (gPC) expansions [4, 8, 19, 20].

The generalized Polynomial Chaos (gPC) transforms the stochastic problem into an equivalent deterministic problem with an efficient low computation time. The accuracy of the results depends on the polynomial order and the problem transformation is in higher dimension.

For the vibroacoustic problem with uncertainties, using the gPC is affronted by the exploded large model size that can be impossible for some industrial cases. To overcome these difficulties, some scientific works in the literatures used a model order reduction technique [7, 14].

In this contest, we propose a parametric uncertainty propagation study based on spectral stochastic finite elements method (SFEM). Four main steps are used to reduce the computational costs. The random design parameters can be chosen based on their sensitivities. A Latin Hypercube Sampling method (LHS) [13, 15] is implemented to generate input sample data. The LHS is used with a normal deviation in the range of Six-Sigma around the nominal value for each random parameter. A Model Order Reduction technique with a controlled accuracy is used [2]. The generalized Polynomial Chaos (gPC) is used and compared to the results obtained by LHS.

The presented methodology's performances are illustrated through a numerical double panel example coupled with fluid cavities. This example is commonly used for experimental and numerical fluid structure system behavior prediction [1, 6].

2 Stochastic Vibroacoustic Damped Model

The fluid structure problem is based on the hypothesis of linear acoustics for a compressible fluid, not viscous, and a small vibration of the fluid flow. For a displacement-pressure formulation (u, p), the damped fluid-structure coupled system with poroelastic material in the rigid walls is based on the following linear equations system:

$$\left(\begin{bmatrix} K_s & -C \\ 0 & K_f \end{bmatrix} - \omega^2 \begin{bmatrix} M_s & 0 \\ \rho_f C^T & M_f \end{bmatrix} + \frac{j\omega}{Z_a(\omega)} \begin{bmatrix} 0 & 0 \\ 0 & A_f \end{bmatrix}\right) \begin{Bmatrix} U \\ P \end{Bmatrix} = \begin{Bmatrix} F_s \\ 0 \end{Bmatrix} \quad (1)$$

where $M_s K_s$ are respectively the mass matrix and the stiffness matrix of the structure, K_s and M_s are respectively the matrices corresponding to the discretization of the kinetic energy and compressibility of the fluid. F_s is the external load applied to the structure and C is the coupling matrix. A_f is the absorbing acoustic matrix, symmetric and depending on the absorbing surface geometry.

In order to damp the fluid cavity resonance, the rigid walls on the source cavity are coated by poroelastic patches characterized by the frequency evolution of the acoustic impedance $Z_a(\omega)$.

In the context of parametric approach of uncertainties and Gaussian distribution, the random design parameters θ can be represented as:

$$\theta = \theta_o + \Delta\theta = \theta_o(1 + \delta_o \xi) \quad (2)$$

where, θ_o is the nominal value of θ, $\Delta\theta$ is the dispersion around θ_o, δ_o is the dispersion level and ξ is a central Gaussian variable.

Consequently, the finite element model in Eq. (1) can be written in the stochastic following compact form of θ:

$$\left([\mathbf{K}(\theta)] - \omega^2[\mathbf{M}(\theta)] + \frac{\mathbf{j}\omega}{\mathbf{Z_a}(\omega)}[\mathbf{A_f}(\theta)]\right)\{\mathbf{Y}(\theta)\} = \{\mathbf{F}\} \quad (3)$$

The symbols (i.e. $\mathbf{K}(\theta)$) are used to define the random variables and $\{\mathbf{Y}(\theta)\}^{\mathbf{T}} = \{\mathbf{U^T}\,\mathbf{P^T}\}$ is the stochastic response.

3 Resolution Methods

3.1 Direct Method

By using the Eq. (1) and the direct inversion of the dynamic stiffness leads to the computation of the following response:

$$\begin{Bmatrix} U \\ P \end{Bmatrix} = \left(\begin{bmatrix} K_s & -C \\ 0 & K_f \end{bmatrix} - \omega^2 \begin{bmatrix} M_s & 0 \\ \rho_f C^T & M_f \end{bmatrix} + \frac{j\omega}{Z_a(\omega)} \begin{bmatrix} 0 & 0 \\ 0 & A_f \end{bmatrix}\right)^{-1} \begin{Bmatrix} F_s \\ 0 \end{Bmatrix} \quad (4)$$

This method is unachievable for large model size. The inversion calculated at each frequency step is extremely time-consuming and still non possible for various cases. The accuracy is the essential advantage for this method. However, it will be considered as a reference for the comparative study.

3.2 Model Order Reduction Method

The resolution of Eq. (1) is performed using a model order reduction technique presented by Ben Smida et al. [2]. The main objective is to build a reduction modal basis that minimizes the numerical computational cost for predefined prediction accuracy.

The computation of the uncoupled modal basis is built using the following structure and fluid uncoupled eigenvalue problem:

$$\begin{cases} (-\omega^2 M_s + K_s)U = 0 \\ (-\omega^2 M_f + K_f)P = 0 \end{cases} \Rightarrow \quad [T_0] = \begin{bmatrix} T_{s0} & 0 \\ 0 & T_{f0} \end{bmatrix} \tag{5}$$

The structure modal basis is truncated following the frequency criterion $f_s = n * f_{\max}$, Where $f_{\max}$ is the maximum frequency of interest. Otherwise there is no truncation criterion for the fluid basis. The equation system (2) is projected on the truncated modal basis:

$$\left(-\omega^2 \begin{bmatrix} \tilde{M}_s & 0 \\ \rho_f \tilde{C}^T & \tilde{M}_f \end{bmatrix} + \begin{bmatrix} \tilde{K}_s & -\tilde{C} \\ 0 & \tilde{K}_f \end{bmatrix}\right) \begin{Bmatrix} q_s \\ q_f \end{Bmatrix} = \begin{Bmatrix} \tilde{F}_s \\ 0 \end{Bmatrix} \tag{6}$$

where $\tilde{K}_s = T_{so}^T K_s T_{so}$, $\tilde{M}_s = T_{so}^T M_s T_{so}$, $\tilde{K}_f = T_{fo}^T K_f T_{fo}$ and $\tilde{M}_f = T_{fo}^T M_f T_{fo}$ $\tilde{C}_s = T_{so}^T K_s T_{fo}$, $\tilde{F}_s = T_{so}^T F_s$, $U = T_{so}^T q_s$ and $P = T_{fo}^T q_f$.

The computation time is significantly reduced using this superposition method. It is obviously used in structural dynamics. Its main inconvenient for a coupled system, is the non-consideration of the coupled modes effect. Consequently, one can obtain a bad convergence of the output result.

The model reduction method used is based on the use of a truncated uncoupled modal basis with a number of chosen n_s^0 structure and n_f^0 fluid modes. The truncation and the kept modes in the basis are selected by an energy criteria presented in the following equations:

$$if \quad E_{s,i} = \frac{T_{so}^T(i) K_s T_{so}(i)}{\sum_{i=1}^{n_s^o} \left(T_{s,sel}^T(i) K_s T_{s,sel}(i)\right)} \geq Tol \quad Then \quad T_{s,i+1}^{sel} = \begin{bmatrix} T_{s,i}^{sel} & T_{so}(i) \end{bmatrix} \tag{7}$$

$$if \quad E_{f,i} = \frac{T_{fo}^T(i) K_f T_{fo}(i)}{\sum_{i=1}^{n_f^o} \left(T_{f,sel}^T(i) K_f T_{f,sel}(i)\right)} \geq Tol \quad Then \quad T_{f,i+1}^{sel} = \begin{bmatrix} T_{f,i}^{sel} & T_{fo}(i) \end{bmatrix} \tag{8}$$

where i represents the current selected mode index.

The basis is accomplished by some vectors to get into account the coupling and the non-considered modes. The vectors $q_{f,i}$ describe the load effects applied by the structure on the fluid. The vectors $q_{s,i}$ describe the load effects applied by the fluid on the structure. They are calculated using the equations:

$$\begin{cases} q_{s,i} = \left(-\omega_c^2 M_s + K_s\right)^{-1}\left(F_s + CT_{f,i}^{sel}\right) \\ q_{f,i} = \left(-\omega_c^2 M_f + K_f\right)^{-1}\omega_c^2 \rho_f C^T T_{s,i}^{sel} \end{cases} \tag{9}$$

The enriched basis T_E is built using the criteria $E_{s,i}$ and $E_{f,i}$ with a *Tol* tolerance condition:

$$if \quad \begin{matrix} E_{s,i} \geq Tol \\ E_{f,i} \geq Tol \end{matrix} \quad then \quad \begin{matrix} \left[T_s^E\right] = \left[T_s^{sel} \quad q_{f,1}\ldots q_{f,i}\ldots\right] \\ \left[T_f^E\right] = \left[T_f^{sel} \quad q_{s,1}\ldots q_{s,1}\ldots\right] \end{matrix} \tag{10}$$

A modal projection is used to estimate the displacement and pressure responses:

$$\begin{Bmatrix} \hat{U} \\ \hat{P} \end{Bmatrix} = T_E \begin{Bmatrix} q_s \\ q_f \end{Bmatrix} \tag{11}$$

The force residuals, so called error loads, are calculated as follow:

$$\begin{cases} R_{Fs} = \left(K_s - \omega_c^2 M_s\right)^{-1}\hat{U} - C\hat{P} - F_s \\ R_{Ff} = \left(K_f - \omega_c^2 M_f\right)^{-1}\hat{P} - \rho_f \omega_c^2 C^T \hat{U} \end{cases} \tag{12}$$

The displacement residuals are estimated using the force residuals:

$$\begin{cases} R_{Ds} = \left(K_s\right)^{-1} R_{Fs} \\ R_{Df} = \left(K_f\right)^{-1} R_{Ff} \end{cases} \tag{13}$$

The bases T_s^E and T_f^E are fulfilled by the displacement residuals vectors R_{Ds} and R_{Df}. A singular value decomposition (SVD) is therefore necessary.

4 The Proposed Uncertainty Analysis Method

The proposed approach is presented in four principal steps. The initial one is to select the random design parameters according to their sources and their significance level. Therefore, the number of samples decrease and the parameters with weak sensitivity will be calculated only by their deterministic value.

The second step uses a Hypercube Latin Sampling (HLS) technique. Compared to the Monte Carlo sampling, the HLS gives the same accuracy with a less number of samples. The uncertainties deviation ranges of Six-Sigma covering 99.99966% of samples improves the prediction accuracy. The results obtained by the HLS technique will be considered as a reference.

The third step uses a model order reduction technique to overcome the difficulties related to the finite element model size and the uncertainty propagation study which are all known as time consuming.

The fourth and final step compares the approximated stochastic responses obtained by a polynomial chaos expansion (PC) to the reference.

5 Numerical Simulation

5.1 Finite Elements Model

The efficiency of the proposed method is illustrated through a numerical simulation double panel example (Fig. 1). It is made by three cavities filled with air and coupled through two flexible panels. The system has uniform dimension in the (x, y) plane: 0.35 × 0.29 m. The depth is 0.14 m for the source cavity, 0.08 m for the middle cavity and 0.2 m for the receiving cavity. Poroelastic patches, characterized by acoustic impedance $Z_a(\omega)$, are stacked on the five rigid walls of the source cavity.

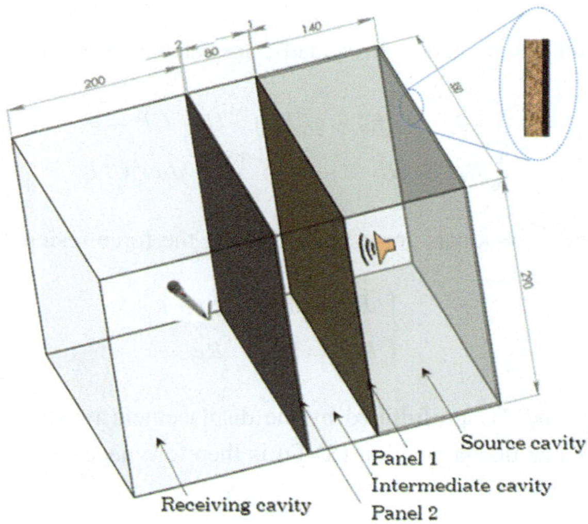

Fig. 1 Geometric model for a three cavities system

The panel 1 is made with epoxy resin and the panel 2 is made with steel. Their physical and geometric properties are presented in the Table 1.

Table 1 Geometric and physical panels properties

Parameters	Young modulus $E(Pa)$	Poisson ratio ν	Density $\rho(\mathrm{Kg/m^3})$	Thickness $h(\mathrm{m})$
Panel 1	4e9	0,4	1200	2e-3
Panel 2	210e9	0,285	7800	1e-3

The finite element model consists of 17 × 14 structural quadrilateral elements and 17 × 14 elements acoustic bricks. The finite element model includes about 9600 DDLs degrees of freedom (DOFs) (3108 DOFs of structure, 6480 DOFs of fluid). The mesh size of the finite elements model is based on a criterion of five to six linear elements per mode shape wavelength.

The structural damping factors introduced respectively for the fluid and structure sub-systems are chosen as $\eta_s = 0.02$ and $\eta_s = 0.002$. An external load exciting the system is located in ((x, y, z) = 0.082, 0.145, 0.12). The vibroacoustic uncertainties analysis is achieved in the frequency range of $[0 - 200Hz]$ containing the first three modes of the coupled system.

The model reduction method is used in the second step; the reduction basis is enriched according to a predefined tolerance $(tol = 10^{-3})$. Singular Value Decomposition (SVD) is useful to guarantee the linear independence of the initial basis modes with the added modes. The model reduction ratio is about 94%.

5.2 Stochastic Response, Results and Discussion

Some mechanical properties are considered as random design parameters with normal Gaussian distribution (Table 2). This selection is based on a preliminary design parameters sensitivities study for the panel 1 considered as a passive sound level control. The deterministic value is used as the nominal value for each variable. The dispersion level is $\delta_o = 20\%$. The central Gaussian variable ξ is calculated using hypercube Latin sampling method in the range of Six Sigma.

Table 2 Random design parameters

	Random parameters	Distribution
Panel 1	Young modulus $E_1\,(Pa)$	Normal Gaussian
	Density $\rho_1\,(\mathrm{Kg/m^3})$	Normal Gaussian

Figure 2 shows the stochastic responses by the mean value of 500 samples for the random design parameters. The acoustic indicators used are the power level for the acoustic cavity (a) and the mean quadratic velocity for the structure panel (b). The uncertainties propagation study is simulated by the proposed strategy using the PC technique compared to the reference one using the HLS. The accuracy of the PC technique depends on the polynomial order that increase the model size.

The acoustic power level in the receiving cavity and the mean quadratic velocity for the panel 2 are predicted with an error level about 10 dB and a relative low simulation time (Table 3). It is well shown that the sensitivity of the stochastic response under the effect of the uncertain parameters is more significant at high frequencies.

Table 3 CPU time (s)

	HLS	PC
CPU (s)	50444	938
Ratio (%)	98.1%	

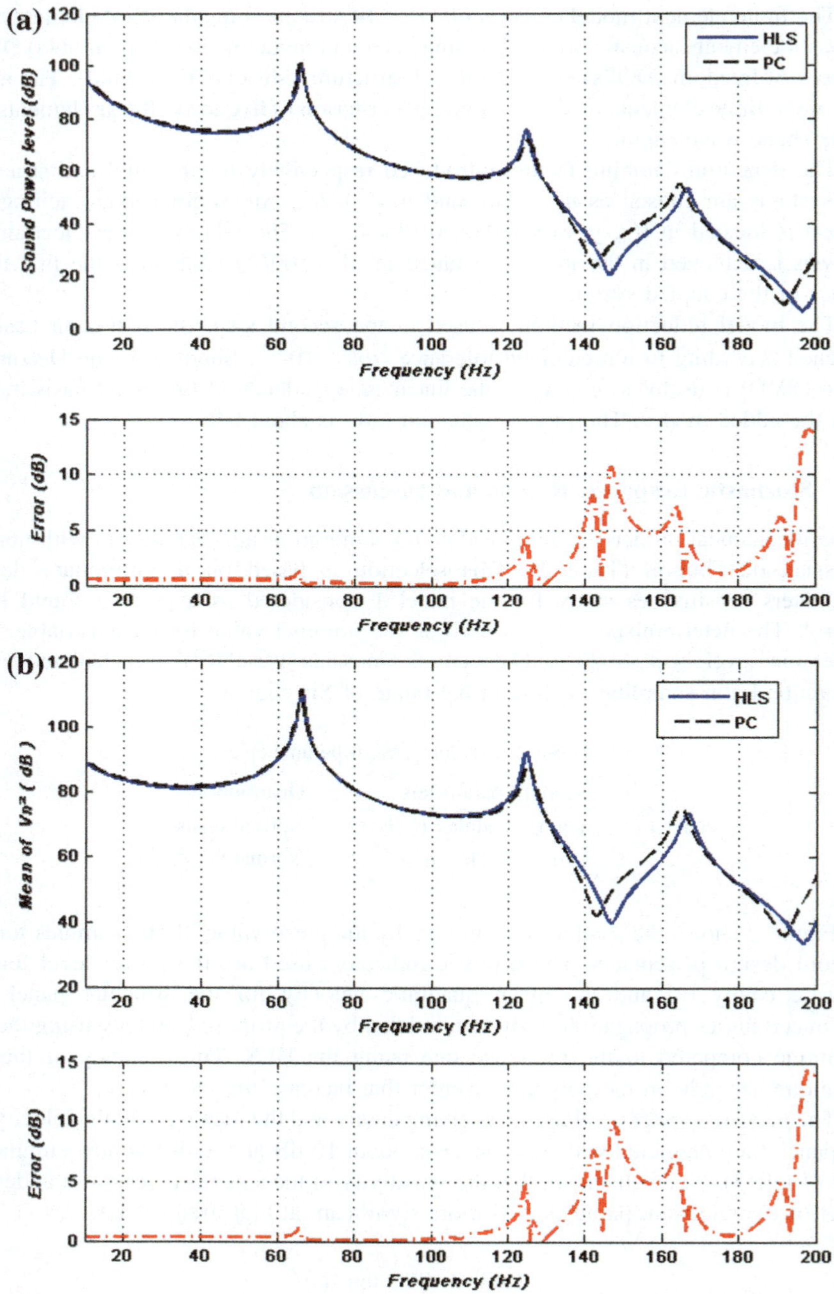

Fig. 2 Vibroacoustic indicators, Mean, relative errors (HLS-PC): **a** acoustic power level, **b** mean of quadratic velocity

6 Conclusion and Outlook

In this paper, uncertainties propagation study for a coupled fluid structure system is proposed. A polynomial chaos expansion (PC) technique is used and compared to the results obtained using Hypercube Latin Sampling (HLS) technique. The performances of the proposed strategy are tested throw double panels coupled fluid structure system.

It is well shown that the simulation cost in CPU time is significantly reduced to 98%. Nevertheless, the accuracy of the acoustic indictors can be improved.

As outlook, a polynomial chaos expansion (PC) technique can be used with a high polynomial order.

References

1. Akrout A, Karra C, Hammami L, Haddar M (2008) Viscothermal fluid effects on vibro-acoustic behavior of double elastic panels. Int J Mech Sci 50:764–773
2. Ben Smida B, Majed R, Bouhaddi N, Ouisse M (2012) Investigations for a model reduction technique of fluid-structure coupled systems. Part C J Mech Eng Sci 226:42–54
3. Caflisch Russel E (1998) Monte Carlo and quasi-Monte Carlo methods. Acta Numerica 7:1–49
4. Dammak K, Koubaa S, El Hami A, Walha L, Haddar M (2019) Numerical modelling of vibro-acoustic problem in presence of uncertainty: application to a vehicle cabin. Appl Acoust 144:113–123.
5. Deb MK, Babuška IM, Oden JT (2001) Solution of stochastic partial differential equations using Galerkin finite element techniques. Comput Methods Appl Mech Eng 190:6359–6372
6. Doutres O, Atalla N (2011) Experimental estimation of the transmission loss contributions of a sound Package placed in a double wall structure. Appl Acoust 72:372–379
7. Ezvan O, Batou A, Soize C, Gagliardini L (2017) Multilevel model reduction for uncertainty quantification in computational structural dynamics. Comput Mech 59:219–246. https://doi.org/10.1007/s00466-016-1348-1
8. Ghanem RG, Spanos PD (1991) Stochastic finite elements: a spectral approach, Springer
9. Kamiński M (2007) Application of the generalized perturbation-based stochastic boundary element method to the elastostatics. Eng Anal Boundary Elem 31:514–527
10. Loève M (1977) Probability Theory, 4th edn. Springer, New York
11. Matthies HG, Brenner CE, Bucher CG, Guedes Soares C (1997) Uncertainties in probabilistic numerical analysis of structures and solids stochastic finite elements. Struct Saf 19:283–336
12. Mingjie W, Huang Qibai H (2016) A new hybrid uncertain analysis method for structural-acoustic systems with random and interval parameters. Comput Struct 175:15–28. https://doi.org/10.1016/j.compstruc.2016.07.001
13. Mulder VL, De Bruin S, Schaepman ME (2013) Representing major soil variability at regional scale by constrained Latin Hypercube Sampling of remote sensing data. Int J Appl Earth Obs Geoinf 21:301–310
14. Ohayon R, Soize C (2017) Computational vibroacoustics in low- and medium-frequency bands: damping, ROM, and UQ modeling. Appl Sci 7:586–625
15. Plischke E, Borgonovo E, Smith C (2013) Global sensitivity measures from given data. Eur J Oper Res 226:536–550

16. Soize C (2000) A non-parametric model of random uncertainties for reduced matrix models in structural dynamics. Probab Eng Mech 15:277–294
17. Soize C (2013) Stochastic modeling of uncertainties in computational structural dynamics-recent theoretical advances. J Sound Vib 332:2379–2395. https://doi.org/10.1016/j.jsv.2011.10.010
18. Van den Nieuwenhof B (2003) Stochastic finite elements for elastodynamics: random field and shape uncertainty modeling using direct and modal perturbation-based approaches. PhD thesis, Université catholique de Louvain
19. Wiener N (1938) The homogeneous chaos. Am J Math 60:897–936
20. Xiu D, Karniadakis GE (2003) Modeling uncertainty in flow simulations via generalized polynomial chaos. J Comput Phys 187:137–167
21. Yamaguchi T, Kurosawa Y, Enomoto H (2009) Damped vibration analysi using finite element method with approximated modal damping for automotive double walls with a porous material. J Sound Vib 325:436–450

Robust 2D-Spatial Fourier Transform Identification of Wavenumber-Space Characteristics of a Composite Plate

Lajili Ramzi[1,2,3(✉)], K. Chikhaoui[3], M.-L. Bouazizi[3,4], and A. Bisharat[4]

[1] Laboratory of Tribology and Dynamics of Systems (LTDS), Ecole Centrale de Lyon, 36 Avenue Guy de Collongues, 69130 Écully, France
ajiliramsis@gmail.com

[2] National School of Engineers of Tunis (ENIT), University of Tunis el Manar, BP 37, Le Belvedere, 1002 Tunis, Tunisia

[3] Preparatory Engineering Institute of Nabeul (IPEIN), Research Unit of Structural Dynamics, Modelling and Engineering of Multi-Physics, 8000, M'rezgua, Nabeul, Tunisia
ajiliramsis, chikhaoui2013@gmail.com, mohamedlamjed@gmail.com

[4] Mechanical Department, College of Engineering, Prince Sattam Bin Abdulaziz University, Al-Kharj, Kingdom of Saudi Arabia
bisharatawni@yahoo.com

Abstract. This paper addresses the issue of wave propagation features in a honeycomb sandwich plate over a broadband frequency range. The special emphasis putted on such materials is due to their growing industrial integration resulting from interesting mechanical and material properties, such as high energy dissipation and resistance/weight ratio. A two-dimensional spatial Discrete Fourier Transform (2D-DFT) is employed with experimentally measured displacement field, as primary input, to identify a complete wave propagation direction-dependent dispersion equation of the sandwich plate. Valuable insights into the wavenumber-space (k-space) profiles, in relation with the structural orthotropic behavior, are highlighted. The 2D-DFT method is proved to be efficient in a deterministic framework. Nevertheless, its robustness against input parameters' uncertainty needs to be evaluated as well to achieve more realistic k-space characteristics' identification. The impact of measurement points' localization's uncertainty on the 2D-DFT identifications is statistically investigated. The obtained results show the large variability of the identified k-space parameters and reveal the important identification sensitivity to such measuring errors involved in the experimental manipulations.

Keywords: Wide-band identification · K-space · Spatial fourier transform · Experiments · Uncertainties · Honeycomb sandwich plate

N. Aifaoui et al. (Eds.): CMSM 2019, LNME, pp. 271–281, 2020.
https://doi.org/10.1007/978-3-030-27146-6_29

1 Introduction

In the wavenumber space (k-space), several identification methods consist on extracting the wavenumber content from experimentally measured or numerically computed data [4, 6, 10]. The most frequently used methods in the literature are those by correlation [6], the Prony series [7], the Mc Daniel method [17], the Inhomogeneous Wave Correlation (IWC) method and its variants [3, 4, 14, 15] and the Fourier Transforms (FTs).

The two-dimensional Discrete Fourier Transform (2D-DFT) is widely used in many engineering areas. It is an extension of the one-dimensional (1D) phase spectrum method, developed by Sachse and Pao [21] for the measurement of the velocity of stress wave, to determine the amplitude and velocity of different modes propagating at the same frequency. Indeed, 2D-DFT identification was proposed in Alleyne and Cawley [1] and Bolton et al. [2] in the context of multimodal analysis. It permits to identify the dispersion curves and the characteristics of multiple modes by transforming the received amplitude-time data to amplitude-wavenumber data at discrete frequencies. Alleyne and Cawley [1] varied frequencies and phase velocities in a single test and measured the amplitudes and velocities of propagating Lamb waves using the 2D-DFT. The authors showed that the computational requirements of the 2D-DFT are fairly modest. Wave propagation in light weight plates with different lattice-shaped core geometries was studied by Kohrs et al. [12, 13]. The authors showed that the 2D-DFT implementation is simple using standard Fast FT (FFT) algorithms, but the method is very demanding in terms of computational time since a complete vibration field must be calculated and a high resolution is required for the FFT. Wave propagation analysis in anisotropic ribbed panels and 1D and 2D honeycomb sandwich structures, over a wide frequency range, was carried by Ichchou et al. [10, 11]. The IWC method was used as an improvement of the classical Fourier Transforms, with an independence of geometry, boundary conditions and source location. The 2D-DFT was also applied by Huang [9] to study the interaction between defects and guided waves in ribbed panels. Moreover, Ruzek et al. [20] applied the 2D-DFT to approximate dispersion curves of 2D structures from measured vibratory field. In [23, 24], the 2D-DFT was used to post-process experimentally measured displacement field of perforated plates under harmonic excitation. The authors proved the efficiency of the method compared to the condensed wave finite element method and the IWC method. The material properties estimation from an experimentally measured vibration field of thin plates was carried out in Roozen et al. [19]. The authors used a wave fitting approach compared to the 2D-DFT.

The 2D-DFT presents two major advantages: bijectivity, which allows inverse transform (IDFT), and rapidity, using FFT algorithm. However, it has mainly three drawbacks: aliasing, leakage and k-space discrimination. For more details, one can refer to works mentioned above.

The main purpose of this paper is to analyze wave propagation features in a honeycomb sandwich plate over a broadband frequency range, through the estimation of the wavenumber-space (k-space) profiles. Besides, addressing such issue in a deterministic framework remains a constraining hypothesis in vibration mechanics.

Therefore, it is inevitable to account for uncertainty to achieve more realistic k-space characteristics' identification. Measuring errors, for instance, could be involved in experimental manipulations and measurement points' coordinates is an example of uncertain parameters. A sample-based uncertainty propagation method is thus included into the identification process: the Latin Hypercube Sampling (LHS) [8, 18]. Statistical investigations could then be performed to quantify the impact of uncertainty on the 2D-DFT identifications. The robustness of these identifications against uncertainty is thus evaluated.

The originality of this work lies thus in proposing a stochastic identification process which combines the 2D-DFT with the LHS method. The proposed process allows identifying the k-space wave propagation features in a honeycomb sandwich plate over a broadband frequency range.

2 Theoretical Backgrounds

Similarly to standard DFT, which explore the spectral content of time signals, spatial DFT permits to explore the k-space content of spatial fields. Practically, a spatial sampling at points on a uniform discrete 2D grid (of dimensions $L_x \times L_y$) is necessary, such as $x_i = i\Delta x, y_j = j\Delta y$, with $0 \leq i \leq N_x - 1$, $0 \leq j \leq N_y - 1$ and $N_x \times N_y$ is the number of equally distributed spatial sampling points.

The 2D-DFT permits to transform the displacement field u as:

$$\hat{u}\left(k_{xp}, k_{yq}\right) = \frac{1}{N_1 N_2} \sum_{i=0}^{N_1-1} \sum_{j=0}^{N_2-1} u\left(x_i, y_j\right) e^{-i\left(k_{xp} x_i + k_{yq} y_j\right)} \tag{1}$$

where $k_{xp} = p\Delta k_x$ and $k_{yq} = q\Delta k_y$ are the wavenumbers' exponentials, with:

$$\Delta k_x = \frac{2\pi}{N_x \Delta x}; \Delta k_y = \frac{2\pi}{N_y \Delta y} \tag{2}$$

Note that the 2D-DFT is $2\pi/\Delta x$ and $2\pi/\Delta y$ periodic. Moreover, its bijectivity property implies that the transform $\hat{u}\left(k_{xp}, k_{yq}\right)$ contains all information contained in the field $u\left(x_i, y_j\right)$, which allows IDFT.

3 K-Space Identification of Sandwich Plate Using Experiment-Based 2D-Spatial DFT

Let consider a sandwich composite plate of dimensions 1 × 1 m^2. The core of the plate is in aluminum honeycomb (Nida Nomex) of thickness 13 mm. The skin of the plate is made of laminated carbon fiber epoxy (Hexcel reference 43199) and is of thickness 0.5 mm each side, with a resin fiber ratio of 50%.

The plate is freely suspended under bending load. It is excited by a point force using an electrodynamic shaker (Brüel & Kjaer, 4810) at the position (x = 0.3 m,

$y = 0.5$ m), in the frequency range [0–3200 Hz], Fig. 1. The normal velocity of the plate is measured by a laser vibrometer (Brüel & Kjaer, 8330) at 1849 measuring points ($\Delta x = 0.0231$, $\Delta y = 0.0230$ m). A force sensor (Brüel & Kjaer Type 8001) permits to collect the injected force and generate the white noise. The fields are sampled and acquired using a Fourier analyzer interfaced with a Hewlett Packard Paragon 35654A sampler.

Fig. 1 Experimental measuring configuration of the displacement field of the honeycomb sandwich plate, excited by an electrodynamic shaker at $x = 0.3$ m and $y = 0.5$ m, using a laser vibrometer

The 2D-DFT is here used to identify the k-space profiles which correspond to the variation of the wavenumber in the k-space with respect to the wave propagation directions. The 2D-DFT permits to identify these characteristics by transforming input displacement fields to output wavenumbers at discrete frequencies. Since displacement fields are measured experimentally, the 2D-DFT could be called experiment-based.

Figures 2 and 3 illustrate the displacement fields measured at two arbitrarily chosen frequencies, 2374 and 3200 Hz, and the associated k-space profiles identified using the 2D-DFT.

Identifying the k-space profile by the 2D-DFT consists on defining the maximums that appear during the decomposition given by Eq. (1). These maximums highlight the profile of the wavenumber variation in the k-space for each frequency and in any wave propagation direction.

Comparing the results at each frequency, one can show the analogy between the variation of the displacement field along the plate (the most important amplitudes of the waves propagating from the excitation point $x = 0.3$ m and $y = 0.5$ m) and the k-space profiles (the maximums of the 2D-DFT corresponding to the wavenumbers transporting

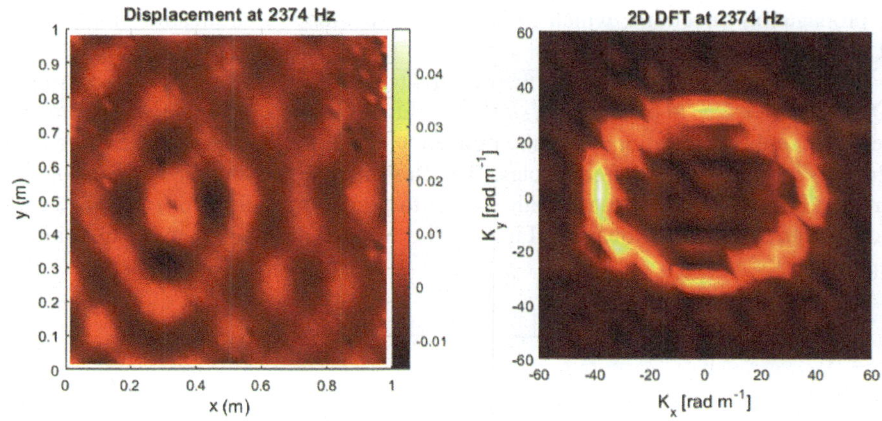

Fig. 2 Deterministic displacement field of the honeycomb sandwich plate and associated k-space profile identified using the 2D-DFT at frequency 2374 Hz

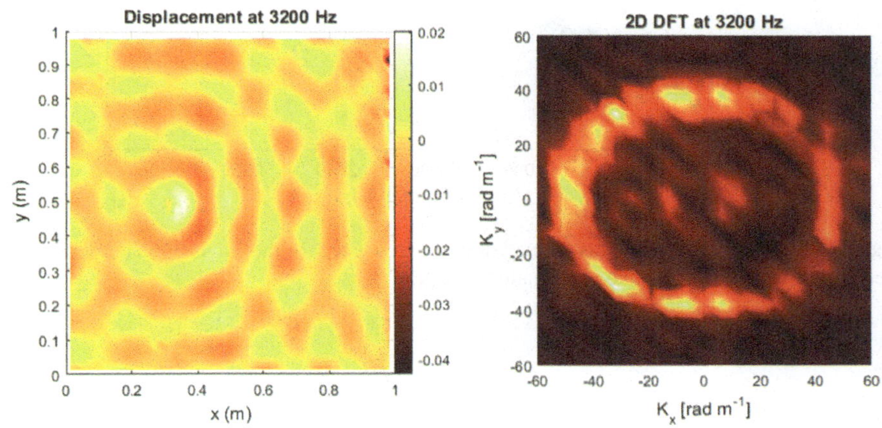

Fig. 3 Deterministic displacement field of the honeycomb sandwich plate and associated k-space profile identified using the 2D-DFT at frequency 3200 Hz

energy). Subsequently, in the mid-high frequency domain particularly, if the excitation point is not known a priori, it can be identified.

Moreover, the results show an elliptical shape of the k-space profiles which is more pronounced by increasing frequency. This points out the elliptic orthotropy of the honeycomb plate. Regarding the shape of the ellipse, an angle of orthotropy near 0° is a priori revealed. A more detailed study of this elliptical orthotropy is subject of works in progress.

The 2D-DFT is typically a mid-high frequency method. At low frequencies, a discontinuity of the k-space profiles is detected. Results are not shown here for the sake of brevity. This discontinuity is due to the fact that energy is not evenly distributed over all propagation directions, which makes the wavenumber identification impossible in all directions.

Comparing the coordinates of the highlighted k-space profiles identified using the 2D-DFT ($K_x = k\ cos\theta$ and $K_y = k\ sin\theta$, respectively), one can deduce the increase of the wavenumber proportionally to the frequency, due to the decrease of the wave length which is infinite at low frequencies. To illustrate the wavenumber variation as a function of frequency, the mean with respect to propagation directions is calculated, at each frequency, and shown in Fig. 4. The wavenumber is computed using the expression:

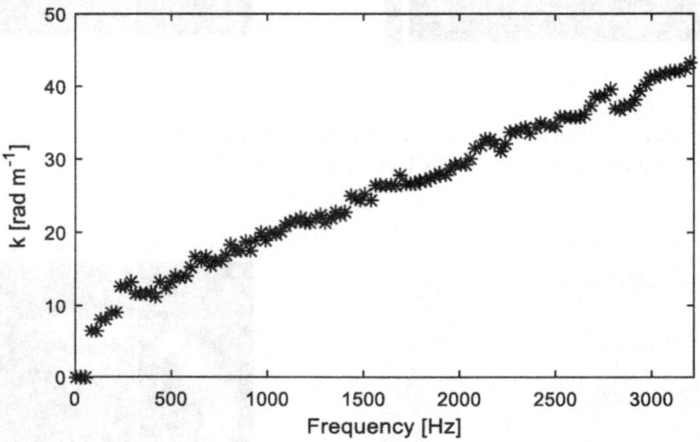

Fig. 4 Deterministic variation, as a function of the frequency, of the wavenumber computed as a mean with respect to the wave propagation direction

$$k = \sqrt{K_x^2 + K_y^2} \tag{3}$$

Based on the works performed by Thite and Ferguson [22] and Egreteau [5], the excitation point location has a significant effect on the 2D-DFT accuracy. The authors showed that the excitation point must be close to, or even centered on, the measurement zone. Indeed, in presence of damping, vibrations (and so wave propagation energy) are significantly reduced which makes the wavenumber identification as difficult as the measurement point is far from the excitation location. On the other hand, it is important to note that the 2D-DFT implementation is simple through a standard FFT algorithm. For more details on the advantages and difficulties encountered when applying this method, the references cited above could be consulted.

In a stochastic framework, measurement points' coordinates are supposed to be uncertain with a probabilistic modeling of their variability. Indeed, the stochastic measurement points' vector x_s is expressed as:

$$x_s = x(1 + \delta_x \xi) \tag{4}$$

where x is the vector containing the mean coordinates of the measurement points, $\delta_x = 2\%$ is the dispersion value and ξ is a normal random variable.

The measurement points' coordinates variability is supposed along both x and y directions, which is more realistic than a simple perturbation along only one axis.

To propagate uncertainties and thus evaluate their impact on identified k-space profiles, a stochastic identification process combines the 2D-DFT with a 1000 samples-based LHS method. Statistical post-processing quantifications are then performed. Figures 5 and 6 illustrate the mean and envelope (extreme statistics) profiles of the k-space at frequencies 2374 and 3200 Hz, respectively.

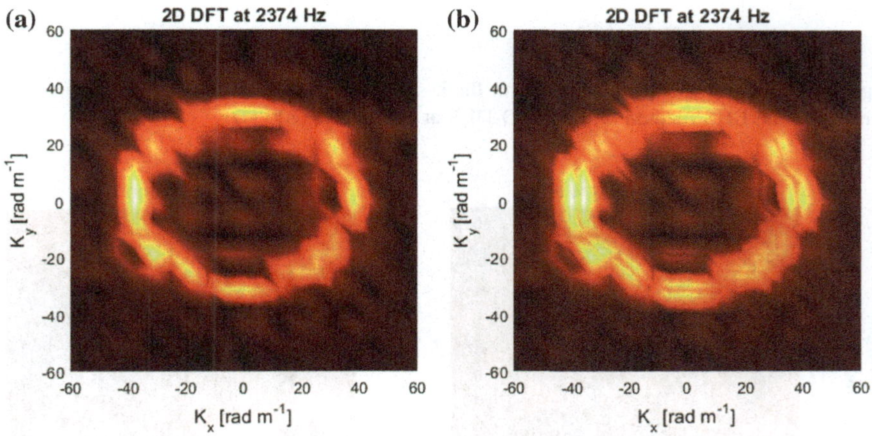

Fig. 5 **a** Mean and **b** envelope profiles of the k-space estimated using the identification process combining the LHS method with the 2D-DFT at frequency 2374 Hz

The envelope profiles of the k-space, Figs. 5b and 6b, show the significant impact of uncertainties on the wavenumber variability. This variability can furthermore be quantified by the dispersion which is computed as the ratio of the standard deviation by the mean, Fig. 7. A dispersion level of nearly 3.35% is obtained, compared to 2% imposed on the measurement points' coordinates.

The elliptic orthotropy of the honeycomb sandwich plate, with an angle of orthotropy close to 0°, allows simplifying the k-space profiles representations to only a quarter of the plane. Standard deviations' representation, for instance, is here simplified as shown in Fig. 7.

The statistical quantifications of the wavenumber computed, over the considered frequency domain, as a mean with respect to the wave propagation direction, is illustrated in Fig. 8. The width of the envelope increases proportionally to the frequency. Since the mean increases similarly, a nearly constant dispersion value of 3.35% is obtained along the frequency band of interest.

Comparing the results obtained here with those of previous works, published in [14, 16], the impact of uncertainty of the measurement points' coordinates on the

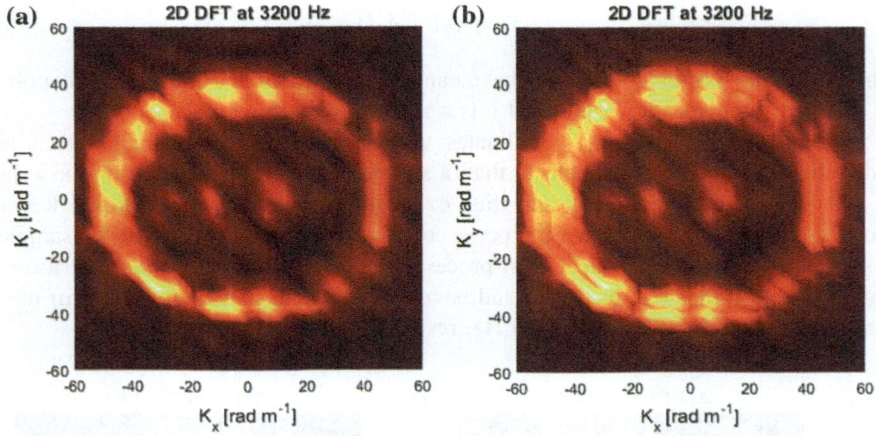

Fig. 6 **a** Mean and **b** envelope profiles of the k-space estimated using the identification process combining the LHS method with the 2D-DFT at frequency 3200 Hz

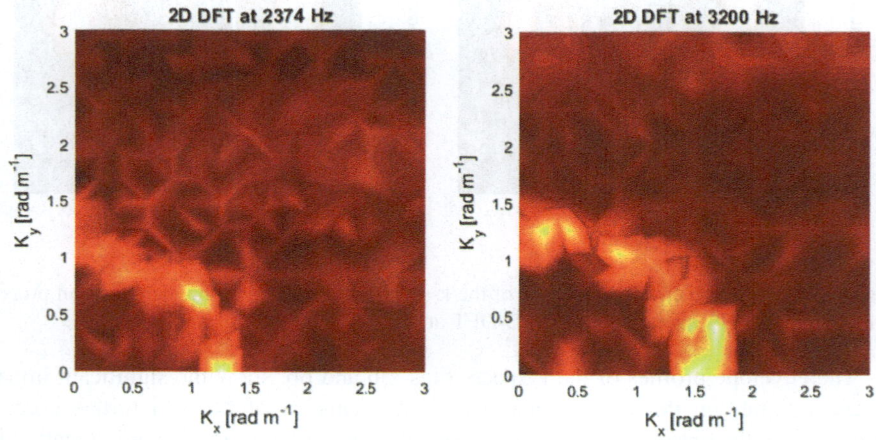

Fig. 7 Standard deviations of the k-space profiles estimated using the identification process combining the LHS method with the 2D-DFT at frequencies 2374 and 3200 Hz, respectively

wavenumber identification in the case of the honeycomb sandwich plate is much more important than that in the case of the honeycomb beam.

4 Conclusion

The present paper presented an experiment-based two-dimensional identification process to estimate frequency and propagation direction dependent k-space characteristics of a honeycomb sandwich plate. The proposed process is based on a two-dimensional

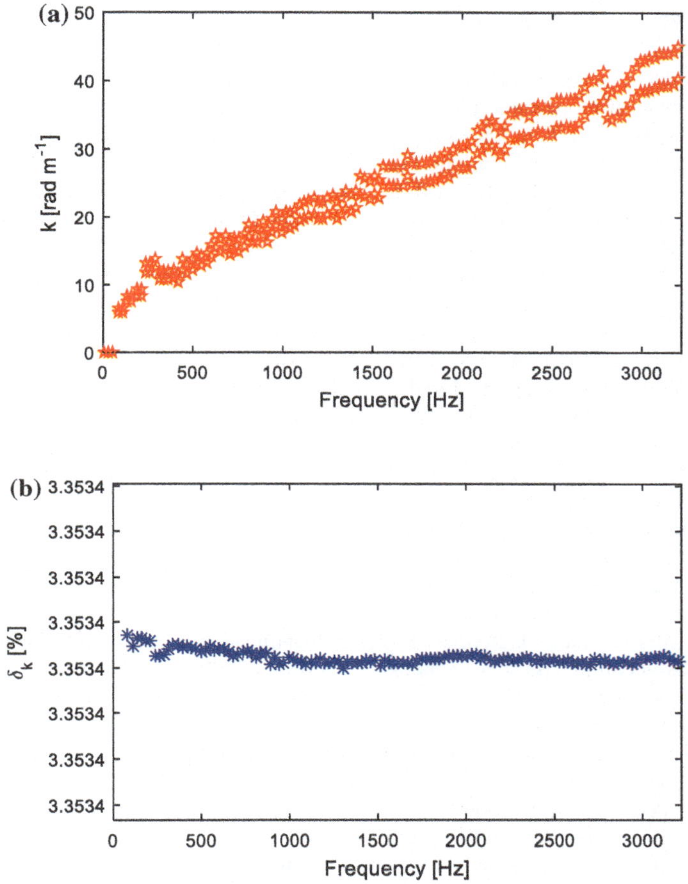

Fig. 8 **a** Envelope and **b** dispersion of the wavenumber computed, over the considered frequency domain, as a mean with respect to the wave propagation direction

Discrete Fourier Transform and used as input an experimentally measured displacement field. Its extension, combining this transform with the Latin Hypercube Sampling method, allowed a robust identification of the k-space profiles in presence of parametric uncertainties on the measurement points' coordinates. Statistical investigations of the uncertainty impact on the 2D-DFT identifications illustrated an output-to-input dispersion ratio of nearly 1.7, which reveals the important identification sensitivity to such measuring errors involved in experimental manipulations.

The works performed here are limited to k-space profiles' identification and could be extended to estimate the spatial damping (wave attenuation) variation with respect to frequency, wave propagation direction and uncertainty.

References

1. Alleyne D, Cawley P (1991) A two-dimensional Fourier transform method for the measurement of propagating multimode signals. J Acoust Soc Am 89:1159–1168
2. Bolton JS, Song HJ, Kim YK, Kang YJ (1998) The wave number decomposition approach to the analysis of tire vibration. In: Proceedings of the noise-conference, vol 98, pp 97–102
3. Berthaut J, Ichchou MN, Jezequel L (2003) Identification in large frequency range of effective parameters of two-dimensional structures by means of wave numbers. Mec Ind 4:377–384
4. Berthaut J, Ichchou MN, Jezequel L (2005) K-space identification of apparent structural behavior. J Sound Vib 280:1125–1131
5. Egreteau T (2011) Développement numérique et implémentation expérimentale d'une méthodologie d'identification de l'équation de dispersion dans les structures aéronautiques. Université de Sherbrooke, Sherbrooke (Québec) Canada
6. Ferguson NS, Halkyard CR, Mace BG, Heron KH (2002) The estimation of wavenumbers in two dimensional structures. In: Proceedings of ISMA II, pp 799–806
7. Grosh K, Williams EG (1993) Complex wave-number decomposition of structural vibrations. J Acoust Soc Am 93:836–848
8. Helton JC, Davis FJ (2003) Latin hypercube sampling and the propagation of uncertainty in analyses of complex systems. Reliab Eng Syst Saf 81:23–69
9. Huang T (2012) Multi-modal propagation through finite elements applied for the control of smart structures, Ph D thesis, École Central de Lyon, France
10. Ichchou MN, Bareille O, Berthaut J (2008) Identification of effective sandwich structural properties via an inverse wave approach. Eng Struct 30:2591–2604
11. Ichchou MN, Berthaut J, Collet M (2008) Multi-mode wave propagation in ribbed plates: part I, wavenumber-space characteristics. Int J Solids Struct 45:1179–1195
12. Kohrs T, Petersson BAT (2007) Wave guiding effects in light weight plates with truss-like core geometries. In: 19th International congress on acoustics Madrid
13. Kohrs T, Petersson BAT (2009) Wave beaming and wave propagation in light weight plates with truss-like cores. J Sound Vib 321:137–165
14. Lajili R, Bareille O, Bouazizi M-L, Ichchou MN, Bouhaddi N (2018) Composite beam identification using a variant of the inhomogeneous wave correlation method in presence of uncertainties. Eng Comput 35(6):2126–2164
15. Lajili R, Bareille O, Bouazizi M-L, Ichchou MN, Bouhaddi N (2018) Inhomogeneous wave correlation for propagation parameters identification in presence of uncertainties. In: CMSM2017, design and modeling of mechanical systems, pp 823–833
16. Lajili R, Chikhaoui K, Bouazizi M-L (2019) Statistical investigations of uncertainty impact on experiment-based identification of a honeycomb sandwich beam. In: ICAV2018, advances in acoustics and vibration II, pp 176–185
17. McDaniel JG, Shepard WS (2000) Estimation of structural wave numbers from spatially sparse response measurements. J Acoust Soc Am 108:1674–1682
18. McKay MD, Beckman RJ, Conover WJ (1979) A comparison of three methods for selecting values of input variables in the analysis of output from a computer code. Technometrics 21:239–245
19. Roozen NB, Labelle L, Leclère Q, Ege K, Gerges Y (2017) Estimation of plate material properties by means of a complex wavenumber fit using Hankel's functions and the image source method. J Sound Vib 390:257–271
20. Ruzek M, Guyader J-L, Pézerat C (2014) Information criteria and selection of vibration models. J Acoust Soc Am 136:3040–3050

21. Sachse W, Pao Y-H (1978) On determination of phase and group velocities of dispersive waves in solids. J Appl Phys 49:4320–4327
22. Thite AN, Ferguson NS (2004) Wavenumber estimation: further study of the corrélation technique and use of SVD to improve propagation direction resolution. University of Southampton, Institute of Sound and Vibration Research, Southampton, UK, p 35
23. Zhou CW (2014) Wave and modal approach for multi-scale analysis of periodic structures, PhD thesis, École Central de Lyon, France
24. Zhou CW, Lainé JP, Ichchou MN, Zine AM (2016) Numerical and experimental investigation on broadband wave Propagation features in perforated plates. Mech Syst Signal Process 75:556–575

Physical Only Modes Identification Using the Stochastic Modal Appropriation Algorithm

Maher Abdelghani[1,2(✉)]

[1] University of Sousse, Sousse, Tunisia
maher.abdelghani@gmail.com
[2] LASMAP, EPT, La Marsa, Tunisia

Abstract. Recently the author proposed a new In-Operation modal identification algorithm, namely the Stochastic Modal Appropriation algorithm (SMA) which identifies the frequency and the damping ratio simultaneously in a single step. The key idea is to rotate parametrically the outputs correlation sequence so as it looks like the system impulse response. We show in this work that SMA rejects automatically harmonics as well as spurious/numerical modes leading therefore to physical-only modes identification. After a mathematical proof, the method is validated on a simulated system.

Keywords: In-operation modal analysis · Modal appropriation · Spurious modes · SMA

1 Introduction

The correlation sequence of the outputs of a structure under its operating conditions is a sum of decaying phased sinusoids. The phase shift is dependent on the damping ratio and has a particular expression which is derived based on a physical background. Suppose there is a noise mode present in the correlation sequence (decaying phased sinusoid). The phase shift does not have the particular expression as for a physical mode. With the SMA algorithm, the correlation sequence is parametrically rotated until one hits a zero phase shift. We get then a pure sinusoid (impulse response for a physical mode and just a pure sine for a noise mode). When the phase angle between the input and output is +/−pi/2 the mode is isolated. This is a necessary condition. The noise mode may also be isolated.

However since the rotation annihilates the phase angle which is related to the damping ratio (the pole) the frequency and the damping ratio will be estimated simultaneously only and only if the mode is physical. This is a consequence of the particular structure of the phase angle in the correlation sequence. A noise mode will not satisfy this property.

In conclusion the physical mode satisfies the following 3 conditions which make it unique:

1. Input/output phase angle +/−pi/2
2. The frequency and damping ratio are estimated
3. The frequency and damping ratio are estimated simultaneously.

N. Aifaoui et al. (Eds.): CMSM 2019, LNME, pp. 282–288, 2020.
https://doi.org/10.1007/978-3-030-27146-6_30

We argue and we prove it, that a mode that satisfies simultaneously these 3 conditions is necessarily a physical mode.

After elaborating our approach on a SDOF system we show, as an example, that our algorithm is robust against harmonics through a simulated example.

2 The SMA Algorithm

Consider a SDOF system excited with unmeasured white noise. The impulse response of the system may be written as [3]:

$$h(t) = C_h e^{-\xi\omega_n t}\sin(\omega_d t) \tag{1}$$

where ξ is the system damping ratio, ω_n is the system natural frequency and ω_d is the damped natural frequency.

The correlation sequence of the system output may be written as [3]:

$$R(t) = C_r e^{-\xi\omega_n t}\sin(\omega_d t - \phi(\xi)) \tag{2}$$

where $\phi(\xi)$ is a phase shift that depends on the damping ratio. Both the impulse response and the correlation sequence can be considered as rotating vectors in the complex plane but with decaying amplitudes (spirals).

In the INOPMA algorithm [1], the correlation sequence was assumed to be an impulse response. This has led to the fact that the mode is isolated at a frequency $\omega^* = \omega_n\sqrt{1-4\xi^2}$ and not the natural frequency ω_n. This may be considered as a limitation of INOPMA. In this work we overcome this limitation and we show that it is still possible to isolate the mode at its natural frequency.

Let $\bar{R}(t,\alpha)$ be the image of $R(t)$ by a linear anti-symmetric function that depends on a certain design parameter α and consider the following sequence:

$$H(t,\alpha) = R(t) + \bar{R}(t,\alpha) \tag{3}$$

$H(t,\alpha)$ is then a rotation and stretching of the sequence $R(t)$. By varying α, it is possible to modify the phase shift as well as the amplitude of the correlation sequence and hence modify the damping ratio leading to a pure sinusoid. At this stage the mode is isolated.

In this work, we consider the following anti-symmetric function:

$$F(R(t),\alpha) = j\alpha R(t) \tag{4}$$

The SMA algorithm is similar to the INOPMA algorithm in the sense that one takes the convolution of the modified correlation sequence with a harmonic force with varying frequency.

In the frequency domain this means that the system transfer function (the Laplace transform of the correlation sequence) is multiplied by a complex factor $(1 + j\alpha)$. It can easily be shown that the transfer function phase angle is zero exactly at the following condition:

$$\begin{cases} \omega = \omega_n \\ \alpha = 2\xi \end{cases} \tag{5}$$

Geometrically this means that when the mode is isolated the vector representing the correlation sequence describes a circle meaning that the conservative part of the system is isolated. The non-conservative part follows immediately. Hence the system natural frequency as well as the damping ratio are identified at the same step.

3 Harmonics Rejection

We prove in this section that the SMA algorithm is robust against harmonics modeled as modes with zero damping.

Consider a SDOF system excited with unmeasured white noise. We show in the sequel that if the damping ratio is zero then the mode can not be identified (the phase angle is never zero) hence rejected.

The SMA algorithm is based on considering the following parametric correlation sequence:

$$H(t, \alpha) = (1 + j\alpha)R(t)$$

The Laplace Transform of this function can be shown to write as:

$$G(s) = (1 + j\alpha)\frac{s + 2\xi\omega_n}{s^2 + 2\xi\omega_n s + \omega_n^2}$$

The Imaginary part of the Frequency Response is:

$$I = (2\xi\alpha\omega_n + \omega)(\omega_n^2 - \omega^2) - (2\xi\omega_n - \alpha\omega)(2\xi\omega\omega_n)$$

While the Real part is:

$$\mathrm{Re} = (\omega_n^2 - \omega^2)(2\xi\omega_n - \alpha\omega) + 2\omega^2\xi\omega_n$$

When the damping ratio is zero the tangent of the phase angle of the Frequency response reduces to:

$$tg = -\frac{\omega(\omega_n^2 - \omega^2)}{\alpha\omega(\omega_n^2 - \omega^2)} = -1/\alpha$$

Which is always different from zero. Consequently for a harmonic the angle between the input and the output is never zero meaning that the harmonic is never identified (no zero crossing).

4 Spurious Modes Rejection

We define in this section a spurious/numerical mode as a mode that is fitted to the system characteristic equation. This leads to a spurious frequency and damping that we still denote in the sequel as wn and zeta. The correlation sequence of the system output is given by [2]:

$$R(t) = e^{-\xi\omega_n t}\left[\cos(\omega_d t) + \frac{\xi}{\sqrt{1-\xi^2}}\sin(\omega_d t)\right] \quad (6)$$

The phase shift in this correlation sequence is given by:

$$tg(\theta) = \frac{\xi}{\sqrt{1-\xi^2}} \quad (7)$$

This particular expression of the phase shift is valid for physical modes only [4]. We propose to show in the sequel that if the phase shift of a correlation sequence enjoys this particular expression, then the mode is necessarily physical.

Consider the following correlation sequence:

$$R_x(t) = e^{-\xi\omega_n t}[\cos(\omega_d t) + x\sin(\omega_d t)]$$

The Laplace Transform of $(1 + j\alpha)R_x(t)$ is:

$$G_x(s) = (1 + j\alpha)\frac{s + \omega_n(\xi + x\sqrt{1-\xi^2})}{s^2 + 2s\xi\omega_n + \omega_n^2}$$

The numerator writes as:

$$(1 + j\alpha)(j\omega + \omega_n(\xi + x\sqrt{1-\xi^2}))(-\omega^2 + 2j\omega\xi\omega_n + \omega_n^2)$$

And the imaginary part writes as:

$$\begin{aligned}\mathrm{Im} = & -\omega^3 + \omega\omega_n^2 - \alpha\omega^2\omega_n(\xi + x\sqrt{1-\xi^2}) + \alpha\omega_n^3(\xi + x\sqrt{1-\xi^2}) \\ & + 2\xi\omega_n^2\omega(\xi + x\sqrt{1-\xi^2}) - 2\xi\omega^2\omega_n\alpha\end{aligned}$$

Under the isolating conditions

$$\begin{cases} \omega = \omega_n \\ \alpha = 2\xi \end{cases}$$

$$\omega_n^3(2\xi(\xi + x\sqrt{1-\xi^2}) - 4\xi^2) = 0$$

Leading to:

$$x = \frac{\xi}{\sqrt{1 - \xi^2}}$$

Which proves that under the SMA isolating conditions the mode is necessarily physical.

5 Simulation Validation

It is not our intention in this section to give a comprehensive study of the performance of the SMA algorithm nor it is to study its limitations. This will be reported in a different study. We will be limited here to validate the proposed theory on a simple SDOF system.

The considered system parameters are taken as m = 2 kg, k = 10000 N/m and c = 8 Ns/m. The excitation is a white noise with unit variance. This leads to the following modal parameters; $\omega_n = 11.254$ Hz and $\xi = 2.83\%$. The output is then simulated using a sampling frequency of Fs = 64 Hz and 2% measurement noise is added to the output. Figure 1 shows the identification results onn this data set.

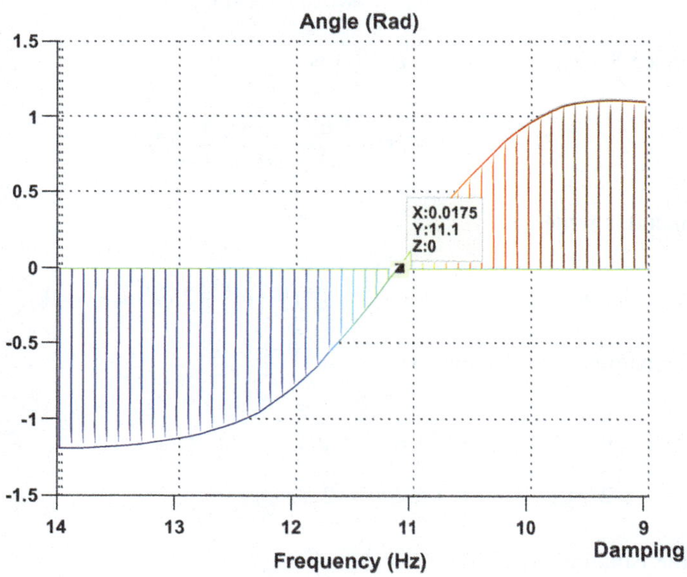

Fig. 1 Phase angle as a function of frequency and alpha

5.1 Harmonics Rejection

We add in this section a harmonic component with frequency 5 Hz and amplitude 0.1 N. Figure 2 shows the identification results and we notice that the harmonic component is rejcted.

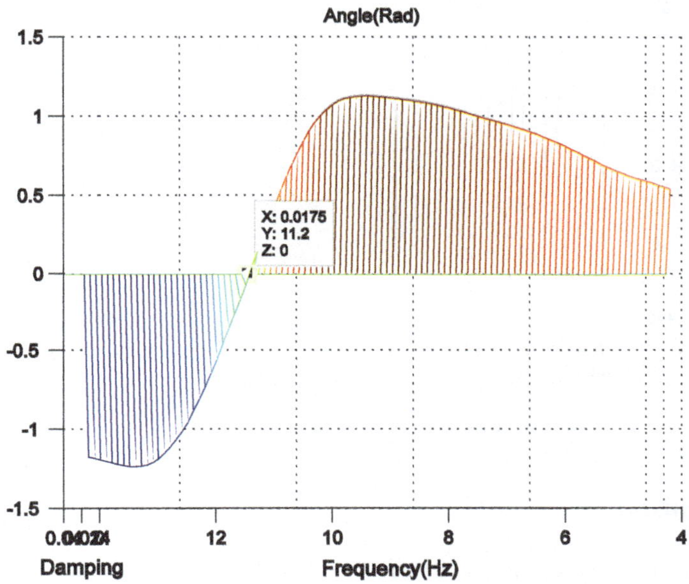

Fig. 2 Hamonics rejection example

5.2 Spurious Modes Rejection

We simulate spurious mode here by considering a colored excitation. We drive a unit white noise through a AR(2) process whose output serves a driving excitation to the system. Figure 3 shows that SMA is robust against spurious modes.

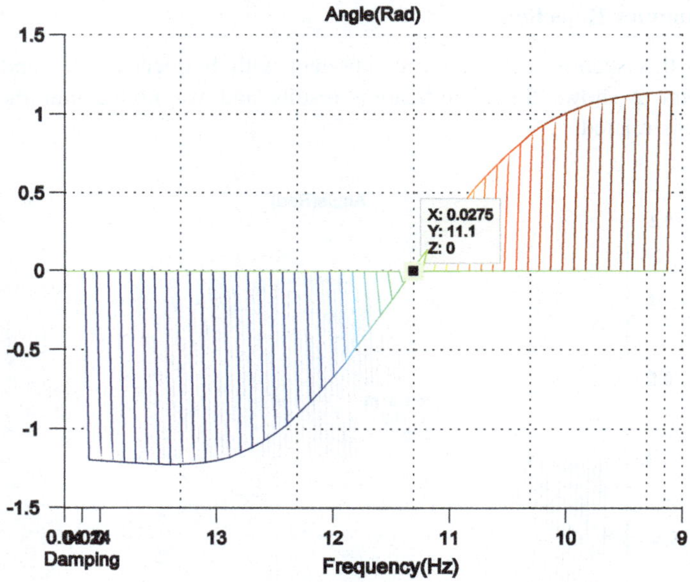

Fig. 3 Spurious modes rejection

6 Conclusion

We showed in this paper that the SMA algorithm rejects automatically spurious modes as well as harmonics. No pre-processing or post-processing is required. This approach has been validated on a simple simulated example. Experimental validation is ongoing.

References

1. Abdelghani M, Inman DJ (2015) Modal appropriation for use with In-Operation modal analysis. J Shock Vib
2. Abdelghani M, Friswell MI (2018) Stochastic modal appropriation (SMA). In: IMAC'2018, USA
3. Balmès E, Chapelier C, Lubrina P, Fargette P (1996) An evaluation of modal testing results based on the force appropriation method. In: International modal analysis conference, Orlando
4. Meirovitch L (1986) Elements of vibration analysis. McGraw-Hill

Comparative Study of Smart Structures Vibrations Under Earthquake Excitations

Houssameddine Chitaoui(✉), Abdellatif Megnounif, and Zahira Benadla

RISAM Laboratory, Civil Engineering Department, University of Tlemcen, Tlemcen, Algeria
houssameddine.chitaoui@univ-tlemcen.dz

Abstract. Due to the evolution of design techniques and material qualities in civil engineering, the structures become lighter. When these structures will be located in environments prone to earthquakes or high winds, this lightness may accentuate the vibration causing major problems to structures such as failure, discomfort, noise… Currently, many researchers are interested in this problem to balance between lightness and vibration resistance. From the review of the literature, several methods are proposed in this field, the most used one is based on the principle of the control of vibration. In this case, the reduction of structural vibrations is done by adding a mechanical system composed of intelligent materials. In this paper, the performance of the control system, such as a passive control using tuned mass dampers TMD and/or an active control using active tendons AT and ATMD on a high building subjected to seismic excitation, is studied. A parametric study is conducted by varying a certain key parameter such as the position, the number and the type of the control systems. Finally, to find the ideal position of the active system, three techniques are proposed: method of modal controllability, controllability index and genetic algorithm. For all the cases, numerical simulations are established at MATLAB and the results are illustrated and compared.

Keywords: Control · Optimization · Modal controllability · Controllability index · Genetic algorithm

1 Introduction

Several researchers have attempted the problem of positioning the actuators/shock absorbers. Pourzeynali et al. [1] have made a comparison between the control with TMD and ATMD. The control force applied on TMD is computed by linear quadratic regulator LQR and a combination between fuzzy logic and genetic algorithm. Rao and Pan [2], used GA to solve the problem of discrete optimal location of the actuators and said their approach produce a global optimal solution or nearby. Liu et al. [3] studied the effect of 18 different excitations of the shake down on the optimal placement of actuators to decrease the greatest displacement of the upper floor and found that the placement is not influenced by the level of earthquake. Rao and Sivasubramanian [4] did a meta-heuristic algorithm multi boot (MSGNS) as simulated annealing. This algorithm

N. Aifaoui et al. (Eds.): CMSM 2019, LNME, pp. 289–299, 2020.
https://doi.org/10.1007/978-3-030-27146-6_31

demonstrated by numerical experiments. However, little effort has been made to date on the optimal placement of actuators and dampers with optimization criteria. Cheng et al. [5] have done an article on optimal placement of actuators and dampers. They proposed formulation of a general optimization problem based on stochastic approach and establishment of a solution procedure. Finally, they proved an optimally applied seismic response control system is much more efficient than a control system without optimal device location, and becomes more practical. This work differs from the previous papers in several points. Most of the articles do not prove the data used, such as the data of the analyzed structures. That is why to do this comparative study in this work, the data of the structure is taken and validated with [1]. Most of the previous relevant researches concentrate to make fix the number of the actuators in control systems. In the present work, the number of actuators in the AT is integrated to see their effect on the greatest displacement of the structure comparing with passive control such as TMD. Most of the papers use GA to find the optimal placement. However, the proposed work make a comparison enter three methods of optimization and their influence in arriving at the optimal positioning of actuators. Two are classical methods which are modal controllability and controllability index [6]. The third is use the GA. In the present work, the LQR algorithm is employed for active control of structure. In the end, we will also show the advantage of the GA to find the optimal position of the different control systems and the influence of making several earthquake excitations.

2 Classical Active Control Algorithm (LQR)

In control theory, the equation of motion written in state-space Eq. (1) is:

$$\{\dot{Z}(t)\} = [A]\{Z(t)\} + [B_u]\{u(t)\} + \{B_r\}\ddot{x}_g(t) \tag{1}$$

The control force {u(t)} is determined by minimizing the quadratic performance index J, given by Cheng et al. [6]:

$$J = \int \left(\{Z(t)\}^T[Q]\{Z(t)\} + \{u(t)\}^T[R]\{u(t)\}\right)dt \tag{2}$$

[Q] and [R] are response weightage matrices for structure's response and control force respectively. J represent the balance between the structural response and the control energy to reduce the response of the structure (Fig. 1) [7].

$$\{u(t)\} = -[G]\{Z(t)\} = [R]^{-1}[B_u]^T[P(t)]\{Z(t)\} \tag{3}$$

where [G] is gain matrix, [P] is determined by solving Riccati's matrix equation. Summarize the method for calculating the control force of an active control:

- Define the matrices [Q] and [R].
- Calculate [P(t)] solution of the Riccati.
- Calculate the gain matrix [G] and finally, calculate the control force {u(t)}.

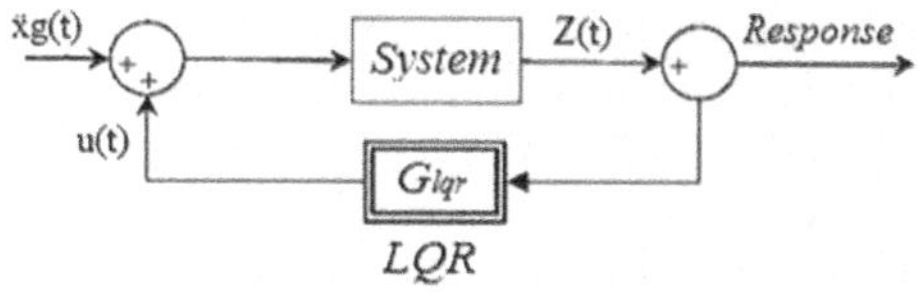

Fig. 1 Block diagram of the system with LQR control

3 Methods of the Optimal Position of the Control Systems

3.1 The Method of Modal Controllability

The modal controllability method is one of the techniques that allow us to define the optimal position of an AT in a given structure [6]. Using the modal analysis of the system, we get:

$$\{x(t)\} = [\Phi]\{q(t)\} \tag{4}$$

$[\Phi]$ and $\{q(t)\}$ are the matrices of modal and modal coordinate mode shapes. By dividing by the term of the mass, we will have for each value of "i", i = (1,…n), n is the number of DOF. The modal equation:

$$\dot{q}_i(t) + 2\xi_i w_1 \dot{q}_i(t) + w_i^2 q_i(t) = \gamma_i u(t) + \delta_i \ddot{x}_g(t) \tag{5}$$

So the efficiency of the control force is proportional to the absolute value of "γ_i". The optimal position of AT corresponds to the max absolute value of "γ_i".

3.2 Controllability Index

The method of the controllability index was developed. For an AT control system, the index of controllability is [6]:

$$\rho(x) = \max\sqrt{\sum_{j=1}^{n}\{(\Delta[\emptyset_j(x)]/\Delta x)Y_j(t)\}^2} \tag{6}$$

where x is a (%) of the total height of the structure at the location of the actuators ($0 \le x \le 1$); $\emptyset_j$ (x) is a form of modes; Y_j (t) is the max value of the response spectrum for "jth" mode; Δ[ ] refers to the spatial difference of the quantity from position x1 to position x2.

The optimal location is defined as the value of (x) for which ρ(x) is maximum.

3.3 Genetic Algorithm

GA is a search algorithm used in mathematical calculation to solve the optimization. It is a stochastic research algorithm based on the mechanics of natural and genetic selection of populations [8]. The GA differs from a classical, derivative-based, optimization algorithm in two main ways, as summarized in the following. In classical algorithm a single point is generated at each iteration and the sequence of points approaches an optimal solution. But in GA a population of points is generated at each iteration and the best point in the population approaches an optimal solution. In this work, we are looking for the floor in which the AT is located in order to obtain the best performance. Also, we should create an objective function. It is only for structure analysis and contain active control program with LQR. It has inputs and output. The inputs are location vector and number of AT. The output is the min of the total displacement of the structure that one can have. After, the duty of GA gives different inputs to the objective function to find the best entry in which the output total displacement is min.

4 Numerical Studies

4.1 Validation of the Model Taken

An eleven story building is used as an example. This building is modeled as a shear frame and the problem is solved in state space. The properties of the structure in Table 1 are taken from [1]:

Table 1 Building structural data

Floors	Mass (kg)	Floors	Stiffness (N/m)
1	2.15e05	1	4.68e08
2–3	2.01e05	2	4.76e08
4	2.00e05	3	4.68e08
5–7	2.01e05	4–7	4.50e08
8–10	2.03e05	8–10	4.37e08
11	1.76e05	11	3.12e08

In [1] the TMD and ATMD are used to compare the results of displacement. We compare and valid our results of displacement using the same property of TMD and ATMD with LQR approach under the El Centro and Northridge earthquake. Let's remember that the controller systems are installed on the top floor. This comparison is to justify the choice of model taken, so the found values have a minimal error with the values of [1]. Also, to apply the position optimization methods those come after.

4.2 Application of Modal Controllability

This method represented in Eq. (7) is the technique that allows us to define the optimal position of an AT in the given structure. On this model, we will apply the active control by some AT to decide in which floor the optimal position have.

$$\left\{\delta_i^j\right\} = \{\Phi_i\}^T\{\gamma\}_j \tag{7}$$

where “i” is number of Eigen modes and “j” is number of floor.

$\left\{\delta_i^j\right\}$ is vector of modal controllability and $\{\Phi_i\}^T$ is an eigen mode.

$\{\gamma\}_j$ is vector of localization of AT control force installed on the ith floor.

Thus, the ideal place of AT corresponds to the maximum absolute of “δ_i^j”.

We note that each mode has its own optimal position of the AT. This method is simple in the execution. It is independent of the earthquake. It has a major disadvantage, because it is really effective in the case of a predominant single mode structure and it is not the case in the intelligent structures, because its can have several significant modes.

According to the most dominant mode of this structure and which corresponds to the max displacement, it is the 9th mode. So we vary the seismic signal and the number of AT with their optimal position to compare the effectiveness of this variation as it is shown in Table 2.

For the 9th mode, modal controllability is:

$$|\delta_9^1| > |\delta_9^2| > |\delta_9^3| > |\delta_9^4| > |\delta_9^5| > |\delta_9^6| > |\delta_9^7| > |\delta_9^8| > |\delta_9^9| > |\delta_9^{10}| > |\delta_9^{11}|$$

Table 2 Comparison of the effect of the different nbrs of AT with their optimal position

Floors	El centro earthquake					Northridge earthquake				
	Max uncontrolled (m)	Controlled to uncontrolled response ratio (reduction ratio)				Max uncontrolled (m)	Controlled to uncontrolled response ratio (reduction ratio)			
		1AT	3AT	5AT	All Floors		1AT	3AT	5AT	All Floors
1	0.020	1.00	0.95	0.90	0.85	0.040	0.97	0.95	0.92	0.83
2	0.040	0.80	0.75	0.75	0.82	0.080	0.93	0.90	0.90	0.88
3	0.059	0.86	0.86	0.86	0.81	0.110	0.90	0.87	0.85	0.83
4	0.074	0.94	0.87	0.87	0.83	0.14	0.92	0.89	0.87	0.86
5	0.09	0.93	0.90	0.83	0.82	0.16	1.00	0.96	0.93	0.86
6	0.10	0.92	0.92	0.83	0.84	0.180	0.94	0.91	0.90	0.88
7	0.12	0.90	0.90	0.86	0.83	0.190	0.94	0.91	0.89	0.86
8	0.12	0.91	0.91	0.83	0.82	0.200	0.95	0.93	0.92	0.90
9	0.13	0.92	0.85	0.84	0.84	0.22	0.95	0.93	0.91	0.89
10	0.14	0.92	0.86	0.79	0.81	0.23	0.95	0.93	0.91	0.90
11	0.145	0.91	0.89	0.82	0.81	0.240	0.95	0.93	0.91	0.88

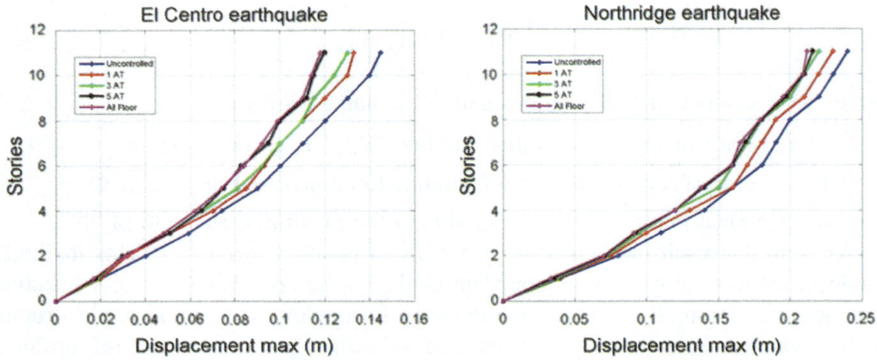

Fig. 2 Comparison of max disp. response of 1/2/5 AT with uncontrolled and fully controlled

According to Table 2 and Fig. 2 we see that: placing the AT in the lower half of the structure greatly reduces the max displacement. However, putting the AT in all floors by comparing with 5AT and 3AT, affects only 1% and 8% respectively of additional reduction for El Centro and 3 to 5% for Northridge earthquakes, which shows that putting the AT in more than half of the floors it has no effect.

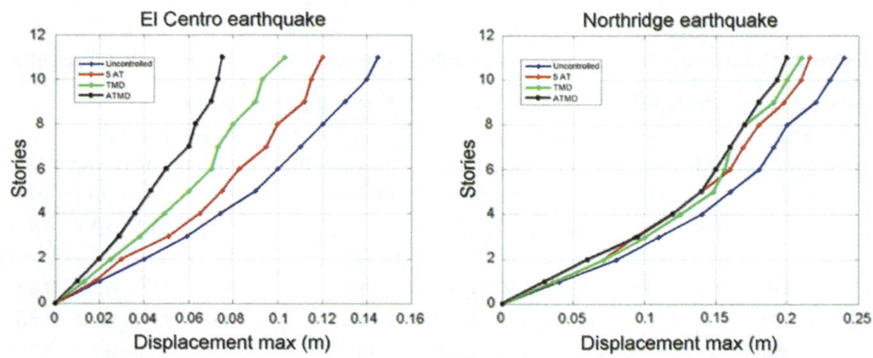

Fig. 3 Comparison of max disp. response of 5 AT with TMD and ATMD

In Fig. 3 to economize the cost of control, we chose to compare the control with 5AT, TMD and ATMD. For El Centro earthquake, we note that there is a great influence of the different control systems on max displacement of the structure from the 3rd to the top level. The effect of TMD is 14% and for ATMD is 34% compared to 5AT. For Northridge earthquake, from the 2nd to the 5th floor the control with AT is better than the other two systems. After, the difference of max displacement stay constant within limits 1%. Makes us think that, the active control with some number of AT can replace at least the TMD. Everything depends on the properties of the AT to apply the wish active control force.

4.3 Application of Controllability Index

Procedure for calculating this index for a controlled structure with more than one AT. The natural frequencies of this building are defined as follows:

$$f_i = [1.04; 3.08; 5; 6.78; 8.42; 9.95; 11.33; 12.60; 13.63; 14.39; 14.91]$$

Using this response spectrum in Fig. 4, drawn by earthquake software [9] to have the max values of the modal response:

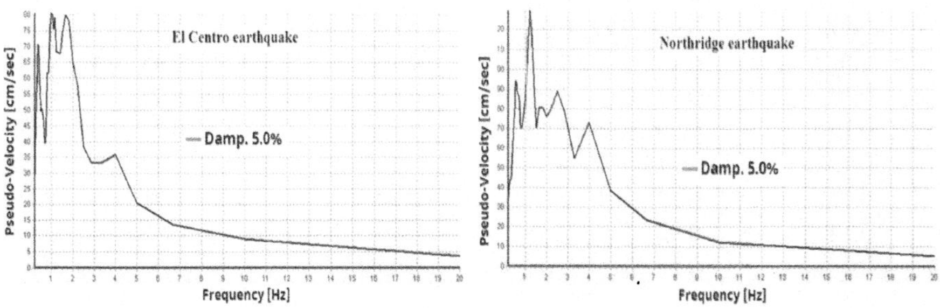

Fig. 4 Response spectra for El Centro and Northridge earthquakes

Assuming that the floors of the structure have the same height: $\Delta x = 0.5$.

Taking the maximum value of the index of controllability to find the optimal position for AT. So this index depends on response spectra of earthquake.

- For El Centro spectra:

$$\rho_4 > \rho_5 > \rho_3 > \rho_6 > \rho_7 > \rho_2 > \rho_8 > \rho_9 > \rho_{10} > \rho_1 > \rho_{11}$$

- For Northridge spectra:

$$\rho_3 > \rho_4 > \rho_5 > \rho_6 > \rho_7 > \rho_8 > \rho_2 > \rho_9 > \rho_{10} > \rho_{11} > \rho_1$$

Applying these positions founded on our structure, by varying the number of AT. The results are in Table 3 and Fig. 5.

Table 3 and its graphical representation in Fig. 5 show that: even by varying the position of the 5AT, their effect on reducing the maximum displacement of the structure is almost like the effect of the fully controlled structure. Just a difference of 7% for El Centro and 1% for Northridge earthquakes. This means that the positions founded by the modal controllability are better than the one founded by the index of controllability if by comparing reduction ratio.

Table 3 Comparison of the effect of the different nbrs of AT with their optimal position

Floors	El centro earthquake					Northridge earthquake				
	Max uncontrolled (m)	Controlled to uncontrolled response ratio (reduction ratio)				Max uncontrolled (m)	Controlled to uncontrolled response ratio (reduction ratio)			
		1AT	3AT	5AT	All Floors		1AT	3AT	5AT	All Floors
1	0.020	1.00	0.95	0.95	0.85	0.040	0.97	0.95	0.90	0.83
2	0.040	1.00	0.95	0.90	0.82	0.080	0.93	0.91	0.90	0.88
3	0.059	0.93	0.91	0.88	0.81	0.110	0.90	0.88	0.86	0.83
4	0.074	0.97	0.93	0.90	0.83	0.14	0.92	0.90	0.85	0.86
5	0.09	0.95	0.91	0.87	0.82	0.16	1.00	0.98	0.87	0.86
6	0.10	0.97	0.93	0.90	0.84	0.180	0.94	0.92	0.88	0.88
7	0.12	0.90	0.90	0.89	0.83	0.190	0.94	0.94	0.89	0.86
8	0.12	0.91	0.91	0.83	0.82	0.200	0.95	0.94	0.90	0.90
9	0.13	0.92	0.92	0.90	0.84	0.22	0.95	0.91	0.89	0.89
10	0.14	0.92	0.91	0.85	0.81	0.23	0.95	0.92	0.91	0.90
11	0.145	0.96	0.91	0.88	0.81	0.240	0.95	0.93	0.89	0.88

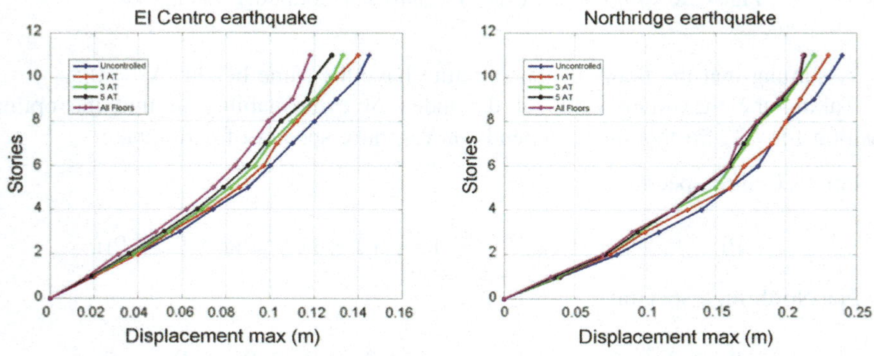

Fig. 5 Comparison of max disp. response of 1/3/5 AT with uncontrolled and fully controlled

Comparing now, the 5AT with the TMD and ATMD which are represented in Fig. 6. So the reduction of the maximum displacement with the active control with 5AT it diminishes a little on the top floor. But still the control with 5AT closer to the TMD and ATMD if we see compared to the Northridge earthquake.

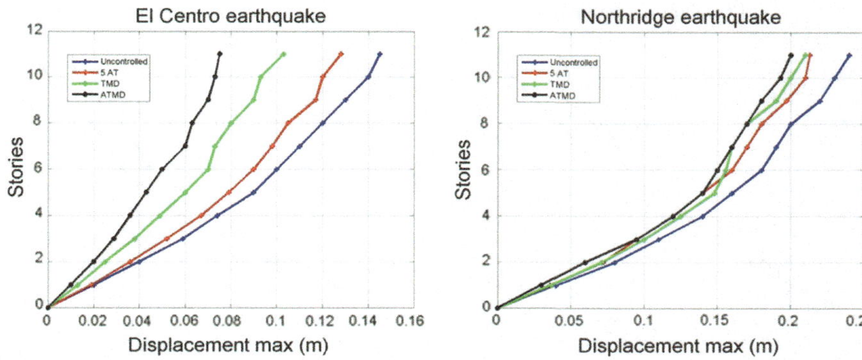

Fig. 6 Comparison of max disp. response of 5 AT with TMD and ATMD

4.4 Application of Genetic Algorithm

In Table 4, we see that the same optimal positions for different number of AT founded by the modal controllability method in mode 9th are founded here by GA.

Table 4 Optimal position for 1/3/5 AT by genetic algorithm

El centro earthquake	Nbrs of AT	Optimal positions (Floors)					Max controlled response (m)	Reduction ratio	Max uncontrolled response (m)
	1 AT	1					0.133	0.91	0.145
	3 AT	1	2	3			0.130	0.89	
	5 AT	1	2	3	4	5	0.120	0.82	
Northridge earthquake	Nbrs of AT	Optimal positions (Floors)					Max controlled response (m)	Reduction ratio	Max uncontrolled response (m)
	1 AT	1					0.230	0.95	0.240
	3 AT	1	2	3			0.225	0.93	
	5 AT	1	2	3	4	5	0.216	0.91	

As can be seen from the Fig. 7, for El Centro earthquake the max displacements of stories of modal controllability and GA are identical which gives a huge reduction approximately 18% and the graph concerning controllability index has 12% compared with uncontrolled structure. For Northridge earthquake the max displacements of stories of the three optimization methods are identical with reduction of 9% for modal controllability and GA and reduction of 11% for controllability index.

From Table 5, it can be seen that the optimal positions of AT varies between 1st and 2nd floor. Concerning the TMD, it optimal positions always on the top half of the structure. For ATMD, it optimal positions stays always in the top floor.

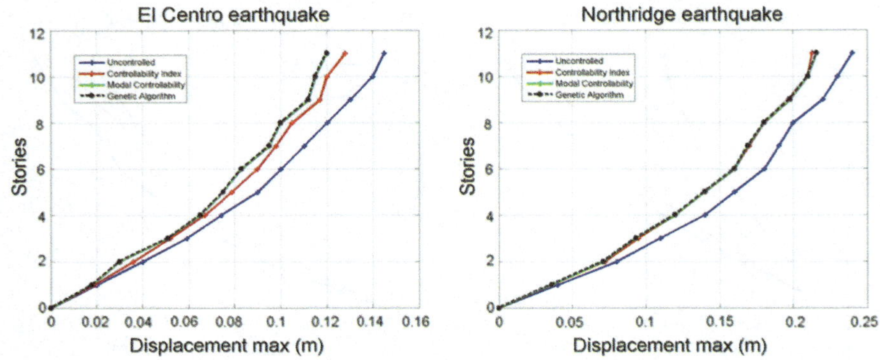

Fig. 7 Comparison of max disp. of uncontrolled and controlled structure with 5AT of the three optimization methods

Table 5 Optimal position for 2AT, 2TMD and 2ATMD under different earthquake

Earthquake name	Max uncontrolled response (m)	Optimal positions of AT		Reduction ratio	Optimal positions of TMD		Reduction ratio	Optimal positions of ATMD		Reduction ratio
Kobe	0.504	4	5	0.91	5	10	0.91	3	11	0.75
Chi Chi	0.231	1	2	0.88	10	11	0.88	10	11	0.65
Loma Prieta	0.123	1	2	0.97	1	2	0.97	1	11	0.87
Kocaeli	0.161	1	2	0.81	2	7	0.81	1	11	0.57
Landers	0.113	1	4	0.84	6	10	0.84	10	11	0.57
Friuli	0.100	1	2	0.86	10	11	0.86	10	11	0.50

5 Conclusion

Control with AT is less expensive and simple in their installation compared to the TMD and ATMD. The results show that AT need to be placed in the lower half of the structure. The position of the AT at the level of the structure stays an important factor in the control. The effect of controlling a structure with a certain amount of AT is equal to controlling the structure completely. Some places in the structure are advantageous. A number of AT with their optimal position can reduce the max displacement of the structure as the reduction carried out by the TMD. The optimal position is determined according to two criteria, first is in relation to the max displacement and second in relation to the AT controls force.

References

1. Pourzeynali S, Lavasani HH, Modarayi AH (2007) Active control of high rise building structures using fuzzy logic and GA. J Eng Struct 29:346–357
2. Rao SS, Pan TS (1991) Optimal placement of actuators in actively controlled structures using GA. AIAA J 29:942–943
3. Liu DK, Yang YL, Li QS (2003) Optimum positioning of actuators in tall buildings using genetic algorithm. Comput Struct 81:2823–2827

4. Rao ARM, Sivasubramanian K (2008) Optimal placement of actuators for active vibration control of seismic excited tall buildings using (MSGNS) algorithm. J Sound Vib 311:133–159
5. Cheng FY, Jiang H, Zhang X (2002) Optimal placement of dampers and actuators based on stochastic approach. Earthq Eng Eng Vib 1671–3664, 02-0237-13
6. Cheng FY, Jiang H, Lou K (2008) Smart structures innovative systems for seismic response control. CRC Press
7. Sugumar R, Kajjam S, Datta TK (2013) Implementation of active seismic control of a building frame using ANSYS, Int J Conceptions Mech Civ Eng. ISSN: 2357–2760
8. Goldberg DE (1989) GA in search, optimization and machine learning. Addison-Wesley Longman Publishing Co., Inc. Boston, MA, USA. ISBN: 0201157675
9. SeismoSignal: Earthquake software for signal processing of strong-motion data

The Influence of Process Parameters on Single Point Incremental Forming: Numerical Investigation

A. Bouhamed[1(✉)], H. Jrad[1], L. Ben Said[1,3], M. Wali[1,2], and F. Dammak[1]

[1] Laboratory of Electromechanical Systems (LASEM), National Engineering School of Sfax, University of Sfax, Route de Soukra Km 4, 3038 Sfax, Tunisia
{abir.bouhamed, Fakhreddine.dammak}@enis.tn, hanen.j@gmail.com, bensaid_rmq@yahoo.fr, mondherwali@yahoo.fr

[2] Department of Mechanical Engineering, College of Engineering, King Khalid University, Abha, Saudi Arabia

[3] Mechanical Engineering Department, College of Engineering, University of Hail, Hail, Saudi Arabia

Abstract. In this paper, elasto-plastic model based on non-associated flow rule is implemented in Abaqus/Explicit software via VUMAT subroutine to study the influence of some process parameters: cone wall angle, tool diameter, and sheet thickness, on the formability, plastic deformation and thinning during single point incremental forming (SPIF) process. The present work presents useful guidance to show the effect of these three parameters in improving the quality of manufactured products and in obtaining a better formability during SPIF process.

Keywords: Single point incremental forming (SPIF) · Non-associated flow rule · Elasto-plastic model · Formability · Thinning

1 Introduction

Single Point Incremental Forming (SPIF) has drawn great attention and increasing interest in manufacturing sheet metal products that have fuelled academic and industrial research in SPIF over the past two decades. The SPIF process is well suited and highly recommended due to its flexibility for small batch and customized production and it is deliberated as a rapid prototyping technique. The principal goal, which stimulates the development of SPIF, is the absence of specific tooling and their greatest formability compared to conventional processes. In SPIF, a spherical headed tool, that follows a predefined tool path conforming to final part geometry, moves over the surface of the sheet and incrementally deforms the flat sheet by a progression of localized deformation.

The SPIF process suffers from poor geometric accuracy of the manufactured parts. In fact, the tool path is one of the principal parameters, which influences the geometric accuracy of the formed part. Hence, many researches works have been carried out on

N. Aifaoui et al. (Eds.): CMSM 2019, LNME, pp. 300–308, 2020.
https://doi.org/10.1007/978-3-030-27146-6_32

the tool path optimization and development, namely [2] who compared between four tool paths strategies developed by a known CAM software (CATIA V5 R17) to see the effect of these strategies on the ISF parameters. Further, [1] gave a parameterized forming strategy to reduce the manufacturing time and homogenize thickness distribution of an asymmetric mechanical piece. Recently, a number of reported review papers have been published an overview on the influence of process parameters in incremental sheet forming such as presented in [6] which gives a particular attention to the ISF process parameters and depicted that the formability and the deformation are significantly influenced by the wall angle during the incremental forming process. Additionally, [8] have controlled the effects of wall angle on Al 3003 strain hardening for parts forming by incremental forming to provide insight into the specific effect of this process on the strain hardening of Al 3003.

Li et al. [11] studied the influence of three parameters including step-down size, sheet thickness, and tool diameter on the thinning and mechanical properties to provide useful guidance in improving the product quality formed by incremental sheet forming.

Different material behaviors proposed for assessing the formability taking into account the effects of process parameters. Ben Said et al. [3] developed an elasto-plastic constitutive model coupled with isotropic ductile damage based on associated flow rule to predict the damage in SPIF process and to study the influence of some process parameters on the damage evolution and forming effort during the SPIF of manufactured parts.

In this paper, elasto-plastic model based on non-associated flow rule is implemented in Abaqus/Explicit software via VUMAT subroutine to depict the effect of some process parameters including wall angle, tool diameter and sheet thickness on the formability during SPIF process. The reason behind using these parameters is to find out their effects on surface quality and thinning of the truncated cone.

2 Constitutive Model

This section aims to present an elasto-plastic constitutive model based on non-associated flow rule and mixed isotropic/kinematic hardening to describe the behavior of the blank sheet during the SPIF process. Simple quadratic constitutive equations are presented in Table 1 and detailed in [4, 5, 7, 10].

Table 1 Elasto-plastic constitutive equations based on non-associated flow rule

a/Yield function	c/Plastic potential function
$f = \varphi_f(\boldsymbol{\sigma}) - \sigma_P \leq 0$ (1)	$F = \varphi_F(\boldsymbol{\sigma}) - \sigma_P + \frac{1}{2}\sum_{k=1}^{M} \frac{b_k}{a_k} \boldsymbol{X}_k : \boldsymbol{X}_k$ (5)
$\varphi_f(\boldsymbol{\sigma}) = \Vert\boldsymbol{\sigma}\Vert_P = \sqrt{\boldsymbol{\sigma}^t \boldsymbol{P} \boldsymbol{\sigma}}$ (2)	$\varphi_F(\boldsymbol{\sigma}) = \Vert\boldsymbol{\sigma}\Vert_{\mathbf{Q}} = \sqrt{\boldsymbol{\sigma}^t \boldsymbol{Q} \boldsymbol{\sigma}}$ (6)
$\sigma_P = \sigma_Y + R(r);\ R(r) = Q\left(1 - e^{-\beta r}\right)$ (3)	d/Kinematic hardening $\boldsymbol{X} = \sum_{k=1}^{M} \boldsymbol{X}_k, \dot{\boldsymbol{X}}_k = a_k \dot{\varepsilon}^p - b_k \dot{\gamma} \boldsymbol{X}_k$ (7)
b/Plastic strain tensor $\dot{\varepsilon}^p = \dot{\gamma} \frac{\partial F}{\partial \boldsymbol{\sigma}} = \dot{\gamma} \boldsymbol{n};\ \boldsymbol{n} = \frac{1}{\varphi_F} \boldsymbol{Q} \boldsymbol{\sigma}$ (4)	e/Isotropic hardening $R = R(r)$ (8)

In Table 1, $\boldsymbol{\sigma}$ is the stress tensor, $\boldsymbol{\alpha}_k$ and $\mathbf{X}_k$ are the kinematic hardening variables, σ_Y is the initial yield stress, $\boldsymbol{\varepsilon}^p$ is the plastic strain tensor, r and R are the isotropic hardening variables, a_k, b_k, β and Q are material parameters $\boldsymbol{P}$ and $\boldsymbol{Q}$ are the Hill'48 anisotropic operators defined the yield function and the plastic potential function respectively.

$$\boldsymbol{P} = \boldsymbol{P}(F,G,H,N,M,L) = \begin{bmatrix} H+G & -H & -G & 0 & 0 & 0 \\ & H+F & -F & 0 & 0 & 0 \\ & & F+G & 0 & 0 & 0 \\ & & & 2N & 0 & 0 \\ & Sym & & & 2M & 0 \\ & & & & & 2L \end{bmatrix} \tag{9}$$

$$\boldsymbol{Q} = \boldsymbol{Q}\left(F',G',H',N',M',L'\right) \tag{10}$$

3 FEM Simulation

Numerical simulations based on the finite element model (FEM) have been carried out with emphasis on assessing and improving the formability in SPIF process. The simulation of the SPIF process was performed using elasto-plastic model based on a non-associated flow rule implemented in Abaqus/Explicit software via VUMAT subroutine, which has the ability to manipulate the contact problem between tool and the part surface as presented in [4]. Furthermore, the blank was pretended to be anisotropy and meshed by S3R shell element with five integration points through the thickness. The mechanical properties of the AA6022-T43 aluminum alloy are shown in Table 2. A Coulomb friction coefficient of 0.1 is assumed between the punch and the blank. In this study, three parameters including wall angle, which means the angle between the deformed sheets to the horizontal plane, sheet thickness, and tool diameter were picked as variable in SPIF process. The reason behind using these different parameters presented in Table 3 is to find out their effects on surface quality and thickness variation on the manufactured part. The forming strategy for three different cone wall angle is illustrated in Fig. 1.

Table 2 Mechanical properties of AA6022-T43 [12]

Young modulus (GPa)	70				
Poisson's ratio	0.33				
Anisotropic coefficients	Yield function coefficients	F	G	H	N
		0.632	0.496	0.504	1.585
	Potential function coefficients	F'	G'	H'	N'
		0.697	0.493	0.507	1.228
Voce law hardening (MPa)	$\sigma_Y + R(r) = 136 + 110\left(1 - e^{-7.5r}\right)$				
Kinematic hardening	$a = 1400\,\text{MPa}$; $b = 20$;				

Table 3 Values of process parameters

Parameter	Value
Wall angle	45°, 55°, 65°
Sheet thickness	1 mm, 1.5 mm, 2 mm
Tool diameter	10 mm, 15 mm, 20 mm

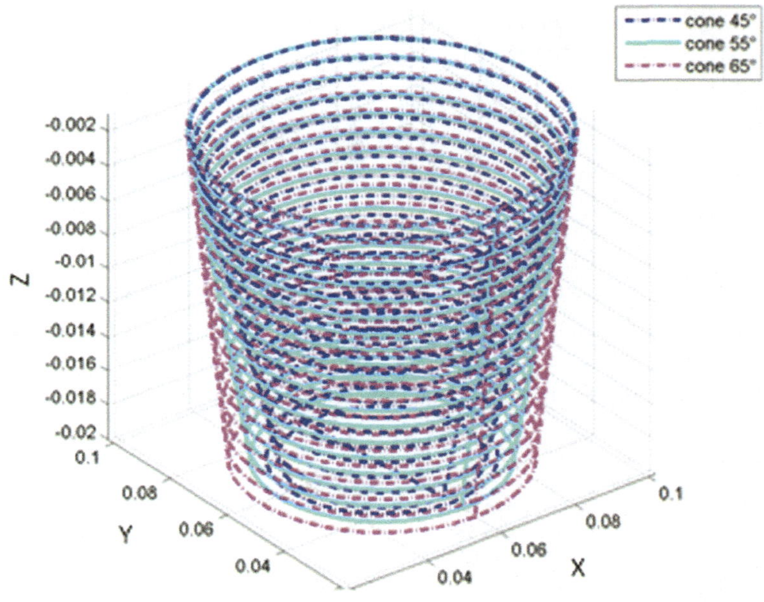

Fig. 1 Tool paths

4 FEM Results

Formability is the capability of the sheet metal to deform without obtaining specific forms of damage. For that, we aim in this section to analyze the influence of some process parameters on forming capability in terms of plastic deformation and thickness distribution. In order to investigate the effect of cone wall angle, comparisons are made between a series of truncated cones were formed with wall angle 45°, 55°, 65° to evaluate the thinning and the formability of AA6022-T43 aluminum alloy. The FE simulations of SPIF process based on the proposed non-associated model which specified by the separation between yield function and plastic potential function.

Figures 2 and 3 illustrate the variation of thickness along the transverse direction of the truncated cone and the forming limit diagram (FLD), which is used to investigate the formability during the SPIF process, respectively. According to these figures, it is noted that the maximum thinning and major principal strain can be reached with wall angle 65°, there were strains up to 80%. It should be emphasize that this thinning can cause the occurrence of the crack and necking as shown in [3]. Therefore, we can conclude that the increasing of the cone wall angle seems to be the major process parameter, which is given a better formability, but it can reach the failure.

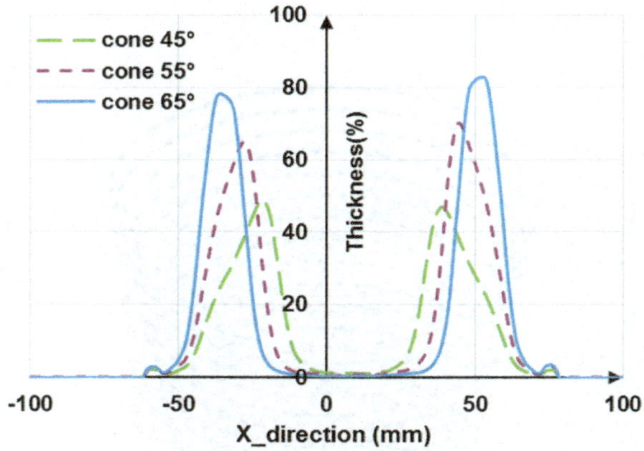

Fig. 2 Thickness strain along the transverse direction

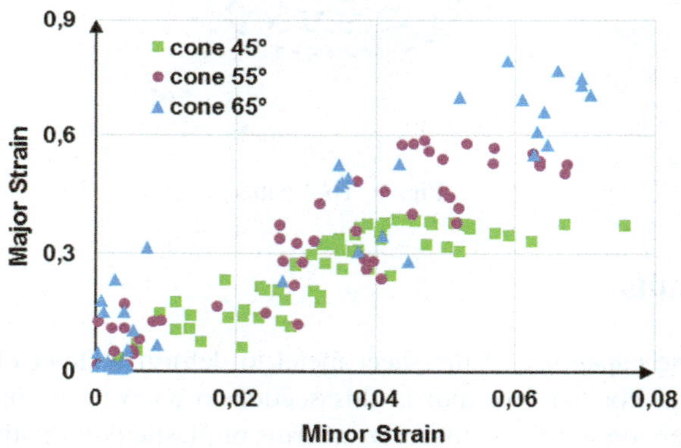

Fig. 3 Numerical FLD diagram

In addition, the rise of tool diameter has no remarkable difference in thinning of the sheet material as shown in Fig. 4. However, it decrease the plastic strain during the SPIF operation (Fig. 5). The maximum of plastic strain is located principally near the tool path. It was shown that very high strains could be achieved for a lower tool diameter. This can be explained by the fact that a highly concentrated zone deformation, caused by small tool diameter, increased the formability as demonstrated in [9].

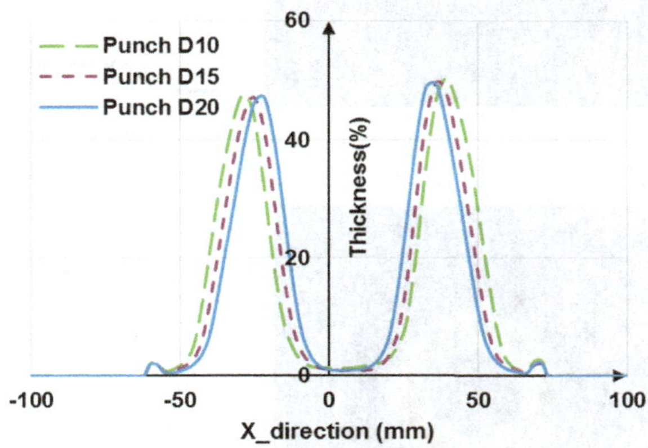

Fig. 4 Thickness evolution

Figure 6 depicts the effect of the sheet thickness and the cone wall angle on the minimal thickness. It shows that the thinning increases when the cone wall angle increases. In addition, the sheet thickness increases, in a nearly linear way with the growth of the thinning.

The higher thinning is achieved with 2 mm sheet thickness and 65° wall angle. This result explains that the deformed part reached damage since the thinning exceeds 100%.

As result, the cone wall angle should be chosen by taking into account of the sheet thickness of the manufactured part to avoid the damage.

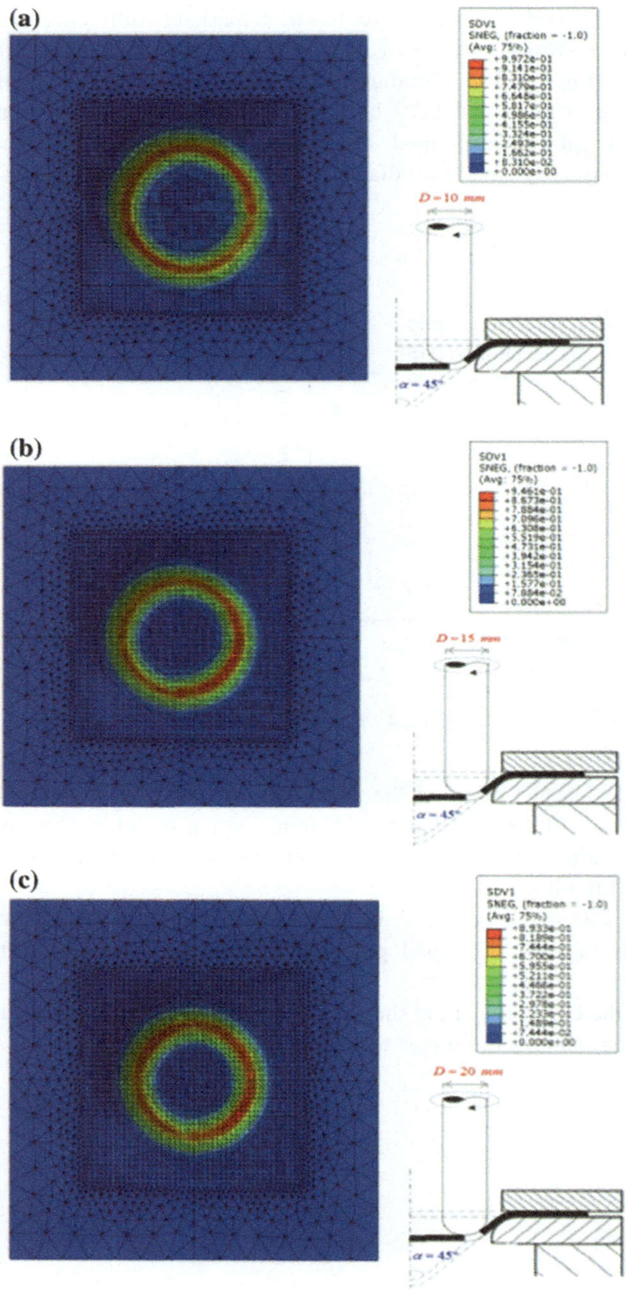

Fig. 5 Plastic strain distribution considering different tool diameter **a** D = 10 mm, **b** D = 15 mm, **c** D = 20 mm

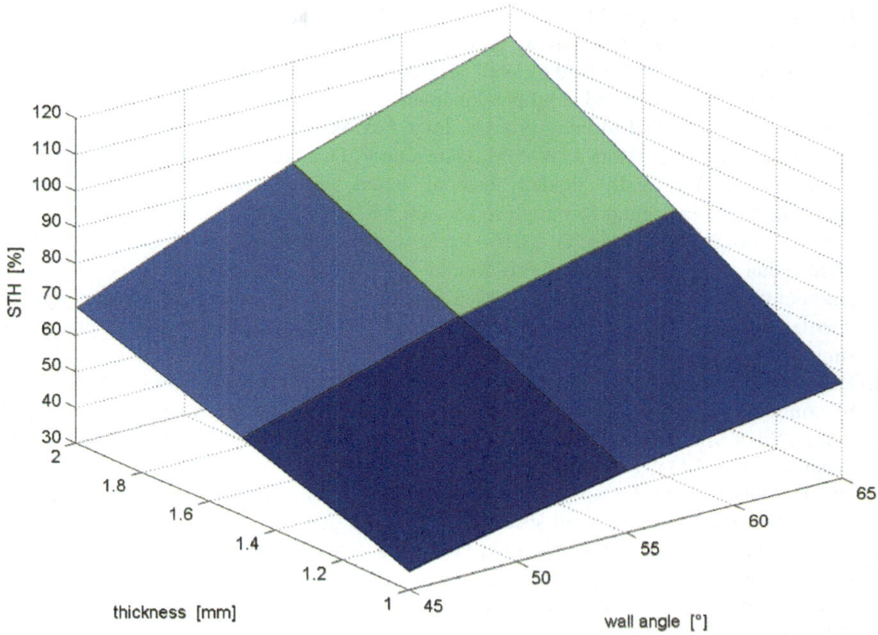

Fig. 6 The relationship of the sheet thickness and the cone wall angle with the minimal thickness

5 Conclusion

In this research work, a finite element model based on non-associated flow rule was implemented on user-defined material subroutine (VUMAT) to simulate the SPIF process of truncated cones manufactured from AA6022-T43 aluminum alloy. The numerical results illustrate the influence of some process parameters, including tool diameter, sheet thickness, and cone wall angle, on thinning, plastic deformation and formability during SPIF operation. This work mainly emphasizes that the rise of cone wall angle increase the formability, but it can reach a failure when the choice of sheet thickness was not optimal. A small effect can be noticed when increasing the tool diameter and the sheet thickness. This work provides additional insight to successfully deform AA6022-T43 aluminum alloy by SPIF process.

References

1. Azaouzi M, Lebaal N (2012) Tool path optimization for single point incremental sheet foring using response surface method. Simul Model Pract Theory 24:49–58
2. Ben Said L, Mars J, Wali M, Dammak F (2016) Effects of the tool path strategies on incremental sheet metal forming process. Mech Ind 17:411

3. Ben Said L, Mars J, Wali M, Dammak F (2017) Numerical prediction of the ductile damage in single point incremental forming process. Int J Mech Sci 131:546–558
4. Bouhamed A, Jrad H, Ben Said L, Wali M, Dammak F (2019) A non-associated anisotropic plasticity model with mixed isotropic–kinematic hardening for finite element simulation of incremental sheet metal forming process. Int J Adv Manuf Technol 100(1–4):929–940
5. Bouhamed A, Jrad H, Mars J, Wali M, Gamaoun F, Dammak F (2019) Homogenization of elasto-plastic functionally graded material based on representative volume element: application to incremental forming process. Int J Mech Sci 160:412–420
6. Gatea S, Ou H, McCartney G (2016) Review on the influence of process parameters in incremental sheet forming. Int J Adv Manuf Technol 87:479–499
7. Ghorbel O, Mars J, Koubaa S, Wali M, Dammak F (2019) Coupled anisotropic plasticity-ductile damage: modeling, experimental verification, and application to sheet metal forming simulation. Int J Mech Sci 150:548–560
8. Hagan E, Jeswiet J (2003) Effect of wall angle on Al 3003 strain hardening for parts formed by computer numerical control incremental forming. Proc Inst Mech Eng 217:1571–1579
9. Isik K, Silva MB, Tekkaya AE, Martins PAF (2014) Formability limits by fracture in sheet metal forming. J Mater Process Technol 214:1557–1565
10. Jrad H, Mars J, Wali M, Dammak F (2019) Geometrically nonlinear analysis of elastoplastic behavior of functionally graded shells. Eng. Comput. 35(3):833–847
11. Li Y, Chen X, Zhai W, Wang L, Li J, Guoqun Z (2018) Effects of process parameters on thickness thinning and mechanical properties of the formed parts in incremental sheet forming. Int J Adv Manuf Technol 98:3071–3080
12. Wali M, Autay R, Mars J, Dammak F (2016) A simple integration algorithm for a non-associated anisotropic plasticity model for sheet metal forming. Int J Numer Meth Eng 107:183–204

Experimental Investigation and Finite Element Modeling on Incremental Forming Process of Aluminum Sheet Alloys

Asma Ben Khalfallah[1,2(✉)], Slim Ben Elechi[1], and Riadh Bahloul[1]

[1] Laboratory of Mechanical Engineering (LGM), National Engineering School of Monastir (ENIM), University of Monastir, Av. Ibn El Jazzar, 5000 Monastir, Tunisia
asmabenkhalfallah92@gmail.com, {sbenelechi,bahloul_riadh}@yahoo.fr

[2] National Engineering School of Sousse (ENISo), Sousse Technological Center, University of Sousse, 4054 Sahloul, Tunisia

Abstract. Incremental sheet metal forming process is a new procedure that forms three-dimensional parts of metal in a thin sheet. In particular, single point incremental forming of sheet metal is considered as a process that forms products without using complex dies and specific forming tool. Through this process, a cylindrical rotating punch with hemispherical end shape follows a predefined continuous or discontinuous trajectory to deform the sheet plastically. This fabrication method is known for its flexibility and the adaptation to complex geometrical shapes [6]. In the present work, the single point incremental forming process (SPIF) has been investigated experimentally and numerically using 3D finite element analysis (FEA). Regarding concerns of the material, the sheets were produced from aluminum alloy. This study focuses on using numerical simulations as a tool to predict and control some mechanical and geometrical responses. In order to understand the effect choice of model constitutive laws, we intend to compare between two relationships of stress-strain hardening behavior, implemented on ABAQUS software, with the experimental results. Based on the obtained findings, a comparison study was presented in this paper between experimental and numerical results. Different outputs responses were extracted such as global geometry (springback error, shape and final achieved section profiles) and thickness distribution. Therefore, the results obtained from the simulation were validated experimentally and good correlations are found, also the process strategies show good agreement with the experiments. Simultaneously, we conclude the most efficient hardening behavior of the material that insures the obtaining of results that are as close as possible to the experimental ones.

Keywords: Single point incremental forming (SPIF) · Aluminum alloy · Finite element analysis (FEA) · Modeling of hardening behavior · Springback · Final thickness distribution

N. Aifaoui et al. (Eds.): CMSM 2019, LNME, pp. 309–317, 2020.
https://doi.org/10.1007/978-3-030-27146-6_33

1 Introduction

Incremental sheet forming (ISF) is a new forming method. It is a process in which a hemisphere-shaped tool is used to shape the sheet into the wanted form. The tool travels over the material causing the sheet to deform plastically, to accomplish the desired shape body [3]. This method is cheap and less expensive when compared with other processes, which require a customized punch for every needed form [4, 5]. The single point incremental sheet forming (SPIF) is the most used type of this procedure, due to its flexibility. However, this method is notorious for the springback phenomenon. This disadvantage occurs due to the accumulation of residual stress during the forming process. After removing the tool and the blank holder, the sheet will go through a relaxation stage, hence, the springback phenomena. The aim of our work is top study the effect of the hardening law models (Swift and Voce) and the resolution methods (explicit and implicit) on different outputs such as final shape, sheet thickness variation and springback.

2 Modeling and Simulation

The geometry considered in this study was a truncated cone, with 60° wall inclination angle, an upper diameter of 160 mm and a depth of 30 mm as shown in Fig. 1. A punch tool with a hemispherical head, with a diameter of 10 mm is used. The desired geometry was first prepared using a commercial 3D CAD-CAM software CATIAV5R21. A discontinuous toolpath with a vertical increment step size of 1 mm was generated. The obtained punch trajectory was converted into a numerical file by using a specific postprocessor and then the different coordinates were implemented in ABAQUS according to a global coordinate system.

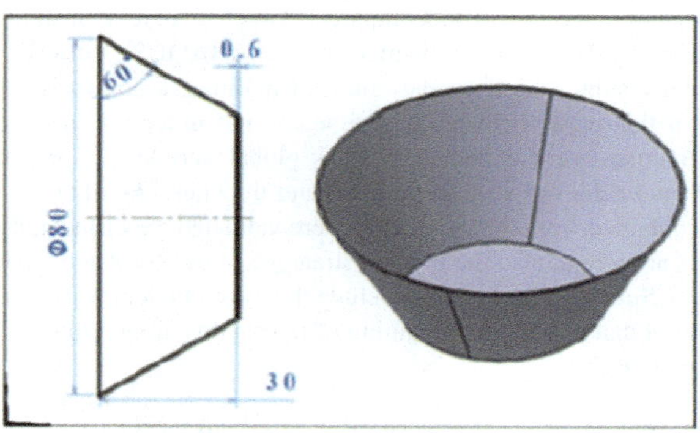

Fig. 1 3D CAD geometry of the formed part by SPIF process

The finite element simulation was carried out using ABAQUS® software. The sheet was modeled as a deformable shell body with 0.6 mm of thickness, it was meshed into S4R shell elements with 5 integration points through the thickness. The tool is considered as a rigid body. The interaction between the two body (punch/sheet) is a surface to surface contact with hard contact and a coulomb's friction value of 0.1.

The material studied in this paper is an aluminum alloy AA1050. The mechanical

Table 1 Mechanical properties of AA1050

Parameters	Symbols	Value
Density (Kg/m^3)	ρ	2700
Young's modulus (GPa)	E	69
Poisson's ratio	υ	0.3
Anisotropy coefficient r_0	–	1.79
Anisotropy coefficient r_{45}	–	2.45
Anisotropy coefficient r_{90}	–	1.7

Table 2 Swift and Voce hardening laws

Law	Function	Material constants
Swift [1]	$\sigma(\varepsilon_p) = \sigma_0(A + \varepsilon_p)^n$	$\sigma_0 = 119.5$ MPa
		A = 0.000142
		n = 0.235
Voce [7]	$\sigma(\varepsilon_p) = \sigma_0(1 - Ae^{-\beta\varepsilon_p})$	$\sigma_0 = 170$ MPa
		A = 0.24
		β = 281.5

properties of the mentioned material are shown in Table 1 (Table 2).

3 Results and Discussion

The aim of this work is to compare the simulation results of two different models based on its material property modeling. In our case, we study two hardening models: Swift and Voce laws. Another objective was to discover the effect of the resolution method on the springback study meaning we are going to compare the springback results issued from an explicit and an implicit calculation. The simulation was carried out on a calculator with an Intel® core^{TM}2 i7-6700 CPU processor and 8 GB of RAM.

We are mainly focusing on the springback phenomena that occurs after the procedure of incremental forming. As mentioned, after releasing the sheet, we remark the occurrence of springback and the change in the final shape of piece caused by the residual stress accumulated during the process. Our goal here is to study this phenomenon through the final shape variation before and after sheet release. Thus, we present two steps: the first one is dedicated for the forming process and the second one is for the springback. We want to study the effect of the resolution methods on the final results especially the springback output. Thus, we propose two sets of tests (Table 3).

Table 3 Simulation steps and strategy

	Forming step	Springback step
Test1	Explicit	Explicit
Test2	Explicit	Implicit

The simulation results are compared with the experimental results conducted by [2] using a 3-axis CNC vertical milling machine.

Since the material is anisotropic, we will start by comparing the results from both models and the experiment results by extracting the final shape of our specimen before springback along X and Y-axis respectively (Fig. 2).

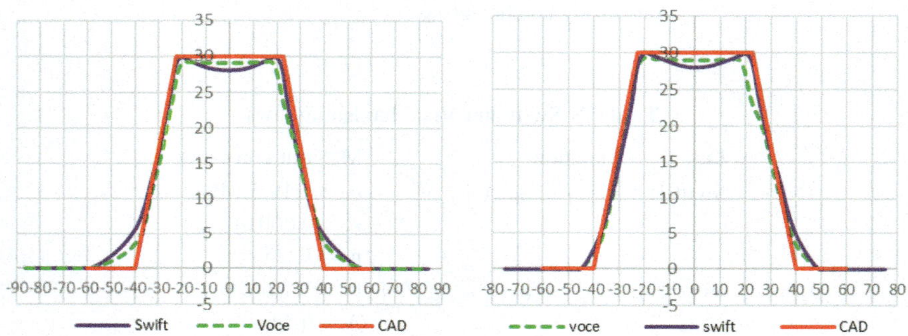

(a) Along X direction (b) Along Y direction

Fig. 2 Shape comparison before the appearance of springback **a** Along X **b** Along Y

From the last two figures, we can see a good agreement between the CAD geometry and the simulation outputs. In the first figure, we can remark also that the Voce model is closer to the theoretical results especially along Y axis. On the other hand, the Swift model present a more important deformation especially on the bottom of the cone, which does not exist in the CAD profile. Since the results along Y axis are more accurate, for Voce and Swift models, alike in the following figures we will only present the results along this axis.

In the following part, we will compare the total deformation of our specimen after releasing the sheet and removing all type of fixation. As mentioned before, the springback study is done through an implicit and an explicit step to see the difference. It should be mentioned that for the springback step, we applied a boundary condition by limiting the displacement of 3 nodes in the upper part of the specimen (see Fig. 3).

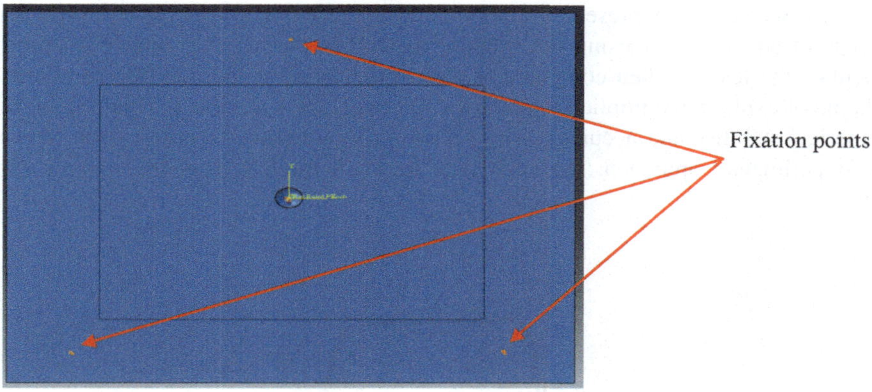

Fig. 3 The disposition of the 3 fixation points

We will start by presenting the final shape results extracted from the explicit-explicit calculation. Figure 4 present a comparison between the experimental profile and the numerical one deduced from the simulation for both Swift and Voce laws.

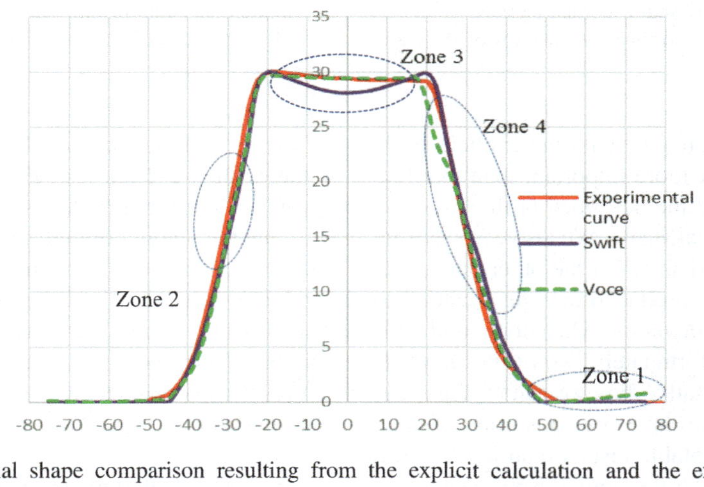

Fig. 4 Final shape comparison resulting from the explicit calculation and the experimental measurements after springback

The first remark is that the Voce model presents a more important springback when compared with the Swift results, especially in the fixation zone (zone 1). Nevertheless, it can be noticed, that both models become closer to the experimental one especially in the wall region, noted (zone 2) in the Fig. 4. For the overall shape, the Voce hardening law still presents a good agreement with the real results, if you took into consideration that the Swift law shows an important deformation on the bottom of the cone, indicated by (zone 3).

Now we move to present the results from the explicit-implicit test. The most important advantage of using the implicit method is that it takes almost no time to calculate the results when compared to the explicit one. In addition, the combination scheme of explicit and implicit is preferred to predict the springback, resulting in more numerical stability and accuracy. The following figure shows a comparison between the experimental approach, the voce and the swift models from the implicit work (Fig. 5).

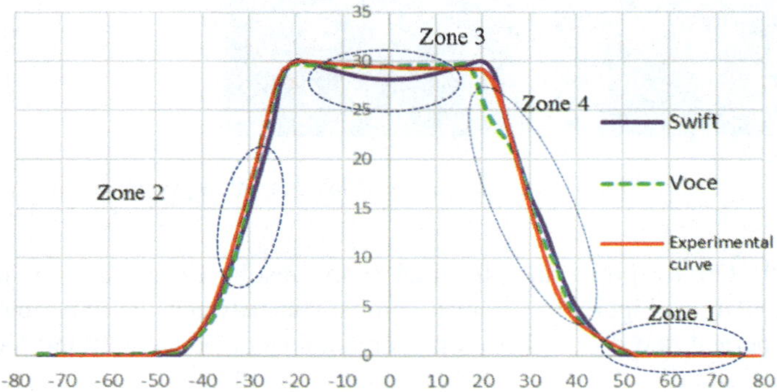

Fig. 5 Final shape comparison resulting from the implicit calculation and the experimental records after springback: profiles of the formed parts along the central cross-section (along the YZ plane)

In accordance with the explicit results, we observe that the Voce model presents a springback more important than that found on the explicit-explicit test. Zone 3 shows clearly that the deformed profile predicted by means of the Swift model is very distant from the real shape. We also can see from zone 1, that the amount of deformation is not very remarkable. On the other hand, it can be observed from the Voce model that the springback produced in zone 4 caused the irregularity to increase and the final shapes presents a deviation when compared with the experimental results, but this difference is small. Unfortunately, we do not have the evolution of the geometry after springback experimentally, thus we cannot conclude which results to consider reliable. However, through the three previous comparison, we can conclude that the Voce model offers more acceptable results than the Swift model.

To make the comparison clearer, we chose in this part of study to make a comparison between the two numerical analysis: implicit and explicit FEM methods applied for prediction of the cross-sectional profiles of the final deformed shapes after springback, for both Swift and Voce. Figure 6a, b illustrate the numerical prediction and the measurement of the geometrical profile along the symmetric axis. We remark that the implicit model gives us better springback results when compared with the explicit one, for both Swift and Voce. We can see that the implicit results are closer to the experimental results when comparing the final profile. In Fig. 6b, the fixation zones presents an important springback, which makes us confirm that the Voce model through the explicit methods could not be taken into account. However, from the implicit-Voce results, we could see good agreements with the real profile. On the other hand, for both the explicit and the implicit methods, the Swift model shows results, which can be described, as not fitted as shown in Fig. 6a.

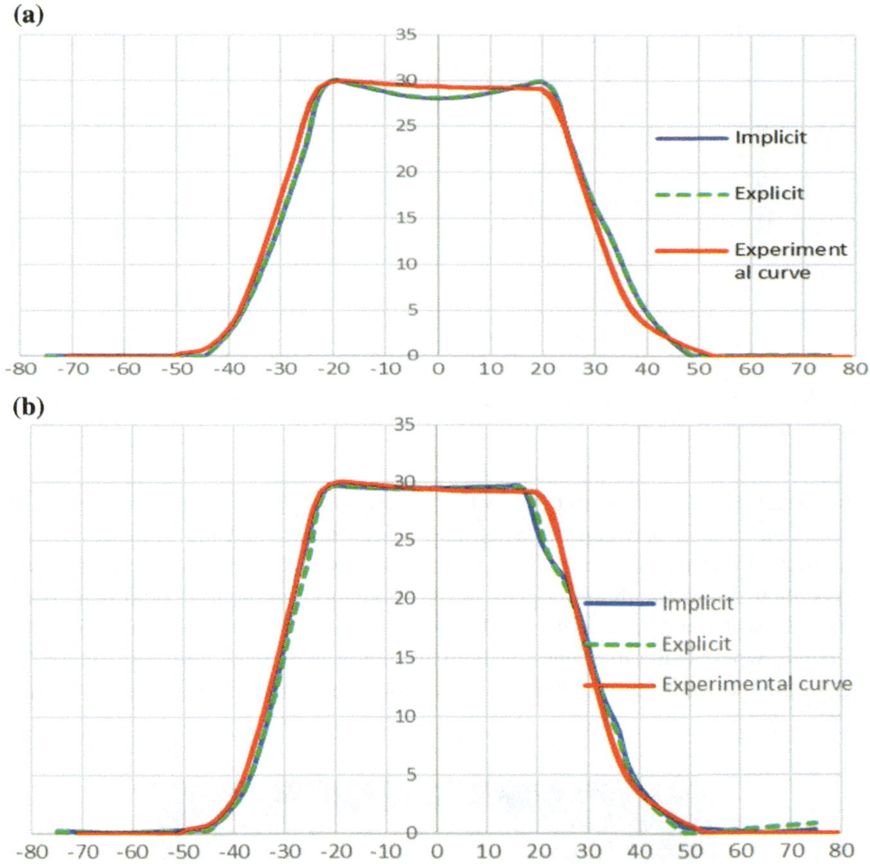

Fig. 6 A comparison of explicit and implicit predictions of the profile along the depth of the final geometry for: **a** the Swift model and **b** the Voce model

Finally, we present the sheet thickness profiles of the cones, which are plotted against the radial dimension. They are deduced from both Swift and Voce models and compared with the theoretical model derived from the so-called sine law and defined by the relation relationship (1).

$$t_f = t_0 \sin(90 - \alpha) \tag{1}$$

In Fig. 7, the evolution of the thickness profiles of two truncated cones formed of an aluminum alloy sheet AA1050 predicted numerically for Voce and Swift and by adopting the implicit schemas.

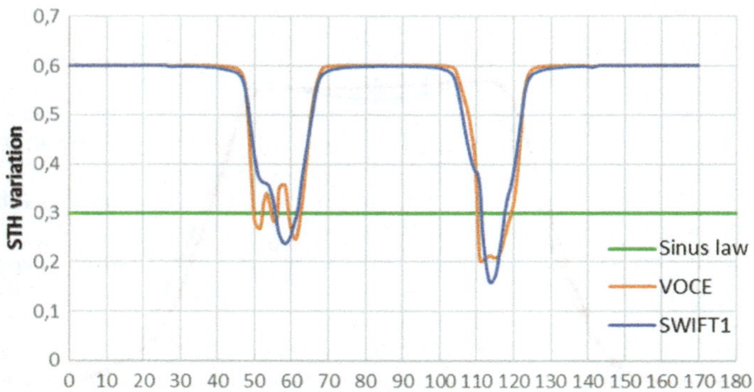

Fig. 7 Numerical prediction of the final sheet thickness evolution along the radial position with combined numerical approach

In our case, the final predicted sheet thickness would be 0.3 mm. The figure above shows a comparison between the Swift and Voce outputs and the Sinus law. We can see a crucial thinning in the right wall of the cone, which could risk the destruction of the specimen. Thus, the thinning peak in the simulation prediction could be explained.

It can also be shown that the thickness at the bottom of the cone remains almost unchanged, whereas the thickness in the wall region is reduced abruptly from 0.6 mm to 0.15 mm and 0.2 mm, respectively for the hardening laws: Swift and Voce models (Fig 8).

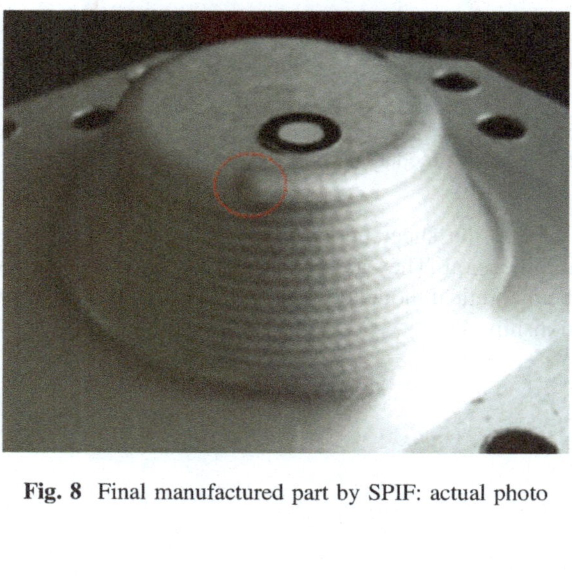

Fig. 8 Final manufactured part by SPIF: actual photo

4 Conclusion

Multiple comparisons have been presented in this work, we studied the simulation results from two proposed hardening models, Swift and Voce, by comparing the final shape obtained with the experiment results. We started by studying the anisotropy of the both models and it was shown that anisotropy has no main effect on the desired results. When comparing both models before springback occurrence, it was concluded that Voce model is closer than the Swift model to the CAD shape. Since Springback is a notorious phenomenon in incremental sheet forming, we carried a comparison between the springback results predicted from an implicit and an explicit calculation method and in the same time, we presented the results issued from both hardening laws, i.e., Swift and Voce. When comparing the simulation profiles with the experimental ones, we could deduce that the explicit method offers an important springback especially for the Voce model. On the other hand, the Voce implicit springback step shows a good agreement with the real shape. Concerning the Swift model, all the results from the implicit and the explicit calculations present a deviation from the experimental shape, especially in the bottom of the cone. As a recapitulation, we can conclude that the Voce model gives as results that are closer to the reality and that obviously the implicit method has the closer profile, but the explicit method, despite being time consuming, gives results that can be acceptable as well. An experimental campaign will be carried out. First to confirm, with a great certainty, the most efficient model for springback, and second to study the effect of different parameters such as cone depth, tool vertical step and fixation method on the springback.

References

1. Ben Ayed L, Ben Robert C, Delam A (2014) Simplified numerical approach for incremental sheet metal forming process. To cite this version: Science Arts & Métiers (SAM). Eng Struct 62:75–86
2. Boudhaouia S, Gahbiche MA, Giraud E (2018) Experimental and numerical study of single point incremental forming for a spiral toolpath strategy. Lect Notes Mech Eng 1007–1015. https://doi.org/10.1007/978-3-319-66697-6_99
3. Edwards WL, Grimm TJ, Ragai I, Roth JT (2017) Optimum process parameters for springback reduction of single point incrementally formed polycarbonate. Procedia Manuf 10:329–338. https://doi.org/10.1016/j.promfg.2017.07.002
4. Nimbalkar DH, Nandedkar VM (2013) Review of incremental forming of sheet metal components. Int J Eng Res Appl 3:39–51
5. Sbayti M, Bahloul R (2015) Numerical study of warm sheet incremental forming process. In: The 6th international congress design and modelling of mechanical systems CMSM'2015
6. Sbayti M, Bahloul R, Belhadjsalah H, Zemzemi F (2018) Optimization techniques applied to single point incremental forming process for biomedical application 1789–1790
7. Shamsari M, Mirnia MJ, Elyasi M, Baseri H (2018) Formability improvement in single point incremental forming of truncated cone using a two-stage hybrid deformation strategy. Int J Adv Manuf Technol 94:2357–2368. https://doi.org/10.1007/s00170-017-1031-5

Ductile Fracture Characterization of an IF Steel Tensile Test by Numerical Simulation

Latifa Arfaoui[1(✉)], Amel Samet[2], and Amna Znaidi[2]

[1] LR-MAI ENIT, Ecole Nationale d'Ingénieurs de Tunis ENIT, Rue Béchir Salem Belkhiria Campus Universitaire, BP 37, 1002 Le Bélvédère, Tunisie
arfaoui_latifa@hotmail.fr

[2] LR-MAI ENIT, Institut Préparatoire aux Etudes d'Ingénieurs d'El Manar IPEIEM, B.P. 244, 2092 Tunis, Tunisie
{amelsamet, amna.znaidi}@laposte.net

Abstract. The automobile chassis is manufactured from various metal formed parts that are joined together mainly by welding. Under same load conditions, each material behaves differently depending on its properties. The knowledge of these properties is crucial and requires experimental data. For an isotropic material model, the data needed can be extracted from a simple tensile test. But, for orthotropic materials, more experimental equipments are necessary. In such cases, the finite element method can be used for the numerical approximation of the problem. In this paper, a numerical modeling of the mechanical response of an Interstitial Free (IF) steel subjected to a low strain rate tension is described, in the particular case of isotropic linear elasticity and orthotropic plasticity under the isotropic hardening assumption. The modeling of the material behavior related to the ductile fracture was based on the approach of Hillerborg et al. (Cem Concr Res 6(6):773–781, 1976, [1]). Firstly, the theory behind the model was explained. Afterwards, the finite element model was described and the related parameters were defined. Finally, the numerical results, performed using the finite element software ABAQUS®, were compared to the experimental data determined through the tensile tests carried out by Cumin et al. (Tech Gaz 23 (1):229–236, 2016, [2]) on the HC260Y steel.

Keywords: Finite element modeling · Orthotropic material · Ductile fracture · IF steel

1 Introduction

Currently, one of the most important steels for the automotive industry in cars fabrication for the external parts is the interstitial free steel (IF) which is combining high strength and great ductility. IF steels contain only a small amount of carbon and nitrogen (less than 50 ppm in weight). They are manufactured through vacuum degassing process and generally stabilized by titanium and/or niobium additions that combined to the soluble carbon and nitrogen form precipitates, resulting in a high cold formability [3].

The determination of the tensile properties of these steels is crucial due to their use mainly in the numerical simulation of sheet metal forming process. In our case, the

N. Aifaoui et al. (Eds.): CMSM 2019, LNME, pp. 318–327, 2020.
https://doi.org/10.1007/978-3-030-27146-6_34

numerical model was developed in the aim of its use later to investigate the tensile properties of the Laser welded IF steel sheets by FEM method.

2 Elastic-Plastic Constitutive Equations

The elastic behaviour is described by the Hooke's isotropic linear elasticity law as follows

$$\sigma = D : \varepsilon^{e} \tag{1}$$

where D is the 4th order elasticity tensor.

By considering the small elastic strains assumption, the total strain value in the structure can be defined as the sum of the elastic and plastic strain

$$\varepsilon = \varepsilon^{e} + \varepsilon^{p} \tag{2}$$

The plastic strain-rate tensor is determined from the associated plastic flow rule as

$$\dot{\varepsilon}^{p} = \lambda h(\sigma, \alpha) \tag{3}$$

where h is a tensorial function, λ is the plastic multiplier and α is an internal hardening variable.

The yield function f has the following form

$$f(\sigma, \alpha) = 0 \tag{4}$$

f defines a yield surface, which is the boundary between elastic and plastic flow. This surface can evolve according to the hardening internal variables. The evolution of those one can be expressed as

$$\dot{\alpha} = \lambda l(\sigma, \alpha) \tag{5}$$

The function l could be scalar or tensorial depending on α.

3 Hill's Yield Criterion for Orthotropic Materials

A yield criterion predicts the beginning of the plastic deformation in materials. The isotropic criteria are used when the plastic deformation obtained for a given stress state is independent of the material orientation. The earliest criteria were proposed by Von Mises and Tresca.

The rolling operations employed for metal sheets production lead to anisotropy, which makes the plastic properties of the samples a function of orientation.

To model a material with initial anisotropy, the key parameters are determined from measurements carried out on tensile specimens cut out at different angles relative to a direction of anisotropy. Uniaxial tensile tests performed on these samples allow to measure quantities like yield stresses and so-called Lankford coefficients R, these coefficients are defined by Meuwissen [4]

$$R_\alpha = \frac{\varepsilon^P_{trans}}{\varepsilon^P_{thick}} \tag{6}$$

where α is the angle from an anisotropy-direction to the tensile-direction of the sample, ε^P_{trans} is the plastic strain in the direction transverse to the tensile direction and ε^P_{thick} is the plastic strain in the thickness-direction of the sample. The Lankford coefficients and the yield stress values depends on the orientation angle α for anisotropic materials.

For metal sheets, the anisotropy type is often assumed to be orthotropic. To describe orthotropic plasticity under general plane stress states, a yield criterion was proposed by Hill. The yield function is given by Meuwissen [4]

$$\Phi = F(\sigma_y - \sigma_z)^2 + G(\sigma_z - \sigma_x)^2 + H(\sigma_x - \sigma_y)^2 + 2L\tau^2_{yz} + 2M\tau^2_{zx} + 2N\tau^2_{xy} = 1 \tag{7}$$

4 Hardening Laws

The hardening rule describes how the yield surface changes (size, shape…) as the result of plastic deformation except for elastic-perfectly-plastic material whose yield surface remains fixed. It also determines when the material will yield again if the loading is continued or reversed.

Two main hardening rules are used to prescribe the evolution of the yield surface which are isotropic hardening and kinematic hardening.

Isotropic hardening is when the yield surface remains the same shape but expands uniformly in all directions with plastic flow. However, kinematic hardening, corresponds to the case when the yield surface remains constant in size and translates in the direction of yielding.

Isotropic hardening models are usually used for large strain simulations which makes them suitable for the FEM modelling of drawing or other metal forming operations. The Hollomon's power law is the simplest and without doubt the most used one. It is defined as follows [5]

$$\sigma = K\varepsilon^n_p \tag{8}$$

where σ is the Cauchy stress, k the strength coefficient, ε_p is the plastic strain and n is the strain-hardening exponent. Another widely used model is the Swift's law, which can be described by the following relation [5].

$$\sigma = K(\varepsilon_0 + \varepsilon_p)^n \tag{9}$$

where ε_0, K and n are constants for a particular material. The identification of different parameters related to a material model, described by its yield criterion, hardening law and evolution law, requires experimental data [6].

5 Ductile Fracture

There are four mechanisms in metals and metal alloys that may lead to failure in a material: intergranular fracture, cleavage fracture, fatigue and ductile fracture.

Ductile rupture of metals can be defined, referring to microscopic observations, as the nucleation, growth and coalescence of micro-cavities or voids (Fig. 1) [7].

The damage process can be interpreted as: (a) nucleation is mainly a result of deformation incompatibilities between metallic matrix and non-metallic inclusions which leads to the formation of voids; (b) void growth as a result of plastic straining and hydrostatic stress.; and (c) Coalescence of the growing cavities that eventually causes the material's failure. This fracture mechanism is typical for dominant tensile loading case.

Two main parameters govern voids nucleation and growth in metals: stress triaxiality and plastic strain. A triaxial stress state takes place when all the three principle stresses have non zero values. Stress triaxiality k is defined as the ratio of the hydrostatic pressure or mean stress σ_H to the Von Mises equivalent stress σ_{eq} [7]

$$k = \frac{\sigma_H}{\sigma_{eq}} = \frac{(\sigma_{xx} + \sigma_{yy} + \sigma_{zz})/3}{\frac{1}{\sqrt{2}}\left[(\sigma_{xx} - \sigma_{yy})^2 + (\sigma_{xx} - \sigma_{zz})^2 + (\sigma_{yy} - \sigma_{zz})^2\right]^{\frac{1}{2}}} \quad (10)$$

where σ_{xx}, σ_{yy} and σ_{zz} are the principle stresses in the x, y and z directions.

Based on the stress triaxiality value, it is possible to find out the state of stress in any point of strained object (Fig. 2) [8].

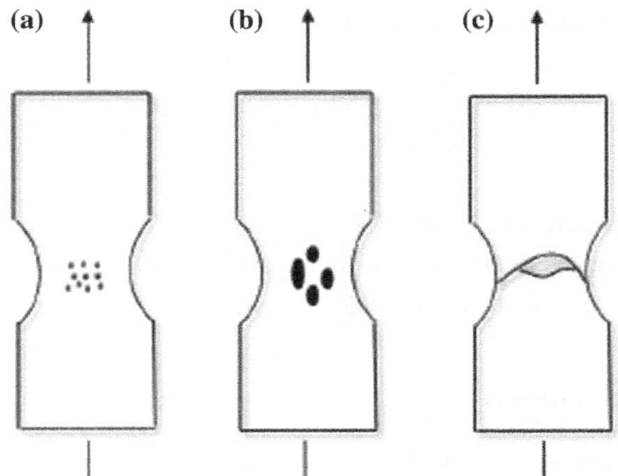

Fig. 1 Schematic representation for damage process: **a** nucleation, **b** growth and **c** coalescence

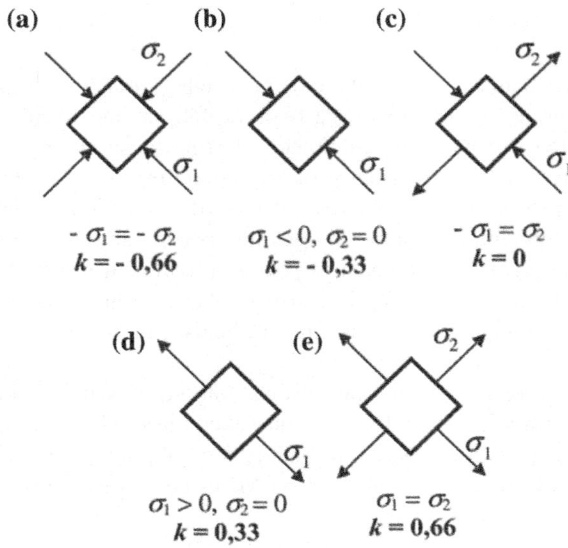

Fig. 2 The *k* factor values for an individual plane stress cases: **a** biaxial compression, **b** uniaxial compression, **c** simple shear, **d** uniaxial tension, **e** biaxial tension

The approach implemented in the finite element software ABAQUS is proposed by Hillerborg [1]. Hillerborg's Fracture Energy, Gf, is defined as the energy required to open a unit area of crack as a material parameter. The softening response after damage initiation is characterized by a stress-displacement response rather than a stress-strain response. The Fracture Energy is defined as

$$G_f = \int_{d\bar{\varepsilon}_0^{pl}}^{d\bar{\varepsilon}_f^{pl}} L\sigma_y d\bar{\varepsilon}^{pl} = \int_0^{\bar{u}_f^{pl}} \sigma_y d\bar{u}^{pl} \tag{11}$$

where L is the characteristic length of the element, $\bar{\varepsilon}_0^{pl}$ is the equivalent plastic strain at the onset of damage, $\bar{\varepsilon}_f^{pl}$ is the equivalent plastic strain at failure and $\bar{u}_f^{pl}$ is the equivalent plastic displacement at failure.

6 Damage Evolution

The concept assumes the growth of voids in the material. The damage, D, is basically a measure for the reduction of the cross section area with respect to the upcoming voids (Fig. 3).

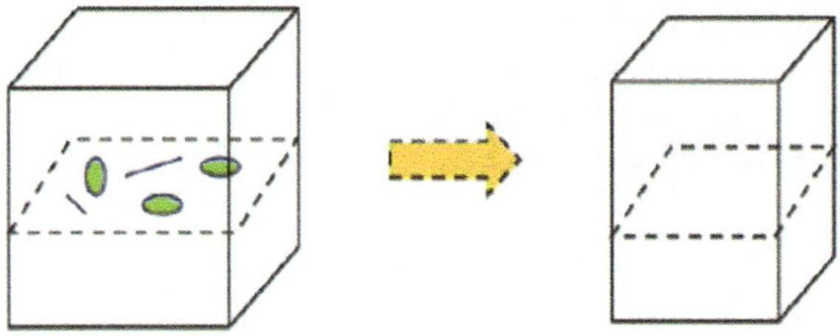

Fig. 3 The current section area and the effective section area

If A represents the cross section area including the area of the voids, and A_{eff} represents the cross section area excluding the area of the voids, [9] define the damage parameter as bellows

$$D = 1 - \frac{A_{eff}}{A} \tag{12}$$

Another formulation of Lemaitre's relation is a basic equation coupling damage to the stresses

$$\sigma_{eff} = \frac{F}{A_{eff}} = \frac{\sigma_{True}}{1 - D} = > \sigma_{True} = (1 - D)\sigma_{eff} \tag{13}$$

The damage model included in ABAQUS, based on the energy release model, is a complete model with a damage initiation criterion and a damage evolution model leading to failure.

7 Finite Element Model

The uniaxial tensile test numerical simulation has been performed using the FE software ABAQUS. The simulations have been carried out on the HC260Y steel which is a high strength IF steel, mainly used in the automotive industry especially for the stamping and forming of automotive panels. The geometry of the model together with the sample dimensions according to the ISO 6892-1-2016 norm are provided in Fig. 4, the thickness value has been chosen equal to 1 mm.

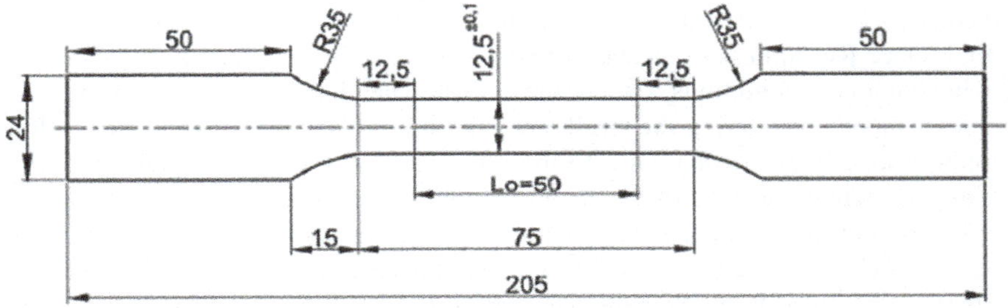

Fig. 4 The geometry of the model drawn in ABAQUS/CAE

The material has been assumed to be isotropic in its elastic properties and following an isotropic strain hardening rule in its plastic properties. The material parameters for elastic strain have been as follows: $E = 200$ GPa, $v = 0.3$. The hardening laws chosen to describe the stress-strain curves were power laws [2], described by the following equations

$$\sigma = 673.1\varepsilon_p^{0.21} \text{ (Hollomon)} \tag{14}$$

$$\sigma = 683.88\left(0.003 + \varepsilon_p\right)^{0.22} \text{ (Swift)} \tag{15}$$

The Hill's yield criterion has been adopted to describe the orthotropic plasticity in the sample. The parameters of the criterion have been calculated using the R values for three different directions of loading in a plane from MD, R_0, R_{45} and R_{90} [10]. Referring to the experimental tests performed by Znaidi [11]. The Lankford coefficients values have been as follows (Table 1).

Table 1 Lankford experimental coefficients

R_{00}	R_{45}	R_{90}
2.19	1.82	2.72

For plane stress condition, the criterion parameters are reduced, they are given by

$$\frac{H}{G} = R_0, \frac{F}{G} = \frac{R_0}{R_{90}}, \frac{N}{G} = \left(R_{45} + \frac{1}{2}\right)\left(\frac{R_0}{R_{90}} + 1\right) \tag{16}$$

The behaviour up to the point of ultimate strength was controlled by the elastic-plastic properties of the material. When damage initiated, the equivalent plastic strain has the same value for the different regions of the sample. After the damage initiation, the inputted plastic properties for the material have been set to rise linearly following the same slope as right before damage was initiated. The damage true stress true strain path σ has been defined by applying a damage parameter D, so that $\sigma = (1 - D)\bar{\sigma}$. The data has been inserted to the material model as a tabular relationship of the damage parameter D to the fracture displacement value $\bar{u}^{pl}$.

In the considered case, a uniaxial tension has been simulated, so that the stress triaxiality factor η value was 0.33. The uniaxial stress's state lasts until the neck is created. Once Rm limit is exceeded, it varies significantly in different specimen zones.

The boundary conditions are chosen to simulate the action of the grips on the sample clamped ends during the tensile test. In one end is applied boundary conditions restraining all degrees of freedom. In the other end, all translations and rotations are restrained except the translation on the y-direction. The last one is kept free to allow the sample straining under a concentrated force applied through a reference point.

The model is meshed using the S4R element provided in ABAQUS. This element is an eight-nodes linear brick element. The option element removal is activated to allow the deletion of failed elements from the mesh once the material stiffness is full fully degraded.

The numerical predictions obtained with Hill's model are compared to the measured data determined from the tensile tests performed by Cumin et al. [2].

The experimental results are described specifically in the following for flat specimens cut from a sheet plate according to a parallel orientation to the rolling direction. The tensile tests were performed employing 7120 N load cell (mean value). True stress vs true strain was plotted in graphs for five different tests (Fig. 5).

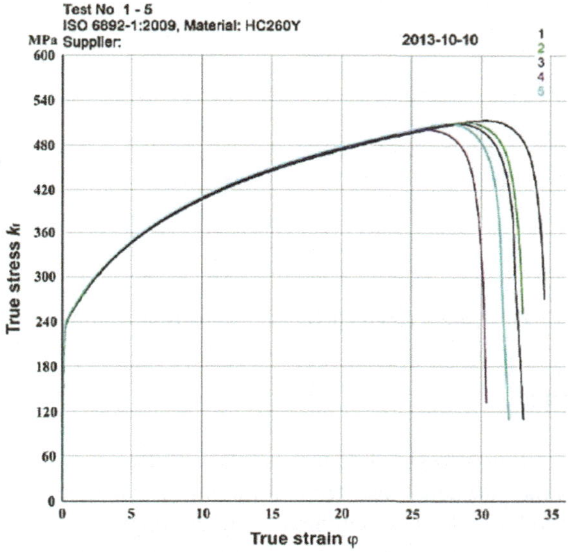

Fig. 5 Results from uniaxial tensile test [2]

The material properties used as input for the numerical model are related to the test 4. The specimen's shape during the tensile test was represented in the Fig. 6 from the initial state until the final fracture.

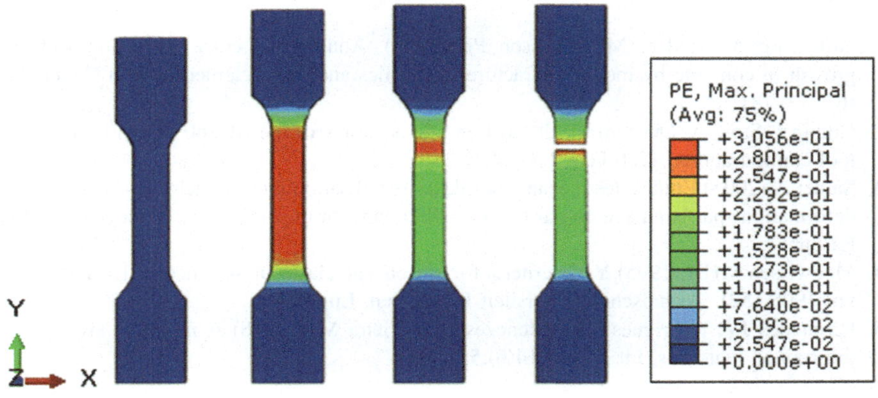

Fig. 6 Specimen's shape during tensile loading

The strain versus stress curve obtained by numerical prediction as well as the experimental curve are shown in the Fig. 7. The comparison of the numerical and experimental curves shows a good agreement especially for the elastic plastic behavior. When damage initiates at an equivalent plastic strain of 0.26, the stress values determined by the numerical predictions seems to be higher than the experimental values. This is may be due to the high mesh dependency of the results. The total failure of the model occurs when the strain reaches 0.3.

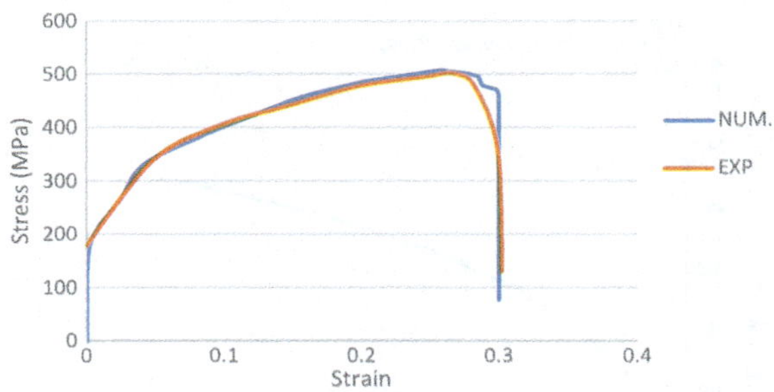

Fig. 7 Experimental stress strain curve and comparison with ABAQUS simulation

8 Conclusions

The aim of this work was to simulate the ductile fracture of an IF steel subjected to a tensile loading in the particular case of isotropic linear elasticity, orthotropic plasticity using the Hill's model under isotropic hardening assumption.

The numerical predictions are in a good agreement with the results of the experiments performed by Cumin et al. [2], even they show a high mesh dependency. This model was developed in the aim of its use later to investigate the tensile properties of the Laser welded IF steel sheets by FEM method.

References

1. Hillerborg A, Modéer M, Petersson PE (1976). Analysis of crack formation and crack growth in concrete by means of fracture mechanics and finite elements. Cem Concr Res 6 (6):773–781
2. Cumin J, Stoić A, Duspara M (2016) Bending accuracy of the HC260Y steel in different v-tool configurations. Tech Gaz 23(1):229–236
3. Samet A (2005) Etudes des mécanismes de recristallisation dans un acier IF-Ti après faibles deformations par laminage ou par traction. PhD thesis at the University of Paris-Sud (Orsay, Essonne)
4. Meuwissen MHH (1995) Yield criteria for anisotropic elasto-plastic metals (DCT rapporten; vol 1995.152). Technische Universiteit Eindhoven, Eindhoven
5. Fernandes JV, Rodrigues DM, Menezes LFI, Vieira MF (1998) A modified swift law for prestrained materials. Int J Plast 14(6):537–550

6. Daghfas O, Znaidi A, Gahbiche A, Nasri R (2018) Identification of the anisotropic behavior of an aluminum alloy subjected to simple and cyclic shear tests. J Mech Eng Sci. https://doi.org/10.1177/0954406218762947
7. Bressan C, Vaz M Jr (2014). Stress-strain behavior and ductile fracture characterization of IF tensile test by 2d and 3d numerical simulations. Congresso Brasileiro de Engenharia e Ciência dos Materiais
8. kut S (2010) A simple method to determine ductile fracture strain in a tensile test of plane specimen's. Metalurgija 49(4):295–299
9. Lemaitre J, Chaboche JL (1994) Mechanics of solid materials. Cambridge University Press
10. Salah HBH, Khalfallah A, Znaidi A, Dogui A, Sidoroff F (2003) Constitutive parameters identification for elastoplastic materials in finite deformation. J de Physique IV. https://doi.org/10.1051/jp4:20030165
11. Znaidi A (2004) Orthotropic plasticity in large deformation. PhD thesis at the Faculty of Mathematical, Physical and Natural Sciences of Tunis

Predictive Modeling and Optimization of Cutting Parameters During the Turning of Inconel 718 Using Taguchi Method

Wassila Frifita[1(✉)], Sahbi BenSalem[1,2], and Mohamed Athmane Yallese[3]

[1] Laboratory of Applied Mechanical Research and Engineering (LR-MAI), National School of Engineers of Tunis(ENIT), ELMANAR University, 1002 Tunis, Tunisia

[2] IPEI Nabeul, Mrazka University Campus, University of Carthage, 8000 Nabeul, Tunisia

[3] Mechanics and Structures Research Laboratory (LMS), May 8th 1945 University, P.O. Box 401, 24000 Guelma, Algeria

Abstract. The present paper focuses on the determination of the optimum cutting conditions leading to minimum cutting force (F_a, F_r, F_z) as well as cutting power in the case of the turning of the Inconel 718 using the tool holder referenced by SVV 2020 K-11 and the carbide double-sided 35° rhombic cutting tool. The optimization is based on the Taguchi method. Furthermore, the orthogonal array, the signal-to-noise ratio (S/N) and the analysis of variance (ANOVA) are respectively exploited to establish the statistical significance of the cutting parameters on different technological ones studied. Three parameters are studied in this paper, namely: feed rate f, nose radius rε and cutting speed V_c. The experimental results revealed that the nose radius is the most factors influencing the three cutting force components followed by the cutting speed for the feed force and the feed rate the two other components. Moreover, a mathematical model relating both of cutting force and cutting power to the cutting parameters were developed.

1 Introduction

Aircraft engines and compressors are parts of severe conditions where temperature and pressure can reach abnormal values. Conventional metals can't work under these conditions. That's why manufacturers come back to refractory super alloys characterized by theirs excellent mechanical properties and their ability to maintain these properties at high temperatures. Among these alloys, nickel-based alloys we cited Inconel 718. It is used in the high pressure part of the compressor and the gas turbine for temperatures goes from 450 °C to 700 °C [1, 2].

Due to its high mechanical proprieties, the Inconel 718 is characterized by lower machinability. This propriety can be the best explication of different machining problems such as surface integrity [3], tool wear, tool life, cutting temperature, cutting forces and cutting power. Thereby, these machining characteristics are directly related to the different cutting conditions. Thus, a great attention should be given to the

N. Aifaoui et al. (Eds.): CMSM 2019, LNME, pp. 328–334, 2020.
https://doi.org/10.1007/978-3-030-27146-6_35

investigation of the impact of process parameters. For that, literature reviewers contain various studies in this context. According to Mishra et al. [4], the feed rate, the of cut and the cutting speed have an important impact on cutting force, feed force and material removal rate (MRR) when turning Inconel 718 using a carbide coated insert with 4 μm coating of TiAlN-TiN and under dry conditions.

Moreover, the use of chamfered inserts had a significant effect on cutting forces. It allowed to decrease cutting forces and then to improve surface integrity [5]. Returning to literature reviews, it was clear that there was a lack of studies that taking the nose radius as factor on response characteristics. Besides, most studies treat the case of dry conditions and little analysis were about wet machining. In addition, all previous studies concerned the laminated Inconel 718. Whereas, in this paper, we focus on the turning of the molded Inconel 718 under wet conditions. Especially, the investigation of the effect of three cutting parameters: cutting speed, feed rate and nose radius on the cutting forces and the cutting power. The cutting speed and the feed rate with three levels whether the nose radius only with two levels. In fact, the workpiece material and the cutting tools where obtained from the collaborated society and there were only two nose radius available.

Based on ANOVA analysis and the S/N ratio, the most significant factors and the optimum conditions were determined.

2 Problem Definition

The main objective of this paper was to determine the optimum cutting condition during the turning of the Inconel 718 and to develop the mathematical model for the three cutting components of cutting force (F_a, F_r, F_z) and the cutting power. Three parameters were taken into consideration: the nose radius with two levels (0.4 and 0.8 mm) the cutting speed and the feed rate with three levels (33/47 and 70 m/min) and (0.08/ 0.11 and 0.14 mm/rev).

3 Experimental Setup

3.1 Material

The experiment tests are carried out on the CNC lathe, SN40C model. The cutting inserts used were the VNMG 120304 and the VNMG 120308 mounting the tool holder SVVNN 2020 K-12F. The cutting forces measurements were obtained directly using a Kistler three-component dynamometer model 9257B Fig. 1a, b.

3.2 Taguchi Method

It is an efficient quality method that permits to determine the most significant parameter (s) for the characteristic performance in question and the optimum combination. For that, it uses the S/N and the ANOVA tools [6]. The OA retained was the mixed L_{18}

(a) SN40C lathe (b) Dynamometer 9257B

Fig. 1 Conventional machine and kistler dynamometer

(two factors with three levels and one factor with two levels) and the adequate category of S/N ratio for this problem was the smaller is the better which expressed by:

$$\frac{S}{N_S} = -10 \times \log(\frac{1}{n}\sum_{i=1}^{n} y_i^2) \tag{1}$$

where y_i: observed data and n: number of observation.

4 Result and Discussion

In the present work, the results treatments were realized under the Minitab17 interface software.

4.1 Signal to Noise Ratio

The cutting forces results are transformed using Eq. (1) to S/N ratio. Then the S/N ratio for feed, thrust and cutting force were calculated. Since it was an orthogonal array, then the effect of each factor at different levels was independent. Thereby, the mean S/N ratio for each level of nose radius, cutting speed and feed rate were calculated. These results are summarized in Table 1.

It's clear from the table above that the cutting forces are depending on the three factors. The analysis of the S/N ratio showed that the three factors feed, nose radius and cutting speed had significant effects on response factors with different degree of signification. For that, Delta (the size of effect) was deducted as the difference between the maximum and the minimum value of the effect for each factor. The factor with the largest value of Delta is the most significant one.

In the present study, the nose radius was the most significant factors for the three cutting forces followed by the cutting speed and the feed rate in the third case.

Table 1 Means S/N ratios responses for cutting forces

	Level	r_ε	V_c	f
Feed force	1	−20,92	−22,64	−21,99
	2	−22,97	−21,91	−21,63
	3		−21,29	−22,22
	Delta	2,05	1,3 5	0,0 6
	Rank	1	2	3
Thrust force	1	−32,73	−35,68	−35,27
	2	−38,07	−35,49	−35,54
	3		−35,03	−35,40
	Delta	5,34	0,65	0,27
	Rank	1	2	3
Cutting force	1	−32,28	−34,59	−33,75
	2	−35,43	−33,80	−33,89
	3		−33,17	−33,92
	Delta	3,15	1,42	0,17
	Rank	1	2	3

4.2 Main Effects Plot for S/N Ratio

The results bellow: mean S/N ratio response given in Table 1 and data means of responses were plotted for feed, thrust and cutting force in Figs. 2, 3 and 4 respectively.

Regarding the main effects plots for means, Figs. 2a, 3a and 4a describe the variation of technological factors with cutting parameters and their sensitivity sense. It's clear that the three cutting forces increase sharply with the increase of nose radius. However, with any increase in cutting speed, cutting forces decrease considerably without exceeding the effect of the nose radius. For the feed rate, it's seen that the major variation was observed for the feed force in comparison with the two other components. Feed force decrease up to 0.11 mm/rev then it increase with increasing feed rate.

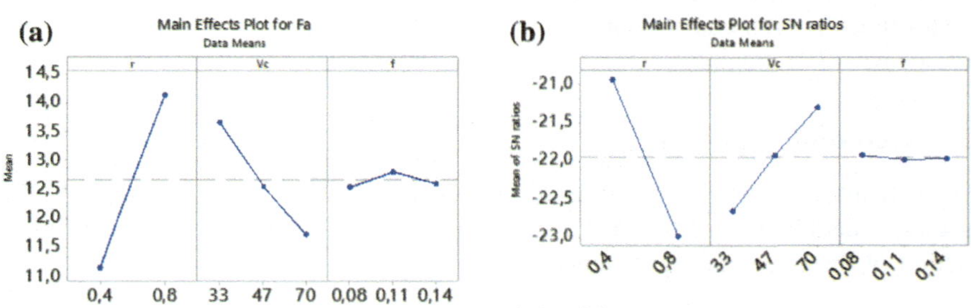

Fig. 2 Main effects plot for S/N ratio and for means responses of F_a

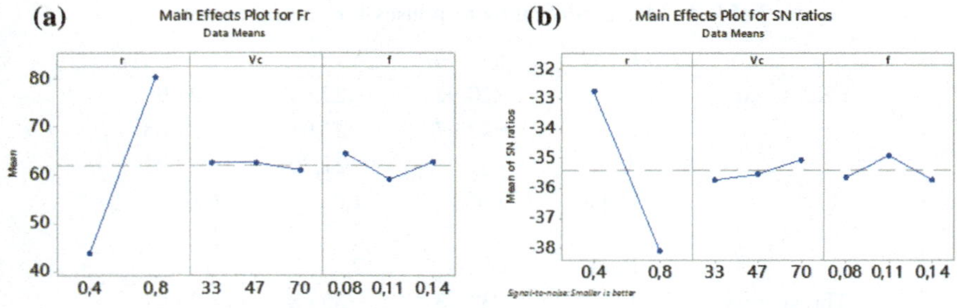

Fig. 3 Main effects plot for S/N ratio and for means responses of F_r

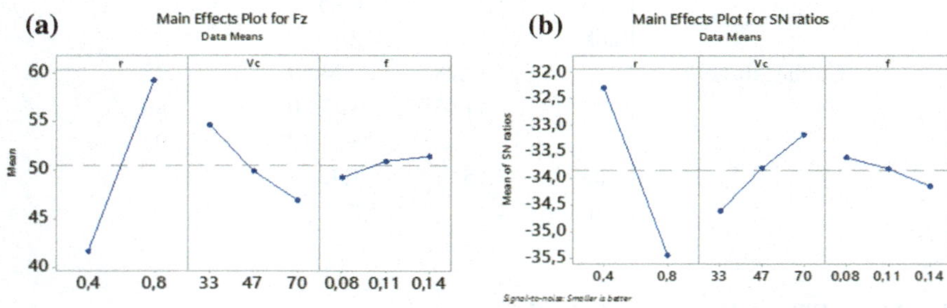

Fig. 4 Main effects plot for S/N ratio and for means responses of F_z

Concerning Figs. 2b, 3b and 4b, these graphs serve not only to present the effect of each parameter on response factors but also to determine directly the optimum combination of process parameters giving an optimum response.

The main objective was to optimize cutting forces. Then, we have an interest to minimize response factors which corresponded to the characteristic category "the smaller is better". The highest S/N ratio corresponds to the smallest noise.

For that, we were referred to figures above to determine the optimum combination of process parameters. For example for the feed force (Fig. 2b), the optimum levels of input factors leading to minimize the feed force were: nose radius of 0.4 mm, cutting speed of 70 m/min and feed rate of 0.11 mm/rev. This combination was the optimum case for the two other components excluding the feed rate optimum level for the cutting force F_z was 0.08 mm/rev.

4.3 Regression Equation

The mathematical models of cutting forces were carried out using the multiple linear regression method under Minitab 17. The equations below present the cutting forces as function of the three cutting parameters.

$$F_a = 6,99 + 11,01 \times r_\varepsilon - 0,06 \times V_c + 46,4 \times f + 0,05 \times r_s \times V_c - 53,2 \times r_s \times f - 0,20 \times V_c \times f \quad (2)$$

$$F_r = 16,0 + 36,9 \times r_\varepsilon - 0,22 \times V_c + 241 \times f + 1,11 \times r_\varepsilon \times V_c - 9 \times r_\varepsilon \times f - 4,40 \times V_c \times f \quad (3)$$

$$F_z = 13,0 + 52,1 \times r_s - 0,13 \times V_c + 301 \times f + 0,39 \times r_s \times V_c - 253 \times r_s \times f - 2,75 \times V_c \times f \quad (4)$$

For the cutting power the analysis was based on the results given above. As it was known that the cutting power and the cutting force component were proportional and related with the following expression:

$$P_c = \frac{F_z \times V_c}{60} \quad (5)$$

Using the cutting forces equations, the optimum value of each response obtained with introducing the optimum process parameters founded in the previous section: F_a opt =9.8N2, F_r opt =38.74 N and F_z opt =36. 2 N. The cutting power was then 42.28 W.

5 Conclusion

The optimization process in this study was carried out using the Taguchi method which seems to be not only the most adequate method but also the most efficient. The present problem was solved with the L18 Taguchi mixed array. The results were carried out basing on the S/N ratio and the main effect plots. The main objective was reached by the determination of the most significant factor for the cutting forces which the nose radius followed by the cutting speed and then the feed rate. The optimum solution that minimizes the cutting forces corresponded to the lowest level of nose radius, the highest one for the cutting speed and the middle level for the feed rate. Finally, a predictive model for response factors was carried out.

Acknowledgements. This work was realized in collaboration with the society "ADI Tunisie" especially Mr. Jouini Saleh as a supervisor and University May 8, 1945 Guelma, LMS laboratory, Algeria espicially Mr Yallese Mohamed Athmane. The authors also gratefully acknowledge the helpful comments and suggestions of the reviewers, which have improved the presentation.

References

1. Zamzemi F (2007) caractérisation de modèles de frottement aux interfaces pièce- outil-copeau en usinage: application au cas de l'usinage des aciers et de l'inconel 718, thèse, école centrale Lyon, France
2. Alexandre F (2004) Aspects probabilistes et microstructuraux de l'amorçage des fissures de fatigue dans l'alliage Inconel 718, thèse, Ecole Nationale Supérieure des Mines de Paris
3. Frifita W et al (2018) Optimization of surface roughness using the method Taguchi, application for workpiece in Inconel 718, Poster, 4th Tunisian Congress of Mechanics. Hammamet, Tunisia
4. Mishra G et al (2016) Optimization of cutting force, feed force and material removal rate (MRR) in turning of Inconel 718. Int J Sci Res
5. Pawade RS et al (2007) an investigation of cutting forces and surface roughness damage in high-speed turning of Inconel 718. Int J Adv Manuf Technol
6. Tsao CC (2009) Grey-Taguchi method to optimize the milling parameters of aluminum alloy. Int J Adv Manuf Technol 40:41–48

Effect of Multi-stage Incremental Formatting Strategy (DDDD) on Sheet Thickness and Profile

Mohamed Fethi Bouzid, Mondher ben Ahmed(✉), Khaled Zid, and Rafik Tarchoun

Higher Institute of Technological Studies, Mahdia, Tunisia
mfbouzid@yahoo.fr, benahmedmondher@yahoo.fr,
zid_khaled70@yahoo.fr, rtarchoun@yahoo.fr

Abstract. Incremental forming of sheet metal is a new sheet forming process. He has shown a variety of applications ranging from the automotive field to the biomedical field. However, sheet metal parts cannot be made in a one-step incremental forming process because the maximum profile angle of the wall that can be formed is limited for a given sheet material and for a given thickness. This limitation can be overcome with the multi-step incremental forming process (MSIF). In the framework of this study, it is proposed to analyze by experimental techniques the thickness and the profile of a truncated cone obtained by several steps forming an incremental MSIF. The study has been carried by forming a conical cup with 85° wall angle in four stages. Experimental tests were performed for specific conditions such as forming strategy, tool path, increment and cutting parameters. ABAQUS numerical simulations were performed to validate the experimental results at the end of each incremental forming step. The numerical and experimental results of the profile sheet metal are the subject of a comparative study between the theoretical, simulated and experimental models.

Keywords: Incremental forming · Finite element simulations · MSIF · Profile sheet metal · Thickness sheet metal

1 Introduction

The principle of the incremental forming process is to deform plastically a sheet locally by means of a small hemispherical tip tool. In most applications of incremental forming, the sheet is embedded on its contour to prevent it from slipping inwards. The final shape of the part is generated by the path of the tool.

This process can be considered in three categories:

- Single point incremental forming (SPIF) [1].
- Two Point Incremental Forming (TPIF) [2].
- Multi-stage incremental forming (MSIF) [3].

In this paper, multi stage incremental forming (MSIF) tests are presented using a CNC machine bench to produce a truncated cone with a wall angle of 85°. Initially, the part

N. Aifaoui et al. (Eds.): CMSM 2019, LNME, pp. 335–344, 2020.
https://doi.org/10.1007/978-3-030-27146-6_36

was formed with a 60° wall angle and increased in the subsequent steps of 70°, 80° and 85° [4].

2 Numerical Simulation

The finite element simulations were performed to study the evolution of the profile in each step to validate the theoretical model as shown in (Fig. 1).

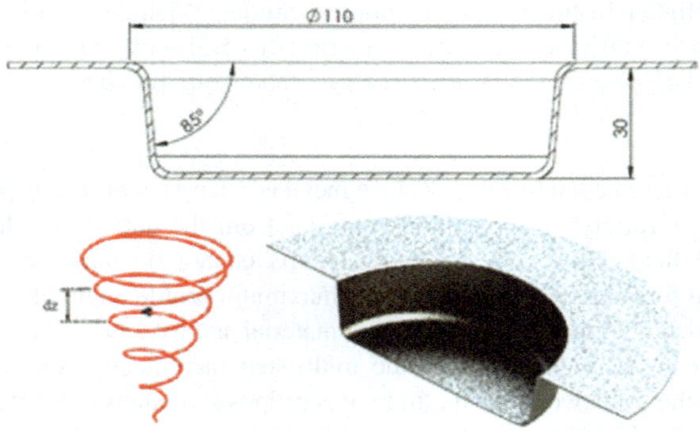

Fig. 1 Theoretical model and trajectory of the tool

Figure 2 shows the steps to perform the numerical simulation of the multi stage incremental forming process (MSIF).

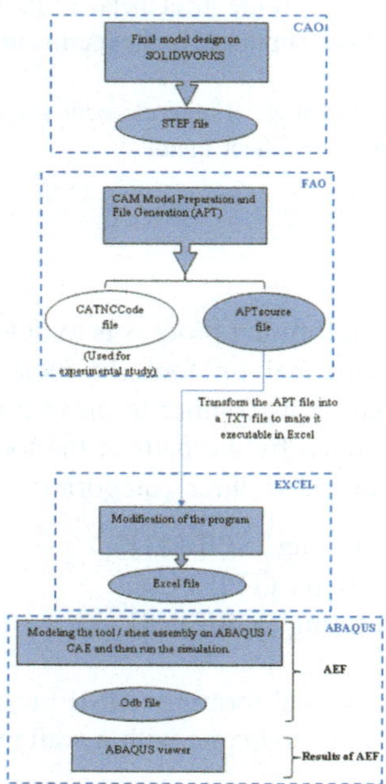

Fig. 2 Digital simulation approach

- **Evolution of the profile**

Figure 3 shows the evolution of the profile. Note that the maximum depth value does not exceed 30 mm.

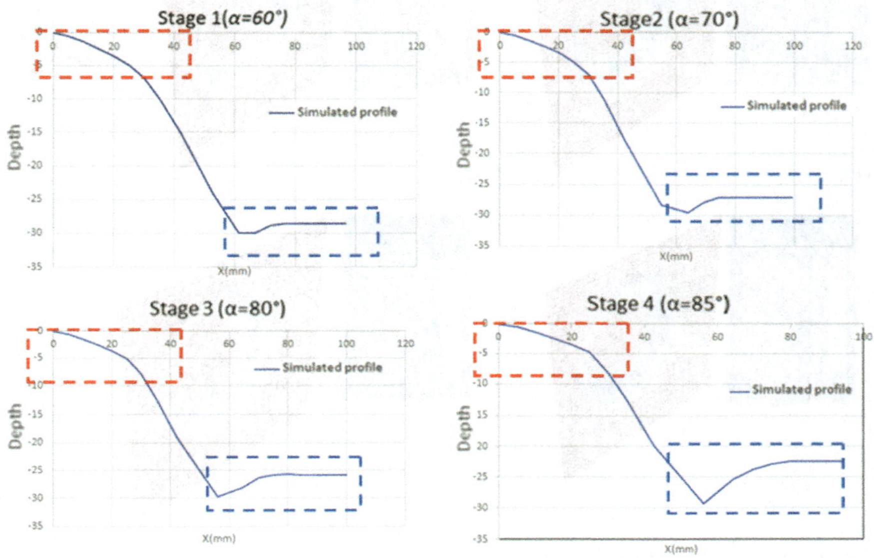

Fig. 3 Evolution of the profile of the sheet

The profiles obtained from the sheet are presented above. The red zone is a deformation that appeared because of the boundary conditions used for the simulation, which is related to the lack of support against the part in the real case.

An elastic return appears in the zone of the bottom of the cone (zone in blue) and it is more remarkable in steps 3 and 4.

- **Results of variation of thickness**

Figure 4 shows the thickness variations for the entire sheet.

The following graphs (Fig. 5) represent for each step the distribution of the thickness obtained by cutting the section perpendicular to the Y direction.

The thickness decreases until the minimum value is reached in step four.

Thickness results, deformed areas on the sheet confirm that the MSIF strategy can form a cone with vertical wall.

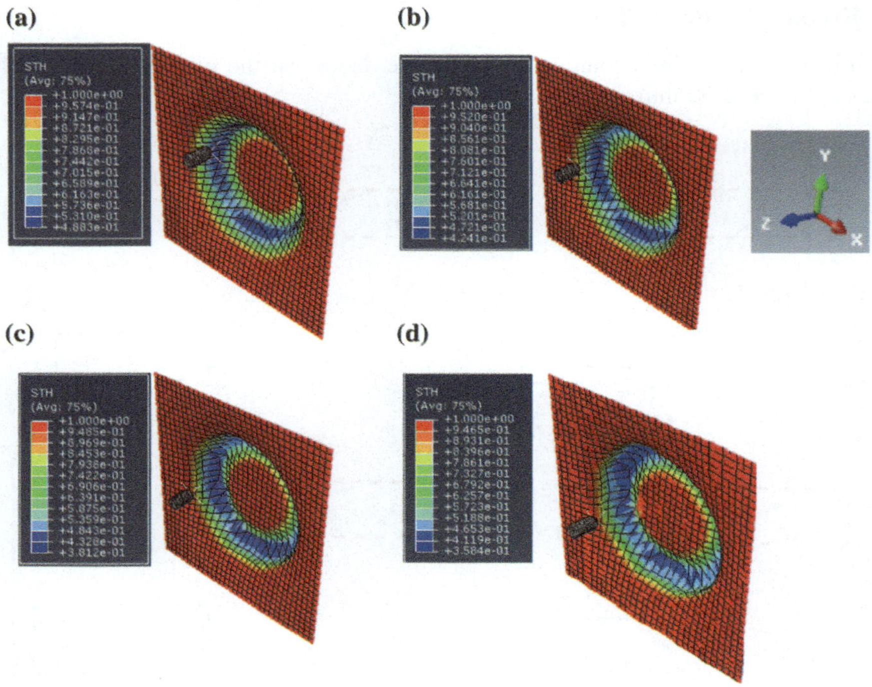

Fig. 4 Thickness distribution in the four steps **a** α = 60°, **b** α = 70°, **c** α = 80°, **d** α = 85

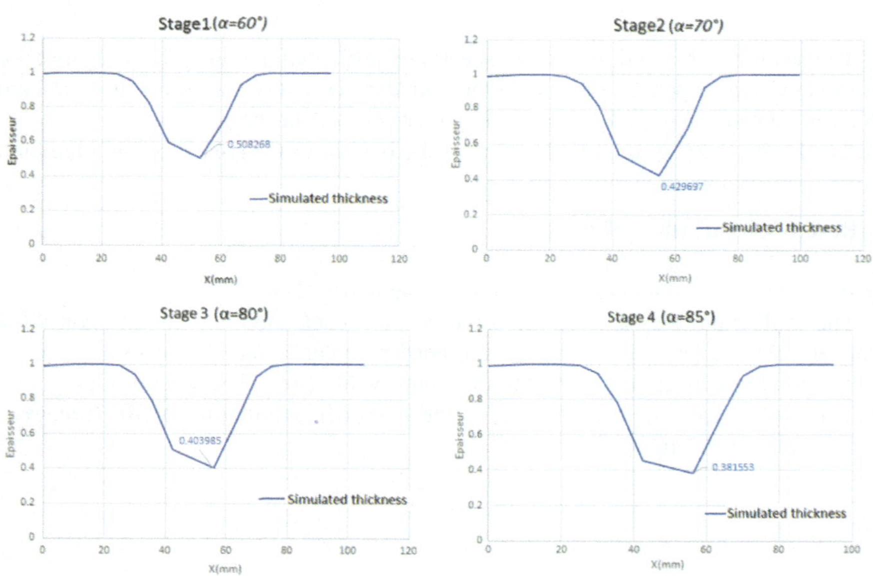

Fig. 5 Evolution of z depth in the four stages

3 Experimental Setup

For this experimental analysis a test bench was used to determine the profile under consideration. The test stand is mounted on a 3-axis SPINNER MVC 850 CNC vertical machining center. The forming tool used is a tool with a hemispherical head diameter = Ø10 mm in 42CrMo4 steel. The test specimen is a square blank of S235 steel sheet in the dimensions are 190 × 190 mm thickness 1 mm.

In Table 1, the multi stage incremental forming test parameters are illustrated.

Table 1 Multi-stage Incremental Forming test parameters

Parameters	Levels
Strategy	DDDD (Dawn, Dawn, Dawn, Dawn)
Path	"Z level" trajectory
Increment Δz	0.5 mm
Tool speed	600 tr/min
Feed rate speed of tool	1000 mm/min
Lubrication	With lubrication

- **Means of profile measurement**

The profile projector has been used for profile measurement and at the same time for thickness checking, are optical measuring devices projecting a profile image of a part on a screen (Fig. 6).

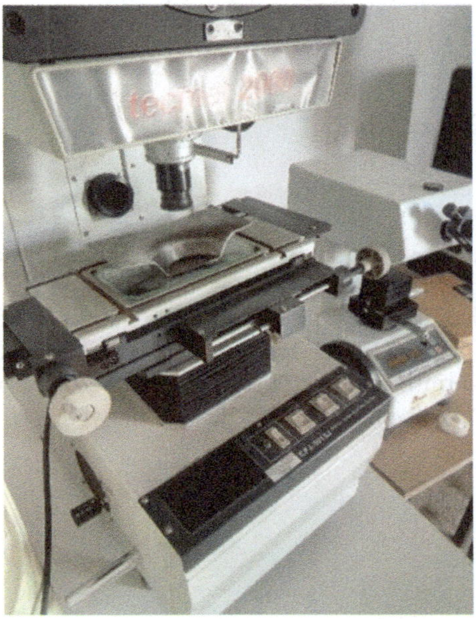

Fig. 6 Vertical profile projector

- **Means for measuring the thickness**

Since the walls of the objects to be measured do not allow the use of traditional calipers, ultrasonic measurement techniques are used; For this we will use a thickness sensor that allows us to measure the variation of the thickness of our room (Fig. 7).

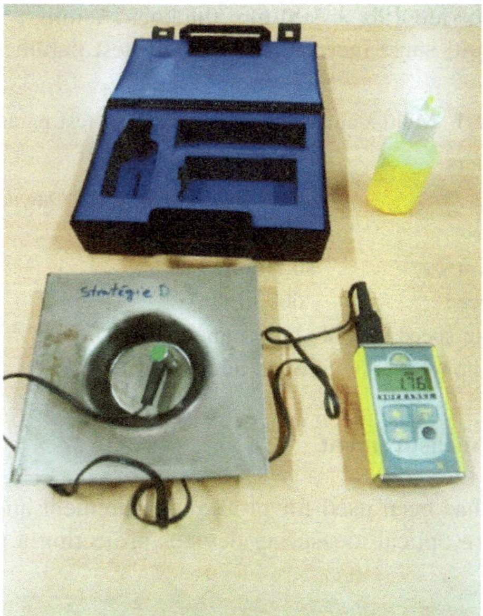

Fig. 7 Thickness measuring sensor

4 Results and Discussion

- Depth evolution

The following, we present an extract of the results of the experimental work bearing on the evolution of the profile presented in (Fig. 8).

It may be noted that the shape obtained is different to that of the theoretical model (CAD) of the part, a degraded form appeared at the bottom of the part (clearer in the stages 3 and 4), this form is thanks to the accumulation of the material by the movement of the tool down, this phenomenon called "pillowing".

Fig. 8 Forms obtained after each stage (stage1: α = 60°) (stage2: α = 70°), (stage3: α = 80°) et (stage4: α = 85°)

In order to collect numerical and experimental results, a comparison was made of the depth of the room between the theoretical, simulated and experimental models presented in (Fig. 9).

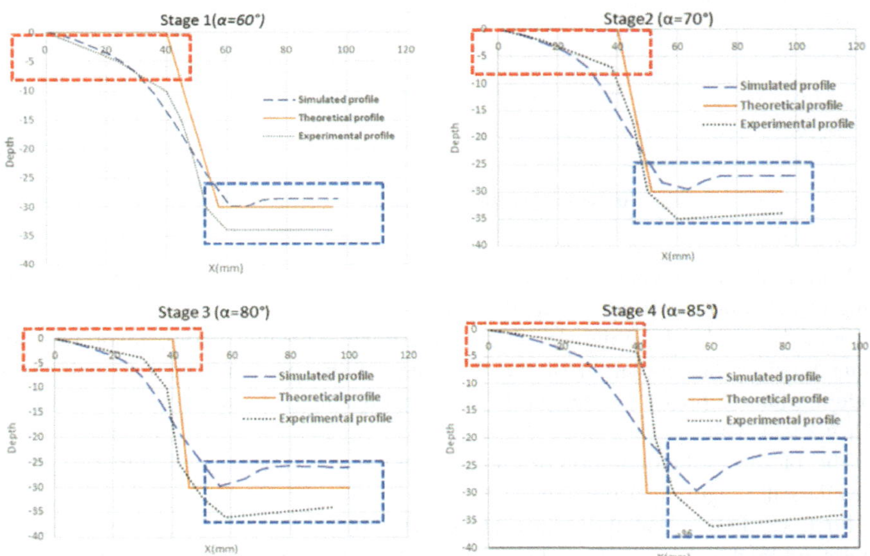

Fig. 9 Presentation of the theoretical, simulated and experimental depths in the four steps

The experimental depth reaches 36 mm in stage 4; it is the consequence of the phenomenon of "pillowing" which increases the desired final depth.

The numerical and experimental profiles have an offset from the theoretical profile at the top of the room. This shift is due to deformation that can be limited with a counter piece.

- **Results of variation of thickness**:

The results obtained with a thickness sensor are shown on (Fig. 10).

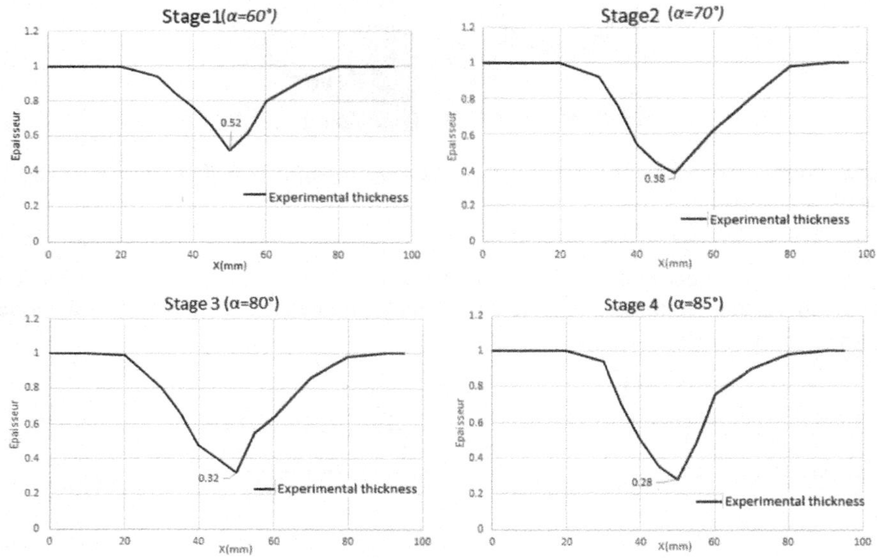

Fig. 10 Thickness distribution in the different steps

Note that the thickness decreased after each pass until waiting for the minimum value in the fourth step which equals 0.28 mm.

In order to collect the numerical and experimental results, a comparison was made at the level of the evolution of the thickness of the piece in order to validate the modeling and to prove that the MSIF strategy capable of producing a cone truncated with vertical wall.

The 11 shows the evolution of thickness in the experimental and numerical studies as well as the theoretical thickness.

First, we notice that the values are close between the numerical results and the theoretical thickness in the four stages. For the experimental thickness, we notice that it exceeds the theoretical value for the steps 2, 3 and 4, so we can say that the prediction of thickness by the theoretical and numerical methods remains difficult.

The results obtained show that multi-pass incremental forming can produce a truncated cone with a vertical wall and overcomes the sine law.

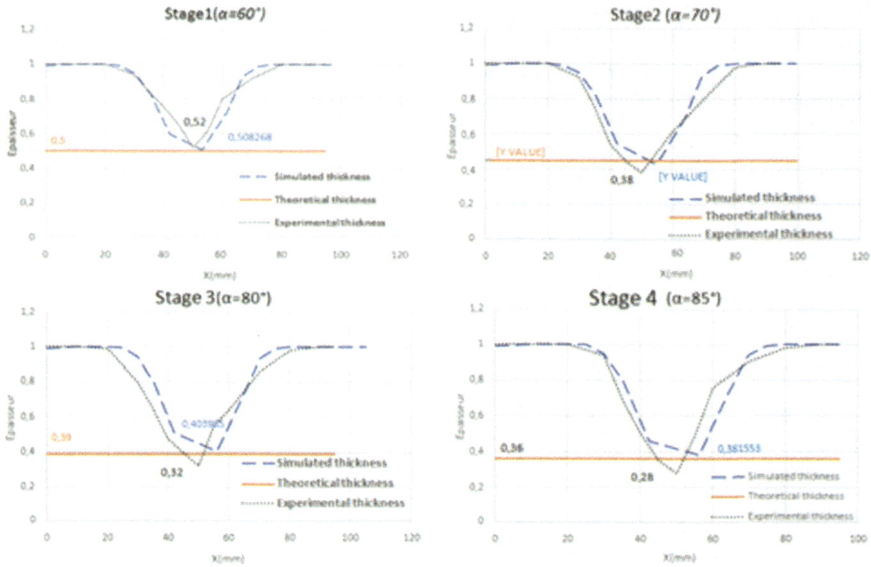

Fig. 11 Simulation and experimental thickness distribution in the four steps with respect to the theoretical thickness

5 Conclusion

The experimental tests carried out using a test bench on the material S235, allowed us to obtain a set of the results of the process of multi-pass incremental forming MSIF.

The work developed in this paper helps to prove that the MSIF strategy is capable of producing a truncated cone with a wall angle very close to 90°.

At the end of this study, you can:

- Avoid deformation of the upper part of the part while adding a support against piece to the test bench.
- Overcome the phenomenon of "pillowing" when forming the sheet while dividing the final depth of the piece on the number of steps.
- Change the "DDDD" strategy by a "DDDU" strategy with a depth greater than the first three movements.

References

1. Silva MB, Skjoedt M, Atkins AG, Bay N, Martins PAF (2008) Single-point incremental forming and formability–failure diagrams. J. Strain Anal 43(1):15–35. https://doi.org/10.1243/03093247JSA340
2. Jeswiet J, Micari F, Hirt G, Bramley A, Duflou J, Allwood J (2005) Asymmetric single point incremental forming of sheet metal. CIRP Ann-Manuf Technol 54(2):88–114

3. Skjødt M, Silva MB, Martins PAF, Bay N (2010) Strategies and limits in multi-stage single-point incremental forming. J Strain Anal Eng Des 45(1):33–44
4. Suresh K, Regalla SP, Kotkundae N (2018) Finite element simulations of multi-stage incremental forming process. Mater Today: Proc 5(2):3802–3810

Human Skills Evaluation to Improve Production Performance: Case of a Company in the Cosmetics Sector

Khalil Tliba[1,2](✉), Thierno M. L. Diallo[2], Olivia Penas[2], Noureddine Ben Yahia[1], and Jean-Yves Choley[2]

[1] Université de Tunis-ENSIT Laboratoire de Mécanique, Productique et Energétique (LR18ES01), Tunis, Tunisie
tlibakhalil@gmail.com, nourdine.benyahia@yahoo.com

[2] Supméca - Institut Supérieur de Mécanique de Paris, Laboratoire QUARTZ (EA 7393), Paris, France
{thierno.diallo, olivia.penas, jean-yves.choley}@supmeca.fr

Abstract. Human resource skills have a significant impact on the production systems performance and competitiveness of companies. In fact, many companies are aware of the human potential in their development, to improve their overall performance and to cope with global economic and technological changes. Hence the need for any company to assess its staff potential and to put into place relevant training plans becomes a crucial challenge, in order to meet its current and future needs. Employees' competencies assessment includes all the activities related to the planning, monitoring, evaluation, recognition and development of individual work performance. In this paper, we present a method to evaluate the human performance based on the job and skills management, in order to improve industrial performance. This assessment relates to the analysis of the data and information gathered from surveys of employees working in a company's cosmetic production departments.

Keywords: Industrial performance · Knowledge and skills assessment · Continuous improvement

1 Introduction

In order to deal with international competition and ensure their sustainability, firms are obliged to innovate and be competitive in terms of cost, quality and time. Innovation and competitiveness can then be considered as the two dimensions of a company's overall performance. These two requirements necessitate the acquisition of some resources and an efficient and effective management of these resources. These resources include tangible assets (machines, equipment, buildings, ICT-information and communication technologies infrastructures…) and intangible assets (human resources mainly). The contribution of intangible assets as knowledge to the overall performance has been reported in several studies [1, 16, 21, 24]. However, even if the link between knowledge and performance is one of the most discussed topics in the knowledge

N. Aifaoui et al. (Eds.): CMSM 2019, LNME, pp. 345–354, 2020.
https://doi.org/10.1007/978-3-030-27146-6_37

management literature [16], the quantification of the knowledge contribution to the overall industrial performance and the causal link between knowledge and performance are difficult to establish [25]. ISO 9001, version 2015 [18], one of the most widely applied management system standards, explicitly mentions the knowledge as a resource to be managed. It requires that companies identify and manage the knowledge needed to achieve the compliance of products and services.

Knowledge management (KM) is "the set of management activities that enable the firm to deliver value from its knowledge asset" [21]. The main KM activities are knowledge sharing, creation, application, storage and identification [15]. Therefore, KM encompasses information and communication technologies (ICT) management [21] but also the human resource management (HRM). KM and HRM are then very closely interrelated as the most important resource from employees relates to their knowledge [21]. KM objectives include the protection against any loss of knowledge related to staff departures or simple forgetfulness of employees in company and the facilitation of the sharing of information necessary for the proper functioning of processes in a logic of risk prevention and continuous improvement. Thus, in order to efficiently exploit existing knowledge assets, in accordance with their present and future needs, companies must design and implement a consistent knowledge management strategy. On this point, strategic knowledge management (SKM) deals with the construction of a KM strategy, the monitoring and evaluation of knowledge resources and the benchmarking of the business environment for their developmental need [16].

The aim of this research work is to support the development of a SKM by assessing the needs in terms of knowledge and skills of production operators and to propose an action plan to the company management in order to satisfy the identified needs in the assessment process. This study was conducted in a cosmetic production company. This paper is organized as follows: the next section introduces the management and the assessment of knowledge and skills based on a literature review. Section 3 presents our research work through the skills assessment survey and its outcomes followed by an analysis for the future SKM. We conclude in Sect. 4.

2 Management and Assessment of Knowledge and Skills

In general, knowledge is classified in most research publications as either tacit or explicit knowledge [14, 29]. Explicit knowledge can be easily acquired, standardized and generalized through manuals, documents and simple operating procedures, whereas tacit knowledge relates to the skills and the know-how of a person that cannot be easily generalized or shared, it is based on the experience, the intellectual level, diploma, and academic training [11, 13]. On this subject, Nonaka proposes and develops a theoretical model to facilitate the dynamic creation of sufficient organizational knowledge through a continuous dialogue between tacit and explicit knowledge [26]. Additionally, (Saint-Onge [27] explains all the aspects and roles of explicit and tacit knowledge in the three elements of intellectual capital in an organization (Human capital, Customer capital and Structural capital). Indeed, the enterprise is considered as a knowledge integration institution. Knowledge has become a vital resource for innovation, performance and competitiveness of organizations in all fields [12]. The real value of organizations depends on their knowledge base, the ideas and insights in the heads of their employees and the performance of their human capital. As a result, some knowledge management activities, as the acquisition, conversion, application, and protection of knowledge, are positively related to the organizational performance [25].

2.1 Knowledge Management

Knowledge Management (KM) is the management of corporate knowledge, skills, and knows-how that can improve firm's performance. According to Wong et al. [28], KM is concerned with the exploitation and development of knowledge (explicit and tacit) of an organization, aligned with the company's objectives. Several KM implementation models are suggested and proven with KM research [3, 9, 17, 23]. These models are developed according to different approaches and perspectives [9]. KM system (KMS) is a set of techniques and strategies to analyze, organize, improve, disseminate [11], maintain and share knowledge and experiences in an organization [26].

In parallel, in the industrial environment, the knowledge management is a novelty of Quality Management System—ISO 9001 in version 2015 [18], where knowledge is described in the seven quality management principles: customer focus, leadership, engagement of people, process approach, improvement, evidence-based decision making and relationship management. In the Chap. 7 "Support", it notably demands the company to identify and manage the knowledge needed to achieve compliance of products and services. The base of knowledge has also been described in the ISO 9000 [19] as "the available set of information (or data) constituting a justified conviction and having a strong certainty of being true" and factual. Finally, as it is now obvious that skills and knowledge management have a direct impact on business performance, the implementation or the improvement of KM must imperatively begin with an evaluation of staff knowledge.

2.2 Knowledge and Skills Assessment

Knowledge has been well recognized as intangible but is an important asset for achieving performance and competitiveness [4]. Then knowledge evaluation is crucial for performance management, as it allows an organization to evaluate, monitor, and improve its human capital. Chen and Chen [7] classified KM performance measurement into eight categories (namely qualitative analysis, quantitative analysis, financial indicator analysis, non-financial indicator analysis, internal performance analysis, external performance analysis, project-oriented analysis, and organizational oriented analysis).

Kuah and Wong [22] covered the most important KM performance measurement approaches and grouped them into two broad categories (qualitative and quantitative) and eight associated models (Balanced scorecard (BSC), Skandia navigator, Intangible asset monitor, Tobin's q ratio, Human resource accounting (HRA), KP^3 methodology, Knowledge management performance index (KMPI), and User-satisfaction-based system (USBS)). Whereas the BSC links learning components and other intangible assets to organizational performance, in order to map KM objectives across the four BSC key areas (financial performance, internal business processes, customer, and learning and growth), this method does not provide an explicit explanation on how to conduct an assessment [20].

The Skandia navigator is an assessment and management tool capable of measuring the hidden dynamic factors of human, customer and structural capital that underpin the visible aspects of a company's buildings and products, thus reflecting more accurately the value of a company [10]. Even if it is easy to implement, the very large number of measures it proposes may cause confusing interpretations. Intangible asset monitor presents a complete measure of organizational success and shareholder value with a knowledge viewpoint. However, this model is more concentrated on management objectives and is inadequate for the purpose of quantifying intangible assets. Tobin's q ratio is an important tool in KM development and for laying the preliminary work for intellectual capital measurement. Unfortunately, the application of Tobin's q in KM does not take into account the replacement costs for intangible assets. HRA is an approach that quantifies the economic value of staffs including cost models, human resource value models and monetary emphasis models, it supports the conversion of employees' knowledge and experience into monetary value [5]. KP^3 (knowledge, product, process, and performance) assesses the contribution of knowledge to business performance by employing product and process as intermediaries. It shows more explicitly the relation between knowledge and business performance [2]. The KMPI is a new metric to assess the KM performance of a firm, which is based on a logistic function including five components used to determine the KCP: knowledge creation, accumulation, sharing, utilization, and internalization. The KMPI method improves the quality of decision-making and establishes and evaluates KCP [6]. Finally, the USBS provides an alternative way to evaluate KM solely based on knowledge users [8].

Additionally, [28] present a review on Knowledge Management (KM) performance measurement in the past two decades and show how KM performance measurement has changed during this period. Therefore, in addition to the previous classifications, they propose to categorize KM performance measurement into six new categories: traditional, advanced, deterministic, stochastic, general result oriented and specific result oriented.

Finally, despite many methods of knowledge assessment, there is no approach to assess both skills and knowledge. For this reason, we propose a qualitative approach to capture both skills and knowledge. In the following paragraph, we will present the survey we conducted to assess the skills and knowledge of employees in the production of a company in the cosmetics sector.

3 Approach and Outcome of the Skills Assessment Survey

This study is based on surveys conducted in the form of individual interviews with company staff. Its aim is to evaluate the know-how and skills of the employees and to measure the gaps with respect to the needs. This skills assessment will help to identify and highlight the strengths and weaknesses for each operator, in the light of expected current or future industrial performance. Therefore, the proposed method consists in first collecting the information and the points of view of all the actors involved in the production system, in order to then make proposals likely to improve both their working environment and the industrial performance.

The survey was carried out after an analysis of historical data related to products and manufacturing equipment and a diagnosis of the production system, in order to measure the current industrial performance according to the measurement methods and performance indicators existing in the company and define failures and thus correlate failures and non-conformities with human factors. The survey is conducted in three stages: Pre-survey, Survey, and Post-survey.

3.1 Pre-survey

This step consists in carrying out the actions prior to the elaboration of the questionnaire: interview with the managers of each company department and retrieving the additional documents (job profile, role description, current position, training followed and qualifications, skills matrix) useful to:

- Define the purpose of the survey: *study of the impact of human factors on production performance by evaluating the current workers' skills and the skills level required by the job*
- Make an inventory of available resources: *examining the time availability, additional documentary sources and various material resources*)
- Define the targeted population and the chosen sample (services concerned, sample size and individuals).

3.2 Survey

To assess the impact of human factors on industrial performance, two different types of media have been developed: the first to audit the performance of the maintenance department which has been completed with the department manager and the second type is a questionnaire completed in the form of an individual interview with the employees selected in the sample.

The survey support consists of two complementary questionnaires. The first contains 52 standard questions for all services; this questionnaire assesses the knowledge and skills of operators within the company and their degree of involvement in the work and is completed in the form of an interview. This standard questionnaire addresses the following points: (i) knowledge of standards, company certifications and Lean Manufacturing tools implemented, (ii) skills, strengths and weaknesses, (iii) training, (iv) future of the employees and their desire to evolve, (v) motivation, degree of involvement in the work and behavior, (vi) relationship with other services (maintenance and quality for example) and working conditions, (vii) relationship with managers and (viii) proposals to improve production. The second is in the form of a self-assessment, which includes questions on the details of the work performed and the operator's know-how. These questions are based on the job description of each position. They assess technical, relational, organizational and adaptation skills. Then the draft questionnaire has been tested by three people (whose characteristics fit those required of members of the survey population), in order to assess the ease of understanding, the degree of acceptance and the ease of interpretation. The test is an absolutely necessary step that must be carried out with rigor. This test resulted in the removal of 11 questions that affected the susceptibility of two of the three people chosen for the test. The objective of these removed questions was to determine the degree of use and familiarization of operators with new technologies in their daily lives. Finally, during the survey, the questionnaires were collected, stored carefully and classified by department and type of population interviewed, before being checked and verified regarding their correct filling.

3.3 Post-survey

The post-survey stage includes the survey counting, the results analysis and finally the formulation of an action plan. In the survey counting step, all the responses of the employees have been summarized in a single document, in order to facilitate the analysis and extraction of important information for the definition of the gap to expectations and the drafting of an action plan. For the closed-ended questions, the ratios for all responses (number of responses by yes or no) for the entire population selected have been calculated. For the open and semi-open questions, the answers considered useful and beneficial for the analysis have been collected (out-of-scope and inconsistent responses with regard to the objective of the study, have been eliminated for the next step. The analysis of the results reveals the following elements. Main operators lack of standards knowledge and certifications, and few apply in practical Lean tools. Operators do not encounter difficulties in the execution of tasks, except for certain organizational problems and the format change which requires excessive physical effort; they have a high mastery of the theoretical and technical knowledge of the profession and a good ability to work in a team. They mention a lack of practical training on the machines and the mechanical subsystems that constitute them. Indeed, the content of current training is not clear or methodical, and as a result is of limited use for their work. In addition, temporary workers and new employees do not benefit from training programs (their only training is provided by production staff). Despite good relations between operators within the same department, the lack of communication is highlighted. Similarly, regarding management, they point to a lack of communication with the manager and inadequate human resources management on the production line, with too much rotation between the different workspaces. Finally, the survey shows a good involvement of operators in their work but not in the company objectives.

After analyzing the results of surveys, an action plan based on four areas of evolution has been prepared. It consists of continuous improvement of maintenance, production, quality and communication. The corresponding improvement areas are respectively detailed in Tables 1, 2, 3 and 4.

Table 1 Production performance improvement areas

Improve the knowledge of the quality management system and Good Manufacturing Practices
Improve the practices of 5S, SMED and TPM
Improve the polyvalence of the operators on the machines by the relevant rotations between the workstations according to: the necessary learning time, the adaptability of each operator to the new workstation, and its complexity
Improve the format changes
Improve the production and stock management system
Involve the production staff in the maintenance actions through awareness and clarification of the problems (causes and effects) during the intervention

Table 2 Quality service performance improvement areas

Increase production by reducing problems related to non-compliance of outsourced products
Provide useful information to company to better understand their subcontractor technical capabilities
Enable the company to further develop its achievements in terms of partnerships with its subcontractors through a better knowledge of their weaknesses and strength
Minimize the cost and time of control throughout the plant
Create a database with the technical characteristics and material resources specific to each supplier
Integrate the operators' skills into this database
Fill this database with traceability and analysis of the problems mentioned above

Table 3 Communication system performance improvement areas

Develop a more reliable internal communication system
Promote listening to the internal client
Foster listening to the external customer
Sensitize and motivate staff
Improve downward communication
Involve the staff and team leaders

Table 4 Proposal of a lifelong learning plan

		Short-term	Mid-term
Training and learning plan for temporary workers	Mastering workstations	X	
Practical SMED training	Controlling the changeover times	X	
Practical training of TPM	Mastering maintenance interventions	X	

(*continued*)

Table 4 (*continued*)

		Short-term	Mid-term
Practical training of 5S	Improving the organization of workplaces	X	
Pedagogical training for internal trainers	Creating competent internal trainers		X
GPEC training to facilitate the implementation of this approach	Establishing a job and skills management system	X	
Industrial and robotic automatic training	Preparing operators for production line automation	X	
Language (English and Spanish)	Facilitating the operator's use of software Facilitating communication		X
More practical training on machines	Mastering production equipment	X	

4 Conclusion

Knowledge and skills are of major importance in terms of competitiveness and innovation. In order to avoid loss of knowledge related to staff departures or forgetfulness, facilitate capitalization and transfer of existing knowledge and to identify knowledge gaps, companies must design and implement a consistent knowledge management strategy. The monitoring and evaluation of knowledge resources is one of the first steps in the implementation of this strategy. This research work addresses the assessment of both knowledge and skills in manufacturing industries. After having reviewed main KM performance measurement and concluded that quantitative methods cannot assess these two intangible assets, we propose and detail a qualitative approach to capture them through a survey decomposed in three steps: pre-survey, survey and post-survey. Based on the obtained results on a cosmetic company case study, recommendations have been proposed to improve its industrial performance. Future work will address the implementation of a SKM, based on these results.

Acknowledgements. This work was financially supported by the "PHC Utique" program of the French Ministry of Foreign Affairs and Ministry of higher education and research and the Tunisian Ministry of higher education and scientific research in the CMCU project number 19G1127.

References

1. Ahmad N, Lodhi MS, Zaman K, Naseem I (2017) Knowledge management: a gateway for organizational performance. J Knowl Econ 8:859–876. https://doi.org/10.1007/s13132-015-0282-3

2. Ahn J-H, Chang S-G (2004) Assessing the contribution of knowledge to business performance: the KP^3 methodology. Decis Support Syst Knowl Manag Tech 36:403–416. https://doi.org/10.1016/S0167-9236(03)00029-0
3. Alavi M, Leidner DE (2001) Review: knowledge management and knowledge management systems: conceptual foundations and research issues. MIS Q 25:107–136. https://doi.org/10.2307/3250961
4. Bontis N (2001) Assessing knowledge assets: a review of the models used to measure intellectual capital. Int J Manag Rev 3:41–60. https://doi.org/10.1111/1468-2370.00053
5. Bontis N, Dragonetti NC, Jacobsen K, Roos G (1999) The knowledge toolbox: a review of the tools available to measure and manage intangible resources. Eur Manag J 17:391–402. https://doi.org/10.1016/S0263-2373(99)00019-5
6. Chang Lee K, Lee S, Kang IW (2005) KMPI: measuring knowledge management performance. Inf Manag 42:469–482. https://doi.org/10.1016/j.im.2004.02.003
7. Chen AP, Chen MY (2005) A review of survey research in knowledge management performance measurement: 1995–2004. In: Proceedings of I-KNOW, pp 87–95
8. Chin K-S, Lo K-C, Leung JPF (2010) Development of user-satisfaction-based knowledge management performance measurement system with evidential reasoning approach. Expert Syst Appl 37:366–382. https://doi.org/10.1016/j.eswa.2009.05.051
9. Costa V, Monteiro S (2016) Key knowledge management processes for innovation: a systematic literature review. VINE J Inf Knowl Manag Syst 46:386–410. https://doi.org/10.1108/VJIKMS-02-2015-0017
10. Edvinsson L (1997) Developing intellectual capital at Skandia. Long Range Plann 30:366–373. https://doi.org/10.1016/S0024-6301(97)90248-X
11. Goh SC (2002) Managing effective knowledge transfer: an integrative framework and some practice implications. J Knowl Manag 6:23–30. https://doi.org/10.1108/13673270210417664
12. Grant RM (1996) Toward a knowledge-based theory of the firm. Strateg Manag J 17:109–122. https://doi.org/10.1002/smj.4250171110
13. Gupta B, Iyer LS, Aronson JE (2013) Knowledge management: practices and challenges. Manag Data Syst, Ind. https://doi.org/10.1108/02635570010273018
14. Haldin-Herrgard T (2000) Difficulties in diffusion of tacit knowledge in organizations. J Intellect Cap 1:357–365. https://doi.org/10.1108/14691930010359252
15. Heisig P (2009) Harmonisation of knowledge management—comparing 160 KM frameworks around the globe. J Knowl Manag 13:4–31. https://doi.org/10.1108/13673270910971798
16. Inkinen H (2016) Review of empirical research on knowledge management practices and firm performance. J Knowl Manag 20:230–257. https://doi.org/10.1108/JKM-09-2015-0336
17. Iskandar K, Jambak MI, Kosala R, Prabowo H (2017) Current issue on knowledge management system for future research: a systematic literature review. Proc Comput Sci 116:68–80. https://doi.org/10.1016/j.procs.2017.10.011
18. ISO (2015) ISO 9001:2015, Quality management systems—requirements
19. ISO (2015) ISO 9000:2015, Quality management systems—fundamentals and vocabulary
20. Kaplan RS, Kaplan RS, Norton DP, Norton DP (1996) The balanced scorecard: translating strategy into action. Harvard Business Press
21. Kianto A, Andreeva T (2012) Does knowledge management really matter? Linking knowledge management practices, competitiveness and economic performance. J Knowl Manag 16:617–636. https://doi.org/10.1108/13673271211246185
22. Kuah CT, Wong KY (2011) Knowledge management performance measurement: a review. Afr J Bus Manag 5:6021–6027. https://doi.org/10.5897/AJBM10.1091

23. Liao S (2003) Knowledge management technologies and applications—literature review from 1995 to 2002. Expert Syst Appl 25:155–164
24. Mardani A, Nikoosokhan S, Moradi M, Doustar M (2018) The relationship between knowledge management and innovation performance. J High Technol Manag Res 29:12–26. https://doi.org/10.1016/j.hitech.2018.04.002
25. Mills AM, Smith TA (2011) Knowledge management and organizational performance: a decomposed view. J Knowl Manag 15:156–171. https://doi.org/10.1108/13673271111108756
26. Nonaka I (1994) A dynamic theory of organizational knowledge creation. Organ Sci 5:14–37. https://doi.org/10.1287/orsc.5.1.14
27. Saint-Onge H (1996) Tacit knowledge the key to the strategic alignment of intellectual capital. Plan Rev 24:10–16. https://doi.org/10.1108/eb054547
28. Wong KY, Tan LP, Lee CS, Wong WP (2015) Knowledge management performance measurement: measures, approaches, trends and future directions. Inf Dev 31:239–257. https://doi.org/10.1177/0266666913513278
29. Yang J, Chyi Lee C (2000) Knowledge value chain. J Manag Dev 19:783–794. https://doi.org/10.1108/02621710010378228

Optimization of FDM Manufacturing Parameters of a Biodegradable Thermoplastic (PLA)

N. Benali(✉), D. Hammami, M. Khlif, and C. Bradai

Electromechanical Systems Laboratory (LASEM), National School of Engineers of Sfax, University of Sfax (ENIS), PB, 1173-3038 Sfax, Tunisia

Abstract. Fused deposition modeling FDM is an additive manufacturing technology (AM) at fast growing and that could enhance manufacturing because of its ability to create functional parts and highly innovative with suitable mechanical properties having a complex geometric shape. Various process parameters used in FDM affect the quality of the prototype. The control of these specific parameters involved during the manufacturing is an important activity. For this, the parameters of the 3D printer must be identified and controlled in order to study and optimize their influence on the mechanical properties of prototypes printed using the Taguchi method. The thickness of the layer and the filling orientation are the most influential on the modulus of elasticity and the tensile strength, respectively. Furthermore, the thermomechanical behavior by DMA tests and the requirements to optimize the factors generated by the FDM technology.

Keywords: Additive manufacturing · FDM · Biodegradable thermoplastic · Optimization · Mechanical properties · DMA

1 Introduction

The environmental impacts of the plastic materials are a critical issue during a technological evolution. The choice we make will have consequences for the future generations. Biopolymers represent an ecological and effective solution when the life cycle is considered. Among these materials, Polylactic acid (PLA) as a biodegradable polymer, non-toxic and manufactured without solvents harmful to nature [3] PLA is a rigid material and probably deformable at temperatures above Tg.

At present, PLA is widely used in additive manufacturing with the technology FDM (Fused Deposition Modeling) [12]. The FDM procedure is an additive manufacturing method used to create functional and robust parts directly from computer generated models [6].

Figure 1 shows the schematic illustration of the FDM process. The material is drawn through a nozzle, where it is heated and is then deposited layer by layer. The nozzle can move horizontally and a platform moves up and down vertically after each new layer is deposited. the process has many parameters that affect the final part

N. Aifaoui et al. (Eds.): CMSM 2019, LNME, pp. 355–362, 2020.
https://doi.org/10.1007/978-3-030-27146-6_38

quality, but its potential and viability is great when these parameters are successfully controlled [1, 2]. This layer-by-layer manufacturing allows an unprecedented freedom in manufacturing complex, with precision and control that cannot be made through traditional manufacturing routes [8, 11].

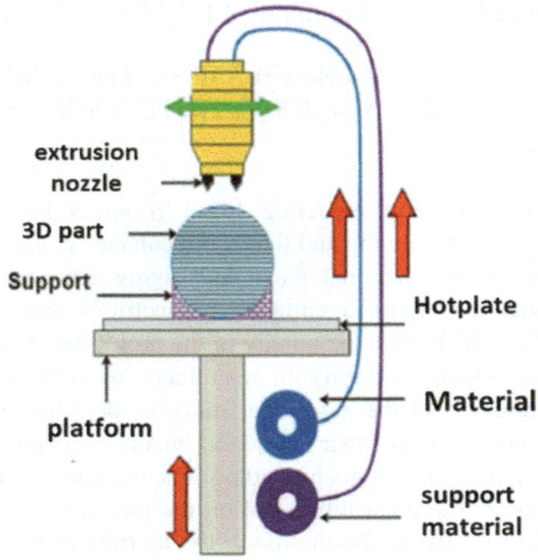

Fig. 1 Schematic representation of the FDM process [6]

FDM manufacturing has been introduced commercially in early 1990s by Stratasys Inc., USA [5], it has been used ever more widely among all additive manufacturing technologies. FDM covers various fields of engineering and industry, such as mold design, aerospace and automotive [10]. FDM is the opposite of subtractive manufacturing, which is cut out and hollowing out a part plastic or metal with for instance a turning machine, without the need for machining, FDM makes it possible to obtain complex and solid part directly from a CAD file. Several thermoplastic materials are commercially available for use with FDM including polylactic acid (PLA), acrylonitrile butadiene styrene (ABS), polycarbonate (PC) and several varieties of the aforementioned materials [4]. However, improved mechanical properties and reduced cost are challenges that the additive manufacturing community is working to overcome [13], so there is a need to optimize the process parameters both from a technological and economic point of view [7].

This article presents a comprehensive review DMA test of PLA polymers, as well as the optimization of FDM process parameters involving statistical design of experiments and optimization techniques.

2 Materials and Methods

2.1 Dynamic Mechanical Analysis of PLA

Specimens of section 2 × 12 mm were prepared in the form of printed plates. These specimens were subjected to sinusoidal shear stresses. The DMA test provides information on the properties of PLA material placed in minor and sinusoidal oscillation depending on the temperature and time, by submitting it at a slight displacement of 5^{-6} m and a low frequency of 10 Hz. The mechanical properties (stiffness K and Coulomb module G) are traced by the machine according to the temperature.

2.2 Design of Experiment Based on Taguchi Method

The second part is devoted to studying the effects of the parameters of the FDM process on the mechanical properties. Tensile strength and modulus of elasticity depend on four important factors such as layer thickness, manufacturing speed, filling orientation and melting temperature. They are defined as follows (Fig. 2):

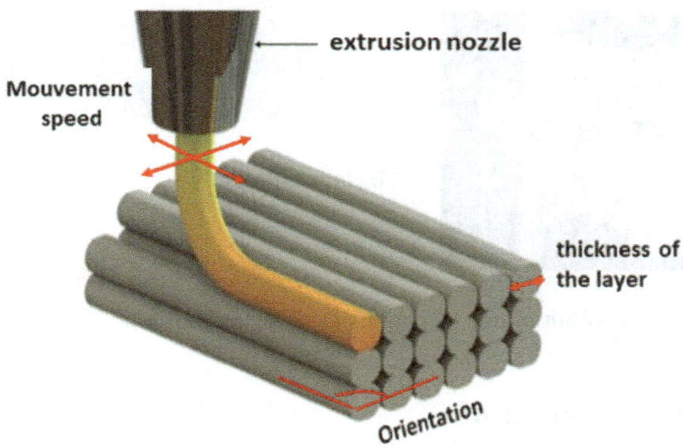

Fig. 2 The different print parameters of a layer [9]

To reduce the number of experiments, consider the design of a fractional Taguchi table L8 (eight experience). The high and low level of each factor is coded as 1 and 2, respectively. Having decided to use two levels by factor, the brainstorming exemplifies the values in Table 1.

Table 1 Selected factors and areas of study

Factor	Level 1	Level 2
Thickness (X1)	0.35 mm	0.2 mm
Speed (X2)	80 mm/s	40 mm/s
Temperature (X3)	200 °C	190 °C
Orientation (X4)	−45/45°	0/90°

In order to manufacture the specimens designed, the prepared STL files have been imported into the FDM machine (Tobeca2) via a software driver (Repetier Host) to start manufacturing (Fig. 3b). To determine the mechanical properties of 3D printed parts and the variability in these properties when different user-controlled printing parameters are used, this investigation looked at the influence of the four parameters on to tensile strength, and Young's modulus. These two properties are used after as two answers to build the Taguchi table. For obtaining the mechanical properties of the 3D specimens, tensile tests were performed using a mechanical testing machine with a 10 kN load cell, following ISO 527 (Fig. 3a–c) standard test and a cross-head speed of 5 mm · min^{-1}.

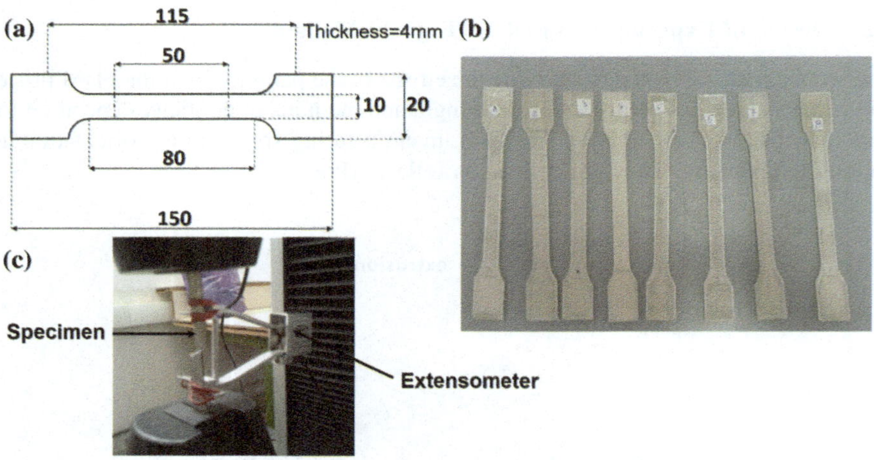

Fig. 3 Elaboration and mechanical characterization of 3D specimens

3 Results and Discussion

3.1 DMA Test of a PLA

The main characteristic of polylactic acid PLA is the transition, associated with the glass transition Tg which is equal to 57 °C, materialized on the DMA curves.

Figure 4a indicates that a decrease in the dynamic stiffness K causes a decrease in the properties of the material with temperatures above 57 °C.

Furthermore, the resulting coulomb modulus (G) and glass transition temperature in term of loss factor of the PLA are shown in Fig. 4b. Note that the coulomb modulus decreased rapidly between 57 and 75 °C due to the glass transition temperature of PLA.

3.2 Influence of FDM Process Parameters

Table 2 presents the effects of the parameters and their levels on the modulus of elasticity E (GPa) and the tensile strength Rm (MPa). The 'average' column of the table is the average of the results for both levels 1 and 2. These results make it possible to

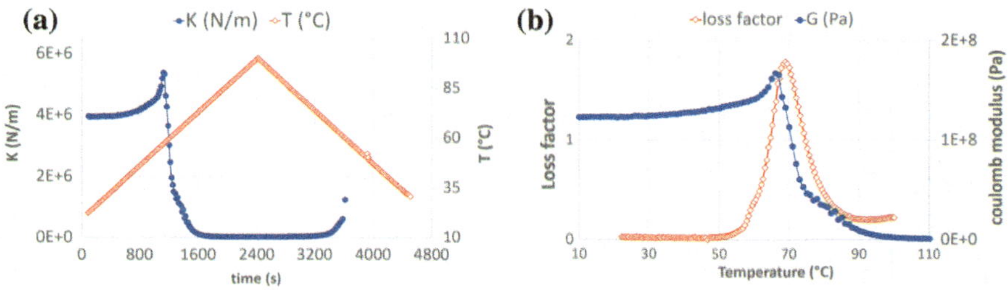

Fig. 4 **a** Variation of stiffness k, **b** coulomb modulus G and loss factor according to the temperature

trace the effects of each factor in percentage. X1 and X2 are the most significant parameters on modulus of elasticity and the tensile strength respectively, as shown in the Table 2.

Table 2 Effects of factors on mechanical properties

		E (GPa)		Rm (MPa)	
Parameters	Level	Average	Effect	Average	Effect
X 1	1	2.62	−0.25	34.56	−2.41
	2	2.87		36.97	
X 2	1	2.83	0.17	34.84	−1.85
	2	2.66		36.69	
X 3	1	2.71	−0.06	37.27	3.02
	2	2.77		34.25	
X 4	1	2.71	−0.07	31.84	−7.84
	2	2.78		39.68	

The percentage effects are shown in Figs. 5 and 6 in the form of a Pareto diagram. For the Young's modulus, the thickness factor comes first with an effect of 30.9%, the speed is in second position, then the orientation and the temperature seem to be the least influential (7.4%) as shown in Fig. 5. For the mechanical strength, Fig. 6 shows that the orientation factor seems to be the most influential in order of importance. We find respectively; an influence of orientation of 41.2 and 15.9% of the temperature. The speed factor seems to be the least influential with 6.6%.

The contour graphs are illustrated in Figs. 7 and 8 allowed to observed the relationship between the adjusted response values and two continuous factors. The darker bands correspond to the higher probabilities of mechanical properties with adjusted response ranges.

Figure 7 shows the analysis of the contour plots for Young's modulus, which revealed that high value (>3 GPa) was obtained with the following combinations: (a) thickness 0.2 mm (Level 2), speed 80 mm/s (Level 1), (b) temperature 190 °C (Level 2) and (c) orientation 0/90° (Level 2). For tensile strength revealed that high

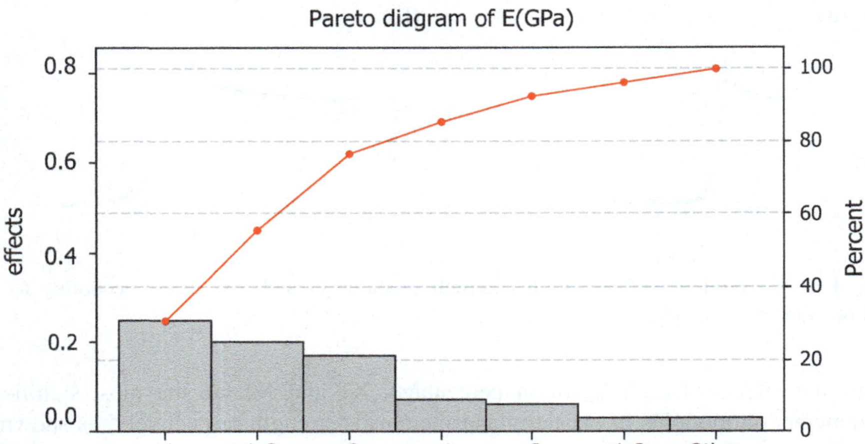

Factors	x1	x1x2	x2	x4	x3	x1x3	Other
Effects	0.25	0.20	0.17	0.07	0.06	0.03	0.03
Percent	30.9	24.7	21.0	8.6	7.4	3.7	3.7
Cum %	30.9	55.6	76,5	85.2	92.6	96.3	100

Fig. 5 Pareto diagram for Young's modulus E (GPa)

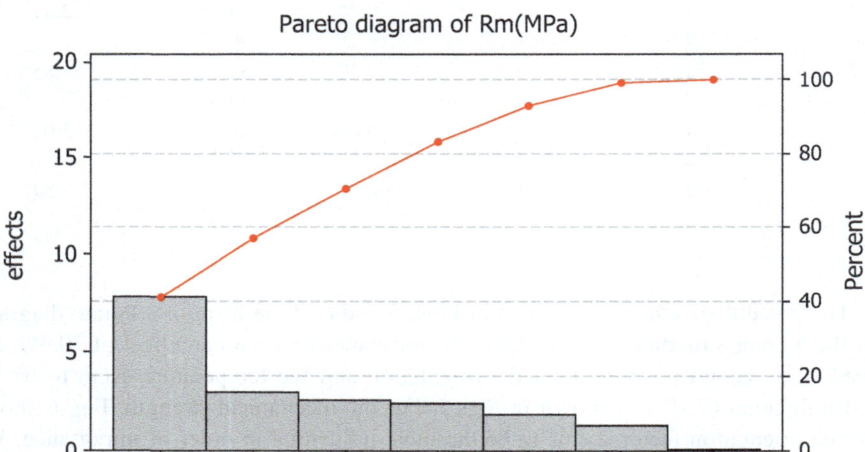

Factors	x4	x3	x1x2	x1	x2	x1x3	Other
Effects	7.84	3.02	2.57	2.41	1.85	1.27	0.09
Percent	41.2	15.9	13.5	12.7	9.7	6.7	0.5
Cum %	41.2	57	70.5	83.1	92.9	99.5	100

Fig. 6 Pareto diagram for tensile strength Rm (MPa)

tensile strength (>40 MPa) was obtained with the following combinations, as shown in Fig. 8: (a) thickness 0.2 mm (Level 2), speed 40 mm/s (Level 2), (b) temperature 200 °C (Level 1) and (c) orientation 0/90° (Level 2).

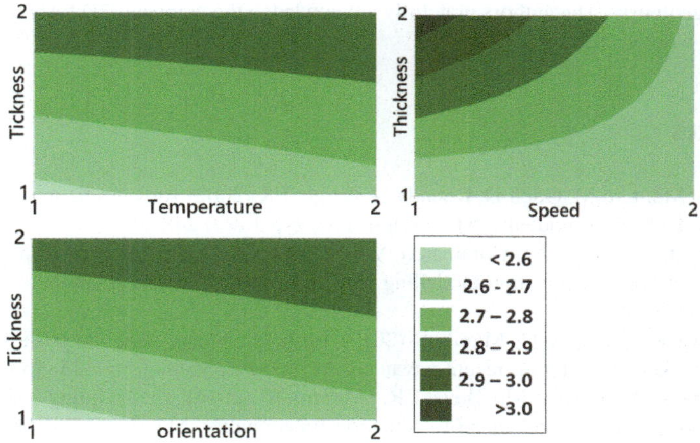

Fig. 7 Contour graph for Young's modulus E (GPa)

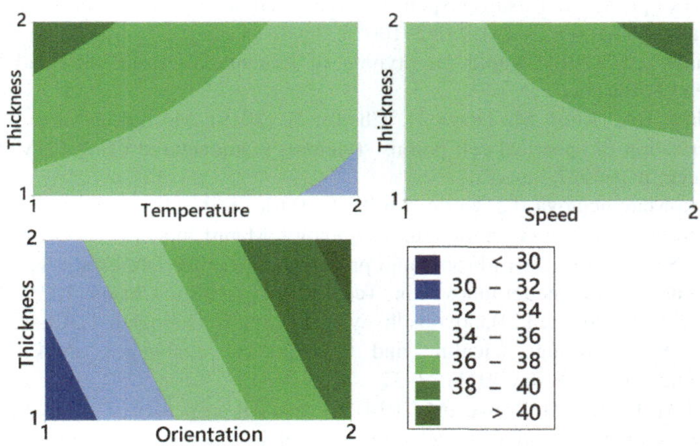

Fig. 8 Contour graph for tensile strength Rm (MPa)

4 Conclusion

In this work, experiments were conducted on a biodegradable polymer (PLA) by dynamic mechanical analysis (DMA) to evaluate the mechanical properties as a function of temperature. We also introduced the Fused deposition Modeling technology (FDM) and critical manufacturing parameters that we specifically identified in this study.

The variation of the parameters has a quantified and significant impact on the mechanical properties. The thickness of the layer, the temperature of the fusion, manufacturing speed and filling orientation are evaluated using the Taguchi statistical method, and that the thickness of 0.2 mm and the orientation 0/90° are the most influential on the modulus of elasticity and the mechanical strength.

Acknowledgements. The authors gratefully acknowledge the company 3D form which helps to improve the work.

References

1. Ahn SH, Montero M, Odell D, Roundy S, Wright PK (2002) Anisotropic material properties of fused deposition modeling ABS. Rapid Prototyp J 8(4):248–257
2. Casavola C, Cazzato A, Moramarco V, Pappalettere C (2016) Orthotropic mechanical properties of fused deposition modelling parts described by classical laminate theory. Mater Des 90:453–458
3. Fatyeyeva K, Chappey C, Marais S (2017) Biopolymer/clay nanocomposites as the high barrier packaging material: recent advances. In: Food packaging, pp 425–463
4. Le Duigou A, Castro M, Bevan R, Martin N (2016) 3D printing of wood fibre biocomposites: from mechanical to actuation functionality. Mater Des 96:106–114
5. Minetola P, Iuliano L, Marchiandi G (2016) Benchmarking of FDM machines through part quality using IT grades. Proc CIRP 41:1027–1032
6. Mohamed OA, Masood SH, Bhowmik JL (2016) Mathematical modeling and FDM process parameters optimization using response surface methodology based on Q-optimal design. Appl Math Modell 40(23–24):10052–10073
7. Narayan R (ed) (2014) Rapid prototyping of biomaterials: principles and applications. Woodhead Publishing
8. Naserddine BA, Khlif M, Dorra H, Chedly B (2019) Mechanical and morphological characterization of spherical cell porous structures manufactured using FDM process. Eng Fract Mech, p 106527
9. Novakova-Marcincinova L, Kuric I (2012) Basic and advanced materials for fused deposition modeling rapid prototyping technology. Manuf Ind Eng 11(1):24–27
10. Rayna T, Striukova L (2016) From rapid prototyping to home fabrication: How 3D printing is changing business model innovation. Technol Forecast Soc Change 102:214–224
11. Tofail SA, Koumoulos EP, Bandyopadhyay A, Bose S, O'Donoghue L, Charitidis C (2018) Additive manufacturing: scientific and technological challenges, market uptake and opportunities. Mater Today 21(1):22–37
12. Tymrak BM, Kreiger M, Pearce JM (2014) Mechanical properties of components fabricated with open-source 3-D printers under realistic environmental conditions. Mater Des 58: 242–246
13. Zierdt P, Theumer T, Kulkarni G, Däumlich V, Klehm J, Hirsch U, Weber A (2015) Sustainable wood-plastic composites from bio-based polyamide 11 and chemically modified beech fibers. Sustain Mater Technol 6:6–14

Investigation of Delamination Factor in High Speed Milling on Carbon Fiber Reinforced Plastics

Adel Abidi[1(✉)] and Sahbi Ben Salem[1,2]

[1] Lab MAI, National School of Engineers of Tunis, Tunis el Manar University, Tunis, Tunisia
adel.abidi8@gmail.com
[2] Preparatory Institute of Engineering Mrezka Nabeul (IPEIN), Nabeul, Tunisia

Abstract. High speed machining is a suitable process to increase the productivity and reduce the cost during manufacturing. Delamination factor (DF) is the main inconvenient while composites drilling. In this work the high speed milling used for drilling carbon fiber reinforced plastic composite (CFRP) with different parameters such as hole size, cutting speed and feed speed. The Response surface methodology (RSM) was used with the Central composite design (CCD) to predict DF in function of different combination of machining factors. The analysis of variance (ANOVA) used for evaluation the effect of the studied factors on the composite. Experimental results showed that the DF at the inter hole is greater than the DF at the exit hole which have a significant effect on the hole size.

Keywords: CFRP · Milling · ANOVA · Delamination · RSM

1 Introduction

Drilling process of Carbon fiber reinforced plastics (CFRP) composites is one of the most challenges for the aerospace industries because of their anisotropic behavior [1] and heterogeneity. Many kinds of damages are happened during this process including delamination; matrix cracking, burring and fiber pull out [2–4]. Delamination is considered as the most severe kind of damage [5] which conducts to significant reduction in the material strength like the degrading of its long term performance.

Several studies [6, 7] have demonstrated that delamination damage correlates closely with cutting forces which are influenced by the cutting conditions.

Non-conventional machining hole processes are being very used to manufacture holes in CFRP composites thanks to its several advantages including low machining cost and high cutting speed leading to low cutting forces, and good machining quality. Among many non-conventional hole manufacturing processes, helical milling is considered as the best alternative choice to conventional drilling [8]. Helical milling is an orbital drilling that it is referred to helical path of the milling tool during its rotation around its axis. Including the advantages of this cutting process is the lower thermal damage and reduction in burr volume as leads to minimize the cutting forces and the

N. Aifaoui et al. (Eds.): CMSM 2019, LNME, pp. 363–374, 2020.
https://doi.org/10.1007/978-3-030-27146-6_39

axial forces [9] to improve the hole quality and the geometrical accuracy [10]. Consequently, helical milling process gives lower delamination and damage in machining of CFRP [11, 12]. According to Geier and Szalay [13], they found during the orbital milling process, that the delamination factor increases slowly while the increasing of cutting speed.

2 Experimental Set Up

2.1 Materials and Methodology

The composite material made by carbon fiber reinforced epoxy resin. It is produced by hand layup process with a woven fabric CFRP laminates which made up of 8 layers, composed of 4 layers oriented as 0/90° and 4 layers with ±45° fiber orientation, with a total thickness of 4.5 mm, as shown in Fig. 1. Coupons were cut in 500 × 500 mm have been used to carry out the experimental tests (Table 1).

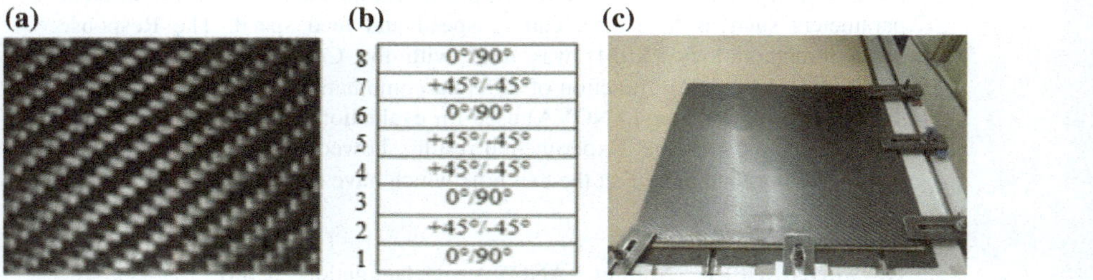

Fig. 1 **a** Carbon Twill weave12K-600 g/m^2, **b** CFRP laminate layup **c** prepared laminate

Table 1 CFRP sheets characteristics

Material	Matrix	Fiber	Dimensions	Thickness	Stacking
CFRP	Epoxy	Carbon fiber	500 × 500 mm	4.5 mm	$[0/90, \pm45, 0/90, \pm45]_S$

The drilling experiments were conducted on a MÉCANUMERIC three-axis CNC high speed cutting machine with a maximum spindle speed of 30.000 rpm. The machine was equipped with a vacuum cleaner for removing material chips while milling operation from the workspace. The work piece was putted on machine table and fixed by individual clamps. Each clamp is in adequate distance to each other to avoid bending the CFRP panel (Fig. 2). The tests were performed without coolant to prevent the laminate contamination with the cutting fluid. A solid uncoated carbide end mill was used with only one cutting edge (one-flute) and its helix angle is 25° with a diameter of 4 mm.

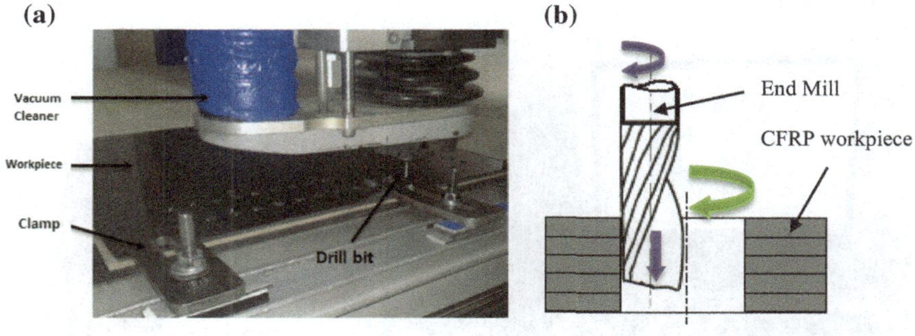

Fig. 2 Experimental setup **a** CNC machine; **b** Tool motion in orbital drilling

2.2 Design of Machining Experiments

The Design of Experiments (DOE) is an analysis method used for the design of the machining experiments plan and modeling and analyzing the effect of control factors on the studied response in the other hand. Furthermore Design of experiments is handled to minimize the experimentation cost in number of tests and time. Among the different tools of DOE, Central composite design (CCD) is the best selected tool in this work according to types and number of factors, their levels and interactions. Then an experimental plan generated by this method provided 15 runs of experiments which consist of the varied combinations of three levels of input cutting parameters which are hole diameter (D), feed speed (F) and spindle speed (N).

In this work, the Design-Expert_11 software was used for the design, modeling, analysis and optimization of the process variables. For obtaining the optimum delamination factor at entry and exit holes, the optimal machining levels of input parameters of the high speed milling process can be found with the use of Response Surface Methodology (RSM). The studied factors are selected based on the literature results. These factors are presented in table below with their different levels (Table 2).

Table 2 Cutting parameters

Factors			Levels		
Code	Designation	Symbol	Low	Medium	High
A	Hole diameter	D (mm)	8	10	12
B	Feed speed	F (mm/min)	360	600	840
C	Spindle speed	N (rpm)	6000	10000	14000

The delamination factors were measured by a Stereo Optical Microscopy M80 (SOM) with HD digital Microscope Camera Leica MC170 HD with interface that provides a full high definition live and a standard capture resolution of 5Mpixels, and were analyzed by Leica software (Fig. 3). The obtained values are resumed in the next Table 3.

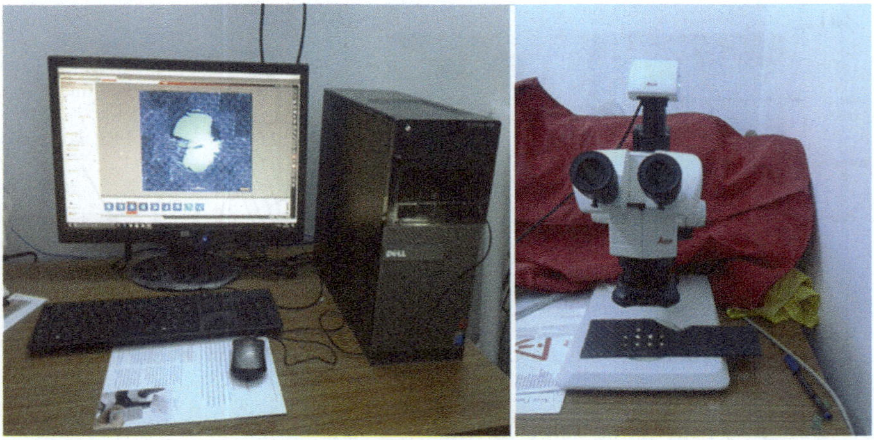

Fig. 3 Instrument used in the experimental study: Microscope Leica M80

Table 3 Design of experiments

Std	Run	D (mm)	F (mm/min)	N (rpm)	DF enter	DF exit
1	3	8	360	6000	1.06558	1.044583
2	4	12	360	6000	1.02053	1.00911
3	12	8	840	6000	1.028625	1.01379
4	5	12	840	6000	1.014444	1.004742
5	7	8	360	14000	1.08046	1.019042
6	1	12	360	14000	1.028333	1.01137
7	14	8	840	14000	1.048958	1.01696
8	9	12	840	14000	1.024306	1.026167
9	2	8	600	10000	1.04887	1.024667
10	10	12	600	10000	1.016	1.01147
11	15	10	360	10000	1.039733	1.021033
12	8	10	840	10000	1.02454	1.01473
13	6	10	600	6000	1.0316	1.020633
14	11	10	600	14000	1.047667	1.015067
15	13	10	600	10000	1.030333	1.01863

3 Results and Discussions

Delamination can be defined by calculating the delamination factor (DF) is defined as the quotient between the maximum diameter of damage (Dmax) and the normal diameter (Dnom). The formula used to calculate the delamination factor [14] is shown in the following Eq. (1):

$$DF = D_{max}/D_{nom} \quad (1)$$

where
D_{max} = maximum diameter of damage (mm)
D_{nom} = nominal diameter (mm).

3.1 Entry Delamination

The response surface methodology (RSM) was used to estimate the delamination factor at the entry hole for the milling process of CFRP composite. According to the model summary statistics results (Table 4), the quadratic model is the most adequate model for modeling of DF. This choice is explained by the high level of determination coefficient ($R^2 = 0.996$). The Predicted R^2 (0.9688) is in high agreement level with the Adjusted R^2 (0.9903) that the difference is less than (0, 2). Comparing to other types of models indicates the best fit for the model. The accuracy of the generated model was identified by analysis of variance (ANOVA) technique giving the following results (Table 5).

Table 4 Model summary statistics of Entry delamination factor (DF)

Source	Std. Dev.	R^2	Adjusted R^2	Predicted R^2	PRESS
Linear	0.007739	0.864595	0.827666	0.695969	0.001479
2FI	0.004844	0.961411	0.932468	0.824654	0.000853
Quadratic	**0.00184**	**0.99652**	**0.990256**	**0.968822**	**0.000152**
Cubic	0.001954	0.999215	0.989016	−1.36658	0.011514

Table 5 ANOVA for response surface quadratic model of entry DF

Analysis of variance table [Partial sum of squares—Type III]						
Source	Sum of	df	Mean	F-value	p-value	
Model	**0.004848**	**9**	**0.000539**	**159.0793**	**1.31926E-05**	**Significant**
A-D	0.002852	1	0.002852	842.1992	9.10453E-07	
B-F	0.000879	1	0.000879	259.6118	1.67778E-05	
C-N	0.000475	1	0.000475	140.3644	7.54344E-05	
AB	0.000425	1	0.000425	125.6466	9.8648E-05	
AC	3.85E-05	1	3.85E-05	11.36757	0.019855419	
BC	7.05E-06	1	7.05E-06	2.082461	0.208590731	
A^2	4.58E-09	1	4.58E-09	0.001354	0.972074317	
B^2	1.69E-07	1	1.69E-07	0.049807	0.832230715	
C^2	0.000135	1	0.000135	39.80858	0.001472463	
Residual	1.69E-05	5	3.39E-06			
Cor total	0.004865	14				

From Table 5, it can be observed that the model entry delamination factor is significant, the P-values for model are less than 5% which means that model terms are significant and the F-value of 159.08 explains the model significance with a confidence more than 95% (p-value = $1.32 \cdot 10^{-5}$). In this case A, B, C, AB, AC, C^2 are significant model terms. It shows also that the interaction terms between the spindle speed and feed speed is not significant (BC: F-Value: 2.082461, P-Value: 0.2086) than the other interaction terms.

The developed model analysis of the entry DF was performed by ANOVA. At first step to compare the use of the ANOVA analysis, it is necessary to evaluate the correlation between the measured and the predicted values and the range of residual between them. To check the accuracy of the model, Fig. 4 shows the normal data distribution. In normal probability plot all points are near to a straight line with a minimum of residual therefor it shows the good correlation that exists between the developed model and the experimental data. Consequently, the relation between the predicted and the experimental values is presented in Fig. 5 which shows a very good correlation.

To check more the suitability and the model correlation with the experimental results, a comparison between measured and predicted values in order of experiment runs are presented in Fig. 6.

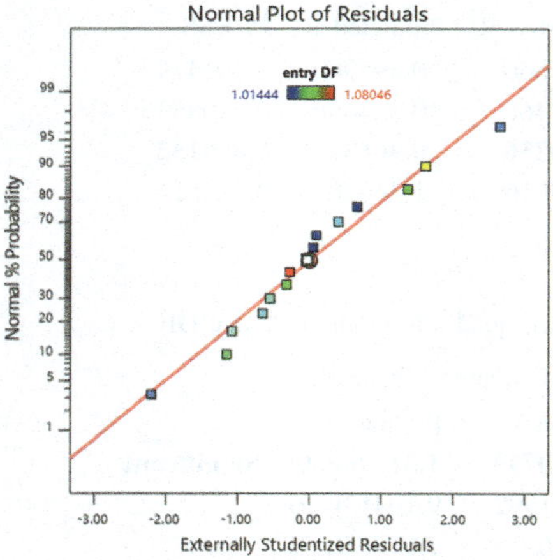

Fig. 4 Normal residuals plot

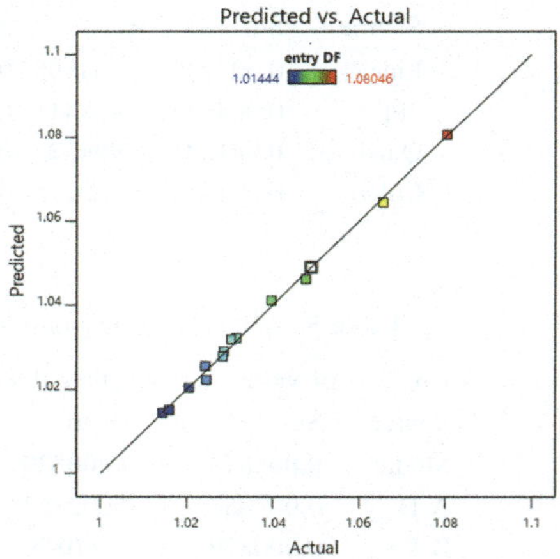

Fig. 5 Predicted versus actual plot of Ra

Figure 7 is displayed the perturbation plot which helps to compare the variation of all the factors at a particular point. The factors C show curvature which indicates that the response is not sensitive to spindle speed. However it decreased while the increasing for both hole size and feed speed (A and B).

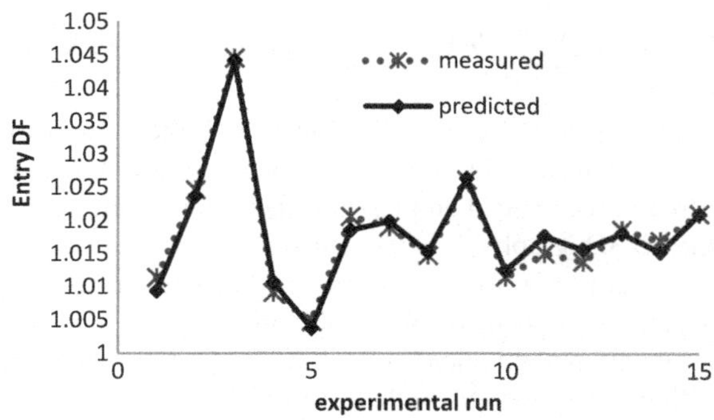

Fig. 6 Comparison between measured and predicted values for *Ra*

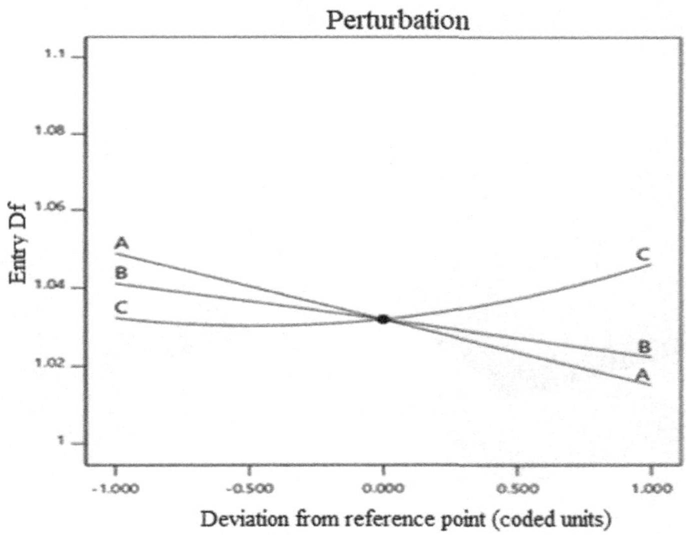

Fig. 7 Perturbation curves of input factors

The response design was generated with RSM. It is a mathematical and statistical technique, which help us to derive an adequate relationship between the desired response and input variables.

Now representing the entry DF is a function of different variables: Hole diameter 'D', tool feed speed 'F' and rotation speed 'N'. The quadratic model used to calculate the variables impact is presented in Eq. (2). Where Y is the predicted response, b_0 is the constant term and the coefficient of linear, square, interaction factors are respectively b_i, b_{ii} and b_{ij} and ε is the random error.

$$\mathbf{Y} = \mathbf{b_0} + \sum\nolimits_{i=1}^{K} b_i X_i + \sum\nolimits_{i,j}^{k} b_{ij} X_i X_j + \sum\nolimits_{i}^{k} b_{ii} X_{ii}^2 + \varepsilon \tag{2}$$

The delamination of drilled holes is developed with the help of the RSM using the CCD model to develop various 3D response surfaces. Figure 8 displays the interaction effects of hole diameter, feed speed and spindle speed on delamination factor. As can it be observed from the 3D graphs that the entry DF increased as the increase of the spindle speed with a nonlinear function. Otherwise it is inversely proportional both to the hole diameter and feed speed for all combinations.

The RSM based on mathematical model developed for predicting delamination during orbital milling, is expressed with the final Eq. (3) in terms of actual factors below:

$$\begin{aligned}\text{Entry DF} = {} & +1.237 - 0.01503 * \text{D} - 0.0002 * \text{F} - 5.2 \cdot 10^{-06} * \text{N}\, 1.52 \cdot 10^{-05} \text{D} * \text{F} - 2.7 \cdot 10^{-07} * \text{D} * \text{N} \\ & + 9.78 \cdot 10^{-10} * \text{F} * \text{N} + 1.06 \cdot 10^{-05} * \text{D}^2 - 4.4 \cdot 10^{-09} * \text{F}^2 + 4.53 \cdot 10^{-10} * \text{N}^2\end{aligned} \tag{3}$$

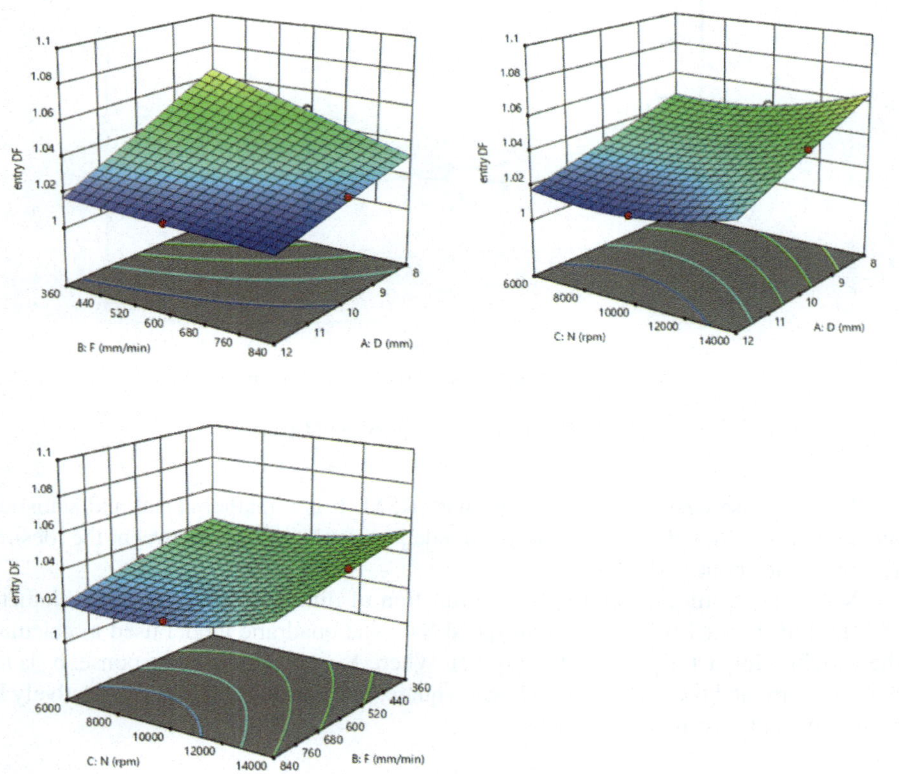

Fig. 8 3D surface response for entry DF

3.2 Exit Delamination

According to the model summary statistics results (Table 6) of exit delamination factor (exit DF), we can observe that the determination coefficient (R^2 = 0.9765) and the Predicted R-squared (0.80179) is in good coincidence with the Adjusted R^2 (0.95896) and with a difference less than (0.15). From these data and comparing to other types of models indicates that the two factors interactions (2FI) model is the most suitable model for modeling exit DF. Then the ANOVA confirmed the significance of this model with the following results (Table 7).

Table 6 Model Summary Statistics

Source	Std.	R^2	Adjusted	Predicted	PRESS	
Linear	0.008605	0.329405	0.146515	−0.65988	0.002016	
2FI	**0.001887**	**0.976547**	**0.958957**	**0.801791**	**0.000241**	**Suggested**
Quadratic	0.002383	0.97663	0.934564	0.649513	0.000426	
Cubic	0.000709	0.999587	0.994214	−0.24663	0.001514	Aliased

Table 7 ANOVA for 2FI model (exit DF)

Source	Sum of	df	Mean	F-value	p-value	
Model	**0.001186**	**6**	**0.000198**	**55.51751**	**4.37E-06**	**Significant**
A-D	0.000316	1	0.000316	88.64225	1.33E-05	
B-F	8.27E-05	1	8.27E-05	23.21141	0.001325	
C-N	1.81E-06	1	1.81E-06	0.508024	0.496256	
AB	0.000234	1	0.000234	65.82312	3.95E-05	
AC	0.000265	1	0.000265	74.45962	2.52E-05	
BC	0.000287	1	0.000287	80.46066	1.9E-05	
Residual	2.85E-05	8	3.56E-06			
Cor total	0.001215	14				

For the ANOVA of exit DF, the model terms of hole size and feed speed are significant are significant that is their p-value less than 0.05. So the model is very sensitive to these factors more than the spindle speed. Furthermore all the interaction between the three factors is significant.

Figures 9, 10, 11 show the accuracy of ANOVA model. For normal probability distribution all points are near to a straight line with a minimum of residuals (Fig. 10) therefore it shows the good correlation that exists between the developed model and the experimental data. Consequently, the relation between the predicted and the experimental values (Fig. 11) confirm the good coincidence between predicted and actual data.

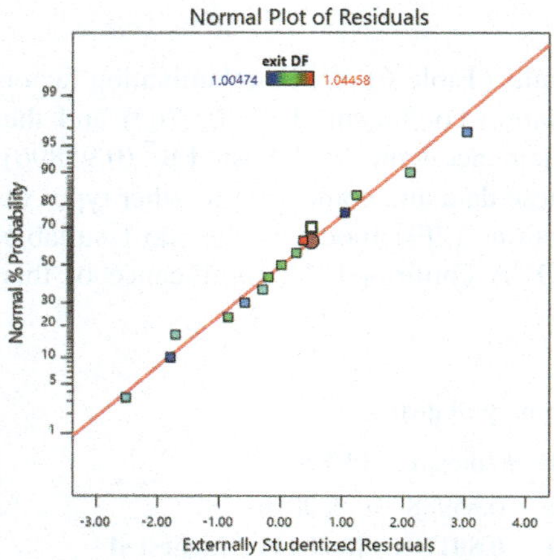

Fig. 9 Normal residuals Plot

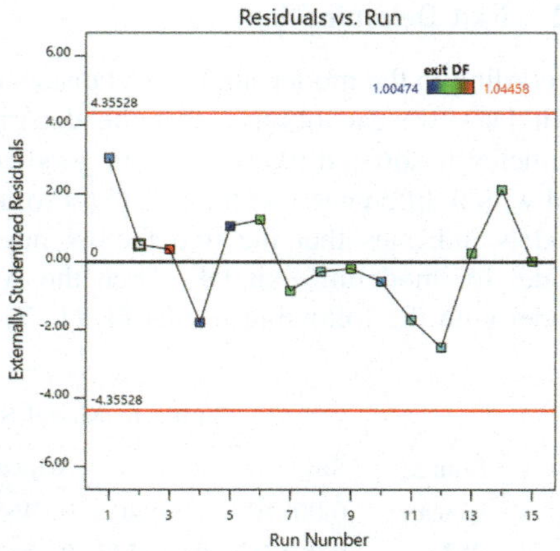

Fig. 10 Residual versus experimental run

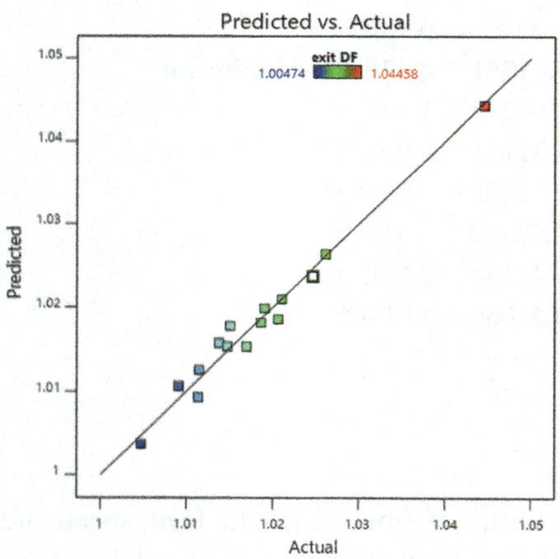

Fig. 11 Predicted versus actual plot of exit delamination

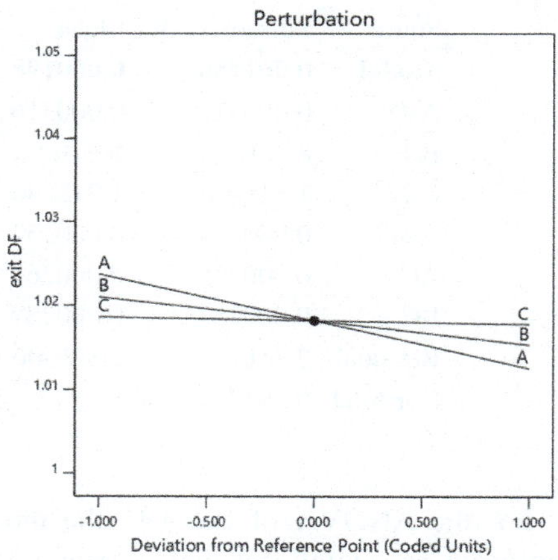

Fig. 12 Perturbation curves of input factors

The Fig. 12 showed the tendency of variation of the input factors which are represented with a linear function; in one hand the delamination factor at exit hole decreased as the increase of hole size and the feed speed levels as observed that it is more sensitive for the hole size. In the other hand the variation of spindle speed effect is negligible which is constant for the different levels of spindle speed that it was not the same case for the entry DF (Fig. 13).

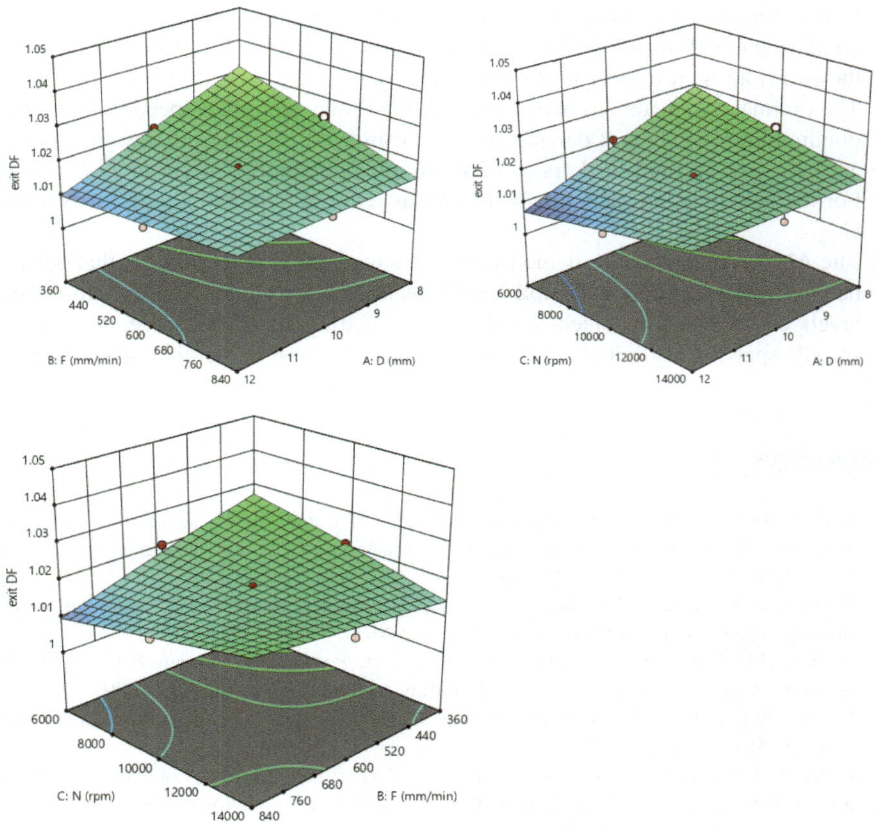

Fig. 13 3D surface response for exit DF

As can it be observed from the 3D graphs that the exit DF increased as the increase of the spindle speed with a nonlinear function. Otherwise it is inversely proportional both to the hole diameter and feed speed for all combinations. The exit DF can be estimated mathematically using ANOVA Eq. (4) in terms of actual factors as below:

$$\begin{aligned}\text{Exit DF} = & +1.2315 - 0.01677\text{D} - 0.00019\text{F} - 1.1 \cdot 10^{-05}\text{N} + 1.13 \cdot 10^{-05}\text{DF} \\ & + 7.2 \cdot 10^{-07}\text{DN} + 6.23 \cdot 10^{-09}\text{F} * \text{N}\end{aligned} \quad (4)$$

4 Conclusion

In the present study, machining experiments were carried out in CFRP laminate using an uncoated mono-flute end mill in order to analyze and optimize cutting conditions. According to this study, the following conclusions can be retained:

- It was found, with using an uncoated carbide end mill with mono flute and a composite CFRP with a twill fabric plies, that delamination factor has a less range at the exit hole with respect to the entry DF.
- The optimal parameters to obtain good quality holes with high speed milling and to obtain minimum level of delamination are estimated according to design expert_11 are hole size 12 mm, feed speed 828 mm/min and spindle speed about 6200 rpm.
- For the delamination range for both cases is reduced comparing with many studies [13].
- The ANOVA results have demonstrated, at a confidence level of 95%, that hole size has the greatest influential impact delamination factor which is minimum for high levels of hole size and feed speed and low cutting speed. However the effect of spindle speed is negligible at exit hole and it is the most influential on the entry DF.

References

1. Teti R (2002) Machining of composite materials. CIRP Ann Manuf Technol 51(1):611–634
2. Shin YC, Dandekar CR (2012) Modeling of machining of composite materials: a review. Int J Mach Tools Manuf 57:102–121
3. Durão LM, De Moura MF, Marques AT (2008) Numerical prediction of delamination onset in carbon/epoxy composites drilling. Eng Fract Mech 75:767–2778
4. El-Hofy MH, Soo SL, Aspinwall DK, Sim WM, Pearson D, Harden P (2011) Factors affecting workpiece surface integrity in slotting of CFRP. Proc Eng 19:94–99
5. Liu D, Tang Y, Cong WL (2012) A review of mechanical drilling for composite laminates. Compos Struct 94:1265–1279
6. Piquet R, Ferret B, Lachaud F, Swider P (2000) Experimental analysis of drilling damage in thin carbon/epoxy plate using special drills. Compos Part A Appl Sci Manuf 31(10):1107–1115
7. Hocheng H, Tsao CC (2005) The path towards delamination-free drilling of composite materials. J Mater Process Technol 167:251–264
8. Pereira RB, Brandão LC, de Paiva AP, Ferreira JR, Davim JP (2017) A review of helical milling process. Int J Mach Tools Manuf 120:27–48
9. Wang GD, Melly SK, Li N (2018) Experimental studies on a two-step technique to reduce delamination damage during milling of large diameter holes in CFRP/Al stack. Compos Struct
10. Meshreki M, Attia MH Sadek A (2012) Characterization and optimization of orbital drilling of woven carbon fiber reinforced epoxy. CIRP Ann Manuf Technol 61(1):123–126
11. Sadek A, Meshreki M, Attia MH (2012) Characterization and optimization of orbital drilling of woven carbon fiber reinforced epoxy laminates. CIRP Ann Manuf Technol 61:123–126. https://doi.org/10.1016/j.cirp.2012.03.089
12. Voss R, Henerichs M, Kuster F (2016) Comparison of conventional drilling and orbital drilling in machining carbon fibre reinforced plastics (CFRP). CIRP Ann Manuf Technol 65:137–140. https://doi.org/10.1016/j.cirp.2016.04.001
13. Geier N, Szalay T (2017) Optimisation of process parameters for the orbital and conventional drilling of uni-directional carbon fibre-reinforced polymers (UD-CFRP). Measurement. https://doi.org/10.1016/j.measurement.2017.07.007
14. Grilo TJ et al (2013) Experimental delamination analyses of CFRPs using different drill geometries. Compos Part B Eng 45(1):1344–1350

The Effect of High-Speed Milling on Surface Roughness of 42CrMo4 Hardened Steel Using a Ball Nose End-Mill Cutter

Sai Lotfi[1(✉)], Belguith Rami[1,2], Baili Maher[2], Dessein Gilles[2], and Bouzid Wassila[1]

[1] Unité de Génie de Production Mécanique et Matériaux, ENIS, Route Soukra Km 3,5, B.P. 1173-3038 Sfax, Tunisie
lotfi_sai@yahoo.com

[2] Laboratoire Génie de Production, Université de Toulouse; INPT/ENIT, 47 Avenue D'Azereix, BP 1629 Tarbes Cedex, France

Abstract. The surface roughness is a decisive criterion for the quality of the machined surface. Many researchers are interested to study the effects of the machining parameters on the surface quality as: the cutting conditions, the machining strategies, the tool geometries and the machining errors. All these studies are developed for the stationary feed rate and neglected the cinematic effects caused by machine deceleration and acceleration when the tool trajectories change a direction. The objective of this research was to investigate the effect of the velocity changes on the surface roughness. A set of machining tests in high-speed end-milling of the 42CrMo4 material by a ball nose end-mill is made. For the same cutting conditions, the roughness is measured on three zones respectively the acceleration, the stationary and the deceleration zone. It was seen that the cinematic change causes a poor surface roughness.

Keywords: Roughness · High-speed milling · 42CrMo4 · Acceleration and deceleration

1 Introduction

High-speed ball end milling is used in mold industries where complex parts of 42CrMo4 hardened steel are machined. The surface roughness is a determinant criterion, its interested by some researchers who studied using a different method and different cutting parameters optimization.

Bouzid et al. [2] studied the influences of the cutting speed, the feed rate and the axial depth of cut on the surface roughness, the Taguchi method is used to determinate the best combination corresponding to the optimum value of the roughness R_a and R_t. The same method was used by [1, 10], they analyzed the effects of the cutting speed, the feed rate and the radial depth of cut on the roughness S_a. The surface roughness is optimized using the nose radius, depth of cut, feed rate and cutting speed. The material studied was the Inconel 718 milled by a cemented carbide tool.

Seculik et al. [3] optimized the roughness using a genetic algorithm method (GA). The same method was used by Jatti et al. [7] to develop a surface roughness model of

N. Aifaoui et al. (Eds.): CMSM 2019, LNME, pp. 375–381, 2020.
https://doi.org/10.1007/978-3-030-27146-6_40

the Aluminum alloys LM6, varying the cutting speed, the feed and the depth of cut. Wojciechowski et al. [4] analyzed the surface roughness parallel to the feed direction of hardness steel after ball end-milling with various surface inclination angle and feed per tooth value. Hossein and Nafis [5] studied the surface roughness as a function of cutter axis Inclination angle, tool diameter, spindle speed, feed rate and depth of cut. The developed model uses response surface methodology (RSM) based on the artificial intelligence method. Dikshit et al. [6] used the same methodology in the case of high-speed ball-end milling, the Mathematical model for surface roughness has been developed in terms of cutting speed, feed per tooth, axial depth of cut and radial depth of cut. An experimental work was conducted by Kurt et al. [8] to study the effects of the tool path strategies and the tool diameter on the surface roughness and the surface errors in the case of free-form surfaces. Quintana et al. [9] studied the influence of the geometric characteristics of ball end mill cut on the surface roughness as function of cutting tool radius and radial depth of cut.

Corner motion is a common and important case in contouring applications. a sharp corner is formed by two consecutive contours with discontinues in their first derivatives [11]. In general, these two linear contours are fed into the Acceleration and deceleration processor one after the other. Two modes are used in contouring applications the first is the exact stop mode which the machine stop in the end of each block, the velocity reached zero and after that start the next block and the second is the continuous mode which the machine starts to execute the second command before the first one is completed executed. The second mode uses a look-ahead algorithm; this method was developed in many researches in terms of optimizing the machining time not a surface roughness.

However, almost all of above methods are limited to constant feed rate but the dynamic behavior of the machine and the acceleration and deceleration process in the end of blocks or in the changing direction of the cut influences the feed value.

In this paper, an experimental study was conducted to analyses the surface profiles. We study the variation of the roughness profiles along one normal-block which started by a zero velocity, accelerate to reach the stationary feed rate, decelerate at the end of the block to reach a zero velocity. The first region is the acceleration zone, the second is called the stationary zone and in the end of the third is the deceleration zone.

2 Experimental Results

The workpiece material used in the experiments was a 42CrMo4 steel; the top surface is a circular part with diameter 100 mm. It is divided on three areas as shown in Fig. 1, two lateral areas for the fixation on the Kistler table dynamometer and the middle surface with size 82 mm × 42 mm is prepared for the cutting tests. The tool path is the "one-way" in feed direction started from the acceleration zone and finished at the deceleration zone.

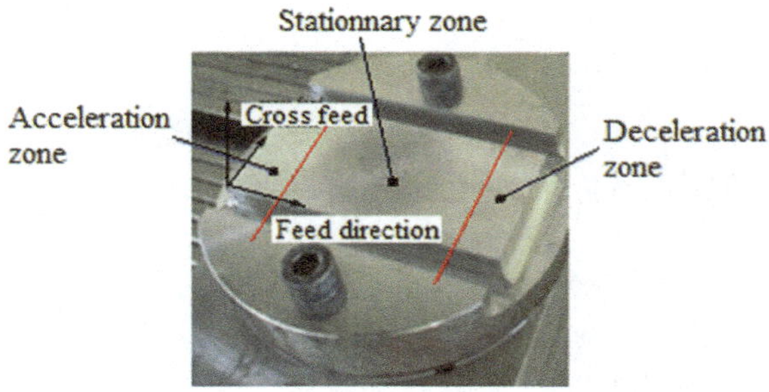

Fig. 1 Prepared workpiece and the 3 zones of studies

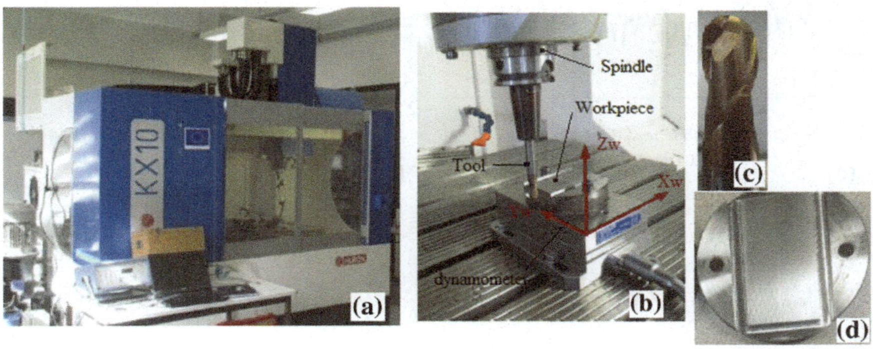

Fig. 2 **a** Milling machine, **b** experimental setup, **c** ball end mill cutter, **d** image of machined surface

The cutting tool used is a ball end mill cutter coated with TiSiN with a diameter $d = 10$ mm, two teeth $N_f = 2$, the helix angle was $i_0 = 30°$, the total length was $L = 100$ mm and the active length was 15 mm as shown in Fig. 2c. The cutter was traveled along a linear tool path. All cutting experiments were conducted in one way milling. The tests were conducted without cutting fluid. The machine used is a CNC HS Milling HURON KX10, Fig. 2a. The maximum acceleration is $A = 3$ m/s^2 and a maximum jerk $J = 50$ m/s^3. Kistler three components dynamometer model 9257B was used for force data acquisition. The surface topography in 3D image is measured by an infinite focus Alicona machine with an extension multiplied by ×20. The tool radial run-out e is measured directly using a dial indicator Holex on the cylindrical part of the tool. The eccentricity ρ location angle was measured from the angular position of the maximum value according to the position of the cutting edge tangent. The modal parameters (M, C, K) are measured using an impact test with the Hammer Meggitt Endevco.

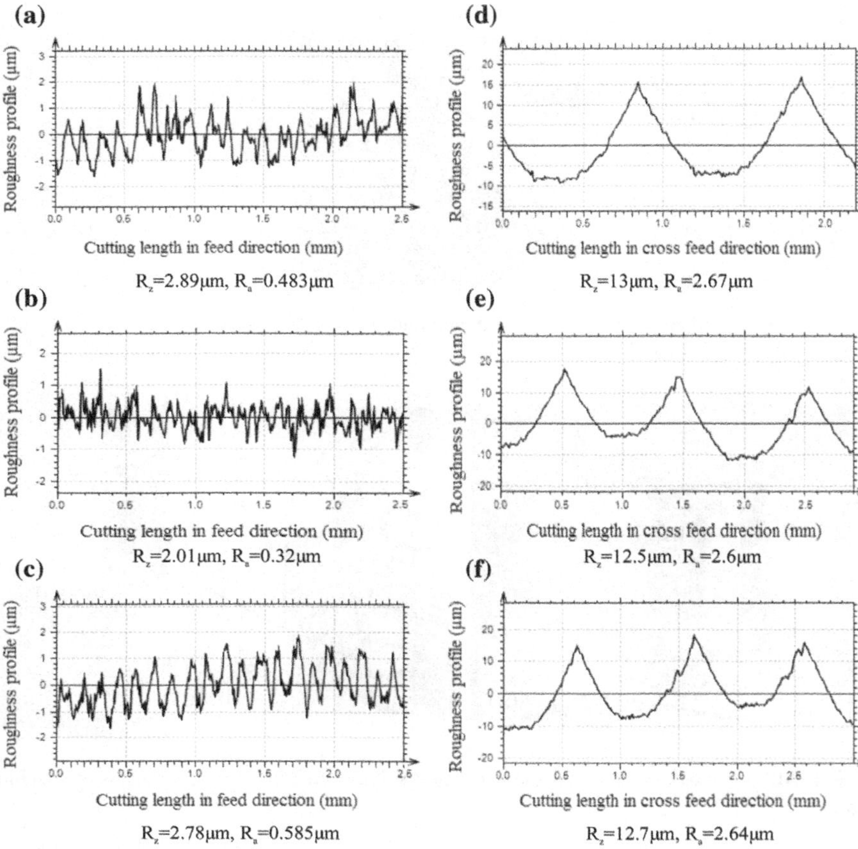

Fig. 3 Experimental measured roughness profiles in feed and cross feed directions: **a–d**: acceleration zone, **b–e**: stationary zone, **c–f**: deceleration zone. N = 15000 rev/min, V_c = 471 m/min, f_z = 0.13 mm/teeth/rev, a_e = 1 mm, e = 6.52 µm and ρ = 53.48°

The roughness profiles are extracted from a 2D topography image using MontainMap softwear. For the roughness in the feed direction, the profiles are measured in the same line traversing the three zones presented in Fig. 1 and located near the tool tip point. For the cross feed direction the profiles are measured at the extreme right and the extreme left respectively for the acceleration and deceleration zones and in the middle of the stationary zone.

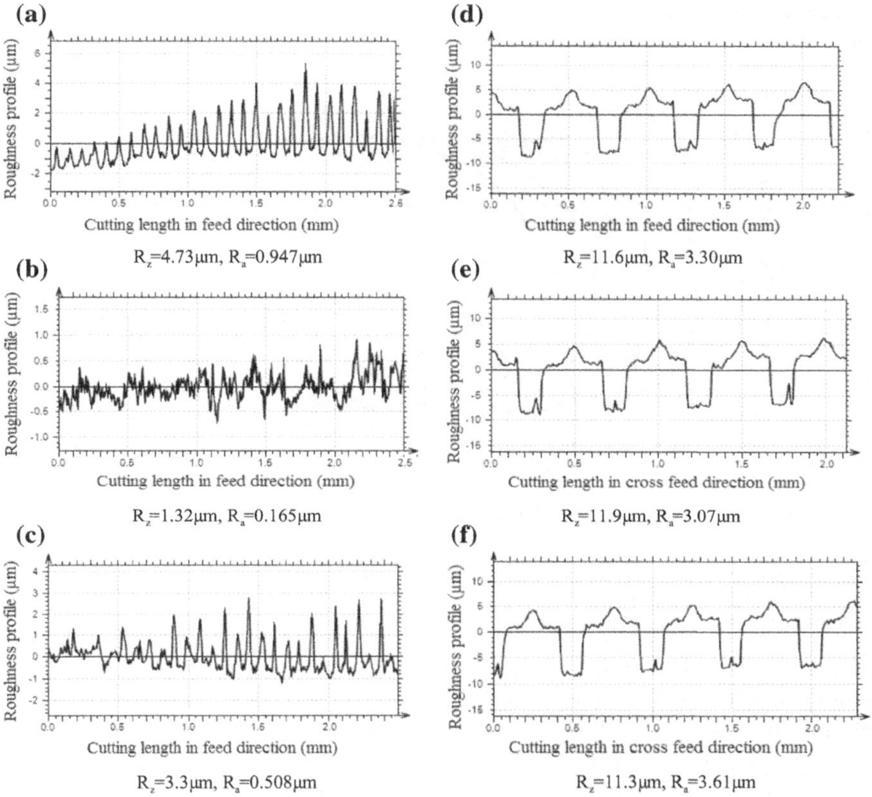

Fig. 4 Experimental measured roughness profile in feed and cross feed directions **a–d**: acceleration zone, **b–e**: stationary zone, **c–f**: deceleration zone. $N = 22000$ rev/min, $V_c = 690.8$ m/min, $f_z = 0.088$ mm/teeth/rev, $a_e = 0.5$ mm, $e = 5.55$ µm and $\rho = 106.26°$

3 Interpretations

The Figs. 3 and 4 show the results of the roughness measured on the acceleration, the stationary and the deceleration zones of the two cutting tests in high speed milling for different parameters ($V_c = 471$ m/min, $f_z = 0.13$ mm/teeth/rev, $a_e = 1$ mm, $e = 6.52$ µm and $\rho = 53.48°$) for the first test and ($V_c = 690.8$ m/min, $f_z = 0.088$ mm/teeth/rev, $a_e = 0.5$ mm, $e = 5.55$ µm and $\rho = 106.26°$) for the second. They prove that the roughness is higher in the acceleration and deceleration zones in both measuring directions.

Based on the machining errors caused by the cutting force, as the vibration and the deflection and the geometrical tool setting errors as the radial runout the tool teeth rotate with an eccentricity around the axis of the spindle. These defects are modeled by an equivalent radius different to the normal effective rotational radius of each tooth around the axis of the spindle. The simulation of the equivalent radius of each tooth using the experimental results is shown in Fig. 5.

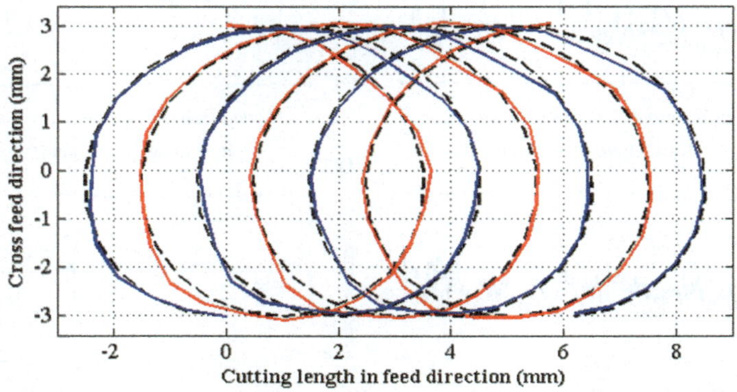

Fig. 5 Effects of the machining errors on the trajectories of the tool teeth

The effects of the dynamic of the machine on the surface roughness profiles in both directions are explicated as follows:

Roughness profiles in feed direction
The roughness R_t and R_a increase in the acceleration and deceleration zones. It can be seen from the Figs. 3 and 4 that the roughness profiles are rougher respectively in the Fig. 3a and c, in Fig. 4a and c than the profiles respectively in Figs. 3b and 4b. theoretically the feed-peak is proportional to the feed per tooth. In the deceleration and acceleration zones the feed-peak height decrease as the feed per tooth decreases for a constant spindle speed. Contrary in the stationary zone the feed-peak height is more important. But it can be seen from the experimental results that the machining errors as the deflection, the vibrations and the runout deteriorate the feed-pick cusp and ameliorate the surface roughness in feed direction.

Roughness profiles in cross-feed direction
In the stationary zone, the cross-feed peak is proportional to the radial depth of cut. The feed rate is at its maximum, the dynamic of the machine and the high-cutting forces cause a tool deflection and vibration, in addition to the runout, these machining errors increase the equivalent radius of the tooth around the spindle axis and decrease the cross-feed-pick value. The roughness R_t and R_a in this second zone as shown in Figs. 3e and 4e are smaller than the roughness in the first and third zones as shown in Figs. 3d and 4d and in Figs. 3e and 4e.

4 Conclusion

In this paper we study the influences of the dynamic of the machine caused by the high speed on the roughness profiles. It can conclude that the acceleration and the deceleration on the changing direction zone increase the roughness R_t and R_a compared to the stationary zone in the both feed and cross feed directions. This experimental study must be concluded from the 3D surface topography study on further works.

Acknowledgements. The work is carried out thanks to the support and funding allocated to the Unit of Mechanical and Materials Production Engineering (UGPM2/UR17ES43) by the Tunisian Ministry of Higher Education and Scientific Research.

References

1. Mersni W, boujelbene M, ben Salem S, Alghamdi A-S (2018) Optimization of the surface roughness in ball end milling of titanium alloy Ti–6Al–4 V using the Taguchi method. Proc Manuf 20: 271–276
2. Bouzid W, Zghal A, Sai L (2004) Taguchi method for design optimization of milled surface roughness. Mater Technol 19–3:159–162
3. Sekulic Pejic M, Brezocnik V, Gostimirović M, Hadzistevic M (2018) Prediction of surface roughness in the ball-end milling process using response surface methodology, genetic algorithms, and grey wolf optimizer algorithms. Adv Prod Eng Manag 13–1:18–30
4. Wojciechowski S, Twardowski P, Chwalczuk T (2013) Surface roughness of hardened steel after ball end milling with various surface inclinations. Arch Mech Technol Autom 33–1:55–64
5. Hossain MSJ, Nafis A (2012) Surface roughness prediction modeling for AISI 4340 after ball end mill operation using artificial intelligence. I J Sci Eng Res 3–5:1–10
6. Dikshit MK, Puri AB, Maity A (2016) Optimization of surface roughness in ball-end milling using teaching-learning-based optimization and response surface methodology. J Eng Manuf 1–12
7. Jatti VS, Sekhar R, Patil RK (2013) Study of ball nose end milling of LM6 Al Alloy: surface roughness optimization using genetic algorithm. I J Eng Tech 5–3:2859–2865
8. Kurt M, Hartomacýoðlu S, Mutlu B, Köklü U (2012) Minimization of the surface roughness and form error on the milling of free-form surfaces using a grey relational analysis. Mater Technol 46–3:205–213
9. Quintana G, Ciurana J, Ribatallada J (2010) Surface roughness generation and material removal rate in ball end milling operations. Mater Manuf Process 25–6:386–398
10. Vishal KM, Pankaj K, Baljeet S (2014) A review of optimization of surface roughness of Inconel 718 in end milling using Taguchi method. J Eng Res Appl 4–12:103–109
11. Chuan S, Peiqing Y (2011) The look-ahead function-based interpolation algorithm for continuous micro-line trajectories. I J Adv Manuf Technol 54:649–668

Multi-optimization of Stellite 6 Turning Parameters for Better Surface Quality and Higher Productivity Through RSM and Grey Relational Analysis

Brahim Ben Fathallah[2](✉), Riadh Saidi[1], Tarek Mabrouki[1], Salim Belhadi[3], and Mohamed Athmane Yallese[3]

[1] Applied Mechanics and Engineering Laboratory (LR-11-ES19), University of Tunis El Manar, ENIT, BP 37, Le Belvédère, 1002 Tunis, Tunisia
{brahim.benfathallah, riadh.saidi, tarek.mabrouki}@enit.utm.tn

[2] Mechanical, Material and Process Laboratory (LR99ES05) ENSIT, University of Tunis, 5 AV Taha Hussein Montfleury, Tunis, Tunisia[3] Mechanics and Structures Research Laboratory (LMS), May 8th 1945 University, 401, 24000 Guelma, Algeria

Abstract. The present paper consists of an experimental study to the effect of turning parameters on surface roughness of Cobalt alloy (Stellite 6) and the optimization of machining parameters based on Grey relational analysis. Taguchi's design of experiments (DOE) is used to carry out the tests. The response surface methodology is successfully applied in the analysis of the effect the turning parameters on surface roughness parameters. Second order mathematical models in terms of machining parameters are developed from experimental results. The experiment is carried out by considering four machining conditions, namely noise radius, cutting depth, cutting speed and feed rate as independent variables and average arithmetic roughness as response variables. It can be seen that the tool noise radius and feed rate are the most influential parameters on the surface roughness. The adequacy of the surface roughness model was established using analysis of variance (ANOVA). An attempt was also made to optimize cutting parameters using a Grey relational analysis to achieve minimum surface roughness and maximum material removal rate.

Keywords: Surface roughness · Optimization · Turning · Difficult-of-cut · RSM · GRA

Abbreviations

ANOVA	Analysis of variance
a_p	Depth of cut (mm)
f	Feed rate (mm/rev)
GRA	Grey relational analysis
HRC	Rockwell hardness
MRR	Material removal rate (cm^3/min)

N. Aifaoui et al. (Eds.): CMSM 2019, LNME, pp. 382–391, 2020.
https://doi.org/10.1007/978-3-030-27146-6_41

R^2	Determination coefficient (%)
Ra	Arithmetic mean roughness (μm)
RSM	Response surface methodology
V_c	Cutting speed (m/min)
α	Clearance angle (degree)
γ	Rake angle (degree)
λ	Inclination angle (degree)
χ_r	Cutting edge angle (degree)
$x_i(k)$	Grey relational generation
$\xi_i(k)$	Grey relational coefficient
γ_i	Grey relational grade
γ_m	Mean value of the grey relational grade.

1 Introduction

The use of cobalt alloys is widely used in the aerospace industry for applications requiring excellent mechanical strength at high temperatures, higher melting temperature, high-temperature corrosion and oxidation resistance [1, 2]. However, these alloys are also known to be difficult to machine [3]. The main problems encountered in the dry machining of cobalt alloys are low material removal rates and very limited tool life.

In the literature, different machining methods and modeling procedures have been studied by different scientists to optimize conditions in order to ultimately predict the machining of the Stellite 6 execution [4, 5] and achieve greater efficiency.

Nevertheless, to our knowledge, it is important to conduct studies on the influence of machining conditions on the changes of surface characteristics of cobalt-based alloys. Bagci and Aykut [6] examined the effects of machining conditions on tool wear, cutting force and chip morphology using the design experimental method (DOE) when symmetric surfacing the Stellite 6 superalloy with carbide tools with or without PVD coating under dry machining. Several researchers [7, 8] had carried out an experimental study on the machining of cobalt-based refractory materials to find the optimal cutting conditions for different cutting conditions. They used various optimization techniques based on the RSM method with sequential quadratic programming algorithm for a constrained problem. For cutting parameters optimization authors have also used [9–11]. Artificial neural networks (ANN), genetic algorithm (GA), genetically optimized neural network system (GONNS) and response surface methodology (RSM) are other modeling techniques that are used to optimize the process in machining operations [10–12]. Sarikaya and Güllü [13] and Folea et al. [14] looked at the effect of machining parameters on tangential force; chip morphology and tool wear during face milling of cobalt alloy (Stellite 6). Bruschi et al. [15] evaluated the contribution of cutting conditions on microstructure, surface integrity and tool wear when turning of the cobalt alloy (CoCrMo) using a carbide insert coated with PVD TiALN under conventional lubrication conditions. They discovered that the feed rate is the parameter that most affects the quality of the cutting surface and tool wear. Bagci and Aykut and Aykut et al. [5, 6] point up that the higher the feed rate and depth of cut, the higher is the arithmetic surface roughness, for all cooling mode and tool types.

A very few research studies on the impact of cutting parameters on the mechanical and microstructural properties of Stellite 6, Zaman et al. [16] and Yingfei et al. [17] have identified residual compressive stress in the axial direction and radial traction in the surface and subsurface direction in all conditions during turning operations. In addition, the affected surface layers are characterized by higher micro-hardness values for all cutting tests [15]. These results were consolidated in the machining of cobalt alloy (Stellite 6) by Yingfei et al. In a goal to attain perfect machining and production efficiency, various modeling methods and techniques have been applied to optimize cutting conditions [4, 13, 18]. Sarikaya and Gullu [13] have inspected the optimization of cooling mode with MQL method (Minimum Quantity Lubrication) when turning of cobalt alloy (haynes 25), by choosing the Taguchi method, which is based on relational analysis of Grey (GRA). For the purpose of comparing the cooling mode (dry, wet, MQL and cryogenic) when turning of cobalt alloy (Stellite 6), Sarikaya and Güllü [4] examined the effect on the clearance wear using Taguchi method and ANOVA method to determine the optimal roughness. Grey relational analysis is useful for treatment of marginal, incomplete and uncertain information. It can be used to successfully to resolve the difficult relationships between multiple performance features. This Grey theory established by Julong [19]. Several researchers are using this technique to optimize cutting conditions. Saha and Mandal [20] investigated multi-response optimization of the turning process for an optimal combination of settings to obtain minimum surface roughness, cutting power and tool vibration frequency using grey relational analysis (GRA). Dubey et al. [21] studied the multiple response optimization of the end milling parameter using the Taguchi gray-based method. They found that spindle speed was the most influential factor to minimize surface roughness.

In this sense, from previous studies in the literature, it appears important to carry out scientific research to find cutting parameters affecting the machining properties of cobalt alloy (Stellite 6). For this reason, the present study focuses on the optimization of the finishing parameters during turning operations of cobalt alloy (Stellite 6). The purpose is to predict the evolution of arithmetic mean roughness (Ra). A contribution to optimizing the cutting parameters when machining cobalt alloy (Stellite 6) is also given. In this regard, the RSM approach and Grey relational analysis were exploited.

2 Experimental Procedure

2.1 Equipment and Materials

The goal of this experimental work was to investigate the effects of cutting conditions on surface roughness, and to establish a correlation between surface quality and productivity. In order for this nose radius tool, cutting speed, feed rate and depth of cut were chosen as process parameters. The straight turning operations are performed on three parts because each part is subdivided on bearings separated by grooves in order to obtain the necessary combinations for the adopted experimental design. The machined parts are based on the cobalt alloy (Stellite 6), with a chemical composition including 28.25% Cr, 3.74% W 1.17% Mo, 1.11% C, 1.89% Fe, 2.32% Ni, 0.57% Mn, 1.20% Si, 0.004% P, 0.001% S, and balance of Co. This material has an average hardness of 41 HRC. The machining parameters are presented in Table 1. Surface roughness measurements were performed by using a Surftest 201 Mitutoyo with a cut-off length of 0.8 mm and sampling length of 5 mm.

Table 1 Experimental conditions

Item	Description
Machine tool	SN 40C, with 6.6 kW spindle power of the Czech company "TOS TRENCIN"
Cutting insert	Coated with PVD (Ti, Al) N_2, CNGG, SGF geometry, (1105 grade) Sandvik ISO
Tool holder	PCLNR 2020K12
Working tool geometry	Cutting edge angle (χ_r = +95°), inclination angle (λ = −6°), rake angle (γ = −6°) and clearance angle (α = 0°) Nose radius: 0.2, 0.4 and 0.8 mm
Process parameters Cutting speed (V_c) Feed rate (f) Depth of cut (a_p) Cooling mode	30, 55 and 80 m/min 0.08, 0.12 and 0.16 mm/rev 0.15, 0.3 and 0.45 mm Synthetic conventional oil mode (20%)

2.2 Design of Experiments

The response surface methodology (RSM) was used to determine the relationships between machining parameters and process characteristics. The relationship between the input parameters (r, Vc, f and ap) and surface roughness (Ra) to them is usually expressed with the aid of the following quadratic polynomial Eq. (1):

$$Y = b_0 + \sum\nolimits_{i=1}^{4} b_i X_i + \sum\nolimits_{i,j}^{4} b_{ij} X_i X_j \sum\nolimits_{i=1}^{4} b_{ii} X_i^2 + \varepsilon \quad (1)$$

Here, Y indicates the estimated Ra, b_0 is the constant term, bi represents the linear effects, b_{ii} represents the pure quadratic effects, b_{ij} represents the second level interaction effects, X_i and X_j indicate the associated coded variables (r, V_c, f and ap) and (ε) represents the error in predicting experimental surface roughness.

3 Effect of Cutting Conditions on Surface Roughness

The effects of the combined machining conditions (r, V_c, f and a_p) in the surface generated when turning the Stellite 6 cobalt alloy on the surface roughness Ra were evaluated. These responses are presented in Table 2.

Table 2 Experimental results when varying cutting conditions

Trail no.	Cutting conditions				Surface roughness Ra (μm)
	r (mm)	$V_{c(m/min)}$	f (mm/rev)	$a_{p\ (mm)}$	
1	0.2	30	0.08	0.15	0.98
2	0.2	30	0.08	0.30	0.95
3	0.2	30	0.08	0.45	1.04
4	0.2	55	0.12	0.15	1.91
5	0.2	55	0.12	0.3	1.78
6	0.2	55	0.12	0.45	1.42
7	0.2	80	0.16	0.15	2.65
8	0.2	80	0.16	0.3	2.1
9	0.2	80	0.16	0.45	2.28
10	0.4	30	0.12	0.15	1.1
11	0.4	30	0.12	0.3	1.25
12	0.4	30	0.12	0.45	1.18
13	0.4	55	0.16	0.15	1.7
14	0.4	55	0.16	0.3	1.83
15	0.4	55	0.16	0.45	1.6
16	0.4	80	0.08	0.15	0.53
17	0.4	80	0.08	0.3	0.68
18	0.4	80	0.08	0.45	0.6
19	0.8	30	0.16	0.15	1.3
20	0.8	30	0.16	0.3	1.12
21	0.8	30	0.16	0.45	1.1
22	0.8	55	0.08	0.15	0.53
23	0.8	55	0.08	0.3	0.36
24	0.8	55	0.08	0.45	0.46
25	0.8	80	0.12	0.15	0.91
26	0.8	80	0.12	0.3	0.64
27	0.8	80	0.12	0.45	0.72

The material removal rate (MRR) is calculated by Eq. (2):

$$MRR = V_c \times f \times a_p \tag{2}$$

The recourse to the RSM method with the aim of minimize the number of cutting tests for design optimization of the effect of cutting parameters (r, V_c, f and a_p) on Ra are given. The aim of ANOVA analysis is to specify which cutting conditions can affect significantly the responses. The results derived from the ANOVA method for the surface roughness variations Ra indicates that the feed rate was the main factor affecting Ra with a contribution of about 50.05%. The other terms (r), (r × f) and (r × r) have a lower contribution with 37.54%, 3.21% and 2.89%, respectively. The relationship between cutting parameters and execution estimations was displayed by quadratic relapse and exhibited by Eq. (3) with coefficients of determination, R^2 are 97.15%. The regression models are obtained using the Design Expert Software.

$$\begin{aligned} Ra = -0.416 - 2.88 \times r + 5.78E - 03 \times V_c + 26.25 \times f + 0.34 \times a_p - 1.9E \\ - 03 \times r \times V_c - 15.66 \times r \times f + 0.35 \times r \times a_p - 0.047 \\ \times V_c \times f - 9.46E - 03 \times V_c \times a_p - 10.16 \times f \times a_p + 3.12 \\ \times r^2 + 1.76E - 05 \times V_c^2 + 1.31 \times a_p^2 \end{aligned} \tag{3}$$

Figure 1a displays the differences between the measurements and the predicted Ra. This comparison clearly proves that the proposed second-order models are able to represent the system in the given experimental field. Figure 1b shows the normal probability diagrams of Ra. These models are adequate as represented by the points falling in a straight line in the normal probability curve. It indicates that errors are normally distributed.

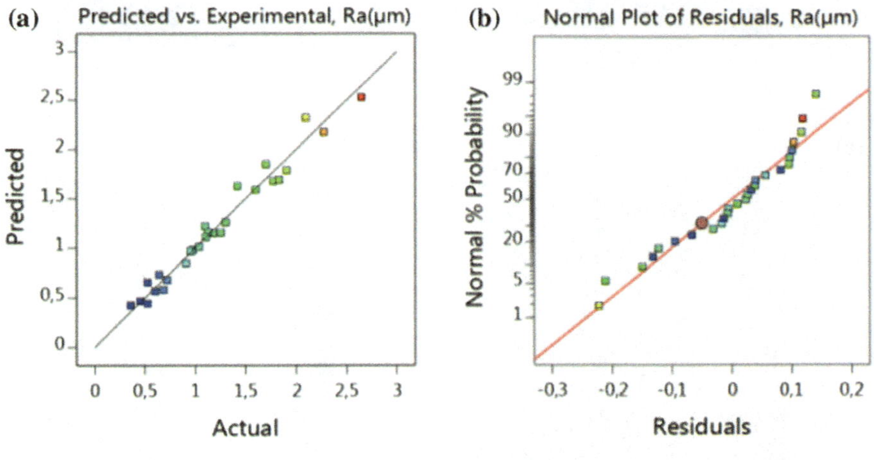

Fig. 1 **a** Comparison between experimental and predicted values of technological parameters and **b** Normal probability plots of Surface roughness

Surface quality is an important measure of machining quality, while surface roughness is one of the most important and fundamental parameters. The effects of cutting conditions (r, Vc, f and ap), during turning operations (straight) of cobalt based alloy (Stellite 6), on Ra are outlined in this section. For commodities, some results are given with 3D graphics and 2D contour illustrations.

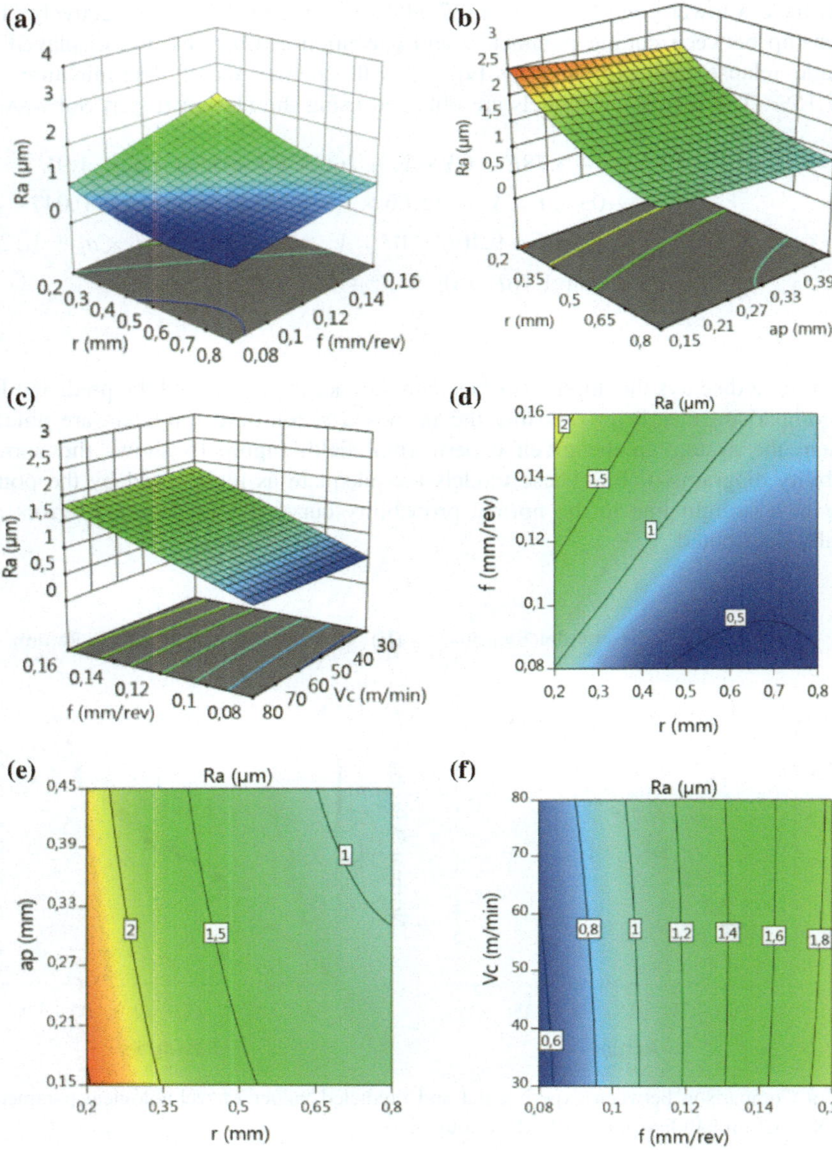

Fig. 2 (a–f) Effect of cutting condition on surface roughness evolution, 3D graphs and 2D contours

Figure 2a, b display that an increase of arithmetic mean roughness can be obtained with the smaller feed rate (f) and higher nose radius (r) for the accompanying cutting conditions: V_c = 80 m/min and a_p = 0.45 mm. The increase in feed rate results in an increase in the residual surface area of the cut and increases roughness. Also it can be mentioned that for higher both cutting depth and nose radius, the lower are surface arithmetic mean roughness values Ra. These are appeared by iso-contours graph (Fig. 2c, d) for $V_c = 80$ m/min and f = 0.16 mm/rev. Figure 2e, f shows with the arithmetic mean surface roughness, Ra when the feed rate and cutting speed vary for a fixed value r = 0.4 mm and $a_p = 0.15$ mm. Always, it can be noted that a smaller feed rate has a greater effect on the improvement of Ra than V_c. A minimal surface roughness is available with a high feed rate and a high cutting speed. This is due to a higher feed rate causing vibrations. This results in an increase in heat production and therefore an increase in surface roughness.

4 Optimization of Cutting Conditions Using Grey Relational Analysis

Multi-optimization of the cutting conditions by considering the various performance characteristics of the cobalt alloy (Stellite 6) using GRA is presented. These characteristics, including Ra and MRR, are selected to evaluate the effects of machining. The cutting conditions that are correlated with the performance characteristics selected in this study are the noise radius, cutting speed, feed rate and depth of cut, respectively. The experimental results are then introduced to calculate the coefficient and grades according to the GRA. The optimized cutting conditions leading simultaneously to higher MRR and lower surface roughness. To obtain optimal cutting conditions, the "smaller-the-better" quality characteristic has been used for minimizing surface roughness. The original sequence can be generalized as follows by Eq. (4):

$$x_i^*(k) = \frac{\max x_i^{(0)}(k) - x_i^{(0)}(k)}{\max x_i^{(0)}(k) - \min x_i^{(0)}(k)} \quad (4)$$

The MRR is the main answer, deciding the machinability of Stellite 6 turning. For the "larger-the-better", the original sequence can be generalized as follows by Eq. (5):

$$x_i^*(k) = \frac{x_i(k) - \min x_i(k)}{\max x_i(k) - \min x_i(k)} \quad (5)$$

where $x_i^*(k)$ is the generating value of Grey relational analysis; $\min x_i(k)$ is the minimum value of $x_i(k)$; $\max x_i(k)$ is the maximum of $x_i(k)$, $k = 1$ for Ra and $k = 2$ for MRR; $i = 1, 2, 3, \ldots, 27$ for experiment numbers 1–27.

Since the experimental design is orthogonal, it is then possible to distinguish the effect of each machining parameter on the grey relational grade at different levels. For example, the mean of the grey relational grade for radius noise at levels 1, 2 and 3 can be calculated by averaging the grey relational grade for the experiments 1–7, 8–18 and 19–27 respectively shown in Table 3.

Table 3 Response table for the mean grey relational grade

Cutting conditions	Level 1	Level 2	Level 3	Main effect (max-min)	Rank
r, mm	0.486	0.523	**0.590**[a]	0.104	1
Vc, m/min	0.487	0.523	**0.589**[a]	0.102	2
f, mm/rev	**0.541**[a]	0.515	0.491	0.050	4
a_p, mm	0.484	0.529	**0.585**[a]	0.101	3

Total mean value of the grey relational grade: $\gamma_m = 0.533$
[a]Levels for optimum grey relational grade

With the Grey relational analysis, it is possible to obtain the optimal conditions of machining to consider simultaneously minimal surface roughness and maximum material removal rate. This technique also allows optimum machining parameters to be obtained for a desired surface roughness and maximum metal removal rate by Grey relational analysis.

5 Conclusion

The present study investigates the use of response surface methodology (RSM) and Grey relational analysis (GRA) to optimize material removal rate (MRR) and arithmetic surface roughness (Ra) for the selected domain of machining parameters (cutting depth (a_p), cutting speed (V_c), feed rate (f) and nose radius (r)). It was concluded that:

- The result of analysis by the ANOVA method shows that the nose radius and feed rate are the most significant factor affecting the arithmetic surface roughness compared to the cutting speed and depth of cut.
- The obtained results of predicted models in terms of Ra values corroborate with the experimental ones in the full design space with correlation coefficient of 97.15%.
- Results coming from ANOVA method applied for Ra variations show that the feed rate presents the most important factor affecting Ra evolution with a contribution about 50.05%. The other terms (r), (r × f) and (r × r) have less contribution with 37.54%, 3.21% and 2.89% significances, respectively.
- The result of Grey relational analysis conduct to higher productivity and best quality in the field of optimal cutting conditions: r = 0.8 mm, Vc = 80 m/min, f = 0.08 mm/rev, and ap = 0.45 mm.

References

1. Davis JR (2000) Nickel, cobalt, and their alloys. ASM international
2. Sato J, Omori T, Oikawa K, Ohnuma I, Kainuma R, Ishida K (2006) Science 312:90–91
3. Ezugwu E (2005) Int J Mach Tools Manuf 45:1353–1367
4. Sarıkaya M, Güllü A (2015) J Clean Prod 91:347–357
5. Aykut S, Kentli A, Gulmez S, Yazicioglu O (2012)

6. Bagci E, Aykut Ş (2006) Int J Adv Manuf Technol 29:940.
7. Schlegel D, Lebaal N, Folea M (2011) Recent Res Manuf Eng Cut: 156–162
8. Schlegel D, Lebaal N, Folea M (2012) Int J Adv Manuf Technol 60:55–63
9. Davim JP, Aveiro P (2016) Design of experiments in production engineering. Springer
10. Fathallah BB, Saidi R, Dekhli C, Belhadi S, Yallese MA (2019) Int J Ind Eng Comput. https://doi.org/10.5267/j.ijiec.2019.3.001
11. Saidi R, Fathallah BB, Mabrouki T, Belhadi S, Yallese MA (2019) Modeling and optimization of the turning parameters of cobalt alloy (Stellite 6) based on RSM and desirability function. Int J Adv Manuf Technol 100(9–12):2945–2968.
12. Bağcı E, Aykut Ş (2014) Arab J Sci Eng 39:8135–8146
13. Sarıkaya M, Güllü A (2014) J Clean Prod 65:604–616
14. Folea M, Schlegel D, Lupulescu N, Parv L (2009) Modeling surface roughness in high speed milling: cobalt based superalloy case study, pp 353–357
15. Bruschi S, Ghiotti A, Bordin A (2013) Effect of the process parameters on the machinability characteristics of a CoCrMo alloy. Trans Tech Publ, pp 1976–1983
16. Zaman HA, Sharif S, Kim D-W, Idris MH, Suhaimi MA, Tumurkhuyag Z (2017) Procedia Manuf 11:563–570
17. Yingfei G, de Escalona PM, Galloway A (2017) J Mater Eng Perform 26:312–326
18. Aykut Ş, Bagci E, Kentli A, Yazıcıoğlu O (2007) Mat Des 28:1880–1888
19. Julong D (1989) J Grey Syst 1:1–24
20. Saha, Mandal N (2013) Int J Ind Eng Comput 4:51–60
21. Dubey, Abhishek, et al (2008) A parametric design study of end milling operation using grey based taguchi method. Int J Emerging Technol Adv Eng 4(4)

Numerical Determination of Cutting Stability Lobes in Orthogonal Milling

Wael Baklouti, Charfeddine Mrad(✉), and Rachid Nasri

Laboratory of Applied Mechanics and Engineering (LMAI), National Engineering School of Tunis (ENIT), University of Tunis el Manar (UTM), BP 37 Le Belvedere, 1002 Tunis, Tunisia
WaelBaklouti@gmail.com,
{Charfeddine.Mrad, Rachid.Nasri}@enit.rnu.tn

Abstract. The vibrations in orthogonal milling are simultaneously due to the friction between the tool and the workpiece and to the flexibility of the cutting system, leading to chip section variation to cause the chatter phenomenon. Indeed, these vibrations induce an increasing machining disorder leading to an unstable cutting to cause a poor surface quality and a rapid tool wear, in addition to different operational risks. This work presents a numerical approach to establish cutting stability lobes in orthogonal metal milling. The Abaqus software is used to perform the simulations, the Johnson-Cook laws are considered for the workpiece material strength and fracture, the Coulomb law is retained for the friction between the workpiece and the tool, the tool is considered flexible, and the thermal conductivities of the workpiece and the tool are introduced. The results at various cutting speeds are obtained for different tool flexibilities: radial, transversal, and combined; and are then compared to conclude on the cutting stability lobes.

Keywords: Orthogonal milling · Chatter phenomenon · Numerical cutting · Stability lobes

1 Introduction

The occurrence of self-excited vibrations due to the friction of the tool against the workpiece leads to the chatter phenomenon. This phenomenon causes a poor surface quality and a rapid tool wear, in addition to various operational risks.

To avoid chatter, two ways are possible: changing the behavior of the cutting system or establishing cutting stability lobes. Cutting stability lobes are easier to obtain.

An example of cutting stability lobes is presented in Fig. 1 [1]. The frontier between a stable cut and an unstable cut has a lobe shape which depends of the cutting speed and the cut depth. The lobes indicate thus the favorable and unfavorable cutting parameters.

The work presents a numerical approach to establish cutting stability lobes in orthogonal metal milling. The Abaqus software is used to perform the simulations, using a descriptive procedure. The results at various speeds are obtained firstly when considering radial tool flexibility, secondly when considering transversal tool

N. Aifaoui et al. (Eds.): CMSM 2019, LNME, pp. 392–398, 2020.
https://doi.org/10.1007/978-3-030-27146-6_42

flexibility, and thirdly when considering combined tool flexibility: radial and transversal. The obtained lobes are finally compared to conclude on the numerical milling process.

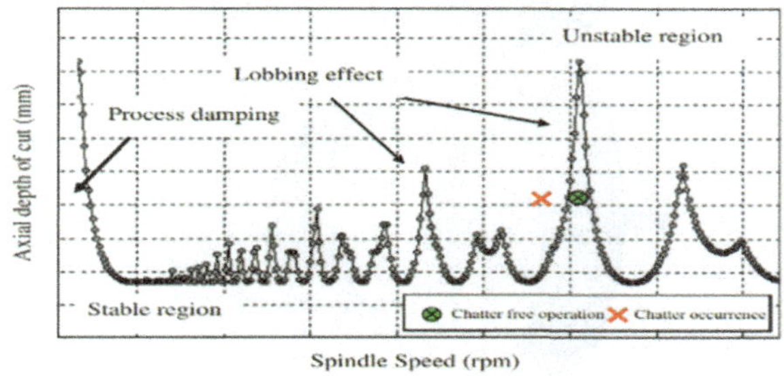

Fig. 1 Cutting stability lobe

2 Approach Description

The Abaqus software is used. The numerical study is done for one tooth, the feed effect can be then neglected. The used model is 2D, to reduce the computation time. The numerical machining is done for a single pass, the other possible passes are similar.

The tool cut depth (ap) is in the axial direction, while the axial cut depth (ae) is in the radial direction. The feed is in the direction perpendicular to both tool cut depth and axial cut depth. The cutting parameters are shown in Fig. 2.

The tool is considered rigid, is rigidly linked to the workpiece and is flexibly linked to the machine. The meshing uses a 2D plane strain solid element CPE4RT: four-node, bilinear displacement-temperature, reduced integration, hourglass control; taking the thermal effect into consideration [2].

The Johnson-Cook laws are considered for the workpiece material strength and fracture, and the Coulomb law is retained for the friction between the workpiece and the tool.

The Lagrangian formulation is used due to its advantages: the chip morphology is conserved, the cutting force is considered, and the surface quality is accurate.

The explicit scheme is considered, the implicit scheme may encounter numerical problems in dynamical applications [3].

3 Numerical Cutting

The tool is considered of different flexibilities: radial, transversal, and combined (transversal and radial). The tool flexibilities are viscoelastic. The radial flexibility is referred to orthogonal turning.

Figures 3 and 4 show respectively, the turning numerical model and the turning stability lobe, considered as reference. The axial cut depth corresponds to the feed rate [4, 5].

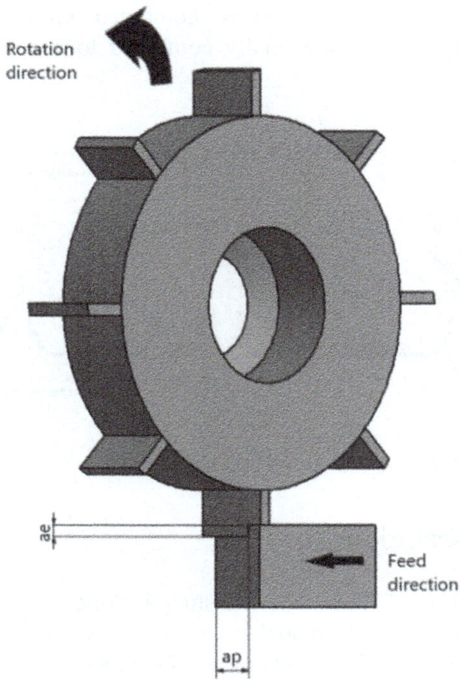

Fig. 2 Cutting parameters

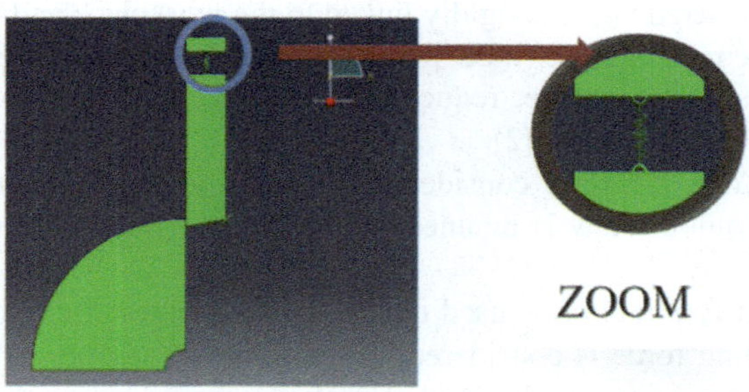

Fig. 3 Turning numerical model

3.1 Radial Flexibility

The tool is modeled as a dynamic system of one degree-of-freedom: mass, spring, damper; in the radial direction (y axis), as shows Fig. 5.

The milling process is taken similar to the turning process to make the comparison possible: The tools are of one degree-of-freedom in the radial direction, the tools have the same parameters and suspension specifications, the materials are the same for the

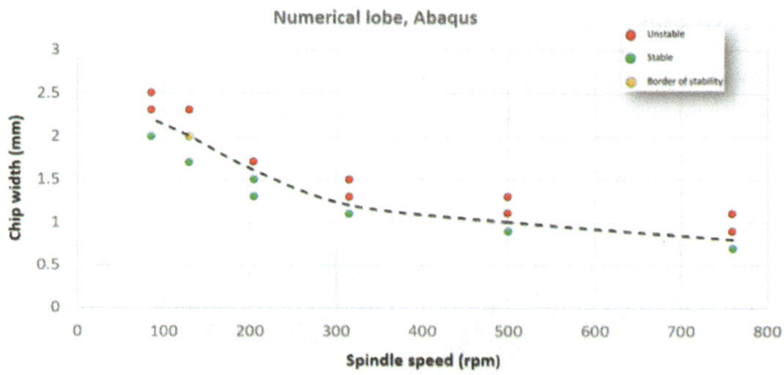

Fig. 4 Turning stability lobe

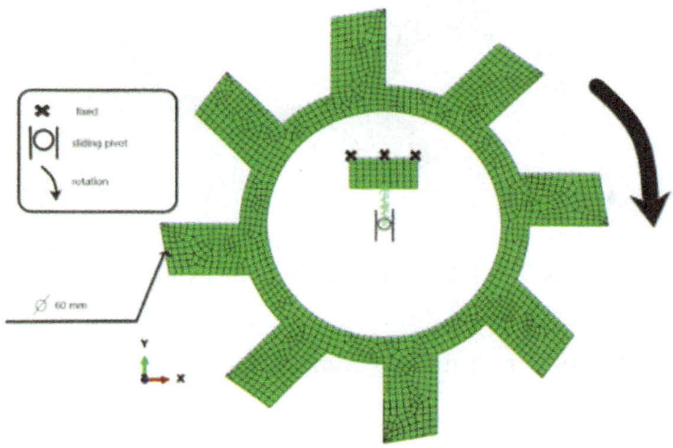

Fig. 5 Milling radial numerical model

workpieces and the tools, the cutting conditions and the friction coefficients are the same, and the instantaneous chip sections are the same.

Figure 6 shows the obtained milling radial stability lobe.

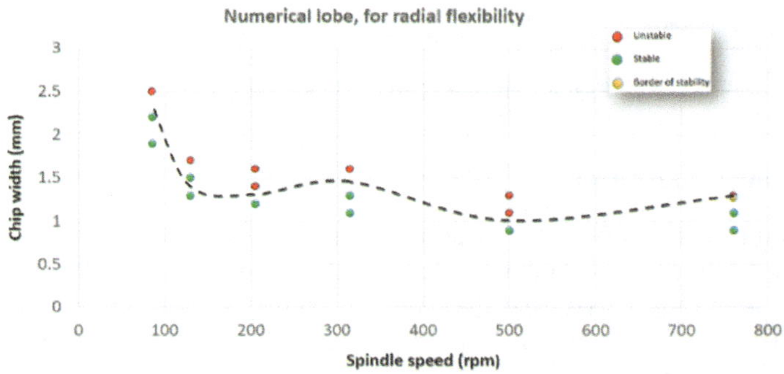

Fig. 6 Milling radial stability lobe

3.2 Transverse Flexibility

The tool is modeled as a dynamic system of one degree-of-freedom: mass, spring, damper; in the transversal direction (x axis), as shows Fig. 7.

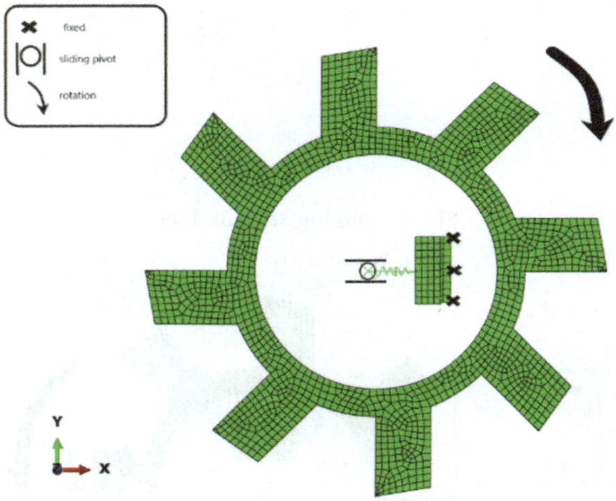

Fig. 7 Milling transversal numerical model

Figure 8 shows the obtained milling transversal stability lobe.

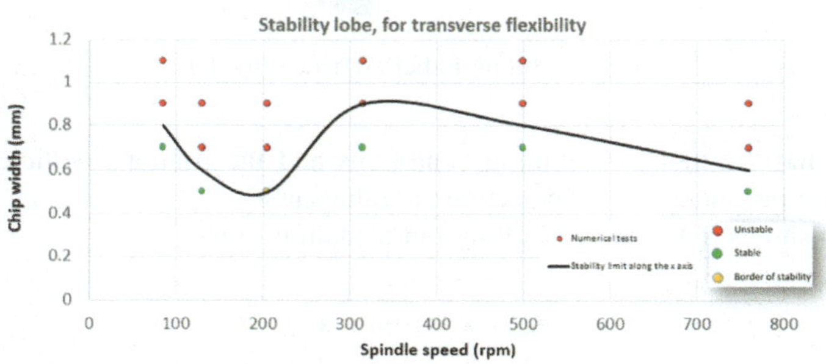

Fig. 8 Milling transversal stability lobe

3.3 Combined Flexibility

The tool is modeled as a dynamic system of two degree-of-freedom: the transverse and the radial flexibilities are simultaneously considered; as shows Fig. 9. The transverse flexibility is along the x axis, and the radial flexibility is along the y axis.

Figure 10 shows the obtained milling combined stability lobes: in the x and y directions.

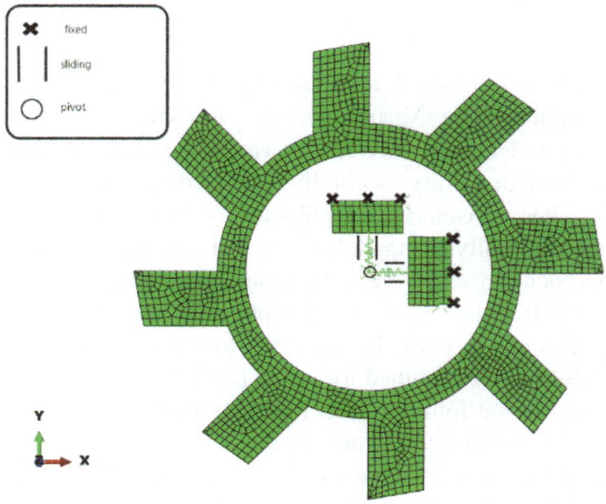

Fig. 9 Milling combined numerical model

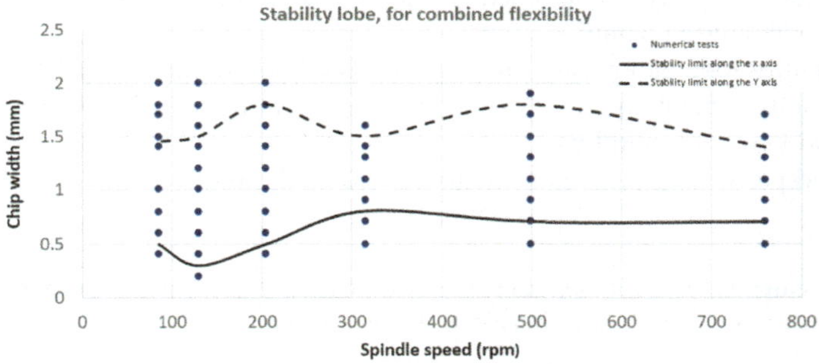

Fig. 10 Milling combined stability lobes

4 Results Comparison

The milling stability lobe obtained for radial flexibility and the turning stability lobe considered as reference, are clearly different despite all the common machining details.

The milling stability lobes obtained for transverse and radial flexibilities are notably different despite all the common direction details.

The milling stability lobe along the x axis is lower than the stability lobe along the y axis, when the flexibility is combined.

The milling stability lobe along the x axis is different from the transversal flexibility lobe and the stability lobe along the y axis is different from the radial flexibility lobe.

5 Conclusion

The work presented a numerical approach to establish cutting stability lobes in orthogonal metal milling. The Abaqus software is used to perform the simulations, using a descriptive procedure. The results at various speeds are obtained firstly when considering radial tool flexibility, secondly when considering transversal tool flexibility, and thirdly when considering combined tool flexibility: radial and transversal. The obtained lobes are finally compared to conclude on the numerical milling process.

The used approach presents the following main advantages: Practical laws of material strength and fracture are adopted, a physical law of friction is used, and the thermal conductivities of the workpiece and the tool are considered.

The milling stability lobe obtained for radial flexibility and the turning stability lobe considered as reference, are found different despite all the common machining details. This confirms that the two processes are distinct.

The milling stability lobes obtained for transverse and radial flexibilities are notably different despite all the common directions details. This confirms that the cutting is asymmetric.

The milling stability lobe along the x axis is lower than the stability lobe along the y axis, when the flexibility is combined. Thus, the stability lobe along the x axis is to be retained for the cutting stability.

The stability lobe along the x axis is different from the transversal flexibility lobe and the stability lobe along the y axis is different from the radial flexibility lobe. This confirms the two axes coupling.

The workpiece roughness is to be examined in the future to confirm the obtained results.

Acknowledgements. The authors gratefully acknowledge the helpful comments and suggestions of the reviewers, which have improved the presentation.

References

1. Quintana G, Ciurana J (2011) Chatter in machining processes: a review. Int J Mach Tools Manuf 51:363–376
2. Hajmohammadi MS, Movahhedy MR, Moradi H (2014) Investigation of thermal effects on machining chatter based on FEM simulation of chip formation. J Manuf Sci Technol 7:1–10
3. Rebelo N, Nagtegaal JC, Taylor LM (1992) Numerical methods in industrial forming processes. Adv Struct Eng Mech 1:99–108
4. Mahnama M, Movahhedy MR (2010) Prediction of machining chatter based on FEM simulation of chip formation under dynamic conditions. Int J Mach Tools Manuf 50:611–620
5. Baklouti W, Mrad C, Nasri R (2018) Numerical study of the chatter phenomenon in orthogonal turning. Int J Adv Manuf Technol 99:755–764

Prediction of Forces Components During the Turning Process of Stellite 6 Material Based on Artificial Neural Networks

Riadh Saidi[1(✉)], Brahim Ben Fathallah[2], Tarek Mabrouki[1], Salim Belhadi[3], and Mohamed Athmane Yallese[3]

[1] Applied Mechanics and Engineering Laboratory (LR-11-ES19), University of Tunis El Manar, ENIT, BP 37, Le Belvédère, 1002 Tunis, Tunisia
{riadh.saidi,tarek.mabrouki}@enit.utm.tn

[2] Mechanical, Material and Process Laboratory (LR99ES05) ENSIT, University of Tunis, 5 AV Taha Hussein Montfleury, Tunis, Tunisia
brahim.benfathallah@enit.utm.tn

[3] Mechanics and Structures Research Laboratory (LMS), May 8th 1945 University, 401, 24000 Guelma, Algeria
belhadi23@yahoo.fr, yallese.m@gmail.com

Abstract. In this study, artificial neural networks (ANNs) were used to study the effects of machinability on cutting parameters during turning of the cobalt alloy (Stellite 6). Cutting forces with three axes (Fx, Fy and Fz) were predicted by changing the tool tip radius (r), cutting speed (V_c), feed rate (f) and cutting depth (a_p) with conventional lubrication. Experimental studies were conducted to obtain training and test data and a feed-forward back-propagation algorithm was used in the networks. The main test parameters are the tool tip radius (r, mm), cutting speed (V_c, m/min), feed rate (f, mm/rev), cutting depth (a_p, mm) and cutting forces (Fx, Fy and Fz, N). r, V_c, f and a_p were used as input data while Fx, Fy and Fz were used as output data. The mean percentage values of root mean square error (RMSE), mean absolute percentage error (MAPE) and mean absolute deviation (MAD) for Fx, Fy and Fz using the proposed models were obtained around 2 and 4.79%, respectively for training and testing. These results show that ANNs can be used to predict the effects of machinability on cutting parameters when cutting Stellite 6 on turning process. The results highlighted the performance of the studied configuration.

Keywords: Cutting force · ANN · Turning · Stellite 6 · Modelling

1 Introduction

The use of cobalt alloys is widely used in the aerospace industry for applications requiring excellent mechanical strength at high temperatures, higher melting temperature, high-temperature corrosion and oxidation resistance [1, 2]. However, these alloys are also known to be difficult to machine [3]. The main problems encountered in the

N. Aifaoui et al. (Eds.): CMSM 2019, LNME, pp. 399–408, 2020.
https://doi.org/10.1007/978-3-030-27146-6_43

dry machining of cobalt alloys are low material removal rates and very limited tool life. The low thermal conductivity and high chemical affinity for many materials often cause the formation of an adhesion layer on the face of the tool, resulting in rapid wear. Also, the low elastic modulus of cobalt alloys is the main cause of vibrations generated during machining.

In the literature, different machining methods and modeling procedures have been studied by different scientists to optimize conditions in order to ultimately predict the machining of the Stellite 6 execution [4–6] and achieve greater efficiency.

Nevertheless, to our knowledge, it is important to conduct studies on the influence of machining conditions on the changes of surface characteristics of cobalt-based alloys.

Several researchers [7, 8] had carried out an experimental study on the machining of cobalt-based refractory materials to find the optimal cutting conditions for different cutting conditions. They used various optimization techniques based on the RSM method with sequential quadratic programming algorithm for a constrained problem. For cutting parameters optimization authors have also used [9, 10]. Artificial neural networks (ANN), genetic algorithm (GA), genetically optimized neural network system (GONNS) and response surface methodology (RSM) are other modeling techniques that are used to optimize the process in machining operations [9–15].

Sarikaya and Gullu [16] and Folea et al. [17] observed that the effect of machining parameters on tangential force, chip morphology and tool wear during face milling of cobalt alloy (Stellite 6) revealed the most prominent factor of largest number of cutting parameters. Bruschi et al. [18] evaluated the contribution of cutting conditions on microstructure, surface integrity and tool wear when turning of the cobalt alloy (CoCrMo) using a carbide insert coated VD with PTiALN under conventional lubrication conditions. They discovered that the feed rate is the parameter that most affects the quality of the cutting surface and tool wear. Bagci and Aykut [5] and Aykut et al. [6] point up that the higher feed rate and depth of cut proportionally with the higher is the arithmetic surface roughness, for all cooling mode and tool types.

In turning and milling operations, the cobalt based-alloy conserves its hardness and strength, which renders machining difficult. This necessitates increasing the tangential forces with the rise of cutting parameters, but this is uncertainly with the cutting speed [19].

A very few research studies on the impact of cutting parameters on the mechanical and microstructural properties of Stellite 6, Zaman et al. [20] and Yingfei G. et al. [21] have identified residual compressive stress in the axial direction and radial traction in the surface and subsurface direction in all conditions during turning operations. In addition, the affected surface layers are characterized by higher micro-hardness values for all cutting tests [18]. These results were consolidated in the machining of cobalt alloy (Stellite 6) by Yingfei et al. [21]. In a goal to attain perfect machining and production efficiency, various modeling methods and techniques have been applied to optimize cutting conditions [4, 22].

In order to predict the results of the cutting process, industrialists and researchers are using artificial neural networks. Toparli et al. [23] used two methods for calculating residual stresses in water-hardened steel bars from 600 °C: the finite element method (FEM) and the ANN method. For the latter, a multi-layer feed-forward network with error gradient retro propagation was used. The results obtained by the ANN method are comparable to those obtained by the FEM. Umbrello D. et al. [24, 25] and Ambrogio et al. [26] have developed theoretical models to predict the cutting conditions required to achieve a given residual stress state using ANNs. Kafkas et al. [27] presented a feed-forward RNA-based approach with a hidden layer to determine residual stresses in PM nickel-based steel (FLN2-4405) with different heat treatments. They used the error gradient retro-propagation algorithm in the learning and training functions of Levenberg-Marquart (LM) and Scaled Conjugate Gradient (SCG). Karataş et al. [28] opted for an RNA with the error gradient back-propagation algorithm in learning. They used the results of experimental measurements as training and test data to determine residual stresses as a function of shot blasting conditions. The results of the mathematical modelling were acceptable.

From previous studies in the literature, it appears important to carry out scientific research to find cutting parameters affecting the machining properties of cobalt alloy (Stellite 6). In this context, we propose a methodology approach for modelling an ANN configuration, adapted to predict the cutting forces during turning of the finishing parameters during turning operations of cobalt alloy (Stellite6). The purpose is to predict the evolution of cutting forces components (Fx, Fy and Fz).

2 Experimental Procedure

The goal of this experimental work was to investigate the effects of cutting conditions on cutting force components. In order for this nose radius tool, cutting speed, feed rate and depth of cut were chosen as process parameters. The straight turning operations are performed on three parts because each part is subdivided on bearings separated by grooves in order to obtain the necessary combinations for the adopted experimental design. The machined parts are based on the cobalt alloy (Stellite 6), with a chemical composition including 28.25% Cr, 3.74% W 1.17% Mo, 1.11% C, 1.89% Fe, 2.32% Ni, 0.57% Mn, 1.20% Si, 0.004% P, 0.001% S, and balance of Co. This material has an average hardness of 41 HRC.

A Taguchi design L27 is applied to the experiment planning. Selected input factors and their levels are shown in Table 1. The ranges of the cutting parameters are chosen according to the recommendations of the manufacturer of the cutting tools (i.e., Sandvik). For cutting forces measurements, a dynamometer (type Kistler 9257B) allowing measurements from range [−5 to 5] KN was used.

Table 1 Experimental conditions

Item	Description
Machine tool	SN 40C, with 6.6 kW spindle power of the Czech company "TOS TRENCIN"
Cutting insert	Coated with PVD (Ti, Al) N2, CNGG, SGF geometry, (1105 grade) Sandvik ISO
Tool holder	PCLNR 2020K12
Insert and tool geometry	Cutting edge angle (χ_r = +95°), inclination angle (λ = −6°), rake angle (γ = −6°) and clearance angle (α = 0°) nose radius: 0.2, 0.4 and 0.8 mm
Process parameters	
Cutting speed (V_c)	30, 55 and 80 m/min
Feed rate (f)	0.08, 0.12 and 0.16 mm/rev
Depth of cut (a_p)	0.15, 0.3 and 0.45 mm
Cooling mode	Synthetic conventional oil mode (10%).

3 Artificial Neural Networks Model

Artificial neural networks are one of the methods of artificial intelligence inspired by the behaviour of the human brain to perform useful functions related to industrial applications such as the assignment of approximation, classification and modelling tasks. It uses a non-linear and bounded algebraic function whose value depends on parameters called coefficients (weight). An ANN model consists of three connected layers, each consisting of one or more neurons. The first layer is the input in which each neuron represents a variable, while the second layer is the hidden layer. The latter receives the information from the input layer and processes it using a transfer function. Finally, the last layer is the output layer which can consist of one or more outputs that represent the numerical values of the responses or dependent variables. The neurons most frequently used are those for which the transfer function (F) is a non-linear function (usually a hyperbolic tangent) of a linear combination of the inputs:

$$F = tanh \sum\nolimits_{i=1}^{n} w_i x_i \tag{1}$$

where (x_i) represent the neuron input variables, (w_i) the adjustable parameters, and (n) the neuron number.

In order to predict the cutting force components Fx, Fy, and Fz, seven multilayer perceptrons (constituted of one input, one hidden, and one output layer) were used. Each perceptron had one output only because ANN models with single outputs give better results compared to those with several outputs. The experimental procedure consisted of 27 tests of which 18 tests were used for network learning and 9 validation tests randomly selected to assess the effectiveness of the network. The learning of the neural network was performed through the application of a retro-propagation algorithm based on a descending gradient. In the present case, a learning rate of $\eta = 0.1$ was selected. Figure 1 presents the adequate ANN architectures to model the Fx. It is the same for ANN architectures for cutting forces Fy and Fz. The experimental and predicted results with ANN approach of cutting force components (Fx, Fy and Fz) as a function of cutting conditions during the turning of Stellite 6 are presented in Table 2.

Table 2 Results of measured and predicted of cutting forces components when turning Stellite 6

Trial	Cutting parameters				Measured	Predicted	Measured	Predicted	Measured	Predicted
	r (mm)	Vc (m/mn)	f (mm/rev)	ap (mm)	Fx(N)	Fx(N)	Fy(N)	Fy(N)	Fz(N)	Fz(N)
1	0.2	30	0.08	0.15	13.0	12.4	38.2	39.4	43.1	44.4
2	0.2	30	0.08	0.30	61.6	61.1	59.3	56.7	93.9	92.7
3	0.2	30	0.08	0.45	110.2	110.4	53.3	56.0	147.4	149.0
4	0.2	55	0.12	0.15	36.9	38.5	44.0	40.7	70.0	69.7
5	0.2	55	0.12	0.30	92.4	99.1	57.9	64.4	131.2	138.4
6	0.2	55	0.12	0.45	174.8	174.7	69.6	70.7	218.3	215.7
7	0.2	80	0.16	0.15	28.0	28.2	60.4	60.7	82.0	85.4
8	0.2	80	0.16	0.30	108.8	100.3	94.4	94.5	171.6	170.4
9	0.2	80	0.16	0.45	201.1	201.3	113.1	110.6	262.4	260.6
10	0.4	30	0.12	0.15	19.6	17.7	39.7	46.9	55.5	51.9
11	0.4	30	0.12	0.30	52.8	53.0	71.0	73.6	109.6	111.2
12	0.4	30	0.12	0.45	106.4	106.2	84.1	84.7	183.9	185.2
13	0.4	55	0.16	0.15	25.7	25.3	57.6	57.5	90.5	86.7
14	0.4	55	0.16	0.30	74.6	74.4	92.9	93.7	152.8	162.8
15	0.4	55	0.16	0.45	166.6	160.1	121.0	118.6	259.2	258.9
16	0.4	80	0.08	0.15	22.4	23.0	47.3	51.2	54.4	54.9
17	0.4	80	0.08	0.30	68.4	68.3	80.1	82.2	96.6	99.5
18	0.4	80	0.08	0.45	147.3	147.3	99.3	95.0	175.6	174.7
19	0.8	30	0.16	0.15	21.1	21.1	52.2	52.3	69.3	70.4
20	0.8	30	0.16	0.30	37.0	36.6	82.1	81.8	118.3	119.3
21	0.8	30	0.16	0.45	90.1	90.0	123.8	124.4	208.5	208.7
22	0.8	55	0.08	0.15	26.2	26.2	71.8	64.5	58.4	59.2
23	0.8	55	0.08	0.30	39.3	44.1	97.4	98.9	98.0	94.2
24	0.8	55	0.08	0.45	97.6	97.4	126.3	130.6	150.3	152.2
25	0.8	80	0.12	0.15	23.6	23.7	73.1	77.5	70.4	83.7
26	0.8	80	0.12	0.30	63.5	42.9	115.9	113.5	137.6	128.9
27	0.8	80	0.12	0.45	115.6	115.8	149.3	154.1	207.0	205.2

The mathematical models obtained by the ANN method for the three criteria Fx, Fy, and Fz are expressed by the following Eqs. (2)–(4):

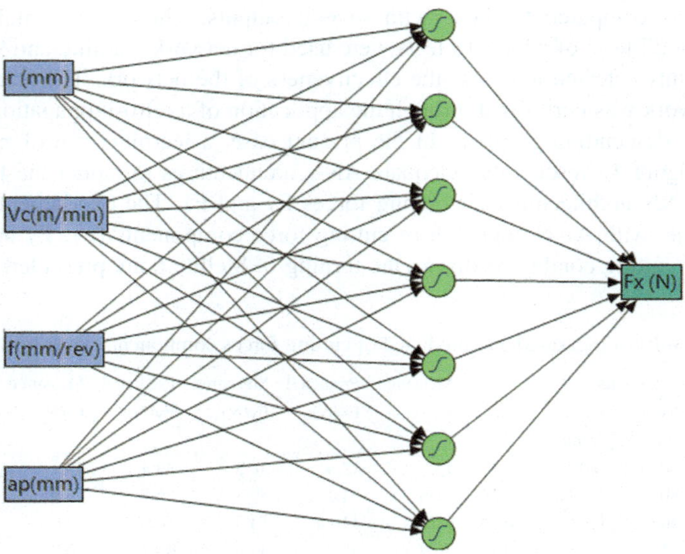

Fig. 1 ANN architecture (4-7-1) to model the Fx

$$
\begin{aligned}
&\text{Fx} = -64.8H1 + 23.88H2 + 13.81H3 + 10.91H4 + 49.3H5 - 19.46H6 - 81H7 + 138.25\\
&\quad \text{With :}\\
&\quad H1 = \text{TanH}(0.5(-4.06r + 0.04V_c + 23.69f + 1.84a_p - 2.31))\\
&\quad H2 = \text{TanH}(0.5(-6.26r - 0.01V_c + 9.59f + 8.965a_p - 0.65))\\
&\quad H3 = \text{TanH}(0.5(0.87r + 0.024V_c - 30.78f + 9.23a_p - 0.28))\\
&\quad H4 = \text{TanH}(0.5(-5.89r + 0.031V_c - 4.97f - 13a_p + 6.92))\\
&\quad H5 = \text{TanH}(0.5(-4.27r + 0.067V_c + 49.44f - 1.94a_p - 6.08))\\
&\quad H6 = \text{TanH}(0.5(-0.8r + 0.02V_c + 8.48f - 14.97a_p + 2.32))\\
&\quad H7 = \text{TanH}(0.5(0.87r - 0.02V_c - 2.7f - 10.52a_p + 5.68))
\end{aligned}
\tag{2}
$$

$$
\begin{aligned}
&\text{Fy} = 132.35H1 - 94.16H2 - 7.03H3 - 123.19H4 + 101.54H5 - 72.68H6 - 02.2H7 + 129.42\\
&\quad \text{With :}\\
&\quad H1 = \text{TanH}(0.5(0.85 + 0.21r - 0.01V_c - 1.94f - 0.76a_p))\\
&\quad H2 = \text{TanH}(0.5(-0.64 + 2.66r + 0.001V_c + 7.77f - 7.58a_p))\\
&\quad H3 = \text{TanH}(0.5(-3.78 + 2.57r + 0.015V_c + 5.73f + 2.01a_p))\\
&\quad H4 = \text{TanH}(0.5(1.11 + 0.59r - 0.004V_c - 6.65f + 0.14a_p))\\
&\quad H5 = \text{TanH}(0.5(-4.12 + 1.28r + 0.024V_c + 5.22f + 2.23a_p))\\
&\quad H6 = \text{TanH}(0.5(3.72 - 4.29r - 0.005V_c - 19.78f + 2.02a_p))\\
&\quad H7 = \text{TanH}(0.5(-1.6 - 0.2r + 0.002V_c + 11f + 1.6a_p))
\end{aligned}
\tag{3}
$$

$$Fz = -60.08H1 + 53.74H2 - 3.63H3 - 23.66H4 + 52.1H5 - 20.46H6 + 3.3H7 + 165.31$$
$$\text{With :}$$
$$H1 = \text{TanH}(0.5(2.09 + 3.54r + 0.01V_c - 7.89f - 9.42a_p))$$
$$H2 = \text{TanH}(0.5(-6.89 - 0.19r + 0.02V_c + 15.11f + 8.94a_p))$$
$$H3 = \text{TanH}(0.5(-5.46 + 8.02r + 0.06V_c - 6.58f - 2.41a_p))$$
$$H4 = \text{TanH}(0.5(-7.37 + 3.43r + 0.03V_c + 44.01f - 8.42a_p))$$
$$H5 = \text{TanH}(0.5(-6.38 + 4.23r + 0.002V_c + 29.41f + 1.08a_p))$$
$$H6 = \text{TanH}(0.5(6.89 + 0.08r - 0.11V_c - 12.5f - 1.04a_p))$$
$$H7 = \text{TanH}(0.5(3.69 - 0.75r - 0.044V_c - 32.72f + 11.78a_p)) \quad (4)$$

The comparison between the experimental values (measured) of Fx, Fy, and Fz, and those estimated by the mathematical models obtained by ANN method is presented in Figs. (2, 3 and 4). they are found very close. However, the models obtained by the ANN method show a better correlation with the experimental data.

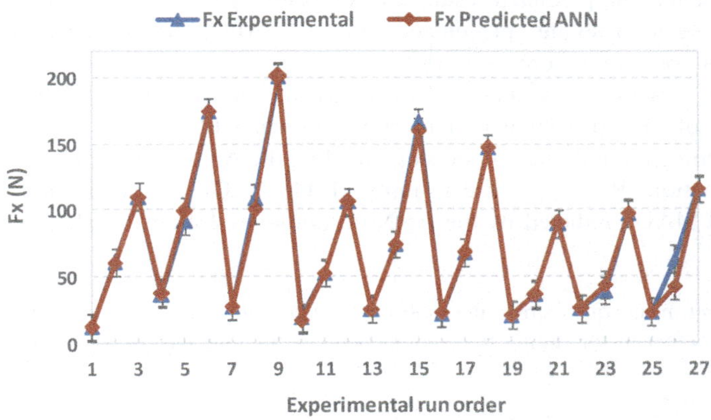

Fig. 2 Comparison between experimental values and predicted by ANN for Fx

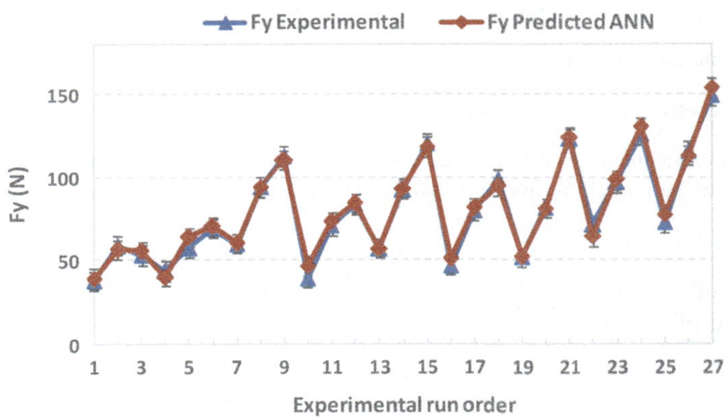

Fig. 3 Comparison between experimental values and predicted by ANN for Fy

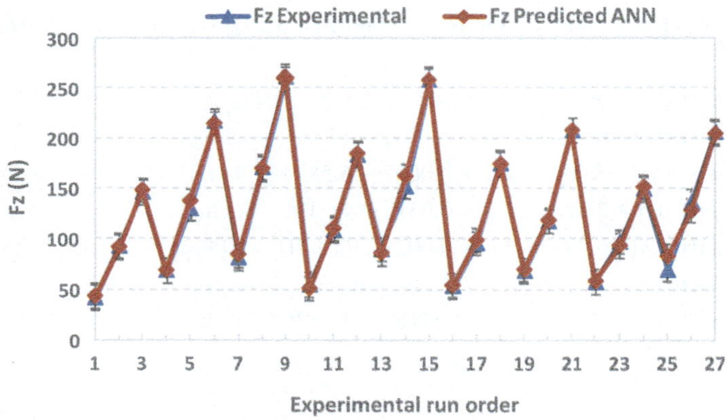

Fig. 4 Comparison between experimental values and predicted by ANN for Fz

In order to properly evaluate the predictive capabilities of the models developed, the experimental and predicted results are compared in terms of root mean square error (RMSE), mean absolute percentage error (MAPE), mean absolute deviation (MAD) and correlation coefficient (R^2).

Table 3 presents the values of RMSE, MAPE, MAD and R^2 obtained from the application of the artificial neural network approach (ANN) for Fx, Fy, and Fz. It should be pointed out that all models obtained by the ANN approach have a coefficient of determination (R^2) very close to unity (Table 3). On the other hand, the RMSE, MAPE and MAD predicted by the ANN do not exceed 4.79% (Table 3).

Table 3 Root mean square error, mean absolute percentage error, mean absolute deviation and correlation coefficient for cutting force components in training and validation steps

Efforts (N)	Fx	Fy	Fz
Root mean square error, $RMSE = \sum_{i=1}^{n} \sqrt{\frac{(E_i-P_i)^2}{n}}$	4.79	3.35	4.29
Mean absolute percentage error (%), $MAPE = \sum_{i=1}^{n} \frac{\lvert (E_i-P_i)/E_i \rvert}{n} \times 100$	3.37	3.84	2.86
Mean absolute deviation, $MAD = \sum_{i=1}^{n} \frac{\lvert E_i-P_i \rvert}{n}$	2.04	2.59	2.88
Correlation coefficient, $R^2 = 1 - \frac{\sum_{i=1}^{n}(P_i-E_i)^2}{\sum_{i=1}^{n}(E_i-Y_e)^2}$	0.9915	0.987	0.9953

n number of experiments, E_i experimental value of the ith experiment, P_i predicted value of the ith experiment by model, Y_e average of the experimentally determined values

The results of simulations with a 7-4-1 architecture obtained clearly show the performance of the proposed neural network configuration.

4 Conclusion

A calculation method based on neural networks has been developed to predict the cutting forces generated by a material removal operation with the cutting tool. An architecture with four inputs which are cutting depth (a_p), cutting speed (V_c), feed rate (f) and nose radius (r) and only one output (cutting forces, Fx, Fy and Fz) was used.

Artificial neural networks (ANN) was successfully applied in modeling of cutting forces components (Fx, Fy and Fz) for the selected domain of the input machining parameters which are cutting depth (a_p), cutting speed (V_c), feed rate (f) and nose radius (r). The obtained results of predicted models in terms of cutting force components values corroborate with the experimental ones in the full design space. The mathematical models developed for Fx, Fy, and Fz using the ANN method are very useful for prediction purposes. A close correlation between the predicted and measured data was found with (R^2) varying between (0.98 and 0.99).

The results show that the optimal cutting conditions can be selected in different machining configurations. The elaborated models served as a tool to calculate the cutting forces components for various sets of input parameters. This study allowed us to provide a digital tool for selecting cutting conditions from a minimum of tests and its effectiveness was successfully tested on four representative configurations.

References

1. Davis JR (2000) Nickel, cobalt, and their alloys. ASM international
2. Sato J, Omori T, Oikawa K, Ohnuma I, Kainuma R, Ishida K (2006) Cobalt-base high-temperature alloys. Science 312(5770):90–91
3. Ezugwu E (2005) Key improvements in the machining of difficult-to-cut aerospace superalloys. Int J Mach Tools Manuf 45(12–13):1353–1367
4. Sarıkaya M, Güllü A (2015) Multi-response optimization of minimum quantity lubrication parameters using Taguchi-based grey relational analysis in turning of difficult-to-cut alloy Haynes 25. J Clean Prod 91:347–357
5. Bagci E, Aykut Ş (2006) A study of Taguchi optimization method for identifying optimum surface roughness in CNC face milling of cobalt-based alloy (stellite 6). Int J Adv Manuf Technol 29:940–947
6. Aykut S, Kentli A, Gulmez S, Yazicioglu O (2012) Robust multiobjective optimization of cutting parameters in face milling
7. Schlegel D, Lebaal N, Folea M (2011) Cutting conditions optimization in a cobalt-based refractory material. Recent Res Manuf Eng Cut:156–162
8. Schlegel D, Lebaal N, Folea M (2012) Cost optimization for the cutting a cobalt chrome refractory material. Int J Adv Manuf Technol 60(1–4):55–63
9. Bağcı E, Aykut Ş (2014) The effects of tool position, coating and cutting parameters on forces, power, MRR and wear in face milling of stellite 6. Arab J Sci Eng 39(11):8135–8146
10. Saidi R, Ben Fathallah B, Mabrouki T et al (2019) Modeling and optimization of the turning parameters of cobalt alloy (Stellite 6) based on RSM and desirability function. Int J Adv Manuf Technol 100:2945–2968

11. Nouioua M, Yallese MA, Khettabi R et al (2017) Investigation of the performance of the MQL, dry, and wet turning by response surface methodology (RSM) and artificial neural network (ANN). Int J Adv Manuf Technol 93:2485–2504
12. Meddour I, Yallese MA, Bensouilah H et al (2018) Prediction of surface roughness and cutting forces using RSM, ANN, and NSGA-II in finish turning of AISI 4140 hardened steel with mixed ceramic tool. Int J Adv Manuf Technol 97:1931–1949
13. Fathallah B, Saidi R, Dakhli C, Belhadi S, Yallese M (2019) Mathematical modelling and optimization of surface quality and productivity in turning process of AISI 12L14 free-cutting Steel. Int J Ind Eng Comput 10(4):557–576
14. Labidi A, Tebassi H, Belhadi S et al (2018) Cutting conditions modeling and optimization in hard turning using RSM, ANN and desirability function. J Fail Anal Prev 18:1017–1033
15. Aouici H, Elbah M, Yallese MA et al (2016) Performance comparison of wiper and conventional ceramic inserts in hard turning of AISI 4140 steel: analysis of machining forces and flank wear. Int J Adv Manuf Technol 87:2221–2244
16. Sarikaya M, Güllü A (2014) The analysis of process parameters for turning cobalt-based super alloy Haynes 25/L 605 using design of experiment. Solid State Phenomena
17. Folea M, Schlegel D, Lupulescu N, Parv L (2009) Modeling surface roughness in high speed milling: cobalt based superalloy case study. In: Proceedings of 1st International Conference on Manufacturing Engineering Quality Production System, pp 353–357
18. Bruschi S, Ghiotti A, Bordin A (2013) Effect of the process parameters on the machinability characteristics of a CoCrMo alloy. In: Key Engineering Materials. Trans Tech Publ, vol 554, pp 1976–1983
19. Aykut Ş, Bagci E, Kentli A, Yazıcıoğlu O (2007) Experimental observation of tool wear, cutting forces and chip morphology in face milling of cobalt based super-alloy with physical vapour deposition coated and uncoated tool. Mater Des 28(6):1880–1888
20. Zaman HA, Sharif S, Kim D-W, Idris MH, Suhaimi MA, Tumurkhuyag Z (2017) Machinability of Cobalt-based and Cobalt Chromium Molybdenum Alloys-A Review. Procedia Manuf 11:563–570
21. Yingfei G, de Escalona PM, Galloway A (2017) Influence of cutting parameters and tool wear on the surface integrity of cobalt-based stellite 6 alloy when machined under a dry cutting environment. J Mater Eng Perform 26(1):312–326
22. Aykut Ş, Gölcü M, Semiz S, Ergür H (2007) Modeling of cutting forces as function of cutting parameters for face milling of satellite 6 using an artificial neural network. J Mater Process Technol 190(1–3):199–203
23. Toparli M, Sahin S, Ozkaya E, Sasaki S (2002) Residual thermal stress analysis in cylindrical steel bars using finite element method and artificial neural networks. Comput Struct 80(23):1763–1770
24. Umbrello D (2005) FE analysis of machining processes: innovative experimental techniques for results assessing. PhD thesis, Mechanical Engineering, University of Calabria
25. Umbrello D, Ambrogio G, Filice L, Shivpuri R (2008) A hybrid finite element method–artificial neural network approach for predicting residual stresses and the optimal cutting conditions during hard turning of AISI 52100 bearing steel. Mater Des 29(4):873–883
26. Ambrogio G, Filice L, Umbrello D, Shivpuri R, Hua J (2006) Application of NN technique for predicting the residual stress profiles during hard turning of AISI 52100 steel. In: Proceedings of the 9th ESAFORM Conference, pp 595–598
27. Kafkas F, Karataş Ç, Sozen A, Arcaklioğlu E, Saritaş S (2007) Determination of residual stresses based on heat treatment conditions and densities on a hybrid (FLN2-4405) powder metallurgy steel using artificial neural network. Mater Des 28(9):2431–2442
28. Karataş C, Sozen A, Dulek E (2009) Modelling of residual stresses in the shot peened material C-1020 by artificial neural network. Expert Syst Appl 36(2):3514–3521

A Finite Element Procedure for Thermal Buckling Analysis of Functionally Graded Shell Structures

S. Trabelsi(✉), S. Zghal, and F. Dammak

Laboratory of Electromechanical Systems (LASEM), National Engineering School of Sfax, University of Sfax, B.P W3038 Sfax, Tunisia
sorourtrabelsi@yahoo.fr, souhirzghal@yahoo.fr, fakhreddine.dammak@enis.tn

Abstract. The purpose of this paper is to explore the buckling problems of functionally graded conical shells due to thermal loadings. The response is obtained for a uniform increase in temperature along the thickness direction of the shell. The equations governing the behavior of the conical shell are written from the modified Reissner-Mindlin formulation. Properties of the shell are estimated using the Voigt rule of mixture via the power function. The temperature dependence of the material constituents is also considered. A comparison study of the obtained results with those available in the literature is presented in order to validate the proposed model. Then, the effects of the power-law exponent and geometrical parameters are examined.

Keywords: Thermal buckling · Functionally graded material · Rule of mixture · Modified Reissner-Mindlin theory

1 Introduction

The conical shell structures have been designed to perform in various engineering applications such as the rocket nozzle liner, the shafts of gas turbine and the aircraft jet engine. In fact, these structures may receive heat and thermo-mechanical loadings which may induce buckling and post-buckling problems. Recently, the Functionally Graded Materials (FGMs) are recognized as materials capable to offer reliable engineering solutions for industrial applications involving high thermo-mechanical loadings. The use of ceramics and metals together and the smooth variation of the microstructure from one material to another can avoid the problems related to stress concentrations and delamination [1]. Based on these unique properties, many attempts have been made for static, dynamic, bending, buckling, vibration and piezoelastic analysis of FG structures [2, 3, 4–6, 13, 4–6].

On the other hand, modeling of the kinematics of these structures can be achieved using three approaches namely: the Classical plate theory (CPT), the Reissner-Mindlin theory (FSDT) and the higher-order theory (HSDT). A number of investigations have been conducted on the buckling and the non-linear deflection of FGM conical shells under thermal loads. Buckling of the FG truncated conical shells under thermal field

N. Aifaoui et al. (Eds.): CMSM 2019, LNME, pp. 409–416, 2020.
https://doi.org/10.1007/978-3-030-27146-6_44

was investigated by Bhangale et al. [7] by applying a semi-analytical finite element procedure based on the FSDT theory. With the same theory, Naj et al. [8] derived an analytical solution to investigate the thermal and mechanical buckling of FGM truncated conical shells. Akbari et al. [9] presented a semi-analytical solution to solve buckling analysis of FGM conical shell using the classical shell theory. Besides, Sofiyev [10, 11] presented some significant findings for thermal buckling analysis of FG conical shells. Furthermore, the thermo-mechanical buckling of FG conical panels were examined by Zhao and Liew [12] with finite element procedure based on the FSDT theory.

Thermal buckling investigations of conical shell structure are restricted in number and most of these investigations have been conducted upon the exact solution. This paper deals with the non-linear buckling analysis of FGM conical shells under thermal loads using a finite element (FE) procedure. The formulation is based on a modified Reissner-Mindlin theory which adopts a quadratic function in the description of the shear stains. The proposed approach represents an amelioration in the Reissner-Mindlin theory in terms of the computational time and the description of large deformations and rotations. In the present study, critical temperatures of FGM conical shells subjected to uniform distributed temperature along the thickness are obtained by solving the eigenvalue problem. The rule of mixture is used to find the thermo-mechanical properties of the FG cones. The present results of the critical buckling temperatures are compared with the literature in order to highlight the efficiency of the modified Reissner-Mindlin theory.

2 Material Properties of FGM Conical Shells

The FGM cone is assumed to be made of metal and ceramic phases. The material properties of these components vary gradually in the thickness direction and they can be evaluated by the rule of mixture, given as follow:

$$P(T,z) = [P_m(T) - P_c(T)]V_m(z) + P_c(T), V_m(z) = \left(\frac{1}{2} + \frac{z}{h}\right)^p \quad (1)$$

where $P_m(T)$ and $P_c(T)$ denote the effective material properties of the metal and ceramic constituents, respectively. $V_m(z)$ is the volume fraction of metal phase and p is the power-law exponent. Moreover, it should be noted that the material properties of the shell can be temperature-dependent (TD) and the dependency may be expressed as follow:

$$P(T,z) = P_0\left(P_{-1}T^{-1} + 1 + P_1T + P_2T^2 + P_3T^3\right) \quad (2)$$

where P_0, P_{-1}, P_1, P_2 and P_3 are the temperature coefficients.

3 Basic Equations of FGM Shell

3.1 Kinematics of the Shell Model and Strain Field

The position vectors of any material point (p) are denoted by bold letters which are given in the initial and current configurations C_0 and C_t, respectively as follows:

$$\mathbf{X}_q = \mathbf{X}_p + z\mathbf{D}, \mathbf{x}_q = x_p + z\mathbf{d} \tag{3}$$

$z \in [-h/2, h/2]$ is the thickness of the structure and $\mathbf{d}$ denote the director vector. In large deformations case, the strain field is defined using the Green-Lagrange strain tensor. The membrane and bending strain vectors are given as:

$$\mathbf{e} = \begin{bmatrix} e_{11} \\ e_{22} \\ 2e_{12} \end{bmatrix}, \chi = \begin{bmatrix} \chi_{11} \\ \chi_{22} \\ 2\chi_{12} \end{bmatrix}, \tag{4}$$

The transverse shear strain vector is defined as:

$$\gamma = f(z)\begin{bmatrix} \gamma_{13} \\ \gamma_{23} \end{bmatrix}, f(z) = 5/4\left(1 - 4(z/h)^2\right). \tag{5}$$

The function $f(z)$ allows a quadratic distribution of the transverse shear strains and constitutes an amelioration in the Reissner-Mindlin theory [13].

The generalized strain field is expressed as follow:

$$\Sigma = [\mathbf{e} \quad \chi \quad \gamma]^T \tag{6}$$

3.2 Weak Form and Constitutive Relations

The weak form of equilibrium equation is expressed as:

$$G = \int_A [\mathbf{N} \cdot \delta\mathbf{e} + \mathbf{M} \cdot \delta\chi + \mathbf{T} \cdot \delta\gamma] dA - G_{ext} = 0 \tag{7}$$

where **N**, **M** and **T** denotes the components of the stress resultants (membrane, bending and shear, respectively). The stress resultant vector is obtained as:

$$\mathbf{R} = \begin{bmatrix} \mathbf{N} \\ \mathbf{M} \\ \mathbf{T} \end{bmatrix} \tag{8}$$

The constitutive relation of the FG conical shell under thermal loadings can be expressed as:

$$\mathbf{R} = \mathbf{H_T\Sigma} - \mathbf{R^{th}}, \mathbf{H_T} = \begin{bmatrix} \mathbf{H_m} & \mathbf{H_{mb}} & \mathbf{0} \\ \mathbf{H_{mb}} & \mathbf{H_b} & \mathbf{0} \\ \mathbf{0} & \mathbf{0} & \mathbf{H_s} \end{bmatrix} \tag{9}$$

$\mathbf{R^{th}}$ represents the thermal resultants and $\mathbf{H_T}$ denotes the elastic modulus, where the components $\mathbf{H_m}, \mathbf{H_{mb}}$ and $\mathbf{H_b}$ are given as follows:

$$(\mathbf{H_m}, \mathbf{H_{mb}}, \mathbf{H_b}) = \int_{-h/2}^{h/2} (1, z, z^2)\mathbf{H}dz, \mathbf{H_s} = \int_{-h/2}^{h/2} f(z)^2 \mathbf{H_\tau} dz \tag{10}$$

$$\mathbf{H} = \frac{E(z)}{1-\nu^2(z)} \begin{bmatrix} 1 & \nu(z) & 0 \\ \nu(z) & 1 & 0 \\ 0 & 0 & \frac{(1-\nu(Z))}{2} \end{bmatrix}, \mathbf{H_\tau} = \frac{E(z)}{2(1+\nu(z))} \begin{bmatrix} 1 & 0 \\ 0 & 1 \end{bmatrix} \tag{11}$$

E(z) and $\nu(z)$ are the Young's modulus and the Poisson's ratio, respectively. The temperature rise ΔT can be considered as uniform or non-uniform through the thickness direction, which equations can be written as follows:

$$\Delta T = T_f - T_i \tag{11}$$

where T_i is the initial temperature of the shell and T_f is the final temperature.

$$\Delta T(z) = T(z) - T_i = (T_m - T_i) + (T_c - T_m)\eta(z) \tag{12}$$

where $\eta(z)$ represents the non-uniform temperature distribution function given by:

$$\eta(z) = \frac{\int_{-h/2}^{z} \frac{dz}{k(Z)}}{\int_{-h/2}^{h/2} \frac{dz}{k(Z)}} \tag{13}$$

3.3 Thermal Buckling Problem

After the derivation of the weak form of equilibrium equations and the constitutive relations, the finite element discretization is solved using 4-node shell elements. For more details, one can be referred to [14]. Then, the determination of the critical buckling temperature is obtained by the resolution of the eigenvalue problem which its expression can be given by:

$$(\mathbf{K_M} + \lambda_{cr}\mathbf{K_G}).\mathbf{U_n} = 0 \tag{14}$$

$\mathbf{K_M}$ and $\mathbf{K_G}$ are the material and the geometric matrices, respectively. λ_{cr} is the critical (eigenvalue) buckling temperature.

4 Results and Discussion

Firstly, the accuracy of the proposed modified Reissner-Mindlin model is tested for thermal buckling analysis of clamped (CC) FGM conical shells made of Alumina (Al_2O_3) and stainless steel (SUS304). The material properties are highly TD and determined from experiments (Reddy and Chin [15], where the temperature coefficients are tabulated in Table 1. The FG cone is subjected to uniform temperature rise and its geometrical properties are given as reported by [7]: $(\alpha = 30^\circ, R_1/h = 215.8026 \text{ and } L/h = 304.7896)$. The conical shell is meshed using S4 FE with 40 elements along the z-direction. The obtained results are compared to those given by [7] and listed in Table 2. The results reveal a good agreement with those reported in the literature. In fact, the relative error for different power coefficients (p) is between 1–5% and this can be explained by the use of the quadratic function in the present formulation. In addition, the continuous material gradient is defined in this work through the variation at Gauss integration points of the element, while [7] adopts a layered model for the computation of materials grading variation.

Furthermore, Table 2 demonstrates that the fully metallic cone ($p = 0$) has the lowest critical temperature and this temperature increases with an increase in ceramic constituents ($p > 1$). Thus, the stiffness of the structure becomes higher with the ceramic reinforcements.

Table 1 Material properties of$\mathbf{Al_2O_3/SUS304}$ [15]

Material	Properties	P_0	P_{-1}	P_1	P_2	P_3
Al_2O_3	E(GPa)	349.55e9	0	−3.853e−4	4.027e−7	−1.673e−11
	ν	0.26	0	0	0	0
	α (K^{-1})	6.8269e−6	0	1.834e−4	0	0
	K(W/mK)	−14.087	−1123	−6.227e−3	0	0
SUS304	E(GPa)	201.04e9	0	3.079e−4	−6.534e−7	0
	ν	0.3262	0	−2.002e−4	3.797e−7	0
	α (K^{-1})	12.33e−6	0	8.086e−4	0	0
	K(W/mK)	15.379	0	0	0	0

Table 2 Critical thermal buckling temperature ΔT_{cri}(K) of clamped $\mathbf{Al_2O_3/SUS304}$ conical shell with TD material properties under different power-law index p

p	Present	Semi-Analytical [7]	P. Error %
0.0	146.77	140.74	4.1
0.5	177.1	170.29	3.8
1	193.22	188.31	2.5
5	231.25	235.68	1.9
10	243.12	251	3.24
15	248.36	257.69	3.75
100	259.89	273	5
1000	262	274.28	4.6

In the following, a parametric study is conducted to examine the influences of the cone angle 'α' and the power law coefficient 'p' on the critical buckling temperature (ΔT_{cri}). Figure 1 depicts the variation of ΔT_{cri} of FG clamped conical shell under uniform temperature rise with different power law coefficient and various value of cone angle. As the numerical results show, the conical shells with the lower cone angle (α = 15°) have the highest thermal stability resistance in comparison with the conical shells with (α = 30°). Clearly, the elastic stiffness of the structures decreases with the increase on the cone angle. The first mode shape of the FG conical shell under uniform temperature rise (UTR) through the thickness with (p = 1) is provided in Fig. 2.

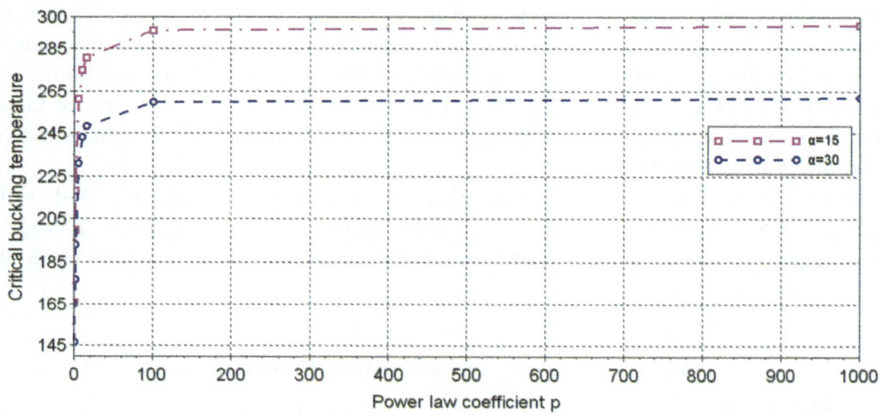

Fig. 1 Effect of cone angle α and the power law index on critical buckling temperature rise ΔT_{cri} of FG conical shells with TD material properties

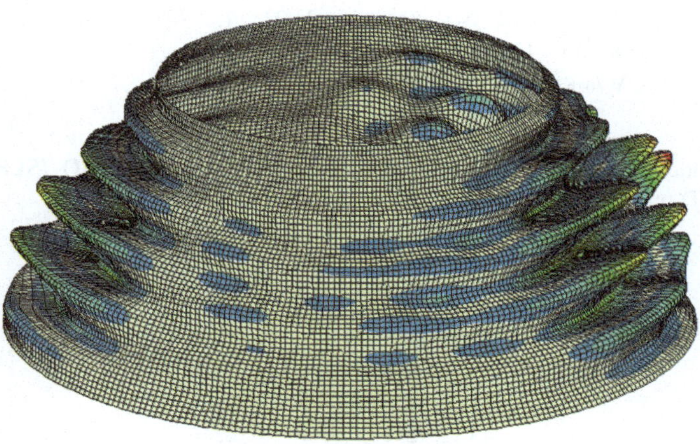

Fig. 2 Thermal buckling mode shape of FG conical shell under UTR with (p = 1)

5 Conclusion

Thermal buckling analysis of FG conical shell is presented. The governing equations are developed using a modified Reissner-Mindlin theory. This model allows the description of a parabolic distribution of the shear deformation along the shell thickness. In fact, the standard FSDT theory used in the literature considers the effects of a constant transverse shear deformation and a shear correction factors are required. The effective material properties of the FG cone are assumed to be graded across the thickness direction and temperature dependent. The numerical results show that the present model provides the assessment of the critical buckling temperature of FG conical shells with the compromise good accuracy/low computational time. Furthermore, the flexural rigidity of the FG cone subjected to thermal loads is improved with the variation of the power-law coefficient.

References

1. Koizumi M (1997) FGM activities in Japan. Compos Part B Eng 28:1–4
2. Zghal S, Frikha A, Dammak F (2017) Static analysis of functionally graded carbon nanotube-reinforced plate and shell structures. Compos Struct 176:1107–1123
3. Wali M, Hajlaoui A, Dammak F (2014) Discrete double directors shell element for the functionally graded material shell structures analysis. Comput Methods Appl Mech Eng 278:388–403
4. Zghal S, Frikha A, Dammak F (2018) Non-linear bending analysis of nanocomposites reinforced by graphene-nanotubes with finite shell element and membrane enhancement. Eng Struct 158:95–109
5. Zghal S, Frikha A, Dammak F (2018) Mechanical buckling analysis of functionally graded power-based and carbon nanotubes-reinforced composite plates and curved panels. Compos Part B 150:165–183
6. Zghal S, Frikha A, Dammak F (2018) Free vibration analysis of carbon nanotube-reinforced functionally graded composite shell structures. Appl Math Model 53:132–155
7. Bhangale R, Ganesan N, Padmanabhan Ch (2006) Linear thermoelastic buckling and free vibration behavior of functionally graded truncated conical shells. J Sound Vib 292:341–371
8. Naj R, Sabzikar M, Eslami MR (2008) Thermal and mechanical instability of functionally graded truncated conical shells. Thin Wall Struct 46(1):65–78
9. Akbari M, Kiani Y, Eslami MR (2014) Thermal buckling of temperature-dependent FGM conical shells with arbitrary edge supports. Acta Mech 226(3):897–915
10. Sofiyev AH (2011) Thermal buckling of FGM shells resting on a two-parameter elastic foundation. Thin Wall Struct 49(10):1304–1311
11. Sofiyev AH, Zerin Z, Kuruoglu N (2017) Thermoelastic buckling of FGM conical shells under non-linear temperature rise in the framework of the shear deformation theory. Compos Part B 108:279–290
12. Zhao X, Liew KM (2011) An element-free analysis of mechanical and thermal buckling of functionally graded conical shell panels. Int J Numer Methods Eng 86:269–285
13. Trabelsi S, Frikha A, Zghal S, Dammak F (2018) Thermal post-buckling analysis of functionally graded material structures using a modified FSDT. Int J Mech Sc 144:74–89

14. Trabelsi S, Frikha A, Zghal S, Dammak F (2019) A modified FSDT-based four nodes finite shell element for thermal buckling analysis of functionally graded plates and cylindrical shells. Eng Struct 178:444–459
15. Reddy JN, Chin CD (1998) Thermomechanical analysis of functionally graded cylinders and plates. J Therm Stresses 21(6):593–626

Thermal Expansion Behavior of Al 2017 Alloy Matrix Composites Prepared by Stir Casting

Mariem Bhouri[1(✉)] and Foued Mzali[1,2]

[1] Laboratory of Thermal and Energetic Systems Studies, Monastir University, National Engineering School of Monastir (ENIM), Monastir, Tunisia
foued.mzali@enim.rnu.tn

[2] Department of Mechanical Engineering, Monastir University, National Engineering School of Monastir (ENIM), Monastir, Tunisia

Abstract. In this present work, an attempt has been made to produce metal matrix composite using Al 2017 alloy as matrix material reinforced with graphite particles using stir casting technique. Initially, Al 2017 alloy charged into a crucible was superheated in the furnace. Besides, preheated graphite particles were dispersed into the vortex of molten 2017A alloy. The mechanical stirring was carried out to improve wettability and distribution. The composite mixture was poured into a non-permanent mold with including the form of normalized specimens tensile. Different Microstructures of Al 2017 alloy matrix composites showed more porosity. Thermal conductivity of the prepared composite was determined before and after the addition of graphite particles to note the extent of improvement.

Keywords: Stir casting · Non-permanent mold · Aluminium matrix composites · Thermal expansion property

1 Introduction

Particle reinforced metal matrix composites (PMMC) are becoming one of the most promising candidate for electronic packaging applications [15] due to their tailorable thermophysical properties, such as thermal conductivity (TC) to dissipate the heat, and low coefficient of thermal expansion (CTE) to decrease the thermal expansion mismatch among the devices. In particular, Aluminium matrix composites reinforced with graphite particles are used because of its high (TC) and low density of the Al metal phase and low (CTE) of the Graphite [1, 4]. Any fluctuation with temperature can induce large thermal stresses due to the CTE mismatch between matrix and reinforcements which, ensuing thermal distortion, thermal fatigue and creep failures. It requires a deep understanding in order to optimize the content of reinforcement providing a good coefficient of thermal expansion (CTE) and thermal conductivity (TC) matching for thermal management.

N. Aifaoui et al. (Eds.): CMSM 2019, LNME, pp. 417–425, 2020.
https://doi.org/10.1007/978-3-030-27146-6_45

Several researchers have studied the thermal expansion behaviour of Aluminium matrix composites containing various types of reinforcement as SiC [7], Al_2O_3 [3] and graphite [1]. They showed that the CTE was affected by several factors such as temperature, amount and distribution of the reinforcement phase, properties of the interfacial phase and thermal history. With increasing the amount of the SiC particles, the CTE of Al 7075 alloy composites decreased [7]. A similar observation was made in respect of the CTE of Aluminum composites reinforced with SiC particles [9]. Deng et al. [2] reported that the CTE of Al 2024 alloy composites reinforced with carbon nanotubes was lower than unreinforced matrix. Elomari et al. [3] showed that the CTE of pre-strained Aluminium composites reinforced with Al_2O_3 increased also with the level of plastic yielding and flow. They attributed this to the thermal stress induced in the composites resulting from the difference in the CTE between the reinforcement and matrix.

In our case, to manufacture composites, stir casting process will be employed because the final distribution of reinforcement particles in the matrix could be controlled [10, 11]. The major advantage of stir casting process is that it offers better matrix-particles bonding due to stirring action of particles into the melt [8].

In this study, stir casting method was used to elaborate Al 2017 alloy composites with different content of graphite. Microstructures and the CTE were analysed while varying the volume fraction of graphite particles within Al 2017 alloy in order to reveal the effect of graphite on the CTE of composites. Furthermore, the thermal expansion of the specimens that were produced has been measured via dilatometry method.

2 Experimental Procedure

Stir casting technique was used to fabricate Al 2017 alloy composites reinforced with graphite particles. The matrix metal was Al 2017 alloy (Composition in wt% Cu 4.01, Si 0.66, Mn 0.64, Mg 0.57 and rest Al). Graphite particles with average size 75 μm was used as reinforcement

Firstly, The Al 2017 alloy was superheated to 800 °C, above the liquidus temperature, to melt completely. Then, to preserve the slurry in the semi-solid state, the melt Al 2017 alloy was cooled below the liquidus temperature. Secondly, 2.5, 5 and 7.5 vol. % of preheated graphite at 500 °C for 30 min were added to the melt Al 2017 alloy in graphite crucible and mixed by using the preheated stirrer (Fig. 1a). Then, the composite slurry was heated again to a liquid state and the mechanical mixing was carried out for 15 min at an average mixing speed of 700 rpm to improve uniform distribution and to promote wettability without inducing turbulence flow or gas entrapment.

The mixture was poured in a non-permanent mould with including the form of standardized specimen's tensile (Fig. 1a). The cast specimens tensile were presented in Fig. 1b).

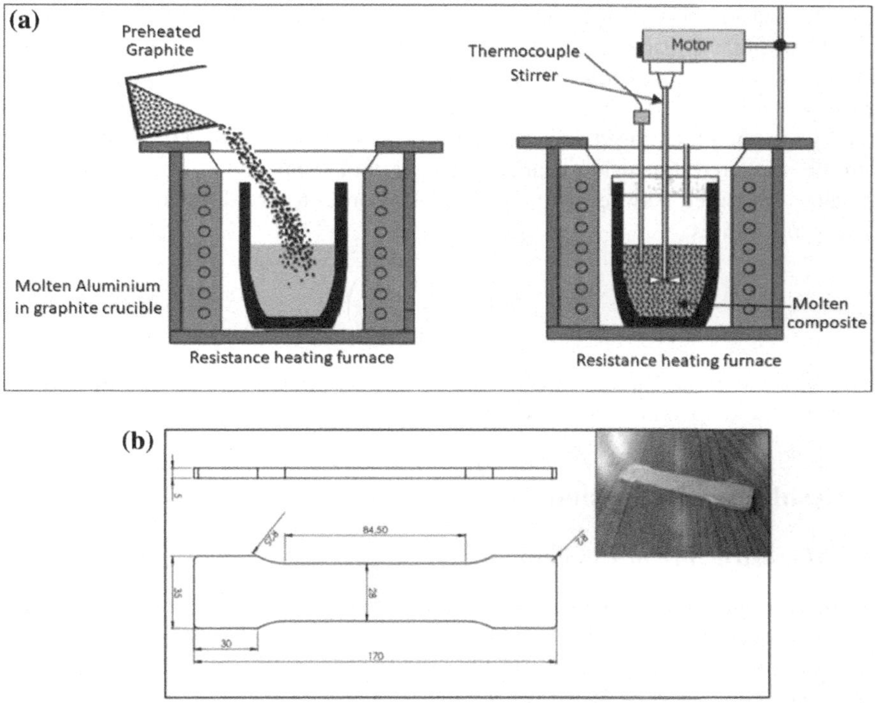

Fig. 1 **a** Stir Casting setup **b** Tensile test specimen of composites

The specimens test were cut and polished in order to characterize the microstructures and the thermophysical properties. Microstructural characterization studies were performed by using optical microscopy (OM) after slight etching with Keller's reagent of composites with different content of graphite.

For density test, the measurement of dimensions and weight of each composites were applied. Then, relative and theoretical densities were calculated following Eqs. (1) and (2), in which the theoretical densities of Al 2017 alloy and graphite were set equal to 2.79 g/cm^3 and 2.24 g/cm^3, respectively.

$$D = \rho_{measured} / \rho_{theoretical} \quad (1)$$

$$\rho_{theoretical} = \rho_{composite} = \rho_m V_m + \rho_p (1 - V_m) \quad (2)$$

where D the relative density, $\rho_{measured}$ the sintered density, $\rho_{theoretical}$ the theoretical density of composite. V is the volume fraction, ρ is the density of the component, and the indices m and p denote the matrix and particles phases, respectively.

CTE measurements for composites were conducted in SETSYS Evolution TMA from 42 to 500 °C under the protective argon gas atmosphere. Composites samples with 8 × 8 mm and 3–4 mm thickness were used. The heating and cooling rates for measurements were 10 °C/min and 30 °C/min respectively were continuously monitored by the computer-based data acquisition system. The linear CTE can be calculated from the measured dimension change as function of temperature (Eq. 3) by averaging the values measured between 250 and 400 °C in order to avoid fluctuations induced at lower temperatures:

$$CTE = \frac{1}{L_0}\left(\frac{\Delta L}{\Delta T}\right) \quad (3)$$

where L_0 is the initial length of the sample at room temperature and ΔL is the change in length over a temperature range ΔT.

3 Result and Discussion

3.1 Microstructure of Composites

The microstructures of composites reinforced with different content of graphite particles are observed (Fig. 2) by conducting the microscopic observation. The visual examinations indicate that after the Keller reagent chemical attack for 30 s, the majority phase (Al) is clear while the phase (Graphite) is dark with finer size. Then, we distinguish the so-called "primary" α (Al) dendrites as well as the eutectic θ (Al_2Cu) present with a "coppery" shade in the matrix (Al). Also, it can be observed that a uniform distribution of particles and also some agglomeration of particles at few places were detected in the composites reinforced with 7.5wt% Graphite (Fig. 2d). In addition, casting defects such as porosity and blisters have been seen also in the micrographs (Fig. 2a). Residual pores were sited at the interface between dendrites of Al 2017 alloy. It was reported that the porosity in cast Al 2017 alloy was instigated by gas entrapment during stirring, and shrinkage during solidification process [6]. Introducing graphite particles into matrix induced also the increase of porosity in composites. In fact, the presence of graphite particles in the melt Al 2017 alloy matrix reduced the liquid metal flow [12].

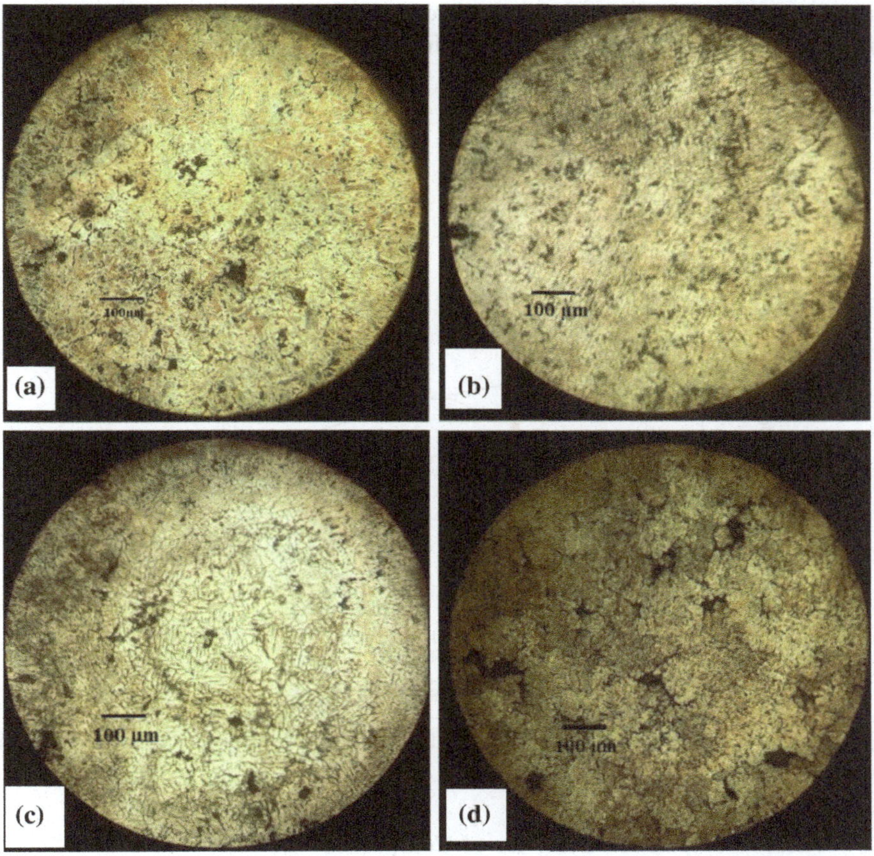

Fig. 2 Optical microstructure of Al 2017 matrix composites by stir casting with (a) 0%, (b) 2.5%, (c) 5% and (d) 7.5% of graphite particles

3.2 Density of Composites

Measured and theoretical densities of cast Al 2017 alloy and its composites are given in Fig. 3. The measured density of cast Al 2017 alloy is lower compared to the theoretical density. This is explained by the presence of pores resulted during stir casting. In addition, it shows that the density of the composite decreases as the amount of graphite particles increases. Because of the less bonding status between graphite particles and Al 2017 alloy, the density of non-unreinforced composites is higher than that of reinforced composites. Besides, Fig. 3 displays also the porosity (Eq. 4) as a function of Graphite particles, it apparently reveals that the porosity of all composites increased markedly with increasing graphite particles volume fraction.

$$Porosity(\%) = [1 - (\rho_{measured}/\rho_{theoretical})]x100 \quad (4)$$

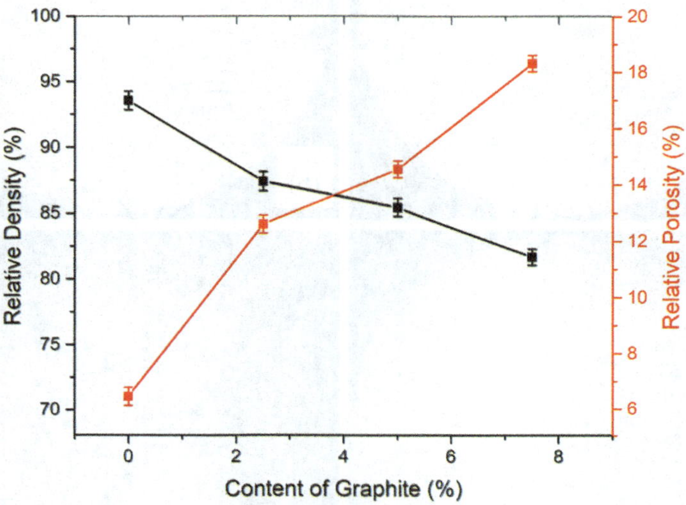

Fig. 3 Relative density and relative porosity of composites with different contents of graphite particles

3.3 Thermal Expansion Behavior of Composites

As mentioned above, coefficient of thermal expansion (CTE) of composites is important property for thermal management applications which were determined following Eq. (3) and compared with estimated values based on theoretical models for composite presented in [14].

Compared to the CTE value of Al 2017 alloy from literature, the obtained CTE (Fig. 4) of cast Al 2017 alloy is low. This is due to the presence of the high defect such as porosities [14].

Figure 4 shows the variation of obtained CTE of composites with increasing volume fraction of graphite particles. The plots in Fig. 4 indicate the calculated coefficient of thermal expansion based on theoretical models presented in [14].

It is noticed that CTE decreases with increasing graphite content. As the amount of graphite increases from 2.5 to 7.5 wt%, there is a slight decrease in CTE from 22.08×10^{-6} to 21.25×10^{-6} °C^{-1}. A gradual decrease in CTE values can be explained by the lower CTE of graphite (2.7×10^{-6} °C^{-1}) than Al 2017 alloy (22.9×10^{-6} °C^{-1}). Thus, the expansion of Al 2017 alloy matrix was constrained by graphite phase.

Similar results were verified by Kuen-Ming Shu and G.C. Tu [13] which they used Copper/silicon carbide composites. They noticed that with the increase of Silicon carbide particles, CTE values decrease. They explained this decrease by the generation of huge amount of residual stress in the composites while the content of reinforcement phase increase due to the large difference between the CTEs of the reinforcement and matrix. Etter et al. [5] used two unfiltered graphite with AlSi7Mg matrix to prove that the CTE of composite decreases at least related to monolithic AlSi7Mg. In addition, the use of 10 wt% MWNT to reinforced Al 2024 alloy allows the drop of the CTE of the composites [2].

Figure 4 compares the experimental CTEs obtained for Al 2017 alloy composites reinforced with different content of graphite particles and those calculated using the models. It is seen that all the models overestimated the CTEs of the composites, which could be due to deficiencies inherent in the original assumptions of the models. For instance, these models did not consider the effects of porosity, matrix plasticity and reactions between two phases.

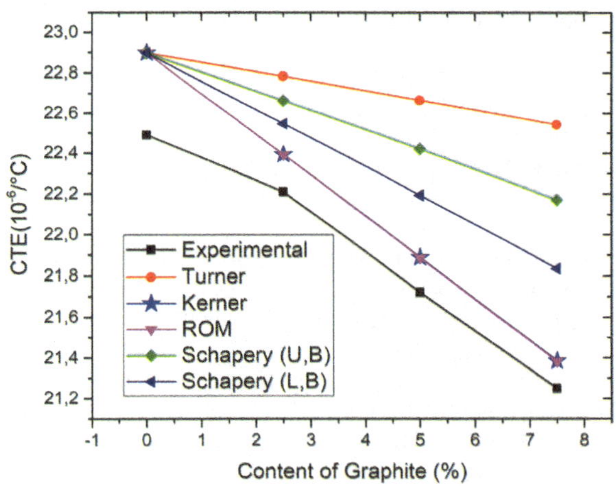

Fig. 4 Comparison of theoretical models and experimental CTE of composites with different contents of graphite particles

4 Conclusion

The present work on preparation of Al 2017 alloy metal matrix composite for different volume fractions of graphite particles by stir casting and evaluation of thermophysical and microstructural properties has led to the following conclusion:

- The composites containing 2.5, 5 and 7.5% of graphite particulates were successfully synthesized by melt stirring method using three stages mixing combined with preheating of the reinforcing phase.
- The optical micrographs of composites produced by stir casting method show the fairly random distribution of graphite particles in the Al 2017 alloy metal matrix.

The microstructure of the composites contained the primary α-Al dendrites as well as the eutectic θ (Al_2Cu).
- During stir casting method, several defects such as porosity and blisters are presented.
- The lower density of graphite avoids the formation of a denser structure of composites and as a result, porosity increases with graphite addition.
- Introducing graphite particles into Al 2017 alloy matrix resulted in the decreased thermal expansion of the composites since graphite has a lower thermal expansion coefficient than that of the Al 2017 alloy matrix.
- Theoretical models could not predict the experimental values of CTE due to the presence of many defects such as porosities during casting process.

References

1. Chen JK, Huang IS (2013) Thermal properties of aluminum-graphite composites by powder metallurgy. Compos. Part B Eng 44(1):698–703. https://doi.org/10.1016/j.compositesb.2012.01.083
2. Deng CF, Ma YX, Zhang P, Zhang XX, Wang DZ (2008) Thermal expansion behaviors of aluminum composite reinforced with carbon nanotubes. Mater Lett 62(15):2301–2303. https://doi.org/10.1016/j.matlet.2007.11.086
3. Elomari S, Boukhili R, Lloyd DJ (1996) Thermal expansion studies of prestrained Al_2O_3/Al metal matrix composite. Acta Mater 44:1873–1882. https://doi.org/10.1016/1359-6454 (95) 00299-5
4. Estrada-Guel I, Carreño-Gallardo C, Mendoza-Ruiz DC et al (2009) Graphite nanoparticle dispersion in 7075 aluminum alloy by means of mechanical alloying. J Alloys Compd 483 (1–2):173–177. https://doi.org/10.1016/j.jallcom.2008.07.190
5. Etter T, Papakyriacou M, Schulz P, Uggowitzer PJ (2003) Physical properties of Graphite/Aluminum composites produced by gas pressure infiltration method. Carbon 41: 1017–1024. https://doi.org/10.1016/s0008-6223(02)00448-7
6. Hashim J, Looney L, Hashmi MSJ (1999) Metal matrix composites: production by the stir casting method. J Mater Process Technol 92–93:1–7. https://doi.org/10.1016/s0924-0136 (99)00118-1
7. Karthikeyan B, Ramanathan S, Ramakrishnan V (2010) Thermo physical property measurement of metal-matrix composites. Mater Des 31:S82–S86. https://doi.org/10.1016/j.matdes.2009.12.007
8. Kok M (2005) Production and mechanical properties of Al_2O_3 particle-reinforced 2024 aluminum alloy composites. J Mater Process Technol 161: 381–387. https://doi.org/10.1016/j.jmatprotec.2004.07.068
9. Lee HS, Hong SH (2003) Pressure infiltration casting process and thermophysical properties of high volume fraction SiC p/Al metal matrix composites. Mater Sci Technol 19: 1057–1064. https://doi.org/10.1179/026708303225004396
10. Narciso J, Alonso A, Pamies A, Cordovilta A et al (1994) Wettability of binary and a ternary alloy of the system Al–Si–Mg with SiC particulates. Scr Metall 31: 1495–1500. https://doi.org/10.1016/0956-716x(94)90063-9
11. Ray S (1995) Casting of metal matrix composites. Key Eng. Mater 104–107: 417–446. https://doi.org/10.4028/www.scientific.net/KEM.104-107.417

12. Sajjadi SA, Ezatpour HR, Torabi Parizi M (2012) Comparison of microstructure and mechanical properties of A356 aluminum alloy/Al_2O_3 composites fabricated by stir and compo-casting processes. Mater Des 34: 106–111. https://doi.org/10.1016/j.matdes.2011.07.037
13. Shu KM, Tu GC (2003) The microstructure and the thermal expansion characteristics of Cu/SiCp composites. Mater Sci Eng A 349: 236–247. https://doi.org/10.1016/s0921-5093(02)00788-8
14. Uju WA, Oguocha INA (2012) A study of thermal expansion of Al-Mg alloy composites containing fly ash. Mater Des 33(1): 503–509. https://doi.org/10.1016/j.matdes.2011.04.056
15. Zweben C (1998) Advances in Composite Materials for Thermal Management in Electronic Packaging. JOM 50(6): 47–51. https://doi.org/10.1007/s11837-998-0128-6

Material and Geometric Nonlinear Analysis of Ceramic/Metal Functionally Graded Cylindrical Shell

H. Jrad[2(✉)], J. Mars[2], M. Wali[1,2], and F. Dammak[2]

[1] Department of Mechanical Engineering, College of Engineering, King Khalid University, Abha, Saudi Arabia
[2] Laboratory of Electromechanical Systems (LASEM), National Engineering School of Sfax, University of Sfax, Route de Soukra Km 4, 3038 Sfax, Tunisia
hanen.j@gmail.com, jamelmars@yahoo.fr, mondherwali@yahoo.fr, Fakhreddine.dammak@enis.tn

Abstract. This paper presents material and geometric nonlinear analysis of ceramic/metal functionally graded cylindrical shell using a user-defined subroutine (UMAT) developed and implemented in Abaqus/Standard. The behavior of the ceramic/metal functionally graded cylindrical shell is assumed elasto-plastic with isotropic hardening according to Ludwik hardening law. Using the Mori–Tanaka model and self-consistent formulas of Suquet, the effective elastoplastic material properties are determined and are assumed to vary smoothly through the thickness of the cylindrical shell. The effects of the geometrical parameters and the material distribution on nonlinear responses are examined.

Keywords: 4-node cylindrical shell · Functionally graded structure · Suquet formulations · Elastoplastic composite

1 Introduction

Functionally graded materials (FGMs) are the advanced materials with varing properties in dimensions. These are made of two or more constituent phases with continuous and smoothly varying composition in preferred directions. The properties of material depend on the spatial position in the structure of material. Functionally graded materials are currently being applied in a number of industries, with a huge potential to be used in other applications in the future.

These materials are gaining attention for aerospace, automobile, biomedical, defence, electrical/electronic, energy, marine, and thermoelectronics owing to extraordinary mechanical, thermal, and chemical properties. Functionally graded materials are also ideal for reducing the mismatch in the thermo-mechanical properties in metal–ceramic bonding that help to prevent debonding and to sustain severe working environments. Many researchers have shown great interest in modelling FG shells, namely [1–5, 8, 10–16, 19–21, 22, 24, 25] using different shells theories.

N. Aifaoui et al. (Eds.): CMSM 2019, LNME, pp. 426–434, 2020.
https://doi.org/10.1007/978-3-030-27146-6_46

In various engineering applications FGM structures, especially areas including the structural and the engineering applications can endure large displacements and withstand large deformations and large rotations. However, a linear analysis of FGM shell structures in areas requiring a high degree of safety is insufficient, FGM element type is not available in the digital software item library.

Further, understanding the elastic-plastic behaviour of FG composites is important, many works defined the relationship between stress and total strain in elasto-plastic materials in UMAT subroutine and implemented it into abaqus, [6, 7, 17].

The goal of this work is to have an accurate modeling of geometry, constitutive laws and an efficient resolution strategy in order to more realistically simulate the response of this type of structures.

In this context, it is proposed to extend finite element modeling using the ABAQUS software to take into account the elasto-plastic response of FGM shells undergoing large displacements and rotations under static loading. The elastoplastic modeling is based on 4-node shell element and a user defined subroutine (UMAT) is implemented in Abaqus/Standard to conduct the material nonlinear analysis of the ceramic/metal functionally graded cylindrical shell.

2 Theoretical Formulation

In this section, the geometry and kinematics of the geometrically non-linear elasto-plastic FGM model are briefly described.

2.1 Material Properties of FG Cylindrical Shell

The material properties of the particle reinforced metal matrix composite cylindrical shell are assumed to vary continuously throughout the thickness direction (z-axis direction), according to the following equation:

$$Y_{FGM}(z) = Y_m + (Y_c - Y_m)\left(\frac{z}{h} + \frac{1}{2}\right)^p \tag{1}$$

where subscripts m and c denote metal and ceramic, respectively. Y designates Young modulus and p is the power-law index. The Poisson's ratio for both metal and ceramic is assumed to be constant. The Mori–Tanaka model is employed to describe the elastic properties of studied FGM composed of elastoplastic metal matrix reinforced by elastic ceramic particles. Accordingly, Poisson's ratio and Young's modulus can be expressed through the thickness as function of The effective bulk modulus $k(z)$ and the effective shear modulus $G(z)$:

$$v(z) = \frac{1}{2}\frac{3k(z) - 2G(z)}{3k(z) + G(z)} \quad ; \quad E(z) = 2G(z)(1 + v(z)) \tag{2}$$

2.2 Kinematic Assumptions, Weak Form and Constitutive Relations

The position vector of any material point (*q*) of the shell structure located at the distance z from the mid surface can be related to the position vector of (p) in both initial and deformed configurations as:

$$\begin{cases} \mathbf{X}_q(S^1,S^2,z) = \mathbf{X}_p(S^1,S^2) + z\mathbf{D}(S^1,S^2) \\ \mathbf{x}_q(S^1,S^2,z) = \mathbf{x}_p(S^1,S^2) + z\lambda(S^1,S^2)\mathbf{d}(S^1,S^2) \end{cases} \tag{3}$$

with $z \in [-h/2; h/2]$, h is the thickness of the structure $\xi = (\xi^1,\xi^2,\xi^3 = z)$ denoting the curvilinear coordinates, λ is the stretching parameter. $\boldsymbol{D}$ and $\boldsymbol{d}$ are the shell director vectors in initial and deformed configurations, respectively. h represents the thickness.

The strain ε can be decomposed in in-plane and transverse shear strains as:

$$\begin{cases} \varepsilon_{\alpha\beta} = e_{\alpha\beta} + z\lambda\chi_{\alpha\beta} & ;\alpha,\beta = 1,2 \\ \gamma_{\alpha 3} = \mathbf{a}_\alpha.\mathbf{d} \end{cases} \tag{4}$$

where are the local orthonormal shell direction in deformed state $e_{\alpha\beta}$, $\chi_{\alpha\beta}$ and $\gamma_{\alpha 3}$ are the components of linearized membrane, bending and shear strains vectors expressed by:

$$e_{\alpha\beta} = \frac{1}{2}(a_{\alpha\beta} - A_{\alpha\beta}) \quad ; \quad \chi_{\alpha\beta} = \frac{1}{2}(b_{\alpha\beta} - B_{\alpha\beta}) \quad ;\alpha,\beta = 1,2 \tag{5}$$

In matrix notation, the membrane, bending, and shear strains vectors are given by

$$\mathbf{e} = \begin{bmatrix} e_{11} \\ e_{22} \\ 2e_{12} \end{bmatrix} \quad ; \quad \boldsymbol{\chi} = \begin{bmatrix} \chi_{11} \\ \chi_{22} \\ 2\chi_{12} \end{bmatrix} \quad ; \quad \gamma = \begin{bmatrix} \gamma_{13} \\ \gamma_{23} \end{bmatrix} \tag{6}$$

The strains expressions provided in Eq. (5) is used in the weak form of equilibrium equations as below:

$$W = \int_A (\delta\mathbf{e}.\mathbf{N} + \delta\boldsymbol{\chi}.\mathbf{M} + \delta\gamma.\mathbf{T})dA - W_{ext} = 0 \tag{7}$$

where the membrane $\boldsymbol{N}$, bending $\boldsymbol{M}$ and shear $\boldsymbol{T}$ stresses resultants can be written in matrix form:

$$\boldsymbol{N} = \int_{-h/2}^{h/2} \begin{bmatrix} \sigma_{11} \\ \sigma_{22} \\ \sigma_{12} \end{bmatrix} dz, \boldsymbol{M} = \int_{-h/2}^{h/2} \lambda z \begin{bmatrix} \sigma_{11} \\ \sigma_{22} \\ \sigma_{12} \end{bmatrix} dz, \boldsymbol{T} = \int_{-h/2}^{h/2} \begin{bmatrix} \sigma_{13} \\ \sigma_{23} \end{bmatrix} dz \tag{8}$$

where σ is the stress tensor.

The generalized resultant of stress $\boldsymbol{R}$ and strain Σ vectors are expressed as:

$$R = [N \quad M \quad T]^T \quad , \Sigma = [e \quad \chi \quad \gamma]^T \tag{9}$$

Using Eqs. (3), (6), (7) and (9), the resultant of stress $\boldsymbol{R}$ is related tothe strain field Σ as below:

$$\mathbf{R} = \mathbf{H}_T \boldsymbol{\Sigma} \tag{10}$$

with $\mathbf{H}_T$ is the linear coupling elastic matrix.

2.3 Constitutive Relations in Elastoplasticity

Elastic and plastic strain components can be additively decomposed by

$$\boldsymbol{\varepsilon} = \boldsymbol{\varepsilon}^e + \boldsymbol{\varepsilon}^p \tag{11}$$

where $\boldsymbol{\varepsilon}^e$ is the elastic reversible part and $\boldsymbol{\varepsilon}^p$ is the plastic irreversible part. The elastic strain rate is related to the stress rate by

$$\dot{\boldsymbol{\sigma}} = \mathbf{D}: \boldsymbol{\varepsilon}^e = \mathbf{D}:(\dot{\boldsymbol{\varepsilon}} - \dot{\boldsymbol{\varepsilon}}^p) \tag{12}$$

It can be noted that constitutive relations of the elastoplastic material are given in terms of the rates of stress and strain. Elastic and plastic strains are separated because unique elastic strain generates stress which can only be calculated by integrating the stress rate over the past load history. The integration is conducted using the implicit Euler method (Backward Euler method), the strain-driven integration algorithm allows to have the stress history from the strain history. The description of how the yield surface changes with plastic deformation is called the hardening law. In the present work uses isotropic hardening illustrated by Ludwik law:

$$\sigma_p = \sigma_Y + K\, r^n \tag{13}$$

where σ_Y, K and n, are effective elastoplastic parameters, evaluated using self-consistent model, Suquet [23] as follows:

$$\sigma_Y = \frac{E}{E_m}\sigma_{Ym} \quad ; \quad K = K_m \frac{1 + V_c}{(1 - V_c)^n} \quad , n = n_m \tag{14}$$

3 Finite Element Resolution

The structure discretization is achieved using a four node shell element, involving three rotational and three translational degrees of freedom per node. Numerical simulations are performed using the commercial software ABAQUS to study the elastoplastic response of ceramic/metal FGM shells taken into account large displacements and finite rotations.

Numerical analysis is conducted using the UMAT subroutine implemented into ABAQUS. It should be mentioned that for each integration point through the thickness of the shell using the integration point number (KSPT), ABAQUS make a call for the UMAT subroutine

4 Numerical Results

Let's consider a FGM cylinder pulled in the middle. It is concerned with the deformation of an open-ended cylinder FGM shell, under the action of two outward forces 180° apart, [18]. The geometry of the studied cylindrical shell geometry is depicted in Fig. 1. Only one half of the structure is modeled owing to symmetry, using 20 × 20, S4 elements. The circumferential periphery is fully clamped and the maximum applied load at the free edge is fixed to $F_{max} = 2 \times 10^6$.

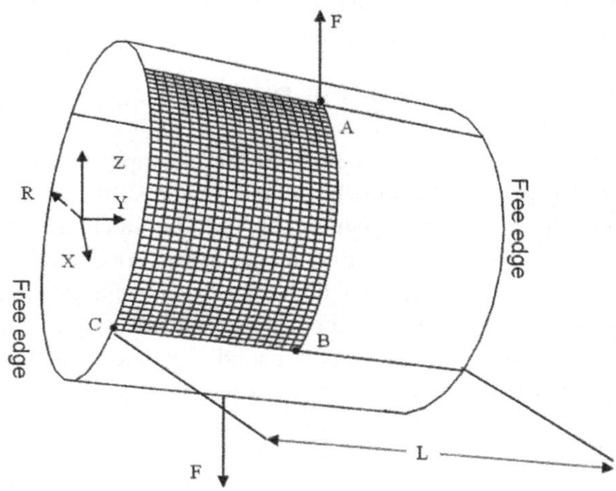

Fig. 1 FGM cylinder shell subjected radial pulling forces, [18]

The geometrical parameters are as follows: The cylinder length is L = 10.35, the radius R = 4.953 and the thickness h = 0.094. The material properties of the metal and ceramic components are given in Tables 1 and 2.

Table 1 Mechanical properties of aluminum Al and SiC materials, Gunes et al. [9]

Materials	Young modulus (GPa)	Poisson's ratio
Al 6061	67	0.33
SiC	302	0.17

Table 2 Elastoplastic properties of the Al 6061 material

Materials	$\sigma_{Ym}(MPa)$	$K_m(MPa)$	n_m
Al 6061	80	237.33	0.3878

Elastic and elastoplastic responses, of load against the radial deflections at points A, B and C, of the open-end FG metal-ceramic shell, are examined for different power index. Figures 2, 3 and 4 show load against elastic and elastoplastic radial deflections at points A, B and C of the studied structure. It can be clearly observed that there are great differences between the elastoplastic solutions and the elastic solutions. Indeed, elastoplastic radial deflections are more important than elastic ones considering the same imposed load.

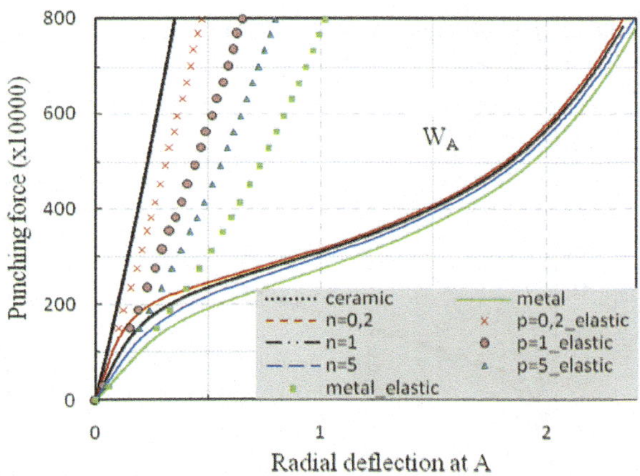

Fig. 2 Elastic and elastoplastic responses of a FGM cylinder shell at A, (F = 2 × 10^6)

The load–displacement curves are located between those of the metal and ceramic shells.

Figure 5 represents the elastoplastic and elastic deformed configurations for the cylinder FGM shells under the maximum load for the same power law index *p* = *1*.

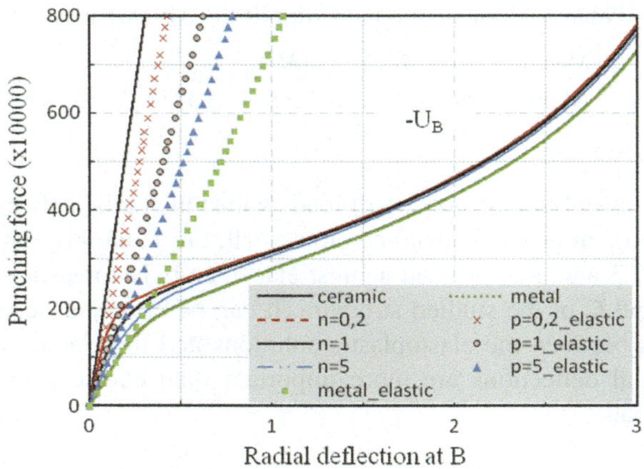

Fig. 3 Elastic and elastoplastic responses of a FGM cylinder shell at B, ($F = 2 \times 10^6$)

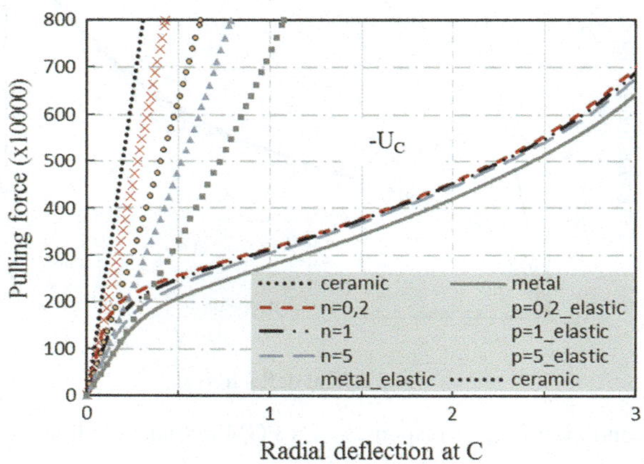

Fig. 4 Elastic and elastoplastic responses of a FGM cylinder shell at C, ($F = 2 \times 10^6$)

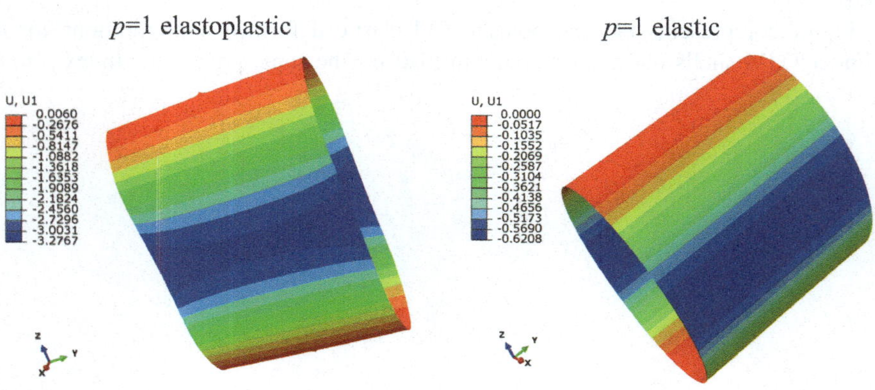

Fig. 5 Elastoplastic and elastic deformed shapes of the FGM cylinder shell configuration for power law index $p = 1$, U_1: radial displacement

5 Conclusion

A geometrically non-linear analysis of elastoplastic cylindrical FGM shell undergoing large deformations is conducted in this paper using 4-node shell finite element. A user-material UMAT subroutine is developed and implemented into ABAQUS in order to introduce the elastoplastic material properties of the FGM shell. Numerical results of loads-deflections curves of the FGM cylinder subjected to large displacements are presented. The effect of power law coefficient on shells elastoplastic responses are also examined. It was found that the power-law index has significant effect on deflection of the studied structure.

References

1. Alizadeh M, Fattahi AM (2019) Non-classical plate model for FGMs. Eng Comput 35 (1):215–228
2. Attia A, Bousahla AA, Tounsi A, Mahmoud SR, Alwabli AS (2018) A refined four variable plate theory for thermoelastic analysis of FGM plates resting on variable elastic foundations. Struct Eng Mech 65(4):453–464
3. Barati MR, Shahverdi H (2016) A four-variable plate theory for thermal vibration of embedded FG nanoplates under non-uniform temperature distributions with different boundary conditions. Struct Eng Mech 60(4):707–727
4. Bouderba B, Houari MSA, Tounsi A, Mahmoud SR (2016) Thermal stability of functionally graded sandwich plates using a simple shear deformation theory. Struct Eng Mech 58 (3):397–422
5. Bouhadra A, Tounsi A, Bousahla AA, Benyoucef S, Mahmoud SR (2018) Improved HSDT accounting for effect of thickness stretching in advanced composite plates. Struct Eng Mech 66(1):61–73
6. Bouhamed A, Jrad H, Said LB, Wali M, Dammak F (2019) A non-associated anisotropic plasticity model with mixed isotropic–kinematic hardening for finite element simulation of incremental sheet metal forming process. Int J Adv Manuf Technol 100(1–4):929–940
7. Bouhamed A, Jrad H, Mars J, Wali M, Gamaoun F, Dammak F (2019) Homogenization of elasto-plastic functionally graded material based on representative volume element: application to incremental forming process. Int J Mech Sci 160:412–420
8. Ghannad M, Nejad MZ, Rahimi GH, Sabouri H (2012) Elastic analysis of pressurized thick truncated conical shells made of functionally graded materials. Struct Eng Mech 43(1):105–126
9. Gunes R, Aydin M, Apalak MK, Reddy JN (2014) Experimental and numerical investigations of low velocity impact on functionally graded circular plates. Compos B Eng 59:21–32
10. Jrad H, Mallek H, Wali M, Dammak F (2018) Finite element formulation for active functionally graded thin-walled structures. C R Mec 346(12):1159–1178
11. Jrad H, Mars J, Wali M, Dammak F (2018) An extended finite element method for modeling elastoplastic FGM plate-shell type structures. Struct Eng Mech 68(3):299–312
12. Jrad H, Mars J, Wali M, Dammak F (2019) Geometrically nonlinear analysis of elastoplastic behavior of functionally graded shells. Eng Comput 35(3):833–847

13. Katariya PV, Hirwani CK, Panda SK (2019) Geometrically nonlinear deflection and stress analysis of skew sandwich shell panel using higher-order theory. Eng Comput 35(2):467–485
14. Mallek H, Jrad H, Algahtani A, Wali M, Dammak F (2019) Geometrically non-linear analysis of FG-CNTRC shell structures with surface-bonded piezoelectric layers. Comput Methods Appl Mech Eng 347:679–699
15. Mallek H, Jrad H, Wali M, Dammak F (2019) Geometrically nonlinear finite el-ement simulation of smart laminated shells using a modified first-order shear defor-mation theory. J Intell Mater Syst Struct 30(4):517–535
16. Mallek H, Jrad H, Wali M, Dammak F (2019) Piezoelastic response of smart functionally graded structure with integrated piezoelectric layers using discrete double directors shell element. Compos Struct 210:354–366
17. Mars J, Said LB, Wali M, Dammak F (2018) Elasto-Plastic modeling of low-velocity impact on functionally graded circular plates. Int J Appl Mech, 1850038
18. Mars J, Koubaa S, Wali M, Dammak F (2017) Numerical analysis of geometrically non-linear behavior of functionally graded shells. Lat Am Solids Struct 14(11):1952–1978
19. Mellouli H, Jrad H, Wali M, Dammak F (2019) Geometrically nonlinear meshfree analysis of 3D-shell structures based on the double directors shell theory with finite rotations. Steel Compos Struct 31(4):397–408
20. Mellouli H, Jrad H, Wali M, Dammak F (2019) Meshfree implementation of the double director shell model for FGM shell structures analysis. Eng Anal Boundary Elem 99:111–121
21. Mellouli H, Jrad H, Wali M, Dammak F (2019) Meshless implementation of ar-bitrary 3D-shell structures based on a modified first order shear deformation theory. Comput Math Appl 77:34–49
22. Rajasekaran S (2018) Analysis of axially functionally graded nano-tapered Timoshenko beams by element-based Bernstein pseudospectral collocation (EBBPC). Eng Comput 34 (3):543–563
23. Suquet P (1997) Effective properties of nonlinear composites. Continuum micromechanics. CISM courses and lecture notes. Springer, Heidelberg 377:197–264
24. Hajlaoui A, Chebbi E, Dammak F (2019) Buckling analysis of carbon nanotube reinforced FG shells using an efficient solid-shell element based on a modified FSDT. Thin-Walled Struct 144:106254
25. Woo J, Merguid SA (2001) Non-linear analysis of functionally graded plates and shallow shells. Int J Solids Struct 38:7409–7421

Buckling Analysis of Carbon Nanotube-Reinforced FG Shells Using an Enhanced Solid-Shell Element

E. Chebbi(✉), A. Hajlaoui, and F. Dammak

Laboratory of Electromechanical Systems (LASEM), National Engineering School of Sfax, University of Sfax, B.P.1173-3038 Sfax, Tunisia
{chebbi.elouni,abdhajlaoui}@gmail.com,
fakhreddine.dammak@enis.rnu.tn

Abstract. This paper presents a higher-order solid-shell element for buckling behavior of carbon nanotubes reinforced functionally graded shells based on higher-order shear deformation concept. Four different types of reinforcement along the thickness are considered. This finite element is used to study the buckling behavior of carbon nanotubes reinforced functionally graded shells and to investigate the influence of same parameters on the buckling behavior.

Keywords: Solid-shell element · FG-CNTRCs · HSDT · Buckling

1 Introduction

Based on the higher-order shear deformation plate theory, Shen and Zhu [13] studied the compressive postbuckling under thermal environments. Using von Kármán-type of kinematic nonlinearity and include plate-foundation interaction, Shen et al. [12] studied the buckling and postbuckling behavior of functionally graded graphene-reinforced composite laminated plates in thermal environments. Using a higher order shear deformation theory with a von Kármán-type of kinematic nonlinearity, Shen [15] investigated the thermal buckling behavior of functionally graded carbon nanotube-reinforced composite cylindrical shells. Shen and Xiang [16] investigated the post-buckling behavior of nanotube-reinforced composite cylindrical shells. Using the harmonic differential quadrature (HDQ) method, Mehri et al. [11] examined the buckling of a pressurized CNT reinforced functionally graded truncated conical shell.

On other hand, shell element based on discrete double directors has been widely used in the literature to model recently FGM and FG-CNTRCs materials and one can be refereed to these publications [17, 1, 2, 18, 3] among others.

This paper presents an improvement of the solid-shell formulation developed in the author's previous works [6, 5, 8, 7, 7, 4]. This improvement is achieved by imposing a parabolic shear strain distribution through the CNTRC shell thickness in the compatible strain part and vanish on top and bottom faces of the shell. Consequently, the shear correction factor is no longer needed as in [5] and [10]. Therefore, a special representation of the transverse shear strains is chosen in the present formulation.

N. Aifaoui et al. (Eds.): CMSM 2019, LNME, pp. 435–442, 2020.
https://doi.org/10.1007/978-3-030-27146-6_47

2 Finite Element Formulation

The position vectors of the initial and the current configuration are denoted by **X** and **x**, respectively. The covariant base vectors in the initial and deformed configuration are given by

$$G_k = \frac{\partial X}{\partial \xi^k}, \quad g_k = \frac{\partial x}{\partial \xi^k}, \quad g_k = \frac{\partial x}{\partial \xi^k} \tag{1}$$

The covariant metric tensor G at a material point ξ, in the initial and deformed configuration are defined by

$$G = [G_i.G_j], \quad g = [g_i.g_j], \quad i,j = 1,2,3 \tag{2}$$

This leads to the following Green-Lagrangean strain tensor

$$E = \frac{1}{2}(g - G), \quad E_{ij} = \frac{1}{2}(g_{ij} - G_{ij}) \tag{3}$$

2.1 The Weak Form

The EAS method is based on the following assumption

$$E = E^c + \tilde{E} \tag{4}$$

where E^c and $\tilde{E}$ are respectively the compatible part and the enhanced part of the Green-Lagrange strain tensor. The variational framework of the EAS method, which is based on the three-field variational functional, in Lagrangean formulation, is written as

$$\Pi(u, \tilde{E}, \tilde{S}) = \int_V [\psi(E) - \tilde{S} : \tilde{E} - u.F_V] dV - \int_{\partial V_f} u.F_S dA = 0 \tag{5}$$

where ψ is the strain energy function and u, $\tilde{E}$ and $\tilde{S}$ are the independent tonsorial quantities which are: displacement, enhanced assumed Lagrange strain and assumed second Piola-Kirchhoff stress fields respectively. Vectors, F_V and F_S, in Eq. (5), are the prescribed body force and surface traction respectively. The weak form of this modified functional may be obtained as

$$W = \int_V [\delta E : S - \delta u.F_V] dV - \int_{\partial V_f} \delta u.F_S dA = 0 \tag{6}$$

where S is the Piola-Kirchoff stress tensor given by $S = \partial\psi / \partial E$.

2.2 Compatible Strains

Then the compatible part of the Green-Lagrange strain tensor becomes as follows:

$$\boldsymbol{E}^c = \boldsymbol{T}^{-T}\begin{bmatrix} \frac{1}{2}(g_{11} - G_{11}) \\ \frac{1}{2}(g_{22} - G_{22}) \\ \sum_{A=1}^{4} \frac{1}{4}(1+\xi_A\xi)(1+\eta_A\eta)\frac{1}{2}(g_{33}^A - G_{33}^A) \\ (g_{12} - G_{12}) \\ \frac{1}{2}\left[(1-\eta)\left(g_{13}^B - G_{13}^B\right) + (1+\eta)\left(g_{13}^D - G_{13}^D\right)\right] \\ \frac{1}{2}\left[(1-\xi)\left(g_{23}^A - G_{23}^A\right) + (1+\xi)\left(g_{23}^C - G_{23}^C\right)\right] \end{bmatrix} \tag{7}$$

where the matrix $\boldsymbol{T}$ is the transformation of the strain tensor from parametric coordinates to the local Cartesian coordinates given in [5].

2.3 Enhanced Green Lagrange Strains

The enhanced Green-Lagrange strain part is related to the vector of the internal strain parameters α as:

$$\tilde{\boldsymbol{E}} = \tilde{\boldsymbol{M}}\,\boldsymbol{\alpha},\ \delta\tilde{\boldsymbol{E}} = \tilde{\boldsymbol{M}}\,\delta\boldsymbol{\alpha},\ \Delta\tilde{\boldsymbol{E}} = \tilde{\boldsymbol{M}}\,\Delta\boldsymbol{\alpha} \tag{8}$$

where $\tilde{\boldsymbol{E}}$, $\delta\tilde{\boldsymbol{E}}$ and $\Delta\tilde{\boldsymbol{E}}$ are total, virtual and incremental enhanced Green Lagrange strain tensor respectively. The crucial assumption of the EAS method is the enforcement of the orthogonality conditions for the assumed stress field $\tilde{\boldsymbol{S}}$ and the enhanced strain $\tilde{\boldsymbol{E}}$. This orthogonality conditions impose the following choice for the interpolation function matrix $\tilde{\boldsymbol{M}}$ to be expressed as follows

$$\tilde{\boldsymbol{M}} = \frac{\det \boldsymbol{J}_0}{\det \boldsymbol{J}}\boldsymbol{T}_0^{-T}\boldsymbol{M}_{\xi\eta\zeta},\quad \int_{-1}^{1}\int_{-1}^{1}\int_{-1}^{1}\boldsymbol{M}_{\xi\eta\zeta}\,d\xi\, d\eta\, d\zeta = 0 \tag{9}$$

where the subscript '0' means evaluation at the center of the element in the natural coordinates, $\boldsymbol{J} = [\boldsymbol{G}_1, \boldsymbol{G}_2, \boldsymbol{G}_3]$ is the Jacobian matrix. The interpolation matrix $\boldsymbol{M}_{\xi\eta\zeta}$, in Eq. (9), is expression in term of the parametric coordinates (ξ, η, ζ). Three choices of matrix $\boldsymbol{M}_{\xi\eta\zeta}$ will be considered with 9 parameters.

$$\boldsymbol{M}_{\xi\eta\zeta}^{9} = \begin{pmatrix} \xi & 0 & 0 & 0 & 0 & 0 & 0 & 0 & 0 \\ 0 & \eta & 0 & 0 & 0 & 0 & 0 & 0 & 0 \\ 0 & 0 & \zeta & 0 & 0 & 0 & 0 & \xi\zeta & \eta\zeta \\ 0 & 0 & 0 & \xi & \eta & 0 & 0 & 0 & 0 \\ 0 & 0 & 0 & 0 & 0 & \frac{1}{5}-\zeta^2 & 0 & 0 & 0 \\ 0 & 0 & 0 & 0 & 0 & 0 & \frac{1}{5}-\zeta^2 & 0 & 0 \end{pmatrix} \tag{10}$$

3 Carbon Nanotube Reinforced Composite Shell

A CNTRC shell structure made from a mixture of SWCNT and an isotropic matrix is considered. In this study, the shells are assumed to have five different patterns of reinforcement over the cross sections, that is, UD is uniformly distributed; FG-V, FG-Λ, FG-O and FG-X designate the other four types of functionally graded distributions of CNTs.

In the present study and due to the simplicity and convenience, the extended rule of mixtures approach which contains the efficiency parameters is employed extensively to extract the elastic properties of the (CNTRCs) shell as follows [14]

$$E_{11} = \eta_1 V_{CNT} E_{11}^{CNT} + V_m E_m, \quad \frac{\eta_2}{E_{11}} = \frac{V_{CNT}}{E_{22}^{CNT}} + \frac{V_m}{E_m}, \quad \frac{\eta_3}{G_{12}} = \frac{V_{CNT}}{G_{12}^{CNT}} + \frac{V_m}{G_m} \tag{11}$$

where (E_{11}, E_{22}) and G_{12}^{CNT} are the Young's and shear modulus of CNT, respectively; E_m and G_m are the Young's modulus and shear modulus of the isotropic matrix; η_1, η_2 and η_3 are the CNT efficiency parameters. On the other hand, the volume fraction of CNTs (V_{CNT}) and matrix (V_m) should equal unity.

Similarly, the effective Poisson ratio and mass density depend weakly on position can be determined in the same way as

$$\nu_{12} = V_{CNT}^{*} \nu_{12}^{CNT} + V_m \nu_m \tag{12}$$

$$\rho = V_{CNT} \rho^{CNT} + V_m \rho^m \tag{13}$$

where $(\nu_{12}^{CNT}, \rho^{CNT})$ and (ν_m, ρ^m) are the Poisson's ratio and mass density of CNT and matrix, respectively. V_{CNT}^{*} represent the total volume fraction of (CNTs) which can be obtained as a function of mass fraction of CNTs (w_{CNT}), mass density of matrix (ρ^m) and mass density of CNTs (ρ^{CNT}) as

$$V_{CNT}^{*} = \frac{w_{CNT}}{w_{CNT} + \frac{\rho^{CNT}}{\rho^m} - \frac{w_{CNT}\rho^{CNT}}{\rho^m}} \tag{14}$$

The uniform and four types of functionally graded distributions of the carbon nanotubes along the shell thickness used in this paper are assumed to be

$$V_{CNT} = \begin{cases} V_{CNT}^{*} & \text{(UD CNTRC)} \\ 2\left(1 - 2\frac{|z|}{h}\right) V_{CNT}^{*} & \text{(FG-O CNTRC)} \\ 4\frac{|z|}{h} V_{CNT}^{*} & \text{(FG-X CNTRC)} \\ \left(1 + 2\frac{z}{h}\right) V_{CNT}^{*} & \text{(FG-V CNTRC)} \\ \left(1 - 2\frac{z}{h}\right) V_{CNT}^{*} & \text{(FG-Λ CNTRC)} \end{cases} \tag{15}$$

4 Numerical Simulations

To verify the present formulation, a comparison is carried out for simply supported CNTRC/PmPV plate under uniaxial compressive pressure (see Fig. 1). The geometrical properties of the plate are shown in Fig. 1. A 16 × 16 × 1 meshing is used for this test. The material properties of both matrixes at room temperature of 300 K and (10; 10) single walled carbon nanotubes (SWCNT) are given according to molecular dynamic (MD) simulation are shown in Table 1. The CNT efficiency parameters (η_i) associated with the given volume fraction (V^*_{CNT}) are taken from same literatures cited previously are presented in Table 2. For this example, we assume that the critical buckling load factors are evaluated by choosing Nx = 10 shape functions. A 20 × 20 × 1 meshing is used for this test.

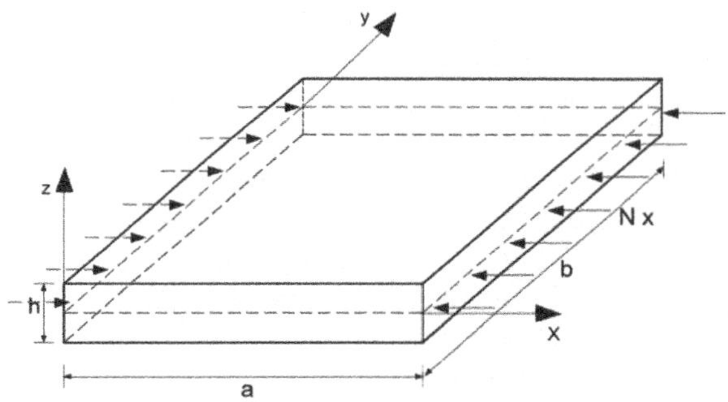

Fig. 1 Geometric of FG-CNTRC plate

Table 1 Material properties of PmPV matrix at room temperature of 300 K and (10; 10) single walled carbon nanotubes (SWCNT)

PmPV	(10,10) SWCNTs
$E_m = 2.1GPa$	$E_{11}^{CNT} = 5.6466\,TPa$, $E_{22}^{CNT} = 7.0800\,TPa$
$\rho^m = 1150\,Kg/m^3$	$G_{12}^{CNT} = 1.9445\,TPa$,
$v^m = 0.34$	$G_{12} = G_{13} = G_{23}$
	$\rho^{CNT} = 1400\,Kg/m^3$
	$v_{12}^{CNT} = 0.175$

Table 2 CNT efficiency parameters for different values of (CNTs) volume fractions

V^*_{CNT}	η_1	$\eta_2 = \eta_3$
0.11	0.149	0.934
0.14	0.150	0.941
0.17	0.149	1.381

In Table 3, the critical buckling load parameters ($\widehat{n}_{cri} = N_{cri}b^2/E_mh^3$) of simply supported FG-CNTRC square plates are determined and compared with the IMLS-Ritz method [19] and the First order shear deformation plate theory [9]. It can be found that the results obtained by the present model are in good agreement with those found in the literature.

Table 3 Uniaxial buckling load factor ($\widehat{n}_{cri} = N_{cri}b^2/E_mh^3$), for simply supported FG-CNTRC square plate

V^*_{CNT}	a/h		Present	$\widehat{n}_{cri}$	
				[9]	[19]
0.11	100	UD	39.4856	39.3391	39.1158
		FG-X	57.2361	57.0671	56.7373
		FG-O	21.5380	21.4639	21.3316
	50	UD	38.1832	38.0600	37.7998
		FG-X	57.3723	54.3867	54.0305
		FG-O	21.1782	21.0904	20.9312
	10	UD	18.8524	18.9825	18.9783
		FG-X	19.5539	22.1655	22.1612
		FG-O	13.9789	13.7938	13.8356
0.14	100	UD	49.5652	–	49.0816
		FG-X	72.1583	–	71.5516
		FG-O	26.6469	–	26.3572
	20	UD	37.3059	–	–
		FG-X	47.6342	–	–
		FG-O	23.0132	–	–

The influence of the length-to-thickness ratio on the uniaxial buckling load factor for various distribution of CNTs (UD, FG-V, FG-O, FG-X) are illustrated in Fig. 2.

In this test, a length $a = 10$ and volume fraction $V^*_{CNT} = 0.14$ will be considered. It can be concluded that the uniaxial buckling load factor of simply supported FG-CNTRC square plates is almost independent of the length-to-thickness ratio (a/h) when the plates become thin $a/h > 50$. It is shown that the highest and lowest uniaxial buckling load factor is obtained for the FG-X and FG-O distribution, respectively. This may be explained by the distribution of the CNT that is higher condensed near the top and bottom surfaces of FG-X-CNTRC, which is not the case for FG-O-CNTRC. On the other hand, since the highest value of uniaxial buckling load factor is observed at the FG-X-CNTRC distribution.

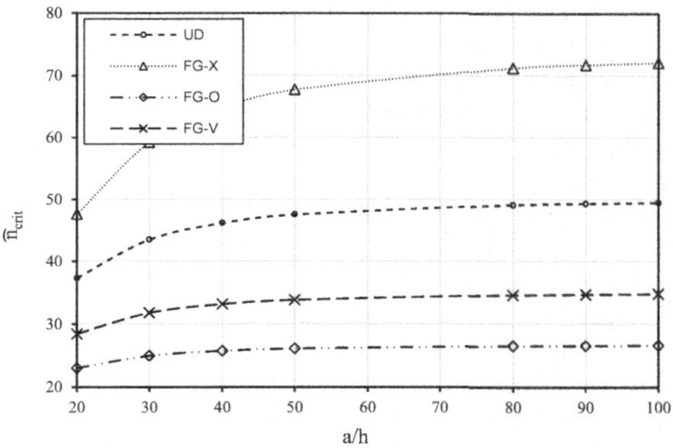

Fig. 2 Influence of the length-to-thickness ratio on the uniaxial buckling load factor $(\widehat{n}_{cri} = N_{cri}b^2/E_m h^3)$ of simply supported FG-CNTRC square plate $(a/b = 1,\ a = 10,\ V^*_{CNT} = 0.14)$

5 Conclusion

In the present paper we presented a higher-order solid-shell element formulation with an imposed parabolic shear strain distribution through the shell thickness in the incompatible strain part. Numerical solution for buckling behavior of CNTRC plate under uniaxial compressive pressure. Comparison studies were performed to verify the accuracy and validity of the present formulation and the results were found to be in good agreement with works taken from the literature. The effects of distribution of CNTs and (CNTs) volume fraction on the buckling loads of CNTRC plates have been studied with various parameters.

References

1. Frikha A, Hajlaoui A, Wali M, Dammak F (2016) A new higher order C0 mixed beam element for FGM beams analysis. Compos Part B Eng 106:181–189
2. Frikha A, Wali M, Hajlaoui A, Dammak F (2016) Dynamic response of functionally graded material shells with a discrete double directors shell element. Compos Struct 154:385–395
3. Frikha A, Zghal S, Dammak F (2017) Finite rotation three and four nodes shell elements for functionally graded carbon nanotubes-reinforced thin composite shells analysis. Comput Methods Appl Mech Eng
4. Hajlaoui A, Chebbi E, Triki E, Dammak F 2018 Post-buckling analysis of functionally graded materials structures with enhanced solid-shell elements. Lecture notes in mechanical engineering
5. Hajlaoui A, Jarraya A, El Bikri K, Dammak F (2015) Buckling analysis of functionally graded materials structures with enhanced solid-shell elements and transverse shear correction. Compos Struct 132:87–97

6. Hajlaoui A, Jarraya A, Kallel-Kamoun I, Dammak F (2012) Buckling analysis of a laminated composite plate with delaminations using the enhanced assumed strain solid shell element. J Mech Sci Technol 26:3213–3221
7. Hajlaoui A, Triki E, Frikha A, Wali M, Dammak F (2017) Nonlinear dynamics analysis of FGM shell structures with a higher order shear strain enhanced solid-shell element. Lat Am J Solids Struct 14:72–91
8. Hajlaoui A, Wali M, Ben Jdidia M, Dammak F (2016) An improved enhanced solid shell element for static and buckling analysis of shell structures. Mech Ind 17
9. Kiani Y (2016) Shear buckling of FG-CNT reinforced composite plates using Chebyshev-Ritz method. Compos Part B Eng 105:176–186
10. Mars J, Koubaa S, Wali M, A FD (2017) Numerical analysis of geometrically non-linear behavior of functionally graded shells. Lat Am J Solids Struct 14:1952–1978
11. Mehri M, Asadi H, Wang Q (2016) Buckling and vibration analysis of a pressurized CNT reinforced functionally graded truncated conical shell under an axial compression using HDQ method. Comput Methods Appl Mech Eng 303:75–100
12. Shen H-S, Xiang Y, Lin F, Hui D (2017) Buckling and postbuckling of functionally graded graphene-reinforced composite laminated plates in thermal environments. Compos Part B 119:67–78
13. Shen H-S, Zhu ZH (2012) Postbuckling of sandwich plates with nanotube-reinforced composite face sheets resting on elastic foundations. Eur J Mech A/Solids 35:10–21
14. Shen HS (2009) Nonlinear bending of functionally graded carbon nanotube-reinforced composite plates in thermal environments. Compos Struct 91:9–19
15. Shen HS (2012) Thermal buckling and postbuckling behavior of functionally graded carbon nanotube-reinforced composite cylindrical shells. Compos Part B Eng 43:1030–1038
16. Shen HS, Xiang Y (2013) Postbuckling of nanotube-reinforced composite cylindrical shells under combined axial and radial mechanical loads in thermal environment. Compos Part B Eng 52:311–322
17. Wali M, Hajlaoui A, Dammak F (2014) Discrete double directors shell element for the functionally graded material shell structures analysis. Comput Methods Appl Mech Eng 278:388–403
18. Zghal S, Frikha A, Dammak F (2017) Static analysis of functionally graded carbon nanotube-reinforced plate and shell structures. Compos Struct 176:1107–1123
19. Zhang LW, Lei ZX, Liew KM (2015) An element-free IMLS-Ritz framework for buckling analysis of FG-CNT reinforced composite thick plates resting on Winkler foundations. Eng Anal Bound Elem 58:7–17

Static Analysis of Carbon Nanotube-Reinforced FG Shells Using an Enhanced Solid-Shell Element

A. Hajlaoui, E. Chebbi(✉), and F. Dammak

Laboratory of Electromechanical Systems (LASEM), National Engineering School of Sfax, University of Sfax, B.P. 1173-3038 Sfax, Tunisia
fakhreddine.dammak@enis.rnu.tn

Abstract. The static behavior of carbon nanotubes reinforced functionally graded shells is studied using the enhanced assumed strain (EAS) solid-shell element based on higher-order shear deformation concept. Four different types of reinforcement along the thickness are considered. Furthermore, the developed solid-shell element allows an efficient and accurate analysis of carbon nanotube-reinforced functionally graded shells under linear static conditions. The influences of some geometrical and material parameters on the static behavior of shell structures are discussed.

Keywords: Solid-shell element · Carbon nanotube · FG-CNTRCs · HSDT

1 Introduction

Stimulated by the concept of FGMs, the pattern of the functionally graded (FG) distribution of reinforcement has been successfully applied for CNT-reinforced composite materials. CNT-based FGMs were first proposed by Shen [13] with CNT distributions within an isotropic matrix designed specifically to grade them with certain rules along the desired directions to improve mechanical properties of the structures.

A series of investigations about FG-CNTRC beam, plate and shell was then conducted to study the mechanical properties of nanocomposites. Based on the first-order shear deformation plate theory, Zhu et al. [17] carried out bending analyses of thin-to-moderately thick composite plates reinforced by singlewalled carbon nanotubes. Based on three-dimensional theory of elasticity, Alibeigloo [1] examined the bending behavior of FG-CNTRC rectangular. Mehrabadi and Sobhani Aragh [12] investigated stresses caused by bending behavior of functionally graded carbon nanotube-reinforced open cylindrical shells subjected to mechanical loads. Using the element-free kp-Ritz method, Zhang et al. [16] examined flexural strength of carbon nanotube reinforced composite cylindrical panels. Jeyaraj and Rajkumar [11] studied static behavior of FG-CNTRC plate under non-uniform elevated temperature fields using the finite element method. Based on discrete double directors shell element, the static, free vibration and geometrical non linearity behaviors of FG-CNTRCs structures was investigated [2–5, 15].

N. Aifaoui et al. (Eds.): CMSM 2019, LNME, pp. 443–451, 2020.
https://doi.org/10.1007/978-3-030-27146-6_48

This paper presents an improvement of the solid-shell formulation developed in the author's previous works [8, 7, 10, 9, 9, 6]. This improvement is achieved by imposing a parabolic shear strain distribution through the CNTRC shell thickness in the incompatible strain part and vanish on top and bottom faces of the shell. Consequently, the shear correction factor is no longer needed as in [7]. Therefore, a special representation of the transverse shear strains is chosen in the present formulation.

2 Finite Element Formulation

The position vectors of the initial and the current configuration are denoted by **X** and **x**, respectively. The covariant base vectors in the initial and deformed configuration are given by

$$\boldsymbol{G}_k = \frac{\partial \boldsymbol{X}}{\partial \xi^k},\ \boldsymbol{g}_k = \frac{\partial \boldsymbol{x}}{\partial \xi^k},\ k = 1, 2, 3 \tag{1}$$

The covariant metric tensor $\boldsymbol{G}$ at a material point ξ, in the initial and deformed configuration are defined by

$$\boldsymbol{G} = \left[\boldsymbol{G}_i.\boldsymbol{G}_j\right],\ \boldsymbol{g} = \left[\boldsymbol{g}_i.\boldsymbol{g}_j\right],\ i,j = 1, 2, 3 \tag{2}$$

This leads to the following Green-Lagrangean strain tensor

$$\boldsymbol{E} = \frac{1}{2}(\boldsymbol{g} - \boldsymbol{G}),\ E_{ij} = \frac{1}{2}\left(g_{ij} - G_{ij}\right) \tag{3}$$

2.1 The Weak Form

The EAS method is based on the following assumption

$$\boldsymbol{E} = \boldsymbol{E}^c + \tilde{\boldsymbol{E}} \tag{4}$$

where $\boldsymbol{E}^c$ and $\tilde{\boldsymbol{E}}$ are respectively the compatible part and the enhanced part of the Green-Lagrange strain tensor. The variational framework of the EAS method, which is based on the three-field variational functional, in Lagrangean formulation, is written as

$$\Pi\left(\boldsymbol{u}, \tilde{\boldsymbol{E}}, \tilde{\boldsymbol{S}}\right) = \int_V \left[\psi(\boldsymbol{E}) - \tilde{\boldsymbol{S}} : \tilde{\boldsymbol{E}} - \boldsymbol{u}.\boldsymbol{F}_V\right] dV - \int_{\partial V_f} \boldsymbol{u}.\boldsymbol{F}_S dA = 0 \tag{5}$$

where ψ is the strain energy function and $\boldsymbol{u}$, $\tilde{\boldsymbol{E}}$ and $\tilde{\boldsymbol{S}}$ are the independent tonsorial quantities which are: displacement, enhanced assumed Lagrange strain and assumed second Piola-Kirchhoff stress fields respectively. Vectors, $\boldsymbol{F}_V$ and $\boldsymbol{F}_S$, in Eq. (5), are the prescribed body force and surface traction respectively. When invoking the classical orthogonality condition, $\int_V \tilde{\boldsymbol{S}} : \tilde{\boldsymbol{E}}\, dV = 0$, the number of independent variables in the original functional is reduce to two $\left(\boldsymbol{u}, \tilde{\boldsymbol{E}}\right)$. The weak form of this modified functional may be obtained as

$$W = \int_V \left[\delta \boldsymbol{E} : \boldsymbol{S} - \delta \boldsymbol{u}.\boldsymbol{F}_V\right] dV - \int_{\partial V_f} \delta \boldsymbol{u}.\boldsymbol{F}_S dA = 0 \tag{6}$$

where $\boldsymbol{S}$ is the Piola-Kirchoff stress tensor given by $\boldsymbol{S} = \partial\psi / \partial \boldsymbol{E}$.

2.2 Compatible Strains

Then the compatible part of the Green-Lagrange strain tensor becomes as follows:

$$\boldsymbol{E}^c = \boldsymbol{T}^{-T} \begin{bmatrix} \frac{1}{2}(g_{11} - G_{11}) \\ \frac{1}{2}(g_{22} - G_{22}) \\ \sum_{A=1}^{4} \frac{1}{4}(1 + \xi_A \xi)(1 + \eta_A \eta)\frac{1}{2}(g_{33}^A - G_{33}^A) \\ (g_{12} - G_{12}) \\ \frac{1}{2}\left[(1 - \eta)\left(g_{13}^B - G_{13}^B\right) + (1 + \eta)\left(g_{13}^D - G_{13}^D\right)\right] \\ \frac{1}{2}\left[(1 - \xi)\left(g_{23}^A - G_{23}^A\right) + (1 + \xi)\left(g_{23}^C - G_{23}^C\right)\right] \end{bmatrix} \tag{7}$$

where the matrix $\boldsymbol{T}$ is the transformation of the strain tensor from parametric coordinates to the local Cartesian coordinates given in [7].

2.3 Enhanced Green Lagrange Strains

The enhanced Green-Lagrange strain part is related to the vector of the internal strain parameters $\boldsymbol{\alpha}$ as:

$$\tilde{\boldsymbol{E}} = \tilde{\boldsymbol{M}}\,\boldsymbol{\alpha}, \delta\tilde{\boldsymbol{E}} = \tilde{\boldsymbol{M}}\,\delta\boldsymbol{\alpha}, \Delta\tilde{\boldsymbol{E}} = \tilde{\boldsymbol{M}}\,\Delta\boldsymbol{\alpha} \tag{8}$$

where $\tilde{\boldsymbol{E}}$, $\delta\tilde{\boldsymbol{E}}$ and $\Delta\tilde{\boldsymbol{E}}$ are total, virtual and incremental enhanced Green Lagrange strain tensor respectively. The crucial assumption of the EAS method is the enforcement of the orthogonality conditions for the assumed stress field $\tilde{\boldsymbol{S}}$ and the enhanced strain $\tilde{\boldsymbol{E}}$. This orthogonality conditions impose the following choice for the interpolation function matrix $\tilde{\boldsymbol{M}}$ to be expressed as follows

$$\tilde{\boldsymbol{M}} = \frac{\det \boldsymbol{J}_0}{\det \boldsymbol{J}} \boldsymbol{T}_0^{-T} \boldsymbol{M}_{\xi\eta\zeta}, \quad \int_{-1}^{1}\int_{-1}^{1}\int_{-1}^{1} \boldsymbol{M}_{\xi\eta\zeta} d\xi d\eta d\zeta = 0 \tag{9}$$

where the subscript ‘0’ means evaluation at the center of the element in the natural coordinates, $\boldsymbol{J} = [\boldsymbol{G}_1, \boldsymbol{G}_2, \boldsymbol{G}_3]$ is the Jacobian matrix. The interpolation matrix $\boldsymbol{M}_{\xi\eta\zeta}$, in Eq. (9), is expression in term of the parametric coordinates (ξ, η, ζ). Three choices of matrix $\boldsymbol{M}_{\xi\eta\zeta}$ will be considered with 9 parameters.

$$M^9_{\xi\eta\zeta} = \begin{pmatrix} \xi & 0 & 0 & 0 & 0 & 0 & 0 & 0 & 0 \\ 0 & \eta & 0 & 0 & 0 & 0 & 0 & 0 & 0 \\ 0 & 0 & \zeta & 0 & 0 & 0 & 0 & \xi\zeta & \eta\zeta \\ 0 & 0 & 0 & \xi & \eta & 0 & 0 & 0 & 0 \\ 0 & 0 & 0 & 0 & 0 & \frac{1}{5}-\zeta^2 & 0 & 0 & 0 \\ 0 & 0 & 0 & 0 & 0 & 0 & \frac{1}{5}-\zeta^2 & 0 & 0 \end{pmatrix} \quad (10)$$

the term $1/5 - \zeta^2$ ensures the parabolic transverse shear distribution in thickness direction. This enhancement requires a numerical integration rule at least $2 \times 2 \times 3$.

3 Carbon Nanotube Reinforced Composite Shell

In the present study and due to the simplicity and convenience, the extended rule of mixtures approach which contains the efficiency parameters is employed extensively to extract the elastic properties of the (CNTRCs) shell as follows [13]

$$E_{11} = \eta_1 V_{CNT} E_{11}^{CNT} + V_m E_m, \; \frac{\eta_2}{E_{11}} = \frac{V_{CNT}}{E_{22}^{CNT}} + \frac{V_m}{E_m}, \; \frac{\eta_3}{G_{12}} = \frac{V_{CNT}}{G_{12}^{CNT}} + \frac{V_m}{G_m} \quad (11)$$

where (E_{11}, E_{22}) and G_{12}^{CNT} are the Young's and shear modulus of CNT, respectively; E_m and G_m are the Young's modulus and shear modulus of the isotropic matrix; η_1, η_2 and η_3 are the CNT efficiency parameters. On the other hand, the volume fraction of CNTs (V_{CNT}) and matrix (V_m) should equal unity.

Similarly, the effective Poisson ratio and mass density depend weakly on position can be determined in the same way as

$$v_{12} = V_{CNT}^* v_{12}^{CNT} + V_m v_m, \; \rho = V_{CNT} \rho^{CNT} + V_m \rho^m \quad (12)$$

where $\left(v_{12}^{CNT}, \rho^{CNT}\right)$ and (v_m, ρ^m) are the Poisson's ratio and mass density of CNT and matrix, respectively. V_{CNT}^* represent the total volume fraction of (CNTs) which can be obtained as a function of mass fraction of CNTs (w_{CNT}), mass density of matrix (ρ^m) and mass density of CNTs (ρ^{CNT}) as

$$V_{CNT}^* = \frac{w_{CNT}}{w_{CNT} + \frac{\rho^{CNT}}{\rho^m} - \frac{w_{CNT}\rho^{CNT}}{\rho^m}} \quad (13)$$

The uniform and four types of functionally graded distributions of the carbon nanotubes along the shell thickness used in this paper are assumed to be

$$V_{CNT} = \begin{cases} V^*_{CNT} & \text{(UD CNTRC)} \\ 2\left(1 - 2\frac{|z|}{h}\right)V^*_{CNT} & \text{(FG-O CNTRC)} \\ 4\frac{|z|}{h}V^*_{CNT} & \text{(FG-X CNTRC)} \\ \left(1 + 2\frac{z}{h}\right)V^*_{CNT} & \text{(FG-V CNTRC)} \end{cases} \tag{14}$$

4 Numerical Simulations

In the present section, the FE formulation is employed to solve numerically the static analysis of two different carbon nanotube-reinforced functionally graded shell structures (FG-CNTRCs): CNTRC/PmPV. PmPV is expressed form of: PmPV: poly {(m-phenylenevinylene)-co-[(2,5-dioctoxy-p-phenylene) vinylene]}.

The (CNTs) are assumed to be of the type of armchair (10,10) single walled carbon nanotube (SWCNTs). The material properties of both matrixes at room temperature of 300 K and (10; 10) single walled carbon nanotubes (SWCNT) are given according to molecular dynamic (MD) simulation are shown in Table 1. These material properties are the same as the model described in literature [14] to provide a comparison with their results. In this study, the CNT efficiency parameters (η_i) associated with the given volume fraction (V^*_{CNT}) are taken from same literatures cited previously are presented in Table 2.

Table 1 Material properties of PmPV matrix at room temperature of 300 K and (10; 10) single walled carbon nanotubes (SWCNT)

PmPV	(10,10) SWCNTs
$E_m = 2.1 GPa$ $\rho^m = 1150\, Kg/m^3$ $\nu^m = 0.34$	$E^{CNT}_{11} = 5.6466\, TPa,$ $E^{CNT}_{22} = 7.0800\, TPa$ $G^{CNT}_{12} = 1.9445\, TPa,$ $G_{12} = G_{13} = G_{23}$ $\rho^{CNT} = 1400\, Kg/m^3$ $\nu^{CNT}_{12} = 0.175$

Table 2 CNT efficiency parameters for different values of (CNTs) volume fractions

V^*_{CNT}	η_1	$\eta_2 = \eta_3$
0.11	0.149	0.934
0.14	0.150	0.941
0.17	0.149	1.381

Among available literatures, there is no comprehensive study illustrate the static behavior of (FG-CNTRC) skew plate. The only exception is the work of Zghal et al. [14], where the authors used a 3d-shell model based on a double directors shell element.

This test aims to examine the applicability of the present formulation to obtain highly accurate approximate for static analysis of CNTRC/PmPV skew plate (Fig. 1) with various skew angles and length-to-thickness ratios. Thus, a comparison of non-dimensional central deflection ($\bar{w} = |w/h|(a/2, a/2)$) of CNTRC/PmPV skew plate obtained by the present method and by the proposed method by [14] (SHO4) is given in Table 3.

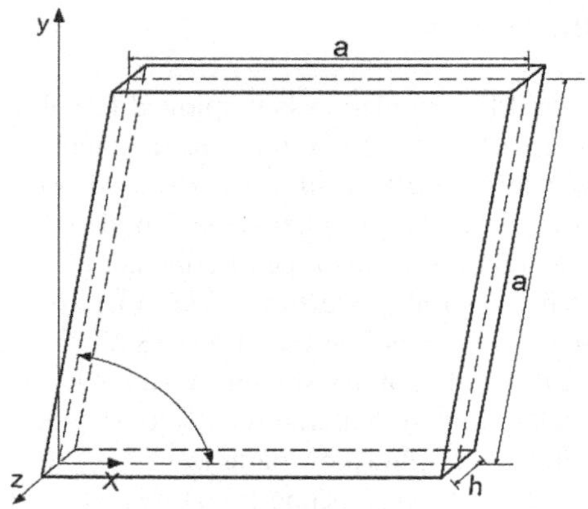

Fig. 1 Geometric of FG-CNTRC skew plate

Figure 1 shows the analysis model for CNTRC/PmPV skew plate. In Fig. 1, a is the length side of the skew plate, h is the total thickness of the skew plate in the z-direction, and α is the skew angle of the skew plate. The CNTRC/PmPV skew plate is subjected to uniformly distributed transverse load equal to $q_0 = -1.0 \times 10^5 N/m^2$. In addition, a set of distribution of CNTs (UD, FG-V, FG-O and FG-X) is examined with various

Table 3 Non-dimensional central deflection of clamped FG-CNTRC skew plate for different skew angles, distributions of CNTs and length-to-thickness ratios. $V^*_{CNT} = 0.17$

a/h		10			
α(°)		90	60	45	30
UD	Present	1.417×10^{-3}	1.220×10^{-3}	8.980×10^{-4}	4.419×10^{-4}
	SHO4	1.249×10^{-3}	1.115×10^{-3}	8.485×10^{-4}	4.273×10^{-4}
FG-X	Present	1.420×10^{-3}	1.210×10^{-3}	8.831×10^{-4}	4.343×10^{-4}
	SHO4	1.161×10^{-3}	1.040×10^{-3}	7.952×10^{-4}	4.057×10^{-4}
FG-V	Present	1.502×10^{-3}	1.279×10^{-3}	9.271×10^{-4}	4.475×10^{-3}
	SHO4	1.336×10^{-3}	1.173×10^{-3}	8.751×10^{-4}	4.302×10^{-4}
FG-O	Present	1.513×10^{-3}	1.290×10^{-3}	9.429×10^{-4}	4.518×10^{-3}
	SHO4	1.577×10^{-3}	1.349×10^{-3}	1.349×10^{-3}	4.617×10^{-4}

length-to-thickness ratios $(a/h = 10)$ and skew angles $(\alpha = 90°, 60°, 45°, 30°)$. A 32 × 32 × 3 meshing is used for this test.

Table 3 list the non-dimensional central deflection of clamped FG-CNTRC skew plate (h = 2 mm). It can be found that the results obtained by the present model are in good agreement with those found in the works of [14].

The following discussion examines the effect of skew angle on non-dimensional stress of clamped FG-O-CNTRC skew plate subjected to uniformly distributed transverse load $q_0 = -1.0 \times 10^5 N/m^2$, with $a/h = 50$ and $V^*_{CNT} = 0.17$. It is interesting that while for all stress plotted in Fig. 2 there is an remarkable effect of skew angle α due to the fiber orientation. Furthermore, the non-dimensional stress increases with the increase in skew angle. On the other hand, the central axial stress distributions $\bar{\sigma}_{xx} = h^2/q_0a^2\sigma_{xx}(a/2, a/2, z)$ are anti-symmetric about the mid-plane and equal to zero at mid-plane, top and bottom surface Fig. 2a. The non-dimensional in-plan shear stress $\bar{\sigma}_{xy} = h^2/q_0a^2\sigma_{xy}(0, 0, z)$ along the non-dimensional thickness are zero at mid-plane and anti-symmetric about the mid-plane Fig. 2b. The non-dimensional shear transverse stress $\left(\bar{\sigma}_{xz} = h^2/q_0a^2\sigma_{xz}(0, a/2, z),\ \bar{\sigma}_{yz} = h^2/q_0a^2\sigma_{yz}(a/2, 0, z)\right)$ through the thickness are symmetric about the mid-plane Fig. 2c and d.

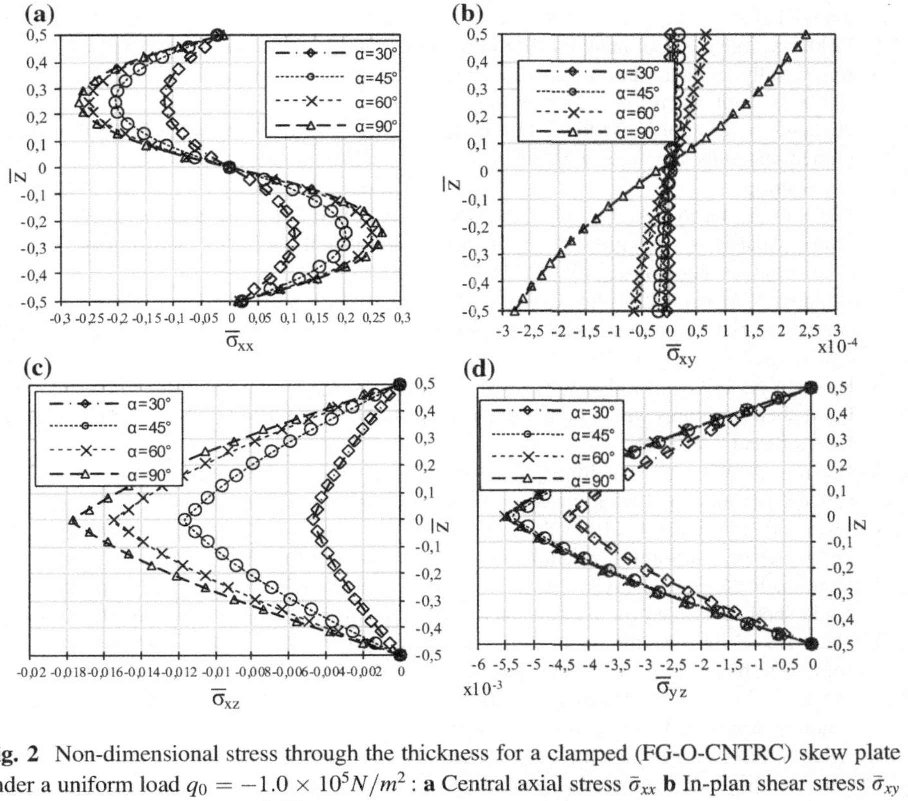

Fig. 2 Non-dimensional stress through the thickness for a clamped (FG-O-CNTRC) skew plate under a uniform load $q_0 = -1.0 \times 10^5 N/m^2$: **a** Central axial stress $\bar{\sigma}_{xx}$ **b** In-plan shear stress $\bar{\sigma}_{xy}$ **c** Shear transverse stress $\bar{\sigma}_{xz}$ **d** Shear transverse stress $\bar{\sigma}_{yz}$

5 Conclusion

In this paper, a solid-shell finite element with enhanced assumed strain is developed to study the static behavior of carbon nanotubes reinforced functionally graded shells. The solid-shell element assures a quadratic distribution of the transverse shear strain. The proposed formulation and that form literature are in good agreement for static analysis of carbon nanotube-reinforced FG structures.

References

1. Alibeigloo A (2013) Static analysis of functionally graded carbon nanotube-reinforced composite plate embedded in piezoelectric layers by using theory of elasticity. Compos Struct 95:612–622
2. Frikha A, Dammak F (2017) Geometrically non-linear static analysis of functionally graded material shells with a discrete double directors shell element. Comput Methods Appl Mech Eng 315:1–24
3. Frikha A, Hajlaoui A, Wali M, Dammak F (2016) A new higher order C0 mixed beam element for FGM beams analysis. Compos Part B Eng 106:181–189
4. Frikha A, Wali M, Hajlaoui A, Dammak F (2016) Dynamic response of functionally graded material shells with a discrete double directors shell element. Compos Struct 154:385–395
5. Frikha A, Zghal S, Dammak F (2018) Finite rotation three and four nodes shell elements for functionally graded carbon nanotubes-reinforced thin composite shells analysis. Comput Methods Appl Mech Eng 329:289–311
6. Hajlaoui A, Chebbi E, Triki E, Dammak F (2018) Post-buckling analysis of functionally graded materials structures with enhanced solid-shell elements, Lecture notes in mechanical engineering
7. Hajlaoui A, Jarraya A, El Bikri K, Dammak F (2015) Buckling analysis of functionally graded materials structures with enhanced solid-shell elements and transverse shear correction. Compos Struct 132:87–97
8. Hajlaoui A, Jarraya A, Kallel-Kamoun I, Dammak F (2012) Buckling analysis of a laminated composite plate with delaminations using the enhanced assumed strain solid shell element. J Mech Sci Technol 26:3213–3221
9. Hajlaoui A, Triki E, Frikha A, Wali M, Dammak F (2017) Nonlinear dynamics analysis of FGM shell structures with a higher order shear strain enhanced solid-shell element. Lat Am J Solids Struct 14:72–91
10. Hajlaoui A, Wali M, Ben Jdidia M, Dammak F (2016) An improved enhanced solid shell element for static and buckling analysis of shell structures. Mech Ind 17
11. Jeyaraj P, Rajkumar I (2013) Static behavior of FG-CNT polymer nano composite plate under elevated non-uniform temperature fields. Procedia Eng 64:825–834
12. Mehrabadi SJ, Sobhani Aragh B (2014) Stress analysis of functionally graded open cylindrical shell reinforced by agglomerated carbon nanotubes. Thin-Walled Struct 80:130–141
13. Shen HS (2009) Nonlinear bending of functionally graded carbon nanotube-reinforced composite plates in thermal environments. Compos Struct 91:9–19
14. Zghal S, Frikha A, Dammak F (2017) Static analysis of functionally graded carbon nanotube-reinforced plate and shell structures. Compos Struct 176:1107–1123
15. Zghal S, Frikha A, Dammak F (2018) Free vibration analysis of carbon nanotube-reinforced functionally graded composite shell structures. Appl Math Model 53:132–155

16. Zhang LW, Lei ZX, Liew KM, Yu JL (2014) Static and dynamic of carbon nanotube reinforced functionally graded cylindrical panels. Compos Struct 111:205–212
17. Zhu P, Lei ZX, Liew KM (2012) Static and free vibration analyses of carbon nanotube-reinforced composite plates using finite element method with first order shear deformation plate theory. Compos Struct 94:1450–1460

Effect of the Type of Binder on Thermal and Mechanical Properties of Mortar with Doum Palm Fiber

Nairi Fatma[1,2](✉), Alleque Lamis[2], Zitoune Redouane[2], and Zidi Mondher[1]

[1] Laboratory of Mechanical Engineering (LGM), National School of Engineers of MONASTIR (ENIM), Monastir, Tunisia
fatma.naiiri@gmail.com
[2] Institut Clément Ader (ICA), CNRS UMR 5312, 133c, av. de Rangueil, 31077 Toulouse Cedex, France

Abstract. This paper present a study of the effect of the type of binder on mortars properties reinforced with Doum palm fiber. Two types of binder used for the production of mortars containing Doum palm fiber. Cement is used as a binder for mortar 1 and plaster is used for mortar 2. The mass content of fiber in the test samples was varied from 0.5% to 1.5% with a step of 0.5%. Thermal characteristics of specimens in term of conductivity and diffusivity were studied. Moreover, the flexural and compressive strength of the samples were evaluated. The measurement of the thermal properties of the specimens shows that plaster has the lowest values of thermal conductivity. A report between densities and mechanical properties of these mortars was realized. These results also report that the density of the specimens has an important role in mechanical properties. The more porous is the material present the lower thermal conductivity. For both mortars, the use of Doum palm fiber improves the mortar ductility without affecting the mechanical requirement for construction materials.

Keywords: Mortar · Doum palm fiber · Cement · Plaster · Mechanical properties

1 Introduction

The use of natural fiber in mortar formulation is justified by the improvement of mechanical and thermal performance.in addition to the evident reduction of weight of the specimen [13]. Portland cement, since his development over 175 years, becomes the dominant binder used in mortar and concrete for construction [7]. Cement production is one of the important contributors to the CO_2 emission.

Different investigation to find a suitable alternative in order to decrease the cost of the construction material and the atmospheric CO_2 concentration has launched a year ago. Nevertheless, there are many alternatives have been analyzed by researchers to replace cement in concrete or mortar [1]. Thus, in cement-based products for civil construction, a study examined the use of paper sludge as an environmentally friendly alternative for the production of eco-efficient cement [5]. Another study tested paper

N. Aifaoui et al. (Eds.): CMSM 2019, LNME, pp. 452–459, 2020.
https://doi.org/10.1007/978-3-030-27146-6_49

waste as a secondary raw material for producing clinker for the mortar [3]. Other studies have indicates that cement incorporated with recycled paper mill residues are high and meet the requirements of building materials [10, 11]. Various types of vegetable fibers have becomes commonly used as reinforcement in composite materials such as jute [15], hemp [12], coir [2] and flax.

Several studies have evaluated the use of different types of natural fibers in mortar and concrete composite. Filho and al suggested the use of coconut fibers as reinforcement of cement composite. Moreover, the compressive strength and modulus of rupture of the composites increase by increasing the contents of coconut until optimum values [14].

The aim of this research is to highlight the effect of the incorporation of Doum palm fibers on the cement and gypsum mortar properties.

2 Experimental Procedure

2.1 Materials

2.1.1 Doum Palm Fiber

Natural fibers used in this study are obtained from petiole of Hyphaene thebaica with common names doum palm. Doum palm is used as a decorative plant in rural and urban areas. Yearly important quantities of old and big leaves are pulled out and discarded. Therefore to reduce the waste of these renewable materials, we opted for exploiting these abandons fibers as mortar reinforcement.

The fibers were extracted from the petiole of Doum palm fiber using a grinder make in a mechanical engineering laboratory.

Sodium hydroxide (NaOH) was used for the surface treatment. It was conducted by immersing the Doum palm fiber in 1% of NaOH solution at 105 °C for 1 h. Furthermore, the fibers were rinsed several times to neutralize them from the residues of NaOH. Finally, the fiber was bleached with sodium chloride and soaked in distilled water.

2.1.2 Binder

Ordinary Portland cement II(32.5R) has been used as a binder for the first type of the composite. The chemical composition of cement has been presented in Table 1.

Table 1 Chemical composition of cement (%)

CaO	SiO_2	Al_2O_3	Fe_2O_3	MgO	K_2O	Na_2O	SO_3
60	17	4	3	5	1	1	3.5

Gypsum has also used as a binder for the second type of composite. It was produced by heating natural gypsum between 150° and 170° according to the reaction.

$$CaSo_4\,2H_2O \rightarrow CaSo_4\,0.5H_2O + 1.5H_2O \quad (1)$$

The technical feature of used gypsum are presented in Table 2 (Fig. 1).

Table 2 Gypsum particle size distribution

Size (μm)	250	180	90	63
Finer (%)	≤0.2	1–2.5	12–17	22–28

2.1.3 Sand

Particle size distribution of the sand determined in accordance with is given in Fig. 2.

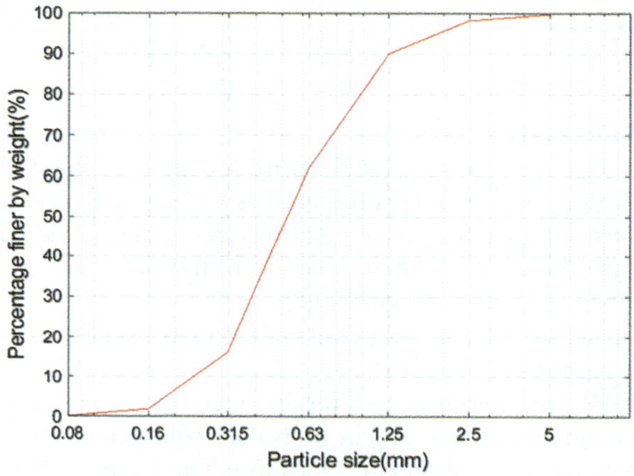

Fig. 1 Grain size distribution

For the two types of composite rolled sand with (0/2), granulomere was used. The physical properties are presented in Table 3.

Table 3 Sand physical properties

MF fineness modulus	The equivalent of sand to the piston ESP (%)	The equivalent of sand visual ESV (%)	Apparent volumic mass (kg/m^3)
3.315	71.91	78.56	1470

2.2 Mixing Procedure and the Preparation of the Composite

Two types of mortar reinforced with Doum palm fiber noted mortar1 and mortar 2 was evaluated. For the production of mortar 1, the binder used is the cement. Gypsum was used as the second type of binder for the production of mortar 2.

For all the composite different weight percentage of Doum fiber from 0.5% to 2% with a step of 0.5% were added (Table 4).

Table 4 Mixing ratio by weight

Designation	Constituent	Ration in the mix
	$\frac{Water}{Plaster+sand}$	0.6
Gypsum mortar	$\frac{Sand}{Plaster}$	0.5
	Cement	1
Cement mortar	Sand	3
	Water	0.5

2.3 Experimental Investigation

The prismatic specimens (40*40*160 mm) were used for the flexural strength tests according to the standard ASTM C348 [4]. Three specimens were examined and average values were saved. The compressive strength test was carried out according to the standard ASTM C349 [9].

The thermal conductivity test was determined using the Hot disk method on specimens with dimension 20 mm × 11mm × 10mm.

3 Results and Discussions

3.1 Fibers Surface

Figure 2 shows the surface morphology of Doum palm fiber. The surface is composed of fibrils bonded together by pectin and other compounds.

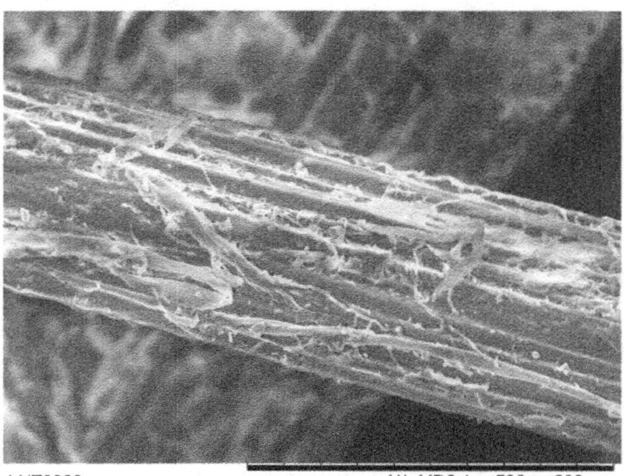

Fig. 2 SEM image of untreated Doum palm fiber

Natural fiber characterized by their chemical composition and particle and especially by the percentage of lignin, cellulose, and hemicellulose. Pectin and lignin present the binding agent in the fiber.

Figure 3 shows the cross-section of the Doum palm fiber underlining the structure of fiber—cells. The diameter of Doum palm fiber after treatment is around 164.2 μm.

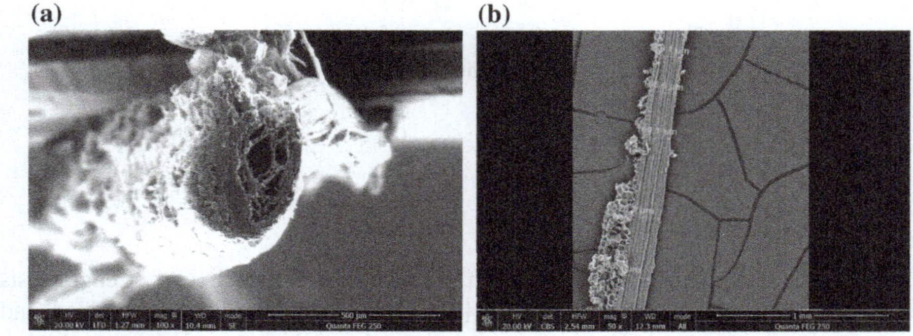

Fig. 3 **a** SEM image of the cross-section of Doum palm fiber **b** Diameter of (**a**) treated Doum palm fiber

3.2 Mechanical Properties

Figure 4 presents the flexural and compressive strength of mortar as a function of fiber ratio. These figures show that cement mortar has greater mechanical properties than those of gypsum mortar. The gap of resistance between different mortar is less than

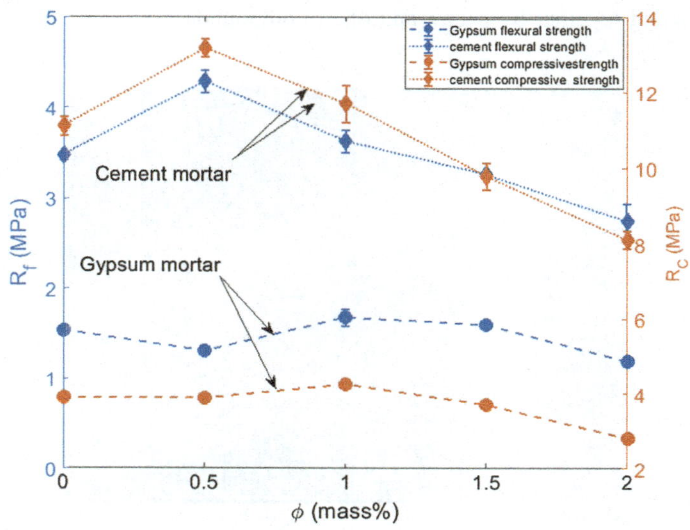

Fig. 4 Mechanical properties of cement and gypsum mortar

43.74%. The improve of resistance by adding 0.5% of Doum fiber for cement mortar is about 18.63 and 23.63% for compressive and flexural strength respectively. However, beyond 1% of fibers content, the compressive strength of cement mortar samples decrease because the cement mortar becomes less manageable and difficult to be used due to the excess of fibers. These results are compliant with other studies [6, 8].

For gypsum mortar, the higher mechanical properties are obtained by adding 1% of Doum palm fiber. Thus, the decrease in mechanical properties beyond this value is a result of the poor fibers/matrix adhesion and the increase of the porosity in the matrix due to the air entertainment.

3.3 Thermal Properties

The dry density of mortars and thermal conductivity are plotted against fiber ratio in Fig. 5.

It is clear that as far as fiber ratio increased lower dry densities were obtained for both mortars. Additionally, densities of gypsum mortar are lower than those of cement mortar. The difference is about 43.83%.

It can be observed that the choice of binder plays a more interesting role on density than the fiber ratio.

It can be observed a coherent relation between thermal conductivity, density and binder type. The thermal conductivity of cement mortar is higher than gypsum mortar as these are denser as explained previously. This report demonstrates the effect of composite porosity on thermal conductivity. Lower is the density (higher porosity), lower is thermal conductivity. These results are coherent with those of Grouly ('resumes_final_impression', no date) who indicated that thermal conductivity is vigorously addicted by the type of binder. So for a specified thermal conductivity, gypsum mortar has a lower density than that of cement mortar.

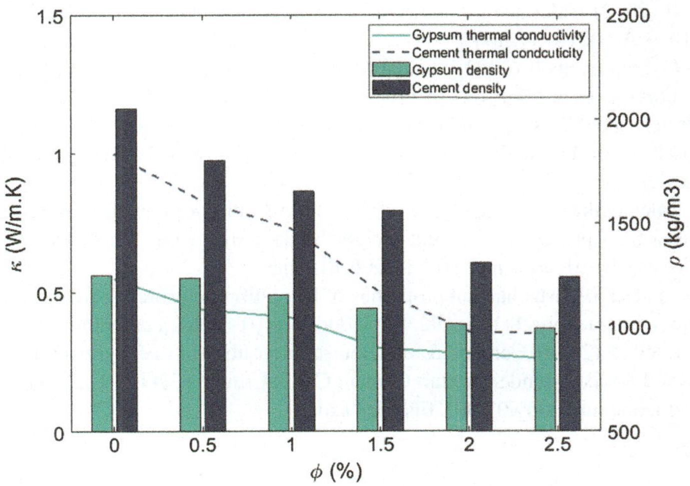

Fig. 5 Thermal conductivity and density of composite as a function of fiber ratio

4 Conclusion

From this study, mechanical and thermal properties of gypsum and cement mortars reinforced with Doum palm fiber were analyzed.

Results report an important decrease in thermal conductivity of the composite adding to an increase of mechanical properties for specific fiber content.

The most interesting results are obtained by adding 0.5% and 1% of fiber for cement mortar and gypsum mortar respectively.

References

1. Amirul M et al (2017) Australian Journal of Basic And Properties of Mortar with Red Gypsum as Cement Replacement Material by Using Industrial Approach Method, 11 November, pp 115–121. https://doi.org/10.22587/ajbas.2017.11.13.13
2. Asasutjarit C et al (2007) Development of coconut coir-based lightweight cement board. Constr Build Mater 21(2):277–288. https://doi.org/10.1016/j.conbuildmat.2005.08.028
3. Buruberri LH, Seabra MP, Labrincha JA (2015) Preparation of clinker from paper pulp industry wastes. J Hazard Mater. Elsevier B.V., 286, pp 252–260. https://doi.org/10.1016/j.jhazmat.2014.12.053
4. Cement H, Statements B (2002) Standard test method for flexural strength of hydraulic-cement mortars 1(4)
5. Frías M, Rodríguez O, De Rojas MIS (2015) 'Paper sludge, an environmentally sound alternative source of MK-based cementitious materials. A Review 74:37–48. https://doi.org/10.1016/j.conbuildmat.2014.10.007
6. Ismail M (2005) Compressive and tensile strength of natural fibre-reinforced cement based composites, July 2005, pp 1–12
7. Juenger MCG et al (2011) Cement and concrete research advances in alternative cementitious binders. Cement Concr Res 41(12):1232–1243. https://doi.org/10.1016/j.cemconres.2010.11.012 Elsevier Ltd
8. Lu J et al (2012) Natural fiber composites for structural. Mater Des 32(1):1–2. https://doi.org/10.1023/A:1021013921916
9. Mortars C, Statements B (2002) Standard test method for compressive strength of hydraulic-cement mortars (Using Portions of Prisms Broken in Flexure) 1, pp 1–4
10. Raut SP et al (2012) Reuse of recycle paper mill waste in energy absorbing light weight bricks. Constr Build Mater 27(1):247–251. https://doi.org/10.1016/j.conbuildmat.2011.07.053 Elsevier Ltd
11. Raut S, Ralegaonkar R, Mandavgane S (2013) 'Utilization of recycle paper mill residue and rice husk ash in production of light weight bricks. Arch Civil Mech Eng 13(2):269–275. https://doi.org/10.1016/j.acme.2012.12.006 Elsevier
12. Sedan D et al (2008) Mechanical properties of hemp fibre reinforced cement: influence of the fibre/matrix interaction 28:183–192. https://doi.org/10.1016/j.jeurceramsoc.2007.05.019
13. Silva LM et al (2010) Cement & concrete composites role of lightweight fillers on the properties of a mixed-binder mortar. Cement Concr Compos 32(1):19–24. https://doi.org/10.1016/j.cemconcomp.2009.07.003 Elsevier Ltd.

14. Tolêdo Filho RD et al (2000) Durability of alkali-sensitive sisal and coconut fibres in cement mortar composites. Cement Concr Compos 22(2):127–143. https://doi.org/10.1016/S0958-9465(99)00039-6
15. Zhou X et al (2013) Materia ls and design fracture and impact properties of short discrete jute fibre-reinforced cementitious composites. J Mater Design 49:35–47. https://doi.org/10.1016/j.matdes.2013.01.029 Elsevier Ltd.

Numerical Investigation of Reverse Redrawing Process Using a Non Associated Flow Rule

Olfa Ghorbel[1(✉)], Sana Koubaa[1], Jamel Mars[1], Mondher Wali[1,2], and Fakhreddine Dammak[1]

[1] Laboratory of Electromechanical Systems (LASEM), National Engineering School of Sfax, University of Sfax, B.P 1173-3038, Sfax, Tunisia
olfa@ghorbel.net, sana.kouba@enis.tn, fakhreddine.dammak@enis.rnu.tn
[2] Department of Mechanical Engineering, College of Engineering, King Khalid University, Abha, Saudi Arabia

Abstract. Deep drawing process is commonly used to produce particular components like aerospace and automotive structural parts. Based on the drawing ratio, it can be performed in a single or a multiple-stage drawing. Due the complexity of this process, finite element simulations are considered as a powerful tool for the both reasons: reducing times and costs, and improving quality and productivity. The current study is conducted to evaluate the performance and the capability of a non associated flow rule (NAFR) approach on numerical results during reverse re-drawing process of DDQ mild steel metal. The adopted model is implemented on ABAQUS software using user interface material subroutine (VUMAT).

Keywords: Non associated flow rule · DDQ mild steel · Reverse re-drawing

1 Introduction

Deep drawing is a difficult process that need a specific choice of dependant parameters related to tools geometry, frictional condition, applied forces, mesh sensitivity, gap value, among others. In the case of a high drawing ratio, re-drawing process is required. When considering this process, the result of the first stage in terms of stress and thickness variation influences the subsequent behaviour. Therefore in order to obtain a successful product, the determination of the suitable combination of the above parameters is a hard task.

Numerical simulation of deep drawing forming process is nowadays an indispensable tool in the automotive and aerospace industry, particulary for a virtual product conception and tools geometries. In fact, the conventional manufacturing based on an empirical procedure is gradually replaced by the finite element simulation. It is a very powerful tool to optimize the process parameters and to solve technological problems (fracture, necking). Many experimental and numerical studies are undertaken in order to more understand the redrawing process and to estimate the influence of the material and process conditions [7, 18]. There is a general agreement that the accuracy of the

N. Aifaoui et al. (Eds.): CMSM 2019, LNME, pp. 460–467, 2020.
https://doi.org/10.1007/978-3-030-27146-6_50

numerical results is basically depending on the constitutive equations that describe the hardening law and the anisotropic behavior of the sheet metal [8, 16, 19, 20].

The objective of the present work is to study the ability of a non associated flow rule combined with isotropic hardening to simulate accurately a reverse re-drawing process of a cylindrical cup, proposed recently by Danckert et al. [9]. The anisotropic Hill'48 yied criterion is adopted [10]. It is a quadratic criterion, widely used in the literature due to its ability to predict the anisotropic behavior and the formability of several materials [1–6, 12–15, 17, 21, 22]. Numerical development is implemented into a user-defined material subroutine (VUMAT) using the commercial finite element code ABAQUS. Good correlation is observed between numerical and experimental in both stages.

2 Constitutive Equations

In this section, the adopted constitutive equations are based on an anisotropic elasto-plastic model, coupled with isotropic-kinematic hardening.

The strain tensor $\boldsymbol{\varepsilon}$ is decomposed in elastic and a plastic part based on infinitesimal approximation, such as

$$\boldsymbol{\varepsilon} = \boldsymbol{\varepsilon}^e + \boldsymbol{\varepsilon}^p \quad (1)$$

The elastic strain is related to the stress tensor $\boldsymbol{\sigma}$ as follows

$$\boldsymbol{\sigma} = \boldsymbol{D} : \boldsymbol{\varepsilon}^e \quad (2)$$

where $\boldsymbol{D}$ is the elastic stiffness matrix.

Using a non-associated elastoplastic model, the yield criterion and the plastic potential are chosen as in Wali et al. [22].

$$f = \sqrt{\frac{3}{2}}\varphi_f(\xi) - (\sigma_Y + R) \leq 0, \xi = \boldsymbol{\sigma} - \boldsymbol{X} \quad (3)$$

$$F = \sqrt{\frac{3}{2}}\varphi_F(\xi) - (\sigma_Y + R) + \frac{1}{2}\sum_{k=1}^{M}\frac{b_k}{a_k}\boldsymbol{X}_k : \boldsymbol{X}_k \quad (4)$$

ξ is the effective stress tensor, $\boldsymbol{X} = \sum_{k=1}^{M} \boldsymbol{X}_k$ is the back-stress tensor, R is the drag stress in isotropic hardening, σ_Y is the initial yield stress and a_k, b_k are material parameters.

In this work, the isotropic hardening function is defined by using Swift hardening law

$$\sigma_Y + R(\kappa) = K\left(\varepsilon_0^p + \kappa\right)^n \tag{5}$$

κ is the isotropic hardening parameter K, ε_0^p, and n are material parameters.

$\varphi_f(\xi)$ and $\varphi_F(\xi)$ are the equivalent stresses, which are continuously and differentiable functions. They can be generally, isotropic, orthotropic, quadratic or non-quadratic. In this contribution, the quadratic anisotropic function in the form Hill's [10] is adopted as follows:

$$\varphi_f(\xi) = \|\xi\|_{\boldsymbol{P}} = \sqrt{\xi^t \boldsymbol{P} \xi}, \varphi_F(\xi) = \|\xi\|_{\boldsymbol{Q}} = \sqrt{\xi^t \boldsymbol{Q} \xi} \tag{6}$$

In the three dimensional form, $\boldsymbol{P}$ and $\boldsymbol{Q}$ are fourth order tensors, obtained by taking

$$\boldsymbol{P} = \frac{2}{3}\boldsymbol{H}, \quad \boldsymbol{Q} = \frac{2}{3}\boldsymbol{H}' \tag{7}$$

$$\boldsymbol{H} = \boldsymbol{H}(F, G, H, N, M, L) = \begin{bmatrix} H+G & -H & -G & 0 & 0 & 0 \\ & H+F & -F & 0 & 0 & 0 \\ & & F+G & 0 & 0 & 0 \\ & & & 2N & 0 & 0 \\ & Sym & & & 2M & 0 \\ & & & & & 2L \end{bmatrix}, \tag{8}$$

$$\boldsymbol{H}' = \boldsymbol{H}\left(F', G', H', N', M', L'\right)$$

Finally, the evolution equations for the present problem are derived as:

$$\dot{\kappa} = \dot{\gamma}, \dot{\boldsymbol{X}}_k = a_k \dot{\boldsymbol{\varepsilon}}^p - b_k \dot{\gamma} \boldsymbol{X}_k \tag{9}$$

$$\dot{\boldsymbol{\varepsilon}}^p = \dot{\gamma}\frac{\partial F}{\partial \boldsymbol{\sigma}} = \sqrt{\frac{3}{2}}\dot{\gamma}\boldsymbol{n}, \boldsymbol{n} = \frac{1}{\varphi_F}\boldsymbol{Q}\xi \tag{10}$$

$$\dot{\boldsymbol{X}} = \sum_{k=1}^{M} \dot{\boldsymbol{X}}_k = \dot{\gamma}\boldsymbol{H}_X, \boldsymbol{H}_X = a\sqrt{3/2}\boldsymbol{n} - \sum_{k=1}^{M} b_k \boldsymbol{X}_k, a = \sum_{k=1}^{M} a_k \tag{11}$$

$\dot{\gamma}$ is the plastic multiplier which is consistent with the loading/unloading condition

$$\dot{\gamma} \geq 0, f \leq 0, \dot{\gamma} f = 0 \tag{12}$$

A return mapping algorithm based on fully implicit backward Euler integration is adopted. The enforcement of the consistency condition and the exactness of the equivalent stresses function φ_F is reduced to the two scalar equations. The unknowns of this system of equations are the plastic multiplier $\Delta\gamma$ and the equivalent stresses function $\varphi_{F,n+1}$, which are solved with the Newton-Raphson method. For more details see [22].

Note that the implementation of the proposed algorithm in large deformation is achieved by replacing tensorial variables in the constitutive equations by their rotational corresponding quantities. A commercial finite element code is used (ABAQUS) by considering hypo-elastic approach (olfa et al. 2019).

3 Numerical Simulations of Reverse Re-drawing Process

All numerical simulations in the present study are based on a reverse deep drawing process, which is proposed as a benchmark at Numisheet'99 conference [11]. It is schematically presented in Fig. 1. All process parameters and tools dimensions are shown in Table 1. The final product is a cylindrical cup performed for DDQ mild steel with initial thickness equal to 0.98 mm. The elasto-plastic material properties of the sheet are listed in Table 2.

Both drawing stages are simulated using ABAQUS software with the user defined material subroutine (VUMAT). The quarter blank sheet is meshed using C3D8R elements. The blank holder (1, 2), the die (1, 2) and the punch (1, 2) are assumed to be discret rigid bodies.

Table 1 Tools dimensions of the reverse re-drawing process

Tools components and process parameters	Description
Punch 1	ØD = 100 mm; R = 5.5 mm
Die 1	ØD = 104.5 mm; R = 8 mm
Blank-holder 1	ØD = 104.5 mm; gap = 1.13 mm
Punch 2	ØD = 73.4 mm; R = 8.5 mm
Die 2	ØD = 78 mm; R = 5.5 mm
Blank-holder 2	ØD = 104.5 mm; R = 7 mm; gap = 1.4 mm
Blank	ØD = 170 mm; e = 0.98 mm
Lubrifiant	Yes

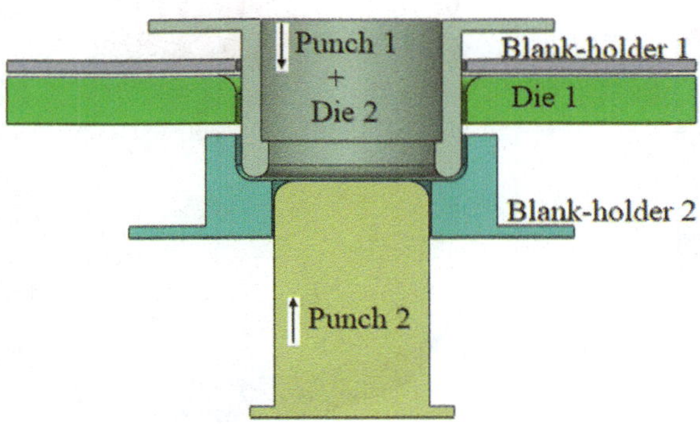

Fig. 1 Reverse deep drawing device (schemmatic view)

Table 2 Elasto-plastic material parameters of DDQ mild steel [16]

Elastic parameters	Hill'48 parameters		Hardening swift law
E = 210000 MPa	F = 0.3224	F' = 0.3137	K = 568.34 MPa
Nu = 0.3	G = 0.3663	G' = 0.3663	n = 0.233
	H = 0.6337	H' = 0.6337	$\varepsilon_0^p = 172\,\text{MPa}$
	L = M=N = 1.4658	L' = M' = N' = 1.1764	

4 Results and Discussions

The equivalent plastic strain prediction with the final cup shape at final stages, using a NAFR_HH are plotted in Fig. 2. Based on this figure, the proposed anisotropic model is in good agreement with the result presented in the work of [20, 16].

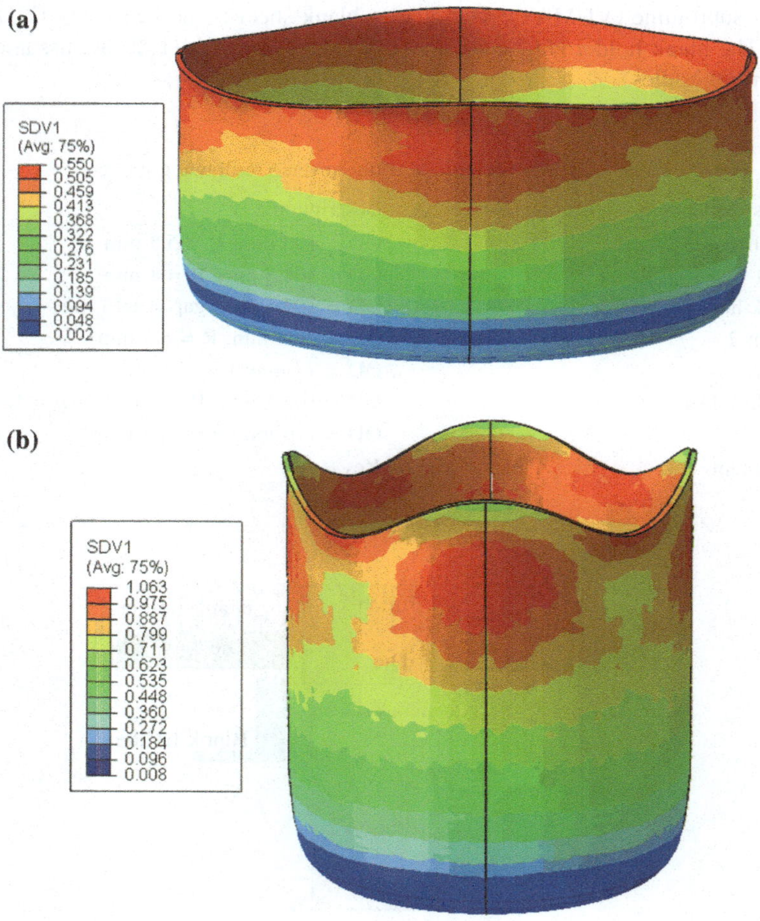

Fig. 2 Equivalent plastic strain (SDV1) of the cylindrical cup shape using NAFR_HH **a** stage 1 **b** stage 2

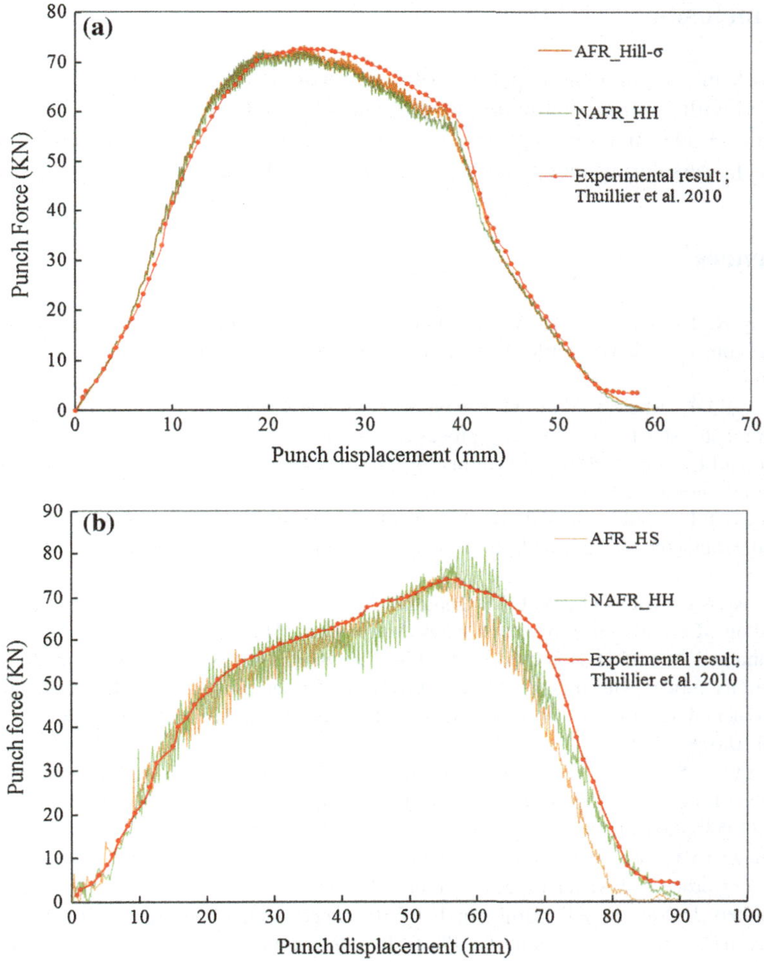

Fig. 3 Force Punch—displacement during reverse redrawing process **a** stage 1 **b** stage 2

Also, numerical simulations of Punch load-displacement in stage (1) and stage (2) are performed and compared with the experimental curves in Fig. 3. Despite the good agreement of the both models (AFR_Hill-σ/NAFR_HH) in the first stage, numerical result using the NAFR_HH reproduce more experiments comparing to associated plasticity model in the second stage. This can be explained by the fact that the NAFR can predict more the thickness distribution of the cup in the end of the first stage, given as an input in the second one. However, there is still a few discrepancies when comparing to the experimental profile. More refinements of the anisotropic model are necessary such as including kinematic hardening and damage parameters.

5 Conclusion

This work investigates the capability of a non associated anisotropic plasticity model combined with isotropic hardening to simulate accurately reverse redrawing processes. Quadratic Hill48 function is adopted for the description of plastic behavior of DDQ mild steel, considering independent functions for yielding and plastic potential.

References

1. Autay R, Koubaa S, Wali M, Dammak F (2017) Numerical implementation of coupled anisotropic plasticity-ductile damage in sheet metal forming process. J Mech 34(04):417–430
2. Ben Said L, Mars J, Wali M, Dammak F (2016) Effects of the tool path strategies on incremental sheet metal forming process. Mech Ind 17(4):411
3. Ben Said L, Mars J, Wali M, Dammak F (2017) Numerical prediction of the ductile damage in single point incremental forming process. Int J Mech Sci 131–132:546–558
4. Belhassen L, Koubaa S, Wali M, Dammak F (2016) Numerical prediction of springback and ductile damage in rubber-pad forming process of aluminum sheet metal. Int J Mech Sci 218–26
5. Belhassen L, Koubaa S, Wali M, Dammak F (2017) Anisotropic effects in the compression beading of aluminum thin-walled tubes with rubber. Thin-Walled Struct 902–10
6. Bouhamed A, Jrad H, Ben Said L, Wali M, Dammak F (2018) A non-associated anisotropic plasticity model with mixed isotropic–kinematic hardening for finite element simulation of incremental sheet metal forming process. Int J Adv Manuf Technol. https://doi.org/10.1007/s00170-018-2782-3
7. Chung SY, Swift HW (1953) An experimental investigation into the redrawing of cylindrical shells. Proc Inst Mech Eng B J Eng Manuf 1:437–447. https://doi.org/10.1177/095440545300100126
8. Cannizzaro L, Micari F, Noto La Diega S (1990) Finite element analysis of the reverse drawing process. J Mater Process Technol 24:441–450
9. Danckert J, Nielsen KB, Hojbjerg P(1990) Experimental investigation of Numisheet'99 benchmark test C. In: Gelin JC, Picart P (eds.), Proceeding of the fourth international conference and workshop on numerical simulation of 3D sheet forming processes (Numisheet'99), Besancon, France, pp 637–642
10. Hill R(1979) Math Proc Camb Philos Soc 85:179–191
11. Koubaa S, Mars J, Wali M, Dammak F (2017) Numerical study of anisotropic behavior of Aluminum alloy subjected to dynamic perforation. Int J Impact Eng 101:105–114
12. Koubaa S, Belhassen L, Wali M, Dammak F (2017) Numerical investigation of the forming capability of bulge process by using rubber as a forming medium. Int J Adv Manuf Technol 92:1839–1848
13. Mars J, Wali M, Jarraya A, Dammak F, Dhiab A (2015) Finite element implementation of an orthotropic plasticity model for sheet metal in low velocity impact simulations. Thin Walled Struct 89:93–100
14. Mars J, Ben Said L, Wali M, Dammak F (2018) Elasto-plastic modelling of low-velocity impact on functionally graded circular plates. Int J Appl Mech 10(04):1850038
15. Mars J, Chebbi E, Wali M, Dammak F (2018) Numerical and experimental investigations of low velocity impact on glass fiber–reinforced polyamide. Compos B 146(1):116–123

16. Neto DM, Oliveira MC, Alves JL, Menezes LF (2014) Influence of the plastic anisotropy modelling in the reverse deep drawing process simulation. Mater Design 60:368–379
17. Ghorbel O, Mars J, Koubaa S, Wali M, Dammak F (2019) Coupled anisotropic plasticity-ductile damage: Modeling, experimental verification, and application to sheet metal forming simulation. Int J Mech Scie 150:548–560
18. Parsa MH, Yamaguchi K, Takakura N, Imatani S (1994) Consideration of the re-drawing of sheet metals based on finite-element simulation. J Mater Process Technol 47:1–2
19. Thuillier S, Manach PY, Menezes LF, Oliveira MC (2002) Experimental and numerical study of reverse re-drawing of anisotropic sheet metals. J Mater Process Technol 125–126:764–771
20. Thuillier S, Manach PY, Menezes LF (2010) Occurrence of strain path changes in a two-stage deep drawing process. J Mater Process Technol 210:226–232
21. Wali M, Autay R, Mars J, Dammak F (2015) A simple integration algorithm for a non-associated anisotropic plasticity model for sheet metal forming. Int J Numer Meth Eng 107 (3):183–204
22. Wali M, Chouchene H, Ben Said L, Dammak F (2015) One-equation integration algorithm of a generalized quadratic yield function with Chaboche non-linear isotropic/kinematic hardening. Int J Mech Sci 92:223–232

Low Velocity Impact-and-Damage Study of DD13 Sheet Metal

Sana Koubaa(✉), Olfa Ghorbel, Jamel Mars, and Fakhreddine Dammak

Laboratory of Electromechanical Systems (LASEM), National Engineering School of Sfax, University of Sfax, B.P 1173, 3038 Sfax, Tunisia
koubaasana@yahoo.fr, jamelmars@yahoo.fr, olfa@ghorbel.net, fakhreddine.dammak@enis.rnu.tn

Abstract. Due to their crystallographic texture, sheet metals generally present a significant anisotropy. The DD13 sheet metal is one of the hot-rolling steels, also called mild steels that is designed for deep and extra deep drawing applications. It is characterized by high fracture elongation and suitable for chassis components and wheel outer borders. A finite element (FE) -implementation of an anisotropic plasticity model coupled with damage and isotropic hardening for the DD13 metal in low velocity impact simulations is performed. The elasto-plastic model includes isotropic elasticity, anisotropic yielding, associated plastic flow and isotropic hardening. The plastic-damage model is implemented into a user-defined material (VUMAT) subroutine for the commercial finite element code ABAQUS/Explicit. Material properties are defined using the experimental results of (Ghorbel et al (2019) Int J Mech Sci 150:548–560, [1]). A parametric study of the effect of anisotropy, -plate thickness and impactor properties is carried out. The numerical results of the impact force history and impact velocity are discussed and validated.

Keywords: Impact · Perforation · DD13 · Coupled anisotropic plasticity-damage

1 Introduction

The investigation of metallic structural impact is of interest in several engineering sectors as transport, offshore [2], naval [3], aerospace, defense and process industries [4] to predict the onset of necking. The use of numerical methods for structural design is a powerful tool to achieve safety structures. Nevertheless, impact problems include plasticity, large deformations-and fracture that require numerical accuracy and robustness [5]. The reliability of the FE impact simulations mainly depends on the appropriate definition of the physical problem and the influencing parameters. In literature, in the low-velocity sheet-impact studies, the parameters investigated are the nose shape of the impactor [6], the impact velocity,-boundary conditions [7], or the sheet properties [8]. The literature on impact loading includes velocity ranges from low [9] to moderate and high impact [10]. According to [11], the low-velocity is up to about 20 ms^{-1} and the moderate velocity is ranged between 20–300 ms^{-1} for metal sheets.

N. Aifaoui et al. (Eds.): CMSM 2019, LNME, pp. 468–476, 2020.
https://doi.org/10.1007/978-3-030-27146-6_51

A particular case of impact consists in the failure induced by perforation. The failure process consists on plug ejection for blunt nose shape projectile [12], necking for hemispherical impactor and petalling process for conical nose shape impactor.

Although various aspects of metallic plate's dynamic impact were studied, most of researches used isotropic behavior and neglect anisotropy, which is a key aspect that should be taken into account. Few studies are interested in anisotropy when studying perforation of sheet metals. Mars et al. [13] discussed the effect of anisotropy on low velocity impact on AA plates. It was revealed that anisotropy has a great effect at low velocity impact. However, in this previous work, damage was considered independently with the constitutive model. Recently, [14] carried out numerical study of the perforation process of target AA5754-O Aluminum plate by integrating 2-equation algorithm of -generalized quadratic yield criterion of Hill, based on the mixed hardening. Comparison between J2 and Hill yield criteria associated with isotropic ductile damage was conducted. It was depicted that anisotropy has a great effect on petalling failure process.

Therefore, to take into account the anisotropic behavior of sheet metal during impact and to accurately predict the failure process, a coupled advanced anisotropic elastoplastic constitutive equations based on CDM when integrating isotropic and non linear hardening is developed. Based up on the work of [14, 15, 16–18] the integration algorithm of the constitutive equation leads to two local scalar equations which are solved by Newton-Raphson method. The DD13 sheet metal is considered which is one of the hot-rolling steels, also called mild steels that is designed for deep and extra deep drawing applications. The mechanical behavior of the DD13 sheet is investigated in previous study [1]. According to [19], material with ferritic crystalline structure display cold forming and cutting performance. Thus, the material can be also suited for off-shore structures mainly subjected to dynamic impact. The results of [19, 20] show that the quasi-static tests provide a reliable reference for impact loading conditions. The present work contributes to the evaluation of the associated anisotropic plasticity fully coupled with ductile damage-to predict the real behavior of DD13 steel material at low impact velocity up to 25 ms^{-1}.

2 Finite Element Simulations

2.1 Constitutive Equations

In this section, the anisotropic elasto-plastic model with non-linear isotropic/kinematic hardening and fully coupled with-isotropic ductile damage is presented based on CDM approach. In the context of thermodynamics of irreversible processes, we use some couples of states variables [21]. So, we define the elastic strain and-stress tensors $(\boldsymbol{\varepsilon}^e, \boldsymbol{\sigma})$ for the elasto-plastic flow; the isotropic hardening (r, R) for the size of the yield stress and kinematic-hardening (α, $\mathbf{X}$) for the displacement of the yield stress; and the internal isotropic ductile damage variable d. Using the hypothesis of the total energy-equivalence, the-effective-state-variables are defined-on-fictive-undamaged state as follows:

$$\tilde{\varepsilon}^e = (1-d)^{0.5}\varepsilon^e, \tilde{\sigma} = \sigma/(1-d)^{0.5}, \tilde{r} = (1-d)^{0.5} r$$
$$\tilde{R} = R/(1-d)^{0.5}, \tilde{\alpha} = (1-d)^{0.5}\alpha, \tilde{X} = X/(1-d)^{0.5} \quad (1)$$

Based on the infinitesimal strain approximation, the Helmholtz free energy is:

$$\psi(\varepsilon^e, \alpha_k, r_k, d) = \psi^e(\varepsilon^e, d) + \psi^p(\alpha_k, r_k, d) \quad (2)$$

The state relations are:

$$\sigma = \frac{\partial\psi}{\partial\varepsilon^e}, X_k = \frac{\partial\psi}{\partial\alpha_k}, R_k = \frac{\partial\psi}{\partial r_k}, Y = -\frac{\partial\psi}{\partial d} \quad (3)$$

σ is the stress tensor, X_k is the associated variable referring to the back-stress tensors, R_k is the drag stress in isotropic hardening and Y is the associated damage variable. Considering the associated effect of damage with isotropic and kinematic hardening, the terms of elastic and plastic state potential are:

$$\psi^e(\varepsilon^e, d) = \frac{1}{2}\tilde{\varepsilon}^e : D : \tilde{\varepsilon}^e = \frac{(1-d)}{2}\varepsilon^e : D : \varepsilon^e \quad (4)$$

$$\psi^p(\alpha_k, r_k, d) = \frac{1}{2}\left(\sum_{k=1}^{N} Q_k\tilde{r}_k^2 + \sum_{k=1}^{M} a_k\tilde{\alpha}_k{:}\tilde{\alpha}_k\right) = \frac{(1-d)}{2}\left(\sum_{k=1}^{N} Q_k r_k^2 + \sum_{k=1}^{M} a_k\alpha_k{:}\alpha_k\right) \quad (5)$$

Q_k and a_k denote the isotropic and the kinematic hardening modulus parameters, respectively. D represents the elastic operator:

$$D = \begin{bmatrix} \lambda+2\mu & \lambda & \lambda & 0 & 0 & 0 \\ & \lambda+2\mu & \lambda & 0 & 0 & 0 \\ & & \lambda+2\mu & 0 & 0 & 0 \\ & & & \mu & 0 & 0 \\ & & & & \mu & 0 \\ Sym & & & & & \mu \end{bmatrix}, \mu = \frac{E}{2(1+\nu)}, \lambda = \frac{E\nu}{(1+\nu)(1-2\nu)} \quad (6)$$

Using Eq. (4), the stress-like variables are:

$$\sigma = (1-d)D : \varepsilon^e, X_k = (1-d)a_k\alpha_k, k = 1:M, R_k = (1-d)Q_k r_k, k = 1,..,N$$
$$Y = \frac{1}{2}\varepsilon^e : D : \varepsilon^e + \frac{1}{2}\left(\sum_{k=1}^{N} Q_k r_k^2 + \sum_{k=1}^{M} a_k\alpha_k : \alpha_k\right) \quad (7)$$

The plastic flow, f, and the plastic potential, F, are:

$$\begin{aligned} f &= \frac{1}{(1-d)^{0.5}}[\varphi(\xi) - R] - \sigma_Y \\ F &= f + \frac{1}{2(1-d)}\left[\sum_{k=1}^{N}\frac{\beta_k}{Q_k}R_k^2 + \sum_{k=1}^{M}\frac{b_k}{a_k}\boldsymbol{X}_k : \boldsymbol{X}_k\right] + F_d(Y,d) \end{aligned} \tag{8}$$

$$\xi = \boldsymbol{\sigma} - \boldsymbol{X}, \boldsymbol{X} = \sum_{k=1}^{M}\boldsymbol{X}_k, R = \sum_{k=1}^{N}R_k = (1-d)\sum_{k=1}^{N}Q_k r_k \tag{9}$$

The function $\varphi(\xi)$ in Eq. (9) is the equivalent stress with a quadratic yield function, given in 3D form by:

$$\varphi = \|\xi\|_{\boldsymbol{H}} = \sqrt{\xi^t \boldsymbol{H}\xi}, \boldsymbol{H} = \begin{bmatrix} H+G & -H & -G & 0 & 0 & 0 \\ & H+F & -F & 0 & 0 & 0 \\ & & F+G & 0 & 0 & 0 \\ & & & 2N & 0 & 0 \\ & \text{Sym} & & & 2M & 0 \\ & & & & & 2L \end{bmatrix} \tag{10}$$

$\boldsymbol{H}$ is -the Hill'48 anisotropic operator. The constant F, G, N, M and L are calibrated from uniaxial tensile tests [1]. The evolution equations of the effective internal variables are:

$$\dot{\boldsymbol{\varepsilon}}^p = \dot{\gamma}\frac{\partial F}{\partial \boldsymbol{\sigma}} = \dot{\gamma}\boldsymbol{n} = \frac{\dot{\gamma}}{(1-d)^{0.5}}\tilde{\boldsymbol{n}}, \tilde{\boldsymbol{n}} = \frac{1}{\varphi}\boldsymbol{H}\xi \tag{11}$$

$$\dot{\boldsymbol{\alpha}}_k = -\dot{\gamma}\frac{\partial F}{\partial \boldsymbol{X}} = \dot{\boldsymbol{\varepsilon}}^p - \frac{\dot{\gamma}}{1-d}\frac{b_k}{a_k}\boldsymbol{X}_k = \dot{\boldsymbol{\varepsilon}}^p - \dot{\gamma}b_k\boldsymbol{\alpha}_k \tag{12}$$

$$\dot{r}_k = -\dot{\gamma}\frac{\partial F}{\partial R_k} = \dot{\gamma}\left[\frac{1}{(1-d)^{0.5}} - \frac{\beta_k R_k}{Q_k(1-d)}\right] = \dot{\gamma}h_k, h_k = \frac{1}{(1-d)^{0.5}} - \beta_k r_k \tag{13}$$

$$\dot{d} = \dot{\gamma}\bar{Y}, \bar{Y} = \frac{\partial F_d}{\partial Y} \tag{14}$$

$$\dot{\boldsymbol{X}}_k = \dot{\gamma}\left[a_k(1-d)^{0.5}\tilde{\boldsymbol{n}} - \left(b_k + \frac{\bar{Y}}{1-d}\right)\boldsymbol{X}_k\right] \tag{15}$$

The enforcement of the consistency condition and the damage equation are reduced to two scalar equations [1]:

$$\begin{cases} f_1(\Delta\gamma, d) = \frac{1}{(1-d)^{0.5}}[\varphi - R] - \sigma_Y = 0 \\ f_2(\Delta\gamma, d) = d - d_n - \Delta\gamma\bar{Y} = 0 \end{cases}, \varphi = \left[\bar{\xi}^T.\boldsymbol{I}_c^{-T}.\boldsymbol{H}.\boldsymbol{I}_c^{-1}.\bar{\xi}\right]^{1/2} \tag{16}$$

The unknowns are the plastic multiplier $\Delta\gamma$ and the damage variable d. The derived equations are solved using the Newton Method.

2.2 Finite Element Model

We consider a clamped circular plate impacted by hemispherical and conical nose shape impactors (see Fig. 1). The diameter of the impactor is 6.35 mm and the DD13 plate has the radius of 60 and 2 mm thickness. The striker is modeled as analytical rigid body. A 3-D Finite Element model is developed using ABAQUS/VUMAT. Simulations are performed by taking into account the geometric symmetry of the circular plate. The plate is meshed with 8100 C3D8R elements (0.4 mm-radius direction, 5 elements in thickness). A hard contact law is applied [22].

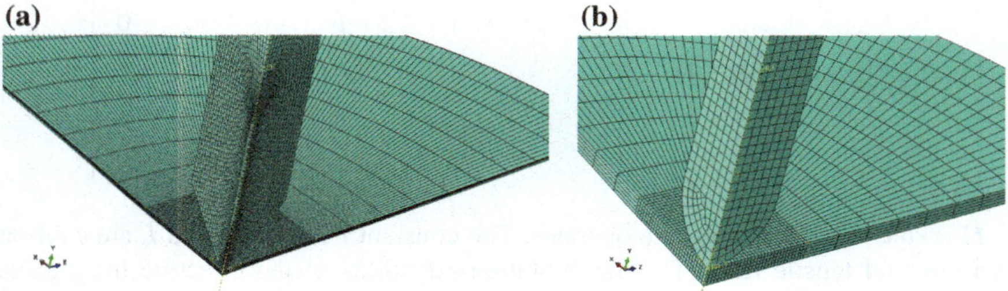

Fig. 1 3D impact computational problem of circular plate with rigid impactor, **a** conical nose shape, **b** hemispherical nose shape

The DD13 is a hot rolled sheet metal -used in several industrial applications like chassis components and offshore structures among others. This sheet metal is a high elongation material with ferritic microstructure. Table 1 shows the chemical composition. To form the fundamentals of constitutive modelling of DD13 sheet metal, a set of material tests was carried out. Uniaxial tension tests provided material parameters related to elasto-plastic with isotropic hardening response and damage evolution by correlating experimental results with numerical simulations [1]. The identified elastic parameters of DD13 sheet metal are listed in Table 2. 3D bending tests were carried out. It was depicted that the contribution of the kinematic parameters on the material behavior is not significant and can be neglected [23].

Table 1 DD13 sheet material chemical-composition

C	S	Mn	P
0.08	0.03	0.4	0.03

Table 2 Hill constants, isotropic hardening-IH and damage parameters identification for DD13, [1]

Elastic parameters	E = 220 GPa	σ_y = 278 MPa		
Hill'48-r	F = −0.43	G = −0.47	H = −0.53	N = −1.49
IH	Q_1 = 490 MPa	Q_2 = 50 MPa	β_1 = 1.35	β_2 = 30
Damage	β = 5	S = 1.5	S = 200 MPa	Y_0 = 15 MPa

3 Simulation Results

Numerical simulations are performed on the perforation of DD13 sheet metal plates using ABAQUS/Explicit. The elasto-plastic constitutive relation fully coupled with isotropic ductile damage and-integrating isotropic hardening was implement in the VUMAT user subroutine. The anisotropy effect is studied by comparing von Mises model J2 coupled with isotropic hardening (J2—IH) and the orthotropic Hill'48 yield criterion coupled with isotropic hardening (Hill-IH). Figure 2 depicts the impact force at an initial velocity $V_0 = 25\ \text{ms}^{-1}$. The plate thickness was fixed to 2 mm. One can notice a variance between-isotropic and anisotropic impact force response. The reduction of force corresponds to the start of the crack propagation and the petal performing. Isotropic model overestimates perforation. This is in a good correlation with the velocity evolution results and with previous study [22].

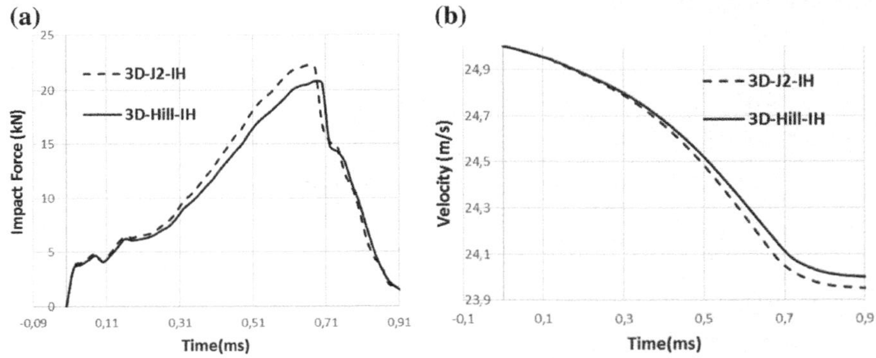

Fig. 2 **a** Forces evolution from simulation, **b** hemispherical projectile velocity evolution

A study on the projectile nose shape influence on the perforation simulation of 2 mm thick circular DD13 plate at an impact velocity $V0 = 19\ \text{ms}^{-1}$ is conducted. One can remark that the projectile shape significantly influences the petalling process and failure modes. The hemispherical projectile brings necking, the conical shape includes petalling due to piercing which is in an agreement with [24]. The damage evolution is depicted in Fig. 3. The residual velocity V_r are plotted versus impact velocity V_i for a series of impact simulations. The analytical equation of the residual velocity developed by [25] is also plotted. One can deduce accurate results of the ballistic limit (Fig. 4).

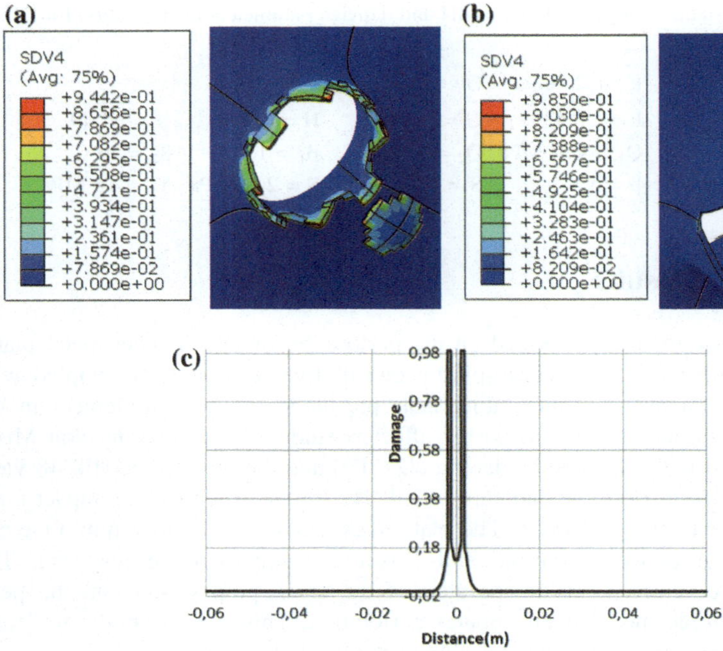

Fig. 3 2mm thick DD13 target impacted normally by: **a** hemispherical projectile; **b** conical projectile; **c** damage evolution (SDV4); Hill-IH

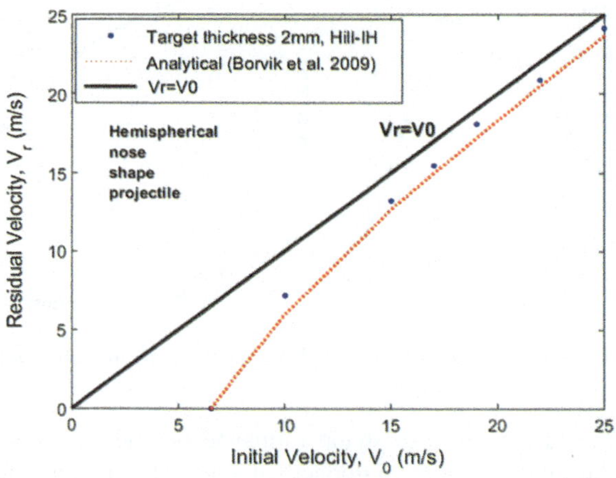

Fig. 4 Residual velocity as a function of initial velocity for 2 mm target thickness impacted by 8 kg striker (Hill-IH)

4 Conclusion

In spite of the large interest in structure impact of sheet metals, limited material database is existing in the literature on the high ductile DD13 ferritic metal regarding anisotropy and damage when subjected to impact and perforation. In this paper, a local approach, based on the CDM fully coupled with thermodynamically constitutive anisotropic elasto-plastic model with mixed hardening is developed to predict the mechanical response of DD13 mild-metal. A parametric study of the effect of anisotropy, plate thickness and impactor properties is carried out. Results of the impact force and velocity evolutions are discussed and validated with literature. The outcomes of this study and the developed mode are suitable for future studies on dual phase stiffened steel plate that may take parts as large structure components in offshore or ship to predict fracture. Also, the present study can be useful for the analysis of clinch-adhesive technique in the automotive industry leading to the proper design of the joint and manufacture of safe and durable fastening, particularly by predicting the energy necessary the hybrid joint dynamic fracture and impact loading.

References

1. Ghorbel O, Mars J, Wali M, Dammak F (2019) Int J Mech Sci 150:548–560
2. Langseth M, Larsen PK (1990) Int J Impact Eng 9(3):289–316
3. Simonsen BC (1997) Ship grounding on rock—I. Theory Marine Struct 10(7):519–562
4. Marciniak Z, Kuczyński K (1967) Limit strains in the processes of stretch-forming sheetmetal. Int J Mech Sci 9(9):609–620
5. Paik JK, Pedersen PT (1996) Modelling of the internal mechanics in ship collisions. Ocean Eng 23(2):107–142
6. Wilkins ML (1978) Mechanics of penetration and perforation. Int J Eng Sci 16:793–807
7. Iqbal MA, Khan SH, Ansari R, Gupta NK (2013) Experimental and numerical studies of double-nosed projectile impact on aluminum plates. Int J lmpact Eng 54:232–245
8. Iqbal MA, Gupta NK, Sekhon GS (2006) Behaviour of thin aluminium plates subjected to impact by ogive-nosed projectiles. Defence Sci J 56:841–852
9. Grytten F, Borvik T, Hopperstad OS, Langseth M (2009) Low velocity perforation of AA5083-H116 aluminium plates. Int J lmpact Eng 36:597610
10. Mohotti D, Ngo T, Raman SN, Ali M, Priyan M (2014) Plastic deformation of polyurea coated composite aluminium plates subjected to low velocity impact. Int J Mater Design 56:696713
11. Jones N, Paik JK (2012) Impact perforation of aluminium alloy plates. Int J lmpact Eng 48:46–53
12. Kpenyibga KM, Jankowiak T, Rusinek A, Pesci R, Wang B (2015) Effect of projectile nose shape on ballistic resistance of interstitial-free steel sheets. Int J lmpact Eng 79:83–94
13. Mars J, Wali M, Jarraya A, Dammak F, Dhiab A (2015) Finite element implementation of an orthotropic plasticity model for sheet metal in low velocity impact simulations. J Thin-Walled Struct 89:93100
14. Koubaa S, Mars J, Wali M, Dammak F (2017) Numerical study of anisotropic behavior of Aluminum alloy subjected to dynamic perforation. Int J Impact Eng 101:105–114
15. Belhassen L, Ben Said L, Koubaa S, Wali M (2018) Effects of using flexible die instead of flexible punch in rubber pad forming process. Lec Note Mech Eng, 259–267

16. Autay R, Koubaa S, Wali M, Dammak F (2017) Numerical implementation of coupled anisotropic plasticity-ductile damage in sheet metal forming process. J Mech 34:417–480
17. Belhassen L, Koubaa S, Wali M, Dammak F (2017) Anisotropic effects in the compression beading of aluminum thin-walled tubes with rubber. Thin walled Struct 119:902–910
18. Belhassen L, Koubaa S, Wali M, Dammak F (2016) Numerical prediction of springback and ductile damage in rubber-pad forming process of aluminium sheet metal 117:218–226
19. Gruben G, Sølvernes S, Berstad T, Morin D, Hopperstad OS, Langseth M (2017) Low-velocity impact behaviour and failure of stiffened steel Plates. Marine Struct 54(2017):73–91
20. Gruben G, Langseth M, Fagerholt E, Hopperstad OS (2016) Low-velocity impact on high-strength steel sheets: an experimental and numerical study. Int J Impact Eng 88:153–171
21. Saanouni K, Chaboche J-L (2003) Computational damage mechanics application to metal forming simulation. In: Milne I, Ritchie RO, Karihaloo B (Eds), Comprehensive structural integrity: numerical and computational methods, Pergamon, Oxford, pp 321–376
22. Koubaa S, Belhassen L, Wali M, Dammak F (2017) Numerical investigation of the forming capability of bulge process by using rubber as a forming mediu. Int J Adv Manuf Technol 92:1839–1848
23. Chongthairungruang B, Uthaisangsuk V, Suranuntchai S, Jiratherearanat S (2013) Spring-back prediction in sheet metal forming of high strength steels. J Mater Design 50:253–266
24. Rodriguez-Martinez JA, Rusinek A, Pesci R, Zaera R (2012) Analysis of the strain induced martensitic transformation in austenitic steel subjected to dynamic perforation. EDP Sci, https://doi.org/10.1051/epjconf/20122604036
25. Borvik T, Forrestal MJ, Hopperstad OS, Warren TL, Langseth M (2009) Perforation of AA5083-H116 aluminium plates with conical-nose steel projectiles Calculations. Int J lmpact Eng 36:426–437

Mechanical Characterization of Thin Films Using Nanoindentation Technique. a Numerical Study

Asma Jellad(✉)

Sousse Mechanical Laboratory (LMS), National Engineering School of Sousse, BP264, Sousse, Tunisia
ja_asma@yahoo.fr

Abstract. Elastic-plastic properties of thin films deposited on elastic-plastic substrate were extracted using numerical nanoindentation tests from the force-displacement curve. In order to limit the substrate effect on measuring thin film elastic-plastic properties, three theoretical models were studied in this work. All these models use a parametric identification method based on dimensional and finite element analyses to extract relationships between the indentation parameters and the elastic-plastic properties of the film and the substrate. The mechanical properties of several thin films were identified using these models.

Keywords: Thin films · Nanoindentation · Finite element method (FEM)

1 Introduction

Thin films are widely used in modern technology and applications such as integrated circuits, metal cutting, thermal barrier coatings and plastic injection moulding due to their high physical properties like thermal and chemical stability, corrosion resistance and also high hardness, strength and wear resistance [1–3]. In order to understand, predict and improve the functionality and reliability of devices containing thin films, it's necessary to evaluate the mechanical characterization of these thin films [4, 5]. Furthermore, nanoindentation technique has been successfully used to measure the mechanical properties of thin films. This method was conducted to investigate the elastic and plastic properties of various thin films deposited on their substrates. However, it was difficult to extract only the film properties due to the small thickness of the films [6]. It was shown that to extract the mechanical properties of thin films from the nanoindentation test, we have to limit the indentation depth less than 10–20% of the film thickness [7].

In this study, we show that the elastic modulus, yield stress and work hardening exponent of the film material could be extracted from the indentation curve and using dimensional analysis and finite element methods. We conducted several numerical indentation tests using ABAQUS code to examine the effectiveness of three analytical models from literature used to determine mechanical properties of thin films. In this work, we identified elastic-plastic film properties using these models and compared them with those obtained in the literature.

N. Aifaoui et al. (Eds.): CMSM 2019, LNME, pp. 477–486, 2020.
https://doi.org/10.1007/978-3-030-27146-6_52

2 Methods Measuring Elastic-Plastic Thin Film's Properties

2.1 Oliver and Pharr Method

Nanoindentation technique was used in order to determine the mechanical properties of thin films. The hardness H and the reduced elastic modulus E_r were obtained from the experimental load-displacement curve using the Oliver and Pharr model [8]. Oliver and Pharr analytical approach was based on Doerner and Nix [9], and Snedden [10] research. The hardness of the material was determined using the maximum loading point of the load-displacement curve. Material's elastic properties were obtained using the elastic part of the unloading curve. The hardness H was defined as the mean pressure under the indenter:

$$H = \frac{P_{\max}}{A_c} \quad (1)$$

where, P_{max} is the maximum applied force obtained directly from the force–displacement curve, and A_c is the projected contact area of the indenter tip with the material. The initial slope of the unloading curve was used to measure the elastic properties of the material. The contact stiffness, S (stiffness of the contact between the indenter tip and the sample material), was used to calculate the elastic reduced modulus of the specimen E_r. The stiffness, S was expressed by [8, 9]:

$$S = \frac{dP}{dh} = \frac{2}{\sqrt{\pi}} E_r \sqrt{A_c} \quad (2)$$

dP/dh is the initial slope of the unloading curve, P is the applied force and h is the indenter displacement (Fig. 1). h_f is the final penetration depth, h_{max} is the maximum penetration depth and P_{max} is the maximum force applied to the indenter (Fig. 1).

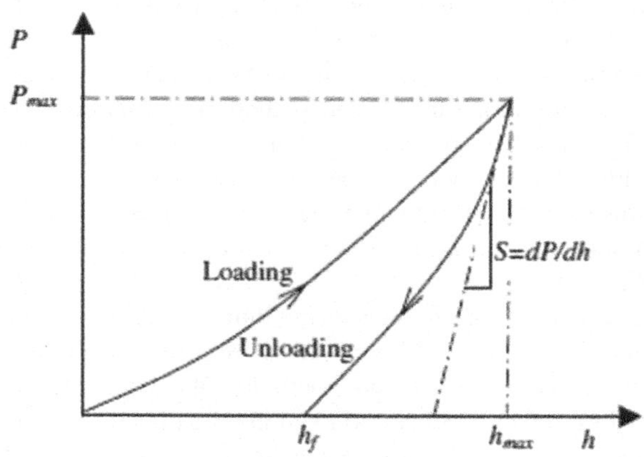

Fig. 1 Experimental load-displacement curve

Therefore, the elastic modulus of the film is given by [8, 9]:

$$E_f = (1 - \upsilon_f^2)\left(\frac{1}{E_r} - \frac{1 - \upsilon_i^2}{E_i}\right)^{-1} \quad (3)$$

where υ_f is the Poisson ratio of the film while E_i and υ_i are the elastic modulus and the Poisson ratio of the indenter.

2.2 Numerical Methods

- **Dimensional analysis**

The indenter was considered as rigid, conical with half angle θ indenting into an elastic-plastic film coated on an elastic-plastic substrate (Fig. 2). Both the film and the substrate materials were characterized by a power law relation defined as [11]:

$$\varepsilon = \begin{cases} \frac{\sigma}{E} & for\, \sigma \leq \sigma_y \\ \frac{\sigma_y}{E}\left(\frac{\sigma}{\sigma_y}\right)^{\frac{1}{n}}, & others \end{cases} \quad (4)$$

where σ_y is the initial yield stress and n is the work hardening exponent, while E and υ are the material elastic modulus and Poisson ratio, respectively.

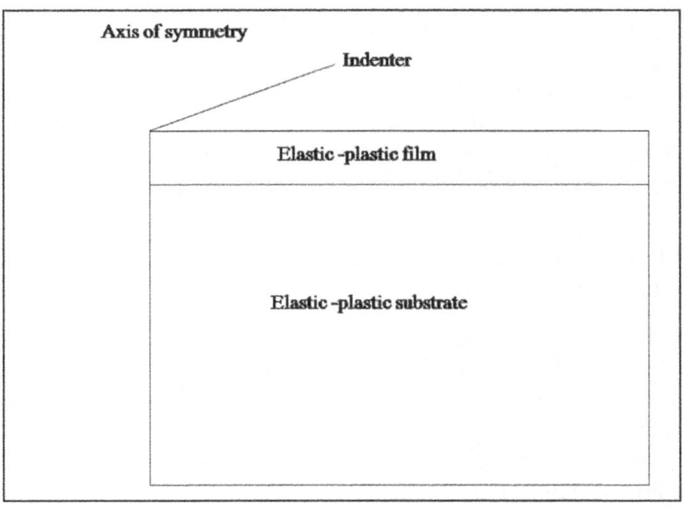

Fig. 2 Geometric model of normal indentation on the film/substrate system

The dimensional analysis permit to have implicit dimensional relationships between film parameters (σ_{yf}, E_f, n_f), substrate parameters (σ_{ys}, E_s, n_s) and the indentation parameters (P_{max}, h_{max}, h_f) using the Buckingham-π theorem [12].

- **Finite element modeling**

The FEM simulations allow to determine explicit relationships between the indentation parameters (P_{max}, h_{max}, h_f) and the elastic-plastic properties of the film (σ_{yf}, E_f, n_f) and the substrate parameters (σ_{ys}, E_s, n_s). The film mechanical properties were extracted by the resolution of the obtained explicit equations.

In this study, the FEM simulations were conducted using the commercial code ABAQUS. The indenter was assumed as a rigid conical with a half angle of 70.3°. The model was considered as axisymmetric to simplify the calculation. The FE simulations require finer meshing near the contact region and a gradually coarser mesh further from the contact region to ensure numerical accuracy (Fig. 3).

The film and the substrate were assumed as isotropic. Generally the initial yield stress and the elastic modulus of the film was about 0.1–2 and 50–250 GPa, respectively and for the substrate was about 0.03–1.1 and 40–210 GPa, respectively [13]. Therefore, several finite element simulations were carried out for combinations of elastic-plastic properties in these wide ranges of film and substrate material parameters.

3 Identification of the Elastic-Plastic Thin Films Properties Using Three Analytical Models

In this work, three analytical models (Jiang, Zhou and Huang model, Liao, Zhou, Huang and Jiang model and Ma, Zhou, Long and Lu model) were used in order to avoid the substrate effect on measuring elastic-plastic film's properties. All these models were based on both dimensional analyses and finite element method (FEM) to extract mechanical properties of the film [11, 13, 14]. These models studied the case of thin elastic-plastic films deposited on elastic-plastic substrates.

3.1 Identification of the Elastic-Plastic Film Properties Using Jiang, Zhou and Huang Model

Numerical indentations tests were carried out on three film- substrate systems (Material 1, Material 2, Material 3) using ABAQUS code by introducing the mechanical properties of the substrate and the films [13]. Figure 4 shows the geometric model and the used boundary conditions.

The film-substrate system was modeled using 3500 elements of the CAX4 type (four axisymmetric bilinear quadratic nodes). The area under the indenter is more refined. The obtained force-displacement curves of the three studied materials are shown in Fig. 5.

The film elastic-plastic properties (Ef, σyf, nf) were then identified using these indentation charge-displacement curves and the explicit equations in Jiang, Zhou and Huang model [13]. The charge-displacement curves were used to extract the maximum

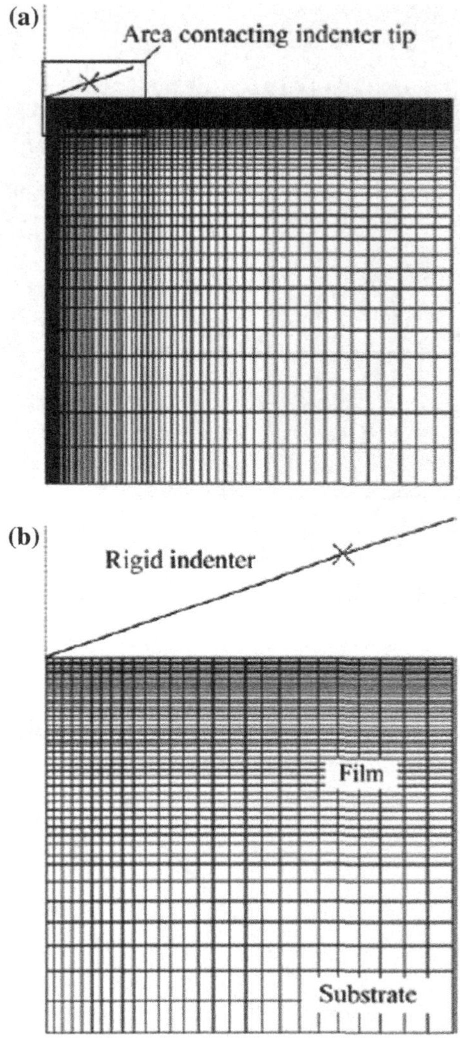

Fig. 3 Finite element mesh of the axisymmetric indentation: **a** Finite element mesh for the total model **b** Finite element mesh near the contact zone

load P_m, the maximum penetration depth h_m and the parameter x witch was defined by the function:

$$P = P_m \left(\frac{h}{h_m} \right)^x \tag{5}$$

This law describes the behavior of the loading curve. The determined values of the parameter x were shown in Table 1.

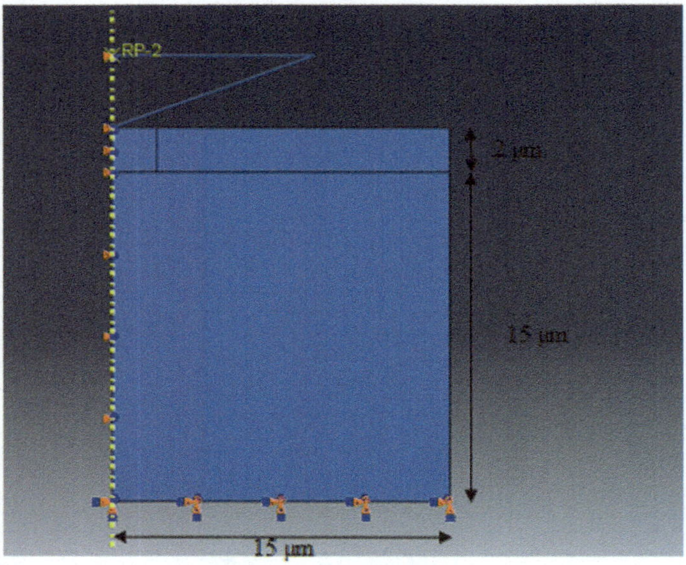

Fig. 4 Geometric model and boundary conditions

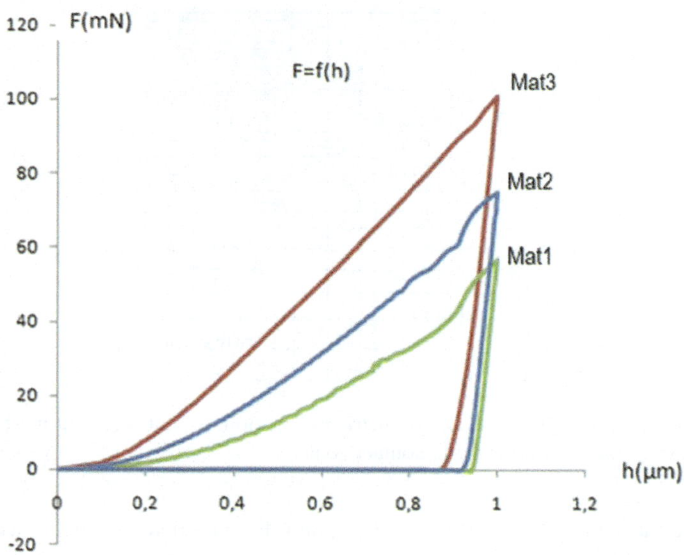

Fig. 5 Force-displacement curves of the studied film/substrate systems under three numerical indentation tests

Table 1 x values of the three studied films

Film material	x value
Material 1	2.0
Material 2	1.5
Material 3	3.0

Table 2 Comparison between the input film properties and the identified film properties

Film material	Input properties			Identified properties (our results)			Identified properties (Jiang et al. results)		
	Ef (GPa)	σyf (MPa)	nf	Ef (GPa)	σyf (MPa)	nf	Ef (GPa)	σyf (MPa)	nf
Material 1	80	500	0.1	79.5	503	0.24	79.5	500	0.1
Material 2	130	1000	0.2	129.1	1020	0.20	129.1	998.6	0.2
Material 3	210	1800	0.4	207.8	1809	0.20	207.8	1799.5	0.4

The explicit equations in Jiang, Zhou and Huang model [13] were then resolved using Matlab software to obtain σyf and nf properties. These equations form a non linear system. Its resolution is done using the "fsolve" function. The input film's properties and those identified are shown in Table 2. The identified film mechanical properties were found to be close to those used in numerical simulation.

3.2 Identification of the Elastic-Plastic Film Properties Using Liao, Zhou, Huang and Jiang Model

Numerical indentations tests were performed on four film- substrate systems (Material 0, Material 1, Material 2, Material 3) by introducing the mechanical properties of the substrate and the films [11]. The indentation geometric model and the used boundary conditions were shown in Fig. 6.

The film-substrate system is modeled using 8215 CAX4 elements (four axisymmetric bilinear quadratic nodes). The area under the indenter is more refined. Figure 7 shows the obtained load-displacement curves of the studied materials.

These load-displacement curves and the explicit relationships used in Liao, Zhou, Huang and Jiang model were used to determine the elastic-plastic properties of the films (Ef, σyf, nf).

The total work *Wt* (the area under the loading curve) and the residual work *Wu* (the area under the unloading curve) were determined using the loading and the unloading

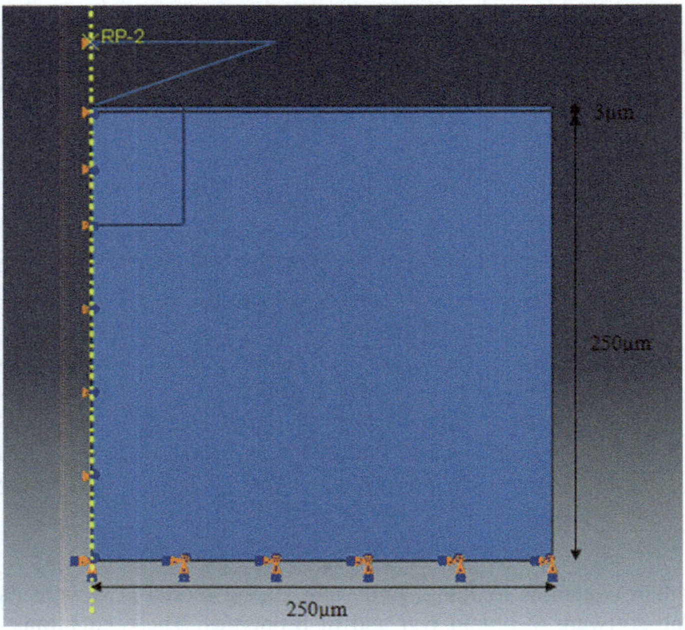

Fig. 6 The indentation geometric model and the used boundary conditions

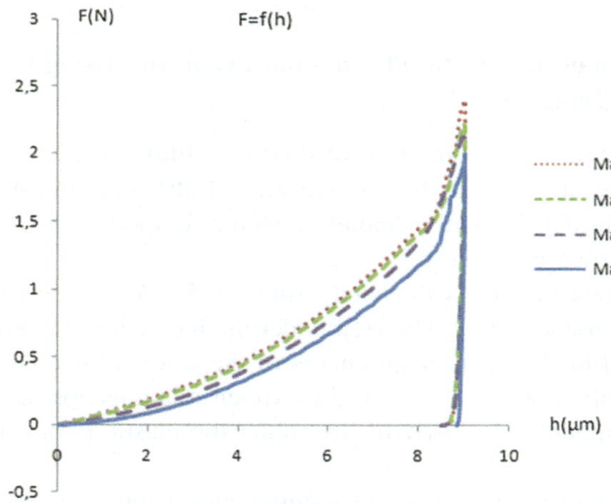

Fig. 7 Load-displacement curves of the studied materials

equations and the TI Interactive software. The computed total and residual work values were show in Table 3.

The Table 4 shows the input properties of the materials and those determined by resolving the explicit equations in Liao, Zhou, Huang and Jiang model using Matlab

Table 3 Wt and Wu values of the studied materials

Film material	Wt (N. μm)	Wu (N. μm)
Material 0	4.826	0.0940
Material 1	4.975	0.0942
Material 2	4.980	0.0960
Material 3	4.540	0.1000

Table 4 Comparison between the input film properties and the identified film properties

Film material	Input properties			Identified properties (our results)			Identified properties (Liao et al. results)		
	Ef (GPa)	σyf (MPa)	nf	Ef (GPa)	σyf (MPa)	nf	Ef (GPa)	σyf (MPa)	nf
Material 0	50	200	0	48	191.5	0.04	50	195.57	0
Material 1	50	1000	0.4	47	1100	0.25	50	976.54	0.4
Material 2	130	2000	0.2	105	2080	0.15	109.6	1999.4	0.2
Material 3	210	1800	0.4	205	1900	0.30	214.6	1799.5	0.4

software. It is seen from Table 4 that the identified film elastic-plastic properties were close to that of the input data.

3.3 Identification of the Film Properties Using Ma, Zhou, Long and Lu Model

To identify the film properties (Ef, σyf, nf), we solve the explicit two equations from Ma, Zhou, Long and Lu model using Matlab software. These equations form a non-linear system. Its resolution is done by the function "fsolve". The following Table 5 shows the input and the identified mechanical properties obtained for the four studied materials (Material 0, Material 1, Material 2 and Material 3) [14]. The identified properties were found to be close to the input values.

Table 5 Comparison between the input film properties and the identified film properties

Film material	Ef (GPa)	Input properties		Identified properties (our results)		Identified properties (Ma et al. results)	
		σyf (MPa)	nf	σyf (MPa)	nf	σyf (MPa)	nf
Material 0	210	500	0.1	441	0.14	514	0.10
Material 1	180	300	0.25	252	0.30	280	0.26
Material 2	70	500	0.12	560	0.16	489	0.13
Material 3	9	300	0.05	315	0.06	320	0.05

4 Conclusion

Numerical indentation tests were conducted on several film/substrate systems to determine the elastic-plastic properties of the films. Therefore, we have shown that several analytical methods could be applied in order to limit the substrate effect on determining film mechanical properties. These methods use two major steps: dimensional analyses to extract implicit relationships between the indentation parameters and the elastic-plastic properties of both film and substrate and the second step was finite element analyses to extract the explicit relationships used to determine the film elastic-plastic properties. In this study, the effectiveness of these analytical methods was verified by the good agreement between the input properties and the identified film elastic-plastic properties.

References

1. Schroeer A, Ensinger W, Wolf GK (1991) Comparison of the corrosion behavior and hardness of steel samples (100cr6) coated with titanium nitride and chromium nitride by different institutions using different deposition techniques. Mater Sci Eng A140:625–630
2. Brandl W, Gendig C (1996) Corrosion behavior of hybrid coatings. Thin Solid Films 290–291:343–347
3. Terrat JP, Gaucher A, Hadj rabah H, Fillit RY (1991) Structure and mechanical properties of reactively sputtered chromium nitrides. Surf Coat Technol 45:59–65
4. Summary E (2003) Workshop on silicon nanoelectronics and beyond. Hillsboro (OR): Intel's Jones Farm Conference Center
5. Kesler O (1998) Measurements of residual stress in plasma-sprayed metallic ceramics and composite coatings. Mater Sci Eng, A 257(2):215–224
6. Huang X, Pelegri AA (2006) Mechanical characterization of thin film materials with nanoindentation measurements and FEM analysis. J Comps Mater 40(15):1393–1406
7. Gamonpilas C, Busso EP (2004) On the effect of substrate properties on the indentation behavior of coated system. Mater Sc and Eng A 380(1–2):52–61
8. Oliver WC, Pharr GM (1992) An improved technique for determining hardness and elastic modulus using load and displacement sensing indentation experiments. J Mater Res 7 (6):1564–1583
9. Doerner MF, Nix WD (1986) A method for interpreting the data from depth-sensing indentation instruments. J Mater Res 1(4):601–609
10. Sneddon IN (1965) The relation between load and penetration in the axisymmetric Boussinesq problem for a punch of arbitrary profile. Int J Eng Sci 3:47–57
11. Liao Y, Zhou Y, Huang Y, Jiang L (2009) Measuring elastiv-plastic properties of thin films on elastic plastic substrates by sharp indentation. Mech of Mater 41:308–318
12. Buckingham E (1914) On physicallysimilar systems. Phys Rev 4:354–376
13. Jiang L, Zhou Y, Huang Y (2010) Elastic-plastic properties of thin films on elastic-plastic substrates characterized by nanoindentation test. Transact Non Ferr Met Soc China 20:2345–2349
14. Ma ZS, Zhou1 YC, Long SG, Lu CS (2012) An inverse approach for extracting elastic-plastic properties of thin films from small scale sharp indentation. J Mater Sci Technol 28 (7):626–635

Numerical Study of SPIF Process of Al–Cu Bimetal Sheet Using Finite Element Analysis: Influence of Process Parameters on the Mechanical and Geometrical Responses

Henia Arfa[1(✉)], Wifak Ben Abdelkader[1,2], and Riadh Bahloul[1]

[1] Laboratory of Mechanical Engineering (LGM), National Engineering School of Monastir (ENIM), University of Monastir, Avenue Ibn Eljazzar, 5019 Monastir, Tunisia
henia_arfa@yahoo.fr, bahloul_riadh@yahoo.fr, benabdelkaderwifak@gmail.com

[2] National Engineering School of Sousse (ENISo), University of Sousse, Sousse Erriadh, BP 264, 4023 Sousse, Tunisia

Abstract. Single point incremental forming (SPIF) which does not require any high capacity press machine nor a set of dies with specific shape, is an emerging sheet metal forming technology, capable of manufacturing complex parts at low cost for small to medium-batch production. This method is explained by the small plastic deformation caused by the force applied by a small punch on a sheet. The main reason to carry out such a process is to take the advantages of materials with different properties, such as high strength, low density, low price, and corrosion resistibility, at the same time and in a single component. The usage, of bimetal sheet is also justified because of the combination of the advantages offered by both materials of which it is composed and to a reduction of the disadvantages presented by each material if taken separately. The study presented in this research paper concerns a numerical investigation conducted on the incremental sheet metal forming process of Al and Cu bimetal composite. The finite element (FE) model validation was performed to compare the results obtained from the numerical simulations with experimental data available in the literature. The effects of process factors, which are considered as input parameters, such as: the tool diameter, wall inclination angle and layer arrangement, were investigated with finite element method (FEM) approaches on the forming time (the required time to forming the sheet), forming forces (the required forces to form the sheets), dimensional accuracy and maximum thickness variation (the maximum difference between initial thickness and formed sample thickness) of a truncated pyramid. The composite sheet behavior in a forming process differs from single-layer sheets and depends on the layers arrangement (contacted layer with the tool) and thickness. In this regard, the effect of layers arrangement on the forming behavior of (Al/Cu) bimetal sheet is of particular interest in the present study. Therefore, several tests were conducted to investigate the

N. Aifaoui et al. (Eds.): CMSM 2019, LNME, pp. 487–497, 2020.
https://doi.org/10.1007/978-3-030-27146-6_53

influences of some other variables, such as layers arrangement on the formability of the sheet material in a single point incremental forming process and the variations of force versus time diagram, defined as outputs.

Keywords: SPIF process · Al/Cu bimetallic composite material · Layer arrangement · Numerical simulation · Input parameters · Output responses

1 Introduction

Incremental forming process is one of the most promising new techniques in metal forming process. A hemispherical tool is pressed over the sheet metal to form the desired shape. The tool path is controlled by a CNC machine. It is based on the methodology of producing the designed shape by progressive movement of the hemispherical tool. Since localized deformation is developed during forming, more stretching occurs than in conventional forming. Due to slow forming process, this can be limited only to small batch production systems. This process is capable of forming sheet metal, which is used in automotive, biomedical and aerospace industries [1]. Figure 1 present the principle of single point incremental forming.

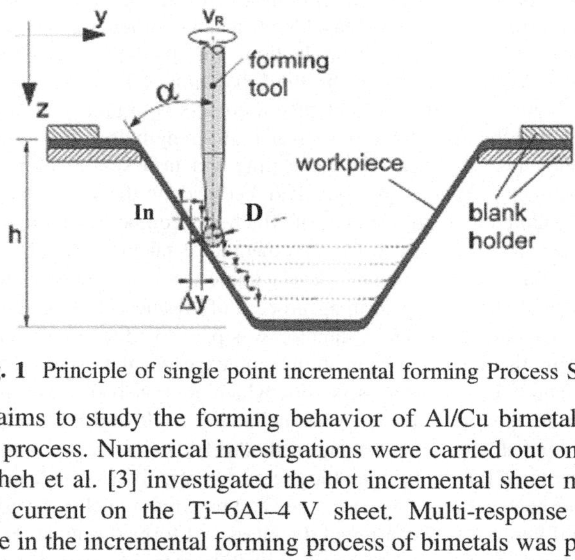

Fig. 1 Principle of single point incremental forming Process SPIF

This paper aims to study the forming behavior of Al/Cu bimetals through incremental forming process. Numerical investigations were carried out on SPIF of bilayer sheet. Honarpisheh et al. [3] investigated the hot incremental sheet metal forming by using electrical current on the Ti–6Al–4 V sheet. Multi-response optimization of hyperbolic shape in the incremental forming process of bimetals was performed by [4]. Honarpisheh et al. [5] investigated dimensional accuracy in the incremental forming process of (Al/Copp) bimetals. Sakhtemanian et al. [6] studied the effect of layers' arrangement of low-carbon steel/commercially pure titanium bimetal sheet in the single point incremental forming process, experimentally and numerically. In this work, truncated square pyramid is formed to study the formability of (Al/Copp) bimetal sheet. The effect of layers arrangement has been investigated in the single point incremental forming process (SPIF) numerically. The main reason to carry out such a process is to take the advantages of materials with different properties, such as high strength, low density and resistibility, at the same time and in a single component. The composite sheet behavior in a forming process differs from single-layer sheets and depends on the layers arrangement and thickness. In this regard, a useful control of the process by determining and monitoring the forces between the punch and the sheet is aimed.

2 Numerical Modeling of SPIF

The Single Point Incremental Sheet Forming (SPIF) process allows the production of the complex three-dimensional shapes from Computer Aided Design (CAD) models without specially designed forming tools. Figure 2a, b illustrated the geometric representation of a square pyramid and the tool path describing its geometry, respectively. The last one was generated by the punch during the forming process of truncated pyramid by adopting a discontinuous trajectory strategy.

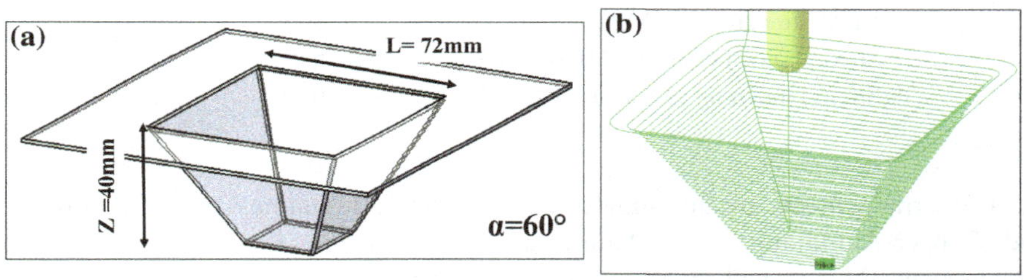

Fig. 2 **a** Geometries and dimensions carried out **b** Discontinuous trajectory for pyramidal model

Representative frustums square pyramids with dimensions of 72 mm × 72 mm and a 60° wall angle were formed. The maximum drawing depth is equal to z = 40 mm. According to the designed equipment, square sheets with a side of 135 mm were utilized during numerical simulations. The initial thickness of the sheet metal before its working is fixed at a value equal to 1.2 mm. The thicknesses division of the couple Al/Copp bimetals in the Al and Copp sheets was 0.8 mm and 0.4 mm, respectively. The Finite Element (FE) study focuses on the SPIF process based on a unidirectional tool path strategy of contours and step-downs to obtain truncated square pyramid.

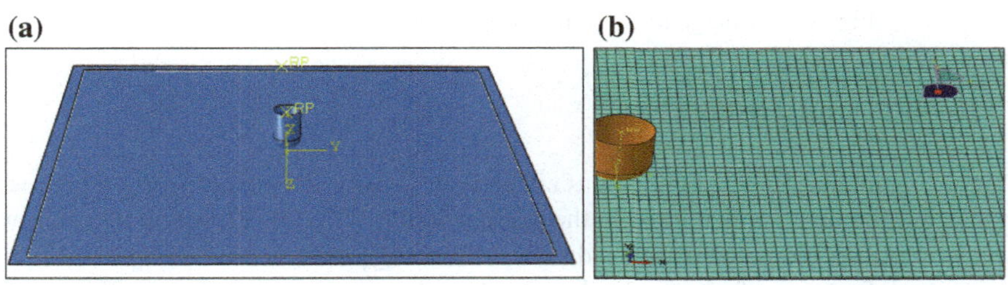

Fig. 3 **a** Three-dimensional numerical simulation of single point incremental forming of sheet metal and **b** The finite element meshing configuration of the initial blank

Different process parameters are considered. The numerical simulation of SPIF has been done with the numerical code Abaqus\Explicit. The tool is modeled as a rigid surface. Due to the little thickness of the bimetal sheet compared to its length and width, quadrilateral shell elements with 4 nodes and 6 degrees of freedom per node (S4R) and five Gaussian reduced integration points through the thickness direction were used. Figure 3 shows the punch tool assembly on the bilayer sheet for the simulation.

Table 1 summarizes the mechanical properties of both materials (copper-C10100 and aluminum-1050) sheets obtained thanks to uniaxial tensile testing technique.

Table 1 Isotropic hardening properties of Aluminum and Copper alloy sheets from tensile test [5]

	Aluminum-1050	Copper-C10100
Density (kg/m^3)	2700	8940
Young's modulus, E (GPa)	70	115
Yield strength, YS (MPa)	122	306
Strength coefficient, K (MPa)	232	540
Strain hardening exponent, n	0.12	0.11

Two arrangements are investigated in this study: Aluminum-Copper (Alum/Copp) and Copper-Aluminum (Copp/Alum), as shown in Fig. 4. In the case of such arrangement, the tool will be in contact with the upper layer.

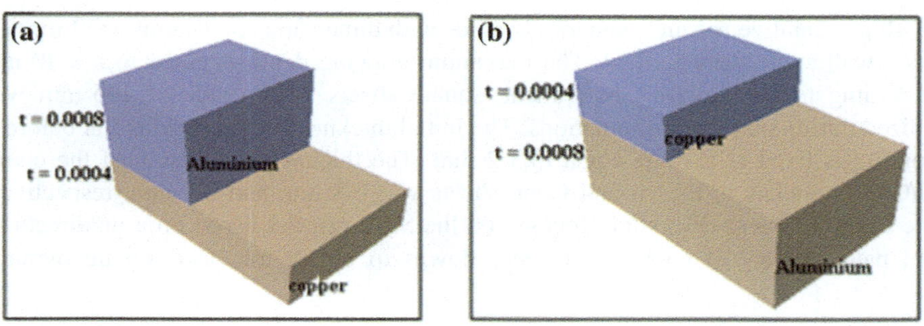

Fig. 4 **a** Arrangement (Alum/Copp), **b** Arrangement (Copp/Alum)

3 Results and Discussion

Finite element analysis is employed as a technique to simulate manufacturing processes or other mechanical applications. In the present study, ABAQUS Explicit software was used to simulate the deformation induced in forming processes including single point incremental forming of sheet metal and to predict the forming force. This section provides information about the results obtained in the frame of the present work, with regards to the influence of different process parameters on the characteristics of the parts produced by incremental sheet forming and the comparison between the results predicted by the numerical model and the ones obtained experimentally [5]. The objectives of these studies are to identify and analyze the effects of the principal geometrical parameters related to the wall angle and the tool diameter on the characterization of the process.

3.1 Forming Force: Comparison Between Experimental and Numerical Results

In order to effectively determine the impact of variables in SPIF, numerical simulations using finite element analysis were carried out. The three factors varied are wall inclination angle, tool size, and layers arrangement. The forming factors are listed in Table 2.

Table 2 The levels of forming factors for numerical experiments

Wall angle	55°, 60°, 65°
Tool diameter	10 mm, 16 mm
Layers arrangement	Aluminium/Copper, Copper/Aluminium

The graphs of Fig. 4 summarize the time plot of punch forces attained during the single point incremental forming process of the layer's bimetal (Copper/Aluminum). The evaluations of vertical forces (Fz) provided by the punch in incremental CNC sheet metal forming process were investigated by applying two approaches: experimental analysis [5] and numerical modelling on forces determination for improving knowledge of a single point incremental forming. Both Fig. 5a_1–a_3 represent the evolution of the Fz component measured and predicted respectively by experimental and numerical approaches throughout the incremental forming process. The geometry chosen as a benchmark represents a truncated pyramid shape with dimensions of 72 mm × 72 mm, a wall inclination angle $\alpha = 60°$ and an inner height of 40 mm. Square sheets (135 mm × 135 mm × 1.2 mm), were employed during numerical tests. The thicknesses of Al/Cu bimetals in the Aluminum and Copper layers were 0.8 mm and 0.4 mm, respectively.

By comparing the force-time curves for both experimental and numerical solution methods, it can be inferred that first a good agreement is obtained between the two approaches. Secondly, these results show that the maximum average force occurs in the case where the tool diameter was d = 16 mm and the layers' arrangement corresponding to the Aluminum when it was used as the top layer and was in contact with the forming tool. Secondly, these results show that the maximum average force occurs in the case where the tool diameter was 16; Aluminum was used as the top layer. The effect of arrangement layers on the forming forces was examined. The results indicate that the average force decreases when the copper is used as the top layer. By tracking this process, yielding starts from outer layer to inner one. The aluminum naturally has lower yield stress than copper. Therefore, as the Aluminum-1050 sheets is the outer

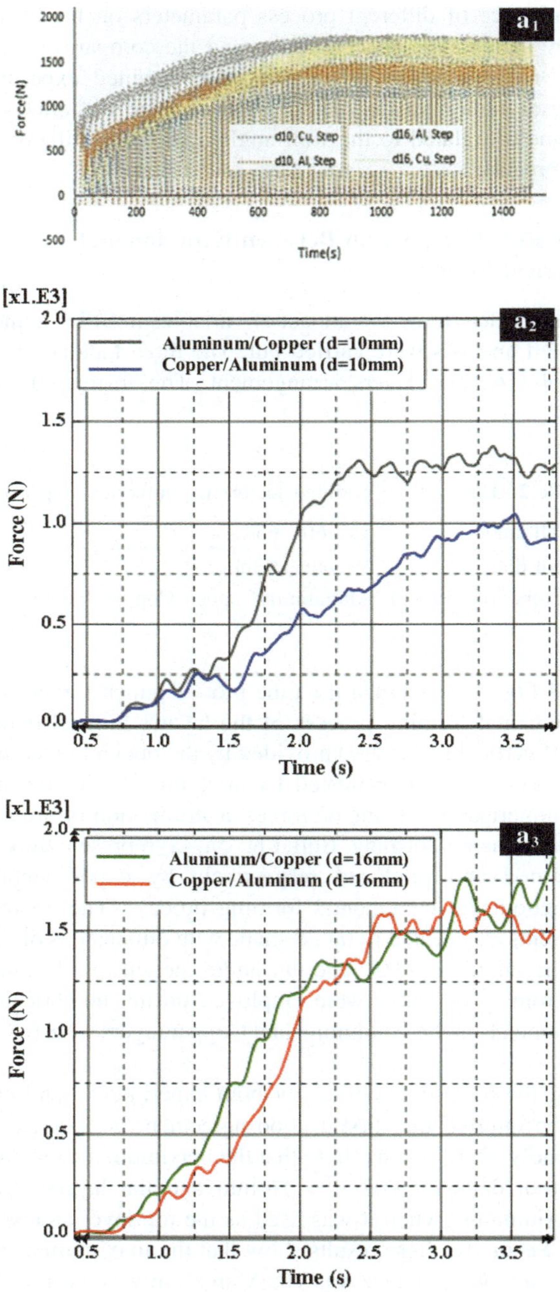

Fig. 5 Variation of forming force for different sheet arrangements $\mathbf{a}_1$ Experimental results $\mathbf{a}_2$ Numerical prediction of the vertical force curves (Fz) for d = 10 mm and $\mathbf{a}_3$ Numerically simulated force curves (Fz) corresponding to tool diameter d = 16 mm

layer, plastic deformation or in other words moving of dislocations starts sooner and as a result, the lower force is needed in order to start the deformation.

3.2 Effect of the Wall Angle on the Distribution of Final Sheet Thickness and Equivalent Plastic Strain

After validating the finite element model and in order to investigate the effects of process factors, the numerical results of the distribution of sheet thickness and equivalent plastic strain are presented in this section. The formed geometry chosen was a pyramid with dimensions of 72 mm × 72 mm with a depth equal to 40 mm. The thicknesses in Aluminum and Copper layers were 0.8 mm and 0.4 mm, respectively. The layers arrangement considered was Al/Copp, it means that the Aluminum sheet was in the upper layer (where the forming tool was in contact with the surface of Aluminum alloy sheet). The tool diameter and the vertical pitch adopted in this series of simulations are 10 mm and 1 mm respectively.

The process analysis based on finite element modelling is required in order to study and analyze the effect of process factors on the characteristics of incrementally shaped parts. This will allow to deepen the understanding of the process itself. In the present section, some numerical simulations were carried out in order to determine the effects of ISF process parameters, such as the wall angle, and other one not presented in this paper, like (initial sheet thickness, forming tool diameter, vertical step size, nature of tool path contours and arrangement mode of bimetal layers) on the final sheet thickness and equivalent plastic strain distribution and on the other geometrical and mechanical responses not considered here. Thus, to see the influence of this geometrical parameter, three truncated pyramid were formed, each having varying the wall angle, corresponding respectively to the wall inclination angles α of 50°, 60° and 70°.

The 3D iso-values shown in Fig. $6a_1$ describe the distribution of final sheet thickness. Figure $6a_1$–a_3 report the simulated incremental forming sequences and highlight the calculated thickness histories of the truncated pyramids with wall inclination angles α of 55°, 60° and 65° respectively. Firstly, it can be seen that the thickness in the central region of the bottom of the pyramid remains unchanged, whereas the thickness in the wall region is visibly reduced. In fact, during the process the sheet is severely stretched along the radial direction. Secondly, the field distribution of thickness changes when the wall angle is different. The deformed slope angle α could be considered as an important factor that influenced the final results provided by SPIF process: with the increase of the wall angle α, the thickness reduction reaches a threshold value about 0.4407 mm from an initial thickness of 1.2 mm, and as a consequence, a broken product is manufactured. At the end of the process simulations, the calculated final thicknesses in the wall region are severely reduced from 1.2 mm to 0.649 mm, 0.524 mm, and 0.440 mm, respectively for the 55°, 60° and 65° wall angle. Therefore, it results in nonuniform localized thinning in the wall region of the formed pyramidal parts.

Deformation behavior of sheet metal in single point incremental forming process is numerically investigated using an explicit finite element code ABAQUS/Explicit. It needed developing a simple finite element model which is capable to give an accurate prediction of some characteristics of the formed parts.

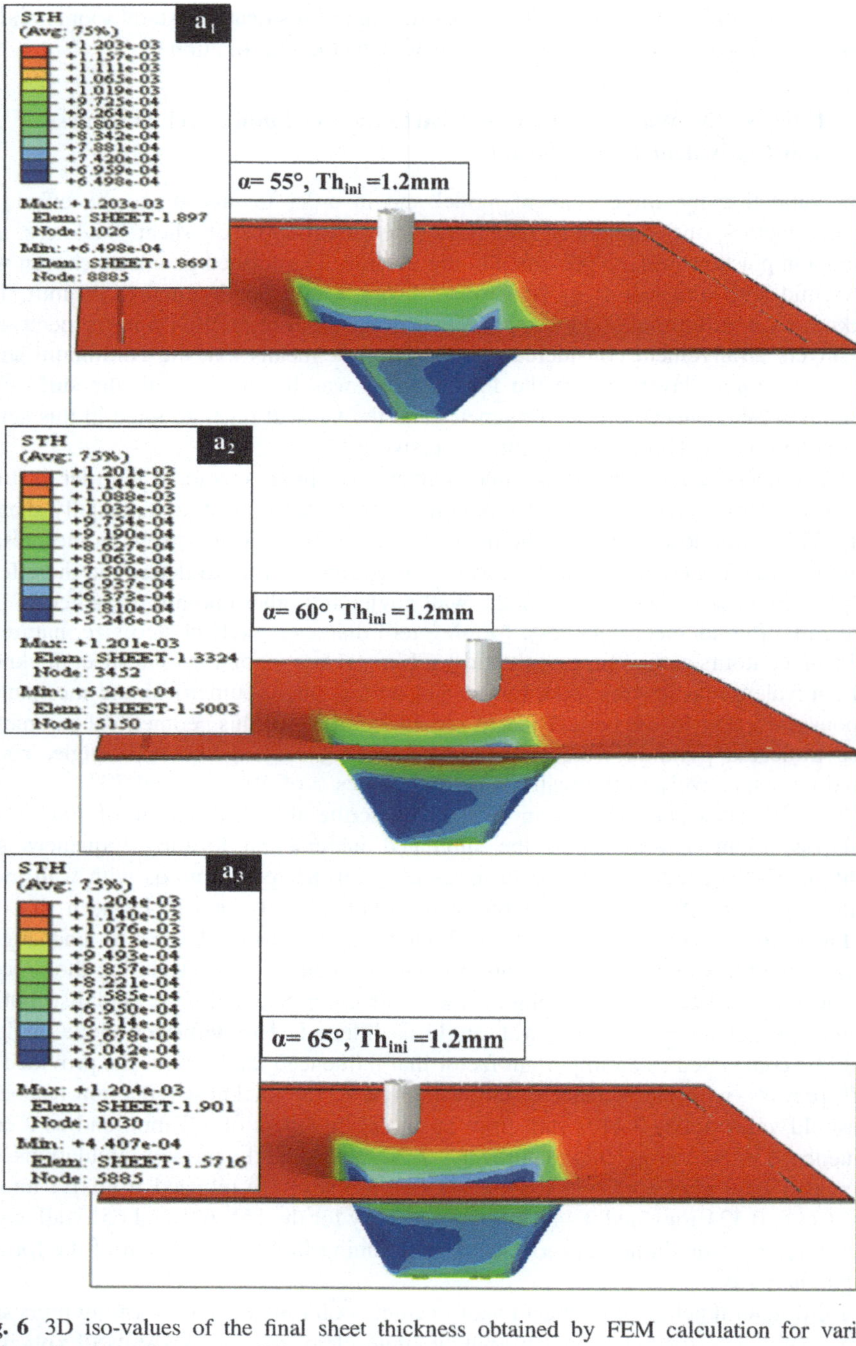

Fig. 6 3D iso-values of the final sheet thickness obtained by FEM calculation for various deformed wall angles **a**$_1$ $\alpha = 55°$ **a**$_2$ $\alpha = 60°$ **a**$_3$ $\alpha = 65°$

During the SPIF process, the deformation takes place locally as the tool moves contour by contour. Throughout one contour, only the material close to the contact area with the tool undergoes limited deformation due to the small vertical step size. This small deformation accumulates over a series of succeeding contours. These observations are summarized in Fig. 7a$_1$–a$_3$ which show the global evolution of the accumulated equivalent plastic strain of the mentioned range of pyramidal shapes having varied slope angles. As mentioned by [2], the strain paths are characterized by a typical step-trend: each strain increment is the direct consequence of the action of the tool as it

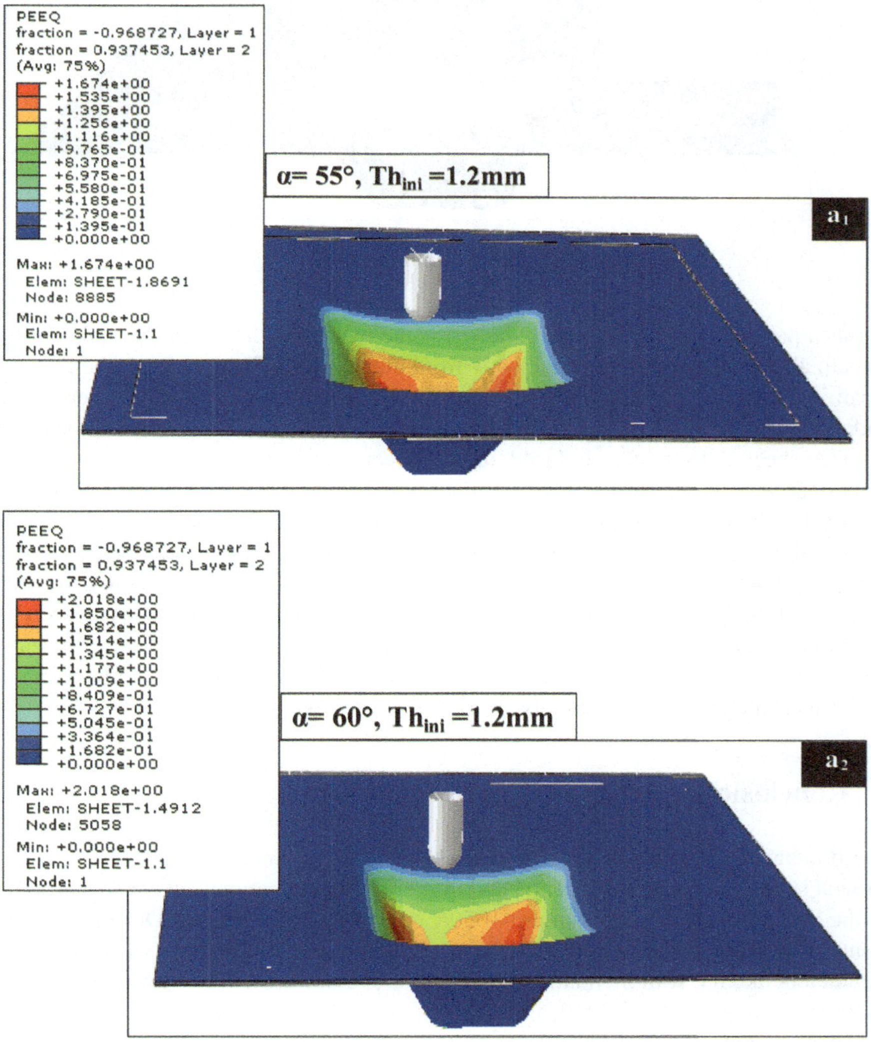

Fig. 7 3D iso-values of equivalent plastic strain distributions obtained by FEM calculation for various deformed wall angles

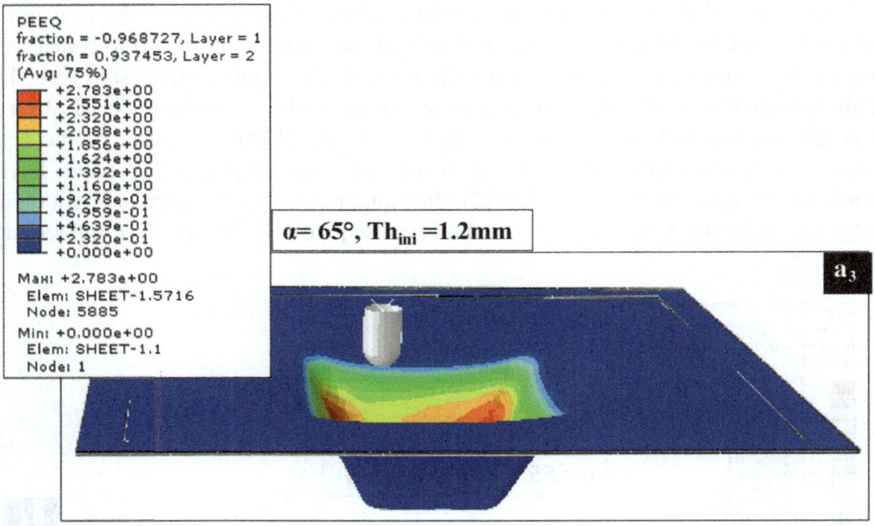

Fig. 7 (*continued*)

passes a particular element. In turn, no strain increment occurs when the tool continues its path along the same contour far away from that element. This confirms the feature of localized deformation that characterizes the SPIF process: a given material point undergoes its strain through progressive, small increments each time the tool passes by.

The sheet deformation induced in the sheet metal which is predictable by the numerical simulations shows that the maximum value of the equivalent plastic strain intended at the end of the process and after the removal of tool, evolve in a proportional way to the increase of the wall inclination angle α. Figure 6a_1–a_3 displays the 3D contour plots of the equivalent plastic strain corresponding to the created truncated pyramid in the incremental forming process and simulated by FEM. Consequently, the maximum values of this response were identified. They are very significant as they can achieve 1.674, 2.018, and 2.783 corresponding to the formed pyramids with an increase of the wall inclination respectively from 55° to 65 ° with an increment of 5°.

4 Conclusions

The bimetal can play an important role in the industry. Since there are few studies on bimetal sheet forming, experimental and numerical investigations have been carried out in the present study on Aluminum-1050/Copper-C10100 bimetal. During the incremental sheet metal forming of Al/Cu multilayer sheet, the influence of some process parameters like the wall inclination angle, the tool diameter and the layers arrangement on the mechanical and geometrical responses, such as the forming forces, the maximum thickness variations and the equivalent plastic strain distributions, have been analyzed. Comparison between numerical and experimental results indicates, firstly the reliability of the proposed model provided and that the predicted values of vertical force in z direction are in good agreement with the experimental one, thus confirming the potentiality of the FEM tool. Secondly, it was observed from this bimetal forming technique that the forming force increases with the increasing of the tool diameter and when the Aluminum material will be used as a top layer (in contact with tool). Also, the equivalent plastic strain distributions and the maximum thickness variations predicted in the used geometry increase with increasing of the wall angle. There is a strong dependence of the deformed sheet thickness on the draw angle α, which can lead to inhomogeneous thickness distributions in the final part.

Following from the above results in SPIF, the process heaviness and the equivalent plastic strains on the sheet surface increase with the increasing of the wall inclination angle, determining larger thinning.

References

1. Ambrogio G, De Napoli L, Filice L, Gagliardi F, Muzzupappa M (2005) Application of incremental forming process for high customized medical product manufacturing. J Mater Process Technol 162:156–162. https://doi.org/10.1016/j.jmatprotec.2005.02.148
2. Ambrogio G, Filice L, Micari F (2006) A force measuring based strategy for failure prevention in incremental forming. J Mater Process Technol 177(1–3):413–416. https://doi.org/10.1016/j.jmatprotec.2006.04.076
3. Honarpisheh M, Niksokhan J, Nazari F (2016) Investigation of the effects of cold rolling on the mechanical properties of explosively-welded Al/St/Al multilayer sheet. Metal Resear Technol 113(1):105. https://doi.org/10.1051/metal/2015049
4. Honarpisheh M, Keimasi M, Alinaghian I (2018) Numerical and experimental study on incremental forming process of Al/Cu bimetals: influence of process parameters on the forming force, dimensional accuracy and thickness variations. J Mech Mater Struct 13(1):35–51. https://doi.org/10.2140/jomms.2018.13.35
5. Honarpisheh M, Mohammadi Jobedar M, Alinaghian I (2018) Multiresponse optimization on single-point incremental forming of hyperbolic shape Al-1050/Cu bimetal using response surface methodology. Int J Adv Manuf Technol 96:3069–3080. https://doi.org/10.1007/s00170-018-1812-5
6. Sakhtemanian MR, Honarpisheh M, Amini S (2018) Numerical and experimental study on the layer arrangement in the incremental forming process of explosive-welded low-carbon steel/CP-titanium bimetal sheet. Int J Adv Manuf Technol 95:3781–3796. https://doi.org/10.1007/s00170-017-1462-z

Effect of Multiple Impacts on Thin Leading Edges of Turbine Blade Treated by Laser Shock Peening Process

Manel Ayeb[1,3](✉), Mounir Frija[1,2], and Raouf Fathallah[1,3]

[1] Unit of Mechanical and Materials Production Engineering (UGPMM/UR17ES43), National Engineering School of Sfax (ENIS), University of Sfax, Sfax, Tunisia
Ayeb.manel2705@gmail.com, Mounir.frija@gmail.com, Raouf.fathallah@gmail.com

[2] Institute of Applied Sciences and Technology of Sousse (ISSATSo), University of Sousse, Sousse, Tunisia

[3] National Engineering School of Sousse (ENISo), University of Sousse, Sousse, Tunisia

Abstract. In this paper, we will study the influence of multiple laser impacts on thin leading edges of a turbine blade. A numerical analysis based on a 3D finite element method of thin leading edge specimens of a turbine blade made of titanium super-alloy (Ti–6Al–4V) is performed using the commercial software ABAQUS. A repetitive time Gaussian increment pressure that is uniformly applied at a square affected region is used to characterize the LSP loading. We apply the visco-elastic-plastic of the Johnson-Cook law coupled with damage in order to develop the treated material behavior law. The objective of this simulation is to predict the mechanical surface modifications generated by the laser shock processing: (i) the residual stresses, (ii) the plastic strains and (iii) the Johnson-Cook superficial damage. These modifications are well analyzed for thin leading edges of a turbine blade treated by a square laser spot that can effectively treat a considerable part with a coverage rate that is below 5% comparing with a circular laser spot.

Keywords: Laser shock peening · Square laser spot · Finite element method Multiple impacts · Thin leading edges

1 Introduction

Laser shock processing or peening (LSP) is an innovative surface treatment that exploits laser energy in order to generate plasticized shock waves in the treated target. This process guarantees an extremely thin process (Fig. 1). The mechanical result of the use of a pulsed laser radiation with a few nanoseconds time period and around a decade of Gigawatts per square centimeter power density engenders a hardening production as a consequence of a shock wave propagation and an in-depth plastic deformations gradient in the superficial layer of a target. Therefore, these plastic strains

N. Aifaoui et al. (Eds.): CMSM 2019, LNME, pp. 498–506, 2020.
https://doi.org/10.1007/978-3-030-27146-6_54

produce residual stresses that are very favorable with the purpose of improving the surface properties of the treated workpiece.

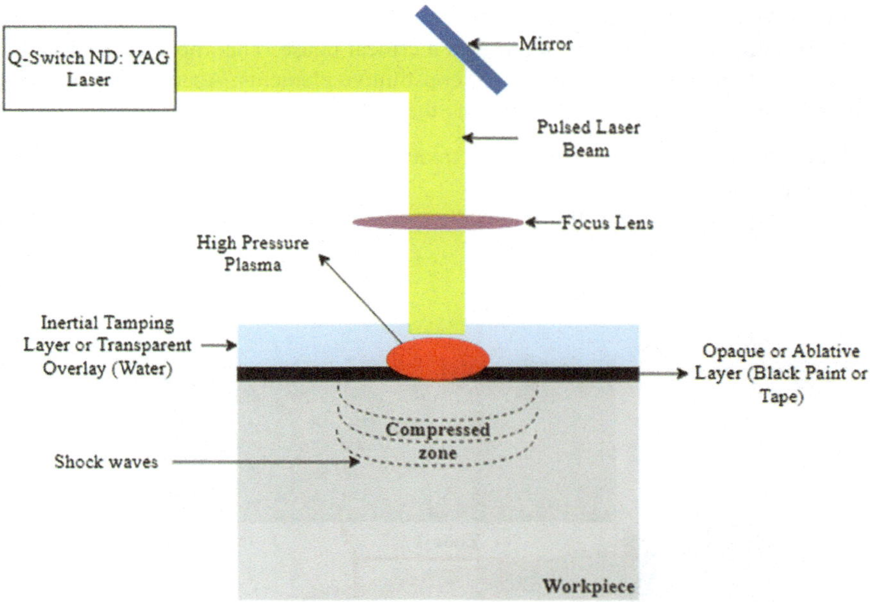

Fig. 1 The experimental mechanism of the LSP treatment

Consequently, the study of the in-depth residual stress, the induced plastic strain and the Johnson-Cook superficial damage fields on the target treated by the LSP treatment is too substantial. Previously, these effects have been already examined with circle laser spot. Besides, some researchers studied the in-depth residual stress field generated by the square laser spot that it is likely to engender the stress hole. As well, the surface morphology made by square laser impacts is slicker than that generated by circular impacts [1].

In this work, we will numerically simulate a square laser spot based on the finite element method (FEM) with the aim of analyzing the effects of multiple impacts on the Ti–6Al–4V thin leading edge surface of a turbine blade. Moreover, the objective of this study is to examine these effects on: the in-depth residual stresses, the induced plastic strains and the Johnson-Cook superficial damage.

2 Finite Element Model Investigation of the LSP Treatment

2.1 Geometric Model

Through our study, we will adopt a square laser spot which is increasingly used in the LSP process to satisfy certain requirements of industrial applications. As well, to model this square laser spot that is applied to the upper surface of the material, we proceed to a

3D model of a representative specimen of a Ti–6Al–4V turbine blade exploiting the finite element code ABAQUS/EXPLICIT (Fig. 2) to examine the multiple impacts effect on the in-depth residual stresses, the equivalent plastic strain and the damage condition generated by the square laser spot.

The optimal mesh selection represents a critical phase. The impacted zone must be constituted of an appropriate number of constitutive elements. Accordingly, the model contains two areas that for each zone has the following mesh (Fig. 2):

- Zone 1 that is subjected to the LSP treatment impacts: 66000 Hex elements with C3D8R type.
- Zone 2 which constitutes the rest of the specimen: 141418 Tet elements with C3D4 type (A 4-node linear tetrahedron).

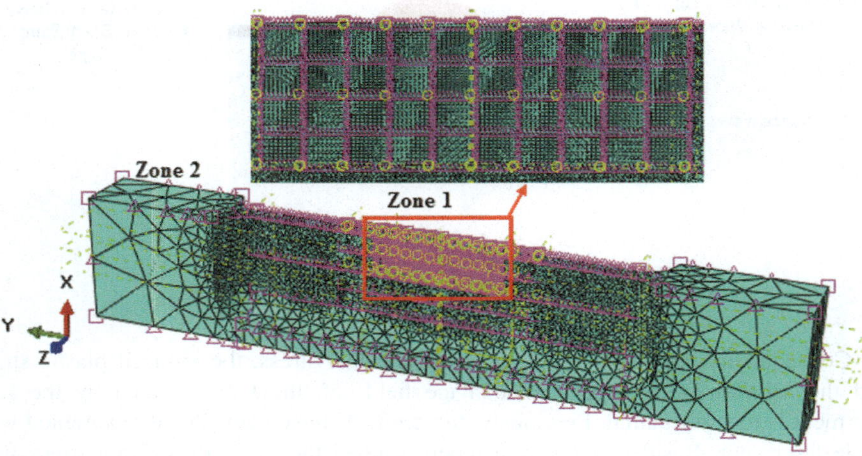

Fig. 2 Mesh details of the 3D model of a representative specimen of Ti–6Al–4V titanium superalloy turbine blade

2.2 General Hypotheses of the Study

So as to realize the LSP FEM implementation, we contemplate the following assumptions:

(i) We neglect the edge effects.
(ii) The pressure is spatially uniform over the part that will be treated.
(iii) The material arrangements are perpendicular to the surface of the massif.
(iv) The heating influence because of the application of an ablative thermo-protective layer.

2.3 Material Representative Properties

We use the Johnson-Cook empirical behavior law [2] to develop the thermo-visco-plastic behavior of the Ti–6Al–4V material. The Johnson-Cook law is frequently employed in extremely dynamic process simulation and used to denote the following Von Mises flow:

$$\sigma_y(\bar{\varepsilon}^p) = [A + B(\bar{\varepsilon}^p)^n].\left[1 + C\ln\left(\frac{\dot{\bar{\varepsilon}}^p}{\dot{\varepsilon}_0}\right)\right]\left[1 - \left(\frac{T - T_0}{T_f - T_0}\right)^m\right] \tag{1}$$

In Eq. (1), the instantaneous strain rate that is described in the second term is between $1\,\mathrm{s}^{-1}$ and $10^6\,\mathrm{s}^{-1}$ (The strain rates created by LSP can exceed $10^6\,\mathrm{s}^{-1}$ inside the specimen [3]).

In our paper, we neglect the heating effect considering that the temperature doesn't increase though the propagation of the shock wave. Consequently, the Johnson-Cook law turns as following:

$$\sigma_y(\bar{\varepsilon}^p) = [A + B(\bar{\varepsilon}^p)^n].\left[1 + C\ln\left(\frac{\dot{\bar{\varepsilon}}^p}{\dot{\varepsilon}_0}\right)\right] \tag{2}$$

The five material constants (Table 1) of the Johnson-Cook law are A, B, C, m and n and characterize the Johnson-Cook parameters.

Table 1 Johnson-Cook plasticity parameters [5]

A [MPa]	B [MPa]	C	m	n	T_0 [K]	T_f [K]
1098	1092	0.014	1.1	0.93	298	1878

Indeed, the damage will be developed by the Johnson-Cook law [3] and given by the following equations:

$$\bar{\varepsilon}^p_{rupture} = \left[d_1 + d_2\exp\left(d_3\frac{\sigma_H}{\sigma_{eq}}\right)\right]\left[1 + d_4\ln\left(\frac{\dot{\bar{\varepsilon}}}{\dot{\varepsilon}_0}\right)\right]\left[1 + d_5\frac{T - T_0}{T_f - T_0}\right] \tag{3}$$

$$D = \sum\frac{\Delta\bar{\varepsilon}^p}{\bar{\varepsilon}_f} = \sum\frac{\Delta\varepsilon^p_{eq}}{\bar{\varepsilon}_f} \tag{4}$$

The material constants d_1, d_2, d_3, d_4 and d_5 are represented in Table 2.

Table 2 Johnson-Cook damage parameters [5]

d_1	d_2	d_3	d_4	d_5
−0.09	0.27	0.48	0.014	3.87

In Eq. (3), the equivalent plastic strain to fracture (rupture) $\bar{\varepsilon}^p_{rupture}$ is a function of the hydrostatic stress σ_H, the Von Mises equivalent stress σ_{eq} and the reference strain $\dot{\varepsilon}_0$ that is equal to $1\ \mathrm{s}^{-1}$.

In Eq. (4), the accumulated damage D is a function of the incremental plastic strain $\Delta\varepsilon^p_{eq}$ and the equivalent plastic strain to fracture (rupture) $\bar{\varepsilon}_f$. The stresses will be set to zero and the crack will appear, when D reaches 1.

2.4 Boundary Conditions and Loading Induced by LSP Treatment

Figure 3 shows the boundary conditions and the loading generated by the LSP.

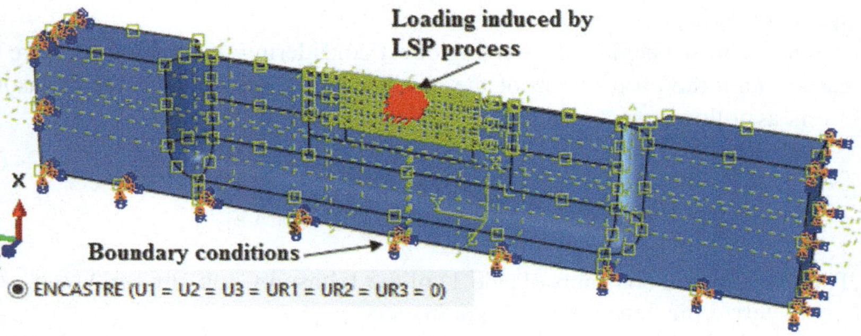

Fig. 3 Loading and boundary conditions generated by the LSP process

The specific boundary conditions that are taken into account in our paper, are as follows:

- The base and the two extreme sides of the workpiece are blocked ($U_r = U_\theta = U_z = UR_r = UR_\theta = UR_z$).
- The vertical central line is defined equivalent to a symmetry line.
- The boundary field temperature is equal to 298 K.

In our work, we will treat the dynamic response of the part subjected to a square laser spot that it consists to apply a temporal law of a spatially constant pressure on any part of the workpiece surface.

According to the Fabbro model [4], the peak pressures expressed in gigapascal (GPa) are estimated for different power densities while the wavelength and the pulse width are respectively equal to 1064 nm and 20 ns and adopted by the following equation:

$$P_{\max} = 0.01\sqrt{\frac{\alpha}{2\alpha+3}}\sqrt{Z}\sqrt{AI_0} \tag{5}$$

In Eq. (5), the peak pressure $P_{\max}$ is a function of the efficiency coefficient α, the equivalent impedance Z that is expressed in gram per square centimeter per second (g/cm^2/s), the absorbing factor A and the laser power density I_0 that is expressed in gigawatt per square centimeter (GW/cm^2).

In this paper, the full width at half maximum (FWHM), the maximum peak pressure (P_{max}) and the dimension of the square laser spot are respectively equal to 20 ns, 5 GPa and 3 × 3 mm^2. Used for a FWHM equal to 20 ns, the normalized applied peak pressure profile is illustrated in Fig. 4.

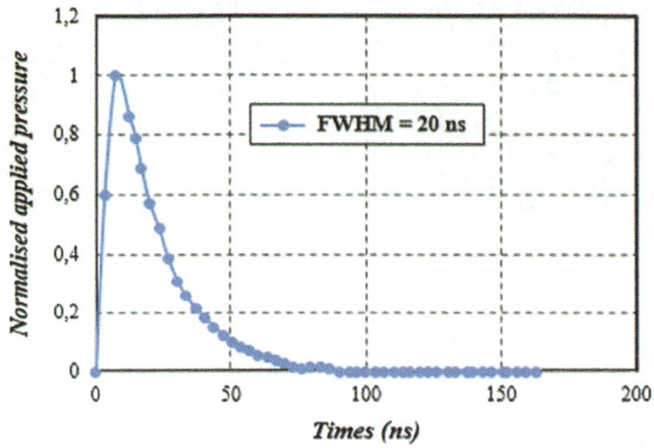

Fig. 4 The normalized applied pressure along time for FWHM = 20 ns

2.5 Principal Calculation Steps

The calculation steps [6–8] that are taken into account in our work will be partitioned into two stages that are:

- The 'CHARG' loading step is applied for a period time of 1000 ns.
- The 'RELAX' relaxation step lasts 0.5 ms.

3 Results and Discussions

The finite element code ABAQUS/EXPLICIT is used to accomplish the analysis, in our work. In fact, the explicit procedure ensures a several robustness. As well, it has an exceptional resolution of rapid impacts. That's why, we use it to simulate our model with the intention of studying the effect of multiple impacts on the in-depth residual stresses, the generated equivalent plastic strain and the Johnson-Cook damage state induced by the square laser spot. The results are represented at depth that is created at 1.5 mm from the upper surface of the treated workpiece (Fig. 5).

We note that the compressive residual stress increases in ultimate value while the superposed impacts number increases. Also, the plastic depth of the compressive stress increases while the number of superposed impacts increases (Fig. 6). Also, we can notice that the residual compressive stresses are very near to the surface treated by LSP, that represents an interest for inhibiting and delaying the cracks propagation.

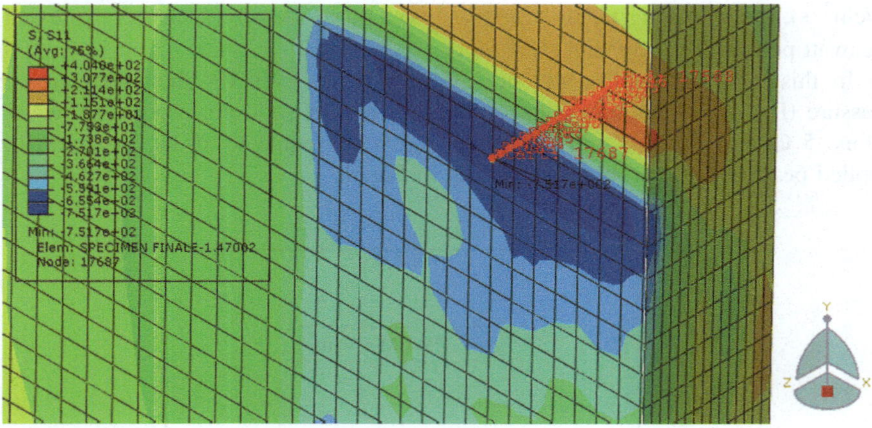

Fig. 5 Path used to plot the results exposed at depth

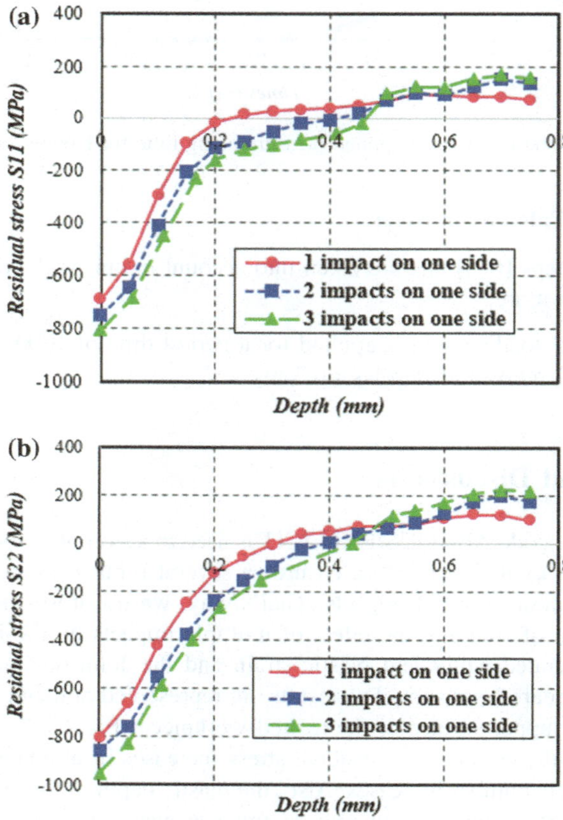

Fig. 6 Influence of the number of superposed impacts on the residual stress **a** (S11) and **b** (S22) at depth in the treated specimen

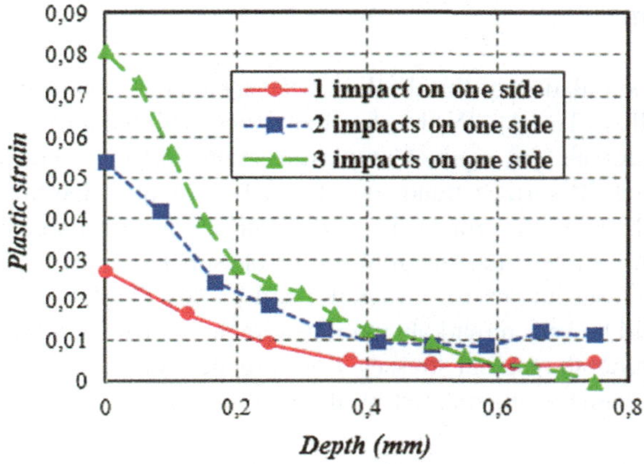

Fig. 7 Effect of the superposed impacts number on the plastic strain at depth in the treated specimen

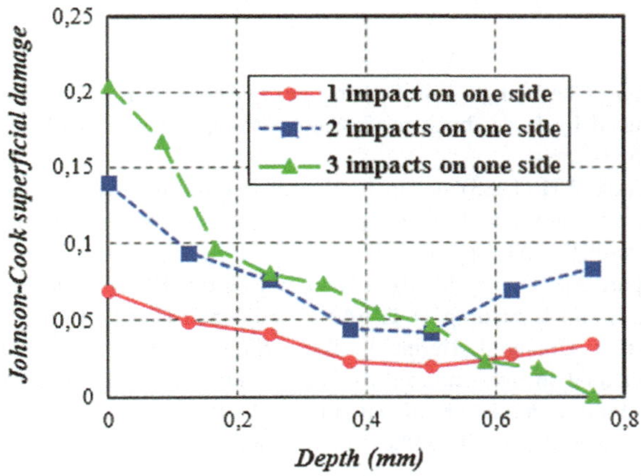

Fig. 8 Effect of the superposed impacts number on the Johnson-Cook superficial damage at depth in the treated specimen

Hence, the equivalent plastic strain is immediately caused by the residual stresses. Thus, it increases at depth by increasing the number of superposed impacts (Fig. 7).

Besides, the Johnson-Cook superficial damage (Fig. 8) increases at depth by increasing the number of superposed impacts. The damage is usually fewer than 1, for a single laser impact.

4 Conclusion

A finite element simulation of thin leading edge specimens of a turbine blade made of titanium super-alloy (Ti–6Al–4V) has been analyzed with the aim of predicting the in-depth residual stresses, the equivalent plastic strain and the Johnson-Cook damage generated by the LSP surface treatment. The effect of the superposed laser impacts number on the Ti–6Al–4V thin leading edge surface are investigated. The obtained results have demonstrated that the Johnson-Cook superficial damage doesn't exceed 70% for a single laser impact for a titanium super alloy (Ti–6Al–4V).

This paper will be very remarkable to illustrate the influence of the superposed laser impacts on the in-depth residual stresses, the equivalent plastic strain and the Johnson-Cook superficial damage engendered by the LSP treatment.

Acknowledgements. This work is carried out thanks to the support and funding allocated to the Unit of Mechanical and Materials Production Engineering (UGPMM/UR17ES43) by the Tunisian Ministry of Higher Education and Scientific Research.

References

1. Wen Cao Z, Gang Che Z, Kun Zou S, Xing Fei Q (2011) Numerical simulation of residual stress field induced by laser shock processing with square spot. J Shanghai Univ (English Edition) 15(6):553–556
2. Johnson GR, Cook WH (1983) A constitutive model and data for metals subjected to large strains, high strain rates and high temperatures. In: Proceedings of the 7th international symposium on ballistics 54, pp 541–547
3. Smith PR, Shepard MJ, Prevéy PS III, Clauer AH (2000) Effect of power density and pulse repetition on laser shock peening of Ti-6AI-4 V. J Mater Eng Perform 9:33–37
4. Peyre P, Sollier A, Chaieb I, Berthe L, Bartnicki E, Braham C, Fabbro R (2003) FEM simulation of residual stresses induced by laser peening. Eur Phys J Appl Phys 23:83–88
5. Kay G (2003) Failure modeling of Titanium 6Al-4V and Aluminum 2024-T3 with the Johnson-Cook material model, Office of Aviation Research Washington, D.C.20591, September 2003
6. Frija M, Ayeb M, Seddik R, Fathallah R, Sidhom H (2018) Optimization of peened-surface laser shock conditions by method of finite element and technique of design of experiments. Int J Adv Manuf Technol 97(1–4):51–69
7. Ayeb M (2018) Développement d'un modèle de prévision de l'état induit du traitement, efocsa, Editions universitaires européennes, ISBN-13:978-613-8-39716-8, 92 p., 13 April 2018
8. Ayeb M, Frija M, Fathallah R (2018) Prediction of residual stress profile and optimization of surface conditions induced by laser shock peening process using artificial neural networks. Int J Adv Manuf Technol

Experimental Study of Immiscible Polymer Blends: Morphology and Rheology

Z. Baccouch[1(✉)], S. Mbarek[2], and M. Jaziri[3]

[1] Laboratory of Electromechanical Systems, National School of Engineers of Sfax, LR99ES36, University of Sfax, Sfax, Tunisia
[2] Mechanical Laboratory of (LMS), LR11ES36, National School of Engineers of Sousse, University of Sousse, Sousse, Tunisia
[3] Materials Department, Laboratory of Electrochimie and Environnement, National School of Engineers of Sfax, University of Sfax, B.P. 3038, Sfax, Tunisia

Abstract. This paper focuses on the relation between the rheological and the morphological properties of immiscible polymer blends. Polystyrene droplets (PS) have been synthesized in the laboratory. The blends of EVA2840/PS at ratios 90/10, 70/30, and 50/50 (wt%/wt%) were prepared by Rheomix HAAKE 600VR internal batch mixer with roller rotor. Rheological tests were made at a temperature of 140 °C. Rheological measurements using a rotational rheometer with parallel-plate geometry are considered in order to lead to a better understanding of the dispersion of the PS droplets. Rheological examinations showed that the storage, the loss moduli, and the complex viscosity of blends have an extra elasticity at lower frequency. As a result, analysis of Han diagrams revealed that EVA2840/PS blend are immiscible.

Keywords: Blends · Rheological properties · EVA · PS

1 Introduction

Blending polymers is a strategy to increase plastic material performance [1]. The procedure is to use polymers and to blend them in the melt in order to accomplish the target properties. Nevertheless, polymer pairs are immiscible and form a multiphase system leading to a more complex rheology [2]. The understanding of the rheological properties of polymer blends is important in order to optimize the processing condition. Flow behavior of the polymers is reliant on the molecular and geometrical characterization as well as processing conditions like shear rate, temperature, flow time etc. But in polymer blends, the flow behavior is concurrently dependent on some others factors such as interfacial adhesion, interfacial thickness, miscibility and morphology of blends [3]. The morphology of blends is directed by the viscosity and phase elasticity of the blend components [4]. Moreover, these parameters are dependent on shear rate and blend composition. On the other hand, at melt state the morphological stability is a result of system thermodynamic and interfacial interactions [5, 6]. In this case the basic

N. Aifaoui et al. (Eds.): CMSM 2019, LNME, pp. 507–513, 2020.
https://doi.org/10.1007/978-3-030-27146-6_55

evidence about the effect of different parameters on viscosity, processing, elasticity is useful in selection of proper polymers under a given set of processing conditions. Therefore. The study of the relationship between rheological, morphological and thermodynamic factors is of agreat importance [7].

Furthermore, the end-use performance of the blend is influenced by final morphology which itself is substantially affected by rheological behavior of the system [8–10]. Most of the studies considered the rheological behavior of immiscible blends [11–13]. In these blends, interfacial tension has a controlling role on both rheology and morphology since it influences the dispersed particle size as well as particle size distribution. The interfacial tension is governed by the structural similarity of the components and blend composition [14].

The objective of this work is a systematic study aiming the simultaneous expertise of the melt state linear viscoelastic behavior of EVA2840 and their blends and its relationship with morphological observations.

2 Materials and Techniques

2.1 Materials

Selected polymers must meet several criteria.

The **ethylene-vinyl acetate copolymer** (EVA) is mainly selected for its melting temperature is lower compared to the transition glass temperature of polysterene inclusions (PS) to gether with its lower thermal variation in the viscosity. To do so, two types of EVA were selected. The ethylene- vinyl acetate (EVA) is a copolymer composed of ethylene and vinyl acetate obtainable by a process of high-pressure radical polymerization. Two poly(ethylene-co-vinyl acetate) (EVA), kindly supplied by Arkema, of different molar masses have been used. The amount of acetate groups contained in these copolymers is 28% by weight.

Polystyrene (PS for short) –(CH2–CH(Ph))n- is the polymer obtained by polymerizing styrene monomers CH2 = CH–Ph. The used PS, in this study, was synthesized in the laboratory in order to meet certain characteristics molar mass weight of about 30000 g/mol). The volume ratio between the styrene monomer used and the water/alcohol synthesis medium is 1/5; and that between the water and the alcohol is 3/5. We used butanol sold by ACROS. The mixture is then placed for 7 h at a temperature of 70 °C, with stirring; the formed polystyrene must be purified and filtered. The polystyrene is, then, presented in the form of powder to be dried, in a vacuum oven, for 24 h at 40 °C to remove the last traces of the solvent.

The temperature of the glass transition for PS was measured at around 85 °C.

2.2 Techniques

All blends were prepared by using Rheomix HAAKE. 640p internal batch mixer with roller rotor. All the experiments were performed under nitrogen atmosphere in order to prevent oxidative degradation. The two components (PS, EVA) were loaded to the mixing chamber simultaneously and compounded at 80 °C for 5 min.

The composite were subsequently compression-moulded using a laboratory press at 80 °C for 3 min into 1 mm thick sheets and then cooled to room temperature.

The rheological measurements are carried out in the dynamic mode with an ARES rheometer equipped with an air-pulsed oven. This thermal environment ensures a temperature control within 0.1 °C. The samples were placed between plate-plate interacting fixtures. The zero gap is set by contact, the thicknesses are thus minima; the error is estimated of +0.010 mm with respect to the indicated value. The temperature were 140 °C.

The morphology of the composite was investigated by scanning electron microscopy (SEM) using a Hitashi S800 model. The samples were fractured in liquid nitrogen.

3 Results and Discussion

Figure 1 show the evolution of the storage modulus of the EVA2840/PS blends schemed together with that of the neat polymers at the same temperature of 140 °C. In the frequency range, the storage modulus of pure EVA2840 is lower than that of the EVA blends. This behavior is known as an interfacial tension effect and has been detailed in the literature in the case of polymer blends even from a theoretical point of view [15]. In fact, in the case of dispersed droplets in a continuous matrix, this effect has been described as an excess of elasticity or as the occurrence of long relaxation times. It is correlated to the deformability of the dispersed phase which gives rise to a shape relaxation. This process is clearly connected to the behavior of the neat polymer but the volume fraction of dispersed phase, the size, the distribution of size and the interfacial tension have the most important effects. The phenomenon is also observed in other morphologies such as fully or partial co-continuous parameter is the interfacial energy, product of the interfacial area with the interfacial tension.

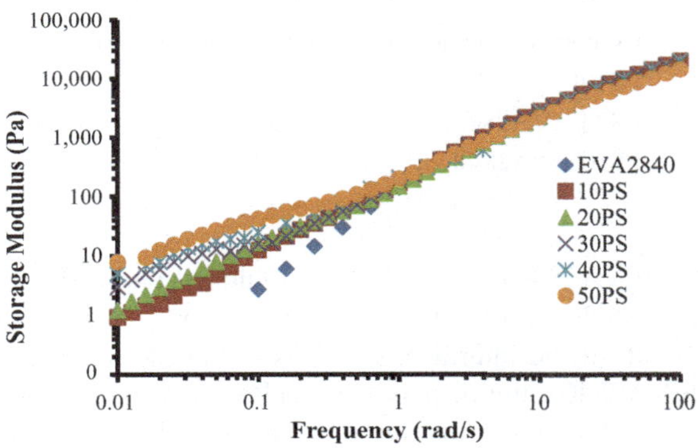

Fig. 1 Storage modulus of EVA2840/PSblends at 140 °C

This phenomena is not observable in the high frequency range since in this case the dynamic spectrometry is investigative scales that are less important than the size of the dispersed phase morphologies.

Figure 2 represent the morphologies of the composite EVA2840/PS, respectively, for the compositions 90/10, 70/30 and 50/50.

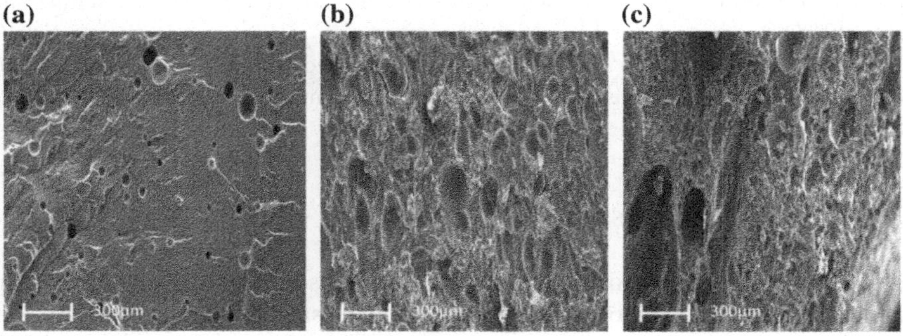

Fig. 2 Photos MEB of EVA2840/PS: **a** PS10 EVA2840, **b** PS 30 EVA2840 and **c** PS50 EVA2840

We observe a nodular morphology in which the PS minority phase is dispersed in the form of particles of different sizes in the EVA matrix 2840 that corresponds to the majority phase. These particles of spherical shape generally have a more or less large size distribution depending on the composition of the composite (Table 1). For these composite, the distribution is very heterogeneous, i.e. the concentration of the particles is different from one location to another at each fracture surface (Fig. 2).

Table 1 Size of the particles of the dispersed phase in the blends EVA284/PS

Composite compositions	Particles diameter (µm)
PS10EVA2840	30–50
PS30 EVA2840	50–80
PS50 EVA2840	70–120

Figure 2a (PS10EVA2840), show that the minority phase PS is presented in the form of spherical particles with a large size dispersion, On the other hand, the PS particles present voids at the interface, which is a typical example of poor adhesion. Figure 2b (PS30EVA2840), the dispersed particles are bigger, they have a generally spherical shape with a broad size distribution. The particles appear to be embedded in the matrix. The enlargement of the size of the particles of the dispersed PS phase could be the result of an agglomeration of PS particules. Concerning composite of intermediate compositions Fig. 2c (PS50EVA2840); we observe an arrangement of the particles in form of a coarse dispersion since we observenodules agglomerates.

Figure 3 show the evoution of the loss modulus of the EVA2840/PS blends and of the neat polymer at 140 °C is plotted. In the frequency range under investigation, theEVA2840 shows the highest values of the loss modulus. The modulus of the blend is between those of the neat components. The resultsare schemed in terms of complex viscosity on Fig. 4. EVA2840 has the highest viscosity and in most cases, the viscosity ofthe blend is also between those of the neat componentsexcept for low amount of PS in EVA2840. It could also be shown that the 50/50 composition displays a high complexviscosity in the low frequency range. Similar effects were already reported by Steinmann and Huitric for blends of PS and PMMA or blends of PE and PA12 [16, 17]. Using a fine analysis of the morphology, they could attribute the observed maximum with the point at which phase inversion occurs.

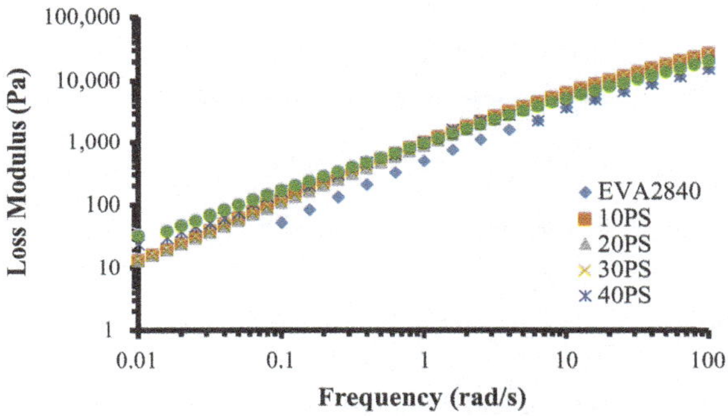

Fig. 3 Loss modulus of EVA2840/PSblends at 140 °C

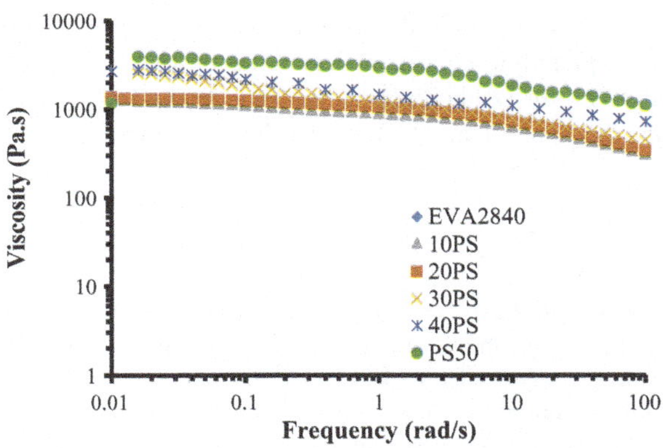

Fig. 4 Complex viscosity of EVA2840/PS blends at 140 °C

In order to investigate the immiscibility and the morphology of the blend via rheological data is making use of the Han diagrams [18]. The storage modulus is plotted versus the loss modulus in the log-log scale Han diagram,. If the blend is miscible, an identical slop can be highlighted for blends with different contents of a particular pair of neat components.

Contrary, the two polymers are immiscible. Figure 5 shows the Han diagram for the EVA2840/PS blends. The results show that these blends, are immiscible thanks to their different gradient.

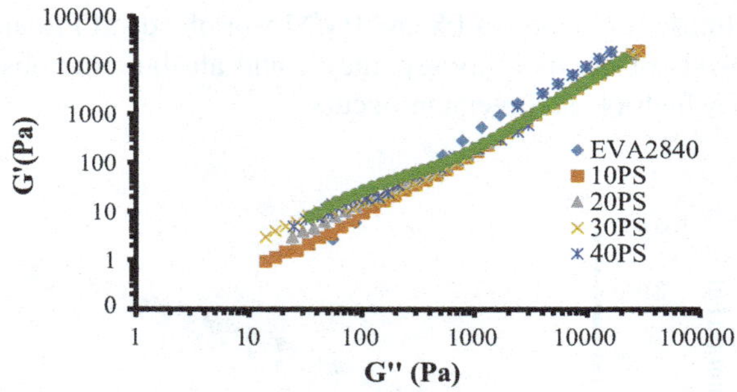

Fig. 5 Han plot of EVA2840/PS blend at 140 °C

4 Conclusion

In conclusion, the the rheological behavior of the EVA2840/PS blends is mainly governed by the EVA. By increasing PS content the storage modulus and viscosity of the blends increase. A higher storage modulus is obtained for the EVA2840 blends as compared to the neat component at low frequencies. this behavior is related to the interfacial tension effect, and in the case of dispersed droplets in a matrix, is attributed to the long relaxation times. Thus, the predominant difference in the elasticity at the low frequencies is explainable energy which is the product of the interfacial area and interfacial tension he observed increase in modulus in the EVA2840/PS blends results in chain stiffening and hence leads to longer relaxation times in such system.

References

1. Baccouch Z, Mbarek S, Jaziri M (2017) Experimental investigation of the effects of a compatibilizing agent on the properties of a recycled poly (ethylene terephthalate)/ polypropylene blend. Polymer Bull 74(3):839–856
2. Carrot C, Mbarek S, Jaziri M, Chalamet Y, Raveyre C, Prochazka F (2007) Immiscible blends of PC and PET, current knowledge and new results: rheological properties. Macromol Mater Eng 292(6):693–706
3. Moly KA, Bhagawan SS, Sabu Thomas (2002) Melt elasticity behaviour and extrudate characteristics of LLDPE/EVA blends: effect of blend ratio, compatibilisation and dynamic cross-linking. Mater Lett 53(4–5):346–352

4. Devaux (2007) Model Studies of transreactions in condensation polymers. Transreactions in Condensation Polymers, pp 125–158
5. Fasce R, Seltzer P, Frontini VJ, Rodriguez Pita EBAV, Pacheco ML (2005) Dias "Mechanical and fracture characterization of 50:50 HDPE/PET blends presenting different phase morphologies", Polymer Eng Sci
6. Kwag H, Rana D, Cho K, Rhee J, Woo T, Lee BH, Choe S (2000) Polym Eng Sci 0:1672–168
7. Han CD (1983) Lem KW Temperature-independent correlation of elasticresponses of viscoelastic liquids. Polym Eng Rev 2:135–165
8. Risse S, Lan Tighzert F, Berzin B (2014) Vergne, Microstructure, rheological behavior, and properties of poly(lactic acid)/poly(butylene succinate)/Organoclay Nanocomposites. J Appl Polymer Sci (12):131
9. Jones R, Pollock HM, Geldart D, Verlinden A (2003) Inter-particle forces in cohesive powders studied by AFM: Effects of relative humidity, particle size and wall adhesion. Powder Technol 132:196–210
10. Leclair A, Favis BD (1996) The role of interfacial contact in immiscible binary polymer blends and its influence on mechanical properties. Polymer 37:4723–8
11. Sadiku-Agboola O, Sadiku ER, Adegbola AT, Biotidara OF (2011) Materials sciences and applications, MSA Rheological Properties of Polymers: Structure and Morphology of Molten Polymer Blends
12. Stockelhuber KW, Svistkov AS, Pelevin AG, Heinrich G (2011) Impact of filler surface modification on large scale mechanics of styrene butadiene/silica rubber composites. Macromolecules 44:4366–4381
13. Lee SS, Jeong HM, Jho JY, Ahn TO (2000) Miscibility of poly(ethylene terephthalate)/poly (estercarbonate) blend Polymer 41:1773
14. MurffS R, Paul DR, Barlow JW (1984) Thermal and mechanical behavior of polycarbonate–poly(ethylene terephthalate) blends. J Appl Polym Sci 29:3231
15. Palierne JF (1990) Linear rheology of viscoelastic emulsions with interfacial tension. Rheol Acta 29:204
16. Steinmann S, Gronski W, Friedrich C (2002) Quantitative rheological evaluation of phase inversion in two-phase polyemer blends with co-continiuos morohology. Rheol Acta 41:77
17. Huitric J, Médéric P, Moan M, Jarrin, J (1998) Influence of composition and morphology on rheological properties of polyethylene/polyamide blends. Polymer 39(20):4849–4856
18. Linder L, Henrichs PM, Hewitt JM, Massa DJ (1985) J Chem Phys 82:1585

Extension of the Jiles–Atherton Hysteresis Model to Characterize the Magneto-Mechanical Behavior: Experimental and Numerical Investigations for Stator Blanking

K. Hergli[1(✉)], H. Marouani[2], and M. Zidi[1]

[1] University of Monastir, LGM, ENIM, Avenue Ibn-Eljazzar, 5019 Monastir, Tunisia
hergli.khaoula@gmail.com, mondher.zidi@enim.rnu.tn
[2] College of Engineering, Muzahimiyah Branch, King Saud University, P.O. Box 2454, Riyadh 11451, Saudi Arabia
hmarouani@ksu.edu.sa

Abstract. Stators of electrical machines are manufactured by ferromagnetic sheet metals blanking. The literature shows that the magnetic efficiency depends on the mechanical deformation/stress states. Thus, the simulation of manufacturing processes leading to the mechanical deformation mapping coupled to a magnetic behavior model can be useful in the design of the rotating machines. This work focuses on a magneto-mechanical coupling approach applied to a stator teeth blanking. An experimental characterization of the magnetic behavior of a Fe–Si alloy under tensile tests is done. The magnetic behavior is described by Jiles–Atherton model. An extended formulation is then proposed to model the magnetic behavior under mechanical deformation. Genetic algorithms are used to identify the corresponding hysteresis parameters. Finite element simulation of teeth blanking is done under Abaqus. This simulation allows us to access to the deformation states on the blanked sheet. A Python code is implemented to extract the plastic equivalent strain for each element mesh from the finite element simulation. Then, according to the extended Jiles–Atherton model, the corresponding magnetic induction is calculated. Finally, a new Abaqus output is created, called magnetic induction, and is injected into the Abaqus file to depict the magnetic parameter distribution. Results show the magnetic induction distribution and signpost that the magnetic degradation reaches 24% near the cut edge. The affected zone is 1.5 mm large.

Keywords: Jiles–Atherton hysteresis model · Ferromagnetic materials · Magneto-mechanical coupling · Finite element modeling · Blanking

N. Aifaoui et al. (Eds.): CMSM 2019, LNME, pp. 514–523, 2020.
https://doi.org/10.1007/978-3-030-27146-6_56

1 Introduction

Stress and strain states are ones of the major factors disturbing magnetic behaviors and properties of ferromagnetic materials. Along with frequency [16], temperature [14, 6, 15, 3] and material properties (grain size, texture, thickness…) [18], an applied stress considerably alters the material magnetic properties. For this reason, ferromagnetic sheet metal forming processes affect the magnetic response and efficiency of rotor and stator electrical steel elements. They lead to local changes of microstructure, harmful residual stress and elasto-plastic deformations mainly at the zone of deformation (in the cut edge) [7, 8, 11, 12, 17]. The importance of the problem has led to several experimental and numerical investigations [1, 2, 7, 8, 13]. Therefore, all the above-mentioned researches have limitations and are non-sufficient to well describe the magnetic behavior evolution with elastoplastic deformations. In fact, they do not take into account the effect of all the cutting parameters or the material parameters. Thus, they cannot define properly the complex state induced by the cutting process studied.

Our work explores the magneto-mechanical coupling describing the degradation of magnetic properties of stator teeth blanking. Firstly, experimental investigations are detailed to introduce the used material, magnetic experimental set-up and the experimental hysteresis curves at different plastic deformation levels. Then, the numerical approaches are described. We focus on the finite element-blanking model and on the Phyton-Abaqus coupling procedure. The main results show the magnetic induction distribution and magnetic induction degradation, which is larger near the cut edge.

2 Experimental Section

In this section, we describe the used material and the experimental procedure adopted to obtain the experimental hysteresis curves.

2.1 Material and Experimental Set up

The material used in this study is a fully process non-oriented Fe-3 wt% Si steel sheet which is widely applied in the main parts of rotors and stators of electrical machines. Test specimens are a 0.35 mm thick strips with a width of 20 mm and a length of 250 mm. Their chemical composition is given in [4] and their mechanical properties are given in Table 1.

The magnetic measurements were carried out at different ranges of plastic strain using a universal testing machine (at room temperature) from 0 to 10% of total strain. The applied stress and the magnetic field were both in the rolling direction. Magnetic measurement apparatus and more details are given in [4].

Table 1 Mechanical properties of material

Young modulus (GPa)	Yield strength (MPa)	Ultimate strength (MPa)	Poisson's ratio (−)	Density (Kg. m^{-3})	Strength coefficient K (MPa)	Strain hardening exponent n (−)
195	310	433	0.29	7800	770	0.26

2.2 Experimental Results

Measured hysteresis curves at different state of plastic deformation are given in Fig. 1. As it is shown, hysteresis loop gets more and more large with deformation. We notice a degradation on induction at saturation B_s and an elevation of hysteresis losses and coercive field H_c.

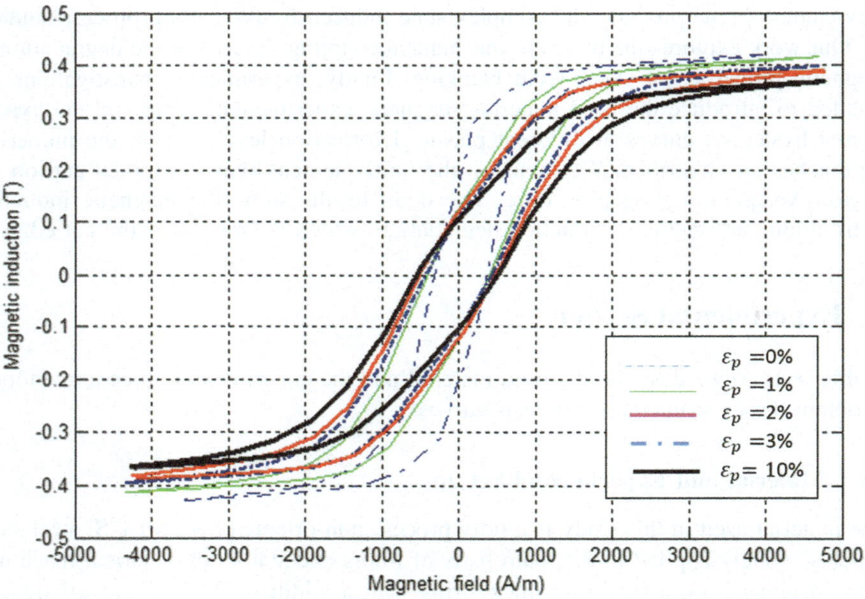

Fig. 1 Measured hysteresis curves for different ranges of plastic deformation [4]

2.3 Hysteresis Model

J-A hysteresis model [5] is expressed by the following differential equation which is function of 5 parameters $\{M_s, a, \alpha, c, k\}$:

$$\frac{dM}{dH} = (1-c)\frac{M_{an} - M_{irr}}{\delta k - \alpha(M_{an} - M_{irr})} + c\frac{dM_{an}}{dH} \quad (1)$$

where M_s is the saturation magnetization, a represents the domain density, α is a domain coupling parameter, c is a coefficient of reversibility ranging from 0 to 1 and k is a pinning factor.

The J-A hysteresis model is extended to better describe the magnetic behavior of the used material under plastic deformation. Among the five parameters of the model, identification procedure and sensitivity study result that the domain density parameter a is the most sensitive to the deformation. So it will be expressed as an analytical function of deformation [4].

$$a = \frac{119 + 650.\varepsilon_p}{1 + \varepsilon_p} \tag{2}$$

The extended model was compared to experimental data and the mean square error between measurements and model data was less than 5%.

3 Numerical Aspect

In this section we present the numerical simulation of stator teeth blanking.

3.1 Finite Element Modelling

Due to the symmetry of the problem, only a half-part is simulated. The specimen, a rectangular sheet with dimensions of 15 × 45 × 0.8 mm, is fixed between a die and a blank holder as it is shown is Fig. 2. Punch, die and blank holder have a fitting radius of 0.2 mm. The radial clearance between the punch and the die, relative to the thickness of the sheet, is 10%. The die and blank holder are fixed. The punch is moving with a speed of 100 mm/s. The Coulomb friction model is used to represent the contact between the sheet and the punch with a coefficient of friction equal to 0.2. The punch, the die and the blank are considered rigid bodies.

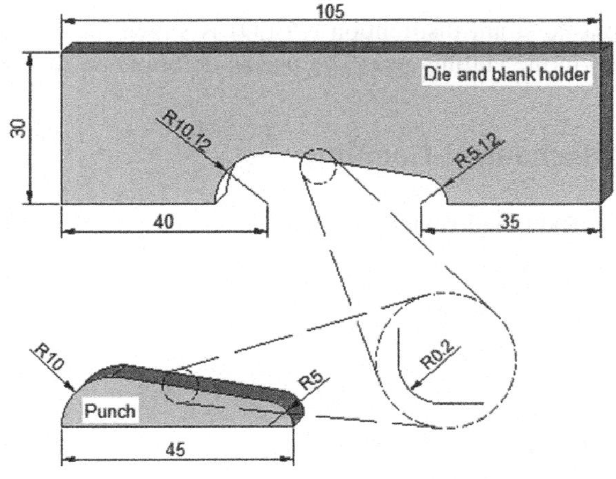

Fig. 2 Schematic description of blanking tools

The blanked sheet is meshed with hexahedral elements with reduced integration (type C3D8R). A finer mesh is made in the zone which undergoes the greatest distortions, in the deformed area. The Eulerian–Lagrangian technique is used in order to improve and guarantee the convergence of the problem. Figure 3 shows the meshed structure.

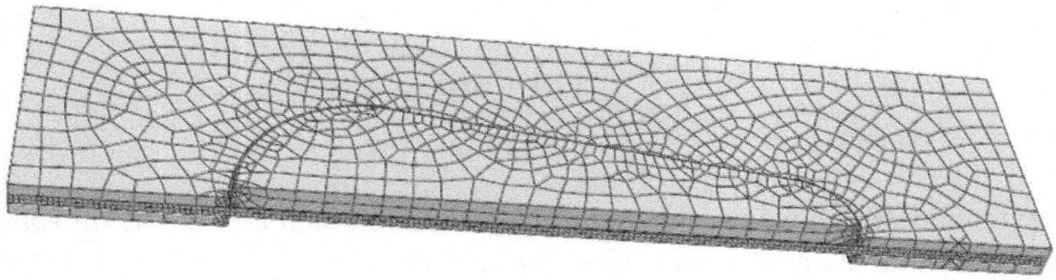

Fig. 3 Finite element mesh of the simulated structure

Finite element simulation of sheet metal blanking is performed using Gurson-Tvergaard-Needleman model to model ductile fracture [9]. The nine Gurson–Tvergaard–Needleman parameters values are summarized on Table 2. f_0 is the initial void fraction, $\{q_1, q_2, q_3\}$ are adjustable material parameters, f_N is the volume fraction of the nucleating void, ε_N is the mean strain for void nucleation, S is the standard deviation, f_c is the critical void volume fraction and f_F is the void volume fraction at failure [8, 10].

Table 2 Gurson–Tvergaard–Needleman parameters

f_0	q_1	q_2	q_3	ε_N	f_N	S	f_c	f_F
0.01	1.5	1	2.25	0.3	0.04	0.1	0.11	0.12

3.2 Finite Element Results

The equivalent plastic strain distribution (PEEQ) is shown in Fig. 4. Stress concentration is localized in the cutting area. The plastic deformation reaches 154.5%.

4 Magneto Mechanical Coupling

In this section, we detail the magneto-mechanical coupling steps and the major results.

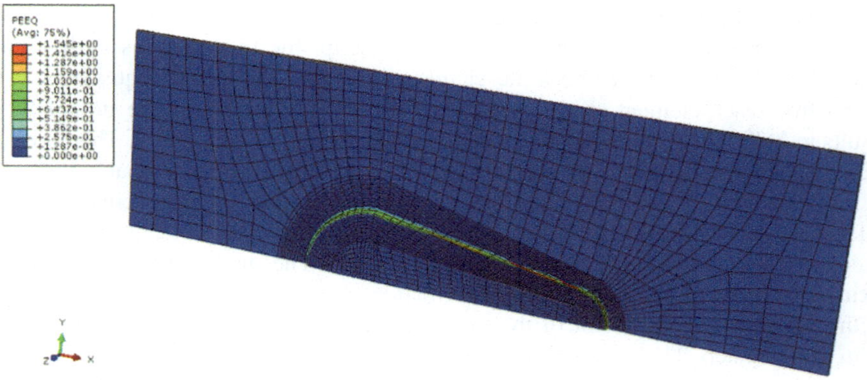

Fig. 4 Equivalent plastic strain PEEQ distribution

4.1 Coupling Procedure

The magneto-mechanical coupling methodology used is based on a weak coupling between the plastic deformation (generated by blanking) and the magnetic field: The mechanical state of the blanked sheet is obtained using Abaqus/Explicit to access the distribution of the plastic deformation. The results will be used to perform the magnetic calculation under Python. The procedure is defined in Fig. 5.

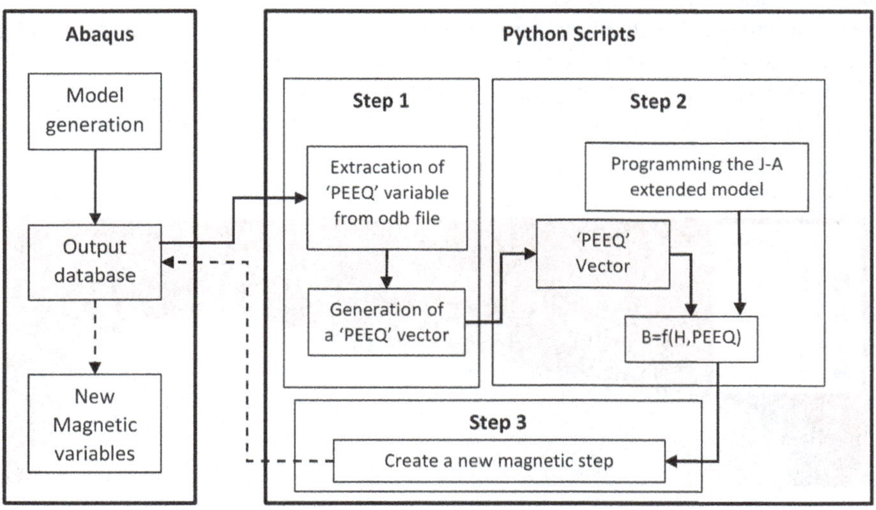

Fig. 5 Flow chart of the coupled approach (Abaqus/Python)

Actually, the coupled magneto-mechanical analysis is split into three separate steps: The first step consists on developing a Python script file that contains Python commands which extract from the finite element simulation the plastic equivalent strain PEEQ from each element mesh of the specimen. This script allows the storage of the results to build tables, make mathematical calculations and draw curves. Therefore, the PEEQ mapping will be converted into a PEEQ 'numpy' array (vector) that will serve as an input for the next step. The size of the vector is the number of the specimen elements (11925 element). Then, an analytical computation of the magnetic hysteresis cycle under Python will be developed in the second step. The modified J-A static model, taking into account the elastoplastic deformation [4], is then implemented. The inputs of the program are the vector of the magnetic field H and the PEEQ vector resulting from the first step. The output is a matrix B = f (H, PEEQ), revealing the evolution of magnetic induction B as well as the degradation of magnetization for each element of the specimen. Finally, in the third step, a new Python script create a new step "Magnetic step" under Abaqus. This new step encloses the magnetic results of the previous calculations and introduce two new variables (i.e. 'Magnetic induction' and 'Degradation').

4.2 Coupling Results

Our coupling approach allows providing the results of the magneto-mechanical modeling and predicting the zones and the magnetization degradation rate for each element of the sample due to a plastic deformation by blanking. The distribution of the magnetic induction B follows the distribution of the equivalent plastic strain PEEQ. A significant fall of the induction is noted when the deformation increases. The magnetic degradation is particularly detected in the area near the cutting edge. Figures 6 and 7 highlight the magnetic flux distribution and the corresponding degradation for a magnetic field H = 1000 A/m. The degradation distribution shows that the magnetic degradation is extreme near the cut edge, it reaches 24%. The affected zone is 1.5 mm large (Fig. 8).

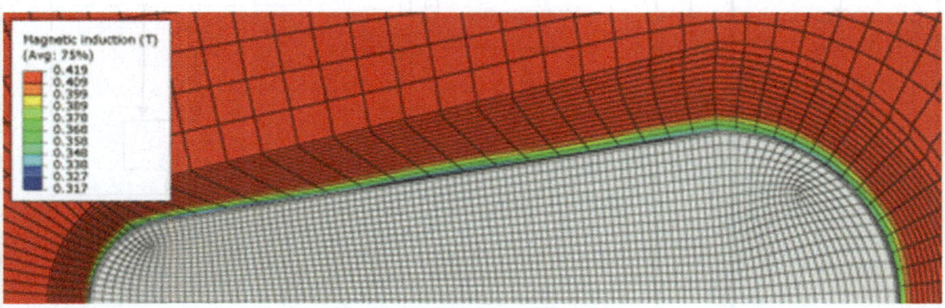

Fig. 6 Magnetic induction B(T) distribution for H = 1000 A/m

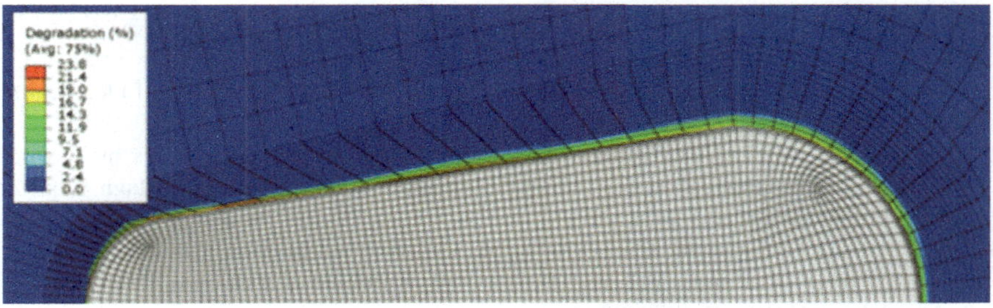

Fig. 7 Degradation of the magnetic induction for H = 1000 A/m

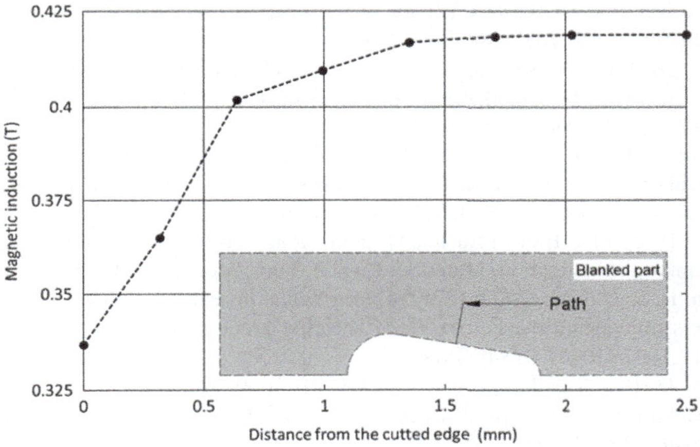

Fig. 8 Magnetic induction evolution among selected path

5 Conclusion

In this paper a numerical study of the magneto-mechanical coupling is developed and applied to stator teeth blanking. The finite element modelling is applied with a Gurson–Tvergaard–Needleman constitutive model and an Eulerian–Lagrangian adaptive mesh to guaranty best convergence of the model. An extended Jiles–Atherton model is used to model the magnetic behavior of a fully process non oriented Fe-3 wt%Si under mechanical deformation. The coupling model consists on a coupling between the plastic deformation induced by blanking process and the magnetic behavior. Abaqus simulations and Python codes were combined. The main results show that the elasto-plastic deformation considerably reduces the magnetization mainly in areas near the cutting edge. The degradation of the magnetic induction reaches 24% for a medium applied field H = 1000 A/m. The coupling model could proof a powerful tool for researchers on elasto-plastic effect on magnetization for electrical steel structures.

References

1. Aydin U, Rasilo P, Martin F et al (2017) Magneto-mechanical modeling of electrical steel sheets. J Magn Magn Mater. https://doi.org/10.1016/j.jmmm.2017.05.008
2. Bernard L, Daniel L (2015) Effect of stress on magnetic hysteresis losses in a switched reluctance motor: Application to stator and rotor shrink fitting. IEEE Trans Magn. https://doi.org/10.1109/TMAG.2015.2435701
3. Chen D, Fang L, Kwon Bi, Bai B (2017) Measurement research on magnetic properties of electrical sheet steel under different temperature, harmonic and dc bias. AIP Adv 7:1–5. https://doi.org/10.1063/1.4979490
4. Hergli K, Marouani H, Zidi M (2017) Numerical determination of Jiles–Atherton hysteresis parameters: Magnetic behavior under mechanical deformation. Phys B Condens Matter 549:74–81. https://doi.org/10.1016/j.physb.2017.09.126
5. Jiles DC, Atherton DL (1986) Theory of ferromagnetic hysteresis. J Magn Magn Mater. https://doi.org/10.1016/0304-8853(86)90066-1
6. Kuo SK, Lee WC, Lin SY, Lu CY (2015) The influence of cutting edge deformations on magnetic performance degradation of electrical steel. IEEE Trans Ind Appl 51:4357–4363. https://doi.org/10.1109/TIA.2015.2440351
7. Leuning N, Steentjes S, Schulte M et al (2016) Effect of elastic and plastic tensile mechanical loading on the magnetic properties of NGO electrical steel. J Magn Magn Mater 417:42–48. https://doi.org/10.1016/j.jmmm.2016.05.049
8. Marouani H, Ben IA, Hug E, Rachik M (2008) Rate-dependent constitutive model for sheet metal blanking investigation. Mater Sci Eng, A. https://doi.org/10.1016/j.msea.2007.10.023
9. Marouani H, Rachik M, Hug E (2012) Experimental investigations and FEM simulations of parameters influencing the Fe-(wt.3%)Si shearing process. Mech Ind 13:271–278. https://doi.org/10.1007/s10461-013-0526-9
10. Marouani H, Ben Ismail A, Hug E, Rachik M (2009) Numerical investigations on sheet metal blanking with high speed deformation. Mater Des 30:3566–3571. https://doi.org/10.1016/j.matdes.2009.02.028
11. Naumoski H, Maucher A, Vandenbossche L, et al (2014) Magneto-optical and field-metric evaluation of the punching effect on magnetic properties of electrical steels with varying alloying content and grain size. In: 2014 4th international electric drives production conference, EDPC 2014—proceedings
12. Naumoski H, Maucher A, Herr U (2015) Investigation of the influence of global stresses and strains on the magnetic properties of electrical steels with varying alloying content and grain size. In: 2015 5th international conference on electrical drives prod EDPC 2015 - Proc 1–8. https://doi.org/10.1109/edpc.2015.7323206
13. Ossart F, Hug E, Hubert O, et al (2000) Effect of punching on electrical steels: experimental and numerical coupled analysis. In: IEEE transactions on magnetics
14. Raghunathan A, Melikhov Y, Snyder JE, Jiles DC (2009) Modeling the temperature dependence of hysteresis based on Jiles–Atherton theory. In: IEEE transactions on magnetics
15. Sixdenier F, Messal O, Hilal A, et al (2015) Temperature dependent extension of a hysteresis model To cite this version : temperature dependent extension of a hysteresis model
16. Szewczyk R, Frydrych P (2010) Extension of the Jiles-Atherton model for modelling the frequency dependence of magnetic characteristics of amorphous alloy cores for inductive components of electronic devices. In: Acta Physica polonica A

17. Xiong X, Hu S, Hu K, Zeng S (2016) Texture and magnetic property evolution of non-oriented Fe-Si steel due to mechanical cutting. J Magn Magn Mater 401:982–990. https://doi.org/10.1016/j.jmmm.2015.10.023
18. Zaizen Y, Omura T, Fukumura M, et al (2016) Evaluation of stress distribution due to shearing in non-oriented electrical steel by using synchrotron radiation. AIP Adv 6, https://doi.org/10.1063/1.4944342

Product Development Process Based on Open Technologies

Mohammed Akerdad(✉), Ahmed Aboutajeddine, and Mohamed El Majdoubi

Laboratory of Mechanical Engineering, Faculty of Science and Technology of Fez, Sidi Mohamed Ben Abdellah University, Fez, Morocco
{mohammed.akerdad, ahmed.Aboutajeddine, mohammed.elmajdoubi}@usmba.ac.ma

Abstract. Nowadays, the world is experiencing many changes, the technology is conquered by open hardware movement and the competition is based mainly on the development of ground-breaking products. These technological and economical changes are pushing companies to reinvent themselves to develop products that are cheaper, better and of good quality than competitors. An appealing avenue for these companies is to use open hardware solutions in the development of their new products. However, product development processes used nowadays are not suited to exploit the full potential of open hardware. This work addresses this issue by first performing an in-depth assessment of the academic literature on product innovation processes and later proposing a product development process based on open technologies. The process introduces a new product development rationale that includes four phases: the first phase is looking for opportunities, which involves generating opportunities and creative ideas. The second phase is the opportunity assessment phase, which consists of selecting and capturing opportunities. The third phase is the validation of opportunities phase through technical feasibility, client desirability and sustainability. The fourth phase is engineering design, which includes concept generation, prototyping testing, and product manufacturing. The developed process hopes to open the way for businesses, especially in emerging countries like Morocco, to be competitive on the market by helping them designing superior products based on open technologies.

Keywords: Product design process · Open technology · Prototyping

1 Introduction

Companies in emerging countries are facing major challenges due to globalization, short products life cycles and limited access to technology. Thus, the use of open technology tools appears as a noteworthy opportunity to overcome these limitations and gain competitive advantage and success. However, this opportunity is restricted by the knowledge and exploitation of factors influencing the success of new products.

Cooper and Edgett consider that an innovation strategy and the ideation practice are among the factors of the success of a product [8]. Other authors consider that a coherent approach to the development of creative ideas is the factor that helps designers develop

N. Aifaoui et al. (Eds.): CMSM 2019, LNME, pp. 524–532, 2020.
https://doi.org/10.1007/978-3-030-27146-6_57

innovative new products [13]. In other words, a good product development process is the main factor of economic progress and competitiveness in the market. In fact, studies have shown that about 80% of product development projects fail even before completion and more than 50% of projects have no return on investment [5]. These studies highlight the need to develop a new product development approach that incorporates new design methods to reduce cost and development time as well as help companies develop products of good quality and being competitive in the market.

Researchers and specialists have studied the product development process (PDP). Chauhan and al have considered it as one of the most critical areas of competence of a company [4]. According to Unger and Eppinger, PDPs are the procedures and methods that companies use to design and market new products. Krishnan and Ulrich, on the other hand, defined a PDP as "transforming a market opportunity and a set of assumptions about product technology into a product available for sale" [19]. A literature assessment on product development processes shows that PDPs vary following authors, which explains why some researchers are focused on identifying best product development practices [1]. This variety expands the choice for companies to use the most suitable PDP to their contexts. However, the key to success for a company usually depends on the upstream phases of product development process [18]. Actually, Cooper considered the pre-development phase as a critical factors for success [18]. Other authors, on the other hand, have suggested that the prototyping phase is the most critical phase of a PDP and that influences the overall results of the design [3].

This paper proposes a new product development approach for designing products based on open technologies to reduce development time, improve quality and minimize errors. It begins, in Sect. 2, with a literature review on product development processes specifically on the front end and prototyping phase. In Sect. 3, we detail our proposed process which includes four key phases: opportunity search, opportunity assessment, opportunity validation and engineering design. Section 4 illustrates the assumptions from our research. Finally, a conclusion is drawn with as prospective work to validate the process.

2 Literature Review of Product Development Process

The PDP is at the heart of any company strategic issues, which explains it importance in the success or failure of product introduction on the market [28]. In the literature, there are several product development processes in different forms: sequential processes, concurrent processes, and spiral processes.

The sequential process has dominated the industry for almost 30 years (Ulrich and Eppinger 2015). This process corresponds to all the steps and activities that a company uses to design and market a product. It begins with a planning phase, directly related to R & D activities. The result of this phase is the project's mission, which becomes the starting point for the concept development phase that will guide the design team in the other phases. The mechanical design process proposed by Ullman proposes six phases for developing products [10]. Each of these phases involves the implementation of generic tasks, and the success of a phase is approved by a design review, which marks its completion. Pahl and Beitz, on the other hand, defined their product development process in four phases. Each phase contains specific activities (Fig. 1) [26].

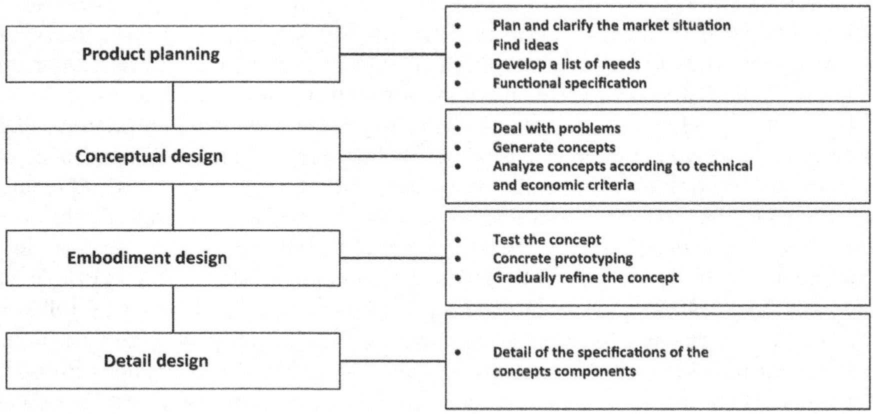

Fig. 1 Product development process of Pahl and Beitz 2007

Many companies are adopting concurrent processes for developing new products. The design phases in these processes take place in a non-linear and iterative way which explains the flexibility of these processes and the reduction of deadlines.

The biggest competitors to concurrent models are the more flexible spiral processes that have been adapted from software development [7]. The Spiral model is based on design cycles, it includes a risk-based approach to product development [15]. This process eliminates unnecessary work and quickly moves towards a finalized product by building a series of steps and looping "build; test; feedback; revise" [7]. Among the disadvantages of spiral processes is its complexity which requires considerable attention from the management [33].

In general, and in recent publications, for example: Benabidin 2014 summarize the design models proposed by design theorists from 1967 to 2013 [2], we find that all the models listed have been divided into six main phases: establishing a need, analysis of task, conceptual design, embodiment design, detailed design and implementation. According to recent studies of several researchers, the phase of pre-development or front-end phase is among the critical phases for the success of a product. In addition, the prototyping phase is a crucial phase for the development of a good quality product. These two phases of PDPs (Front end, Prototyping) are the subject of the next part.

2.1 Fuzzy Front End

In 1991, Smith and Reinerlsen are the first to give the name of "front end" to this phase. They described it as an unpredictable, unstructured and chaotic process. The term "front end" describes the first stage of developing an idea [30]. Others defined it as the work that is done to develop a product before it enters the formal product development system [20]. Some authors consider that this phase includes demonstrations of technical feasibility, financial viability, business plan preparation, business model development and preliminary market analysis [21]. Thus, this phase offers significant benefits in terms of performance, business and technical risk reduction, and cost reductions [9]. Smith and Reinertsen suggest including the Front-End Phase in the Product Development Process. It is worth noting that some product development companies call this stage: frontal innovation [17].

Authors in the area of innovation management offer different models for managing front-end innovation. Wheelwright et al. 1992 developed a "funnel development" model which consists in generating ideas then progressively refining them in order to select the best one [34]. Koen, Bertels and Kleinschmidt have developed "the new concept development process" [18]. This model is circular in shape and is divided into three parts. The part that illustrates the front-end phase is a wheel that has five elements: identifying opportunities, analyzing opportunities, generating ideas, selecting ideas, and defining concepts. In addition, Frishammar and his colleagues proposed a six-phase process for the development of radical ideas and concepts [11]. This framework provides key objectives, key activities and results for each phase.

Measuring the performance of these models has proved to be complex because of the different ways of measuring success in the literature. A recent study has been done; to analyze the impact of the various activities of the Front-End phase; it uses three measures to achieve successful front-end action: efficiency, creativity, and reducing uncertainty [32]. Others consider inter-functional interaction as a critical factor on new product performance [31].

2.2 Prototyping Phase

Prototyping is a key activity for both convergence (validation) and divergence (exploration) in a design space [3]. As Wall says, "Prototyping is one of the most critical activities in new product development" [23]. According to Otto & Wood in 2001, a prototype is an artifact that approximates a feature (or multiple features) of a product, service, or system [35].

Studies have shown that designers frequently use physical prototypes to facilitate and simplify the process [14]. These physical models help clarify customer requirements [12], reduce errors, and can improve design decision-making. Prototypes are also useful to communicate concepts within the design team. Authors note that companies need to use a guided prototyping approach to quickly develop viable and desirable new products, using fewer resources [24]. Prototyping is then an important activity that must be used in every steps of the design process.

3 New Approach of Product Development Process

Design process is the procedures that companies use to develop new products and market them. Technological change, competition and the rise of startups are forcing companies to frequently develop new products. For this, we propose a product development process. This model was based on a literature review of a product type innovation. Figure 2 illustrates the steps of the open hardware-based design process.

This process aims at designing innovative products that are viable for a company. This process can shorten development time, use early and frequent testing with prototyping, strong customer integration, make design flexible, and seek out new or unknown markets.

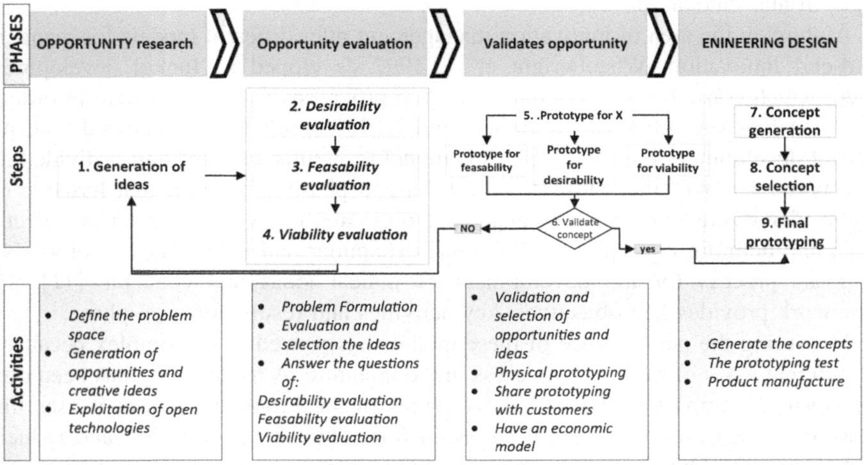

Fig. 2 New product development process based open technologies

This process presents a roadmap for innovation. It uses prototyping tools to ensure the success of products on the market. More details on the three phases will be discussed in the following sections.

3.1 Opportunity Research

The starting point for all good products is the idea [11], which is an opportunity, an assumption about value creation. This first phase focuses on identifying opportunities and project ideas. For some designers, it is very abstract to propose new ideas. That is why we are offering designers the opportunity to harness the great potential of open source technology to build on the insights and opportunities that exist in the marketplace by visiting communities and online platforms like Crowdfunding, Kickstarter and Arduino forums and others [29]. The credo of these communities is to share new ideas and new technologies as well as needs that exist in the markets [25].

3.2 Opportunity Evaluation

To evaluate the ideas found in the first phase, we propose a new theoretical method for evaluating ideas that is inspired by design thinking frameworks and ideas tunnel, called opportunity evaluation. This new method consists of going through a convergent tunnel that contains three opportunity assessment filters: desirability assessment, viability assessment and feasibility assessment. Members of the development team can come together to evaluate opportunities by answering questions related to desirability, feasibility and sustainability (Table 1). If the result of the first evaluation is affirmative, the designers can proceed to the second evaluation until the last evaluation. Otherwise, the idea or opportunity is rejected from the tunnel and another idea is tested until a relevant idea is found. The chosen idea is moved then to the next stage: opportunity validation phase.

Table 1 Questions to evaluate opportunities

Steps of the evaluation	Questions
Desirability evaluation	• Does the idea found meet a customer need? • Is the idea valuable to customers
Feasibility evaluation	• Is the idea technologically feasible? • Is the idea technically feasible?
Viability evaluation	• Can we integrate into the market? • Can we have a return on investment?

3.3 Validates Opportunities

The third phase is devoted to validating ideas or opportunities. Based on previous research on the impact of prototyping methods on PDPs, we propose to apply the "prototyping for X" (PFX) method. This method is intended to help designers focus efforts and resources to create products that test hypotheses and provide product information [24]. As shown in Fig. 2, PFX is based on three essential lenses, according to [22]:

- Desirability prototyping: consists of creating prototypes to validate that the product has a desirability on the market.
- Feasibility prototyping: consists of creating prototypes to validate that the product is technically feasible and functional.
- Viability prototyping: consists of creating prototypes that test the probability of adapting to time and budget constraints.

To make these prototypes, we can use open source hardware tools. This open hardware creates more flexible products at a low cost. In other words, the new technological revolution in open hardware has given designers the tools they need to turn creative ideas into innovative and viable products. This new technological trend is accessible and usable by all designers who share the results of their work and who collaborate and cooperate with different people around the world [16]. Our product development process relies on open hardware to develop opportunities quickly and at low cost. Thus, help designers and creators test concepts with rapidly evolving prototyping [6]. For example: Collaborative practice and shared design reduce capital costs by 90% to 99% compared to conventional costs for scientific equipment [27].

3.4 Engineering Design

We have validated the concepts and ideas in the third phase. The fourth phase of "Technical Design" consists of following the steps of conventional product development processes, which begin with the generation of concepts and culminate in final prototyping. Designers can also use open hardware tools such as open source 3D printers and open software to build virtual prototypes and simulations. This phase allows designers to obtain the final product manufactured according to the technical specifications and customer requirements.

4 The Assumptions Derived from Processes

After a thorough analysis of our process, we have made assumptions about the process:

- **Hypothesis 1**: Idea generation allows companies and designers to generate promising ideas.
- **Hypothesis 2**: Opportunity assessment allows filtering ideas and finding the ones that are relevant.
- **Hypothesis 3**: Prototyping allows designers to obtain feedback from their concepts on the technical feasibility, timeliness and viability of the concept.
- **Hypothesis 4**: Open technologies reduce the manufacturing costs of prototypes.

To test these hypotheses, a case study will be developed in future work. This case study aims to validate our process and improve it in order to know the limits and constraints encountered during the execution of the process.

5 Conclusion and Future Work

This work has been devoted to the development of a new product development process that aims to help companies and designers design cheaper and better products. Our process is based on four essential phases: the search for opportunities, the evaluation of opportunities, the validation of opportunities and the engineering design. Our process takes into account the factors of success of the product: technical feasibility, customer desirability and viability of the company. Our future goal is to improve tools and methods used in the process and apply the process to a specific case study to validate the process and our hypothesis.

References

1. Barczak G, Griffin A, Kahn KB (2009) Perspective: trends and drivers of success in NPD practices: results of the 2003 PDMA best practices study*. J Prod Innov Manag 26:3–23
2. Benabid Y (2014) Contribution à l'amélioration du processus de conception des produits innovants: Développement d'outils d'aide au choix des processus. ENSAM, Paris

3. Camburn BA, Jensen D, Crawford R, Otto K, Wood K (2015) Evaluation of a strategic method to improve prototype performance with reduced cost and fabrication time. In: 80–4 Proceedings 20th international conference on engineering design ICED 15 Vol 4, pp 27–30
4. Chauhan AS, Nepal B, Soni G, Rathore APS (2018) Examining the state of risk management research in new product development process. Eng Manag J 30:85–97
5. Cooper RG (2001) Winning at new products: accelerating the process from idea to launch
6. Cooper RG (2014) What's next?: After stage-gate. Res-Technol Manag 57:20–31
7. Cooper RG, Edgett SJ (2008) Maximizing productivity in product innovation. Res Technol Manag 51:47–58
8. Cooper RG, Edgett SJ (2012) Best practices in the idea-to-launch process and its governance. Res Technol Manag 55:43–54
9. Cooper RG, Kleinschmidt EJ (1994) Determinants of timeliness in product development. J Prod Innov Manag 11:381–396
10. David GU (2011) The mechanical design process (McGraw-Hil)
11. Frishammar J, Dahlskog E, Krumlinde C, Yazgan K (2016) The front end of radical innovation: a case study of idea and concept development at prime group. Creat Innov Manag 25:179–198
12. Gordon VS, Bieman JM (1995) Rapid prototyping: lessons learned. IEEE Softw 12:85–95
13. Heininger K, Chen H-E, Jablokow K, Miller SR (2018) How engineering design students' creative preferences and cognitive styles impact their concept generation and screening. (Am Soc Mech Eng), p V007T06A032–V007T06A032
14. Hess T, Summers JD (2013) Case study: evidence of prototyping roles in conceptual design
15. Hicks C, McGovern T (2009) Product life cycle management in engineer-to-order industries. Int J Technol Manag 48:153–167
16. Jackson A (2014) Makers: the new industrial revolution. J Des Hist 27:311–312
17. Jensen ARV (2017) Front end innovation: navigating situated spaces of actors and models. Aalborg Universitetsforlag
18. Koen PA, Bertels HMJ, Kleinschmidt E (2014) Managing the front end of innovation—Part I: results from a three-year study. Res Technol Manag 57:34–43
19. Krishnan V, Ulrich KT (2001) Product development decisions: a review of the literature. Manag Sci 47:1–21
20. Markham SK (2013) The impact of front-end innovation activities on product performance. J Prod Innov Manag 30:77–92
21. Markham SK, Ward SJ, Aiman-Smith L, Kingon AI (2010) The valley of death as context for role theory in product innovation. J Prod Innov Manag 27:402–417
22. Menold J, Simpson TW, Jablokow KW (2016) The Prototype for X (PFX) framework: assessing the impact of PFX on desirability, feasibility, and viability of end designs. V007T06A040
23. Menold J, Jablokow K, Simpson T, Seuro R (2017) Evaluating the discriminatory value and reliability of ideation metrics for their application to concept development and prototyping. (Am Soc Mech Eng), p V007T06A034-V007T06A034
24. Menold J, Simpson TW, Jablokow K (2018) The prototype for X framework: exploring the effects of a structured prototyping framework on functional prototypes. Res Eng, Des
25. Oberloier S, Pearce JM (2018) General design procedure for free and open-source hardware for scientific equipment. Designs 2:2
26. Pahl G, Beitz W (2013) Engineering design: a systematic approach (Springer Science & Business Media)
27. Pearce JM (2016) Return on investment for open source scientific hardware development. Sci Public Policy 43:192–195

28. Perrin-Bruneau F (2005) Proposal of an integration strategy of new design methods: elements for definition of the "methods" integrator role. Theses. Arts et Métiers ParisTech
29. Raasch C, Herstatt C, Balka K (2009) On the open design of tangible goods. RD Manag 39:382–393
30. Reid SE, Brentani UD (2004) The fuzzy front end of new product development for discontinuous innovations: a theoretical model. J Prod Innov Manag 21:170–184
31. Russell RK, Tippett DD (2008) Critical success factors for the fuzzy front end of innovation in the medical device industry. Eng Manag J 20:36–43
32. Schweitzer F, Gabriel I (2012) Action at the front end of innovation. Int J Innov Manag 16:1240010
33. Unger D, Eppinger S (2011) Improving product development process design: a method for managing information flows, risks, and iterations. J Eng Des 22:689–699
34. Wheelwright SC, Clark KB (1992) Revolutionizing product development: quantum leaps in speed, efficiency, and quality (Simon and Schuster)
35. Wood KL, Jensen D, Bezdek J, Otto KN (2001) Reverse engineering and redesign: courses to incrementally and systematically teach design. J Eng Educ 90:363–374

Failure Mechanism of Sandwich Panels Under Three-Point Bending

Raja Ouled Ahmed Ben Ali(✉) and Sami Chatti

LMS, Ecole Nationale d'Ingénieurs de Sousse, Sousse, Tunisie
ouledahmed_raja@live.fr, sami.chatti@enim.rnu.tn

Abstract. Sandwich materials are potential candidates instead of traditional materials in several fields as aerospace, civil engineering and automotive because of their mechanical properties and especially their high ratio bending stiffness to weight. Three-point bending is a frequent process for forming sandwich panels before usage. This study presents an analysis of the damage of the sandwich panels during quasi-static tests in three-point bending. Experimental tests leading to the failure of the core of the sandwich material were carried out. Finite element analysis was also conducted for the numerical prediction of observed damage. Also, analytical Gibson's modified model is considered to obtain the critical loads leading to the failure of the sandwich panels. This allows constructing a mode map for failure modes of sandwich panels in three points bending process.

Keywords: Three points bending · Thick sandwich panel · Numerical simulation · Damages · Failure map

1 Introduction

Sandwich plates are increasingly used in a wide range of industrial products [1] varying from automobiles and airplanes to simple home appliances due to the properties such as lightweight, vibration reduction, acoustic noise damping, and heat insulation [2]. In addition, the sandwich panel metal/polymer/metal have been tested in standard or specialized tests such as shearing, three point bending, four point bending, and indentation [3–5]. The most common failure mode of the foam core sandwich structures is core shearing, followed by local indentation collapse and face yielding [3, 5, 6]. The risk of mechanical buckling, and decohesion between the skins and the core constitute the main weaknesses of sandwich panels. Under bending, a sandwich panel undergoes various modes of degradation classified in several categories by [7–9]. Kim and Hwang [10] studied theoretically and experimentally the effect of decohesion between skin and core on the stiffness of sandwich panels. Idriss et al. [11] have shown that the crack propagation occurs in three stages: decohesion between the core and the upper skin, core shear and debonding between the lower skin and the core. However, among these studies, there is a lack of systematical research on the steel/polyurethane/steel sandwich structures' failure mechanism. Based on the aforementioned literature, limited study was conducted on the modeling of damage propagation in polyurethane foam cores. For the above-mentioned reasons, in this paper the

N. Aifaoui et al. (Eds.): CMSM 2019, LNME, pp. 533–544, 2020.
https://doi.org/10.1007/978-3-030-27146-6_58

failure mechanism of sandwich panel in three point bending test are investigated by applying the cellular solids theory [12]. In addition, experiments and numerical simulations are carried out to validate the theoretical prediction.

2 Analytical Analysis

Several failure modes have been identified for sandwich panels in three-point bending [12]: (a) face yielding; (b) wrinkling of the compressive face; (c) core shearing, (d) face debonding, (e) indentation. The last mode of failure occurs when the loading is extremely localized and can be avoided by increasing the loading area. To analyze the failure mechanism of the panels, it is necessary to characterize the normal stress and the shear stress acting on the skins and the core.

The maximum stress occurs in the cross section which has the maximum moment. In the load case of three-point bending, the maximum moment and the shear can be easily acquired in terms of the concentrated load P, that:

$$M = \frac{PL}{4} \tag{1}$$

Considering the shear stress as linear through the faces and constant through the core since the faces are much stiffer and thinner than the core, the normal and shear stresses can be expressed as [13]:

$$\sigma_{p\max} = \pm \frac{PL}{4e_p bd} \tag{2}$$

$$\sigma_{c\max} = \pm \frac{PLe_c}{8D} E_c \tag{3}$$

$$\tau_{c\max} = \frac{P}{2bd} \tag{4}$$

where D is the equivalent flexural rigidity, which can be expressed as [13]:

$$D = \frac{E_p e_p d^2 b}{2} + \frac{E_p t_p^3 b}{6} + \frac{E_c e_c^3 b}{12} \tag{5}$$

2.1 Modified Gibson's Model

Instead of a flat loading head and supports used in the experiments and analysis of [3], a cylindrical loading head and supports are used in the present experiments and analysis. Hence some modifications are required. Three main failure modes of sandwich panel steel/polyurethane/steel (face yielding, wrinkling face and core shearing) are considered in this study.

Face yielding

This mode occurs when the maximum normal stress in faces reaches the yield strength of the face material

$$\sigma_p = \sigma_{yp} \quad (6)$$

where

σ_{yp}: yield strength of the face material

The critical load for the face yield mode is given by

$$P_{cr1} = \frac{4be_p e_c}{L}\sigma_{yp} \quad (7)$$

Wrinkling face

This mode occurs when the maximum normal stress in faces reaches the local elastic instability stress, in this case:

$$\sigma_p = \sigma_{wp} \quad (8)$$

σ_{wp} is wrinkling stress in the face material which can be expressed as follow [13]:

$$\sigma_{wp} = \frac{3E_p^{1/3}E_c^{2/3}}{\left[12(3-\upsilon_c)^2(1+\upsilon_c)^2\right]^{1/3}} \quad (9)$$

where υc: Poisson's ratio of foam material

The critical load of wrinkling face mode is given as [14]:

$$P_{cr2} = \frac{4be_p e_c}{L}E_p^{1/3}E_s^{2/3}\left[0.28\left(\frac{\rho_c}{\rho_s}\right)^{4/3} + 0.2\left(\frac{\rho_c}{\rho_s}\right)^{2/3}\right] \quad (10)$$

where ρ_s, E_s: Density and Young's modulus of foam's cell-wall material, respectively.

Core shearing

The shear force is carried mainly by the foam core when the sandwich panel is subjected to a transverse shear force. If the shear stress in the foam core reaches the shear strength of the foam core material, the initial failure will be in the foam core. The critical load of core shear mode is written as [14]:

$$P_{cr3} = 2be_c\left[2.32\left(\frac{\rho_c}{\rho_s}\right)^{3/2} - 0.28\frac{\rho_c}{\rho_s}\right]\sigma_{ys} \quad (11)$$

where σ_{ys} Yield strength of foam's cell-wall material

Table 1 gives the mechanical and geometrical characteristics of sandwich panels Steel/Polyurethane/Steel (Table 1).

Table 1 Mechanical and geometrical characteristics of sandwich panels Steel/Polyurethane/Steel

Panel's geometry (mm)			Skin material (MPa)		Core material (Kg/m^3MPa)			
b	e_p	e_c	E_p	σ_{yp}	ρ_c	ρ_s	E_s	σ_{ys}
50	0.5	40	200000	400	40	1170	1600	53.4

Figure 1 plots the sandwich panel failure loads of different failure modes in terms of the length between supports L. From this figure it is observed that the critical loads of the first two failure modes decrease with increasing length between the supports L, while the critical load of the shear failure of the core is constant. In addition, it is observed that the panels suffer a failure according to the mode of core shear for lengths between weak supports. While, for wide distances between supports, it is the folding mode of the skins which intervenes. In addition, the skin yielding failure mode is less likely to occur since the corresponding failure limit load is much higher than the other two modes.

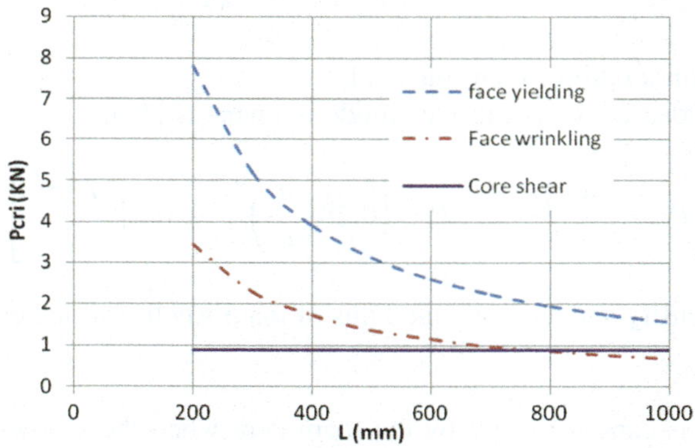

Fig. 1 Failure loads of sandwich panel steel/polyurethane/steel

2.2 Failure Mode Map

The failure mode map can be constructed from Eqs. (7) (10) and (11), with dimensionless parameters the relative density of foam ρc/ρs and ratio of skin thickness to span length ep/L as the coordinates. The diagram is divided into three regions. Within each region one failure mechanism is dominant. The regions are separated by three transition lines, which represent the panel designs for which two mechanisms have the same failure load. The three transition lines are governed by Eqs. (12), (13) and (14), respectively, skin folding and skin lamination, skin wrinkling and core shear and that between core shear and plasticization of skin. These equations are obtained for equal critical loads for two particular failure modes. It is clear that these transition lines depend mainly on the strength of face and core materials.

$$\frac{\rho_c}{\rho_s} = \left(\frac{\sigma_{yp}}{0.26\,E_p^{1/3}\,E_S^{2/3}}\right)^{3/2} \tag{12}$$

$$\frac{\rho_c}{\rho_s} = \left(\frac{0.85epE_p^{1/3}\,E_s^{2/3}}{0.62\,\sigma_s L}\right)^{4} \tag{13}$$

$$\frac{\rho_c}{\rho_s} = \left(\frac{1.13\sigma_p e_p}{0.65\,\sigma_s L}\right)^{3/4} \tag{14}$$

An initial failure mode map according to the geometry and material properties of sandwich panels steel/polyurethane/steel is predicted in Fig. 2.

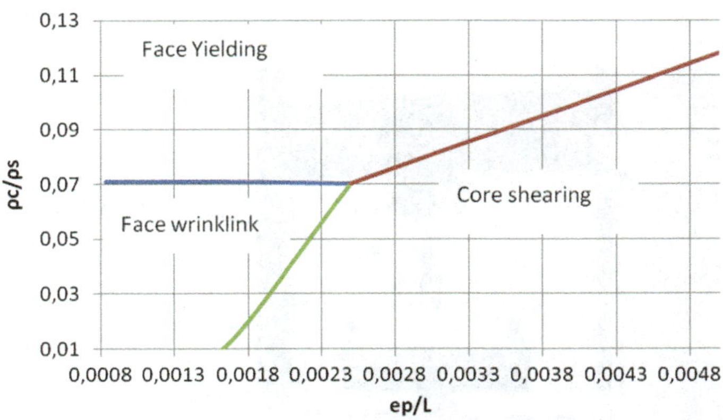

Fig. 2 Failure mode map of sandwich panel steel/polyurethane/steel

3 Experimental and Numerical Procedures

3.1 Experimental Procedures

The core of the sandwich specimens used in this study consists of polyurethane PUR is closed-cell rigid foam plastic. The foam properties could be obtained from uniaxial compression tests according to ASTM C 365-57 standard. The skin used in this study consists of galvanized steel with high specific strength and stiffness. The skin properties could be obtained from tensile tests according to NF EN 10002-1 standard. The mechanical properties of the polyurethane foam core and the steel skin were experimentally obtained and reported in Table 2.

Table 2 Mechanical properties of the steel skin and the polyurethane foam core

	Steel skins	Foam core
Density ρ [kg/m^3]	7800	40
Yield stress σ_0 [MPa]	440	0.41
Young's modulus E [MPa]	200000	3.31
Poisson's ratio ν	0.3	0.4
Strength R_m [MPa]	453	0.53

Quasi-Static three-point bending tests were conducted with the MTS testing machine to acquire the load–displacement curves (Fig. 3). All the specimens were obtained from sandwich panels composed by a polyurethane foam core and steel skins. Subsequent tests are performed with a displacement rate of 10 mm/min. The specimens were tested to failure.

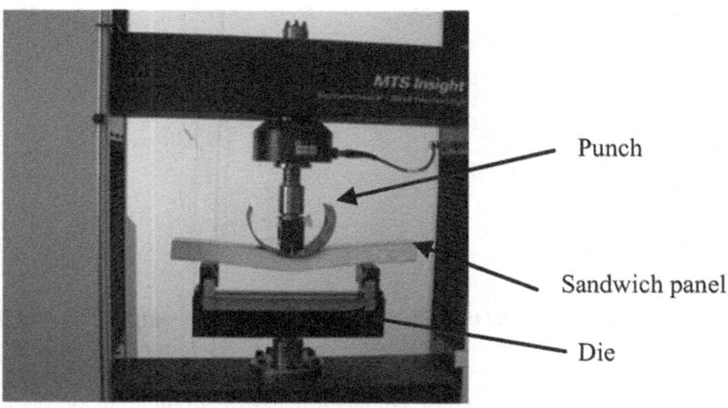

Fig. 3 Experimental set-up of the three-point bending test

3.2 Numerical Simulations

The FEM software package ABAQUS/Explicit was used to simulate the three-point bending of sandwich sheets. Figure 4 gives the two-dimensional geometric modeling with Abaqus software with a mesh size sufficiently refined to ensure precise results and adequate boundary conditions. Due to the material symmetry, only a half of the section of the sandwich panel was considered. The core was modeled as foam of an elastic–plastic material using hardening curves obtained from compression tests. The skin sheet was modeled as elastic-plastic material.

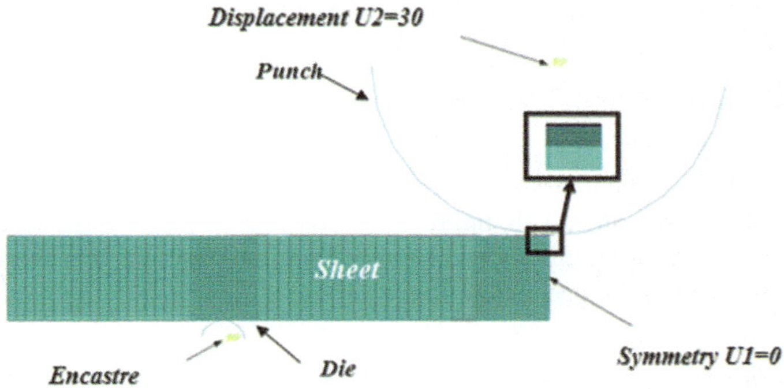

Fig. 4 FE model of three-point bending process

Damage initiation in the foam core was modeled by a shear damage criterion.

The shear criterion assumes that the equivalent plastic strain at the onset of the damage. $\bar{\varepsilon}_S^{pl}$ is a function of the shear ratio and strain rate:

$$\bar{\varepsilon}_S^{pl}\left(\theta_S, \dot{\bar{\varepsilon}}^{pl}\right) \tag{15}$$

where $\theta_S = (q + k_s p)/\tau_{max}$ is the shear stress ratio, τ_{max} is the maximum shear stress, and k_s is a material parameter. The damage onset is occurred when:

$$w_S = \int \frac{d\bar{\varepsilon}^{pl}}{\bar{\varepsilon}_S^{pl}\left(\theta_S, \dot{\bar{\varepsilon}}^{pl}\right)} = 1 \tag{16}$$

where ω_S is a state variable which increases with the equivalent plastic strain.

4 Results and Discussion

Figure 5a shows a shear failure of the core in three-point bending. Since foams have generally lower mechanical properties than skins, they will be affected by damage initiation. Figure 5b shows another mode of fracture: debonding skin/core that spreads under the skin several millimeters along the length of the panel. This fracture is due to the presence of defects in the junction between the skins and the foam.

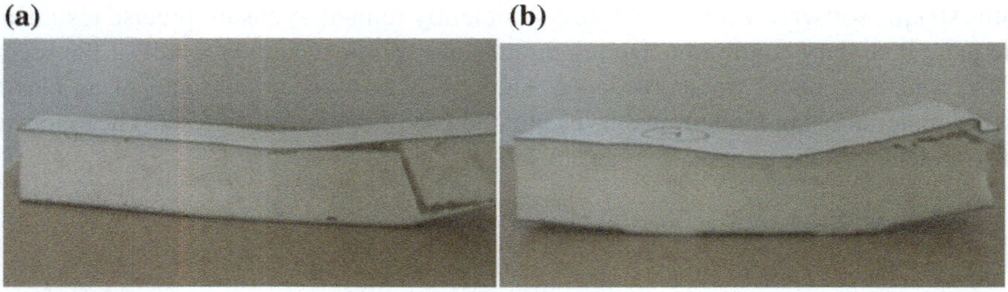

Fig. 5 Failure modes of sandwich panels, **a** Shear failure of the core; **b** debonding skin/core

Figure 6 shows the curves of the stored energies ALLKE, ALLIE and ALLKE/ALLIE versus time. This figure shows that the kinetic energy (ALLKE) does not exceed 5% of the total energy (ALLIE), which demonstrates that the influence of inertial force is within the acceptable range.

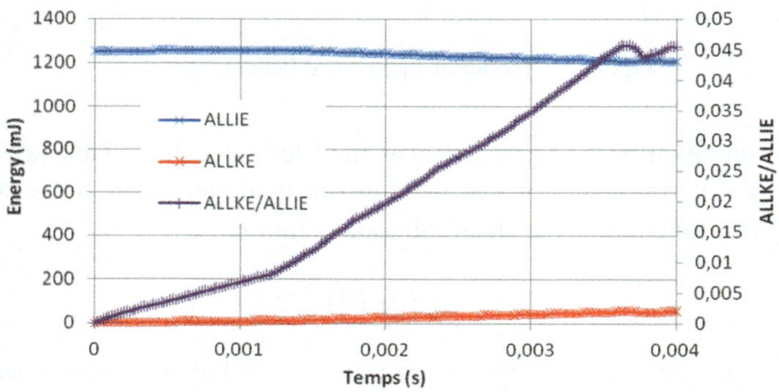

Fig. 6 ALLKE, ALLIE and ALLKE/ALLIE ratio versus time curves

Figures 7 and 8 show experimental and numerical load-displacement curves of sandwich panel in three-point bending test. The obtained curves can be divided into three regions. In the first region, a linear trend is observed with a small deformation. The second region corresponds to a nonlinear behavior in which the maximum load is reached, significant drop of the peak load is observed for all sandwich structures. This sudden drop is due to the foam cracking. In the third region, a plateau was observed until failure with small evolution and the specimen continued to sustain the load but never exceeded the previous peak load. It can be seen that a satisfactory agreement was found between the experimental result and the FEA result. The peak load in numerical results was slightly higher than the experimental ones. This is due to initial defects in sandwich composite which is not considered in FEM analysis. Figure 7 plots the load-displacement curves by considering the effect of the distance between supports. From this figure it can be seen that the failure parameters (loads and displacements) increase with the decrease of the distance between supports as can be expected. Figure 8 shows that the failure parameters increase with the increase of the foam thickness. Values of failure loads, failure displacements and stiffness are presented in Tables 3 and 4.

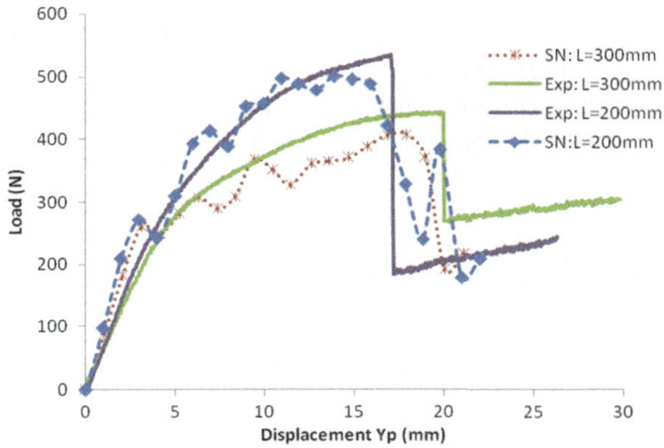

Fig. 7 Bending curves for different lengths between supports

Table 3 Static characteristics of sandwich panel for different length between supports

	L = 200 mm	L = 300 mm
Failure load (N)	514	411
Failure displacement (mm)	17.7	19.3
Stiffness (N/mm)	82	61

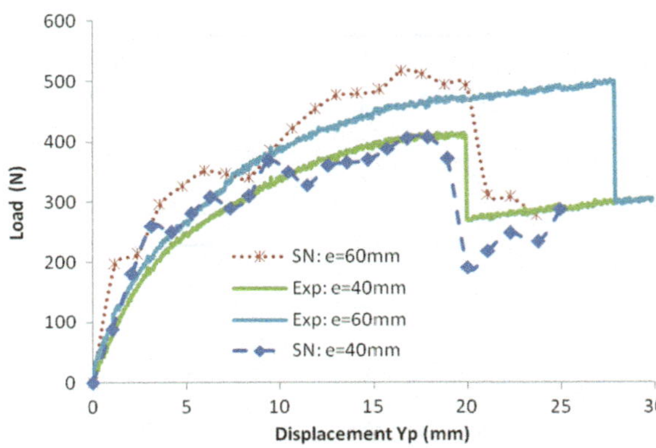

Fig. 8 Bending curves for different foam thickness

Total equivalent plastic strain occurred in the sandwich plates are gathered from the finite elements analyses and illustrated in Fig. 9. As can be seen, the maximum equivalent plastic strain is located in the foam under the punch and in the lower part of the foam inclined at about 45° to the punch/panel contact. Notice that failure was initiated at these locations.

Table 4 Static characteristics of sandwich panel for different foam thickness

	e = 40 mm	e = 60 mm
Failure load (N)	411	476
Failure displacement (mm)	19.3	22.2
Stiffness (N/mm)	61	73

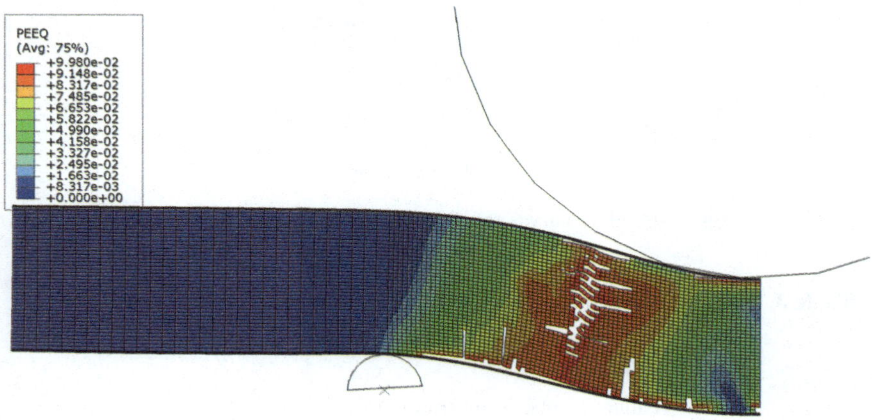

Fig. 9 Equivalent plastic strain distribution

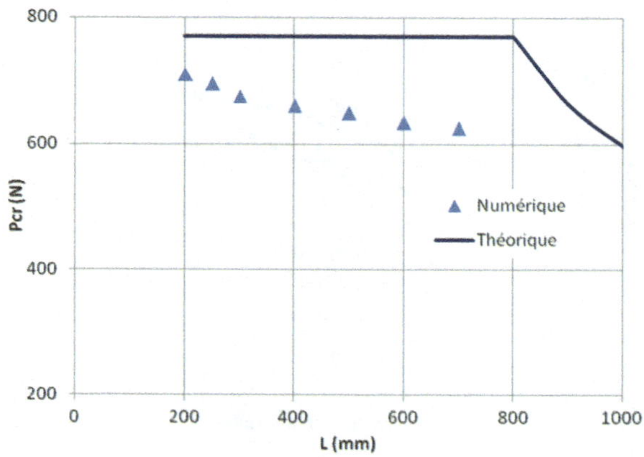

Fig. 10 Numerical results of the failure loads of the sandwich panel steel/polyurethane/steel

Figure 10 gives a comparison between the critical loads obtained from the theoretical studies, concerning the two modes of damage skin wrinkling and core shearing, and the loads obtained numerically. From this figure we observe that the maximum error is 21% which can be explained by the simplifying assumptions used in obtaining the theoretical failure criteria.

5 Conclusion

The purpose of this paper is to investigate the damage behavior of sandwich panel steel/polyurethane/steel by considering the effect of the variation of several geometrical parameters during quasi-static tests in three-point bending. Experimental tests leading to the failure of the core of the sandwich material were carried out. The FE model was validated by comparing the load–displacement curves of the sandwich panels between experimental and FE analysis, the comparisons show a satisfactory agreement between the experimental and numerical results. Also, analytical Gibson's modified model is considered to obtain the critical loads leading to the failure of the sandwich panels. This allows constructing a mode map for failure modes of sandwich panels in three points bending process.

References

1. Kim KJ, Rhee MH, Bik C (2009) Development of application technique of aluminium sandwich sheet for automotive hood. Int J Precis Eng Manuf 10:71–75
2. Link TM (2001) Formability and performance of steel-plastic-steel laminated sheet materials. In: SAE technical paper, 01-0079
3. Zenkert D, Burman M (2011) Failure mode shifts during constant amplitude fatigue loading of GFRP/foam core sandwich
4. Ferreira JAM, Costa JDM (1998) Static behaviour of PVC foam composite sandwich panel. J Cell Polym 17(3):177–192
5. Gimenez I, Farooq MK, Mahi AEl, Kondratas A, Assarar M (2004) Experimental analysis of mechanical behaviour and damage development mechanisms of PVC foams in static tests. Mater Sci 10(1)
6. Mamalis A, Spentzas K, Manolakos D, Ioannidis M, Papapostolou D (2008) Experimental investigation of the collapse modes and the main crushing characteristics of composite sandwich panels subjected to flexural loading. Int J Crashworthiness 13(4):349–362
7. Andrews EW, Moussa NA (2009) Failure mode maps for composite sandwich panels subjected to air blast loading. Int J Impact Eng 36:418–425
8. Daniel IM, Gdoutos EE, Wang KA, Abot JL (2002) Failure modes of composite sandwich beams. Int J Damage Mech 11:309–334
9. Gdoutos EE, Daniel IM, Wang KA (2003) Compression facing wrinkling of composite sandwich structures. Mech Mater 35:511–522
10. Kim HY, Hwang W (2002) Effect of debonding on natural frequencies and frequency responses functions of honey sandwich beams. Compos Struct 55:51–62 92, 2703–2711
11. Idriss M (2013) Analyse expérimentale et par éléments finis du comportement statique et vibratoire des matériaux composites sandwich sains et endommagés. Thèse de doctorat

12. Gibson LJ, Ashby MF (1997) Cellular solids: structure and properties. Cambridge University Press
13. Allen G (1969) Analysis and design of structural sandwich panels. Pergamon Press, Oxford, p 1969
14. Yu JL, Wang EH, Li JR, Zheng ZJ (2008) Static and low-velocity impact behavior of sandwich beams with closed-cell aluminum-foam core in three-point bending. Int J Impact Eng 35:885–894

Analysis on the Dependence of the Fracture Locus on the Pressure and the Lode Angle

Nouira Meriem[1,2], Oliveira Marta[3], Khalfallah Ali[1,4](✉), Alves José[5], and Menezes Luís[3]

[1] Laboratoire de Génie Mécanique, Ecole Nationale d'Ingénieurs de Monastir, University of Monastir, Av. Ibn El Jazzar, 5019 Monastir, Tunisia
nouirameriem93@gmail.com, ali.khalfallah@gmail.com
[2] Ecole Nationale d'Ingénieurs de Sousse, University of Sousse, 4003 Sousse, Tunisia
[3] Department of Mechanical Engineering, CEMMPRE, University of Coimbra, Polo II Pinhal de Marrocos, 3030-788 Coimbra, Portugal
{marta.oliveira, luis.menezes}@dem.uc.pt
[4] Institut Supérieur des Sciences Appliquées et de Technologie de Sousse, University of Sousse, 4003 Sousse, Tunisia
[5] Department of Mechanical Engineering, CMEMS, University of Minho, Campus de Azurém, 4800-058 Guimarães, Portugal
jlalves@dem.uminho.pt

Abstract. Classical fracture models assume that the stress triaxiality is the key parameter controlling the magnitude of the fracture strain. However, recent works shown the influence of other parameters that characterize the stress state on the prediction of fracture strains. In this work, two uncoupled fracture models, Mae and Wierzbicki [8] and Xue and Wierzbicki [9], were analysed using finite element models. These models define a ductile fracture *locus* formulated in the 3D space of the stress triaxiality, Lode angle parameter and the equivalent fracture strain. The material selected was a cast A356 aluminium alloy for which the model parameters were previously defined. Two groups of tests are analysed in order to provide additional information on the material ductility. The first corresponds to plane strain tests carried out on flat plates with different grooves. The second corresponds to uniaxial tension tests applied on smooth and notched round bars, which were designed with different notch radii. These specimens allow covering a wide range of stress triaxiality. The present work extracts the evolution of the equivalent plastic strain at fracture, the stress triaxiality and the Lode angle parameter in order to evaluate the possibility of using either smooth and notched bars tests or smooth and notched bars tests and grooved plates to evaluate the 3D *locus* for high values of stress triaxiality. In this context, a new function is proposed to describe the equivalent plastic strain at fracture based on the stress triaxiality and the Lode angle parameter.

Keywords: Uncoupled fracture models · 3D fracture *locus* · Equivalent fracture strain · Stress triaxiality · Lode angle

N. Aifaoui et al. (Eds.): CMSM 2019, LNME, pp. 545–557, 2020.
https://doi.org/10.1007/978-3-030-27146-6_59

Nomenclature

E	Young's modulus
k	Exponent in curvilinear Lode angle dependence function
m	Damage exponent
n	Strain hardening exponent
p	Mean pressure
p_{lim}	Limiting pressure below which no damage occurs
q	Exponent in pressure dependence function
$\eta = \sigma_m/\sigma_{eq}$	Stress Triaxiality
$\bar{\varepsilon}_f$	Equivalent fracture strain
$\bar{\varepsilon}_{f0}$	Reference equivalent fracture strain
$(\bar{\varepsilon}_{\mathbf{f,t}})$	Effective fracture strain under uniaxial tension
$(\bar{\varepsilon}_{\mathbf{f,s}})$	Effective fracture strain under pure shear
γ	Ratio of fracture strains
ξ, θ_l	Third invariant of the deviatoric stress tensor, Lode angle
μ_p	Pressure dependence function
μ_θ	Lode angle dependence function
ϑ	Poisson's ratio
$\sigma_{1,2,3}$	Principal components of the Cauchy stress tensor
σ	Cauchy Stress tensor
σ_m	Mean stress
$\sigma_{eq}, \bar{\sigma}$	Equivalent stress
$J_3 = \mathrm{s}_1\mathrm{s}_2\mathrm{s}_3$	Is the third stress invariant
D	Damage accumulation
y_0, k and n	Swift law hardening parameters.

1 Introduction

Although it has been extensively studied, the prediction of ductile fracture of metallic materials still presents considerable challenges when resorting to numerical tools to analyse the mechanical behaviour of structural components under various loading conditions. The models proposed in literature for ductile fracture prediction, can be divided in two groups: micromechanical models and continuum damage models. The first relies on the mathematical description of the mechanisms of void nucleation, growth and coalescence. The second group is based on the definition of a damage variable, which acts as a softening mechanism. This type of models can also include phenomenological laws, which can be implemented using a coupled or an uncoupled approach. The Johnson and Cook's uncoupled model [1] integrated the effect of stress triaxiality, strain rate, and temperature [2]. Bao and co-worker [3, 4] designed and performed tests on several type of specimens to calibrate the fracture *locus* in a wide range of stress triaxiality. They showed that the fracture strain does not have to be a monotonically decreasing function of the stress triaxiality [5]. Xue and co-workers [6] introduced the Lode angle parameter in the definition of the 3D fracture *locus*. This model is similar to the one proposed by Wilkins [7] in the sense that the fracture *locus* is constructed in the 3D space, which defines an equivalent strain to fracture, based on the stress triaxiality and the third invariant of the deviatoric stress tensor.

In this work, the grooved plane strain plates specimens with different notches and tensile tests on smooth and notched round bars will be examined using two uncoupled phenomological fracture models: Mae and Wierzbicki [8] and Xue and Wierzbicki [9].

2 Ductile Fracture Models

A number of ductile fracture models have been proposed to predict failure. In this work, two phenomenological uncoupled models are presented and discussed. They take into account different material parameters to assess the effective fracture strain, for various loading conditions. The mean pressure, the equivalent stress and the stress triaxiality are expressed by the following equations, since the material is assumed to have isotropic plastic behaviour, described by the von Mises yield criterion:

$$p = -\sigma_m = -\frac{1}{3}tr(\sigma) = -\frac{1}{3}(\sigma_1 + \sigma_2 + \sigma_3) \tag{1}$$

$$\bar{\sigma} = \sigma_{eq} = \sqrt{\frac{1}{2}}\sqrt{\left[(\sigma_1 - \sigma_2)^2 + (\sigma_2 - \sigma_3)^2 + (\sigma_3 - \sigma_1)^2\right]} \tag{2}$$

$$\eta = -\frac{p}{\sigma_{eq}} = \frac{\sigma_m}{\sigma_{eq}} \tag{3}$$

2.1 Mae and Wierzbicki [8]

Mae and Wierzbicki [8] indicated that the ductile fracture *locus* consists of three branches in the whole range of the stress triaxiality, which define the equivalent strain to fracture ($\overline{\varepsilon_f}$) as follows:

$$\left\{\begin{array}{ll} \overline{\varepsilon_f} = \frac{D_1}{1+3\eta}, & -\frac{1}{3} \leq \eta \leq 0 \\ \overline{\varepsilon_f} = \overline{\varepsilon_{f,t}} + (\overline{\varepsilon_{f,t}} - \overline{\varepsilon_{f,s}})(3\eta - 1), & 0 \leq \eta \leq \frac{1}{3} \\ \overline{\varepsilon_f} = D_2 e^{D_3 \eta} + D_4, & \eta \geq \frac{1}{3} \end{array}\right\} \tag{4}$$

Table 1 Material parameters for ductile fracture characterization of the cast Aluminum Alloy for the model of Mae and Wierzbicki [8]

D1	D2	D3	D4
0.2733	0.1417	−1.545	0.0

where D_1, D_2, D_3, D_4 are the model parameters that must be evaluated for each material. Table 1 presents the parameters determined for a cast aluminium alloy A356, as shown in Mae and Wierzbicki [8].

2.2 Xue and Wierzbicki [9]

Xue and Wierzbicki [9] assume that two other variables play an important role in the evaluation of the equivalent strain to fracture: the mean pressure and the Lode angle (θ_l). The ductile fracture envelope assumes the following form:

$$\overline{\varepsilon_f} = \varepsilon_{f0}.u_p(p).u_\theta(\theta_l) \tag{5}$$

where the function $u_p(p)$ adopts a logarithmic form:

$$u_p(p) = \begin{Bmatrix} 1 - q\log(1 - \frac{p}{p_{\lim}}), & p \geq p_{\lim}(1 - \exp(\frac{1}{q}) \\ 0 & p \leq p_{\lim}(1 - \exp(\frac{1}{q}) \end{Bmatrix} \tag{6}$$

and **:

$$u_\theta(\theta_l) = \gamma + (1 - \gamma)(\frac{6\,|\theta_l|}{\pi})^k \tag{7}$$

where $(\theta_l \in [-\frac{\pi}{6}, \frac{\pi}{6}])$. The Lode angle is one of the several parameters that are commonly used to demote the azimuthal angle on an octahedral plane in the principal stress space. It can be defined by:

$$\theta_l = \tan^{-1}(\frac{1}{\sqrt{3}}\frac{2\sigma_2 - \sigma_1 - \sigma_3}{\sigma_1 - \sigma_3}) \quad \text{or by} \quad \theta_l = -\frac{1}{3}\sin^{-1}(\frac{27J_3}{2\overline{\sigma}^3}) \tag{8}$$

$$\xi = \frac{27J_3}{2\overline{\sigma}^3} \tag{9}$$

The parameters $\varepsilon_{f0}, \gamma, p_{\lim}, q, k, m$ need to be identified for each material. Table 2 presents the set of the parameters determined for an aluminium alloy A356.

Table 2 Material constants for ductile fracture characterization of the cast Aluminum Alloy for the model of Xue and Wierzbicki [9]

ε_{f0}	$p_{\lim}$	q	γ	k	m
1.20	800.0 MPa	1.5	0.6	1.0	1.0

2.3 Damage Evolution

The models previously described define the equivalent plastic strain at fracture in function of the stress state variables used to characterize the stress state. These models allow a direct prediction of the equivalent plastic strain at fracture, for a monotonic stress state. However, when there are changes in the stress path it is necessary to evaluate the impact of those changes in the predicted equivalent plastic strain at fracture. This is commonly done using a damage parameter, D. Ductile failure occurs when the damage parameter reaches the critical damage value of 1.0. The evolution of damage is defined by [9]:

$$D = \int_{0}^{\varepsilon_c} f(\bar{\varepsilon}_p, \bar{\varepsilon}_f) d\bar{\varepsilon}_p = 1 \tag{10}$$

where ε_c is the equivalent strain at fracture.

For the case of Mae and Wierzbicki [8] model, the damage evolution can be represented by Eq. (11).

$$D = \int_{0}^{\bar{\varepsilon}^p} \frac{1}{\bar{\varepsilon}_f} d\bar{\varepsilon}_p \tag{11}$$

For Xue and Wierzbicki [9] model, the damage accumulation is expressed in terms of the ratio between the current plastic strain and the equivalent strain to fracture. The damage plasticity model can be represented by the following expressions [9],

$$\dot{D} = m\left(\frac{\bar{\varepsilon}_p}{\bar{\varepsilon}_f}\right)^{m-1} \frac{\bar{\varepsilon}_p}{\bar{\varepsilon}_f} \tag{12}$$

In this work, the exponent value of damage m is equal to one, i.e. the damage rule corresponds to a linear damage function.

3 Numerical Simulations

Based on the literature review, geometries of standardized and non-standardized test specimens were identified, as shown in Fig. 1. The numerical simulations were performed with the DD3IMPsolver [10, 11]. The commercial software GiD was used as pre and post-processor. Thus, after building the CAD models of the specimens (see Fig. 1), a three-dimensional finite element mesh was built in GID, using eight-node solid finite elements. A selective reduced integration (SRI) technique is employed, with eight and a single GP for the deviatoric and hydrostatic parts of the velocity field gradient, respectively.

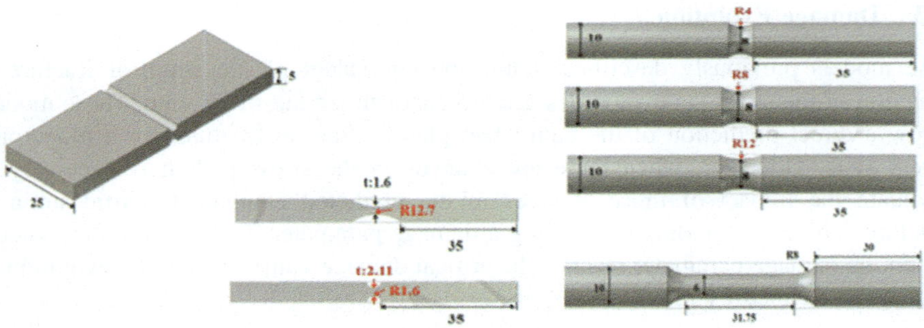

Fig. 1 A sketch of flat-grooved plane strain specimens (left), and smooth and notched round bars specimens [12]

Table 3 Material properties used in the numerical simulation [13]

Material parameters				
Swift law + Linear kinematic hardening			Isotropic elastic behaviour	
y_0	k	n	ϑ	E
200.167	556.06	0.2010	0.33	86 GPa

The numerical simulations were performed considering the two fracture uncoupled models using the cast aluminium alloy. The Swift law (isotropic hardening) is adopted:

$$Y = k(\varepsilon_0 + \bar{\varepsilon})^n \tag{13}$$

Y represents the yield stress and its evolution during deformation. The initial yield stress y_0 can be written as a function of k, ε_0 and n as follows: $y_0 = k\varepsilon_0^n$. Table 3 present the material parameters.

3.1 Effect of Models Parameters

In this section, the prediction of ductile fracture is analysed for the two groups of classical tests carried out:

- **Flat-grooved Plate Plain Strain Specimens**

Eight different ratio of t/R were considered. All specimens had a constant thickness t = 1.2 and 2.11 mm, respectively at the groove, but the radii of the grooves are equal to 1.6, 4, 8 and 12.7 mm, respectively. The loading condition is simple tension. The notation adopted for these tests is "Tx Rymm", were x is the thickness of the plate and y is the groove radius.

- **Smooth and Notched Round Bars Specimens**

Four values of a/R were assigned to specimens a/R = 0 (smooth round bars), a/R = 1/2, 1 and 2/3 (notched round bars). The notation adopted for theses tests is "R = ymm", were x is the radius of the notch.

Table 4 summarizes the results obtained with both models. Note that the model proposed by Mae and Wierzbicki [8, 13] only takes the stress triaxiality into account. On the other hand, the model proposed by Xue and Wierzbicki [9] assumes the influence of the mean pressure and the Lode angle. The value of $\bar{\varepsilon}_f$ is the one obtained by the model, assuming a constant value for the variables that characterize the stress state, equal to the one predicted at the onset of ductile fracture. On the other hand, $\bar{\varepsilon}_f^p$ corresponds to the numerically predicted equivalent plastic strain at the location where fracture is predicted by the model. Therefore, the results shown in Table 4 highlight the importance of the damage accumulation variable, since it is clear that there are higher differences between both values when adopting the Xue and Wierzbicki [9] model.

Table 4 A summary of numerical results of fracture strains, stress triaxialities and Lode angle parameters for a cast A356 aluminium alloy

	Tests	$\bar{\varepsilon}_f^p$	$\bar{\varepsilon}_f$	η	θ_l
Mae and Wierzbicki [8]	T1.6 R1.6 mm	0.033	0.032	0.967	−0.05
	T1.6 R4 mm	0.047	0.047	0.713	−0.02
	T1.6 R8 mm	0.054	0.053	0.642	−0.02
	T1.6 R12.7 mm	0.058	0.057	0.597	−0.09
	T2.11 R1.6 mm	0.032	0.031	0.994	−0.13
	T2.11 R4 mm	0.045	0.045	0.747	−0.02
	T2.11 R8 mm	0.053	0.052	0.651	−0.04
	T2.11 R12.7 mm	0.056	0.055	0.614	−0.06
	Smooth round bar	0.085	0.084	0.339	0.999
	R = 4 mm	0.064	0.064	0.519	0.93
	R = 8 mm	0.068	0.067	0.483	0.999
	R = 12 mm	0.073	0.071	0.453	0.999
Xue and Wierzbicki [9]	T1.6 R1.6 mm	0.271	0.182	1.21	−0.01
	T1.6 R4 mm	0.365	0.262	0.996	0.09
	T1.6 R8 mm	0.370	0.281	0.919	0.04
	T1.6 R12.7 mm	0.382	0.288	0.878	0.02
	T2.11 R1.6 mm	0.292	0.223	1.080	0.005
	T2.11 R4 mm	0.349	0.267	0.934	−0.01
	T2.11 R8 mm	0.378	0.298	0.857	0.02
	T2.11 R12.7 mm	0.388	0.296	0.860	0.02
	Smooth round bar	0.673	0.337	0.942	0.996
	R = 4 mm	0.541	0.385	0.929	0.999
	R = 8 mm	0.605	0.478	0.782	0.999
	R = 12 mm	0.647	0.483	0.763	0.999

Although not shown here, this is also related with the fact that the stress triaxiality follows a more stable evolution during the deformation process than the mean pressure and the Lode angle. Also, the mean pressure and the Lode angle evolutions are more sensitive to the mesh descritization adopted.

The analysis of Table 4 also shows that, when considering a tensile loading, all specimens considered present a positive value for the stress triaxiality for the point where ductile fracture is predicted by both models. When considering the Mae and Wierzbicki [8, 13] the flat-grooved plates present a negative value for the Lode angle, while for the Xue and Wierzbicki [9] model, ductile fracture is predicted either for positive or negative values of this parameter. Nevertheless, the Lode angle parameter presents a value close to zero for the flat-grooved plates and a value close to 1.0 for the round bars. Moreover, the Xue and Wierzbicki [9] model is predicting the occurrence of ductile fracture for higher values of tensile displacement, i.e. equivalent plastic strain. Note that this results from the fact that, the parameters shown in Table 2 are recommended by the authors to be used in a coupled implementation of the model, which is not the one adopted in the current work.

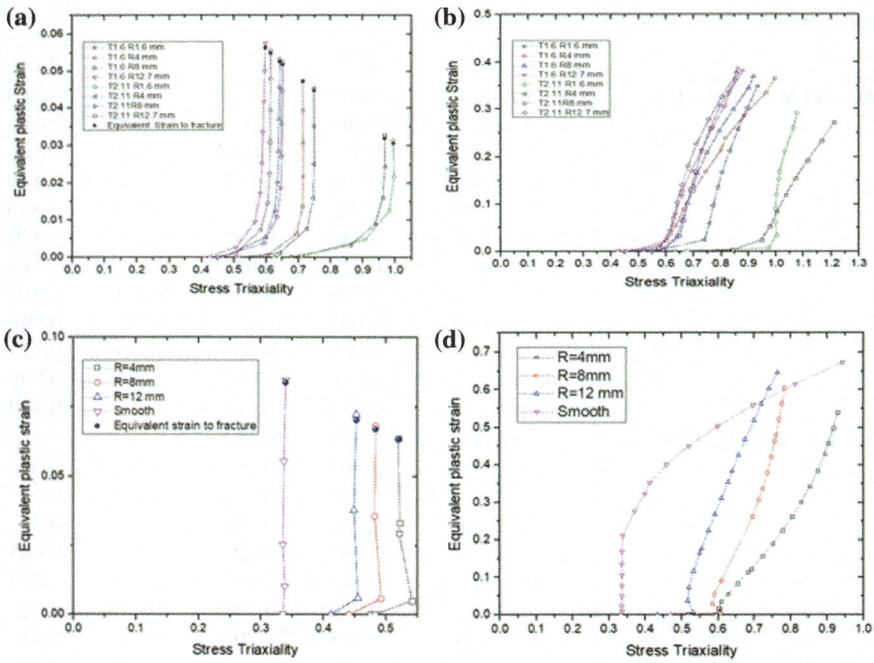

Fig. 2 Equivalent plastic strain versus stress triaxiality in the center of flat-grooved plates (**a**), (**b**) and smooth and notched round bars (**c**), (**d**) from numerical simulations using the fracture models: **a**, **c** Mae and Wierzbicki [8] and **b**, **d** Xue and Wierzbicki [9]

Figure 2 shows the evolution of the equivalent plastic strain with the stress triaxiality, obtained for the central element, with the two fracture uncoupled models. Note that the different scale used for both models results from the fact that the Xue and Wierzbicki [9] model always predicts the onset of ductile fracture for much higher values of equivalent plastic strain. The use of uncoupled models always assures that the evolution of the variables that characterize the stress state is equal. The fact that the stress triaxiality suffers an abrupt increase for the central element denotes the onset of necking.

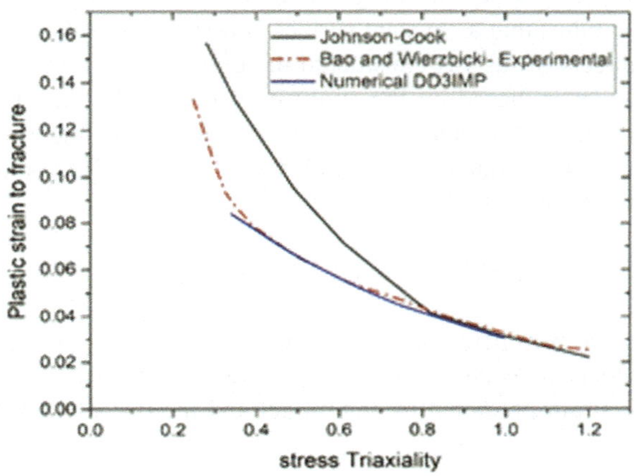

Fig. 3 Fracture loci of the cast Aluminum Alloy [8]

Both Table 4 and Fig. 2 show the effect of stress triaxiality on the predicted fracture strain, under a fixed Lode angle parameter. As the stress triaxiality increases, the equivalent fracture strain and, consequently, the equivalent plastic strain at fracture decreases, for both models. This is an important feature of ductile fracture models as noted by many authors including Johnson-Cook [1]. In fact, theoretical analysis [14, 15] and numerous experimental studies [1, 3, 8, 13] have proved that the fracture strain increases when the stress triaxiality increases. Moreover, since the stress triaxiality is quite constant in the beginning of the tests (see Fig. 2), the damage accumulation follows a linear trend, meaning that the numerically predicted equivalent plastic strain at fracture is almost equal to the one predicted by the the Mae and Wierzbicki model [8]. In fact, as shown in Fig. 3 the results show a good agreement with the experiments performed by Bae and Wierzbicki [8, 16] for the high range of stress triaxiality.

3.2 3D Fracture Locus

Based on the analysis of the results, the two sets of test enable the definition of two boundary limits, for positive values of stress triaxiality:

- $\xi = 0$ corresponding to plane strain, $\bar{\varepsilon}_f^{(0)}$
- $\xi = 1$ corresponding to axial symmetry in deviatoric tension, $\bar{\varepsilon}_f^{(+)}$

This enables the construction of a 3D fracture *locus* that defines the equivalent plastic strain to fracture, based on the stress triaxiality and Lode angle parameter, such as $(\eta, \xi, \bar{\varepsilon}_p^f)$.

If the effect of the Lode angle parameter on the fracture *locus* is neglected, only one set of tests should be considered. In this case, the expression adopted for the range of high values of stress triaxiality is the one corresponding to the third branch of Eq. (4) of Mae and Wierzbicki [8] model, meaning that three parameters need to be identified, D_2, D_3 and D_4. This emplies that a constant surface is assumed for all values of Lode angle parameter. If both sets of tests is considered the effect of the Lode angle parameter is also included. However, based on the analysis of the results, the mean pressure is a variable that with the deformation and is quite sensitive to the mesh discretization adopted. Therefore, a new function is adopted to describe the 3D fracture *locus* for high values of stress triaxiality, as follows [17]:

$$\bar{\varepsilon}_f = D_1 e^{D_2 \eta} - \left(D_1 e^{D_2 \eta} - D_3 e^{D_4 \eta}\right)\left(1 - |\xi|^{\frac{1}{n}}\right)^{n} \tag{14}$$

In these case five parameters need to be identified, D_1, D_2, D_3 and D_4 and n.

An objective function is chosen in order to minimize the average error between the equivalent plastic strain predicted by the model and the one of each test, as follows:

$$\underset{(D_1,D_2,D_3,D_4)}{Min}(Error) = \underset{(D_1,D_2,D_3,D_4)}{Min} \sum_{i=1}^{N} \left|\bar{\varepsilon}_f, i_{Num} - \bar{\varepsilon}_{f,i\,Calc}\right| \tag{15}$$

where test i listed in Table 4 and N is the total number of tests, $\bar{\varepsilon}_{f,i\,Num}, \bar{\varepsilon}_{f,i\,Calc}$ refer to the numerical determined fracture strain and the calculated fracture strain, respectivelty.

Figure 4 shows the 3D fracture *locus* calibrated using the numerical results obtained with Mae and Wierzbicki [8] model and assuming no influence of the Lode angle parameter. The values obtained for the model parameters are also shown in the Fig. 4. The comparison of the values obtained with the ones used in the numerical simulations (see Table 2) confirms that the optimization procedure recovers the proposed values. This validates the proposed procedure based on the results shown in Fig. 3. Figure 5 shows the 3D fracture *locus* presented in Eq. (14), calibrated using the numerical results obtained with Xue and Wierzbicki [9] model. The set of parameters obtained was the following: $D_1 = 1.9286, D_2 = 1.795, D_3 = 0.9664, D_4 = 1.366$, $n = 0.2$. Moreover, Fig. 5 shows the results for the two boundary limits.

The models previously described define the equivalent plastic strain at fracture in function of the stress state variables used to characterize the stress state. Figures 4 and 5 show a comparison of the 3D surfaces obtained with each model. These surfaces allow a direct prediction of the equivalent plastic strain at fracture, for a monotonic stress state. The calibrated 3D fracture locus can give us a visualized overall view of the models. The shape of the 3D fracture *locus* predicted by Eqs. (4) and (14) are clearly different. In the case of Eq. (4), for $\gamma = 1$, (γ is the ratio of Lode angle function for

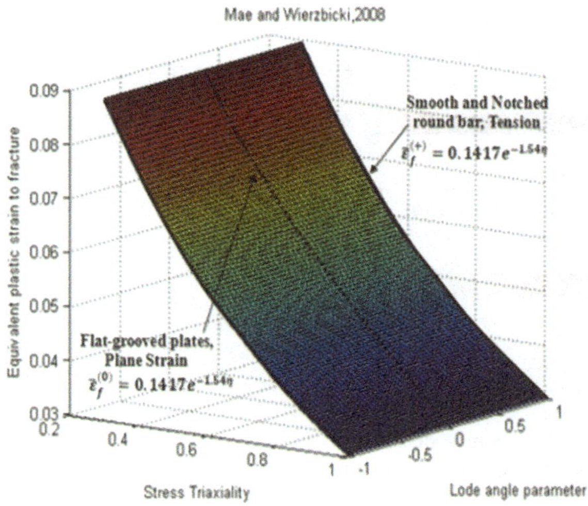

Fig. 4 The 3D Fracture locus of the cast Aluminum Alloy, Mae and Wierzbicki [8]

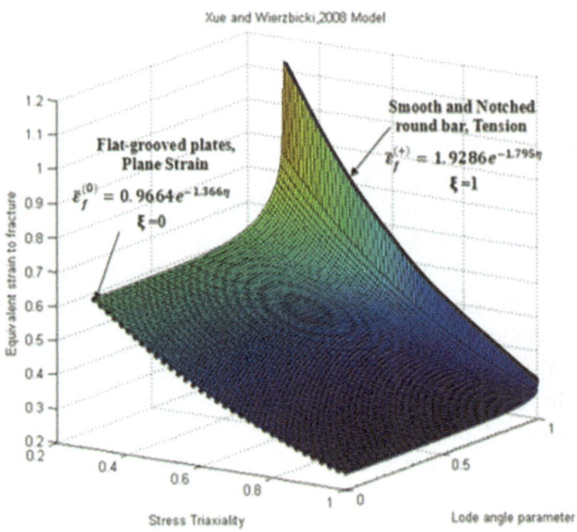

Fig. 5 The 3D Fracture locus of the cast Aluminum Alloy, Xue and Wierzbicki [9]

$\theta_l = 0$ and $\theta_l = -\pi/6$) the predicted fracture locus flattens out and for $0 \prec \gamma \prec 1$, the shape is a concave, similar to that shown in the Xue and Wierzbicki [8, 9, 14], model. Thus, the shape of the 3D fracture locus is sensitive to the Lode angle parameter.

Both models are based on an uncoupled phenomenological approach between fracture and the variables that characterize the stress state; the stress triaxiality, the Lode angle parameter. Therefore, the evolution of the stress and strain distributions is only dictated by the plasticity model adopted and is the same whatever the fracture

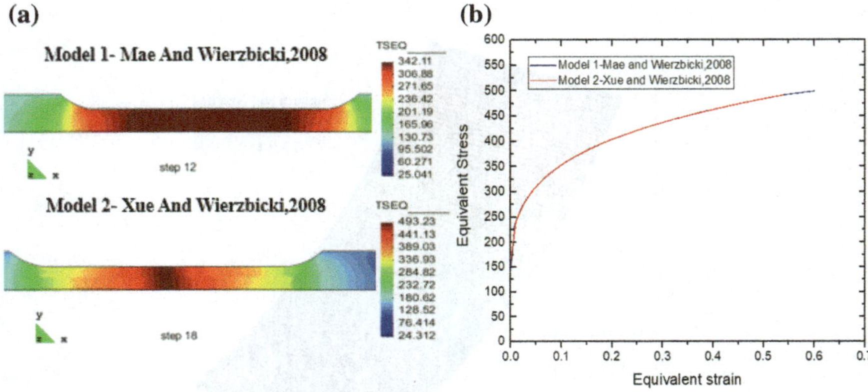

Fig. 6 **a** The equivalent Stress distribution in specimen, **b** the equivalent Stress versus the equivalent Strain, for the tensile test on smooth round bar

model adopted. This means that fracture strain is predicted for different displacement has a direct result of the fracture locus adopted to analyze the results. Figure 6 indicated that the model adopted, for the tensile test on smooth round bar, has no influence on the stress-strain distribution.

4 Conclusion

In this work two uncoupled models: Mae and Wierzbicki [8] and Xue and Wierzbicki [9], are sued to evaluate the occurrence of ductile fracture, considering two sets pf experimental test. These sets are characterized by presenting a wide range of positive values of stress triaxiality, for two values of Lode angle parameter. The results are used to test a procedure that enables the definition of the fracture *locus*, $(\eta, \xi, \bar{\varepsilon}_p^f)$, following the same approach used by several authors p to construct the fracture *loci* based on experimental results. The results show that the use of tensile tests on round and notched round bars and grooved plates specimens enables the identification of the fracture *locus*, for positive values of stress triaxiality and Lode angle parameter.

References

1. Johnson GR, Cook WH (1985) Fracture characteristics of three metals subjected to various strains, strain rates, temperatures and pressures. Eng Fract Mech 21(1):31–48
2. Teng X, Wierzbicki T (2006) Evaluation of six fracture models in high velocity perforation. Eng Fract Mech 73(12):1653–1678
3. Bao Y (2003) Prediction of ductile crack formation in uncracked bodies. PhD thesis, Massachusetts Institute of Technology, Cambridge
4. Bao Y, Wierzbicki T (2004) On fracture locus in the equivalent strain and stress triaxiality space. Int J Mech Sci 46(1):81–98

5. Bao Y, Wierzbicki T (2005) On the cut-off value of negative triaxiality for fracture. Eng Fract Mech 72(7):1049–1069
6. Bai Y, Wierzbicki T (2008) A new model of metal plasticity and fracture with pressure and lode dependence. Int J Plast 24(6):1071–1096
7. Wilkins ML, Streit RD, Reaugh JE (1980) Cumulative-strain-damage model of ductile fracture: simulation and prediction of engineering fracture tests. Lawrence Livermore Laboratory, Technical Report No. UCRL-53058
8. Teng X, Mae H, Wierzbicki T (2008) Calibration of ductile fracture properties of a cast aluminum alloy. Mater Sci Eng, A 459:156–166
9. Xue L, Wierzbicki T (2008) Damage accumulation and fracture initiation in uncracked ductile solids subject to triaxial loading. Int J Solids Struct 44(16):5163–5181
10. Menezes LF, Teodosiu C (2000) J Mater Process Technol 97:100–106
11. Oliveira MC, Alves JL, Menezes LF (2008) Arch Comput Methods Eng 15:113–162
12. Jinwoo L, Se-Jong K, Daeyong K (2017) Metal plasticity and ductile fracture modeling for cast aluminum alloy parts. J Mater Proc Technol, https://doi.org/10.1016/j.jmatprotec.2017.12.040
13. Mae H, Teng X, Comparison of ductile fracture properties of aluminum castings: Sand mold vs. metal mold. Int J Solids Struct 45:1430–1444
14. Rice JR, Tracey DM (1969) On the ductile enlargement of voids in triaxial stress fields. J Mech Phys Solids 17:201–217
15. McClintock FA (1968) A criterion of ductile fracture by the growth of holes. ASME J Appl Mech 35:363–371
16. Teng X,Wierzbicki T, Bai Y (2008) On the application of stress triaxiality formula for plane strain fracture testing. J Eng Mater Technol https://doi.org/10.1115/1.3078390
17. Antonin P, Jan R, Miroslav S (2013) Identification of ductile damage parameters. In: Simulia community conference

Finite Element Analysis of Single Point Incremental Forming Process of Metallic Composite Sheet: Application to Titanium-Steel Bimetal Sheet Forming

Wifak Ben Abdelkader[1,2(✉)], Henia Arfa[1], and Riadh Bahloul[1]

[1] Laboratory of Mechanical Engineering (LGM), National Engineering School of Monastir (ENIM), University of Monastir, Avenue Ibn Eljazzar, 5019 Monastir, Tunisia
benabdelkaderwifak@gmail.com,
henia_arfa@yahoo.fr, bahloul_riadh@yahoo.fr
[2] National Engineering School of Sousse (ENISo), University of Sousse, BP 264 Sousse Erriadh, 4023 Sousse, Tunisia

Abstract. Incremental sheet forming (ISF) process is a new manufacturing process where no dedicate dies are needed, neither are specific tools. In fact, it consists on deforming a sheet plastically in progressive way. This plastic deformation is caused by applying a vertical force generated by hemispherical punch piloted with a CNC machine into a thin sheet. This process has been applied to metals and polymers. However, few researches are made in the field of composite materials. In fact, this type of material shows higher properties than each material if considered independently. Recently, some research works are published, in particular those related to bimetals sheets. These findings open a new area of scientific researches. In other hand, the application of single point incremental forming (SPIF) on bilayers materials gain a lot of interest. In this study, we present a finite element analysis (FEA) of SPIF applied to Titanium-Steel bimetal sheet (Ti/St). The present work aims to predict the effect of some parameters such as layer arrangement and step size on the forming forces evolution and the thickness distribution of a truncated pyramid. Precisely, layer arrangement constitutes one of the major factors influencing the material formability and the load applied by the punch. Results show that higher forming forces are obtained when the upper layer is steel.

Keywords: Incremental sheet forming process (ISF) · Finite element analysis · Layer arrangement · Forming forces · Thickness distribution

1 Introduction

A metallic sheet could be formed by a variety of manufacturing process starting from hammering to the use of complexes and high technologies. The most used methods in industries to produce sheet metal component are stamping and deep drawing. Unfortunately, they are not suitable for small batches and prototypes due to their high cost of equipment. In the other hand, a new manufacturing process has known a great growth

N. Aifaoui et al. (Eds.): CMSM 2019, LNME, pp. 558–566, 2020.
https://doi.org/10.1007/978-3-030-27146-6_60

in the last decade named incremental sheet forming (ISF). It is considered as a low-cost alternative process for deep drawing and stamping [1, 2]. It consists on forming 3D shape by applying a progressive and local plastic deformation. The tool moves on a predefined path into a thin sheet, which reminded fixed along the process. In fact, a hemispherical punch piloted by a numerical commanded milling machine generates a local force able to create a standing deformation. By getting into the sheet at each increment along z-axis, we create the final shape of the piece. Besides the low cost, this manufacturing process has a great formability and high flexibility. It is also able to produce complex shapes. Those benefits attract research's attention. Nowadays, incremental forming process knows a great development. Many studies are conducted in order to better understand this new forming technique carried out on a one metal sheet. However, few efforts are made to investigate the forming behavior of composites. Recently, bimetal sheet attracts researcher's attention. Those layered materials offer many advantages. In fact, this type of composite shows higher properties than each material if considered independently. By way of illustration, they combine higher strength with low density. In addition, they demonstrate two different behaviors confer by each side. Layered materials are made of dissimilar layer linked to each other by different method such as explosive welding [3]. Bimetallic material constituted with titanium and steel aims to enhance the mechanical property and reduce cost of equipment made entirely from titanium. Lately, studying the forming behavior of bilayer material in the ISF gains a lot of interest. Ashouri and Shahrajabian [4] examined the formability of bilayer hybrid brass/steel sheets experimentally through single point incremental forming. They compared the deformation behavior of two modes of arrangement brass/steel and steel/brass. They studied the effect of punch diameter, feed rate and vertical increment on strain, fracture angle and fracture height. Zahedi et al. [5] compared the SPIF process of bilayer sheet aluminum/steel numerically and experimentally. Honarpisheh and Alinaghian [6] carried out an experimental approach and a numerical study on SPIF process of aluminum/copper bimetallic sheet in order to elucidate the influence of process parameters on the forming force, dimensional accuracy and thickness variations. In the same context of composite forming technologies, Sakhtemanian et al. [7] realized a finite element analysis and experimental study of steel/titanium bilayer sheet in incremental sheet forming process of a truncated pyramid of 10 mm height. They explained the effect of layer arrangement and step size on the forming forces, as well as the effect of step increment on microstructure properties. In the present paper, a finite element simulation of steel/titanium bimetallic composite during single point incremental forming is presented. The aim of this work is to analyze the effect of SPIF process parameters such as the layer arrangement and the step size on the forming force and the thickness distribution.

2 Finite Element Analysis

In the current study, the commercial ABAQUS/Explicit software is used to simulate the deforming behavior of steel/titanium composite during single point incremental forming process. Two modes of layers arrangement were exanimated. The first mode is

(St/Ti) referred to Steel-Titanium, where the steel layer is in contact with the tool. The second mode is (Ti/St) referred to Titanium-Steel.

2.1 Sheet Materials and Pyramid Geometry

In this study, we aim to investigate the behavior of titanium (grade 2) and low carbon steel bimetal composite sheet. The initial blank sizes of each material are (200 × 200 × 1) mm. The mechanical properties of each metal are given in Table 1.

Table 2 summarizes the process parameters and the geometry specifications of the formed truncated pyramid. A discontinuous path was used as a tool trajectory.

Table 1 Mechanical properties of titanium and low carbon steel [7]

Material	Titanium (CP-Ti)	Low carbon steel (St-12)
Density (kg m^{-3})	$\rho = 4510$	$\rho = 7800$
Young's modulus (GPa)	E = 163	E = 208
Poison's ratio	$\nu = 0.3$	$\nu = 0.29$
Yield stress (MPa)	$\sigma_y = 350$	$\sigma_y = 232$
Ultimate tensile strength (MPa)	UTS = 460	UTS = 400

Table 2 Process parameters and geometry specifications of a truncated pyramid

Process parameters	Values	Geometrical parameters	Values
Initial thickness (mm)	2	Wall angle (°)	45
Punch diameter (mm)	10	Major base (mm)	108 × 108
Vertical step (mm)	0.25, 0.5, 0.75	Height (mm)	25

2.2 Description of the FE Model

The equipment required for incremental sheet forming is composed of a punch with a hemispherical end, the sheet blank and the backing plate. Those parts should be modeled in the FE simulation. In addition, the tool trajectory should be defined. Besides, all the interactions and the boundary conditions must be integrated. We choose to simulate the behavior of the layered composite during SPIF of a truncated pyramid. To do so, firstly we design the geometry with CAD software, such as a commercial 3D CAD-CAM software CATIAV5R21. Secondly, the punch trajectory was generated and converted into a numerical file (APT file) by using a specific postprocessor and then integrated into ABAQUS. This file defines the control tool movement. The punch moves gradually with an increment Δz along z-axis until it reaches the final depth and traces a series of contour lines. In our simulation, the punch was modelled by adopting the assumption of an analytical rigid body hypothesis, with a diameter of 10 mm. A node is designed as a reference point. The movement of the tool is described by time series (t, x), (t, y) and (t, z). Those time series constitute the boundary conditions for the punch. It is a fundamental step to define all the boundary conditions applied also to the sheet and the fixture. Those conditions reflect the experimental procedure and the real set up conditions. In order to model the clamping effect of the backing plate, an encastre constraint is used as a boundary conditions applied to the outer edges of the sheet. Moreover, the backing plate is also modeled as an analytical body. An encastre constraint is used as a boundary condition applied on its reference point. For a coherent modeling, we should also define the interaction between the tool and the sheet, and the interaction between the fixture and the sheet. To simulate the friction between the tool and the composite sheet, a surface-to-surface contact model is adjusted with a penalty method and a predefined friction's coefficient. A surface-to-surface contact model with a hard contact method is associated to the interaction between the bilayer sheet and the fixture. The bimetallic composite sheet has a small thickness compared to its width and length. Hence, to mesh the part, a S4R shell elements was selected. To affect the material property, a composite layup option is utilized. Consequently, the sheet is subdivided into two plies steel and titanium. Figure 1 illustrates the boundary conditions and the finite element meshing configuration.

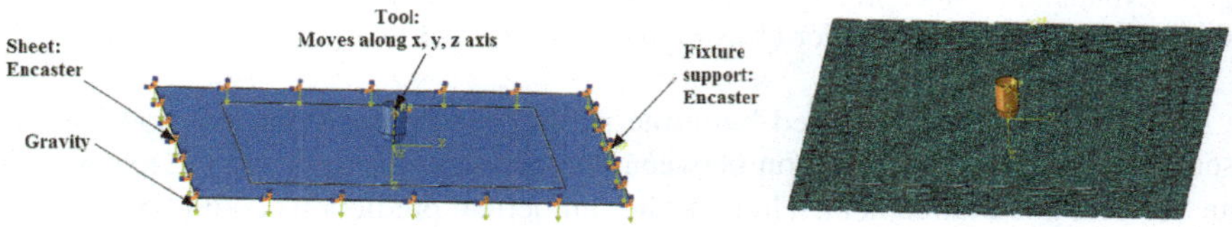

Fig. 1 The finite element-meshing configuration of the initial blank and boundary conditions

The simulation aims to predict the forming forces provided by the punch during the single point incremental forming. Two modes of layer arrangement are investigated: steel/titanium indexed (St/Ti) and titanium/steel (Ti/St), as shown in the following Fig. 2. The upper layer is the layer in contact with the tool. In another words, if we consider the St/Ti mode, the layer in contact with the tool is the steel.

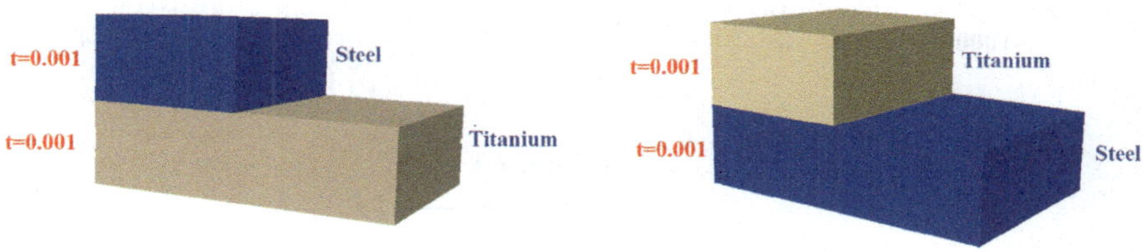

Fig. 2 Layers's arrangement: St/Ti and Ti/St

3 Results and Discussion

3.1 Validation of the FE Model

Figure 3 presents the evolution of the forming force in x, y and z directions according to the following conditions: (St/Ti) mode, a programmed internal depth of the pyramid truncated at 10 mm, 45° wall angle and a vertical step size equal to 0.3 mm. Those forces were measured experimentally and they are taken from literature [7].

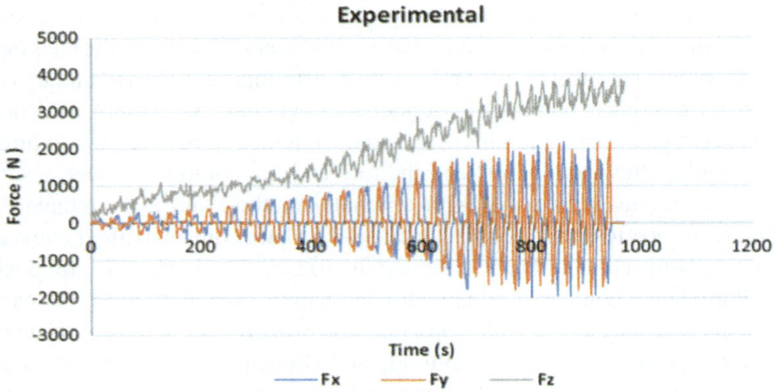

Fig. 3 Evolution of three forming forces components (Fx, Fy and Fz) experimentally measured and exerted on the sheet metal for (Ti/St) mode and $\Delta z = 0.3$ mm [7]

The forming forces obtained by numerical approach using our FE model are represented by Fig. 4. A comparison between the simulation results and the experimental data shows a good agreement. Thus, further numerical predictions could be established using this FE model.

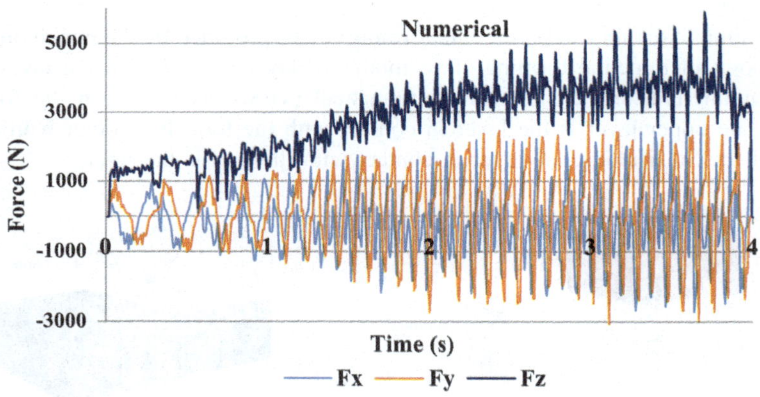

Fig. 4 Numerical prediction of the forming forces diagram in case of (Ti/St) mode of layers arrangement and an incremental step size $\Delta z = 0.3$ mm

3.2 Influence of Layers' Arrangement and Different Vertical Steps Down on the Variations of Forming Force Versus Time Diagram

In this part of study, we aim to investigate the effects of layer arrangement in the bimetallic composite and the step size variation on how the mechanical resistance of Titanium-Steel bimetal sheet (Ti/St) properties during single point incremental forming process changed. The effect of layer positioning and vertical steps down on the forming forces evolution were examined in Fig. 5. The curves presented in Fig. 5a, b, c show

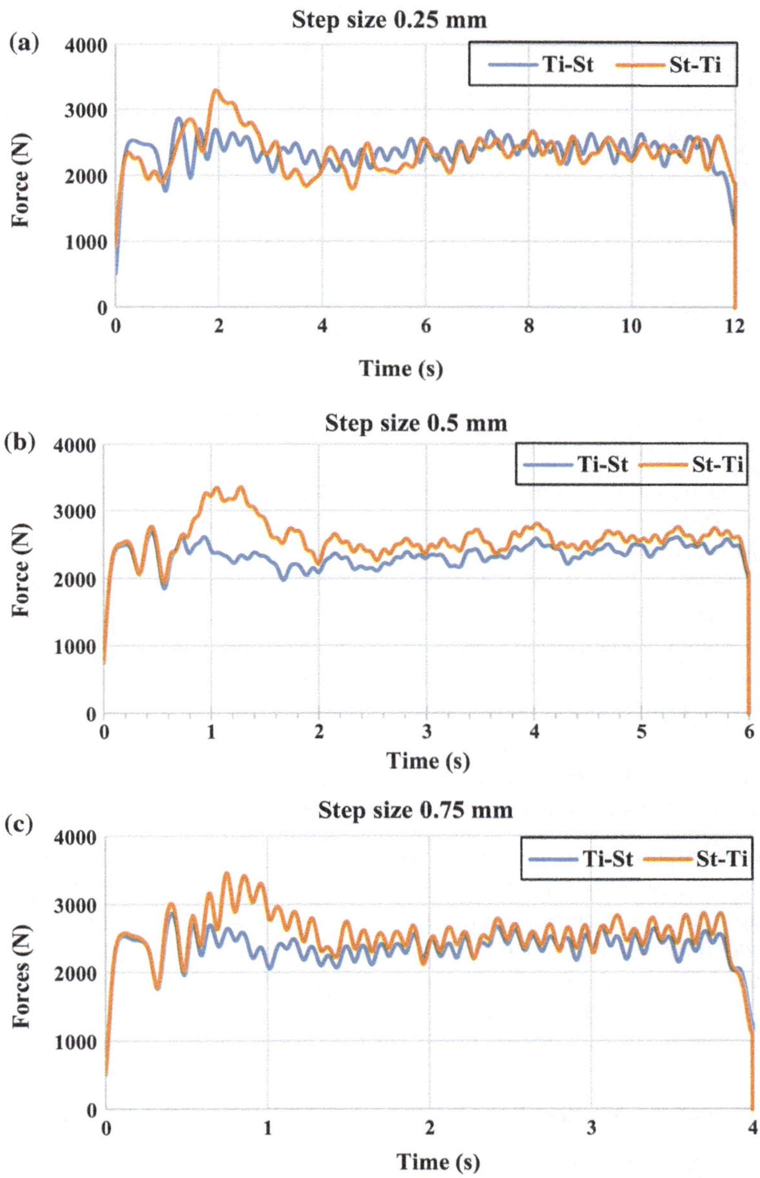

Fig. 5 Effect of layer arrangement on the forming force for different step size: **a** Δz = 0.25 mm, **b** Δz = 0.5 mm and **c** Δz = 0.75 mm

the evolution of the resultant forces versus time corresponding to the two types of layer arrangements, and for parts formed by using 0.25, 0.5 and 0.75 mm vertical size of tool displacement, respectively. The various curves in the three graphs make it possible to deduce two observations. The first is related to the mechanical resistance of the part

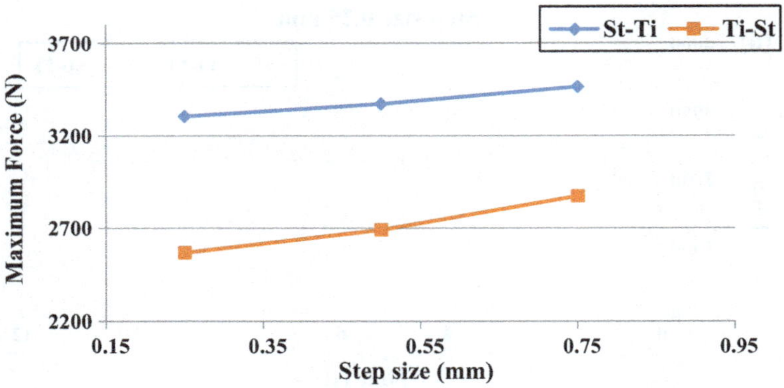

Fig. 6 The effect of step size on the maximum force in two modes of arrangements St/Ti and Ti/St

where the resultant of the forces increases considerably with the increase in the vertical increment step size as can be seen in Fig. 5. The second observation concerns the comparison between the simulation results of the predicted total forces. Similar trends in the evolution of force curves when the step size value was varied can be observed for various arrangement of layers. As it is shown in Fig. 5, higher forming forces are obtained when the low carbon steel is in contact with the punch. In fact, (St/Ti) mode shows higher force than the (Ti/St) mode for all different vertical increment during SPIF process of the bilayer composite.

3.3 The Effect of Vertical Pitch Size on the Peak Magnitude of the Forming Loads Acting on the Punch

Figure 6 illustrates the influence of vertical step on the maximum forming force of bimetallic composite for the (St/Ti) and (Ti/St) mode. As it has been observed from the following figure, an increase in the vertical pitch size has a substantial impact on the magnitude of the force required for successful operation in the forming. It can also be concluded that the numerical simulation results show that an increase in the vertical step also causes a considerable increase in the maximum value of the resultant forming force, and this could be clearly observed for a mode of layers' arrangement of type (St/Ti).

3.4 Effect of the Vertical Step Down on the Thickness Distribution

In this section, we aim to investigate the effect of vertical increment step size Δz on the thickness distribution of final product for both cases of layer sequence. In Fig. 7 are pictured the 3D iso-values of the final thickness for various step size values and different contacted layer. Results show that by increasing the vertical pitch, the forming force increases and the thickness distribution is more uniform, which means that a thicker SPIF part can be achieved. As shown in Fig. 7, the maximum thinning is achieved when the top layer is the low carbon steel and for both cases of the step size $\Delta z = 0.5$ mm and $\Delta z = 0.75$ mm.

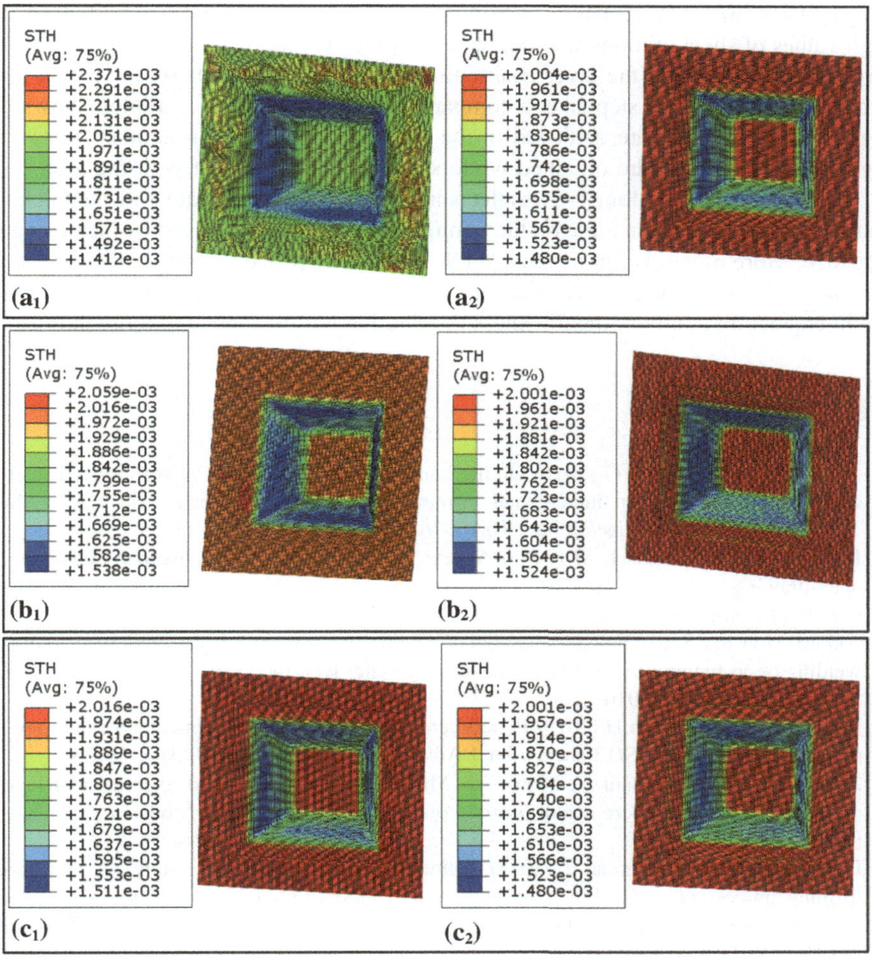

Fig. 7 Iso-values of the final sheet thickness distribution obtained by FEM for various step size: $\mathbf{a_1}$ $\Delta z = 0.25$ mm (Ti-St), $\mathbf{a_2}$ $\Delta z = 0.25$ mm (St-Ti), $\mathbf{b_1}$ $\Delta z = 0.5$ mm (Ti-St), $\mathbf{b_2}$ $\Delta z = 0.5$ mm (St-Ti), and $\mathbf{c_1}$ $\Delta z = 0.75$ mm (Ti-St) and $\mathbf{c_2}$ $\Delta z = 0.75$ mm (St-Ti)

4 Conclusion

In the current investigation, the influence of step size and the effect of layer arrangement of a bimetallic composite sheet on the forming force and the thickness distribution were studied. Firstly, numerical results were confronted to experimental one for the following parameters: step size equal to 0.3 mm, Ti/St mode and a programmed inner drawing depth of 10 mm. This comparison shows a good agreement, both by numerical and experimental approaches. Then, some process parameters were changed in order to understand their effect on the forming force and the thickness distribution during the SPIF process. The following conclusions are confirmed through this study.

The layer arrangement has an important consequence on thickness distribution. The high values of sheet thinning are obtained for the St/Ti mode. It was also demonstrated through this study, that the layer positioning has an important role on the punch loads evolution. For different step size, a comparison of the tool forces variation shows that higher forming forces are achieved in the case of St/Ti mode. In another words, the higher forming efforts are obtained while steel layer is in contact with the tool.

The increment size along z-axis affects in a more or less important way the thickness distribution. In fact, an increase on the thinning behavior was observed by increasing the step size. More precisely, an increase on the vertical step size induces a decrease of the thickness. By increasing the vertical step size, the forming force also increases in both arrangements. In addition, the thickness distribution will be more and more uniform.

References

1. Al-ghamdi KA, Hussain G (2016) On the comparison of formability of roll-bonded steel-Cu composite sheet metal in incremental forming and stamping processes. Int J Adv Manuf Technol 87:267–278. https://doi.org/10.1007/s00170-016-8488-5
2. Hirt G, Ames J, Bambach M (2005) A new forming strategy to realise parts designed for deep-drawing by incremental CNC sheet forming. Steel Res Int 76(2–3):160–166. https://doi.org/10.1002/srin.200505989
3. Akbari Mousavi SAA, Farhadi Sartangi P (2009) Experimental investigation of explosive welding of cp-titanium/AISI 304 stainless steel. Mater Des 30(3):459–468. https://doi.org/10.1016/j.matdes.2008.06.016
4. Ashouri R, Shahrajabian H (2017) Experimental investigation of incremental forming process of bilayer hybrid brass/St13 sheets. Int J Adv Des Manuf Technol 10(3):127–135
5. Zahedi A, Mollaei-Dariani B, Morovvati MR (2015) Numerical and experimental investigation of single point incremental forming of two layer sheet metals. Modares Mech Eng 14 (14):1–8
6. Honarpisheh M, Alinaghian I (2018) Numerical and experimental study on incremental forming process of Al/Cu bimetals: Influence of process parameters on the forming force, dimensional accuracy and thickness variations. J Mech Mater Struct 13(1):35–51. https://doi.org/10.2140/jomms.2018.13.35
7. Sakhtemanian MR, Honarpisheh M, Amini S (2018) Numerical and experimental study on the layer arrangement in the incremental forming process of explosive-welded low-carbon steel/CP-titanium bimetal sheet. Int J Adv Manuf Technol 95(9–12):3781–3796. https://doi.org/10.1007/s00170-017-1462-z

Bending Fatigue Behavior of Flax and Carbon Fiber Reinforced Epoxy Resin

M. Ben Ameur[1,2(✉)], A. El Mahi[1], J. L. Rebiere[1], M. Beyaoui[2], M. Abdennadher[2], and M. Haddar[2]

[1] Laboratoire d'acoustique de l'université du Maine, (LAUM UMR CNRS 6613), Le Mans Université, Av. O. Messiaen, 72085 Le Mans Cedex 9, France
abderrahim.elmahi@Univ-lemans.fr,
Mariem.Ben_Ameur.Etu@univ-lemans.fr

[2] Laboratoire de recherche de Mécanique, Modélisation et Production (LA2MP), Ecole Nationale D'Ingénieurs de Sfax, Département Génie Mécanique, Route Soukra, 3038 Sfax, Tunisie
moez.beyaoui@yahoo.fr, moezabd@yahoo.fr,
Mohamed.haddar@enis.rnu.tn

Abstract. The purpose of the present work is to investigate experimentally the mechanical fatigue behavior of unidirectional flax and carbon fiber reinforced composites. Static and fatigue bending tests were realized on laminates made of flax and carbon fibers impregnated with an epoxy resin. Displacement-controlled bending fatigue tests were conducted on standard specimens with a mean displacement of 50% of the ultimate failure displacement. Specimens were subjected to an applied displacement level of 70% until 10^4 cycles. A comparison of the fatigue properties of the flax and carbon fiber composites was made. The stiffness degradation curves show that carbon composites have higher resistance than flax composites. Hysteresis cycles, dissipated energy and loss factor evolution were measured and discussed. The results obtained show that the flax fiber composites have a damping property higher than the carbon fiber reinforced composites under fatigue bending tests.

Keywords: Flax fiber · Carbon fiber · Static behavior · Bending fatigue behavior

1 Introduction

Composite materials reinforced with conventional fibers, particularly with carbon, are knowing a continuous development in all industrial fields such as marine, aeronautics, sports, etc. Nowadays, researchers have become interested in the use of bio-based composites. Composites reinforced with natural fibers are ecological, biodegradable, their availability is unlimited and they have a relatively interesting specific property. Therefore, plant fibers can be an alternative to synthetic ones [1]. Flax fiber reinforced composites are the most interesting vegetal-based composites that have been studied. Their mechanical characteristics are the subject of the work of [2, 3]. In addition to the mechanical properties, dynamic properties have become an important factor for high

N. Aifaoui et al. (Eds.): CMSM 2019, LNME, pp. 567–575, 2020.
https://doi.org/10.1007/978-3-030-27146-6_61

performance applications. [4] studied the mechanical and dynamic properties of glass, flax and carbon fiber composites. They found that flax fiber composites have higher performance for damping behavior, but poor mechanical properties compared to carbon fiber composites. Furthermore, these studies are not sufficient because of the use of plant fiber reinforced composite materials in high performance and structural applications requires further information on their dynamic properties such as fatigue behavior as well as damping. Indeed, the problem of the dissipated energy is an interesting factor in the mechanical design of structures. Sufficient damping is needed to reduce vibration of structures and avoid fracture fatigue.

This paper presents an experimental study to investigate the mechanical behavior of flax fiber reinforced epoxy (FFRE) and carbon fiber reinforced epoxy (CFRE). The FFRE and CFRE specimens were subjected to static and fatigue loading under three-point bending tests. The static bending tests were carried in order to determine the ultimate parameters. The cyclic fatigue tests were carried out to analyze the damage and to quantify the stiffness degradation. Hysteresis loop was used to evaluate the energy dissipation and the loss factor for the two types of laminates.

2 Experimental Procedures

2.1 Materials

FFRE and CFRE specimens are made of unidirectional flax tape fiber and unidirectional carbon fiber fabric, with areal weights of 200 and 300 g/m^2, respectively. These materials are supported in a SR 1500 epoxy resin mixed with 33% of an SD 2505 hardener. The two types of materials are unidirectional specimens with 0° fiber directions and consist of 12 layers. The flax fibers were dried in an oven for one hour at 110 °C before manufactoring. The composite plates were manufactured using a vacuum molding process by the so-called "bag" technique. They were cured under a pressure of 0.5 bar and at room temperature (20 °C) for 7 h. The plates obtained are exposed to room temperature for at least two days to ensure complete polymerization of the resin. Rectangular test specimens having a length of 120 mm and a width of 15 mm were cut using a high-speed diamond band saw. No lubricating fluid was used when cutting specimens to avoid moisture absorption.

2.2 Bending Tests

In order to determine the ultimate failure displacement, static three-point bending tests were performed on the two types of specimens (Fig. 1). Five replicates have been tested in each sample. The span lengths l (distance between supports) was chosen so as to keep the same ratio (l/h) between the flax laminates and the carbon laminates. The dimensions used for the test specimens are presented in Table 1. The experimental three-point bending tests were carried out on a standard INSTRON hydraulic machine connected to a dedicated computer for control and data acquisition. The tests were performed at room temperature according to the ASTM D790-86 standard with a test speed of 2 mm/min.

Table 1 Dimensions of FFRE and CFRE specimens

Laminates	Total length L (mm)	Width b (mm)	Thickness h (mm)	Span length l (mm)
FFRE	120	15	6.0	100
CFRE	120	15	3.3	55

Bending fatigue tests, under displacement control have been conducted on FFRE and CFRE specimens with identical testing conditions to the static tests. The waveform of the signal is sinusoidal (Fig. 2) with a constant frequency rate of 5 Hz. The fatigue tests were conducted with a constant mean displacement d_{mean} equal to 50% of the ultimate failure displacement d_{rup} and an applied displacement level r_d ($r_d = d_{max}/d_{rup}$) equal to 70% where d_{max} the maximum applied displacement. During these tests, displacement and load were recorded up to 10000 cycles.

Fig. 1 Three-point bending test

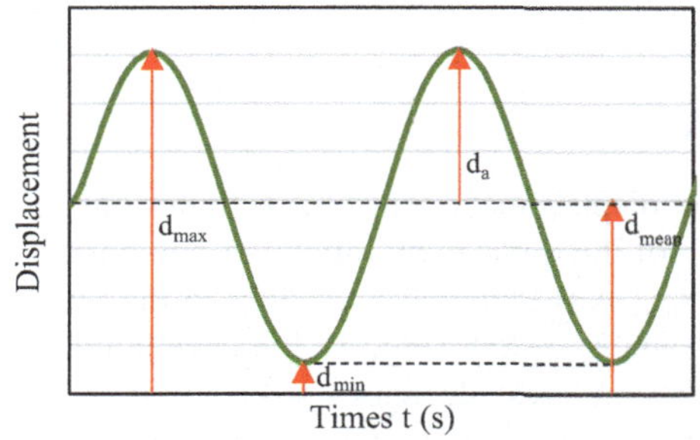

Fig. 2 Sinusoidal loading for fatigue testing

3 Results and Discussion

3.1 Static Results

Static three-point bending tests were carried on unidirectional FFRE and CFRE specimens to investigate the mechanical behavior such as stiffness, ultimate failure stress, ultimate failure displacement and flexural modulus. The stress-deflection curves for FFRE and CFRE specimens are given in Fig. 3. These curves allow us to distinguish different behavior. For CFRE specimens, the curve shows a linear behavior up to final fracture of test specimens which is of brittle type. However, in the case of FFRE specimens, the curve presents a linear part followed by a nonlinear one until the maximum stress. This behavior is due to the significant contribution of the flax fibers. The flexural characteristics of the composites are reported in Table 2.

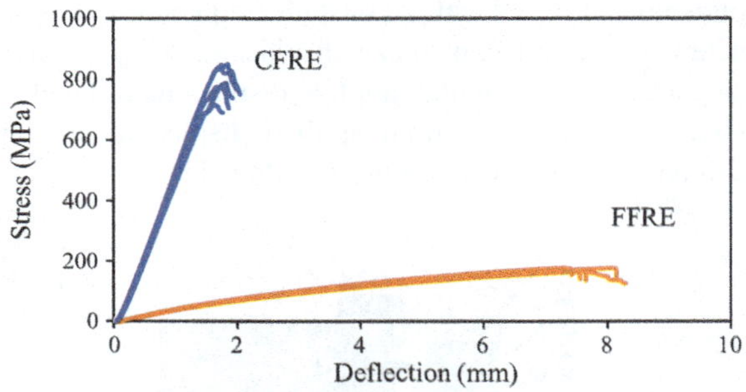

Fig. 3 Quasi-static responses in bending loading for carbon laminates and flax laminates

Table 2 Flexural characteristics for FFRE and CFRE composites

	Stiffness ($N\ mm^{-1}$)	Failure stress (MPa)	Failure displacement (mm)	Flexural modulus (GPa)
FFRE	170	170	7.6	13
CFRE	1100	750	1.75	85

3.2 Fatigue Results

The stiffness loss under displacement control constitutes one of the most used methods to follow the evolution of the damage by fatigue of composites [5]. During these tests, the maximum stress σ_{max} was recorded according to the number of cycles N, for the applied displacement level $r_d = 70\%$ (Fig. 4). For each laminate (FFRE or CFRE), it is observed that the maximum stress decreases slowly with the cycles number.

Figure 5 presents the decrease in the maximum stress $\sigma_{max}/\sigma_{0max}$ according to the number of cycles using a logarithmic scale, where σ_{0max} is the value of the maximum stress at the first cycle. The results obtained for the two types of laminates were reported in the same plot. For each laminate, the stiffness loss of the specimen can be observed in two phases:

- In the first time, a rapid decrease of the ratio $\sigma_{max}/\sigma_{0max}$ appear since the first few cycles. This effect can be associated to the initiation of damage and multiplication of cracks in matrix or fibers.

- In the second phase, a progressive stress reduction. This phase presents the stable propagation of damage and cracks during test until 10000 cycles.

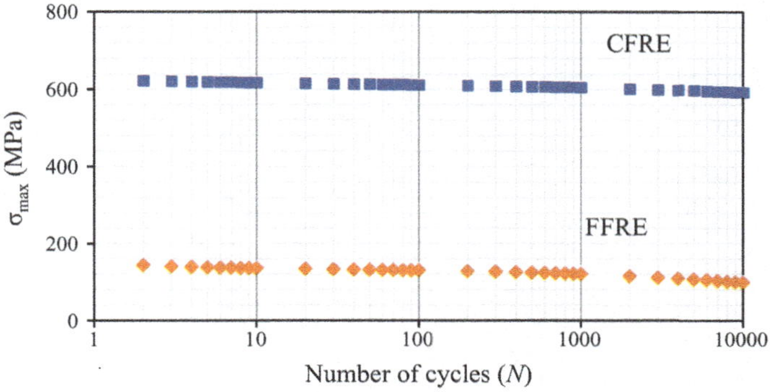

Fig. 4 Evolution of the maximum stress versus number of cycles for carbon and flax laminates at the applied displacement level $r_d = 70\%$

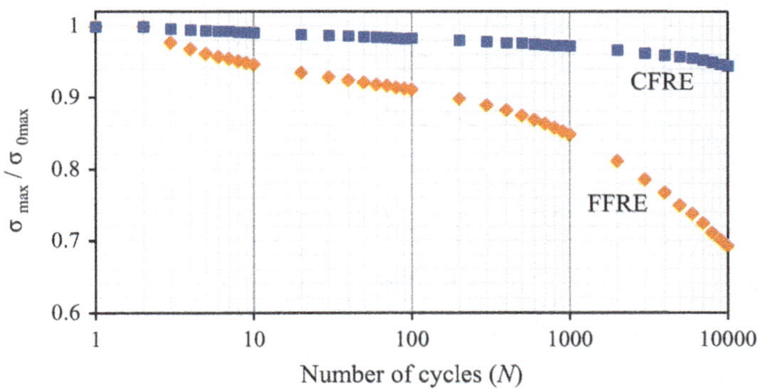

Fig. 5 Loss of the normalized stress ($\sigma_{max}/\sigma_{0max}$) versus number of cycles using a logarithmic scale for carbon and flax laminates

Finally, at 10000 cycles, the decrease is about 5% for the carbon laminates, while it is about 30% for the flax fiber laminates. This result showed that the FFRE laminates have a strength lower than that of CFRE laminates.

For each cycle, 200 experimental points were recorded during these tests. The hysteresis loops are obtained from the load vs displacement curves. The evolution of the hysteresis loops is shown in Fig. 6 at given cycles: 10, 100, 1000, 5000 and 10000 cycles. The hysteresis loops in these figures are translated along the displacement axis to better clarify the figure. For each specimen, it can be seen that the area under the hysteresis loops and the peak of the load decrease with the increase of the cycles number. This decrease is due to the rigidity loss, which is governed by the damage development in the specimens. For the carbon laminates, the area in the hysteresis loops is small as compared to the flax laminates. Also, we clearly observe that the maximum load loop is high for the CFRE in comparison with the FFRE. This difference can be explicated by the intrinsic behavior of the flax fiber.

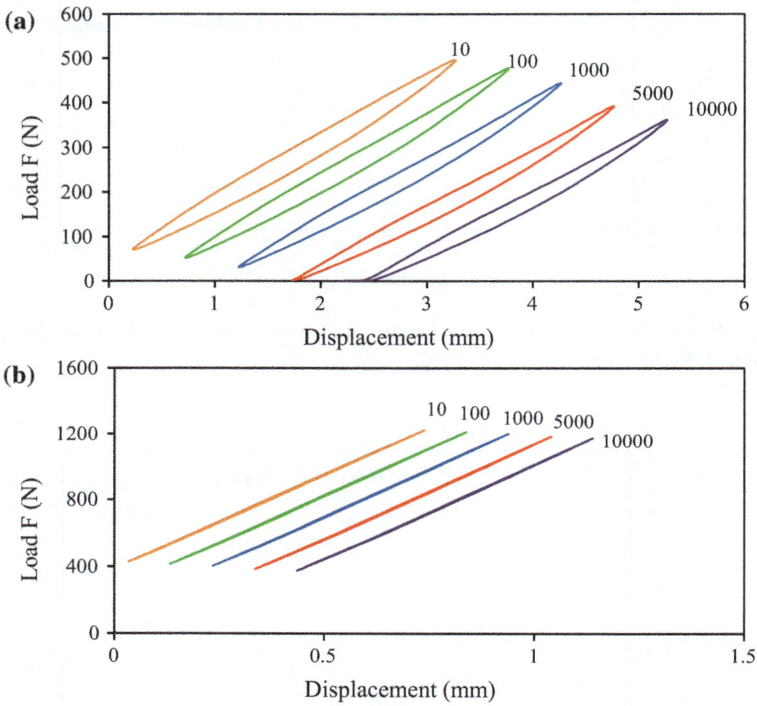

Fig. 6 Hysteresis loops for different cycles: **a** FFRE and **b** CFRE

The evaluation of the dissipated energy in one cycle in the material can be a useful mean to know about the material. Two types of energy could be reported from the hysteresis loop. The area inside the hysteresis curves represents the dissipated energy E_d and the area under the top of the hysteresis curves represents the maximum potential energy E_p [6] Fig. 7. For each cycle, the dissipated E_d and potential E_p energies are given by a simple trapezoidal summation as follows:

$$E_d = \frac{1}{2}\sum_{i=1}^{n}(d_{i+1} - d_i)\{[f(d_{i+1}) + f(d_i)] - [g(d_{i+1}) + g(d_i)]\}. \tag{1}$$

$$E_p = \frac{1}{2}\sum_{i=1}^{n}(d_{i+1} - d_i)[f(d_{i+1}) + f(d_i)]. \tag{2}$$

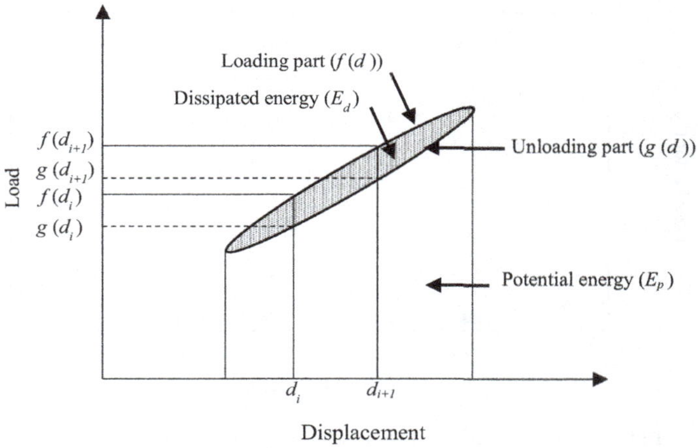

Fig. 7 Dissipated and potential energy

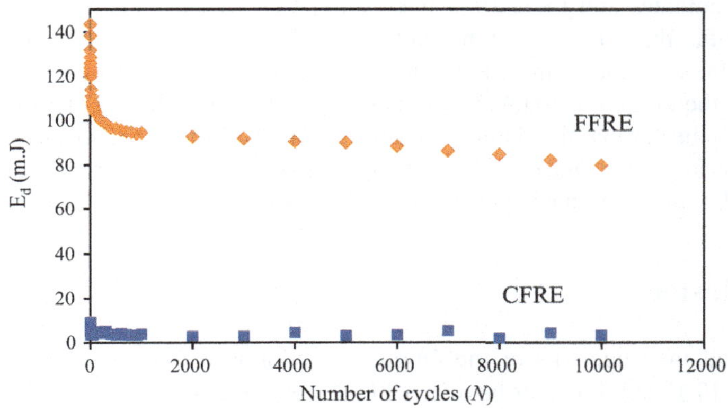

Fig. 8 Evolution of the dissipated energy (m.J) versus the number of cycles

The dissipated energy as a function of cycles number for the FFRE and CFRE are presented in Fig. 8. For each laminate, the dissipated energy decreases with the increase of the number of cycles. This decrease is shown in two phases: First part with rapid decrease in first cycles and second one with very slow decrease. Also, it is clearly observed that the flax laminates dissipated more energy that the carbon laminates. In fact, in the first phase the difference is about 130 m.J but at final cycle the difference is about 75 m.J.

To evaluate the damping (the loss factor) of the studied laminates, the dissipated energy per cycle is compared to the maximum potential energy:

$$\eta = \frac{E_d}{2\pi E_p}. \tag{3}$$

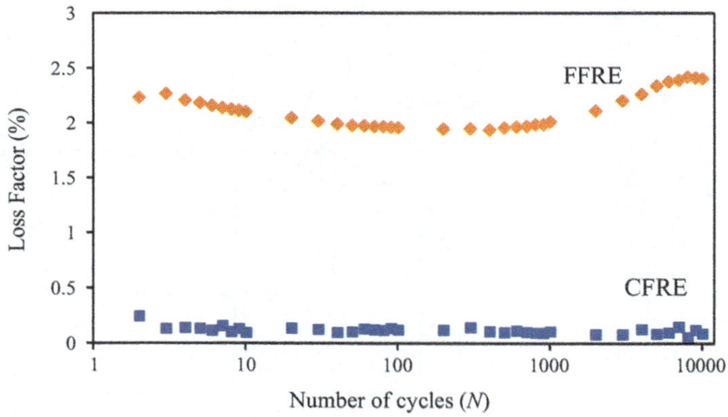

Fig. 9 Evolution of the loss factor versus the number of cycles

Figure 9 presents the evolution of the loss factor versus the number of cycles for the flax fiber laminates and the carbon fiber laminates. The loss factors decrease at first cycles and after that it becomes constant for the FFRE and CFRE specimens. For the flax laminates, the loss factor increases after stabilization and this behavior is due to the increase of the amount of damage and confirmed by the high loss of rigidity in comparison with the carbon fiber laminates. In addition, the FFRE specimens present a high value of loss factor in range of 2–2.5% comparing with the CFRE which is around 0.1%. Thus, the flax fiber laminates present a good damping behavior under fatigue tests.

4 Conclusion

Experimental investigations of the fatigue behavior were conducted on carbon fiber laminates CFRE and flax fiber laminates FFRE. Specimens were subjected to static and fatigue bending tests with displacement control. At first, the properties at failure of the studied specimens were determined from the static bending tests. The stiffness evolution, the hysteresis loops, the energy dissipation and the loss factor were studied for specimens subjected to 3-point bending fatigue tests. The damage propagation was studied using the stiffness degradation method and the evolution of the loss factor. The results obtained prove that the flax fiber reinforced epoxy present a good damping behavior than that of carbon fiber reinforced epoxy under the fatigue bending tests.

References

1. Wambua P, Ivens J, Verpoest I (2003) Natural fibres: can they replace glass in fibre reinforced plastics? Compos Sci Technol 63:1259–1264
2. Monti A, EL Mahi A, Jendli Z, Guillaumat L (2018) Quasi-static and fatigue properties of a balsa cored sandwich structure with thermoplastic skins reinforced by flax fibres. J Sandwich Struct Mater, 1–24

3. Haggui M, El Mahi A, Jendli Z, Akrout A, Haddar M (2018) Static and fatigue characterization of flax fiber reinforced thermoplastic composites by acoustic emission. Appl Acoust
4. Duc F, Bourban PE, Manson JAE (2014) Damping of thermoset and thermoplastic flax fibre composites. Compos Part A 64:115–123
5. El Mahi A, Bezazi A (2009) Describing the flexural behaviour of cross-ply laminates under cyclic fatigue. Appl Compos Mater 16:33–53. https://doi.org/10.1007/s10443-008-9076-0
6. Idriss M, El Mahi A, Assarar M, El Guerjouma R (2013) Damping analysis in cyclic fatigue loading of sandwich beams with debonding. Composites: Part B 44:597–603

Quasi-static Properties of a Bio-Based Sandwich Structure with an Auxetic Core

Khawla Essassi[1,2(✉)], Jean-Luc Rebiere[1], Abderrahim El Mahi[1], Mahamane Toure[1], Mohamed amine Ben Souf[2], Anas Bouguecha[2], and Mohamed Haddar[2]

[1] Acoustics Laboratory of Maine University (LAUM) UMR CNRS 6613, Maine University, Av. O. Messiaen, 72085 Le Mans cedex 9, France
khawlaessassi@gmail.com, {Jean-Luc.Rebiere, abderrahim.elmahi, Mahamane.Toure.Etu}@univ-lemans.fr
[2] Laboratory of Mechanics Modeling and Production (LA2MP), National School of Engineers of Sfax, University of Sfax, BP N° 1173, 3038 Sfax, Tunisia
{khawlaessassi, bensouf.mohamedamine}@gmail.com, anas.bouguecha@gmx.de, mohamed.haddar@enis.rnu.tn

Abstract. Auxetic structure became the center of attention in the recent years because of their abnormal properties such as negative Poisson's ratios. These materials exhibited many benefits like strength, higher stiffness and energy absorption, in comparison with conventional ones. This article presents the mechanical characterization of a bio-based sandwich structure with an auxetic core. The structure has been manufactured using additive manufacturing (3D printing technology). The core and the skins are made from the same biological material which is the polylactic acid. Thus, the sandwich structure is 100% bio based which makes it recyclable. Different density of the auxetic structure was studied in order to evaluate their influence on the mechanical properties of the sandwich. Uniaxial loading was performed on the skins in order to obtain the mechanical properties of the materials and on the core to obtain the Poisson's ratios. Then, the sandwiches were tested in three points bending. The flexural and shear rigidity were measured for each beam. The results obtained showed that the auxetic core density plays a major role on the mechanical characteristic of the sandwich structure.

Keywords: Bending · Poisson's ratio · Auxetic structure · Bio-based sandwich · 3D printing

1 Introduction

Re-entrant honeycomb structures have become the center of attention because of the benefits they offer. These materials present a negative Poisson's ratio, higher stiffness, strength and energy absorption capacity than their counterparts [1, 2]. Therefore, auxetic structure are widely used in different domain such as aerospace, automotive and numerous other applications. Thanks to the benefits stated above, these structures are used as a core material in sandwich composites to increase their shear stiffness and their ability to absorb energy [3]. Different experiences were developed to study the tensile

N. Aifaoui et al. (Eds.): CMSM 2019, LNME, pp. 576–585, 2020.
https://doi.org/10.1007/978-3-030-27146-6_62

[4] and bending [5] behaviour of these meta-materials. The complex geometry of the architecture structure induces to developed new approaches to manufacture them. Despite the complexity of these structures, additive manufacturing, also known as 3D printing technologies, is considered one of the most efficient way to manufacture them.

Because of the environmental challenges we're facing recently, bio-composite have become the center of attention of many researches. These materials have shown promising characteristics such as their biodegradability, recyclability and lower cost. Bio-based composite materials present an interesting alternative to their synthetic counterparts [6, 7].

In this context, a sandwich composite with auxetic core using a bio-based material is manufactured using the 3D printing technique. Different auxetic density were tested. The mechanical properties of the auxetic structure under tensile tests and of the sandwich composite under three-point bending tests were determined. Moreover, the effect of the auxetic density on the flexural stiffness and shear stiffness were measured.

2 Materials and Method

2.1 Materials and Manufacturing

The auxetic structure and the sandwiches beams are manufactured using the polylactic acid (PLA) 1.75 mm in diameter provided by a NANOVIA. It is a bio-based, biodegradable and renewable material dedicated to additive manufacturing techniques. The 3D printing used here is the MakerBot Replicator2 Desktop (Fig. 1). It consists of an extrusion head, filament rolls, building platform, X-Y motion mechanism, etc. The topography was designed using a CAD software. Considering the layer-by-layer fabrication process of the 3D printing, the layer orientation was selected in a way that improves the mechanical properties of the sandwich. As a consequence, the same orientation was adopted to print all specimens.

Fig. 1 The MakerBot replicator2 desktop

The geometry of the auxetic structure using as a core material of the sandwiches is given in Fig. 2. l is the original length of the inclined cell walls and h is the length of the vertical ones. θ is the initial angle between the inclined walls and the X axis. t is the cell wall thickness. The thickness of the auxetic structure is 5 mm. The width of the specimens is set to be equal to 25 mm. It is discretized to a specific number of unit cells. Each number of cells selected along the width gives us a unique core density. From Fig. 2, H = 2(h + lsinθ) and L = 2lcosθ are the specimen length on X and Y axes, respectively. The relative density [5] of the auxetic core is calculated by:

$$\frac{\rho}{\rho_s} = \frac{t/l(h/l+2)}{2\cos\theta(h/l+\sin\theta)} \tag{1}$$

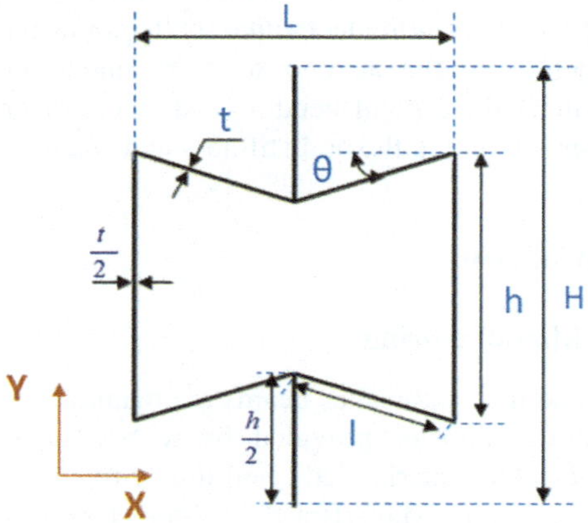

Fig. 2 Design of unit cell of auxetic structure

The dimensions of the different unit cell and the relative densities of each configuration are given in Table 1.

Table 1 Design parameters of the auxetic structure

Cell number in width	ρ/ρ_s (%)	l (mm)	h (mm)	θ (degree)	t (mm)
2 cells	16.7	6.65	8.52	−20	0.6
3 cells	25.1	4.43	5.68	−20	0.6
4 cells	33.5	3.32	4.26	−20	0.6

2.2 Tensile Test

To determine the properties of the auxetic structure and the sandwich composite, it is essential to characterize the base PLA material in advance. Dogbone coupons, in agreement with ASTM standard D638 [8], were printed with PLA materials. Then, the tensile tests were conducted using a standard hydraulic machine INSTRON with a load cell of 10 kN and rate of 1 mm/min. Also, the auxetic structures were printed and tested in order to determine their Poisson's ratio. The tested beams have a length of 100 mm, a width of 25 mm and a thickness of 5 mm. To avoid damaging the auxetic structure while clamping the beams, the clamping was set on a bloc of 25 mm length printed with beams. The displacement was measured by mean of a transversal and a longitudinal extensometer. The failure modes of the different samples were analyzed (Fig. 3).

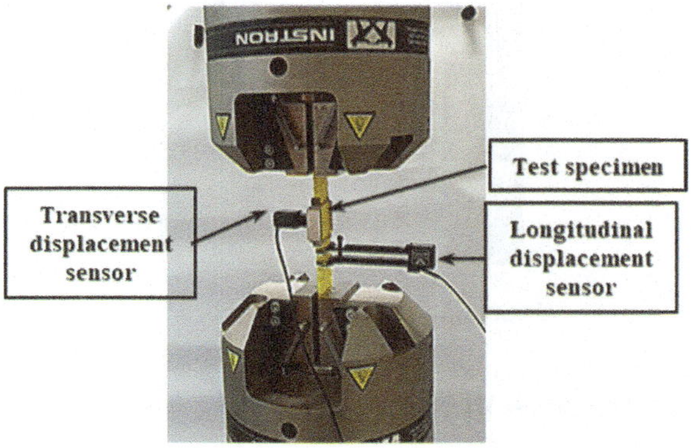

Fig. 3 Set-up of the tensile test

2.3 Three-Point Bending Test

Three-point bending tests were performed on the different sandwich configurations according to the ASTM C393 standard test methods [9], as shown in the Fig. 4. The quasi-static load was applied at a displacement rate of 5 mm/min with 10 kN force sensor. Three 3D printed sandwiches with different core density were considered. Tests were performed to evaluate the flexural stiffness and the shear stiffness of the sandwiches. Thus, bending tests were performed on the sandwichs considering different span length, from 100 to 240 mm, in their linear domain. Also, samples with 120 mm in length, 25 mm in width and 7 mm in thickness (2 mm for the two skins and 5 mm for the core) are tested until breaking. These tests make it possible to analyze the properties at break of the sandwich structures. For each boundary condition, tests were repeated in order to take into account the variability of the results due to the experimental conditions.

Fig. 4 Set-up of the three-point bending test

3 Results and Discussion

3.1 Material Properties

To evaluate the material properties, tensile tests are performed on 3D printed PLA dogbone coupons. The experimental tensile stress versus strain curves obtained is presented in Fig. 5. Three specimens are used to perform uniaxial tensile tests. The averaged experimental properties of the studied material are reported in Table 2. Different static properties are obtained (Young's modulus, Poisson's ratio, strain to failure and ultimate strength). Under maximum loads, the tested material does not fail directly. They maintain some residual strength until final failure.

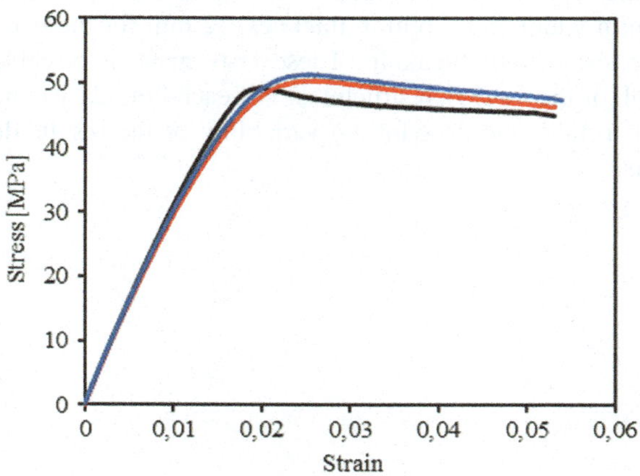

Fig. 5 Engineering stress-strain curves under tensile load of 3D printed PLA dogbone coupons

Table 2 Experimental properties of 3D printed PLA dog-bone coupons

Average mechanical properties	Young's modulus E (GPa)	Ultimate strength σ_u (MPa)	Yield strain ε_y	Plastic strain at break ε_p	Poisson's ratio ν^*
PLA material	3.15	51.36	0.026	0.053	0.3

3.2 Poisson's Ratio of the Auxetic Structure

Tensile tests were performed on the auxetic specimens for different relative density. The Poisson's ratio is determined in the linear-elastic deformation region. Figure 6a–c shows the Transversal-Longitudinal Strain curves of PLA auxetic structure with different relative density. The Poisson's ratio can be evaluated as:

$$\nu_{LT} = -\frac{\varepsilon_T}{\varepsilon_L} \tag{2}$$

According to this equation, it is possible to evaluate the Poisson's ratio of the auxetic structure. Results are given in Table 3. These materials have exhibited the unusual property of getting fatter when stretched. That is why values are negative.

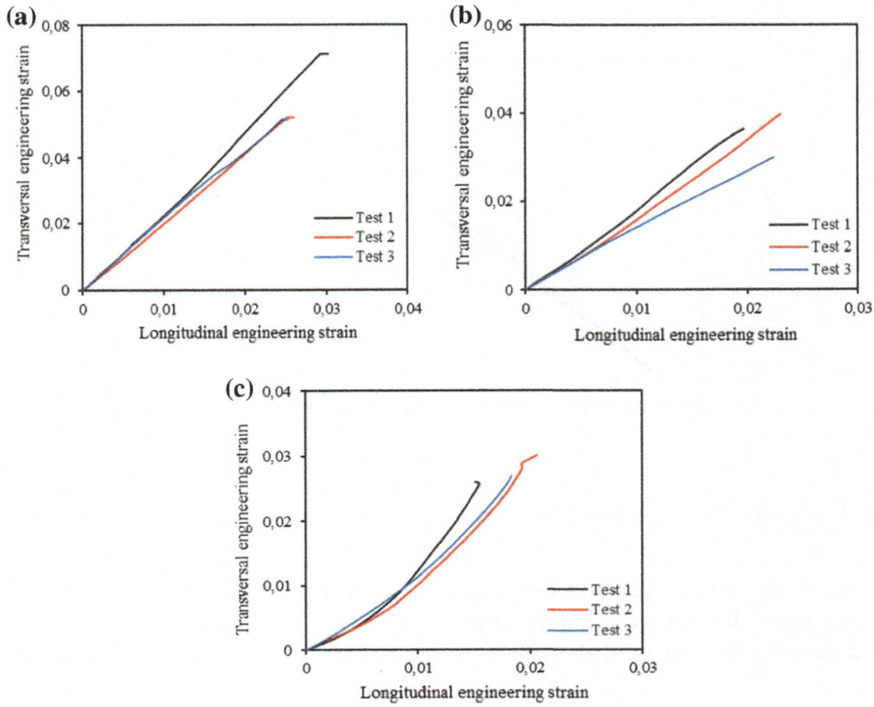

Fig. 6 Transversal Strain-Longitudinal Strain curves under tensile load of PLA auxetic structure with relative density: **a** 16.7%, **b** 25.1%, **c** 33.5%

Table 3 Poisson's ratio of the auxetic structure

Configurations tests	Poisson ratio ν_{LT}			
	Test 1	Test 2	Test 3	Average
2 cells	−2.45	−2.08	−2.07	−2.2
3 cells	−1.75	−1.92	−1.33	−1.66
4 cells	−1.43	−1.69	−1.45	−1.52

3.3 Bending Performance of Sandwiches Composites

The 3D printed sandwichs beams with different relative core density were tested under three-point bending obtained by experimental tests with a span length of 110 mm. The effect of the core density on the bending behavior is examined. Figure 7 presents the load–deflection curves obtained for different core density. Results show that the specific core density has a significant effect on the load-deflection curves. The sandwich composite with the higher relative core density has the higher loading forces, stiffness and the largest bending deflection. Specimens exhibit the same behaviour. Firstly, a linear elastic domain extends up to a deflection of about 5 mm for the different configurations studied. Then, the beams exhibit a shorter nonlinear behaviour until the break.

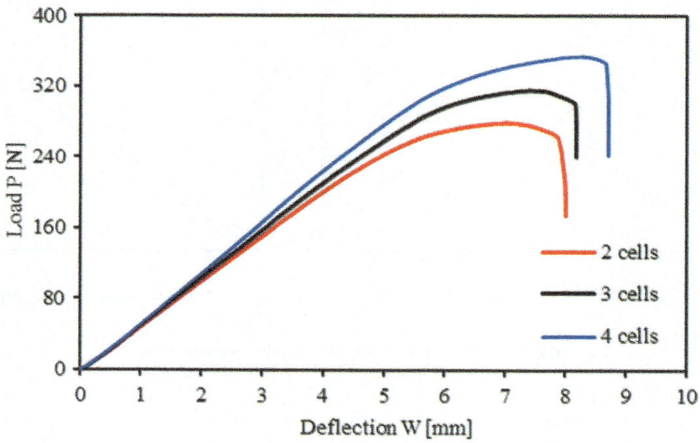

Fig. 7 Three-point bending tests performed on sandwich beams with three different core density

An elastic analysis was affected on the sandwichs beams with different core density to determine the mechanical properties of the skins and core. Under three-points bending tests, the relation that connect the applied load P to the deflection W is given by:

$$\frac{W}{Pd} = \frac{d^2}{48D} + \frac{1}{4N} \tag{3}$$

where d is the span length, D the flexural stiffness and N the shear stiffness of the beam.

Therefore, beams were tested in their elastic domains and the span length was varied from 100 to 240 mm. Figure 8 presents the evolution of W/(Pd) as a function of the square of the span length (d^2). The experimental plots are then fitted by a linear equation. According to Eq. (3), the flexural stiffness D is deduced from the slope of the linear fitting curve and the shear stiffness N is deduced from the intercept.

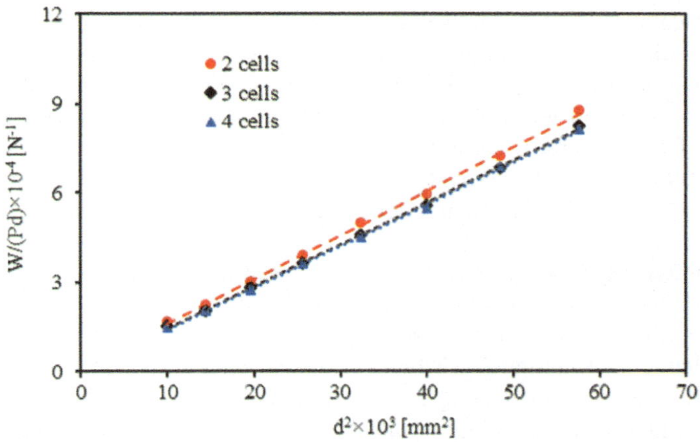

Fig. 8 Evolution of the ratio (w/pd) with respect to the span length in three-point bending under elastic deformations

Also, the flexural and shear stiffness can be calculated as:

$$D = \frac{E_f t_f h^2 b}{2} \tag{4}$$

$$N = bhG_c \tag{5}$$

Using these equations, the Young's modulus of the skins and the shear modulus of the auxetic core can be calculated. Table 4 presents the flexural stiffness D, the shear stiffness N, the Young's modulus of the skins E_f and the shear modulus of the auxetic core G_c for every sandwich configuration. The Young's modulus of the skins is very close to the Young's modulus measured for the PLA material.

Table 4 Elastic and ultimate properties of the skins and the auxetic core

Relative core density (%)	Flexural stiffness D (N mm^2)	Shear stiffness N (N)	Young's modulus of the skins E_f (MPa)	Shear modulus of the auxetic core G_c (MPa)
16.7	1407758	15296	3128	102
25.1	1480389	50722	3289	338
33.5	1493336	53348	3319	356

4 Conclusion

An experimental evaluation of the quasi-static properties of a bio-based sandwich composite with an auxetic core is presented. This structure is produced using the additive manufacturing technique (3D printing). Three different core density out of PLA were printed and tested under tensile tests and quasi-static three-point bending conditions. The structural responses, Poisson's ratio, flexural stiffness, shear stiffness, Young's modulus of the skins and the shear modulus of the auxetic core were calculated. That's why, different span lengths were considered in the three-point bending tests.

It is found that the core density of the sandwich structures has a significant effect on their failure mechanism and mechanical properties. Under tensile tests, re-entrant honeycombs display the expected behaviour. Yet, when stretched, they exhibited the unusual behaviour of getting fat. Moreover, under bending, sandwich with high core density have shown an interesting shear stiffness comparing to that of a lower core density. Considering the previous results, a numerical model can be developed and used to validate the mechanical properties found. This model can be used also to study the effect of the core topology and geometrical parameters of the auxetic structures.

References

1. Hou S, Liu T, Zhang Z, Han X, Li Q (2015) How does negative Poisson's ratio of foam filler affect crashworthiness? Mater Des 82:247–259
2. Ren X, Das R, Tran P, Ngo TD, Xie YM (2018) Auxetic metamaterials and structures: A review. Smart Mater Struct 27(2):023001
3. Liu T, Hou S, Nguyen X, Han X (2017) Energy absorption characteristics of sandwich structures with composite sheets and bio coconut core. Compos B Eng 114:328–338
4. Zhang J, Lu G, Wang Z, Ruan D, Alomarah A, Durandet Y (2018) Large deformation of an auxetic structure in tension: Experiments and finite element analysis. Compos Struct 184:92–101
5. Li T, Wang L (2017) Bending behavior of sandwich composite structures with tunable 3D-printed core materials. Compos Struct 175:46–57
6. Faruk O, Bledzki AK, Fink HP, Sain M (2012) Biocomposites reinforced with natural fibers: 2000–2010. Prog Polym Sci 37(11):1552–1596

7. Lefeuvre A, Bourmaud A, Morvan C, Baley C (2014) Elementary flax fibre tensile properties: correlation between stress–strain behaviour and fibre composition. Ind Crops Prod 52: 762–769
8. Standard A. D638 (2010) Standard test method for tensile properties of plastics. ASTM International, West Conshohocken, PA. https://doi.org/10.1520/d0638-10
9. ASTM C393/C393M (2016) Standard test method for core shear properties of sandwich constructions by beam flexure. ASTM, West Conshohocken, PA

Characterization of CrN/CrAlN/Cr_2O_3 Multilayers Coatings Synthesized by DC Reactive Magnetron Sputtering

K. Aouadi[1(✉)], B. Tlili[2], C. Nouveau[1], A. Besnard[1], and M. Chafra[3]

[1] LaBoMaP, Arts et Métiers ParisTech, Rue Porte de Paris, Cluny 71250, France
aouadikhalil@hotmail.com,
{corinne.NOUVEAU, aurelien.BESNARD}@ensam.eu
[2] Université de Tunis, El-Manar Ecole Nationale d'Ingénieurs de Tunis, LR-11-ES19, Laboratoire de Mécanique Appliquée et Ingénierie (LR-MAI), 1002 Tunis, Tunisia
tlilii_brahim@yahoo.fr
[3] Laboratoire de Systèmes et de Mécanique Appliquée, 2078 La Marsa, Tunis, Tunisie
chafra_moez@yahoo.fr

Abstract. The CrN/CrAlN/Cr_2O_3 multilayer coatings were deposited by reactive magnetron sputtering DC on 90CrMoV8 stainless steel under various oxygen flow rates. The structure and crystalline phases are characterized by the x-ray diffractometer. Through SEM, a dense and coherent is revealed in CrN/CrAlN/Cr_2O_3 multilayer coatings. The friction and wear behaviors obtained with the ball-on-disc test show that all multilayer films exhibit a good wear resistance, especially the one with an oxygen flow rate of 10 sccm. Nevertheless, in sea water the film without a top layer of Cr_2O_3 have the lowest coefficient of friction. This behavior is attributed to the interfacial strengthening and the existence of the upper passivation layer Cr_2O_3. Adding to that, the film obtained under an oxygen flow rate of 10 sccm show the lowest grain size and the maximum hardness and elastic modulus could respectively, 45 and 417 GPa.

Keywords: Multilayer coatings · Microstructure · Hardness · Wear

1 Introduction

Oxide thin films coatings are widely used in different applications. They are known by their good oxidation resistance, good corrosion resistance, good wear resistance and good thermal properties [1, 6, 13]. However, when applied as individual coatings, oxides films haven't good properties.

To enhance the hardness of by developing super-lattice [8], some researchers studied an oxide/oxide structure. Nevertheless, in some cases, hardness has not been improved specially for Al_2O_3/ZrO_2 super-lattice [23]. This is explained either by the fact that the shear modulus of the two oxides is similar, or by the formation of an amorphous phase. In fact, Chang et al. [6] deposited an electrolytic ZrO_2/Al_2O_3 double layer coating by varying temperature. They note that the maximum hardness is about

N. Aifaoui et al. (Eds.): CMSM 2019, LNME, pp. 586–594, 2020.
https://doi.org/10.1007/978-3-030-27146-6_63

7.2 GP under a temperature of 300 °C. Likewise, they observed that ZrO_2/Al_2O_3 multilayer coating has a very good corrosion behavior with a corrosion resistance which is 10 times higher than that of uncoated steel. Recently nitride/oxide super-lattice with high hardness has been successfully developed. Studies have shown that these thin films have a good hardness which can reach 45 GPa. However, these super-lattices have relatively low deposition times. Many researchers have been made to increase the deposition time of these super-lattices using reactive magnetron sputtering technique. Yue et al. [25] developed VN/SiO_2 multilayers with SiO_2 thicknesses ranging from 0.45 to 1.70 nm. They showed that when the thickness of SiO_2 is less than 1 nm, multilayer films crystallized according to the VN matrix and the two layers exhibit epitaxial growth. The hardness of this coating increases to 34 GPa. Li et al. [15] developed $TiAlN/SiO_2$ films by reactive magnetron sputtering. They found that hardness reaches his maximum (37 GPa) when the thickness of SiO_2 is about 0.6 nm. However, the variation of the thickness of the VN layer has no effect on the mechanical and structural properties.

Chromium oxide, Cr_2O_3, is an oxide which has been used as a protective coating of cutting tools due to its outstanding wear and corrosion resistance. It's known by its high hardness and melting temperature, good resistance to wear and high oxidation temperature [5]. In order to improve the properties of chromium oxides, several researchers have studied the effect of increasing the oxygen content. Luo et al. [19] developed chromium oxides by RF magnetron sputtering by varying oxygen flow. They showed that as oxygen flow increases from 2.3 to 3.22 sccm, the hardness increases rapidly from 8 to 30 GPa. Hones et al. [10] deposited Cr_2O_3 films by RF magnetron sputtering. They showed that the grain size of chromium oxides increases linearly with the deposition temperature. However, with the increase of the oxygen level, the grain size increases initially and then remains constant for a percentage of 20% at. Similarly, they showed that the hardness significantly depends on the O_2 content since it reaches a maximum value of 31 GPa for an oxygen content between 15 and 25 at.%. Nevertheless, for an O_2 content of 30% at. The hardness is very low (2.5 GPa). Liu et al. [17] developed Cr_2O_3 films by plasma glow discharge with different oxygen flow rates (5, 10, 15 and 20 sccm) over a chromium nitride film as a top layer. They studied the corrosion resistance in an aqueous 3.5% NaCl solution. They observed that the application of a thin layer of Cr_2O_3 on the stainless steel greatly improves its corrosion resistance. Ho et al. [9] deposited Cr_2O_3 films on a CrN layer as a top layer. The adhesion test shows the appearance of cracks which is probably due to the improved hardness of Cr_2O_3. The addition of a top layer of Cr_2O_3 also lowered the coefficient of friction.

In this work we present results multilayers films $CrN/CrAlN/Cr_2O_3$ were synthesized through using DC magnetron sputtering technique. The effect of oxygen rate on the properties of the top layer Cr_2O_3 is studied. In this study, the chemical composition, microstructure, mechanical, and tribological properties of the multilayer coatings $CrN/CrAlN/Cr_2O_3$ coatings were investigated.

2 Experimental Procedures

The CrN/CrAlN/Cr_2O_3 multilayer coatings were deposited by DC reactive magnetron sputtering (KENOSISTEC-KS40V).

Before deposition, substrates were in situ etched under argon plasma at −700 V for 10 min to ensure a better adhesion of the coatings. The base pressure prior to sputtering was lower than 2.10^{-5} Pa and heated at 300 °C. During deposition, the working pressure was 0.5 Pa.

For the deposition process, chromium and aluminum target with purity of 99.95% was used. To deposited CrN and CrAlN monolayers The Cr and Al target power was set respectively at 1500 W and 1000 W and flow rates of Ar and N_2 were 68.8 and 33.3 sccm, respectively. For the Cr_2O_3 thin films, the Cr target power was set at 1500 W. The flow rate of O was determined with the hysteresis experimental. After this experimental process, the flow rate of oxygen chosen is 0, 10 and 20 sccm. A substrate bias voltage of −500 V is applied to all coatings during deposition. Before deposited CrN/CrAlN/Cr_2O_3 multilayer film, a thin layer about 120 nm Of Cr is deposited to ensure the adhesion of the coating.

The microstructure and surface morphology of the multilayer coatings were observed by SEM field emission (JEOL JSM 7610F). The friction coefficient is determined by rotative ball-on-disk tests. The applied fixed load was 5 N and the sliding speed was 3 cm/s. The counterpart used is an alumina ball (Al_2O3) 6 mm of diameter. The distance slide by the ball is about 200 m. The wear volume of different coatings was calculated from 3D optical profilometer (VEECO, Wyko NT-1100) profiles. The wear volume was calculated from profilometer profiles. To be more precise, eight sections of depth in the wear track are chosen. The film surfaces topography was analyzed by atomic force microscope (AFM) (type XE Park 70) using the tapping mode. An area of 25 μm^2 (5 × 5) was scanned. Additional analyses with the Gwyddion software were performed to determine the RMS roughness of the multilayer coatings. The coating hardness was determined by nanoindentation tests using a MTS XP Nano indenter equipped with a Vickers indenter and using the Rahmoun's model [22]. The adhesion tests were carried out by means of a micro-scratch tester (Scratch Tester Millennium 200) equipped with Rockwell spherical diamond indenter. To calculate an average adhesion value, each test is repeated three times. Figure 1 illustrates a descriptive scheme of multilayer film.

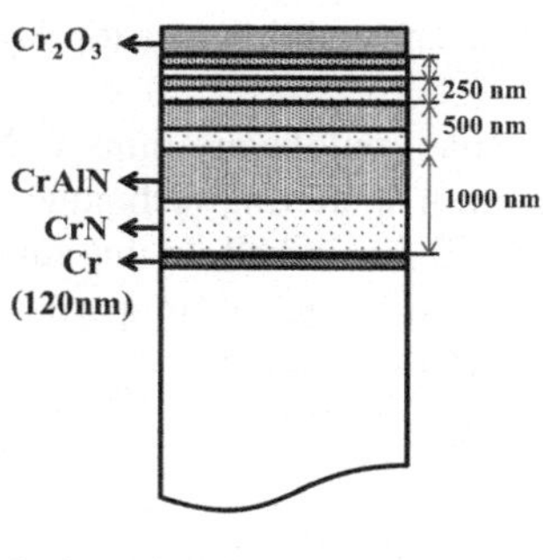

Fig. 1 Scheme of the multilayer systems

3 Results and Discussion

3.1 Microstructure and Morphology

The observation of the SEM images of the cross-section and surface morphology of the multilayer coatings CrN/CrAlN/Cr_2O_3 was carried out using a high-resolution FEG SEM. Figure 2 shows the cross-section of the multilayer films without and with a top layer of Cr_2O_3.

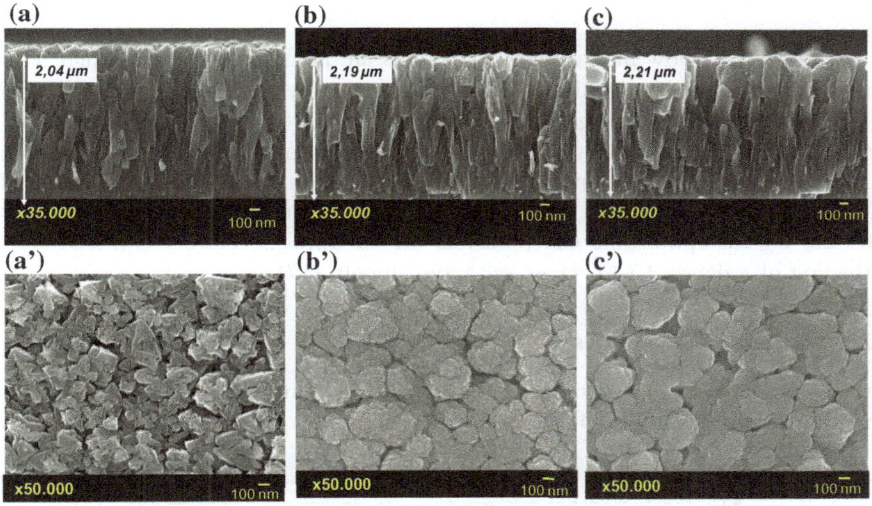

Fig. 2 Cross-sectional SEM image and surface morphology of multilayer films with a top layer of Cr_2O_3 at oxygen rate: (**a**, **a'**) 0 sccm (**b**, **b'**) 10 sccm (**c**, **c'**) 20 sccm

All coatings represent a columnar and dense structure. The observations of the images of the surface morphology of the different multilayer coatings (Fig. 1a'–c') show that the application of a top layer of oxide leads to a variation of the shape of the columns. The film without a top layer of oxide present pyramid-like surface features. The presence of nano-pores between the columns is very visible. According to the Mahieu's model structure zones [20], this structure corresponds to the I_c zone. By adding an oxide layer with a small amount of O_2 (10 sccm), a rounded and assembled columnar tops appear and form a cauliflower-like structure. Always, according to the Mahieu's model, this structure corresponds to the zone II. This area is denser and more homogeneous than Ic area. We notice that the number of nano-pores decreases compared to the multilayer coating without top layer. By increasing the oxygen flow rate to 20 sccm, the surface is always smooth with very small columns which seem less granular than those obtained with an oxygen flow rate of 10 sccm.

The atomic force microscopy (AFM) images surface of multilayer films are presented in Fig. 2.

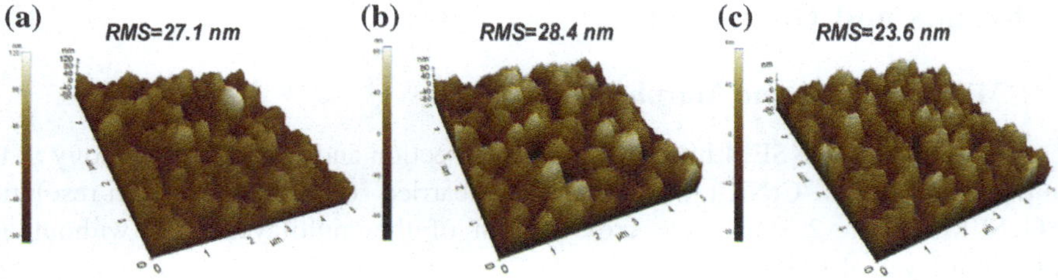

Fig. 3 Three-dimensional AFM of multilayer films with a top layer of Cr_2O_3 deposited on silicon substrate at oxygen rate: **a** 0 sccm **b** 10 sccm **c** 20 sccm

AFM images show that all multilayer films are homogeneous, dense and weakly porous. The morphology of these coatings appears to be a set of continuous mounds.

The roughness of coatings is also shown in Fig. 3. The films with a top layer of Cr_2O_3 obtained with an oxygen flow rate of 20 sccm, has the lowest roughness (23.6 nm). Generally, the growth of Cr_2O_3 films is determined by the diffusion current of Cr^{3+} ions and the oxygen rate adsorption. Also, with the increase in oxygen flow rate, the O_2 adsorption rate can moderate the Cr diffusion current and leads to the formation of a denser Cr_2O_3 layer and more smooth [17].

3.2 Mechanical Properties

The hardness and Young modulus of the multilayers films as a function of the oxygen flow rate are shown in Fig. 4.

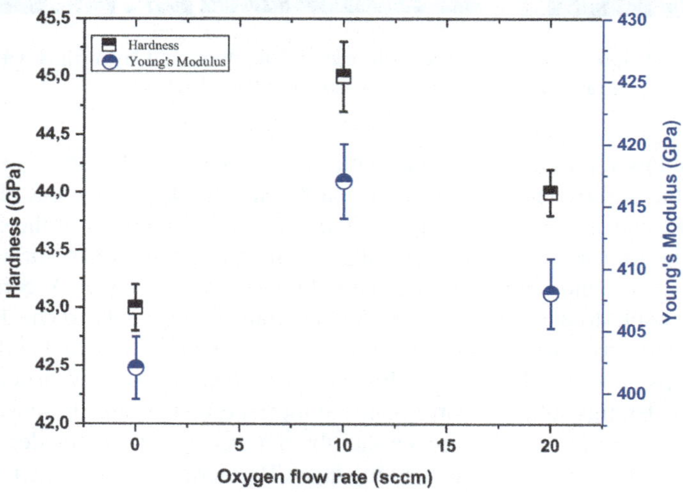

Fig. 4 Hardness and Young's modulus of multilayer films CrN/CrAlN/Cr_2O_3

Figure 4 show that the addition of a top layer of Cr_2O_3 obtained with 10 sccm of oxygen flow rate causes an increase in hardness and Young's modulus and it reaches values of 45 and 417 GPa. However, for the coatings without top layer the hardness is about 43 GPa and the Young's modulus is 402 GPa. This improvement can be attributed to the increase in the number of interfaces between the film of Cr_2O_3 and the CrN/CrAlN coating. Indeed, the interfaces are considered as barriers to dislocation which prohibit the growth of columnar grains through the layers. Blocking dislocations due to column-to-column discontinuity contributes to improved hardness [21]. Also, increasing hardness may be attributed to the existing of hard oxide coating as a barrier to plastic deformation [2]. Likewise, good hardness of the coatings with top layer is due to the morphology of these films. In fact, the multilayer coating without top layer presents pores, however these pores disappear in films with Cr_2O_3 top layer (Fig. 2) [7]. Indeed, pores are known by their detrimental effect on hardness. Also, the strain effect and hall-Petch strengthening can explain the increasing of the hardness [14]. By increasing the oxygen flow rate to 20 sccm, hardness decrease to 44 GPa. Similar results have been shown by Barshilia et al. [3]. This decrease can be attributed to the decrease in hardness of the Cr_2O_3 layer. In fact, Luo et al. [18] developed Cr_2O_3 films by magnetron sputtering. They varied the flow of oxygen from 2 to 2.3 sccm and showed that the hardness of chromium oxide decreases with this increase of the oxygen rate from 14 to 7 GPa. Barshilia et al. [4] attribute the decreasing of hardness to the voids appears during the growing films which influence the mechanical properties.

Figure 5 shows the Lc_1 and Lc_2 critical loads of the CrN/CrAlN/Cr_2O_3 multilayers films as a function of the oxygen flow rate.

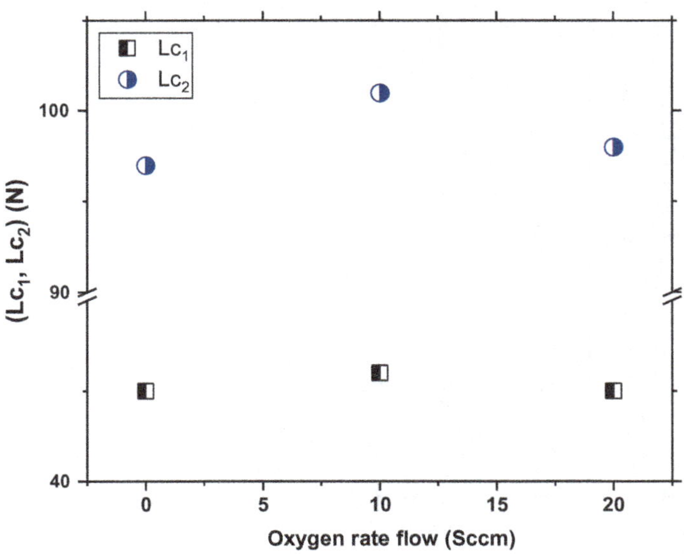

Fig. 5 Critical load of CrN/CrAlN/Cr_2O_3 multilayer coatings

From Fig. 5 we have noticed that the addition of a top layer of oxide (Cr_2O_3) to the CrN/CrAlN multilayers hasn't any effect on the force corresponds to the development of cracks (Lc_1). For Lc_2, it increases when a layer of Cr_2O_3 is deposited as a top layer on the multilayer film and reaches a value of 102 N, which is a more than respectable result for layers that should be used in mechanical applications. This improvement in adhesion is attributed to the increase in the number of interfaces [11]. Indeed, the interfaces can delay the propagation of cracks. In addition, the deposition of a small thickness of a layer Cr_2O_3 can form a coherent interface with CrAlN films resulting in a significant increase of adhesion. This improvement can also be due to the improvement of the microstructure. Indeed, when comparing Fig. 2a' with Fig. 2b', c', it can be shown that there exist less pores. For the CrN/CrAlN/Cr_2O_3 multilayer film obtained with a flow rate of oxygen of 20 sccm Lc_2 slightly decrease. This may be due to the decrease of hardness (Fig. 4). Indeed, many researchers have reported that the critical load increases linearly with the hardness of coatings [12, 24] and this confirm our results.

3.3 Coefficient of Friction

Figure 6 shows the coefficient of friction (COF) of the CrN films after sliding against Al_2O_3 balls (normal load of 5 N) as a function of oxygen flow rate.

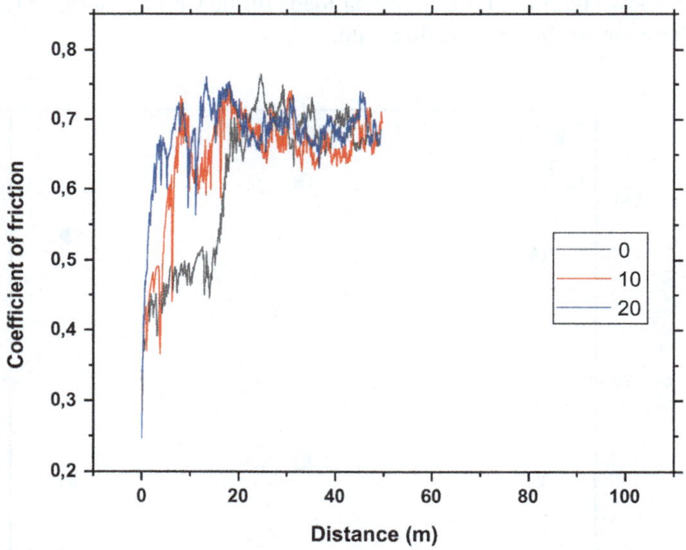

Fig. 6 Coefficient of friction of CrN/CrAlN/Cr_2O_3 multilayer coatings

From Fig. 6 we note that the addition of a layer of Cr_2O_3 hasn't any influence on the friction behavior. The coefficient of friction of the various multilayers is about 0.69. Ho et al. [9] have shown that the deposition of a Cr_2O_3 layer on a CrN layer has reduced its coefficient of friction from 0.4 to 0.26. This is not the case for our films.

This may be due to the thinness of our oxide top layer. In addition, the friction of the ball on the surface of the film can increase the amount of oxide due to the local increase in temperature. This new oxide can weaken the bond strength between the top layer of Cr_2O_3 and the layers below. In fact, if the oxygen atoms (O) are sufficiently incorporated in the upper layer, it becomes more fragile and has cracks for a critical quantity of oxygen [16].

4 Conclusion

The main objective of the present work was to investigate the effect of a top layer of chromium oxide (Cr_2O_3) in the properties the multilayer CrN/CrAlN/Cr_2O_3. All films were deposited by DC reactive magnetron sputtering and the influence of the oxygen flow rate was studied. The cross section SEM image shows that all coatings present a columnar structure. For the surface morphology, multilayer without top layer exhibits pyramid-like surface features. Whereas, films with a layer of Cr_2O_3 represent a form of a cauliflower-like structure. Also, the application of a top layer improves mechanical properties. In fact, by adding a layer of Cr_2O_3, the hardness and the adhesion increase. For tribological properties, we noted that, the application of a top layer hasn't any influence. This may be attributed to the small thickness of the layer of Cr_2O_3.

References

1. Alirezaei S, Monirvaghefi SM, Salehi M, Saatchi A (2007) Wear behavior of Ni–P and Ni–P–Al_2O_3 electroless coatings. Wear 262:978–985. https://doi.org/10.1016/j.wear.2006.10.013
2. Alirezaei S, Monirvaghefi SM, Salehi M, Saatchi A (2004) Effect of alumina content on surface morphology and hardness of Ni–PAl_2O_3(α) electroless composite coatings. Surf Coat Technol 184:170–175. https://doi.org/10.1016/j.surfcoat.2003.11.013
3. Barshilia HC, Surya Prakash M, Poojari A, Rajam KS (2004) Corrosion behavior of nanolayed TiN y NbN multilayer coatings prepared by reactiveer direct current magnetron sputtering process. Thin Solid Film 460:133–142. https://doi.org/10.1016/j.tsf.2004.01.096
4. Barshilia HC, Rajam KS (2008) Growth and characterization of chromium oxide coatings prepared by pulsed-direct current reactive unbalanced magnetron sputtering. Appl Surf Sci 255:2925–2931. https://doi.org/10.1016/j.apsusc.2008.08.057
5. Celik E, Tekmen C, Ozdemir I, Cetinel H, Karakas Y, Okumus SC (2003) Effects on performance of Cr_2O_3 layers produced on Mo/cast-iron materials. Surf Coat Technol 175:1074–1081. https://doi.org/10.1016/S0257-8972(03)00361-X
6. Chang C, Yen SK (2004) Characterization of electrolytic ZrO_2/Al_2O_3 double layer coatings on AISI 440C stainless steel. Surf Coat Technol 182:242–250. https://doi.org/10.1016/S0257-8972(03),00862-4
7. Gaillard Y, Rico VJ, Jimenez-Pique E, Gonzalez-Elipe A (2009) Nanoindentation of TiO_2 thin films with different microstructures. J Phys D Appl Phys 42:145305–145314. https://doi.org/10.1088/0022-3727/42/14/145305
8. Helmersson U, Todorova S, Barnett SA et al (1987) Growth of single-crystal TiN/VN strained-layer superlattices with extremely high mechanical hardness. J Appl Phys 481:1–5. https://doi.org/10.1063/1.339770

9. Ho Wei-Yu, Huang D-H, Huang L-T, Hsu C-H, Wang D-Y (2004) Study of characteristics of Cr_2O_3/CrN duplex coatings for aluminum die casting applications. Surf Coat Technol 178:172–177. https://doi.org/10.1016/j.surfcoat.2003.06.017
10. Hones P, Diserens M, Lévy F (1999) Characterization of sputter-deposited chromium oxide thin films. Surf Coat Technol 121:277–283. https://doi.org/10.1016/S0257-8972(99),00384-9
11. Huang S, Chen S, Kuo Y et al (2011) Mechanical and tribological properties evaluation of cathodic arc deposited CrN/ZrN multilayer coatings. Surf Coat Technol 206:1744–1752. https://doi.org/10.1016/j.surfcoat.2011.10.029
12. Ichimura H, Rodrigob A (2000) The correlation of scratch adhesion with composite hardness for TiN coatings. Surf Coat Technol 126:152–158. https://doi.org/10.1016/S0257-8972(00),00541-7
13. Kim H, Koh Y, Kim H (2000) Densification and mechanical properties of B_4C with Al_2O_3 as a sintering aid. J Am Ceram Soc 65:2863–2865. https://doi.org/10.1111/j.1151-2916.2000.tb01647.x
14. Kong M, Wu X, Huang B, Li G (2009) Epitaxial growth and superhardness effect of TiN/AlON nanomultilayers synthesized by reactive magnetron sputtering technology. J Alloy Compd 485:435–438. https://doi.org/10.1016/j.jallcom.2009.05.133
15. Li W, Liu P, Wang J et al (2011) Microstructure and mechanical properties of TiAlN/SiO_2 nanomultilayers synthesized by reactive magnetron sputtering. Mater Lett 65:636–638. https://doi.org/10.1016/j.matlet.2010.11.073
16. Lin C, Tsai Y, Duh J (2010) Effect of grain size on mechanical properties in CrAlN/SiNx multilayer coatings. Thin Solid Films 518:7312–7315. https://doi.org/10.1016/j.tsf.2010.04.100
17. Liu H, Tao J, Xu J, Chen Z, Gao Q (2009) Corrosion and tribological behaviors of chromium oxide coatings prepared by the glow-discharge plasma technique. Surf Coat Technol 204:28–36. https://doi.org/10.1016/j.surfcoat.2009.06.020
18. Luo F, Gao K, Pang X, Yang H, Qiao L, Wang Y (2008) Characterization of the mechanical properties and failure modes of hard coatings deposited by RF magnetron sputtering. Surf Coat Technol 202:3354–3359. https://doi.org/10.1016/j.surfcoat.2007.12.020
19. Luo F, Pang X, Gao K, Yang H, Wang Y (2007) Role of deposition parameters on microstructure and mechanical properties of chromium oxide coatings. Surf Coat Technol 202:58–62. https://doi.org/10.1016/j.surfcoat.2007.04.066
20. Mahieu S, Depla D (2009) Reactive sputter deposition of TiN layers: modelling the growth by characterization. J Phys D Appl Phys 42:053002. https://doi.org/10.1088/0022-3727/42/5/053002
21. Mori T, Fukuda S, Takemura Y (2001) Improvement of mechanical properties of Ti/TiN multilayer film deposited by sputtering. Surf Coat Technol 140:122–127. https://doi.org/10.1016/S0257-8972(01),01021-0
22. Rahmoun K, Iost A, Keryvin V, Guillemot G, Chabane Sari NE (2009) A multilayer model for describing hardness variations of aged porous silicon low-dielectric-constant thin films. Thin Solid Films 518:213–221. https://doi.org/10.1016/j.tsf.2009.07.040
23. Sproul WD (1996) Reactive sputter deposition of polycrystalline nitride and oxide superlattice coatings. Surf Coat Technol 87:170–176. https://doi.org/10.1016/S0257-8972(96),02977-5
24. Wang Z, Zhang D, Ke P, Liu X, Wang A (2015) Influence of substrate negative bias on structure and properties of TiN coatings prepared by hybrid HIPIMS method. J Mater Sci Technol 31:37–42. https://doi.org/10.1016/j.jmst.2014.06.002
25. Yue J, Liu Y, Li G (2008) Microstructure and mechanical properties of VN/SiO_2 nanomultilayers synthesized by reactive sputtering. Mater Lett 62:1621–1623. https://doi.org/10.1016/j.matlet.2007.09.040

Mechanical Characterization of Composite GRC Under Different Solicitations

Marwa Gaaloul[1(✉)], Ated Ben Khalifa[1], Walid Harizi[2], Mondher Zidi[1], and Abdelouhed Laksimi[2]

[1] Mechanical Engineering Laboratory, National Engineering School of Monastir, Tunisia University of Monastir, Ibn El Jazzar Street, 50001 Monastir, Tunisia
Marwagaaloul0@gmail.com

[2] Roberval Laboratory, Compiegne Technology University, Compiègne, France

Abstract. This work involves discovering an experimental study in the mechanical behavior of Glass Fiber Reinforced Cement (GRC), which is a composite material consisting of Portland cement and chopped alkali-resistant glass fibers. For that, a series of bending and compressive tests were carried out on GRC specimens. In order to identify the mechanisms of damage and detect the cracks of these structures, microscopic observations have been made using Environmental Scanning Electron Microscope (ESEM). These observations show that the breakdown of the matrix and fibers are the dominant damage mode.

Keywords: Composite GRC · Bending tests · Compression tests · ESEM · Damage mode

1 Introduction

Recently, the construction industry is oriented towards the use of the composite material in order to solve requirements unsatisfied by using traditional materials such as the fragility of cementitious matrices. Indeed, the reinforcement of the matrix by the fibers prevents the microcracks to enlarge.

The reinforcement of cement by glass fiber is known as GRC composite (Glass fiber Reinforced Cement), which is a material made of a white cement matrix in which chopped glass fibers are randomly dispersed. It has been widely used in the construction industry for non-structural elements, especially in façade panels, thanks to its lightness, ductility and good mechanical strength compared to concrete [1].

The use of GRC in building should promote corrosion resistance, water permeability, sound insulation and fire resistance.

The GRC composite is generally manufactured either by the spraying process: This method is used for large items such as cladding panels. By this process the cement-sand mixture and the glass fibers are simultaneously projected in a mold by a spray gun which also ensures the cutting of fiber. The GRC composite is sprayed in thin layers until it reaches the desired thickness, normally 10–15 mm. Or by pre-mix technic which used to create small objects, during this process the cement and the sand are

N. Aifaoui et al. (Eds.): CMSM 2019, LNME, pp. 595–602, 2020.
https://doi.org/10.1007/978-3-030-27146-6_64

mixed dry, then adding the water to the mixture. This type of implementation requires the use of short fibers (≈13 mm) [2].

One of the most problems of GRC composite is the durability: the glass fibers become brittle over time because of the alkalinity of the cement mortar. To get rid of this problem several industrialists have developed a new type of alkali-resistant glass fibers AR (special resistant alkali glass fibers) and mortar additives to prevent the chemical and physical processes that lead to embrittlement of the GRC [3, 4].

In order to determine the mechanical properties of the composite GRC different tests were carried out on samples of the composite, for example: Alejandro et Al. have performed a series of tensile and four-bending tests on GRC specimens. Results have shown that the GRC composite has a ductile behavior [3, 5]. Arabi has done a series of compressive tests. The results of these tests indicate that the addition of fibers causes a slight decrease in the mechanical properties [6].

In this study to obtain the optimized material properties and characterize the production conditions, experimental tests were carried out to determine the mechanical characteristics and the stress-strain diagrams under static loading.

2 Mechanical Characterizations

GRC composite specimens are produced by the spray method by Aglo Pierre company. The composite consists of Portland cement, AR (Cem Fil) short fibers (34 mm) which are randomly distributed, silica, sand, water and plasticizer.

2.1 Specimens Preparation

A series of tests were carried out on GRC composite specimens. For bending tests a plate 600×600 mm^2 of dimension and 10 mm of thickness was constructed by the simultaneous projection of the mortar and AR fibers in a wooden mold. After 28 days of mixing, the plate was cut into rectangular test samples of $276 \times 50 \times 10$ mm^3 of dimensions using a guided saw. Then, the test samples are ready to undergo the bending test. For the compression tests the mortar and fibers are sprayed in a cylindrical PVC mold of dimensions 100×200 mm^2.

2.2 Bending Tests

The bending tests were conducted according to ASTM C947 [7] on the prismatic test specimens, cut from a mother plate. These tests were achieved on the LLOYD machine, occupied with a maximum load equal to 20 KN with a speed of 3 mm min^{-1}. The specimen is based on two supports spaced by a distance L = 160 mm and the load F is applied to the center of the sample until rupture, as shown in Fig. 1.

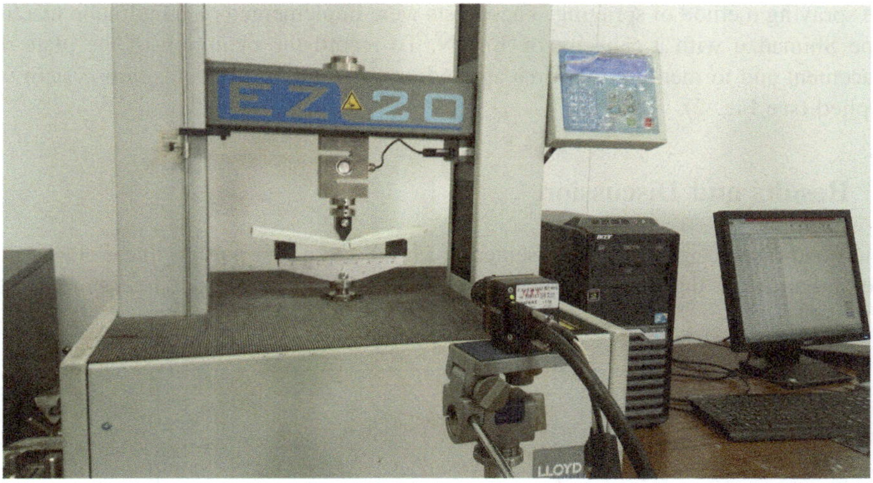

Fig. 1 bending test set up

2.3 Compression Tests

The compression tests are carried out on cylindrical specimens of D 100 L200 of dimension according to the concrete standard BS EN 12390 [8] and manufactured by

Fig. 2 Compressive test set up

the spraying method of spraying. These tests were implemented on a hydraulic machine type Shimadzu with a capacity of 30 KN. To record the evolution of the plate displacement and to measure the force applied on the sample, an acquisition system was applied (see Fig. 2).

3 Results and Discussion

The Load-displacement curve presented in Fig. 3, has a first portion linear elastic until the initiation of the first crack. Followed by another nonlinear slice showing the

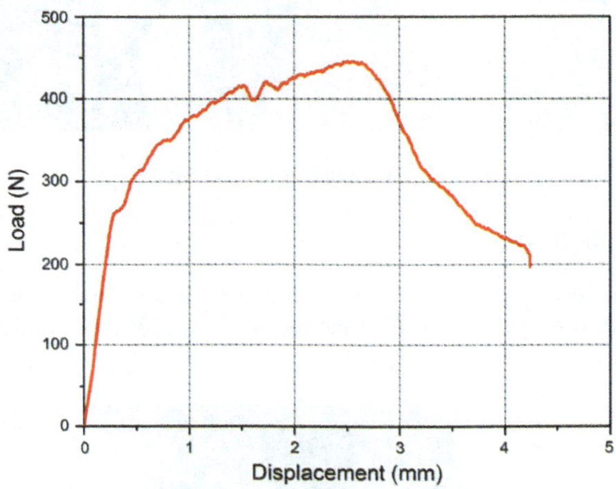

Fig. 3 Stress strain curve after bending test 0

beginning of the damage in the form of microcracks.

The Fig. 4 shows that GRC, after compressive test, retains its shape despite the appearance of cracks. The stress-strain curve presented in Fig. 5 confirm that the GRC specimens is not totally damaged after undergoing compression test. It's clear that there is no strain drop at the end of the test.

These two results illustrate that the GRC composite has a ductile behavior.

The values of bending and compressive strengths of GRC composite obtained after bending and compressive tests carried out on three prismatic samples for flexure tests and three cylindrical specimens for compression test, are presented in Table 1.

Fig. 4 GRC specimen after compressive test

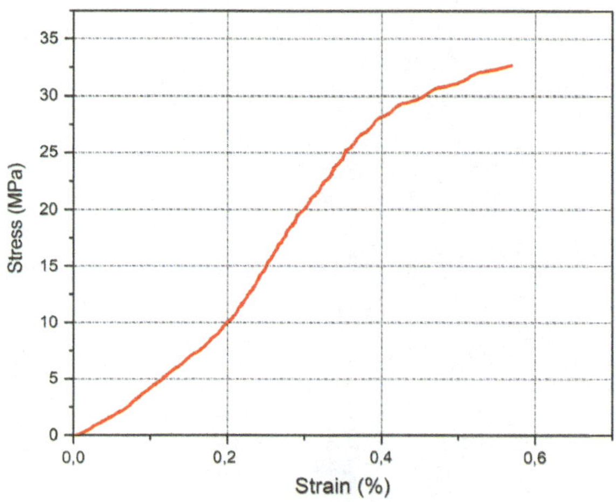

Fig. 5 Stress strain curve of GRC composite after compressive test

Table 1 Values of breaking stress

Parameters	Values (MPa)
Bending strength	10.75
	12.26
	11.09
	Average = 11.37 ± 0.79
Compressive strength	20.66
	32.74
	20.53
	Average = 24.64 ± 7.01

4 Observations

In order to identify the damage mechanisms of GRC composite, a microstructural observation was done on fracture zone of GRC composite specimens after bending tests using SEME (see Figs. 6 and 7).

These observations highlight that the most dominant damage mechanisms are the matrix micro-cracks and fibers breakage.

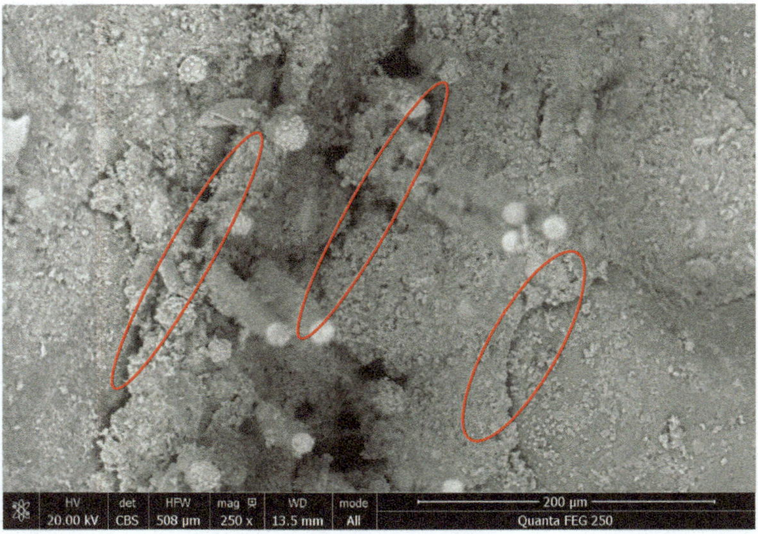

Fig. 6 Matrix microcracks

Fig. 7 Fiber breakage

5 Conclusion

This paper presents mechanical tests (bending and compression) carried out on prismatic and cylindrical specimens of GRC composite containing 2.21% glass fiber mass fraction. It is revealed that the GRC has a ductile behavior.

The ESEM observations illustrate that the monitoring damage mechanisms are the matrix microcracks and fibers breakage.

Acknowledgements. Authors are thankful to the whole staff of the AgloPierre Company for their great collaboration. We are also gratefully acknowledge the helpful comments and suggestions of the reviewers, which have improved the presentation.

References

1. Enfedaque A, Gálvez JC, Suárez F (2015) Analysis of fracture tests of glass fibre reinforced cement (GRC) using digital image correlation. Constr Build Mater 75:472–487. https://doi.org/10.1016/j.conbuildmat.2014.11.031
2. AgloPierre Fiche technique, 655–657
3. Diaz AE, Paradela LS, Sánchez-Gálvez V (2010) The effect of silica fume and metakaolin on glass-fibre reinforced concrete (GRC) ageing. Mater Construcción 60:67–82. https://doi.org/10.3989/mc.2010.52009
4. Bentur A, Ben-Bassat M, SCHNEIDER D (1985) Durability of glass-fiber-reinforced cements with different alkali-resistant glass fibers. J Am Ceram Soc 68:203–208. https://doi.org/10.1111/j.1151-2916.1985.tb15298.x

5. Enfedaque A, Cendón D, Gálvez F, Sánchez-Gálvez V (2010) Analysis of glass fiber reinforced cement (GRC) fracture surfaces. Constr Build Mater 24:1302–1308. https://doi.org/10.1016/j.conbuildmat.2009.12.005
6. Arabi N (2006) Contribution à l'étude du comportement mécanique du composite Ciment-Verre et durabilité des fibres de verre
7. C78/C78M (2010) Standard test method for flexural strength of concrete (Using Simple Beam with Third-Point Loading) 1. Hand C78-02:1–4. https://doi.org/10.1520/c0078
8. BE (2003) Standard test method for compressive strength of cylindrical concrete specimens 04:1–5. https://doi.org/10.1520/c0039

Model Parameters Identification of Adhesively Bonded Composites Tubes Under Internal Pressure

Sonia Braiek[1,2](✉), Ated Ben Khalifa[1], Redouane Zitoune[2], and Mondher Zidi[1]

[1] Université de Monastir, ENIM, LGM, Rue Ibn El Jazzar, 5000 Monastir, Tunisia
soniabraiekl@gmail.com
[2] Université Toulouse-III-Paul-Sabatier, ICA, INSA, UPS, Mines Albi, ISAE, 3 Rue Caroline Aigle, Toulouse, France
redouane.zitoune@iuttlse3.fr

Abstract. The present work aims to numerically investigate the pressure behaviour of adhesively bonded interlocked composites tubes. To this end, a finite element model has been investigated based on the meso-model (interface and ply) concept. The identification of the model parameters has been carried-out through fracture mechanical tests and cyclic tensile tests on [±45] and [±80] specimens. The mathematical models as well as the experimental tests of both interface and ply are presented. It has been proved through numerical outcomes that the joined tube outer diameter is proportional to the applied pressure.

Keywords: Composite tube · Adhesive bonding · Internal pressure · Finite element model

1 Introduction

Composites tubes are being increasingly used in tremendous industrial fields especially for high-pressure containers owing to their good corrosion resistance and high strength-to-weight ratio [1, 2]. Consequently, the assembly of these tubular structures became highly required to obtain a piping system used for several industrial applications. In fact, the adhesive bonding has been known to be a good technique for assembling composites structures comparing to riveting or bolting assembly method [3].

In this context and with the collaboration of CTRA Company, filament wound joined composites tubes have been manufactured for liquid storage and transportation. In spite of the innumerable advantages of adhesively bonded tubes, their use for high pressure applications is still restricted due to the important gaps in the joined tubes behaviour understanding.

Several research works [4, 5] have been devoted to investigate the pressure behaviour of single composites tubes. Others works [6, 7] have focused on the pressure behaviour of joined composite tubes with different assembly configuration. Guoqiang et al. [8] investigated the pressure behaviour of adhesively bonded tube using hybrid

N. Aifaoui et al. (Eds.): CMSM 2019, LNME, pp. 603–613, 2020.
https://doi.org/10.1007/978-3-030-27146-6_65

technique. Mertiny and Ellyin [9] focused their research on the biaxial behaviour of joined tubes using cylindrical sleeve coupler. Vallée et al. [10] established a comparison between the mechanical behaviour of bolted and bonded joint tubes.

The present work denotes an insight into a numerical investigation of the joined interlocked composites tubes pressure behaviour. To this end, a finite element model based on the meso-model concept has been developed in order to predict the pressure behaviour of adhesively bonded tubes. The identification of the model parameters has been also conducted through fracture mechanical test as well as cyclic tensile tests on [±45] and [±80] specimens.

2 Finite Element Model

In order to predict the pressure behaviour of adhesively bonded composite tube, a finite element model was developed. To this end, a non linear analysis based on the meso-model concept is conducted. At the meso-scale, the composite tube is modeled as a laminated structure (ply). The adhesive layer is considered as a two dimensional entity using a cohesive elements (interface). The two meso-constituents ply and interface depend on a number of parameters that are required to be identified to be introduced in the numerical model.

2.1 Interface Model

The delamination phenomenon of the adhesively bonded composite tube along the internal pressure test is modeled through the interface model. Indeed, the aforementioned model is used since it predicts the damage initiation and propagation in the bonding zone. It is important to mention that the interface is modeled as a two dimensional entity between the two tubes.

The strain energy of the interface is given by the following form

$$E_{D,interface} = \frac{1}{2}\left[\frac{\langle\sigma_{33}\rangle_-^2}{k_3} + \frac{\langle\sigma_{33}\rangle_-^2}{k_3(1-d_1)} + \frac{\sigma_{32}^2}{k_2(1-d_2)} + \frac{\sigma_{31}^2}{k_1(1-d_3)}\right] \quad (1)$$

where

k_1, k_2 and k_3 are the interface elastic characteristics.

d_1, d_2 and d_3 are the damage variables.

$$k_1 = \frac{2G_{13}}{e} \quad (2)$$

$$k_2 = \frac{2G_{23}}{e} \tag{3}$$

$$k_3 = \frac{E_3}{e} \tag{4}$$

where

e: the interface thickness

$G_{13}, G_{23}\ et\ E_3$: the ply elastic characteristics.

The thermodynamic forces related to the damage variables is written in the following form

$$Y_{d1} = \frac{\sigma_{31}^2}{2k_3(1-d_1)^2} \tag{5}$$

$$Y_{d3} = \frac{\langle\sigma_{33}\rangle^2}{2k_3(1-d_3)^2} \tag{6}$$

$$Y_{d2} = \frac{\sigma_{32}^2}{2k_2(1-d_2)^2} \tag{7}$$

The interface damage evolution is written in function of the thermodynamic forces by the following form:

$$Y_{interface} = [(Y_{d3})^\alpha + (\delta_1 Y_{d1})^\alpha + (\delta_2 Y_{d2})^\alpha]^{\frac{1}{\alpha}} \tag{8}$$

$$\overline{Y}_{interface} = \sup_{\tau \le t}(Y_{interface}) \tag{9}$$

where

δ_1, δ_2 are the coupling parameters

α, n are the material parameters.

The damage evolution law is written in the following form:

$$W(Y) = \left(\frac{n}{1+n}\frac{\langle \overline{Y}_{interface} - Y_{0,interface}\rangle}{Y_{c,interface} - Y_{0,interface}}\right)^n \tag{10}$$

$$\begin{aligned} d_1 = d_2 = d_3 &= 1 \qquad if\, d < 1 \\ d_1 = d_2 = d_3 &= W(Y) \quad otherwise \end{aligned}$$

where

$Y_{0,interface}$ is the threshold value

$Y_{c,interface}$ is the crictical value.

2.2 Ply Model

The ply is modeled as an orthotropic material accommodating two damage mechanisms namely (i) fiber/matrix debonding and (ii) matrix microcracking.

The ply strain energy is expressed in the following form

$$E_D = \frac{1}{2}\left[\frac{\langle\sigma_{11}^2\rangle_+}{E_1^0} + \frac{\varphi(\langle\sigma_{11}\rangle_-)}{E_1^0} - \frac{2\upsilon_{12}^0\sigma_{11}\sigma_{22}}{E_1^0} + \frac{\langle\sigma_{22}\rangle_+^2}{E_2^0(1-d')} + \frac{\langle\sigma_{22}\rangle_-^2}{E_2^0} + \frac{\sigma_{12}^2}{G_{12}^0(1-d)}\right] \tag{11}$$

where

d and d′: scalar damage variables.

An evolution law is defined for each damage mechanisms in the present form:

$$Y_d = \frac{\partial E_D}{\partial d} = \frac{1}{2}\frac{\sigma_{12}^2}{G_{12}^0(1-d)^2} \tag{12}$$

$$Y_{d'} = \frac{\partial E_D}{\partial d'} = \frac{1}{2}\frac{\sigma_{22}^2}{E_2^0(1-d')} \tag{13}$$

The ply damage evolution is then given by the following form

$$Y(t) = \sqrt{Y_d(t) + bY_{d'}(t)} \tag{14}$$

$$\bar{Y}(t) = \sup_{\tau\leq t}(Y(t)) \tag{15}$$

where

b represent the coupling parameter

Y_d *and* $Y_{d'}$ are the thermodynamic forces.

The ply damage development laws is written in function of the thermodynamic forces as follow

$$d = \frac{\langle \overline{Y} - Y_0 \rangle_+}{Y_c} \quad if\, d < 1 \quad d' = \frac{\langle \overline{Y} - Y_0' \rangle_+}{Y_c'} \quad if\, d'$$
$$d = 1 \qquad otherwise \qquad\qquad d' = 1 \qquad otherwise \tag{16}$$

where

Y_c, Y_c', Y_0 et Y_0' are the material characteristics

In order to consider the anelastic strain resulting from the matrix plasticity, a damage-plasticity model is presented based on the effective stress $\tilde{\sigma}$:

$$\tilde{\sigma} = \begin{bmatrix} \sigma_{11} \\ \frac{\langle \sigma_{22} \rangle}{1-d'} \\ \frac{\sqrt{2}\sigma_{12}}{(1-d)} \end{bmatrix} \tag{17}$$

The plasticity development is expressed in the following form:

$$f = \sqrt{\tilde{\sigma}_{12}^2 + a^2 \tilde{\sigma}_{22}^2} - R(p) - R_0 \tag{18}$$

where

a represent the coupling coefficient.

$$a^2 = \frac{\dot{\varepsilon}_{22}^p (1 - d')^2 \sigma_{12}}{2\dot{\varepsilon}_{12}^p (1 - d)^2 \sigma_{22}} \tag{19}$$

p is the accumulated plastic strain.

3 Parameters Identification

The modeling of the composite tube under internal pressure loading depends on several parameters that are required to be identified through experimental tests.

3.1 Interface Parameters

The interface model depends on the elastic and damage parameters. Indeed, the elastic parameters k_1, k_2 and k_3 are drawn through the ply elastic parameters (Eqs. 2, 3 and 4). The damage parameters G_I, G_{II} and G_{III} are identified from fracture mechanical tests (mode I, mode II and mode III) (Fig. 1). These tests were conducted on adhesively bonded E glass/VE 411 composites plates following ASTM D5528 [11] and ASTM D7905 [12] standards.

Table 1 summarizes the interface parameters.

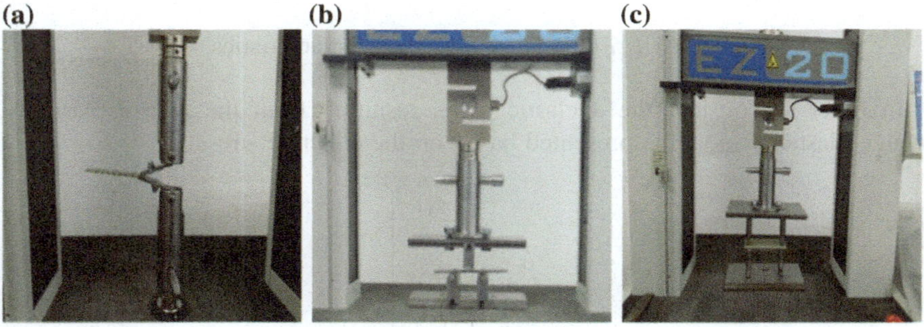

Fig. 1 Fracture mechanical tests. With **a** DCB, **b** ENF, **c** ECT

Table 1 The interface model parameters

Parameters					
Elastic parameters	K_1(MPa/mm)	K_2(MPa/mm)	K_3(MPa/mm)		
Value	2400	4074	5296		
Damage parameters	G_I	G_{II}	G_{II}	α	n
Value	3	6	1	0.9	0.5

3.2 Ply Parameters

Except the elastic parameters $E_1^0, E_2^0, E_3^0, G_1^0, G_2^0, G_3^0, \upsilon_{12}^0, \upsilon_{13}^0, \upsilon_{23}^0$ the ply model depends on the damage development laws parameters. The aforementioned parameters were identified by means of cyclic tensile tests on [±45°] and [±80°] E glass/VE411 composites specimens. In fact, the cyclic tensile tests carried-out on [±45°] samples allow the determination of the ply shear behaviour. Although, the tests performed on [±80°] specimens allow the identification of the coupling parameters between the transverse and shear behaviour.

3.2.1 [±45°] Specimens

Figure 2 illustrates the tensile load-unload curve of E glass/VE411 laminates. For each load/unload i, these values are identified:

ε_i^e: Elastic strain; ε_i^p: Plastic strain; d_i: Damage variable $d = \frac{G_{12-}^0 G_{12}^i}{G_{12}^0}$ and Y_i: Thermodynamic force $Y_i = \frac{\sigma_{12}^i}{(1-d_i)\sqrt{2G_{12}^0}}$.

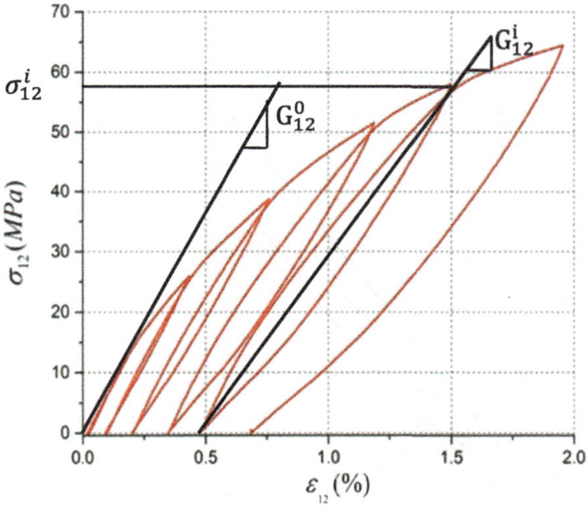

Fig. 2 Tensile load-unload curve of E glass/Vinylester [±45°] composite laminate

In order to determine Y_0 and Y_c, the shear damage curve (Fig. 3) is plotted.

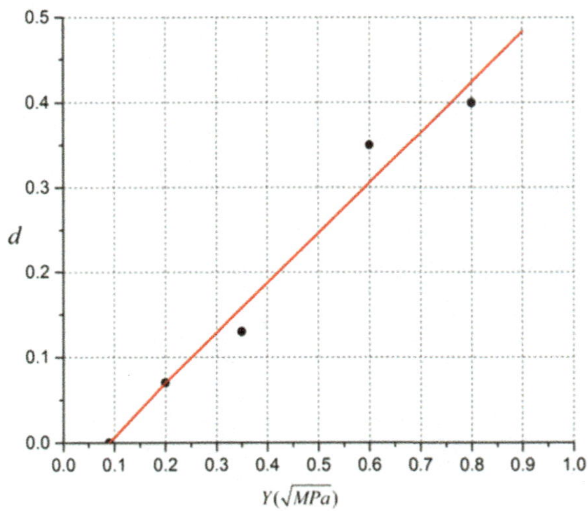

Fig. 3 Shear damage master curve of E glass/Vinylester 411 composite

3.2.2 [±80°] Specimens

As for the previous tests on [±45°] samples, cyclic tensile tests were carried-out on E glass/VE411 [±80°] specimens to determine Y_0' and Y_c' (cf. Figs. 4 and 5).

Referring to the damage parameters evolution d and d', the coupling parameters b is defined in the following form: $d' = bd$.

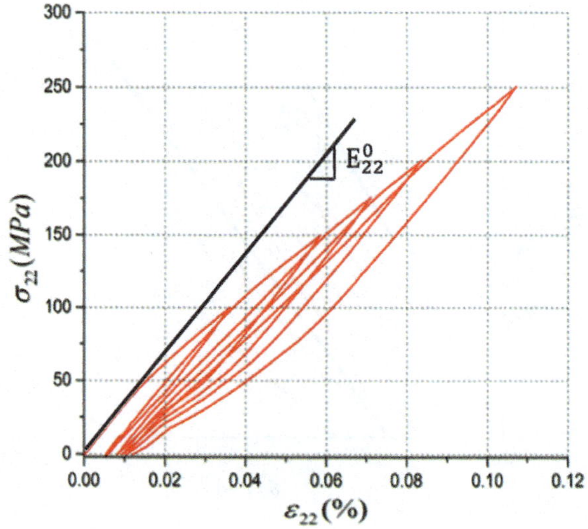

Fig. 4 Load-unload curve of E glass/Vinylester 411 [±80°] composite

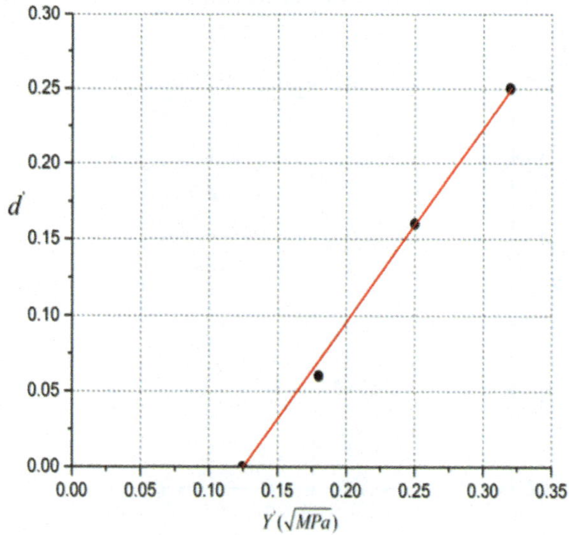

Fig. 5 Transverse damage master curve of E glass/Vinylester 411 composites specimens

3.2.3 Plasticity Parameters

The threshold values ($R + R_0$) and the accumulated plastic strain p were directly obtained from the shear test performed on [±45°] composites specimens.

$$R + R_0 = \frac{\sigma_{12}}{(1-d)}$$

$$p = \int_0^{\varepsilon_{12}^p} 2(1-d)d\varepsilon$$

The plot $R + R_0 = f(p)$ is witnessed in Fig. 6. In fact, this curve is described by the following form $R = Kp^{\gamma}$. The ply parameters are shown in Table 2.

Table 2 The Ply model parameters

Parameters						
Elastic parameters	E_1 (MPa)	$E_2 = E_3$ (MPa)	$G_{12} = G_{13}$ (MPa)	G_{23} (MPa)	$\upsilon_{12} = \upsilon_{13}$	υ_{23}
Value	7010	5296	1200	2037	0.28	0.3
Damage parameters	Y_0 (MPa)	Y_c (MPa)	Y'_0 (MPa)	Y'_c (MPa)	b	
Value	0.01	3.24	0.014	1	4.4	
Plasticity parameters	R_0	K (MPa)	a	γ		
Value	11.6	1194	0.9	0.57		

Based on the meso-model concept, a finite element model of adhesively bonded composite tube under internal pressure loading was developed using the experimental identified parameters.

Figure 6 illustrates the pressure evolution versus radial displacement of the tube outer diameter. The analysis of this curve denotes that the pressure increase inside the tube leading to the outer diameter rise until reaching 1.8 mm. At this stage, the leakage takes place. In fact, the leakage can be explained by the fact that the delamination phenomenon takes place between the two adherends and the adhesive layer.

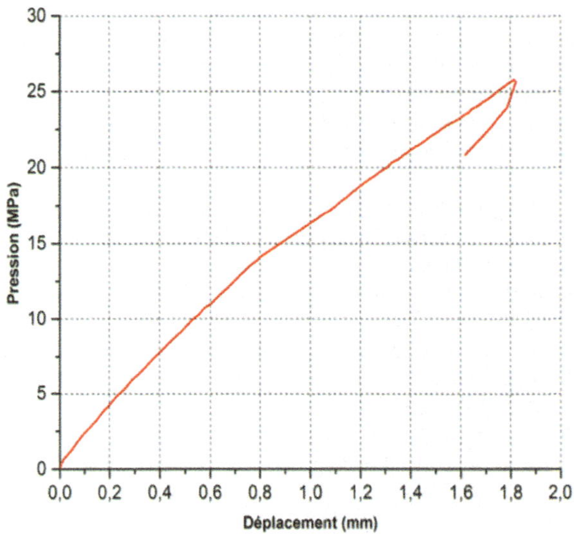

Fig. 6 Numerical result of E glass/Vinulyster 411 composite tube under internal pressure

4 Conclusion

The pressure behaviour of adhesively bonded composites tubes was numerically investigated. The following conclusions can be drawn:

- The proposed model was based on the meso-model concept.
- The parameters of the meso-constituents ply and interface were experimentally identified to be introduced in the developed model. In fact, the aforementioned parameters were determined through fracture mechanical tests and cyclic tensile tests on [±45] and [±80] specimens.
- It has been revealed that the tube outer diameter is proportional to the applied pressure.
- The leakage formation is the result of the delamination phenomenon between the two adherends and the adhesive layer.

Acknowledgements. Authors are thankful to Monastir University and the Tunisian Ministry of Higher Education and Scientific Research for their support (LGM: LAB-MA-05). We are also grateful to the whole staff of the CTRA TUNISIA Company for their great collaboration.

References

1. Bai J, Seeleuthner P, Bompard P (1997) Mechanical behaviour of ±55° filament wound glass-fibre/epoxy-resin tubes: I. Microstructural analyses, mechanical behaviour and damage mechanisms of composite tubes under pure tensile loading, pure internal pressure, and combined loading. Compos Sci Technol 57:141–153. https://doi.org/10.1016/S0266-3538(96)00124-8
2. Martins LAL, Bastian FL, Netto TA (2012) Structural and functional failure pressure of filament wound composite tubes. Mater Des 36:779–787. https://doi.org/10.1016/j.matdes.2011.11.029
3. Saleem M, Zitoune R, El Sawi I, Bougherara H (2015) Role of the surface quality on the mechanical behavior of CFRP bolted composite joints. Int J Fatigue 80:246–256. https://doi.org/10.1016/j.ijfatigue.2015.06.012
4. Onder A, Sayman O, Dogan T, Tarakcioglu N (2009) Burst failure load of composite pressure vessels. Compos Struct 89:159–166. https://doi.org/10.1016/j.compstruct.2008.06.021
5. Samanci A, Avci A, Tarakcioglu N, Şahin ÖS (2008) Fatigue crack growth of filament wound GRP pipes with a surface crack under cyclic internal pressure. J Mater Sci 43:5569–5573. https://doi.org/10.1007/s10853-008-2820-x
6. Skoryi LPT, Skoryi IA (1973) Stresses in bonded joints of thin cylindrical shells. Plenum Publ Corp, pp 108–111
7. Apalak MK (2006) Stress analysis of an adhesively bonded functionally graded tubular single lap joint subjected to an internal pressure. Sci Eng Compos Mater 13:183–211. https://doi.org/10.1007/s11581-016-1887-7
8. Li G, Davis D, Stewart C, Peck J, Pang SS (2003) Joining composite pipes using hybrid prepreg welding and adhesive bonding. Polym Compos 24:697–705. https://doi.org/10.1002/pc.10063

9. Mertiny P, Ellyin F (2006) Joining of fiber-reinforced polymer tubes for high-pressure applications. Polym Compos 27:99–109
10. Vallée T, Tannert T, Meena R, Hehl S (2013) Dimensioning method for bolted, adhesively bonded, and hybrid joints involving Fibre-Reinforced-Polymers. Compos Part B Eng 46:179–187. https://doi.org/10.1016/j.compositesb.2012.09.074
11. ASTM D 5528 (2004) Standard test method for Mode I interlaminar fracture toughness of unidirectional. Annu B ASTM 3:1–12. https://doi.org/10.1520/d5528-01r07e03.2
12. ASTM D 7905 (nd) Standard test method for determination of the mode II interlaminar fracture toughness of unidirectional fiber-reinforced polymer matrix composites. Annu B ASTM. https://doi.org/10.1007/s00125-004-1663-5

Influence of the Nitrided Layers Microstructure on the Fatigue Life Improvements of AISI 4140 Steel

Rafik Bechouel[1,2](✉), Nabil Laalai[1], and Mohamed Ali Terres[1]

[1] Mechanical, Material and Processes Laboratory (LR99ES05), ENSIT, University of Tunis, 5 Avenue Taha Hussein Montfleury, 1008 Tunis, Tunisia
Bechouel.rafik@gmail.com,
mohamedali.terres@esstt.rnu.tn

[2] National Engineering School of Sousse, ENISO, University of Sousse, Sousse, Tunisia

Abstract. Nitriding is an important industrial process to improve the mechanical properties of components, especially by producing compressive residual stresses. Gas and ion nitriding has become a popular thermo-chemical surface treatment, which is being used to develop thermal/mechanical fatigue and wear characteristics of steels. In this study, the gas and ion nitriding of AISI 4140 steel was carried. The micro-structure, the micro-hardness, the residual stresses distribution and the crack resistance of the hardened steel are determined. These analysis and characterization are carried out using optical microscopy, scanning electronic microscopy, X-ray diffraction and mechanical measurements (micro hardness and residual stresses) of treated material. The results are intended to contribute in defining and optimizing the adequate choice of treatments for this type of steel in industrial conditions. The gains, expressed in term of endurance limit, brought by these treatments are established by three-points bending fatigue tests and discussed in relation to the residual stresses evolution under the cyclic loading conditions. The fatigue fracture resistance is analyzed by methods of fracture mechanisms. This reveals that the gain provided by the gas nitriding (50%) is about 8% against 32% for the ion nitriding. This is primarily allotted to a high level of compressive residual stresses for ionic nitrided state compared to the gas nitrided state.

Keywords: Gas nitriding · Ion nitriding · Microstructure · Compressive residual stresses · Fatigue life · AISI 4140 steel

1 Introduction

Majority of failures in engineering materials such as fatigue fracture, wear and corrosion, are very sensitive to the structure and properties of the material surface. These characteristics and, in particular, the magnitude of the tensile surface residual stress were found by machining to affect significantly the fatigue life of mechanical components that have by accelerating the initiation and the propagation of the fatigue cracks. It is well established that the fatigue strength of mechanical components can be

N. Aifaoui et al. (Eds.): CMSM 2019, LNME, pp. 614–623, 2020.
https://doi.org/10.1007/978-3-030-27146-6_66

enhanced by characteristics of the material surface. On the other hand, it was found that the near-surface compressive residual stresses usually extend the fatigue life [1–6]. Manufacturing surface having high near-surface mechanical properties (hardness, compressive residual stresses) is one of the major challenges for increasing the lifetime of mechanical parts. Surface mechanical treatments like shot peening, or thermo chemical techniques like nitriding, lead to compressive residual stresses which prevent the initiation of cracks. Nitriding is a thermo chemical process largely applied to increase the surface hardness, fatigue life of iron-based alloys. Several surface modification techniques are already well known and established, such as ion nitriding [1–5], gas nitriding [6–8].

Ion nitriding is one of the most studied and industrially adopted processes, which involves the introduction of nitrogen in the structure of the material through diffusional mechanisms. Gas nitriding has become a popular thermo-chemical surface treatment, which is being used to develop thermal/mechanical fatigue and wear characteristics of iron and steels. Theoretically, microstructure of the nitrided layer can be controlled by adjusting the gas nitriding process parameters such as ammonia decomposition rate and nitriding duration. During the nitriding treatment two zones are produced in low-alloy steels, namely, the so-called compound layer and the diffusion zone [5, 8–10]. Therefore, characterization of the relationship between nitriding process parameters, and microstructure and fatigue characteristics of the nitrided layers is crucial to ensure stable service of the nitrided components under different conditions. Fatigue strength of mechanical components can be greatly enhanced by generating compressive residual stress, increasing the hardness and reducing the grain size.

In this study, the effect of 4 technological variants, such as untreated, ion nitriding and gas nitriding (35 and 50%) on the fatigue limit of a low-alloy steel is investigated Micro-structural observation, micro-hardness and XRD measurement of residual stress were performed. Three point bending fatigue tests were conducted on nitrided specimens to evaluate the endurance limit improvements at 10^6 cycles. The fatigue limit of all series was experimentally determined and compared with the as-received specimens. The mechanisms of fatigue crack initiation and propagation were investigated based on scanning electron microscope (SEM) observations of the fatigue fracture surface of the tested specimens.

2 Material and Experimental Procedures

The material used in this study was low alloyed steel AISI 4140, for which the chemical composition and the mechanical properties are given in Tables 1 and 2. These samples have been nitrided using the ionic and gas nitriding techniques. The AISI 4140 steel samples were ionic nitrided (IN State) at 520 °C for a total duration of 20 h in a gas mixture of 20% N_2 + 80% H_2 at 2 mbar and gas nitriding at 525 °C for 24 h under NH_3 = 35% (GN35 State) and NH_3 = 50% (GN50 State) to get nitrided layer.

Table 1 Chemical composition of studied steel

C	Mn	Si	S	Cr	P	Mo	Ni	Fe
0.41	0.77	0.28	0.028	1.02	0.019	0.16	0.16	Bal.

Table 2 Mechanical characteristics of untreated 42CrMo4 steel

Tensile characteristics				Hardness
$Rp_{0.2}$ (MPa)	Rm (MPa)	At (%)	E (GPa)	$HV_{0.1}$
978	1050	16.5	201	356

After nitriding, the samples were cross-sectioned, polished, and then etched with Nital 4% used to examine the microstructure of the nitrided layers formed with an optical microscope. A Study of the morphology of the surface of the layers was carried out using a SEM. Changes in the mechanical properties of the nitrided surfaces of the samples were characterized by micro hardness tests. The micro hardness tests were performed on a Vickers micro hardness tester with a load of 50 g (HV0.05). The micro hardness depth profile was determined by taking the micro hardness measurements from the edge towards the core of the cross-sectioned samples.

The residual stresses on the nitride surface were measured by X-ray diffraction with a SET-X residual stress diffractometer. Fatigue tests were conducted using a MTS machine tester to investigate the effect of the harder surface layers on the fatigue crack initiation and the fatigue strengths of specimens at room temperature. These fatigue tests were conducted by three-points bending on notched specimens (Kt = 1.6) and were carried out at a frequency of 15 Hz with a two stresses ratios of R1 = 0.1. The objective of examining the effects of 4 technological variants by comparing nitrided and non-nitrided materials, Therefore, 15 samples were used to construct S–N curves. The fatigue-fractured surfaces were observed by SEM. The applied stress, the crack initiation position were then analyzed.

3 Results and Discussion

3.1 Microstructural Characterization

Ion nitriding and gas nitriding techniques produced different nitride layers in the AISI 4140 steel surfaces in terms of morphology and thickness. The cross-sectional micrographs of the 4 technological variants can be seen in Fig. 1. Two distinct layers were formed during each of the nitriding times, where the outermost layer was the compound or white layer, and the innermost was the diffusion layer. As observed in previous investigations with low alloy steel [8–12], the nitrided layer composition depends on three parameters: time, temperature, and composition of the gaseous mixture. As can be seen, their thicknesses differ only slightly between each other. The metallographic analysis showed that the nitriding process produced a relatively thick nitrided layer consisting of a white layer at the top and an adjacent nitrogen diffusion layer on the 4140 steel surfaces. The white layer formed in ion nitriding had a thickness of 5 μm, the layer produced at gas nitriding 35% showed a thickness of 10 μm, while the layer produced at gas nitriding 50% showed a thickness between 10 and 15 μm.

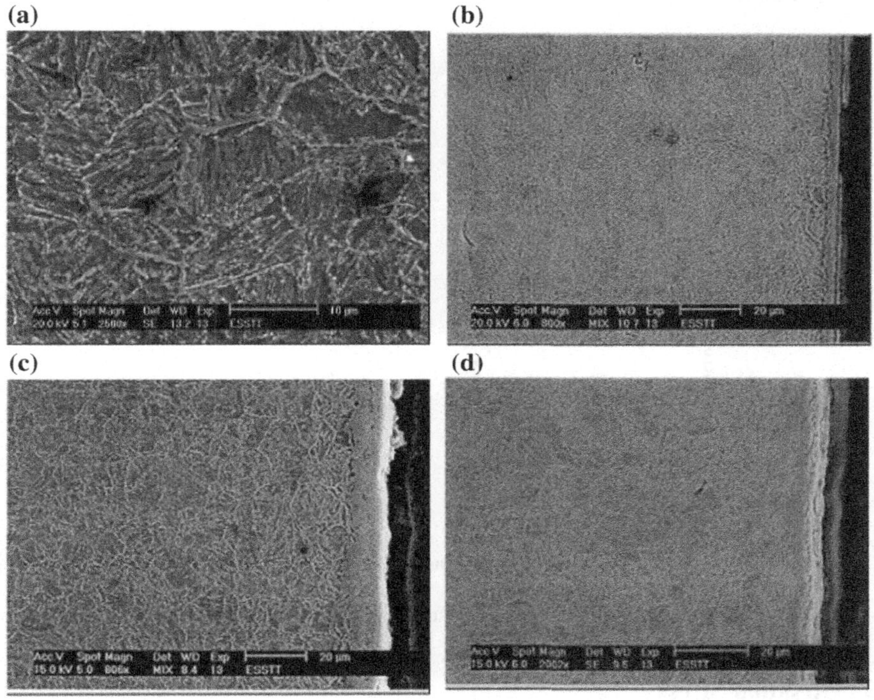

Fig. 1 Cross-sectional views of the microstructure produced at untreated sample (**a**), at ion nitriding (**b**), at gas nitriding (35%) (**c**) and at gas nitriding (50%) (**d**)

3.2 Work Hardening

The micro hardness was measured along the depth on the cross section of the nitrided samples. The hardness profiles in Fig. 2 show a strong hardening effect with the ion and gas (35%) nitriding. This is associated with the nitrogen compositional profile through the nitrided surface, developed as a consequence of the nitriding mechanism [2, 5, 7, 11]. The maximum hardness value measured from the ion nitrided surface was observed to be 1110 HV0.05, which is about 3 times as hard as the untreated material. This dramatic increase in surface hardness was due to a fine and homogeneous coherent chromium nitride precipitation of Cr N type [2, 5, 6]. As discussed in the literature [2, 8, 10], alloys with a high chromium content tend to exhibit a strong interaction between the Cr and N.

The pronounced hardness gradient, in the treated samples, results from the formation of iron nitrides (γ'-Fe_4N and ε-$Fe_{2-3}N$) and Cr N, characterizing a crystalline structure different from the substrate. These create a higher hardness gradient between surface and nucleus and therefore cause higher residual stresses than in samples where the layer depths are higher. In this case, however, with the restriction of the substrate, higher residual stresses can be created when compared with a gradual decrease in the interstitial element quantity, such as observed in deeper layers.

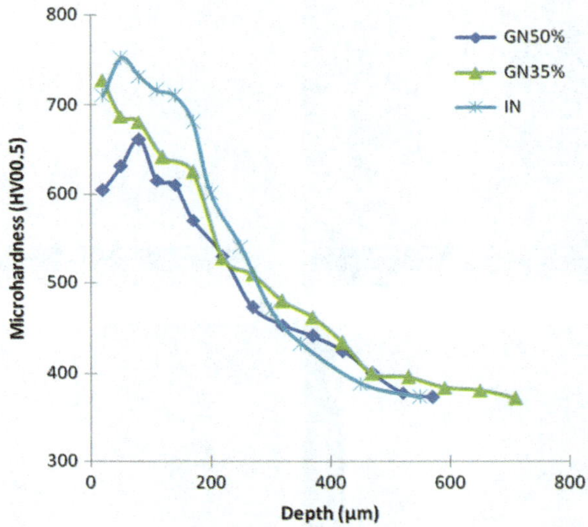

Fig. 2 Micro hardness profiles measured along depth on the cross-section

3.3 Residual Stress

These structure transformations are characterized by the formation of nitrides and carbonitrides in the compound layer, whereas the micro distortions related to nitrogen insertion in the diffusion zone are at the principal cause of the compressive residual stresses. These compressive stresses level decreases from the sample surface to attain approximately 0.5 mm of depth in under layer, until very low values or tensile slightly (Fig. 3). The analysis of this study results show the significant hardening of the diffusion layer that can be charged to the precipitation supported by the chromium contents. The compressive residual stresses, like the work-hardening, are controlled primarily by the induced metallurgical transformations during the nitriding. It should be noted that the residual stresses distribution, established during nitriding evolves during cooling at the end of the treatment cycle [3, 5, 8, 11].

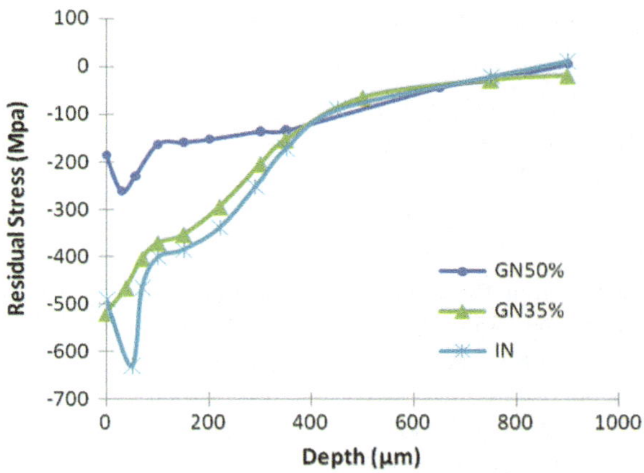

Fig. 3 Residual stress profiles for the nitrided layers of the AISI 4140

3.4 Fatigue Behaviour

The three points bending fatigue tests results obtained from the notched specimens for $R_1 = 0.1$, are reported in the Wöhler diagrams. It is shown in Fig. 4 that the ion nitriding improves the fatigue strength in the high cycle fatigue domain (HCF). The surface work-hardened layer and compressive residual stresses increase the stress of crack initiation and propagation, therefore, the bending-fatigue life of ion-nitrided specimens, can be increased significantly. The endurance limits at 10^6 cycles, determined by the staircase method are presented in Table 3. The fatigue strength is improved at a ratio of 32% for 42CrMo4 steel after ion nitriding.

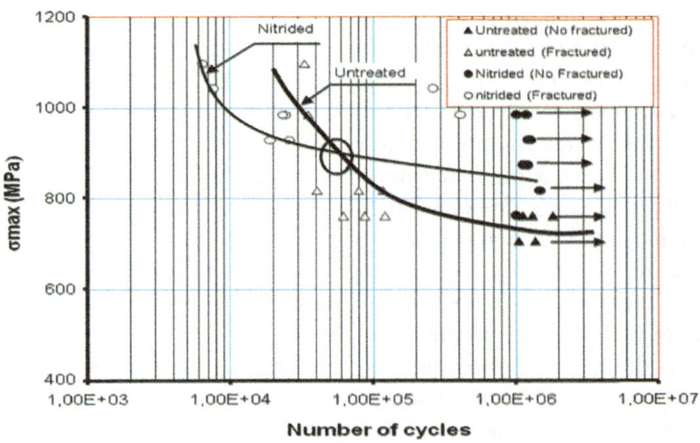

Fig. 4 Fatigue lifetime improvements by nitriding of the AISI 4140

Table 3 Endurance limits at 10^6 cycles for R = 0.1

State	σ_D (MPa)	Variation rate of σ_D (%)
Untreated	718 ± 35	–
Ion nitrided	950 ± 35	32
Gas nitrided 35%	860 ± 35	20
Gas nitrided 50%	775 ± 35	8

It has been proved that the compressive residual stresses resulting from the ion nitriding and gas nitriding (35%) have a favourable effect on the fatigue strength and then improve the fatigue resistance of the nitrided layers [6, 8, 10, 12]. During work component, the beneficial compressive residual stresses at the surface are often imposed to a cyclic loading with positive mean stress. In this case, it has been found that the rate of residual stress relaxation can be drastic in the early stages of fatigue cycling. These levels are experimentally measured during fatigue tests, experimental investigations carried out on nitriding component of 42CrMo4 steel subjected to bending at load controlled (R_1 = 0.1), show that significant relaxation of residual stresses is observed on the compressed surface.

3.5 Fractographic Analysis of Fracture Surfaces

The high hardness and the associated high level of compressive residual stress in nitrided layers makes it difficult to begin and to propagate cracks on the surface by delaying this process. The observations of nitrided surface fatigue fracture with low magnification under the SEM, enable clearly the distinction of two characteristic zones of different fracture modes (Fig. 5): (i) Zone punt, typical to prompt propagation of the fatigue cracks in the nitrided layer, where there are many nucleation localised, which characteristic the stress concentration effects created by specimen notched; (ii) Zone with relief, typical to relatively ductile propagation of the fatigue cracks.

The examinations with high magnification of the fracture zone reveal nucleation and fatigue cracks, starting from the compound layer of the all nitrided states, in both domains. These nucleation's are definitely visible at the specimen surface, in the notched zone close to the rupture, where it appears micro-fractures along the direction perpendicular to the maximum tensile stress generated by the requests of bending fatigue. The fatigue crack propagation occurred in the material was initially characterized by a local plastic deformation, then crack propagates under cyclic loading around the grain boundaries, which leads to the intergranular fracture. For both specimens and in the region of final static failure, the crack propagates by transgranular ductile fracture with dimple morphology (Fig. 6).

The results of this study show the fatigue resistance improvement by nitriding and it appears, closely related to the fatigue crack initiation on the surface. The fatigue cracks starts from surface independently from the loading level, as well as for the nitrided state, which starts from the compound layer. It is observed two distinguish fields from fatigue behaviour (LCF and HCF) of the nitrided layer compared to basic material. It comes out from it that the fatigue behaviour indicated in the HCF region

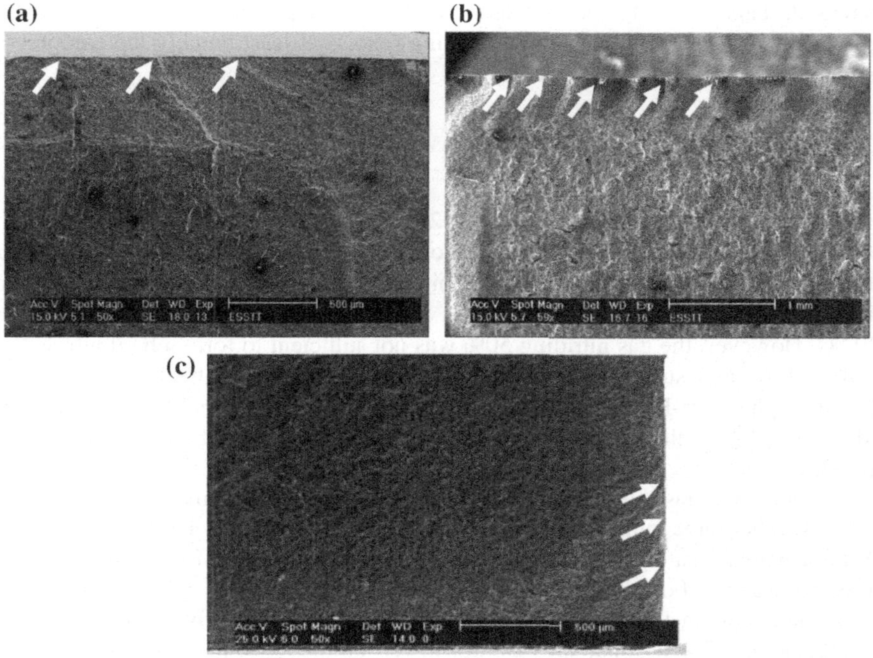

Fig. 5 Fracture facet showing fatigue crack nucleation sites in nitrided layers: ion nitriding (**a**), gas nitriding (35%) (**b**) and gas nitriding (50%) (**c**)

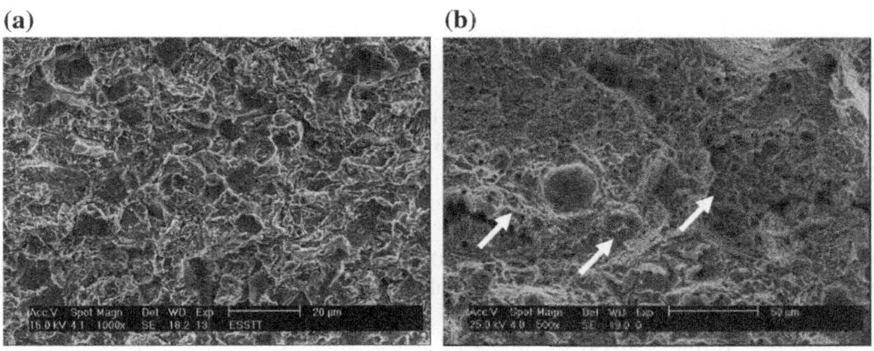

Fig. 6 Fatigue fracture propagation of nitrided state: Brittle aspect in the diffusion zone (**a**) and Ductile aspect in the basic material (**b**)

increases the fatigue life, compared to that of the core material. However, in the constant amplitude LCF region, nitriding can result in an order of magnitude decrease in fatigue life.

This is attributed to the crack initiation site on the surface in the LCF region, and as the case layer has a lower ductility compared to the core, the fatigue resistance is decreased. This is due to the fact that ductility plays an important role in the LCF region [5, 8, 10, 12]. The plastic overloading relaxes at least quickly the compressive residual stresses, produced by nitriding.

4 Conclusion

In this work, the effect of ion nitriding at 520 °C, gas nitriding (35%) and gas nitriding (50%) techniques on the fatigue resistance of AISI 4140 steel was investigated. Various nitriding techniques produced different nitride layers in terms of morphology and thickness. The maximum case depth was achieved on the ionic nitrided samples at 520 °C. However, the gas nitriding 50% was not sufficient to form a hard nitride layer on the 4140 steel surfaces. The nitride layers formed for all treatment times have crystalline phases γ'-Fe_4N and ε-$Fe_{2-3}N$, and Cr N. The compound layer was irregular and consequently the measurements of its thickness and phase quantities were damaged.

The ion and gas (35%) nitriding processes significantly increased the surface hardness. The ionic nitrided sample had the highest surface hardness with 1100 HV0.05, which is a significant increase of approximately 300% if compared to the untreated material. For higher fatigue lives, the effect of different nitriding condition is more significant. In this case, a higher percentage of fatigue life is related to higher surface compressive residual stresses. For lower fatigue lives, there is also a reduction in the difference between fatigue resistance for samples with and without treatment. By curves inclination examination it is observed that nitriding treatment has low influence for lower fatigue lives (higher stresses).

References

1. De la Cruz P, Odén M, Ericsson T (1998) Influence of plasma nitriding on fatigue strength and fracture of a B-Mn steel. Mater Sci Eng A 242:181–194
2. Manisekaran T, Kamaraj M, Sharrif SM, Joshi SV (2007) Erosion studies on surface modified 13Cr–4Ni steels: effect of angle of impingement and particle size. JMEPEG 16:567–572
3. Terres MA, Ben Mohamed S, Sidhom H (2010) Influence of ion nitriding on fatigue strength of low-alloy (42CrMo4) steel: experimental characterization and predictive approach. Int J Fatigue 9:42–56
4. Souza SD, Olzon-Dionysio M, Miola EJ, Paiva-Santos CO (2004) Plasma nitriding of sintered AISI 316L at several temperatures. Surf Coat Technol 84:176–181
5. Gontijo LC, Machado R, Miola EJ, Casteletti LC (2004) Characterization of plasma-nitrided iron by XRD, SEM and XPS. Surf Coat Technol 183:10–17
6. Terres MA, Sidhom H, Ben Cheikh Larbi A, Ouali I, Lieurade HP (2001) Influence de la nature de la couche de combinaison sur la résistance à la fatigue des composants nitrurés. Matériaux Techniques 9:42–56
7. Tokaji K, Ogawa T, Shibata H (1999) The effects of gas nitriding on fatigue behaviour in titanium alloys. J Mater Eng Perform 8:159–167

8. Terres MA, Sidhom H (2010) Influence de la nitruration gazeuse sur la tenue en fatigue flexion des composants nitrurés en acier 42CrMo4. Matériaux Techniques 98:171–184
9. Ribeiro KJB, deSousa RRM, deAraújo FO (2008) Industrial application of AISI 4340 steels treated in cathodic cage plasma nitriding technique. Mater Sci Eng A 479:142–147
10. Ahangarani SH, Mahboubi F, Sabour AR (2006) Effects of various nitriding parameters on active screen plasma nitriding behavior of a low-alloy steel. Vacuum 80:1032–1037
11. Celik A, Efeoglu I, Sakar G, Turkey E (2001) Microstructure and structural behaviour of ion-nitrided AISI 8620 steel. Mater Charact 46:39–44
12. Terres MA, Bechouel R, Ben Mohamed S (2017) Low cycle fatigue behaviour of nitrided layer of 42CrMo4 steel. Int J Mater Sci Appl 6:18–27

Characterization of the Surface Roughness of a GFRP by a 3D Profilometer After Trimming

Meher Azouzi[1(✉)], Ated Ben Khalifa[2], Anne Collaine[3], and Michel Tourlonias[3]

[1] Laboratoire de Physique et Mécanique Textiles (LPMT), ENSISA, Université de Haute Alsace, Alsace, Mulhouse, France
meher.azouzi@uha.fr
[2] Laboratoire de Génie Mécanique (LGM), ENIM, Université de Monastir, Monastir, Tunisie
[3] LPMT, ENSISA, Mulhouse, France
{anne.collaine,michel.tourlonias}@uha.fr

Abstract. The aim of this study is to characterize the link between the cutting conditions used while trimming a composite material glass fiber reinforced polymer (GFRP) and the surface quality by characterizing the area roughness of the machined surfaces. In the experimental evaluation, a central composite design with 20 combinations was used to study cutting parameters (cutting speed (V_c), radial engagement (a_e) and tooth feed (f_z)). The area roughness parameters (Sa, Sq, Sz) were measured by a KEYENCE VHX-6000 3D profilometer. Response Surface Methodology (MSR) were used to determine mathematical models using experimental data.

Keywords: GFRP · Composite trimming · Area roughness · 3D profilometer · Central composite design · MSR

1 Introduction

The composite materials industry is a sector that has made a great deal of progress over many years [1, 2]. Today, the manufacturing of this material is mainly divided between glass fiber reinforced composites (GFRP) and carbon fiber reinforced composites (CFRP). Although they are less efficient, GFRP composites now represent 95% by volume of the composites produced in Europe [3]. The process of obtaining a composite material part involves two stages: a first stage consists of manufacturing and a second stage called finishing. Trimming is one of the most common finishing processes used. Obtaining sufficient quality in terms of dimensional accuracy, surface finish and absence of residual fibers are the main difficulty in machining composites while at the same time assuring a certain productivity: speed of production, tool costs, etc. In view of the difficulties in obtaining finished quality workpieces, various studies are interested in proposing evolutions of the machining finishing process in order to improve the results [4, 5]. The combined effect of cutting parameters on surface quality or tool

N. Aifaoui et al. (Eds.): CMSM 2019, LNME, pp. 624–633, 2020.
https://doi.org/10.1007/978-3-030-27146-6_67

degradation in GFRP composite drilling operations has been the subject of extensive study [6–9]. In the case of milling GFRP composites, the influence on the surface quality of the machining direction in relation to the fiber orientation is known [10]. The same can also be said for the influence of cutting parameters [11]. However, several research works haven't shown a complete and global study on the trimming of GFRP composites combining tool choice and parameters choice and results, both in terms of quality and productivity. The definition of behaviors, generic theoretical principles are studied but the experiments carried out concern carbon fiber reinforced composites [12, 13]. In the present work, the roughness of the machined surface states is investigated. Indeed, we are interested in the use of area roughness parameters to describe the quality of the machined surface.

2 Response Surface Methodology

The principle is to model the experimental response surface. Among the many types of plans to build response surfaces, we will use a central composite design. Indeed, this technique has the advantage of being suitable for the sequential conduct of a study and requiring a relatively few experiments [14]. RSM allows to evaluate not only the linear effects but also quadratic effects as well as the effects of interaction between the different operating variables, the model used is the quadratic model with interactions:

$$Y = a_0 + \sum_i a_i X_i + \sum b_{ij} X_i X_j + \sum_j c_j X_j^2 \tag{1}$$

where a_0, a_i, b_{ij} and c_j are the regression coefficients of the model and X_i, X_j are the explanatory variables of the model.

3 Materials and Methods

3.1 Material

The composite material was made with Vetrotex 136 tex continuous glass fibers, reference ZTW EC13 136 tex TD22C with a diameter of 13 μm. The matrix is initially liquid thermoset matrix with two components: epoxy/amine system CY219 and Renlam HY5160 (Huntsmann). The test pieces are made by a method of filament winding to obtain a unidirectional composite. The equipment of the manufacturing composite plates is shown in Fig. 1. Indeed, the fiberglass filament is wound on a parallelepiped mold animated by a rotational movement managed by a motor whose speed is controlled. The filament slides in a guiding system whose movement is controlled by the rotation of a worm drive, and is driven by the rotation of the mold. The combination of the rotational speed of the mold and that of the worm drive ensures the implantation of the glass fibers on the mold in the form of parallel lines with two possible orientations: 0° and 90°. The rotation of the worm drive is reversed at each end. At each inversion, the system is stopped manually to reassure manual impregnation with the resin.

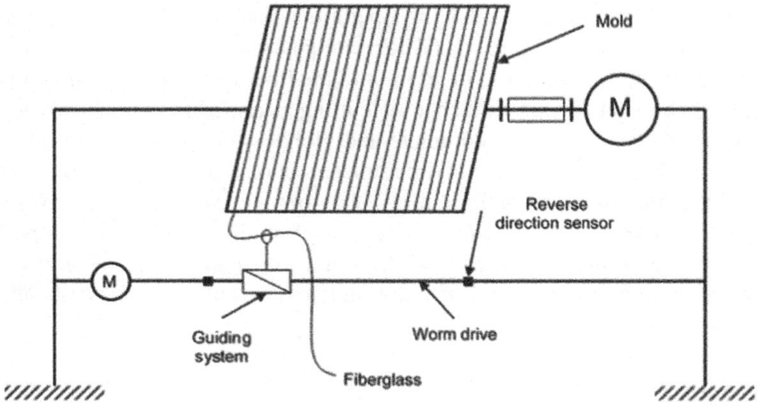

Fig. 1 Device for manufacturing composite plates

3.2 Machine and Tool Used

The machining tests were carried out on a Charly 2U machine equipped with a spindle turning up to 24000 tr/min, allowing feed rates up to 100 mm/s and with a maximum spindle power of 1.1 kW. For the cutting tool (Fig. 2), our choice fell on a one-piece SECO carbide milling cutter with a diameter d = 6 mm whose main characteristics are shown in Table 1.

Table 1 Characteristics of the tool used

Manufacturer's reference	Teeth number	Coating	Specifications
840060R050Z4.0- DURA	4	DURA	Double helix to avoid delamination

Fig. 2 Cutting tool

3.3 Machining Strategy

During machining the composite plate is mounted between two flanges tightened by 8 screws (Fig. 3). During the trimming operation, the machining strategy adopted is climb milling. Indeed, the tool is driven by a translational movement with a feed rate V_f (mm/min) and a rotation speed N (tr/min). As part of this work, all the tests were carried out with a tool overhang D_{ep} = 1 mm and a width of plate a_p = 4 mm. The milling cutter goes out of the spindle about 25 mm.

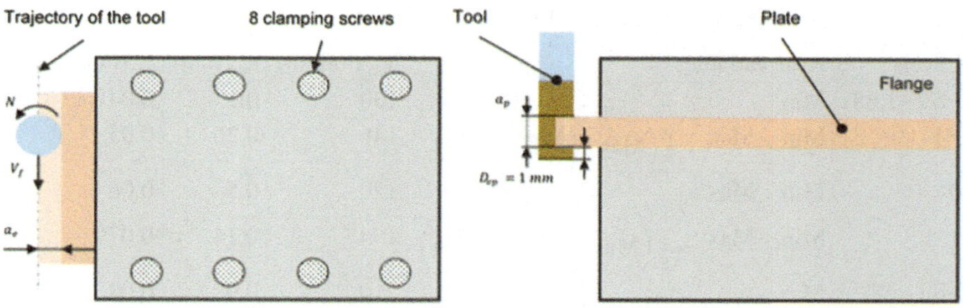

Fig. 3 Trimming strategy

4 Design of Experiments

A central composite design with 20 combinations (Fig. 4) was chosen to study the influence of the three selected cutting parameters (cutting speed V_c, radial engagement a_e and tooth feed f_z). The values of parameters were established in accordance with the recommendations provided by the SECO cutting tool manufacturer.

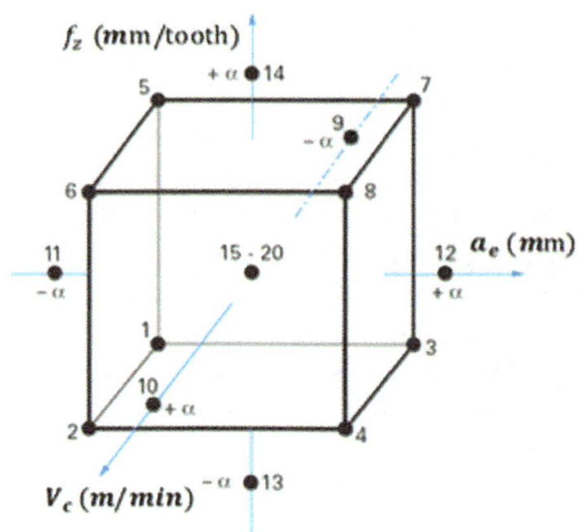

Fig. 4 Central composite design

The experimental design consists of 20 tests: 8 tests corresponding to the orthogonal plane, 6 tests for the axial points and 6 tests at the central point (0, 0, 0) to check the repeatability. The domain of variation of the coded factors is brought back to the interval [Min; Max] divided into 5 levels: −δ, −1, 0, 1, +δ which are calculated according to the expressions given in Table 2, where: $\delta = \sqrt{4}N$ and N is the number of fractional plan tests. Here N = 8, so: $\delta = \sqrt{48} = 1.68$.

Table 2 Variation of the cutting parameters and their levels

Levels	Real variables	V_c (m/min)	a_e (mm)	f_z (mm/tooth)
$-\delta$ (-1.68)	Min	130	0.1	0.04
-1	$\frac{\text{Min} + \text{Max}}{2} - \frac{1}{\delta}\left(\text{Max} - \frac{\text{Min} + \text{Max}}{2}\right)$	146	0.26	0.05
0	$\frac{\text{Min} + \text{Max}}{2}$	170	0.5	0.06
1	$\frac{\text{Min} + \text{Max}}{2} + \frac{1}{\delta}\left(\text{Max} - \frac{\text{Min} + \text{Max}}{2}\right)$	194	0.74	0.07
$+\delta$ (1.68)	Max	210	0.9	0.08

5 Analysis Method of Area Roughness

We have at our disposal a 3D KEYENCE VHX-6000 digital microscope that will allow us to measure area roughness parameters. This microscope is essentially characterized by: advanced lighting and focus, multiple illumination and direct depth composition. With digital microscope, the operator can tilt the lens and camera at a maximum angle of 90°, eliminating the need to manipulate the object by hand and thus facilitate control. The XY stage allows a displacement range of 100 × 100 mm so observation and image assembly are faster. For magnification we can go from 20× to 2000× with one lens. The aim is to characterize the machined surface with area roughness parameters. The microscope used allows us to scan the entire machined surface to calculate the desired parameters. The scanning operation of the entire surface to do the 3D reconstruction is only done if the lens and the camera are perpendicular. In our case we cannot satisfy this condition because of the length of the machined samples. For this reason, we have designed a support inclined at 60° to the vertical to measure all the machined plates. In order to properly characterize the machined surface, we chose to use a 1000× magnification and to average 8 evenly distributed measurement points (Fig. 5). For each "point" we defined a square measurement zone of 250 µm of side and recorded the values of Sa, Sq and Sz by setting the filters to the following values: S-Filter = 2 µm, L-Filter = 0.25 mm. This strategy is based on the instructions in ISO 25178-2.

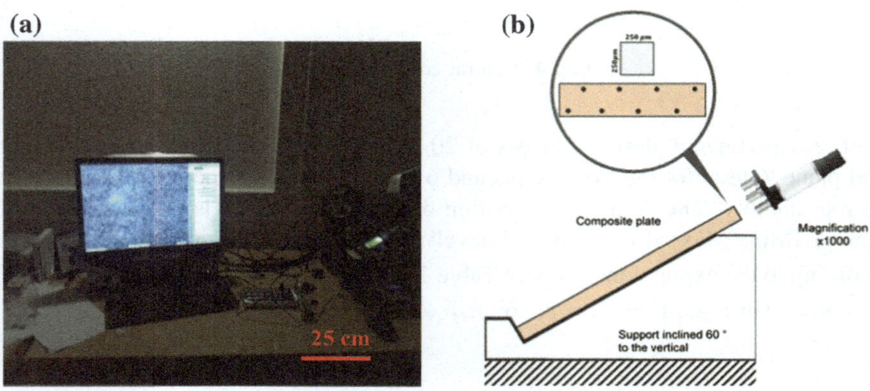

Fig. 5 **a** Area roughness measurement instrument, **b** Area roughness measurement strategy

6 Results and Discussion

The analysis of the machined surface state according to the experimental procedure detailed above, allowed us to recover several area roughness parameters. Among these parameters, the choice was fixed on two parameters which have a mean of measurement which are Sa and Sq and a parameter which describes a maximum height which is Sz. Indeed, these parameters are defined as follows:

$$S_a = \frac{1}{A}\iint_A |Z(x,y)| dx\, dy, \tag{2}$$

$$S_q = \sqrt{\frac{1}{A}\iint_A Z^2(x,y) dx\, dy}; \tag{3}$$

$$S_z = \max_A Z(x,y) + \left|\min_A Z(x,y)\right| \tag{4}$$

6.1 Results

The different results of the area roughness parameters are presented in the following Table 3.

The recovered results show a coincidence at the level of the trend of evolution for the Sa and Sq responses and this one is entirely logical considering the similarity of these two parameters. The found values of the area roughness are in line with the recommendations of the manufacturer VHX.

For the area roughness response Sz, it has the same trend of evolution as the area roughness response Sa with a difference in values as shown in Fig. 6.

Table 3 Experimental matrix and area roughness responses

Test no.	V_c (m/min)	a_e (mm)	f_z (mm/tooth)	Sa (μm)	Sq (μm)	Sz (μm)
1	146	0.26	0.05	1.9	2.9	26.2
2	194	0.26	0.05	2.2	3.6	33.0
3	146	0.74	0.05	2.7	4.1	34.5
4	194	0.74	0.05	2.5	3.9	37.5
5	146	0.26	0.07	2.8	4.1	37.2
6	194	0.26	0.07	3.1	4.5	34.9
7	146	0.74	0.07	3.1	4.8	42.8
8	194	0.74	0.07	3.1	4.7	41.0
9	130	0.5	0.06	2.2	3.1	29.1
10	210	0.5	0.06	3.0	4.4	36.3
11	170	0.1	0.06	2.2	3.2	29.6

(continued)

Table 3 (*continued*)

Test no.	V_c (m/min)	a_e (mm)	f_z (mm/tooth)	Sa (µm)	Sq (µm)	Sz (µm)
12	170	0.9	0.06	2.8	3.9	29.3
13	170	0.5	0.04	3.7	5.3	36.3
14	170	0.5	0.08	3.3	5.0	38.8
15	170	0.5	0.06	1.6	2.4	23.3
16	170	0.5	0.06	2.3	3.5	34.7
17	170	0.5	0.06	2.3	3.4	31.8
18	170	0.5	0.06	1.7	2.3	21.1
19	170	0.5	0.06	3.3	5.2	45.2
20	170	0.5	0.06	4.0	5.7	42.1

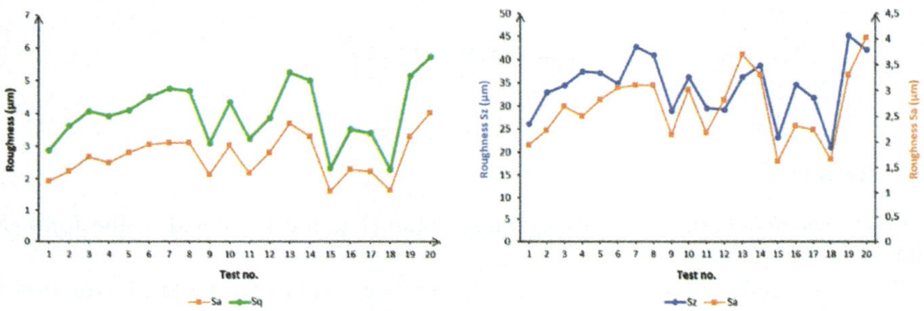

Fig. 6 Variation of the area roughness parameters Sa, Sq and Sz

6.2 Results Study with "Minitab"

For the study of surface roughness results with Minitab, we will limit ourselves only to the study of the area roughness response Sa.

Analysis of the Empirical Model

The studied model takes into account the linear, interaction and quadratic effects. The regression coefficients of this model make it possible to express the relation of the area roughness Sa according to the three parameters studied:

$$S_a = 6.9 + 0.011\,V_c + 6.1\,a_e - 262 f_z - 0.000005\,V_c \times V_c - 0.55\,a_e \times a_e + 2333 f_z \times f_z - 0.0160\,V_c \times a_e + 0.07\,V_c \times f_z - 35\,a_e \times f_z \quad (5)$$

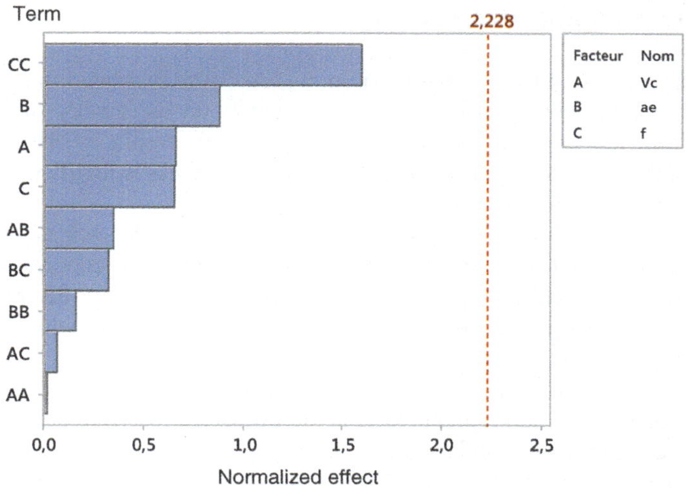

Fig. 7 Pareto diagram of the area roughness response Sa

The Fig. 7 shows that the tooth feed-tooth feed interaction is the most influential. It is followed by radial engagement, cutting speed and tooth feed.

Analysis of Response Surfaces

The relationship between the different operating variables and the response studied is illustrated in the three-dimensional representations of the response surfaces:

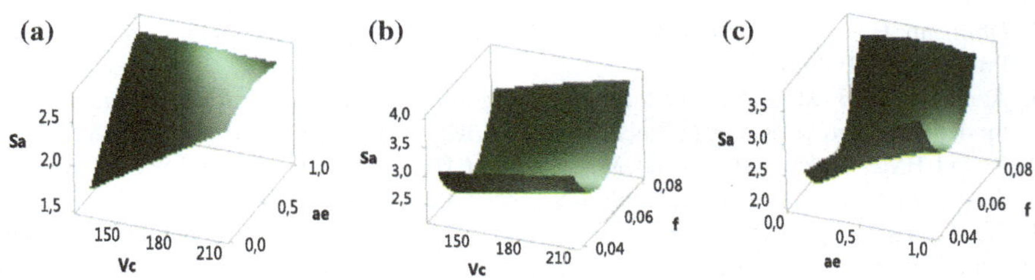

Fig. 8 Response surface results: **a** f_z = 0.006 mm/tooth, **b** a_e = 0.5 mm, **c** V_c = 170 m/min

- The Fig. 8a illustrates the effect of the interaction (V_c, a_e) on the area roughness Sa, more a_e increases more Sa is high.
- The Fig. 8b illustrates the effect of the interaction (V_c, f_z) on the area roughness Sa, the minimum value of Sa corresponds to the central values of V_c and f_z.
- The Fig. 8c illustrates the effect of the interaction (a_e, f_z) on the area roughness Sa, the influence of a_e is more important than f_z.

7 Conclusions and Perspectives

In this work, various GFRP composite trimming tests were presented in order to study the influence of the cutting parameters V_c, a_e and f_z on the quality of the surfaces obtained. From the study of the surface quality and the results of the experimental designs we can conclude that the machined surface roughness is highly sensitive to the variation of the cutting parameters.

Many approaches can be considered to progress in this work, such as improving the experimental protocol, testing other tool engagements (small pass depth and high radial engagement), investigating larger variation ranges of f_z and V_c, testing parameter variations to see if the influence is the same, completing the study with work on tool lifespan …

While the first results are encouraging and the prospects numerous, they highlight the need to quantify vibrations and to have at our disposal a means of characterizing the vibratory behavior of the workpiece during trimming. Vibration measurement will help to better define the operating ranges of a tool in relation to cutting conditions.

References

1. Les composites: des matériaux d'avenir-Partie 2: Marché et évolutions
2. (2017) 2016: a good year for composites. JEC Compos Mag (111):21–23
3. Witten E, Kraus T, Kuhnel M (Nov 2016) Composites market report 2016-market developments, trends, outlook and challenges
4. Mehbudi P, Baghlani V, Akbari J, Bushroa AR, Mardi NA (2013) Applying ultrasonic vibration to decrease drilling-induced delamination in GFRP laminates. Procedia CIRP 6:577–582
5. Uhlmann E et al (2016) Machining of carbon and glass fibre reinforced composites. Procedia CIRP 46:63–66
6. Azmi AI (2015) Monitoring of tool wear using measured machining forces and neuro-fuzzy modelling approaches during machining of GFRP composites. Adv Eng Softw 82:53–64
7. Kavad BV, Pandey AB, Tadavi MV, Jakharia HC (2014) A review paper on effects of drilling on glass fiber reinforced plastic. Procedia Technol 14:457–464
8. Parida AK, Routara BC, Bhuyan RK (2015) Surface roughness model and parametric optimization in machining of GFRP composite: Taguchi and Response surface methodology approach. Mater Today Proc 2:3065–3074
9. Palanikumar K (2011) Experimental investigation and optimisation in drilling of GFRP composites. Measurement 44:2138–2148
10. Venu Gopala Rao G, Mahajan P, Bhatnagar N (2007) Machining of UD-GFRP composites chip formation mechanism. Compos Sci Technol 67:2271–2281
11. Parida A-K, Das R, Sahoo AK, Routara BC (2014) Optimization of cutting parameters for surface roughness in machining of GFRP composites with graphite/fly ash filler. Procedia Mater Sci 6:1533–1538

12. Boudelier A, Ritou M, Garnier S, Furet B (2013) Investigation of CFRP machining with diamond abrasive cutters. J Compos Adv Mater 23:1–X
13. Haddad M, Zitoune R, Eyma F, Castanie B (2014) Study of the surface defects and dust generated during trimming of CFRP: influence of tool geometry, machining parameters and cutting speed range. Compos Part A
14. Box GEP, Draper NR (1987) Empirical model building and response surface. Wiley, New York

Mechanical Behavior of Titanium Aerospace Alloy: TA6V ($TiAl_6V_4$) Obtained Through an Identification Strategy Using CPB06 and Barlat Yield91 Criteria

Rym Harbaoui[1(✉)], Amna Znaidi[2], and Rachid Nasri[1]

[1] Laboratory of Applied Mechanics and Engineering LR-MAI, University Tunis El Manar, ENIT BP37- Le belvédère, 1002 Tunis, Tunisia
rym.harbaoui@gmail.com, rachid.nasri@enit.rnu.tn
[2] IPEIEM: Preparatory Institute for Engineering Studies of El Manar, University Tunis El Manar, BP244, 2092 Tunis, Tunisia
amna.znaidi@laposte.net

Abstract. The applications of titanium alloys are traditionally present in the field of transportation, basically in aircraft engines and turbo-reactors. The major advantage of these titanium alloys is linked to their exceptional mechanical behavior presenting a good mechanical resistance in different conditions of solicitations, an important duration of life associated to a low density; these are capital parameters for this kind of applications. Therefore, titanium (mainly TA6V alloy) is now widely used in the conception of high precision manufactured products such as airplane's engine and airframes skin including fan blades for turbojet engine demanding such advantages. The present work considers the study of the plastic behavior of TA6V as an aerospace material alloy. To achieve such finality, a modeling study followed by an identification strategy is imposed. Thus, the use of constitutive law taking into consideration the microstructural state of titanium is essential to have a reliable model for a further implementation in finite element software. The results found by this modeling study and the identification using CPB06 and Barlat Yield 91 as plastic criteria, will serve to study the anisotropic mechanical behavior of this material under several solicitations.

Keywords: Constitutive laws · Identification · Anisotropy · Titanium

1 Introduction

In the field of aerospace, titanium has been applied for many years. Ti-6Al-4V titanium alloy as represented by (TA6V) is mainly used for the airframe and the engine parts respectively. Since the study of the microstructure of a material enables the

N. Aifaoui et al. (Eds.): CMSM 2019, LNME, pp. 634–642, 2020.
https://doi.org/10.1007/978-3-030-27146-6_68

identification of the physical phenomena responsible for the plastic behavior, an adequate modeling of titanium by non-quadratic criteria taking into account the complex structure of this material is needed. This paper investigates the plastic behavior of titanium as represented by TA6V (grade 5) using the CPB06 and Barlat Yield 91 criteria in two different allotropic states.

2 Titanium in Aerospace

Titanium and titanium alloys are excellent candidates for aerospace applications [1] owing to their high strength to weight ratio and excellent corrosion resistance [2]. Titanium utilization is, however, strongly limited by its higher cost relative to competing materials, primarily aluminum alloys and steels (Fig. 1).

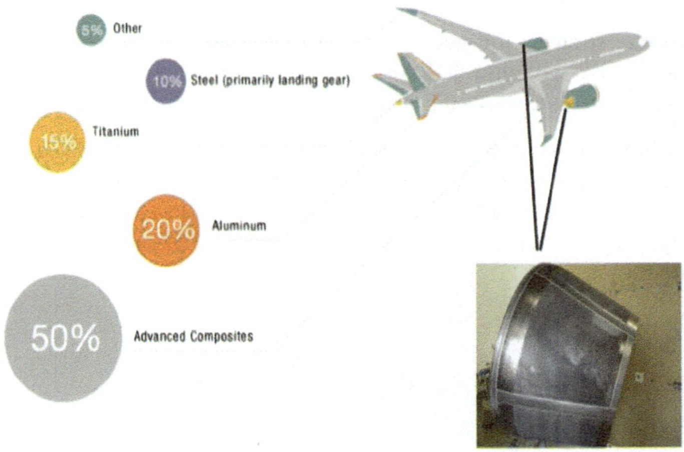

Fig. 1 Titanium alloys in aerospace

3 Titanium Alloys Allotropic Status

Titanium can exist in two crystalline forms which differ in the nature of the crystal lattice with reversible transformation at 882 °C. Titanium can crystallize in the Hexagonal Compact system (HC). Titanium is the alpha (α Ti) which is the most stable form at room temperature. It may also crystallize in the cubic system centered (CC); is beta titanium (Ti β) which exists at elevated temperatures up to melting point. The α and β forms are two allotropic varieties β being lighter than α (Fig. 2).

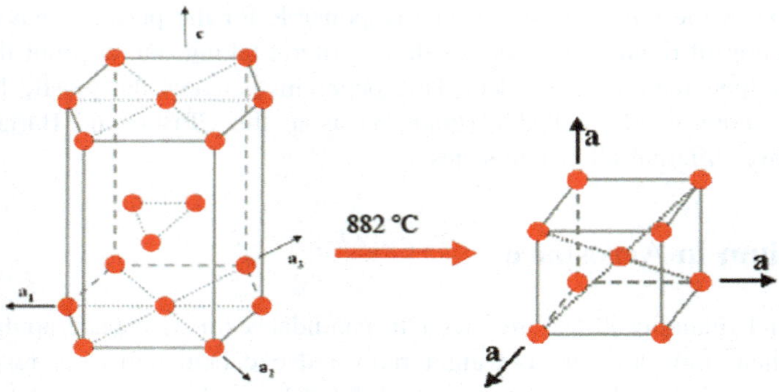

Fig. 2 Elementary mesh in both allotropic states (Revil2010)

4 Identification Model

In this work, the strategy followed takes into consideration: the anisotropic material behavior, the law of evolution and the equivalent stress.

For the identification phase the following assumptions have been adopted:

- Identification through "small perturbations" process, the tests used are treated as homogeneous tests, we neglect the elastic deformation; the behavior is considered rigid plastic incompressible, the plasticity surface evolves homothetically (isotropic hardening) and all tests are performed in the plane of the sheet resulting in a plane stress condition.

The yield criterion is then written as follows:

$$f(\overline{\sigma}^D, \alpha) = \sigma_c(\overline{\sigma}^D) - \overline{\sigma_s}(\alpha) \tag{1}$$

where σ^D is the deviator of the Cauchy stress tensor (incompressible plasticity) and σ_c is the equivalent stress.

The function $\sigma_c(\overline{\sigma}^D)$ satisfies the following condition for all a > 0:

$$\sigma_c(a\overline{\sigma}^D) = a\sigma_c(\overline{\sigma}^D) \tag{2}$$

The evolution of the surface charge is represented in the space deviators of stresses which are defined as follows:

$$\begin{cases} \overline{X}_1 = |\sigma^D| \cos\theta \\ \overline{X}_2 = |\sigma^D| \sin\theta \cos 2\psi \\ \overline{X}_3 = |\sigma^D| \sin\theta \sin 2\psi \end{cases} \tag{3}$$

Using the special setup of the space deviators, the general form of the equivalent orthotropic plan stress is thus

$$\sigma_c(\overline{\sigma}^D) = \sigma_c(\overline{X}_1, \overline{X}_2, \overline{X}_3) = |\overline{\sigma}^D| f(\theta, 2\psi) \tag{4}$$

Any type of criterion can be written in the form Gronostajsk [3]:

$$f(\theta, 2\psi) = |\overline{\sigma}^D| \sigma_s(\alpha) \tag{5}$$

where θ is the angle that defines the test and ψ the off-axis angle.

We will model a law that allows the use of non-quadratic criteria for the identification of anisotropic parameters in HC and CC structure and using Eqs. (1), (2), (3) and (4) we obtain the function that gives the laws of evolution based on anisotropic coefficients [4, 5] (Table 1).

Table 1 Values of θ for several tests

Test	Expansions equibiaxe (E.E)	Simple traction (S.T)	Large traction (L.T)	Simple shear (C.S)
θ	0	π/3	π/6	π/2

5 Identification Procedure

To identify a material, an identification of hardening curves is required (adequate analytical law is chosen). Thus, the identification of parameters that define the anisotropy of the material is required [6]. For this step the non-quadratic yield criterion CPB06 is chosen as yield surface and analytic law Ludwick is used and it's written as follows:

$$\sigma_s = \sigma_0 + k\varepsilon^n \tag{6}$$

Cazacu et al. developed a yield criterion for compact hexagonal materials [7, 8]. The plastic behavior is no longer considered balanced in tension and compression [9].

The criterion is expressed as follows:

$$\sigma^c = \left(\sum_{i=1}^{3} (|q_i| - kq_i)^a \right)^{1/a} \tag{7}$$

To ensure the convexity of the plasticity surface, we fix a $\geq$ 1, 'a' is the degree of homogeneity also known as the shape coefficient.

k: is the material parameter, expressible solely in terms of the ratio between, σ_T the uniaxial yield in tension, and σ_C the uniaxial yield in compression,

$$1 \leq \mathrm{k} \leq 1.$$

where q_1, q_2 and q_3 are the eigenvalues of a modified stress deviator tensor q defined as follows

$$q = C : \sigma^D \tag{8}$$

σ^D is the deviator of the Cauchy stress tensor (incompressible plasticity). The fourth order tensor C carries the anisotropy by 6 coefficients

$\sigma_s(\varepsilon^p)$: Isotropic hardening function; where ε^p is the equivalent plastic strain.

This criterion will be used if the titanium is considered (Ti-α). During the stabilization phase, we can also consider that during the conception of nozzles and some engine parts a deviation of temperature transforms the material to its second allotropic state (Ti-β); hence the use of a second non-quadratic criterion Barlat expressed as follows:

$$\sigma^c = \left(\sum_{i,j=1}^{3} \left(|q_i - q_j| \right)^a \right)^{1/a} \tag{9}$$

6 Results and Discussion

For the identification procedure the detailed model is used in a simplex algorithm.while all the assumptions have been respected. This operation consists in choosing the model coefficients while minimizing the squared difference between the theoretical and experimental results [10].

We notice good identification of hardening curves (see Fig. 3) (same experimental and theoretical results).

Our second phase for the identification is to determine the coefficients of anisotropy (F, G, H, N) and the shape coefficient "a" (Table 3), taking into account the criterion of Barlat with a = 6 for the identification of constitutive equations using materials of centered cubic structure (cc). As for CPB06 we represent the anisotropic coefficients of the forth order tensor and the material parameter k, for a fixed degree of homogeneity (shape coefficient) a = 2 (Table 2).

The identified model allows studying the load surface for each test (see Fig. 4). It is noted that while using Barlat criterion, this material is resistant vis-à-vis the simple shear when compared to simple traction. In contrast, the plastic flow in large traction (the sample length is comparable to its width) is rapidly reached. As for CPB06 criterion, the load surface evolves homothetically the same way for the different tests.

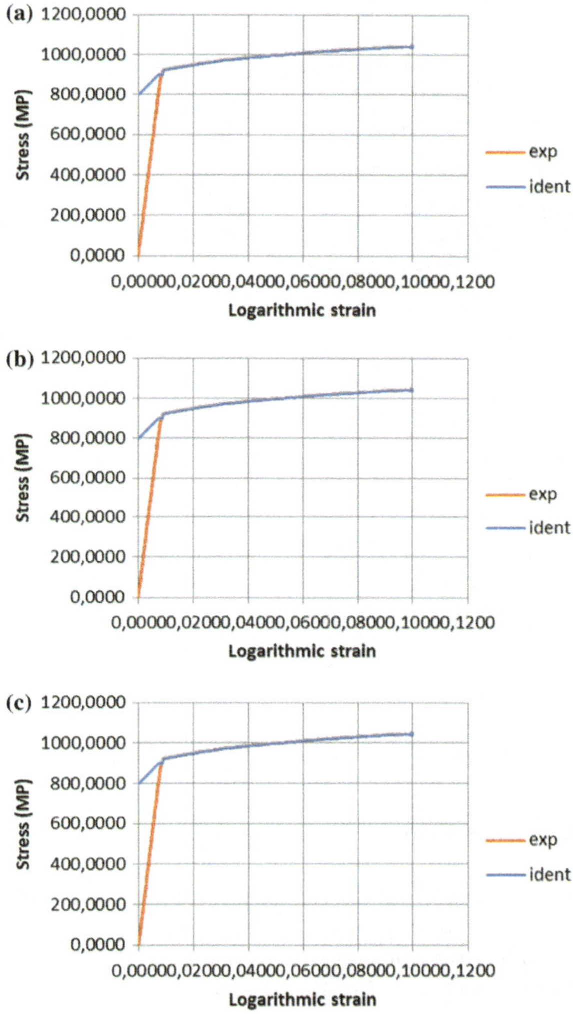

Fig. 3 Identification of hardening curves: **a** $\psi = 00°$ **b** $\psi = 45°$ **c** $\psi = 90°$

Table 2 Identification of the coefficients of anisotropy for a = 2 by CPB06

k	C_{11}	C_{12}	C_{13}	C_{22}	C_{23}	C_{33}	Error
−0.2	1	0.3296	0.4925	−0.5985	−0.9344	1.8839	1.0349e−9
0.002	1	1.1049	−0.4276	−0.8409	1.1157	−0.3414	1.0678e−8

$C_{44} = C_{55} = C_{66} = 1$

Table 3 Identification of the coefficients of anisotropy and the shape coefficient ‘a’ by Barlat Yield 91

F	G	H	N	a
0.2854	0.2064	0.3335	0.8921	6.9584

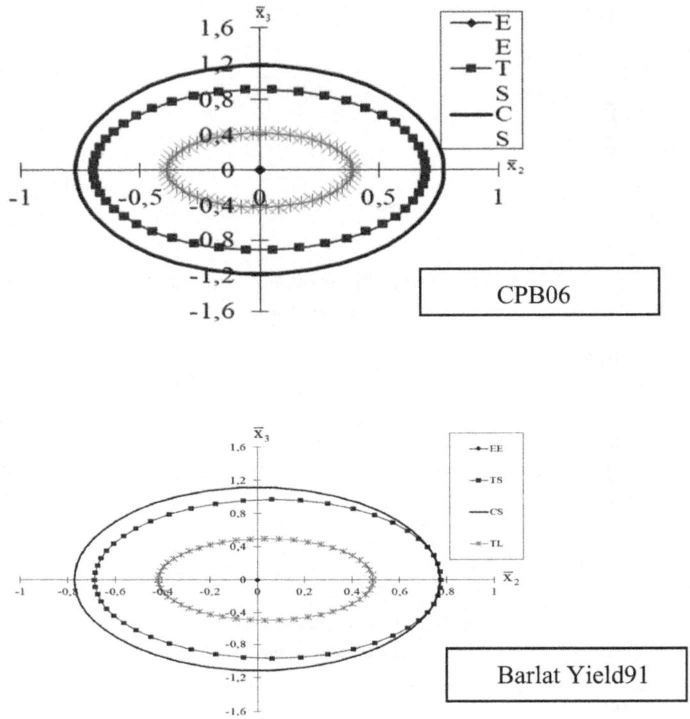

Fig. 4 Evolution of the load surface in the deviators plan (X_2, X3)

7 Numerical Simulation

Numerical simulation approach is developed to evaluate the anisotropic behavior of Boeing's 777 nozzle made of TA6V when subjected to pressure force. This approach consists on implementing the different stress-strain data and the Lankford coefficients for the three different directions of loading in-plane (00°, 45° and 90° to the rolling direction) (Fig. 5).

We notice that the von mises stress is located in the upper part of the nozzle while there's no important difference between the three different directions of loading in-plane (00°, 45° and 90° to the rolling direction).

The numerical convergence of the method would have been verified by implementing the CPB06 criterion into the Finite-element model [11]. Therefore, an explicit implementation of the adequate plastic criterion is needed to have valid results.

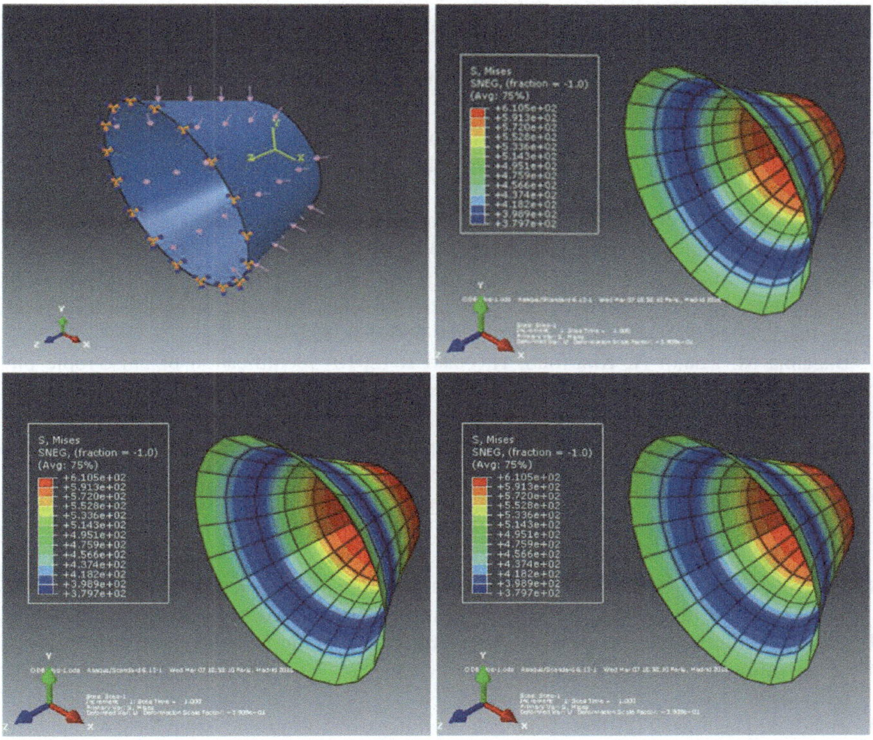

Fig. 5 Von mises stress results shown for three different directions of loading in-plane (00°, 45° and 90° to the rolling direction)

8 Conclusion

In this work, starting from an experimental database we developed an identification strategy that focused on the parameters of plasticity taking into consideration the allotropic state of the titanium material (Tiα and Tiβ) the constitutive law and the coefficients of Lankford.

Thus, for each alloy: the plastic model of behavior, the analytical law and the criteria should be presented. The different parameters of anisotropy for our studied material were identified by CPB06 and Barlat91 criteria. A validation by comparing the model with the experimental database was performed.

The evolution of load surface for the various tests was represented at the end of this identification. This strategy has allowed us to study more precisely the anisotropy of titanium. By a forthcoming integration of the two criteria considered for each allotropic state of titanium in software finite element we reach a high precision manufacturing.

References

1. Boyer RR (1996) An overview on the use of titanium in the aerospace industry. https://doi.org/10.1016/0921-5093(96)10233-1
2. Inagaki I et al (2014) Nippon steel, Sumitomo application and features of titanium for the aerospace industry metal technical report No. 106. http://www.nssmc.com/en/tech/report/nssmc/pdf/106-05.pdf
3. Gronostajsk Z (2000) The constitutive equations for fem analysis Journal of material processing technology. https://doi.org/10.1016/S0924-0136(00)00635-X
4. Daghfas O, Znaidi A, Gahbiche A et al (2017) Plastic behavior of 2024-T3 under uniaxial shear tests. In: International conference design and modeling of mechanical systems. Springer. https://doi.org/10.1007/978-3-319-66697-6_102
5. Baganna M, Znaïdi A, Kharroubi H, Nasri R (2010) Identification of anisotropic plastic behavior laws for aluminum 2024 after heat treatment materials from off-axis testing. In: International conference of plasticity, Toulouse. https://doi.org/10.3221/IGF-ESIS.38.18
6. Hill R (1948) A theory of the yielding and plastic flow of anisotropic metals. Proc Roy Soc Lond A 193:281–297. https://doi.org/10.1098/rspa.1948.0045
7. Cazacu O, Plunkett B, Barlat F (2008) Orthotropic yield criterion description of anisotropy in tension and compression of sheet metals. Int J Plast 24:847–866. https://doi.org/10.1016/j.ijplas.2007.07.013
8. Cazacu O, Plunkett B, Barlat F (2006) Orthotropic yield criterion for hexagonal closed packed metals. Int J Plast 22:1171–1194. https://doi.org/10.1016/j.ijplas.2005.06.001
9. Cazacu O, Ionescu, IR, Yoon JW (2009) Orthotropic strain rate potential for the description of anisotropy in tension and compression of metals. Int J Plast. https://doi.org/10.1016/j.ijplas.2009.11.005
10. Znaidi A, Daghfas O (2016) Identification strategy of anisotropic behavior laws application to thin sheets of aluminum A5. J Theor Appl Mech. https://doi.org/10.15632/jtam-pl.54.4.1147
11. Revil B (2009) Simulation de la mise en forme d'alliage de titane à froid, 19 ème Congrès Français de Mécanique Marseille. https://hal-mines-paristech.archives-ouvertes.fr/hal-00509374

Development of Sustainable Soft Flooring Material to Reduce Fall Injuries

Roohul Abad Khan[1](✉), Rachida El Morabet[2], and Amadur Rahman Khan[3]

[1] Department of Civil Engineering, King Khalid University, Abha, Saudi Arabia
rakhan@kku.edu.sa
[2] Department of Geography, Hassan II University of Casablanca, Casablanca, Morocco
rachidaelmorabet@yahoo.fr
[3] Faculty of Architecture, Aligarh Muslim University, Aligarh, India
amadarch@gmail.com

Abstract. Fall injuries are experienced worldwide ranging from fatalities to hospitalization especially in Elderly population. The available soft flooring options are restricted to specific buildings and location. The study developed soft flooring material incorporating crumb rubber from waste tyres to provide cushion effect in flooring to induce cushion effect. The concrete specimens were developed in form of tiles/slab with varying thickness of 10, 20 and 30 mm. In total 36 specimens were tested with 3 × 2 blank reference specimen without crumb rubber. The specimens were tested in accordance to the force required to cause fracture in human leg bone (192 N) while the testing energy was 198 N. The height of drop was kept 45 cm. The softness was determined with respect to contact time of the impact to the surface while deflection was obtained using dial gauges of 0.01 mm accuracy. The test results of specimen showed two specimens absorbing more than 50% of impact energy. Thereby reducing the chances of fatal/major injuries by 50%. The developed material also solves the big problem of waste rubber tyre disposal in a sustainable manner. The developed material can be successfully applied various location and buildings. Hence, the developed material is green and sustainable both for humans and environment.

Keywords: Fall injuries · Soft flooring · Crumb rubber · Impact energy · Sustainable

1 Introduction

Elder people often fall. From among the population of age 65 and above 28–35% fall annually [4]. This is a major public health concern incurring significant effect in terms of healthcare costs. Falls threaten the independence of elder people and is commonly experienced geriatric syndromes. Additionally, falls are the prime cause of fatal injuries and disabilities for people of 65 years age and above. This renders falls as the most complex and expensive health issue concerning older population worldwide [2]. Each day children are hospitalized due to fall related injuries which also calls for promoting their safety during their stay in hospitals [5]. Children suffering from head injuries due

N. Aifaoui et al. (Eds.): CMSM 2019, LNME, pp. 643–649, 2020.
https://doi.org/10.1007/978-3-030-27146-6_69

to fall might experience effect in motor and cognitive skills as a long term effect [7]. Even the hospitalized patients are not safe from fall injuries which range between 1.4 to 17.9/1000 owing to health services and patients conditions. The fall injuries can reach 30%, with major injuries (fractures, bleeding or death) from 4 to 6% [11]. Injuries contribute around 12% of the worldwide disease burden. Falls account for 424,000 deaths out of 5 million deaths, leading it to become second cause of unintentional injury death globally only second to road traffic injuries [8]. The main culprit of falls and fall injuries are flooring material not because they cause it always but because of their hard and tough material properties. If the flooring material can be softer than conventional concrete, stone, tile etc. the chances of injuries can be reduced. Here is where the demand for soft flooring material brings crumb rubber into the scenario.

Waste rubber tyre are source of crumb rubber which are cheap, inexpensive and widely available. Crumb rubber is being investigated for various properties of concrete for past few decades. Crumb rubber has been employed to study structural properties of concrete like fresh and hardened properties of SCC [6], Compressive stress, Li et al. [15], mechanical and durability [10], bonding [14], Mechanical Properties of high fly ash concrete [1], toughness [16], longitudinal shear resistance [13] as well as Non-structural Applications [3], acoustical properties [9, 12]. The number of research work on influence of crumb rubber on various properties of resultant concrete matrix is high but when it comes to shock absorption capacity it is lacking.

This study attempts to investigate the impact/shock absorbing capacity of the crumb rubber concrete material. The objective of the study is to evaluate the capacity of the developed specimen to absorb impact energy successfully. The energy absorbed will directly correspond with reduction in percentage of injuries due to fall. Additionally study will convert waste rubber tyre into a resource for building material rendering material as green and sustainable.

2 Methodology

In total 30 specimens were casted in form of concrete rubber tiles. Additionally, 3 control specimens with concrete no rubber crumb rubber were also casted. Crumb rubber were originated from waste tyre retreading company. The crumb rubber used was 100% waste from treading company with no processing whatsoever to be employed in the concrete tile. The thickness of concrete was varied from 10, 20 and 30 mm to evaluate its effectiveness in absorbing impact energy. The specification of Concrete is given in Table 1.

Stainless steel ball of 50 mm diameter was used for testing concrete tile. The steel ball was dropped (free fall) from a height of 45 cm. The height was selected as a standard height of bed and chair from where usually fall may occur.

Table 1 Concrete and its materials specification

Parameters	Value
Grade	M30
Proportion of aggregate	0.62
Proportion of sand	0.38
Specific gravity of cement	3.12
Specific gravity of aggregate	2.74
Specific gravity of sand	2.72
Nominal size of aggregate	10 mm
Workability	75 mm
Water to cement ratio	0.45

3 Energy Calculation for Causing Fracture in Bone

For considering effectiveness of developed material in reducing energy in terms of energy absorbed, it is necessary to calculate energy required to cause fracture in bone. This study considered leg of bones to calculate energy needed to cause fracture. Assuming bones as physical body, Hooke's law holds at small deformation,

$$Y = \sigma/\epsilon \tag{1}$$

where Y is young modulus, σ is stress and € is strain. Based on Hooke's law equation study derived following equation to calculate ultimate energy required to cause fracture in bone.

$$U = lS\sigma_B^2/2Y \tag{2}$$

where, U is Ultimate energy required to cause fracture in bone, l is average length of bone taken as 90 cm, s is area of bone 6 cm^2, Y is Youngs modulus 14 × 105, σB is breaking stress of bone 104 N/sm^2.

The calculated amount of force needed to cause fracture in bone comes out to be 192.5 N or J/m. This force was calculated for a person of weight 70 kg.

4 Result and Discussion

The tested specimens were evaluated in terms of energy absorbed based on deflection and contact time. The softness of material was determined based on the contact time as energy dissipates on impact due to increased contact time.

From Fig. 1 it can be derived that in plain concrete with no crumb rubber cracks at failure are linear, uni-directional and wide. While in specimen incorporating crumb, rubber displays multiple non-linear cracks but with reduced width. The linear and wide crack displays sudden failure typical property of concrete while incorporating crumb

Fig. 1 Crack development on surface

rubber displays multiple cracks inducing tensile properties in concrete specimen. The crumb rubber concrete specimens were tested to observe first crack and no. of blows were continued till ultimate failure. The fig above depicts the crack at ultimate failure thus the cracks appear to be much wider than at initial failure. This is significant since the injury due to fall will occur at first failure and rebound impact is most unlikely to occur in case of fall. Thus, it can de deduced that crumb rubber can induce tensile properties of concrete which can be successfully applied to be used as soft flooring material. However, application for structural application needs further research and investigation.

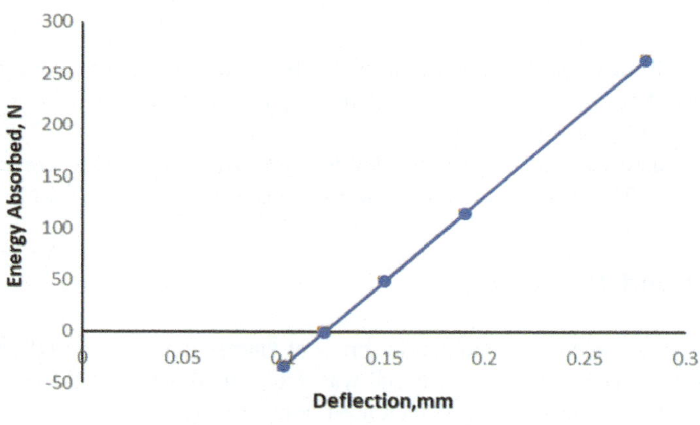

Fig. 2 Energy absorbed versus deflection, 10 mm concrete top thickness

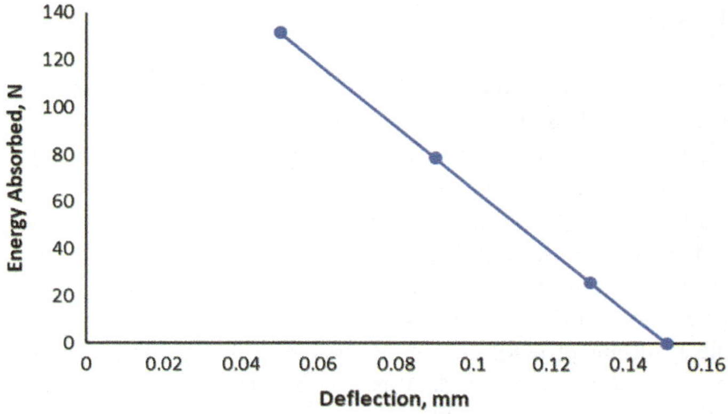

Fig. 3 Energy absorbed versus deflection, 20 mm concrete top thickness

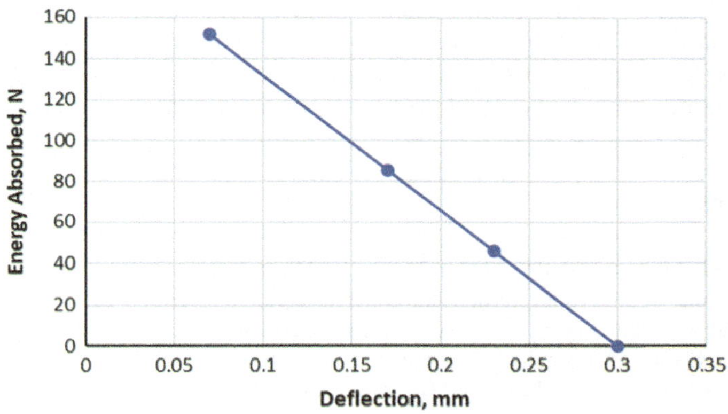

Fig. 4 Energy absorbed versus deflection, 30 mm concrete top thickness

The increased contact time is directly proportional to softness of material. The increased contact time dissipates the energy over longer period of time giving soft and cushion effect to object in contact. In terms of Contact time it can be said that all specimen provides soft feeling or cushion effect. But when specimen was analyzed based on deflection it was discovered that 20 mm (Fig. 3) and 30-mm (Fig. 4) thickness provided softness effect and absorbed energy and were in agreement with the results of contact time theory. But when 10 mm thickness sample Fig. 2 was analyzed it was discovered that only one specimen absorbed impact energy while other four specimens displayed spring effect. The impact increased contact time but it instead of energy dissipation it caused energy accumulation.

The developed material can successfully absorb impact energy and thus can reduce injuries. The specimen with 20 and 30 mm thickness depicted chances of injury by more than 50%.

5 Conclusion

The injuries due to fall are a major health concern especially in elder population. The injuries are not only health issue but also need expensive treatment along with increased stress of the family concerned. The soft flooring material available is mainly restricted to hospital buildings or special care units that too in very low percentage. Adding to which the increased cost of special soft flooring material again limits its application. The study successfully developed a material which can successfully overcome shortcoming of existing material. The finding of the study suggest that crumb rubber can be successfully incorporated in flooring material in combination with concrete. The resultant product can be applied in most of the building floors as the thickness of the material can be varied as per the loading condition of the floor. The energy absorption capacity was achieved above 50% successfully. Hence developed material can effectively reduce fall injuries by absorbing impact energy reducing probability of injury by more than 50%. The material converts waste into resource by incorporating crumb rubber as building material. This solves a major problem of waste tyre disposal and its environmental impacts. Hence, developed material is green and sustainable to use. The future work can employ other material to increase durability. Sensors can be used to derive more variables influencing the performance of the material. The material can be developed evolved as per field performance.

References

1. Adamu M, Mohammed BS, Shahir M (2018) Mechanical properties and performance of high volume fly ash roller compacted concrete containing crumb rubber and nano silica. Constr Build Mater 171:521–538. https://doi.org/10.1016/j.conbuildmat.2018.03.138
2. Al-faisal W (Oct 2006) Falls prevention for older persons, 1–7
3. Aliabdo AA, Elmoaty A, Elmoaty MA, Abdelbaset MM (2015) Utilization of waste rubber in non-structural applications. Constr Build Mater 91:195–207. https://doi.org/10.1016/j.conbuildmat.2015.05.080
4. Almegbel FY, Alotaibi IM, Alhusain FA, Masuadi EM, Al Sulami SL, Aloushan AF, Almuqbil BI (2018) Period prevalence, risk factors and consequent injuries of falling among the Saudi elderly living in Riyadh, Saudi Arabia: a cross-sectional study. BMJ Open 8(1):2–10. https://doi.org/10.1136/bmjopen-2017-019063
5. AlSowailmi BA, AlAkeely MH, AlJutaily HI, Alhasoon MA, Omair A, AlKhalaf HA (2018) Prevalence of fall injuries and risk factors for fall among hospitalized children in a specialized children's hospital in Saudi Arabia. Ann Saudi Med 38(3):225–229. https://doi.org/10.5144/0256-4947.2018.225
6. Aslani F, Ma G, Law D, Wan Y, Xuan V, Le T (2018) Experimental investigation into rubber granules and their effects on the fresh and hardened properties of self-compacting concrete. J Clean Prod 172:1835–1847. https://doi.org/10.1016/j.jclepro.2017.12.003
7. Assiry KA, Abdulmutali HA, Alqahtani AA, Alyahya AA, Elawad ME (2014) Traumatic head injuries in children: experience from Asir, KSA. Online J Med Med Sci Res 5(5):44–47
8. Bhatti JA, Farooq U, Majeed M, Khan JS, Razzak JA, Khan MM (2015) Fall-related injuries in a low-income setting: results from a pilot injury surveillance system in Rawalpindi, Pakistan. J Epidemiol Glob Health 5(3):283–290. https://doi.org/10.1016/j.jegh.2015.01.003

9. Bianca-maria Ş, Nastac D, Vasile O, Bratu M (2016) Sound absorbing materials made by embedding crumb rubber waste in a concrete matrix. Constr Build Mater 124:755–763. https://doi.org/10.1016/j.conbuildmat.2016.07.145
10. Bisht K, Ramana PV (2017) Evaluation of mechanical and durability properties of crumb rubber concrete. Constr Build Mater 155:811–817. https://doi.org/10.1016/j.conbuildmat.2017.08.131
11. Flimban MA, Abduljabar DF, Dhafar KO, Deiab BA, Gazzaz ZJ, Bansuan AU, Suliman MI (2016) Analysis of patient falls among hospitalised patients in Makkah region. J Pak Med Assoc 66(8):994–998. https://doi.org/10.3305/nh.2015.31.3.8292
12. Flores N, Flores-medina D, Hernández-olivares F (2016) Influence of fibers partially coated with rubber from tire recycling as aggregate on the acoustical properties of rubberized concrete. Constr Build Mater 129:25–36. https://doi.org/10.1016/j.conbuildmat.2016.11.007
13. Holmes N, Dunne K, Donnell JO (2014) Longitudinal shear resistance of composite slabs containing crumb rubber in concrete toppings. Constr Build Mater 55:365–378. https://doi.org/10.1016/j.conbuildmat.2014.01.046
14. Kashani A, Ngo TD, Hemachandra P, Hajimohammadi A (2018) Effects of surface treatments of recycled tyre crumb on cement-rubber bonding in concrete composite foam. Constr Build Mater 171:467–473. https://doi.org/10.1016/j.conbuildmat.2018.03.163
15. Li D, Zhuge Y, Gravina R, Mills JE (2018) Compressive stress strain behavior of crumb rubber concrete (CRC) and application in reinforced CRC slab. Constr Build Mater 166:745–759. https://doi.org/10.1016/j.conbuildmat.2018.01.142
16. Li Y, Li Y (2017) Experimental study on performance of rubber particle and steel fiber composite toughening concrete. Constr Build Mater 146:267–275. https://doi.org/10.1016/j.conbuildmat.2017.04.100

Comparative Evaluation of Natural Rubber Properties Blended with Almond Shells Powder with and Without Addition of New Bio-binary Accelerator System

Moez Kamoun[1], Marwa Allouch[1](✉), Jamel Mars[1], Mondher Wali[1,2], and Fakhreddine Dammak[1]

[1] Laboratory of Electromechanical Systems (LASEM), National Engineering School of Sfax, University of Sfax, 3038 Sfax, BP, Tunisia
kamoun_moez2@yahoo.fr, jamelmars@yahoo.fr, mondherwali@yahoo.fr, {marwa.allouch, fakhreddine.dammak}@enis.tn
[2] Department of Mechanical Engineering, College of Engineering, King Khalid University, Abha, Saudi Arabia

Abstract. The mechanical properties, curing characteristics, and swelling behavior of vulcanized natural rubber (NR) blended with almond shells waste powder (ASWP) with a binary accelerator system are investigated. Results indicate that the mechanical properties were improved. Cross linking density of vulcanized natural rubber was measured by the equilibrium swelling method. As a result, the binary accelerator was found to be able to improve both cure rate and cross-linking density. Using the numerical analysis of test interaction between binary accelerator and operational modeling of vulcanization-factor experiments, it can be concluded that the interaction (Garlic oil, Cystine) was significant and the optimum value of bio-binary accelerator was suggested.

Keywords: Characterization · Natural rubber · Almond shell powder
Bio-binary accelerator system

1 Introduction

The last years have seen an increase of rubber consumption unlike many other polymers. In fact, natural rubber (NR) was widely used as the matrix in the composite materials for most engineering applications. A lot of works were conducted on the improvement of the rubber behavior via different approaches. These researches highlight that the addition of the accelerator system in rubber vulcanization reactions gives superior properties [2, 5, 7]. Nevertheless, the choice of an accelerator in the sulfur vulcanization define the kind of network structure created and consequently the specific material properties. In fact, the type, the distribution and the density of cross links have an important effect on the physical properties of cross-linked rubber [4, 8].

In our previous studies, we investigate experimentally the rheological and mechanical properties of the vulcanizate cured with the binary accelerator system based

N. Aifaoui et al. (Eds.): CMSM 2019, LNME, pp. 650–654, 2020.
https://doi.org/10.1007/978-3-030-27146-6_70

on Cystine (Cys) and N-cyclohexyl-2-benzothiazyl sulfonamide (CBS) in conventional vulcanization system of natural rubber (NR) [6].

The purpose of the present study is to evaluate the effect of a new bio-binary system based on Natural Garlic Powder (NGP) and Cystine (Cys) in conventional vulcanization system of natural rubber (NR) blended with almond shell powders.

2 Materials and Methods

2.1 Material

Almond Shell powder and Natural Garlic Powder (NGP) were obtained from Lab-Ltd. NGP was a mixture of a garlic-rich organosulfur compound (OSCs) complex extracted from cloves of garlic. All other chemical ingredients were commercial grade. The based formulation of the mixes used is given in Table 1.

Table 1 Formulations for rubber compounds in phr

Materials (phr)	M0	M0	M0	M0	M4
Natural rubber	100	100	100	100	100
Almond shell powder	10	10	10	10	10
Zinc oxide	5.00	5.00	5.00	5.00	5.00
Stearic acid	2.00	2.00	2.00	2.00	2.00
Cystine (Cys)	0.80	0.80	0.80	0.80	0.80
NGP	0.00	0.25	0.50	0.75	1.00
Sulfur	2.50	2.50	2.50	2.50	2.50

phr, parts per hundred parts rubber

2.2 Composites Preparation

The mixture of natural rubber (NR) was blended using a two-roll mill at 70 °C. The mixing procedure was carried out in accordance with ASTM D 3182.

After the preliminary blending, the mixes were compression molded in an electrically heated hydraulic press at 150 °C for the optimum cure time.

3 Experimental Characterization

3.1 Crosslinking Density Measurement

The cross link density of NR compounds was determined through the equilibrium swelling method. 0.2 g was cut from the hot compacting sample and immersed in pure benzene at 25 °C until equilibrium for 5 days. After the conditioned period, the swelling was stopped and the samples were removed from the liquid. The adhered liquid was rapidly removed from the sample using a filter paper and the swollen weight was measured immediately. The swelling ratio is defined as,

$$R = (m_s - m_0)/m_0 \quad (1)$$

where m_0 is the weight of the initial specimen and m_s is the weight of the swollen specimen.

Then, the cross linking density is obtained by the following relation [1, 3]:

$$\gamma = -[\ln(1 - \psi_r) + \psi_r + \chi\psi_r]/Vs\left(\psi_r^{1/3} - \psi_r/2\right) \quad (2)$$

where V_s was the molar volume of the solvent, χ was the parameter characteristic of polymer–solvent interaction and ψ_r was the volume friction of rubber in the swollen NR vulcanizate.

3.2 Mechanical Properties

Five specimens were cut from each compacted sheets with a Wallace die cutter. Tensile tests were carried out according to ASTM D 412 using an Instron 4465 Universal machine. A cross-head speed of 500 mm/min was used and the tests were performed at ambient temperature.

The Shore A durometer was used for measuring the hardness of a material according to ASTM D 2240.

4 Results and Discussion

Figure 1 gives cross-linking density value measured by the equilibrium swelling method. The result indicates that cross-linking density increase proportionally with NFP content in elaborated samples. This increase is important and very recommended in elastomers. Networks containing high cross-link density is characterized by excellent behavior and their proprieties are acceptable.

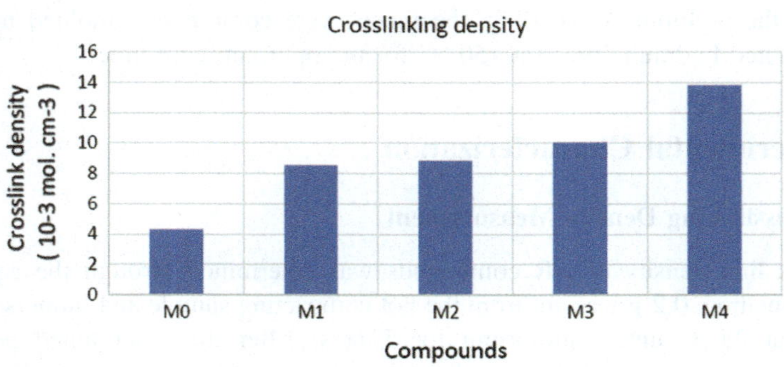

Fig. 1 Cross-linking density values versus NGP content

Figure 2 illustrates the increase of tensile strength with the addition of NGP and it reaches a maximum at a concentration of 0.5 phr of this accelerator. This improvement can be explained by the increase in cross-link density. On the other hand, the elongation at break values shows a decrease as the flexibility of materials decrease resulting from the increase in cross-link density (Fig. 3). The addition of bio-binary accelerator system (NGP, Cys) had also a positive effect on the hardness of samples (Fig. 4). The ideal properties were shown by mix of M2.

These results highlight that a concentration of 0.5 phr of NGP represents the optimum for the vulcanization system. This could be correlated with the amounts of cross-link density of the vulcanizate.

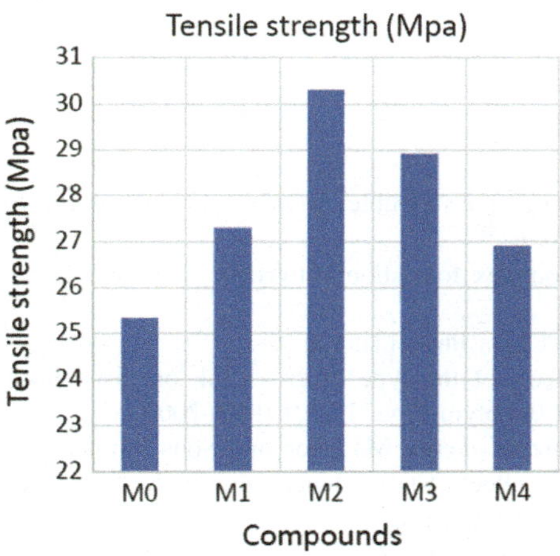

Fig. 2 Tensile strength versus NGP content

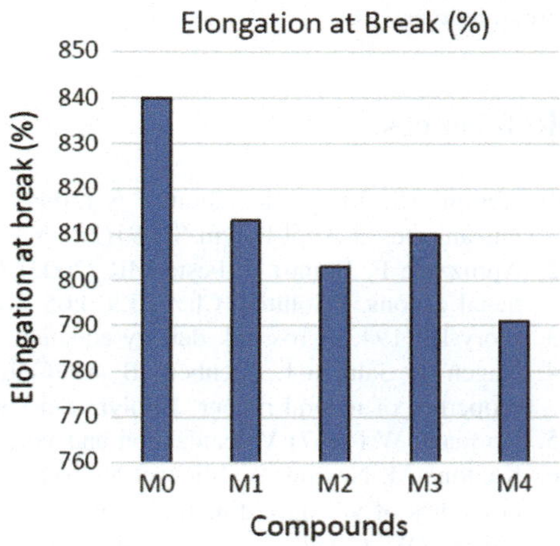

Fig. 3 Elongation at Break versus NGP content

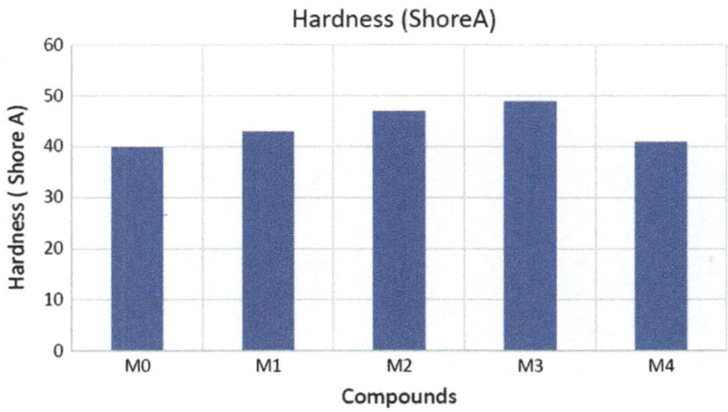

Fig. 4 Hardness determination versus NGP content

5 Conclusion

In the present study, the effect of the new binary accelerator system based on Natural Garlic Powder (NGP) and Cystine (Cys) has been investigated giving emphasis to the processing characteristics and mechanical properties. It is found that a NGP concentration of 0.5 phr improve the production rate and increase significantly cross-link density of the vulcanizates.

The tensile and hardness properties were better than or comparable to those of the reference mix.

Acknowledgements. This work is partially supported by .... The authors also gratefully acknowledge the helpful comments and suggestions of the reviewers, which have improved the presentation.

References

1. Aprem AS, Joseph K, Thomas S (2004) Studies on double networks in natural rubber vulcanizates. J Appl Polym Sci 91(2):1068–1076
2. Apruzzese F, Bottari E, Festa MR (2003) Complex formation between cystine and alkali-metal cations. J Solution Chem 32(1):65–76
3. Flory PJ (1943) Crosslink density equation. J Chem Phys 11:5120
4. Hagen R, Salmén L, Stenberg B (1996) Effects of the type of cross-link on viscoelastic properties of natural rubber. J Polym Sci Part B Polym Phys 34(12):1997–2006
5. Hofmann W (1967) Vulcanization and vulcanizing agents. Maclaren and Sons Limited
6. Kamoun M, Nassour A, Michael N (2009) The effect of novel binary accelerator system on properties of vulcanized natural rubber. Adv Mater Sci Eng
7. Layer RW (1987) A postcrosslinking accelerator system for natural rubb based on thiocarbamyl sulfenamides. Rubber Chem Technol 60(1):89–101
8. Marykutty CV, Mathew EJ, Thomas S (2008) Studies on a new binary accelerator system in sulfur vulcanization of natural rubber. Elastomers Plast, 383–389

Corrosive Wear Resistance of TiO_2 Coatings by the Electrophoretic Deposition Process

Dhiflaoui Hafedh(✉), Khlifi Kaouther, and Ben Cheikh Larbi Ahmed

Laboratoire de Mécanique, Matériaux et Procédés LR99ES05, Ecole Nationale Supérieure d'Ingénieurs de Tunis, Université de Tunis, 5 Avenue Taha Hussein, Montfleury, 1008 Tunis, Tunisia
dhafedh@gmail.com

Abstract. In this study, a TiO_2 coating was prepared by an electrophoretic technique on 316L stainless steel. The corrosion morphology of the coating primarily included pitting, micro-cracks and NaCl attachment. The friction coefficient averages of the TiO_2 coating treated at 850 °C before and after immersion in NaCl medium under different applied loads and at three slip rates are investigated. the results show that the average coefficient of friction of the coatings rises with an increase in the sliding speed and with the increase of the applied load. These results show an increase in the coefficient of friction after immersion in the NaCl solution. The value varies from 0.22 to 0.31 for a slip speed of 100 μm/s and from 0.38 to 0.51 for 300 μm/s.

Keywords: TiO_2 coatings · Scratch test · Tribo-corrosion

1 Introduction

TiO_2 coatings have shown a good protective character against wear. However, the wear of a material is characterized by the chemical aggressiveness of the medium and in parallel, the corrosion of the part can be aggravated by the friction. The interaction between wear and corrosion in an aqueous medium can cause a synergistic degradation effect for the material. The phenomena of wear and corrosion can then be treated independently of one another.

In the literature, tribo-corrosion is defined as the study of the process of damage to the material under the effects of friction and corrosion. Tribo-corrosion is evaluated primarily through the total wear W_t which is easily accessible by measuring mass loss or by profilometry of the used volume. However, tribo-corrosion is often described as wear due to several contributions. Their determination, which is essential for understanding the mechanisms of tribo-corrosion, varies according to the methodologies used.

The first approach developed by Uhlig, for the fretting conditions, and then taken over by Mischler for slip conditions on alloys [7], the total wear is due to two contributions, the wear of the mechanical W_m and the other is accelerated corrosion by wear W_c. The total wear is written in this case:

N. Aifaoui et al. (Eds.): CMSM 2019, LNME, pp. 655–660, 2020.
https://doi.org/10.1007/978-3-030-27146-6_71

$$W_t = W_m + W_c \quad (1)$$

Many studies seek to quantify these two contributions [1–3, 6]. For the majority, the methodology followed consists of integrating the current measured under friction. The resulting charges are converted to $W_{C \cdot A \cdot U}$ (accelerated corrosion by wear) with Faraday's law while W_m is determined by difference:

$$W_m = W_t - W_{C \cdot A \cdot U} \quad (2)$$

In this method, W_m is therefore not considered as a reference quantity and varies according to the applied potential. A study, following this approach, determines W_m in tribo-corrosion tests in distilled water and uses this quantity to determine $W_{C \cdot A \cdot U}$ by difference with W_t [1].

A second approach is based on the fact that the wear of a material subjected to tribo-corrosion is greater than the sum of that observed under friction in non-aggressive medium W_m and that obtained in corrosive medium without mechanical action W_{diss} This additional wear W_{syn} results from the synergy between mechanical action and dissolution [8, 9]. The different volumes are then linked by the relation:

$$W_t = W_m + W_{diss} + W_{syn} \quad (3)$$

The dissolution component is obtained, in the absence of friction, from the integration of the dissolution current during the immersion of the sample in the corrosive solution or the measurement of the loss of mass at the end immersion.

2 Experimental Protocol

Thin TiO_2 films were deposited on 316L stainless steel substrates using the EPD technique. TiO_2 powder from Aldrich was used (Sigma-Aldrich Corporation, St. Louis, MO, USA) with 99.7% of the particles having dimensions inferior to 25 nm. The electrolyte solution was made up of a mixture of 0.06 g of TiO_2 nano-powder with 60 ml of absolute ethanol pure till $\geq$99.8%. An aluminum plate with a dimension of 20×15 mm^2 was used as a counter electrode (anode). In fact, each electrode was put in the media with a 1 cm distance. Then, the substrate material, with 15 mm diameter and a 3 mm thickness, was cut. Samples were automatically polished from 120 to 1000 grit silicon carbide (SiC) papers. This polishing was accompanied by a diamond pastes in order to obtain a mirror finish. Afterward, the electrodes were washed thoroughly by running distilled water. They were rinsed and degreased ultrasonically using acetone. Finally, they were dried for 10 min to remove any residual grease.

Each electrode was put into the media at temperature equal to 25 °C. During the AC-EPD tests, (HP 3314A) generator was utilized to produce the asymmetric AC signal. The latter was monitored with a digital oscilloscope (Tektronix). Optimized EPD was performed for 5 min at an applied voltage of 30 V.

The wear tests were performed with an MST micro-scratch equipped with a diamond Rockwell indenter with a tip radius of 50 μm and a cone angle of 60°. A mode of wear has been used where the number of cycles, the applied load and the sliding speed can be set. All tests were carried out on 316L stainless steel substrates 15 mm in diameter and 3 mm thick coated with TiO_2. The test environment was identical for each test: in air with a humidity of 46%, a temperature of 22 °C and without lubrication. The reciprocating motion is given to the indenter with a 3 mm deflection, moving speeds of 100, 200, and 300 μm/s and applied normal forces of 1.1.5, 2, 2.5 and 3 N. After the same samples, were immersed in a 3.5% solution of NaCl. After 24 h, the samples are removed from the medium and subjected to the same wear tests.

The thicknesses of the coatings were 11 μm. We confirmed this result by cross-section observation SEM (Fig. 1).

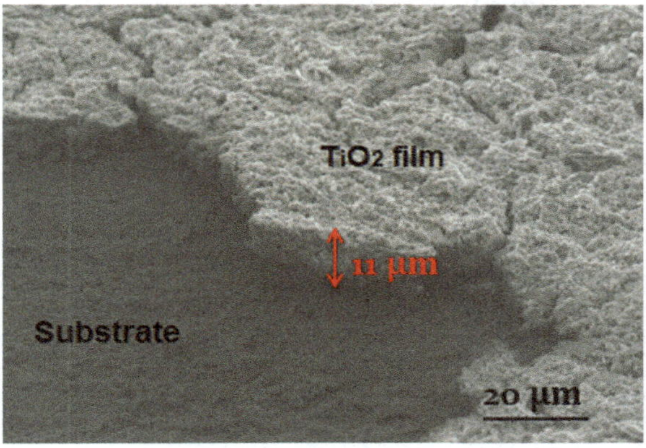

Fig. 1 SEM cross-sectional image

3 Results

The state of TiO_2 deposition surface after immersion in a chloride medium, shows the formation of a precipitate that can be attributed to a dissolution of the TiO_2 layer and thus passage of iron-chromium oxides from the surface to the aggressive medium. This is accompanied by a loss of mass.

The friction coefficient averages of the TiO_2 coating treated at 850 °C before and after immersion in NaCl medium under different applied loads and at three slip rates are reported in Fig. 2. The average coefficient of friction of the coatings rises with an increase in the sliding speed and with the increase of the applied load. These results show an increase in the coefficient of friction after immersion in the NaCl solution. The value varies from 0.22 to 0.31 for a slip speed of 100 μm/s and from 0.38 to 0.51 for 300 μm/s. The fact that the post-immersion coefficients of friction were higher than those before most likely immersion resulted from the synergistic effect between wear, which accelerates the degradation of the material by blistering and corrosion, which leads to the penetration of water and ions between the coating and the substrate [4].

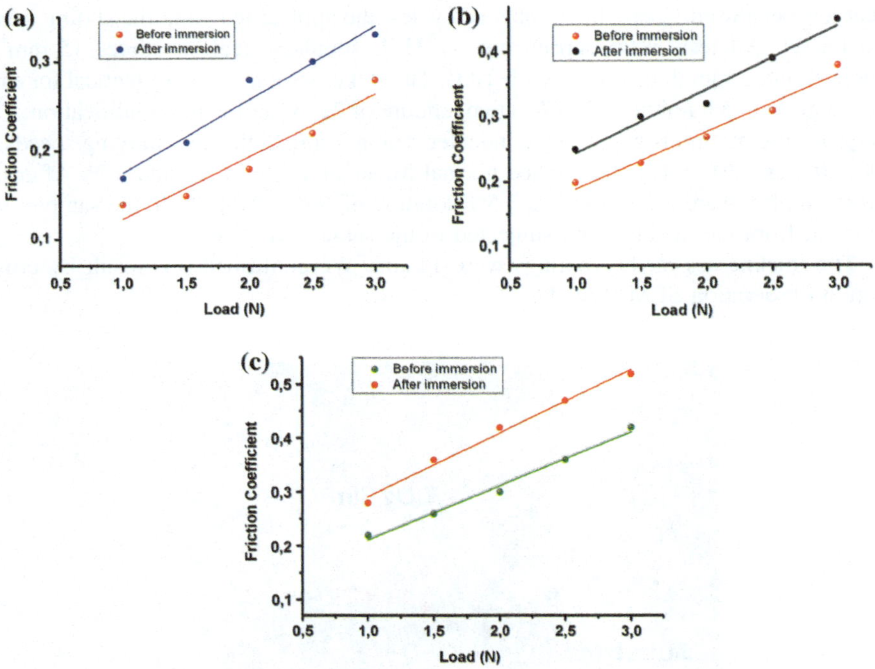

Fig. 2 Contribution of the two wear components in total wear for tribocorrosion tests: **a** 100 μm/s, **b** 200 μm/s and **c** 300 μm/s

The various quantities relating to the wear of the coatings studied are summarized in Table 1.

Table 1 Evolution of wear as a function of the normal force applied

F (N)	W_t after immersion			W_m dried		
	Sliding speed (μm/s)					
	100	200	300	100	200	300
1	3.08	3.21	3.67	2.21	2.25	2.75
1.5	4.83	5.06	5.30	3.65	3.94	4.22
2	7.37	8.50	8.77	5.33	6.19	7.17
2.5	8.58	9.67	10.7	6.20	6.68	7.74
3	9.36	11.12	12.6	8.43	8.85	9.36

The results obtained for wear, presented in this table show that:

(a) In condition of the increasing normal load and whatever the sliding speed:

- Total wear increases;
- Corrosive wear increases;
- Mechanical wear is also increasing.

(b) For constant normal force and slip speed increases:

- Total wear is growing stronger;
- Corrosive wear increases;
- The mechanical wear believes.

In both cases (dry or after immersion), wear volumes showed a similar trend. They increase considerably with the increase of the applied load. The wear behavior of a coating is related to its structure, its mechanical properties and also to its adhesion quality [5, 10]. So to explain the increase in the volume used in the coating we thought to measure the adhesion of the coatings have and after immersion.

The results obtained are shown in Fig. 3 and Table 2. Experiments have shown that critical loads (LC1, LC2 and LC3) have decreased after immersion. The LC3 critical load exhibited a sign of total flaking of the film decreased from 7.99 N dry to a value of 5.32 N after immersion. This result proves a decrease in the level of adhesion of the layer.

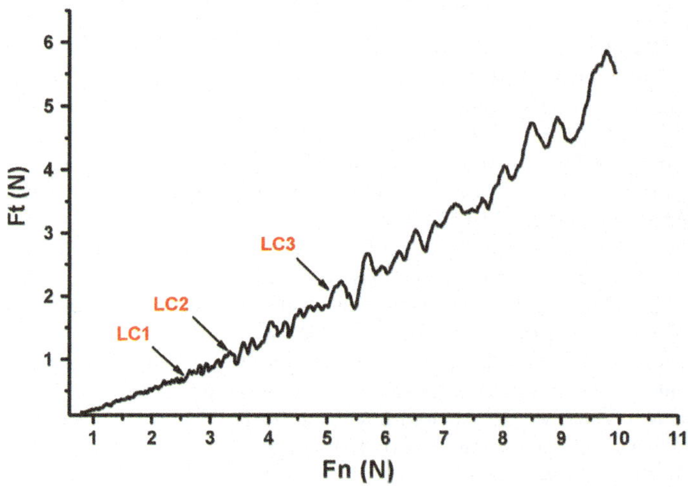

Fig. 3 Evolution of the tangential force as a function of the normal load after immersion

Table 2 Evolution of wear as a function of the normal force applied

Critical load (N)	Dry	Before immersion
LC1	3.22	2.55
LC2	5.08	3.51
LC3	7.99	5.32

4 Conclusion

During this work, the tribological properties of the different deposits were studied according to two distinct protocols; on the one hand in accordance with a “pin on plan” friction test, which makes it possible to evaluate the wear resistance of the layers and, on the other hand, by means of tribo-corrosion tests (immersion in an NaCl medium followed by a friction test) whose purpose is to identify the contribution of each component of wear. For the tribo-corrosion tests, the coating calcined at 850 °C was immersed in NaCl medium for 24 h followed by a friction test with conditions similar to that of dry friction. The total volume of wear increases due to accelerated wear by the degradation of the coating.

Acknowledgements. This work is partially supported by the Laboratoire de Mécanique, Matériaux et Procédés. The authors also gratefully acknowledge the helpful comments and suggestions of the reviewers, which have improved the presentation.

References

1. Benea L, Ponthiaux P, Wenger F, Galland J, Hertz D, Malo J (2004) Tribocorrosion of stellite 6 in sulphuric acid medium: electrochemical behaviour and wear. Wear 256:948–953
2. Dahm KL (2007) Direct observation of the interface during sliding tribo-corrosion. Tribol Int 40:1561–1567
3. Favero M, Stadelmann P, Mischler S (2006) Effect of the applied potential of the near surface microstructure of 316L steel submitted to tribocorrosion in sulfuric acid. J Phys D Appl Phys 39:3175–3183
4. Kim WJ, Kim JG, Kim YS, Ozdemir I, Tsunekawa Y (2007) Wear-corrosion of cast iron thermal spray coatings on Al alloy for automotive. Comp Met Mater 13
5. Lathabai S (1998) An sem study of mild-to-severe wear transition in grey cast iron during dry sliding. Scripta Mater 38:1557–1562
6. Lgried M, Liskiewicz T, Neville A (2012) Electrochemical investigation of corrosion and wear interactions under fretting conditions. Wear 282–283:52–58
7. Mischler S (2008) Tribo electrochemical techniques and interpretation methods in tribocorrosion: a comparative evaluation. Tribol Int 41:573–583
8. Sun Y, Haruman E (2011) Tribocorrosion behaviour of low temperature plasma carburised 316L stainless steel in 0.5 M NaCl solution. Corrosion Sci 53:4131–4140
9. Watson SW, Friedersdorf FJ, Madsen BW, Cramer SD (1995) Methods of measuring wear-corrosion synergism. Wear, Part 2, 181–183:476–484
10. Wu H, Zhao J, Xia W, Cheng X, He A, Yun JH, Wang L, Huang H, Jiao S, Huang L, Zhang S, Jiang Z (2017) A study of the tribological behaviour of TiO_2 nano-additive water-based Lubricants. Tribol Int 109:398–408

Numerical Study of Mechanical Behavior of Agave Fibers Reinforced Composites

Yosra Glouia[1(✉)], Asma El Oudiani[1], Imen Maatoug[1], Rim Zouari[2], and Slah Msahli[1]

[1] Laboratory of Textile Engineering, University of Monastir, Ksar Hellal 5070, Tunisia
gyosral@yahoo.fr
[2] Mechanical Engineering Laboratory, University of Monastir, Monastir, Tunisia

Abstract. An investigation was carried out on local natural cellulosic fibers which have gained interest in the composite field due to their superior specific properties. A multi scale finite element (FE) model of unidirectional natural fiber composite materials with reasonable dimensionality was developed. The mechanical behavior of the composite at macro scale as well as meso-scale was simulated. In particular the response to tensile and three-points bending test was studied. Linear material properties are obtained by using pure strains assumptions in the implicit analysis of the composite, while the non-linear behavior and viscoelastic parameters require the explicit dynamic analysis. Simulation is performed thanks to Abaqus finite element software. Comparison of Experimental and FEM tensile and three-point bending Strength shows very good agreement.

Keywords: Finite element analysis · Natural fibers · Biocomposites · Mechanical behavior · Viscoelasticity · Plasticity · Abaqus

1 Introduction

Composites are the multi-phase material and wonder materials becoming an essential part of today's materials due to their advantages such as low weight, corrosion resistance and high fatigue strength [1]. They are extensively used as materials in making aircraft structures, automotive industry, electronic, packaging, medical equipment, space vehicle and homebuilding [2].

Despite their interesting performance, these materials are polluting and harmful to nature. And considering the increase in environmental constraints such as Kyoto protocol, it was necessary to think of ecofriendly materials as substituents, for instance natural fibers [3]. Natural fibers made polymer composites are gaining more interest in engineering applications due to appreciable structural properties at a relatively low cost which finds its attraction towards the field of research. Composite microstructures are determined by the physical and mechanical properties of the individual materials. Analytical methods provide reasonable prediction for relatively simple configurations of the phases. Complicated geometries, loading conditions and material properties often do not yield analytical solutions, due to complexity and the number of equations.

N. Aifaoui et al. (Eds.): CMSM 2019, LNME, pp. 661–670, 2020.
https://doi.org/10.1007/978-3-030-27146-6_72

In this case, numerical methods are used for approximate solutions, but they still make some simplifying assumptions about the inherent microstructures of heterogeneous multiphase materials, one such method is finite element analysis. Literature reveals that researchers have not been interested till now to viscoelastoplastic behavior law of composite materials. In this context, the present work aims to simulate the mechanical behavior of natural cellulosic fibers reinforced composites using finite element modeling method.

2 Materials and Methods

2.1 Published Data

To be able to model and simulate the mechanical behavior of any material with Abaqus finite element package, several essential parameters must be available especially material geometry and dimensions as well as behavior law. In literature, several researchers (Tunisia, France, Germany…) have worked on natural fibers reinforced composites. Chaabouni studied composites made from thermosetting matrix and Agave fiber [4], Ghali worked on polyester luffa composites [5], Ben Brahim studied tensile properties of unidirectional Alfa-polyester composite [6], Allègue worked on mechanical properties of Posidonia oceanica fibers reinforced cement 5 Allègue [7].

Following this bibliographic synthesis and published data, we notice that we have the maximum of information and data (Curves, the max of tests… etc.) for the Agave fibers, the polyester and epoxy matrices and the composite resulting from these latter components [4]. To this is added the researches of [8] and [9] in which they conducted respectively a physical characterization (diameter, length, morphology…) and a rheological study on Agave fibers. Thanks to these three researches, it is possible to perform a finite element analysis of unidirectional aligned Agave-polyester and Agave-epoxide composite.

Based on [4] and Msahli [8] researches, Agave Americana L. fibers density is equal to 1.36 which is considered as a low weight compared to other hard vegetable fibers. For these reasons, Agave Americana L. fibers are very interesting. We resume in Table 1 elastic parameters (Young's modulus and Poisson coefficient) as well as tensile mechanical characteristics (resistance, elongation) of the performed composites with diverse ratios of fibers.

Table 1 Elastic parameters of Agave based composites [4] and Carma [10]

Matrix	Agave fraction (%)	E (GPa)	ν	Stress (MPa)	Strain (%)
PES	15	1.48	0.32	28	4.8
PES	24	1.07	0.32	31	9
PES	31	1.164	0.32	33	8.5
EPX	10	2.34	0.333	48	3.2
EPX	20	2.317	0.333	24	1.1
EPX	30	2.307	0.333	22	1.1

2.2 Finite Element Modeling

The choice of the length scale, together with the constitutive material models and the related parameters is a key point especially in the fields of composites, which are made of different materials. Several approaches were used and developed during the last years and are reported in the literature. In particular, two different techniques of modelling have been employed in the literature, one technique does not reproduce all the fibers and the matrix but considers the layer as a unique homogenous mean. This approach most likely does not allow the detailed investigation of complex phenomena. The Meso heterogeneous model [11–18] instead reproduces every single fiber and the matrix, as well as the interaction among these. Modeling of the mechanical behavior of fiber-reinforced polymer matrix composites is presented by the example of Tunisian Americana. L Agave fibers in a polyester and epoxide based matrix respectively. Our study contains two different scales macro and meso scale models because of the heterogeneity in composite materials when analyzing them at numerous scales.

3 Results and Discussion

3.1 Macro Scale Modeling, Tensile Test Simulation

Unidirectional Agave reinforced composite is considered as a three-dimensional plate (rectangular shape) when analyzing it at macro scale. Dimensions of the plate are: $100 \times 20 \times 2\ \text{mm}^3$.

Concerning plastic behavior, we used experimental data and we calculate true stress and true strain thanks to the following formulas:

$$\sigma\, true = \sigma nom \times (1 + \varepsilon nom) \tag{1}$$

$$\varepsilon true = \ln(1 + \varepsilon nom) \tag{2}$$

$$\varepsilon plastic = \varepsilon total\, true - \varepsilon elastic\, true \tag{3}$$

$$\varepsilon elastic\, true = \frac{\sigma\, true}{E} \tag{4}$$

The following figures represent the results of the simulation of the tensile test of Agave polyester composites as well as Agave epoxide composites using various weight ratios. We draw for each composite the numerical and the experimental curve. Then, we conclude from the two curves the error generated by the model (Fig. 1).

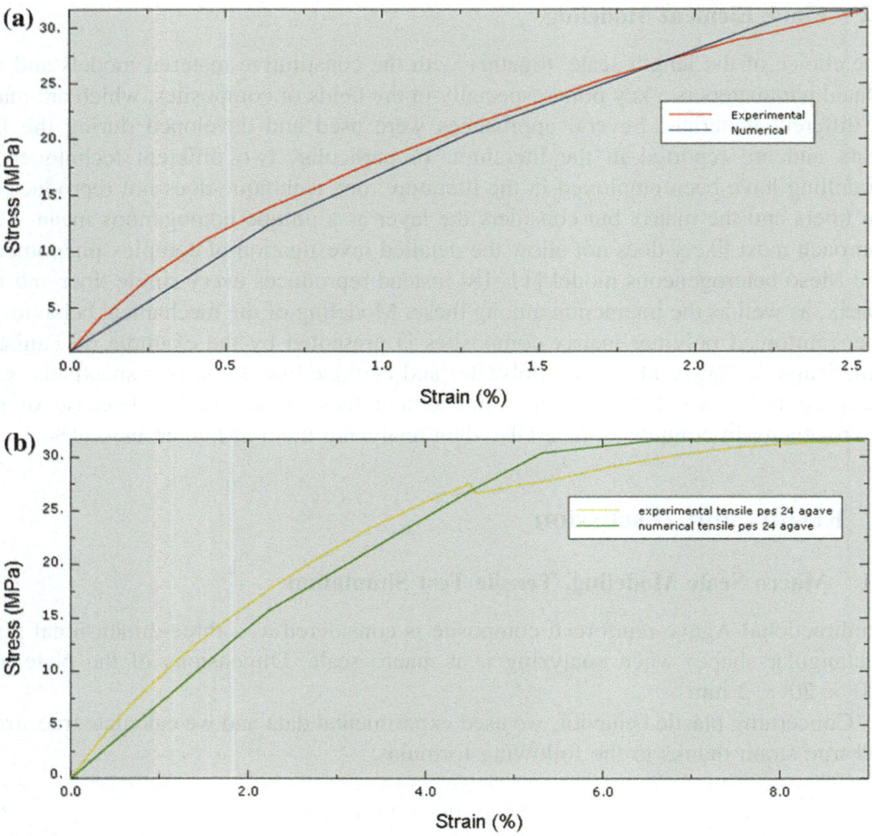

Fig. 1 Comparison between experimental and numerical stress-strain response under tensile test of respectively 15% Agave polyester composite (**a**) and 24% Agave polyester composite (**b**)

For these composites, experimental and numerical curves present the same shape with an error in the range of 5–15%. This error may be due to the assumption of homogeneity of composite material which does not exist concretely. We summarize the error value for each composite in Table 2.

Table 2 Error (%) between experimental and numerical results

Composite	Error (%)
PES 15% Agave	7.62
PES 24% Agave	7.39
EPX 10% Agave	16.25
EPX 30% Agave	5.52

Error value is acceptable since it doesn't exceed 16%. We conclude thus that model converge and may be validated.

3.2 Macro Scale Modeling, 3 Points Bending Test Simulation

We consider the same constitutive law for the model used in the simulation of the tensile test since we are working with the same materials already described.

Numerical flexural modulus was calculated applying the following Formula (5).

$$MOE = \frac{P_1 L^3}{4be^3 y_1} \text{ (MPa)} \tag{5}$$

where P_1 is the maximum load (N), L the range (mm), e the specimen thickness (mm), b the specimen width (mm) and y_1 indicates the defection (mm).

Results are summarized in Table 3.

Table 3 Experimental and numerical values of flexural modulus (MOE) of Agave based composites under 3 points bending test

Composite	Numerical MOE (GPa)	Experimental MOE (GPa)	Error (%)
PES 15% Agave	1.4	1.6	12
PES 24% Agave	2.12	2.5	15
PES 31% Agave	2.4	2.6	7
EPX 10% Agave	2.34	2.4	6
EPX 20% Agave	2.4	2.5	4
EPX 30% Agave	2.7	2.8	3

Comparison of Experimental and FEM tensile and three-point bending Strength shows very good agreement since error doesn't exceed 15%.

3.3 Meso Scale Modeling, Tensile Test Simulation

The material id modelled and analyzed for mechanical properties with finite element method software (ABAQUS 6.14). Our model is a three-dimensional deformable body. We created a rectangle that is then extruded to form a parallelepiped. The modeling of the fibers is created by their partition in the matrix in order to ensure a perfect adhesion. Thus, we can model the composite on the mesoscopic scale, without having to create different pieces. The number of circles corresponds to the mass fraction considered in the study. Dimensions of sample are $100 \times 20 \times 2$ mm^3. Fiber's diameter is about 0.263 mm. Elastic parameters of Agave fibers and matrices are summarized in Table 4.

Table 4 Fibers and matrix properties [4] and Carma [10]

	E (MPa)	ν
Agave fibers	4000	0.33
Polyester matrix	870	0.37
Epoxide matrix	1924	0.386

In order to apply the boundary conditions, we have coupled the two faces z = 0 and Z = 100 mm to a rigid body with well-defined reference points (RP1 and RP2). First, we perform a static analysis and we develop a model in which the composite material is assigned as unidirectional composite by assuming the fiber and matrix materials are linearly elastic. The numerical simulation results appear in Figs. 2 and 3.

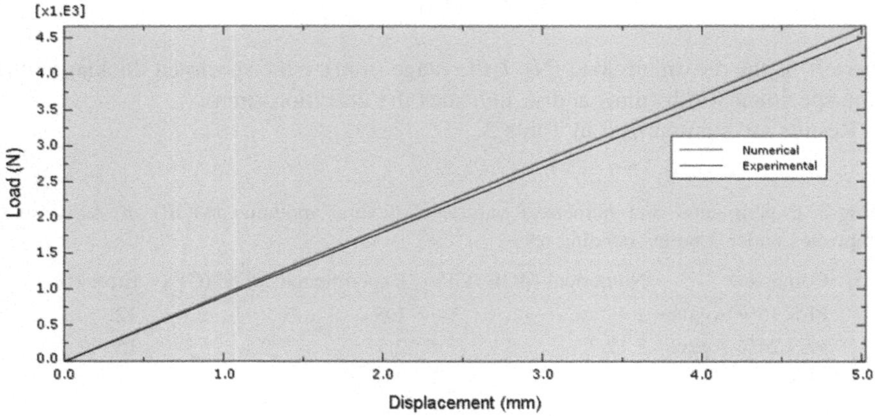

Fig. 2 Comparison between experimental and numerical response under tensile test considering elastic behavior law for epoxide 10% Agave composite

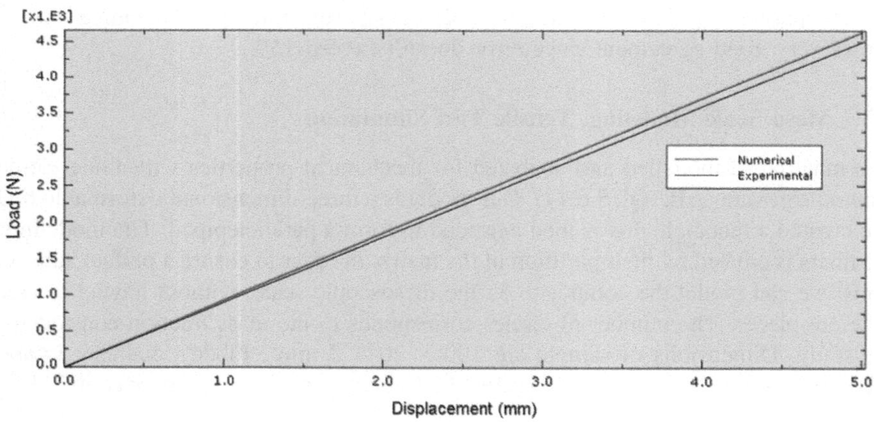

Fig. 3 Comparison between experimental and numerical response under tensile test considering elastic behavior law for polyester 15% Agave composite

The curves in Figs. 2 and 3 present the same shape with the presence of slight difference. The difference between experimental and numerical simulated curves is equal to 14% for the case of polyester/Agave composite. While, for epoxide Agave composite, the difference between numerical and experimental curves is about 2%. Hence the model converges. We calculate also numerical Young's modulus and we compare it with the experimental value as mentioned in the Table 5.

Table 5 Numerical Young's modulus versus Experimental Young's modulus

Composite	Esimulation (GPa)	Eexperimental (GPa)
PES 15 Agave	1.2	0.96 [4]
EPX 10 Agave	2.25	2.2 [4]

Numerical and experimental results have very good agreement.

Although the previous model shows good results, it doesn't reflect the real behavior law of Agave fibers as well as matrices employed in this study. That is why, we propose a second model in which we consider a viscoelastic behavior law for Agave fibers and an elastoplastic behaviour for polyester and epoxide matrix as it is mentioned in literature [19].

The viscoelastic law of fibers is identified by the load–extension test added with the relaxation test. These tests have been then implemented in the ABAQUS finite element package via Prony series method. They define shear moduli and volumetric compression moduli over time as illustrated in Formula (6):

$$g_R(t) = \frac{G_R(t)}{G_0} = 1 - \sum_{i=1}^{N} g_1 \left[1 - e^{\frac{-t}{\tau_i}}\right] \tag{6}$$

where $G_R(t)$ is the shear modulus over time, G_0 the shear modulus at the very beginning of the test, $g_R(t)$ is the ratio between the shear modulus at the beginning of the relaxation test, τ_i is a relaxation time corresponding to the increment i, t is the time during the relaxation test and g_i is the shear modulus corresponding to the increment i and corresponds to the relaxation time τ_i.

Abaqus allows to calculate Prony series from the results of the dimensioned relaxation tests.

In this method, we conduct an explicit analysis. The mechanical properties values are the same mentioned in Table 4. Results of simulation under tensile test for Agave polyester and Agave epoxide composites are illustrated in Fig. 4.

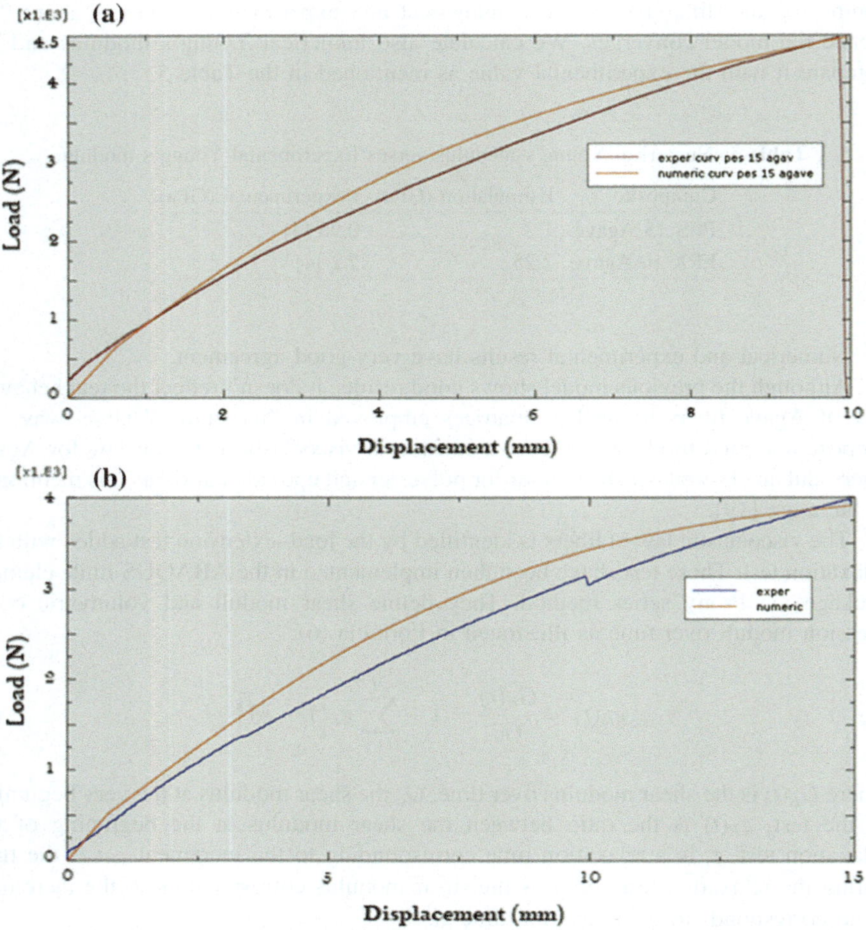

Fig. 4 Comparison between experimental and numerical response under tensile test using explicit analysis and viscoelastoplastic behavior law of respectively polyester 15% Agave composite (**a**) and epoxide 10% Agave (**b**)

The curves of Fig. 4 are very similar and present the same shape with a slight difference that may be due to many factors (presence of porosity, discontinuity, presence of voids in fibers and matrix, anisotropic behaviour of Agave fibers….). Difference between experimental and numerical simulated curves is equal to 5.88% for the case of Agave PES composite and 9% for the case of Agave EPX composite. Hence the model converges. Numerical and experimental results have very good agreement. The error value is smaller for the case of the second model, considering a viscoelastoplastic behavior law.

4 Conclusion

Finite element analysis is performed in two different scales: macro and meso (scale of fibers). For macro scale modeling, we have predicted the behavior and the response of polyester Agave composites as well as epoxide Agave composites. Numerical and experimental results were globally in agreement even so the presence of a slight difference which varied depending on the material studied. This error was in the interval of 5 and 22% concerning tensile test simulation results.

Moreover, we have simulated the behavior of composites under three points bending test. In the one hand, we have calculated flexural modules. And in the other hand, we showed the influence of fibers addition to polyester and epoxide matrices on flexural properties of composite materials. Simulation results were in very good agreement with literature.

Concerning the meso scale modeling, we have predicted the behavior and the response of polyester Agave composites as well as epoxide Agave composites taking into account the properties of matrix and fibers. We started by a simple elastic behavior law for Agave fibers, polyester and epoxide. Even though, this model was in concordance with experimental results, it didn't reflect accurately the behavior law of the materials.

A second model in which we consider a viscoelastic behaviour for Agave fibers and an elastoplastic behaviour for polyester and epoxide matrices was thus performed to remedy to deficiencies of the first simulation model.

The error percentage for the second model does not exceed 9%, which is evaluated as a very satisfying result. The difference in the results, obtained from Finite element analysis and experimental analysis, could be explained by the presence of porosity, discontinuity and presence of voids in fiber and matrix and also due to the anisotropic behavior of Agave fiber [13].

The error between experimental and numerical results of Agave fiber composites could be also attributed to the non-uniformity and heterogeneity of the fibers.

References

1. Wilson MJ (2003) Finite element analysis of glass fibre reinforced thermoplastic composites for structural automotive components. Dissertation, University of Nottingham
2. Berreur L, Maillard B, Nösperger S (2002) L'industrie française des matériaux composites. Strategic study report by Nodal Consultant. 13th may
3. Wambua P, Ivens J, Verpoes, I (2003) Natural fibres: can they replace glass in fibre reinforced plastics? Comp Sci Tech 63(9):1259–1264
4. Chaabouni Y (2005) Microstructure characterisation of Agave Americanan L. fibre; study of Agave fibre reinforced composites. Dissertation, University of Mulhouse
5. Ghali L, Msahli S, Zidi M, Sakli F (2011) Effects of fibre weight ratio, structure and fibre modification onto flexural properties of Luffa-polyester composites. Adv Mater Phys Chem 1(3):78–85
6. Ben Brahim S, Ben Cheikh R (2007) Influence of fibre orientation and volume fraction on the tensile properties of unidirectional Alfa-polyester composite. Comp Sci Tech 67(1): 140–147

7. Allègue L (2014) Mechanical properties of posidonia oceanica fibers reinforced cement. J Compos Mater 49(5):509–517
8. Msahli S, ChaabouniY, Sakli F, Drean JY (2007) Mechanical behavior of Agave Americana L. Fibres: correlation between fine structure and mechanical properties. J Appl Sci 7(24):3951–3957
9. El Oudiani A (2003) Study of the mechanical behavior of Agave fibers Americana L. Dissertation, University of Monastir
10. Carma (2006) http://stibeziers.fr/tsipm/spip_tsipm/IMG/pdf/composite3.pdf
11. Ghering F (2013) Mechanical behaviour and damage study of short hemp fibre reinforced thermoplastics: experimental approach and modelling. Dissertation, University of Lorraine
12. Kumar J, Praveen DN, Thara R, Irfan G (2016) Experimental & finite element analysis of Sisal fibre reinforced composites. Int J Recent Trends Eng Res 2(7):155–160
13. Prasad G, Venkatachalam Akshat, Rathi Rajakuma S (2014) Finite element analysis of jute fibre made hybrid polymer matrix composite. Appl Mech Mater 592(594):363–367
14. Rajesh M, Srinag T, Phani Prasanthi P, Venkataraovenkatrao K (2016) Finite element analysis of coir/banana fiber reinforced composite material. Int J Adv Res Mech Eng Tech 2 (4):29–33
15. Houshyar S, Shanks RA, Hodzic A (2009) Modelling of polypropylene fibre-matrix composites using finite element analysis. EXPRESS Polym Lett 3(1):2–12
16. Sarvana Bavan D, Mohan Kumar GC (2012) Finite element analysis of a natural fiber (Maize) composite beam. J Eng 2013:1–7
17. Da Silva LJ, Panzera TH, Christoforo AL, Durão LMP, Rocco Lahr FA (2012) Numerical and experimental analyses of biocomposites reinforced with natural fibres. Int J Mater Eng 2 (4):43–49
18. Sun CT, Vaidya RS (1996) Prediction of composite properties from a representative volume element. Comp Sci Tech 56(2):171–179
19. Perrot Y, Baley C, Davies P (2006) Influence of low styrene emission polyester resins on the aging behavior of composites in a marine environment. Appl Compos Mater 13(1):1–22

A 3D Numerical Analysis of the Chip Segmentation Mechanism and the Side Burr Formation During the Ti_6Al_4V Alloy Machining

M. Yaich[1,2(✉)], Y. Ayed[2], Z. Bouaziz[1], and G. Germain[2]

[1] Université de Sfax, Ecole Nationale d'Ingénieurs de Sfax, Laboratoire de Mécanique des Fluides Appliquées, Génie des Procédés et Environnement, 3035 Sfax, Tunisia
Mariem.yaich@enis.tn, Zoubeir.bouaziz@gmail.com
[2] Arts et Métiers ParisTech, LAMPA, 2 bd du Ronceray, 49035 Angers Cedex, France
{Yessine.ayed, Guenael.germain}@ensam.eu

Abstract. A 3D finite element modeling of the orthogonal turning process was curried out in the current study. It aims to carefully investigate the mechanisms controlling the chip segmentation and the crack propagation direction in the case of the Ti_6Al_4V machining. Coupled temperature-displacement numerical simulations were performed in the software Abaqus®/Explicit, under different cutting conditions. The instantaneous distribution of numerical thermomechanical variables along the width of cut was investigated. High plastic strains, temperatures and damage were predicted in the median plane of the workpiece, mainly in the shear bands around the tool tip vicinity. Whereas, a reduction of their values was noted while moving towards the chip sides and its upper surface. The 3D numerical simulations pointed out that the orthogonal machining resulted in an increase of the chip width, in addition to the material flow along the X and Y directions. The quantitative analysis of the side burr formation highlighted its sensitivity to the cutting conditions. The definition of high feed rates resulted in pronounced material flow in the workpiece edges, thus the modeling of wider chip. The present study concluded that the chip segmentation is a 3D mechanism. In addition, it pointed out the limitations of the 2D numerical simulations, as well as the inadequacy of the plain strain hypothesis, even in the case of the orthogonal machining.

Keywords: Machining · Abaqus · Ti_6Al_4V · Numerical analysis · Chip · Side burr

N. Aifaoui et al. (Eds.): CMSM 2019, LNME, pp. 671–680, 2020.
https://doi.org/10.1007/978-3-030-27146-6_73

1 Introduction

The machining process of materials with poor machinability, like the titanium alloys [1], is still problematic for both industrials and researchers. Significant material deformation is heavily encountered under high strain rates and it induces a temperature rise in the cutting zones. In addition, a concentrated and important heating is generated in the cutting zones due to the low conductivity of machined material and its pronounced chemical affinity with the cutting tools. Indeed, an accelerated tool wear is generally emphasized [2]. Despite the use of equipment with high precision, the experimental investigation of several instantaneous and local physical phenomena, taking place in very thin cutting zones, is still very expensive and it is not precise enough [3]. The experimental observations made by Pottier et al. [4] have allowed to divide the chip formation process into three main sub-processes: compression, strain localization/failure and sliding. However, despite the adoption of advanced experimental devices (high frame rate camera and optical microscopy) and carrying out in situ and post-mortem observations, the accurate determination of the direction of the deformation/crack propagation within the chip has been not allowed. The authors have required the definition of numerical analysis for valuable information.

In the case of the Ti_6Al_4V alloy machining, several investigations of the literature [5–7] have highlighted the formation of serrated chip, even when low cutting conditions have been defined. The significant thermomechanical coupling has heavily limited the efficiency of experimental tests. In fact, an enhanced understanding of the several phenomena following the chip segmentation has been required to increase the material machinability. The definition of numerical approaches, in addition to the experimental analysis, was heavily important to provide valuable information, thus to deal with the severe loading conditions following the machining of difficult to-cut materials [8]. These approaches has enabled the investigation of transient phenomena and pronounced nonlinearity involved during the chip formation. Furthermore, the valuable advances in the numerical simulations, giving rise to several modeling techniques (finite element (FE) method, discrete element method, smoothed particle hydrodynamics, material point method), have encouraged their use. The availability of powerful commercial codes (Abaqus®, Deform®, AdvantEdge®, Forge®, etc.) has led to the widespread adoption of the FE method, based on the special discretization of the model geometry, in the last decades [7–10].

Indeed, the current paper focuses on numerically investigating the phenomena controlling the chip serration. A set of 3D finite element simulations was set up in the software Abaqus®/Explicit to model the orthogonal machining of the Ti_6Al_4V alloy under several cutting conditions. The distribution of temperatures, plastic strains and damage was investigated. The numerical results corresponding to two different planes of the workpiece, the medium and the side planes, were used to determine the sub-processes of the chip segmentation. The sensitivity of the side burrs modeled in the workpiece edges to the cutting conditions was also studied.

2 Numerical Model

A 3D numerical model was set up to simulate the Ti_6Al_4V orthogonal machining. The Lagrangian formulation integrated in the FE software Abaqus®/Explicit was defined. A fully coupled temperature-displacement analysis was performed. The model geometry was discretized with the 8-node 3D thermally coupled continuum elements (C3D8RT) with reducer integration. The geometry of the tungsten carbide cutting tool and its physical properties are grouped in Table 1 and Table 2, respectively. To reduce the mesh distortion mainly encountered in the cutting zones, because of the severe contact conditions induced in the chip-tool-workpiece interfaces, the workpiece was divided in three tied parts (see Fig. 1): the uncut chip part (P_1), the tool passage part (P_2) and the workpiece support (P_3).

Table 1 Cutting conditions and tool geometry

Cutting conditions	
Cutting speed V_c (m/min)	45–75
Uncut chip thickness f (μm)	0.15–0.3
Width of cut a_p (mm)	3
Tool geometry	
Rake angle corresponding to the orthogonal plane γ_0 (°)	6
Clearance angle corresponding to the orthogonal plane α_0 (°)	7
Edge inclination angle λ_s (°)	0
Edge entering angle κ_r (°)	90
Cutting edge radius r_b (μm)	20

Table 2 Physical properties of the cutting tool and the workpiece [9]

Physical parameters	Tool	Workpiece
Density ρ (kg/m^3)	15,000	4430
Specific heat C_p (J/kg/K)	203	580
Thermal conductivity λ (W/m/K)	46	7.3
Thermal expansion α_p (μm · m/K)	4.7×10^{-6}	8.6×10^{-6}
Elastic modulus, E (GPa)	–	113.8
Poisson's ratio, v	–	0.342
Inelastic heat friction, η_p	–	0.9
Room temperature, T_{room} (K)	293	293
Melting temperature, T_{melt} (K)	–	1943

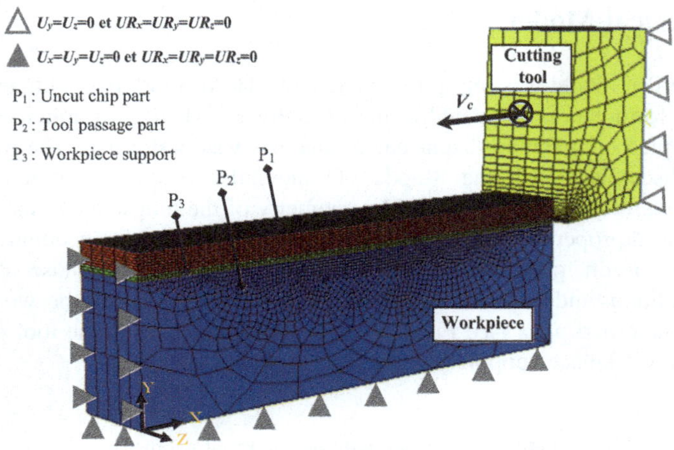

Fig. 1 Model geometry and boundary conditions

To predict the nonlinear ductile behavior of the Ti_6Al_4V alloy, the Johnson-Cook (JC) constitutive models [11, 12] were defined to the workpiece. These criteria have been commonly used the fact that they take into account the effects of the strain hardening, the viscosity, the temperature distribution and the damage initiation on the machined material (see Eqs. 1–3). The damage evolution laws (Eqs. 4–5) were used to predict the progressive degradation of elements mesh, which is induced in the strain localization zones, while minimizing the dependency of the FE solution on the mesh. We note that a rigid cutting tool was considered for all numerical simulations. The Coulomb-Tresca friction criterion (see Eq. 6) was used to model the mechanical contact conditions taking place between the cutting tool and the workpiece. Only the effects of the friction and the plastic deformation of the workpiece material on the heat generation, thus on the temperature rise, were considered in the current study.

$$\sigma = \left(A + B(\varepsilon_p)^n\right) \times \left(1 + C\ln\left(\frac{\dot{\varepsilon}_p}{\dot{\varepsilon}_0}\right)\right) \times \left(1 - \left(\frac{T - T_{room}}{T_{melt} - T_{room}}\right)^m\right) \qquad (1)$$

where, A, B, n, C and m are the JC plasticity coefficients. Their corresponding values are grouped in Table 3. T_{room} and T_{melt} are the reference and the melting temperatures respectively. $\dot{\varepsilon}_0$ is the reference strain rate.

Table 3 JC constitutive coefficients [10, 13]

JC plasticity coefficients				
A (MPa)	*B* (MPa)	*n*	*C*	m
1119	8386	0.473	0.019	0.643
JC damage coefficients				
D_1	D_2	D_3	D_4	D_5
–0.09	*0.25*	*–0.5*	*0.014*	*3.87*

$$w_{JC} = \sum \frac{\Delta \varepsilon_p}{\varepsilon_{i0}} \tag{2}$$

where, $\Delta\varepsilon_p$ is the accumulated plastic strain. ε_{i0} is the plastic strain at the damage initiation and it is computed as follow:

$$\varepsilon_{i0} = \left(D_1 + D_2 \exp\left(D_3 \frac{\sigma_h}{\sigma_{vm}}\right)\right) \times \left(1 + D_4 \ln\left(\frac{\dot{\varepsilon}}{\dot{\varepsilon}_0}\right)\right) \times \left(1 + D_5\left(\frac{T - T_{room}}{T_{melt} - T_{room}}\right)\right) \tag{3}$$

where, D_{1-5} are the JC damage coefficients and they are given by Table 3. σ_h and σ_{VM} are the hydrostatic and the Von Mises stresses, respectively.

$$(D_{ev})_{Linear} = \frac{\bar{u}_p}{\bar{u}_f} = \bar{u}_p \times \frac{\tilde{\sigma}}{2G_f} \tag{4}$$

$$(D_{ev})_{exponential} = 1 - \exp\left(-\int_0^{\bar{u}_p} \left(\frac{\tilde{\sigma}}{G_f}\right) \times d\bar{u}_p\right) \tag{5}$$

where, $(D_{ev})_{Linear}$ and $(D_{ev})_{exponential}$ are the damage variables for linear and exponential failure evolutions, respectively. $\bar{u}_f$ is the plastic displacement at failure, G_f is the fracture energy and $\tilde{\sigma}$ is the flow stress at the damage initiation.

$$\tau_f = \begin{cases} \mu \times \sigma_n & (if\, \mu \times \sigma_n < \tau_{\max}) \\ \tau_f = \tau_{\max} = m_{Tresca} \times k & (if\, \mu \times \sigma_n \geq \tau_{\max}) \end{cases} \tag{6}$$

where, τ_f, μ, σ_n, m_{Tresca} and k are the shear stress, the Coulomb's coefficient, the normal friction stress, the Tresca factor and the yield stress, respectively.

3 Results and Discussion

Figure 2 illustrates the chip morphology predicted for a cutting speed and a feed rate of 45 m/min and 0.3 mm/rev, respectively. A material flow along the different directions was obtained when the cutting tool penetrated in the workpiece. For both modeled sections ($Z = 0$ mm or $Z = \pm 0.25$ mm), high equivalent plastic strains were predicted in the shear bands, mainly close to the inner chip surface that was in contact with the tool rake face. In addition, a slight mismatch between the chip morphology predicted in the two sections was highlighted. Less serrated chip was modeled in the workpiece sides ($Z = \pm 0.25$ mm), where the removed material was also deformed along the width of cut direction and it resulted in the side burrs formation.

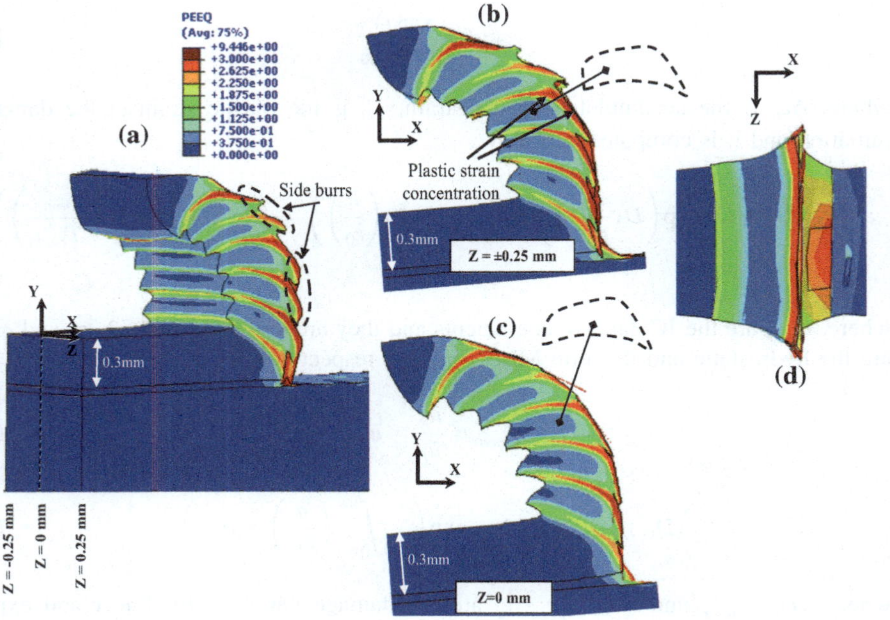

Fig. 2 Chip morphology and equivalent plastic strain distribution in different sections: **a** 3D chip, **b** chip sides, **c** median plane and **d** XZ plane (V_c = 45 m/min and f = 0.3 mm/rev)

To explain these preliminary observations, a detailed analysis of mechanisms controlling the chip segments formation was performed. The instantaneous evolution of temperatures was investigated. Figure 3 shows a temperature rise in the chip, mainly in its inner surface in contact with the tool edge radius, where a heat concentration in the median plane ($Z = 0$ mm) was underlined. The temperatures furthest from this zone were the lowest. With the progressive penetration of the cutting tool in the workpiece, the computed temperatures were propagated in the shear band until reaching the upper surface of the chip and its sides. The non-uniform temperature distribution along the shear plane and the width of cut was mainly due to the contact friction conditions.

For important cutting time, the computed temperature was accentuated and it reached 1000 K (see Fig. 4c, d). Moreover, a significant heat propagation towards the upper chip surface was noted. A thermally extended affected zone (TAZ) was predicted. For high feed rates, its thickness was significantly increased. In addition, more pronounced material flow along the workpiece sides was pointed out. Figure 5 illustrates a significant increase of the average width of the side burr for f = 0.3 mm/rev, under both investigated cutting speeds. Its width was about 40% the modeled feed rate.

The instantaneous and local distribution of the damage variable (*SDEG*) in the chip was studied to better understand the chip segmentation process. Numerical chip illustrated in Fig. 6 underlined that, for both investigated planes, the segments formation was controlled by three main sub-processes:

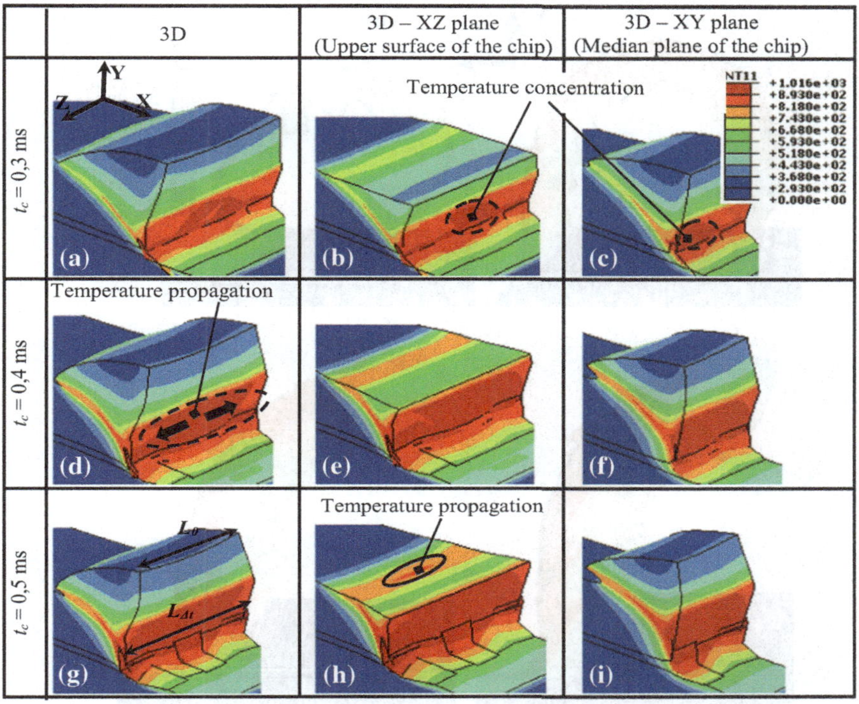

Fig. 3 Instantaneous temperature distribution in the chip for different observation planes (V_c = 45 m/min and f = 0.3 mm/rev)

- Material compression: The machined material was initially compressed without being damaged. An increase of the damage variable was noted in the tool tip vicinity, while almost undamaged material was modeled away from this contact interface.
- Localization and spread of thermomechanical loads: More accentuated and extended sticking contact was involved in the tool-chip interface, following the progressive penetration of the cutting tool in the workpiece. This explains the important local heating of the bottom chip surface, which was in contact with the tool rake face, illustrated by Fig. 3. For these thermomechanical conditions, the material damage was taking place. The most important and concentrated *SDEG* were reached in the median plane, notably in the primary shear band just around the tool tip. Therefore, more pronounced sliding of the chip along the shear plane was obtained, giving rise to slight distinction in the segment geometry from one modeled plane to another.
- Ejection: The chip slid upwards on the tool rake face and a totally formed segment was modeled.

However, the comparison of numerical results illustrated in Fig. 6a and e highlighted that, for the same cutting time (t_c = 0.3 ms), the formation of the first shear

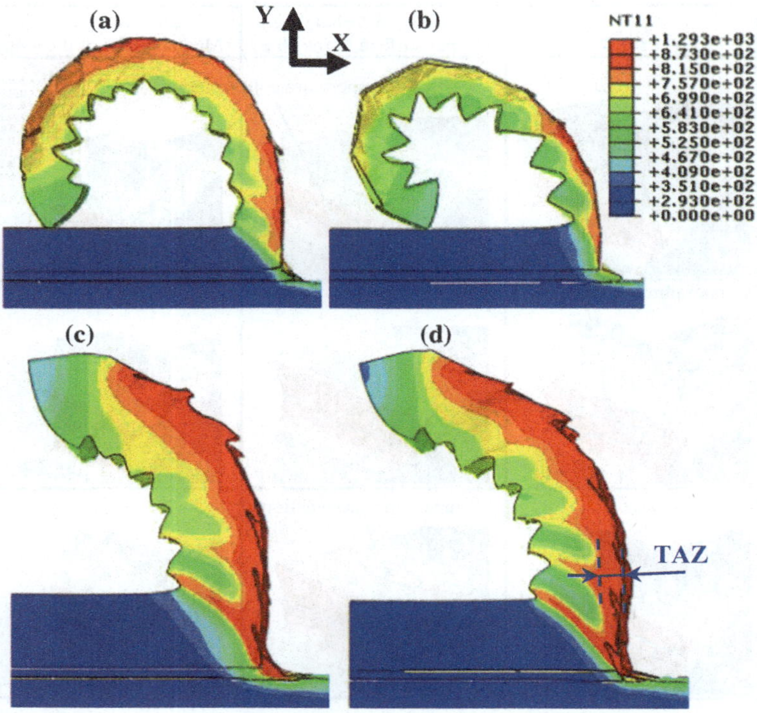

Fig. 4 Effect of the cutting conditions on the temperature distribution: **a** $V_c = 45$ m/min and $f = 0.15$ mm/rev, **b** $V_c = 75$ m/min and $f = 0.15$ mm/rev, **c** $V_c = 45$ m/min and $f = 0.3$ mm/rev and **d** $V_c = 75$ m/min and $f = 0.3$ mm/rev

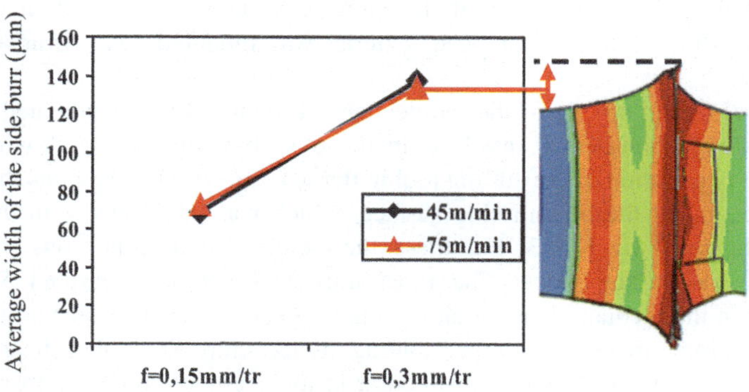

Fig. 5 Effect of the cutting conditions on the numerical side burr

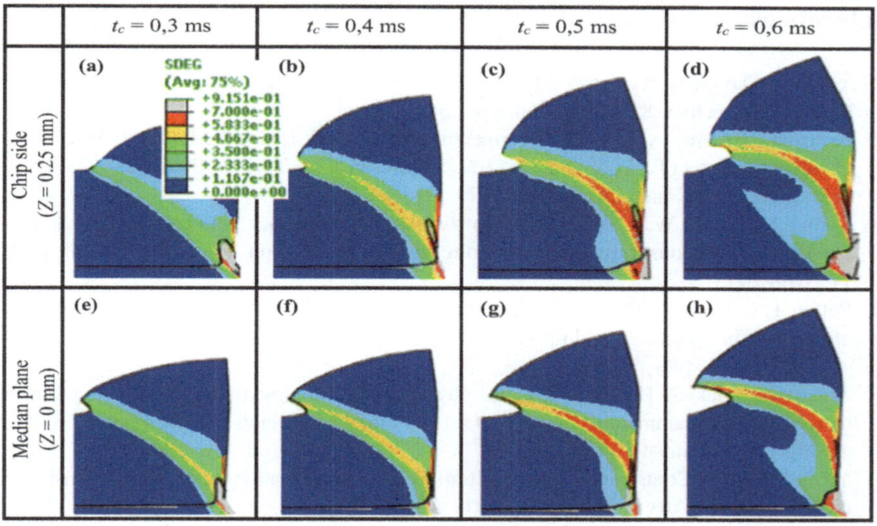

Fig. 6 Instantaneous damage distribution in the chip as a function of the observation plane (V_c = 45 m/min and f = 0.3 mm/rev)

band was initiated in the median plane, while only a compressed removed material was modeled at the chip sides. Similar delay was noted for the other cutting times. This confirmed the hypothesis made by Pottier et al. [4], who supposed a damage initiation from the median plane to the chip sides.

4 Conclusion

The 3D modeling of the Ti_6Al_4V orthogonal machining, carried out in this investigation, allowed to better understand the multi-physical mechanisms controlling the chip formation. A noticeable mismatch in the distribution and the levels of numerical plastic strains, temperatures and damage variables, were pointed out between the median plane of workpiece and its sides. A pronounced thermal softening was predicted in the median plane. It resulted in the modeling of more concentrated and important damage within, mainly in the narrow shear bands, which emphasized the efficiency of the 3D FE modeling. Indeed, the current numerical analysis allowed to underline the inability of the 2D FE modeling to predict the chip formation mechanisms, even in the case of orthogonal machining.

In the current study, an increase of the chip width, due to the material flow along the Z-direction was highlighted. The modeling of several cutting conditions highlighted the sensitivity of side burrs to the feed rate. They became more pronounced, giving rise to additional thermomechanical loads applied to the cutting tool. Although the low intensity of this kind of material flow, compared to that obtained in thickness, the consideration of its effect on the machining was required.

References

1. Veiga C, Davim JP, Loureiro AJR (2013) Review on machinability of titanium alloy: The process perspective. Rev. Adv Mater Sci 34:148–164
2. Arrazola PJ, Garay A, Iriarte LM, Armendia M, Marya S, Le Maître F (2009) Machinability of titanium alloys (Ti_6Al_4V and Ti555.3). J Mater Process Technol 209:2223–2230. https://doi.org/10.1016/j.jmatprotec.2008.06.020
3. Daoud M, Jomaa W, Chatelain JE, Bouzid A (2015) A machining-based methodology to identify materialconstitutive law for finite element simulation. Int J Adv Manuf Technol 77:2019–2033
4. Pottier T, Germain G, Calamaz M, Morel A, Coupard D (2014) Sub-millimeter measurement of finite strains at cutting tool tip vicinity. Exp Mech 54:1031–1042. https://doi.org/10.1007/s11340-014-9868-0
5. Nouari M, Makich H (2014) On the physics of machining titanium alloys: interactions between cutting parameters, microstructure and tool wear. Metals (Basel) 4:335–358. https://doi.org/10.3390/met4030335
6. Yaich M (2017) Contribution à la fiabilisation de la modélisation numérique de l'usinage de pièces en titane. Arts et Métiers Paritech, Angers
7. Zang J, Zhao J, Li A, Pang J (2017) Serrated chip formation mechanism analysis for machining of titanium alloy Ti-6Al-4V based on thermal property. Int J Adv Manuf Technol. https://doi.org/10.1007/s00170-017-0451-6
8. Ali MH, Ansari MNMNM, Khidhir BA, Mohamed B, Oshkour AA (2014) Simulation machining of titanium alloy (Ti-6Al-4V) based on the finite element modeling. J Braz Soc Mech Sci Eng 36:315–324. https://doi.org/10.1007/s40430-013-0084-0
9. Ducobu F, Rivière-Lorphèvre E, Filippi E (2016) Material constitutive model and chip separation criterion influence on the modeling of Ti_6Al_4V machining with experimental validation in strictly orthogonal cutting condition. Int J Mech Sci 107:136–149. https://doi.org/10.1016/j.ijmecsci.2016.01.008
10. Zhang YC, Mabrouki T, Nelias D, Gong YD (2011) Chip formation in orthogonal cutting considering interface limiting shear stress and damage evolution based on fracture energy approach. Finite Elem Anal Des 47:850–863. https://doi.org/10.1016/j.finel.2011.02.016
11. Johnson GR, Cook WH (1985) Fracture characteristics of three metals subjected to various strains, strain rates, temperatures and pressures. Eng Fract Mech 21:31–48
12. Johnson GR, Cook WH (1983) A constitutive model and data for metals subjected to large strains, high strain rates and high temperatures. In: Proceedings of 7th international symposium Ballistics, vol 547, pp 541–547
13. Nemat-Nasser S, Guo WG, Nesterenko VF, Indrakanti SS, Gu YB (2001) Dynamic response of conventional and hot isostatically pressed Ti-6Al-4V alloys: experiments and modeling. Mech Mater 33:425–439. https://doi.org/10.1016/S0167-6636(01)00063-1

A Modified FSDT Model for Static Analysis of Smart Functionally Graded Shells

H. Mallek[2(✉)], H. Mellouli[2], H. Jrad[2], M. Wali[1,2], and F. Dammak[2]

[1] Department of Mechanical Engineering, College of Engineering, King Khalid University, Abha, Saudi Arabia
mondherwali@yahoo.fr
[2] Laboratory of Electromechanical Systems (LASEM), National Engineering School of Sfax, University of Sfax, Route de Soukra km 4, 3038 Sfax, Tunisia
{hanen.mallek,hana.mellouli,Fakhreddine.dammak}@enis.tn, hanen.j@gmail.com

Abstract. This paper investigates static analysis of multilayered shells with integrated piezoelectric materials. An efficient 4-node shell element is developed to solve piezoelastic response of functionally graded structure with embedded piezoelectric actuators and sensors. A modified First order Shear Deformation Theory (FSDT) is introduced in the present method to remove the shear correction factor and improve the accuracy of transverse shear stresses. The properties of subtract material are assumed to be graded through the thickness by the power law distribution while the electric potential is assumed to be a linear function through the thickness of each active sub-layer. Accuracy and convergence of the present model is validated by comparing the numerical results with the published numerical solutions in the literature.

Keywords: Modified FSDT · 4-node shell · Sensors/Actuators · Piezoelectric material · Functionally graded structure

1 Introduction

Smart materials are characterized by a sufficiently strong coupling between the electric field and the elastic deformation, so that they can be used in sensors as well as in actuator applications. These categories of materials are widely used in order to attenuate the acoustical noise [4], control the free vibration [3] and control the deflection control of structural elements [12, 5, 13, 18].

Most of the composite shell FE formulations are based on three different approaches which are: the Kirchhoff theory (CST) [3, 8], the First order Shear Deformation Theory (FSDT) [11, 17] and the High order Shear Deformation Theory (HSDT) [10, 20]. The inefficiency of the Kirchhoff hypothesis appears with neglecting the effects of transverse shear and normal strains of the structure and using the FSDT, shear correction coefficients (equal to 5/6 for isotropic material) should be included to adjust the transverse shear stiffness. The high-order shear deformation theory was established in order to assume a parabolic distribution of the transverse shear strain to correct the constant shear stress in the Mindlin-Reissner theory and to get closer a realistic

N. Aifaoui et al. (Eds.): CMSM 2019, LNME, pp. 681–689, 2020.
https://doi.org/10.1007/978-3-030-27146-6_74

distribution of the transverse shear strain through the thickness [14, 16, 25]. Despite HSDT theories provide a refined approximation of the displacements and deformations of the structure, the number of used kinematic variables is mainly high which leads to a high CPU effort. Hence, attention has turned to modified FSDT shell type finite elements [15, 19, 21, 22]. This new theory is carried out to handle the linear distribution of the shear strain by imposing a new parabolic shear strain distribution across the shell thickness and a zero shear strain on the top and bottom faces of the shell.

In this work, the modified FSDT is introduced to predict the piezoelastic response of Functionally Graded Material (FGM) structures with integrated smart layers using 3D-piezoelectric shell model. The analysis is based on the finite element method by using 4-node shell element and the electric potential is assumed to be linear through the thickness of the piezoelectric layer. Numerical results of piezoelastic behavior of FGM plate under double sinusoidal load without and with piezoelectric effect are presented. Effects of geometrical parameters and power-law index on the response of the piezo-laminated structure are also examined.

2 Theoretical Formulations

In this section, the geometry and kinematics of the modified FSDT model are briefly described.

2.1 Material Properties of FG Plates

In order to model the properties of FGM structures, a description is needed to show the global distribution of material properties. The top surface is "Alumina-rich", whereas the bottom surface is "Aluminum-rich". In this study, using a polynomial material law, the distribution of young modulus Y_{FGM} in terms of shell thickness is assumed to follow [1, 6, 7, 24]:

$$Y_{FGM}(z) = Y_m + (Y_c - Y_m)\left(\frac{z}{h} + \frac{1}{2}\right)^n \tag{1}$$

where $Y_m = 70$ GPa and $Y_c = 380$ GPa are the Young modulus of the metal and ceramic components, respectively and n is the power-law index. The Poisson's ratio for both metal and ceramic is assumed to be constant.

2.2 Kinematic Assumptions, Weak Form and Constitutive Relations

The position vector of any material point (q) of the shell structure located at the distance z from the mid surface can be related to the position vector of (p) in both configurations C_0 and C_t as:

$$\begin{cases} X_q(\xi^1,\xi^2,z) = X_p(\xi^1,\xi^2) + zD(\xi^1,\xi^2) \\ x_q(\xi^1,\xi^2,z) = x_p(\xi^1,\xi^2) + zd(\xi^1,\xi^2) \end{cases}; \quad z \in [-h/2, h/2] \tag{2}$$

with $\boldsymbol{\xi} = (\xi^1, \xi^2, \xi^3 = z)$ denoting the curvilinear coordinates. $\boldsymbol{D}$ and $\boldsymbol{d}$ are the shell director vectors in reference and current configurations, respectively. h represents the thickness.

The strain $\boldsymbol{\varepsilon}$ can be decomposed in in-plane and transverse shear strains as:

$$\begin{cases} \varepsilon_{\alpha\beta} = e_{\alpha\beta} + z\,\chi_{\alpha\beta} \\ 2\varepsilon_{\alpha 3} = \gamma_\alpha \end{cases}; \quad \alpha, \beta = 1, 2 \tag{3}$$

where $e_{\alpha\beta}$, $\chi_{\alpha\beta}$ and γ_α are the components of linearized membrane, bending and shear strains vectors expressed by:

$$\begin{cases} \delta e_{\alpha\beta} = 1/2(\boldsymbol{A}_\alpha \cdot \delta\boldsymbol{x}_{,\beta} + \boldsymbol{A}_\beta \cdot \delta\boldsymbol{x}_{,\alpha}); \quad \delta\gamma = \boldsymbol{A}_\alpha \cdot \delta\boldsymbol{d} + \delta\boldsymbol{x}_{,\alpha} \cdot \boldsymbol{d} \\ \delta\chi_{\alpha\beta} = 1/2(\boldsymbol{A}_\alpha \cdot \delta\boldsymbol{d}_{,\beta} + \boldsymbol{A}_\beta \cdot \delta\boldsymbol{d}_{,\alpha} + \delta\boldsymbol{x}_{,\alpha} \cdot \boldsymbol{d}_{,\beta} + \delta\boldsymbol{x}_{,\beta} \cdot \boldsymbol{d}_{,\alpha}); \quad \alpha, \beta = 1, 2 \end{cases} \tag{4}$$

In matrix notation, the membrane, bending, and shear strains vectors are given by

$$\boldsymbol{e} = \begin{Bmatrix} e_{11} \\ e_{22} \\ 2e_{12} \end{Bmatrix}, \quad \boldsymbol{\chi} = \begin{Bmatrix} \chi_{11} \\ \chi_{22} \\ 2\chi_{12} \end{Bmatrix}, \boldsymbol{\gamma} = \begin{Bmatrix} \gamma_1 \\ \gamma_2 \end{Bmatrix} \tag{5}$$

The electric field $\boldsymbol{E}$, assumed to be constant over an element of the active layer, varies linearly through the thickness of this layer. It can be expressed in terms of the electric potential φ as:

$$\boldsymbol{E} = -\varphi_{,\alpha}; \quad \alpha = 1, 2, 3 \tag{6}$$

The strains and electric field expressions shown in Eqs. (5) and (6) will be used in the weak form of equilibrium equations in order to obtain the numerical solution as below:

$$G = \int_A (\boldsymbol{N} \cdot \delta\boldsymbol{e} + \boldsymbol{M} \cdot \delta\boldsymbol{\chi} + \boldsymbol{T} \cdot \delta\gamma + \bar{q} \cdot \delta\boldsymbol{E}) dA - G_{ext} = 0 \tag{7}$$

where the membrane $\boldsymbol{N}$, bending $\boldsymbol{M}$ and shear $\boldsymbol{T}$ stresses resultants and the electric displacement resultant $\bar{q}$ can be written in the form:

$$\boldsymbol{N} = \int_{-h/2}^{h/2} \begin{bmatrix} \sigma_{11} \\ \sigma_{22} \\ \sigma_{12} \end{bmatrix} dz, \; \boldsymbol{M} = \int_{-h/2}^{h/2} z \begin{bmatrix} \sigma_{11} \\ \sigma_{22} \\ \sigma_{12} \end{bmatrix} dz, \; \boldsymbol{T} = \int_{-h/2}^{h/2} f(z) \begin{bmatrix} \sigma_{13} \\ \sigma_{23} \end{bmatrix} dz, \; \bar{\boldsymbol{q}} = \int_{-h/2}^{h/2} \boldsymbol{q}\, dz \tag{8}$$

where $f(z) = \frac{5}{4}\left(1 - \frac{4z^2}{h^2}\right)$ that insure a parabolic shear strain through the thickness and $\boldsymbol{\sigma}$ is the stress tensor.

The generalized resultant of stress $\boldsymbol{R}$ and strain $\boldsymbol{\Sigma}$ vectors are expressed as:

$$\boldsymbol{R} = \begin{bmatrix} \boldsymbol{N} & \boldsymbol{M} & \boldsymbol{T} & \bar{\boldsymbol{q}} \end{bmatrix}^T_{11\times1}, \boldsymbol{\Sigma} = \begin{bmatrix} \boldsymbol{e} & \boldsymbol{\chi} & \boldsymbol{\gamma} & -\boldsymbol{E} \end{bmatrix}^T_{11\times1} \tag{9}$$

The linear constitutive equations of piezoelasticity expressing the coupling between the elastic and electric fields relevant to present problem can be defined as:

$$\begin{bmatrix} \boldsymbol{\sigma} \\ \boldsymbol{q} \end{bmatrix} = \begin{bmatrix} \boldsymbol{C} & \boldsymbol{p}^T \\ \boldsymbol{p} & -\boldsymbol{k} \end{bmatrix} \begin{bmatrix} \boldsymbol{\varepsilon} \\ -\boldsymbol{E} \end{bmatrix} \tag{10}$$

in which $\boldsymbol{C}$ is the matrix of elasticity, p is the piezoelectric matrix and $\boldsymbol{k}$ is the matrix of dielectric permittivity.

Using Eqs. (3), (6), (8), and (10), the resultant of stress $\boldsymbol{R}$ is related to the strain field $\boldsymbol{\Sigma}$ as below:

$$\boldsymbol{R} = \boldsymbol{H}_T\boldsymbol{\Sigma}, \boldsymbol{H}_T = \begin{bmatrix} \boldsymbol{H}_{11} & \boldsymbol{H}_{12} & 0 & \boldsymbol{H}_{14} \\ & \boldsymbol{H}_{22} & 0 & \boldsymbol{H}_{24} \\ & & \boldsymbol{H}_{33} & \boldsymbol{H}_{34} \\ Sym & & & \boldsymbol{H}_{44} \end{bmatrix}, \tag{11}$$

with $\boldsymbol{H}_T$ is the linear coupling elastic and electric matrix expressed as:

$$\begin{cases} (\boldsymbol{H}_{11}, \boldsymbol{H}_{12}, \boldsymbol{H}_{22}) = \int_{-h/2}^{h/2} (1, z, z^2)\boldsymbol{C}\, dz \\ \boldsymbol{H}_{33} = \int_{-h/2}^{h/2} (f(z))^2\, \boldsymbol{C}_\tau\, dz \\ (\boldsymbol{H}_{14}, \boldsymbol{H}_{24}) = \int_{-h/2}^{h/2} (1, z)\boldsymbol{p}_1^T\, dz \\ \boldsymbol{H}_{34} = \int_{-h/2}^{h/2} f(z)\,\boldsymbol{p}_2^T dz \\ \boldsymbol{H}_{44} = \int_{-h/2}^{h/2} \boldsymbol{k}\, dz \end{cases} \tag{12}$$

where $\boldsymbol{C}$ and $\boldsymbol{C}_\tau$ are in plane and out-of-plane elasticity sub-matrices. $\boldsymbol{p}_1^T$, $\boldsymbol{p}_2^T$ and $\boldsymbol{k}$ represent the in plane and out-of-plane piezoelectric sub-matrices and dielectric permittivity matrix, respectively.

2.3 Finite Element Approximation

The structure discretization is achieved using a four node shell element. The mechanical strains and the electric field have to be approximated by using bilinear shape functions. The interpolation of the shear strains are obtained by using the assumed natural strain (ANS) method in order to avoid shear locking. For more details, one can be referred to Mallek et al. [9]. Therefore, the discrete form of Eq. (7) leads to the discretized static linear piezoelastic equilibrium equation for the structure.

3 Numerical Results

An FGM square plate with two piezoelectric G-1195N layers bonded on the top and the bottom surfaces (Fig. 1) is analyzed in this section in order to outline the performance and the efficiency of the proposed model. The global thickness and the side-to-thickness ratio of the structure are denoted h and a/h, respectively. The thickness of the smart square plate is taken as $0.1 \times h$ for the top and bottom active layers and $0.8 \times h$ for the FGM layer. The structure is subjected to doubly sinusoidal distributed load q expressed as $q = q_0 \sin(\pi x/a)\sin(\pi y/a)$. The material properties of the metal and ceramic components used for the calculation are $Y_m = 70$ GPa, $v_m = 0.3$, $Y_c = 380$ GPa and $v_c = 0.3$ respectively. The material and piezoelectric properties of G-1195N are: $Y = 62$ GPa, $v = 0.31$ $e_{11} = e_{22} = -20.16\,\text{C/m}^2$ and $k_{33} = 3.3610^{-10}$F/m. The deflection and the shear stresses are computed in normalized form as given:

$$\bar{w} = \frac{10Y_c h^3}{a^4 q_0} w\left(\frac{a}{2}, \frac{a}{2}\right), \bar{\sigma}_{xz} = \frac{h}{a\, q_0} \sigma_{xz}\left(0, \frac{a}{2}, 0\right) \tag{6}$$

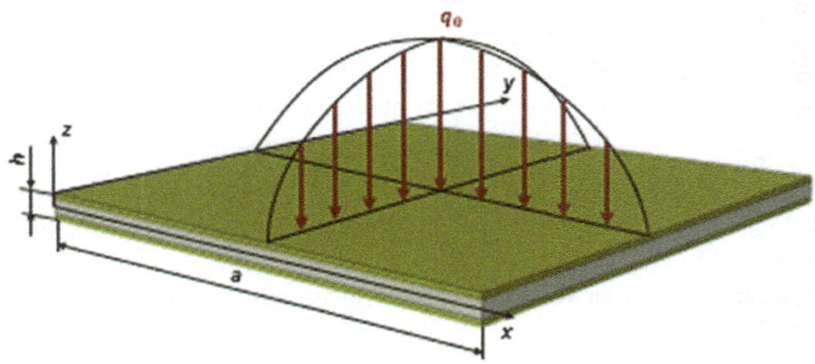

Fig. 1 A simply supported square FGM plate integrated active layers under sinusoidal load

Only investigation on FGM square plate under doubly sinusoidal load without active layers are available in literature. Hence, before proceeding to the numerical results of piezoelastic response of smart FGM and in order to examine the validity and accuracy of the present model, comparisons of results obtained by developed model and those available in literature [2, 23] are presented, where the G-1195N piezoelectric/FGM/G-1195N piezoelectric square plate reduces to a one layer of FGM by setting the thicknesses of piezoelectric layers to be zero, so that, piezoelectric effects vanish.

Figure 2 shows the dimensionless deflection at center of square plate for different values of power law exponent n with and without piezoelectric effect considering a/h = 10. Results obtained by the present model (in the case of FGM plate without active layers) are compared with analytical solutions by Carrera et al. [2]. The present approach shows very close results to Carrera's analytical solution [2]. Moreover, the increase in the material power index *n* leads to an increase in the deflection of the structure for both cases. This is because that as increasing the value of material power index the percentage of aluminum phase will rise. By reason of aluminum has a lower elastic material property that makes such FGM shells less stiff. It is clearly seen from Fig. 2 that deflection center of square plate with integrated piezoelectric layers is higher than that without active layers which is expectable. In fact, due to combined electrical and mechanical interaction for piezoelectric materials, the structure undergo predominant bending.

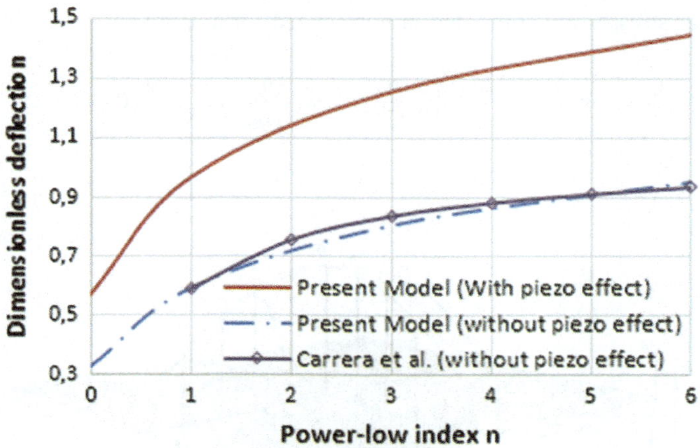

Fig. 2 Dimensionless deflection versus power low index n for a simply supported square plate under sinusoidal load (a/h = 10)

It is important to investigate the influence of the plate thickness on the deformation of the structure with and without piezo active layers. Figure 3 shows the effect of ratio a/h on the deflection at center of the square plate under doubly sinusoidal load for volume fraction index n = 6. As may be seen from Fig. 3 that the results of developed model correlated to solution obtained by Wali et al. [23], who used 3d-shell model based on a double director shell element.

Furthermore, it is found that dimensionless center-point deflection of the structure decreases as the ratio rises for both cases: with and without piezoelectric effects and it becomes almost constant when (a/h > 50). This fact is due to the stiffness which decreases with the increase of the ratio a/h and the plate becomes thinner and then deforms easily. Figure 3 also shows that the thickness of active layers causes a significant effect on the piezoelastic linear behavior of the active structure. Indeed,

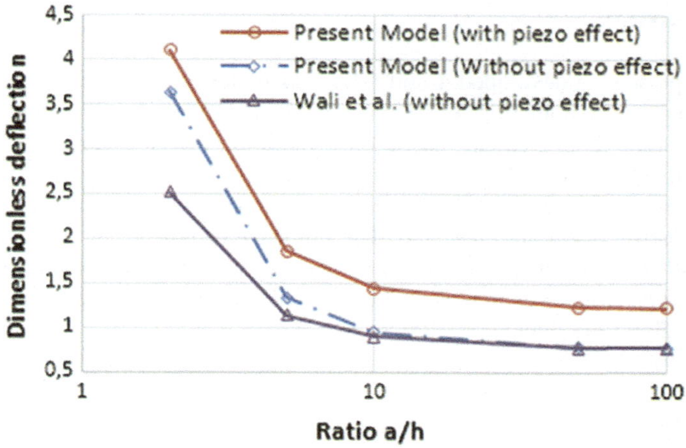

Fig. 3 Dimensionless deflection versus ratio a/h for a simply supported square plate under sinusoidal load (n = 6)

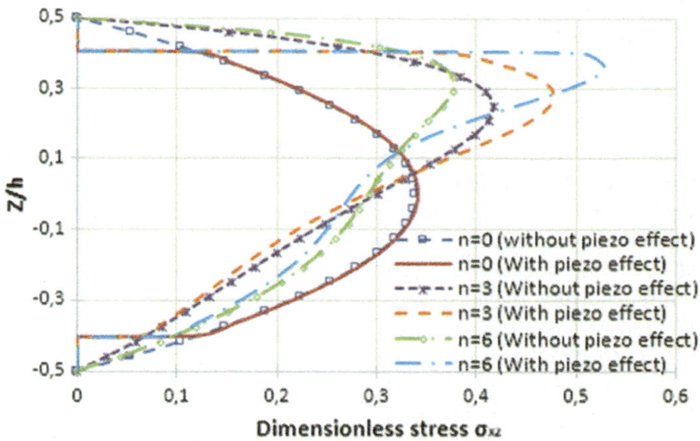

Fig. 4 Dimensionless shear stresses $\bar{\sigma}_{xz}$ for a simply supported square plate under sinusoidal load (a/h = 10)

deflection at center of smart plate is more than 1.5 times of that without piezoelectric layers. Thus, such sensitive geometrical parameter should be carefully considered for the design of smart structures with integrated piezoelectric layers.

The distribution of shear stresses at the center of transversal edge of the plate with and without piezoelectric effect for different volume fraction indexes considering a/h = 10 are depicted in Fig. 4. It can be observed that the piezoelectric layers have a significant effect on the in-plane shear in the layer of the substrate (FGM layer) caused by the static sinusoidal load. In fact, sensors (active layers) tend to increase transverse shear stresses in the substrate layer.

4 Conclusion

This paper presents a developed 4-node shell finite element that can be gainfully used for modeling and simulation of the behavior of smart structure with integrated sensors/actuators. This element, based on a modified FSDT theory, is developed to include stiffness and the electromechanical coupling of the active layers. Numerical results of piezoelastic behavior of FGM plate under double sinusoidal load without and with piezoelectric effect are presented. The influence of power law coefficient and the piezoelectric effect on shells are also examined. It was found that the power-law index and the piezoelectric layer thickness have significant effects on deflection and stress distribution along the thickness.

References

1. Bouhamed A, Jrad H, Jamel M, Wali M, Gamaoun F, Dammak F (2019) Homogenization of elasto-plastic functionally graded material based on representative volume element: application to incremental forming process. Int J Mech Sci 160:412–420
2. Carrera E, Brischetto S, Robaldo A (2008) Variable kinematic model for the analysis of functionally graded material plates. AIAA J 46:194–203
3. Gabbert U, Koppe H, Seeger F, Berger H (2002) Modeling of smart composite shell structures. J Theor Appl Mech 40(3):575–593
4. Gabbert U, Duvigneau F, Ringwelski S (2017) Noise control of vehicle drive systems. Facta Univ Ser Mech Eng 15(2):183–200
5. Jrad H, Mallek H, Wali M, Dammak F (2018) Finite element formulation for active functionally graded thin-walled structures. CR Mec 346(12):1159–1178
6. Jrad H, Mars J, Wali M, Dammak F (2018) An extended finite element method for modeling elastoplastic FGM plate-shell type structures. Struct Eng Mech 68:299–312
7. Jrad H, Mars J, Wali M, Dammak F (2019) Geometrically nonlinear analysis of elastoplastic behavior of functionally graded shells. Eng Comput 35:833–847
8. Mallek H, Jrad H, Algahtani A, Wali M, Dammak F (2019) Geometrically non-linear analysis of FG-CNTRC shell structures with surface-bonded piezoelectric layers. Comput Methods Appl Mech Eng 347:679–699
9. Mallek H, Jrad H, Wali M, Dammak F (2019) Geometrically nonlinear finite element simulation of smart laminated shells using a modified first-order shear deformation theory. J Intell Mater Syst Struct 30(4):517–535
10. Mallek H, Jrad H, Wali M, Dammak F (2019) Piezoelastic response of smart functionally graded structure with integrated piezoelectric layers using discrete double directors shell element. Compos Struct 210:354–366
11. Marinković D, Marinković Z, Petrović G (2012) On efficiency of a single-layer shell element for composite laminated structures. Facta Universitatis Ser: Mech Eng 10(2):105–112
12. Marinković D, Rama G (2017) Co-rotational shell element for numerical analysis of laminated piezoelectric composite structures. Compos B Eng 125:144–156
13. Marinković D, Rama G, Zehn M (2019) Abaqus implementation of a corotational piezoelectric 3-node shell element with drilling degree of freedom. Facta Univ Ser Mech Eng 17(2):269–283

14. Mellouli H, Jrad H, Wali M, Dammak F (2019) Meshfree implementation of the double director shell model for FGM shell structures analysis. Eng Anal Boundary Elem 99:111–121
15. Mellouli H, Jrad H, Wali M, Dammak F (2019) Meshless implementation of arbitrary 3D-shell structures based on a modified first order shear deformation theory. Comput Math Appl 77:34–49
16. Mellouli H, Jrad H, Wali M, Dammak F (2019) Geometrically nonlinear meshfree analysis of 3D-shell structures based on the double directors shell theory with finite rotations. Steel Compos Struct 31:397
17. Rama G (2017) A 3-node piezoelectric shell element for linear and geometrically nonlinear dynamic analysis of smart structures. Facta Univ Ser Mech Eng 15(1):31–44
18. Rama G, Marinković D, Zehn M (2018) Efficient three-node finite shell element for linear and geometrically nonlinear analyses of piezoelectric laminated structures. J Intell Mater Syst Struct 29(3):345–357
19. Shi G (2007) A new simple third-order shear deformation theory of plates. Int J Solids Struct 4443:99–4417
20. Sudhakar AK, Kamal M (2003) Finite element modeling of smart plates/shells using higher order shear deformation theory. Compos Struct 62:41–50
21. Trabelsi S, Frikha A, Zghal S, Dammak F (2018) Thermal post-buckling analysis of functionally graded material structures using a modified FSDT. Int J Mech Sci 144:74–89
22. Trabelsi S, Frikha A, Zghal S, Dammak F (2019) A modified FSDT-based four nodes finite shell element for thermal buckling analysis of functionally graded plates and cylindrical shells. Eng Struct 178:444–459
23. Wali M, Hajlaoui A, Dammak F (2014) Discrete double directors shell element for the functionally graded material shell structures analysis. Comput Methods Appl Mech Eng 278:388–403
24. Zenkour AM (2006) Generalized shear deformation theory for bending analysis of functionally graded plates. Appl Math Model 30:67–84
25. Zghal S, Frikha A, Dammak F (2017) Static analysis of functionally graded carbon nanotube-reinforced plate and shell structures. Compos Struct 176:1107–1123

Experimental Investigation of Mechanical Behavior of NiTi Arch Under Cycling Loading and Cathodically Hydrogen Charging

Riheme Sarraj[1(✉)], Tarek Hassine[1], and Fehmi Gamaoun[1,2]

[1] Laboratory of Mechanics of Sousse, National Engineering School of Sousse, University of Sousse, Sousse, Tunisia
Riheme.Sarraj@eniso.rnu.tn,
tarekhassine007@gmail.com, Fehmi.Gamaoun@insat.rnu.tn
[2] College of Engineering, King Khalid University, Abha, Saudi Arabia

Abstract. The aim of the present study is to predict the NiTi behavior under cyclic loading with hydrogen charging. To achieve this goal, a series of experimental tests have been carried out. First, samples have been cycled until having an imposed deformation of 2.1, 4.8, 7.6 and 8.2% till 50 cycles. Second, orthodontic specimens are submitted to cyclic loading with various strain rates of $5 \times 10^{-3}\ s^{-1}$, $10^{-3}\ s^{-1}$, $5 \times 10^{-4}\ s^{-1}$ and $10^{-4}\ s^{-1}$ under an imposed strain of 7.6%. Finally, arch wires are charged by hydrogen in 0.9% NaCl an aqueous solution at room temperature with a current density of 10 A/m^2 for 2, 3, 4 and 6 h and are aged for 7 days in air. Throughout cyclic loading, a significant degradation of material performance is observed (the critical stress for the start and the end of the martensite transformation, the residual strain and the dissipated energy evolving). This evolution becomes more significant with a higher strain rate and with hydrogen charging rather then without it. Thus, via this work, we can assume that the embrittlement is due to the diffusion of hydrogen and the generation of dislocations after aging.

Keywords: Shape memory alloys · Cyclic effect · Hydrogen · Embrittlement

1 Introduction

NiTi wires have extensively been considered in the medical field [1]. Its use such an alloy is mainly justified by its shape memory effect, biocompatibility and superelasticity [2]. Furthermore, this shape memory alloy has shown excellent mechanical properties and a high resistance to corrosion and abrasion for long time applications. In the oral cavity, the superelastic wire is applied a constant force during the treatment time to control the teeth movement. However, a fracture in the wire will appear after few months of the service life. This embrittlement is due to the superposition of two factors: first, orthodontists manipulate the wire by applying a few number of cycles, which will alter the mechanical behavior of this sample [3]. Second, the NiTi shape memory alloy shows a significant sensitivity to hydrogen absorption, which is caused by the diffusion of an amount of hydrogen impurities due to the presence of fluoride in the toothpaste [4, 5].

N. Aifaoui et al. (Eds.): CMSM 2019, LNME, pp. 690–698, 2020.
https://doi.org/10.1007/978-3-030-27146-6_75

Under cyclic loading, residual strains are accumulated in the structure owing to the residual martensite, lattice defects and dislocations [6]. The hysteresis loop changes its size until reaching a saturated state [7]. The critical stress of the forward and inverse transformation decreases with the growing number of cycles [8].

The degradations of the NiTi arch wire's mechanical behavior depend on the strain rate. This dependence is attributed to the released latent heat during the martensitique transformation and the production of the variant of the martensite in the structure. Gamaoun et al. [1, 9] studied the effect of the strain rate of NiTi wire's mechanical behavior. Their work highlighted that the critical stress and plateau slope went up with the rising of the strain rate, which caused the germination of the new domains. The increase in the temperature inside the wire made the heat release faster, the stress required for the growth of the new domains rise and the slope of the plateau go up during the forward transformation.

Regarding the hydrogen embrittlement, several studies have been developed to investigate this effect on the mechanical behavior of the NiTi wires. Sheriff et al. [10] showed the effects of 10–80 wppm of hydrogen on the behavior of the superelastic NiTi alloy. Their work highlighted that the fatigue life would decrease with the increasing hydrogen quantity. The increase in the concentration of the hydrogen would affect the fatigue life for an imposed strain less than 1.4% and up to 80 ppm of hydrogen. Lachiguer et al. [11] examined the influence of the hydrogen concentration on the thermomechanical behavior of Shape Memory Alloy (SMA). It was proven also that the hysteresis loop area of mechanical loading-unloading would depend on the immersion time in the hydrogen solution. It was shown that the critical stress of austenite-martensite transformation went down from 500 MPa for the as-received sample to 570 MPa after 72 h of hydrogen charging. Besides, the plateau slope increase suggest that with the absorption of hydrogen, more energy would be required for the transformation.

The aim of the present work is to investigated the cyclic effect at different imposed strains, the strain rate effect and the hydrogen effect of the superelastic NiTi orthodontic alloy. It presents a contribution to studying the mechanical behavior of the orthodontic arch wire when it is introduced in the bracket glued on the tooth in the oral cavity.

2 Experimental Procedure

For all experimental tests, NiTi specimens, with dimensions of $20 \times 0.43 \times 0.64$ mm^3, have been used. All the specimens have been submitted to a cyclic loading using an uniaxial-tensile loading Instron 5566-type machine with a load cell of 10 kN. The cyclic tensile tests have been performed at room temperature with a constant strain rate of 10^{-4} s^{-1} with an imposed strain of 2.1, 4.8, 7.6% and 8.2%.

In order to study the strain rate effect on the NiTi wires, cyclic loading tests have been carried out corresponding to an imposed strain of 7.6% with a various strain rate of 5×10^{-3} s^{-1}, 10^{-3} s^{-1}, 5×10^{-4} s^{-1}, and 10^{-4} s^{-1}. Scanning Electron Microscope (SEM) observations using a JEOL JSM-5400 microscope have been made to investigate the fracture surfaces of the cyclic samples with different strain rates.

To investigate the effect of hydrogen under cyclic loading, the specimens have been electrolytically charged by immersion in a 0.9% NaCl solution and using a platinum anode. Specimens have been charged following these conditions: current density of 10 A/m^2 and charging time of 2, 3, 4 and 6 h. The arch wires have been aged for 7 days at room temperature in air after charging.

3 Experimental Results and Discussion

3.1 Effect of Imposed Strain

Figure 1 shows the cyclic stress-strain of the NiTi arch wires obtained in the cyclic loading-unloading tests with various imposed strain of 2.1, 4.8, 7.6 and 8.2%. Some common features of the cyclic loading can be made from these experimental results. An accumulated residual strain is formed and goes up with the number of cycles and the value of the imposed strain. As noted, the critical stress for inducing the martensite and the austenite exhibits a marked decrease in the twenty first cycles until reaching a stabilized state. Moreover, the hysteresis loops area fall with the growth of the number of cycles [3]. The reduction in the critical stress is due to the fatigue, which causes the degradation of the mechanical properties. This effect is considered to be associated with the introduction of small levels of localized slip and build-up dislocations, which results in lower values of the critical stress to introduce martensite and austenite [12]. Previous studies indicated that the increase in the residual strain with cyclic loading was owing to the generation of dislocation during the cyclic loading [6]. Besides, the residual strain would typically stabilize after few cycles [13].

Added to that, the dislocation density tended to growth due to the shear deformations at the martensite–austenite phases, which altered some cyclic properties such as the ability to dissipate energy and the decrease in the hysteresis loop area [14].

3.2 Effect of Strain Rate

Figure 2 depicts the stress–strain curves under cyclic loading under an imposed strain of 7.6% with a various strain rate of $5 \times 10^{-3}\ s^{-1}$, $10^{-3}\ s^{-1}$, $5 \times 10^{-4}\ s^{-1}$ and $10^{-4}\ s^{-1}$. It is remarkable that the residual strain, the critical stress for inducing the martensite and the austenite and the hysteresis loop area evolve with the growing number of cycles. Moreover, these parameters reach a saturated state after about 20 cycles. The most important outcome is the effect of the strain rate on the degradation of the mechanical behavior. In fact, the critical stress of introducing the martensite increase from 400 MPa for a strain rate of $10^{-4}\ s^{-1}$ to 430 MPa for a strain rate of $5 \times 10^{-3}\ s^{-1}$. Furthermore, when the strain rate is equal to $5 \times 10^{-3}\ s^{-1}$, the residual strain is about 1.18% after 50 cycles, but when the strain rate is equal to $10^{-4}\ s^{-1}$, it reaches 0.7% after 50 cycles. The hardening-like slope of the plateau goes up for the higher strain rates due to the rise in the number of domains, which leads to the increase in the temperature in the structure, and consequently the release of heat will be fast.

Accordingly to the previous study, the slope of the plateau of the forward transformation rises with the increasing strain rate owing to the interaction between the

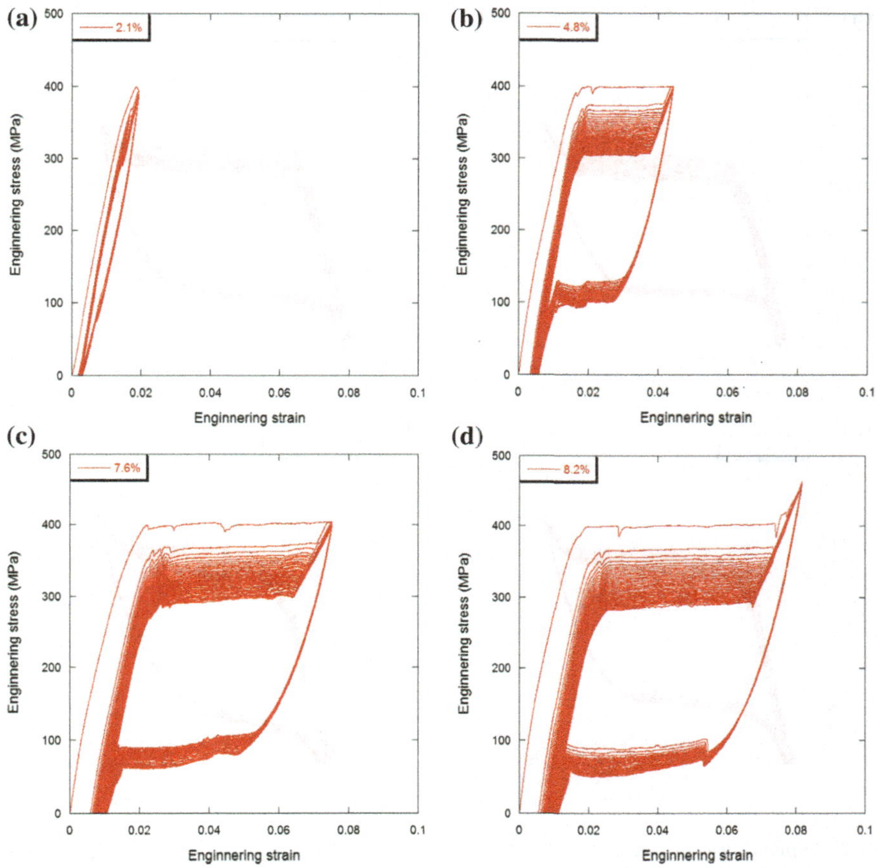

Fig. 1 Typical strain cycling curve after 50 cycles for deformed specimen at various imposed strains: **a** 2.1%, **b** 4.8%, **c** 7.6%, and **d** 8.2%

thermal, mechanical and transitional phenomena [15]. In fact, the internal heat production is fast when the strain rate is important. Yet, because the reduced conduct time, a quantity of heat could not be extracted out if the NiTi wire. Consequently, the temperature in the structure is increased and the critical transformation stresses would grow with the rise in the temperature, and the hardening-like slope of the plateau would go up [16, 17] (Fig. 2).

The SEM fractographs after a tensile test until the rupture of the NiTi arc wire cycled at an imposed strain of 7.6% with a strain rate of 10^{-3} s^{-1} and 10^{-4} s^{-1} are depicted in Fig. 3a and c. Figure 3b and d show the magnified (10×) portion of the central region. Both surfaces of the NiTi wire exhibit "cup and cone" fractures. It consists of small and large dimples. The ductile behavior is present in both samples.

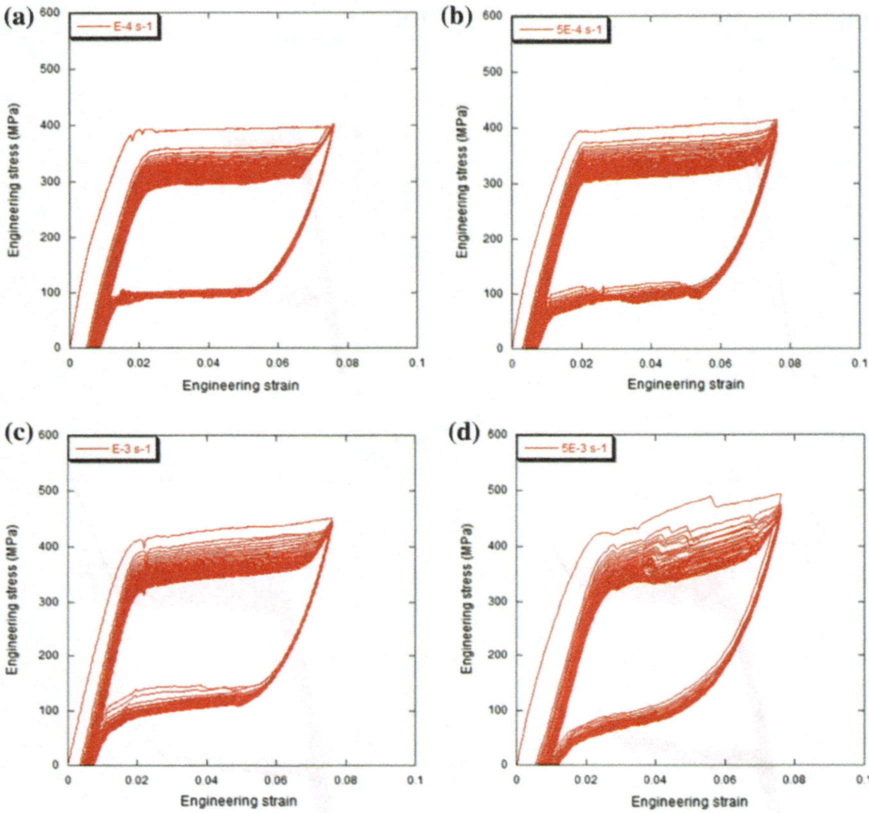

Fig. 2 Typical strain cycling curve after 50 cycles for deformed specimen with an imposed strain of 7.6% and at various strain rates: **a** E^{-4} s^{-1}, **b** $5E^{-4}$ s^{-1}, **c** E^{-3} s^{-1}, and **d** $5E^{-3}$ s^{-1}

3.3 Effect of Hydrogen Charging

Figure 4 represents the stress-strain curves of the charged samples with a current density of $10A/m^2$ after cyclic loading with an imposed strain of 2.1, 4.8, 7.6 and 8.2%. The obtained results indicate that the critical stress of the forward transformation of the charged specimen rises by 40 MPa compared to the as-received ones. Ogawa et al. [18], Yokoyama et al. [19] indicated that this behavior is due to the diffused hydrogen. This absorption delay the martensite transformation and lead to the increase in the critical stress. In addition, the hysteresis loops show a decrease in the size of superelastic plateau and the shape change with the rise of the number of cycles. This behavior is accentuated with the imposed strain of 7.6%, it leads to a sudden fracture of the NiTi wires which it lose their superelastic effect. In addition, the residual strain of the charged NiTi arch wire is more important than the accumulated strain in the as-received NiTi alloy. Furthermore, the plateau of the forward and the inverse transformation rises

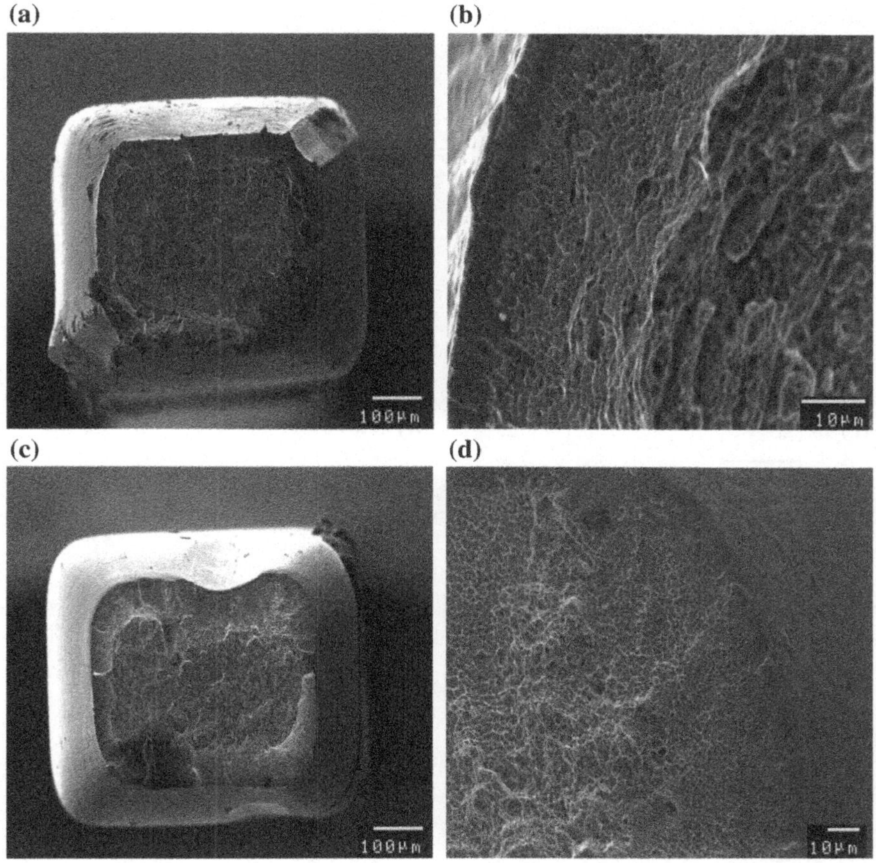

Fig. 3 SEM observation of surface after tensile test until rupture of NiTi arc wire cycled at imposed strain of 7.6% with various strain rate of: **a** 10^{-3} s^{-1}, **b** magnified view of dimples in (**a**), (**c**) 10^{-4} s^{-1} and (**d**) magnified view of dimples in (**c**)

after hydrogen charging. This hardening is explained by the generation of dislocations in the structure during charging.

The embrittlement of the orthodontic arch wire is due to the hydrogen diffusion in the structure under cyclic loading. In fact, the hydrogen diffuses from the surface to the inside of the wire and will be trapped by the dislocation sites existing in the structure. This trapped hydrogen obstructs the martensitic transformation, which leads to a stress rise and hence a delayed fracture of the specimen will be observed. Hydrogen embrittlement of a NiTi alloy is justified by the interactions of hydrogen with dynamic processes: the martensite transformation and dislocation movement [20, 21]. Therefore, a residual stress is introduced in the structure and increased with the number of cycles. The generation of the residual strain is due to the residual martensite [19].

Moreover, in the present study, the cathodic charging of hydrogen is applied at a low strain rate equal to 10^{-4} s^{-1}, i.e., low number of deformation cycles, for low strain

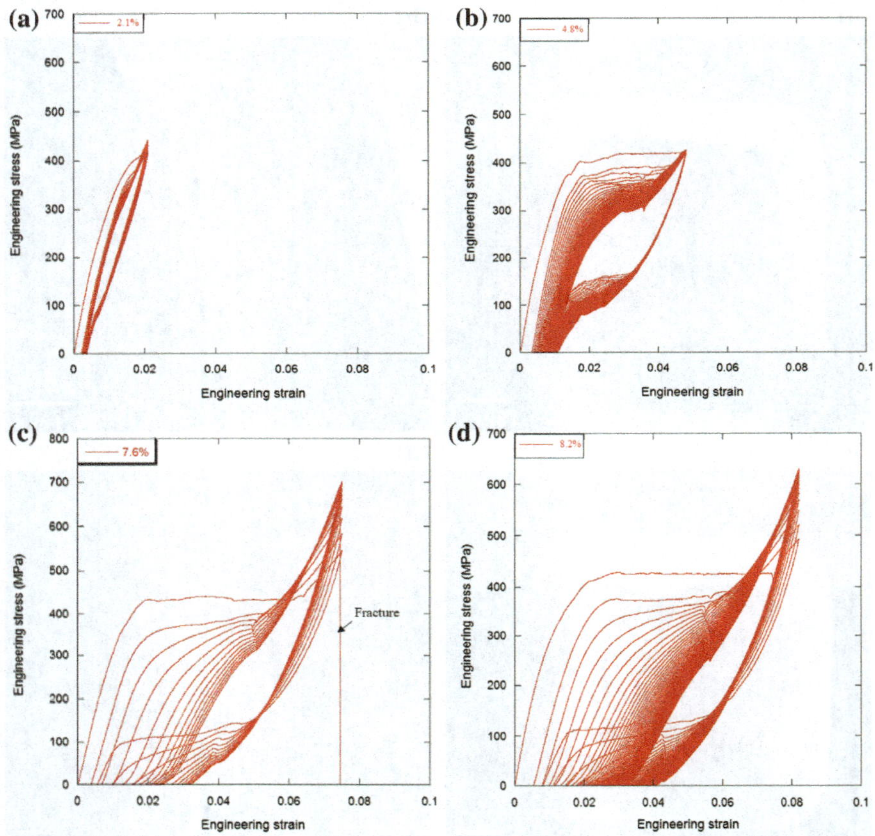

Fig. 4 Typical strain cycling curve after 50 cycles with an imposed strain of 7.6% and at a strain rates equal to 10^{-4} s^{-1} for charged specimen with a current density of 10 A/m^2 for: **a** 2 h, **b** 3 h, **c** 4 h and **d** 6 h and were aged for 7 days in air

rates, the number of the nucleated bands was very limited, compared to the high strain rates. This thermomechanical aspect is attributed to the heat conduction transfer due to the released latent heat of the domain-fronts during sustained loading. During cyclic loading, embrittlement has been observed after 6 h of hydrogen charging and ageing for 7 days for a lower strain rate (10^{-4} s^{-1}). Whereas, previous research [9] shown that, for the same charging and ageing condition with a strain rate equal to 10^{-2} s^{-1} depicts a superelastic behavior. According to these results we can highlight that for the as-charged specimens, the lower the strain rate is low, the more the embrittlement occurs after a shorter hydrogen charging time. This embrittlement is attributed to the interaction between the higher diffused quantity of hydrogen in the volume of the NiTi SMA and the thermo-mechanical aspect of the nucleation and growth of martensitic bands.

4 Conclusion

This article discusses the behavior of NiTi orthodontic arch wire in the oral cavity. To achieve this aim a series of experimental tests (the effect of cyclic loading with different imposed strain, the effect of strain rate and hydrogen effects under the cyclic loading) has been carried out. The results obtained in this experimental tests can be summarized as follows:

During cyclic loading, the results have despite an accumulation of the residual strain. The generation of dislocations leads to a decline in the size of the hysteresis loops of the stress-strain curves. Moreover, the critical stress for introducing martensite and austenite falls with the number of cycles. As the number of cycles rises, all this parameters evolve until reaching stable states.

The degradation of the NiTi arch wire's mechanical behavior depends on strain rate. Indeed, this degradation is owing to the growth of the nucleated domains, which raises with the increasing applied strain rate.

For the charged wires, the critical stress of the martensitique transformation rises up compared to the as-received specimens. Besides, the hysteresis loops decreases and the shape changes with the growth in the cycle number. These results are justified by the high amount of hydrogen tripped by the dislocations generated in the structure during cyclic loading.

References

1. Gamaoun F, Hassine T, Bouraoui T (2014) Strain rate response of a Ni–Ti shape memory alloy after Hydrogen charging. Philos Mag Lett 94(1):30–36
2. Yokoyama K et al (2015) Inhibition of localized corrosion of Ni–Ti superelastic alloy in NaCl solution by Hydrogen charging. J Alloys Comp 639:365–372
3. Sarraj R, Hassine T, Gamaoun F (2018) Mechanical behavior of NiTi arc wires under pseudoelastic cycling and cathodically Hydrogen charging. Mater Res Exp 5(1):15704
4. Elkhal Letaief W, Hassine T, Gamaoun F (2017) A coupled model between hydrogen diffusion and mechanical behavior of superelastic NiTi alloys. Smart Mater Struct 26(7): 75001
5. Letaief E, Wissem TH, Gamaoun F (2017) Tensile behaviour of superelastic NiTi alloys charged with hydrogen under applied strain. Mater Sci Tech (UK) 33(13):1533–1538
6. Miyazaki S, Imai T, Igo Y, Otsuka K (1986) Effect of cyclic deformation on the pseudoelasticity characterizations of Ti–Ni alloys. Metall Trans A, Phys Metall Mater Sci 17 A(1):115–120
7. Tanaka K et al (1996) Phenomenological analysis of plateaus on stress-strain hysteresis in TiNi shape memory alloy wires. Mech Mater 24(1):19–30
8. Strnadel B et al (1995) Cyclic stress-strain characteristics of TiNi and TiNiCu shape memory alloys. Mater Sci Eng, A 202(1–2):148–156
9. Gamaoun F, Hassine T (2015) Ageing effect and rate dependency of a NiTi shape memory alloy after Hydrogen charging. J Alloy Compd 615(S1):S680–S683
10. Sheriff J, Pelton AR, Pruitt LA (2004) Hydrogen effects on NITINOL fatigue. In: Proceedings of the international conference on shape memory and superelastic technologies, pp 111–116

11. Lachiguer A et al (2016) Modeling of Hydrogen effect on the superelastic behavior of Ni-Ti shape memory alloy wires. Smart Mater Struct 25(11):1–11
12. Desroches R, Asce M, Mccormick J, Delemont M (2004) Cyclic properties of superelastic shape memory alloy wires and bars. J Struct Eng 130(1):38–46
13. Brinson L Catherine, Schmidt I, Lammering R (2004) Stress-induced transformation behavior of a Polycrystalline NiTi shape memory alloy: micro and macromechanical investigations via in situ optical microscopy. J Mech Phys Solids 52(7):1549–1571
14. Maletta C et al (2014) Fatigue properties of a Pseudoelastic NiTi alloy: strain ratcheting and hysteresis under cyclic tensile loading. Int J Fatigue 66:78–85
15. Grandi D, Maraldi M, Molari L (2012) A macroscale phase-field model for shape memory alloys with non-isothermal effects: influence of strain rate and environmental conditions on the mechanical response. Acta Mater 60(1):179–191
16. Morin C (2011) A constitutive model for shape memory alloys accounting for thermomechanical coupling. Int J Plast 27(5):748–767
17. Zhang X et al (2010) Experimental study on rate dependence of macroscopic domain and stress hysteresis in NiTi shape memory alloy strips. Int J Mech Sci 52(12):1660–1670
18. Ogawa T, Takashima K, Asaoka K (2007) Hydrogen embrittlement of Ni–Ti superelastic alloy aged at room temperature after hydrogen charging. Mat Sci Eng A 466:106–113
19. Yokoyama K, Tomita M, Sakai J (2009) Hydrogen embrittlement behavior induced by dynamic martensite transformation of Ni–Ti superelastic alloy. Acta Materialia 57(6): 1875–1885
20. Gamaoun F, Ltaief M, Bouraoui T, Zineb TB (2011) Effect of Hydrogen on the tensile strength of aged Ni–Ti superelastic alloy. Art J Intell Mater Syst Struct 22(17):2053–2059
21. Kaneko K, Ogawa T, Moriyama K (2005) Hydrogen embrittlement of work-hardened Ni–Ti alloy in fluoride solutions. Biomaterials 26:101–108

The Effect of Surface Treatment on the Fatigue Behavior of NiTi Alloys

Maha Rokbani[1(✉)], Luc Saint-Sulpice[2], Shabnam Arbab-Chirani[2], and Tarak Bouraoui[1]

[1] ENIM, LGM, University of Monastir, Monastir, Tunisia
rokbani_maha@hotmail.fr, Tarak.bouraoui@gmail.fr
[2] ENIB, FRE CNRS 3744, IRDL, 29200 Brest, France
{Luc.Saint-Sulpice, Shabnam.arbab}@enib.fr

Abstract. NiTi alloys have been widely used as biomaterials especially for the realization of dental devices because of their biocompatibility and superelastic behavior. However, these tools can break frequently during clinical use. The susceptibility of these alloys to cyclic loadings in the presence of hydrogen can be one of major parameters in the degradation of fatigue properties. It is admitted that the state of surface is determining to know the influence of the hydrogen in the fatigue properties. In order to study the impact of some surface treatments and the effect of hydrogen on the fatigue behavior of NiTi alloys at high number of cycles, we propose to use the self-heating method. This methodology has the advantage of being faster and cheaper than traditional fatigue tests and allows to estimate the endurance limit of the material using empirical approach. In this case, the temperature variation is considered as a relevant parameter to predict the fatigue resistance of these alloys. The results of this study showed that after a surface electropolishing treatment, the hydrogen does not affect the fatigue properties of these alloys. However, after a mechanical polishing, the effect of hydrogen is more pronounced leading to a decrease in the fatigue life of NiTi alloys with a high increasing in temperature.

Keywords: NiTi alloys · Surface treatment · Electroploshing · Fatigue · Self-heating

1 Introduction

NiTi alloys are considered as promising materials in the realization of dental tools. However, their fatigue life may be limited [1] and sudden failures can be observed during clinical use [2]. Most of previous works has attributed the decrease of performance of these alloys to the absorption of hydrogen in the surrounding environment of use [3]. Indeed, the finished NiTi alloys used as dental devices are often processed and presented in different environments containing hydrogen. Thus, the ability of these materials to absorb this chemical entity is strongly related to its surface finish and may affect the fatigue properties of these alloys.

The aim of this work is to study the effect of different manufacturing process parameters such as mechanical polishing and electroplishing associated with the effect

N. Aifaoui et al. (Eds.): CMSM 2019, LNME, pp. 699–706, 2020.
https://doi.org/10.1007/978-3-030-27146-6_76

of hydrogen on the fatigue properties of NiTi alloys at high number of cycles (HCF). The choice of these parameters is related in a few steps to the process of manufacturing or using of dental tools. Indeed, some appropriate surface treatments are often applied to dental tools in order to improve their flexibility, their resistance to cyclic fatigue and to reduce the inherence surface defect. The electropolishing is one of the most used processes for finishing instruments in NiTi alloys. Although, the process of manufacturing NiTi alloy differ according to the supplier, it has been proven that in most manufacturing process, hydrogen can be introduced into NiTi alloys.

As a result, since the protective layer on the surface of NiTi alloys may interfere with hydrogen penetration by cathodic hydrogen charging, a mechanical polishing operation was considered in this study to remove this layer and to accelerate the process of introducing hydrogen into the material. In fact, in the oral environment, it has been demonstrated that the surface layer will degrade in the presence of corrosive environments or because of repetitive cyclic stress. Then, in order to study the influence of hydrogen on the fatigue properties of these alloys undergoing these different surface treatments, the self-heating method have been used. Although self-heating tests are rarely performed on NiTi alloys, they are successfully validated in very recent studies [4, 5].

2 Materials and Experimental Procedures

2.1 Material of the Study

The material of this study is a NiTi quasi-equiatomic alloy of nickel and titanium considered for the elaboration of endontic instruments. The specimen is "diabolo" shaped with a length of 6 cm and a diameter of 1.5 mm with a notch of 1 mm (Fig. 1). This geometrical form was chosen to be better adapted for fatigue tests. We have considered the material in three states: Standard (Virgin) and non-standard (Electropolished and polished). Some specimens have been Electropolished for 60 s. Other specimens undergo mechanical polishing with abrasive paper followed by immersion in the acetone for 20 min.

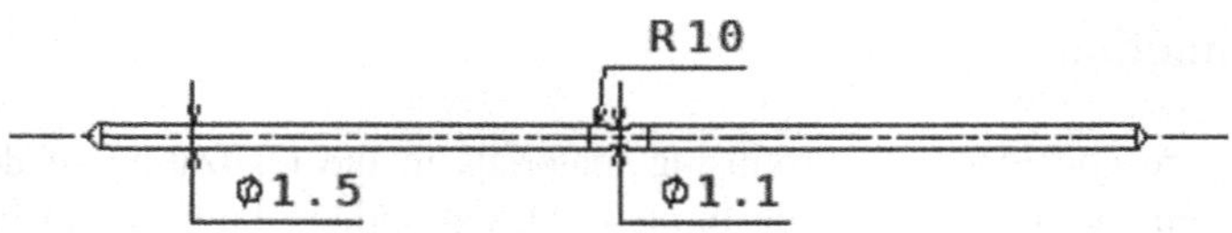

Fig. 1 Geometry of fatigue specimen

2.2 Experimental Procedures

A device of electrolytic hydrogen charging have been used at ambient temperature (20 °C). The device consists on a platinum anode, a NiTi specimen immersed into NaCL 0.9%. The hydrogen charging was made during 14 h under a current density of 20 A/m^2 for different states of NiTi specimens (virgin, mechanically polished and electropolished). Then, a mechanical characterization of the study material with different surface conditions was then carried out by tensile tests at ambient temperature. All tests are performed with a deformation rate of 10^{-4} s^{-1}. Hydrogen charged specimens were tested a few minutes after being removed from the solution. Self-heating tests have been carried out on a Bose-Electroforce test machine (Fig. 2a). This method consists on applying blocks of 3000 cycles on the same specimen at loading levels ranging from 100 to 800 considering increments of 50 MPa between 100 and 300 and 25 MPa for the other blocks.

This sequence is characterized by its load ratio R = 0 and a frequency of 30 Hz. This method consists on measuring the mean temperature of a specimen at successive loading levels. For each level, we have noted the evolution of the temperature of the specimen using an infrared camera. The principle of this method is shown in Fig. 2b.

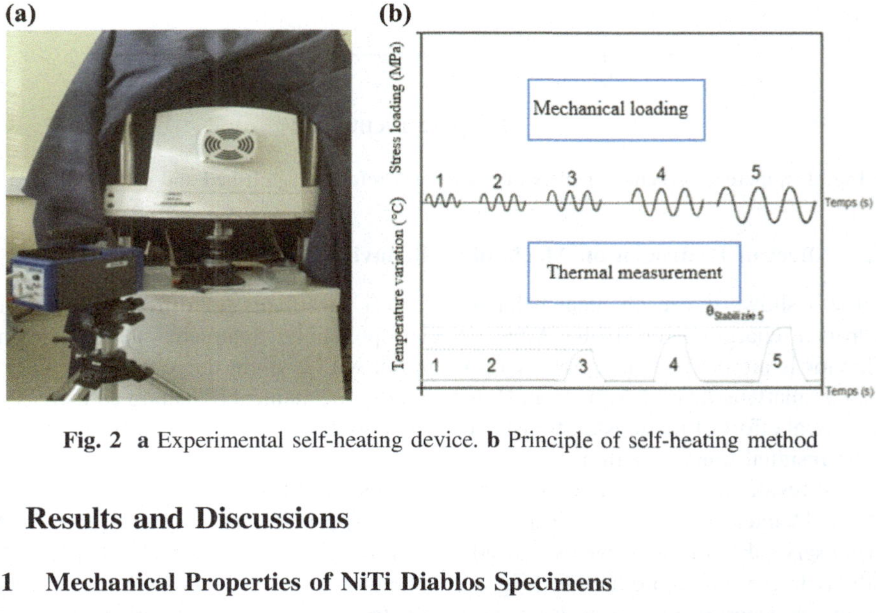

Fig. 2 **a** Experimental self-heating device. **b** Principle of self-heating method

3 Results and Discussions

3.1 Mechanical Properties of NiTi Diablos Specimens

Figure 3 shows the tensile properties (Load-Displacement) of a NiTi diabolo specimen in the reference state. The critical stress of martensitic transformation is approximately evaluated to be 600 MPa for the virgin specimen. From the results observed in the case of electropolished specimens, we can see that the critical stress of martensitic transformation is slightly lower with a high variation in displacement. This variation is more pronounced in the case of mechanical polished specimens, which justifies that mechanical polishing had a significant impact in the modifying of the mechanical response of NiTi alloys.

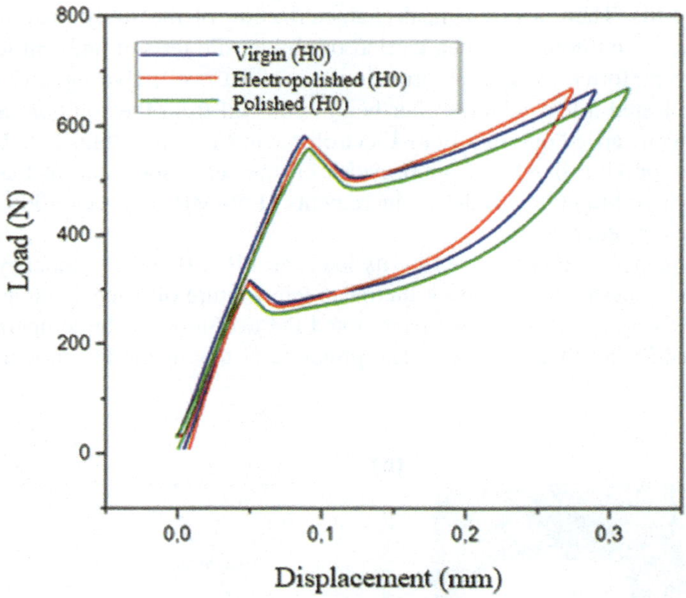

Fig. 3 Superelastic behavior of NiTi specimens: reference, polished and electropolished

3.2 Effect of Hydrogen on Mechanical Behavior of NiTi Specimens

Figure 4 shows the mechanical behavior of NiTi specimens for different states after hydrogen charging for 14 h. After an eletropolishing treatment, the specimen's behavior is almost the same for a virgin one but with a slight decrease in the critical stress of martensitic transformation. However, after mechanical polishing an increase in the critical effort of martensitic transformation is observed accompanied by an increase in the residual transformation.

The results of a cyclic loading test of the superelastic behavior at low number of cycles (14 cycles) for a NiTi virgin and polished specimen are showed in Fig. 5. We can observe that the first cycles carried out on the virgin specimen charged for 14 h with hydrogen are reproducible (Fig. 5a). Therefore, we can conclude that a cyclic loading is accompanied by a decrease in the stress applied as a function of the displacement during cycling. However, after hydrogen charging of polished specimen for 14 h, an appearance of residual stress is more pronounced during the discharge and an increase in the critical stress of martensitic transformation is observed (Fig. 5b).

This failure clearly shows that the impact of hydrogen is more pronounced for polished specimen. After polishing, the surface protective layer is removed which promotes the penetration of hydrogen during electrolytic charging.

Hydrogen introduced represents an obstacle to direct martensitic transformation A–M that increases the critical constraint of transformation and can lead to the fracture of the specimen after a small number of cycles following a cyclic load.

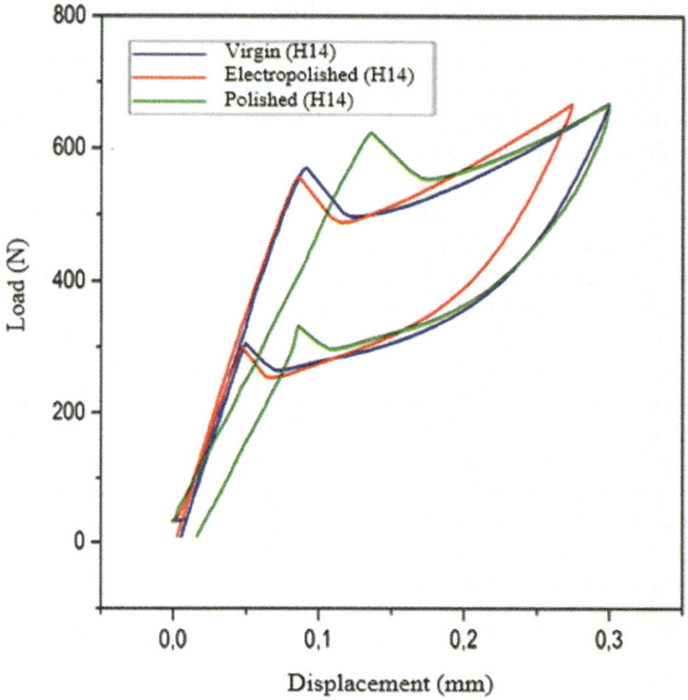

Fig. 4 Superelastic behavior of NiTi specimens charged with hydrogen for 14 h

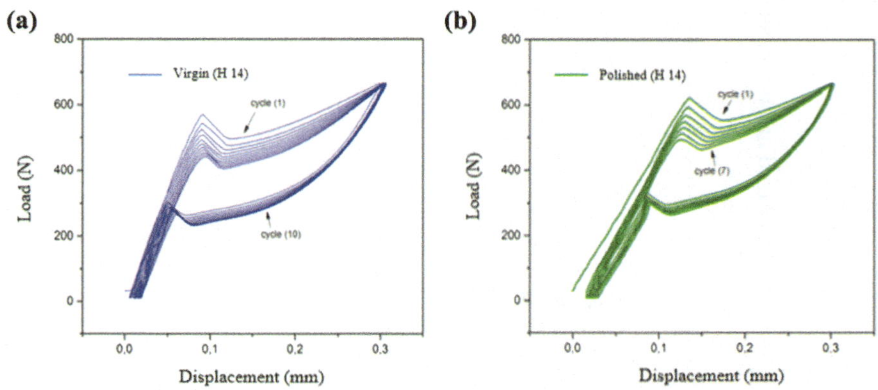

Fig. 5 Cyclic loading curve for NiTi specimen after hydrogen charging. **a** Virgin specimen. **b** Polished specimen

3.3 Determination of Fatigue Properties by Self-heating Method

The Fig. 6 represents the evolution of the stabilized mean temperature as a function of the stress amplitude obtained for different test conditions. The mean endurance limit (σ_{D}) can be determined through an empirical approach. The intersection between the asymptotic behavior of the last points of the self-heating curve and the abscise axis as shown in Fig. 6 is evaluated at about 750 MPa.

We can observe that after a certain level of loading applied, the stabilized average temperature increases significantly what is due to the appearance of dissipative mechanisms. The same procedure have been used to determine the average endurance limit of electropolished and polished specimens (Table 1). These results showed that virgin specimens have a surface that resist well to crack initiation. However, after the electropolishing treatment, we can see that the electropolished specimens presented a low temperature variation, which shifts the curve to the right. It is known that the electropolishing treatment, which has the role of improving the mechanical characteristics of the material, resulted in a slight increase in the value of the average endurance limit in our study and improved subsequently fatigue strength at a large number of cycles (HCF). For the mechanically polished specimen, since the protective layer that previously prevented crack propagation is eliminated, fractures are more likely to occur and therefore the endurance limit value of this specimen is evaluated at about 705 MPa.

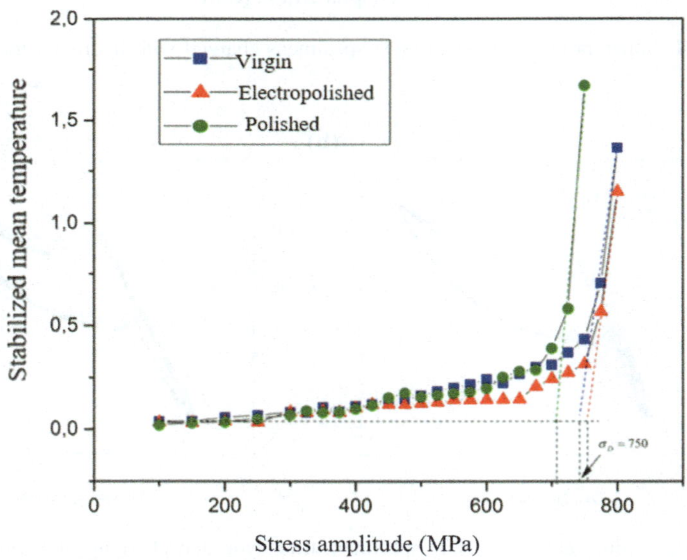

Fig. 6 Self-heating curve for virgin, electropolished and mechanically polished specimens

This method based on extracting temperature elevation from self-heating measurements is used to predict the effect of hydrogen on the different specimens (Fig. 7).

The endurance limit values presented in Table 1 show a significant decrease in the fatigue life of NiTi alloys after hydrogen charging.

From these results, we can conclude that an electropolishing treatment allows preserving the fatigue resistance of NiTi alloys at high number of cycles. Indeed, we can see that the material's endurance does not vary in the presence of hydrogen. As is recognized, an electropolishing treatment alters surface morphology and eliminate surface irregularities and defects that can serve as stress concentration points and favorable hydrogen sites. However, due to mechanical polishing the fatigue life of NiTi alloys is affected and it is significantly reduced (680 MPa) compared to the virgin specimen (750 MPa). Indeed, the rise of the temperature and the deformation during mechanical treatments can cause changes of the structures thus the introduction of the impurities as the inclusions of the abrasives, which can generate corrosion later, and premature fractures. These surface defects appears as favorable sites for the absorption of hydrogen and the initiation of cracks, which explain the reduction of fatigue life.

Table 1 Prediction of the endurance limit of different specimens without and with hydrogen charging

Specimens	Value (MPa)
Virgin (H0)	750
Electropolished (H0)	760
Polished (H0)	705
Virgin (H14)	750
Electropolished (H14)	760
Polished (H14)	680

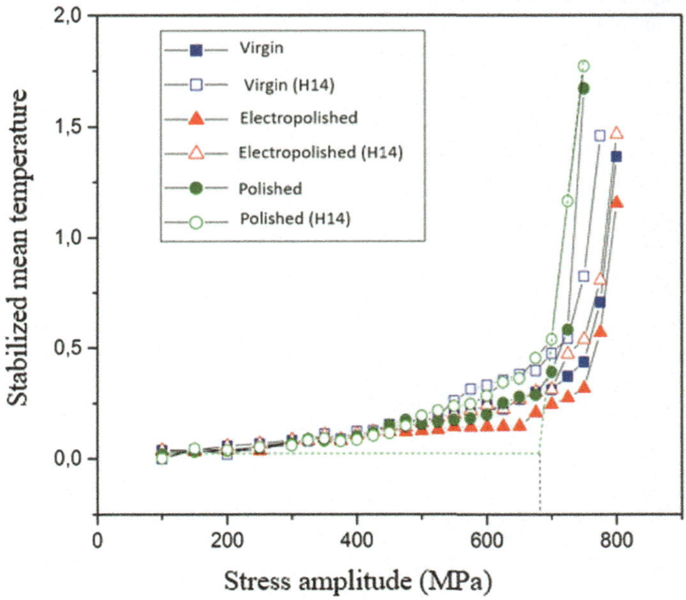

Fig. 7 Self-heating curve for virgin, electropolished and mechanically polished NiTi specimen after hydrogen charging

4 Conclusion

In conclusion, the self-heating method has been successfully applied to NiTi superelastic alloys in order to characterize their fatigue behavior and especially to predict the effect of the interaction between surface conditions and hydrogen in the fatigue resistance of these alloys at high number of cycles (HCF). It should be noted that surface that have been exposed to either electrochemical or mechanical treatment are fundamentally different and undergoes changes in their original properties and can thus affect the fatigue properties of NiTi alloys.

References

1. Mc Kelvey A, Ritchie R (2015). Fatigue-crack propagation in Nitinol: a shape-memory and superelastic endovascular stent material. J Biomed Mater Res 47(3):301–308. https://doi.org/10.1002/(sici)1097-4636(19991205)47
2. Kaneko K, Yokoyama K, Moriyama K, Asaoka K, Sakai J, Nagumo M (2003) Delayed fracture of beta titanium orthodontic wire in fluoride aqueous solutions. Biomaterials 24 (12):2113–2120. https://doi.org/10.1016/S0142-9612(02)00642-7
3. Yokoyama K, Hamada K, Moriyama K, Asaoka K (2001) Degradation and fracture of Ni–Ti superelastic wire in an oral cavity. Biomaterials 22:2257–2262. https://doi.org/10.1016/s0142-9612(00)00414-2
4. Legrand V, Arbab Chirani S, Calloch S (2013) Shape memory alloys fatigue and self-heating of NiTi. In: The 10th international conference on multiaxial fatigue & fracture, vol 9, pp 1–8. https://doi.org/10.1007/978-1-4614-6555-3_11
5. Rokbani M, Saint-Sulpice L, Chirani SA, Bouraoui T (2017) Hydrogen effects on Ni–Ti fatigue performance by self-heating method 26(10):105016–105027. https://doi.org/10.1088/1361-665x/aa86f4

Micro-Scale Investigations on Belt-Finishing Wear Mechanisms and Residual Stresses by Scratch Test: Numerical Study

Naoufal Bouktib(✉) and Abdeljabar Khellouki

Advanced Mechanics and Industrial Applications, Ecole Nationale Supérieure d'Arts et Métiers, Moulay Ismail University, Al Mansour, 15290 Meknès, BP, Morocco
Bouktib.naoufal@gmail.com,
Khellouki_abdeljabar@yahoo.fr

Abstract. The belt-finishing is a new superfinishing process which interests the industrialists in the automotive and aeronautics field. Belt-finishing is known by its ability to improve the surface topography and increase the wear resistance and the fatigue strength of the mechanical parts by inducing a strong residual compressive stress field. The aim of this work is to understand the belt-finishing process at the microscopic scale through a numerical study by ABAQUS software. This study investigates the abrasive wear mode, the influence of the scratching velocity and the variation of the friction coefficient on the residual stresses induced by a scratch test between a rigid abrasive grain and a hard steel work-piece AISI 52100. It has been shown that, for the parameters considered constant throughout the study (penetration depth h = 1 μm, attack angle of the indenter θ = 30° and the grain size D = 30 μm) the wear mode is always ploughing. Moreover, the residual compressive stresses become stronger for lower scratching velocity. Furthermore, the residual compressive stresses show significant values for lower friction coefficient. This situation favors a good mechanical resistance to fatigue of the mechanical parts.

Keywords: Belt-finishing · Scratch test · Abrasive wear mode · Residual stresses · Scratching velocity · Friction coefficient

1 Introduction

Abrasive machining processes are part of the broad field of manufacturing processes including belt-finishing, belt-grinding, lapping, polishing, etc. These methods are manufacturing techniques which use a very hard granular particles to improve the surface integrity of a mechanical part undergoes a hard-turning process. In the specific case, for a belt-finishing process, The abrasive belt, which consists of geometrically calibrated alumina grains, is applied to a rotating part with a pressure, an axial oscillation and a belt feed rate (Fig. 1a).

Belt-finishing is a new superfinishing process in the industry, which makes research in this domain at their beginning. However, efforts have been made in the recent years to understand the mechanisms of this process [1–5].

N. Aifaoui et al. (Eds.): CMSM 2019, LNME, pp. 707–715, 2020.
https://doi.org/10.1007/978-3-030-27146-6_77

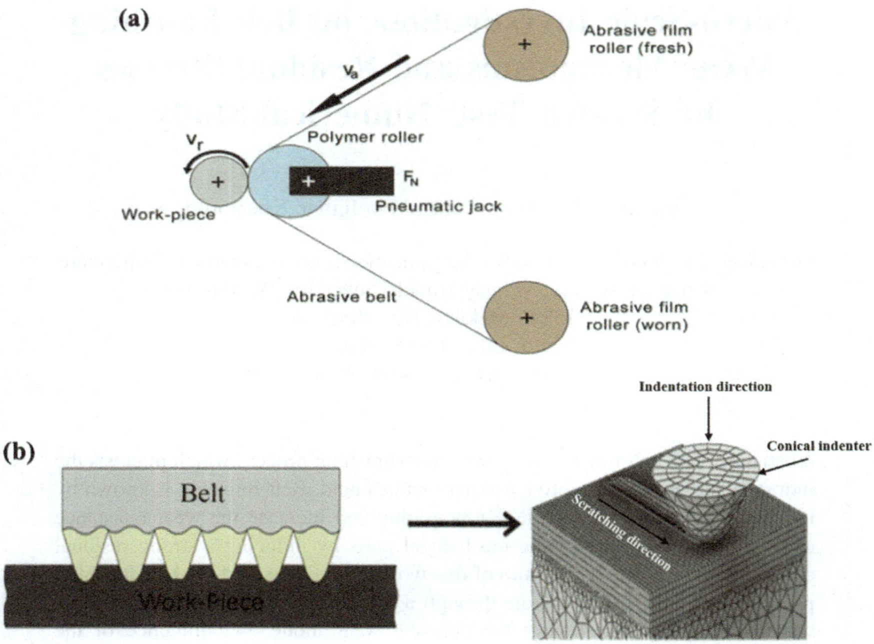

Fig. 1 **a** The belt-finishing process. **b** Scratch test with a rigid conical indenter

Experimental studies at the macroscopic scale have been carried out to measure the residual stresses during the belt-finishing process [4–6]. These studies have shown that this process induces residual compressive stresses up to −1000 MPa over a depth of 6 μm. This situation leads to a significant improvement in the surface integrity which favors a good mechanical resistance to fatigue of the work-piece.

Despite this research, the mechanism of action between the work-piece and the abrasive tool at the local scale is not completely understood, which is due to the complexity of the study. Indeed, the movements and the shape of the abrasive grains and their distribution on the belt are complicated and almost random. Therefore, the microscopic study must take into account some assumptions concerning the shape of the grains as well as their movements, in order to simplify the investigation. This study is usually done through numerical or analytical models that simulate the contact between a grain and a work-piece (Fig. 1b) which aims to examine all the phenomena that occur at the locale scale. Therefore, the wear mechanism and the residual stresses field induced by a scratch test are the phenomenon considered in this paper.

Among the phenomena that appear in the microscopic scale is the wear mechanism. The investigation of abrasive wear mode constitutes an interesting field of research. The latter can be found in different superfinishing process. According to [7], there are three wear modes during a scratch test: cutting, ploughing and wedge formation and the controlling factor is the attack angle and the shear strength.

Regarding the residual stresses, the numerical study of the distribution of residual stress field induced by superfinishing processes did not take all its importance in the literature. However, some researches have tried to numerically study this aspect, such as [4, 6, 8, 9], which has examined the influence of certain parameters such as the penetration depth of the grain, the applied load or the work-piece hardness.

The main objective of this paper is a contribution to the understanding of the mechanisms of the belt-finishing process through a finite element numerical study of the influence of some key-parameters on the wear mode and the distribution of the residual stresses during a scratch test between a rigid abrasive grain and a hard steel work-piece AISI 52100. This material was chosen because it is widely used in the manufacture of rotating elements of car engines (bearing, crankshafts, camshafts, etc.). The study focuses in particular on a characterization of the wear mode and the effect of some parameters that have been marginalized in the previous researches such as the scratching velocity which has a very important effect on the plastic deformation as well as the friction coefficient which is related to the lubrication mode [2].

2 Numerical Study

In order to evaluate the residual stresses and the wear mechanism during a belt-finishing process at the micro-scale, a finite element model of scratch test between a grain and a smooth surface of the workpiece has been proposed. To simplify the study, while being close to the reality of the belt-finishing process, a grain is modeled as conical indenter with spherical tip, while the movement of the grain is a straight line at a constant velocity. This study is focused on two parts. Starting with a characterization of the wear mode using a three-dimensional finite element numerical modeling Fig. 2b, then the effect of the scratching velocity and the friction coefficient on the residual stresses using a two-dimensional finite element numerical modeling Fig. 2a.

2.1 The Description and the Conditions of the Modeling

The finite element analysis is performed using ABAQUS/Standard and ABAQUS/Explicit software. The quasi-static movement of the indenter for all the simulations is carried out to achieve the loading according to two steps, the first one consists in a vertically penetration into the sample by a constant penetration depth h = 1 μm. And the second one allows the scratching of the surface following a horizontal movement with a constant scratching velocity. The horizontal and vertical displacements have been described based on the previous numerical and experimental research of [4], who assumed that 550, is the number of abrasive grains of size 30 μm, which are simultaneously in contact for an applied normal load of 340 N. The latter is close to the optimum found by [3]. Therefore, the local load acting on one grain is estimated to 0.6 N, which correspond to a penetration depth of 1 μm.

The indenter is supposed to be rigid and conical with a spherical tip of diameter D = 30 μm and attack angle $\alpha = 30°$. These dimensions are chosen to be close to the standards used during belt-finishing process [1–4], the work-piece is a deformable hard steel AISI 52,100 of size 1 mm × 0.3 mm for the two-dimensional model Fig. 2b and

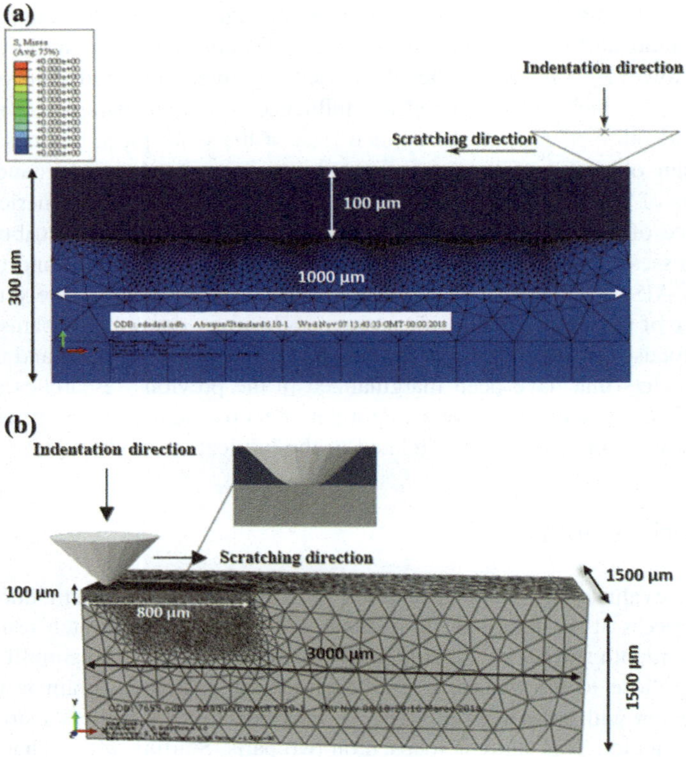

Fig. 2 **a** 2D modeling of the scratch test. **b** 3D modeling of the scratch test

3 mm × 1.5 mm × 1.5 mm for the three-dimensional model Fig. 2a, these sizes was chosen sufficiently large to obtain a steady state of the residual stresses and the deformation in order to ensure their stability without having a non-convergence problem. Table 1 shows the mechanical properties and cutting conditions of the AISI 52100 according to the Johnson-Cook plasticity model [10]. The boundary conditions applied to the work-piece for both models are a fixation of the lower surface and the side surfaces, while the upper surface remains free.

The mesh in the two-dimensional model has 50,000 quadrilateral elements (CPE4RT) of size 0.5 µm in the contact region. While, in the three-dimensional model consists of a (C3D8R) element type of size 1 µm in the contact zone.

The contact between the indenter and the upper surface of the work-piece in the two-dimensional model is made using surface-to-surface method. While in the three-dimensional model a General-Contact was chosen as formulation method with an arbitrary Lagrangian Eulerian (ALE) adaptive method to ensure the convergence of the simulation.

Table 1 The mechanical properties and the cutting conditions of the AISI 52100

Parameters	Values
Density ($kg \cdot m^{-3}$)	7870
Young's modulus (GPa)	210
Poisson coefficient	0.3
Yield stress, A (GPa)	2.49
Hardening modulus, B (GPa)	1.5
Hardening coefficient, n	0.19
The strain rate sensitivity coefficient, C	0.027
The reference plastic strain rate, $\dot{\varepsilon}_0\ (s^{-1})$	1

3 Results and Discussion

3.1 Characterization of the Wear Mode

To characterize the wear mode during the scratch test, a numerical and analytical examination of the contact between the grain and the work piece are considered.

The numerical simulation is done using a fixed scratching velocity of 0.5 m/s and by assuming that the contact is frictionless. The results are plotted in Fig. 3b.

While Fig. 3a shows the wear mode diagram which characterized the abrasive wear mode (ploughing, wedge forming and cutting) as a function of degree of penetration, D_P, and the shear strength at the contact interface, f.

The corresponding degree of penetration, D_p, and shear strength at the contact interface, f, are calculated based on [7, 11] models with Eqs. (1) and (2).

$$D_p = h\sqrt{\frac{\pi H_v}{2W}} \tag{1}$$

$$f = \frac{\left(\left(\frac{6}{\pi}\right)\tan\alpha\right)}{\sqrt{1+\left(\left(\frac{6}{\pi}\right)\tan\alpha\right)^2}} \tag{2}$$

where h is the penetration depth of the indenter, α is the attack angle, W is the load and H_v the Vickers hardness of the workpiece.

As it can be seen from Fig. 3b, the numerical results show that most of the removed materials are pushed to the side, meaning that the dominant wear mechanism is plastic deformation called ploughing, which is in good agreement with the analytical examination Fig. 3a. Moreover, the scratch section is high compared to the wedge section which can be explained by the work-hardening phenomenon (Fig. 4).

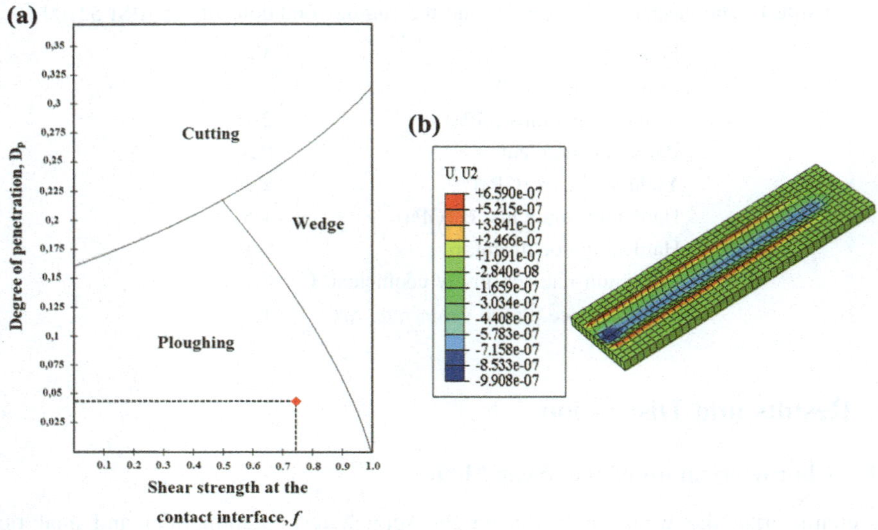

Fig. 3 **a** Analytical examination of wear mode during a scratch test. **b** Numerical study of wear mode during a scratch test (α = 30°, h = 1 μm, D = 30 μm, Frictionless)

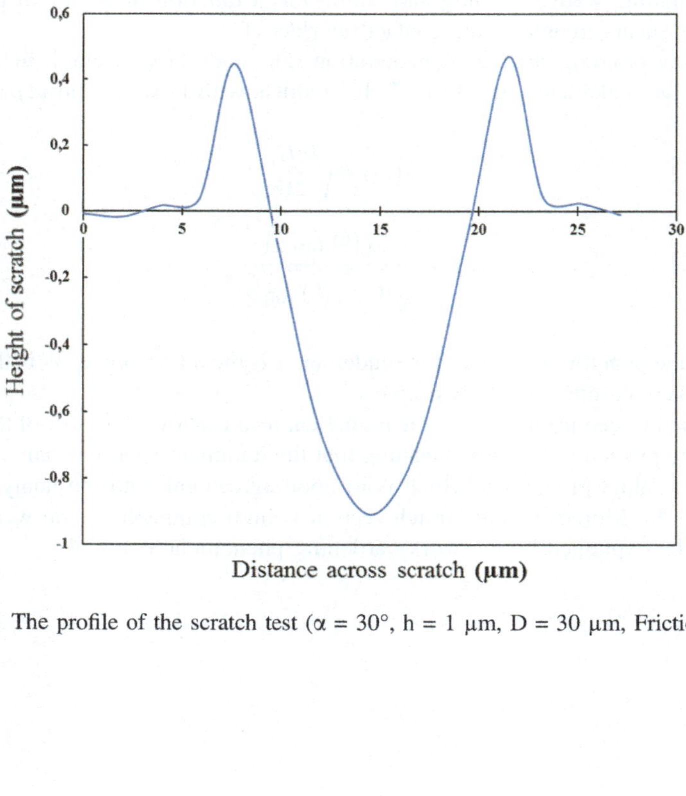

Fig. 4 The profile of the scratch test (α = 30°, h = 1 μm, D = 30 μm, Frictionless)

3.2 The Effect of Scratching Velocity on the Residual Stresses

To understand the relationship between scratching velocity and the residual stress field, four simulations with different scratching velocity values ranging from 0.1 to 2.5 m/s were taken into account.

The contact between the abrasive grain and the workpiece is frictionless which corresponds to a perfect lubrication during a superfinishing process. The results are plotted in Fig. 5.

Whatever the scratching velocity, the residual stress field is modified over a depth of 20 μm, while a maximum compressive stress around 2000 MPa is located between a depth of 3 and 4 μm whatever the scratching velocity. Moreover, it can be seen that the maximum scratching velocity of 2.5 m/s leads to a tensile residual stress in the surface, which is due to the high local temperature in the contact zone.

Furthermore, the lower is the scratching velocity; the higher is the compressive residual stress in the outer-layer. A better result is obtained with a scratching velocity of 0.1 m/s. We can say that in order to have a compressive residual stress during a belt-finishing process which favors the mechanical resistance to fatigue, the feed rate must be chosen as low as possible.

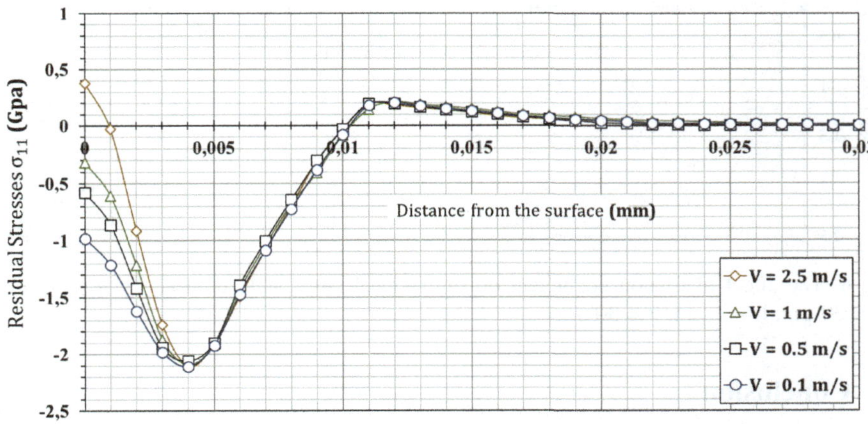

Fig. 5 The effect of scratching velocity on the residual stress field induced by a scratch test (θ = 30°, h = 1 μm, D = 30 μm)

3.3 The Effect of Friction Coefficient on the Residual Stresses

In order to study the effect of friction between the grain and the workpiece on the residual stresses, four simulations were performed by choosing different values of friction coefficient. The friction coefficient μ is chosen between 0 (which corresponds to an abundant lubrication preventing any friction) and 0.3 (which corresponds to a rough contact without lubrication). The scratching velocity used during the different simulations is fixed at 0.5 m/s. The results are plotted in Fig. 6.

Here again, the residual stress field is modified over a depth of 20 μm, while a maximum compressive stress is reached at a depth that varies from 4 to 5 μm, depending on the value of the friction coefficient.

The main result during this simulation is that, the higher is the friction coefficient (bad lubrication); the lower is the compressive stress in the sub-layer.

The tensile residual stress that appear for a maximum friction coefficient of 0.3, are here related to the high plastic deformation of the material leading to a strong release of heat in the contact zone. This state is the consequence of a lack or insufficiency of lubrication.

A minimization of friction phenomena between the abrasive belt and the part during a belt-finishing process, by choosing an appropriate lubrication, is therefore of great importance to have residual compressive stresses which leads to an improvement in the mechanical resistance to fatigue of the mechanical parts.

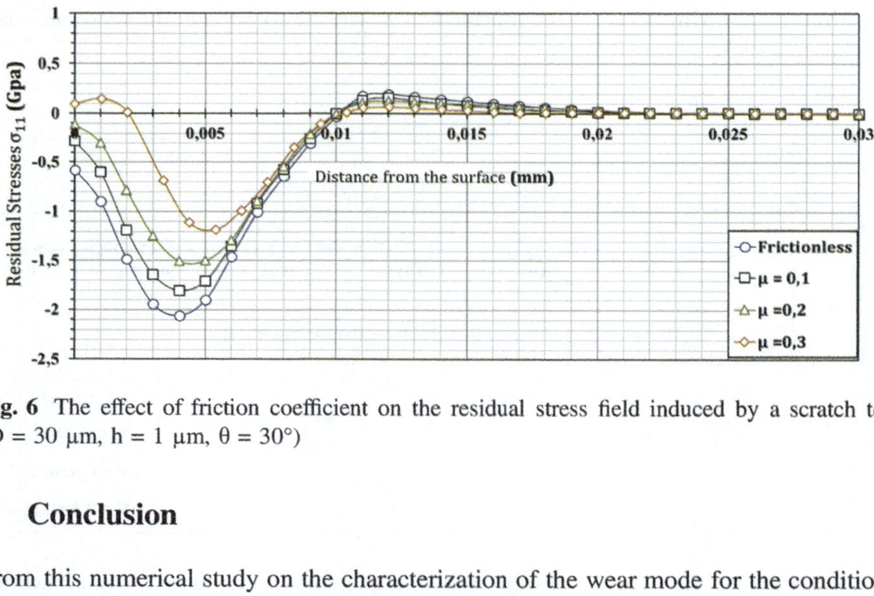

Fig. 6 The effect of friction coefficient on the residual stress field induced by a scratch test (D = 30 μm, h = 1 μm, θ = 30°)

4 Conclusion

From this numerical study on the characterization of the wear mode for the conditions considered constant throughout the paper and the effect of scratch velocity and coefficient of friction on the field of the residual stresses induced during a scratching of 100Cr6 hard steel, we can conclude the following points:

- For a penetration depth of 1 μm and an attack angle of 30°, the wear mode is ploughing. Moreover, the scratch section is always higher than the wedge section due to the work-hardening phenomenon.
- The application of the belt-finishing process with a low feed rate of the abrasive film leads to the formation of compressive residual stresses, which improves the fatigue strength of the mechanical parts.

- Good lubrication in belt-finishing process is a priority condition which must be taken into account in order to reduce the local friction coefficient and to promote compression stresses in the outer layer.

References

1. Khellouki A, Rech J, Zahouani H (2007) The effect of abrasive grain's wear and contact conditions on surface texture in belt finishing. Wear 263(1–6):81–87
2. Khellouki A, Rech J, Zahouani H (2010) The effect of lubrication conditions on belt finishing. Int J Mach Tools Manuf 50(10):917–921
3. Khellouki GA, Rech J, Zahouani H (2007) Influence of the belt-finishing process on the surface texture obtained by hard turning. Proc Inst Mech Eng Part B: J Eng Manuf 221 (7):1129–1137
4. Rech J, Kermouche G, Grzesik W, Garcıa-Rosales CC, Khellouki A, Garcıa-Navas V (2008) Characterization and modeling of the residual stresses induced by belt-finishing on a AISI52100 hardened steel. J Mater Process Technol 208(1–3):187–195
5. Wenxi WANG, Ferdinando SALVATORE, Joël RECH (2018) Effects of belt's adhesive wear on surface integrity in dry belt finishing. Procedia CIRP 71:31–34
6. Kermouche G (2013) Étude de traitements mécaniques des surfaces reposant sur des sollicitations de rayure (superfinition, polissage, galetage). Matériaux & Techniques 101:308
7. Hokkirigawa K, Kato K (1988) An experimental and theoretical investigation of ploughing, cutting and wedge formation during abrasive wear. Tribol Int 21(1):51–57. https://doi.org/10.1016/0301-679x(88)90128-4
8. Kermouche G, Aleksy N, Loubet JL, Bergheau JM (2009) Finite element modeling of the scratch response of a coated time-dependent solid. Wear 267:1945–1953
9. Kermouche G, Rech J, Hamdi H, Bergheau J (2010) On the residual stress field induced by a scratching round abrasive grain. Wear 269:86–92
10. Arrazola PJ, Ozel T (2008) Numerical modelling of 3D hard turning using arbitrary Lagrangian Eulerian finite element method. Int J Mach Mach Mater 4(1):14–25
11. Bowden FP, Tabor D (1966) Friction, lubrication and wear: a survey of work during the last decade. Br J Appl Phys 17(12):1521–1544. https://doi.org/10.1088/0508-3443/17/12/301

Micromechanical Modeling and Simulation of the Elastoplastic Behavior of Composite Materials

Zoubida Sekkate(✉), Ahmed Aboutajeddine, Mounia Bennoura, and Abbass Seddouki

Mechanical Engineering Laboratory, Faculty of Science and Technology, University of Sidi Mohamed BenAbdellah, Fes, Morocco
sekkat.zoubida@gmail.com

Abstract. Composite materials are the preeminent drivers for the significant enhancement to products that meet recent industrial requests. Nevertheless, their variability and the complexity of the innovative products introduce a great challenge in predicting the mechanical response fundamental for the composite design engineering. Then, to pledge a faster and more cost-effective development product, the experimental practice is switched to virtual simulation tools using advanced multi-scale analyses techniques. Indeed; developing predictive micromechanical models that enable to derive the composites' macroscopic behavior in realistic environment based on the microscopic behavior of each constituent is obviously defiance. Therefore, to evaluate the micromechanical response of complex composite materials which are mainly nonlinear, a new micromechanical model is proposed to meticulously typify the elastoplastic behavior of composite materials and determine the optimized design. The model is developed following the composite design perspectives using a Hill-type incremental formulation and the classical J2 plasticity theory to derive the tangent operators in each phase. To assess the predictive accuracy of the estimation, the analogy to the experimental data and the exact finite element solution is proved.

Keywords: Composite · Micromechanical models · Multiphase · Elastoplasticity

1 Introduction

Composite materials have inherently the ability, by mixing different constituents, to cope with the demand of fast-evolving exigencies of innovative materials in different industries. Nevertheless, the anisotropic nature of composite materials has made the use of this type of material limited due to the enormous amount of tests required to find the right mix of materials and processes that meet product requirements and production cycles. Recently, the experimental practice of trial-error based on the selection approach is evolving to a more rational, organized and systematic one through a composite material design process from the atomic scale to the system scale. An important building block of this process is the efficiency of micromechanical models

N. Aifaoui et al. (Eds.): CMSM 2019, LNME, pp. 716–724, 2020.
https://doi.org/10.1007/978-3-030-27146-6_78

that enable to predict the interaction between the microstructure and the effective properties of the product and for which the sample testing is performed only in the final stage for verification. Until recently, the linear homogenization realm has a relatively well-established and stabilized modeling framework. According to the assessment of their predictive capabilities accomplished by [1], the double inclusion model initially proposed by [2, 3] and improved by [4] which overcomes the drawbacks of many existing models in the literature and allows deriving the overall moduli for multiphase composites in a closed-form, can be seen as a more general model compared to other on the literature.

Nevertheless, to avoid overdesign, build better products and reduce costs, non linear analyses are essential [5].

In fact, when we handle nonlinear homogenization, local properties of the phases become position and time dependent. Then, an extension of the homogenization scheme to address elasto-plastic composites is tackled based on the linearization of the constitutive laws and the treatment of the intra-phase's heterogeneity under the concept of the fictitious heterogeneous linear elastic or thermo-elastic comparison composite (LCC) [6]. The linearization is then defined around a reference strain or stress state; to assume that the mechanical characteristic's (The stiffness or the compliance tensors) are uniform per phase and evolve only with the overall load; to finally exploit the mean field homogenization valid in linear case. The starting point of the whole range of linearization schemes, were those of Sachs [7] and Taylor [8]. Thereafter, on the basis of the self consistent formulation developed by Kröner [9], Hill [10] has developed an incremental formulation for nonlinear composite materials. The scheme deals with the linearization of the local constitutive relations of each phase through the use of an elasto- plastic tangent operator and compute the macroscopic response within an incremental step by step homogenization procedure.

Despite the relevance of the incremental formulation to capture in a good manner the visco- elastoplastic behavior of heterogeneous materials, in its original form, the properties are supposed uniform leads to a too stiff macroscopic response [11–13]. With regard to that limitation, many routes have been attempted in order to improve its predictive capabilities in elasto-plastic framework [14, 15]. Jiang and Shao [13] have proved that the use of the isotropic part of local tangent operator to compute only the eshelby tensor give the better predictions compared to the method using isotropization in the computation of Hill tensor or in the different steps to evaluate the macroscopic tangent operator. The method of using isotropic part of tangent moduli was also used by Pettermann et al. [16] to propose an incremental MT formulation under thermal strains condition enabling to take into account the Bauschinger phenomenon. In another study, Teng [17] has established the DI of Nemat Nasser in an incremental form, which accurately captures the Bauschinger phenomenon and provides rigorous predictions at only moderate concentration. The uniqueness of the method is reflected in the examination of the matrix phase as two-phase material consisting of an elastic and plastic region.

Later, Hutchinson [11] has developed a secant procedure to characterize the stady-state-creep behavior for which an incremental form is not needed. The latter method is a total deformation theory, which is similar to a non linear elastic relation and restricted to monotonic and proportional loading. Molinari et al. [18] have proved also that incremental form rather than secant is the practical procedure to obtain a softer material response. The main reason for the weakness of the secant formulation is that to consider a uniform secant modulus per-phase, the latter is generally defined around the average strain or stress in the considered phase.

Consequently, several routes have been investigated to accurately consider the local field fluctuations in each phase. The first one is due to [19] which derive, for incompressible materials, the effective moduli of particle-reinforced composites based on the shear energy in the matrix. The second general procedure was proposed by Suquet [12], which defined the LCC based on the second-order moment.

Anterior to the modifications developed to improve the predictions of the secant formulation, Molinari et al. [20] has developed a tangent procedure as a mix between the incremental and secant formulation in which the constitutive laws led to a linear thermo-elastic problem. Afterward, a new affine scheme Masson et al. [21] has been developed on the basis of the later procedure.

In the present work, an attempt of developing a mean field homogenization scheme for rate- independent elasto-plastic composites, which is suited for general loading and provided accurate results at the different levels and for different volume fractions of heterogeneities. To satisfy the cited criteria, an extension of the double inclusion of [4] is formulated using the incremental linearization procedure. We invoke also in the present work the tangent moduli computation and the issue of isotropization as defined by [14]. The Elasto-plasticity theory in each phase is illustrated through the J2 flow theory. Finally, we derive the effective tangent stiffness tensor and the macro stress increment.

2 Double Inclusion Model

The double inclusion model developed initially by [2] to assess the different type of interactions in two-phase composites. The formulation used to derive the macroscopic moduli of the composite is based on certain assumptions on the mathematical solution of the local fields relying on the Lippman–Schwinger type equation in terms of strain fields:

$$\varepsilon(r) = \varepsilon^0 - \int_V \Gamma(r - r') : \delta C(r') : \varepsilon(r') dr' \tag{1}$$

where ε^0 is the uniform strain field induced by far–field loads in the reference medium and $\Gamma(r - r')$ is the modified green tensor.

Thus, the estimation does not hold accurately the matrix-inclusion interaction. The same other have generalized the model to multi-inclusions models [3] but the estimation of the average field quantities of the extended model was developed in a nonclosed-form solution. Consequently, to surmount these limitations and achieve predictions that are physically acceptable, Aboutajeddine and Neal [4] have reformulated the model to obtain a closed-form solution of the effective elastic moduli for multiphase composite; through the mixing of the non-interacting and the self-consistent approximations.

Based on the Lippmann-Schwinger integral and the Eshelby's inclusion concept, the authors have subdivided the strain concentration tensor into two-levels characterizing the different type of interactions such as:

$$A^{I} = A^{I,DI} : A^{DI} \tag{2}$$

Here the tensor $\boldsymbol{A}^{I,DI}$ is defined at the level of the ***DI*** relating the average strain of the inclusion to that of the double inclusion and $\boldsymbol{A}^{DI}$ relates the average strain of the double inclusion to the over-all strain average.

As mentioned by Aboutajeddine and Neal, the expression of the tensor $\boldsymbol{A}^{DI}$ is defined as:

$$A^{DI} = \left[I + P^{e}_{DI} : (C^{M,e} + f^{I,DI} C^{I,M} : A^{I,DI}\right]^{-1} \tag{3}$$

The strain concentration tensor can be obtained through only the determination of the tensor $\boldsymbol{A}^{I,DI}$ defined by:

$$A^{I,DI} = \left[I + f^{M} \delta P^{I}_{M} : C^{I,M}\right]^{-1} \tag{4}$$

where:

$$\delta P^{I}_{M} = P^{DI}_{M} - (1/f^{M}) \Delta P^{I}_{M} \tag{5}$$

The predictions of the effective elastoplastic response using the DI configuration of Aboutajeddine and Neal over each load increment the linearization is performed through the use of the matrix tangent moduli.

3 Incremental Formulation of Homogenization Model

The pioneer work to tackle the problem of inelastic composites was provided by Hill. The incremental formulation makes it possible to produce a stepwise examination which, at each step, is presented as the treatment of a problem of linear homogenization. For this purpose stress and strain are related in incremental form using the tangent operator:

$$\dot{\sigma}(x,t) = C^{tg}_{r}(\varepsilon(x,t),t) : \dot{\varepsilon}(x,t) \tag{6}$$

where $\mathbf{C}^{\mathbf{tg}}_{\mathbf{r}}$ are the local tangent moduli and $r = 0, 1$ for a two-phase composite. The tangent moduli are not uniform, because at each time it depends on the local stress or strain states. Indeed, in a given state the real nonlinear medium is assimilated to a fictitious reference matrix having uniform tangent moduli $\hat{\mathbf{C}}^{\mathbf{tg}}_{\mathbf{r}}(\mathbf{t})$. This concept has been introduced by [6, 22] defining a linear comparison material (LCC) which consists in replacing the original RVE with an RVE having the mechanical properties varying only with the time:

$$\dot{\sigma}(x,t) = \hat{C}_r^{tg}(t) : \dot{\varepsilon}(x,t) \tag{7}$$

4 Elasto-Plastic Tangent Operator

For Elasto-plastic material, the anisotropic tangent moduli have an important impact in numerical simulations. Two tangent operators can be defined: First from the rate constitutive equation (continuum or elasto-plastic), and second from a discretization in time interval (consistent or algorithmic) [14],

$$\begin{cases} \dot{\sigma} = C^{ep} : \dot{\varepsilon} \\ \delta\sigma_{n+1} = C^{alg} : \delta\varepsilon_{n+1} \end{cases} \tag{8}$$

For simplicity, the algorithmic and consistent operators are derived using the J2 flow theory. Following Doghri [23], the continuum and algorithmic tangent operator are given by:

$$\begin{cases} C^{alg} = C^{ep} - (2\mu)^2(\Delta p)\frac{\sigma_{eq}}{\sigma_{eq}^{trial}}\frac{\partial N}{\partial \sigma}, \quad \frac{\partial N}{\partial \sigma} = \frac{1}{\sigma_{eq}}\left(\frac{3}{2}I^{dev} - N \otimes N\right) \\ C^{ep} = C^{el} - \frac{(2\mu)^2}{h}N\,N\;,\; h = 3\mu + \frac{dR}{dp} > 0 \end{cases} \tag{9}$$

Here, σ_{eq}^{tr} denotes the trail elastic predictor of σ_{eq}. Δp and σ_{eq}^{tr} are the unknown internal variables needed to determine the overall composite behavior. To illustrate the elastoplastic behavior for small deformations, the J2 theory written in a radial return mapping algorithm (RMP) is used. However, the extension to other yield criteria can also be derived, for instance, for a cyclic test; the chaboche's model can be used to characterize the Bauschinger effect. Otherwise, even though the constitutive model is isotropic the tangent operators are anisotropic and the computation of Eshelby's tensor can be performed only numerically as proposed by [24]. To defeat such limitation the matrix phase is assumed isotropic or transversely isotropic. In other words, the use of the isotropic part of the algorithmic tangent operator allows obtaining the analytical expressions of Eshelby's tensor for various shapes of inclusions and gives results that are much softer than those computed using the algorithmic tangent operator. The way to extract the isotropic part of anisotropic operator is not unique [25]. In this work the isotropic tangent operator is defined in a form-identical to the Hook's operator in isotropic linear elasticity:

$$C^{iso} = 3K_t I^{vol} + 2\mu_t I^{dev} \tag{10}$$

where K_t and μ_t are the isotropic tangent bulk and shear moduli which will be extracted using the General method [15]. Then, the Eshelby's tensor can be expressed in a closed-form.

5 Incremental Algorithm Steps

This procedure is derived for the case of double inclusion model consisting of an inclusion I surrounded by a coating of a matrix material and embedded in the overall composite. At the beginning of the scheme $t_{n,}$ the macro strain ε_n and the internal variables are known. If the algorithm is strain driven; for a given strain increment, we compute the stress increment. Accordingly, the derivation of a strain increment in the inclusions can be made from a voigt assumption. The stress increment is then derived from an averaging technique. These steps enable the computation of the instantaneous algorithmic moduli of both inclusions and matrix to characterize the Elasto-Plastic response.

6 Numerical Results and Discussion

In order to elucidate the robustness of the new extension, the previously steps are applied for a Matrix Metal material studied before by [12, 14]. The matrix is an elastoplastic aluminum alloy with a power law isotropic hardening $R(p) = k * p^m$ with k = 416 MPa, m = 0.3895 and the following properties $E_0 = 75$ GPa; $\vartheta_0 = 0.3$; $\sigma_y = 75$ MPa. While a Ceramic inclusions are considered elastic ($E_1 = 400$ GPa; $\vartheta_0 = 0.2$) (Fig. 1).

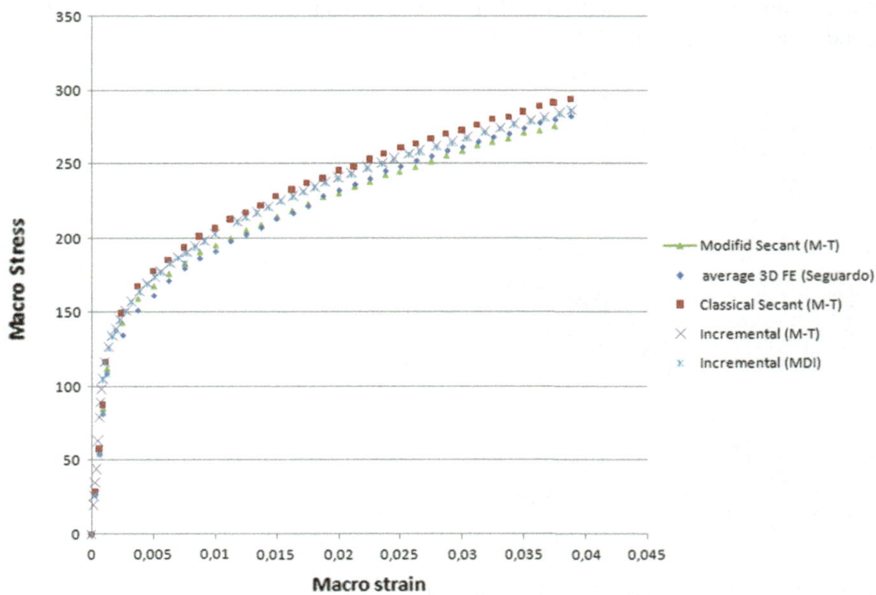

Fig. 1 MMC (v = 30%) under macro-tension. Comparison of different predictions: (1) Modified Secant MT, (2) Average 3D FEs (Segurado), (3) Classical secant MT, (4) Incremental MT and the incremental modified double inclusion (this work)

As Llorca and Segurado [26] have proved that the FE estimations on 2D axisymmetric unit cells for composites with periodic microstructure are in reasonable agreement in an average sense with predictions on 3D FE unit cells for composite with random dispersion of inclusions. This deduction substantiates the comparison done previously.

The new extension compared to the results obtained in [27], gives excellent predictions of the elastic and plastic response even if the reference strain state is computed only using the average-strain. Additionally, as pointed out by [28] the incremental and the second-order secant estimations are adjacent to the target numerical results; which approve the use of the incremental approach to define the LCC. The discrepancies between our formulation and the incremental M-T extension may be due to the use of the MT scheme which is still remaining short even for elastic behavior.

In this section, the reinforcement is a short glass fiber embedded in a polyamide matrix. The matrix is elastoplastic described by an exponential-linear hardening law: R (p) = $k_1 * p + k_2 * (1 - \exp(-m * p))$ with $k_1 = 139$ MPa, $k_2 = 32.7$ MPa, m = 319.4 and $\sigma_y = 29$ MPa. The elastic properties are defined by: $E_0 = 2.1$ GPa, $\vartheta_0 = 0.3$. The fibers are assumed elastic ($E_1 = 72$ GPa, $\vartheta_1 = 0.22$).

The Fig. 2 shows numerical predictions of the new estimate under a macro tension test. The incremental formulation of the double inclusion scheme [4] provide excellent estimations in investigating the elastoplastic behavior of a two phase composite materials and is in good agreement with the second order incremental M-T scheme and the FE results. Therefore, for a future study a linearization of our approach around a second moment reference state may lead to better results than that provided using the second order incremental M-T formulation.

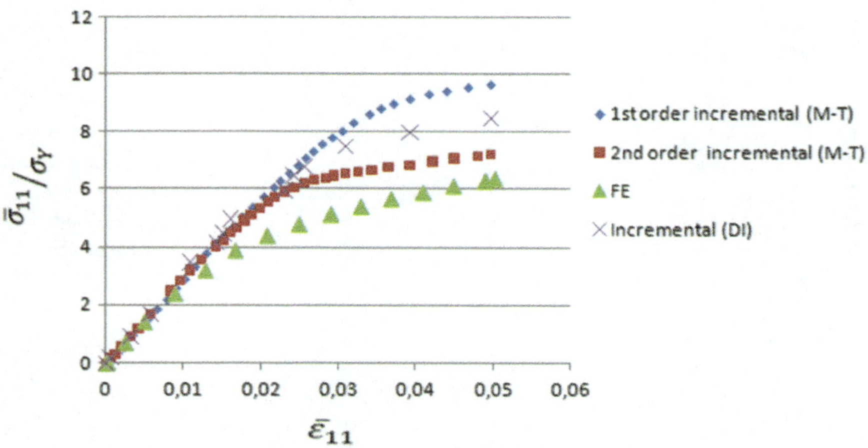

Fig. 2 (vl = 15.7%) under uniaxial tension. The aspect ratio $\alpha = 15$: (1) The first order incremental MT, (2) The second order incremental MT, (3) The FE results and (4) The Incremental modified Double inclusion (this work)

7 Conclusion

In this paper, the modified double inclusion was extended to determine the effective Elasto-plastic response of two-phase composite material for various volume fractions of inclusions.

The extension of the scheme is based on Hill incremental formulation, dealing with rate-independent small-strain composite materials, with the matrix characterized by the Von Mises yield criterion (J2 Elastoplasticity) reinforced with the isotropic and elastic inclusions.

With reference to the stiffness comparison established by [14] of various tangent moduli, we use in our computation the algorithmic (Consistent) tangent moduli. Two hypotheses define the basis of the model: (1) the reference moduli in each phase are computed with the average strain and the increment strain average quantities. (2) The Eshelby's tensor is computed with the isotropic part of the algorithmic moduli. The implicit Euler method is used because of its good accuracy to integrate the incremental stress-strain relation along the loading path. The efficiency of the estimation was tested by considering a two phase's composite MMC running under macro tensile test. The comparison of the proposed incremental double-inclusion predictions with the first and second incremental Mori Tanaka scheme and finite element results exhibits that the extended model provides accurately the nonlinear behavior of Elasto-plastic deformation.

References

1. Bennoura M, Aboutajeddine A (2016) Predictive capabilities of micromechanical models for composite materials. J Reinf Plast Compos 35(14):1115–1125. https://doi.org/10.1177/0731684416640365
2. Hori M, Nemat-Nasser S (1993) Double-inclusion model and overall moduli of multi-phase composites. Mech Mater 14(3):189–206
3. Hori M, Nemat-Nasser S (1999) On two micromechanics theories for determining micro–macro relations in heterogeneous solids. Mech Mater 31(10):667–682. https://doi.org/10.1016/S0167-6636(99)00020-4
4. Aboutajeddine A, Neale KW (2005) The double-inclusion model: a new formulation and new estimates. Mech Mater 37(2):331–341. https://doi.org/10.1016/j.mechmat.2003.08.016
5. Solidworks (2010) Understanding nonlinear analysis
6. Ponte Castañeda P (1991) The effective mechanical properties of nonlinear isotropic composites. J Mech Phys Solids 39:45–71. https://doi.org/10.1016/0022-5096(91)90030-R
7. Sachs G (1928) Plasticity problems in metals. Z Ver Dent Ing
8. Taylor GI (1938) Plastic strain in metals. J Inst Metals 62
9. Kroner E (1961) Zurplastischenverformung des vielkristalls. Acta Metall 9:155, 307–324
10. Hill R (1965) Continuum micro-mechanics of elastoplastic polycrystals. J Mech Phys Solids 13:89, 72:734
11. Hutchinson JW (1976) Bounds and self-consistent estimates for creep of polycrystalline metals. Proc Roy Soc Lond A348:101–127. https://doi.org/10.1098/rspa.1976.0027
12. Suquet P (1996) Overall properties of nonlinear composites: remarks on secant and incremental formulations. https://doi.org/10.1007/978-94-009-1756-9_19

13. Jiang T, Shao JF (2009) On the incremental approach for nonlinear homogenization of composite and influence of isotropization. Comput Mater Sci 46(2):447–451. https://doi.org/10.1016/j.commatsci.2009.03.032
14. Doghri I, Ouaar A (2003) Homogenization of two-phase elasto-plastic composite materials and structures: study of tangent operators, cyclic plasticity and numerical algorithms. Int J Solids Struct 40(7):1681–1712. https://doi.org/10.1016/S0020-7683(03)00013-1
15. Doghri I, Friebel C (2005) Effective elasto-plastic properties of inclusion-reinforced composites: study of shape, orientation and cyclic response. Mech Mater 37(1):45–68. https://doi.org/10.1016/j.mechmat.2003.12.007
16. Pettermann HE, Planskensteiner AF, Bohm HJ, Rammerstorfer FG (1999) A thermo-elasto-plastic constitutive law for inhomogeneous materials based on an incremental Mori–Tanaka approach. Comput Struct 71:197–214. https://doi.org/10.1016/S0045-7949(98)00208-9
17. Teng H (2014) A new incremental formulation of elastic–plastic deformation of two-phase particulate composite materials. J Appl Mech 81(6):061006. https://doi.org/10.1115/1.4026557
18. Molinari A, Ahzi S, Kouddane R (1997) On the self-consistent modeling of elastic-plastic behavior of polycrystals. Mech Mater 26:43–62
19. Qiu YP, Weng GJ (1992) A theory of plasticity for porous materials and particle-reinforced composites. J Appl Mech 59:261–268. https://doi.org/10.1115/1.2899515
20. Molinari A, Canova GR, Ahzi S (1987) A self-consistent approach of the large deformation polycrystal viscoplasticity. Acta Metall 35(12):2983–2994. https://doi.org/10.1016/0001-6160(87)90297-5
21. Masson R, Bornert M, Suquet P, Zaoui A (2000) An affine formulation for the prediction of the effective properties of nonlinear composites and polycrystals. J Mech Phys Solids 48(6–7):1203–1227
22. Tandon GP, Weng GJ (1988) A theory of particle-reinforced plasticity. J Appl Mech 55:126–135. https://doi.org/10.1115/1.3173618
23. Doghri I (2013). Mechanics of deformable solids: linear, nonlinear, analytical and computational aspects. Springer
24. Gavazzi AC, Lagoudas DC (1990) On the numerical evaluation of Eshelby's tensor and its application to elastoplastic fibrous composites. Comput Mech 7(1):13–19
25. Bornert, M., Bretheau, T., Gilormini, P., (Eds.) 2001. Homogénéisation en mécanique des matériaux, 2. Matériaux aléatoires élastiques et milieux périodiques. HERMES Science, Paris, 2001
26. Llorca J, Segurado J (2004) Three-dimensional multiparticle cell simulations of deformation and damage in sphere-reinforced composites. Mater Sci Eng A 365(1–2): 267–274
27. Gonzalez C (2004) Numerical simulation of elasto-plastic deformation of composites: evolution of stress microfields and implications for homogenization models. J Mech Phys Solids 52(7):1573–1593
28. Pierard O, González C, Segurado J, LLorca J, Doghri I (2007) Micromechanics of elasto-plastic materials reinforced with ellipsoidal inclusions. Int J Solids Struct 44(21):6945–6962

Impact of Injection Parameters on Gloss Properties of Grained Polypropylene Parts

Souad Mbarek[1(✉)], Zaineb Baccouch[2], Didier Perrin[3], Olivier Eterradossi[3], Bernard Monasse[4], Helene Garay[3], and Jean-Christophe Quantin[3]

[1] Mechanical Laboratory of (LMS), LR11ES36, National School of Engineers of Sousse, Sousse, Tunisia
[2] Laboratory of Electromechanical Systems, LR99ES36, University of Sfax, National School of Engineers of Sfax, Sfax, Tunisia
[3] Centre des Matériaux des Mines d'Alès (C2MA), Ecole des mines d'Alès, Alès, France
[4] Centre de Mise en Forme des Matériaux (CEMEF), Ecole des mines de Paris, Paris, France

Abstract. This paper deals with the optical and colorometric properties of the recycled polymer during numerous internal reprocess. The effects of the number of grinding-injection cycles, three process parameters (material temperature, mold temperature, and injection rate) were investigated. One most limit for this kind of study is the large number of experiments that requires long time and significant investments. The idea is to vary five injection parameters (Tmaterial, Tmold, injection rate for five injection cycles using statistical approach. The five variables were investigated at three industrial used levels. The number of recycling varies from cycle 0 to cycle 4 at five levels. The complete matrix for screening was designed using D-optimal quadratic design. The experimental design was generated with the statistical software MODDE 10.1-Umetrics. A set of 42 experiments was carried out to determine the influence of injection parameters, polluant and recycling on the appearance properties of smooth and grained surfaces. The statistical software package Nemrodw® version 2007, LPRAI (Marseille, France) was used to analyze the experimental design.

Keywords: Recycling · Gloss properties · PP · Experimental design

1 Introduction

Nowadays, appearance properties of injected objects are one of the most defies for automotive industries. The necessity of recycling was also added for quality commitment. In fact, recycled polymer improve cost performance [1]. Nevertheless, this recycling process often brings unfavorable effects, e.g., an increase in appearance properties. In order to accomplish this compromise between recycling and good esthetic of produces, studies on the effect of recycling on the appearance properties of polymers are re-quired [2]. The effect of recycling on optical properties of polymer films has been few analyzed in the literature [3–5] Among other plastics, polypropylene

N. Aifaoui et al. (Eds.): CMSM 2019, LNME, pp. 725–731, 2020.
https://doi.org/10.1007/978-3-030-27146-6_79

(PP) is a commodity polymer products and used in large quantities for automative applications [6, 7]. The main reasons for the success of PP are its quite good price/performance ratio, its excellent mechanical properties, and suitable optical characteristics [8–10]. In fact, the massive consumption of this polymer makes its recycling strategically very important for the environmental policy of industry [11] Conversely, appearance properties of polymer products are mostly fixed by the processing parameters. A low cavity surface temperature lead to a precipitately polymer melt freeze. Subsequently, a frozen layer will be formed during filling process at the inter-face between the hot polymer melt and the cold mould cavity. This phenomena conducts to a series of defects of the ultimate moulded parts, such as flow mark [12] weld mark [13], swirl mark, roughness [14], low gloss [11], and low replication accuracy [13].

The surface quality of plastic parts can be improved significantly by increasing mold cavity temperature [15]. A better replication of the mold texture can be achieved at a lower melt viscosity at higher shear rates and higher mold temperatures. This provides a higher gloss in smooth regions. The gloss properties had a significant effect on the color: all the factors lead to an increase in gloss showed concomitant effects of increasing the color coordinate b* and of decreasing the lightness L*. Studies were published showing that the processing parameters may affect the gloss of injection molded parts [9], the mold temperature commonly being considered the more important parameter to be controlled. It was shown that the mold temperature is the more relevant parameter and that a clear interaction exists between that parameter and the holding pressure.

This quantitative descriptor of appearance (gloss and color) is the consequence of a psychophysical phenomenon of visual perception correlated with a situation in which the light reflected from the surface of an opaque sample is either predominantly in the specular direction (gloss) or diffuse in all directions (color) [7]. the color depends on the illumination conditions, the observation angle, the optical characteristics of the material, the amount of the colorant present, the surface topography [6], and the gloss [3]. Work on polymeric surfaces has shown a decrease in gloss with increased roughness [15, 17, 18].

In this paper, the effect of process conditions (material temperature, mold temperature, injection rate) are investigated. The impact of recycling on the colorimetric properties of polypropylene containing 2 wt% of pigment using statistical approach were evaluated.

2 Materials and Techniques

2.1 Materials

The PP homopolymer SABIC® PP 575P produced by SABIC Europe was used in the experiments. SABIC® PP 575P is an homopolymer for injection molding. A masterbach supplied by Clariant was used to obtain a gray-beige complexion. The masterbach was characterized with EDS analysis by scanning electron microscopy. The analyze show the presence of: $CaCO_3$ (calcium carbonate), TiO_2 (titanium dioxide) and Sb_2O_3 (Antimony trioxide). Blending of PP with 2 wt% of pigments (masterbatch) was performed in the PEP "plastics technical center" with a co-rotating intermeshing twin-screw extruder (Clextral EVOLUM HT 32, diameter 32 mm, centerline distance 21 mm) with a length to diameter ratio L/D 44. The screw profile is made of conveying and kneading elements, also using opposite pitch to ensure melting, mixing, shearing, and a good dispersion of the components. The extrusion process was carried out with a screw rotation speed of 350 rpm, at a temperature of 210 °C and a throughput of 30 kg · h^{-1}. The colored pellets were also injection-molded to obtain samples for study.

2.2 Techniques

2.2.1 Polymer Processing

The mold was supplied by CFO company specialized on "Manufacturing Design Tools". The mold design was realized in C2MA to obtain samples of 100 mm 100 mm with a thickness of 2 mm. the samples show two different faces, a polished mirror surface and a rough surface. All specimens were prepared on a Krauss Maffei KM50-180CX injection-molding machine. The maximum clamping force is 50 t. During the course of experiment, corresponding to each set of experimental parameters, the 5 first injected samples are thrown away to ensure that the process was stable. PP was injected from zero to five steps of recycling. Samples are grinding with a Cutting Mill SM 300 to be reinjected.

In a purpose of understanding the effect of injection parameters and recycling on the appearance properties of polymers, several injection parameters and cycles numbers must be tested. One most limit for this kind of study is the large number of experiments that requires longtime and significant investments. The idea is to vary three injection parameters temperature of material (Tmaterial), Temperature of the mold (Tmold) and the injection rate for five injection cycles. The work material selected for the study was polypropylene because of its large use and the need to recycle it with the maintaining of good aspect properties.

The effects of four factors known to influence injection conditions, material temperature, mold temperature, injection flow and number of injection cycles were studied using a statistical approach. The three variables (materiel temperature, mold temperature and injection flow) were investigated at three industrial used levels (materiel temperature: 220, 240, 260 °C, mold temperature: 30, 50, 80 °C and injection flow: 10, 20, 40 cm^3/s). The number of recycling varies from cycle 0 to cycle 4 at five levels. The complete matrix for screening was designed using D-optimal quadratic design. The experimental design was generated with the statistical software MODDE 10.1-Umetrics. A set of 45 experiments was carried out to determine the influence of injection parameters and recycling on the appearance properties of samples. The statistical software package Nemrodw® version 2007, LPRAI (Marseille, France) was used to analyze the experimental design.

The response approach involving a D-optimal quadratic design was adopted for studied the effect of injection parameters and recycling on the final response (colorimetric and gloss). Each variable (material temperature, mold temperature, injection rate) was studied at three different levels (1, 2, and 3). The four recycling cycles was studied with four levels (1, 2, 3, and 4). All variables were taken at a central coded value of zero. The minimum and maximum range of variables investigated and the full experimental plan with respect to their actual and coded forms. The analysis of the data was carried out to obtain an empirical model defining the response Y (Eq. 1), which is the value of colorimetric (L*, a*, b*) or gloss (R20, R60, R85).

$$\begin{aligned} Y = b_0 + b_{1A} \cdot X_{1A} + b_{1B} \cdot X_{1B} + b_{2A} \cdot X_{2A} + b_{2B} \cdot X_{2B} + b_{3A} \cdot X_{3A} \\ + b_{3B} \cdot X_{3B} + b_{4A} \cdot X_{4A} + b_{4B} \cdot X_{4B} + b_{4C} \cdot X_{4C} + b_{4D} \cdot X_{4D} \end{aligned} \quad (1)$$

2.2.2 Colorometric Properties

CIELAB was intended to present a standard, approximately uniform color space which could be used by everyone so that color values could be simply compared. In a uniform color scale, the differences between points plotted in the color space correspond to visual differences between the colors plotted. The L* axis is the light-dark axis, and thus gives a measure of the relative brightness of the sample ranging from total black (L* = 0) to the white (L* = 100). The a* axis is the red-green axis (positive a* = red, negative a* = green), and the b* axis is the blue-yellow axis (positive b* = yellow, negative b* = blue) [15, 12] (Fig. 1).

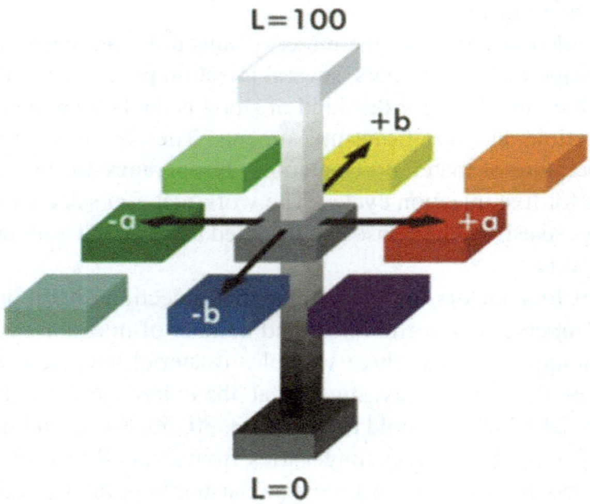

Fig. 1 (L*, a*, b*) representing CIELab color space

3 Results and Discussion

The results of colorimetric responses are analyzed with the statistical software package Nemrodw-version 2007, LPRAI (Marseille, France) and shown respectively in Fig. 2. The reference is taken on the value of the highest level (black bar). The influence of the injection temperature (220, 240 and 260 °C), the mold temperature (30, 50 and 80 °C), the injection rate (10, 20 and 40 cm^3/s) and the number of injection cycle (cycle 0 to cycle 5) on the gloss of different PP samples with grained surface is shown in Fig. 2.

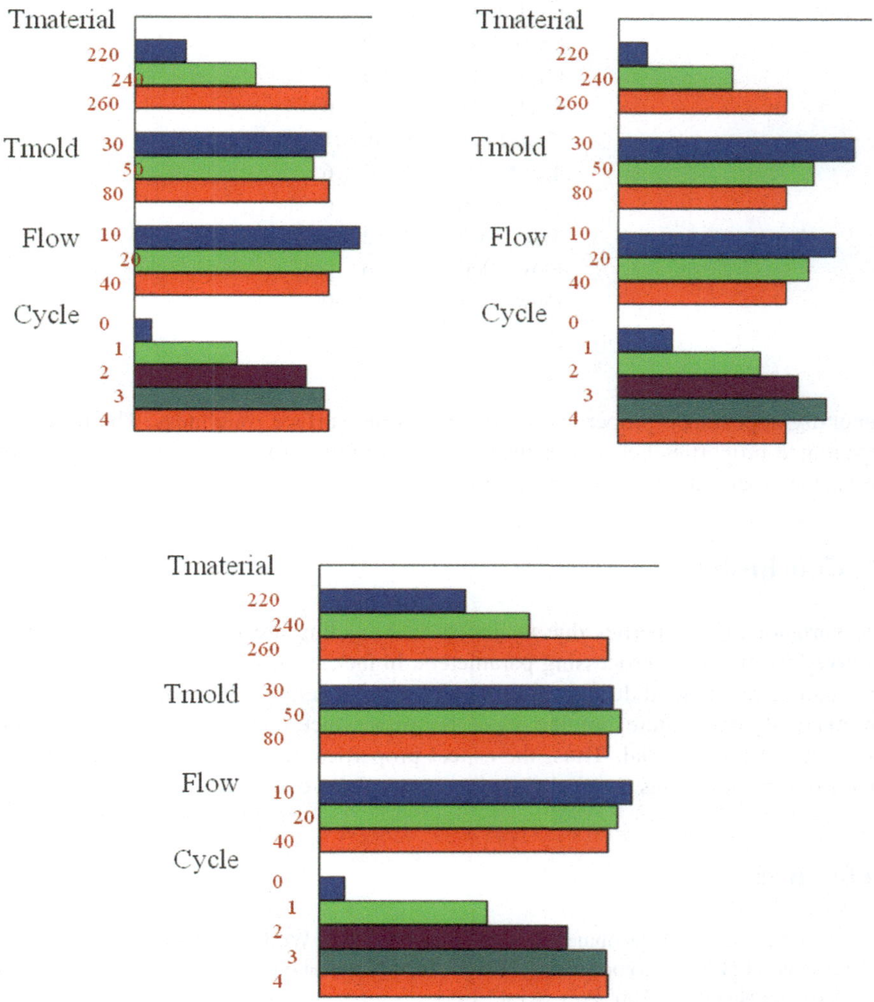

Fig. 2 Graphic representation of the difference in weight of the factors on total effect for the colorimetric response. **a** L*, **b** a*, **c** b*

The colorimetric property depends on variations in injection conditions and number of recycling cycles. Thus, the recycling and material tempera-ture have the most important effect on colorimetric. The variations of colorimetric parameters are obtained by changing the number of cycle and the Table 1.

The appearance properties of the injection molded samples showed that variations in injection parameters and number of cycles had a significant impact on colorimetric.

Table 1 The effect of injection parameters and number of cycles on the colorimetric (Pareto individual effect from response analyses with Nemrodw)

	L*	a*	b*
Cycle 0/4	**42**	24	**57**
Cycle 1/4	12	0	13
Cycle 2/4	0	0	1
Cycle 3/4	0	3	0
T_{mat}220/260	**35**	**46**	21
T_{mat}240/260	9	7	6
T_{mold} 30/80	0	11	0
T_{mold} 50/80	0	2	0
Flow 10/40	2	6	0
Flow 20/40	0	0	0

Generally appearance properties are related to the surface roughness. The differences appearance properties between samples were presumed to arise from the variation in the surface topography in ac-cording to the literature [9, 16].

4 Conclusion

The variations of properties due to numerous recycling are in the same range that is observed by changing processing parameters. In fact, the optical properties of recycled parts can be maintained during numerous internal recyclings. The number of cycle and the material temperature are the most important factors changing the colorimetric properties of the material. Thus, the aspect properties of recycled parts can be maintained during numerous internal recycling up to three cycles.

References

1. Allen NS, Hardy SJ, Jacobine AF, Glaser DM, Yang B, Wolf D, Catalina F, Navaratnam S, Parsons BJ (1991) Photochemistry and photopolymerization activity of perester derivatives of benzophenone. J Appl Poly Sci 42:1451
2. Baccouch Z, Mbarek S, Jaziri M (2017) Experimental investigation of the effects of a compatibilizing agent on the properties of a recycled poly (ethylene terephthalate)/ polypropylene blend. Polym Bull 74:839
3. Dalal EN, Natale-Hoffman KM (1999) Color Res Appl 24:369
4. Dawkins E, Engelmann P, Horton K, Monfore M (1998) J Inj Mold Technol 1:1
5. Donald B, Mathew R (1988) SPE Antec Tech Paper 34:18
6. Huff K (1994) Visual assessment and practical colorimetry in the plastic industry. Bayer AG, Leverkusen, Germany
7. Hunter RS, Harold RW (1987) The measurement of appearance. Wiley, New York

8. Lange D (1981) Color difference measuring instrument. MicroColor, Operating Instructions. Edition 4 BDA 163
9. Mbarek S, Baccouch Z, Perrin D, Eterradossi O, Monasse B, Garay H, Quantin JC (2019) Effect of recycling and injection parameters on gloss properties of smooth colored polypropylene parts: contribution of surface and skin layer. Polym Eng Sci 59:1288–1299
10. Menyhárd A, Gahleitner M, Varga J, Bernreitner K, Jääskeläinen P, Oysd H, Pu-kánszky B (2009) Eur Polym J 45:31–38
11. Vilaplana F, Karlsson S (2008) Quality concepts for the improved use of recycled polymeric materials: a review. Macromol Mater Eng 293:274
12. Tredoux L, Satoh I, Kurosaki Y (1999) Investigation of wave-like flow marks in injection molding: flow visualization and micro-geometry. Polym Eng Sci 39:2233–2241
13. Xie L, Ziegmann G (2008) A visual mold with variotherm system for weld line study in micro injection molding. Microsyst Technol Micro-Nanosyst Inform Storage Process Syst 14:809–814
14. Yoon JD, Hong SK, Kim JH, Cha SW (2004) Cell Polym 23:39
15. Wang L, Huang T, Kamal MR, Rey AD, Teh J (2000) Surface topography and gloss of polyolefin blown films. Polym Eng Sci 40:747–760
16. Baccouch Z, Mbarek S, Perrin D, Eterradossi O, Monasse B, Garay H, Quantin JC (2019) Investigation on the effects of recycling and injection parameters on gloss properties of smooth polypropylene parts. Adv Mater Res Mech Manuf, 223–231
17. Yoshii M, Kuramoto H, Kawana T (1996) The observation and origin of micro flow marks in the precision injection molding of polycarbonate. Polym Eng Sci 36:819–826
18. Zhang A, Zhao G, Guan Y (2014) Effects of mold cavity temperature on surface quality and mechanical properties of nanoparticle-filled polymer in rapid heat cycle molding. J Appl Polym Sci 132(6)

Dynamic Calibration Method for Copper Crusher Gauges Based on Split Hopkinson Pressure Bars Technique and Finite Element Modeling

L. Elkarous[1,2(✉)], A. Nasri[1], and R. Nasri[1]

[1] Ecole Nationale d'Ingénieurs de Tunis, LR-MAI-ENIT, Université de Tunis El Manar, 1002 Tunis, Tunisia
elkarouslamine@gmail.com, abdallah.nasri@enit.utm.tn, rachid.nasri@enit.rnu.tn

[2] Académie militaire, Fondouk Jedid, 8012 Nabeul, Tunisia

Abstract. In the field of interior ballistics, crusher gauges are still widely used for the gas peak pressure measurement within a fired ammunition. Under the effect of the gas chamber pressure, a copper cylinder deforms plastically and records its maximum amplitude. The plastic deformation is after what converted to a peak pressure through the so-called pressure/deformation conversion table which was established generally by a compression test at constant strain rate. From our point of view, this quasi-static calibration does not consider particularly the dynamic behavior of the copper crusher especially required for ballistic dynamic measurements. This might explain the difference of up to 20% in peak pressures between the crusher values and that measured by a piezoelectric pressure transducer, currently the standardized technique for pressure measurement in ballistic proof testing. The present study falls within this topic with the aim of developing a dynamic calibration method for crusher gauges. Testing was conducted through the use of the split Hopkinson pressure bars (SHPB) to investigate the mechanical behavior of the crusher gauges at different high strain rates. First of all, the stress/strain behavior was approached by a constitutive model. Then, the associated model parameter was determined through the use of the SHPB technique under different high strain rates. Finally, the pressure/deformation curve was then established through finite element modeling using Abaqus/Explicit where the gas peak chamber pressure measured by a piezoelectric transducer were taken as reference value.

Keywords: Interior ballistics · Pressure measurement · Piezoelectric transducer · Copper crusher gauge · Dynamic calibration · High strain rate

1 Introduction

Ballistics is in general the study of a projectile's motion in three-dimensional space, particularly those fired by weapon systems. It is subdivided into three branches, namely internal ballistics, external ballistics and terminal ballistics. The aim of interior ballistics is to obtain a required projectile muzzle velocity to hit the target and inflict a

N. Aifaoui et al. (Eds.): CMSM 2019, LNME, pp. 732–742, 2020.
https://doi.org/10.1007/978-3-030-27146-6_80

desired damage to it. The achievement of this parameter is closely related to the work of combustion gases which are issued from the burn of the propellant inside the chamber. Therefore, peak pressure inside the fired ammunition shell must not exceed the yield strength of the made-steel barrel to ensure a permanent safe use of the weapon.

Gas chamber pressure measurement is one of the most demanding measurements in ballistic testing. Knowledge of its maximum value (peak pressure) has become paramount in applications, such as weapon systems development, investigation on ballistic performances of ammunition and safety problems. Unfortunately, the measurement of this variable depends on the used measurement technique. Until the mid of 1960, crusher gauge technique was the widely known and used method for gas peak pressure measurement. A copper or lead made cylinder is compressed by a steel piston placed in contact with combustion gases inside the barrel chamber. Under the effect of the gas chamber pressure, the crusher cylinder deforms plastically and records its maximum amplitude. The plastic deformation is converted to a peak pressure through the so-called pressure/deformation conversion table. Since 1960, piezoelectric transducers have superseded crusher gauges when charge amplifiers have been developed by W.P. Kistler in the 1950s [1]. This technique was rapidly standardized for gas chamber pressure measurement.

Despite its drawbacks, the crusher gauge technique [2] is still widely used nowadays. It is simple (there is no further instrumentation needed), cheap and accurate enough for a rapid estimation of the chamber peak pressure, e.g. in ammunition testing and weapon proofs. Although it is used for dynamic pressure measurements, crusher gauge suppliers continue to provide conversion tables mostly established by quasi-static calibration at very low strain rate (10^{-4}–10^{-2} s^{-1}). As it is widely known that behavior of materials depends strongly onto the strain rate, a difference of up to 20% in peak pressures between the crusher values and that measured by a piezoelectric transducer.

Initially developed for dynamic testing of materials at high strain rates (10^{2}–10^{4} s^{-1}), recent research results demonstrated that the Kolsky bars setup [3], also widely known as split Hopkinson pressure bars (SHPB), is a versatile technique which may also be employed for dynamic calibration of force [4] and pressure sensors [5]. The present study has proved that the SHPB is also a promising technique for dynamic calibration of crusher gauges. This new calibration method has permitted to reduce measurement uncertainty since it has allowed to take into account the dynamic behavior of copper crusher.

First of all, the stress/strain behavior was approached by the Johnson-Cook constitutive model. Then, the associated model parameter was determined through the use of the SHPB technique under different high strain rates. Finally, the pressure/deformation curve was then established through finite element Abaqus/Explicit software [6] where the gas chamber pressure measured by a piezoelectric transducer Kistler type 6215 were taken as reference.

2 Split Hopkinson Pressure Bars Apparatus and Data Analysis

Kolsky was the first person to extend the Hopkinson bar technique to measure the stress/strain response of materials under impact loading conditions. A general Kolsky compression bars apparatus consists of three major components: a loading device, the bars, and a data acquisition and evaluation unit. As schematically shown in Fig. 1, the SHPB setup at the ballistic laboratory of the Royal Military Academy disposes of 2 m long incident and transmission bars with a diameter of 30 mm. They are made out of high strength maraging steel [7].

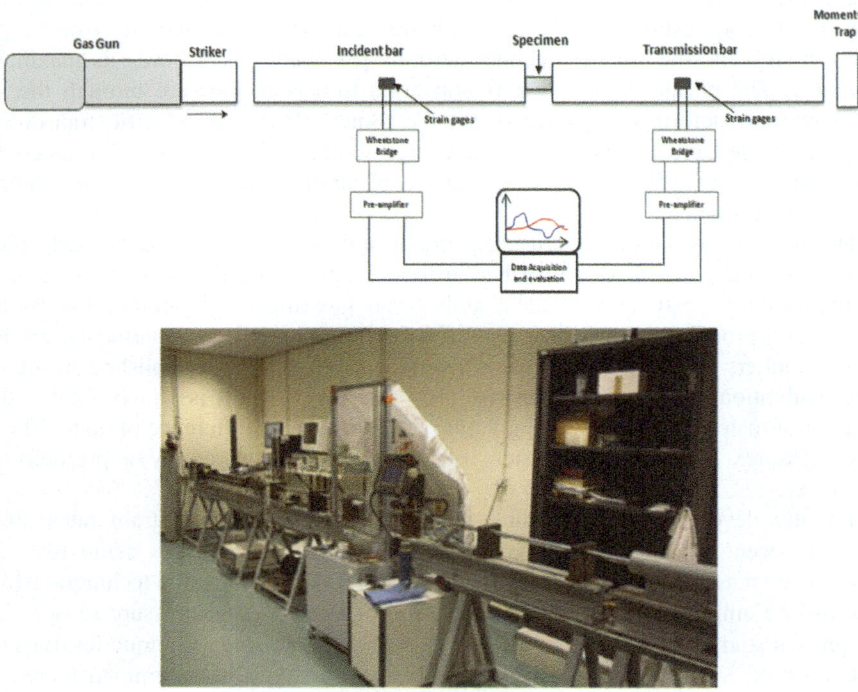

Fig. 1 The SHPB apparatus at the ballistic laboratory of the Royal Military Academy

The striker of 400 mm is launched by a sudden release of the compressed air in a gas gun and accelerated in a long barrel until it impacted on the end of the incident bar. Proper alignment of the SHPB setup was checked by performing a bar-to-bar experiment (without specimen in between the bars). The one-dimensional longitudinal elastic wave velocity was determined by performing a single-bar experiment (original Hopkinson setup).

The cylindrical specimens, used for the experiments and sandwiched one by one between the bars, were made out of the same material as the original copper crushers used in the ballistic experiments, but had slightly different dimensions in order to have a good signal-to-noise ratio for the SHPB testing. The specimens had respectively a height and diameter of 10 and 7.2 mm. They had an initial temperature corresponding to room temperature.

A double set of strain gages was attached symmetrically to the incident and the transmission bar surfaces to eliminate any influence of parasitic bending of the bars. The signals coming from the strain gages were conditioned with a Wheatstone bridge. They were amplified because such signals produced low-amplitude voltage. Finally, the signals were sampled at 10 MHz and recorded using a digital acquisition system.

The impact of the striker on the first incident bar generates a compression longitudinal wave, $\varepsilon_I(t)$. Once this wave reaches the interface between the incident bar and the specimen, part of it is reflected back into the incident bar as wave of tension, $\varepsilon_R(t)$, while the rest transmits into the specimen, $\varepsilon_T(t)$, and gets reflected back and forth inside the specimen due to wave impedance mismatch between the specimen and bars.

Assuming equilibrium conditions in the Kolsky-bar test, strain rate $\dot{\varepsilon}_s$, strain ε_s and stress σ_s inside the specimen material are computed using the following equations (Fig. 2):

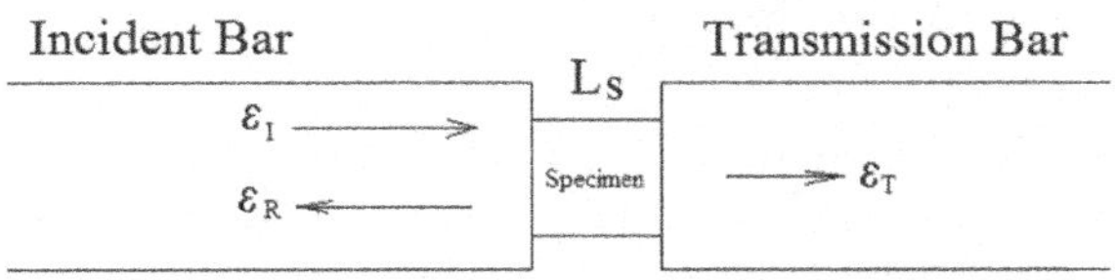

Fig. 2 Testing section of SHPB

$$\dot{\varepsilon}_s = -\frac{2C_0}{L_s}\varepsilon_R \tag{1}$$

$$\varepsilon_s = -\frac{2C_0}{L_s}\int_0^t \varepsilon_R dt \tag{2}$$

$$\sigma_s = E\frac{A}{A_s}\varepsilon_T \tag{3}$$

where C_0 is the longitudinal sound velocity in the bars, L_s is the length of the specimen, A et A_s are respectively the cross-sections of the bar and the specimen and E is the Young's modulus of the bar material.

3 Modeling of the Crusher Gauge Material

In order to investigate the behavior of the crusher gauge, two strength models were considered for crusher material. The first considered model was the Zerilli-Armstrong model [8], which in its version for face-centered cubic materials (FCC) like copper is given by:

$$\sigma = c_0 + c_2\sqrt{\varepsilon_p}e^{-C_3T + C_4\ln(\dot{\varepsilon})T} \tag{4}$$

where, σ is the stress, ε_p is the plastic strain, $\dot{\varepsilon}$ is the strain rate, T is the temperature, and finally C_0, C_2, C_3, C_4 are the material parameters fitted to the real material behavior.

The second considered strength model was the Johnson-Cook model [9], which was originally developed to model the mechanical behavior of body-centered cubic materials (BCC), as steels for instance. Nevertheless, the Johnson-Cook model is a very versatile model that has often been used outside its original scope, typically with satisfactory results. Its mathematical expression is given by:

$$\sigma = \left[A + B\varepsilon_p^n\right]\left[1 + Cln\left(\frac{\dot{\varepsilon}}{\dot{\varepsilon}_0}\right)\right]\left[1 - \left(\frac{T - T_r}{T_m - T_r}\right)^m\right] \tag{5}$$

where A, B, C, n and m are the material parameters to be determined, and $\dot{\varepsilon}_0$ is the reference strain rate (chosen as 1 s^{-1}). T_m and T_r are respectively the melt temperature and the reference temperature of the considered material. The melt temperature T_m was fixed to 1357 K considering the melt temperature of pure copper while the reference temperature T_r was equal to 294 K (room temperature).

The parameters of the strength models were determined using a global optimization approach (all parameters are determined simultaneously) and the least squares method, applied to all eleven experimentally determined stress-strain curves at once. The adiabatic heating of the samples was taken into account by calculating the conversion of the external work into heat and assuming a Taylor-Quinney factor β of 0.9, as is common for copper-based materials [10]. The temperature increase ΔT associated with an elementary deformation step from a stress-strain condition $(\sigma_1, \varepsilon_1)$ to a stress-strain condition $(\sigma_2, \varepsilon_2)$ is then (linearly) approximated by:

$$\Delta T = \frac{\beta \cdot \frac{1}{2}(\sigma_1 + \sigma_2)}{\rho \cdot C_p}(\varepsilon_2 - \varepsilon_1) \tag{6}$$

where, ρ is the density of the material (8960 kg m^{-3}) and C_p is the heat capacity (414 J kg^{-1} K^{-1}).

4 Experimental Dynamic Test Results

First of all, and before performing dynamic SHPB testing, we tried to reproduce the supplier's pression/deformation conversion data. To do this, compression tests were carried out on crusher samples using a universal testing machine (UTM) to determine the compressive strength and then the stress-strain curve of the crusher material as shown in Fig. 3.

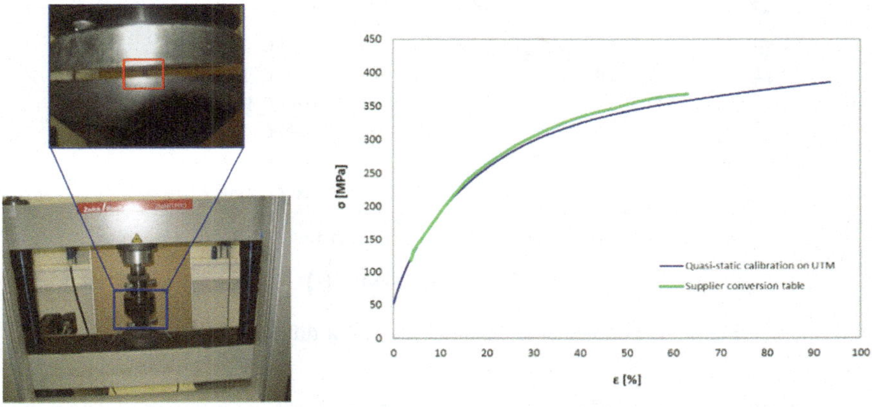

Fig. 3 Supplier conversion data versus quasi-static calibration result

From the examination of the strain-stress curves, obtained results showed clearly that supplier's data were issued from a calibration performed at constant strain rate.

In total, eleven dynamic compression tests were performed in a strain rate range of approx. 650 s^{-1} up to 1400 s^{-1}. The final strains varied from approximately 0.1 up to 0.25, comparable to the strains achieved in the copper crushers during a typical ballistic pressure measurement. The final stress-strain curves of each test were obtained through custom data analysis software implemented in a LabVIEW environment [11] (Fig. 4).

A representative comparison between an experimental stress-strain curve and the modeled curves is given in Fig. 5 as well. It represents a typical true stress/true strain curve for a strain rate of 1181 s^{-1}. The pseudo-elastical part of the curve was removed by applying to the 0.2% offset criterion as used for quasi-static tensile and compression testing.

The resulting parameters for both material models are given in Table 1 and compared with literature values for pure copper (OFHC) [12].

It can be observed that the crusher material shows a much more pronounced strain hardening than OFHC copper, which could most likely be explained by the crusher material being a copper alloy (instead of the initially assumed pure copper material). As in the case of the Johnson-Cook model the temperature sensitivity exponent, m, did not converge to any meaningful physical value (which can be attributed to the fact that all testing was done at room temperature), its value was set arbitrarily to one.

Even if it was developed for BCC materials, the Johnson-Cook model fits the experimental curve better than the Zerilli-Armstrong model, especially for the

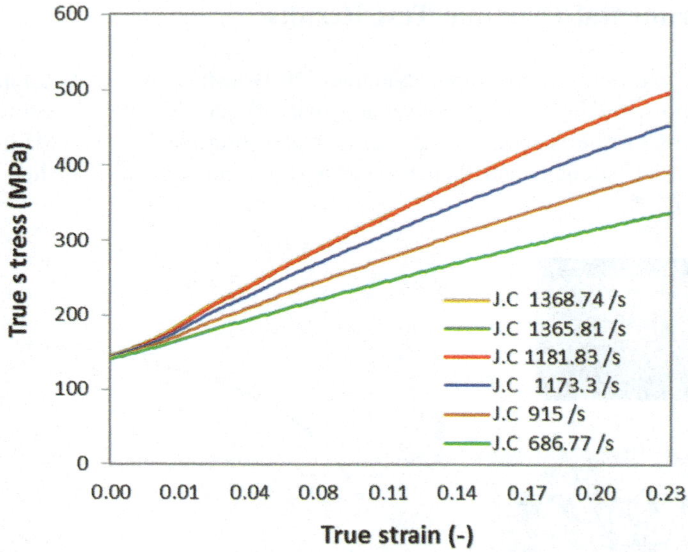

Fig. 4 Experimental stress-strain curves at different strain rates

beginning and the end of the curves. The Johnson-Cook strength model was hence preferred to the Zerilli-Armstrong model for the subsequent finite element modeling. This is also reflected in the total cumulative square errors (which are respectively 7.95×10^5 and 1.20×10^6).

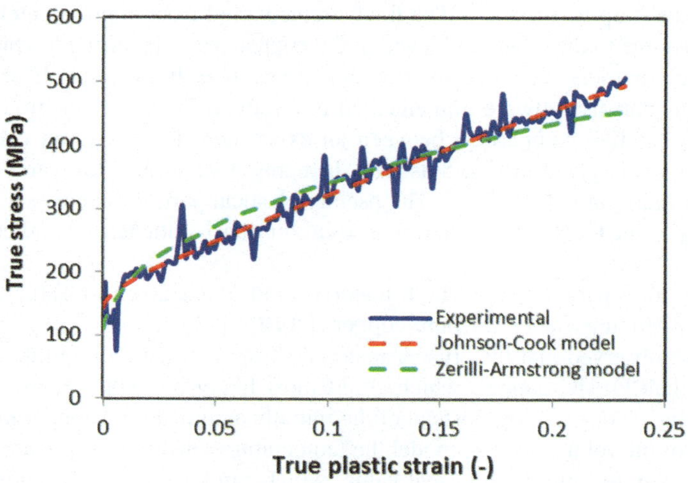

Fig. 5 Comparison between an experimental stress-strain curve and the modeled results

Table 1 Johnson-Cook and Zerilli-Armstrong model parameters

Johnson-Cook model					
	A (MPa)	B (MPa)	C (–)	n (–)	m (–)
SHPB	119	935	0.049	0.89	1
Literature	90	292	0.025	0.31	1.09
Zerilli-Armstrong model					
	C_0 (MPa)	C_2 (MPa)	C_3 (–)	C_4 (–)	
SHPB	107	853	0.0054	0.000681	
Literature	65	890	0.0028	0.000115	

5 Numerical Modeling

The finite element modeling was carried out with the FE software Abaqus/Explicit with a total geometry. The stainless-steel piston and crusher cylinder were modeled as 3D deformable part and discretized with hexahedral C3D8R Lagrangian elements. For the piston, a linear elastic constitutive behavior was also assumed which is defined by E = 197 GPa, ν = 0.272 and ρ = 7800 kg/m^3. The behavior of crusher material was defined with a fitted Johnson-Cook model as the strength model. The contact algorithm surface-to-surface with a frictional coefficient of 0.2 which was applied between Lagrangian parts.

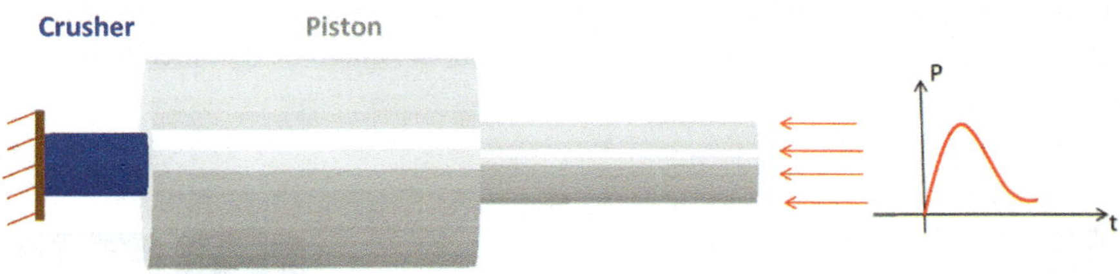

Fig. 6 Modeling of the experimental set-up with the FE Abaqus/explicit software

As shown by Fig. 6, the measured pressure-time curve obtained by a Kistler type 6215 piezoelectric pressure transducer was taken as the input reference pressure. Firing tests have been performed with an instrumented 12.7 mm barrel manufactured according to military standards and mounted on a universal ballistic breech with an interchangeable barrel (BMCI). The piezoelectric transducers and the copper crusher gauge were installed in the same mounting position in respect to barrel length.

The copper crusher gauge had a nominal height of 4.91 mm and a nominal diameter of 3.0 mm. The 12.7 × 99 ammunition filled with the gun powder type 0.50 inch (WC 860) have been used to achieve the firing tests. A full metal jacket projectile type M33 of 42.5 g is also used for all experiments. The cartridge cases were drilled with a hole of 2.5 mm diameter according to military standards. The triggering was realized using a muzzle flash detector which detects the flash at the exit of the projectile from the muzzle. Figure 7 gives an overview the experimental setup.

Fig. 7 Experimental set-up for pressure measurement

The permanent simulated deformation of the crusher Δl_{sim} has been compared to the reference permanent deformation Δl_{ref} corresponding to the initially applied peak pressure P_{peak} (using the conversion table), and to the experimentally obtained average residual deformation Δl_{exp}. The results of the conducted simulations and experiments were confronted to evaluate the capability of the numerical model regarding its robustness and accuracy. If the simulated permanent deformation corresponded to the reference permanent deformation, the developed model could be assumed to be correct (Fig. 8).

Fig. 8 Contour plots of von Mises equivalent stress

6 Results and Discussion

Firing tests were performed using respectively 12, 14 and 16 g of powder charges. Table 2 gives the average the experimental peak pressure and the average of permanent deformations respectively for the piezoelectric Kistler type 6215 and the copper crusher gauges when FNB ammunitions were fired with three different gunpowder masses.

The peak pressure given by the crusher was always lower than the measured by the piezoelectric transducer. The difference between values may reach 20% as for the 12 g powder mass but remains was less than 10% for the higher 16 g powder mass.

Table 2 Reference, simulated and experimental permanent deformations

Powder mass (g)	P_{max} (MPa)	Δl_{ref} (mm)	Δl_{exp} (mm)	$P_{crusher}$ (MPa)	Δl_{sim} (mm)	P_{sim} (MPa)	Rel. difference (%)	
12	159.15	0.69	0.49	127.65	0.64	150.3	19.8	5.5
14	250.15	1.35	1.1	218.4	1.29	243.1	12.7	2.8
16	319.6	1.80	1.6	292	1.73	310	8.6	3

At the examination of the Table 2, the residual deformation rises when powder mass was increased as expected. It was also clear that the simulated residual deformations corresponded better to the reference deformations than to the average experimental deformations. Moreover, the peak pressure values obtained with the simulated deformation became quite close to the reference peak pressure given by the piezoelectric transducer. The difference did not exceed 5% which can be a proof of the accuracy of the dynamic calibration by the SHPB method and thereby the used behavior model. The reason of the remaining difference is certainly due the choice of the required strain rate to carry out the SHPB dynamic tests. Its exact value for crusher cylinder deformation is still unknown. One of the possible solutions consisted of the estimation of this parameter from the pressure-time curve [13].

7 Conclusion

Gas chamber pressure measurement is one of the most frequently performed measurements in ballistic testing. Crusher gauges are still widely used to measure this physical variable. As the accuracy of the measurements depends strongly on the used calibration method, this work was dedicated to elucidate the problem of calibration in order to develop a reliable dynamic calibration technique for copper crusher gauges based on the split-Hopkinson pressure bar technique. Compared to pressure measurement results obtained with the quasi-static calibration, those given by crusher gauges were improved and became close to the piezoelectric measurements when applying the developed calibration method. The accuracy of the model further could be improved to ameliorate the consistency of the behavior model and to overcome the dependence on the strain rate in the dynamic testing of crusher material.

References

1. KIAG (2006) Technical documents, Kistler Instrumente AG
2. NATO (2005) Pressure Measurement by Crusher Gauges NATO Approved Tests for Crusher Gauges AEP-23 Ed. 2, NATO Standardization Agency

3. Kolsky H (1949) An investigation of the mechanical properties of materials at very high loading rates. Proc Phys Soc Lond B62:676
4. Kumme R, Mack O, Bill B, Gossweiler C, Haab HR (2002) Dynamic properties and investigations of piezoelectric force measuring devices. In: Proceedings of the 18th IMEKO TC3 conference, no 1685, Celle. VDI-Berichte, pp 161–171
5. Zhang Y, Zu J, Zang HY (2012) The dynamic calibration method of high-pressure transducer under based on quasi-static function excitation source. Measurement 45:1981–1988
6. Dassault Systèmes SIMULIA (2014) Abaqus 6.14: Abaqus/CAE User's guide, pp 680–694. http://abaqus.software.Polimi.it/v6.14/index.html. Accessed 19 Dec 2018
7. Elkarous L, Coghe F, Pirlot M, Golinval JC (2013) Experimental techniques for ballistic pressure measurements and recent development in means of calibration. J Phys: Conf Ser 459
8. Zerilli F, Armstrong R (1990) Description of tantalum deformation behavior by dislocation mechanics based constitutive relations. J Appl Phys 68:1580–1591
9. Johnson GR, Cook WH (1983) A constitutive model and data for metals subjected to large strains, high strain rates and high temperatures. In: Proceedings of the 7th international symposium on ballistics, The Hague, The Netherlands, pp 541–547
10. Bai Y, Dodd B (1992) Adiabatic shear localization. Pergamon Press, UK
11. National Instruments. LabVIEW (2012) Getting started with LabVIEW (2012), pp 42–67. http://www.ni.com/pdf/373427h.pdf. Accessed 19 Dec 2018
12. Banerjee B (2005) An evaluation of plastic flow stress models for the simulation of high temperature and high-strain-rate deformation of metals. Department of Mechanical Engineering, University of Utah, USA
13. Kuokkala VT, Ramo J, Vuoristo T (2004) Calibration of crusher pressure gauges by high strain rate testing. Tampere University of Technology, Institute of Materials Science, Tampere, Finland, P.O. Box 589, FIN-33101

Improvement of the Quality of Aeronautical Products Stelia Tunisia

Safa Mathlouthi(✉)

General Direction of Technological Studies, High Institute of Technological Studies of Kelibia, Nabeul, Tunisia
mathlouthi.safa@yahoo.fr

Abstract. The major challenge of any company nowadays is to master Quality, Cost and Delivery QCD of its products by optimizing its manufacturing process. In this context, STELIA Tunisia specialized in the aeronautical field has launched this project to improve the quality of its manufactured products in the assembly unit(Panels and frames of lower fuselage and front fuselage). Through a diagnosis, we identified the production line then the nonconformities with the highest Cost Of Poor Quality COPQ. the assembly line of lower fuselage 13–14 is concerned, more particularly nonconformity type "rejects". Following a clear root causes analysis, we concluded that these nonconformities are mainly due to a lack of supervision, quality control and organization. Hence the need to deploy a new strategy for monitoring its processes through the implementation of process monitoring and control plans.

Keywords: Process monitoring plan · FMECA process · Checkpoints · 5S

1 Introduction

STELIA Tunisia is a company specialized on the assembly of aero-structures and fuselage parts. In 2016, it recorded an important deviation of COPQ. This high cost was mainly in the assembly process of lower fuselage 13–14.

To hunt poor quality costs, the company decided to launch a project aimed at resolving this problem. First, a diagnosis was developed to identify its causes and define a solid action plan to be implemented.

2 Diagnosis

The process diagnosis is a first mandatory step. So we must present the product "lower fuselage 13–14" and its assembly process.

2.1 Assembly Process of Lower Fuselage 13–14

Lower fuselage 13–14 is the lower part of an airplane as shown in the Fig. 1:

It is composed of several parts:

- EPC: This is the baggage carrier of the plane

N. Aifaoui et al. (Eds.): CMSM 2019, LNME, pp. 743–751, 2020.
https://doi.org/10.1007/978-3-030-27146-6_81

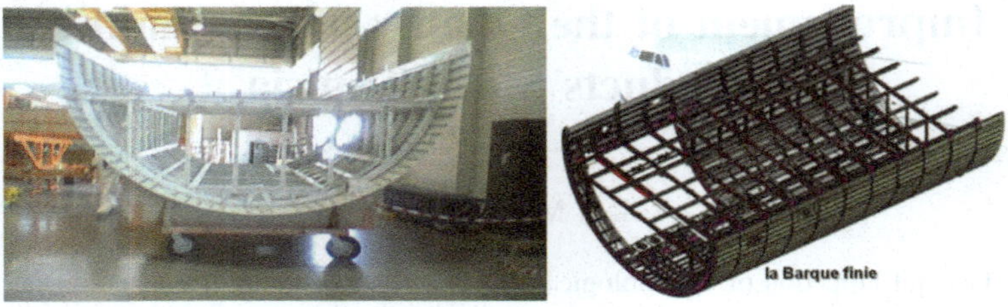

Fig. 1 Lower fuselage 13–14 of an airplane

- PID: Right lower panel is an element for the fuselage of the aircraft
- PIG: left lower panel is an element for the fuselage of the aircraft
- PLD: Right Side panel is a fuselage element of the aircraft
- PLG: Left Side panel is a fuselage element of the aircraft

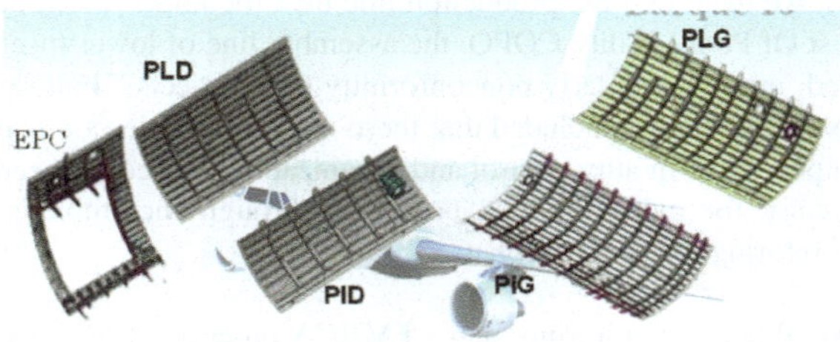

Fig. 2 Structure of the parts: EPC, PID, PIG, PLD and PLG

- ½ BD: Right lower side of the fuselage
- ½ BG: Left lower side of the fuselage
- Floor

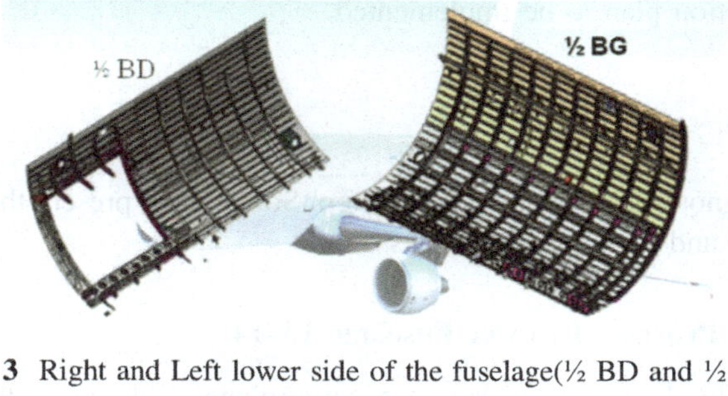

Fig. 3 Right and Left lower side of the fuselage(½ BD and ½ BG)

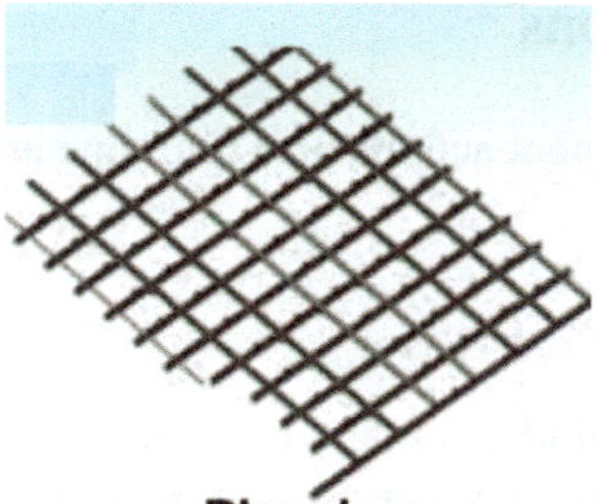

Fig. 4 Floor

Each part has a defined assembly process. To assemble the lower fuselage 13–14, we must go through 9lines: 8 parts assembly lines and a line for the final product (Fig. 2).

The assembly process is carried out manually for the majority of the operations except the riveting which is done semi-automatically on a machine "PRECA". The parts assembly is particularly delicate. Indeed a number of holes must be drilled accurately to accommodate the rivets fixing the frames and stringer (Fig. 3).

A control is made by the quality inspector at the end of each assembly process. He decides the conformity of the obtained products (Fig. 4).

All parts and even the final product pass, almost, by the same process steps:

- Preparation: it consists on preparing tools
- Riveting: Riveting is done on the PRECA machine
- Structure: it concerns the assembly of each part, or the assembly of final product. Generally it consists on doing manual milling, drilling or riveting operations.
- Surface Finishing: This is the last phase of the process.

2.2 Diagnosis and Action Plan

A monthly COPQ tracking during the year 2016 has shown an important deviation versus target and last years results with an upward aspect, which has a big impact on customer satisfaction in term of delivery.

Statistic analysis has shown that "rejects" is the non-conformities type which presents the Pareto diagram head.

Through an Ishikawa and 5 Why analysis, we demonstrated root causes:

- A lack of process monitoring
- A lack of control since current control is done, only, at the end of process
- A lack of tools organization (non identification, errors)

Hence the proposal of the following improvement actions:

- Implementation of process monitoring plans to anticipate defects appearances
- Control at each process step to lighten the control task and improve defects detection.
- Tools organization via the application of the 5S method.

3 Improvement Actions

We will implement improvement actions on a pilot line and then we will duplicate it on the other lines.

3.1 Constitution of the Work Team

The first step is to form a multidisciplinary team in order to involve all the levels, from the user to the manager, in the project as well as to guarantee the implementation of improvement actions.

3.2 Choice of the Pilot Line

The assembly process of the lower fuselage 13–14 has 9 lines. Implementing improvements on this important number of lines on the same time by the same work team seems to be difficult. So it's easier to initiate the project on a pilot line and proceed with the duplication.

The interest of this pilot line is also to demonstrate the feasibility and the gains before extending the improvements on all lines. It is a showcase to attract the attention of those who are not yet involved and who refuse change. We have chosen as a pilot line the line "right lower side of the fuselage" being characterized by high skilled operators as well as their ability to change.

3.3 Creating Process Flow Chart

Before creating the monitoring plan we have identified the production synoptic also called the process Flow Chart for "right lower side of the fuselage" process.

The Process Flow Chart is used to describe the assembly process of "right lower side of the fuselage" from the raw material to the final product.

3.4 Creating the Monitoring Plan

The monitoring plan is the document that allows setting up the monitoring and controlling strategy for a product on the manufacturing process. It ensures compliance with certain criterias through the measurement of indicators.

In order to develop the monitoring plan, we have first conducted a risk assessment using a brainstorming with the team to indicate, for each step of process, the possible risks and the processes to control.

An extract from the risk analysis is presented in the Fig. 5.

Analyse de risque à minima			
	Etape du processus	Risques	Processus à contrôler
10	Préparation des bandes Anti-crique	Inversion outillages de perçage des BAC	Correspondance Bandes Anti-criques / Outillages de perçage (Référence marquée sur pièce Vs Référence gravée sur outillage).
		Utilisation d'un outillage dégradé	Etat des outillages (Aspect, état des sauterelles et état des canons)
		Jeu NC, position des perçages NC et déformation de la pièce.	Serrage et épinglage de la bande Anti-crique en utilisant les sauterelles et les agrafes à vis.

Fig. 5 Risk assesment extract of the assembly process of "right lower side of the fuselage"

This plan relies heavily on the established risk assessment. In fact, for each risk we have indicated the following informations:

- The criterion value (example: conformity of the tools, plating parts, rules for drilling …)
- The reference document (example: the codification of reference documents such as N53C2DR …)
- Frequency
- The monitoring method (visual V or with a monitoring tool)
- Monitoring tools (example, control sheets or OK start.)
- Supervision by (the person responsible for this control)
- Reaction plan (reaction to be established in case of non-conformity)

Below is an extract from the assembly process monitoring plan of "right lower side of the fuselage" (Fig. 6).

Analyse de risque à minima				Plan de surveillance						
	Etape du processus	Risques	Processus à contrôler	Valeur critère	Doc de référence	Fréquence / Taille	Méthode de surveillance	Outil de surveillance	Surveillé par	Plan de réaction
10	Préparation des bandes Anti-crique	Inversion outillages de perçage des BAC	Correspondance Bandes Anti-criques / Outillages de perçage (Référence marquée sur pièce Vs Référence gravée sur outillage).	Correspondance outillage Vs pièce	N53C2DR B68A-1-FI1-010	100%	V	Check-Point	LP	
		Utilisation d'un outillage dégradé	Etat des outillages (Aspect, état des sauterelles et état des canons)	Conformité outillage	Plan outillage	100%	V	Check-Point	IQ	Elimination outillage + Alerte préparateur
		Jeu NC, position des perçages NC et déformation de la pièce.	Serrage et épinglage de la bande Anti-crique en utilisant les sauterelles et les agrafes à vis.	Plaquage pièces	N53C2DR B68A-1-FI1-010	100%	V	Check-Point	LP	Correction immédiate avant perçage

Fig. 6 Process monitoring plan of the assembly process of "right lower side of the fuselage"

3.5 FMECA Process

FMECA is the acronym for Failure Modes, Effects and Criticality Analysis [1].

The monitoring plan presented the different risks as well as the processes to be controlled in order to avoid it.

Nevertheless the risks do not have the same degree of criticality and to apply this monitoring plan it is necessary to improve the current situation.

Hence the goal of carrying out an FMECA Process in order to evaluate the criticality of each risk and to decide on the improvements to be made either by preventive or corrective actions in order to reduce criticality [1].

Thus, for each risk we had:

- Classified it according to 5M (method, machine, labor, environment, material)
- Defined its modes and effects
- Identified its causes using the 5 why method
- Calculated its criticality. Criticality C is the product of 3 quotations ($C = G \times O \times D$).

It should be noted that G relates to the severity of the effect of the risk, O concerns the occurrence or the frequency of the cause and D relates to the non detection of the failure mode.

STELIA has set a limit of criticality Clim = 70

So, the principle is to compare the calculated criticality C to the criticality Clim:

- In case of C > Clim, corrective action is to be considered
- In case of C is lower but just close to Clim, a preventive action is expected.
- In case C <⋘< Clim, nothing to predict

For corrective actions, a follow-up is necessary in order to recalculate the new criticality and compare it to the original one. It allows judging the effectiveness of the established actions.

3.6 Checkpoints

One of the identified problems during the diagnosis phase was the lack of control. At this level, it is necessary to improve risk control in order to guarantee the detection of nonconformities in the concerned station on time. We are trying to move from a control at the end of the process to a control by station.

For this reason, we will use the Checkpoints method.

Checkpoints is a simple and effective tool that exhaustively lists operations and tasks to be controlled.

Following the control result, the principle is:

- If the checkpoints indicate "OK" then the task has been checked successfully and no fault detected.
- If the checkpoints indicate "NOK" then the task has been checked but there is a detection of a defect which must be corrected.

The checkpoints includes:

- The choice of the work shift
- The concerned line, the date and MSN (item code)
- A list of figures, at the top of the document, including the control steps.
- A table at the bottom of the document that includes: the control steps, the supervisor (for example the Quality Inspector QI), the choice of indications OK or NOK. In case of NOK indication, the gap must be noted.

The following figure shows an extract from the checkpoints of the assembly process of "right lower side of the fuselage" (Fig. 7).

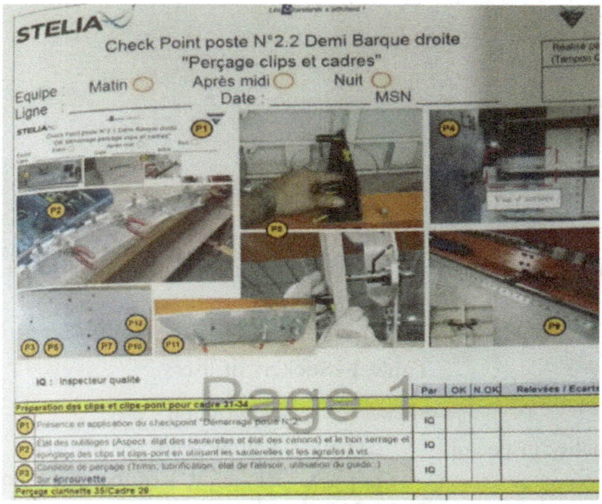

Fig. 7 Checkpoints extract of the assembly process of "right lower side of the fuselage"

3.7 Target Operation Method

The interest of the method "target operation" is to standardize the allocated time for each control task in the implemented Checkpoints.

The following Fig. 8 shows an example of 'target operation' of "right lower side of the fuselage".

3.8 Points Display Method

This method aims to reduce the control time.

It consists on putting stickers on the objects to control or the place where the operation is carried out. Indeed we will facilitate the observation of points for quality inspectors and post leaders.

These stickers contain the number of the operation according to the corresponding Checkpoints.

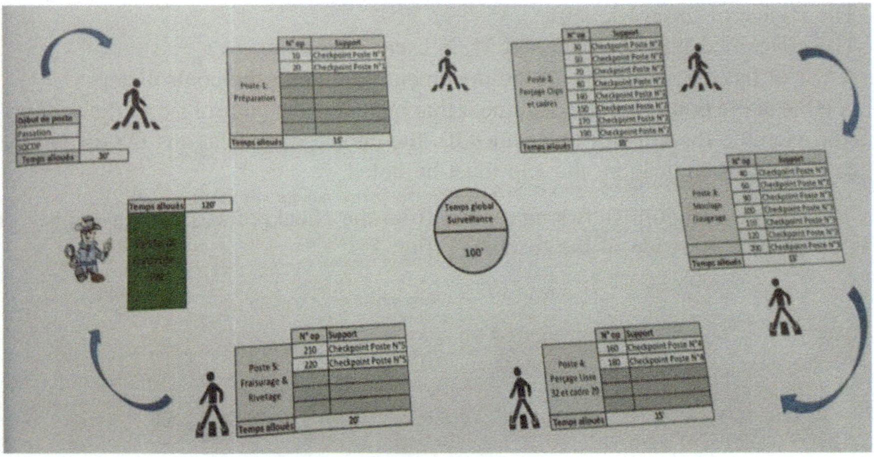

Fig. 8 Target Operation "right lower side of the fuselage"

Table 1 is an example of stickers for a quality inspector or for a team leader.

Table 1 Stickers of users

User	Sticker	Description
Quality inspector	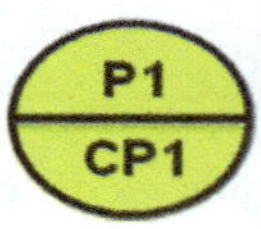	Yellow color for inspector **P1**: point 1 of the check-point **CP1**: Checkpoint 1
Team leader		Purple color for team leader **P1**: point 1 of the check-point **CP1**: Checkpoint 1

An example is presented in this Fig. 9:

Fig. 9 Points display method on "right lower side of the fuselage"

3.9 The 5S Method

The 5S method comes from 5 verbs (Sort, Set in order, Shine, Standardize, Sustain) of Japanese actions summarizing the essential tasks to carry out to organize a specific area [2].

During this project we noticed:

- Defects (breakage, scratching…) in the tooling
- Lack of standards for the storage of tools

Hence the need to set up a 5S method in order to:

- Ensure the good condition of the tools through good packaging
- To highlight the dysfunctions (any defect in the tools)
- Provide the necessary tooling requirements and make them closer to the user
- Prevent the mess
- Deploy a user-friendly standard of storage and organization.

4 Conclusion

The purpose of this project was to improve the quality of the most expensive product in terms of COPQ. We initiated our project with a diagnosis that revealed that ''rejects'' present the major type of nonconformity causing this high COPQ.

Through a root causes analysis, we concluded a lack of monitoring and control as well as a lack of organization on the shop floor.

To resolve these problems we deployed a new control and monitoring strategy by the implementation of the process monitoring plan and the Checkpoints. A 5S method was also implemented to provide good working conditions.

As a result, improvements were achieved on two sides: technical and managerial. Technically, COPQ has decreased, after 2 months, by 20% with a downward aspect. Regarding the managerial aspect, we noted an improvement in people involvement and standard's respect which guarantees the mainting of results. As planned we duplicated the project on all concerned lines of the assembly unit.

In perspective, we proposed to implement KAIZEN concepts encouraging continuous improvements.

References

1. Kelada J, L'Amdec (1994) Ecole des hautes études commerciales, centre d'études en qualité totale
2. Hohmann C (2010) Guide pratique des 5S et du management visuel, 2nd edn

Investigation of the Effects of the Pre-hole Geometrical Parameters on the Countersinking Process

Mosbah Hassen[1(✉)], Attyaoui Slimen[2], and Nasri Rachid[3]

[1] LR- MAI-11-ES19-ENIT, Faculté des Sciences Gafsa (FSG), Université Tunis El Manar, Tunis, Tunisia
hassen.mosbah@yahoo.fr

[2] LR-MAI-11-ES19-ENIT, El Manar, Ecole Nationale d'Ingénieurs de Carthage (ENICAR), Université Tunis, Tunis, Tunisia
attyaouislimen@gmail.com

[3] LR-MAI-11-ES19-ENIT, Ecole Nationale d'Ingénieurs de Tunis (ENIT), Université Tunis El Manar, Tunis, Tunisia
rachid.nasri@enit.rnu.tn

Abstract. The countersinking process is consisted to enlarge a pre-hole using a conical punch in order to allow rivets and screws to sit flush with the surface of the assembled sheets. The geometry of the pre-hole had many effects on this process. The main goal of this study is to characterize the effects of the pre-hole geometrical parameters on the obtained countersunk hole. Their effects on the forming kinematics and the loads were investigated. A finite element model was developed using ***ABAQUS/Standard***. The adopted model was performed with an elasto-plastic behavior and an isotropic hardening rule for the workpiece. A configuration with a maintained blank holder has been adopted for this study. The results from simulations analysis provided one good explanation for the complexity of the observed forming kinematics. The comparison between the simulations and the experimental results confirmed the validity of the adopted finite element model for predicting the loads and the final shape of the countersunk workpiece.

Keywords: Countersinking · Geometric parameters · Forming kinematics · Loads · Finite element · Main effect

1 Introduction

The countersinking process using forming tools became one of the interesting processes for the joining plates in aeronautic and many other fields [5].

In the literature, many studies were about the countersunk hole and joints in the composite plates. The aim researches were focused on cracking, fatigue and surface treatment. Chul and Alten [2] focused their works on the cracking behavior in the countersunk fastener holes. They studied the load transfer in the joints with fasteners holes. A generic workpiece was adopted by referring to the previous works in the same field. Three-dimensional finite element analysis was used. This model was meshed in

N. Aifaoui et al. (Eds.): CMSM 2019, LNME, pp. 752–761, 2020.
https://doi.org/10.1007/978-3-030-27146-6_82

PARTRAN® and the simulations of the crack and the fasteners were studied in Abaqus®. The experiments were carried out by the use of both optical and scanning electron microscopes. The results identified the same locations of the cracks obtained with the simulations. Rana et al. [6] investigated the effects of the geometry shape of the countersunk hole on the fatigue life. They carried out experiments for two cases with two different shapes of the countersunk hole. As consequence, it was found that the rounding of the corner at both the countersunk hole and the rivet head enhances the mechanical resistance of the riveting joint. Therefore, the fatigue lives of the workpiece, with rounded countersunk hole and rounded rivet head, were longer than that workpiece with the usual shapes of rivet.

A modified equation for the stress concentration factor in the countersunk hole was presented by Darwish et al. [3]. In their work, a three-dimensional finite element analysis was implemented in Ansys®. A parametric study was carried out to investigate the main effect of different factors on the stress concentration in a countersunk hole. As a result, they determined new equations for modeling the studied factors factor of the stress concentration.

On the other hand, [4] presented a finite element analysis for a 2D axisymmetric model of countersinking to investigate the forming kinematic during the process. In their work, they presented the parameters of the countersinking. A washer was used as the workpiece for the analysis. They studied some phenomena of the forming kinematics and they plotted the evolutions of some geometrical parameters and loads during the process. In addition, they studied the effects of the inner radius of the workpiece on the forming kinematics during the process. All the results were summarized in one diagram which couples the different geometrical parameters and loads of the countersinking. This diagram was useful to determine the suitable parameters to obtain a standard countersunk hole and the necessary loads for the process.

Mosbah et al. [5] investigated the countersinking process and they validated the model by comparing the numerical and the experimental results. In this works, the effect of the fillet radius of the inner pre-hole was performed to determine the suitable value for the adopted numerical model. It was found that many new phenomena were observed such as the gaps and the expansions. New geometrical parameters were adopted to investigate the forming kinematic phenomenon.

Regarding the previous studies, it can be seen there are few works about the effects of the geometrical parameters. The aim objective of this study was focused on the study of the effects of the pre-hole geometrical parameter on the obtained countersunk hole. In other hand, it was found interesting to investigate the effect of these parameters on the punch load. A parametric study was performed to analyze these effects. An axisymmetric finite element model was carried out using an isotropic hardening rule. The experiments were conducted to validates the numerical results.

2 Material Behavior

The adopted model was built by considering the punch, the die, and the blank-holder as rigid bodies. The workpiece was assumed deformable sheet with an elasto-plastic behavior. The large deformation was taken on account. A monotonous compression test was used to characterize the material behavior as described in the works of Alves et al. [1], Mosbah et al. [5], Zhao and Gary [7]. Thus, the stress-strain law developed in the previous work [5] was adopted for this present study. For the FEM, this law was integrated as a multi-linear model. The mechanical properties of the steel sheet are listed in Table 1.

Table 1 Mechanical properties of the metal

Young's modulus (E)	210 GPa
Poisson's ratio (υ)	0.3
Elastic yield (σ_e)	235 MPa

3 Parameters

To succeed in the countersinking process, the geometrical parameters must be carefully represented and controlled. The Fig. 1 illustrated the adopted parameters for the tools and the workpiece before and after the deformation. The nomenclature of all these parameters was listed in the Table 2.

Table 2 Workpiece parameters

Parameters tools	Symbols	Values
Punch radius	r_p	5 mm
Punch angle	Φ_p	90°
BH displacement	δ_b	0 mm
Blank-Holder inner radius	r_b	5 mm
Fillet of the BH hole	r_s	0.2 mm
Punch travel	δ_p	1.4 mm
Initial shape of the workpiece		
Initial internal radius	r_i	2.5–3 mm
Fillet radius of the initial hole	r_t	0.2–0.5 mm
Initial external radius	r_e	15 mm
Thickness	e	2.5–3 mm
Final shape of the workpiece		
Final thickness	e_f	–
Inferior gap	G_i	–
Final internal radius	r_{fi}	–
Final inferior hole radius	r_{inf}	–
Final external radius	r_{fe}	–
superior radius of countersink	r_c	–
height of countersink	h_c	–

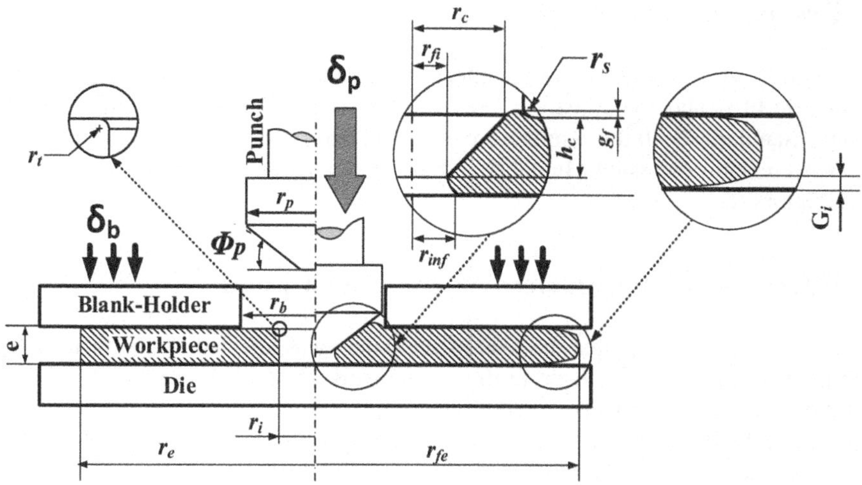

Fig. 1 Workpiece and tools parameters before and after countersinking, Hassen et al. [5]

4 FE Model

In this present work, an FE model was developed using the ***Abaqus/Standard*®**. The same model developed on the previous work of Mosbah et al. [5], was adopted. A 2D axisymmetric model was performed. The die, the punch, and the blank holder were assumed to be rigid. The blank-holder was maintained above the workpiece to prevent the raising up of the workpiece during the countersinking process.

On the other hand, the workpiece was modeled as deformable-body. Thus, it was meshed using 3-nodes linear axisymmetric hybrid mesh CAX3H. A remeshing technique was used to avoid the distortion of the mesh elements.

The contacts between the tools and the workpiece were modeled with the Coulomb's friction law. To describe reasonable contact conditions, the friction coefficient was fixed equal to $\mu = 0.12$.

5 Experiments Tools

Attention must be paid to the experiment to succeed the process and the results. In order to respect the FE model, an experiment tool was manufactured as mentioned in the works of Mosbah et al. [5]. The maintained configuration of the blank-holder was succeeded by applying a slight tightening when using the screw-nut system.

6 Results and Discussions

In this study, the investigation of the effect of the geometrical parameters was based on parametric FEA and static analysis. Several numerical models were performed using Abaqus/Standard®. All the results were extracted using python scripting. The gathered results were analyzed using Minitab® and the sensitivity analysis was performed using three factors (e, r_t, r_i).

7 Effects of the Parameters on the Punch Load

In Fig. 2a, it can be seen that the load of the punch was affected by the factor r_t more than the thickness factor e. By referring to the work of Mosbah et al. [5], it was clear that the value of the r_t affects clearly the F_p. The choice of the adequate value was the objective to befit the numerical evolution of the F_p with the experimental load. It was found that the value $r_t = 0.2$ mm led to the acceptable FE model. Concerning the factor e, it had a significant effect on the load. The material flow under the punch is complicated thus the static analysis was a useful tool to show its effect on the load F_p. The reduced value of the thickness e decreased the developed load F_p and the opposite was true. On the other hand, the initial radius r_i has no significant effect on F_p. As showed in Fig. 2b, all the curves had approximately the same evolution path and they had a slight difference between the final values of F_p for each used radius r_i. The maximum difference was assumed equal to 5% for the case using a thickness $e = 3$ mm and fillet radius $r_I = 0.2$ mm.

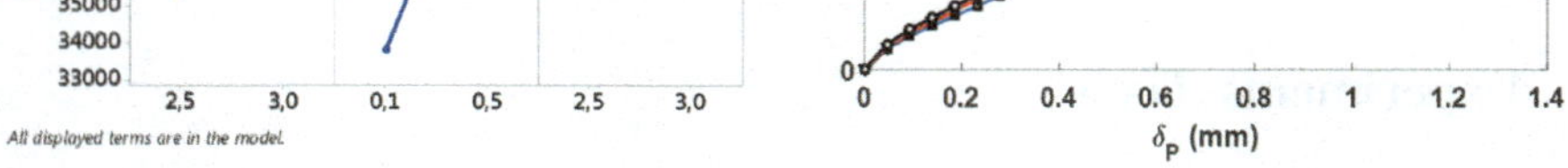

Fig. 2 **a** The main effects of e, r_t and r_i on F_p. **b** The evolution of the F_p versus δ_p ($e = 3$ mm, $r_t = 0.2$ mm)

8 Effects of the Parameters on the Countersunk Hole Shape

The final inner shape of the countersunk hole was the aim objective in this study. To investigate the effect of the factors (e, r_t, r_i) on these shape parameters, their diagrams of the main effects were dressed as shown in Fig. 3.

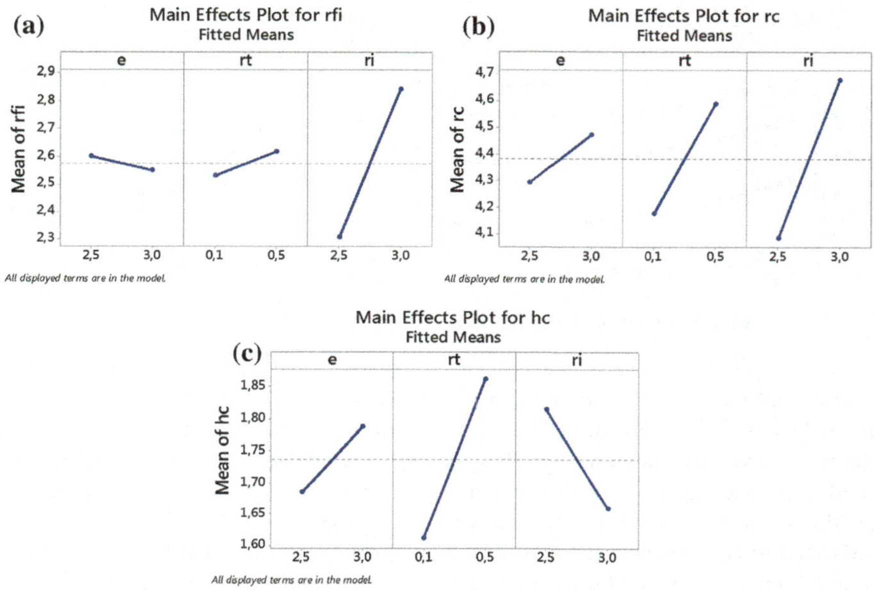

Fig. 3 The main effects of e, r_t and r_i on: **a** r_{if} **b** r_c **c** h_c

The final internal radius r_{fi}, was considered as the most important parameter for the countersunk hole. Regarding to the Fig. 3a, it can be seen that the initial radius of the pre-hole r_i was the main factor that had a significant effect on the retraction r_{fi}. And the other factors e and r_t had no significant effect on r_{fi}. Therefore, attention must be paid for this factor to control the final retraction. It was clear that the reduced value of r_i lead to minimal retraction and the opposite was true. The evolutions of r_{fi} for each curve are approximately the same.

A separate FE analysis was performed using parametric value for r_i. The evolution of the retraction r_{fi} for each value of r_i was dressed on the Fig. 4a, by fixing $e = 3$ mm and $r_t = 0.2$ mm. all the curves had approximately the same evolution. In fact, the retraction was defined as $r_{fi}-r_i$, the Fig. 4b showed precisely its evolution and its effects by r_i. It was found that all the pre-hole radius followed a retraction during the punch travel. After a specific value of the punch travel δ_p, an expansion was observed. For the case of $r_i = 2.5$ mm, the expansion began at $\delta_p = 1.2$ mm where the retraction reached its maximum value -0.25 mm. After this phenomenon, the final value of the retraction was equal to -0.19 mm. Compared to the case $r_i = 3$ mm, it was clear that the expansion of the pre-hole appeared at $\delta_p = 0.8$ mm and the final value of the retraction was greater and equal to -0.23 mm. It was concluded that the reduction of r_i ensures the phenomenon of the retraction. The big value of r_i may lead to the expansion of the pre-hole without any appearance of retraction. Therefore, the retraction and the final radius r_{fi} were affected by this parameter. Thus, attention was paid for it to control the countersinking process and the obtained results.

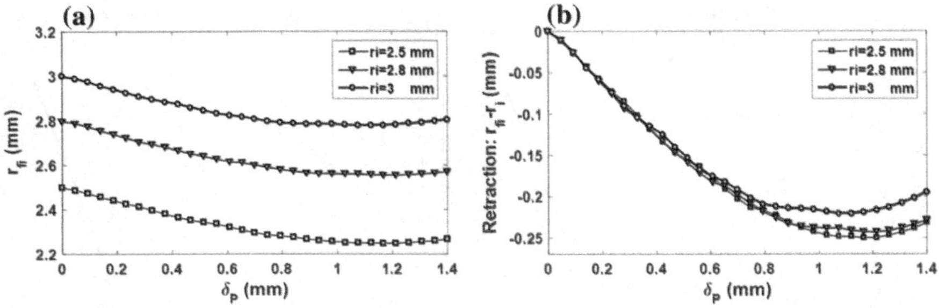

Fig. 4 Effect of the initial radius r_i on: **a** r_{fi} **b** the retraction r_{fi}–r_i

The final shape of the countersunk hole was defined by the h_c and r_c beside the retracted radius r_{fi}. To illustrate the effects of these parameters, their main effect diagram was plotted as shown in Fig. 3b, c. It was found that the thickness *e* did not affect h_c and r_c as much as the two other parameters. For the case of h_c, it was clear that the r_t was the most significant factor when its value decreased the obtained h_c and r_c decreased too. That is why, the choice of the value of r_t was important to succeed in the countersinking process and the obtained results. As explained in section of the analysis of the punch load, this parameter became the most important factor besides the factor $r_{i.}$

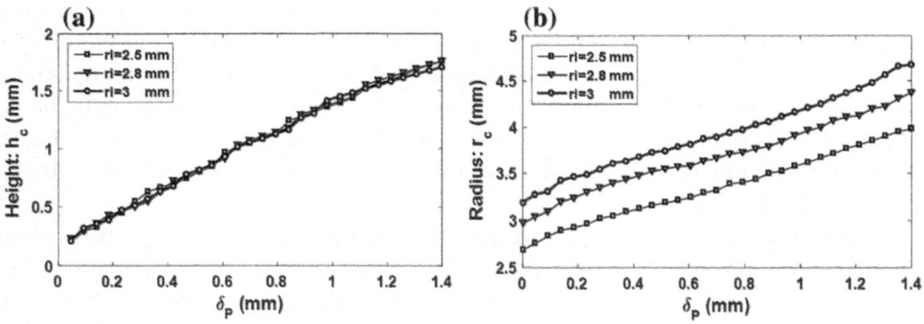

Fig. 5 Effect of the initial radius r_i on: **a** h_c, **b** r_c

Concerning the effect of $r_{i,}$ it was observed that it had a significant effect on r_c more than h_c as shown in Fig. 3b, c. By seeing the Fig. 5a, it was clear that h_c had approximately the same evolution whatever the value of r_i which led to neglecting the effect of this factors. On the contrary, Fig. 5b showed the sensibility of r_c when varying $r_{i.}$ As a consequence, this factor showed the most significant effect on the final shape of the countersunk hole.

9 Experimental Final Shape

To validate the numerical results, the experiments were conducted to obtain the final shape of the countersunk workpiece. The used workpieces were manufactured carefully with $r_i = 2.8$ mm and $r_e = 15$ mm with $e = 3$ mm. The experiment data was compared to the numerical one. The Fig. 6a, b showed the superposition of the numerical and the experimental shapes. A good agreement was obtained which led to adopting the numerical results and so the analysis of the parameter effects. By referring to the work of Hassen et al. [5] it was obvious to conclude the results were suitable whatever the used thickness e.

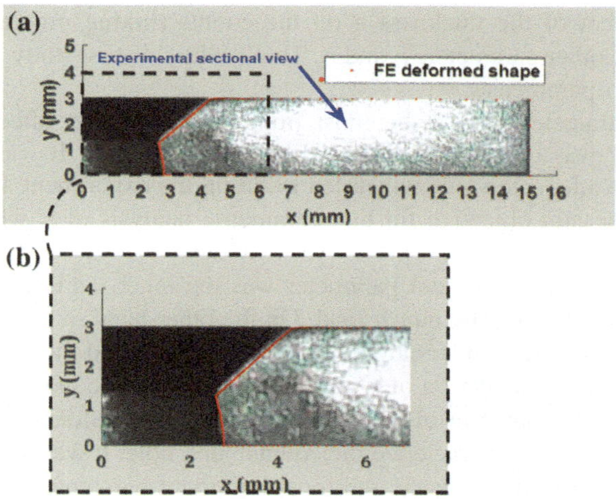

Fig. 6 Superposition of the section view of the workpiece(e = 3 mm, $r_i = 2.8$ mm, $r_t = 0.2$ mm) with the numerical shape

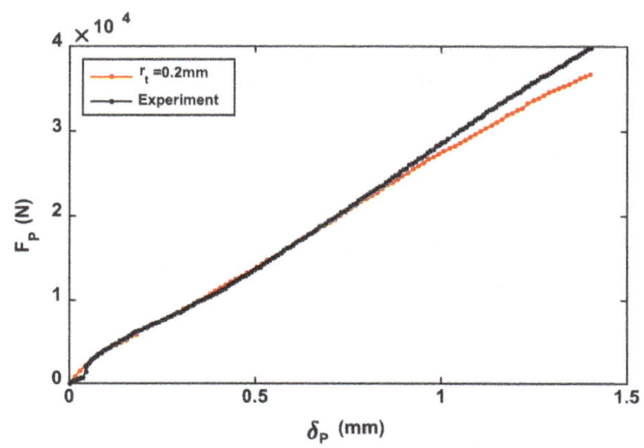

Fig. 7 Punch load Fp evolution versus the punch load travel δ_p

Moreover, the punch load F_p was dressed in the Fig. 7 and it was fitted to the numerical load when using $r_t = 0.2$ mm which confirms the attention paid for this parameter after the inner radius r_i.

10 Conclusion

This study deals with the analysis of the parameter effects on the final shape of the countersunk hole and the developed punch load. The static analysis was a useful tool to investigate the effect of each parameter. The diagram of the main effect was dressed carefully and it gave reliable results.

It was found that the adopted geometric parameters had a different effect on the process. The effect of the thickness e on the countersinking process was neglected comparing to the other parameters r_i and r_t. That's why this first study was concentrated on these last two parameters.

Since the parameter r_t was the most important parameter affecting the load, a parametric study was conducted using a different value of it. It was found that $r_i = 0.2$ mm was the adequate value which led to fitting the experiment and the numerical results. It was the aim objective for this parametric analysis of r_t which explains the attention paid for it.

Moreover, the most important parameter was the inner radius r_i. this one did not have a significant effect on the punch load. On the other hand, it had an obvious effect on the final shape of the countersunk hole. Exactly, it affected the retraction clearly. It was concluded that the reduction of r_i conducted to maximize the retraction and delay of the expansion phenomenon of the countersunk hole. In addition, the parameter rc and hc were important to characterize the countersunk hole. It was remarked they were affected by r_i too. As consequence, the most important parameter affecting the countersunk shape was the inner radius r_i.

The experiment results were compared to the numerical one in order to validate the numerical model and exactly the static analysis. A good agreement was fount for the punch load analysis when using $r_t = 0.2$ mm.

Concerning the future work, it was interesting to investigate all the process parameters to obtain more accuracy for the results and to obtain more explanation about the effects of the parameters and their interactions.

References

1. Alves LM, Nielsen CV, Martins PAF (2011) Revisiting the fundamentals and capabilities of the stack compression test. J Exp Mech 51:1565–1572
2. Chul YP, Alten F (2007) Effect of load transfer on the cracking behavior at a countersunk fastener hole. Int J Fatigue 29:146–157
3. Darwish F, Tadhtoush G (2012) Stress concentration analysis for countersunk rivet holes in orthotropic plates. Eur J Mech A/Solids 37:69–78
4. Jalouli I, Krichen A (2011) Finite element analysis of countersinking process. Int J Adv Manuf Technol 55(5–8):641–648

5. Mosbah H, Jallouli I, Krichen A (2017) Advanced analysis for the countersinking process. Int J Adv Manuf Technol 90(9–12):3473–3481
6. Rana M, Makabe C (2009) The effect of hole shape on the extent of fatigue life improvement by cold expansions. J Eng Failure Anal 16:2081–2090
7. Zhao H, Gary G (1996) The testing and behaviour modelling of sheet metals at strain rate from 10^{-4} to 104 s^{-1}. J Mater Sci Eng 207:46–50

Meshfree Modeling of 3D-Shell Structures Using the Modified First Order Shear Deformation Theory

H. Mellouli[2(✉)], H. Mallek[2], H. Jrad[2], M. Wali[1,2(✉)], and F. Dammak[2]

[1] Department of Mechanical Engineering, College of Engineering, King Khalid University, Abha, Saudi Arabia
mondherwali@yahoo.fr
[2] Laboratory of Electromechanical Systems (LASEM), National Engineering School of Sfax, University of Sfax, Route de Soukra km 4, 3038 Sfax, Tunisia
{hana.mellouli,hanen.mallek, Fakhreddine.dammak}@enis.tn, hanen.j@gmail.com

Abstract. This work develops a meshfree method for the analysis of 3D-shell structure based on the modified first order shear deformation theory. The present meshfree method is based on the radial point interpolation method (RPIM) for the construction of the shape functions with Delta function property using arbitrarily distributed nodes in the support domains. The first order shear deformation theory is improved in this work in order to correct the constant shear strains with the Mindlin-Reissner theory and to get closer to its realistic distribution through the thickness with parabolic curves. The accuracy and convergence of the proposed model is compared to results presented in the literature.

Keywords: Meshfree method · 3D-shell structures · The modified FSDT · RPIM

1 Introduction

In attempts to overcome several existing shortcomings of the FEM, meshfree methods which are based on a set of discrete nodes instead of meshes, have been applied in shell analysis due to the ease node placement and accuracy of computed results. There are many meshfree methods such as the smooth particle hydrodynamics (SPH) [17], The reproducing kernel particle (RKPM) [9], and the element free galerkin based on the global weak form Krysl and Belytschko [7].

The Radial Point Interpolation Method (RPIM) [21] is an interpolator meshfree method that employs both polynomial and radial basis functions (RBFs) to construct the shape functions. The RPIM shape functions possess the Kronecker delta function property due to the use of the radial basis function.

Shell structures are widely used in various mechanical structures, civil engineering, aerospace and naval. Its analysis, therefore, is of great importance for practical

N. Aifaoui et al. (Eds.): CMSM 2019, LNME, pp. 762–769, 2020.
https://doi.org/10.1007/978-3-030-27146-6_83

engineering problems. According to the open literature, it can be carried out with three different theories: the Kirchhoff-love theory [4, 8], the first-order shear deformation theory (FSDT) [1, 5, 6] and the high order-shear deformation theory (HSDT) [2]. The inefficiency of the Kirchhoff-Love hypothesis appears with neglecting the effects of transverse shear and normal strains of the structure. In FSDT, transverse shear is assumed to be constant through the shell thickness and thus requires the computation of shear correction coefficients. The high-order shear deformation theory includes the consideration of realistic parabolic variation of transverse shear stress through the shell thickness to avoid the use of transverse shear correction coefficients. The modified first order shear deformation theory [12, 15, 18, 19] appears in order to get parabolic distribution of the transverse shear stress with the ad of a value of 5/4 for the shear function to correct the strain energy terms and to get rid of the shear correction factors calculation. Its origin is taken from the double directos shell theory [3, 13, 14, 16, 20, 22].

The objective of this work is to present a meshfree method for the analysis of 3D shell structure using the modified first order shear deformation theory. With this approach, the quadratic distribution of the shear strain is satisfied and the RPIM is considered as a meshfree approximation method. Numerical example of a cylindrical shell structure with border moment using the meshfree method is examined in this work in order to highlight the accuracy and performance of the present model.

2 The Modified First Order Shear Deformation Theory Kinematic Assumptions

This section develops the basic formulations of the free modified first order shear deformation theory for shell structures. For convenience of presentation, the Cartesian coordinate system ($\mathbf{E}_i$), i = 1, 2, 3, is adopted to describe the shell geometry in the 3D space. To distinguish the initial configuration C_0 from the deformed C_t, capital letters (respectively lowercase letters) are used for quantities relative to the configuration C_0 (respectively C_t). Vectors will be denoted by bold letters.

2.1 Displacement Field and Strains of the Shell Model

Applying the modified first order shear deformation theory, all material points of the shell are defined using parameterizations in terms of curvilinear coordinates $\boldsymbol{\xi} = (\xi^1, \xi^2, \xi^3 = z)$. The pair ($\boldsymbol{X}_p$, $\boldsymbol{d}$) defines the position of an arbitrary point 'q' of the shell, $\boldsymbol{X}_p$ gives the position of a point 'p' on the shell midsurface and $\boldsymbol{d}$ is the director unit vector. With assumption of the first order shear deformation theory, the position vector of the point q in the deformed configuration is given by:

$$\boldsymbol{x}_q(\xi^1, \xi^2, z) = \boldsymbol{x}_p(\xi^1, \xi^2) + z\boldsymbol{d}(\xi^1, \xi^2) \tag{1}$$

The virtual membrane, bending and shear strains in the reference state C_0 are expressed as: $(\alpha, \beta = 1, 2)$:

$$\begin{cases} \delta e_{\alpha\beta} = \frac{1}{2}\left(\boldsymbol{A}_{\alpha} \cdot \delta \boldsymbol{x}_{,\beta} + \boldsymbol{A}_{\beta} \cdot \delta \boldsymbol{x}_{,\alpha}\right) \\ \delta \chi_{\alpha\beta} = \frac{1}{2}\left(\boldsymbol{A}_{\alpha} \cdot \delta \boldsymbol{d}_{,\beta} + \boldsymbol{A}_{\beta} \cdot \delta \boldsymbol{d}_{,\alpha} + \delta \boldsymbol{x}_{,\alpha} \cdot \boldsymbol{d}_{,\beta} + \delta \boldsymbol{x}_{,\beta} \cdot \boldsymbol{d}_{,\alpha}\right), \quad \boldsymbol{d} = \boldsymbol{D} \\ \delta \gamma_{\alpha} = \boldsymbol{A}_{\alpha} \cdot \delta \boldsymbol{d} + \delta \boldsymbol{x}_{,\alpha} \cdot \boldsymbol{d} \end{cases} \tag{2}$$

Equation (2) can be rewritten in matrix notation [11] as:

$$\boldsymbol{e} = \begin{Bmatrix} e_{11} \\ e_{22} \\ 2e_{12} \end{Bmatrix}, \quad \boldsymbol{\chi} = \begin{Bmatrix} \chi_{11} \\ \chi_{22} \\ 2\chi_{12} \end{Bmatrix}, \quad \boldsymbol{\gamma} = \begin{Bmatrix} \gamma_1 \\ \gamma_2 \end{Bmatrix} \tag{3}$$

2.2 The Meshfree Global Weak Form

The weak form of equilibrium equations, used for the numerical solutions with the meshfree method, is given as:

$$\boldsymbol{G} = \int_A (\boldsymbol{N}.\delta\boldsymbol{e} + \boldsymbol{M} \cdot \delta\boldsymbol{\chi} + \boldsymbol{T} \cdot \delta\boldsymbol{\gamma})dA - \boldsymbol{G}_{ext} = 0 \tag{4}$$

where $\boldsymbol{G}_{ext}$ is the external virtual work and $\boldsymbol{N}$, $\boldsymbol{M}$ and $\boldsymbol{T}$ represent respectively the membrane, bending and shear stress resultants, expressed as:

$$\boldsymbol{N} = \int_{-h/2}^{h/2} \begin{bmatrix} \sigma_{11} \\ \sigma_{22} \\ \sigma_{12} \end{bmatrix} dz, \quad \boldsymbol{M} = \int_{-h/2}^{h/2} z \begin{bmatrix} \sigma_{11} \\ \sigma_{22} \\ \sigma_{12} \end{bmatrix} dz, \quad \boldsymbol{T} = \int_{-h/2}^{h/2} f(z) \begin{bmatrix} \sigma_{13} \\ \sigma_{23} \end{bmatrix} dz \tag{5}$$

Considering the assumption of the modified first order shear deformation theory, the shear function $f(z)$ is expressed as

$$f(z) = \frac{5}{4}\left(1 - 4\left(\frac{z}{h}\right)^2\right) \tag{6}$$

The generalized stress resultants $\boldsymbol{R}$ and strain $\boldsymbol{\Sigma}$ are defined as:

$$\boldsymbol{R} = \{\boldsymbol{N} \quad \boldsymbol{M} \quad \boldsymbol{T}\}^T_{8\times1}, \boldsymbol{\Sigma} = \{\boldsymbol{e} \quad \boldsymbol{\chi} \quad \boldsymbol{\gamma}\}^T_{8\times1}, \tag{7}$$

In the case of an elastic isotropic constitutive model, the generalized stress resultant R_{φ} is related to the generalized strain resultant $\boldsymbol{\Sigma}$, using the material tangent modulus $\boldsymbol{H}_T$, as:

$$\boldsymbol{R} = \boldsymbol{H}_T\boldsymbol{\Sigma}, \quad \boldsymbol{H}_T = \begin{bmatrix} H_{11} & H_{12} & H_{13} \\ & H_{22} & H_{23} \\ sym & & H_{33} \end{bmatrix} \tag{8}$$

$$(\boldsymbol{H}_{11}, \boldsymbol{H}_{12}, \boldsymbol{H}_{22}) = \int_{-h/2}^{h/2} (1, z, z^2)\boldsymbol{H}\, dz \tag{9}$$

$$(\boldsymbol{H}_{13}, \boldsymbol{H}_{23}, \boldsymbol{H}_{33}) = \int_{-h/2}^{h/2} (0, 0, (f(z))^2)\boldsymbol{H}_{\tau} dz \tag{10}$$

where $\boldsymbol{H}$ and $\boldsymbol{H}_{\tau}$ represent respectively the in plane and out-of-plane linear elastic sub-matrices.

2.3 The RPIM Using for the Meshfree Approximation of the Modified First Order Shear Deformation Theory

In this section, a brief introduction of the construction of the radial point interpolation method (RPIM) shape functions is given. The RPIM shape functions combine a radial basis function (RBF) $R^{\mathrm{I}}(X)$ with a polynomial basis function $P^{\mathrm{J}}(X)$. Thus, for a point of interest $X = (x, y)$ located in the support domain, the approximation of the displacement vector is defined as:

$$\boldsymbol{U}(x) = \sum_{\mathrm{I}=1}^{N} R^{\mathrm{I}}(X)a_{\mathrm{I}} + \sum_{\mathrm{J}=1}^{M} P^{\mathrm{J}}(X)b_{\mathrm{J}} = \boldsymbol{R}^T(X)\boldsymbol{a} + \boldsymbol{P}^T(X)\boldsymbol{b} \tag{13}$$

where a_{I} and b_{J} represent respectively the non-constants coefficients of $R^{\mathrm{I}}(X)$ and $P^{\mathrm{J}}(X)$. N is the nodal number in the support domain and M denotes the number of monomial terms with M < N. In matrix form, the displacement vector combines the radial moment matrix $\boldsymbol{R}$ and the polynomial moment matrix $\boldsymbol{P}$ with the vector of coefficients for RBFs $\boldsymbol{a}$ and the vector of coefficients for polynomial matrix $\boldsymbol{b}$, as:

$$\boldsymbol{U} = \boldsymbol{R}\boldsymbol{a} + \boldsymbol{P}\boldsymbol{b} \tag{14}$$

The Gaussian radial basis function approximation is defined at a point of interest X and a defining point X_{j}, as [10]:

$$R^{\mathrm{J}}(X) = \exp(-c^2 \|X_{\mathrm{J}} - X\|^2) \tag{15}$$

where $\|X_{\mathrm{J}} - X\|$ is the Euclidean norm and c is the optimal shape parameter of an RBF. The shape parameter c characterizes the average nodal spacing for all nodes in the local support domain in order reach accurate results.

3 Numerical Example and Results

This section develops a numerical example describing a cylindrical shell structure with border moment in order to outline the performance and the efficiency of the proposed model. The structure of the cylinder is clamped in an edge and subjected to a moment load in the opposite edge, as seen in Fig. 1. Material and geometrical properties of the cylinder

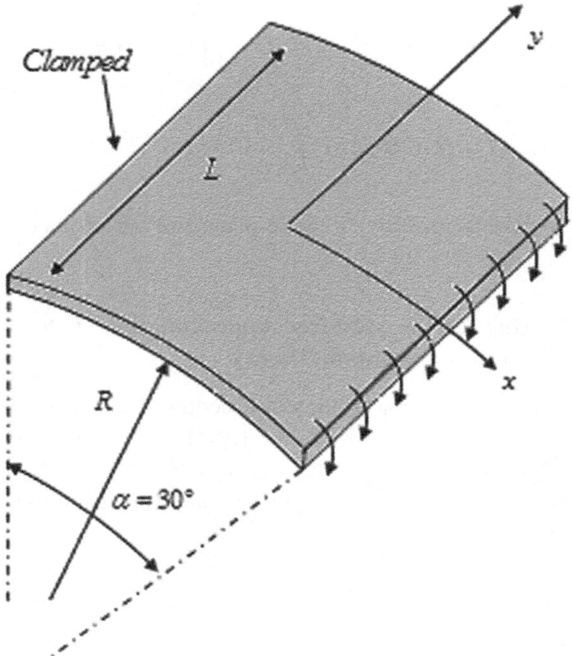

Fig. 1 The geometry of the cylinder with border moment

are given as: elastic modulus $E = 2.1 \times 10^9$ Pa, Poisson's ratio $\nu = 0.0$, radius $R = 2.0$, length $L = 1.0$, and the moment load $m = 240$. For the RPIM, the Gaussian radial basis function approximation is used in this test, where the shape parameter c is equal to 7.

Total nodes of the structure are used to form the global background cells performing the numerical integrations in the meshfree global weak form method, compared to the finite element method. The performance of the present meshfree model is evaluated considering numerical integration with 2 × 2 Gaussian quadrature on the background cells with quadratic polynomial basis.

An analytical solution cited by Costa et al. [1] and equal to $\theta_y = \frac{12mR\alpha}{Eh^3}$, is used to validate and normalize the present results compared with those obtained by Costa et al. [1] using a linear meshfree shell model with the Reissner-Mindlin kinematics based on an inextensible director. In order to highlight the effect of the number of elements per side, comparison between normalized deflections measured in points where the moment loads are applied, is represented in Fig. 2 with variation of the ratio R/h. Figure 3 presents the variation of the shear stress through the structure thickness using 9 nodes per side and a ratio $R/h = 0.1$ since the analytical solution is reached with the present model using these conditions.

As seen in Fig. 2, the present results closely match with the reference results for the case of ratio $R/h = 0.1$ and are good for $R/h = 0.01$. Figure 3 shows that the variation of the shear stress through the structure thickness proves its parabolic distribution

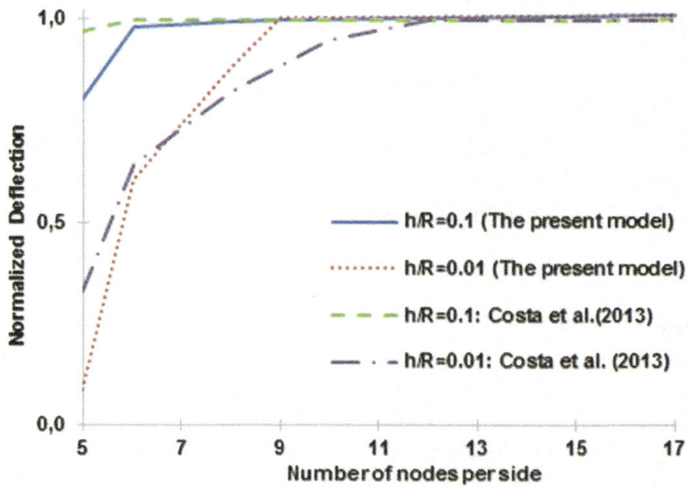

Fig. 2 Results of deflections versus the number of nodes per side for two cases of the ratio R/h

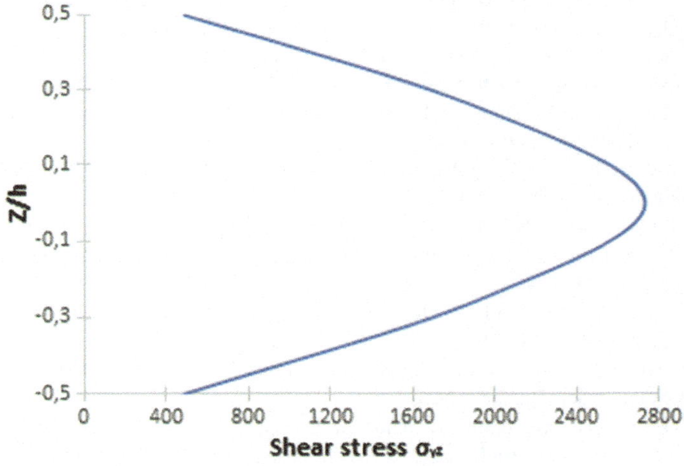

Fig. 3 The variation of the shear stress σyZ through the cylinder thickness

which specifies the use of the modified FSDT. The condition of zero shear stress on top and bottom of structure is released in this work (mainly seen in Fig. 3) in order to remove the C^1 continuity, contrarily of the finite element method.

4 Conclusion

In this study, a linear meshfree method based on the modified first order shear deformation theory is investigated. The radial point interpolation method is used to construct the meshfree shape functions for distributed nodes in the support domain. The numerical example exhibit good agreement to the reference solutions, thus the present meshfree model provides accurate and competitive solution to the linear analysis of shell structures.

References

1. Costa JC, Tiago C, Pimenta PM (2013) Meshless analysis of shear deformable shells: the linear model. Comput Mech 52(4):763–778
2. Ferreira AJM, Batra RC, Roque CMC, Qian LF, Martins PALS (2005) Static analysis of functionally graded plates using third-order shear deformation theory and a meshless method. Compos Struct 69(4):449–457
3. Frikha A, Wali M, Hajlaoui A, Dammak F (2016) Dynamic response of functionally graded material shells with a discrete double directors shell element. Compos Struct 154:385–395
4. Jrad H, Mallek H, Wali M, Dammak F (2018) Finite element formulation for active functionally graded thin-walled structures. CR Mec 346(12):1159–1178
5. Jrad H, Mars J, Wali M, Dammak F (2018) An extended finite element method for modeling elastoplastic FGM plate-shell type structures. Struct Eng Mech 68:299–312
6. Jrad H, Mars J, Wali M, Dammak F (2019) Geometrically nonlinear analysis of elastoplastic behavior of functionally graded shells. Eng Comput 35:833–847
7. Krysl P, Belytschko T (1996) Analysis of thin shells by the element-free Galerkin method. Int J Solids Struct 33(20–22):3057–3080
8. Li W, Nguyen-Thanh N, Zhou K (2018) Geometrically nonlinear analysis of thin-shell structures based on an isogeometric-meshfree coupling approach. Comput Methods Appl Mech Eng 336:111–134
9. Liu WK, Jun S, Zhang YF (1995) Reproducing kernel particle methods. Int J Numer Methods Eng 20(6):1081–1106
10. Liu GR, Gu YT (2005) An introduction to meshfree methods and their programming. Springer Science and Business Media
11. Mallek H, Jrad H, Algahtani A, Wali M, Dammak F (2019) Geometrically non-linear analysis of FG-CNTRC shell structures with surface-bonded piezoelectric layers. Comput Methods Appl Mech Eng 347:679–699
12. Mallek H, Jrad H, Wali M, Dammak F (2019) Geometrically nonlinear finite element simulation of smart laminated shells using a modified first-order shear deformation theory. J Intell Mater Syst Struct 30(4):517–535
13. Mallek H, Jrad H, Wali M, Dammak F (2019) Piezoelastic response of smart functionally graded structure with integrated piezoelectric layers using discrete double directors shell element. Compos Struct 210:354–366
14. Mellouli H, Jrad H, Wali M, Dammak F (2019) Meshfree implementation of the double director shell model for FGM shell structures analysis. Eng Anal Bound Elements 99: 111–121

15. Mellouli H, Jrad H, Wali M, Dammak F (2019) Meshless implementation of arbitrary 3D-shell structures based on a modified first order shear deformation theory. Comput Math Appl 77:34–49
16. Mellouli H, Jrad H, Wali M, Dammak F (2019) Geometrically nonlinear meshfree analysis of 3D-shell structures based on the double directors shell theory with finite rotations. Steel Compo Struct 31(4):397–408
17. Monaghan JJ (1988) An introduction to SPH. Comput Phys Commun 48(1):89–96
18. Trabelsi S, Frikha A, Zghal S, Dammak F (2018) Thermal post-buckling analysis of functionally graded material structures using a modified FSDT. Int J Mech Sci 144:74–89
19. Trabelsi S, Frikha A, Zghal S, Dammak F (2019) A modified FSDT-based four nodes finite shell element for thermal buckling analysis of functionally graded plates and cylindrical shells. Eng Struct 178:444–459
20. Wali M, Hajlaoui A, Dammak F (2014) Discrete double directors shell element for the functionally graded material shell structures analysis. Comput Methods Appl Mech Eng 278:388–403
21. Wang JG, Liu GR (2002) A point interpolation meshless method based on radial basis functions. Int J Numer Methods Eng 54(11):1623–1648
22. Zghal S, Frikha A, Dammak F (2017) Static analysis of functionally graded carbon nanotube-reinforced plate and shell structures. Compos Struct 176:1107–1123

An ABAQUS Implementation of a Solid-Shell Element: Application to Low Velocity Impact

A. Chaker[1,2(✉)], S. Koubaa[1], J. Mars[1], A. Vivet[2], and F. Dammak[1]

[1] Electro-Mechanical System's Laboratory (LASEM), National Engineering School of Sfax, University of Sfax, B.P 1173-3038 Sfax, Tunisia
aidacheker811@gmail.com, koubaasana@yahoo.fr, jamelmars@yahoo.fr, fakhreddine.dammak@enis.tn

[2] Normandie University, ENSICAEN, UNICAEN, CEA, CNRS, CIMAP, 14000 Caen, France
alexandre.vivet@unicaen.fr

Abstract. The main objective of this work is to develop an hexahedral solid shell finite element in order to resolve the numerical locking effects that may have occurred when employing the conventional solid and shell finite elements. The developed formulation relay on the coupling between the Assumed Natural Strain (ANS) and Enhanced Assumed Strain (EAS) methods. The FE model was implemented into the user element (UEL) interface of the FE commercial code ABAQUS by using the UEL FORTRAN subroutine. The robustness and performance of this element are proven using dynamic contact metal sheet behavior at low velocity impact. The obtained results are validated with finding from the literature. The developed solid shell element is efficient from a computational view point as the thickness direction is discretized adopting only single element layer.

Keywords: Finite element · Solid shell element · Low velocity impact · Locking problems

1 Introduction

The main challenge in mechanical engineering is to develop a robust locking-free finite element. Nonetheless, conventional solid and shell finite elements present a source of various locking problems as transverse shear, membrane and volumetric locking problems. To overcome these numerical phenomena, many researchers proposed to develop a specific finite element named solid shell element. Klinkel et al. [10] and (Vu-Quoc and Tan [19]) developed a solid shell element by combining assumed natural strain (ANS) and enhanced assumed strain (EAS) methods. Hajlaoui et al. [9] developed a hexahedral solid shell element with optimal EAS parameters in order to predict the effect of buckling solicitations to functionally graded material (FGM) structures.

Referring to the literature, the solid shell element is widely implemented into home codes [6–8, 16]. But, the simulation of this element via the commercial code ABAQUS is scarcely conducted. Chalal and Abed-Meraim [3] implemented in the user element (UEL) interface of ABAQUS the SBH5, SBH15 and SBH20 hexahedral solid-shell

N. Aifaoui et al. (Eds.): CMSM 2019, LNME, pp. 770–777, 2020.
https://doi.org/10.1007/978-3-030-27146-6_84

elements to simulate FGM benchmark tests. Reinoso and Blázquez [17] implemented an hexahedral solid shell element in the UEL interface of ABAQUS to simulate FG carbone nanotube benchmark tests. Nevertheless, in these references, the solid shell element was applied for only static problems. Besides, to the knowledge of authors, the simulation via ABAQUS of low velocity impact contact problems adopting solid shell element is not yet conducted.

The low velocity impact characterizes the resistance of structures under a brief contact. This phenomenon was widely investigated numerically [1, 12, 13, 20]. In particular, in case of aluminum circular plates, various numerical studies were developed: Chen applied the finite difference method to predict the dynamic response of elastoplastic circular plate subjected to low velocity impact (less than 50 m/s) and to correlate results with experimental study [4]. Mars predicted the effect of low velocity impact to an anisotropic aluminum circular plates with an elastoplastic behavior [15]. The finite element model was discretized by 10 conventional solid hexahedral elements C3D8R with reduced integration through the thickness direction.

In this paper, a robust free-locking solid shell element SS is derived with mixed formation based on the combination between ANS and EAS methods. This developed element is implemented into the UEL interface of the commercial code ABAQUS using FORTRAN user subroutine. The application field of this implemented element is expanded in order to include elastoplastic, dynamic and contact low velocity impact problems on aluminum structures. Our contribution consists on the discretization of the circular plate with only single solid shell element through the thickness direction.

2 Finite Element Formulation

The Green-Lagrange strain tensor $\mathbf{E}$ is decomposed on two strain tensors as:

$$\mathbf{E} = \mathbf{E}^c + \widetilde{\mathbf{E}}, \tag{1}$$

In which $\mathbf{E}^c$ and $\widetilde{\mathbf{E}}$ are respectively the compatible and incompatible strain tensors.

The three field Hu-Washizu (FHW) variational principle is expressed as:

$$\Pi(\mathbf{u}, \widetilde{\mathbf{E}}, \widetilde{\mathbf{S}}) = \int_V \left(\psi(\mathbf{E}^c + \widetilde{\mathbf{E}})\right) dV - \int_V (\widetilde{\mathbf{S}} : \widetilde{\mathbf{E}}) dV - \int_V \mathbf{F}_V \cdot \mathbf{u} dV - \int_{\partial V_f} \mathbf{F}_S \cdot \mathbf{u} dA \tag{2}$$

$\mathbf{u}, \widetilde{\mathbf{S}}$ and ψ are respectively the displacement field, the independent part of the second stress tensor of Piola-Kirchhoff and the Helmhotz strain energy function. While $\mathbf{F}_V$ and $\mathbf{F}_s$ are respectively the body and the surface forces.

The weak form obtained by the minimization of Π is written on the following expression:

$$G(\mathbf{u}, \widetilde{\mathbf{E}}) = \int_V \mathbf{S} : (\delta \mathbf{E}^c + \delta \widetilde{\mathbf{E}}) dV - \int_V \mathbf{F}_V \cdot \delta \mathbf{u} \, dV - \int_{\partial V_f} \mathbf{F}_S \cdot \delta \mathbf{u} \, dA = 0 \tag{3}$$

2.1 Assumed Natural Strain Method

In order to overcome transverse shear locking, [5] developed the ANS method. This method consists to modify the E^c_{13}, E^c_{23} and E^c_{33} as:

$$\begin{bmatrix} 2E^c_{13} \\ 2E^c_{23} \end{bmatrix} = \begin{bmatrix} (1-\eta)E^B_{13} + (1+\eta)E^D_{13} \\ (1-\xi)E^A_{23} + (1+\xi)E^C_{23} \end{bmatrix} \tag{4}$$

$$E^c_{33} = \sum_{P=1}^{4} \frac{1}{4}(1+\xi_P\xi)(1+\eta_P\eta)E^P_{33} \tag{5}$$

where A, B, C, D and P = A_i (i = 1, 2, 3, 4) are the interpolation points as presented in Fig. 1.

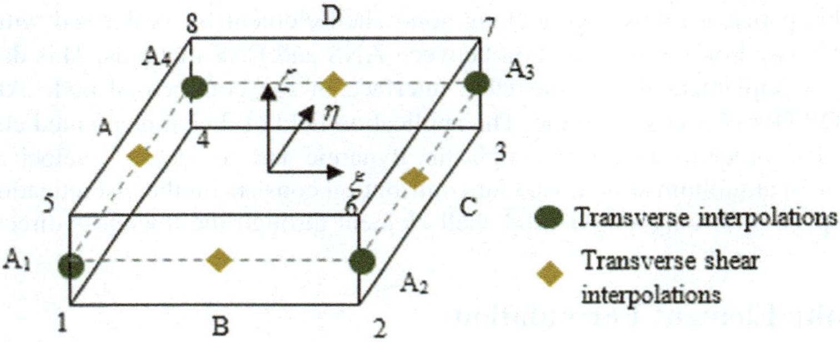

Fig. 1 Transverse shear and transverse interpolation

2.2 Enhanced Assumed Strain Method

To overcome thickness and volumetric locking, (Simo and Refai [18]) derived the EAS method. This method consists to define an enhancing strain tensor as:

$$\tilde{\mathbf{E}} = \frac{\det \mathbf{J}_0}{\det \mathbf{J}} (\mathbf{T}^0_S)^{-T} \tilde{\tilde{\mathbf{E}}} \tag{6}$$

J is the Jacobian matrix, $\mathbf{T}^0_S$, $\mathbf{J}_0$ are the transformation from parametric coordinates to cartesian coordinates and the Jacobian matrix at the center of element. $\tilde{\tilde{\mathbf{E}}}$ is the strain incompatible tensor in the covariant base.

2.3 Finite Element Resolution

Using the finite element method, the displacement field is approximated as:

$$\mathbf{u} = \sum_{I=1}^{n} \mathrm{N}_I \cdot \mathbf{U}_I; \quad \mathrm{N}_I = \frac{1}{8}(1 + \xi_I \cdot \xi) \cdot (1 + \eta_I \eta) \cdot (1 + \eta_I \eta) \tag{7}$$

$\mathbf{U}_I$ is the field displacement at the node I of the element.

The weak form (Eq. 3) presents nonlinear terms. To solve this non-linearity, Newton-Raphson method is adopted. Applying the static condensation, the elementary residual vector and stiffness matrix are written at the following form:

$$\mathbf{R}_e = \mathbf{L}^T \mathbf{H}^{-1} \mathbf{h} + \mathbf{f}^{ext} - \mathbf{f}^{int}; \quad \mathbf{K}_e = \mathbf{K} - \mathbf{L}^T \mathbf{H}^{-1} \mathbf{L} \tag{8}$$

3 Elastoplastic Constitutive Equations

In this work, we assume that the problem of low velocity impact on aluminum circular plates is at the level of small deformations.

The strain tensor $\boldsymbol{\varepsilon}$ is expressed as function of elastic and plastic strain tensor:

$$\boldsymbol{\varepsilon} = \boldsymbol{\varepsilon}^e + \boldsymbol{\varepsilon}^p \tag{9}$$

The stress tensor also is decomposed additively on spherical tensor $\sigma_m \boldsymbol{I}$ and deviatoric tensor $\boldsymbol{S}$ as:

$$\boldsymbol{\sigma} = \sigma_m \boldsymbol{I} + \boldsymbol{S}; \quad \sigma_m = \frac{1}{3} trace(\boldsymbol{\sigma}) \tag{10}$$

Adopting the Von-Mises criterion J_2 with isotropic hardening, the plastic surface f is expressed as:

$$f = \sqrt{3/2} \|\boldsymbol{S}\| - \sigma_y \leq 0 \tag{11}$$

where σ_y is the yield stress.

In this study, Euler implicit numerical scheme is considered. The plastic strain field at the (n + 1) configuration is defined as [2, 11, 14]:

$$\boldsymbol{\varepsilon}^p_{n+1} = \boldsymbol{\varepsilon}^p_n + \sqrt{3/2}\, \Delta\gamma \frac{\boldsymbol{S}_{n+1}}{\|\boldsymbol{S}_{n+1}\|} \tag{12}$$

While γ is the plastic multiplier.

The equation to solve is defined as:

$$f = \sqrt{\frac{3}{2}} \|\boldsymbol{S}^{trial}\| - 3\Delta\gamma\mu - \sigma_y = 0 \tag{13}$$

Finally, considering that the tangent modulus is consistent and applying Newton-Raphson method, the tangent modulus takes the following form:

$$\partial_{\varepsilon}\sigma_{n+1} = K\mathbf{1}\otimes\mathbf{1} + 2\mu(1-\sqrt{3/2}\Delta\gamma\frac{2\mu}{\|S^{trial}\|})(\boldsymbol{I} - 1/3\boldsymbol{1}\otimes\boldsymbol{1}) \\ + 2\mu\sqrt{3/2}\left(\Delta\gamma\frac{2\mu}{\|S^{trial}\|} - A\right)\boldsymbol{n}^{trial}\otimes\boldsymbol{n}^{trial} \tag{14}$$

4 Numerical Results

The aim of this section is to use the solid shell element to predict the impact response on aluminum circular plate. The simulation was conducted via the commercial code ABAQUS/Standard.

The impact model is made of a clamped circular plate with radius $R = 60$ mm and thickness $h = 1$ mm. The plate is impacted by a cylindrical impactor with hemi-spherical nose shape.

The projectile is assumed rigid with radius $R_p = 6.35$ mm and mass $m_p = 54.4$ g with an initial velocity V0 = 29.9 m/s. However, the aluminum circular plate is considered with an elastic perfectly plastic behavior. The material properties are presented in Table 1. This plate is discretized with 4888 SS5 solid shell elements with five independent parameters. The meshing is refined through the contact zone and only one element along the thickness direction is adopted.

Table 1 Circular plate material properties

	Young modulus E (GPa)	Poisson's ratio	Density ρ (kg/m^3)	Yield stress σ_y (MPa)
Aluminum plate	69	0.3	2600	290

The contact between the plate and the projectile is defined as 'node to surface' and 'hard' contact. Owning to the impact model symmetry, only one quarter of this model is studied (see Fig. 2).

The temporal responses of impact force history and projectile velocity obtained by the developed model are presented in Figs. 3 and 4. Present results are compared to the experimenal works of [4] and numerical data of [15] using C3D8 elements in ABAQUS.

We observe that numerical results of the present model are well correlated with those predicted experimentally by [4] and numerically with [15]. Other validation tests are conducted and depicted in Table 2 by varying the impactor mass and velocity. Three cases are cases are considered as:

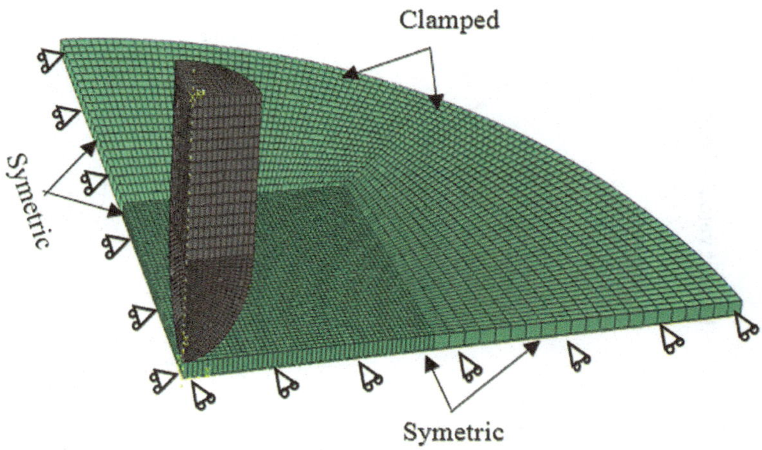

Fig. 2 Completed impact model

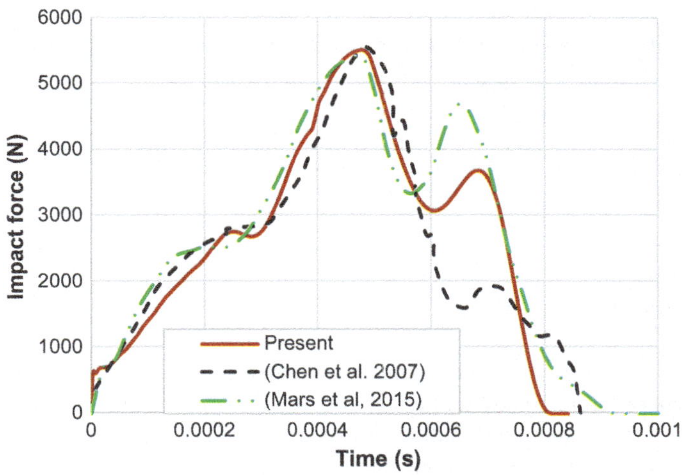

Fig. 3 Dynamic response of impact force history

- **a**/m_p = 23.5 g, R_p = 6.35 mm, V_0 = 49.1 m/s
- **b**/m_p = 35.5 g, R_p = 6.35 mm, V_0 = 41.4 m/s
- **c**/m_p = 54.5 g, R_p = 6.35 mm, V_0 = 29.9 m/s

Referring to the experimental results conducted by [4], can notice that the present numerical model is well validated.

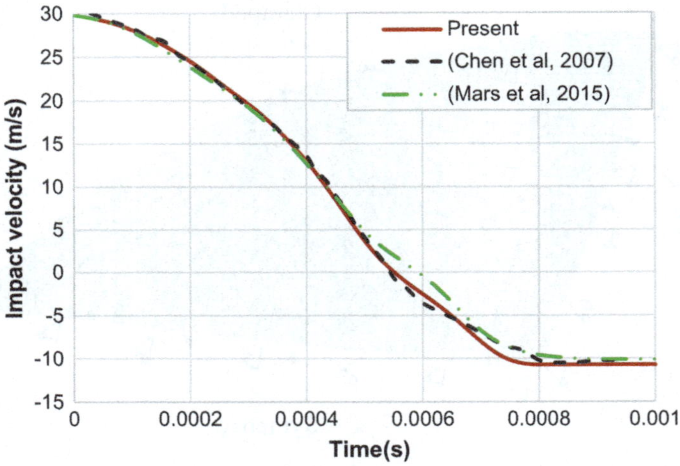

Fig. 4 Dynamic response of impact velocity

Table 2 Comparison between numerical and experimental results

	Peak force (KN)			Separation velocity (m/s)			Impact time (ms)		
	a/	b/	c/	a/	b/	c/	a/	b/	c/
Chen et al. [4]	5.45	5.92	5.5	−16.5	−11.3	−10.2	0.58	0.6	0.86
Present	5.81	6.2	5.52	−11.7	−9.8	−10.6	0.48	0.73	0.81

5 Conclusion

In this paper, a solid shell finite element is implemented into the user element (UEL) of ABAQUS in order to avoid numerical locking problems. The formulation of this element is based on the combination between ANS and EAS methods. The present contribution consists to apply a single element along the thickness direction to simulate the impact problems on aluminum clamped circular plate. The obtained numerical results are correlated with those funding from the literature.

References

1. Abida M, Mars J, Gehring F, Vivet A, Dammak F (2018) Anisotropic visco-elastoplastic modeling of quasi-unidirectional flax fiber reinforced epoxy behavior: an investigation on low-velocity impact response. J Renew Mater 6(5):464–476
2. Bouhamed A, Jrad H, Said LB, Wali M, Dammak F (2019) A non-associated anisotropic plasticity model with mixed isotropic–kinematic hardening for finite element simulation of incremental sheet metal forming process. Int J Adv Manuf Technol 100(1–4):929–940

3. Chalal H, Abed-Meraim F (2018) Quadratic solid-shell finite elements for geometrically nonlinear analysis of functionally graded material plates. Materials 11(6):1046
4. Chen LB, Xi F, Yang JL (2007) Elastic–plastic contact force history and response characteristics of circular plate subjected to impact by a projectile. Acta Mech Sin 23:415–425
5. Dvorkin EN, Bathe KJ (1984) A continuum mechanics based four-node shell element for general non-linear analysis. Eng Comput 1(1):77–88
6. Flores FG (2016) A simple reduced integration hexahedral solid-shell element for large strains. Comput Methods Appl Mech Eng 303:260–287
7. Hajlaoui A, Jarraya A, Kallel-Kamoun I, Dammak F (2012) Buckling analysis of a laminated composite plate with delaminations using the enhanced assumed strain solid shell element. J Mech Sci Technol 26:3213–3221
8. Hajlaoui A, Triki E, Frikha A, Wali M, Dammak F (2017) Nonlinear dynamics analysis of FGM shell structures with a higher order shear strain enhanced solid-shell element. Latin Am J Solids Struct 14:72–91
9. Hajlaoui A, Jarraya A El, Bikri K, Dammak F (2015) Buckling analysis of functionally graded materials structures with enhanced solid-shell elements and transverse shear correction. Compos Struct 132:87–97
10. Klinkel S, Gruttmann F, Wagner W (2006) A robust non-linear solid shell element based on a mixed variational formulation. Comput Methods Appl Mech Eng 195(1–3):179–201
11. Koubaa S, Mars J, Dammak F (2019) Efficient numerical modelling of functionally graded shell mechanical behavior. J Appl Comput Sci 15(1):84–94
12. Koubaa S, Mars J, Wali M, Dammak F (2017) Numerical study of anisotropic behavior of Aluminum alloy subjected to dynamic perforation. Int J Impact Eng 101:105–114
13. Mars J, Chebbi E, Wali M, Dammak F (2018) Numerical and experimental investigations of low velocity impact on glass fiber-reinforced polyamide. Compos B Eng 146:116–123
14. Mars J, Koubaa S, Wali M, Dammak F (2017) Numerical analysis of geometrically non-linear behavior of functionally graded shells. Lat Am J Solids Struct 14(11):1952–1978
15. Mars J, Wali M, Jarraya A, Dammak F, Dhiab A (2015) Finite element implementation of an orthotropic plasticity model for sheet metal in low velocity impact simulations. Thin-Walled Struct 89:93–100
16. Rah K, Van Paepegem W, Habraken AM, Degrieck J (2012) A mixed solid-shell element for the analysis of laminated composites. Int J Numer Meth Eng 89(7):805–828
17. Reinoso J, Blázquez A (2016) Geometrically nonlinear analysis of functionally graded power-based and carbon nanotubes reinforced composites using a fully integrated solid shell element. Compos Struct 152:277–294
18. Simo JC, Rifai MS (1990) A class of mixed assumed strain methods and the method of incompatible modes. Int J Numer Meth Eng 29(8):1595–1638
19. Tan XG, Vu-Quoc L (2005) Efficient and accurate multilayer solid-shell element: non-linear materials at finite strain. Int J Numer Meth Eng 63(15):2124–2170
20. Wali M, Abdennadher M, Fakhfakh T, Haddar M (2011) Dynamic analysis of an elasto-plastic sandwich subjected to low velocity impact. Multidiscipline Model Mater Struct 7:184–206

Forced Vibration Analysis of Functionally Graded Carbon Nanotubes-Reinforced Composite Plates with Finite Element Strategy

S. Zghal(✉), S. Trabelsi, A. Frikha, and F. Dammak

Laboratory of Electromechanical Systems (LASEM), National Engineering School of Sfax, University of Sfax, B.P W3038 Sfax, Tunisia
{souhirzghal, sorourtrabelsi, frikhaahmed}@yahoo.fr, fakhreddine.dammak@enis.tn

Abstract. This paper deals with forced vibration analysis of functionally graded carbon nanotubes reinforced composite (FG-CNTRC) plates. The equations of motion are derived using a finite element strategy which is based on a high order distribution of the displacement field. The material properties of FG-CNTRCs are assumed to be aligned in the axial direction and functionally graded in the thickness direction where uniform (UD) and three graded distributions called FG-V, FG-O and FG-X are taken into account in the analysis. The proposed method incorporates the effect of transverse shear deformations and verifies that the shear stresses vanish on the top and bottom surfaces of the structure. To illustrate the efficiency and the reliability of the present model, temporal deflections curves are provided leading hence to show the ability of the present model in the prediction of vibrational behavior of FG-CNTRC plates with good accuracy.

Keywords: Carbon nanotubes · Functionally graded · Vibration · Finite element strategy

1 Introduction

Recently, the discovery of carbon nanotubes (CNTs) has been regarded as a new opportunity for both engineers and material scientists to investigate high performances qualities. In fact, these materials are recognized by their superior thermal, electrical and mechanical characteristics as mentioned by many researchers [2, 14]. Thus make them promising candidates for the composite reinforcements. Furthermore, functionally graded materials (FGMs) are the advanced inhomogeneous composites which are characterized by the gradually variation of their constituents in the thickness direction. Applying the ideas of FGMs to the CNT-reinforced composites leads to a new class of materials known as functionally graded carbon nanotubes-reinforced composites (FG-CNTRC). Regarding to the great potential of these advanced materials for both industrial and academic fields, the study of the vibrational behavior of these structures has attracted the attention of many scientists in recent years. In the following, the major works related to this subjected are briefly described.

N. Aifaoui et al. (Eds.): CMSM 2019, LNME, pp. 778–785, 2020.
https://doi.org/10.1007/978-3-030-27146-6_85

With the element-free kp-Ritz method which is based on the FSDT theory, Zhang et al. [7, 21] presented elastodynamic analysis of FG-CNTRC plates. Based on the FSDT theory, Malekzadeh et al. [8], Malekzadeh and Heydarpour [9] studied the vibrational behavior of FG-CNTRC plates and cylindrical panels subjected to moving load. Moreover, Kiani presented various analyses on dynamic behavior of FG-CNTRC cylindrical and conical panels [5, 6] within the FSDT framework. Duc and Pham [1] presented the linear dynamic responses of FG-CNTRC truncated conical shells resting on elastic foundations using the classical shell theory. Nonlinear forced vibration analysis of FG-CNTRC cylindrical shells are also examined by Hasrati et al. [4]. In this study, the authors employ the FSDT theory in combination with the Von-Karman hypothesis. Recently, [3, 15–20] carried out several studies on linear static, free vibration, buckling, non-linear analysis of FGM and FG-CNTRC shells using Love-Kirchhoff, Mindlin and high order theories. Active and piezoelastic responses of FGM structures can be found in these publications [10, 11].

Accordingly to this overview about vibrational behavior of FG-CNTRC structures, the major papers employ the FSDT theory with Von-Karman hypothesis which remains restricted to membrane forces and small deformations. Besides, the number of papers related to this subjected is also limited. Thus motivated us to investigate the vibrational behavior of FG-CNTRC plates using a finite element strategy based on double directors finite shell element. This model induces a high order distribution of the displacement field and imposes a zero condition of the shear stresses on the top and bottom surfaces. In this paper, the temporal responses of FG-CNTRC plates subjected to sudden transverse load are performed. Uniformly (UD) and three others distributions labeled: FG-V, FG-X and FG-O are considered. The material properties are estimated through the extended rule of mixture with some efficiency parameters to take into account the size dependency scale of CNTs. The effect of CNTs profiles as well as their volume fractions is also examined.

2 Finite Element Strategy

The used finite element procedure is presented here which is based on a double directors finite shell element model. This model induces a high order distribution of the displacement field and takes into account the effect of shear deformations. It verifies also the zero condition of the shear stress at the top and bottom surfaces. In the following, the geometry and kinematics of the shell structure are briefly described. The initial and deformed configurations are denoted by C_0 and C_t respectively.

2.1 Parameterization of the Geometry and Deformations

The parameterization of the shell geometry is achieved in terms of convective coordinates $(\xi^1, \xi^2, \xi^3 = z)$ where the expression of the position vector of any material point (q) is given by:

$$\mathbf{X}_q(\xi^1,\xi^2,z) = \mathbf{X}_p(\xi^1,\xi^2) + z\mathbf{D}(\xi^1,\xi^2) \quad (1)$$

$$\mathbf{x}_q(\xi^1,\xi^2,z) = \mathbf{x}_p(\xi^1,\xi^2) + f_1(z)\mathbf{d}_1(\xi^1,\xi^2) + f_2(z)\mathbf{d}_2(\xi^1,\xi^2) \quad (2)$$

p represents the normal projection point on mid-surface of the material point q and $z \in [-h/2, h/2]$ is the total thickness of the shell. $\mathbf{D}$ is the reference shell director vector and $\mathbf{d}_1$ and $\mathbf{d}_2$ represent the first and the second shell directors vectors, respectively. f_1 and f_2 are the functions which indicate the high distribution of the displacement field and can be expressed as follows:

$$f_1(z) = z - f_2(z); \quad f_2(z) = \frac{4z^3}{3h^2} \quad (3)$$

The constitutive relations of the FG-CNTRC plate can be written as follows:

$$\mathbf{R} = \mathbf{H}_T\boldsymbol{\Sigma}, \quad \mathbf{H}_T = \begin{bmatrix} H_{11} & H_{12} & H_{13} & \mathbf{0} \\ H_{12} & H_{22} & H_{23} & \mathbf{0} \\ H_{13} & H_{23} & H_{33} & \mathbf{0} \\ \mathbf{0} & \mathbf{0} & \mathbf{0} & H_{44} \end{bmatrix} \quad (4)$$

where $\mathbf{R} = \begin{Bmatrix} \mathbf{N} \\ \mathbf{M}_1 \\ \mathbf{M}_2 \\ \mathbf{T}_1 \end{Bmatrix}$ and $\Sigma = \begin{Bmatrix} \mathbf{e} \\ \chi^1 \\ \chi^2 \\ \gamma^1 \end{Bmatrix}$ represent the generalized resultant of stress and strain vectors, respectively. $\mathbf{H}_T$ is the linear elastic constitutive matrix for an orthotropic FG- CNTRC plate. Its components can be expressed in function of f_1 and f_2 as:

$$(\mathbf{H}_{11}, \mathbf{H}_{12}, \mathbf{H}_{13}, \mathbf{H}_{22}, \mathbf{H}_{23}, \mathbf{H}_{33}) = \int_{-h/2}^{h/2} (1, f_1, f_2, f_1^2, f_1 f_2, f_2^2)\mathbf{H}dz,$$
$$\mathbf{H}_{44} = \int_{-h/2}^{h/2} (f_1')^2 \mathbf{H}_\tau dz \quad (5)$$

$\mathbf{H}$ and $\mathbf{H}_\tau$ denote the in-plane and out-of-plane linear elastic sub-matrices and are given by:

$$\mathbf{H} = \mathbf{T}_1^T \mathbf{H}_L \mathbf{T}_1; \quad \mathbf{H}_\tau = \mathbf{T}_2^T \mathbf{H}_{\tau L} \mathbf{T}_2 \quad (6)$$

$\mathbf{H}_L$ and $\mathbf{H}_{\tau L}$ are the constitutive material matrices of the FG-CNTRC structure; $\mathbf{T}_1$ and $\mathbf{T}_2$ denote the transformation matrices from the orthotropic system to Cartesian system for the FG-CNTRC structure. Their expressions are given by:

$$\mathbf{H}_L = \begin{bmatrix} H_{LL} & H_{LT} & 0 \\ H_{LT} & H_{TT} & 0 \\ 0 & 0 & G_{LT} \end{bmatrix}; \quad \mathbf{H}_{\tau L} = \begin{bmatrix} G_{Lz} & 0 \\ 0 & G_{Tz} \end{bmatrix}; \quad T_1 = \begin{bmatrix} c^2 & s^2 & cs \\ s^2 & c^2 & -cs \\ -2cs & 2cs & c^2 - s^2 \end{bmatrix};$$
$$\mathbf{T}_2 = \begin{bmatrix} c & -s \\ -s & c \end{bmatrix}; \quad c = \cos(\theta); \quad s = \sin(\theta) \tag{7}$$

The subscript L and T indicate the longitudinal and transverse directions for an orthotropic FG-CNTRC plate in the principal material coordinates axes (L-T-N). The components H_{LL}, H_{LT}, H_{TT} and G_{LT} are functions of engineering constants [12] which their forms stick with the material properties of the FG-CNTRC structure.

2.2 Material Properties of FG-CNTRC Plates

We consider here a FG-CNTRC plate of length a, width b and thickness h which is reinforced by single walled carbon nanotubes (SWCNTs) as depicted in Fig. 1.

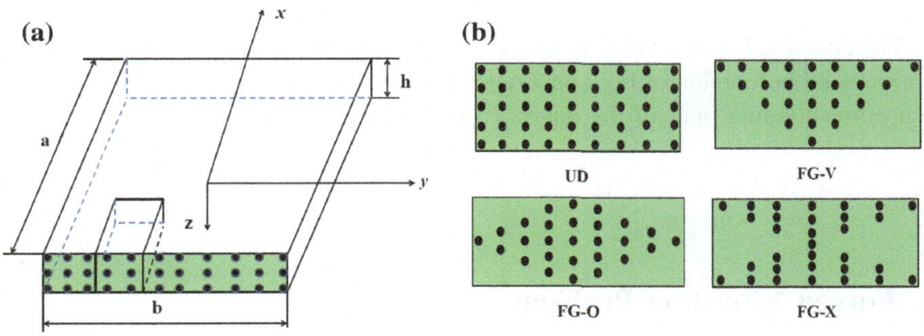

Fig. 1 FG-CNTRC plate. **a** Geometry definition and **b** different CNTs distributions

The volume fraction of CNTs noted, $V_{CNT}(z)$, can be written for each from as:

$$V_{CNT}(z) = \begin{cases} V^*_{CNT}(UD) \\ \left(1 + \frac{2Z}{h}\right) V^*_{CNT}(FG - V) \\ 2\left(1 - \frac{2|Z|}{h}\right) V^*_{CNT}(FG - O) \\ 2\left(\frac{2|Z|}{h}\right) V^*_{CNT}(FG - X) \end{cases} \tag{8}$$

with:

$$V^*_{CNT} = \frac{w_{CNT}}{w_{CNT} + (\rho^{CNT}/\rho^m) - (\rho^{CNT}/\rho^m) w_{CNT}} \tag{9}$$

w_{CNT} represents the mass fraction of CNTs; ρ^m and ρ^{CNT} are the densities of matrix and CNTs, respectively. The effective material properties of the FG-CNTRC plate are estimated using the extended rule of mixture with some efficiency parameters and are given by:

$$E_{11} = \eta_1 V_{CNT} E_{11}^{CNT} + V_m E_m,$$
$$\frac{\eta_2}{E_{22}} = \frac{V_{CNT}}{E_{22}^{CNT}} + \frac{V_m}{E_m}, \tag{10}$$
$$\frac{\eta_3}{G_{12}} = \frac{V_{CNT}}{G_{12}^{CNT}} + \frac{V_m}{G_m}.$$

In Eq. (10), E_{11}^{CNT}, E_{22}^{CNT} and G_{12}^{CNT} are the Young's and shear moduli of CNTs, respectively while E_m and G_m are the corresponding properties of the isotropic matrix. To capture the size dependency of CNTs, efficiency parameters $\eta_j (j = 1, 2, 3)$ are introduced into Eq. (10). The volumes fractions of the CNTs, V_{CNT}, and the matrix, V_m can be given by the following relation:

$$V_m + V_{CNT} = 1. \tag{11}$$

The effective Poisson ratio, as well as the mass density of the FG-CNTRC plate can be expressed in function of Poisson's ratio and mass density of CNTs (υ_{12}^{CNT}, ρ^{CNT}) and their corresponding ones of the matrix (υ_m, ρ^m) phases as follows:

$$\upsilon_{12} = V_{CNT}^{*} \upsilon_{12}^{CNT} + V_m \upsilon_m, \quad \rho = V_{CNT} \rho^{CNT} + V_m \rho^m. \tag{12}$$

3 Forced Vibration Problem

After the derivation of the equations of motion via the variational principle which is based on the weak form of equilibrium, the discretization of the displacement and stain fields is achieved within a four nodes finite shell element. For more details, see [18, 19, 20]. The construction of the stiffness and mass matrices is carried out via the discretization of the internal virtual work and the kinetic energy, respectively. Thereby, the equations of motion related to forced vibration problem of FG-CNTRC plates can be given as:

$$\mathbf{M}\ddot{\mathbf{U}}_{\mathbf{n}} + \mathbf{K}\mathbf{U}_{\mathbf{n}} = \mathbf{F} \tag{13}$$

where **M** and **K** are the mass and stiffness matrices of the FG-CNTRC plate, respectively and **F** is the external load vector. The resolution of Eq. (13) in time domain is achieved using the Newmark's algorithm.

4 Results and Discussion

With the proposed finite element strategy, the vibration analysis of FG-CNTRC plates in terms of temporal deflections is presented. For that, the Poly(m-phenylenevinylene)-co-[(2,5-dioc-toxy-p-phenylene) vinylene] is selected for matrix phase which its material properties at room temperature ($T_0 = 300$ K) are given as indicated by Shen [13]: $E_m = 2.1$ GPa; $\rho^m = 1150$ kg/m^3; $\upsilon_m = 0.34$. Furthermore, the (10, 10) SWCNTs are considered as reinforcements where its material properties are: $E_{11}^{CNT} = 5.6466$ TPa; $E_{22}^{CNT} = 7.0800$ TPa; $G_{12}^{CNT} = 1.9445$ TPa; $\rho^{CNT} = 1400$ kg/m^3; $\upsilon_{12}^{CNT} = 0.175$. We further assume that:$G_{12} = G_{23} = G_{13}$. The values of the efficiency parameters $\eta_j(j = 1, 2, 3)$ for each CNT volume fraction are listed in Table 1. Firstly, the efficiency and reliability of the present finite element strategy is performed. In fact, the temporal responses of a fully simply supported (SSSS) square FG-CNTRC plate of length side a = 25 cm and thickness h = 5 cm are compared to those given by Lei et al. [7] for UD distributions. As well as, the effect of CNT volume fractions ($V_{CNT}^* = 0.11, 0.14, 0.17$) is examined. In addition, the whole of plate is subjected to a transverse sudden dynamic load of value $q_0 = 1 \times 10^5$ N/m^2. The plate is meshed using 16 × 16 finite elements. The obtained results are illustrated in Fig. 2.

Table 1 The different values of efficiency parameters of CNTs

V_{CNT}^*	η_1	η_2	η_3
0.11	0.149	0.934	0.934
0.14	0.150	0.941	0.941
0.17	0.150	1.381	1.381

Figure 2 reveals that the present results are in good agreement with those of Lei et al. [7] which verify the aptitude of the present model in the prediction of temporal responses of FG-CNTRC plates with good accuracy. On other hand, Fig. 2 shows that the amplitude of the displacement w(t) decreases as the CNT volume fraction V_{CNT}^* increases and this is can be explained by the stiffness of the strucutre which becomes higher with the CNTs reinforcements.

Next, the effect of different profiles of CNTs on temporal responses of FG-CNTRC square plates with apothem-to-thickness (r/h = 10) and CNT volume fraction $V_{CNT}^* = 0.11$ is performed. It should be noted that the apothem geometrical parameter refers to the mi-length of the square plate (r = a/2) in the present study and it is used for comparison purpose between the present results and those provided by [21]. An excellent agreement between the obtained results and those of the literature is revealed in Fig. 3. Furthermore, it can be observed that FG-O form induces the highest value of vibration amplitude while the FG-X form provides the lowest one. The amplitude values corresponding to UD and FG-V forms are located between these extreme cases. Therefore, the distributions close to top and bottom surfaces can improve the vibrational behavior of the FG-CNTRC plates more than those near the mid-surface or uniformly distributed.

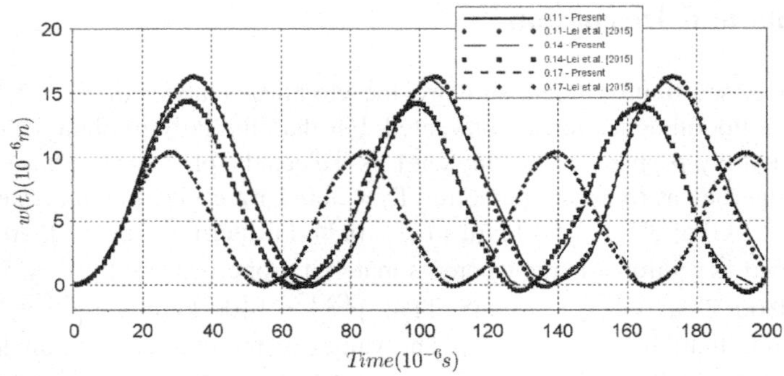

Fig. 2 Effect of CNT volume fraction on temporal responses of UD-CNTRC square plate

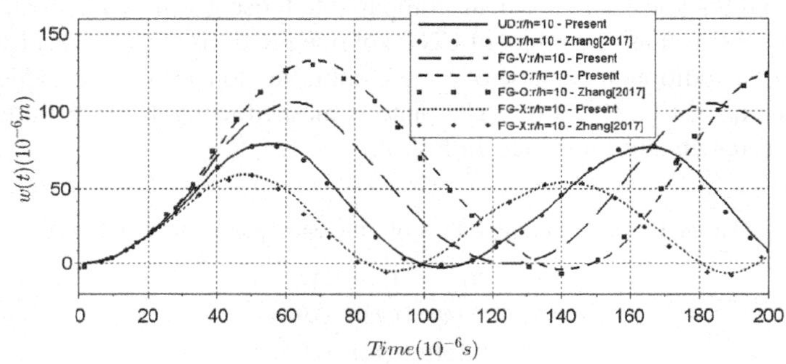

Fig. 3 Effect of different profiles of CNTs on temporal responses of FG-CNTRC square plate with an apothem-to-thickness ratio (r/h = 10) and = 0.11

5 Conclusions

In this paper, the forced vibration analysis of FG-CNTRC plates with a finite element strategy is presented. The governing equations are established using a double directors finite shell model based on the high order distribution of the displacement field. The zero condition of the transverse shear stresses on the top and bottom surfaces is also imposed. The effective material properties are approximated using the extended rule of mixture with some efficiency parameters to take into account the size dependency scale of CNTs. The results reveal that the temporal responses of such strucutres can be enriched as a functionally graded distribution of CNTs via an optimal choice of CNTs.

References

1. Duc ND, Pham DN (2017) The dynamic response and vibration of functionally graded carbon nanotube-reinforced composite (FG-CNTRC) truncated conical shells resting on elastic foundations. Materials 10:1194

2. Esawi AMK, Farag MM (2007) Carbon nanotube reinforced composites: potential and current challenges. Mater Des 28:2394–2401
3. Frikha A, Zghal S, Dammak F (2018) Finite rotation three and four nodes shell elements for functionally graded carbon nanotubes-reinforced thin composite shells analysis. Comput Methods Appl Mech Eng 329:289–311
4. Hasrati E, Ansari R, Torabi J (2017) Nonlinear forced vibration analysis of FG-CNTRC cylindrical shells under thermal loading using a numerical strategy. Int J Appl Mech 09:1750108
5. Kiani Y (2017) Dynamics of FG-CNT reinforced composite cylindrical panel subjected to moving load. Thin-Walled Struct 111:48–57
6. Kiani Y (2017) Analysis of FG-CNT reinforced composite conical panel subjected to moving load using Ritz method. Thin-Walled Struct 119:47–57
7. Lei ZX, Zhang LW, Liew KM (2015) Elastodynamic analysis of carbon nanotube-reinforced functionally graded plates. Int J Mech Sci 99:208–217
8. Malekzadeh P, Dehbozorgi M, Monajjemzadeh SM (2015) Sci Eng Compos Mater 22:37–55
9. Malekzadeh P, Heydarpour Y (2012) Response of functionally graded cylindrical shells under moving thermo-mechanical loads. Thin-Walled Struct 58:51–66
10. Mallek H, Jrad H, Wali M, Dammak F (2019) Piezoelastic response of smart functionally graded structure with integrated piezoelectric layers using discrete double directors shell element. Compos Struct 210:354–366
11. Mellouli H, Jrad H, Wali M, Dammak F (2019) Meshfree implementation of the double director shell model for FGM shell structures analysis. Eng Anal Boundary Elem 99:111–121
12. Reddy JN (1997) Mechanics of laminated composite plates: theory and analysis, 2nd.ed. Florida, New York
13. Shen HS (2009) Nonlinear bending of functionally graded carbon nanotube-reinforced composite plates in thermal environments. Compos Struct 91:9–19
14. Thostenson ET, Ren Z, Chou TW (2001) Advances in the science and technology of carbon nanotubes and their composites: a review. Compos Sci Technol 61:1899–1912
15. Trabelsi S, Frikha A, Zghal S, Dammak F (2018) Thermal post-buckling analysis of functionally graded material structures using a modified FSDT. Int J Mech Soc 144:74–89
16. Trabelsi S, Frikha A, Zghal S, Dammak F (2019) A modified FSDT-based four nodes finite shell element for thermal buckling analysis of functionally graded plates and cylindrical shells. Eng Struct 178:444–459
17. Zghal S, Frikha A, Dammak F (2017) Static analysis of functionally graded carbon nanotube-reinforced plate and shell structures. Compos Struct 158:95–109
18. Zghal S, Frikha A, Dammak F (2018) Free vibration analysis of carbon nanotube-reinforced functionally graded composite shell structures. Appl Math Model 53:132–155
19. Zghal S, Frikha A, Dammak F (2018) Non-linear bending analysis of nanocomposites reinforced by graphene-nanotubes with finite shell element and membrane enhancement. Eng Struct 158:95–109
20. Zghal S, Frikha A, Dammak F (2018) Mechanical buckling analysis of functionally graded power-based and carbon nanotubes-reinforced composite plates and curved panels. Compos Part B 150:165–183
21. Zhang LW, Liu WH, Xiao LN (2017) Elastodynamic analysis of regular polygonal CNT-reinforced composite plates via FSDT element-free method. Eng Anal Bound Elem 76:80–89

Evolution of Mean Velocity and Temperature Field of Variable Density Turbulent Rectangular Jet

Amel Elkaroui[1(✉)], Amani Amamou[1], Mohamed Hichem Gazzah[2], Nejla Mahjoub Saïd[3], and Georges Le Palec[4]

[1] LGM, National Engineering School of Monastir, University of Monastir, Monastir, Tunisia
amel.karoui@hotmail.fr, amani.amamou@yahoo.fr
[2] Quantum and Statistical Physics Laboratory, Faculty of Sciences of Monastir, University of Monastir, Monastir, Tunisia
hichem.gazzah@fsm.rnu.tn
[3] LGM, Preparatory Institute for Engineering Studies, University of Monastir, Monastir, Tunisia
nejla.mahjoub@fsm.rnu.tn
[4] CNRS, IUSTI, Aix Marseille University, Marseille, France
georges.lepalec@univ-amu.fr

Abstract. This chapter investigates numerically the effect of density variation on the mean and fluctuating flow properties of a turbulent rectangular jet using the Reynolds Stress Model (RSM). Two cases are tested: isothermal and heated jets. Predicted results are compared to the existing experimental data and a good agreement between them is found. Axial evolution of mean and fluctuating velocity profiles are investigated in this paper. The decrease of the jet density produces faster jet decay, leading to enhance the mixing of the gas.

Keywords: Jet · Numerical modeling · Reynolds Stress Model (RSM) · Velocity · Density variation

1 Introduction

Jets have been widely studied due to their important application in industry. Many jet flow configurations are encountered in literature as free jets [8, 16], jet in a co-flow [4, 5, 7, 9, 10, 12], jet in a crossflow [13, 14] and jet in a counterflow [1–3]. Among these configurations, free jets are the most studied, especially for round geometry of the nozzle exit. For that, this study is devoted to the configuration of a turbulent rectangular jet in a quiescent stream.

The development of a jet flowing into a stagnant environment is characterized by two main regions: the Zone of the Flow Establishment (ZFE) and the Zone of Established Flow (ZEF). (ZFE) or the potential core region, located near the jet exit, is marked by constant proprieties. In the far field region (ZEF), the flow becomes self-similar in both mean and turbulent properties. Amielh et al. [4] provided an analysis of the effect of density variation of turbulent jets on the velocity near field. Also, Sarh [16]

N. Aifaoui et al. (Eds.): CMSM 2019, LNME, pp. 786–794, 2020.
https://doi.org/10.1007/978-3-030-27146-6_86

studied the influence of density variation on mean and fluctuating velocity and temperature quantities, using the Laser Doppler Anemometry (LDA) and the cold wire anemometry techniques. He concluded that the mixing efficiency increases when the density ratio between the emerging jet and the external environment decreases. In the same context, Elkaroui et al. [8] validated their numerical model, developed with a computer code, for a free turbulent plane jet based on comparison with Sarh's [16] experimental data. Hitchman et al. [11] and Deo et al. [6] performed an investigation of turbulent jets issuing from rectangular nozzles with and without confinement.

Numerical study of a turbulent rectangular air jet at rest is investigated in this paper. The influence of the jet density variation on the velocity field is conducted through two jet situations; isothermal case and heated case. The Reynolds Stress Model (RSM) is used in computational simulations to ensure the comparison between predicted results and available experimental data. The longitudinal centerline velocity, the turbulent intensity and the momentum half-width are analyzed in this study.

2 Computational Procedure

2.1 Governing Equations

To model the turbulent flow, the Reynolds-averaged Naviere Stokes equations are numerically solved. For an incompressible fluid, the governing equations including the continuity (1), and the momentum (2) conservation equations, are considered in the cartesian coordinates system as follow:

$$\frac{\partial}{\partial x_i}(\rho \tilde{U}_i) = 0 \tag{1}$$

$$\frac{\partial}{\partial x_j}(\rho \tilde{U}_i \tilde{U}_j) = -\frac{\partial P}{\partial x_i} + \frac{\partial}{\partial x_j}\left[\mu\left(\frac{\partial \tilde{U}_i}{\partial x_j} + \frac{\partial \tilde{U}_j}{\partial x_i}\right) - \overline{\rho u'_i u'_j}\right] + \rho g_i \tag{2}$$

where $i,j = 1,2$ refer to x, y, respectively. The mean velocity U_i P, g_i, ρ and μ represent the pressure, the gravity in i direction, the fluid density and the dynamic viscosity, respectively. $(\overline{\rho u'_i u'_j})$ is the Reynolds stresses.

Due to the additional unknown, a closure turbulence model is required to solve this system. In this paper, the second order turbulence model, Reynolds Stress Model (RSM) model is used for the closure of this system of equation.

The Reynolds stress equation is written as follow [17]:

$$\frac{\partial}{\partial x_k}\left(\tilde{U}_k \overline{\rho u'_i u'_j}\right) = P_{ij} + G_{ij} + D_{ij} + \Pi_{ij} - \frac{2}{3}\rho \varepsilon \delta_{ij} \tag{3}$$

where:

$$\frac{\partial}{\partial x_k}\left(\tilde{U}_k \overline{\rho u_i' u_j'}\right) \text{ is the convective term} \tag{4}$$

P_{ij} is the production term due to mean strain and is written as follow:

$$P_{ij} = -\left(\overline{\rho u_i' u_k'}\right)\frac{\partial \tilde{U}_j}{\partial x_k} - \left(\overline{\rho u_j' u_k'}\right)\frac{\partial \tilde{U}_i}{\partial x_k} \tag{5}$$

G_{ij} is the production due to buoyancy effects which is given by:

$$G_{ij} = -\beta\left(g_i \overline{\rho u_j' t'} + g_j \overline{\rho u_i' t'}\right) \tag{6}$$

D_{ij} is the diffusion term, is given by:

$$D_{ij} = C_s \frac{\partial}{\partial x_k}\left(\frac{k}{\varepsilon}\left(\overline{\rho u_k' u_l'}\right)\frac{\partial\left(\overline{\rho u_i' u_j'}\right)}{\partial x_l}\right) \tag{7}$$

Π_{ij} is the pressure-strain correlation, is shown by:

$$\Pi_{ij} = \overline{p'\left(\frac{\partial u_i'}{\partial x_j} + \frac{\partial u_j'}{\partial x_i}\right)} \tag{8}$$

2.2 Flow Configuration and Boundary Conditions

The flow configuration of a rectangular air jet flowing vertically in a stagnant ambient is sketched in Fig. 1 The computational domain is rectangular with a domain size 60H × 25H, with H is the actual dimension of the slot width. Two cases of jet configuration are investigated: isothermal and heated jet. The flow development is characterized by the Reynolds number $\mathrm{Re} = \frac{U_j H}{\nu}$ and the densimetric Froude number $Fr = \frac{\rho_j U_j^2}{\left(gH\left(\rho_{co} - \rho_j\right)\right)}$ of the jet flow. At the nozzle exit, initial conditions are recapitulated in the Table 1:

Boundary conditions considered in this chapter are the following:

$$x = 0: \begin{cases} -H/2 < y < H/2: U = U_j; V = 0; T = T_j; k = 10^{-3} U_j^2; \varepsilon = \frac{C_\mu k^{3/2}}{0.03H} \\ H/2 \le y < L_y/2H: U = 0; V = 0; T = T_a; k = 0; \varepsilon = 0 \\ -H/2 \le y < -L_y/2H: U = 0; V = 0; T = T_a; k = 0; \varepsilon = 0 \end{cases}$$

$$x > 0: \begin{cases} y = 0: V = 0; \frac{\partial \Phi}{\partial y} = 0 \ where\ \Phi = U, T, k, \varepsilon \\ y = L_y/2H \ and\ y = -L_y/2H: U = 0; \frac{\partial V}{\partial y} = 0; T = T_a; k = 0; \varepsilon = 0 \end{cases}$$

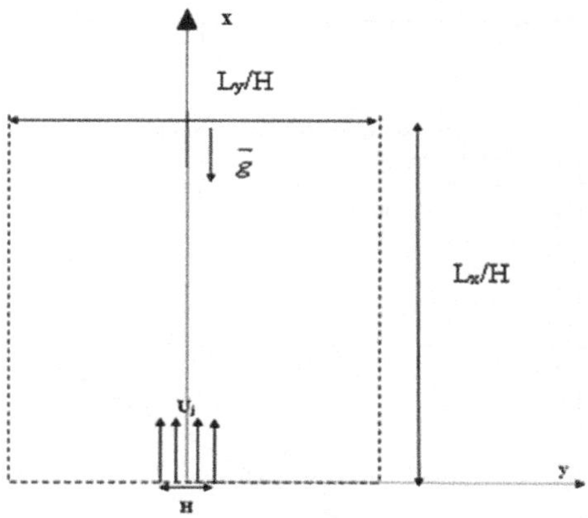

Fig. 1 Flow configuration

Table 1 Flow parameters for a rectangular jet

	Re	Fr	U_j(m/s)	T_j(K)	Mach
Isothermal jet	5000	254	15.5	295	0.045
Heated jet	5000	216	51	595	0.106

$$x{=}L_x/H : \frac{\partial \Phi}{\partial x} = 0 \, where \, \Phi = U, V, T, k, \varepsilon$$

2.3 Numerical Method

The resolution of the governing equations is carried out using the finite volume method (FVM). The pressure–velocity coupling is performed by the SIMPLE algorithm. The system of algebraic equations is solved line by line using the TDMA (Tri-Diagonal Matrix Algorithm) described by Patankar [15]. The used mesh is not uniform and gradually extends according to the longitudinal and transversal directions. A two-dimension grid is elaborated for computation with 60 × 60 cells. This latter is chosen after a careful mesh independence investigation.

3 Results and Discussion

3.1 Centerline Velocity

The variation of the normalized centerline velocity of the jet $\left((U_j - U_\infty)/(U_c - U_\infty)\right)^2$ along the downstream axial distance (x/H), is presented, in Fig. 2, for isothermal and heated jets. U_j U_∞ and U_c is the initial jet velocity, the ambient velocity, and the centerline mean axial velocity respectively.

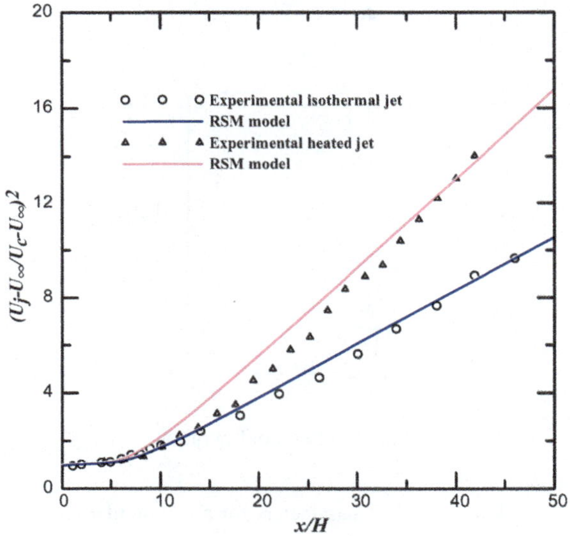

Fig. 2 Axial evolution of the centerline longitudinal velocity linear representation

Numerical simulation results are compared with experimental data of Sarh [16] for isothermal and heated jet. A good agreement is found between the numerical and experimental data. From Fig. 2 the different flowfield regions can be identified. In the near-field region ($0 \le x/H \le 10$), known as the potential core region, the fluid leaves the jet with a constant value which equals to the initial jet velocity. In the zone of established flow ($x/H \ge 10$), the centerline velocity seems to decrease with $x^{1/2}$ for the two jet cases. This law of decay satisfies the following relationship: $\left(\frac{U_j - U_\infty}{U_c - U_\infty}\right)^2 = K_1\left(\frac{x}{H} - C_1\right)$ where K_1 is the velocity decay rate and C_1 is the virtual origin of the jet.

From Table 2, it is seen that the effect of the jet density reduce is manifested by an increase of the velocity decay rate K_1 and the virtual origin C_1, for both heated and isothermal jets. The decay rate for the heated jet is most important than the isothermal jet. This is implies that the rising of the temperature of injection enhances the flow mixing.

Table 2 Velocity decay rate and virtual origin of rectangular jets

	K_1	C_1
Isothermal jet	0.1966	0.1833
Heated jet	0.3224	0.3753

Figure 3 features the axial decay of the longitudinal velocity $((U_c - U_\infty)/(U_j - U_\infty))$ as a function of (x/H) in logarithmic coordinates, to determine if the centerline velocity decay follows a simple power low. The centerline velocity decay can be represented by the following relationship: $\frac{(U_c - U_\infty)}{(U_j - U_\infty)} = A_u \left(\frac{x}{H}\right)^n$.

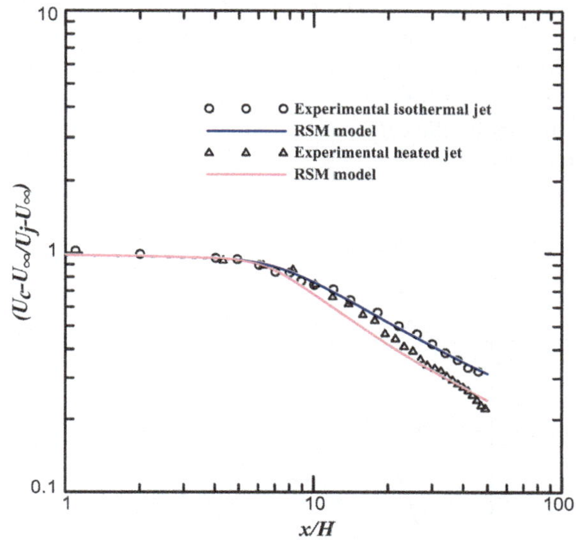

Fig. 3 Axial evolution of the centerline longitudinal velocity logarithmic representation

It is shown that these curves have the same shape starting from the unity in the potential core region, and then decreases progressively. It is evident from this Figure that for the isothermal jet three regions are observed: the first one is situated between $(x/H = 1)$ and $(x/H = 6)$, with n = −0.038, the second one is located between $(x/H = 9)$ and $(x/H = 45)$, with n = −0.5 and where A_u is equal to 2.55. The third extends beyond $(x/H = 45)$ with n = −0.991. It is found that three classical regions that rectangular jets present are observed. Similarly, the value of A_u found for the characteristic region falls within the values cited in the literature and which are between 2.2 and 2.7. However, when the flow is heated a fourth region appears and a very sensitive modification of the exponent n. The extension of these different regions, the value taken by the exponent n in each of them, and the value of A_u when n = −0.5 are given below. The first region $(1 \prec x/H \prec 5)$ with n = −0.043. The second region $(6 \prec x/H \prec 16)$ with n = −0.5 and $A_u = 2$. The third region $(16 \prec x/H \prec 44)$ the exponent n = −0.706. The forth region when $(x/H \succ 44)$ n = −1.21.

The logarithmic representation shows clearly that the heated jet decays more rapidly than the isothermal one. It is worth noting that this decay cannot be represented with a simple power low.

3.2 Centerline Turbulent Intensity

Figure 4 features the evolution of the normalized turbulent intensity $(u'/(U_c - U_\infty))$ along the jet axis (*x/H*). Predicted fluctuating velocities for both isothermal and heated jets are compared to experimental data of Sarh [16]. The RSM model overall reproduces experimental data especially for the isothermal case and in the ZEF. In fact, the fluctuation of velocity increases rapidly in the potential core region to reach an asymptotic value equals to 0.21 and 0.23 for the isothermal and heated jets, respectively. Further, downstream the centerline turbulent intensity remains constant with the longitudinal coordinate.

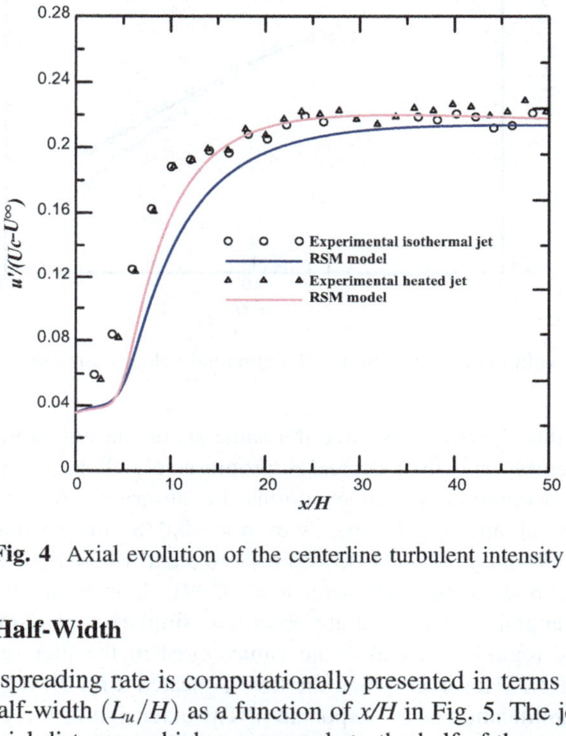

Fig. 4 Axial evolution of the centerline turbulent intensity

3.3 Dynamic Half-Width

The dynamic jet spreading rate is computationally presented in terms of the evolution of the dynamic half-width (L_u/H) as a function of *x/H* in Fig. 5. The jet dynamic half-width L_u is the axial distance, which corresponds to the half of the centerline velocity.

The predicted jet half-width obtained by the second-order RSM model agrees with experimental Sarh's (1990) data for isothermal and heated jets. It is worth noting that, in the vicinity of the nozzle exit, the dynamic half-width remains constant for the isothermal as far the heated jet. In this region, the density variation has no effect and the two jets behave in the same way. Then, the dynamic half-width increases almost linearly with x for the two jet cases. It is clear that the spreading rate for the isothermal jet is higher than that for the heated jet.

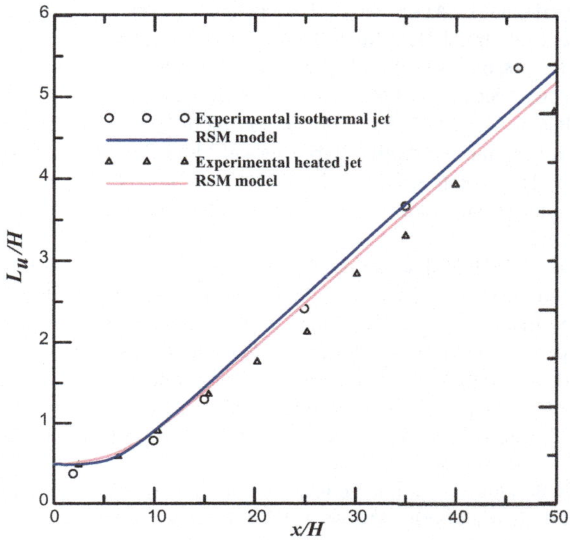

Fig. 5 Axial evolution of the dynamic half-width

4 Conclusion

A computer code is developed to investigate the dynamic behavior of a turbulent rectangular jet in a stagnant environment. Numerical simulations are performed using the second-order closure model (RSM). Two jet cases (isothermal and heated jet) are investigated and predicted results are compared with experimental data deduced from literature.

A good agreement between experimental data of Sarh [16] and numerical results is obtained. The centerline velocity decays increases more rapidly in the case of heated jet. The spreading rate for the isothermal jet is found to be higher than the heated jet spreading. Therefore, the heated jet tends to mix more quickly with the surrounding fluid than isothermal jet.

References

1. Amamou A, Habli S, Mahjoub Saïd N, Bournot Ph, Le Palec G (2015) Numerical study of turbulent round jet in a uniform counterflow using a second order Reynolds stress model. J Hydro-environ Res 9:482–495
2. Amamou A, Habli S, Mahjoub Saïd N, Bournot Ph, Le Palec G (2015) Computational study of mixing behavior of a turbulent jet issuing in a uniform counterflow at low velocity ratios. J Turbul 17(2):237–251
3. Amamou A, Habli S, Mahjoub Saïd N, Bournot Ph, Le Palec G (2016) Computational study of mass and heat transport in a counterflowing turbulent round jet. Appl Thermal Eng 105:724–736

4. Amielh M, Djeridane T, Anselmet F, Fulachier L (1995) Velocity near-field of variable density turbulent jets Int. J Heat Mass Transf 39:2149–2164
5. Ben Haj Ayech S, Habli S, Saïd NM, Bournot P, Le Palec G (2016) A numerical study of a plane turbulent wall jet in a coflow stream. J Hydro-Environ Res 12:16–30
6. Deo RC, Nathan GJ, Mi J (2007) Comparison of turbulent jets issuing from rectangular nozzles with and without sidewalls. Exp Thermal Fluid Sci 32:596–606
7. Elkaroui A, Ben Haj Ayech S, Gazzah MH, Mahjoub Saïd N, Le Palec G (2018) Numerical study of local entropy generation in a heated turbulent plane jet developing in a co-flowing stream 62:605–628
8. Elkaroui A, Gazzah MH, Mahjoub Saïd N, Bournot Ph, Le Palec G (2018) Entropy generation concept for a turbulent plane jet with variable density. Comput Fluids 168:328–341
9. Gazzah MH, Belmabrouk H (2013) Local entropy generation in co-flowing turbulent jets with variable density. Int J Numer Methods Heat Fluid Flow 24:1679–1695
10. Gazzah MH, Belmabrouk H (2014) Directed co-flow effects on local entropy generation in turbulent heated round jets. Comput Fluids 105:285–293
11. Hitchman GJ, Strong AB, Slawson PR, Ray G (1990) Turbulent plane jet with and without confining walls. AIAA J 28:700–1699
12. Imine B, Saber-Bendhina A, Imine O, Gazzah MH (2005) Effects of a directed co-flow on a non-reactive turbulent jet with variable density. Heat Mass Transfer 42:39–50
13. Kalifa RB, Habli S, Mahjoub Saïd N, Bournot H, Le Palec G (2014) Numerical and experimental study of a jet in a crossflow for different velocity ratio. J Braz Soc Mech Sci Eng 36(4):743–762
14. Mahjoub Saïd N, Habli S, Mhiri H, Le Palec G, Bournot Ph (2007) Flow field measurement in a crossflowing. J Fluid Eng 129:551–562
15. Patankar SV (1980) Numerical heat transfer and fluid flow. Hemisphere Publishing, Washington DC
16. Sarh B (1990) Contribution à l'étude des jets turbulents à masse volumique variable et des flammes turbulentes de diffusion, thèse de doctorat à l'Université Pierre et Marie Curie Paris
17. Wilcox DC (2006) Turbulence modeling for CFD, 3rd ed. DCW Industries Inc.

Enhanced Efficiency of InGaN/GaN MQW Solar Cell by Applying Stress

Bilel Chouchen[1(✉)], Mohamed Hichem Gazzah[1], and Hafedh Belmbrouk[2,3]

[1] Quantum and Statistical Physics Laboratory, Faculty of Sciences of Monastir, University of Monastir, 5019 Monastir, Tunisia
bilelchouchen06@gmail.com

[2] Electronics and Microelectronics Laboratory, Faculty of Science of Monastir, University of Monastir, 5019 Monastir, Tunisia

[3] Department of Physics, College of Science AlZulfi, Majmaah University, Al Zulfi 11932, Saudi Arabia

Abstract. With enhancements in the performance of photovoltaic solar cells, the InGaN/GaN multiple quantum wells have been considered as a very promising structure to improve the mechanism of carrier collection and hence the efficiency of conversion. The basic processes for the operation of a solar cell are the generation of electron-hole pairs, the recombination of these carriers into external circuits, and crucial step here is the generation of the electron-hole pairs. The piezoelectric charges induced by external stress generated at the InGaN and GaN interfaces induce an improvement of the electronic properties and the electrical parameters of the InGaN-MQW SC. Here, we demonstrate by a new numerical modeling self-consistent model coupled by the electrical parameters of cells, that the electronic properties and efficiency of conversion InGaN quantum wells SC have been improved by external stress. This study proves that the piezo-phototronic effect modulates the quantum photovoltaic device but also offers a great promise to maximize the use of solar energy in the current energy revolution.

Keywords: Solar cell · Piezo-phototronic effect · InGaN/GaN multiple quantum wells · Electronic proprieties · Conversion efficiency

1 Introduction

In recent decades, InGaN alloys have recently shown considerable interest in many applications in the optoelectronics field. In fact, several physical properties make this alloy a good candidate for photovoltaic applications such as; high radiation resistance, high thermal conductivity, high absorption coefficient (about 10^5 cm^{-1}) Asgari and Razi [1]. In addition, a modulable band gap (ranging from 0.62 to 3.42 eV) covering the maximum of the solar spectrum [2]. However, several factors can improve the photovoltaic energy conversion and the different parameters of the InGaN/GaN MQW's solar cell. Many groups have focused on the optimization of this type of cell, by modifying some geometric parameters such as the quantum well thickness Redaelli

N. Aifaoui et al. (Eds.): CMSM 2019, LNME, pp. 795–803, 2020.
https://doi.org/10.1007/978-3-030-27146-6_87

et al. [3], the thickness of the barrier Wierer et al. [4], and the indium fraction Lai et al. [5]. The piezo-phototronic effect tends to improve the electronic, mechanical and optical proprieties of InGaN/GaN heterostructures Huang et al. [6]. This effect is ensured by the application of a piezoelectric stress Han et al. [7]. In this respect, several numerical and experimental works have been interested in this factor Belghouthi et al. [8], Zhu et al. [9]. These authors have shown that the application of a piezoelectric stress can also affect open-circuit voltage and cell efficiency. These elements are the key parameters that can help us evaluate the performance of the solar cell. In fact, we can be effective by playing on the sign and the value of this strain for modulated and optimized of the electric performance. In this paper, the numerical self-consistent model, based on the finite difference method, consists of analyzing the impact of the piezo-electric strain. In fact, the coupling of the electronic properties and parameters of the InGaN/GaN MQW solar cell in the N face configuration, is presented in Sect. 2. The impact of the variation of the piezoelectric stress of the electronic properties on the cell performance is discussed in Sect. 3. Finally, the Sect. 4 is the conclusion.

2 Numerical Method

2.1 Self-consistent Model

Taking into account the polarization effects, the band edges of the heterostructures being studied can be obtained by self-consistent resolution. This model is coupling between Schrodinger-Poisson equations.

The one dimensional Schrödinger Equation Saidi et al. [8, 10]:

$$-\frac{\hbar^2}{2}\frac{d}{dz}\left(\frac{1}{m^*(z)}\frac{d\psi_{v,k_z}(z)}{dz}\right)+E_c(z)\psi_{v,k_z}(z)=\varepsilon_v\psi_{v,k_z}(z) \tag{1}$$

where v is the subband index, k_z is the vector momentum parallel to the z-direction. $m^*(z)$, $E_c(z)$ are respectively the electron effective mass and the total potential energy. $E_c(z)$, the following terms are included:

$$E_c(z)=V_B(z)+V_H(z)+V_{ex}(z)+V_P(z) \tag{2}$$

$V_B(z)$ is the heterojunction band gap discontinuity, $V_H(z)$ is the effective Hartree potential due to the screened ionized donors, $V_{ex}(z)$ is the exchange-correlation potential resultant from the several body effects and the $V_P(z)$ is the potential energy induced by the polarization charges. The Poisson equation is solved for a potential energy $V_H(z)+V_P(z)$ as:

$$\varepsilon_0\frac{d}{dz}\left(\varepsilon_r(z)\frac{d(V_H+V_P)}{dz}\right)=e^2\left(\frac{\sigma}{e}\delta(z-z_0)+N_D-n(z)\right) \tag{3}$$

where $\varepsilon_r(z)$ the local dielectric constant of the medium is, z_0 is the interface position, $n(z)$ is the sheet concentration of confined electrons, and N_D is the total density of ionized donors.

The local density of 2DEG is calculated as follows:

$$n(z) = \sum_{v} n_v |\psi_{v,k}(z)|^2 \tag{4}$$

with:

$$n_v(z) = \frac{m^* k_B T}{\pi h^2} \log\left[1 + \exp\left(\frac{E_F - E_v}{k_B T}\right)\right] \tag{5}$$

where E_F is the Fermi level.

The polarization developed at the InGaN/GaN interfaces depends on the indium's concentration x within the InGaN layer. It is generally calculated using model developed by Vurgaftman et al. [11], leading to the following Eqs. (6), (7) and (8):

$$P^{SP}_{In_xGa_{1-x}N} = -0.042x - 0.034(1-x) + 0.038x(1-x) \tag{6}$$

$$P^{PZ}_{InN} = -1.373\,\varepsilon + 7.559\varepsilon^2 \tag{7}$$

$$\begin{aligned} P^{PZ}_{GaN} &= -0.918\,\varepsilon + 9.541\,\varepsilon^2 \\ P^{PZ}_{In_xGa_{1-x}N} &= xP^{PZ}_{InN}[\varepsilon(x)] + (1-x)P^{PZ}_{GaN}[\varepsilon(x)] \end{aligned} \tag{8}$$

where P_{sp} and P_{pz} are the spontaneous and piezoelectric polarization, respectively, and ε the strain defined according to the lattice parameter of the substrate a_s and epilayer a_e expression as follows: $\varepsilon(x) = \left[\frac{a_s - a_e(X)}{a_e(X)}\right]$

The dependency of the piezoelectric polarization with the indium's content x within the alloy can be described by a Vegard's law as:

$$P^{PZ}_{In_xGa_{1-x}N} = xP^{PZ}_{InN}[\varepsilon(x)] + (1-x)P^{PZ}_{GaN}[\varepsilon(x)]$$

Accordingly, the total polarization at the interface is given as follows:

$$P = \left[P^{SP}_{In_xGa_{1-x}N} - P^{PZ}_{In_xGa_{1-x}N}\right] - P^{SP}_{GaN} \tag{9}$$

The discontinuity of P between the GaN and InGaN layers develops a polarization charges with an interfacial density σ_s equals to:

$\rho_p = \frac{P}{W_z}$; $\sigma_s = P$ for $0 < z < W_{pz}$ the polarization charges are distributed in all the i-region interface InGaN/GaNMQW of thickness W_{pz}.

2.2 Electrical Parameters of the $In_xGa_{1-x}N/GaN$ MQW Solar Cell

Allow to calculating the electrical parameters of the $In_xGa_{1-x}N$/GaN–MQW solar cell. The current-voltage relationship for a MQW solar cell is expressed as follows by Anderson [12]

$$J(V) = J_{QWO}[1 + r_R\beta]\left[\exp\left(\frac{qV}{k_BT}\right) - 1\right] - J_{ph} \tag{10}$$

β is a parameter related to the intrinsic region,r_R is the radiative enhancement ratio represents the fractional increase in recombination in the i-MQW region

$$J_{QWO} = j_0 \exp\left(\frac{q\rho_p w_{pz}^2}{2\varepsilon V_{th}}\right) \tag{11}$$

The saturation current density is given by:

$$J_0 = \frac{qD_n}{L_n} n_{p0} + \frac{qD_p}{L_p} p_{n0}$$

with

$$n_{p0} = n_i \exp\left(\frac{E_i - E_F}{k_b T}\right) \tag{12}$$

n_{p0}, p_{n0}, is the thermal equilibrium electron in n-type semiconductor.

n_i is the intrinsic carrier density E_i is the intrinsic Fermi level. E_F is the Fermi level calculated by the self-consistent model with polarization effect.

V_{oc} is the open circuit voltage and obtained by Belghouthi et al. [13]:

$$V_{oc} = \frac{k_b T}{q}\left\{ Ln\left(\frac{J_{ph}}{J_{QWO}}\right) - \left(\frac{q\rho_p W_{PZ}^2}{2\varepsilon V_{th}}\right)\right\} \tag{13}$$

η is the efficiency of the photovoltaic cell is given by:

$$\eta = \frac{P_{max}}{P_{in}} = \frac{V_{oc} FF J_{ph}}{P_{in}}$$

3 Results and Discussion

3.1 Description of the Model

Our reference structure is a solar cell p-GaN-$In_{0.2}Ga_{0.8}N$ (MQW)-nGaN is shown in Fig. 1. The polarization charges are assumed to spread in the interface region

InGaN-QW/GaN-QB. This cell is characterized by; 250 nm is the thickness of n-GaN type layer and 200 nm is the thickness of p-GaN layer. The intrinsic region is characterized by five QW-nGaN with; 3 nm is the width of GaN-QB barrier and 10 nm is the width of the well.

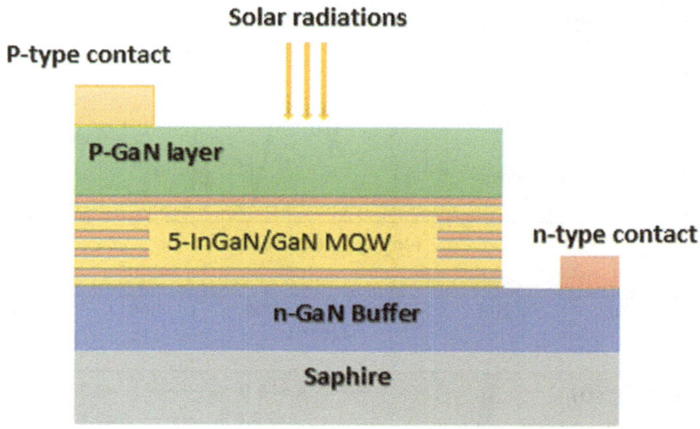

Fig. 1 Model structure InGaN/GaNMQW's solar cells

Figure 2a depicts the energy diagram of the conduction band as a function of the growth direction z. In fact, the conduction band energy and the Fermi level were calculated by the self-consistent numerical by applied the strain. Under of this the effect the conduction band is modified. We notice a significant variation in Fermi level. Based on Eq. 12, the difference of the Fermi energy under various external stress will affect the different photovoltaic characteristics Zhang et al. [8, 14]. Figure 2b shows a zoom at the last i-$In_{0.2}Ga_{0.8}N$ quantum well. It is clear, that the increase and the decrease of the different strain value caused a shift of the conduction band respectively up wards or down wards. The variation of the Fermi energy by the change of strain is shown in Fig. 3. On the one hand, we remark a reduction of the E_F in the case where the tensile/strain is applied ranging from 0 to 0.9%. On the other hand, in the case where the compressive/strain is applied ranging from of 0 to −0.9%, an increase in the Fermi energy is observed. This variation of the E_F will play an important role in the generation- recombination process of the carriers at the intrinsic region InGaN-MQW and the cell performance.

In this work, two types of strain were simulated. In fact, for the case of positive deformation in the proposed model, the piezoelectric positive value indicates under tensile/strain, which results in a strong accumulation of charges in the intermediate region i-InGaN MQW's. In contrary, for a negative value of the application

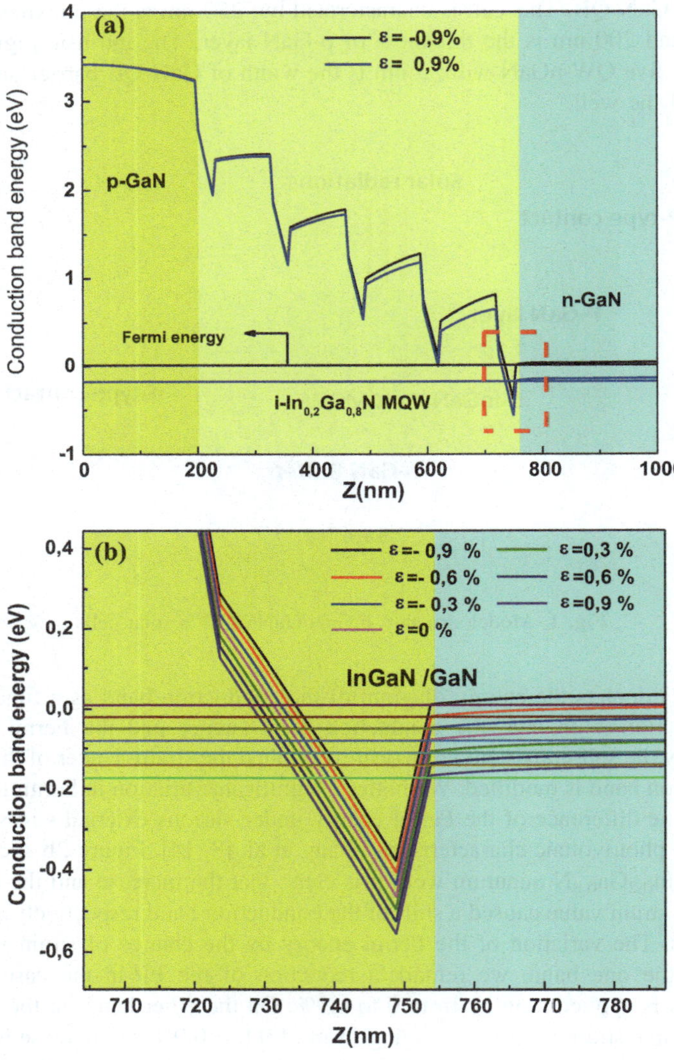

Fig. 2 a, b Conduction band energy along the growth direction z for the 5-periods-$In_{0\cdot2}Ga_{0\cdot8}N$/GaN–MQW on different strain

compressive/strain, the negative piezoelectric charges are created in the i-region. This effect is illustrated in Fig. 3. The dependance in open circuit voltage Voc is illustrated in Fig. 4. The value of Voc is reduced by 4.23% when the strain/compressive is applied ranging from 0 to −0.9%. We notice, that under tensile/strain conditions the Voc is increased of 2.81%. This effect is mainly due to a good confinement of the piezoelectric charges at the interface and the reduction of the Fermi energy. The evolution of the

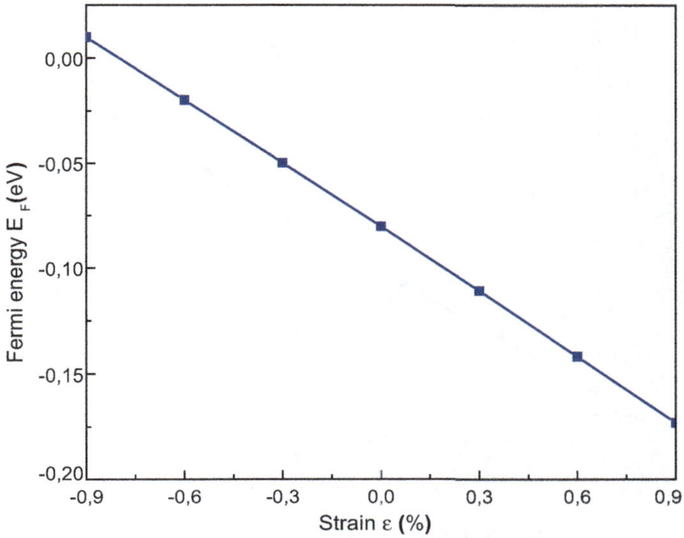

Fig. 3 The Fermi energy $\mathbf{E_F}$ as function of the strain

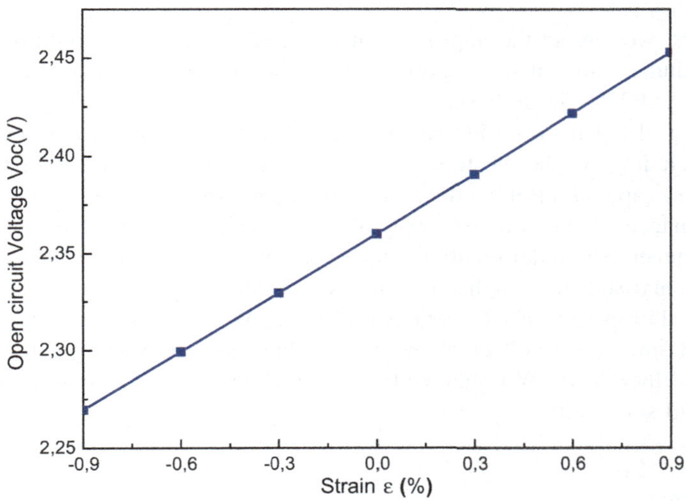

Fig. 4 The open circuit voltage as function of the strain

efficiency η as function of the piezoelectric strain is presented in Fig. 5. We found, that the efficiency is improved of 4.25% at each increasing on the tensile/strain in the range 0 to 0.9%. But, the efficiency is reduces of 3.72% of compressive/strain is applied. The piezo-phototronic effect is a good solution to improve the efficiency of the InGaN-MQW-based solar cell Jiang et al. [15].

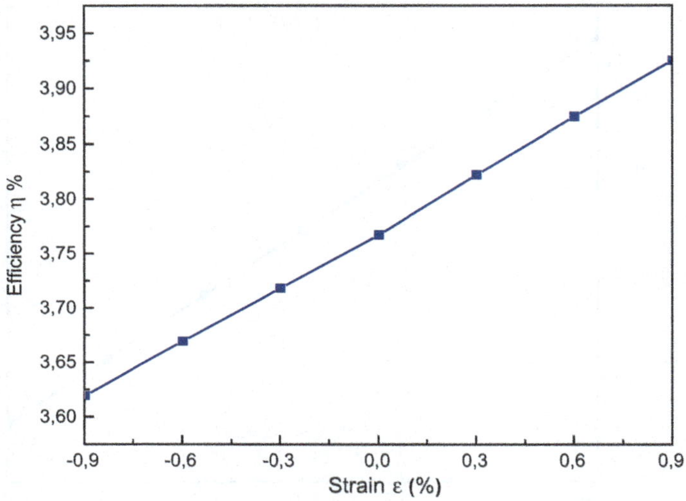

Fig. 5 Efficiency as function of the strain

4 Conclusion

In this work, we present the improvement of the efficiency of the multi-quantum well solar cell under different constraints. In fact, the conversion efficiency is relatively improved by 4.25%. In addition, a numerical model was developed to illustrate the performance of the III-V nitride-based solar cell, and offered many important clarifications, including on the electronic properties such as the structure of the band of energy, band gap and offset conduction band which also have a very important role on the performance of the cell. By applying an external mechanical mechanism in an efficient manner. Strain/deformation which leads to a series of optoelectronic reaction phenomena and scientific applications that will further improve the electronic, optical and electrical properties of InGaN/GaN MQW SC. We have found that the application of tensile/strain is a very effective and recoverable solution to improve the conversion efficiency of InGaN-MQW's solar cells, and which has great potential in the design of nitride-based solar cells.

References

1. Asgari A, Razi S (2010) High performances III-Nitride quantum dot infrared photodetector operating at room temperature. Opt Express 18:14604–14615. https://doi.org/10.1364/OE.18.014604
2. Jeng MJ, Su TW, Lee YL, Chang YH, Chang LB, Lin RM, Jiang JH, Lu YC (2010) effect of Silicon doping on performance of 30-Pair InxGa1-xN/GaN quantum well solar cells. Japanese J Appl Phys 49:052302. https://doi.org/10.1143/JJAP.49.052302
3. Redaelli L, Mukhtarova A, Valdueza-Felip S, Ajay A, Bougerol C, Himwas C, Faure-Vincent J, Durand C, Eymery J, Monroy E (2014) Effect of the quantum well thickness on

the performance of InGaN photovoltaic cells. Appl Phys Lett 105:131105–131101-131105–131104, https://doi.org/10.1063/1.4896679

4. Wierer JJJ, Koleske DD, Lee SR (2012) Influence of barrier thickness on the performance of InGaN/GaN multiple quantum well solar cells. Appl Phys Lett 100:111119. https://doi.org/10.1063/1.3695170
5. Lai KY, Lin GJ, Lai Y-L, Chen YF, He JH (2010) Effect of indium fluctuation on the photovoltaic characteristics of InGaN/GaN multiple quantum well solar cells. Appl Phys Lett 96:081103. https://doi.org/10.1063/1.3327331
6. Huang X, Du Ch, Zhou Y, Jiang Ch, Pu X, Liu W, Hu W, Chen H, Wang ZL (2016) Piezo-Phototronic effect in a quantum well structure. ACS Nano 5:5145–5152. https://doi.org/10.1021/acsnano.6b00417
7. Han X, Chen M, Pan C, Wang ZL (2016) Progress in piezo-phototronic effect enhanced photodetectors. J Mater Chem C 4:11341–11354. https://doi.org/10.1039/C6TC04029B
8. Belghouthi R, Salvestrini JP, Gazzeh MH, Chevallier C (2016) Analytical modeling of polarization effects in InGaN double hetero-junction p-i-n solar cells. Superlattices Microst 100:168–178. https://doi.org/10.1016/j.spmi.2016.09.016
9. Zhu L, Wang L, Pan C, Chen L, Xue F, Chen B, Yang L, Su L, Wang ZL (2017) Enhancing the efficiency of silicon-based solar cells by the piezo-phototronic effect. ACS Nano 11:1894–1900. https://doi.org/10.1021/acsnano.6b07960
10. Saidi I, Bouzaïene L, Gazzah MH, Mejri H, Maaref H (2006) Back doping design in delta-doped AlGaN/GaN heterostructure field-effect transistors. Solid State Commun 140:308–312. https://doi.org/10.1016/j.ssc.2006.08.026
11. Vurgaftman I, Meyer JR, Ram-Mohan LR (2001) Band parameters for III–V compound semiconductors and their alloys. J Appl Phys 89:5815–5875. https://doi.org/10.1063/1.1368156
12. Anderson NG (1995) Ideal theory of quantum well solar cells. J Appl Phys 78:1850–1861. https://doi.org/10.1063/1.360219
13. Belghouthi R, Taamalli S, Echouchene F, Mejri H, Belmabrouk H (2015) Modeling of polarization charge in N-face InGaN/GaN MQW solar cells. Mater Sci Semicond Process 40:424–428. https://doi.org/10.1016/j.mssp.2015.07.009
14. Zhang Y, Yang Y, Wang ZL (2012) Piezo-phototronics effect on nano/microwire solar cells. Energy Environ Sci 5:6850–6856. https://doi.org/10.1039/C2EE00057A
15. Jiang Ch, Jing L, Huang X, Liu M, Du Ch, Liu T, Pu X, Hu W, Wang ZL (2017) Enhanced solar cell conversion efficiency of ingan/gan multiple quantum wells by piezo-phototronic effect. ACS Nano 9:9405–9412. https://doi.org/10.1021/acsnano.7b04935

Analysing 2D Elastic and Elastoplastic Problems with the Element Free Galerkin Method

Z. Sendi(✉) and H. BelHadjSalah

LGM/ENIM, Laboratory of Mechanical Engineering, National Engineers School of Monastir, University of Monastir, Monastir, Tunisia
zohrasendii@yahoo.fr, hedi.belhadjsalah1@gmail.com

Abstract. The paper is devoted to numerical modelling elastic and elastoplastic problems using the Element Free Galerkin method (EFG) based on the Moving Least Square approximation (MLS). Numerical calculations are done for a number of beams and simulated using both the Finite Element Method (FEM) and the Element Free Galerkin one (EFG). Using both methods, the displacements, strains and stresses are compared. The results of this study are presented in the forms of figures and tables.

Keywords: Element free galerkin · Moving least square · Elastoplasticity · Dynamique explicit

1 Introduction

The Finite Element Method (FEM) is, currently, the most used method for the numerical simulation of forming processes. It has a very solid theoretical foundation and continuous improvement through the years. This method is also known by the simplicity of its implementation.

In spite of the success of this method (FEM), it continues to confront some difficulties. Effectively, to track discontinuities (propagation of cracks, damage), for example, remeshing of the structures is usually inevitable. Moreover, advanced behavioral models coupled with temperature or damage become very badly measured using the FEM. In these cases, it is fundamental to use the non-local formulations where the notion of neighborhood is important. But, this is in contrast with the FEM basis where the use of elements is essential.

To solve this problem, Meshless methods have been developed since 70s. These methods are more suitable in the applications where the FEM has problem.

These methods avoid the use of Mesh. So, the famous problem associated to the element distortion will disappear and the structure is, only, represented by a set of nodes. Therefore, the formulation based, simply, on nodes is more suitable to solve problems concerning very large deformations.

For several years the Meshless group has made successive developments of different Meshless methods.

N. Aifaoui et al. (Eds.): CMSM 2019, LNME, pp. 804–814, 2020.
https://doi.org/10.1007/978-3-030-27146-6_88

One of the primary Meshless methods, SPH method (Smooth Particle Hydrodynamics) developed by Lucy [1], Gingold and Monaghan [2] to solve problems in astrophysics and, soon after, the same method was used in fluid dynamics by Monaghan [3].

While SPH and its adapted versions were based on strong form, other methods developed in the 1990s were based on the global weak form.

One of these methods known as the Element Free Galerkin (EFG), which developed in 1994 by Belytschko et al. [4], was improved, then, and called IEFG [5].

After that, the method Reproducing Kernel Particle (RKPM) was developed by Liu et al. [6].

Another group of mesh free methods based on local weak forms had appeared. The most popular method is called Meshless Local Petrov-Galerkin (MLPG), see Atluri and Zhu [7]. The main difference between this MLPG method and both EFG and RKPM is that in the MLPG, the local weak forms are generated on the overlapping subdomains. Although in both EFG and RKPM, the integration of the local weak form is carried out in local subdomains.

In this study, we utilize the Element Free Galerkin (EFG) method by means of the Moving Least Square (MLS) approximation. The most important tasks of this work include two aspects. Primary, an analytical study using the EFG method is developed. Next, the elastic and elastoplastic simulations are performed to reveal the accuracy of the aforementioned method to predict the distribution of the different mechanical quantities (stresses, deformations, plastic deformation …) throughout the structure and at each moment.

The results obtained from the EFG (MLS) simulations are qualitatively and quantitatively compared with those issued from analytical solutions and FEM to examine the applicability and the accuracy of the Meshless Method.

2 Moving Least Square (MLS) Approximation

In this work, the shape functions in the Element Free Galerkin (EFG) method, developped by Belytschko et al. [4] are constructed using the Moving Least Square (MLS) approximation introduced by Lancaster and Salkauskas [8].

In the MLS approximation, the trial function around the point x is:

$$u^h(\mathbf{x}) = \sum_{i=1}^{m} p_i(\mathbf{x}) a_i(\mathbf{x}) = \mathbf{p}^T(\mathbf{x})\mathbf{a}(\mathbf{x}) \tag{1}$$

where $\mathbf{p}$ $(\mathbf{x})$ is a basic functions vector and m is the number of terms of monomials.

The vector of coefficients $\mathbf{a}(\mathbf{x})$ is to be determined using the MLS approximation.

In order to precisely obtain the local approximation of the function, the difference between the approximation and the local nodal values must be minimized by a weighted least-squares method.

Define a functional

$$J = \sum_{i=1}^{n} w(x - x_i)(u^h(\mathbf{x}) - u(x_i))^2 = \sum_{i=1}^{n} w(x - x_i)(\mathbf{p}^T(x_i)\mathbf{a}(\mathbf{x}) - u_i)^2 \quad (2)$$

where w(x–x_i) is a weight function defined on a domain Ω_X called domain of influence (see Liu and Gu [9]) and n denotes the number of particles in Ω_X. From (2), we obtain the vector form as follows:

$$J = (\mathbf{Pa} - \mathbf{u})^T \mathbf{W}(\mathbf{x})(\mathbf{Pa} - \mathbf{u}) \quad (3)$$

where

$$\mathbf{u}^T = (u_1, u_2, \ldots, u_n) \quad (4)$$

$$\mathbf{P} = \begin{bmatrix} p_1(x_1) & p_2(x_1) & \ldots & p_m(x_1) \\ p_1(x_2) & p_2(x_2) & \ldots & p_m(x_2) \\ \ldots & \ldots & \ldots & \ldots \\ \ldots & \ldots & \ldots & \ldots \\ p_1(x_n) & p_2(x_n) & \ldots & p_m(x_n) \end{bmatrix} \quad (5)$$

$$\mathbf{W}(\mathbf{x}) = \begin{bmatrix} w(x - x_1) & 0 & \ldots & 0 \\ 0 & w(x - x_2) & \ldots & 0 \\ \ldots & \ldots & \ldots & \ldots \\ \ldots & \ldots & \ldots & \ldots \\ 0 & 0 & \ldots & w(x - x_n) \end{bmatrix} \quad (6)$$

To minimize the function J, we impose ∂J/∂**a** equal to zero. Therefore, we get the following system to solve:

$$\begin{cases} \frac{\partial J}{\partial a_1} = 0 \Leftrightarrow \sum_{i=1}^{n} 2w(x - x_i)p_1(x_i)[\mathbf{p}^T(x_i)\mathbf{a}(x) - u_i] = 0 \\ \frac{\partial J}{\partial a_2} = 0 \Leftrightarrow \sum_{i=1}^{n} 2w(x - x_i)p_2(x_i)[\mathbf{p}^T(x_i)\mathbf{a}(x) - u_i] = 0 \\ \ldots \\ \ldots \\ \frac{\partial J}{\partial a_m} = 0 \Leftrightarrow \sum_{i=1}^{n} 2w(x - x_i)p_m(x_1)[\mathbf{p}^T(x_i)\mathbf{a}(x) - u_i] = 0 \end{cases} \quad (7)$$

Or also, we can write the following matrix form,

$$\mathbf{W}(\mathbf{x})\mathbf{P}\mathbf{P}^T\mathbf{a}(x) = \mathbf{W}(\mathbf{x})\mathbf{Pu} \quad (8)$$

Then we obtain,

$$\mathbf{a}(x) = \mathbf{A}^{-1}(\mathbf{x})\mathbf{B}(\mathbf{x})\mathbf{u} \quad (9)$$

where

$$\mathbf{A}(\mathbf{x}) = \mathbf{W}(\mathbf{x})\mathbf{P}\mathbf{P}^{T} = \sum_{i=1}^{n} w(x - x_i)\mathbf{p}(x_i)\mathbf{p}^{T}(x_i) \tag{10}$$

and

$$\mathbf{B}(\mathbf{x}) = \mathbf{W}(\mathbf{x})\mathbf{P} = [w(x - x_1)\mathbf{p}(x_1)w(x - x_2)\mathbf{p}(x_2)\ldots\ldots\ldots\ldots w(x - x_n)\mathbf{p}(x_n)] \tag{11}$$

Finally, we substitute the Eqs. (9), (10) and (11) in (1) to get the approximation as,

$$u^{h}(\mathbf{x}) = \sum_{i=1}^{n} \phi_i(\mathbf{x})u_i = \mathbf{\Phi}^{T}(\mathbf{x})\mathbf{u} \tag{12}$$

Let $\mathbf{\Phi}(x)$ denote the matrix of shape functions given by the following expression

$$\mathbf{\Phi}^{T}(x) = \mathbf{p}^{T}(x)\mathbf{A}^{-1}(\mathbf{x})\mathbf{B}(\mathbf{x}) \tag{13}$$

3 Imposition of Boundary Conditions

Imposing boundary conditions in the MLS approximation is not evident as in the Finite Element Method (FEM). Therefore, many techniques of imposing essential boundary conditions have been developped, starting with the Lagrange multiplier method [10], then the Penalty method [10], Nitsche's method [10].

The difference between varieties of imposing boundary conditions methods have been well discussed in Duflot [11].

4 Numerical Integration

In Meshless methods, we do not have the concept of mesh used to build shape functions. Then, the numerical evaluation of integrals is more difficult than in the FEM. Thus, different methods of integration have been developped, see Duflot [11].

5 2D Linear Elasticity by MLS

5.1 Two-Dimensional Plate

In this subsection, we evaluate the performance of the MLS approximation by solving a linear elastic problem.

Therefore, we start with a simple example represented by a two-dimensional plate $[0, 1] \times [0, 1]$ with two symmetry conditions, as shown in Fig. 1, and the displacement $U^{d} = 0.1$ mm is imposed along the y-axis.

While working in the case of plane stresses, we take $E = 2.10^{5}$ MPa and $\nu = 0.3$.

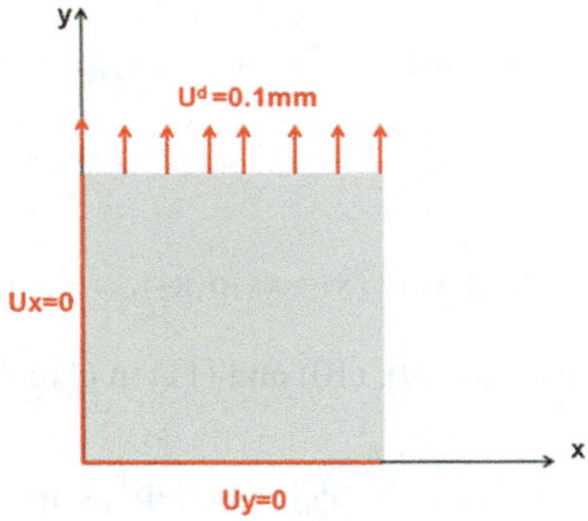

Fig. 1 Dimensions and boundary conditions of the plate

After simulating the plate by the finite element method (FEM) and by the EFG method based on the MLS approximation, we plotted the evolution of the displacements Ux and Uy in several paths of the structure.

Referring to Table 1, the results in Fig. 2a, b approve good efficiency and convergence of the MLS method.

Table 1 The coordinates of paths

	Path1x	Path2x	Path3x	Path4x	Path5x	Path1y	Path2y	Path3y	Path4y	Path5y
x-axe (mm)	x = X	x = X	x = X	x = X	x = X	x = 0	x = 0.2	x = 0.5	x = 0.9	x = 1
y-axe (mm)	y = 0	y = 0.2	y = 0.5	y = 0.9	Y = 1	y = Y	y = Y	y = Y	y = Y	Y = Y

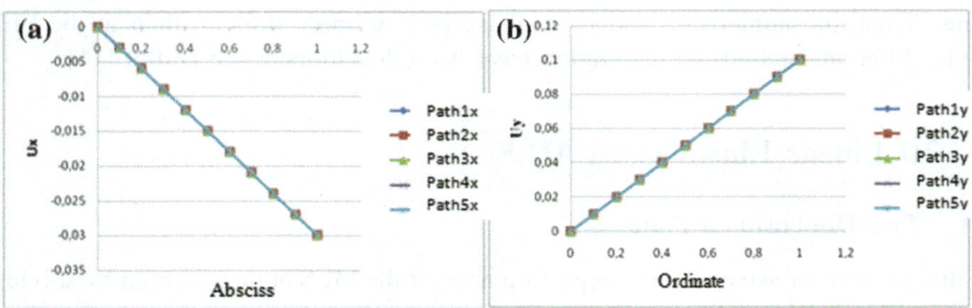

Fig. 2 Displacements Ux **a** and Uy **b** in several paths by MLS

5.2 Timoshenko Beam

In this second elastic linear example, we present the results obtained from the simulation of a Tymoshenko beam (Fig. 3) subjected to a parabolic traction expressed by the equation:

$$t_y(y) = -\frac{P}{6EI}(\frac{D^2}{4} - y^2); I = \frac{D^3}{12} \tag{14}$$

The exact solutions in displacements in plane stresses are given by the following expressions:

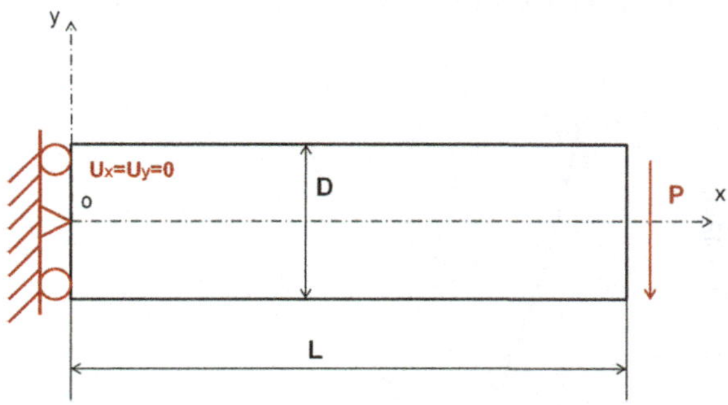

Fig. 3 Dimensions and boundary conditions of the Timoshenko beam

$$u_x(x,y) = -\frac{Py}{6EI}\left[(6L - 3x)x + (2 + \nu)(y^2 - \frac{D^2}{4})\right] \tag{15}$$

$$u_y(x,y) = \frac{P}{6EI}\left[3\nu y^2(L - x) + (4 + 5\nu)\frac{D^2 x}{4} + (3L - x)x^2\right] \tag{16}$$

As well as the exact solutions in constraints:

$$\sigma_{xx}(x,y) = -\frac{P(L - x)y}{I}; \sigma_{xy}(x,y) = -\frac{P}{2}I(\frac{D^2}{4} - y^2); \sigma_{yy}(x,y) = 0 \tag{17}$$

We consider the case of plane stresses and the properties of the material such that $E = 3.10^7$ MPa, ν = 0.3. As for the dimensions, we took D = 12 mm and L = 48 mm. The applied force P is equal to 1000 N.

We also mention that the distribution of the nodes was taken regularly.

The stresses at the center of the beam (x = L/2, yε [−D/2, D/2] were evaluated, plotted and compared with those of the exact solutions given by Eq. (17).

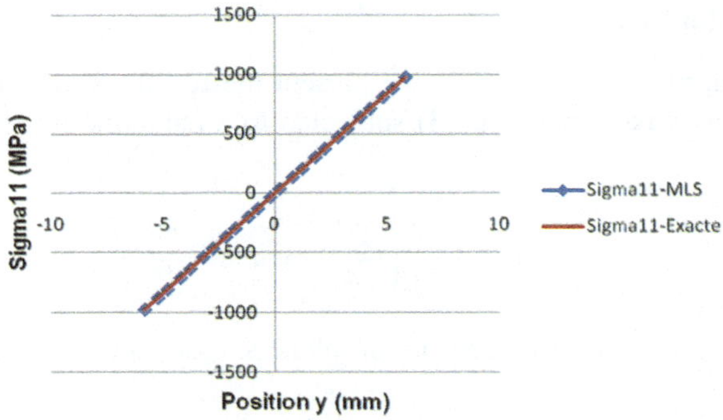

Fig. 4 Comparison of Sigma11 Evolution of x = L/2

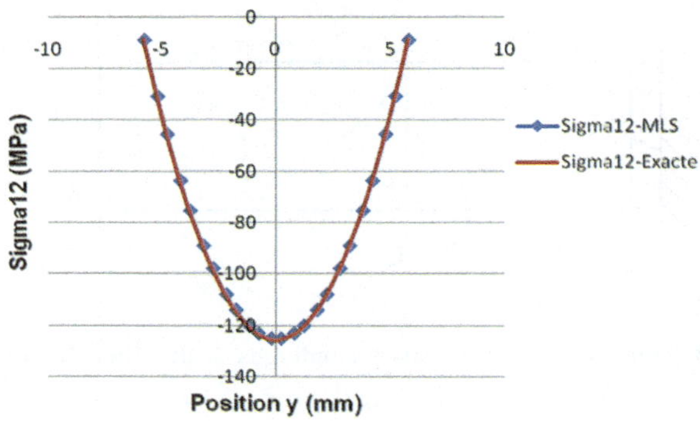

Fig. 5 Comparison of Sigma12 Evolution of x = L/2

Figures 4 and 5 show good agreement between MLS results and exact constraints. To better illustrate this agreement, we calculate the relative error norm between the analytical solutions and the MLS results.

Table 2 Error between analytical and MLS solutions for different node densities

Number of nodes	Relative error norm	
	Displacement	Constraints
33 × 9	2.14×10^{-3}	7.02×10^{-2}
49 × 13	$5{,}61 \times 10^{-4}$	3.52×10^{-3}
65 × 17	$8{,}34 \times 10^{-5}$	1.42×10^{-3}

According to Table 2, the relative error did not exceed the value of 2.10^{-3} in displacements and 7×10^{-2} in stresses. It is also clear that by increasing the number of nodes, the error decreases. Then, we have to optimize the Precision/Time ratio. And this depends on the type of the studied problem and the degree of precision desired.

6 2D Elastoplasticity by MLS

6.1 MLS Discretization and Explicit Scheme

The used method is well presented and explained in Sendi et al. [12].

6.2 Numerical Examples and Discussions

We have modeled a tensile test of a 2D perforated plate in the case of plane stresses. The dimensions as well as the boundary conditions are given in Fig. 6.

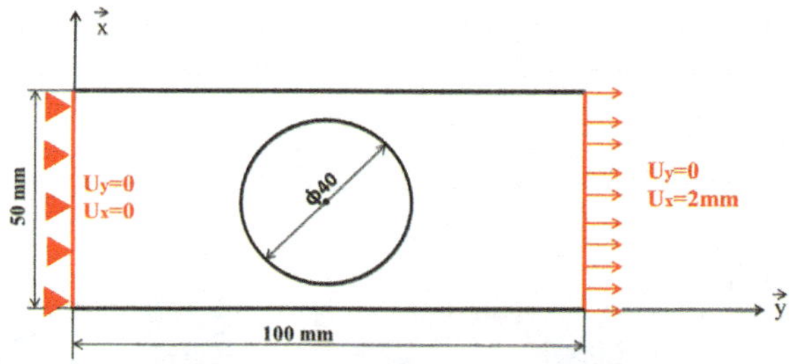

Fig. 6 Dimensions and boundary conditions of the 2D perforated plate

The parameters of the material are summarized in Table 3.

Table 3 Material parameters

Parameters	Symbol	Value
Young module (MPa)	E	2.1×10^5
Poisson coefficient	ν	0.3
Density (kg/m3)	ρ	7850

First, we used the Abaqus software to model and simulate the specimen based on an explicit resolution scheme (see Fig. 7a). Then, we modeled the same test using the EFG (MLS) method implemented under the Matlab software (see Fig. 7b).

In this latter method, we used quadratic weight functions with circular supports having as radius of influence r = 1.4 * dx (dx is the distance between two successive nodes). Thus, we have chosen the penalty method, already explained in Fernandez-Mendez and Huerta [10], to impose boundary conditions.

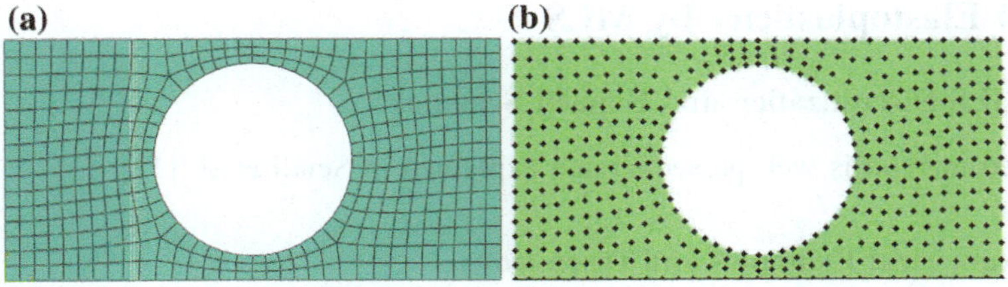

Fig. 7 Mesh used in Explicit Abaqus **a** and Node Distribution for EFG (MLS) **b** in the specimen with central hole

According to Fig. 7a, b and in order to better confirm our conclusions, we note that we used the same distribution of the nodes in the two methods (Table 4).

Table 4 Comparisons of isovalues of two methods EFG (MLS) and FEM

	EFG(MLS)-Matlab	FEM-Abaqus
Von-mises stress (Pa)	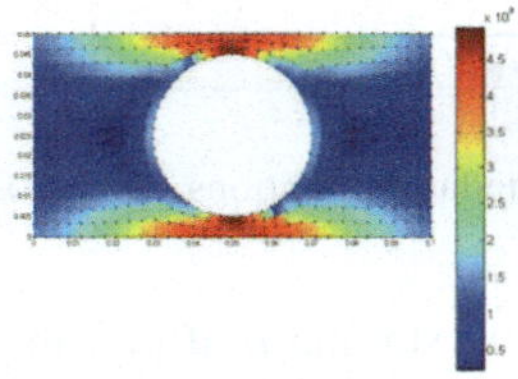	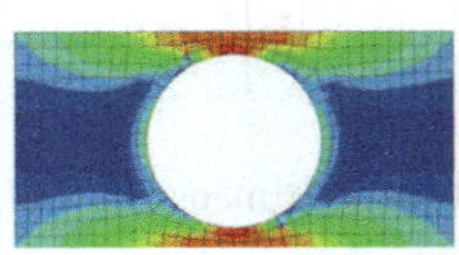
Equivalent plastic strain	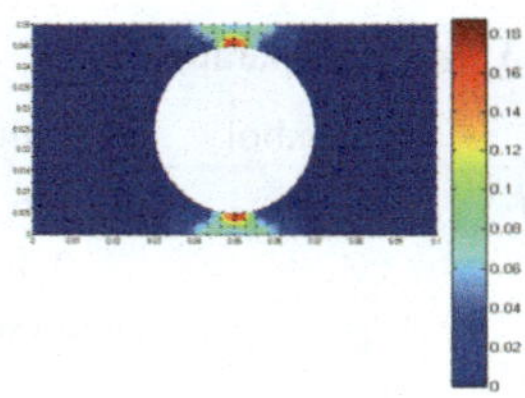	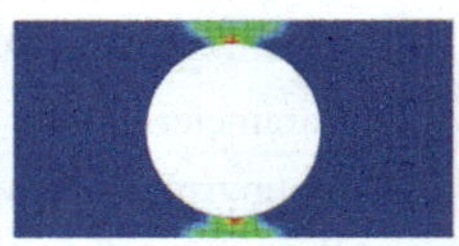

The previous table describes the distribution of physical quantities (deformations, stresses) in the deformed specimen after imposing a displacement of 2 mm. By comparing the results from the Abaqus software (FEM) and those derived from the EFG (MLS) method, we can already see not only quantitative but also qualitative similarity between the different components of different physical quantities. We also notice that plastic flow is well localized in critical zones with stress concentrations. This was clearly modeled by the two methods (EFG and FEM).

According to this example, we can validate the integration of the elastoplastic model. Moreover, it has been used to show that the EFG (MLS) method solves a mechanical problem introducing non-linearity.

7 Conclusions

Using the EFG method and the MLS approximation, two-dimensional elasto-plasticity problems are presented in this work.

The Galerkin weak form is employed to get the equations system, the penalty method is applied for the essential boundary conditions.

Some selected numerical examples are sited to prove that the EFG method has high computational precision and competence.

In addition, this paper presents an evaluation on the efficiency of this method to capture the initiation of the plasticity saturation and its propagation for many steps of time.

Validation of the method has been recognized through comparison of the simulated results with those of analytical solutions and from the Finite Element Method.

Consequently, both qualitative and qualitative conformity between the two methods can be found, signifying that the used Meshless method can reasonably predict the propagation of physical quantities (displacements, deformations, stresses, etc.) until the saturation.

Finally, we conclude that the codes, developed, are ready for use in the modeling and simulating of other mechanical tests.

Therefore, to take advantage the main benefits for which the EFG method has been developed, this study needs to be completed to more complex geometries presenting material discontinuities.

In that case, we intend to simulate and modelize 3D cases presented, essentially, by forming processes with the Element Free Galerkin method.

References

1. Lucy LB (1977) A numerical approach to the testing of the fission hypothesis. Astron J 82:1013–1024
2. Gingold RA, Monaghan JJ (1977) Smoothed particle hydrodynamics: theory and application to non-spherical stars, Monthly Notices R. Astron Soc 181:375–389
3. Monaghan JJ (1988) An introduction to SPH. Comput Phys Commun 48:89–96
4. Belytschko T, Lu YY, Gu L (1994) Element-free Galerkin methods. Int J Numer Methods Eng 37:229–256
5. Zhang Z, Liew KM, Cheng Y (2008) Coupling of the improved element-free Galerkin and boundary element methods for two-dimensional elasticity problems. Eng Anal Boundary Elem 32:100–107
6. Liu WK, Jun S, Zhang YF (1995) Reproducing kernel particle methods. Int J Numer Methods Eng 20:1081–1106
7. Atluri SN, Zhu T (2000) The meshless local Petrov-Galerkin (MLPG) approach for solving problems in elasto-statics. Comput Mech 25:169–179
8. Lancaster P, Salkauskas K (1981) Surfaces generated by moving least squares methods. Math Comput 37:58–141
9. Liu GR, Gu YT (2005) An introduction to Meshfree methods and theirs programming. Springer, Berlin

10. Fernandez-Mendez S, Huerta A (2004) A comparison of two formulations to blend finite elements and mesh-free methods. Comput Methods Appl Mech Eng, 193:1257–1275
11. Duflot M (2004) Application des méthodes sans maillage en mécanique de la rupture, thesis in applied science, University of Liège
12. Sendi Z, Belhadjsalah H, Labergere C, Saanouni K (2016) An element free Galerkin method for an elastoplastic coupled to damage analysis. https://doi.org/10.1051/matecconf/20168007005

Finite-Element Determination of the Equivalent Thermal Conductivity of Hollow Blocks Masonry Wall

Houda Friaa(✉), Myriam Laroussi Hellara, and Abdelwaheb Dogui

Mechanical Engineering Laboratory, LR99ES32, National Engineering School of Monastir, University of Monastir, 5019 Monastir, Tunisia
houda.frea@hotmail.com

Abstract. In the light of the energy crisis, energy saving has become an important topic by every country over the world. Energy consumption of buildings usually takes up to 30–40% of the human's livelihood energy consumption. Hence, the determination of their thermal properties represents an essential task for energy computation. In this paper, a numerical approach is used in order to compute the effective thermal properties of an heterogeneous masonry structure. The structure to be studied is formed by hollow blocks with cavities filled by air, joined periodically with head and bed joints. Those hollow blocks are broadly used due to the good thermal and noise insulation. The thermal conductivities of the solid part of the hollow blocks and of the mortar are the main material input parameters. Moreover, the convection and radiation are taken into account in the cavities. The effective equivalent thermal conductivity tensor is then determined and the effect of radiation and convection is studied.

Keywords: Finite element · Hollow block · Conduction · Convection · Radiation · Equivalent thermal conductivity tensor

1 Introduction

The buildings sector significantly contribute to the total energy consumption all over the world [1, 5, 8]. Many countries have adopted energy-saving policies to reduce their energy consumption. Utilization of energy-efficient building envelope may be a good mean for improving the building energy efficiency. In recent years, hollow block or brick, a new kind of prefabricated building construction material, is becoming more and more popular. This block has good thermal insulation, lightweight and acoustic insulation [12]. Recently, more interest has been made on the determination of the thermal behavior of this complex building structure. Some studies focus on the thermal properties of this structure by using theoretical [2], numerical [4] or experimental methods [3, 9]. These methods play an important role in determining the thermal performance and optimizing the configurations of the hollow block. The related results and conclusions provide valuable references for the engineering applications of the hollow block. One should notice that for experimental method, the financial means necessary could present prohibitive factors for many laboratories. Hence, modeling and

N. Aifaoui et al. (Eds.): CMSM 2019, LNME, pp. 815–822, 2020.
https://doi.org/10.1007/978-3-030-27146-6_89

numerical simulations are an appropriate alternative. Studies on heat transfer characteristics in hollow block walls are very limited. Most of the existing studies neglect the effect of the mortar on the thermal performance of the wall and neglect the effect of multi-mode heat transfer. To overcome this drawback, few papers involve the effect of the multi-mode heat transfer process and heat-insulating property of the configuration of the hollow brick [6, 10]. They conducted a 3D simulation about some kinds of light concrete hollow brick walls and light concrete multi-holed bricks by the finite element method, resulting in complex models with high number of elements and therefore long computational times. In order to solve this problem, another approach has been adopted by [1, 10, 11]. Actually, they simplify the whole wall by using a computational domain which can be regarded as a periodic unit which can reproduce the totality of the building wall by simple translations. In this paper, for the first time, the equivalent thermal conductivity is determined for a masonry wall formed by 500 mm (length) 200 mm (width) and 200 mm (height) hollow concrete block widely applied in the present building constructions all over the world. The finite element software ABAQUS is used. The computational method is employed in order to avoid the limitation for cavity radiation in ABAQUS/standard [7]. A comprehensive investigation on the effect of enclosure configurations on equivalent thermal conductivity is conducted to determine the contributions of every heat transfer mode, i.e., conductivity, natural convection, and radiation.

2 Numerical Study

The investigation of an equivalent thermal conductivity for a hollow blocks masonry wall Ω is carried out on three-dimensional model within a steady state thermal analysis. The computational method is adopted to predict a thermal conductivity of a concrete block wall as well as to determine the effect of various factors that may influence heat transfer through this structure.

2.1 Geometry and Modeling

The masonry wall, as well as the hollow blocks, are shown in Fig. 1. The dimensions of the latter one are summarized in Table 1. This type of blocks is commonly used because of its low cost, its durability and its fireproof. It is made of concrete classification B40, in accordance with NF P 14-402. It is composed of a complex shape with six ($146.67 \times 75.5 \times 180$ mm^3) rectangular holes filled with air. Those cavities present almost 60% of the whole volume.

The modeling of the studied structure consists in adopting a unit cell. The latter is formed by a hollow concrete block joined with mortar on its upper face and one lateral face.

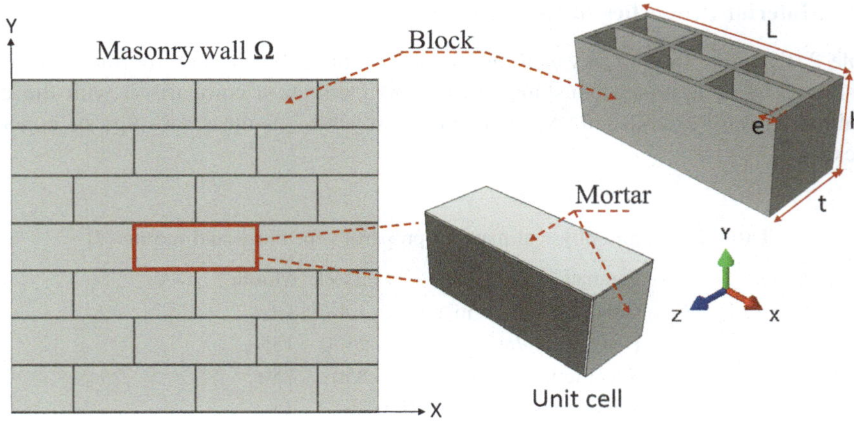

Fig. 1 Masonry wall and the hollow blocks and the unit cell

Table 1 Dimensions of the hollow block

Symbol	Description	Value (mm)
L	Length	500
H	Height	200
t	Thickness	200
e	Partition	15

2.2 Mesh Generation

Quadratic heat transfer finite elements (DC3D20 in ABAQUS) have been used for both block and mortar. In order to guarantee the reliability of the FEM results, a mesh convergence study is performed. This led to a total number of 69,557 finite elements as shown in Fig. 2.

Fig. 2 Finite elements mesh of the chosen unit cell

2.3 Material Properties of the Constituents

Table 2 lists the thermo-physical properties of both block and mortar. All input parameters were assumed constant, in order to facilitate a comparison with the data obtained by well-established experimental approaches on the same type of concrete blocks.

Table 2 Thermo-physical material properties for block and mortar [7]

Parameters	Block	Mortar
Conductivity (W/m°K)	1.4	1.6
Density (Kg/m^3)	1500	1850
Specific heat (J/(Kg °K))	850	750
Film coefficient (W/m^2°K)	25	25
Emissivity	0.9	0.85

2.4 Boundary Conditions

The thermal loading consists in applying a temperature gradient across two faces perpendicular to each of the three spatial directions while the remaining faces are insulated. These boundary conditions generate a heat flow towards the cooler surface. The thermal conductivity in the direction i of the applied temperature gradient can be calculated using Fourier's law:

$$\lambda_i = \frac{-q_i}{S_i}\frac{d_i}{\Delta T} \tag{1}$$

where S_i is the cross-sectional area of the model perpendicular to the heat flow direction, Δd_i is the distance between the two faces across which the temperature gradient $\Delta T = (T_{hot} - T_{cold})$ is applied. In this study, T_{hot} and T_{cold} were arbitrarily set as 400 °C and 100 °C. Whereas q_i represents the sum of the nodal heat flux on the cooler face. It is given by:

$$q_i = \sum_{j=1}^{N} q_j S_j \tag{2}$$

N is the number of elements on the surface, S_j is the surface area of jth element and q_j is the average of the nodal flux values for the jth element. The value of N is dictated by the mesh size, which is arrived after testing different meshes for convergence of the resulting solutions for temperature and flux.

2.5 Heat Transfer in Hollow Blocks

In the solid part of the block and in mortar, the only mode of heat transfer is conduction. Actually, the concrete body itself is porous but the dimensions of the pores are so small that convection and radiation in the pore space can be neglected. However, in the cavities complex heat transfer occurs including conduction, convection and radiation simultaneously. Figure 3 summarizes those different type of heat transfer.

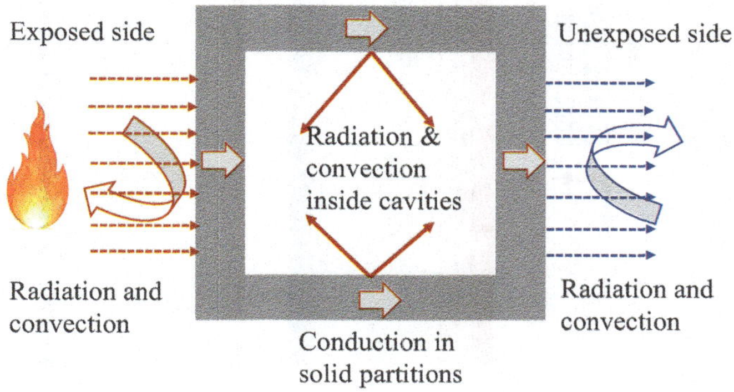

Fig. 3 Schematic diagram showing the mechanism of heat transfer within a cavity

3 Results and Discussions

3.1 Effect of Radiation

In order to evaluate the influence of radiation in the cavities, two cases are simulated. In the first case, the simulations are carried out considering only thermal conductivity in the cavities which is taken into account as 0.026 W/m k. Whereas the second one radiation is also taken into account with an emissivity is 0.9. In both models, the following parameters are also considered: a 4 W/m^2 K film coefficient adopted for the cold surface, a 25 W/m^2 K external film coefficient and a 293 °K ambient temperature.

The temperature distribution in section Y = 200 mm (on the upper face of the unit cell) is presented in Fig. 4 for the two studied cases. One can see that, due to the different thermal conductivities of the block, of the air, and of the mortar, the isotherm is unevenly distributed. One can also notice in Fig. 4b that when taking into account radiation, the temperature gradient in air layers is higher than the one without radiation as shown in Fig. 4a; while it is evenly distributed at the position of block material. One can notice that the temperature distribution depends strongly on the mortar conductivity: if the mortar and block conductivities increase, the insulation of the wall decreases. Mortar, despite its low volume fraction (3.5%), significantly increased the through-thickness thermal conductivity of the wall by nearly 25%. Thus, using insulating mortar with low thermal conductivity can reduce heat transfer through the mortar. In particular, replacing ordinary concrete mortar with insulating mortar with a low thermal conductivity.

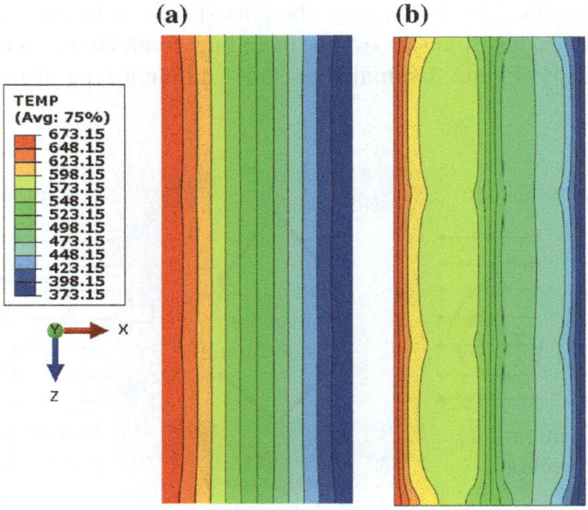

Fig. 4 Typical temperature distribution (**a**) without radiation (**b**) with radiation

Moreover, Fig. 5 shows the temperature variation along the unit cell width at Y = 200 mm. In the case of only conduction and convection, it can be noted that the temperature varies only a little near the boundary faces. Then, the temperature drops very quickly and almost linearly away from this area. For the radiation model, the curve of temperature is also decreasing along the path but no more linearly. We can actually observe several points of inflection on this curve. This is caused by the existence of diffusion regimes (conduction, convection, and radiation), occurring at two different speeds. These numerical results show the important role played by radiation in the heat transfer process in a hollow concrete block wall exposed to fire as shown in Fig. 5. Radiative exchanges accelerate the propagation of heat, but in a non-uniform way because of, in particular, form factors.

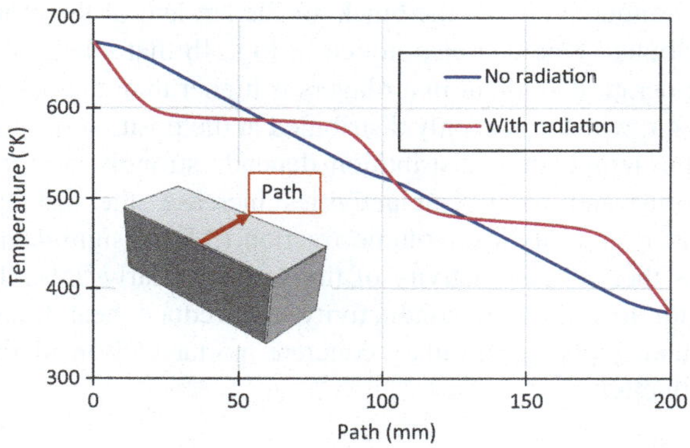

Fig. 5 Thermal field distribution along a chosen path across the block

3.2 Resulting Effective Thermal Conductivity

The masonry structure represents a composite material characterized by an orthotropic behavior which arises from the spatial organization of its constituents. Hence the need for effective thermal conductivity determination per direction giving rise to an effective thermal conductivity tensor. Steady-state boundary conditions are applied to all the models in the X, Y, and Z directions. As evident from the results, the cavities shapes significantly influence the equivalent thermal conductivity of the block wall. In the X directions, the equivalent thermal conductivity is the highest because of the presence of more cavities in this direction as shown in Fig. 6.

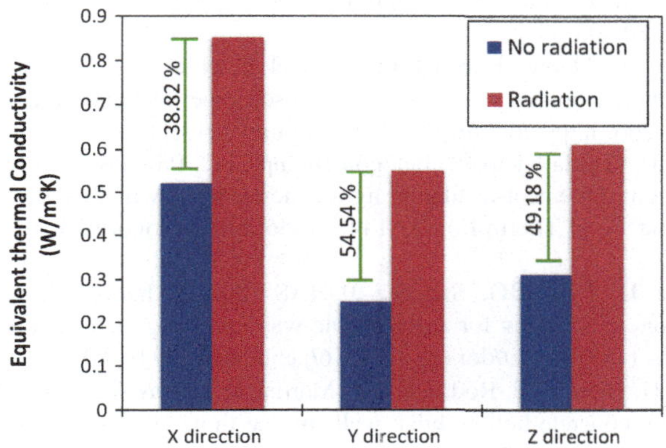

Fig. 6 Equivalent thermal conductivities of the block wall within the three directions

4 Conclusion

A thermal finite element analysis has been performed in three-dimensions in order to study the heat transfer of a concrete hollow blocks wall in different directions. The three models of heat transfer have been taken into account conduction in the solid parts, convection and radiation in the air cavities. This led to the following conclusions:

(1) Equivalent Thermal Conductivity strongly depends on the combined effect of the heat conduction in concrete material, the natural convection and the radiation within the cavities.
(2) The heat radiation inside the cavities increases considerably the equivalent thermal conductivity of the wall; furthermore, the amount and arrangement of holes played important roles on equivalent thermal conductivity.
(3) The equivalent thermal conductivity depends also on both the block and the mortar thermal conductivities.

It must be noted that the thermo-physical material properties for block and mortar are assumed constant in this work which presents an initial step on the special geometry of the studied structure for performing equivalent thermal conductivity tensor. This study will be extended to the determination of this tensor of thermal conductivity when the thermo-physical properties depend on the temperature in further work.

References

1. Alghamdi AA, Alharthi HA (2017) Multiscale 3D finite-element modelling of the thermal conductivity of clay brick walls. Constr Build Mater 157:1–9. https://doi.org/10.1016/j.conbuildmat.2017.09.081
2. Bouchair A (2008) Steady state theoretical model of fired clay hollow bricks for enhanced external wall thermal insulation. Build Environ 43:1603–1618. https://doi.org/10.1016/j.buildenv.2007.10.005
3. del Coz Díaz JJ, Álvarez-Rabanal FP, Gencel O et al (2014) Hygrothermal study of lightweight concrete hollow bricks: a new proposed experimental-numerical method. Energy Build 70:194–206. https://doi.org/10.1016/j.enbuild.2013.11.060
4. del Coz Díaz JJ, García Nieto PJ, Betegón Biempica C, Prendes Gero MB (2007) Analysis and optimization of the heat-insulating light concrete hollow brick walls design by the finite element method. Appl Therm Eng 27:1445–1456. https://doi.org/10.1016/j.applthermaleng.2006.10.010
5. del Coz Díaz JJ, Niet PJG, Sanchez JDHAS (2009) Thermal design optimization of lightweight concrete blocks for internal one-way spanning slabs floors by FEM. Energy Build 41:1276–1287. https://doi.org/10.1016/j.enbuild.2009.08.005
6. del Coz Diaz JJ, Nieto PJG, Rodriguez M, Martinez-Luengas AL (2006) Non-linear thermal analysis of light concrete hollow brick walls by the finite element method and experimental validation. Appl Therm Eng 26:777–786. https://doi.org/10.1016/j.applthermaleng.2005.10.012
7. Keelson H (2018) Fire resistance quantification of non-loadbearing masonry walls-numerical study. Department of Civil and Environmental Engineering Carleton University Ottawa, Ontario
8. Kočí J, Maděra J, Černý R (2015) A fast computational approach for the determination of thermal properties of hollow bricks in energy-related calculations. Energy 83:749–755. https://doi.org/10.1016/j.energy.2015.02.084
9. Kus H, Özkan E, Göcer Ö, Edis E (2013) Hot box measurements of pumice aggregate concrete hollow block walls. Constr Build Mater 38:837–845. https://doi.org/10.1016/j.conbuildmat.2012.09.053
10. Li LP, Wu ZG, Li ZY et al (2008) Numerical thermal optimization of the configuration of multi-holed clay bricks used for constructing building walls by the finite volume method. Int J Heat Mass Transf 51:3669–3682. https://doi.org/10.1016/j.ijheatmasstransfer.2007.06.008
11. Tang DL, Li LP, Song CF et al (2015) Numerical thermal analysis of applying insulation material to holes in hollow brick walls by the finite-volume method. Numer Heat Transf Part A 37–41. https://doi.org/10.1080/10407782.2014.986396
12. Zhang Y, Du K, He J et al (2014) Impact factors analysis on the thermal performance of hollow block wall. Energy Build 75:330–341. https://doi.org/10.1016/j.enbuild.2014.02.037

Formulation of the Dynamic Stiffness Matrix of Prestressed Cross-Ply Laminated Circular Cylin-Drical Shell Subjected to Distributed Loads

Imen Harbaoui[1(✉)] and Mohamed Amin Kadimallah[2]

[1] Laboratory of Applied Mechanics and Engineering, LR-MAI University Tunis El Manar-ENIT BP37- Le Belvédère, 1002 Tunis, Tunisie
imene.harbaoui8@gmail.com

[2] College of Engineering, Civil Engineering Department, Prince Sattam Bin Abdulaziz University, BP 655, Al-Kharj 11942, Saudi Arabia
aminekh6@yahoo.fr

Abstract. This paper describes a procedure for taking into account distributed loads in the calculation of the harmonic response of a cross-ply laminated circular cylindrical shell subjected to internal pressure using the dynamic stiffness method. Based on the first order shear deformation theory founded on love's first approximation theory the dynamic stiffness matrix has been built from which natural frequencies are easily calculated. The vibration analysis is then validated with numerical examples to determine the performance of this model and the effect of presetress on the frequency spectrum. The response of the system is determined with applied equivalent loads on element boundaries. The described approach has many advantages compared to the finite element method, such as reducing the computing time with a minimum model size and higher precision.

Keywords: Continuous element method · Distributed loads/internal pressure · Cross-ply laminated composite shell · Dynamic stiffness method · Harmonic response

1 Introduction

The vibrations of composite prestressed shells have been the subject of many researches. These researches are very interesting since these composite prestressed shells are widely used as elements for various engineering structures such as pressure vessels, pipes, storage tanks, ducts in many engineering applications particularly in the medical field, nuclear and marine. Many papers investigate the free vibrations of a prestressed cylindrical shell. We can say that the first investigation in this field was that of Fung in [1], he studied the effect of pressure on the vibration frequencies for cylinder with freely supported ends. He proved theoretically, that for thin cylinder the internal pressure has a very significant effect on the frequency of the different modes. Theses theoretical results are validate by the experimental in 1957 by Fung et al. [2]. Later Tj et al. investigated the vibrations of pressurized shells using finite element method by considering a cylindrical

N. Aifaoui et al. (Eds.): CMSM 2019, LNME, pp. 823–832, 2020.
https://doi.org/10.1007/978-3-030-27146-6_90

shell partially buried on radial elastic foundation [8]. More recently Hu and Redekop employed the differential quadrature method to investigate prestressed vibration analysis of a cylindrical shell with an oblique end [3, 4] investigate the free vibration and dynamic response of multilayered composite circular cylindrical shells under static axial loads and internal pressure using first order shear deformation theory. In the literature Leissa [5] discussed the nonlinear vibration behavior of prestressed shells.

In the present paper, the dynamic stiffness formulation of cross-ply laminated circular cylindrical shell as defined in previous studies by Casimir et al. [6] and by Harbaoui et al. [7] is developed to deal with prestressed cross-ply laminated circular cylindrical shell based on first order shear deformation theory. In the present formulation both rotatory inertia and shear deformation effects are taken into account. In this study we will show the influence of preload on the shape of the frequency spectra. Knowledge of these spectra is necessary for this purpose to avoid possible resonances in engineering applications.

2 Geometry

We consider cross-ply laminated cylindrical shell of length L, radius R and thickness h as shown in Fig. 1. The cylindrical shell consists of N orthotropic layers which are perfectly bound together.

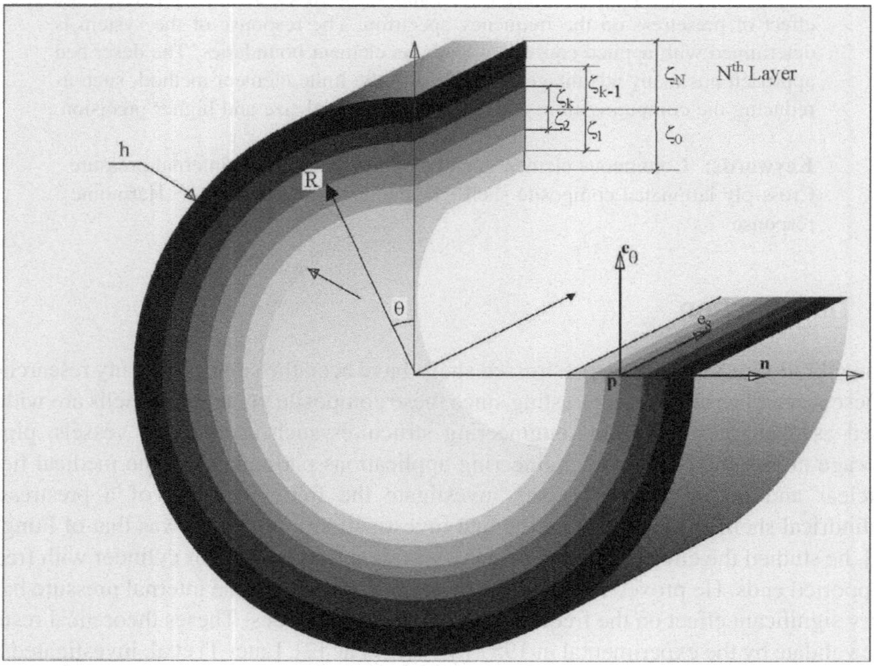

Fig. 1 Cross-ply laminated cylindrical shell [6]

3 Kinematic Assumptions

According to the Mindlin–Reissner assumption, the components of the harmonic displacement are given by Eq. (1):

$$\begin{cases} u_s(s,\theta,\zeta) = U_s(s,\theta) + \zeta\phi_\theta(s,\theta) \\ u_\theta(s,\theta,\zeta) = U_\theta(s,\theta) + z\phi_s(s,\theta) \\ u_\zeta(s,\theta,\zeta) = U_\zeta(s,\theta) \end{cases} \tag{1}$$

where u_s,u_θ and u_ζ are displacement components in the e_s, e_θ and **n** directions respectively. U_s,U_θ and U_ζ are in-plane amplitude displacement components of the shell in the middle surface. ϕ_s, ϕ_θ are the rotations of the middle surface of transverse normal about e_s and e_θ, respectively.

Considering the assumption of small displacements, the strain– displacement relationship of laminated composite cylindrical shell of radius R can be written as:

$$\begin{cases} \varepsilon_{ss} = \frac{\partial u_s}{\partial s} = \frac{\partial U_s}{\partial s} + \zeta\frac{\partial \phi_\theta}{\partial s}, \quad \varepsilon_{\theta\theta} = \frac{1}{R+\zeta}\left(\frac{\partial U_\theta}{\partial \theta} + \zeta\frac{\partial \phi_s}{\partial \theta} + U_\zeta\right) \\ \gamma_{s\theta} = 2\varepsilon_{s\theta} = \frac{1}{R+\zeta}\left(\frac{\partial U_s}{\partial \theta} + \zeta\frac{\partial \phi_\theta}{\partial \theta}\right) + \frac{\partial U_\theta}{\partial s} + \zeta\frac{\partial \phi_s}{\partial s} \\ \gamma_{s\zeta} = 2\varepsilon_{s\zeta} = \phi_\theta + \frac{\partial U_\zeta}{\partial s}, \quad \gamma_{\theta\zeta} = 2\varepsilon_{\theta\zeta} = \frac{1}{R+\zeta}\left(\frac{\partial U_\zeta}{\partial \theta} - U_\theta - \zeta\phi_s\right) + \phi_s \end{cases} \tag{2}$$

4 Lamina Constitutive Relations

Consider a composite shell composed of N orthotropic layers of uniform thickness with the the constitutive orthotropic material axis of the kth layer is oriented with an angle θ_k. The stress–strain relations of the k_{th} layer are reduced to the following equations by neglecting the transverse normal strain and stress:

$$\begin{Bmatrix} \sigma_{ss} \\ \sigma_{\theta\theta} \\ \sigma_{\theta\zeta} \\ \sigma_{s\zeta} \\ \sigma_{s\theta} \end{Bmatrix} = \begin{bmatrix} Q_{11}^k & Q_{12}^k & 0 & 0 & 0 \\ Q_{12}^k & Q_{22}^k & 0 & 0 & 0 \\ 0 & 0 & Q_{44}^k & 0 & 0 \\ 0 & 0 & 0 & Q_{55}^k & 0 \\ 0 & 0 & 0 & 0 & Q_{66}^k \end{bmatrix} \begin{Bmatrix} \varepsilon_{ss} \\ \varepsilon_{\theta\theta} \\ \gamma_{\theta\zeta} \\ \gamma_{s\zeta} \\ \gamma_{s\theta} \end{Bmatrix} \tag{3}$$

where Q_{ij} are the transformed stiffnesses of the kth lamina [9]. The expressions of Q_{ij} are given in Appendix A.

5 Behaviour Equations of Composite Shell

Relations between internal forces and moments resultants are given by considering the internal forces per unit length. See Thinh [9]

$$(N_{ss}, N_{\theta\theta}, N_{s\theta}, N_{\theta s}) = \int_{-h/2}^{h/2} \left[\sigma_{ss}\left(1+\frac{\zeta}{R}\right), \sigma_{\theta\theta}, \tau_{s\theta}\left(1+\frac{\zeta}{R}\right), \tau_{s\theta}\right] d\zeta \tag{4}$$

$$(M_{ss}, M_{\theta\theta}, M_{s\theta}, M_{\theta s}) = \int_{-h/2}^{h/2} \left[\zeta\sigma_{ss}\left(1+\frac{\zeta}{R}\right), \zeta\sigma_{\theta\theta}, \zeta\tau_{s\theta}\left(1+\frac{\zeta}{R}\right), \zeta\tau_{s\theta}\right] d\zeta \tag{5}$$

and

$$(Q_s, Q_\theta) = K\int_{-h/2}^{h/2} \left[\tau_{s\zeta}\left(1+\frac{\zeta}{R}\right), \tau_{\theta\zeta}\right] d\zeta \tag{6}$$

6 The Dynamic Equilibrium Equations

Herein, we use the first order shell theory gives the following for a circular cylindrical shell subjected to harmonic distributed forces [6] and internal pressure. The effect of the internal pressure and the change of membrane curvatures have to be incorporated in the equations of equilibrium according to Timoshenko's theory [1], so the equilibrium equations of the prestressed cylindrical shell in the case of a harmonic regime are as follows:

$$\begin{cases} \frac{\partial N_{ss}}{\partial s} + \frac{1}{R}\frac{\partial}{\partial\theta}\left(N_{s\theta} - \frac{1}{2R}M_{s\theta}\right) - \frac{\overline{N}_\theta}{R}\left(\frac{\partial^2 U_\theta}{\partial s\partial\theta} - \frac{\partial U_\zeta}{\partial s}\right) + f_s = I_0\frac{\partial^2 U_s}{\partial t^2} + I_1\frac{\partial^2 \phi_\theta}{\partial t^2} \\ \frac{\partial}{\partial s}\left(N_{s\theta} + \frac{1}{2R}M_{s\theta}\right) + \frac{\partial N_\theta}{R\partial\theta} + \frac{Q_\theta}{R} + \overline{N}_s\left(\frac{\partial^2 U_\theta}{\partial s^2}\right) + f_\theta = I_0\frac{\partial^2 U_\theta}{\partial t^2} + I_1\frac{\partial^2 \phi_s}{\partial t^2} \\ \frac{\partial Q_s}{\partial s} + \frac{1}{R}\frac{\partial Q_\theta}{\partial\theta} - \frac{N_{\theta\theta}}{R} + \overline{N}_s\left(\frac{\partial^2 U_\zeta}{\partial s^2}\right) + \frac{\overline{N}_\theta}{R}\left(\frac{\partial U_\theta}{R\partial\theta} - \frac{\partial^2 U_\zeta}{R\partial\theta^2}\right) + f_n = I_0\frac{\partial^2 U_\zeta}{\partial t^2} \\ \frac{\partial M_{ss}}{\partial s} + \frac{1}{R}\frac{\partial M_{s\theta}}{\partial\theta} - Q_s = I_1\frac{\partial^2 U_s}{\partial t^2} + I_2\frac{\partial^2 \phi_\theta}{\partial t^2} \\ \frac{\partial M_{s\theta}}{\partial s} + \frac{1}{R}\frac{\partial M_{\theta\theta}}{\partial\theta} - Q_\theta = I_1\frac{\partial^2 U_\theta}{\partial t^2} + I_2\frac{\partial^2 \phi_s}{\partial t^2} \end{cases} \tag{7}$$

It is assumed that the pressure remains usually in its initial direction it's not influenced by the deformation of the shell during vibration. The axial prestress caused by the pressure acting upon the ends of the tank is neglected compared to the circumferential prestress.

Stresses were those due to internal pressure:

$$\begin{aligned} &\overline{N}_\theta = PR, \overline{N}_s = \frac{PR}{2} \\ &\overline{N_{s\theta}} = \overline{T_s} = \overline{T_\theta} = \overline{M_s} = \overline{M_\theta} = \overline{M_{\theta s}} = 0 \end{aligned} \tag{8}$$

$$I_i = \sum_{k=1}^{N} \int_{\zeta_k}^{\zeta_{k+1}} \rho^k \zeta^i d\zeta (i = 0, 1, 2) \tag{9}$$

ρ^k is the material mass density of the kth layer and f_s, f_θ and f_n are distributed loads described on the directions e_s, e_θ and n respectively.

7 Dynamic Stiffness Relation of Thick Laminated Prestressed Cylindrical Shells Subjected to Distributed Loads

7.1 Dynamic Transfer Relation and Sate Vector

Thinh et Nguyen [9] described the strategy for calculating the dynamic stiffness relation. This procedure based on a Fourier expansion of each component of the state vector E is adopted for the case of cross-ply laminated prestressed cylindrical shells.

Where the sate vector is defined as following:

$$E = (u, \mathrm{v}, w, \beta, \beta_\theta, N_s, N_{\theta s}, T_s, M_s, M_{\theta s})$$

According to the 2π periodicity of the circular cylindrical shells with respect to the variable θ, we obtain two uncoupled symmetric and antisymmetric systems

The symmetric solution is such that:

$$^sE = \sum_{m=0}^{\infty} {}^sE_m^{(1)} \cos(m\theta) \quad {}^aE = \sum_{m=0}^{\infty} {}^aE_m^{(2)} \sin(m\theta) \tag{10}$$

The antisymmetric solution is such that:

$$^sE = \sum_{m=0}^{\infty} {}^sE_m^{(2)} \sin(m\theta) \quad {}^aE = \sum_{m=0}^{\infty} {}^aE_m^{(1)} \cos(m\theta) \tag{11}$$

Introducing Fourier expansions in the derivations of all components of state vector with respect to variable s gives a system of equations as follows:

$$\begin{aligned}
&\frac{dU_{sm}^{(1)}}{ds} = f_1, \frac{dU_{\theta m}^{(2)}}{ds} = f_2, \frac{dU_{\zeta m}^{(1)}}{ds} = f_3, \frac{d\phi_{\theta m}^{(1)}}{ds} = f_4, \frac{d\phi_{sm}^{(2)}}{ds} = f_5 \\
&\frac{dN_{ssm}^{(1)}}{ds} = f_6 + \frac{\overline{N}_\theta}{R}\left(\frac{\partial^2 U_{\theta m}^{(2)}}{\partial s \partial\theta} - \frac{\partial U_{\zeta m}^{(1)}}{\partial s}\right) - f_{sm}^{(1)}, \frac{dN_{s\theta m}^{(2)}}{ds} = f_7 - \overline{N_s}\left(\frac{\partial^2 U_{\theta m}^{(2)}}{\partial s^2}\right) - f_{\theta m}^{(2)} \\
&\frac{dQ_{sm}^{(1)}}{ds} = f_8 - \overline{N_s}\left(\frac{\partial^2 U_{\zeta m}^{(1)}}{\partial s^2}\right) - \frac{\overline{N}_\theta}{R}\left(\frac{\partial U_{\theta m}^{(2)}}{R\partial\theta} - \frac{\partial^2 U_{\zeta m}^{(1)}}{R\partial\theta^2}\right) + f_{nm}^{(1)}, \frac{dM_{sm}^{(1)}}{ds} = f_9, \frac{dM_{s\theta m}^{(2)}}{ds} = f_{10}
\end{aligned} \tag{12}$$

The expressions of functions f_i (i = 1–10) are given in Appendix B. The two systems of differential equations are written in the matrix according to Eq. (13):

$$\frac{\partial E_{im}}{\partial s} = \Delta_{im}(s,\omega).E_{im} + f_{im} \quad \text{for} \quad i = 1, 2 \tag{13}$$

where $f_{1m} = (0,0,0,0,0,-f_{1sm},-f_{1\theta m},-f_{1nm},0,0)$ and $f_{2m} = (0,0,0,0,0,-f_{2sm},-f_{2\theta m}-f_{2nm},0,0)$.

$$E_{im}(L) = T_{im}(\omega)E_{im}(0) + {}^{c}E_{im}(0) \tag{14}$$

with: $T_{im}(\omega) = e^{\Delta_{im}(\omega)L}$ represent the dynamic transfer matrices and ${}^{c}E_{im}$ are additional part vectors formulated from the distributed load acting on the surface of the cylindrical shell

7.2 Dynamic Stiffness Relation of Presressed Composite Shell

The dynamic stiffness matrix related the external forces acting on the two edges of the shell to the displacement of these edges. This dynamic stiffness matrix is obtained from transfer matrix after applying several matrix manipulations and by partitioning the transfer matrix into four blocks. It is expressed according to [4] as follows:

$$K_{im}(\omega).U_{im} = F_{im} + {}^{c}F_{im} \text{ for i } \in \{1,2\} \tag{15}$$

where:

$$K_m^{(i)} = \begin{bmatrix} {}^{12}T_m^{(i)^{-1}}.{}^{11}T_m^{(i)} & {}^{12}T_m^{(i)^{-1}} \\ -{}^{12}T_m^{(i)^{-T}} & {}^{22}T_m^{(i)}.{}^{12}T_m^{(i)^{-1}} \end{bmatrix} \tag{16}$$

${}^{c}F_{im}$ represent the force vectors of additional forces computed from the distributed forces acting between the two edges.

8 Numerical Validation

8.1 A Cross-Ply Laminated Prestressed Cylindrical Shell

To validate the present formulation, a cross-ply laminated prestressed shell (CPSE) element was developed with computer program to solve different numerical examples on harmonic response of composite prestressed shells under distributed loads. The results obtained with continuous elements are compared with those obtained with finite elements.

The dynamic stiffness matrix of composite circular cylindrical shell subjected to distributed loads used in this study has been validated previously by Casimir et al. [6]. The object of the present paper is to validate the model under internal static pressure P. In the current case of shell with both ends free the only considered presressed resultant is the circumferential $\overline{N}_{\theta} = PR$

The dimensions of the circular cylindrical shell are:

$$L = 0.1\,\text{m},\ R = 0.07\,\text{m},\ h = 0.025\,\text{m}$$

The shell is composed of four-layer cross-plies from the outer to the inner layer. All layers have the same thickness and material properties. Used material is graphite/epoxy having the following characteristics:

$$E_1 = 138\,\text{GPa},\ E_2 = 8.96\,\text{GPa.},\ G_{12} = G13 = 7.1\,\text{GPa},\ G_{23} = 3.45\,\text{GPa},\ \nu_{12} = 0.3,$$
$$\nu_{12} = \nu_{12}\frac{E_2}{E_1} = 0.019,\ \rho = 1645\,\text{kg/m3}.$$

8.2 Distributed Radial Load

The cylindrical shell shown in Fig. 2 is subjected to Distributed load applied along a line at the outer surface of the cylinder.

The following expression of the Fourier expansion of the force vectors are involved in Eq. (17).

$$f_0^{(1)} = \left(0,0,0,0,0,0,0,-\frac{f}{2\pi},0,0\right)^T \quad \forall m \neq 0, f_m^{(1)} = \left(0,0,0,0,0,0,0,-\frac{f}{2\pi},0,0\right)^T$$
$$\forall m, \quad f_m^{(2)} = 0 \tag{17}$$

20 terms in the Fourier expansions are required to obtain full convergence at frequencies up to 20,000 Hz. The harmonic responses are compared with the results obtained with a commercial FE software application. This comparison is given in Fig. 2. There is a convergence of results obtained by the two models for an internal pressure of 8e7Pa. The advantage of the present method compared to FEM is the calculation time which is preserved in this case is 10 times faster as well as the significant reduction in number of meshes. This method is interesting to overcome the problem of enormous number of elements required by FEM (Fig. 3).

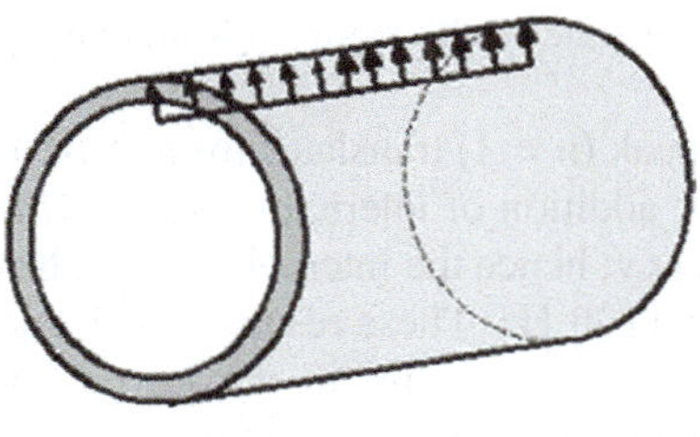

Fig. 2 Distributed load along a line

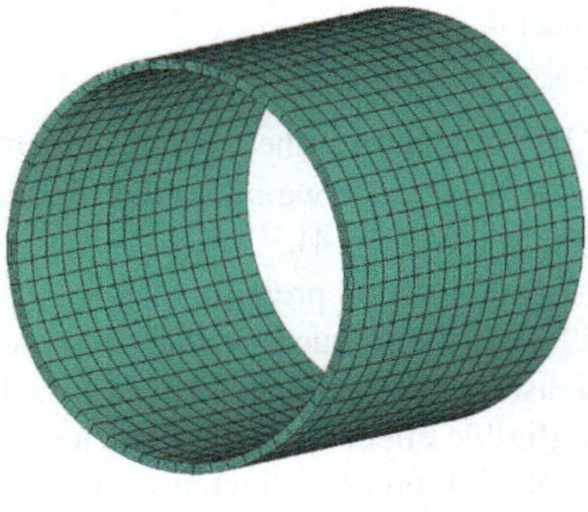

Fig. 3 Finite element model (elements 60 × 15)

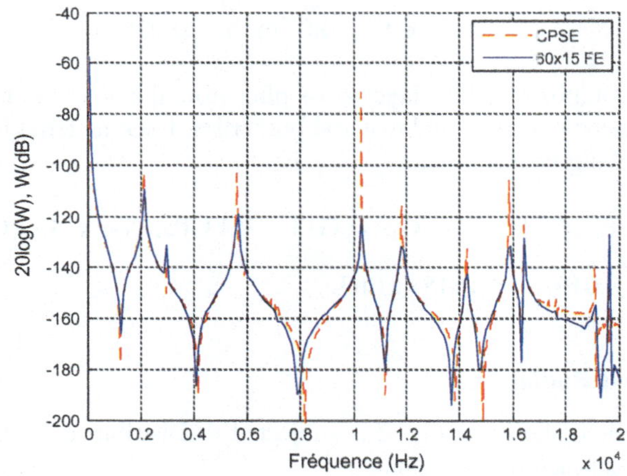

Fig. 4 Harmonic response of a prestressed cylindrical shell (internal pressure P = 8 × 10^7 Pa)

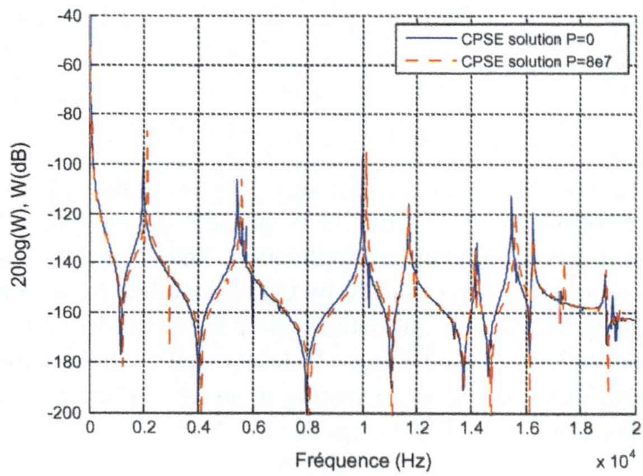

Fig. 5 Comparison of the harmonic response of the unpressurized cylinder and the prestressed cylinder

By comparing the spectra obtained with an unpressurized and a pressurized (8 × 10^7) cylinder we noted that frequency peaks translate to the right with increasing pressure (see Fig. 4).

For an internal pressure of 8 × 10^7 Pa, the first peak (n = 1) translates from 1910 to 2111 Hz. The frequency is slightly increased by the addition of internal pressure. This increase in frequency diminishes for a higher frequency; hence the internal pressure has a negligible effect when frequencies are higher than 1106 Hz. These results shows that the internal pressure increases the natural frequencies, especially in the lower frequency. In other words, the internal pressure tends to stabilize the structure (Fig. 5).

9 Conclusion

The method of dynamic stiffness matrix applied to a prestressed cross-ply laminated composite cylindrical shell subjected to distributed radial loads is described in this paper which has allowed us to take into account the effect of the static pressure in the formulation of a continuous element of axisymmetric composite shell. The introduction of the "internal effect of the pressure and the change of the curvatures of the membranes were incorporated into the equilibrium equations according to Timoshenko theory. This procedure is used for the case of a prestressed composite cylindrical shell subject to radial load acting on their edges. The comparison between the curves obtained by continuous elements and those obtained by finite elements prove the effectiveness of the proposed method in view of the high accuracy of the results as well as the speed of calculation especially in the medium and high frequency ranges.

Appendix A

$$\begin{cases} Q_{11} = \frac{E_1}{1-\nu_{12}\nu_{21}}\cos^4\theta + \frac{E_2}{1-\nu_{12}\nu_{21}}\sin^4\theta \\ Q_{22} = \frac{E_2}{1-\nu_{12}\nu_{21}}\sin^4\theta + \frac{E_2}{1-\nu_{12}\nu_{21}}\cos^4\theta \\ Q_{12} = \frac{\nu_{12}E_2}{1-\nu_{12}\nu_{21}} \\ Q_{44} = G_{23}\cos^2\theta + G_{13}\sin^2\theta \\ Q_{55} = G_{13}\cos^2\theta + G_{23}\sin^2\theta \\ Q_{66} = G_{12} \end{cases}$$

Appendix B

$$\begin{cases} f_1 = mc_4 U_{\theta m} + c_4 U_{\zeta m} + mc_5 \phi_{sm} + \frac{D_{11}}{c_1} N_{sm} - \frac{B_{11}}{c_1} M_{sm} \\ f_2 = \frac{m}{R} U_{sm} - \frac{D_{66}}{c_{10}} N_{s\theta m} + \frac{B_{66}}{c_{10}} M_{s\theta m} \\ f_3 = -\phi_{\theta m} + \frac{1}{kA_{55}} Q_{sm} \\ f_4 = mc_2 U_{\theta m} + c_2 U_{\zeta m} + mc_3 \phi_{sm} - \frac{B_{11}}{c_1} N_{sm} + \frac{A_{11}}{c_1} M_{sm} \\ f_5 = -\frac{m}{R} \phi_{\theta m} + \frac{B_{66}}{c_{10}} N_{s\theta m} - \frac{A_{66}}{c_{10}} M_{s\theta m} \\ f_6 = -I_0\omega^2 U_{sm} - I_1\omega^2 \phi_{\theta m} - \frac{m}{R} N_{s\theta m} + \frac{m}{2R^2} M_{s\theta m} \\ f_7 = \left[\frac{3kA_{44}}{2R^2} + m^2\left(c_6 - \frac{c_8}{2R}\right) - I_0\omega^2\right] U_{\theta m} + m\left(c_6 + \frac{3kA_{44}}{2R^2} - \frac{c_8}{2R}\right) U_{\zeta m} \\ \quad + \left[\frac{I_2\omega^2}{2R} - I_1\omega^2 - \frac{3kA_{44}}{2R} + m^2\left(c_7 - \frac{c_9}{2R}\right)\right] \phi_{sm} + m\left(\frac{c_5}{2R} - c_4\right) N_{sm} + m\left(\frac{c_4}{2R} - c_2\right) M_{sm} \\ f_8 = m\left(c_6 + \frac{kA_{44}}{R^2}\right) U_{\theta m} + \left(c_6 + \frac{kA_{44}m^2}{R^2} - I_0\omega^2\right) U_{\zeta m} + m\left(c_7 - \frac{kA_{44}}{R}\right) \phi_{sm} - c_4 N_{sm} - c_2 M_{sm} \\ f_9 = -I_1\omega^2 U_{sm} - I_2\omega^2 \phi_{\theta m} + Q_{sm} - \frac{m}{R} M_{s\theta m} \\ f_{10} = \left(m^2 c_8 - I_1\omega^2 - \frac{kA_{44}}{R}\right) U_{\theta m} + m\left(c_8 - \frac{kA_{44}}{R}\right) U_{\zeta m} + \left(m^2 c_9 + kA_{44} - I_2\omega^2\right) \phi_{sm} \\ \quad - mc_5 N_{sm} - mc_4 M_{sm} \end{cases}$$

and

$$
\begin{aligned}
c_1 &= A_{11}D_{11} - B_{11}^2, c_2 = (A_{12}B_{11} - A_{11}B_{12})/Rc_1 \\
c_3 &= (B_{12}B_{11} - A_{11}D_{12})/Rc_1, c_4 = (B_{12}B_{11} - A_{11}D_{11})/Rc_1 \\
c_5 &= (D_{12}B_{11} - D_{11}B_{12})/Rc_1, c_6 = (A_{12}c_4 + c_2B_{12} + A_{22}/R)/R \\
c_7 &= (A_{12}c_5 + c_3B_{12} + B_{22}/R)/R, c_8 = (B_{12}c_4 + c_2D_{12} + B_{22}/R)/R \\
c_9 &= (B_{12}c_5 + c_3D_{12} + D_{22}/R)/R, c_{10} = B_{66}^2 - A_{66}D_{66}
\end{aligned}
$$

References

1. Fung YC (1955) On the vibration of thin cylindrical shells under internal pressure. The Ramo-Wooldridge Corporation - guided missile research division; Report No AM 5–8
2. Fung YC, Sechler EE, Kaplan A (1957) On the vibration of thin cylindrical shells under internal pressure. J Aer Sci 24(9):650–660
3. Hu XJ, Redekop D (2004) Prestressed vibration analysis of a cylindrical shell with an oblique end. J Sound Vib 277:429–435
4. Khalili SMR, Azarafza R, Davar A (2009) Transient dynamic response of initially stressed composite circular cylindrical shells under radial impulse load Compos Struct 89(2):275–84
5. Leissa AW (1973) Vibration of shells, washington DC: NASA SP-288
6. Casimir JB, Khadimallah MA, Nguyen MC (2016) Formulation of the dynamic stiffness of a cross-ply laminated circular cylindrical shell subjected to distributed loads. Comput Struct 166:42–50
7. Harbaoui I et al (2018) A new prestressed dynamic stiffness element for vibration analysis of thick circular cylindrical shells. Int J Mech Sci 140:37–50
8. Tj GH, Mikami T, Kanie S, Sato M (2006) Free vibration characteristics of cylindrical shells partially buried in elastic foundations. J Sound Vib 290:78593
9. Thinh Nguyen (2013) Dynamic stiffness matrix of continuous element for vibration of thick cross-ply laminated composite cylindrical shells. Compos Struct 98(2013):93–102

Finite Element Modelling of the Functionally Graded Shells Mechanical Behavior

Sana Koubaa[1(✉)], Jamel Mars[1], Mondher Wali[2], and Fakhreddine Dammak[1]

[1] Laboratory of Electromechanical Systems (LASEM), National Engineering School of Sfax, University of Sfax, B.P 1173-3038 Sfax, Tunisia
sana.kouba@enis.tn, jamelmars@yahoo.fr, fakhreddine.dammak@enis.rnu.tn

[2] Department of Mechanical Engineering, College of Engineering, King Khalid University, Abha, Saudi Arabia
mondherwali@yahoo.fr

Abstract. In this study, numerical analysis of the static bending response of FGM is carried out with different combinaisons of geometries, boundary conditions and mechanical loading. The material properties according to the coordinates of the integration points are defined using the UMAT subroutines in ABAQUS software. The FSDT is used for thin and moderately thick FG shells analysis. The performance of the developed work is illustrated through the solution of several non trivial structure problems from literature. The numerical simulation depicts very close results to solutions in literature which assess the accuracy of the implementation. The proposed solution procedure is significantly efficient from the computational point of view.

Keywords: Functionally graded shells · Finite element · UMAT abaqus

1 Introduction

Functionally graded materials (FGMs) are special kind of composite with a gradual transition of material properties from one material to another. They are made to overcome the problems associated to the discontinuity in conventional composite. The most known FGMs in literature are mainly composed of transition alloys from metal at one surface and ceramic at the opposite surface. Unlike composites, which possess mechanical properties that vary gradually with location, the FGM exhibits a smooth and continuous gradient in both compositional profile and material properties in order to optimize some function of the overall FGM. As the use of FGM increases in several field, it is crucial to develop new methodologies and techniques to characterize them.

Several works are available in literature to analyze the linear mechanical behavior of FG shell structures [1–3]. Most of authors used the classical plate theory CPT based on the Kirchhoff hypothesis; the First Order Shear Deformation theory (FSDT) developed by Mindlin; the Higher-Order Shear Deformation Theory (HSDT). The classical theory is inaccurate as it neglects the effects of transverse shear and normal strains of the structure. However, Reissner-Mindlin theory provides a correct overall

N. Aifaoui et al. (Eds.): CMSM 2019, LNME, pp. 833–841, 2020.
https://doi.org/10.1007/978-3-030-27146-6_91

assessment. In fact, shear correction factors should be incorporated to adjust the transverse shear stiffness. Then, the accuracy of solutions is strongly dependent on the correction factors. Contrarily to FSDT, there is no need of a shear correction factor when using a HSDT. Nevertheless, equations of motion are more complicated to obtain [4]. Hence, the FSDT is widely used to investigate analytically the FG shells response.

The theoretical formulation is typically coupled with FE method to provide the numerical solution of material inhomogeneity effect on FG plates deformation. In most studies, the solution procedure can be implemented into home codes [5, 6, 7]; other authors obtained the numerical solution using the commercial FE commercial softwares [8, 9]. In this regard, using the commercial finite element software ABAQUS, layers are divided into a sufficient number of slices to approximate the gradual variations of the material properties [10]. The main limitation of this method is the expensive CPU time and the non-continuous segmented distribution of material properties.

To the best knowledge of the authors, there are no further accessible documents in literature on ABAQUS implementation of static response of FG shells when taking into account the continuity of material point distribution. The main contribution of this paper consists on introducing an alternative method to define the material properties according to the coordinates of the integration points when considering one layer to avoid an expensive CPU time. the material properties are defined using both interfaces, UMAT and USDFLD subroutines, that were implemented in ABAQUS software. Due to its high efficiency and simplicity, FSDT was used for analyzing thin FG shells. The accuracy and performance of the developed model is illustrated through the solution of several FG structures problems taken from the literature. The proposed solution procedure is significantly efficient from the computational point of view.

2 Material

The FGM shell structure with polynomial material law is adopted (Fig. 1):

$$P_{FGM}(z) = P_m + (P_c - P_m)\left(\frac{z}{h} + \frac{1}{2}\right)^n \tag{1}$$

$P_{FGM}(z)$, P_m and P_c denote, respectively, the effective material property, the metal properties and the ceramic properties. $E_{FGM}(z)$ designs the Young modulus and $\rho_{FGM}(z)$ is the density. h denotes the structure thickness and z is the coordinate measured along the thickness direction. n is the power-law index.

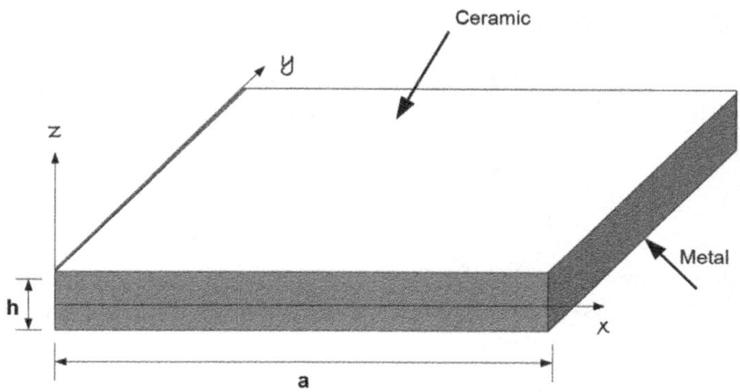

Fig. 1 Geometry of a functionally graded plate

3 Constitutive Equations

We consider a first-order-shear-deformation shell model with a thickness h, made of a FGM with material properties described in Sect. 2. All formulations are developed under the assumption of a linear elastic behavior and small deformations of material. According to the FSDT, the displacement components are:

$$\begin{cases} u_x = u + z\varphi_y \\ u_y = v - z\varphi_x \\ u_z = w \end{cases} \tag{2}$$

where φ_x and φ_y are the rotations of the transverse normal about the Cartesian axis x and y, respectively; u, v and w are the in-plane displacements and deflection of the mid-plane, respectively. The generalized displacement vector u is then

$$\boldsymbol{u} = \left[u,\ v, w,\ \varphi_x,\ \varphi_y\right]^T \quad z \in \left[-\frac{h}{2}, \frac{h}{2}\right] \tag{3}$$

Z is the thickness coordinate of the shell. The state of deformation can be decomposed in in-plane and transverse shear strains as:

$$\begin{cases} \varepsilon_{\alpha\beta} = e_{\alpha\beta} + z\chi_{\alpha\beta} \\ \gamma_{\alpha 3} = 2\varepsilon_{\alpha 3} \end{cases}, \quad \alpha,\ \beta = 1,\ 2 \tag{4}$$

where $\varepsilon_{\alpha\beta}$, $\chi_{\alpha\beta}$ and $\gamma_{\alpha 3}$ refer to the membrane, bending and transverse shear strains, respectively. In matrix notation, the strain vectors are:

$$\boldsymbol{\eta}_m = \begin{bmatrix} e_{11} \\ e_{22} \\ \gamma_{12} \end{bmatrix}, \quad \boldsymbol{\eta}_b = \begin{bmatrix} \chi_{11} \\ \chi_{22} \\ \chi_{12} \end{bmatrix}, \quad \boldsymbol{\eta}_s = \begin{bmatrix} \gamma_{13} \\ \gamma_{23} \end{bmatrix} \tag{4}$$

The constitutive equations can be written as:

$$\begin{aligned}
&\boldsymbol{\sigma} = \boldsymbol{H}(z)\boldsymbol{\eta} \\
&\boldsymbol{\tau} = \boldsymbol{G}(z)\boldsymbol{\eta}_s \\
&\boldsymbol{\eta} = \boldsymbol{\eta}_m + z\boldsymbol{\eta}_b,\ \boldsymbol{\sigma} = [\sigma_{11}, \sigma_{22}, \tau_{12}] \\
&\boldsymbol{\tau} = [\sigma_{13}, \sigma_{23}]
\end{aligned} \tag{5}$$

H(z) and G(z) are the in-plane and out-of-plane linear elastic constitutive matrices, depending on z, expressed as a function of the Young's modulus and the Poisson's ratio:

$$\begin{aligned}
&\boldsymbol{H} = \frac{E(z)}{1 - \nu^2(z)} \begin{bmatrix} 1 & \nu(z) & 0 \\ \nu(z) & 1 & 0 \\ 0 & 0 & (1 - \nu(z))/2 \end{bmatrix}, \\
&\boldsymbol{G}(z) = \frac{E(z)}{2(1 + \nu(z))} \begin{bmatrix} 1 & 0 \\ 0 & 1 \end{bmatrix}
\end{aligned} \tag{6}$$

Let $\boldsymbol{R} = \begin{bmatrix} \boldsymbol{N} \\ \boldsymbol{M} \\ \boldsymbol{T} \end{bmatrix}$, and $\boldsymbol{\Sigma} = \begin{bmatrix} \boldsymbol{\eta}_m \\ \boldsymbol{\eta}_b \\ \boldsymbol{\eta}_s \end{bmatrix}$ be the generalized stress and strain. where the in-plane membrane and bending and transverse shear stresses resultants components can be written in the following form:

$$N_{\alpha\beta} = \int_{-h/2}^{h/2} \sigma_{\alpha\beta}\, dz,\ M_{\alpha\beta} = \int_{-h/2}^{h/2} z\sigma_{\alpha\beta}\, dz,\ T_{\alpha 3} = \int_{-h/2}^{h/2} \tau_{\alpha 3}\, dz \tag{7}$$

One can deduce the constitutive equation:

$$\boldsymbol{R} = \boldsymbol{H}_T\boldsymbol{\Sigma},\quad \boldsymbol{H}_T = \begin{bmatrix} \boldsymbol{H}_m & \boldsymbol{H}_{mb} & 0 \\ \boldsymbol{H}_{mb} & \boldsymbol{H}_b & 0 \\ 0 & 0 & \boldsymbol{H}_s \end{bmatrix} \tag{8}$$

$$(\boldsymbol{H}_m, \boldsymbol{H}_{mb}, \boldsymbol{H}_b) = \int_{-h/2}^{h/2} (1, z, z^2)\boldsymbol{H}(z)\, dz \quad , \quad \boldsymbol{H}_s = \kappa \odot \int_{-h/2}^{h/2} \boldsymbol{G}(z)\, dz$$

Numerical simulation is performed using the commercial software ABAQUS. To avoid stress discontinuity, UMAT interface is implemented in order to define the material properties according to the coordinates of the integration points.

4 Numerical Results

4.1 Benchmark Tests

The performance of the proposed numerical simulation is assessed with several structure problems (i) Bending of a simply supported rhombic plate, (ii) pinched hemispherical shell with 18° hole, (iii) pinched cylinder with end diaphragms (Fig. 2). Results are compared to the findings in Wali et al. [11] in which a 3d-shell model based on a discrete double directors shell element (DDDSE) was introduced. 7 DOF per node are used with FE subroutine.

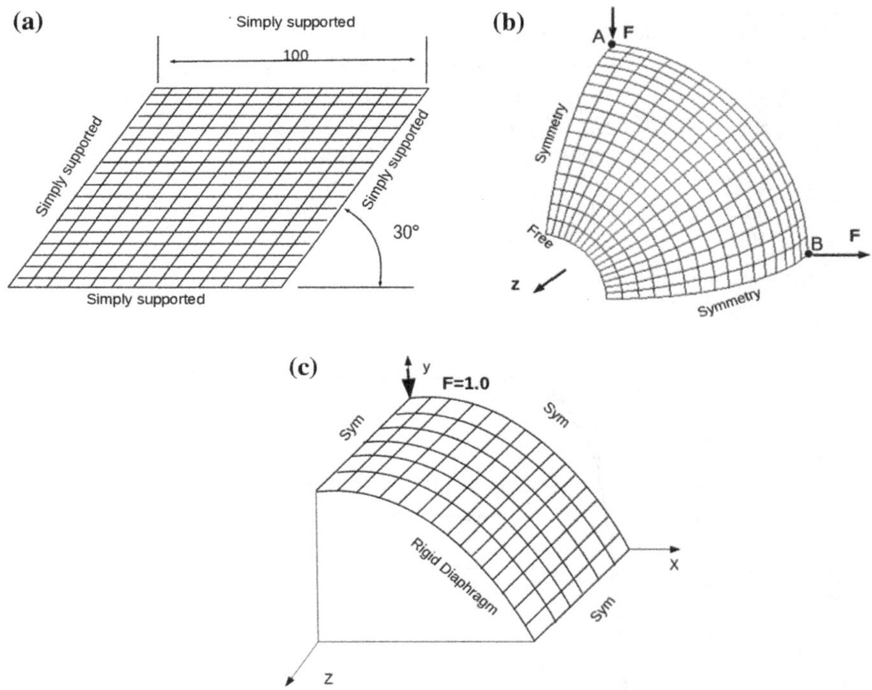

Fig. 2 **a** bending of a simply supported rhombic plate, **b** pinched hemispherical shell, **c** pinched cylinder with end diaphragms

The different shells are modeled with the standard quadrilateral 4-nodes element with three rotational and three translational degrees of freedom per node. Results, based on the FSDT of shell elements are obtained. The FGM structure properties are the triplet (Em, Ec, ν) = (70 GPa, 380 GPa, 0.3) for the metal and ceramic components, respectively. To avoid stress discontinuity at the interfaces, the material properties according to the coordinates of the integration points are defined using the UMAT subroutine in ABAQUS software. All material and geometrical properties are given in a coherent system of units. The power-law index is n = 6 for all cases. The normalized results of deflection and displacement are obtained with the FE solution using 100 elements per side. The FE results of the normalized center-point deflection of rhombic plate, displacement of pinched hemispherical shell and pinched cylinder, are gathered in Table 1. Comparison between present model and elements developed by Wali et al. [11] are in a very good correlation for the three considered cases. We plot in Fig. 3 the normalized displacement with the FE converged solution of 4.746 10 − 4 using 100 elements per side. It can be depicted that the present FE model produces the close results to those of Wali et al. [11] elements. Therefore, it can be concluded that the proposed technique exhibits high performance.

Table 1 Comparison results of present simulations

Node per side	Rhombic plate	Hemispherical shell (10^{-2})
	Present Wali et al. [11]	Present Wali et al. [11]
3	1.843 7.846	4.532 5.190
5	2.761 5.186	5.153 5.398
9	2.889 3.695	5.181 5.269
17	3.073 3.348	5.204 5.220
33	3.189 3.275	5.223 5.229
100	3.278 3.268	5.237 5.232

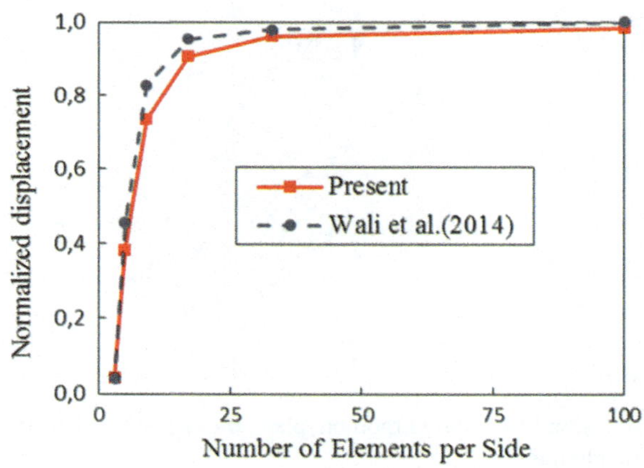

Fig. 3 Normalized displacement of the pinched cylinder with end diaphragms

4.2 Square Plate Under Doubly Sinusoidal Load

In this section, the accuracy of the present model is evaluated for isotropic square FGM plate under doubly sinusoidal load (Fig. 4). The plate is discretized using the element *S4 with a* 20 by 20 mesh. The various non-dimensional parameters are according to [11].

$$\bar{w} = \frac{10E_c h^3}{a^4 q_0} w\left(\frac{a}{2}, \frac{a}{2}\right), \quad \bar{\sigma}_x = \frac{h}{a q_0} \sigma_x \left(\frac{a}{2}, \frac{a}{2}, \frac{h}{2}\right)$$
$$\bar{\sigma}_y = \frac{h}{a q_0} \sigma_y \left(\frac{a}{2}, \frac{a}{2}, \frac{h}{3}\right), \quad \bar{\tau}_{xy} = \frac{h}{a q_0} \sigma_x \left(0, 0, \frac{-h}{3}\right)$$

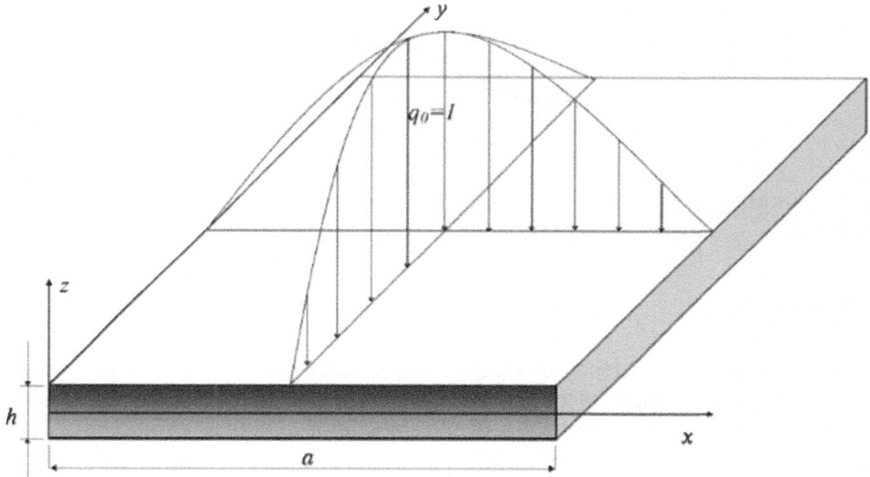

Fig. 4 A simply supported square FGM plate under sinusoidal load

In Fig. 5, we plot the evolution of the dimensionless shear stress across the thickness direction for various power-law index n. Comparisons with [11, 12] are conducted. In [12], a 3D hyperbolic sine plate theory is presented for the case of a simply supported square plates subjected to a bi-sinusoidal mechanical load. 91 mathematical layers were simulated to model the continuous variation of properties across the thickness direction leading to significant computational effort and virtual storage of data. In the present work, the proposed FE simulation presents very close results to literature, which assess the accuracy of the implemented subroutine. Its also of interest to note that the proposed solution procedure is significantly efficient from the computational point of view.

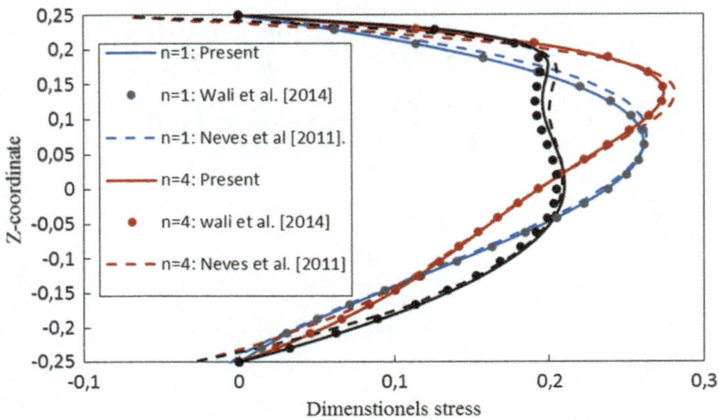

Fig. 5 Dimensionless shear stress $\bar{\tau}_{xz}$ for a simply supported square plate under sinusoidal load ($a = h = 4$)

5 Conclusion

The analysis and the efficiency of FE simulation based on Abaqus-UMAT subroutine for the FGM shell structures is presented in this paper. The material properties are defined according to the coordinates of the integration points to avoid stress discontinuity at the interfaces. By numerical investigation, the proposed model presents a good performance and high accuracy to predict the static behavior of shell structures when comparing with available published results.

References

1. Tornabene F (2009) Free vibration analysis of functionally graded conical, cylindrical shell and annular plate structures with a four-parameter power-law distribution. Comput Methods Appl Mech Eng 2911–2935
2. Ferreira AJM, Roque CMC, Neves AMA, Jorge RMN, Soares CMM, Liew KM (2011) Buckling and vibration analysis of isotropic and laminated plates by radial basis functions. Compos: Part B: 592–606
3. Zghal S, Frikha A, Dammak F (2017) Static analysis of functionally graded carbon nanotube-reinforced plate and shell structures. Compos Struct 176:1107–1123
4. Frikha A, Wali M, Hajlaoui A, Dammak F (2016) Dynamic response of functionally graded material shells with a discrete double directors shell element. Compos Struct 154:385–395
5. Frikha A, Hajlaou A, Wali M, Dammak F (2016) A new higher order c0 mixed beam element for fgm beams analysis. Compos B 106:181–189
6. Frikha A, Zghal S, Dammak F (2018) Dynamic analysis of functionally graded carbon nanotubes-reinforced plate and shell structures using a double directors finite shell element. Aerosp Sci Technol 78:438–451
7. Zghal S, Frikha A, Dammak F (2018) Free vibration analysis of carbon nanotube-reinforced functionally graded composite shell structures. Appl Math Model 53:132–155

8. Nie G, Zhong Z (2007) Axisymmetric bending of two-directional functionally graded circular and annular plates. Acta Mech Solida Sin 20:289–295
9. Alipour MM, Shariyat M (2012) An elasticity-equilibrium-based zigzag theory for axisymmetric bending and stress analysis of the functionally graded circular sandwich plates, using a maclaurin-type series solution. Eur J Mech A/Solids 34:78–101
10. Mao Y, Fu Y, Ai S, Fang D (2013) Interfacial damage analysis of shallow spherical shell with fgm coating under low velocity impact. Int J Mech Sci 30–40
11. Wali M, Hajlaoui A, Dammak F (2014) Discrete double directors shell element for the functionally graded material shell structures analysis. Comput Methods Appl Mech Eng 278:388–403
12. Neves AMA, Ferreira AJM, Carrera E, Cinefra M, Roque CMC, Jorge RMN, Soares CMM (2012) A quasi-3d hyperbolic shear deformation theory for the static and free vibration analysis of functionally graded plates. Compos Struct 94:1814–1825

Design and Modeling of a Mechatronic Power System of an Electric Vehicle

A. Guizani[1(✉)], H. Trabelsi[1], M. Hammadi[2], J. Y. Choley[2], M. Barkallah[1], and M. Haddar[1]

[1] Mechanics, Modelling and Production Research Laboratory, National Engineering School of Sfax, 3038 Sfax, Tunisia
amir.guizani@live.fr, hassen.trabelsi@outlook.fr, bark_maher@yahoo.fr, mohamed.haddar@enis.rnu.tn
[2] QUARTZ EA7393, SUPMECA-Paris, 3 Rue Fernand Hainaut, 93407 Saint-Ouen, France
{moncef.hammadi, jean-yves.choley}@supmeca.fr

Abstract. Electric vehicles are considered as the cars of the future, given the advantages they offer compared to conventional cars, which are equipped with a combustion engine: economy of use, reliability, silence of operation, overall environmental impact, promotes the development of renewable energies and the stability of networks. However, they still have critical problems to solve as high cost, low autonomy and long charging time. The majority of these problems are totally related to the battery package. Thus, the battery package must contain enough energy to have sufficient power capacity for the accelerations and decelerations tests. Also, to have a certain driving autonomy. So, in this study, a mechatronic power system model of an electric vehicle is proposed. It is adapted to estimate the amount of energy needed to travel a certain distance. The modeling and simulation of the system were performed using the object-oriented language Modelica. While, the parametric studies for estimating energy consumption were carried using Model Center software. The developed model ensures that the performances related to the acceleration test and driving distance are respected.

Keywords: Mechatronic power system · Modeling · Simulation · Energy estimate

1 Introduction

The electric car remains a mobility solution of the future. It is still evolving and getting closer step by step in performance terms of the thermal car [10]. However, it still has critical problems to solve. In regards to the economic field, the biggest problem is related to the battery. Its cost is high and its maintenance is expensive [2].

Furthermore, the electric car also has difficulties at the recycling level. It is much more complex and less competitive compared to the thermal car. Indeed, it requires special treatment of batteries. The latter is so expensive for the user that it is generally supported in the manufacturer's warranty.

N. Aifaoui et al. (Eds.): CMSM 2019, LNME, pp. 842–852, 2020.
https://doi.org/10.1007/978-3-030-27146-6_92

Moreover, the electric car is not 100% ecological taken into account the production of electricity necessary for its operation. The electricity used to recharge the battery must be produced by truly renewable energies, such as solar or hydraulic energy.

In terms of autonomy, the majority of available vehicles have autonomy in the order of 100–150 km [2]. This autonomy is largely sufficient for an urban user. However, it is mediocre for a user making only long distances.

In conclusion, we can say that in the case of an electric production made from renewable energies, the electric vehicle seems more beneficial for the environment than the thermal car. In fact, it can significantly reduce gas emissions and air pollution, especially in urban environment where its implementation will be privileged.

The preliminary design of electric cars can be achieved using different multi-domain modeling tools such as Matlab/Simulink [4, 5] and VHDL-AMS [11].

Several studies have been developed in the literature to improve and optimize the performances of electric cars in order to obtain a more efficient and more reliable system. For instance, Schaltz concentrated his work on the design and modeling of a battery package [12]. This package must contain enough power to have a certain driving autonomy. Guizani et al. developed, in the preliminary design phase, a simple model of the electric vehicle [6]; in order to optimize the propulsion system characteristics while respecting the performance requirements related to the different test cases: the maximum speed and the acceleration test.

The focus in this paper will be on the design and modeling of a mechatronic power system of an electric vehicle to estimate the necessary amount of energy power to travel a certain distance.

This paper is structured as follows: after the introduction, Sect. 2 describes the mathematical model used in this study. Section 3 gives the implementation of the presented model in Modelica/Dymola. Section 4 gives the simulation results in order to verify the developed model, Sect. 5 provides parametric studies to estimate the energy consumption of the electric vehicle during an imposed profile and Sect. 6 presents the conclusion remarks.

2 Description of the Mathematical Model

The battery is the energy source of the vehicle. It consists of the series and/or parallel association of elementary cells. The equivalent diagram of the battery is shown in Fig. 1.

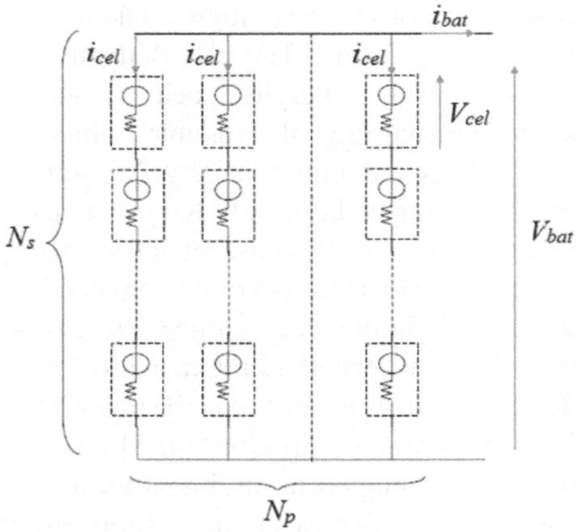

Fig. 1 Equivalent circuit diagram of the battery

The total voltage, the current and the mass of the battery can be calculated by the following expressions [8]:

$$\begin{cases} V_{bat} = N_s.V_{cel} \\ i_{bat} = N_p.i_{cel} \\ M_{bat} = N_s.\,N_p.\,M_{cel} \end{cases} \tag{1}$$

where:

- N_s [−]: the number of cells in series,
- N_p [−]: the number of cells in parallel,
- V_{cel} [V]: the voltage of a cell,
- i_{cel} [A]: the current in a cell,
- M_{cel} [Kg]: the mass of a cell,
- V_{bat} [V]: the voltage of the battery,
- i_{bat} [A]: the current in of the battery,
- M_{bat} [Kg]: the mass of the battery.

In order to evaluate the behavior of the battery, it is necessary to determine the variation of its state of charge *SoC*. The *SoC* corresponds to the amount of charge *C* that can be restored by the battery with respect to its nominal capacity C_O and corresponding to *SoC* = 100% [7]:

$$\begin{cases} \mathrm{SoC} = \mathrm{SoC_{init}} - 100\dfrac{\mathrm{C}}{\mathrm{C_0}} \\ C = \dfrac{1}{3600}\displaystyle\int i_{bat}.dt \end{cases} \tag{2}$$

C and C_O are expressed in *A.h*.

After the mathematical development of the battery, the model is modeled using the object-oriented language Modelica [3], as indicated in the following section.

3 Modeling of the Battery with Modelica

The objective of this study is to develop a Modelica model of the battery able to simulate its real behavior taking into account the consequent decrease of its *SoC*.

There are many technologies of battery. If their operation is similar, the materials used react differently and therefore have different performance. The main Battery Pack's technologies most used by the automotive industry are three: lead-acid batteries, nickel batteries and lithium batteries.

In this article, the battery package is based on a type AMP 20 Lithium Ion Cell[1] because of the advantages that confers compared to other technologies:

- Higher energy density than others,
- A recharge time less important,
- Low risk of pollution.

The external view of the battery model is shown in Fig. 2.

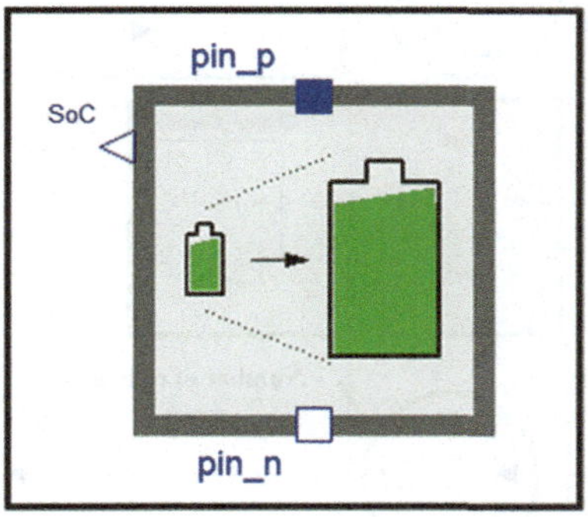

Fig. 2 External view of the battery model developed with Modelica

Going deep into the internal architecture of the model as shown in Fig. 3, the electrical behavior of the battery is described only by blocks chosen from the Modelica library. Before describing the role of each block, it is important to specify that this model operate in cycle. It generate a voltage source between the two ports *pin_p* and *pin_n* of the battery.

[1] http://www.a123systems.com/prismatic-cell-amp20.htm.

The current I_{Bat} generated by the voltage U_{Bat} and measured by the *current Sensor* is compared with the permissible current I_{max}. The minimum current obtained is divided by the *gain1*, which represents the number of cells in parallel N_p, in order to obtain the current circulating in each cell *i*. This current flows thereafter in the *SoC* block.

The other input of this block is the capacity of the battery C_0 connected to the *gain2*. It perform a unit conversion from *A.s* to *A.h*. This block calculates the instantaneous value of the state of charge of a cell by Eq. (2). The calculated value is an input of a $V_{Cel\text{-}SoC}$ table, which provides the voltage of a cell according to the state of charge. The voltage value obtained at the output of the component $V_{Cel\text{-}SoC}$ is multiplied by the *gain3,* which represents the number of cells in series N_s, is used as a reference to generate the battery voltage. To evaluate the behavior of different battery technologies, the designer can modify the V_{Cel_SoC} table at the output of the *SoC* block.

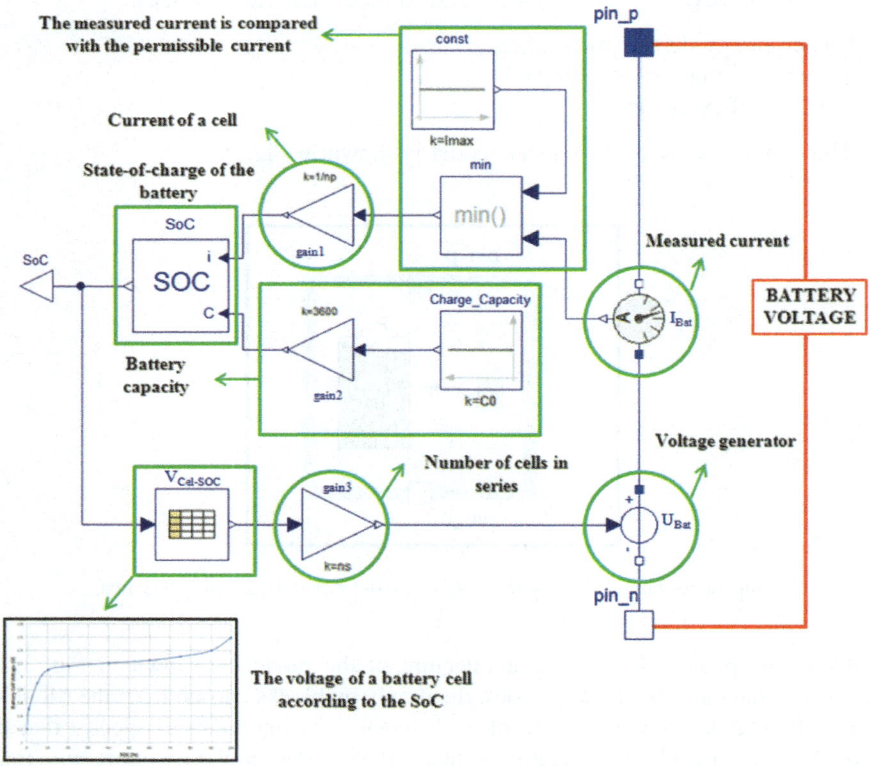

Fig. 3 Internal view of the battery model developed with Modelica

4 Simulation Results

In this part, the analysis model used to perform parametric studies in the following part is made up of the battery in series with equivalent resistance that represents the rest of the electric vehicle model. The Modelica model of the equivalent resistance was developed in [9].

The simulation results of the vehicle according to the New European Driving Cycle NEDC [1], are given by Figs. 4 and 5. The NEDC cycle is used to simulate and verify the model of the battery developed in the previous part.

Figure 4 shows a comparison between the real speed of the vehicle measured at the output of the model and the NEDC cycle imposed on the input. The obtained result confirms that the measured speed follows the profile of the imposed road.

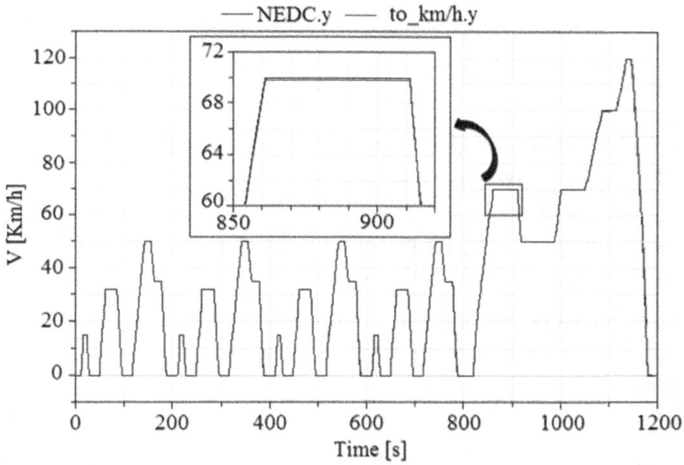

Fig. 4 Comparison between the measured speed of the vehicle and the imposed cycle NEDC

Figure 5 represent the variation of the state of charge along the imposed profile. The final state of charge of the battery is about 67%, which means that the proposed model has lost 16% of its capacity. With the same values of N_s and N_p and a minimum state of charge equal to 20%, the model can repeat 4.5 times the imposed cycle.

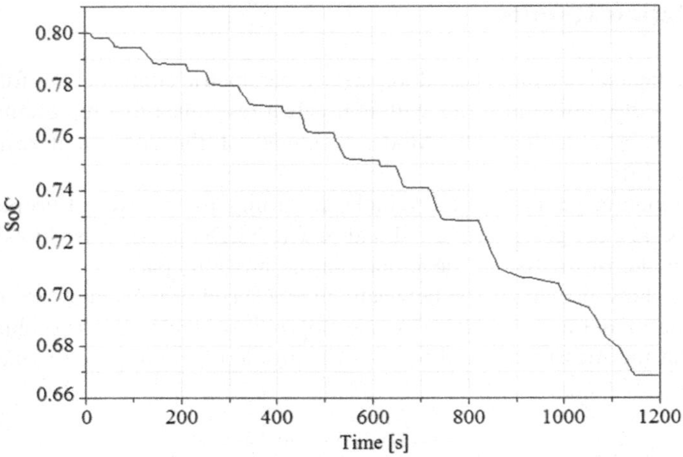

Fig. 5 Variation of the state of charge of the battery during the NEDC profile

5 Parametric Study of the Energy Consumption of the Battery with ModelCenter

The objective of this part is to estimate the amount energy of the battery needed to travel a certain distance. Therefore, the following requirements are taken into account:

- The imposed speed on the input is a constant.
- The road is horizontal, the grade angle α is equal zero.
- The wind speed is zero.
- The battery fully charged.

The model developed in the previous section was used to calculate the value of the energy required by the battery for each kilometer traveled by the vehicle. The only difference, is that the *NEDC* cycle is replaced by a constant speed. It is the reference speed that the vehicle must reach. In addition, other parameters are added within the model, as shown in Fig. 6. These parameters are defined by the following equations [8]:

- *Energy*, is the total energy consumed by the vehicle during the imposed cycle;

$$Energy = \frac{1}{3600}\int (powerSensor.Power)dt \tag{3}$$

- $Range_{Km}$, it is the distance traveled by the vehicle in kilometers;

$$Range_{km} = \frac{1}{1000}\int (speedSensor.speed)dt \tag{4}$$

- $Energy_{Km}$, it is the energy required by the vehicle to travel one kilometer with a constant speed;

$$Energy_{km} = \frac{Energy}{Range_{km}} \tag{5}$$

Figure 7 shows the results of a parametric study performed with ModelCenter[2] to determine the $Energy_{Km}$ function. In this study, the input variables are the total number

```
 // Total Energy
 Real Energy;
 // Energy to travel one kilometer at a constant speed
 Real Energy_Km "Wh/km";
 // Range
 Real Range_km(start=0) "in km";
equation
  der(Energy) = powerSensor.power/3600;
  der( Range_km*1000) = speedSensor.v;
  Energy_Km = if time > 1 then  Energy/Range_km else 0;
```

Fig. 6 Modelica code for estimating the energy required by the battery

of cells, the reference speed to be achieved and the output parameter (response) is the $Energy_{Km}$ function.

Parametric study operates in iterative loop. By setting each time the number of cells, $Energy_{Km}$ is calculated by changing the reference velocity between a minimum value *startingvalue* and a maximum value *endingvalue*, as shown in Fig. 8.

From the results shown in Fig. 8, it is possible to plot the curves in Fig. 9, which show the variation of $Energy_{Km}$ as a function of the speed to be achieved for a certain number of cells.

These graphs can be used as a reference by the designer to calculate the total energy required by an electric vehicle by fixing certain requirements such as:

- The distance to cover,
- The speed of the vehicle,
- The number of cells.

Example Requirements:

- The distance to be covered = 100 km,
- The speed = 100 km/h,

[2] Phoenix Integration. PHX ModelCenter, from http://www.phoenix-int.com/software/phx-modelcenter.php.

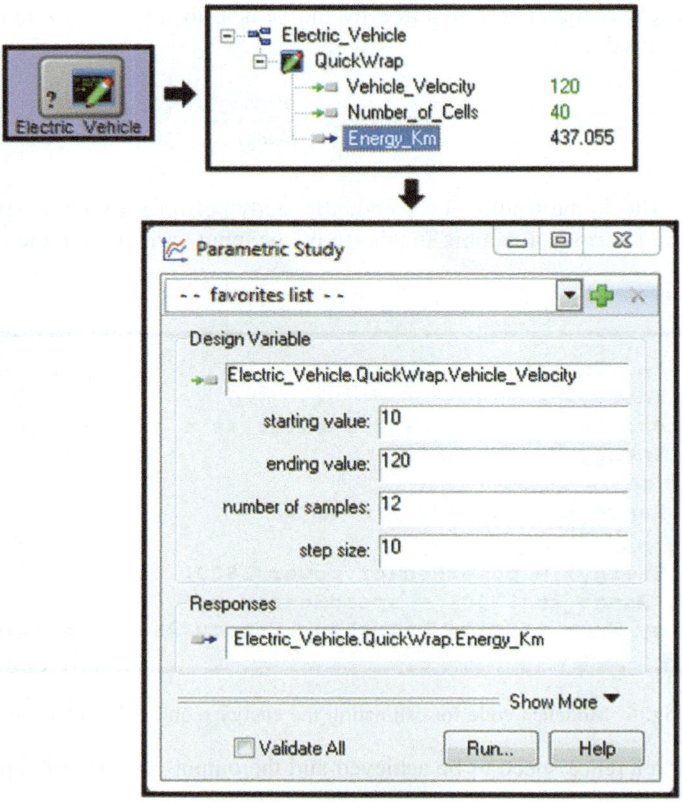

Fig. 7 Graphic interface of the parametric study with ModelCenter

- The number of cells = 300.

The designer must select on the *x* axis the required speed which is equal to 100 km/h, after that, using the curve of 300 cells, he must read on the *y* axis the necessary value of $Energy_{Km}$. If the distance to be traveled is equal to 100 km, considering a cycle of discharge of the battery. Then, total energy *TotalEnergy* is given by the following relation [8]:

$$TotalEnergy = 100.k.Energy_{km} \tag{6}$$

Where *k* is a factor that takes into consideration that the battery needs to be recharged when its *SoC* is below a minimum value (typically SoC_{min} = 20%).

$$k = \frac{1}{SoC_{\text{int}} - SoC_{\text{min}}} \tag{7}$$

Vehicle Velocity	Number of Cells						
	200	250	300	350	400	450	500
10	204.1528	247.0976	290.1852	333.4155	376.7885	420.304	463.963
20	228.289	273.469	318.896	364.571	410.493	456.661	503.077
30	243.546	289.745	336.238	383.025	430.105	477.48	525.148
40	256.811	303.631	350.776	398.244	446.037	494.154	542.595
50	270.615	317.998	365.736	413.828	462.275	511.076	560.232
60	287.893	336.055	384.613	433.567	482.918	532.666	582.81
70	307.261	356.201	405.581	455.401	505.662	556.363	607.505
80	328.379	378.068	428.241	478.899	530.042	581.671	633.784
90	351.243	401.647	452.58	504.042	556.034	608.556	661.609
100	376.01	427.108	478.78	531.027	583.849	637.247	691.222
110	402.923	454.686	507.106	560.148	613.814	668.105	723.02
120	433.201	484.722	537.819	591.716	646.29	701.542	757.472

Fig. 8 Parametric study performed inside Model Center to estimate the energy consumption

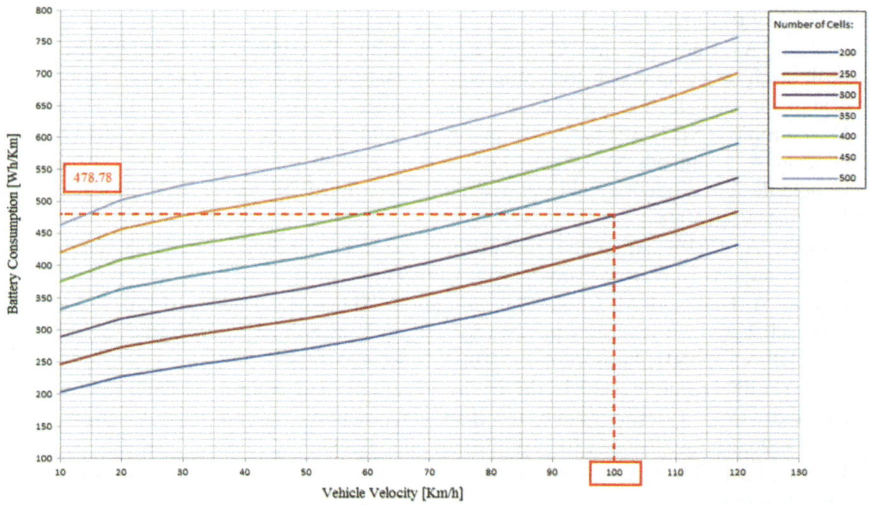

Fig. 9 Variation of energy according to the target speed and the number of cells

6 Conclusion

Our main contribution in this paper was a proposition of a mechatronic power system able to simulate the dynamic behavior of the battery taking into account the consequent decrease of its state of charge.

The presented system has been used to perform parametric studies using ModelCenter software to estimate the amount of energy required to travel a certain distance.

According to the obtained results, the efficiency of the proposed model was tested and confirmed. Futures works will be focus on the optimization of different design variables in order to get a more reliable car that can properly fulfill its mission.

References

1. Barlow TJ, Latham S, McCrae I, Boulter P (2009) A reference book of driving cycles for use in the measurement of road vehicle emissions
2. Chan C, Chau K (2001) Modern electric vehicle technology. Oxford University Press, p 47
3. Fritzson P (2010) Principles of object-oriented modelling and simulation with Modelica 2.1, Wiley https://doi.org/10.1002/9780470545669
4. Gao DW, Mi C, Emadi A (2007) Modeling and simulation of electric and hybrid vehicles. Proc IEEE 95:729–745. https://doi.org/10.1109/JPROC.2006.890127
5. Guizani A, Hammadi M, Choley JY, Soriano T, Abbes MS, Haddar M (2014a) Multidisciplinary optimization of mechatronic systems: application to an electric vehicle Mechatronic systems: theory and applications, Springer, pp 1–14. https://doi.org/10.1007/978-3-319-07170-1_1
6. Guizani A, Hammadi M, Choley JY, Soriano T, Abbes MS, Haddar M (2014b) Multidisciplinary approach for optimizing mechatronic systems: application to the optimal design of an electric vehicle. In: International conference on advanced intelligent Mechatronics (AIM), IEEE/ASME, pp 56–61. https://doi.org/10.1109/AIM.2014.6878046
7. Guizani A, Hammadi M, Choley JY, Soriano T, Abbes MS, Haddar M (2016) Electric vehicle design, modelling and optimization. Mech Ind EDP Sci 17:405. https://doi.org/10.1051/meca/2015095
8. Guizani A (2016) Approche multi-agents pour la conception optimale des systèmes mécatroniques, Université Paris-Saclay, Université Paris-Saclay
9. Guizani A, Hammadi M, Choley JY, Soriano T, Abbes MS, Haddar M (2017) Multi-agent approach based on a design process for the optimization of mechatronic systems. Mech Ind EDP Sci 18:507. https://doi.org/10.1051/meca/2016080
10. Hori Y (2004) Future vehicle driven by electricity and control research on four-wheel-motored" UOT Electric March II" IEEE Transactions on Industrial Electronics. IEEE 51:954–962. https://doi.org/10.1109/TIE.2004.834944
11. Jaber K, Fakhfakh A, Neji R (2011) Modeling and simulation of high performance electrical vehicle powertrains. In: VHDL-AMS, electric vehicles modeling and simulations, p 25. https://doi.org/10.5772/16650
12. Schaltz E (2011) Electrical vehicle design and modeling. Intech Open Access Publisher. https://doi.org/10.5772/20271

Sizing Models and Performance Analysis of Waste Heat Recovery Organic Rankine Cycle System for Internal Combustion Engine

Boughattas Nejmiddin[1(✉)], Hadj Salah Wafa[2], Derbel Aymen[2], and Timoumi Yousef[2]

[1] IPEIK, Kairouan, Tunisia
nejmiddin@yahoo.fr
[2] ENIM, Monastir, Tunisia
Youssef.timoumi@enim.rnu.tn

Abstract. About one third of the energy available on internal combustion engines is actually converted into effective power. The recovery of the dissipation heat and their conversion into electricity is an effective way to increase the efficiency of these engines and therefore reduced their consumptions. Among the current and potentially adopted technologies for this valorisation is the Organic Rankine Cycle system. Thus, the various components constituting the ORC system (motor-pump, heat exchanger and expander) were modelled and dimensioned from the parameters or data provided by the manufacturers' catalogues, and then selected with imperative to optimize the cycle.

Keywords: ORC system · ICE · Expander · Heat exchanger · Motor-pump

Nomenclature

A	Heat exchanger Area m^2
b	Channel spacing of plate m
C_p	Specific heat $J{\cdot}Kg^{-1}{\cdot}K^{-1}$
h	Enthalpy $J{\cdot}Kg^{-1}$
L	Length of plate m
N_{rot}	Rotational speed rpm
$\dot{m}$	Mass flow rate $kg{\cdot}s^{-1}$
N_p	Total number of plate
P	Pressure $N{\cdot}m^{-2}$
$\dot{Q}$	Heat flow W
s	Entropy $J{\cdot}Kg^{-1}{\cdot}K^{-1}$
T	Temperature K
Vs	Displacement m^3tr^{-1}
$\dot{V}_s$	Ideal volume flow rate m^3s^{-1}
U	Heat transfer coefficient $Wm^{-2}K^{-1}$
$\dot{W}$	Power W
W	Width of plate m

N. Aifaoui et al. (Eds.): CMSM 2019, LNME, pp. 853–865, 2020.
https://doi.org/10.1007/978-3-030-27146-6_93

Greek symbols

η	Efficiency
v	Specific volume $m^3 \cdot kg^{-1}$

Subscripts

ex	Exhaust
el	Electrical
em	Electromechanical
exp	Expander
l	Liquid
lm	Log mean
p, pp	Pump
s	Isentropic
sf	Secondary fluid
su	Supply
tp	Two-phase
v	Vapour
wf	Working fluid

Acronyms

EES	Engineering Equation Solver
HP	Horse Power
ICE	Internal Combustion Engine
ORC	Organic Rankine Cycle
WHRS	Waste Heat Recovery System.

1 Introduction

The thermal efficiency of internal combustion engines (ICE) is very low, only about one third of fuel energy is transformed into mechanical energy. The remaining energy is dissipated into the atmosphere as waste heat through the cooling system, and exhaust [7].

The recovery of the dissipation heat and their conversion into electrical or mechanical power is an effective way to increase the efficiency of these engines and therefore reduced their consumptions, as well as minimise the contribution to global warming. Among the current and potentially adopted technologies for this valorisation is the Organic Rankine Cycle system. Its main advantages are the easiness and the availability of its components.

The interest in ORC for waste heat recovery system WHRS on ICE has grown significantly for the past few years [2, 10, 11]. This interest is justified by reductions in consumption expected between 5 and 10% depending on the system and the road cycle considered [8, 10].

The components of the ORC system include typically a motor-pump, an evaporator, an expander, and a condenser (Fig. 1). This system uses organic substances as working fluids, and is a reverse of refrigeration cycle. There is no commercial system today, but several prototypes have been developed since the 1970s, following the first petrol crisis.

Guillaume [1] has developed detailed models for a variety of each component of the ORC system for the mobile application under MATLAB environment. Then he carried a multi-objective optimization (thermodynamic and economic performance) but he doesn't considered electric motor modelisation.

The most of the existing works dedicated to the ORC system were consacred to the solar energy [4, 5]. In addition, the level of detail of the models is most of the time reduced [10], Yang et al. [11].

This work takes into account the stationary aspect of the ORC cycles and precisely those using volumetric scroll expander for the transport applications.

The realization of a detailed steady-state model seems to be the main objective, thus making it possible to get a realistic simulation of the Rankine cycle. It will then make possible to carry out an optimization and dimensioning or sizing the different components of the waste heat recovery organic Rankine cycle system devoted to be coupled to an ICE.

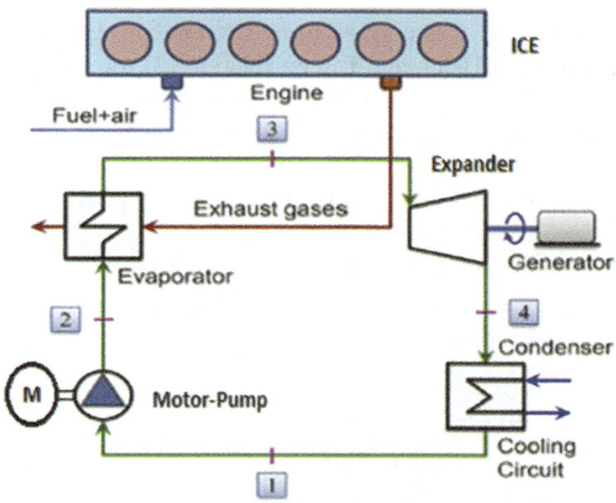

Fig. 1 Schematic diagram of ICE-ORC combined system

2 2 Mathematical Model

The following section focuses on the detailed modelling of the heat exchangers, motor-pump, and expansion machines. All the models are implemented using the EES platform. Then the WHRS model is achieved by interconnecting the components models together. The working fluid considered is the R123.

2.1 Heat Exchanger

The brazed plate heat exchangers are modelled using the log mean temperature difference (LMTD) method, where the evaporator and the condenser includes liquid, two-phase, and vapor zones. This method (LMTD) involves solving the following system of equations for each zone:

$$\dot{Q} = \dot{m}_{wf}\left(h_{wf\_ex} - h_{wf\_su}\right); \dot{Q} = \dot{m}_{sf} C_p \left(T_{sf\_su} - T_{sf\_ex}\right) \quad (1)$$

$$\dot{Q} = UA\Delta T_{lm}; \frac{1}{U} = \frac{1}{h} + \frac{1}{h_{sf}} \quad (2)$$

$$\Delta T_{lm} = \frac{\left(T_{sf\_ex} - T_{wf\_su}\right) - \left(T_{sf\_su} - T_{wf\_ex}\right)}{\ln\left(\frac{\left(T_{sf\_ex} - T_{wf\_su}\right)}{\left(T_{sf\_su} - T_{wf\_ex}\right)}\right)} \quad (3)$$

Heat transfer coefficients U is derived from Thonon correlation [9] for single-phase flow (liquid or vapour) and Hsieh and Lin correlation [3] for two-phase flow (liquid + vapour). The pressure drops is neglected (typically below 0.1 bar for these exchangers).

The total exchange surface of the exchanger is given by:

$$A_{tot} = A_l + A_{tp} + A_v; A_{tot} = \left(N_p - 2\right) L\, W \quad (4)$$

Given the inlet conditions of both fluids, the dimension of the heat exchanger, and that the heat transfer for all zones must also be equal between fluids, it is possible to solve for the outlet conditions.

2.2 Expander-Generators

The expansion stage is done through a scroll hermetic expander, which is characterized by its filling factor Ø and isentropic efficiency $\eta_{s,exp}$ [5].

$$\varphi = \frac{\dot{\mathrm{m}}.v_{su}}{\dot{V}_s} = \frac{\dot{\mathrm{m}}.v_{su}}{\mathrm{V_s}.N_{rot}}; \eta_{s,\mathrm{exp}} = \frac{\dot{W}_{el}}{\dot{W}_s} = \frac{\dot{W}_{el}}{\dot{m}.(h_{su} - h_{ex,s})} \quad (5)$$

They are respectively defined as the ratio of real to ideal mass flow rate and the ratio of real to ideal power generated.

The semi-empirical model of the expander is proposed by Lemort et al. [6]. A polynomial fits were created to correlate the filling factor and the isentropic efficiency as a function of supply pressure P_{su}, and pressure ratio r_p $(P_{wfexp,su}/P_{wfexp,ex})$

$$\sum_{i=0}^{n-1}\sum_{j=0}^{n-1} a_{ij}.\ \ln(r_p)^i.\ \ln(p_{su})^j + a_{n0}.\ \ln(r_p)^n + a_{0n}.\ \ln(p_{su})^n = f(r_p, p_{su}) \tag{6}$$

where the parameters a_{ij} of the polynomial fit are provided in Table 1

Table 1 Coefficient of the polynomial model of the expander

$\eta_{s,exp}$						
j						
i	0	1	2	3	4	5
0	6.34831061E+3	−2.07325125E+3	272.015067E0	−17.9964322E0	602.747139E-3	−8.20388944E-3
1	−4.6222660E+3	1.18102574E+3	−111.050112E0	4.54486911E0	−67.9837592E-3	
2	5.18926734E+3	−1.40315596E+3	141.445478E0	−6.30866773E0	105.088614E-3	
3	−2.7193129E+3	765.497652E0	−80.6286745E0	3.77077331E0	−66.0896654E-3	
4	486.736446E0	−139.912567E0	15.0486978E0	−718.767884E-3	12.8647910E-3	
5	53.1888731E-3					

$\varnothing$			
j			
i	0	1	2
0	4.798	−0.6231	0.02523
1	−0.06549	0.006766	
2	−0.00494		

2.3 Motor-Pump

Since the motor and pump are not lodged into the same shell like the hermetic expanders, the internal and electromechanical losses can be separated. Thus, the global isentropic efficiency is defined by:

$$\eta_p = \eta_{em,p}.\,\eta_{s,p} = \frac{\dot{w}_s}{\dot{w}_{el}} = \frac{\dot{m}_{wf}.(h_{ex,s} - h_{su})}{\dot{w}_{el}} \approx \frac{\dot{m}_{wf}.v_{su}.(p_{ex} - p_{su})}{\dot{w}_{el}} \tag{7}$$

The pumps considered are Hypro plunger pumps (2230B-P, 2351B-P and 2413B-P), and the electric motors are Leeson brand (1HP, 0.5HP) low voltage (48 V DC).

The pump (2351B-P) isentropic efficiency is defined by:

$$\eta_{s,p} = 0.234247552 + 0.220591434\left(\frac{p_{ex}}{p_{nom}}\right) - 0.0179094791\left(\frac{p_{ex}}{p_{nom}}\right)^2 \tag{8}$$

The equation below is derived from manufacturer data, $\eta_{s,p}$ is correlated as a function of normalized outlet pressure. The nominal pressure, P_{nom}, is set to 30 bars. Figure 2 shows the goodness of fit, which results in an error of 1.25%.

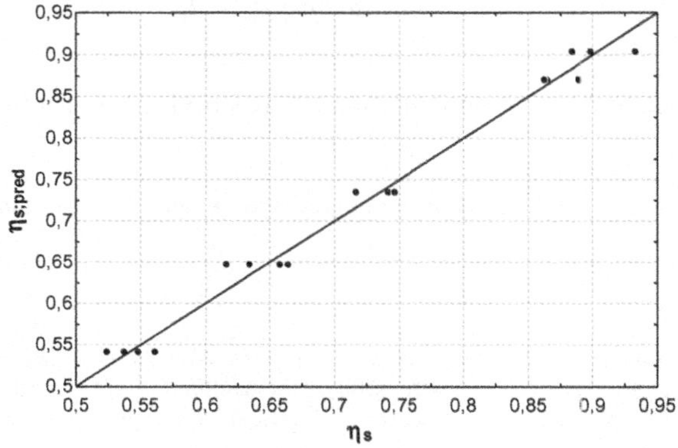

Fig. 2 Predicted versus measured pump 2351B isentropic efficiency

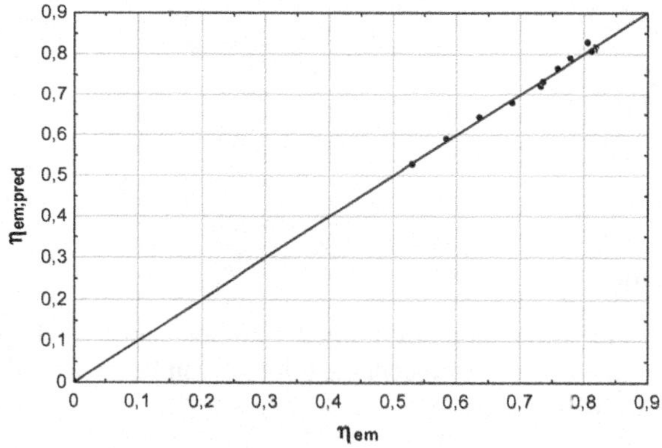

Fig. 3 Predicted versus measured motor 1HP efficiency

The motor (1HP) efficiency is also derived from the manufacturer's data and correlated with fraction of rated mechanical power as follows:

$$\eta_{em,p} = \sum_{k=0}^{6} b_k.\left(\frac{\dot{W}_m}{\dot{W}_{nom}}\right)^k \tag{9}$$

where the rated mechanical power $\dot{W}_{nom}$ is 746 W (1 HP). Table 2 provides the coefficients b_k, and Fig. 3 depicts the goodness of fit, which giving an error of 0.17%.

2.4 Modeling the Complete ORC System

The global model of the cycle is obtained by interconnecting the component models described above, as shown in Fig. 4.

Table 2 Motor (1HP) correlation coefficients for Eq. 9

Coefficient	Value
b_0	2.170250E-03
b_1	4.468185E0
b_2	−9.374727E0
b_3	9.750974E0
b_4	−5.351966E0
b_5	1.474668E0
b_6	−1.608160E-01

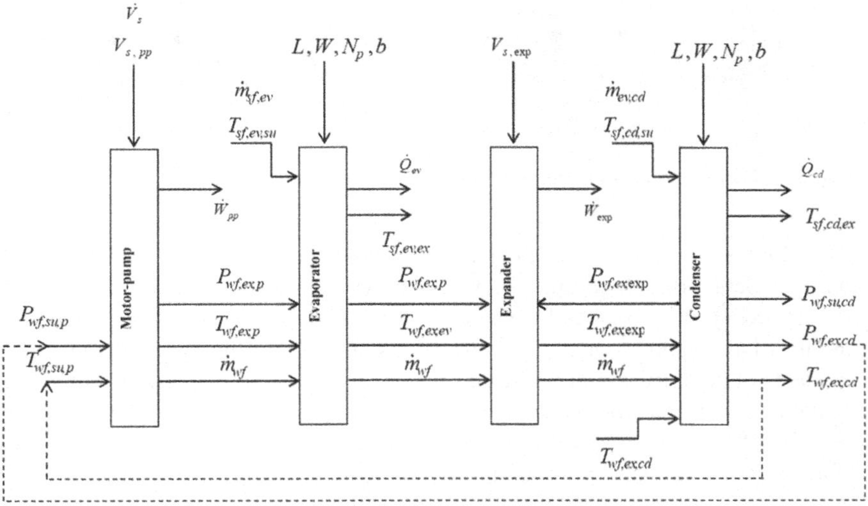

Fig. 4 Schematics of the ORC system model

3 Results and Discussion

The ICE considered in this paper is WP12.480 from Weichai Power CoLtd, whose main parameters are listed in Table 3. The exhaust mass flow rate and temperature depend on the engine torque and the rotational speed, Fig. 5 [11].

Table 3 Parameters of WP 12.480 engine

Parameters	Value
Cylinder number	6
Cylinder bore (mm)	126
Stroke (mm)	155
Displacement (L)	$(\pi.126^2/4).155.6 = 11.6$
The rated power (KW)	353
The rated speed (RPM)	2100

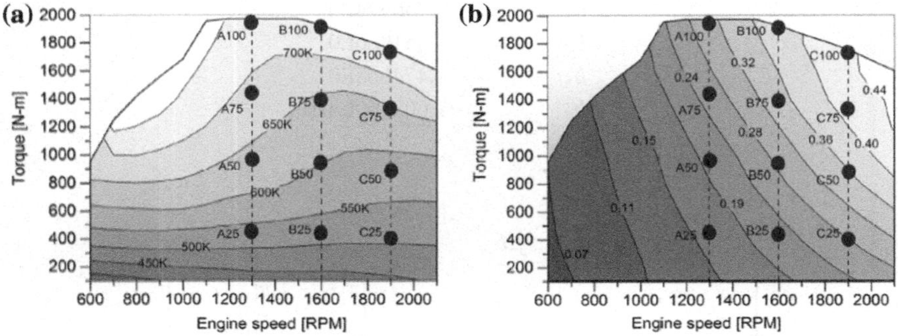

Fig. 5 **a** Engine exhaust temperature map (K); **b** engine exhaust mass flow map (kg/s)

The choice of the electric motor and the volumetric pump is drawn through a numerical calculation. Operating point A50 (1300 rpm and 50% load) is considered (characterized by $\dot{m}_{sf} = 0,205$ kg/s and $T_{sf,su,ev}$ = 364 °C). This choice is made at the beginning of the sizing study under the pretext that motor-pump constitute the key components of the ORC system. The optimized parameters are the powers delivered by the expander and by the ORC system. The expander used in the simulations (ZR48K3E-PFJ) has the highest displacement to guarantee the lowest rotational speed possible. Substantial that the rotational speed of the expander, must not exceed 2900 tr/mn according to the manufacturer's recommendations.

Two models of electric motors (Leason 1HP and 0.5 HP) and three plunger pumps (Hypro 2200B-P, 2300B-P and 2400B-P) have been chosen, which implies six possible combinations. Table 4 lists these combinations as well as the numerical results.

It is revealed from Table 4 that the efficiency of the 1HP electric motor is greater than that of 0.5 HP and that the isentropic efficiency of the 2200B-P pump is significantly higher than the other categories (2300B-P, 2400B-P).

Also, and despite the fact that the combination 1HP + 2351B-P does not provide the best performance and does not deliver the highest power, it was chosen and adopted for this study, on the one hand to respect the constraint of rotation of the expander and secondly to have a heat exchange surface whether for the evaporator or the condenser, reasonable.

Table 4 Results of the simulations of the ORC for different motor-pump combinations

	$\dot{W}_{exp}$ [KW]	$\dot{W}_{orc}$ [KW]	$\dot{W}_{pp}$ [KW]	$\dot{W}_{em}$ [KW]	P_{ev} [bar]	$V_{wf\ pp}$ [L/mn]	$V_{s\ pp}$ [m^3]	η_{orc} [%]	η_{em} [%]	η_{sp} [%]	η_{pp} [%]	A_{ev}	A_{cd}	N_{exp} [rpm]
1P+2230B-P	2.82	2.46	0.29	0.36	19.5	7.44	7.2 10^{-6}	7.51	78.3	67.3	52.7	0.88	5.30	3461
0.5HP +2230B-P	2.61	2.29	0.24	0.32	18.2	6.94	7.2 10^{-6}	7.55	75.7	67.0	50.0	0.77	4.92	3531
1HP +2351B-P	2.68	2.02	0.54	0.66	21.9	7.00	8.7 10^{-6}	6.53	82.0	39.0	31.6	0.86	5.02	2801
0.5HP +2351B-P	2.46	1.83	0.46	0.63	19.9	6.48	8.7 10^{-6}	6.47	73.0	37.2	27.2	0.74	4.64	2946
1HP +2413B-P	6.69	5.06	1.26	1.63	20.0	17.6	16.7 10^{-6}	6.57	77.0	37.0	28.8	1.46 10^{6}	3.9 10^{4}	7934
0.5HP +2413B-P	6.22	5.21	0.63	1.01	18.6	16.5	16.7 10^{-6}	7.23	62.0	64.0	40.0	6.37 10^{5}	2.6 10^{5}	8134

The dimensions of the heat exchangers (evaporator and condenser) depend on the flow rate and the temperature of the secondary fluid, whether for heating or cooling as shown in Figs. 6 and 7.

The plate heat exchangers discussed in Table 4 are Brazepak branded. Two models are discussed (BP 415 and BP 422).

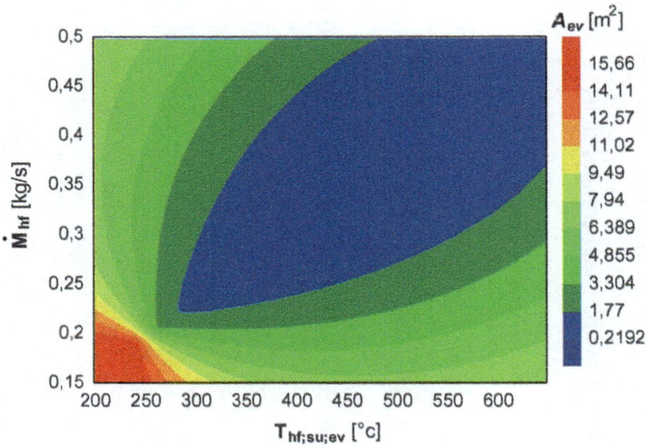

Fig. 6 Evaporator exchange surfaces

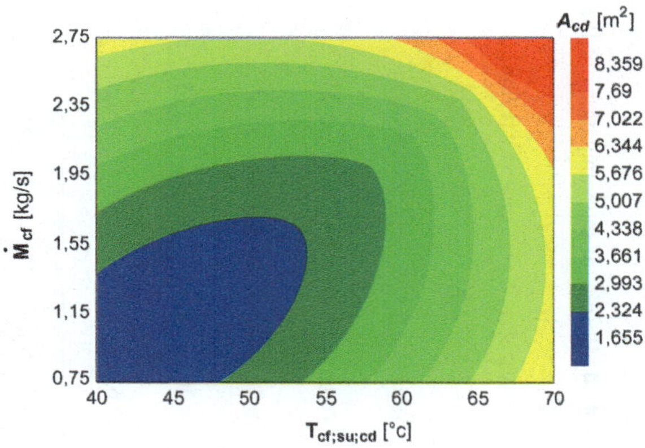

Fig. 7 Condenser exchange surfaces

Engine operating point A25 (1300 rpm and 25% load) is considered (characterized by $\dot{m}_{sf,su,ev} = 0.16$ kg/s and $T_{sf,su,ev} = 267$ °C). The Cooling water feeding temperature is 62 °C, and $\dot{m}_{sf,su,cd} = 1.24$ kg/s.

Our choice corresponds to the BP 422 exchanger, which allows the lowest number of plates possible and therefore less space for the evaporator and the condenser.

Finally, and concerning the expander (COPELAND brand), it must have rotational speed less than 2900 rpm, according to the manufacturer's catalogue (Table 5).

Table 5 Number of plate of heat exchanger

Plate heat exchanger model	Surface area (m^2)	A_{ev} (m^2)	A_{cd} (m^2)	$N_{P,ev}$	$N_{P,cd}$
BP 415	0.053	15.45	5.025	294	97
BP 422	0.099	15.45	5.025	158	53

Based on Fig. 8, we see that the value of 2900 rpm corresponds to a displacement limit equal to 2.15 10–5 m^3/tr. Therefore, we will choose an expander that has a displacement greater than this limit displacement. Consulting the catalogue of the manufacturer (Table 6), we realizes that our choice is fixed on the variant Zr48k3-PFJ, since it is the only one to satisfy the condition stated before.

The optimization of the ORC system was conducted based on the power delivered by the expander and the complete cycle, as well as the overall efficiency. The power of the ORC system and the global efficiency of the cycle are given by:

$$\dot{W}_{orc} = \dot{W}_{\exp} - \dot{W}_{el}; \eta_{orc} = \frac{\dot{W}_{orc}}{\dot{Q}_{ev}} \tag{10}$$

The only parameter to be varied and which can influence the results, is the evaporation pressure or the supply pressure of the expander. The calculation results revealed through Fig. 9, that the optimized evaporation pressure at the inlet of the expander is 17.67 bar. Similarly, if we bases for the optimization on the overall efficiency of the

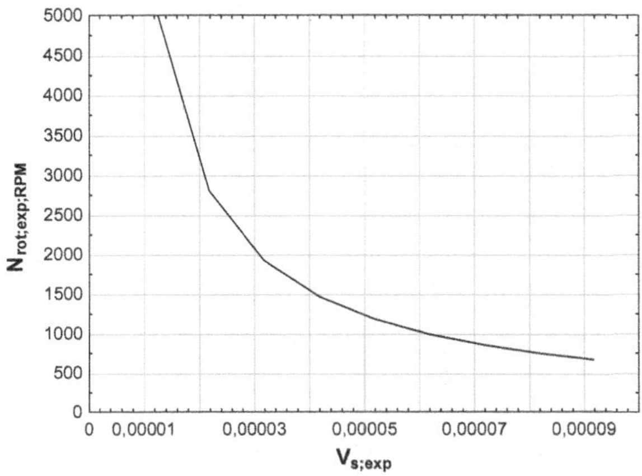

Fig. 8 Expander rotational speed versus displacement

Table 6 Expander displacement

ZR	Displacement (m^3/tr)
ZR42K3-PFJ	1.905×10^{-5}
ZR45k3-PFJ	2.035×10^{-5}
ZR47k3-PFJ	2.110653×10^{-5}
ZR48k3-PFJ	2.183303×10^{-5}

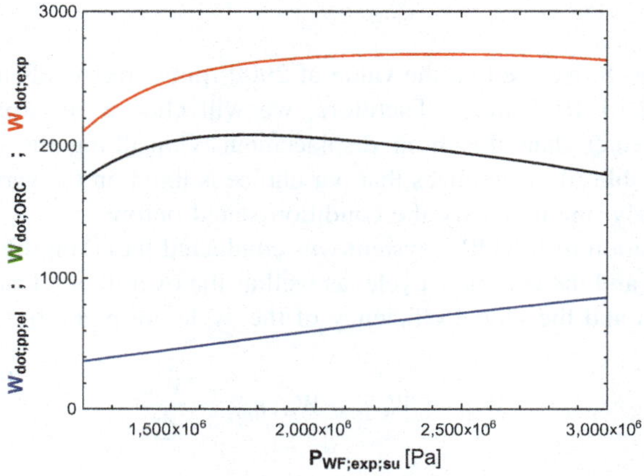

Fig. 9 Power generated by the ORC system

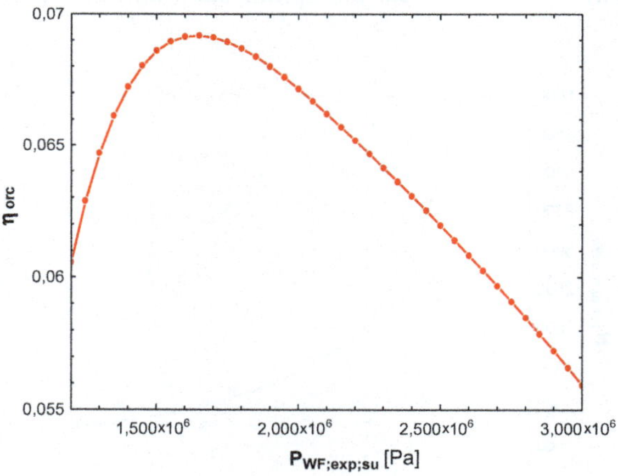

Fig. 10 Overall efficiency of the ORC system

ORC cycle (Fig. 10), we finds an evaporation pressure which is equal to 16.42 bar. But since the input energy is free (exhaust gas). So, despite that the efficiency of the diesel engine will increase, but we will get less energy.

4 Conclusion

The modelisation, optimization and sizing of the different components of WHR ORC system for ICE were investigated.

The results show that:

- The combination of the motor-pump 1HP+2351B-P is adequate for our case
- The Brazepak BP 422 plate heat exchanger is chosen
- The expander ZR48k3-PFJ is adopted
- The optimal evaporating pressure is equal to 17.67 bar

References

1. Guillaume L (2017) On the design of waste heat recovery organic Rankine cycle systems for engines of long-haul trucks. PhD Thesis University of Liège
2. Horst T, Rottengruber HS, Seifert M, Ringler J (2013) Dynamic heat exchanger model for performance prediction and control system design of automotive waste heat recovery systems. Appl Energy 105:293–303
3. Hsieh YY, Lin TF (2002) Saturated flow boiling heat transfer and pressure drop of refrigerant R-410a in a vertical plate heat exchanger. Int J Heat Mass Transfer 45:1033–1044
4. Ireland MK (2014) Dynamic modeling and control strategies for a micro-CSP plant with thermal storage powered by the Organic Rankine Cycle. PhD Thesis MIT
5. Quoilin S (2011) Sustainable energy conversion through the use of organic Rankine cycles for waste heat recovery and solar applications. PhD Thesis University of Liège
6. Lemort V, Declaye S, Quoilin S (2012) Experimental characterization of a hermetic scroll expander for use in a micro-scale Rankine cycle. In Part A. J Power Energy 226:126–136
7. Rody EC, Denis C (2005) Combined cycle for hybrid vehicles. SAE Paper, No. 2005-01-1171
8. Skarke P, Midlam-Mohler S, Canova M (2012) Waste heat recovery from internal combustion engines: feasibility study on an organic rankine cycle with application to the ohio state ecocar phev. In ASME Internal Combustion Engine Fall Technical Conference
9. Thonon B (1995) Recent research and developments in plate heat exchangers. Fuel and Energy Abstracts. 36:361–1995
10. Wei M, Fang J, Ma C, Danish SN (2011) Waste heat recovery from heavy-duty diesel engine exhaust gases by medium temperature ORC system. Sci China Tech Sci 54:2746–2753
11. Yang C, Xie H, Zhou SK (2016) Overall optimization of Rankine cycle system for waste heat recovery of heavy-duty vehicle diesel engines considering the cooling power consumption. Sci China Tech Sci 59:309–321

Entropy Generation Minimization Concept Evaluating Mixing Efficiency Through, Variable Density, Isothermal, Free Turbulent Jet

Nejmiddin Boughattas(✉)

IPEIK, Kairouan, Tunisia
nejmiddin@yahoo.fr

Abstract. This paper is devoted to the numerical and theoretical study of axisymmetric variable density jet discharging into a co-flowing stream. The results reveal that k–ε models give satisfactory agreement with experimental data. The results show also that entropy generation due to mass transfer is very higher than that corresponds to fluid friction. In addition, it was found that Entropy Generation Minimization concept can be used as an indicator to evaluate mixing efficiency and mixedness.

Keywords: Free turbulent jet · Entropy generation · Mixing

Nomenclature

C	Concentration mol·m^{-3}
C_μ, C_ε	Turbulence model constants
C_p	Specific heat. J·Kg^{-1}·K^{-1}
d	Nozzle diameter m
D	Mass diffusivity. m^{2}s^{-1}
G	Production term of k…kg·m^{-1}·s^{-3}
I	Mixing efficiency
k	Turbulent kinetic energy…m^2·s^{-2}
P	Pressure N·m^{-2}
R	Ideal gas constant JKg^{-1}K^{-1}
r	Radial distance m
S_{gen}	Entropy generation rate… WK^{-1}
S'''_{gen}	Local volumetric entropy generation rate Wm^{-3}K^{-1}
T	Temperature K
u, v	Velocity components in x, y direction ms^{-1}
x, y	Cartesian coordinates m
Y	Mass fraction
Z	Mixture fraction

N. Aifaoui et al. (Eds.): CMSM 2019, LNME, pp. 866–881, 2020.
https://doi.org/10.1007/978-3-030-27146-6_94

Greek symbols

ε	Dissipation rate of k $m^2 \cdot s^{-3}$
μ	Dynamic viscosity kg· $(m \cdot s)^{-1}$
ρ	Density $kg \cdot m^{-3}$
λ	Thermal conductivity…$Wm^{-1}k^{-1}$
σ_k, σ_ε	Turbulence model constants
σ_θ	Schmidt or Prandtl number (0.7)
θ	Scalar variable (*C*-*T*-*Y*-*Z*)
θ'^2	Scalar variance
χ	Scalar dissipation rate s^{-1}

Subscripts

a	Air, ambient fluid
0	Nozzle condition
eff	Effective
c	Condition at the jet centreline
in	Value at the jet exit
k	Species index (propane)
t	Turbulent

Acronyms

EGM	Entropy Generation Minimization
STD	Standard

1 Introduction

The Entropy Generation Minimization EGM, is a huge method for optimization design of actual processes and devices. The main advantage of this approach is that can santimize flow irreversibility locally and globally.

Bejan [1] was the first who illustrate the thermodynamic characteristics of isothermal turbulent jet. He has brought to light the thermodynamic aspect for a pure fluid mechanics analysis. Chu and Liu [2] investigated the entropy generation in a high temperature confined jet flow. They indicate that the total entropy generation decreases when the jet Reynolds number increase. Magherbi et al. [8] has formulated the expression of the dimensionless local entropy generation due to viscous effect, heat transfer and diffusion in the case of perfect gas mixture. Then they reports numerically the entropy generation in an inclined enclosure with heat and mass diffusive. Hashiehbaf and Romano [7] experimentally used entropy production concept to evaluate the mixing efficiency in free isothermal jet of water, without taking into account the mass transfer irreversibility. They found that large entropy generation is equivalent to large mixing. Elkaroui et al. [3] as well as Gazzah and Belmabrouk [5] studied the entropy generation in turbulent plane jet and turbulent round jet respectively, considering only viscous and heat irreversibility. They showed that the entropy generation rate grows progressively to reach an asymptotic value along the flow direction as the inlet temperature increases. Many other researches performed studies on the entropy generation and second law analysis in various flow cases under various conditions. Most of them has been focused on entropy generation due to fluid friction and heat transfer, but does not take account mass transfer. In this context, the present paper reports the numerical determination of the entropy generation due to fluid friction and mass transfer, with the aim to evaluating mixing efficiency.

To validate the prediction results, we use as bases of comparison the experimental results of Schefer [13] (turbulent, non-premixed, non-reacting propane-jet flow). Then, we investigate the concept of entropy production to assess the sources of irreversibility. Finally, the analysis of the influence of entropy generation on mixing efficiency and mixedness is done.

2 Mathematical Model

The equations that apply to the turbulent flow are the standard equations of fluid dynamics that are: the conservation of mass, momentum, scalar and scalar fluctuation. In steady regime, these equations constitute a set of coupled partial differential equations that can be written in the cylindrical coordinate in general form as:

$$\frac{\partial}{\partial \mathrm{x}}(\rho\, u\Phi)+\frac{1}{\mathrm{r}}\frac{\partial}{\partial \mathrm{r}}(\mathrm{r}\rho\, v\Phi)=\frac{\partial}{\partial \mathrm{x}}\left[\Gamma_{\Phi}\left(\frac{\partial \Phi}{\partial \mathrm{x}}\right)\right]+\frac{1}{\mathrm{r}}\frac{\partial}{\partial \mathrm{r}}\left[\mathrm{r}\Gamma_{\Phi}\left(\frac{\partial \Phi}{\partial \mathrm{r}}\right)\right]+S_{\Phi} \quad (1)$$

where Φ, Γ_{Φ} and S_{Φ} are respectively, the transportable quantity, the diffusion coefficient and the source term and are revealed in Table 1.

Table 1 Conservation equations

	Φ	Γ_{Φ}	S_{Φ}
Continuity	1	0	0
Axial momentum	u	$\mu_{eff}=\mu+\mu_t$	$-\frac{\partial P}{\partial x}+\frac{\partial}{\partial x}\left(\mu_{eff}\frac{\partial u}{\partial x}\right)+\frac{1}{r}\frac{\partial}{\partial r}\left(r\mu_{eff}\frac{\partial v}{\partial x}\right)$
Radial momentum	v	$\mu_{eff}=\mu+\mu_t$	$-\frac{\partial P}{\partial r}+\frac{\partial}{\partial x}\left(\mu_{eff}\frac{\partial u}{\partial r}\right)+\frac{1}{r}\frac{\partial}{\partial r}\left(r\mu_{eff}\frac{\partial v}{\partial r}\right)-2\mu\frac{v}{r^2}$
Scalar	θ	$\frac{\mu_{eff}}{\sigma_\theta}$	0
Scalar variance	θ'^2	$\frac{\mu_{eff}}{\sigma_\theta}$	$2\frac{\mu_t}{\sigma_\theta}\left(\frac{\partial \theta}{\partial x_j}\right)^2-\rho\chi$

The unknown terms in Table 1 are the turbulent viscosity μ_t and the scalar dissipation rate χ, which require respectively turbulence model and scalar model.

2.1 First-Order *K*-ε Turbulence Model

The *k*-ε model is the mainly common model used in CFD and remained until nowadays the most popular turbulence model. The eddy viscosity μ_t and the scalar dissipation rate χ are related to *k* and to ε as:

$$\mu_t = \rho\, C_\mu \frac{k^2}{\varepsilon} \quad ; \quad \chi = 2\frac{\varepsilon}{k}\, \theta'^2 \tag{2}$$

The modeled equations of *k* and ε are as follows:

$$\frac{\partial(\rho\, U_j k)}{\partial x_j} = \frac{\partial}{\partial x_j}\left[\left(\frac{\mu + \mu_t}{\sigma_k}\right)\left(\frac{\partial k}{\partial x_j}\right)\right] + S_k \quad \text{with} \quad S_k = G - \rho\,\varepsilon \tag{3}$$

$$\frac{\partial(\rho\, U_j\, \varepsilon)}{\partial x_j} = \frac{\partial}{\partial x_j}\left[\left(\frac{\mu + \mu_t}{\sigma_\varepsilon}\right)\left(\frac{\partial \varepsilon}{\partial x_j}\right)\right] + S_\varepsilon \quad \text{with} \quad S_\varepsilon = C_{\varepsilon 1}\frac{\varepsilon}{k} G - C_{\varepsilon 2}\,\rho \frac{\varepsilon^2}{k} \tag{4}$$

The empirical constants for the *k*–ε model are given in Table 2, including the Rodi correction for round jet.

Table 2 Empirical constants of the *k*–ε model

C_μ	$C_{\varepsilon 1}$	$C_{\varepsilon 2}$	σ_k	σ_ε
0.09–0.04 *f*	1.44	1.92–0.0667 *f*	1.0	1.3

2.2 Reynolds Tensor

The first-order turbulence model predicts the turbulent kinetic energy *k*, but does not give any precise information on its repartition according to the directions. There are several ways to better estimate the anisotropic structure of turbulence. One way is to use the Reynolds Stress Model RSM, which is difficult to manage due to the stability problem.

An intermediate solution is to estimate the Reynolds stress from algebraic relations deduced from a second order closure [11]. These relations supplement the first-order turbulence models and are defined by:

$$\overline{u_i' u_j'} = k\left[\frac{2}{3}\delta_{ij} + \frac{1}{(G - \varepsilon + c_1'\,\varepsilon)}\left[(1-\gamma_1)\left(P_{ij} - \frac{2}{3}\delta_{ij} G\right) - \gamma_2 k\left(\frac{\partial U_i}{\partial x_j} + \frac{\partial U_j}{\partial x_i}\right) - \gamma_3\left(D_{ij} - \frac{2}{3}\delta_{ij} G\right)\right]\right] \tag{5}$$

where *k*, ε, *G* and the gradients are calculated with the aid of the first-order turbulence model. Tensors P_{ij} and D_{ij} are given by:

$$P_{ij} = -\left[\overline{u'_j u'_k}\frac{\partial U_i}{\partial x_k} + \overline{u'_i u'_k}\frac{\partial U_j}{\partial x_k}\right] \quad ; \quad D_{ij} = -\left[\overline{u'_j u'_k}\frac{\partial U_k}{\partial x_i} + \overline{u'_i u'_k}\frac{\partial U_k}{\partial x_j}\right] \tag{6}$$

Constants c'_1, γ_1, γ_2 and γ_3 are respectively given by 1.8, 0.76, 0.18 and 0.2 [10].

2.3 Equation of State

The mean density of the mixture can be obtained from the mean mixture fraction using the equation of state. With constant pressure, this leads to:

$$\frac{1}{\rho} = \frac{\mathrm{Y_k}}{\rho_k} + \left(\frac{1 - \mathrm{Y_k}}{\rho_a}\right) \tag{7}$$

2.4 Entropy Generation Rate

In the case of turbulent non-reactive mixture, the local entropy generation rate S'''_{gen} ($\mathrm{Wm^{-3}K^{-1}}$) per unit volume in two-dimensional polar coordinate with single diffusing specie of concentration (C) can be written as [8]:

$$\begin{aligned} \mathrm{S'''_{gen}} = {} & \frac{\mu_{eff}}{T_0}\left\{2\left[\left(\frac{\partial u}{\partial x}\right)^2 + \left(\frac{\partial v}{\partial r}\right)^2 + \left(\frac{v}{r}\right)^2\right] + \left(\frac{\partial u}{\partial r} + \frac{\partial v}{\partial x}\right)^2\right\} \\ & + \frac{\lambda_{eff}}{T_0^2}\left[\left(\frac{\partial T}{\partial x}\right)^2 + \left(\frac{\partial T}{\partial r}\right)^2\right] \\ & + \frac{RD_{eff}}{C_0}\left[\left(\frac{\partial C}{\partial x}\right)^2 + \left(\frac{\partial C}{\partial r}\right)^2\right] \\ & + \frac{RD_{eff}}{T_0}\left[\left(\frac{\partial T}{\partial x}\right)\left(\frac{\partial C}{\partial x}\right) + \left(\frac{\partial T}{\partial r}\right)\left(\frac{\partial C}{\partial r}\right)\right] \end{aligned} \tag{8}$$

where C_0 and T_0 are respectively the bulk concentration and the bulk temperature.

λ_{eff}, and D_{eff} are the effective thermal conductivity and the effective mass diffusivity and expressed as:

$$\lambda_{eff} = \frac{\mu_{eff} Cp}{\mathrm{Pr}} \quad ; \quad D_{eff} = \frac{\mu_{eff}}{\rho_k\, Sc} \tag{9}$$

The first term in Eq. (8) is due to fluid friction, the second is due to pure heat transfer, the third is due to pure mass transfer and the fourth is due to the coupling between thermal and mass transfer.

Since we trait an isothermal jet, only the first term and the third term are considered. Consequently, the entropy generation depends functionally on the local values of velocity and concentration in the domain of interest.

The total entropy generation rate S_{gen} (WK^{-1}) in the jet over the volume can be calculated as follows:

$$S_{gen} = \int_V S'''_{gen} dV \quad \text{Where} \quad dV = r\,dr\,d\theta\,dz \tag{10}$$

3 Boundary Conditions

The boundary conditions are:

- At the inlet, for $x = 0$ and $0 \leq y < d/2$; and in order to overcome as much jet emission influence as possible, the axial velocity profile was calculated from the following relations: $u_{in} = u_{cO}(1-2y/d)^{1/5}$; The turbulent kinetic energy was taken as $k_{in} = u_{cO}(0.04 + 0.1y/d)^2$; the dissipation as $\varepsilon_{in} = (C_\mu\, k_{in}^{1.5})/L_m$; the length scale as $L_m = 0.005d$; $v = 0$; $Z_{in} = 1$; $z' \approx 0$.
  For $x = 0$ and $y \geq d/2$; $u = u_{coflow}$; $k = (0.004\, u_{coflow})^2$; $\varepsilon = (C_\mu\, k^{1.5})/0.002$; $v = 0$; $Z = 0$; $z' \approx 0$; $k_p = k_t = 0.5\, k$ and $\varepsilon_p = 0.5\varepsilon_t = 0.5\varepsilon$
- In the free boundary parallel to the axis, we consider the following conditions: $u = 0$; $\partial v/\partial y = 0$; $k = 0$; $\varepsilon = 0$; $Z = 0$; $z' = 0$
- On the symmetry axis, the radial velocity and radial gradients of all variables are set to zero: $(\partial \Phi/\partial y)_{r=0} = 0$ and $v = 0$; $\Phi = u, k, \varepsilon, Z$ or z'
- At the outflow boundary, the gradient of dependent variables in the axial direction is set to zero: $(\partial \Phi/\partial x)_{\text{outlet}} = 0$; $\Phi = v, k, \varepsilon, Z$ or z'

4 Numerical Procedure

The calculation procedure is based on the TEACH code [6]. The Finite Volume Method (FVM) with a staggered grid as described by Patankar [9] is used for the numerical resolution of the transport equations. The formulation is elliptic convection-diffusion and the pressure velocity coupling is achieved through the SIMPLE algorithm (Semi Implicit Method for Pressure Linked Equation). The convergence of the numerical solution was based on the sum of the absolute residual sources and must be less than 0.3%. The diffusion and the convection coefficients are discretized using a Hybrid scheme (Upwind scheme for Peclet number higher or equal than 2 and Central Differencing scheme for Peclet number less than 2). The system of algebraic equations is solved line by line using the TDMA algorithm (Tri Diagonal Matrix Algorithm) with Gauss-Seidel iterative method. To prevent divergence of the coupled set of equations, an under-relaxation method is employed. Calculation is carried out with a structured, orthogonal, expanding grid, with an expansion factor of 1.05. In fact, the mesh in both directions becomes wider when one moves away from the emission section of the jet. The considered calculation domain was 2000 mm in the streamwise and 450 mm in the radial direction. Through grid refinement studies, it was established that the results are independent of numerical influences for grids finer than 100 × 60 meshes. The system of Eqs. (5), (6) is solved using a Predictor-Corrector Algorithm as recommended by Picart et al. [10].

5 Results and Discussion

The flow configuration is shown in Fig. 1 and consists of a propane jet ejected from a round nozzle issuing vertically downward from a pipe into a co-flowing air. The geometric and flow parameters used in the present study are the same as those in the experimental works of Schefer [13] and are summarized in Table 3.

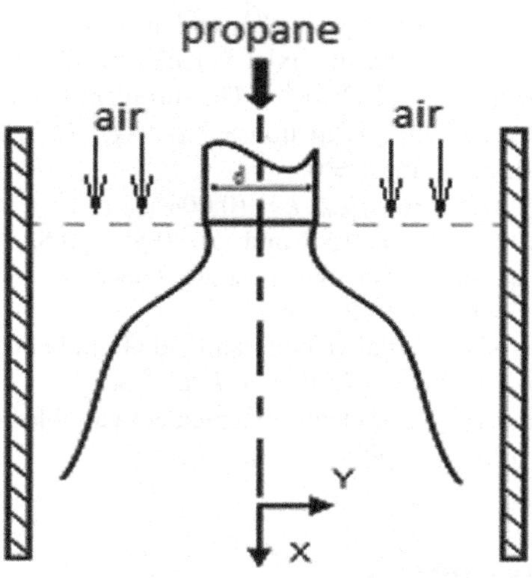

Fig. 1 Simulation test case

Table 3 Operating conditions

U_O (m/s)	69
U_{coflow} (m/s)	9.2
d (mm)	5.26
Re	68000
ρ_k (Kg/m^3)	1.882
ρ_a (Kg/m^3)	1.202
T_k (K)	294
T_a (K)	294

Figure 2 shows the mean longitudinal velocity decay as a function of the normalized distance x/d. Figure 3 represents the axial evolution of the mean mixture fraction in the jet. The predicted results obtained agree reasonably with the experimental findings [13].

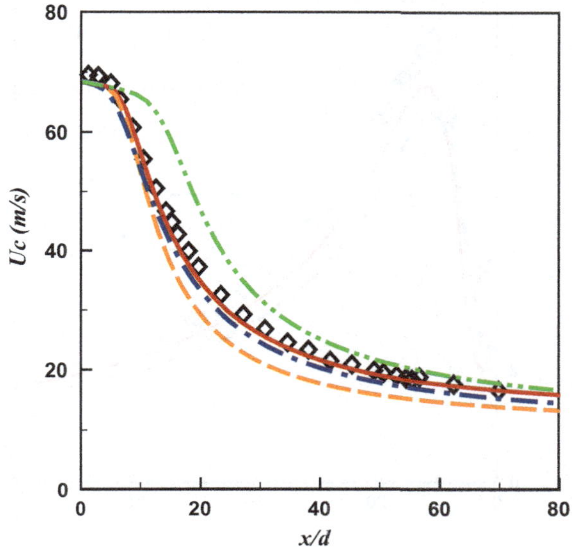

Fig. 2 Centerline profiles of axial velocity

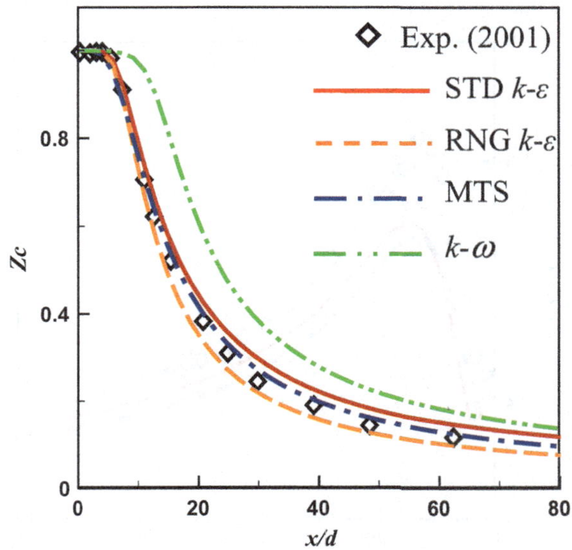

Fig. 3 Centerline profiles of Zc

The evolution of the streamwise u'_c and spanwise v'_c velocity fluctuation along the centreline is shown in Figs. 4 and 5. As already mentioned above and because of the isotropy assumption of the first order turbulence models, the normal and tangential Reynolds stresses are computed using algebraic relation which supplement first order turbulence models [11].

The simulation results under predict slightly the experimental data.

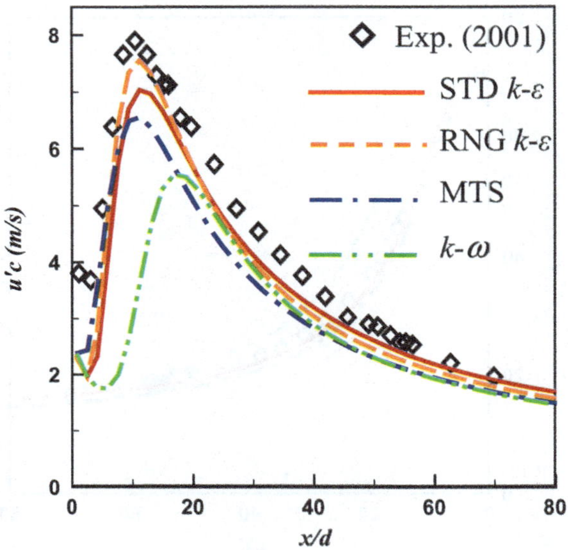

Fig. 4 Centerline profiles of u'c

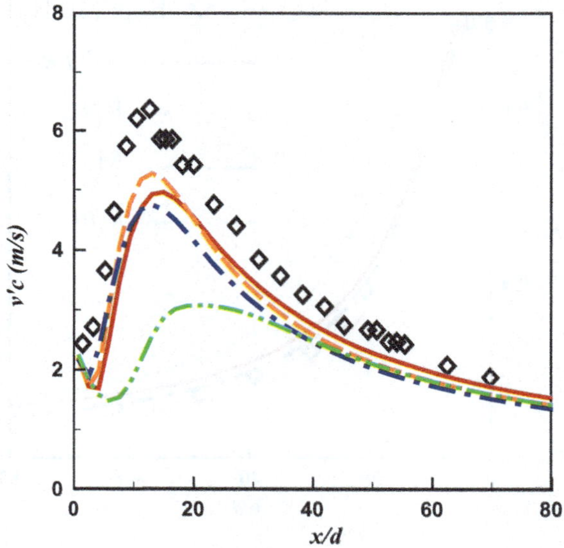

Fig. 5 Centerline profiles of v'c

Transversal evolution of the mean and the fluctuating mixture fraction at the same downstream section from the injection $x/d = 15$ are presented in Figs. 6 and 7. The comparison between experimental measurement and numerical results shows a good concordance for the mean mixture fraction; while for the fluctuating mixture fraction, STD k-ε turbulence model overestimate the value of experimental data.

Figures 8 and 9 show respectively the radial profiles of the mean and fluctuating mixture fraction in the downstream section $x/d = 50$. It is clear that numerical results of STD k-ε turbulence model agree reasonably with the experimental data.

Figure 10 summarizes the local entropy generation contours for the different source of irreversibility. In all cases, the local entropy contours are very dense and have high values near the nozzle exit. This indicates that the viscous dissipation and the gradient concentration are important near the nozzle exit and becomes negligible in the affinity region of the jet. Indeed, the mixing layer expands along the direction of the jet.

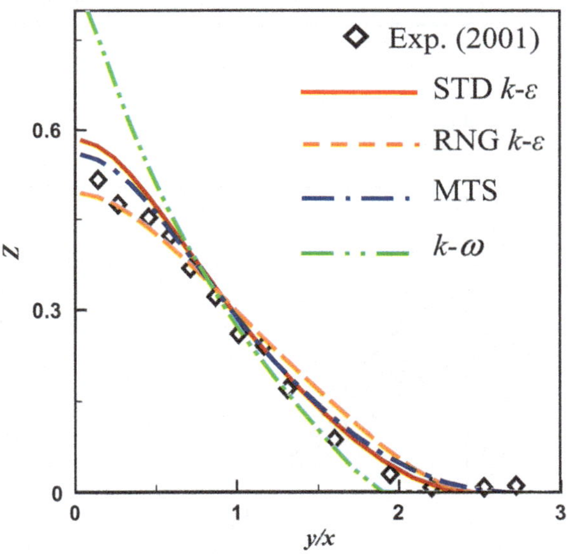

Fig. 6 Radial profiles of Z at $x/d = 15$

Through a careful comparison of the regions in the vicinity of the nozzle exit related to these cases, it appears that entropy generation due to mass transfer is very higher than that corresponds to fluid friction. The high value of the local total entropy generation is 40 $\text{wm}^{-3}\text{k}^{-1}$.

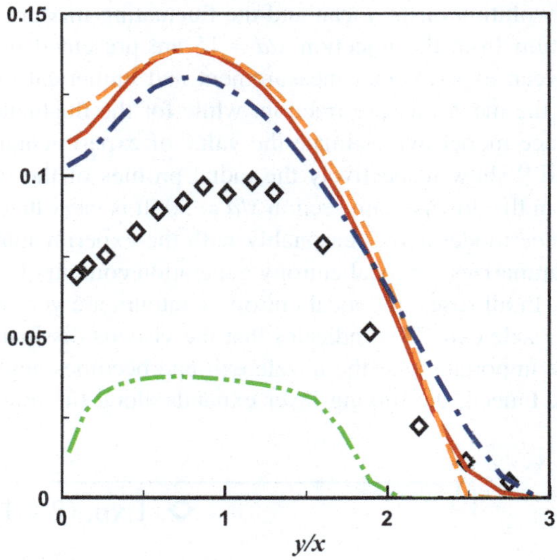

Fig. 7 Radial profiles of z' at $x/d = 15$

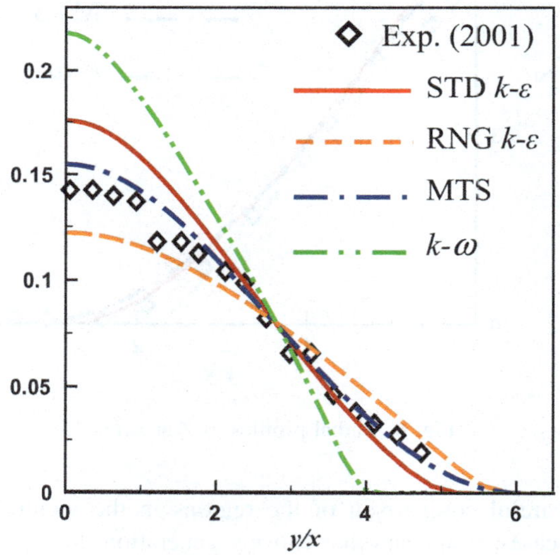

Fig. 8 Radial profiles of Z at $x/d = 50$

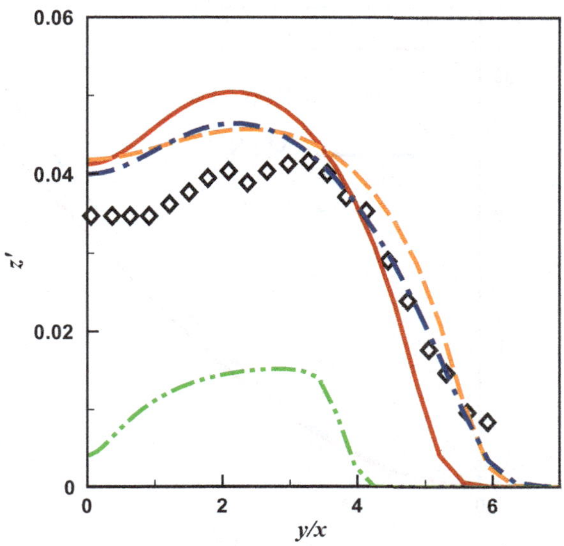

Fig. 9 Radial profiles of z' at $x/d = 50$

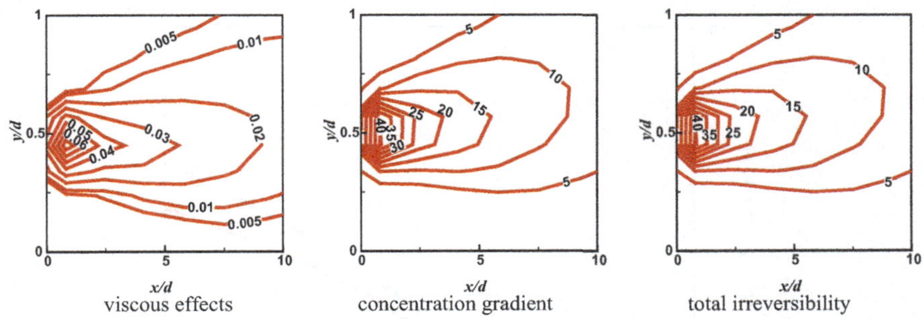

Fig. 10 Local entropy generation contour

Figure 11 represents the axial evolution of the mixing efficiency I defined as [12]:

$$I = \frac{1}{\rho_k d^3} \int_0^\infty \bar{\rho}\tilde{\theta}\, r dA \quad \text{with} \quad dA = r\, dr \tag{11}$$

The mixing efficiency determines the quality of propane injected at the jet nozzle and found away from the symmetry axis. It is shown that the mixing efficiency grows as x/d increases.

The evolution of the scalar based on the mixture fraction half-width along the axial direction is plotted in Fig. 12. The same pace with the evolution of the mixing efficiency is observed.

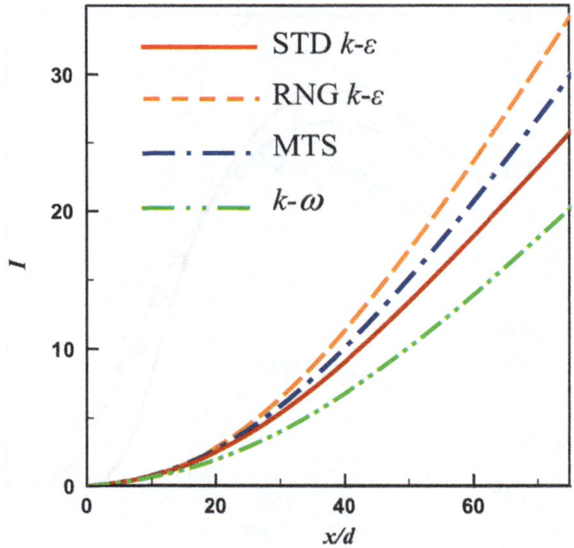

Fig. 11 Evolution of the axial mixing efficiency

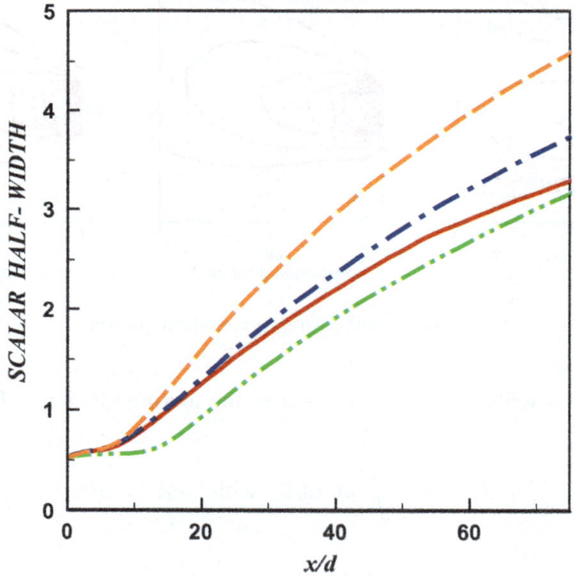

Fig. 12 Centreline values of the mixture fraction half-width

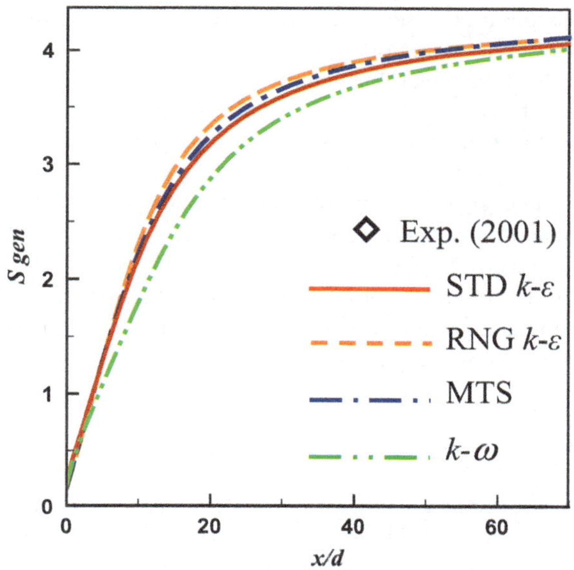

Fig. 13 Axial evolution of the total entropy generation rate

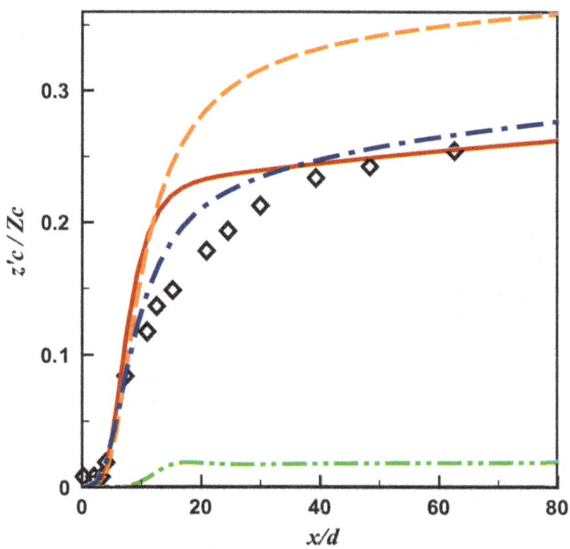

Fig. 14 Centreline values of the scalar fluctuation intensity

Figure 13 represents the axial evolution of the total entropy generation rate in the jet. It is shown that the entropy generation grows as *x/d* increases. Moreover, there is a sudden increase in the entropy generation in the flow region where *x/d* is lesser then 20 in which the jet is more unsteady and has the more distorted profiles, right before it reaches the self-preservation region.

The axial profile of the scalar fluctuation intensity or mixedness on the jet centerline is shown on Fig. 14. The profile increase with *x/d* and then it reach an asymptotic value, as well as the behavior of total entropy generation rate. This is due to the expression of entropy production (Eq. 8) which is approximated by the concentration gradient. Thus, the entropy is well approximated by the RMS fluctuation divided by the mean concentration [4], and not only by the mean square velocity derivatives, as Hashiehbaf and Romano [7] stated.

From Figs. 11, 12, 13 and 14, it is clearly that the highest total entropy production rate gives the best mixedness and the best mixing efficiency. So we can state that large entropy production is equivalent to best mixing.

6 Conclusion

In the present paper, results of a numerical investigation of the behaviour of a turbulent round, variable density, jet discharged in a coflowing ambient stream is presented. Numerical results are compared with experimental data reported in the literature. The main conclusions can be summed up as follows:

- Mass transfer irreversibility approximates the total entropy production in the case of isothermal, variable density jet.
- Entropy generation and mixing efficiency are intimately proportional.

References

1. Bejan A (1991) Thermodynamics for an isothermal flow: the two-dimensional turbulent jet. Int J Heat Mass Transfer 34:93–114
2. Chu SX, Liu LH (2009) Entropy generation analysis of two dimensional high temperature confined jet. Int J Thermal Sci 48:998–1006
3. Elkaroui A, Gazzah H, Mahjoub N, Le BP, Palec G (2018) Entropy generation Concept for a turbulent plane jet with variable density. Comput Fluids 168:328–341
4. Everson D, Manin L, Sirovic M, Winter M (1995) Quantification of mixing and mixing rate from experimental observations. 33rd Aerospace Science Meeting and Exhibition NY
5. Gazzah MH, Belmabrouk H (2014) Local entropy generation in co-flowing turbulent jets with variable density. Int J Numer Meth Heat Fluid Flow 24:1679–1695
6. Gosman AD, Ideriah FJK (1976) Teach T. Imp College Dept Mech Eng London
7. Hashiehbaf A, Romano GP (2014) An experimental investigation on mixing enhancements in non-circular sharp-edged nozzles using the entropy production concept. J Turbul 15:411–428
8. Magherbi M, Abbasi H, Hidouri N, Ben brahim A (2006) Second law analysis in convective heat and mass transfer. Entropy 8:1–17
9. Patankar SV (1980) Numerical heat transfer and fluid flow. Series in computational methods in mechanical and thermal sciences. Hemisphere/Mc Graw-Hill, New York

10. Picart A, Berlemont A, Gouesbet G (1986) Modelling and predicting turbulence fields and the dispersion of discrete particles transported by turbulent flows. Int J Heat and Mass Transfer 12:237–261
11. Rodi W (1979) turbulence models for environmental problems. Von Karman Institute for fluid dynamics: Lecture series 1979–2, Prediction methods for turbulent flows, 15–19 January
12. Sanders H, Sarh B, Gökalp I (1996) Numerical prediction of the structure of high temperature axisymmetric turbulent jets. Int J Therm Sci 35:232–242
13. Schefer RZ (2001) Data base for turbulent, non-premixed, non-reacting propane-jet flow. Combustion Research Facility Sandia National Laboratories Livermore CA

Soft Underwater Robots Imitating Manta Actuated by Dielectric-Elastomer Minimum-Energy Structures

Eiji Hasegawa[1,2(✉)], Moncef Hammadi[1], Jean-Yves Choley[1], and Aiguo Ming[2]

[1] SUPMECA, 3 rue Fernand Hainaut, 93400 Saint-Ouen, France
hasegawa1994.uec@gmail.com, moncef.hammadi@supmeca.fr
[2] Ming Lab, UEC, 1-5-1 Chofugaoka, Chofu, Tokyo 182-8585, Japan

Abstract. Biomimetic robots imitating the characteristics of animals are being developed to help humans understand related phenomena in order to engineer new devices. Particularly, soft robotics are being incorporated to help imitating the natural flexibility of animals. Hence, Dielectric-Elastomer Minimum-Energy Structures (DEMES) are used to fabricate soft robots. The advantages of this solution is that the manufacturing procedure is simple. However, high voltages are required for driving them. The challenge is to use this technology at relatively lower voltages. In this research, we aim to realize Soft Underwater robots driven at low voltage to imitate the swimming of Manta. Therefore, we made a DEMES actuator usable in water by sandwiching the carbon as the electrode with two dielectric elastomers. Our robot swims faster at a lower voltage compared to the soft underwater robot that we made in a previous study.

Keywords: Biomimetic robotics · Soft actuators · Dielectric-elastomer minimum energy structures (DEMES)

1 Introduction

The ocean provides a lot of resources, and exploration for this is the focus of today's world. As a result, the demand for underwater robots has increased over a wide range, including ecological investigation of marine organisms, seabed survey, maintenance inspections of facilities.

A biomimetic robot can be considered as a mechanism exhibiting high capability. Living organisms existing in nature have evolved diversely and adapt to the environment and survival competition, and have improved their abilities. In recent years, focusing on the features of living organisms and elucidating their functions and structures, development of high-performance biomimetic robots by imitation and application has been advanced [1, 2]. In addition to that, in consideration of the flexibility of living organisms in conventional robotics engineering, the development of biomimetic robots incorporating soft robotics, which uses flexible materials not only for structures but also for actuators, has been actively developed.

N. Aifaoui et al. (Eds.): CMSM 2019, LNME, pp. 882–891, 2020.
https://doi.org/10.1007/978-3-030-27146-6_95

An underwater robot can be cited as an application field of biomimetic soft robot. Most underwater robots that are in practical used are screw type, but there are problems such as propulsive efficiency, noise, large weight, involvement of aquatic life [3–6]. As a solution to this problem, research and development of underwater robots and soft underwater robots that mimic the propulsion of fins and body of fishes has been actively conducted.

As actuators used for soft robots, ionic polymer metal composites (IPMC) [7, 8], dielectric elastomers (DEs) [9, 10], shape memory alloys (SMA) [11], piezoelectric ceramics [12], electroactive polymers [13, 14] have been developed. These actuators are widely used in the field of biomimetic as compared with conventional motor actuators. These power and accuracy are not comparable to conventional motors, but there are many merits such as easy fabrication and miniaturization.

In this study, we used dielectric elastomers among soft actuators. This actuator has a feature that large distortion can be obtained, both elastomeric materials and stretchy electrodes are easy to obtain, we can produce arbitrary structure because of its simple structure. A disadvantage is that it is necessary to apply a high voltage of several thousand volts to drive it [15].

The dielectric elastomer actuator has a simple structure in which an elastomer as a dielectric is sandwiched between stretchable electrodes. By applying a voltage to the dielectric elastomer, Maxwell stress p(1) works on the electrodes on both sides (Fig. 1). The actuator is thereby extended. When quitting the voltage application resumes. This motion is used as an actuator. Here, p is the stress, ε_0 is the dielectric constant of the vacuum, ε_r is the dielectric constant of the dielectric, E is the electric field, V is the applied voltage, and z is the thickness of the dielectric.

$$P = \epsilon_0 \epsilon_r E^2 = \epsilon_0 \epsilon_r \left(\frac{V}{z}\right)^2 \tag{1}$$

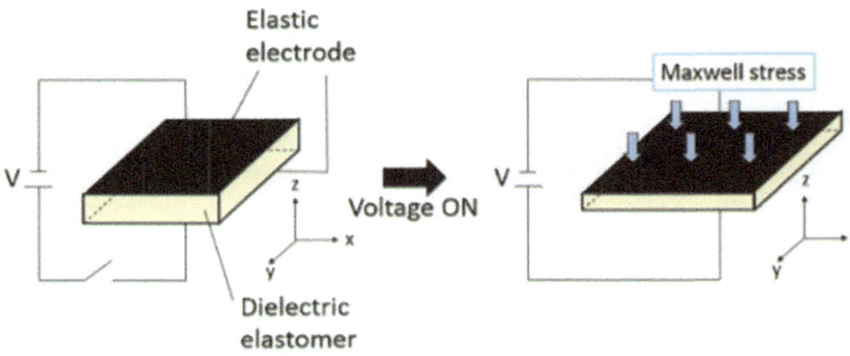

Fig. 1 Elongation of dielectric elastomer

This paper is organized in the following order. In Sect. 2, swimming behavior of manta is analyzed. In Sect. 3 describes the structure of newly created actuators and robots. In Sect. 4 describes the results of performance tests of actuators and robots. In Sect. 5, we will conclude.

2 Swimming Method Classification

The swimming movement of aquatic organisms can be roughly divided into two categories, BCA (Body/Caudal Actuation) and MPA (Median/Paired Actuation), depending on the part and size used for propulsion (Fig. 2) [16]. In BCA, they perform movement using tail bill and body, classifying tuna, eel, and sharks. In MPA, behaviors using back billets, buttocks and chest burials are performed, and rays and ostraciidae are classified. Furthermore, it is classified into Oscillatory and Undulatory respectively based on the frequency at which organisms operate. Manta is classified as MPA-O and can swim much more efficiently than underwater moving objects of artificial objects is there. The swimming form due to the fluttering motion of manta is characterized by a relatively low flapping frequency and large flapping amplitude. Therefore, it is possible to swim with excellent quietness and not disturb surrounding environment. Moreover, it has high maneuverability in the vertical direction and the horizontal direction and it can make quick direction change and vertical movement.

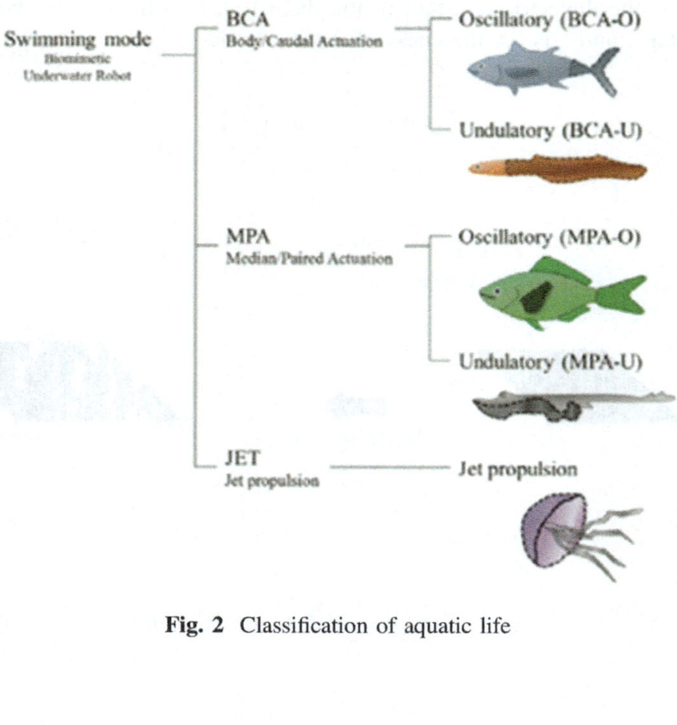

Fig. 2 Classification of aquatic life

As an advantage of incorporating manta into a robot, a fish robot having a chest fin spreading to the left and right has easy control and stable operability. In addition, unless precise movement or velocity is required, it is possible to spread the fins and glide in the water, further noise reduction and efficient movement can be expected (Fig. 3).

Fig. 3 The biological system to be bioinspired (Manta ray)

3 Actuator and Robot Structure

3.1 Actuator Structure

A structure called Dielectric Elastomer Minimum Energy Structure (DEMES) has been developed by using the elongation principle of dielectric elastomer [17]. As an elastomer, acrylic adhesive tape 3 MVHB is used. First, elongate the elastomer. This pre-stretch exerts a restoring force on the elastomer (Fig. 4). Attach the elongated dielectric elastomer to a flexible frame made of very thin plastic (Fig. 5). At this time, the frame is cut out into an elliptical shape. Releasing the elastomer from pre-stretch generates elastic energy to bend the frame. At this time, the elastomer assumes a curved surface having the smallest area. This establishes a minimum-energy structure.

Applying electrodes (carbon grease) on both sides of the part where the PET frame is cut off and the dielectric elastomer is exposed (Fig. 6). By applying a voltage, this actuator is bent. By adjusting the size of the ellipse to be cut, the rigidity also changes, and an ideal bending motion is realized even with a relatively small voltage. As a feature of DEMES, it is possible to realize an ideal bending of a dielectric elastomer actuator requiring a high voltage for driving even at a relatively low voltage.

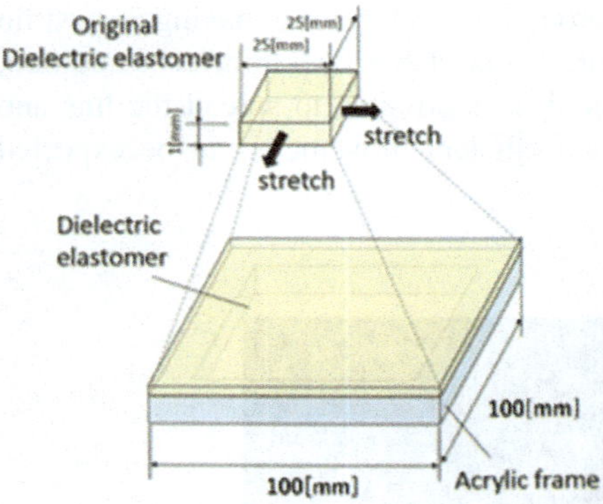

Fig. 4 Pre-stretch of dielectric elastomer

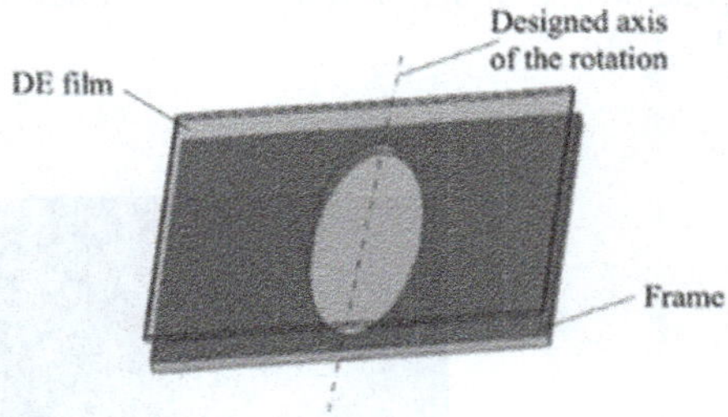

Fig. 5 Structure of conventional DEMES

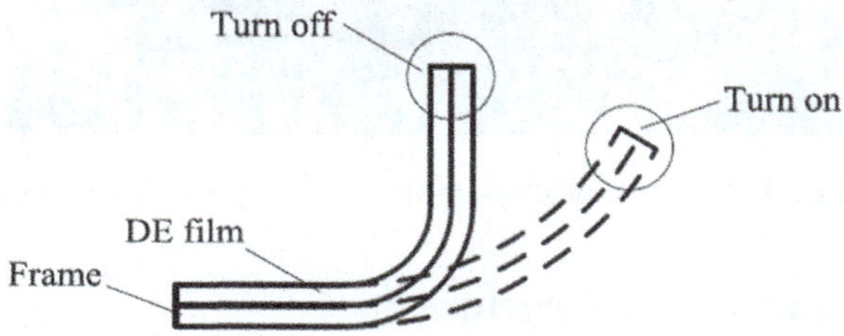

Fig. 6 Bending and extension of DEMES

However, DEMES cannot be used in water because the positive part of the electrode is outside. It can be used by adding an insulator to the positive side of the electrode in water, but this will lose flexibility and degrade performance. Therefore, we sealed the electrode with two elastomers and used a method that does not use an insulator by setting the surrounding water as the ground [18]. In this research, we newly designed and fabricated an actuator with two concepts of DEMES which realizes ideal flexion at low voltage and used ground as water (Figs. 7 and 8). We used 3 M VHB 1 mm thick for elastomer and stretched 400% during pre-stretching.

3.2 Robot Structure

We made a robot consisting of a body part and two fins (Figs. 9 and 10). The body part is the actuator itself. Fins are reproduced with 0.5 mm thick silicon, and a 0.5 mm CFRP is affixed to the tip. The total length was 55 mm, the width was 183 mm, and the mass was 13.8 g. The figure shows the whole dimensions and the real thing. Floats are arranged forward and backward.

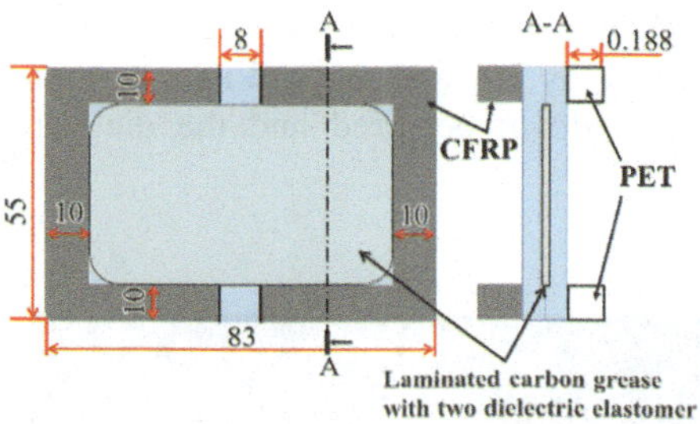

Fig. 7 Newly designed DEMES

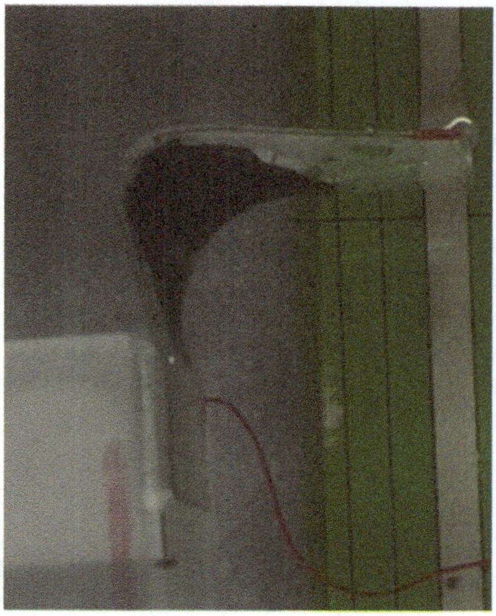

Fig. 8 The fabricated model of DEMES

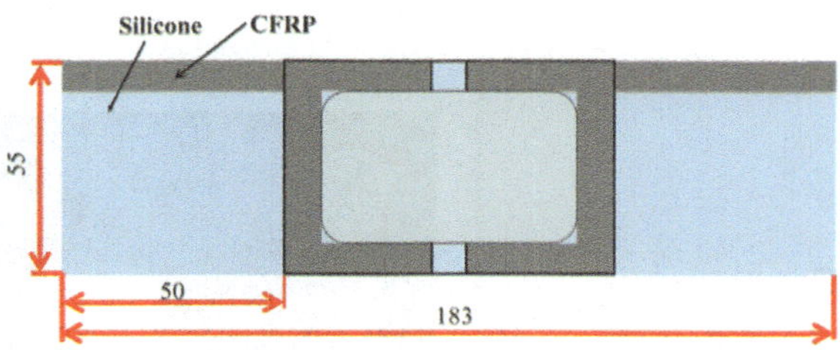

Fig. 9 Robot dimensions

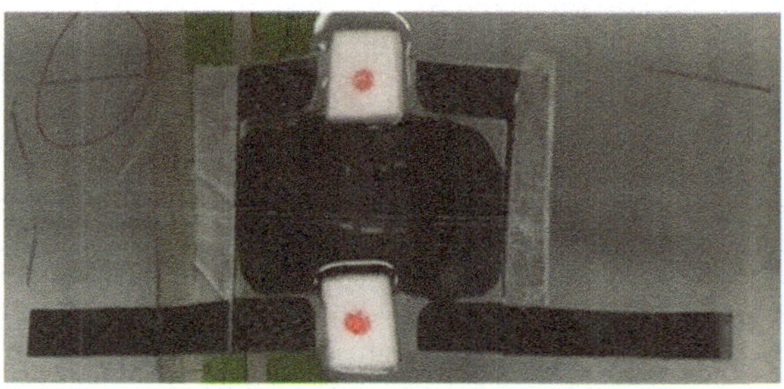

Fig. 10 The fabricated Robot

4 Measurement Results

4.1 Actuator

We put the actuator in the aquarium as shown in Fig. 11 and carried out the test. The temperature of water was 16 °C. When the applied voltage to the actuator was raised by 0.1 kV from 2.0 kV, DEMES could be driven even in water. The bending angle, which is the difference between the initial angle and the angle after driving, was measured by video analysis (Fig. 12). The Fig. 13 shows the bending angle of the actuator when a voltage is applied to the actuator. The bending angle increased as the applied voltage was increased. When applying 2.4 kV, it was bent by 30° and reached its maximum. Also, when 2.5 kV was applied, dielectric breakdown occurred and the dielectric elastomer broke.

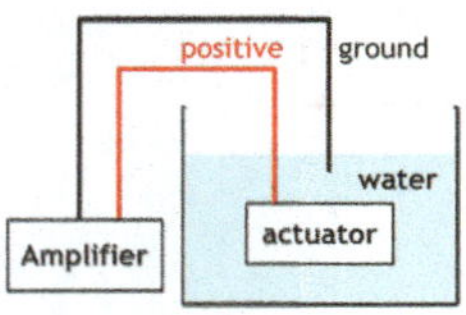

Fig. 11 Overall view of apparatus

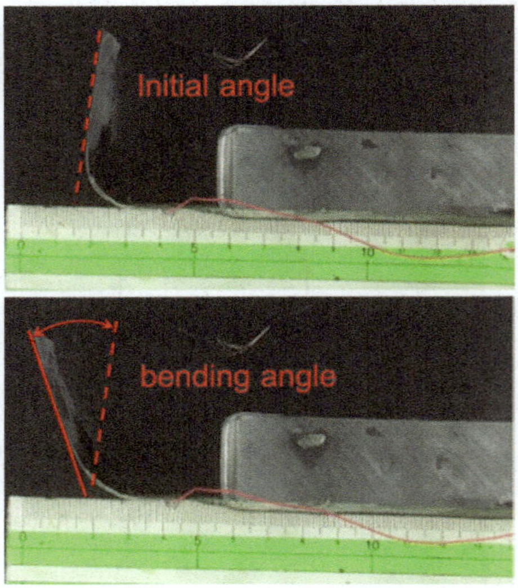

Fig. 12 Measurement method of bending angle

4.2 Robot

We verified the robot's performance. A square wave with a voltage of 2.3 kV was applied and increased from 1 Hz by 1 Hz (Fig. 14). It became 44 mm/s of the peak at 3 Hz, and it decreased from then. We thought that this is because the frequency of the input voltage exceeds the response of the dielectric elastomer and the bending angle of the actuator is reduced. Since the velocity was peak at the frequency 3 Hz, the frequency was fixed at 3 Hz, the applied voltage of the rectangular wave was increased by 0.1 kV from 2.0 kV (Fig. 15). The Fig. 15 shows the relationship between applied voltage and velocity. The maximum velocity at an input voltage of 2 Hz was

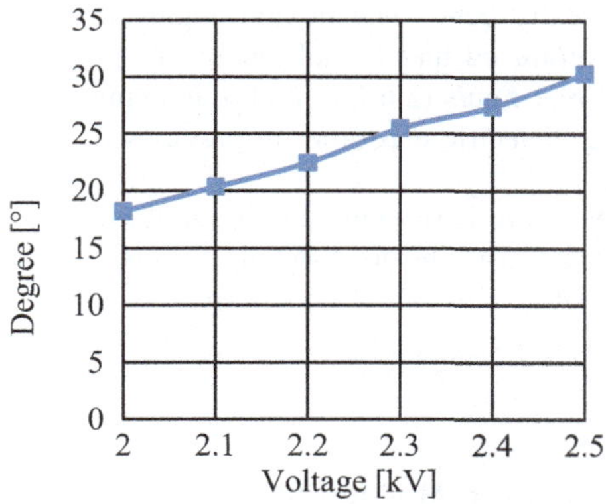

Fig. 13 Relation between voltage and bending angle

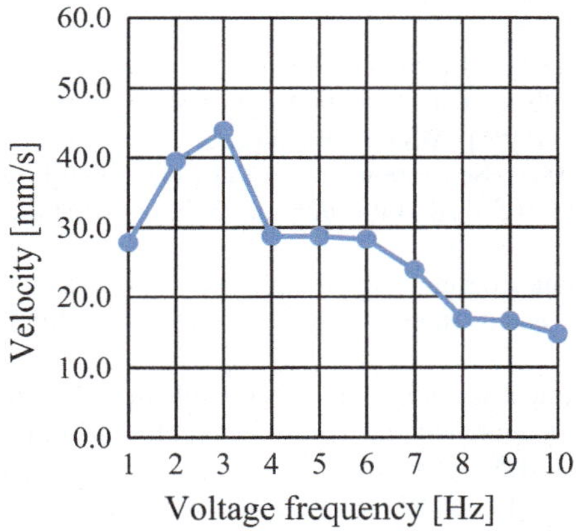

Fig. 14 Relation between voltage frequency and swimming velocity

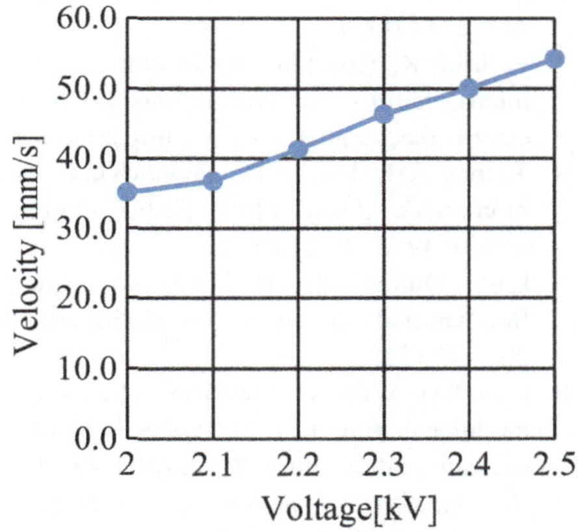

Fig. 15 Relationship between voltage and swimming velocity

54.1 mm/s. The dielectric breakdown occurred at 2.6 kV. In the previous study [18], the swimming velocity at the time of application of a rectangular wave (3 Hz, 7 kV) voltage is about 40 mm/s. Therefore, our robot can swim at lower voltage than previous research.

5 Conclusion

In this research, we aimed to realize a soft underwater robots imitating manta activated by dielectric elastomers, which swim at low voltage. We manufactured newly and used DEMES as an actuator usable in water. We realized this actuator by sandwiching the electrode carbon as two dielectric elastomers and adjusting the stiffness of the PET frame so that it can ideally bend. The flexion angle of the actuator was 30° at the maximum. The robot produced had a total length of 55 mm, a mass of 8.4 g, and a swimming velocity of 39.2 mm/s (2.6 kV, 2 Hz). It swims at lower voltage than a soft underwater robot using dielectric elastomer in previous research.

Acknowledgements. This work is supported by Quartz Lab-Supmeca in France, and MING Lab in UEC-Tokyo in Japan. The authors gratefully acknowledge the helpful comments and suggestions of the reviewers.

References

1. Phamduy P, LeGrand R, Pori M (2015) Robotic fish design and characterization of an interactive iDevice-controlled robotic fish for informal science education. IEEE Robot Autom Mag 22:86–96. https://doi.org/10.1109/MRA.2014.2381367
2. Marchese AD, Onal CD, Rus D (2014) Autonomous soft robotics fish capable of escape maneuvers using fluidic elastomer actuators. Soft Robot 1:75–87. https://doi.org/10.1089/soro.2013.0009
3. Perlrine R, Kornbluh R, Joseph J, Chiba S (1997) Electrostriction of polymer films for microactuators. In: Proceedings IEEE The tenth annual international workshop on micro electro mechanical systems. https://doi.org/10.1109/MEMSYS.1997.581811
4. Techet AH, Hover F, Triantafyllou MS (2003) Separation and turbulence control in biomimetic flows. Flow Turbul Combust 71:105–118. https://doi.org/10.1023/B:APPL.0000014923.28324.87
5. Liu J, Dukes I, Hu H (2005) Novel mechatronics design for a robotic fish. In: IEEE/RSJ International conference on intelligent robots and systems. https://doi.org/10.1109/IROS.2005.1545283
6. Low KH, Willy A (2005) Development and initial investigation of NTU robotic fish with modular flexible fins. In: Proceedings of the IEEE international conference on mechatronics and automation. https://doi.org/10.1109/ICMA.2005.1626681
7. Yamakita M, Kamamichi N, Kozuki T, Asaka K, Luo Z-W (2005) A snake-like swimming robot using IPMC actuator and verification of doping effect. In: IEEE/RSJ International conference, pp 2035–2040. https://doi.org/10.1109/iros.2005.1545485
8. Firouzeh A, Ozmaeian M, Alasty A (2012) An IPMC-made deformable-ring-like robot. Smart Mater Struct 21:1–11. https://doi.org/10.1088/0964-1726/21/6/065011
9. Zhao X, Suo Z (2008) Method to analyze programmable deformation of dielectric elastomer layers. Appl Phys Lett 93:251902. https://doi.org/10.1063/1.3054159
10. Hines L, Petersen K, Sitti M (2016) Inflated soft actuators with reversible stable deformations. Adv Mater 28:3690–3696. https://doi.org/10.1002/adma.201600107
11. Seok S, Onal CD, Cho K-J, Wood RJ, Rus D, Kim S, Meshworm (2013) A peristaltic soft robot with antagonistic nickel titanium coil actuators. IEEE/ASME Trans Mechatron 18:1485–1497. https://doi.org/10.1109/TMECH.2012.2204070
12. Saito Y, Takao H, Tani T, Nonoyama T, Takatori K, Homma T, Nagaya T, Nakamura M (2004) Lead-free piezoceramics. Nature 432:84–87. https://doi.org/10.1038/nature03028
13. Carpi F, Menon C, De-Rossi D (2010) Electroactive elastomeric actuator for all-polymer linear peristaltic pumps. IEEE/ASME Trans Mechatron 15:460–470. https://doi.org/10.1109/TMECH2009.2028884
14. Kornbluh RD, Pelrine R, Pei Q, Heydt R, Stanford S, Oh S, Eckerle J (2002) Electroelastomers: applications of dielectric elastomer transducers for actuation, generation,

and smart structures. In: SPIE 4698 Smart Structures and Materials 2002: Industrial and Commercial Applications of Smart Structures Technologies, pp 254–270. https://doi.org/10.1117/12.475072
15. Shintake J, Shea H, Floreano D (2016) Biomimetic underwater robots based on dielectric elastomer actuators. In: Intelligent robots and systems IEEE/RSJ international conference. https://doi.org/10.1109/IROS.2016.7759728
16. Chu W-S, Lee K-T, Song S-H (2012) Review of biomimetic underwater fast robots using smart actuators. Int J Precis Eng Manuf 13:1281–1292. https://doi.org/10.1007/s12541-012-0171-7
17. Zhao J, Wang S, McCoul D, Xing Z, Huang B, Liu L, Leng J (2016) Bistable dielectric elastomer minimum energy structure. IOP Sciehnce, Smart Mater Struct 25. https://doi.org/10.1088/0964-1726/25/7/075016
18. Li T, Li G, Liang Y, Cheng T et al (2017) Fast-moving soft electronic fish. Am Assoc Adv Sci https://doi.org/10.1126/sciadv.1602045

A Novel in-Pipe Robot Design with Helical Drive

Houbab Abid[1(✉)], Ajmi Houidi[2], and Abdel Fattah Mlika[3]

[1] ISET Sousse, Sousse, Tunisia
houbababid@yahoo.fr
[2] ISSAT Sousse. LMS, Sousse, Tunisia
ajmi.hdi@gmail.com
[3] ENISO. LMS, Sousse, Tunisia
abdelfattah.mlika@gmail.com

Abstract. Liquid and gas pipelines are all around us in today's society. The frequent inspection and maintenance of such pipeline grids is very important. Recently, many pipeline inspection robot systems have been developed. In this paper, a novel in-pipe robot design is presented. The proposed design consists of two modules. The first one is the guiding module. It is formed by a driving motor and is guided along the pipe by a set of wheels moving parallel to the axis of the pipe. The second is the driving module. It is forced to follow a helical motion thanks to tilted wheels rotating around the axis of the pipe. Furthermore, the proposed design has much better mobility turning a bend due to its flexible systems supporting wheels. It has the capability to cross a bend without loses of balance and using a single drive motor. Afterwards the control of the system will be simple. Problem of robots with helical drive in the bend, new design, together with a comparative study between theoretical and simulation results will be presented as well.

Keywords: In pipe robot · Helical drive · Bend pipe

1 Introduction

Papers Pipelines have been used as major tools for transporting or delivering gas, water, oil, or other fluidic materials. Over a period of operation, various problems may occur inside pipelines due to ageing and external impact. Some of the most common failure modes are corrosion, pitting, crack, dents, ovalization or deformation. In order to extend the life expectancy of pipelines and to avoid unexpected accidents caused by such problems, regular inspection inside pipes is strongly recommended and is in fact practiced by many companies involved in gas, water, or oil supplies. Whenever a problem is spotted, rehabilitation or repair jobs must be done inside pipes. Due to the small space and hazardous environment inside pipes, the application of robots for such jobs is considered as the most attractive available solution.

Because of the importance of pipeline transported substance (expensive, poisonous or necessity), the variety of pipeline application background (pipelines may vary in terms of material, size, grid complexity, and applications), and the massive use of

N. Aifaoui et al. (Eds.): CMSM 2019, LNME, pp. 892–901, 2020.
https://doi.org/10.1007/978-3-030-27146-6_96

pipelines for gas and fluidic material transportation, a lot of research works have been conducted and reported on developing methods and machines to carry out various jobs inside pipelines.

Roh [1] classified the in-pipe robots according to their mechanisms of displacement. We distinguish:

- The pig is one of the most well-known commercial one that is passively driven by fluid pressure inside oil pipelines and employed for the inspection of pipelines with large diameters [2];
- The wheel type is similar vehicles with wheels [3, 4];
- The crawler type robot with caterpillars instead of wheels [5];
- The wall press type denotes the robot with flexible mechanism for pressing the wall whatever means that has advantage in climbing vertical pipelines [6];
- walking type possesses articulated legs and it can produce various motions [7];
- The Screw type [8, 9].

Other types of wheel mechanisms proposed, such as the dual-wheel [10], dual-wheel caster drive [11], orthogonal-wheel [12] and double-wheel drive [13]. A comprehensive overview of the development and classification of different kinds of wheeled mobile robots can be found elsewhere [14–20].

2 The Design Concept

Figure 1 a shows the design of a basic module. It is mainly composed of a set of driving wheels, guiding wheels. Both driving wheels and guiding wheels press firmly onto to the internal surface of pipe when in operation. The driving wheels are titled angle. When forced to rotate against the pipe wall, the driving wheels will follow a helical path, thereby generate a thrust force along the pipe axial direction. The thrust force is the locomotive force for the robot.

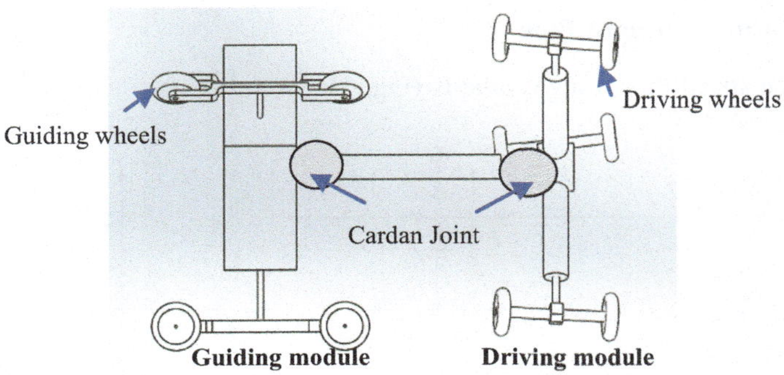

Fig. 1 Concept of a basic module

3 Advance Speed

3.1 Nomenclature

See Fig. 2.

R	Curved pipe radius
$\dot{\varphi}$	Angular Velocity of robot along G_1Z1_i axis
α	The Wheel tilt angle
x, y, z	Coordinate of center of gravity G_1
l	Distance between point G_1 and point G_{3i} along G_1X_{2i} axis
a	Distance between point G_{3i} and point G_{4i}
r	Wheel radius
$\dot{\gamma}_{4i}$	Angular velocity of wheel (4i)
$\dot{\beta}$	Angular velocity of robot at axis x_0

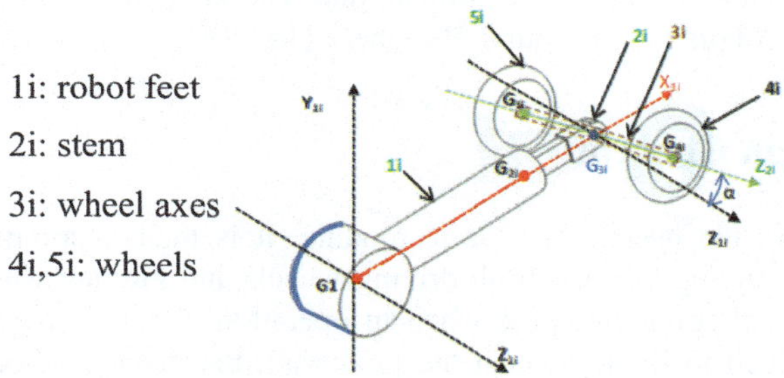

Fig. 2 Parameter of the driving module

3.2 Robot in a Straight Pipe

The advance speed in a straight tube is (Fig. 3):

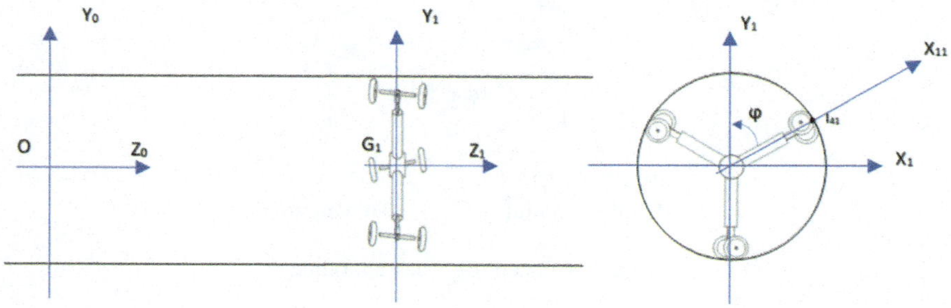

Fig. 3 Robot in a straight pipe

$$\overrightarrow{V_{G1/Ro}} = \frac{d(\overrightarrow{O_0G_1})}{dt}\bigg|_{/Ro} = \dot{z}\overrightarrow{z}_1$$

The advance speed of the point of contact between wheel and tube (I_{41}).

$$\overrightarrow{V_{I41/Ro}} = \begin{pmatrix} a\sin(\alpha)\dot{\varphi} \\ (r+l)\dot{\varphi} - r\dot{\gamma}_{41}\cos(\alpha) \\ \dot{z} - \dot{\gamma}_{41}.r.\sin(\alpha) \end{pmatrix}_{x1.y1.z1}$$

condition using the rouling without slip condition we have:

$$\overrightarrow{V_{I41/Ro}} = 0$$

From where the speed of advance of the robot in rectilinear tube is:

$$\dot{z} = (r+l)tg(\alpha)\dot{\varphi}$$

3.3 Robot in Pipe Bend

Under the effect of the springs the wheels always remain in contact with the inner surface of the tube and the tilt angle of the wheels imposes a constant advance of the centers of the wheels along $\vec{z}_1$.

The constant advance speed is:

$$\dot{z} = (r+l)tg(\alpha)\dot{\varphi}$$

From where a rotation speed around the bend center around the axis $\overrightarrow{x}_1$ is (Fig. 4):

$$\dot{\beta} = \frac{(l+r)tg(\alpha)\dot{\varphi}}{R + l\sin(\varphi)}$$

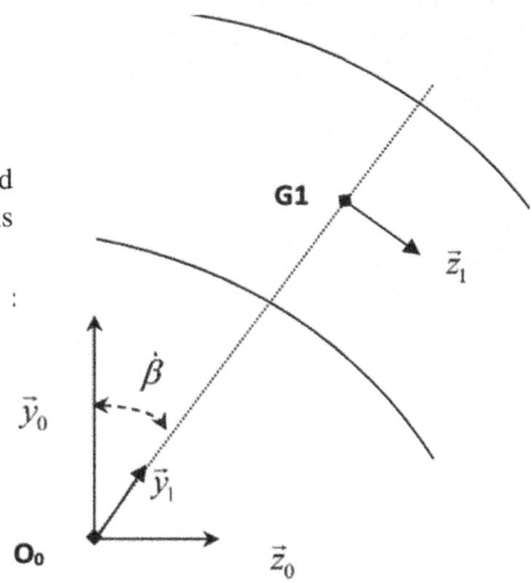

Fig. 4 Robot in pipe bend

3.4 Problem of the Robot in a Bend

When the robot enters into a bend, with a constant speed of the wheel centers, this will necessarily cause the inclination of the platform of the robot, this motion taken by the robot will give it a tilt and thus completely lose contact with the tube. This behavior is illustrated in Fig. 5.

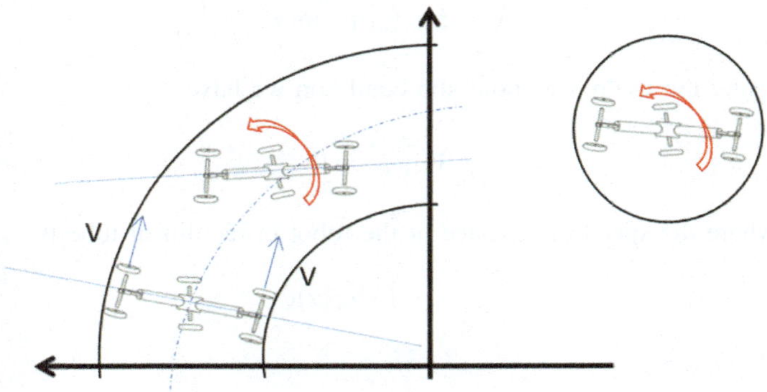

Fig. 5 Tipping of the robot

To avoid this problem, it is necessary that the rotation speed of the robot around the axis x be constant. Therefore, when the robot passes through the bend, the rotational speed of the inner wheel should be slowed down, while the rotational speed of the outer wheel should be increased as shown in the Fig. 6.

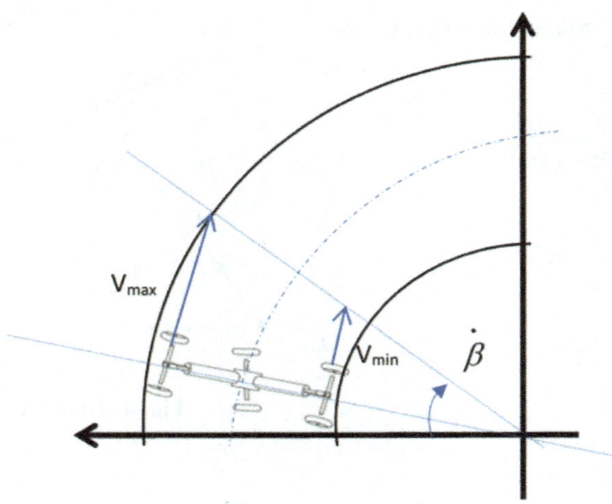

Fig. 6 Advances without tipping

The rotation speed is:

$$\dot{\beta} = \frac{(l+r)tg(\alpha)\dot{\varphi}}{R+l\sin(\varphi)} = cst$$

Which leads us to the following expression:

$$tg(\alpha) = \frac{(R+l\sin(\varphi))\dot{\beta}}{(r+l)\dot{\varphi}}$$

Thus, the tilting angle varies in a way similar to:

$$\alpha = \alpha_0 + k\sin(\varphi), \quad k = cst$$

Az a solution one can use actuators on the level of the three feet of the robot in order to control their tilt angles. The variation of the angle depends on several fixed and variable parameters such as the bend radius the robot dimensions the tube radius and also depends on the angular position of the wheel in the tube which makes this solution very delicate, since we cannot determine the actual angular position of the wheels in the tube. To overcome this problem, we present in this work a new design that allows an advance without problem in the bend pipe.

4 The New Design

In this work we propose a new robot structure with helical movement. The difference in speeds between the outer and inner wheels we added an elastic translation between the stem and the pair wheel axe as shown in the Fig. 7b.

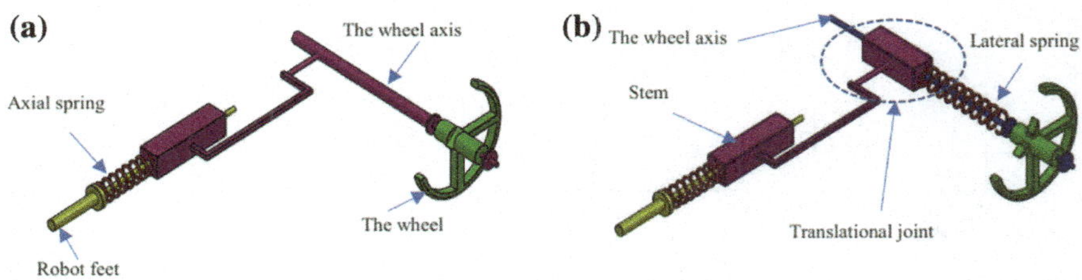

Fig. 7 **a** Old architecture **b** New architecture

With this new solution we added three degrees of freedom for the body which are translations along the three wheels axes as shown in the Fig. 8.

With these three degrees of freedom the wheels speeds will be independent and the speed of advance of the body will no longer be controlled by a single constant speed like the old architecture and with the three new lateral springs the body can advance in the bend with an average speed without losing the balance.

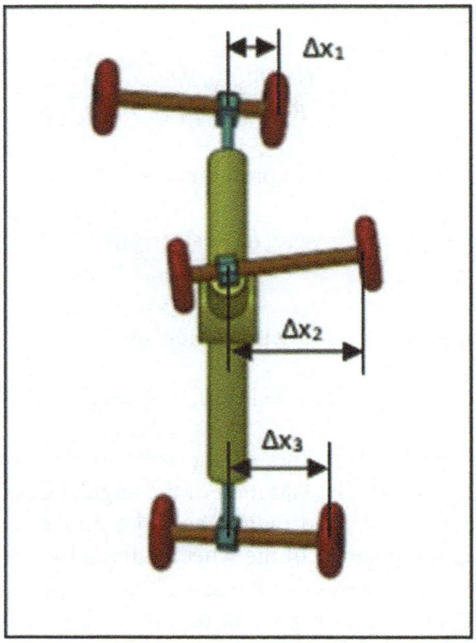

Fig. 8 The three degrees of freedom

In Fig. 9 we present the variation of one of the three translation mobility in a bend. This new flexibility the wheels advance in the curve independent of the body.

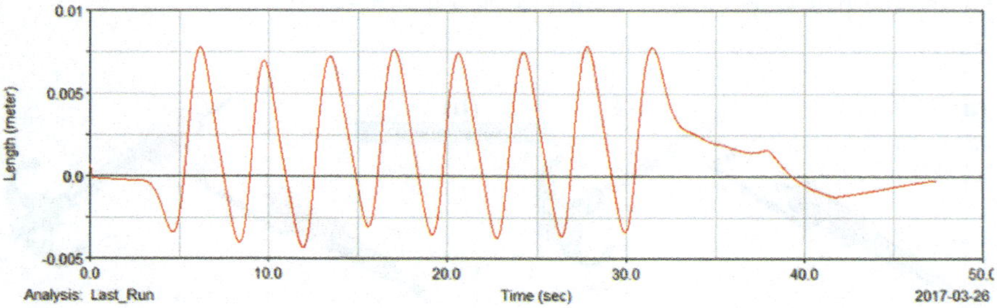

Fig. 9 The variation of the new degree of freedom (radius pipe 156 mm, radius bend 200 mm)

5 Expression of the Advance Speed in a Bend

In a bend the extremity of the body follows three independent rotation speeds (Fig. 10):

$$\dot{\beta}_1 = \frac{(l+r)tg(\alpha)\dot{\varphi}}{R + l\sin(\varphi)}$$

$$\dot{\beta}_2 = \frac{(l+r)tg(\alpha)\dot{\varphi}}{R + l\sin(\varphi + \pi/3)}$$

$$\dot{\beta}_3 = \frac{(l+r)tg(\alpha)\dot{\varphi}}{R + l\sin(\varphi + 2\pi/3)}$$

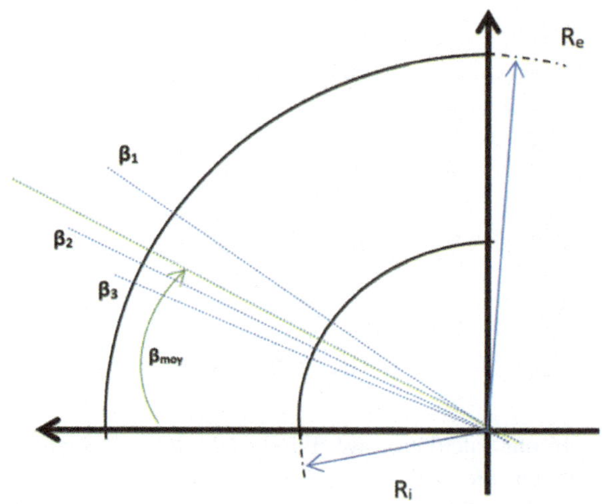

Fig. 10 Advance speed in a bend

The body of the robot under the effect of the lateral springs will be positioned in the middle to counterbalance. Hence the rotation speed of the center of the body is:

$$V = R\dot{\beta}_{moy} = R(\beta_1 + \beta_2 + \beta_3)/3$$

Figure 11 presents the advance speed of the body center in a bend.

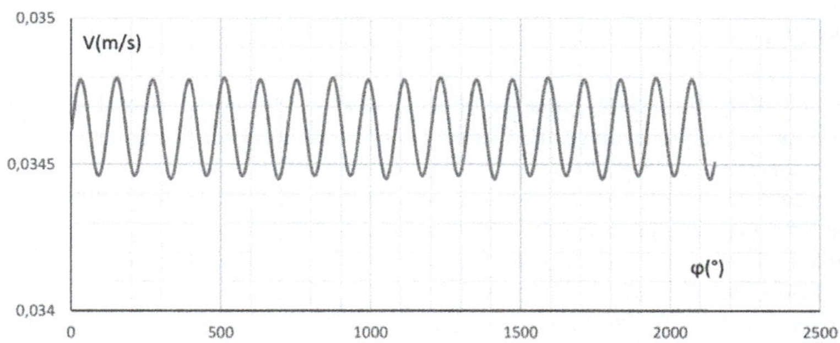

Fig. 11 The advance speed of the body center

6 Simulation

After simulation the robot crosses the bend without tilting, in the following graph we present the advance speed in a bend (Fig. 12).

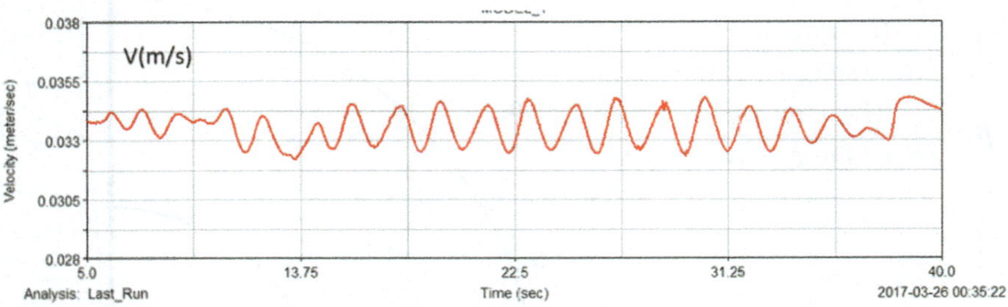

Fig. 12 The advance speed of the body center

In our calculation we did not take into consideration the effects of vertical springs as well as the load of the drive system: motor and the two drive shafts with two Hooke joints. We take into account these effects we can conclude that the theoretical results and simulation are very close.

7 Conclusions

We presented in this paper a new architecture of a helical motion robot able to circulate in a small radius bend. Thus, we can eliminate the slip phenomenon considerably and avoid any jamming or tilting of the robot. As perspective we will work on an optimization approach in order to adjust the different spring stiffness with an object of increasing the circulation speed.

References

1. Roh SG, Shoy HR (2002) Strategy of navigation inside pipelines with differentiel-drive inpipe robot. In: Proceedings IEEE international conference on Robotics and Automation (ICRA 2002) pp 2575–2580
2. Okamoto J Jr, Adamowsk JC, Tsuzuki M, Buiochi F, Camerini C (1999) Autonomous system for oil pipelines inspection. Mechatronics 9:731–743
3. Okada T, Sanemori T (1987) A three-wheeled sel-adjusting vehicle in a pipe, FERRET-1. Int J Robot Res 6(4):60–75
4. Mharamatsu M, Namiki N, Koyoma U, Suga Y (2000) Autonomous mobile robot in pipe for piping operations. In: IEEE/RSJ international conference of intelligent robots and systems (IROS) 3:2366–2 171
5. Roman HT, Pelligrino BA, Sigrist W (1993) Pipe crawling inspection robots: an overview. IEEE Trans Energy Convers, 576–583

6. Kwon Y-S, Lee B, Whang, I-C, Yi B-J (2010) A pipeline inspection robot with a linkage type mechanical clutch. In: The 2010 IEEE/RSJ international conference on intelligent robots and systems, October 18–22, Taipei, Taiwan
7. Fukuda T, Hosokai H, Uemura M (1989) Rubber gas actuator driven by hydrogen storage alloy for in-pipe inspection mobile robot with flexible structure. In: IEEE International conference on robotics and automation (ICRA), vol 3, pp 1847–1852
8. Hayashi I, Iwatsuki N, Iwashima S (1995) The running characteristics of a screw principle microrobot in a small bent pipe. In: Proceedings of international symposium on micro machine and human science, pp 225–228
9. Iwashima S, Hayashi I, Iwatsuki N, Nakamura K (1994) Development of in-pipe operation micro robots. In: Proceedings of international symposium on micro machine and human science, pp 41–45
10. Leow YP, Angeles J, Low KH (2000) A comparative mobility study of three-wheeled mobile robots. In: Proceedings of 6th international conferenceon control, automation, robotics and vision, Singapore
11. Wada M, Takagi A, Mori S (2000) Caster drive mechanisms for holonomic and omni-directional mobile platforms with no over constraint. In: Proceedings IEEE international conference on robotics and automation, San Francisco, CA, pp 1531–1538
12. Killough SM, Pin FG (1992) Design of an omni-directional and holonomic wheeled platform prototype. In: Proceedings IEEE international conference on robotics and automation, nice, pp 84–90
13. Ferrière L, Raucent B (1998) Rollmobs, a new universal wheel concept. In: Proceedings international conference on field and service robotics, Leuven, pp 1877–1882. Carlisle B (2000) Robot mechanisms. In: Proceedings IEEE international conference on robotics and automation, San Francisco, CA, pp 701–708
14. Dudek G, Jenkin M (2000) Computational principles of mobile robotics. Cambridge University Press, New York, NY. McKerrow PJ (1991) Introduction to robotics. Addison-Wesley, Singapore
15. Meystel A (1991) Autonomous Mobile Robots—Vehicles with Cognitive Control. World Scientific, Singapore
16. Campion G, Bastin G, D'Andréa-Novel B (1996) Structural properties and classification of kinematic and dynamic models of wheeled mobile robots. IEEE Trans Robot Automat 12:47–62
17. Ostrovskaya S (2001) Dyanamics of quasiholonomic and nonholonomic reconfigurable rolling robots. PhD Thesis, McGill University, Montreal
18. Angeles J (2002) The robust design of parallel manipulators. In: Proceedings 1st international controlled robotic systems for handling and assembly, Braunschweig, pp 9–30
19. Jones JL, Flynn AM (1993) Mobile robots: inspiration to implementation. Peters, Wellesley, MA
20. Qingyou Liu, Tao Ren, Chen Yonghua (2013) Charateristic analysis of a novel in-pipe driving robot. Elsevier Mechatron 23:419–428

Cable-Driven Parallel Robot (Eight Cables): Motors Command in Position and in Velocity

Hajer Ben Amor(✉), Sami Bennour, Houssein Lamine, Abdelbadiâ CHAKER, and Abdelfattah Mlika

Mechanical Laboratory of Sousse (LMS), National Engineering School of Sousse, University of Sousse, 4054 Sousse, Tunisia
hajourabenamor@gmail.com, houssein.lamine@gmail.com, abdelfattah.mlika@gmail.com, {sami.bennour.meca, abdelbadia.chaker}@eniso.u-sousse.tn

Abstract. In this paper, the inverse kinematic modeling of a cable driven parallel robot made out of eight cables is developed in order to control the robot's moving platform according to a desired trajectory. An example of a Cartesian desired trajectory is fed to the kinematic model and thereby the control variables, i.e., angular position and velocity of each motor are determined. Due to the limited resources, Experiments was carried out using one motor. A comparison between the theoretical and the experimental results is given and discussed.

Keywords: Cables driven robot · Motor control · Position · Trajectory

1 Introduction

Cable-Driven Parallel Robots (CDPR) are mechanisms composed of a base and a moving-platform connected by a set of cables in a given number of points. CDPRs have attracted the attention of researchers and industrials due to their several advantages compared to the serial robots. In fact, they are characterized by a large workspace, high dynamics, an important payload capacity, an easy reconfiguration… [1–4]. Thereby, several studies have proposed prototypes for different applications, e.g., the famous Skycam robot [5].

This paper investigates the Inverse Kinematic Model of an eight-cable CDPR. The developed model is used to generate a specific trajectory of actuators in relation to a desired motion of the robot platform in the space. Indeed, the angular position and velocity of each actuator are calculated in order to satisfy a desired profile motion, i.e., a desired position trajectory in function of time. To illustrate, case studies are addressed and presented.

N. Aifaoui et al. (Eds.): CMSM 2019, LNME, pp. 902–909, 2020.
https://doi.org/10.1007/978-3-030-27146-6_97

2 Notations and Kinematic Modeling

In this paper a CDPR is composed of moving plate-form which is connected to motors shafts through pulleys by eight cables. Motors are fixed on a base. Figure 1 shows an example of CDPR: 7 cables falcon robot [6].

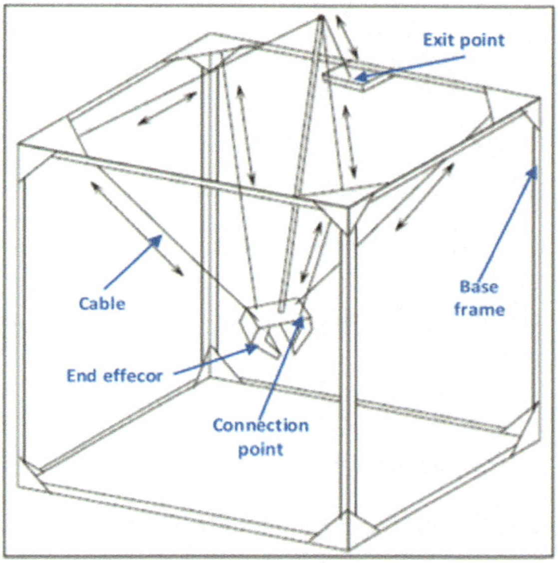

Fig. 1 Example of CDPR: 7 cables falcon robot

As shown in the Fig. 2, the moving plate-form is defined by the $B_i(i = 1, \cdots, 8)$ points and a base defined by the $A_i(i = 1, \cdots, 8)$ points. The reference frames for the moving plate-form and the base are: R_P = (P, x, y, z) and R_b = (O, x_O, y_O, z_O), respectively.

The displacement of the moving platform is established by the variation of the cables lengths $L_i(i = 1, \cdots, 8)$.The kinematic modeling consists in developing the motors angular position $q_i(i = 1, \cdots, 8)$ as a function of the platform pose.

The pose vector $\mathbf{X} = [x \quad y \quad z \quad \alpha \quad \beta \quad \gamma]^T$ denotes the position and the orientation of the platform in relation to R_b, in which $[\alpha \quad \beta \quad \gamma]^T$ is the Euler angles vector.

Consequently, the rotation matrix **Q** can be written in term of **X-Y-Z** Euler angels as follows [7]:

$$Q = \begin{pmatrix} c\beta c\gamma & -c\beta s\gamma & s\beta \\ c\alpha s\gamma + s\alpha s\beta c\gamma & c\alpha c\gamma - s\alpha s\beta s\gamma & -s\alpha c\beta \\ s\alpha s\gamma - c\alpha s\beta c\gamma & s\alpha c\gamma + c\alpha s\beta s\gamma & c\alpha c\beta \end{pmatrix} \tag{1}$$

where s and c denote, respectively, sin and cos functions.

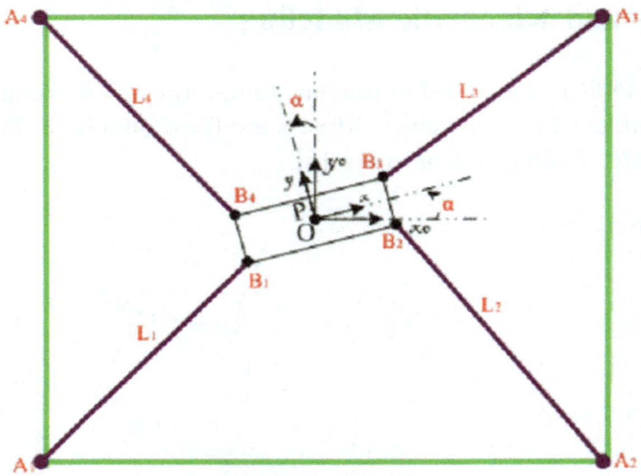

Fig. 2 Kinematic modeling of the CDPR

On the other hand, the motors angular position can be written as:

$$q_i = \frac{\Delta L_i}{r_i} = \frac{L_i - L_{i0}}{r_i} \quad i = 1, \cdots, 8 \tag{2}$$

where $L_{i0}(i = 1, \cdots, 8)$ is the initial length for each cable and $r_i(i = 1, \cdots, 8)$ is the pulley radius.

Moreover, the cable length vector can be written as follows:

$$\boldsymbol{l_i} = L_i\boldsymbol{u_i} = \boldsymbol{r} - \boldsymbol{Q}\boldsymbol{r_{b_i}} - \boldsymbol{r_{a_i}} \quad i = 1, \cdots, 8 \tag{3}$$

where

$$\begin{aligned} \boldsymbol{u_i} &= \frac{[\boldsymbol{A_iB_i}]_{R_b}}{\|\boldsymbol{A_iB_i}\|} \\ \boldsymbol{r} &= [\boldsymbol{OP}]_{R_b} \\ \boldsymbol{r_{b_i}} &= [\boldsymbol{B_iP}]_{R_P} \\ \boldsymbol{r_{a_i}} &= [\boldsymbol{OA_i}]_{R_b} \end{aligned} \quad i = 1, \cdots, 8 \tag{4}$$

Therefore, the motors angular velocity is given as follows [7]:

$$\dot{q}_i = \boldsymbol{R^{-1}J(X)\dot{X}} \quad i = 1, \cdots, 8 \tag{5}$$

where $\dot{q}_i = [\dot{q}_1, \ldots\ldots, \dot{q}_8]$ is the motors angular velocity vector, is a diagonal matrix containing the r_i coefficients, $\dot{X}$ is the end-effector angular velocity and $J(X)$ is the Jacobian matrix of the robot.

$$R^{-1} = diag(\frac{1}{r_1}, \cdots, \frac{1}{r_8}) \tag{6}$$

$$J(X) = \begin{pmatrix} u_1^T & (Qr_{b_1} \times u_1)^T \\ \vdots & \vdots \\ u_8^T & (Qr_{b_8} \times u_8)^T \end{pmatrix} \tag{7}$$

3 Point-to-Point Motion

3.1 Method

In this section, the developed kinematic model is employed in order to control the motors to move the end-effector from an initial pose $\mathbf{X_i}$ to a desired pose $\mathbf{X_f}$ during a given time t_f. The end-effector movement is called Point-to-Point motion when no interpolation constraints its motion, in fact, only the motors motion is smoothed.

To produce a smooth motion, we require that the motors velocity starts and ends at zero. The choice of a third order polynomial function for q_i represents a valid solution to satisfy the desired conditions [8].

The desired profiles are illustrated in Figs. 3 and 4.

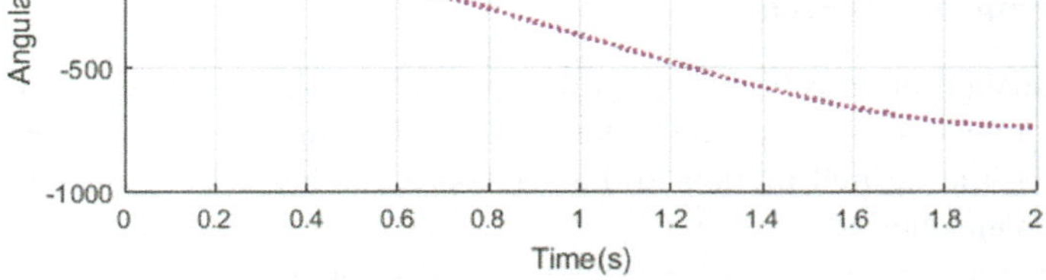

Fig. 3 Motors angular positions profiles

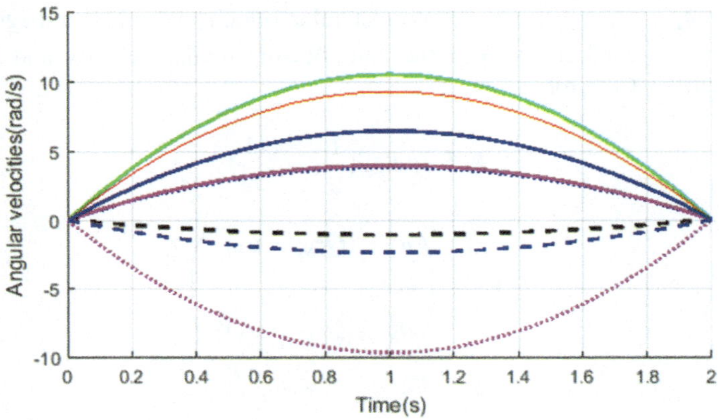

Fig. 4 Motors angular velocities profiles

Therefore, the motion timing for the angular position qi is then given by:

$$q_i(t) = a_0 + a_2t^2 + a_3t^3 \quad i = 1, \cdots, 8 \tag{8}$$

where

$$\begin{cases} a_0 = q_{0i} \\ a_2 = \frac{3(q_{fi} - q_{0i})}{t_f^2} \quad i = 1, \cdots, 8 \\ a_3 = \frac{-2(q_{fi} - q_{0i})}{t_f^3} \end{cases} \tag{9}$$

where q_{fi} *and* q_{0i} present, respectively, the final and the initial angular positions which are calculated by the application of the inverse kinematic model.

3.2 Experimental Results

The experimental architecture of the motor control is shown in Fig. 5, in which the actual motor velocity and position are calculated using the encoder feedback. The generation of the desired motion to the servo-drive is achieved using Arduino Mega 2560 microcontroller.

This section presents the experimental results for an example of Point-to-Point motion. Due to the limited resources, experimentation is carried out using only one actuator. In fact, we imposed a linear Y-axis motion; the platform moves from $\mathbf{X_i}$ (0 0 0 0 0 0) to $\mathbf{X_f}$ (0 0.1 0 0 0 0) in 2 s. The CDPR is to be used for a medical application (Gait rehabilitation) and (Lusardi, 2003) shows the average speed of walking for adult [9]. For this experiment the desired speed is 0.05 m/s.

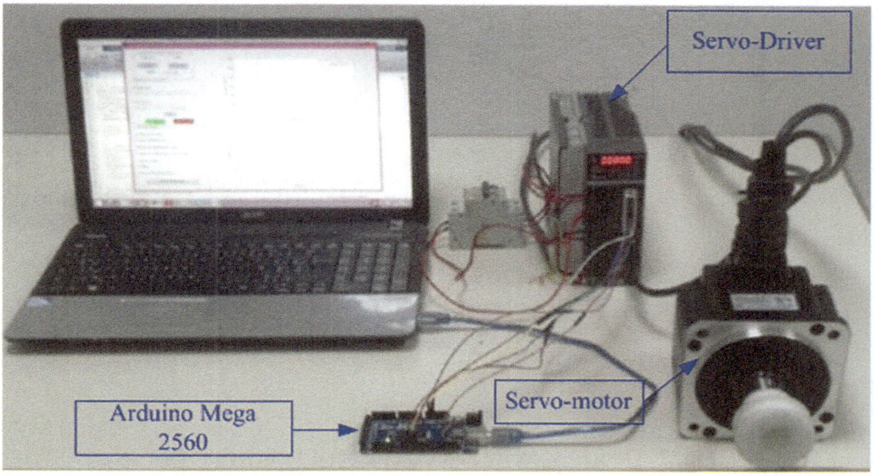

Fig. 5 Motors control architecture

Figures 6 and 7 show the results in term of position and velocity of the trajectory planning of the motor.

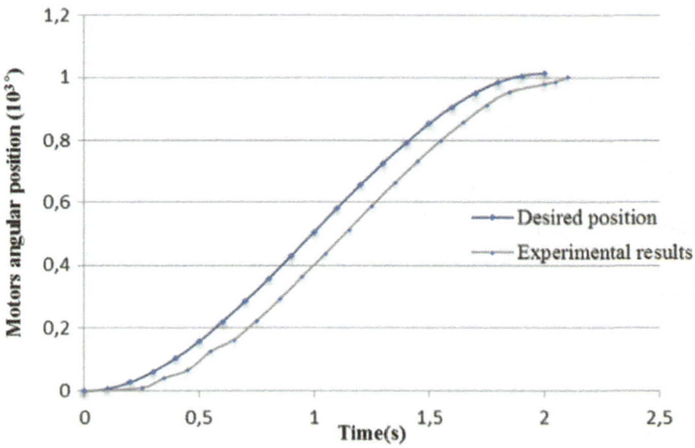

Fig. 6 Desired position and experimental position profiles

We can observe that the measured motor angular position and velocity are similar to the desired ones. Also, a shift in the time axis can be remarked; the theoretical trajectory time is 2 s while the resulting trajectory time is 2.13 s.

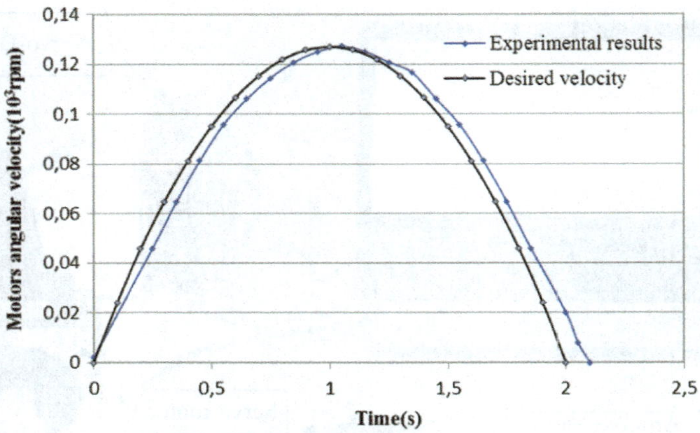

Fig. 7 Desired velocity and experimental velocity profiles

It is interesting to point out that this time error can be explained by the delay due to the data transmission issue. This difference can be tolerated in this phase of work since a compiled code will be generated for all motors assembled and implemented on a real time core in the future work.

4 Conclusions

This paper has introduced a generic inverse kinematic model and a joint path planning strategy of a CDPR robot. The objective is to compute the motors position profile in order to move the end-effector for a desired trajectory. The Inverse kinematic modeling is employed to retrieve the joint position from a desired end-effector pose. In the following, the joint trajectory is smoothed using a polynomial function, i.e., a Point-to-Point motion is generated. An example of a Cartesian motion was given to conduct a comparison between theoretical and experimental results.

References

1. Antoine M, Stéphane C, Philippe C (2018) Design of a cable-direct-driven robot with grasping device. Procedia 290–295
2. Lorenzo P, Paolo G (2014) Cable-direct-driven-robot (CDDR) with a 3-link pas-sive serial support. Robot Comput-Integr Manuf 265–276
3. Saeed B, Amir K (2006) Cable-based Robot manipulators with transla-tional degrees of freedom. In: Industrial robotics. Modelling and control, pp 211–236
4. Trevisani A (2010) Underconstrained planar cable-direct-driven robots: a trajectory planning method ensuring positive and bounded cable tensions. Mechatronics 113–127
5. Cone L (1985) Skycam: an aerial robotic camera system. Byte 10(10)
6. Lamine H, Bennour S, Romdhane L (2016) Design of cable-driven parallel manipulators for a specific workspace using. Adv Robot 585–594

7. Lamaury J (2014) Contribution à la commande des robots parallèles à redondance d'actionnement. *Archives ouvertes HAL*
8. Craig JJ (2005) Introduction to robotics mechanics and control Third edn. Pearson Education International
9. Lusardi M (2003) Functional performance in community living older adults. J Geriatr Phys Ther 14–22
10. Williams RL, Gallina P, Vadia J (2003) Planar translational cable-direct-driven robots. J Robot Syst 107–120

Bio-Inspired CPG Based Locomotion for Humanoid Robot Application

Mohamed A. Sayari[1(✉)], Neila Masmoudi[1], and Riadh Zaier[2]

[1] Electromechanical Systems Laboratory (LASEM), National Engineering School of Sfax University of Sfax, BP 1173 3038 Sfax, Tunisia
sayari.mohamed.a@gmail.com, neila.masmoudi@enis.tn
[2] Sultan Qaboos University, Muscat, Oman
zaier@squ.edu.om

Abstract. Considerable attention has been paid to the development of stable walking robots. Indeed, biped locomotion becomes a broadcast area where various research topics such as artificial intelligence, control theory and neuroscience cope to enhance the abilities of the robots. In this paper, Central Pattern Generator (CPG) which are neural circuits that generates oscillations for rhythmic patterns are used to control the humanoid. The choice of using a CPG as a motion generator is motivated by the naturality of the generated pattern. Moreover, CPG offers the possibility to control the gait speed ensuring an easy modulation of the walking speed. Here, the Zero Moment point is used to measure the stability of the humanoid while walking. Thus, in this paper, a humanoid robot's walking model is presented with a strong emphasis on stability and representativity of actual human walking. Furthermore, the methodological considerations in the implementation of a CPG controller for a humanoid robot application are also presented.

Keywords: Biped locomotion · Zero moment point · Central pattern generator

1 Introduction

The Biped Locomotion research became in the few decades one of the most exciting topics in the robotics research framework. Indeed, one of the motivation behind the use of a biped configuration comparatively to the wheeled and quadruped one is that the environment where humans live is adapted to the biped architecture.

In this framework, many techniques have been developed to measure the stability of a walking gait. One among them is the Zero Moment Point (ZMP) trajectory tracking [1, 2]. According to this criterion, a walk is said stable, if for a given walking trajectory, the projection of ZMP on the ground plane remains within the support polygon. Considerable attention has been paid to this approach for the possibility that it offers for reducing the various degrees of freedom present in a humanoid robot to a simple ZMP trajectory tracking.

N. Aifaoui et al. (Eds.): CMSM 2019, LNME, pp. 910–918, 2020.
https://doi.org/10.1007/978-3-030-27146-6_98

Previous research has documented various models for ZMP problem formulation. A widely used model is the one proposed by Kajita in [3] it consists of a simple balancing cart table system. To control this plant, many algorithms have been proposed. Harada et al. [4] derived an analytical expression for the COM trajectory given a piecewise polynomial ZMP trajectory. Meanwhile, Kajita proposed a combination of a PID controller and a preview control approach for ZMP trajectory tracking [5].

In this paper, the locomotion control of a planar biped robot is presented. The motion generation is ensured by a central pattern generator called Matsuoka Oscillator. The aim of this work is to emphasis the benefits of the use of central pattern generator for biped locomotion in term of stability and smoothness of the pattern. To ensure a high similarity with human walking, the full dynamics of the planar robot is presented as well as the inverse kinematics.

2 Central Pattern Generator

CPG shows a great deal of interest when we deal with robots that perform rhytmic movement. Many types of CPG have been used for motion generation in walking. One of the most commonly used GPG is the one proposed by Matsuoka [6]. It consists of two neurons in mutual and self inhibition [6].

One of the advantages of CPG are that oscillators naturally adapt themselves to the dynamic of the robot and optimize the energy of the actuation and can generate rthymic output without feedback of motor and sensor from limbs and other muscle targets. Matsuoka oscillator is widely used in locomotion for its behavior inspired by the biology. For the Matsuoka Oscillator, Each neuron is governed by two equations and is able to generate rhythmic signal with no external input.

$$\tau_1 \dot{x}_1 = c - x_1 - \beta v_1 - \gamma [x_2]^+ - \sum h_i [g_j]^+ \tag{1}$$

$$\tau_2 \dot{v}_1 = [x_1]^+ - v_1 \tag{2}$$

$$\tau_2 \dot{x}_2 = c - x_2 - \beta v_2 - \gamma [x_1]^+ + \sum h_i [g_j]^- \tag{3}$$

$$\tau_1 \dot{v}_2 = [x_2]^+ - v_2 \tag{4}$$

$$Y_{out} = [x_1]^+ - [x_2]^+ \tag{5}$$

where $[x_1]^+ = \max(x_1, 0)$ and $[x_1]^- = \min(x_1, 0)$.

The figure below depicts the output and the state of the Matsuoka oscillator (Fig. 1).

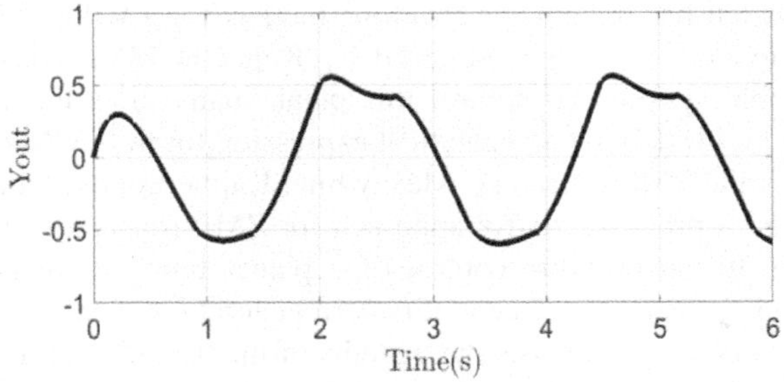

Fig. 1 Output of the matsuoka oscillator

3 Problem Formulation

Many publications appeared documenting the use advanced models to represent the human locomotion. Among those one, some models focused on emphasis the likelihood with human anatomy by increasing the number of degrees of freedom utilized while modelling the biped robot [7, 8]. On the other hand, various researches supported the use of simplified models like the inverted pendulum model and the cart table model, which encapsulate the overall dynamics while reducing the computation [9] (Fig. 2).

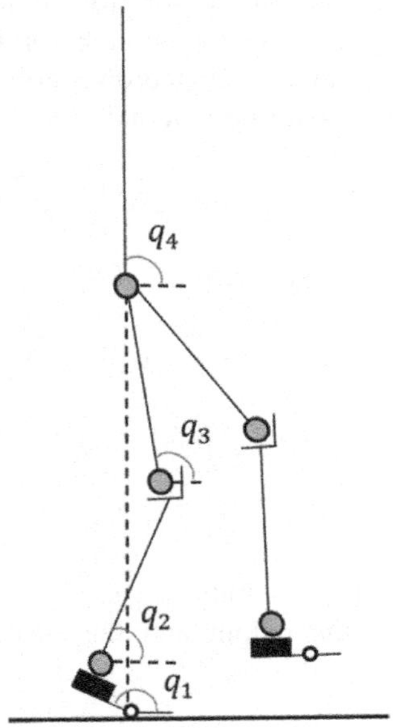

Fig. 2 Schematic representation of the biped robot

In order to fully understand the mechanics underlying human locomotion, both the Kinematics and dynamical formulation of the biped robot are presented in this section.

$$\emptyset = \begin{bmatrix} 1 & 0 & 0 & 0 & 0 & 0 & 0 \\ -1 & 1 & 0 & 0 & 0 & 0 & 0 \\ 0 & -1 & 1 & 0 & 0 & 0 & 0 \\ 0 & 0 & -1 & 1 & 0 & 0 & 0 \\ 0 & 0 & 0 & -1 & 1 & 0 & 0 \\ 0 & 0 & 0 & 0 & -1 & 1 & 0 \\ 0 & 0 & 0 & 0 & 0 & -1 & 1 \end{bmatrix}$$

$$\theta = \emptyset q + \emptyset'$$

$$\emptyset = \begin{bmatrix} \alpha - \pi & \frac{\pi}{2} - \alpha & 0 & 0 & 0 & 0 & \alpha \end{bmatrix}^T$$

where α is the angle between the anklebone and the toe and $\emptyset$ the transition matrix between the absolute angular position and the relative angular position.

The different steps involved in biped locomotion process are depicted here below (Fig. 3).

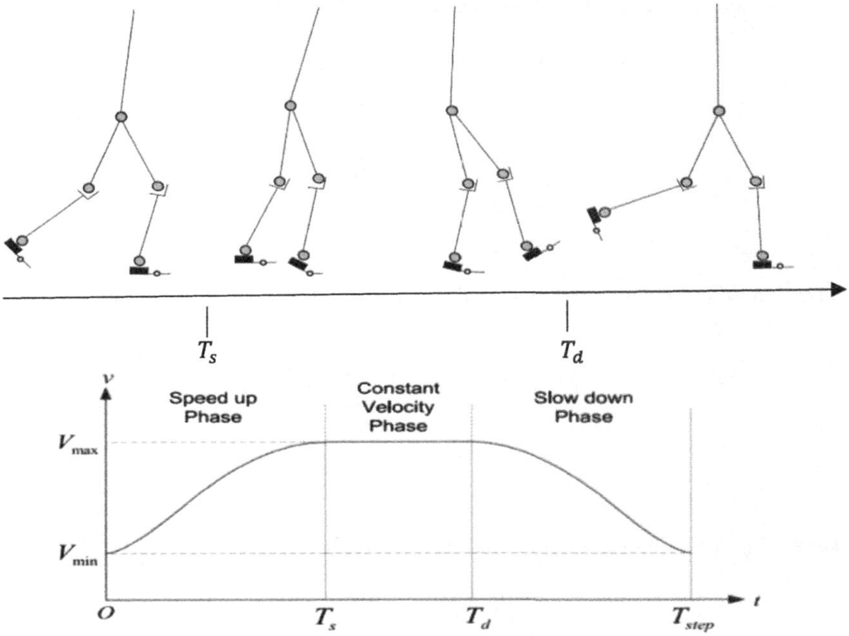

Fig. 3 Walking phases and leg extension velocity [7]

The kinematic analysis of the planar biped is shown here below. In fact, here the inverse kinematic of the all-angular positions are calculated based on the hip position (X_h, Y_h)

$$\theta_1 = q_1 - \pi + \alpha$$

$$\theta_2 = \varphi - \beta - \frac{\pi}{2} + \theta_1$$

$$\theta_3 = \text{a cos}\left(\frac{(X_h - l_1C_1)^2 + (Y_h - l_1S_1)^2 - l_2^2 - l_3^2}{2l_2l_3}\right)$$

$$\theta_4 = \frac{\pi}{2} - \varphi + \beta - \theta_1 - \theta_3$$

$$\varphi = atan2(Y_h - l_1S_1, X_h - l_1C_1)$$

$$\beta = \text{a cos}\left(\frac{(X_h - l_1C_1)^2 + (Y_h - l_1S_1)^2 + l_2^2 - l_3^2}{2l_2\sqrt{(X_h - l_1C_1)^2 + (Y_h - l_1S_1)^2}}\right)$$

where l_1, l_2 and l_3 are the length of the different segments.

$$X_h = l_1C_1 + l_2C_2 + l_3C_3$$

$$Y_h = l_1S_1 + l_2S_2 + l_3S_3$$

$$S_k = \sin(q_k)$$

$$C_k = \cos(q_k)$$

The dynamics of the robot is calculated using Lagrange equations as follow

$$U = mgY_c = -mg\sum_{k=1}^{3} l_kS_k$$

$$K = \frac{1}{2}m(X_c^2 + Y_c^2)$$

Here U is the potential energy and K the kinetic energy

$$\frac{d}{dt}\frac{\partial K}{\partial q_i} - \frac{\partial K}{\partial q_i} + \frac{\partial U}{\partial q_i} = T_i$$

The Equation of motion are shown here below

$$\tau = M\ddot{q} + B\dot{q}^2 + G$$

$$M = \begin{bmatrix} ml_1^2 & ml_1 l_2 C_{2-1} & ml_1 l_3 C_{3-1} \\ ml_1 l_2 C_{2-1} & ml_2^2 & ml_2 l_3 C_{3-2} \\ ml_1 l_3 C_{3-1} & ml_2 l_3 C_{3-2} & ml_3^2 \end{bmatrix}$$

$$B = \begin{bmatrix} 0 & -ml_1 l_2 S_{2-1} & -ml_1 l_3 S_{3-1} \\ -ml_1 l_2 S_{2-1} & 0 & -ml_2 l_3 S_{3-2} \\ -ml_1 l_3 S_{3-1} & -ml_2 l_3 S_{3-2} & 0 \end{bmatrix}$$

$$G = \begin{bmatrix} -mgl_1 C_1 \\ -mgl_2 C_2 \\ -mgl_3 C_3 \end{bmatrix}$$

$$S_{k-l} = \sin(q_k - q_l)$$

$$C_{k-l} = \cos(q_k - q_l)$$

4 ZMP Formulation

Several publications have appeared in recent years documenting the use of CPG for ZMP profile generation. These studies have been pushed by the fact that CPG naturally adapt themselves to the dynamic of the robot, due to their biological inspired behavior [3].

Since human walk in manner that minimizes the total energy ensuring the minimum of external control is critical when dealing with biped locomotion.

Moreover, knowing that human locomotion is a rhythmic pattern, the use of the oscillator shows a great deal. Along with the ability of oscillator to adapt themselves to the dynamics of the robot, CPG offers the possibility of coupling multiple oscillator to ensure a harmonized locomotion pattern with the minimum use of external control.

To highlight the efficiency of the proposed combination, the Matsuoka oscillator' parameters were tuned in order to generate similar output as the artificial signal that mimics the human joint motion.

The Figs. 4 and 5 depict the output of the Matsuoka oscillator for different parameters choice.

As shown in Figs. 4 and 5, changing the parameters of the oscillator updates both the amplitude and the frequency of the oscillator. This possibly enable the control of the speed of the robot. Indeed decreasing the γ parameter while increasing the value of the β parameters would increase the frequency of the oscillation.

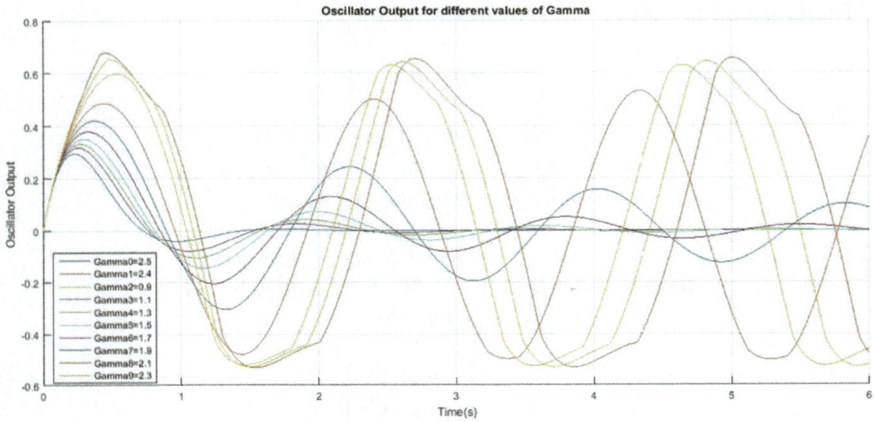

Fig. 4 Oscillator output for different γ values

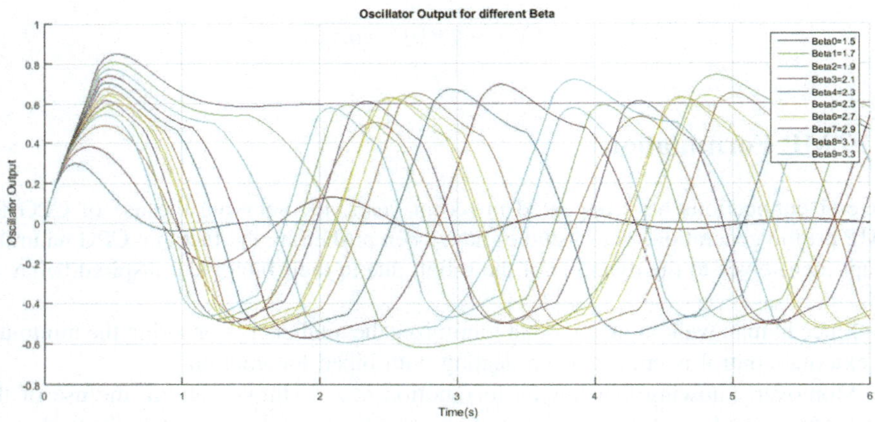

Fig. 5 Oscillator output for different β values

To demonstrate the stability of the generated pattern for the planar model proposed, the ZMP position generated using Matsuoka oscillator as input compared to an COM trajectory generated by the artificial joint motion.

The ZMP position is calculated using the equation below [10]

$$x_{zmp} = \frac{\sum_{i=1}^{n} m_i(\ddot{z}_i + g)x_i - \sum_{i=1}^{n} m_i \ddot{x}_i z_i - \sum_{i=1}^{n} I_{iy}\ddot{\Omega}_{iy}}{\sum_{i=1}^{n} m_i(\ddot{z}_i + g)}$$

In the proposed model is planar model no inertia matrix is inputted in formula of the ZMP. The resulting trajectory of the ZMP is shown in the Fig. 6.

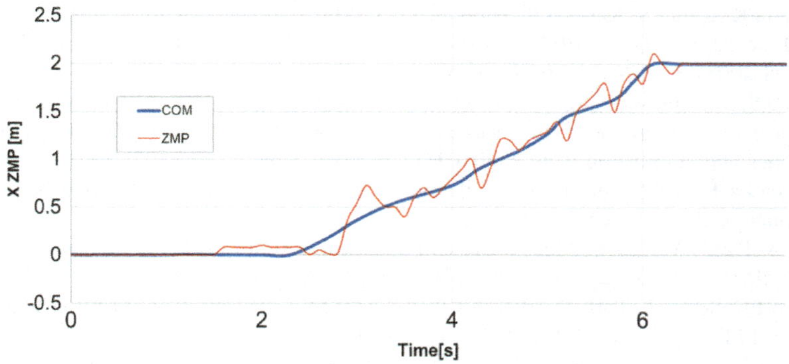

Fig. 6 Trajectory of the ZMP

The output of the simulation demonstrates the stability of the walking pattern generated by the CPG. Here, the major advantage of proper tuning of the central patter generator is the possibility to reduce the use of external control. Indeed, classical control strategies which constrains the robot joints motion reduce the energy efficiency and the smoothness of the pattern generated. While a proper understanding of the effect of each parameter not only enable the generation of a stable pattern but also speed control.

5 Conclusion

In this paper, a bio-inspired approach for motion generating commonly used in biped locomotion has been presented. Matsuoka oscillator has been chosen to be the optimized pattern generator used to perform a stable walk. Here, the Zero Moment point is used to measure the stability of the humanoid while walking. Thus, in this paper, a humanoid robot's walking model is presented with a strong emphasis on stability and representativity of actual human walking.

References

1. Huan TT, Van Kien C, Anh HPH, Nam NT (2018) Adaptive gait generation for humanoid robot using evolutionary neural model optimized with modified differential evolution technique. Neurocomputing 320:112–120
2. Sabaapour MR, Hairi Yazdi MR, Beigzadeh B (2016) Passive dynamic turning in 3D biped locomotion: an extension to passive dynamic walking. Adv Robot 30:218–231
3. Kajita S, Kanehiro F, Kaneko K, Fujiwara K, Harada K, Yokoi K et al (2003) Biped walking pattern generation by using preview control of zero-moment point. In: 2003 IEEE international conference on robotics and automation (Cat. No.03CH37422), vol 2, pp 1620–1626

4. Harada K, Kajita S, Kaneko K, Hirukawa H (2003) ZMP analysis for arm/leg coordination. In: Proceedings 2003 IEEE/RSJ international conference on intelligent robots and systems (IROS 2003) (Cat. No.03CH37453), vol 1, pp 75–81
5. Kajita S, Matsumoto O, Saigo M (2001) Real-time 3D walking pattern generation for a biped robot with telescopic legs. In: Proceedings 2001 ICRA. IEEE international conference on robotics and automation (Cat. No.01CH37164), vol 3, pp 2299–2306
6. Matsuoka K (1985) Sustained oscillations generated by mutually inhibiting neurons with adaptation. Biol Cybern 52:367–376
7. Lee S, Park I-W (2016) Mechanical and electrical design of a biped humanoid which has multiple motors on each lower body joint. Intell Serv Robot 9:49–61
8. Moro FL, Sentis L, Park J, Atkeson CG, Gienger M, Goswami A et al (2017) Whole-Body Control [TC Spotlight]. IEEE Robot Autom Mag 24:12–14
9. Yahmedi ASA, Sayari MA (2014) Efficient walking of a simple biped with a torso. 2nd Middle East conference on biomedical engineering, p 382–384
10. VukobratoviĆ M, Borovac B (2004) Zero-moment point—thirty five years of its life. Int J Humanoid Robot 01:157–173

Analysis and Modeling of a Variable Capacity and an Accelerometer Using MEMS-RF Technology

Agengui Ilyes[1(✉)] and Jabri Ihssen[2]

[1] Institut Supérieur des Études Technologiques de Sousse, ISET Sousse, Sousse, Tunisia
Ilyeslagengui@gmail.com
[2] Université de Tunis El Manar, ENIT-LSE, Tunis Le Belvédère, LR11ES15 Tunis, Tunisia
j.ihssen@hotmail.fr

Abstract. This paper focus on the modeling and the designing of a reconfigurable antenaa LWA with variable capacity using MEMS-RF devices. The variable capacity MEMS can be built using an on surface micro-manufacturing in CPWG. The armatures of the MEMS–RF varactor are separated by an air gap. The choice of the materials and the adequate topology define the performance of the antenna wished in the range of frequency about 30 GHz. Simulation results of the proposed varactor are performed using the MOMENTUM tools of the ADS software. The main objective of this part is to build a variable capacity based on serial plate's forms. The variation of the capacity is reached when the height between these plates varies. This type of varactor is in the range of pF for a frequencies band between 1 and 50 GH. A proposed Accelerometer is also analyzed, modeled and simulated denoted the distribution of the electrical flux and the power losses a function of separation distance referring to ANSYS Maxwel software.

Keywords: MEMS-RF · Varactor · Antenna · Accelerometer · MOMENTUM·ADS · Ansys maxwel

1 Introduction

The Micro Electrical Mechanical System, MEMS, is a microelectronic integrated system composed of analog and digital electronic elements as well as electromechanical elements designed to perform sensor or actuator functions.

The micro mechanical device embedded with electronics/electrical system fabricated through a mix of integrated circuit manufacturing and micro-machining process. The combination of silicon microelectronics with micro-machining technology has made possible the realization of complete systems-on-chips (SoC). MEMS technology enables the development of intelligent components that benefit both a very cost-effective performance/cost ratio because of their ability to be mass-produced by collective manufacturing processes. Thanks to their low losses, low power consumption, high linearity, and high silicon compatibility, MEMS components are good candidates

N. Aifaoui et al. (Eds.): CMSM 2019, LNME, pp. 919–930, 2020.
https://doi.org/10.1007/978-3-030-27146-6_99

for manufacturing circuits in frequencies beyond 30 GHz. The intrinsic qualities of MEMS-RF components can advantageously replace PIN diodes, FET transistors and can be used in phase shifters which are needed for modern telecommunication systems or military systems. They can be implemented also in reconfigurable circuits with low losses (impedance matching network, variable filters…) or in high performance instrumentation systems. These small systems have several advantages (lightness, portability and integrality) within more complex complete systems; the MEMS consume little and are able to achieve high performance due to a very high sensitivity and accuracy.

The Ka band is specifically used for underground network antenna applications. The wavelength in this narrow band is 10 mm. The design of the LWA [1] with variable capacitance is obtained by MEMS-RF capacitors creating a variation of the propagation constant and allowing the scanning of the antenna beam.

The variable capacitors MEMS can be constructed easily using a micromachining surface in CPWG with a central conductor strip and two plates that are separated by an air gap. An electrostatic force caused by the voltage of polarization translates into motion which will be ends when the electrostatic force and mechanical force are equal. A variation of the propagation constant β caused by the movement of the movable plate capacitor allows the variation of the antenna beam direction.

The main purpose of this paper is brought as follows. The first part is dedicated to outline the mechanical and electrical properties of MEMS-RF components and their advantages and disadvantages in terms of packaging and reliability. We will present the mechanical properties, especially the various forces implemented during the actuation as well as the equation describing the movement of the mobile membranes. Then, we will describe the electrical properties and the equivalent diagrams of the most ordinary structures taking into account the cutoff frequency and holding power. The second part deals with the modeling and the designing of a variable capacitor based on MEMS-RF technology. Finally, simulation results of a varactor and an accelometer are performed.

2 Mechanical and Electrical Characteristic

Many physical phenomena negligible at the macroscopic scale become specific in the case of microstructures such as viscous damping, surface friction, and bending and torsion phenomena. In addition, many sources of nonlinearities are involved in the behavior of flexible microstructures. They may be due to the material itself, or to the complex mechanical phenomena that result from elastic deformation. Most microsystems are made of mechanical components such as beams or membranes, which is why many models are built around these two structures and generally, it is not unusual to find approximations that bring the model back to these basic formats.

2.1 Mecancials Properties

The electrostatic force applied to the mobile beam, F_e is directly related to the amount of charge Q, presented in the beam and the fixed electrode by the expression below:

$$F_e = \frac{1}{2}\mathrm{QE} = \frac{1}{2}\frac{\varepsilon_r\varepsilon_0 A V^2}{(d_0 - x)^2} \tag{1}$$

When E is the electrostatic field created by the applied difference potential. ε_0: the permittivity of the vacuum (8.854•10^$^{(-12)}$ F/m). $\varepsilon_r = 1$: Relative permittivity, A: The common surface between the beam and its fixed electrode. V: The potential difference applied between the frames. d_0: The distance initially separating the two electrodes (the gap). x: Moving the moving beam towards the electrode.

The edge effect of the electric field depends on the size of the electrodes. It is generally not necessary to take into account the variation of edge effects as a function of $\mathbf{d_0}$, since these effects hardly change when the beam moves less than $\mathbf{d_0}$/3.

The expression of the mechanical force of return is given by:

$$\mathrm{F_m} = k \cdot x \tag{2}$$

In the case of a bridge made on a coplanar line the stiffness constant is for expression (hypothesis of the Bernoulli beam)

$$\mathrm{K} = \mathrm{K'K''} = \frac{32EWt^3}{l^3}\left(\frac{1}{2 - \left(2 - \frac{W}{l}\right)\left(\frac{W}{l}\right)^2}\right) + \frac{8\sigma(1-\gamma)Wt}{l}\left(\frac{1}{2 - \frac{W}{l}}\right) \tag{3}$$

where $E(GPa)$: Young's module,$l(m)$: Length of the membrane $t(m)$: Thickness of the membrane $w(m)$: Width of the membrane $W(m)$: The length of the lower actuation electrode, γ: The Poisson's ratio σ: The bi-axial residual stress.

The part k' is due to the geometry of the beam and the materials of which it is composed as the Young's modulus $E(GPa)$ and the moment of inertia $I(m^4)$.

The part k'' as a result of the manufacture is thanks to the biaxial residual stress σ (Pa) in the beam and is valid only in the case of a membrane in tension.

The calculated constant at the extremity of the beam is expressed as

$$K = \frac{\sigma(2\ \mathrm{E\ W\ w\ t^3})}{(8\mathrm{t^3W} - 6\mathrm{l^2w^2} + \mathrm{w^4})} \tag{4}$$

As illustrated in Fig. 1, the variation between the stiffness constant for a gold beam as a function of $\frac{t}{l}$ where the force is distributed at the center (k'_c) and at the end of the beam (k'_e) is given by these equations:

$$k'_c = 32Ew\left(\frac{t}{l}\right)^3 \cdot \frac{1}{8\left(\frac{x}{l}\right)^3 - 20\left(\frac{x}{l}\right)^2 + 14\left(\frac{x}{l}\right) - 1} \quad k'_e = 4Ew\left(\frac{t}{l}\right)^3 \cdot \frac{1}{\left(\frac{x}{l}\right)\left(1 - \left(\frac{x}{l}\right)\right)^2} \tag{5}$$

The resolution of the equation which translates the equality between the mechanical force of return and the electrostatic force makes it possible to determine the zone of stability.

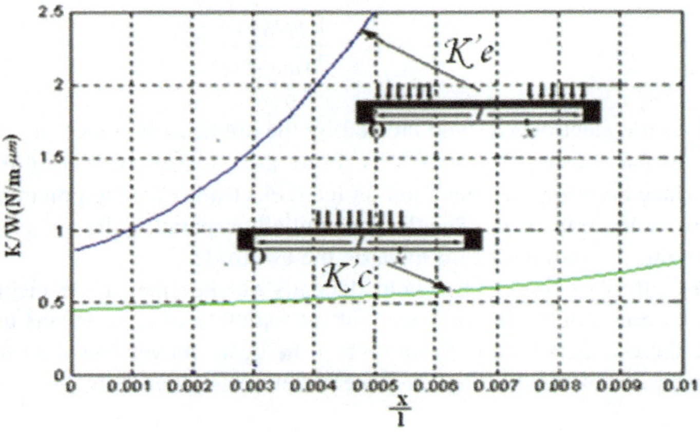

Fig. 1 Load distributed at the center and at the end of a beam

The graphical resolution of the equation $F_m = F_x$ indicates that the stable zone extends from x = 0 to x = d0/3, is one-third of the initial inter-electrode distance.

$$F_m = F_x \Rightarrow -kx + \frac{\varepsilon_o \varepsilon_r A V^2}{2(d-x)^2} = 0 = > \mathrm{X} = \mathrm{d}/3 \tag{6}$$

The stability limit is reached for a voltage equal to

$$\mathrm{V}_P = \sqrt{\frac{8kd^3}{27\varepsilon_0 \mathrm{A}}} \tag{7}$$

As shown in Fig. 2, the voltage drop-down menu is plotted as a function of the distance where the maximum stable deviation of the movable plate is equal to d0/3 (Fig. 2).

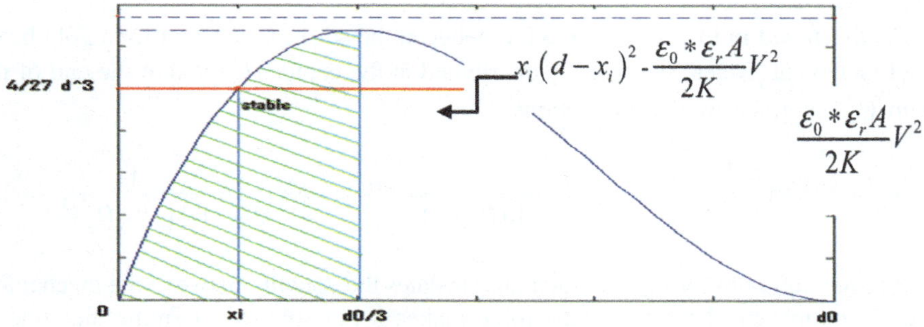

Fig. 2 variation of the mobile armature according to the height of GAP

The damping force does not only depend on the geometry of the beam, but also on the environment in which the structure is located, it can be written as

$$F_a = \zeta \cdot V \tag{8}$$

where ζ: damping coefficient, V: the speed at which the membrane moves.

For a rectangular beam, which has a small amplitude displacement (in the order of the initial height of the beam), the damping coefficient can be written

$$\zeta = \frac{3\,\mu\, l\, w}{2\,\pi\, d_0} \tag{9}$$

where μ is the viscosity of the gas, for air (μ = 1.8.10 − 5 kg/m^3).

The VAN DER WAALS forces and repulsive contact forces occur when the membrane comes into contact with a material in case of descent. These forces can be written as [1]

$$F_{vdw} + F_{cont} = \frac{c_1 \cdot A_{cont}}{(d_0 - x)^3} \cdot \frac{c_2 \cdot A_{cont}}{(d_0 - x)^{10}} \tag{10}$$

A_{cont}: The surface facing the contact level, d_0: The initial height of the beam
c_1: The surface energy due to VAN DER WAALS forces
c_2: Distance between the beam and the contact electrode surface.

These two numbers are strongly depending on the surface state of the microstructure, conditions. A mathematical function specially developed by FLAVIIS and COCCIOLI is used to model the contact forces of micromechanical devices.

$$F_c = \left[\exp^{-[\alpha(Z-t_d)]^\beta} - 1\right] \cdot U(t_d - Z) \tag{11}$$

where α et β: Numbers used to express the rate of variation of the forces as a function of the height, $U(x)$ is the step function.

In the case of small amplitude displacements, the mechanical movement of the membrane can be modeled by the following equation

$$\mathrm{m}\frac{d^2z}{dt^2} + \xi\frac{dz}{dt} + \mathrm{kz} = \mathrm{F}_{el} \tag{12}$$

The Laplace transform of this equation makes it possible to obtain the mechanical transfer function of the device

$$\mathrm{H} = \frac{Z(jw)}{F_{el}(jw)} = \frac{1}{k}\frac{1}{1 - (\frac{w}{w_{mec}})^2 + \frac{jw}{Q_{meca} w_{meca}}} \tag{13}$$

2.2 Electricals Properties

The capacitance value for micro-electromechanical component can be computed analytically by the following expression

$$C = \frac{\varepsilon_0 A}{g + \frac{t_d}{\varepsilon_r}} \tag{14}$$

The equivalent model of an electro-statically actuated suspended beam component is composed of a capacitance, inductance and resistance that are mounted in series or in parallel depending on the structure considered. Figure 3 shows the equivalent electrical model of the beam according to the serial configuration. The model is simulated using ADS software.

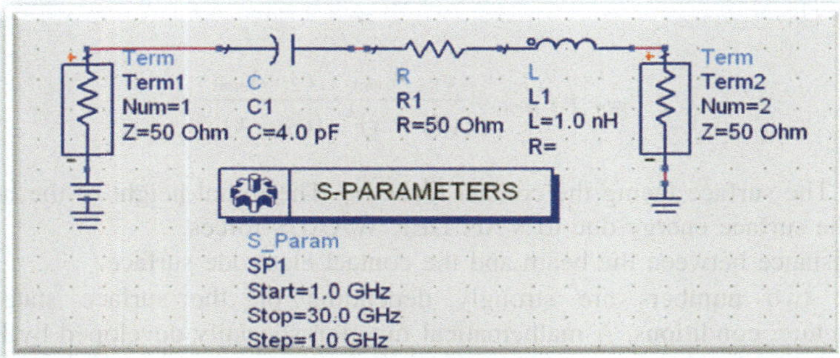

Fig. 3 Capacitive serial device

The device admits a maximum power that should not be exceeded to ensure its proper operation. This power value depends on the geometry of the considered structure. The maximum power allowed is

$$P_{max} = \mathrm{Re}\left(\frac{V_p^2}{Z_0}\right) = \frac{V_r^2}{Z_0}(c\, w\, Z_0)^2 \quad \mathrm{for}(c\, w\, Z_0) > > 1 \tag{15}$$

where V_r is the release voltage, and C the capacitance of the component in the low state.

3 Modeling of the Proposed Varactor

Figure 4 illustrates the S-parameters of the simulated device as well as the phase shift that it generates. The four distinct frequency peaks show good properties up to 30 GHz for (S11 and S22) and 55 GHz for (S12 and S21).

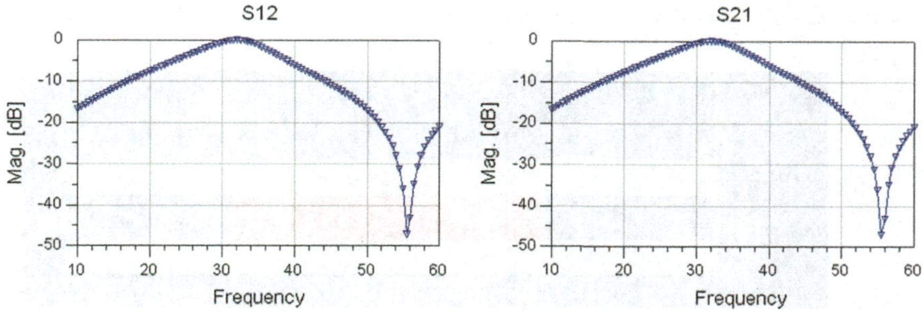

Fig. 4 S-parameter of the serial device

Figure 5 demonstrates the distribution of the field's lines in two directions and Fig. 6 illustrates the Full WAVE simulation. Figure 7 shows the schematic of the varactor and Fig. 8 presents the 3D diagram of the capacity and the mobile armature. Figure 9 regroups the simulation results obtained using MOMENTUM software on both planes E and H. The results (a), (b) and (c) are for air blade heights of 10, 8 and 6.67 μm respectively.

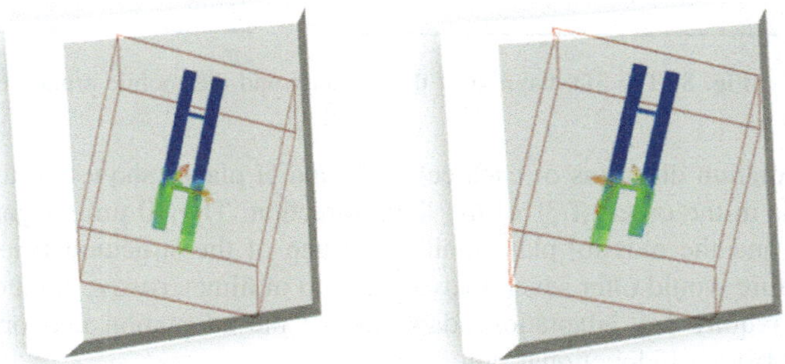

Fig. 5 Distribution of the field's lines in two directions

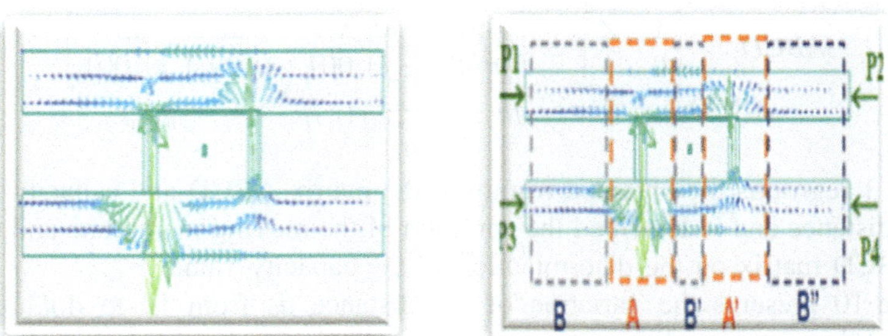

Fig. 6 Full WAVE simulation

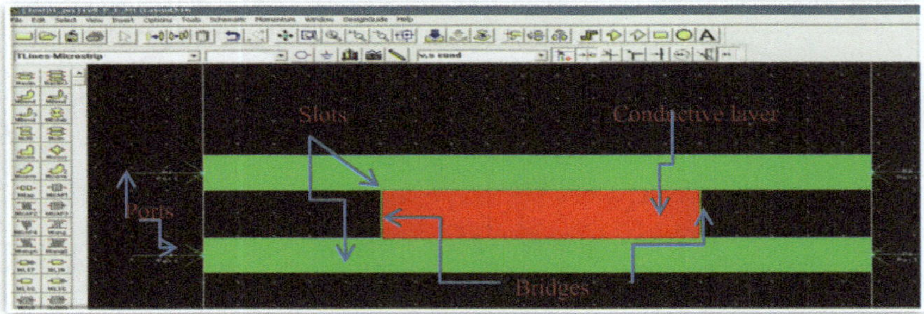

Fig. 7 Full WAVE simulation

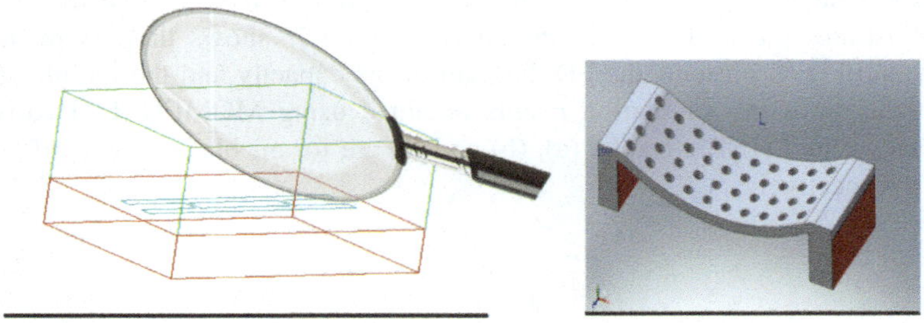

Fig. 8 The 3D diagram of the capacity and the mobile armature

The Radiation diagrams of each cell in E and H planes show that the variation of the radius is in the order of 3° of the lined direction. The 10 μm air gap between the fixed plate and the moving plate limited the gap of the structural beam. The planar LWA structure would offer several advantages to millimeter-wave frequencies. In fact, it will not require any adaptation adaptation of the adaptation network and will be operational from a bidirectional unit.

The capacity is obtained by varying the distance between the two fixed and movable plates. The ADS software makes it possible to measure this capacity by passing through the matrix ABCD. The expression of the matrix ABCD is given by

$$[ABCD] = \begin{bmatrix} 1 & Z \\ 0 & 1 \end{bmatrix} \simeq \begin{bmatrix} 0,99 \pm j0,001 & Real - j\beta \\ 0,001 \pm j0,001 & 0,99 \pm j0,001 \end{bmatrix}$$
$$ABCD_{12} = \mathrm{Re}al - j\beta \Rightarrow -\frac{j}{cw} = -j\beta \Rightarrow C = \frac{1}{\beta * 2\pi f} \qquad (16)$$

Table 1 regroups the obtained results of the matrix ABCD and capacity value for different distance and it described the variation of the moving membrane and the effect of the ABCD matrix on the determination of the capacity value.

Figure 10 presents the variation of the distance d_0 from 10 to do/3, and as a function of the corresponding capacity for each displacement of the moving armature.

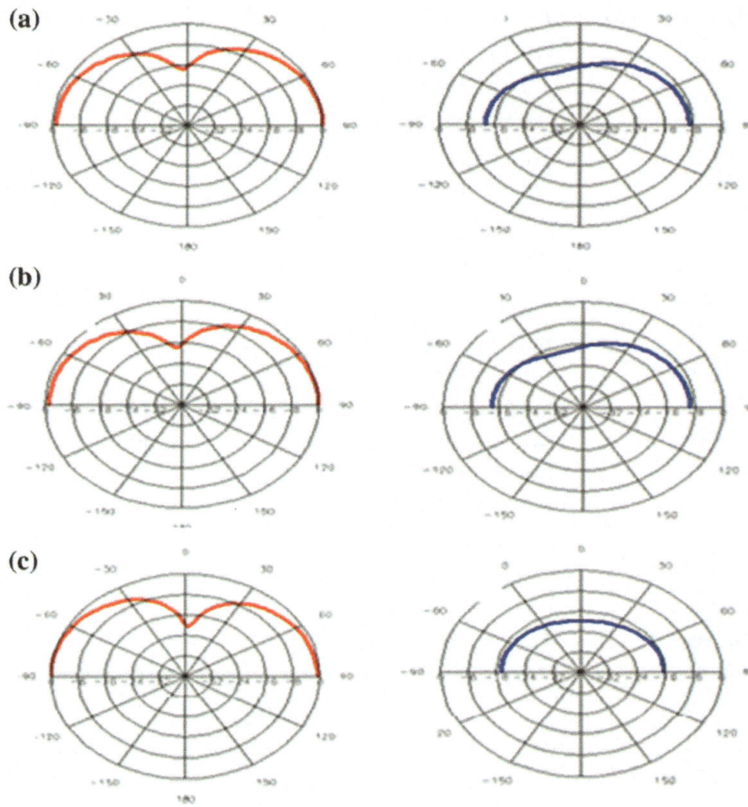

Fig. 9 Radiation diagrams on E and H planes

Figure 11 presents the modeling of the varactor using the electromagnetic simulation software ANSYS Maxwell, which demonstrate the variations of the electric field at the terminals of the proposed varactor with an excitation as a function of the separation distance between the two owners. Figure 11 illustrate the Power loss for the varactor.

Figure 12 illustrates the 3D power loss and the 3D electrical flux distribution for the proposed varactor as a function of the separating distance of the two frames. As the distance between the two frames increases with the insulating effect of the air, the loss of energy increases (Fig. 12).

4 Modeling of the Proposed Accelerometer

The accelerometer is a sensor used to measure linear acceleration on 3 axes. The principle of most accelerometers is based on the fundamental law of dynamics. More precisely, it consists of the equality between the inertia force of the seismic mass of the

Table 1 Obtained results of ABCD matrix and capacity value for various distance

$d_0(\mu m)$	*ABCD(1,2)*	*C(pF)*
10	777.516–j2.031 E3	2,613
9,75	775.433–j2.028 E3	2,617
9,5	772.911–j2.025 E3	2,621
9,25	770.661–j2.023 E3	2,623
9	768.316–j2.020 E3	2,627
8,75	765.973–j2.017 E3	2,631
8,5	763.434–j2.014 E3	2,635
8,25	760.799–j2.011 E3	2,639
8	758.078–j2.008 E3	2,643
7,75	755.223–j2.004 E3	2,648
7,5	752.244–j2.001 E3	2,652
7,25	749.140–j1.99 E3	2,657
7	745.931–j1.993 E3	2,663
6,75	742.539–j1.989 E3	2,668
6,5	738.985–j1.984 E3	2,675

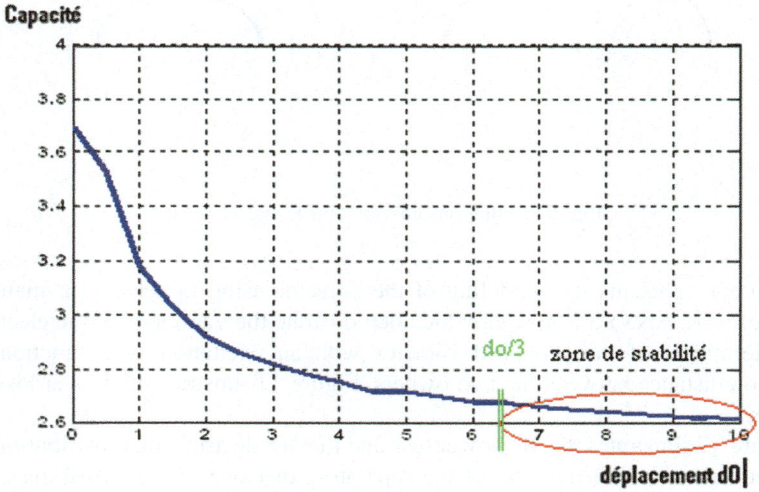

Fig. 10 The value of the capacity following the displacement of the mobile armature

sensor and a restoring force applied to this mass. There are two major families of accelerometers: non-controlled accelerometers and servo accelerometers. The sensor consists of micro-structures: micro-beams and micro-lamella. The general dimensions of these structures vary from 1 to 100 μm. These micro-structures are part of 2 categories: the fixed part and the mobile part. The simulation of the proposed Accelerometer is performed using on ANSYS Maxwell software in order to consults

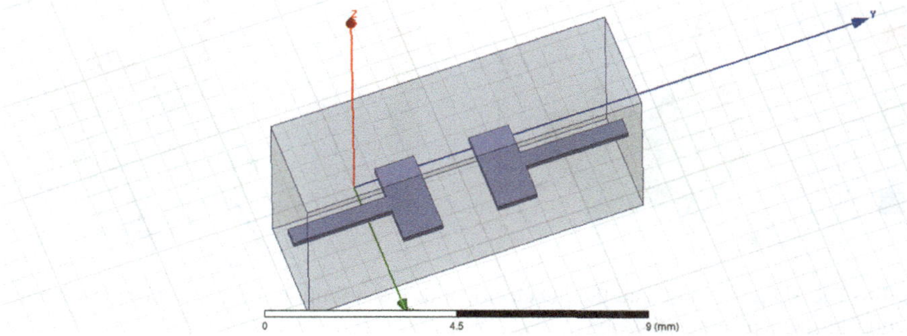

Fig. 11 Proposed accelerometer using ANSYS software

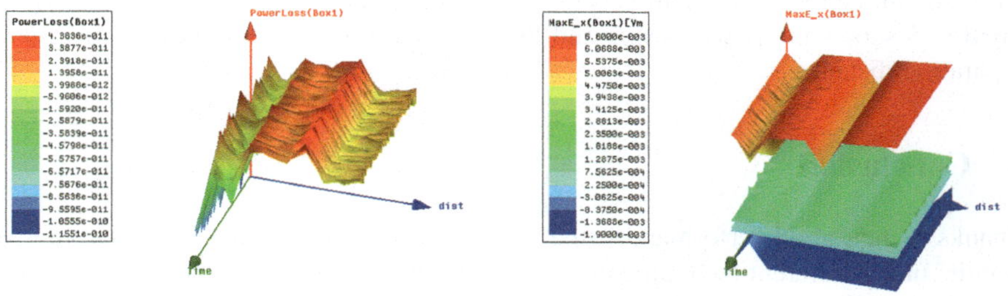

Fig. 12 The 3DPower loss and the 3DElectrical flux distribution for the proposed varactor as a function of the separating distance

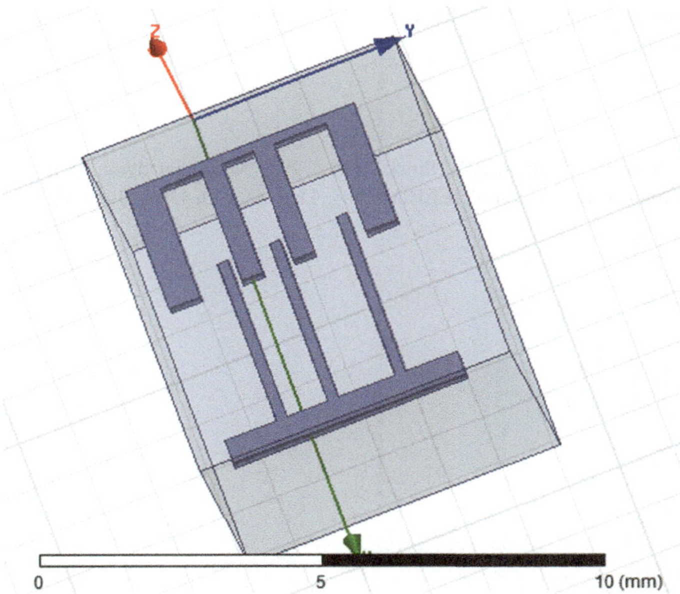

Fig. 13 Proposed accelerometer using ANSYS software

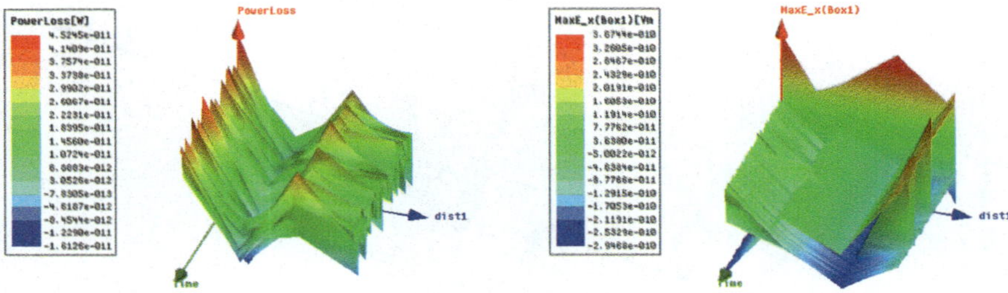

Fig. 14 The 3D power loss and the 3D electrical flux distribution for the proposed varactor as a function of the separating distance

the electrical behavior as a function of the separation distance between the blades and the value measured at the terminals of the fixed and mobile part as shown in Fig. 13. Figure 14 shows the power losses and the electrical flux distribution as a function of separation distance.

5 Conclusion

Thanks to their good performances, the MEMS-RF components are integrated in many circuits under different technologies. Based on these components variable capacitance is designed allowing the variation of the antenna beam (reconfigurable antennas) in a telecommunications field. The modeling of the varactor is performed using ADS software and ANSYS Maxwel tools. A proposed Accelerometer is also been analyze, modeled and simulated denoted the distribution of the electrical flux and the power losses a function of separation distance.

Reference

1. Chan EK, Kan, EC, Dutten, RW, Pinsky PM (1997) Nonlinear dynamic modeling of micro machined microwave switches. IEEE MTT-S, Denver, CO, USA, June 1997

Co-simulation Study of a Two Wheeled Vehicle Equipped with an ABS System

Aymen Khadr[1(✉)], Ajmi Houidi[2], and Lotfi Romdhane[1,3]

[1] Laboratory of Mechanics Sousse (LMS), National Engineering School of Sousse, University of Sousse, Sousse, Tunisia
khadr.aymen@yahoo.fr, lotfi.romdhane@gmail.com
[2] LMS, Higher Institute of Applied Sciences and Technology of Sousse, University of Sousse, Sousse, Tunisia
ajmi.houidi@issatso.rnu.tn
[3] College of Engineering, The American University of Sharjah, Sharjah, UAE

Abstract. The Anti-lock Braking System (ABS) is an active safety system for two wheeled vehicles (TWV). It is used to control the dynamics of a TWV during an emergency braking phase and to improve its driving safety. The main objective of this study is to establish a virtual model of a TWV equipped with a control scheme for an ABS system. The developed model will be used to investigate the reliability of an ABS system when used with a TWV. For this purpose, firstly a TWV multi-body dynamic model was constructed using ADAMS/View. Furthermore, a control model of the ABS system was created using MATLAB/Simulink. Then, a co-simulation approach is established using MATLAB/Simulink and ADAMS/Control. The performance of the co-simulation model is assessed by simulations for different initial speeds and for different road conditions. The obtained results show the benefits of using a co-simulation approach in studying such a complex systems. Moreover, the effects of the active safety system (ABS) on the dynamic of a TWV, during emergency braking, are studied.

Keywords: Antilock braking system (ABS) · Two wheeled vehicle
Co-simulation · Active safety

1 Introduction

Modeling and simulation computer tools bring significant benefits in the automotive field especially in the early stages of the process in order to reduce development costs. Indeed, these tools particularly allow the design of security systems such Anti-lock Braking Systems (ABS) and understanding the braking performance of a vehicle in general and particularly the performance of a two-wheeled vehicle (TWV). To simulate the braking dynamics of a TWV, we are going to design a co-simulation environment formed by a multibody model of the TWV interacting with the road and designed with the ADAMS software. The second part of the co-simulation environment consists in a simplified ABS controller which will be implemented using Simulink software in order to regulate the braking torque in slip conditions.

N. Aifaoui et al. (Eds.): CMSM 2019, LNME, pp. 931–939, 2020.
https://doi.org/10.1007/978-3-030-27146-6_100

2 Multibody Model of a Two Wheeled Vehicle

In this study, the full multibody model used to describe the dynamics of the TWV is adopted from the references [1, 2]. It's constructed using ADAMS/View and it's composed of six articulated rigid bodies with 11 degrees of freedom (Fig. 1): the frame including the rider body, the front wheel, the rear wheel, the sprung part of the fork including the handlebars, the unsprung part of the fork and the unsprung part of the rear wheel. The suspensions are taken into account in this model: the front suspension is between the sprung and unsprung part of the fork and the rear suspension is between the frame and the unsprung part of the rear wheel. To consider the interaction between tire and road this contact is modeled by the PAC-MC tire model of ADAMS based on Magic-Formula [4].

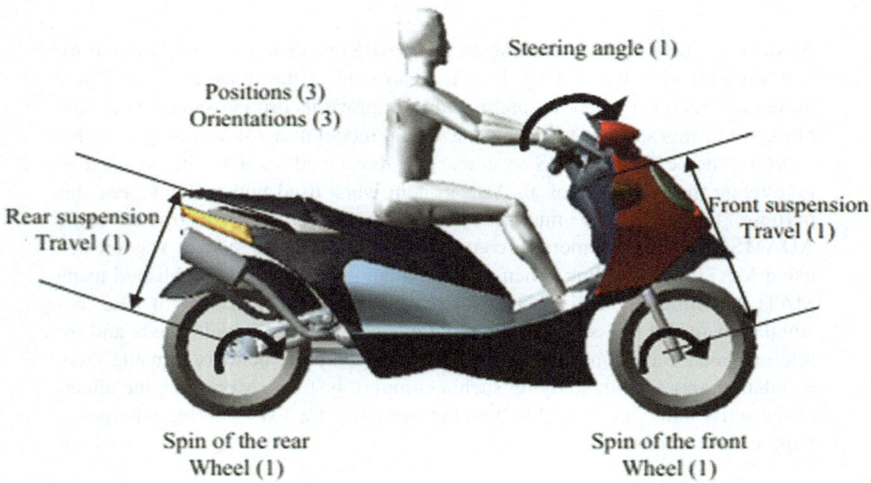

Fig. 1 Multibody model of the TWV

3 Anti-lock Braking System (ABS)

3.1 Braking Dynamics and ABS Principle

In this study the TWV moves on a straight line, so only the longitudinal force F_x is taken into account Fig. 2a. This force is given by the magic formula in terms of the longitudinal slip ratio κ_x. This quantity is calculated by the following relation:

$$\kappa_x = -\frac{V_x - \Omega R}{V_x} \tag{1}$$

where Ω, R, and V_x denote the wheel angular velocity, the wheel rolling radius, and the longitudinal speed of the contact point, respectively.

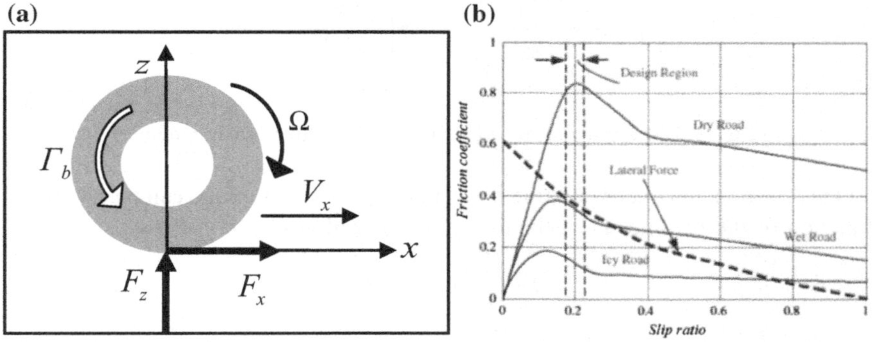

Fig. 2 **a** Forces acting on a single wheel **b** Coefficient of friction versus wheel slip [6]

If the slip ratio is positive the longitudinal force acts as a tractive force $F_x > 0$ else, it acts as a braking force $F_x < 0$. When the wheel is locked ($\Omega = 0$) we obviously have $\kappa_x = -1$. Consequently the braking distance increases, the steering is disabled and the rider can lose the controllability of the vehicle. So the main role of an ABS is to adjust the applied braking force in order to avoid the lock of wheels and to maintain the steering and the stability of the TWV. This means that the wheel slip should be kept within a range for which the friction coefficient is near its peak value (see Fig. 2b). As shown in Fig. 2b, it is apparent that for all types of road surfaces the value of the coefficient of friction is optimal when the wheel slip rate is about 0.2. This value is taken in most ABS control strategies such Wang et al. [7], Poursamad [5], Meymand and Gharaveisi [3].

3.2 ABS Controller

In this study, a simple hydraulic line and pressure build-up will be modeled: assuming there is an input pressure commanded P_d, which would travel through a hydraulic line and terminated by a piston actuator, where pressure might build up to P_b inducing a normal force F_b on the disc brake caliper and finally a braking torque Γ_b acts on the wheel.

In our ABS, we will adapt a bang-bang (on-off) controller. It acts upon the error signal (e) between the actual slip and the desired slip. Then it returns a value of the input pressure commanded at the entrance to the hydraulic line P_d based on the sign function described by the following rule:

$$P_d = sign(e) = \begin{cases} +1 & \text{if} \quad e > 0 \\ 0 & \text{if} \quad e = 0 \\ -1 & \text{if} \quad e < 0 \end{cases} \tag{2}$$

Figure 3 shows the block diagram used for modeling the hydraulic brake dynamics in Simulink. Note that, a saturation block is included in order to limit the brake pressure to a maximum value.

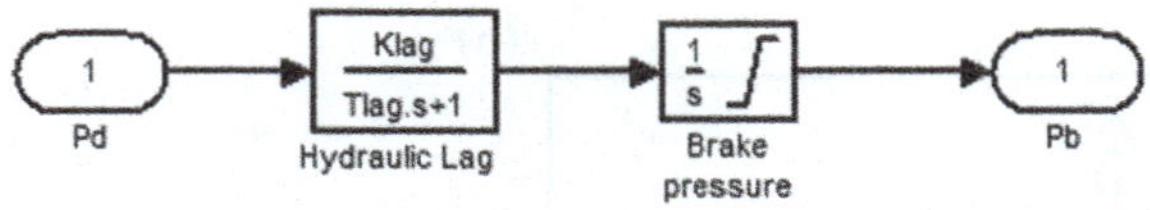

Fig. 3 Block diagram for modeling the hydraulic brake dynamics

Once the brake cylinder pressure is obtained the braking torque can be calculated as a function of the parameters of the disc brake as follows:

$$\Gamma_b = 2R_{eff}\mu_{pad}P_bA_b \tag{3}$$

where R_{eff}, μ_{pad} and A_b are the effective pad radius, the pad friction coefficient and the cylinder cross area, respectively.

4 Co-Simulation ADAMS/Simulink

The developed co-simulation consists in the interaction between ADAMS and Simulink; firstly we establish the multibody model of the TWV with ADAMS/View (Fig. 1). Next we identify its inputs (the front and the rear braking torques) and its outputs (the front and rear longitudinal slips) then we export the TWV model in the form of a block named **adams_sub** (Fig. 4).

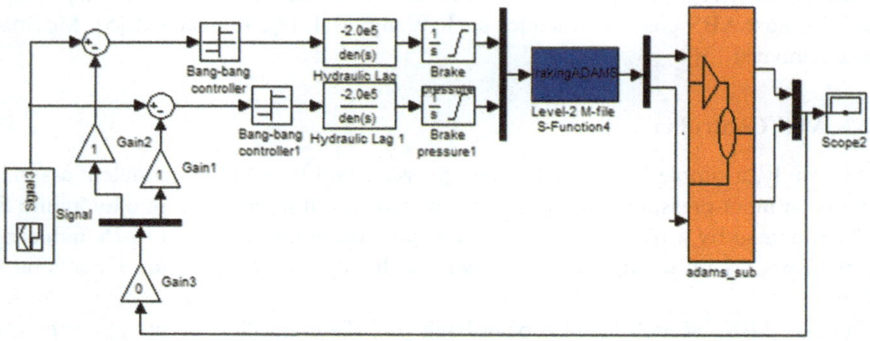

Fig. 4 Control scheme in Simulink

Finally Simulink is used to build a closed loop control scheme, which aims to control of the longitudinal slip of the two wheels of the TWV and to regulate the braking torques during an emergency braking as shown in Fig. 4.

5 Simulation and Results

To simulate the braking performance of the TWV with and without an ABS control system under braking in a straight line, the following conditions are considered:

- The TWV is initialized with a speed of 60 km/h (16.668 m/s),
- Simulations are performed for two types of road surface: on a dry road and on a wet road.
- The desired wheel slip rate is taken −0.2 (a negative value in the case of a braking force).

5.1 Simulation on a Dry Road

The following Figs. 5, 6, 7 and 8 show the TWV behavior during a braking on a dry road. By examining the results obtained without the ABS controller, one note that the longitudinal slips (Fig. 5) varies from (0) to (−1) since the application of the brake until the moment when the wheels stop. The wheel lock is obtained in 0.81 s for the front wheel and in 0.72 s for the rear wheel (Fig. 7).

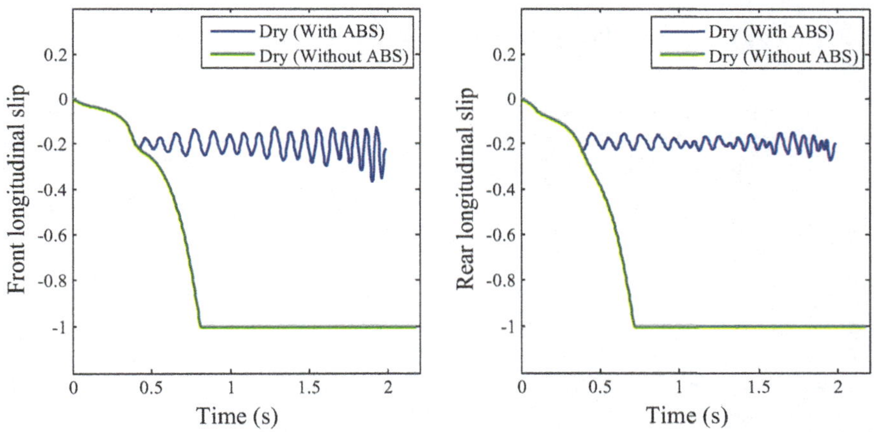

Fig. 5 Front and rear longitudinal slips

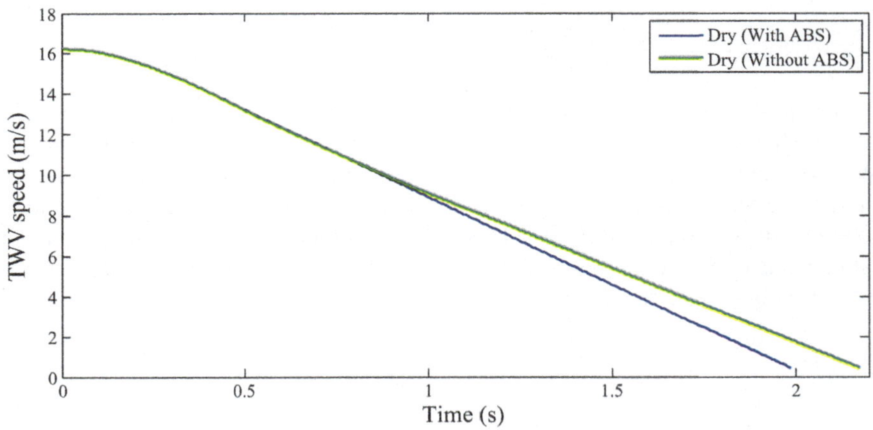

Fig. 6 Linear speed of the TWV

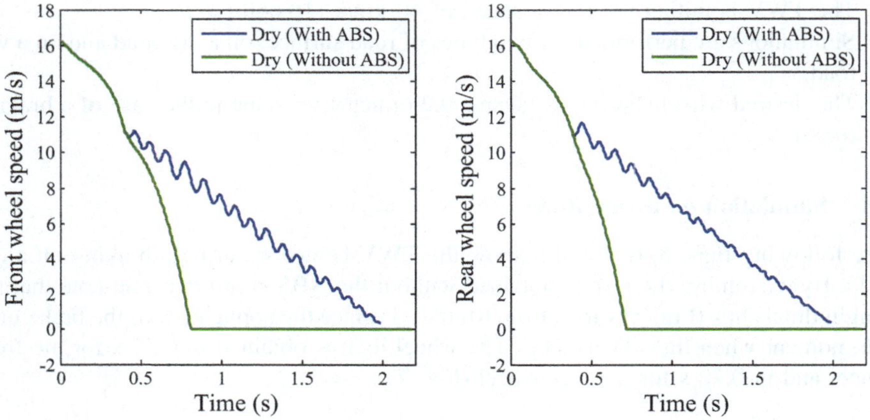

Fig. 7 Front and rear wheels speeds

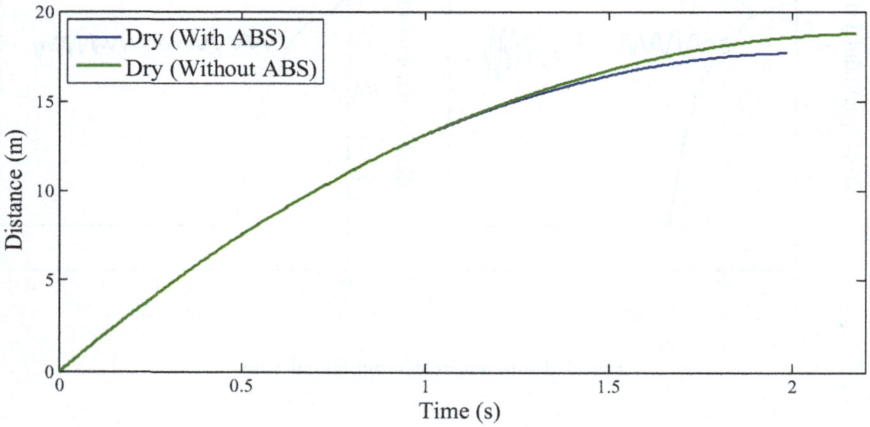

Fig. 8 Distance traveled by the TWV

The stopping distance is 18.79 m and occurred after 2.17 s (Fig. 8). This indicates that the wheels have been locked before the TWV stops. Therefore the control of the TWV was lost due to wheel lock. Through the use of an ABS control system (Bang-Bang), it is shown that: the stopping distance is slightly reduced to 17.72 m and occurred after 1.98 s instead of 2.17 s. The longitudinal slips oscillate around the desired value (−0.2) so there is no wheel lock before the TWV stops. Consequently, the ABS controller keeps the TWV under control.

5.2 Simulation on a Wet Road

The next Figs. 9, 10, 11 and 12 show the TWV behavior during a braking on a wet road.

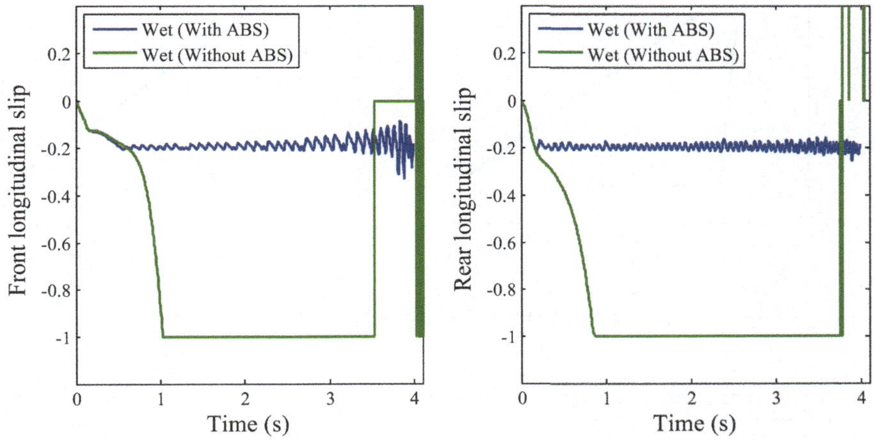

Fig. 9 Front and rear longitudinal slips

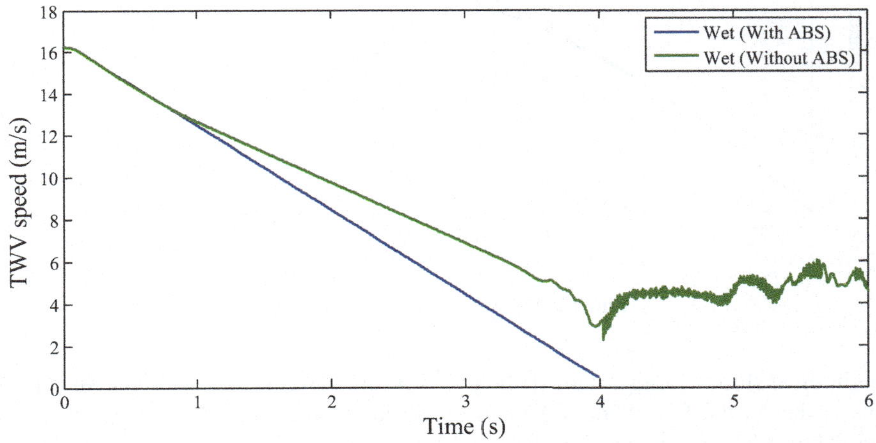

Fig. 10 Linear speed of the TWV

We have the same behavior as that obtained for a dry road. We can note that without an ABS controller the wheel lock is obtained in 1.02 s for the front wheel and in 0.85 s for the rear wheel (Fig. 11). Consequently the control of the TWV was lost.

The wheels remain locked and the longitudinal slips are equal to (−0.1). At (t = 3.52 s) for the front wheel and at (t = 3.75 s) for the rear wheel the values of the

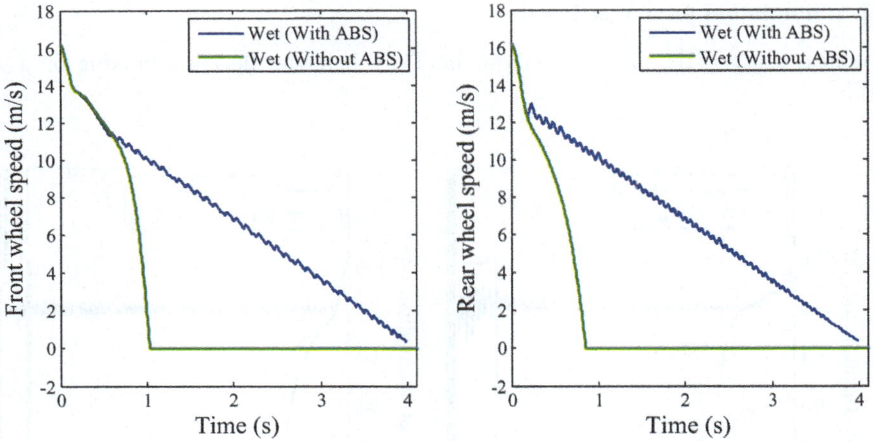

Fig. 11 Front and rear wheels speeds

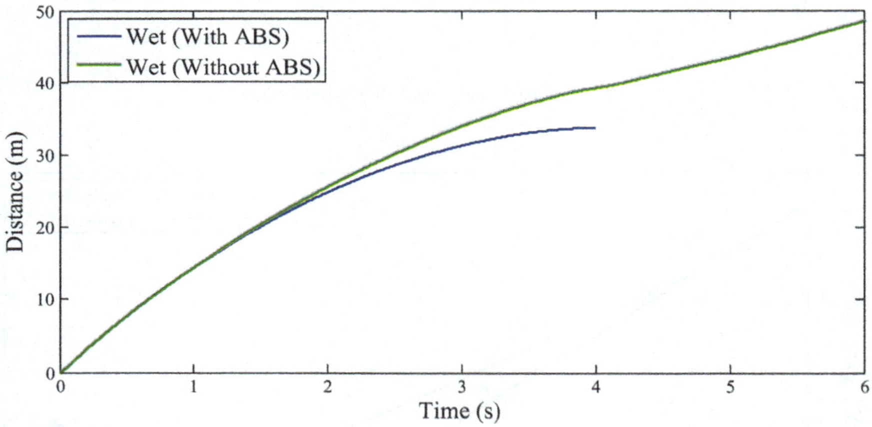

Fig. 12 Distance traveled by the TWV

longitudinal slips no longer correspond to (−0.1) and have a complex behavior (Fig. 9). This is due to the fall of the TWV observed during the simulation. This fall can be verified also by examining Figs. 10 and 12 so it is clear that the speed of the TWV is not canceled and the stopping distance cannot be determined exactly because the TWV remains in motion when the TWV fell.

By using the ABS controller, it is shown that the longitudinal slips oscillate around the desired value (−0.2) so there is no wheel lock before the TWV stops. Consequently, the ABS controller keeps the TWV under control and there is no falling.

Compared to the results obtained for the dry road one can notice that the speed of the TWV was canceled after a longer time (3.98 s) instead of 1.98 s for the dry road. The stopping distance has been increased to 33.85 m instead of 17.72 m for the dry road.

As shown on the previous curves, by comparing the results obtained with an ABS controller with those obtained without an ABS controller we can deduce the bang-bang controller shows the best results in terms of braking distance and ensures that the wheel will never lock. Consequently the stability of the TWV is maintained and is still controlled by the driver.

6 Conclusion and Perspectives

In this paper, an environment of co-simulation has been developed, by incorporating an ABS controller into a full TWV model. This co-simulation aims to study the braking performance of such type of vehicle in a straight line by controlling the longitudinal slips and regulating the braking torques acting on wheels. The obtained results show that the developed ABS controller acts to avoid wheel lock for two types of road surface (wet and dry). Therefore, our co-simulation is able to predict the braking behavior by applying ABS and the simulations carried out showed that the developed ABS controller (Bang-Bang) ensures that the TWV can maintain steering and stability during braking.

Future work will use this environment of co-simulation to study and analyze the braking behavior under turn maneuvers and use more complex controllers. Also it's interesting to valid the developed ABS controller by experimental tests.

References

1. Houidi A, Khadr A, Romdhane L (2017) Dynamic modeling and handling study of a two-wheeled vehicle on a curved track. Mech Ind 18(4):409
2. Khadr A, Houidi A, Romdhane L (2013) Development of co-simulation environment with ADAMS/simulink to study maneuvers of a scooter. In: Design and modeling of mechanical systems. Springer, Berlin, pp 37–43
3. Meymand NE, Gharaveisi A (2014) Optimization of Bang-of-Bang TS-fuzzy based via DARLA technique for ABS system. J Adv Comput Res 5(3)
4. MSC Software (2010) PAC MC ADAMS/Tire Help
5. Poursamad A (2009) Adaptive feedback linearization control of antilock braking systems using neural networks. J Mechatron 19:767–773
6. Sharkawy AB (2010) Genetic fuzzy self-tuning PID controllers for antilock braking systems. Eng Appl Artif Intell 23:1041–1052
7. Wang WY, Chen GM, Tao CW (2003) Stable anti-lock braking system using output-feedback direct adaptive fuzzy neural control. In: International conference on systems, man and cybernetics, pp 3675–3680

Optimization of a Flexible Multibody System Design Variables Using Genetic Algorithm

Mohamed Amine Ben Abdallah[1,2(✉)], Imed Khemili[3], and Nizar Aifaoui[1]

[1] Laboratoire de Génie Mécanique, Ecole Nationale d'Ingénieurs de Monastir, Université de Monastir, Monastir, Tunisie
med.amie.b.abdallah@gmail.com
[2] Ecole Supérieure Privée d'Ingénieurs et Technologies, Ariana, Tunisie
[3] Laboratoire de Mécanique de Sousse, Ecole Nationale d'Ingénieurs de Sousse, Université de Sousse, Sousse, Tunisie
khemili_imed@yahoo.fr

Abstract. The dynamic behavior of multibody systems has been widely studied. Thus, effects of imperfections such as clearance, friction and flexibility on the dynamic behaviour are dealt with various tremendous works. For a given dynamic response, the mechanism design variables needs to be defined. This identification approach is known as the mechanism synthesis. Despite all these imperfections, the mechanism should describe a precise workspace traduced by the trajectory path of the effector component. In this work, the dynamic synthesis for a multibody system with imperfections is presented. A demonstrative slider crank mechanism with a flexible connecting rod has been used for the algorithm validation. The identification approach is based on its dynamic responses such as: the slider velocity and acceleration and the transversal deflection of the flexible connecting rod. A genetic algorithm has been developed to identify its design variables. This algorithm is implemented under Matlab(c). The presented results are in great agreement with the real mechanism dimensions.

Keywords: Flexible multibody system · Genetic algorithm · Dynamic synthesis · Design variables · Imperfections

1 Introduction

The multibody systems are required to be very accurate in several applications such as welding, assembly, and medical robots. Furthermore, the optimization of the mechanism design variables is necessary in order to achieve the desired path. Moreover, many constraints depending on the mechanism and the applications field could be faced. In some applications, the mechanism should describe a defined path. In general, the mechanisms are designed with the assumptions of rigid bodies and perfect joints. However, the real dynamic responses are far away from the desired one. In this work, an genetic algorithm is developed in order to identify the mechanism design variables

N. Aifaoui et al. (Eds.): CMSM 2019, LNME, pp. 940–952, 2020.
https://doi.org/10.1007/978-3-030-27146-6_101

based on its dynamic responses. The algorithm gives the optimal design variables for the mechanism.

The identification approaches have been dealt with various tremendous works. Laribi et al. have studied the path generation for a four-bar mechanism [1]. An enhanced genetic algorithm incorporating a fuzzy logic loop has been presented. The results confirm the ability the developed algorithm has to retrieve the mechanism design variables giving the desired trajectories path.

Erkaya et al. has been interested in artificial intelligence to the synthesis and identification of mechanism design variables. An algorithm based on a neural network is used for a slider crank mechanism optimization [2]. More recently, the authors have used a genetic algorithm optimization for a four-bar mechanism [3]. Then, the authors have coupled the neural network and the genetic algorithm (NN-GA) [4–6] to develop a hybrid optimization algorithm to synthesize multibody systems with clearance in joints.

Jorge Ambrosio has presented an optimization methodology for a composite flexible multibody system [7]. A demonstrative slider crank mechanism with a flexible connecting rod has been presented.

Saeed Ebrahimi has treated the dimensional synthesis of the path generated for a four-bar mechanism [8]. The optimization problem is solved using the Imperialist Competitive Algorithm (ICA). A comparison of the ICA with a set of other heuristic algorithms has been presented.

W Y Lin has established a spatial synthesis of a four-bar mechanism using the Genetic Algorithm and Differential Evolution Hybrid Algorithm (DEHA) in order to optimize the path generation [9, 10]. More recently, the author has presented a combined mutation strategy for the Differential Evolution Algorithm (DEA) for a spatial synthesis of a five-bar mechanism [11].

In this work, a dynamic synthesis of a flexible slider crank mechanism is presented. The mechanism synthesis is established using the genetic algorithm optimization method. Thus, based on a dynamic response such as the slider velocity, the slider acceleration and the transversal deflection of the flexible connecting rod, the proposed algorithm is able to recognize the optimal desired design variables dealing with the given dynamic response. An acceptable error between the dynamic response for the proposed mechanism design variables and the reference dynamic response constitutes the stop condition for the algorithm.

2 Modeling the Mechanism

The dynamic resolution of the motion equation for multibody systems is one of the stiffest problems due to the highly non-linear equations system. Based on parameters depicted in Fig. 1, an analytical study has been established for the flexible slider crank mechanism with perfect joints. The dyad finite element method [12] has been adopted for the elastic slider crank mechanism.

$$\begin{bmatrix} M & \Phi_q^T \\ \Phi_q & 0 \end{bmatrix} \begin{bmatrix} \ddot{q} \\ \lambda \end{bmatrix} = \begin{bmatrix} Q_e + Q_v \\ Q_c \end{bmatrix} \quad (1)$$

$$\Phi(\mathrm{q}, \mathrm{t}) = 0 \quad (2)$$

where, M is the system mass matrix, Φ(q, t) the vector of kinematic constraints, Φ_q the jacobian matrix of Φ(q,t), Φ_q^T the transpose matrix of Φ_q, $\ddot{q}$ the state of accelerations vector, $Q_c = -\Phi_{tt} - (\Phi_q\dot{q})_q\dot{q} - 2\Phi_{qt}\dot{q}$ the right hand side vector of acceleration equations, or vector of quadratic velocity terms, Φ_{tt} the double partial derivative of kinematic constraint equations with respect to time, the Φ_{qt} derivative of Jacobian matrix with respect to time, q the vector that contains the state of positions, or vector of generalized coordinates, $\dot{q}$ the vector that contains the state of velocities, λ the vector of Lagrange multipliers associated with constraints, Q_e the total applied forces, Q_v the total constrained forces.

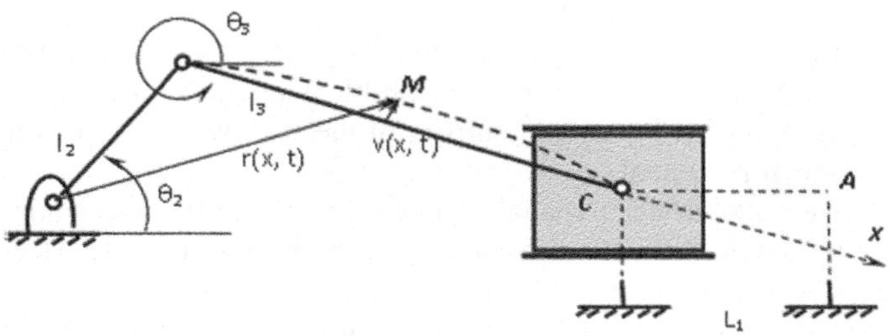

Fig. 1 The slider crank mechanism parameters

The constraint equation for the slider crank mechanism, which is a system of one degree of freedom, with holonomic constraints based on general coordinates, goes as follows:

$$\Phi(q, t) = \begin{pmatrix} \varphi_1(q, t) \\ \varphi_2(q, t) \end{pmatrix} = \begin{pmatrix} l_2 \cos\theta_2 + l_3 \cos\theta_3 - x_c \\ l_2 \sin\theta_2 + l_3 \sin\theta_3 \end{pmatrix} = \begin{pmatrix} 0 \\ 0 \end{pmatrix} \quad (3)$$

The mechanism parameters are detailed in Table 1.

Table 1 Mechanism parameters simulation

Mechanism parameters simulations	
Crank length (mm)	50
Connecting rod length (mm)	350
Slider mass (kg)	0.25
Angular velocity (rad/sec)	28.16

For a given mechanism dynamic response such as the velocity and the acceleration of the slider and the mid-point transversal deformation of the connecting rod, a synthesis method based on Genetic Algorithm (GA), able to identify its design variables will be presented in the next section.

3 Identification Approach

The Genetic Algorithm (GA) [14] is one of the most prominent optimization tools used for the last few decades. It is mainly inspired from natural phenomena. In fact, it is based on nature law, which assigns the best probability or chance for the strongest individual to survive. Otherwise, the low quality or weak individuals have a high death probability. Figure 2 shows the flowchart of the GA which is mainly divided in the following steps, wherein P_C is the crossover and P_m is the mutation probability.

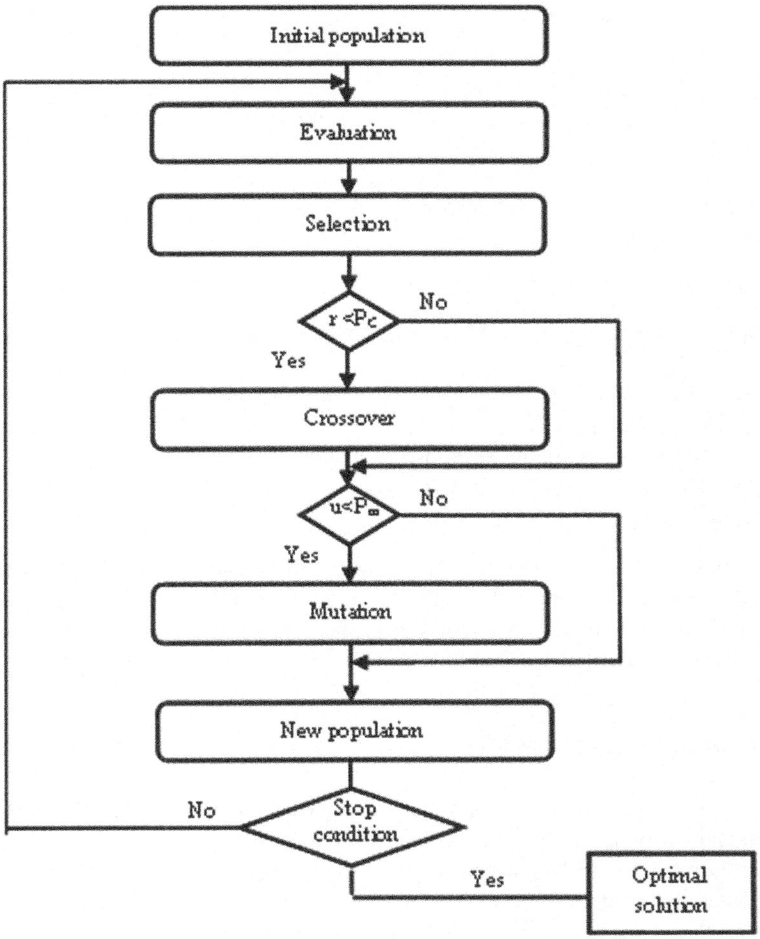

Fig. 2 The Genetic Algorithm flowchart

3.1 Initial Population Choice

In this step, an initial population is chosen with a defined number of chromosomes. Every chromosome contains its own allele. For the presented optimization problem, the number of alleles depends on the mechanism parameters required. In this work, two design variables are considered, mainly the crank length and the connecting rod length. Mass and inertia will be optimized consequently due to their relation to the length of each part.

3.2 Fitness Function

In order to classify the chromosomes, the evaluation of each one in the initial population, is necessary. This evaluation will foster the selection of the best individual for the next steps of the genetic algorithm process. The evaluation is based on the performance of each individual, in the following objective function:

$$F = \min\left(error\right)$$

$$error = \sqrt{\frac{1}{n}\sum_{i=1}^{n}\left(\left(X_{iDV} - X_{it\arg et}\right)/X_{it\arg et}\right)^2} \tag{4}$$

$$Subjet\ to\ :\begin{cases} lb_2 \le l_2 \le ub_2 \\ lb_3 \le l_3 \le ub_3 \end{cases} \tag{5}$$

With:

lb_1, ub_1 : are respectively the lower and upper bounds of the search interval for the crank length.
lb_2, ub_2 : are respectively the lower and upper bounds of the search interval for the connecting rod length.
X_{iDV} : is the design variable response at instant i.
$X_{it\arg et}$: The target response at instant i.
n : Response curve total number of points.

Throughout this step, a selection probability will be assigned to each chromosome. Then, the selection operation will be similar to a casino wheel operation [15], wherein the best individual evaluation has a higher probability to be retained. Nevertheless, the chromosomes selection with a low performance remains always possible.

4 Simulations and Results

The main results of the numerical simulations are discussed in this section. Three types of mechanism responses are used for the dynamic synthesis of the mechanism design variables: the velocity and the acceleration of the slider and the transversal deformation of the connecting rod. It is noticed that the GA greatly depends on the following parameters: the generation number, the initial population size, and the probabilities Pc and Pm. Therein, the effect of each parameter has been studied in order to highlight the ability of the developed algorithm to retrieve the desired mechanism design variables.

4.1 Effects of Generation Number

The dynamic synthesis of the flexible slider crank mechanism based on its dynamic responses deploying a genetic algorithm optimization proposes a set of design variable parameters. The effect of the generation number has been studied with an initial population size of 10 individuals. It is noticed that the crossover and mutation probabilities are fixed respectively to 0.9 and 0.3.

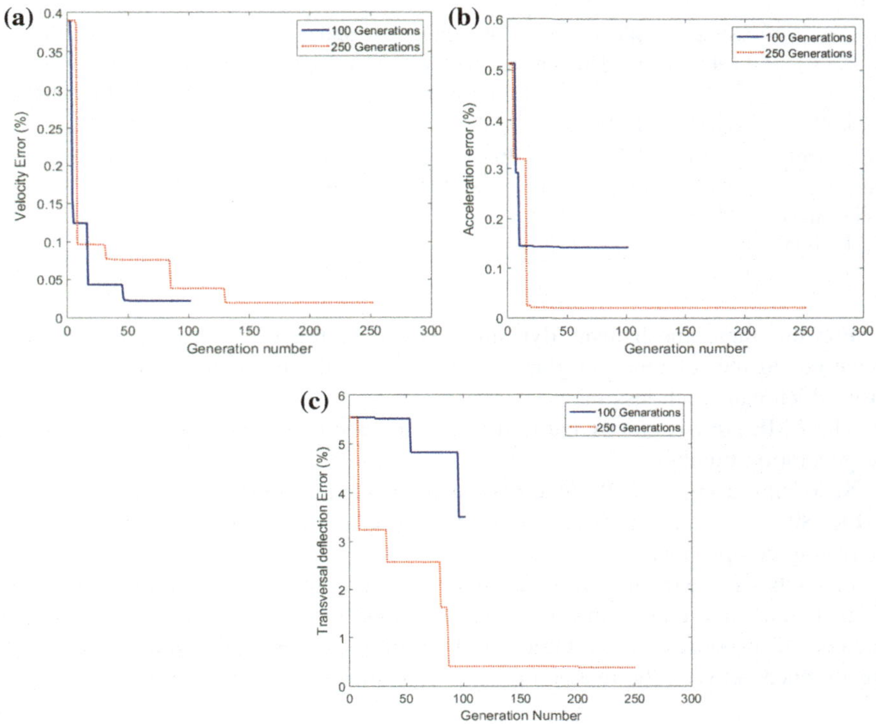

Fig. 3 The error evolution, **a** the velocity error, **b** the acceleration error, **c** the transversal deflection error

Figure 3a–c show respectively the slider velocity and acceleration and connecting rod transversal deformation errors according to the generation numbers. These errors represent the Root Min Square (RMS) values between the reference mechanism response and the proposed response given by the GA optimization algorithm.

It is noticed that, for the three types of dynamic responses the estimated error greatly decreases when increasing the generation number. However, the GA needs a much more CPU time to converge. All these aspects are summarized in Table 2.

Table 2 Generation number effects

Generation number (10 chromosomes)						
100				250		
	Slider velocity	Slider acceleration	Connecting rod transversal deformation	Slider velocity	Slider acceleration	Connecting rod transversal deformation
Error %	0.021	0.142	3.504	0.019	0.0194	0.382
CPU (sec)	329.4	387.3	363.4	798.28	901.72	794.88
Retained mechanism (l_2, l_3) mm	(49.9, 348.5)	(49.2, 345)	(50.8, 353.3)	(50.1, 50.9)	(50.1, 351)	(49.8, 350.5)

For the three mechanism dynamic responses, the GA optimization algorithm converges to the reference mechanism (50; 350) with an acceptable minimum RMS error which can reach less than 0.02%.

The RMS error is greatly reduced (can reach about 10 times less) when increasing the generation number.

Scarifying a higher CPU time computation, for generation number of 250, (from 330 to 800 s) to have a greater error reduction of up to 10 times less is a very interesting compromise.

In synthesis, the retained mechanism design variables correspond for a crank length of 50, 1 mm and a connecting road length of 350, 9 mm. Figure 4 shows the retained mechanism responses vs. the target mechanism responses, which illustrates the good concordance between the retained and the reference mechanisms responses.

4.2 Effects of Population Size

This section is devoted to the focus on the initial population size effects on the GA proposed solutions. In fact, the initial population size has been extended to 20 individuals, which are randomly chosen. The generation number, the crossover and mutation probabilities are maintained respectively to 250, 0.9 and 0.3.

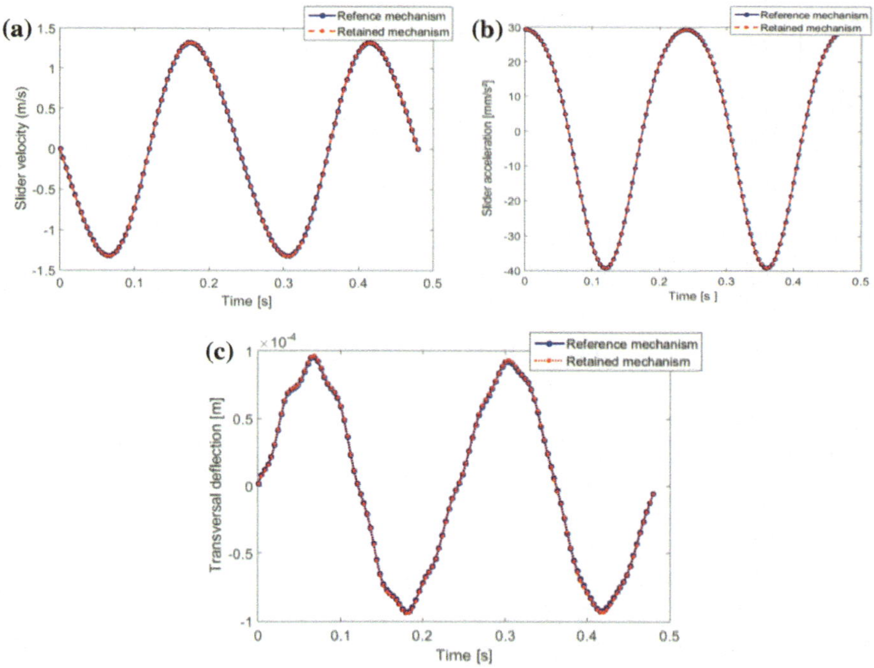

Fig. 4 Retained mechanism responses: **a** Slider Velocity, **b** Slider acceleration, **c** Connecting rod transversal deformation

Figure 5a–c depict the GA evolution error for both initial population sizes of 10 and 20 individuals.

It is noticed that, for the three types of dynamic responses, the estimated error greatly decreases when increasing the initial population size. Nevertheless, the GA needs a much more CPU time to converge. The most results are summarized in Table 3.

In synthesis, after 250 iterations and for an initial population size of 20 individuals, the algorithm proposes a couple of design variables of (50, 349.9) mm for a CPU time of 879 s and an estimated error of 0.0004%. This proposed couple of design variables provides a better accuracy compared to the results provided by the algorithm using a 10-individuals initial population, which seems perfectly logic. However, the computation time increases twice more to give an estimated error reduction 100 times less.

For the retained mechanism design variables, corresponding to a crank length of 50 mm and a connecting rod length of 349.9 mm, the retained mechanism responses agree perfectly with the reference mechanism response (Fig. 6).

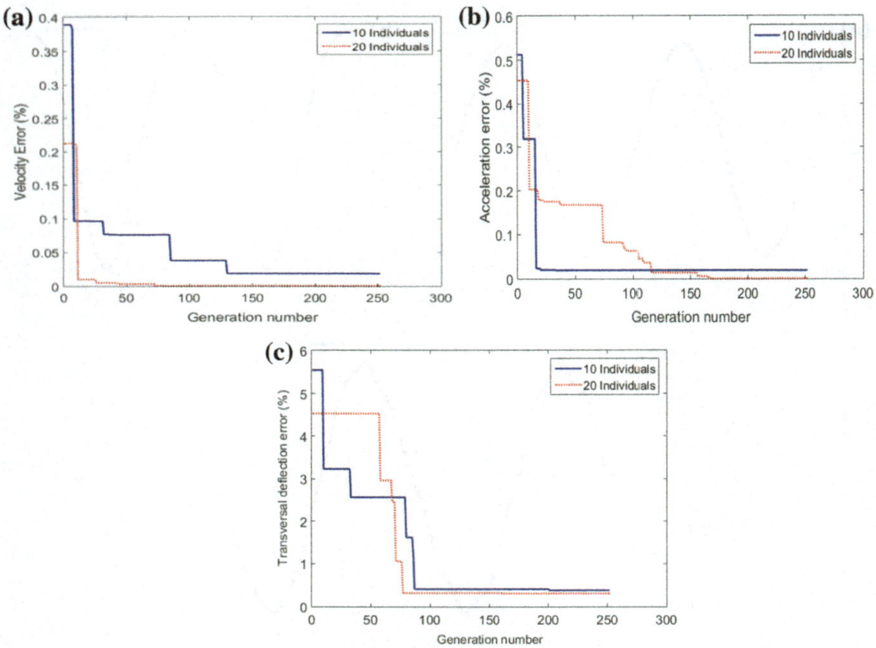

Fig. 5 The error evolution, **a** the velocity error, **b** the acceleration error, **c** the transversal deflection error

Table 3 The iteration number effects

Generation number (20 chromosomes)						
100				250		
	Slider velocity	Slider acceleration	Connecting rod transversal deformation	Slider velocity	Slider acceleration	Connecting rod transversal deformation
Error %	0.0207	0.042	0.382	0.0012	0.00043	0.307
CPU (sec)	350.12	401.93	374.37	788.53	879.41	817.91
Retained mechanism (l_2, l_3) mm	(49.9, 348.7)	(49.8, 347.3)	(49.8, 350.5)	(50, 350.3)	(50, 349.9)	(49.8, 350.4)

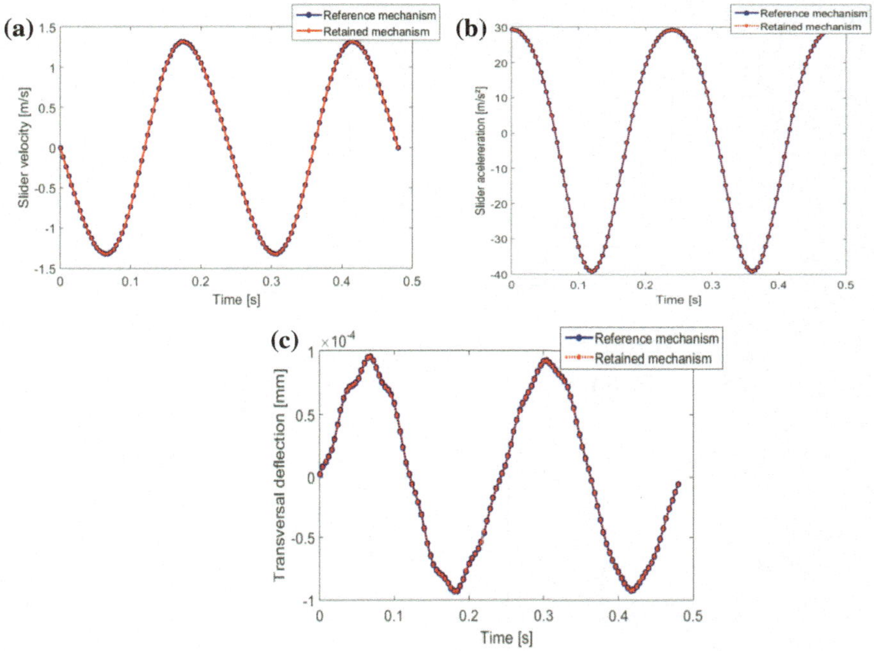

Fig. 6 The retained mechanism responses, **a** the slider velocity, **b** the slider acceleration, **c** the transversal deflection

4.3 Effects of Crossover and Mutation Probabilities

For all the previously presented results obtained for a respectively crossover and mutation probabilities of 0.9 and 0.3, these probabilities values are mostly used in the literature [14, 15].

These results are realized for a 20-individual initial population and a generation number of 250. It is noticed that for the three types of dynamic responses the estimated error significantly declines when increasing the crossover and mutation probabilities. In fact, as shown in Fig. 7, for a fixed mutation probability of 0.1, the minimal RMS error is obtained for a crossover probability of 0.9.

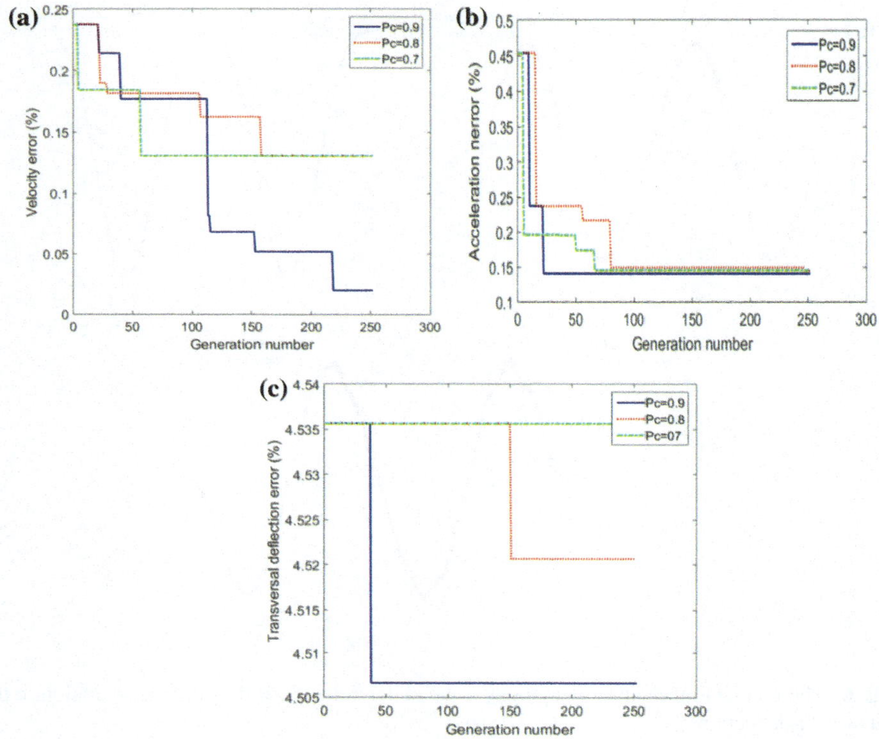

Fig. 7 Results for Pm = 0.1 with different Pc values: **a** Slider velocity, **b** slider acceleration, **c** connecting rod transversal deformation

Figure 8 illustrates the mutation probability variation effects for a chosen crossover probability of 0.9. The simulations results confirm that a 0.3 mutation probability is the most appropriate for the best convergence of the GA optimization algorithm.

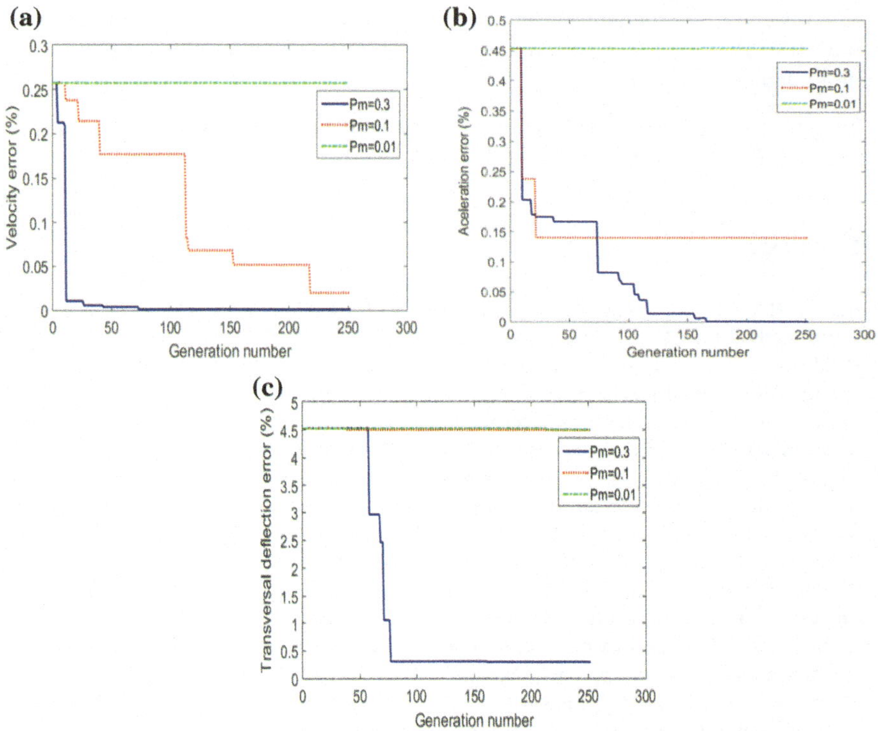

Fig. 8 Results for Pc = 0.9 with different Pm values: **a** Slider velocity, **b** slider acceleration, **c** connecting rod transversal deformation

5 Conclusion

In this work, a dynamic synthesis for a flexible slider crank mechanism has been developed. The identification approach is based on an GA optimization algorithm. Using three mechanism dynamic responses, the algorithm is able to give optimal design variables that satisfy perfectly the required mechanism response accuracy.

It has been demonstrated that:

- Based on the three mechanism dynamic responses, the GA optimization algorithm converges to the reference mechanism with a minimum RMS error.
- The RMS error is greatly reduced when increasing the generation number and an enlarged initial population size.
- For a high generation number and an enlarged initial population size, the CPU time computation increases. It is not considered an inconvenience given the important gain in the mechanism accuracy traduced by a minimum RMS error compared to the reference mechanism.

– The optimum GA parameters are: respectively the crossover and mutation probabilities of 0.9 and 0.3, generation number of 250 and initial population size of 20 individuals.

In future works, the synthesis of a mechanism including material characteristics with multiple joint clearances should be studied.

References

1. Laribi MA, Mlika A, Romdhane L, Zeghloul S (2004) Combined genetic algorithm-fuzzy logic method (GA-FL) in mechanisms synthesis. Mech Mach Theory 39:717–735
2. Yildirim Ş, Erkaya S, Su S, Uzmay I (2005) Design neural network model for transmission angle of a modified mechanism. J Mech Sci Technol 19(10):1875–1884
3. Erkaya S, Uzmay I (2009) Determining link parameters using genetic algorithm in mechanism with joint clearance. Mech Mach Theory 44(1):222–234
4. Erkaya S, Uzmay I (2008) A neural_genetic (NN-GA) approach for optimizing mechanism having joints with clearance. Multibody Syst Dyn 20(1):69–83
5. Erkaya S, Uzmay I (2009) optimization of transmission angle for slider crank mechanism with joint clearances. Struct Multidisc Optim 37(5):493–508
6. Erkaya S (2012) Prediction of vibration characteristics of a planar mechanism having imperfect joints using neural network. Mech Sci Technol 26(5):1419–1430
7. Erkaya S (2013) Trajectory optimization of a walking mechanism having revolute joints with clearance using ANFIS approach. Nonlinear Dyn 71(1–2):75–91
8. Ambrósio JAC, Neto MA, Leal RP (2007) Optimization of a complex flexible multibody systems with composite materials. Multibody Syst Dyn 18(2):117–144
9. Ebrahimi S, Payvandy P (2015) Efficient constrained synthesis of path generating four bar mechanism based on the heuristic optimization algorithm. Mech Mach Theory 85:189–204
10. Lin WY, Hsiao KM (2016) A new differential evolution algorithm with a combined mutation strategy for optimum synthesis of path generating. Proc Ins Mech Eng Part C: J Mech Eng Sci 231(14):275–2690
11. Lin WY (2012) Optimum path synthesis of a four link mechanism with rolling contacts 2010. Proc. Inst Mech Eng Part C: J Mech Eng Sci 226(2):544–551
12. Romdhane L, Dhuibi H, Hadj Salah H (2003) Dynamic analysis of planar elastic mechanisms using the dyad method. Proc Inst Mech Eng Part K: J Multi-Body Dyn 217 (1):1–14. https://doi.org/10.1243/146441903763049397
13. Lin WY (2013) Optimum path generation of a geared five bar mechanism. Adv Mech Eng 5, ID 757935
14. Goldberg DE (1989) Genetic algorithms in search, optimization, and machine learning, Addison-Wesley
15. Kheder M, Trigui M, Aifaoui N (2015) Disassembly sequence planning based on a genetic algorithm. Proc. Inst Mech Eng Part C: J Mech Eng Sci 229(12):2281–2290

Water-Hammer Control in Pressurized Pipe Flow Using Dual (LDPE/LDPE) Inline Plastic Sub Short-Sections

Mounir Trabelsi[1,2](✉) and Ali Triki[2]

[1] Department of Mechanics, National Engineering School of Sfax, University of SFAX, B.P. 1173, 3038 Sfax, Tunisia
mounirtrabelsi.enis@gmail.com

[2] Research Unit: Mechanics, Modelling Energy and Materials M2EM, National Engineering School of Gabès, University of Gabès, Gabès, Tunisia
ali.triki@enis.rnu.tn

Abstract. A dual technique-based inline strategy was explored in this research to enhance the conventional technique skill with regard to the limitation of wave oscillation period spreading. Instead of the single short section employed by the latter technique, the former utilizes a couple of two sub-short sections made of low density polymeric material (**LDPE**). Numerical computations were performed using the Method of Characteristics for the discretization of **1-D** unconventional water-hammer model embedding the Vitkovsky and Kelvin-Voigt formulations. The dual technique efficiency was considered for an operating event involving the onset of cavitating flow. Results evidenced the reliability of the proposed technique for mitigating excessive hydraulic-head drop and rise, and demonstrated that the (**LDPE/HDPE**) plastic sub-short section combination provided an acceptable trade-off between hydraulic-head attenuation and transient wave oscillation period spreading. Ultimately, a sensitivity analysis of the wave amplitude attenuation and wave period spreading to the employed plastic sub-short sections lengths and diameters was reported to estimate the near-optimal values of the sub-short section dimensions.

Keywords: Cavitation · Design · Dual · Inline · Kelvin-Voigt · LDPE · Plastic material · Method of characteristics · Viscoelasticity · Vitkovsky · Water-Hammer

1 Introduction

Water-hammer control constitutes a major concern for hydraulic researchers and designers in order to protect hydraulic utilities from damage and to ensure the global economic efficiency and safety of hydraulic utilities, besides providing an adequate service level. Water hammer control is the most important feature governing the serviceability, integrity and safety of pressurized-piping utilities. An effective design of industrial pressurized piping systems does require the mitigation of unacceptable conditions onset; while ensuring the adequate service level.

N. Aifaoui et al. (Eds.): CMSM 2019, LNME, pp. 953–961, 2020.
https://doi.org/10.1007/978-3-030-27146-6_102

From a design side, there are a large variety of design tools available to control severe water-hammer surge impacts (e.g.: surge tanks, air valves, or pressure vessels) [2, 7, 9]; [3, 5, 6, 10–14]; [15, 18].

In this context, the inline design strategy was recognized as being an effective tool for water-hammer control in pressurized-pipe flow. In particular, Triki [10, 11] explored the efficiency of an inline-based design strategy to upgrade existing steel-piping systems face to both up-and down-surge water-hammer severe effects. Precisely, the inline strategy, applied by the author, consists in substituting a short-section of the transient sensitive region of the existing steel-piping system by another one made of plastic pipe-wall material. Namely, the author made use of the high-and low-density polymeric material types ((**HDPE**) or (**LDPE**)). The author proved that such a technique could successfully be employed to attenuate excessive hydraulic-head rise and drop. Though interesting, this strategy exhibits a wave oscillation period spreading; which may affect negatively the operational procedure of the hydraulic system; such as increase of critical time of valve closure.

Overall, results evidenced a close dependency between hydraulic-head attenuation and wave period spreading. Specifically, the author observed that the case using an (**LDPE**) plastic short-section allowed more important hydraulic-head attenuation and a larger period spreading as compared with the case employing an (**HDPE**) short-section. Based on this result, it may be concluded that, as the wave speed of the short-section pipe-wall material decreases, the hydraulic-head attenuation and the wave oscillation period spreading increase. Physically, this result has obvious explanations. Indeed, the reduced modulus and the viscoelastic mechanical behavior of plastic materials, used for the short-section pipe-wall, results in a reduced wave speed (which may attenuate the surge wave amplitude), and a retarded strain of the pipe-wall (which may expand the wave fluctuation period) [4, 12, 13].

Alternatively, this paper addresses an improved (**LDPE/LDPE**) dual technique based on substituting up-and down-stream sub short sections of the existing steel-piping system by other ones made of (**LDPE**) plastic materials. This idea is intended to address the conventional technique drawback, mentioned above, by reducing the wave reflection time throughout the piping system; besides profiting from the important hydraulic-head attenuation provided by the forgoing plastic material type, of the system and limits the influence of the pressure wave.

The (**1-D**) unconventional water-hammer solver used to approximate the flow parameters are given in the next section.

2 Materials and Methods

To account for unsteady friction losses and pipe-wall viscoelastic behavior, the (**1-D**) unconventional water-hammer embedding the Vitkovsky et al. and the Kelvin-Voigt formulations is widely used in the literature [1, 16]; Szymkiewicz and Mitosek 2013):

$$\frac{\partial h}{\partial t} + \frac{a_0^2}{gA}\frac{\partial q}{\partial x} + 2\frac{a_0^2}{g}\frac{d\varepsilon_r}{dt} = 0 \tag{1}$$

$$\frac{1}{A}\frac{\partial q}{\partial t}+g\frac{\partial h}{\partial x}+g\left(h_{f_s}+h_{f_u}\right)=0 \tag{2}$$

where, h is the hydraulic-head; q is the flow discharge; A is the pipe cross sectional area; g is the gravity acceleration; a_0 is the wave speed; h_{f_s} is the quasi-steady head-loss component per unit length; h_{f_u} is the unsteady friction losses evaluated using the Vitkovsky et al. [16] formula: $h_{f_u}=(k_v/gA)\{(\partial q/\partial x)+a_0\,Sgn(Q)|\partial q/\partial x|\}$, in which, $k_v=0.03$ is a decay coefficient, x and t are the coordinates along the pipe axis and time, respectively.

The retarded radial strain ε_r may be expressed basing on the linear-viscoelastic Kelvin-Voight formulation [1]:

$$\varepsilon_r(x,t)=\sum_{k=1}^{N_{kv}}\varepsilon_{rk}=\sum_{k=1}^{N_{kv}}\frac{\alpha D}{2e}\rho g\int_0^s[h(x,t)-h_0(x)]\frac{J_k}{\tau_k}e^{-\frac{s}{\tau_k}}ds \tag{3}$$

where, J_0 designates the elastic creep compliance, J_k and $\tau_k(k=0\cdots n_{kv})$ denote the creep-compliance and the retardation-time coefficients associated with k^{th} Kelvin-Voigt element, respectively, n_{kv} is the number of Kelvin-Voigt elements.

The Method of Characteristics (**MOC**) procedure selected for transient flow predictions is next outlined briefly (a detailed derivation of the general algorithm is reported in e.g.: [10, 11, 15].

The compatibility equations performed by the **MOC** procedure are given by:

$$\mathbf{C}^{j\pm}:\frac{dh}{dt}\pm\frac{a_0^j}{gs^j}\frac{dq}{dt}+\frac{2a_0^2}{g}\left(\frac{\partial\varepsilon_r}{dt}\right)\pm a_0^j h_f^j=0\,\text{along}\,\frac{\Delta x^j}{\Delta t}=\pm\frac{a_0^j}{c_r^j} \tag{4}$$

in which, the superscript j refers to the pipe number ($1\le j\le np$; np is the number of pipes), Δt denotes the time-step increment and c_r designates the Courant number associated with the spatial-discretization of the jth pipe.

Basing on Eq. (4), the relationships between the hydraulic-head and the discharge may be written as follows:

$$\mathbf{C}^{\pm}:\begin{cases}q_{i,t}^j=c_p^j-c_{a-}^j h_{i,t}^j\\ q_{i,t}^j=c_n^j+c_{a+}^j h_{i,t}^j\end{cases}\text{along}\frac{\Delta x^j}{\Delta t}=\pm\frac{a_0^j}{c_r^j} \tag{5}$$

where, $c_p^j=\left(q_{i-1,t-1}^j+(1/B^j)h_{i-1,t-\Delta t}^j+c_{p1}^{\prime\prime j}+c_{p1}^{\prime\prime\prime j}\right)/\left(1+c_p^{\prime j}+c_{p2}^{\prime\prime j}+c_{p2}^{\prime\prime\prime j}\right)$;
$B=a_0/(gA)$;$c_n^j=\left(q_{i+1,t-1}^j+(1/B^j)h_{i+1,t-\Delta t}^j+c_{n1}^{\prime\prime}+c_{n1}^{\prime\prime\prime}\right)/\left(1+c_n^{\prime j}+c_{n2}^{\prime\prime j}\right)$;
$c_{a+}^j=1+c_{p2}^{\prime\prime\prime j}/\left(B^j\left(1+c_{p2}^{\prime j}+c_{p2}^{\prime\prime j}\right)\right)$; $c_p^{\prime j}=R^j\Delta t\left|q_{i-1,t-1}^j\right|$; $c_n^{\prime j}=R^j\Delta t\left|q_{i+1,t-1}^j\right|$;
$R^j=f^j/2D^jA^j$;
$c_{p_1}^{\prime\prime j}=k_v\theta\,q_{i,t-1}^j-k_v(1-\theta)\left(q_{i-1,t-1}^j-q_{i-1,t-2}^j\right)-k_v\mathbf{sgn}\left(q_{i-1,t-1}^j\right)$
$(q_{i,t-1}^j-q_{i-1,t-1}^j)$;

$c''^{j}_{n_1} = k_v\theta\, q^j_{i,t-1} - k_v(1-\theta)\left(q^j_{i+1,t-1} - q^j_{i+1,t-2}\right) - k_v\mathbf{sgn}\left(q^j_{i+1,t-1}\right)$
$(q^j_{i,t-1} - q^j_{i+1,t-1})$;

$c'''^{j}_{p_1} = -c'''^{j}_{n_1} = -2a^j_0 A^j \Delta t \sum_{k=1}^{n_{kv}} \left[\varepsilon^j_{rk}(x,t)/\partial t\right]$;

$c'''^{j}_{p_2} = c'''^{j}_{n_2} = 2a^j_0 A^j c_0 \gamma \sum_{k=1}^{n_{kv}} J^j_k\left(1 - e^{-(\Delta t/\tau_k)}\right)$;

$\varepsilon^j_{rk,i,t-\Delta t} = J^j_k c_0 \left\{ \left[h^j_{i,t-\Delta t} - h^j_{i,0}\right] - e^{-(\Delta t/\tau_k)}\left[h^j_{i,t-2\Delta t} - h^j_{i,0}\right] - \tau_k \right.$
$\left. (1 - e^{-(\Delta t/\tau_k)})\left[h^j_{i,t-\Delta t} - h^j_{i,t-2\Delta t}\right]/\Delta t \right\} \quad + e^{-(\Delta t/\tau_k)}\varepsilon_{rk,i,t-2\Delta t}$

$c''^{j}_{p_2} = c''^{j}_{n_2} = k_v\theta$, ($\theta = 1$ is a relaxation coefficient); and $c_0 = \alpha\gamma D^j/2e^j$.

It is worth noting that the above **MOC** algorithm is established for a one-phase flow regime. For a cavitating flow regime, the discrete gas cavity procedure (**DGCM**) may be included into the conventional **MOC** solution.

Basically, the **DGCM** procedure assumes that void cavities are lumped at the computing sections.

The discretization of the perfect gas law for an isothermic evolution of each gas cavity leads to [19]:

$$\forall^j_{g\,i,t}\left(h^j_{i,t} - z^j_i - h_v\right) = \left(h_0 - z^j_i - h_g\right)\alpha_0 A \Delta t \tag{6}$$

in which, h_0 is the hydraulic-head reference, α_0 the void fraction at h_0, z^j_i the pipe axis elevation and h_g the gauge hydraulic-head of the liquid.

The discretization of the continuity equation applied for the cavity control volume leads to the expression of the cavity volume:

$$\forall^j_{g\,i,t} = \forall^j_{g\,i,t-2\Delta t} + \left[\psi\left(q^j_{d\,i,t} - q^j_{u\,i,t}\right) - (1-\psi)\left(q^j_{d\,i,t-2\Delta t} - q^j_{u\,i,t-2\Delta t}\right)\right]\sqrt{2} \tag{7}$$

where, q_u and q_d are the flowrates, computed at the upstream and downstream sides of the cavity interface [19].

It is interesting to point out here that the flow regime is regarded as cavitating type for: $\forall^j_{g\,i,t} \prec 0$. Otherwise, it is considered as a one-phase type.

Series connection of multi-pipes:

The discharge and the hydraulic-head at the series connection may be expressed, assuming no flow storage common hydraulic grade-line elevation [10, 17, 19]:

$$q^{j-1}_{ns^j,t} = q^j_{1,t} \text{ and } h^{j-1}_{ns^j,t} = h^j_{1,t} \tag{8}$$

in which, the right and left hands of Eq. (8) designate the hydraulic parameter values at the upstream and downstream sides of the junction.

Next section is devoted to assess the reliability of the dual technique to control water-hammer waves involving a cavitating flow onset.

3 Applications, Results and Discussion

The case study concerns a sloping steel pipe (length: $L = 100$ m; internal diameter: $D = 50.6$ mm; pipe-wall thickness: $e = 3.35$ mm and elastic-wave-speed: $a_0 = 1369.7$ m/s, up-and down-stream axis elevations $z_d = 0$ m and $z_u = 2.03$ m, respectively) connecting two pressurized-tanks (Fig. 1a). The gauge saturated hydraulic-head of the liquid is $h_g = -10.29$ m. The initial steady-state regime is established for a constant flow velocity: $V_0 = 0.3$ m/s and a constant hydraulic-head maintained at downstream pressurized-tank: $h_0^{T_2} = 21.4$ m. The transient regime is provoked by the sudden and full closure of the upstream valve:

$$q_{|x=0} = 0\, and\, h_{|x=L} = h_0^{T_2}(t \succ 0) \tag{9}$$

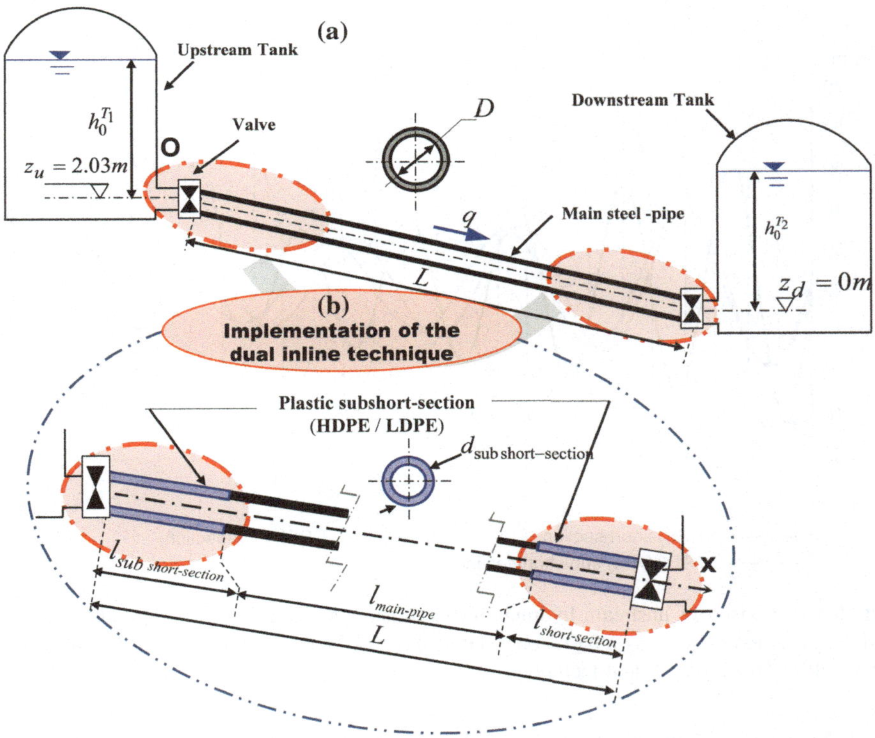

Fig. 1 Schematic implementation of the dual-inline technique

For this transient situation, the dual technique consists in substituting an up-and down-stream sub short-section of the original steel-piping system by a couple of plastic sub short section made of (**LDPE**) material (Fig. 1b), whereas the conventional technique handles only the upstream extremity of the original piping system using the same procedure. The creep compliance coefficients of the generalized Kelvin-Voigt model associated to (**LDPE**) material are: $J_0 = 1.54\,\text{GPa}^{-1}$; $\{J/\tau\}_{1\cdots5}(\text{GPa}^{-1}/\text{s}) = \{7.54/(89 \times 10^{-6}); 10.46/0.022; 0.262/1.864\}$ [8].

As a first investigation step, the sub short-sections lengths and diameters values, used in the dual technique, $l^{\text{dual}}_{\text{sub short-section}} = 5\,\text{m}$ and $d^{\text{dual}}_{\text{sub short-section}}(= D) = 50.6\,\text{mm}$, respectively; however, the corresponding short-section length and diameter used in the conventional technique, are: $l^{\text{conventional}}_{\text{short-section}} = 10\,\text{m}$ and $d^{\text{conventional}}_{\text{short-section}} = 50.6\,\text{mm}$, respectively.

Figure 2, illustrates the hydraulic-head signals predicted into the original system case along with their counterpart involved by the protected systems cases using the (**HDPE**) or (**LDPE**) short-section-based conventional technique or the (**LDPE/LDPE**) sub short-sections–based dual technique. Jointly, the main characteristics of the wave curves, illustrated in Fig. 2, are enumerated in Table 1.

A key observation from Fig. 2 relates to the cavitation onset in the original system

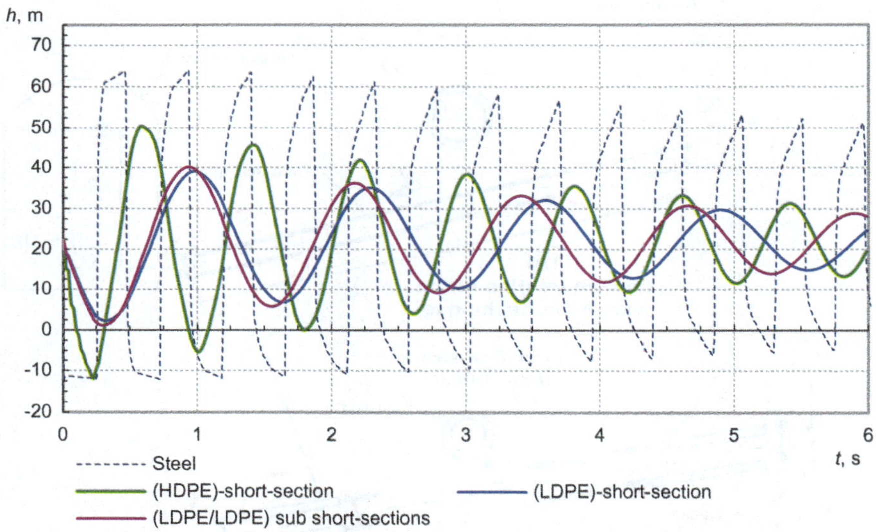

Fig. 2 Comparison of upstream hydraulic-head signals predicted into the original system case and the protected system cases based on a (HDPE) or (LDPE)–based conventional technique and the (LDPE/LDPE)–based dual technique

case. As per Fig. 2 and Table 1, succeeding the upstream valve closure, the hydraulic-head first falls to the gauge value $h_{\min} = -10.2\,\text{m}$, and subsequently rises to a first hydraulic-head peak value: $h_{\max} = 64.2\,\text{m}$. In other words, the magnitude values of hydraulic-head-drop and-rise, computed referring to the initial steady-state benchmark, are $\Delta h^{-}_{\textbf{steel-pipe}} = 32.6\,\text{m}$ and $\Delta h^{+}_{\textbf{steel-pipe}} = 41.8\,\text{m}$, respectively.

Nonetheless, Fig. 2 proves that the cavitating flow regime may be avoided if the hydraulic system is controlled using the (**LDPE**) conventional technique or the

(LDPE/LDPE) dual technique–based inline strategy. For instance, Fig. 2 (and Table 1) depicts a low hydraulic-head drop magnitude equal to: $\Delta h^{-}_{\mathbf{LDPE}} = 19.6\,\text{m}$ or $\Delta h^{-}_{\mathbf{LDPE/LDPE}} = 20.7\,\text{m}$, associated with **(LDPE)** short-section–based conventional technique or **(LDPE/LDPE)** sub short-sections–based dual technique, respectively. Specifically, these configurations allow a significant attenuation of first hydraulic-head crest as compared with those depicted into the original system case: $\delta h^{-}_{\mathbf{LDPE}} = \Delta h^{-}_{\mathbf{LDPE}} - \Delta h^{-}_{\mathbf{steel}} = 12.98\,\text{m}$ or $\delta h^{-}_{\mathbf{LDPE/LDPE}} = \Delta h^{-}_{\mathbf{LDPE/LDPE}} - \Delta h^{-}_{\mathbf{steel}} = -11.85\,\text{m}$, respectively. Similarly, the **(LDPE)** setup of the conventional technique or the **(LDPE/LDPE)** setup of dual technique involves a attenuation of first hydraulic-head peak $\delta h^{+}_{\mathbf{LDPE}} = \Delta h^{+}_{\mathbf{LDPE}} - \Delta h^{+}_{\mathbf{steel}} = 16.5\,\text{m}$ or $\delta h^{+}_{\mathbf{LDPE/LDPE}} = \Delta h^{+}_{\mathbf{LDPE/LDPE}} - \Delta h^{+}_{\mathbf{steel}} = 17.7$ m, respectively, as compared with the original system case.

Furthermore, the dual technique based on a **(LDPE/LDPE)** sub short-sections induces a spreading of the wave oscillations period equal to: $\delta T^{1}_{\mathbf{LDPE/LDPE}} = \left|T^{1}_{\mathbf{steel}} - T^{1}_{\mathbf{LDPE-LDPE}}\right| = 0.763\,\text{s}$ as compared with that corresponding to the original system case. Similarly, a lower period spreading is induced by the **(LDPE/LDPE)** sub short-sections–based dual technique relatively to the **HDPE** short-section–based conventional technique ($\delta' T^{1}_{\mathbf{LDPE-LDPE}} = \left|T^{1}_{\mathbf{HDPE}} - T^{1}_{\mathbf{LDPE-LDPE}}\right| = 0.437\,\text{s}$). Nevertheless, the **(LDPE/LDPE)** configuration of the dual technique augmented slightly the period spreading relatively to that induced into the **(LDPE)** configuration of the conventional technique ($\delta'' T^{1}_{\mathbf{LDPE/LDPE}} = \left|T^{1}_{\mathbf{LDPE}} - T^{1}_{\mathbf{LDPE-LDPE}}\right| = 0.075\,\text{s}$).

Table 1 Characteristics of hydraulic-head waves in Fig. 2

Parameters		Steel main-pipe	Plastic (sub) short-section		
			(HDPE)	(LDPE)	(LDPE/LDPE)
h_{max}: 1st hydraulic-head peak	[m]	64.2	50.2	38.9	40.1
h_{min}: 1st hydraulic-head crest	[m]	−10.2	−10.2	2.8	1.6
T^1: period of the 1st cycle of wave oscillation	[s]	0.472	0.798	1.31	1.235

From the above discussions, it may be confirmed that the **(LDPE/LDPE)** configuration of the dual technique provides an acceptable trade-off between the attenuation of hydraulic-head peak (and crest) and the limitation of spreading of hydraulic-head oscillation period.

Additional task concerns the sensitivity analysis of the first upstream hydraulic-head peak or crest values and wave oscillation period value to the sub short-section diameter and length. For completeness, this analysis is illustrated in Fig. 3a and b, respectively. Figure 3a and b show that the reduction in hydraulic peak or crest magnitude and the augmentation of the period of hydraulic-head oscillation are principally driven by an increase in upstream sub-short-section length and diameter. Besides, small attenuation (or amplification) effects associated with hydraulic-head

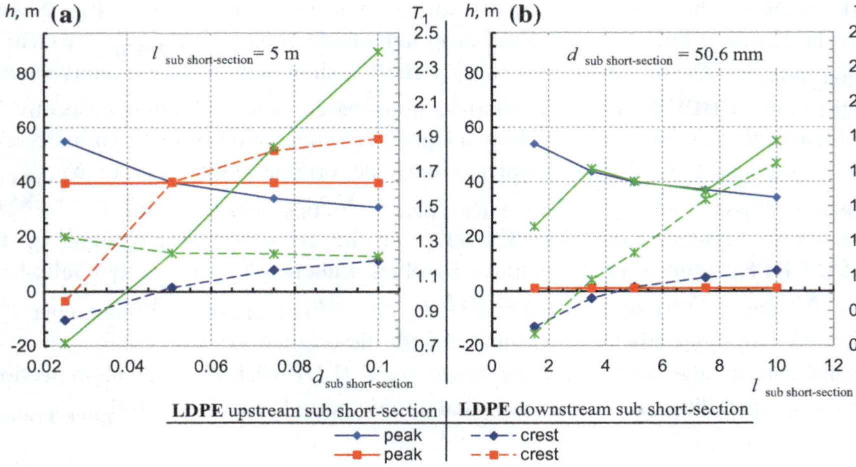

Fig. 3 Variation of hydraulic-head peak and crest depending on the short-section: **a**—diameter and **b**—length

peak or crest and wave oscillation period are observed beyond the sub-short-section diameter and length values: $l^{\text{downstream}}_{\text{sub short-section}} = 2.5\,\text{m}$ and $d^{\text{downstream}}_{\text{sub short-section}} = 0.0506\,\text{m}$.

Hence, the foregoing length and diameter values may be considered as the near optimal values for both up-and-down-stream sub short-sections.

4 Conclusion

To sum up, the (**LDPE/LDPE**) sub short-sections–based dual-technique provided significant attenuation of hydraulic-head peak and crest. Furthermore, this strategy differs in interesting ways from the conventional technique based inline strategy with regard to limitation of the spreading hydraulic-head oscillation period. Additionally, the parametric study of hydraulic-head peak or crest values with respect to the sub-short-section length and diameter identified the near-optimal value for dimensioning the dual sub short-sections.

Though the dual technique-based inline control strategy is numerically tested on a single pipeline system, experimental investigations on the presented strategy may be considered as motivating research perspectives of this study.

References

1. Aklonis JJ, MacKnight WJ, Shen M (1972) Introduction to polymer viscoelasticity. Wiley-Interscience-John Wiley & Sons, Inc
2. Bergant A, Simpson AR, Tijsseling A (2006) Waterhammer with column separation: a historical review. J Fluids Struct 22(2):135–171. https://doi.org/10.1016/j.jfluidstructs.2005.08.008

3. Besharat M, Tarinejad R, Ramos H (2015) The effect of water hammer on a confined air pocket towards flow energy storage system. J Water Supply Res Technol-Aqua 65(2):116–126. doi.:https://doi.org/10.2166/aqua.2015.081
4. Ferry JD (1970) Viscoelastic properties of polymers, 2nd edn. John Wiley & Sons, New York
5. Fersi M, Triki A (2018) Investigation on redesigning strategies for water-hammer control in pressurized-piping systems. J Press Vessel Technol–Trans ASME. https://doi.org/10.1115/1.4040136
6. Fersi M, Triki A (2019) Alternative design strategy for water-hammer control in pressurized-pipe flow. In: Fakhfakh T, Karra C, Bouaziz S, Chaari F, Haddar M (eds) Advances in acoustics and vibration II. ICAV 2018. Applied condition monitoring, vol 13, pp 135–144. Springer, pp 157–165. https://doi.org/10.1007/978-3-319-94616-0_16
7. Ghilardi P, Paoletti A (1986) Additional viscoelastic pipes as pressure surge suppressors. Proceedings of the 5th international conference on pressure surges. Cranfield (UK), pp 113–121
8. Güney MS (1983) Water-hammer in viscoelastic pipes where cross-section parameters are time dependent. In: Proceedings of the 4th international conference on pressure surges. BHRA, Bath, U.K, pp 189–209
9. Massouh F, Comolet R (1984) Étude d'un système anti-bélier en ligne-Study of a water-hammer protection system in line. La Houille Blanche 5:355–362. https://doi.org/10.1051/lhb/1984023
10. Triki A (2016) Water-hammer control in pressurized-pipe flow using an in-line polymeric short-section. Acta Mech 227(3):777–793. https://doi.org/10.1007/s00707-015-1493-13
11. Triki A (2017) Water-Hammer control in pressurized-pipe flow using a branched polymeric penstock. J Pipeline Syst-Eng Pract–ASCE 8(4):04017024. doi.: https://doi.org/10.1061/(asce)ps.1949-1204.0000277
12. Triki A (2018-a) Further investigation on water-hammer control inline strategy in water-supply systems. J Water Suppl Res Technol-AQUA 67(1):30–43. https://doi.org/10.2166/aqua.2017.073
13. Triki A (2018-b) Dual-technique based inline design strategy for Water-Hammer control in pressurized-pipe flow. Acta Mech 229(5):2019–2039. doi.org/https://doi.org/10.1007/s00707-017-2085-z
14. Triki A, Fersi M (2018) Further investigation on the Water-Hammer control branching strategy in pressurized steel-piping systems. Int J Press Vessels Pip 165(C):135–144. https://doi.org/10.1016/j.ijpvp.2018.06.002
15. Triki A, Chaker MA (2019) Compound technique-based inline design strategy for water-hammer control in steel pressurized-piping systems. Int J Pres Ves Pip 169C:188–203. https://doi.org/10.1016/j.ijpvp.2018.12.001
16. Vitkovsky JP, Lambert MF, Simpson AR, Bergant A (2000) Advances in unsteady friction modelling in transient pipe flow. In: The 8th international conference on pressure surges BHR the hague the Netherlands
17. Wan W, Huang W (2018) Water hammer simulation of a series pipe system using the MacCormack time marching scheme. Acta Mech 229(7):3143–3160. https://doi.org/10.1007/s00707-018-2179-2
18. Wan W, Zhang B, Chen X (2019) Investigation on water hammer control of centrifugal pumps in water supply pipeline systems. Energies 12(1):108. doi.:https://doi.org/10.3390/en12010108
19. Wylie EB, Streeter VL (1993) Fluid transients in systems. Prentice Hall, Englewood Cliffs NJ